# Mammals and Birds as Bioindicators of Trace Element Contaminations in Terrestrial Environments

Elżbieta Kalisińska
Editor

# Mammals and Birds as Bioindicators of Trace Element Contaminations in Terrestrial Environments

## An Ecotoxicological Assessment of the Northern Hemisphere

Springer

*Editor*
Elżbieta Kalisińska
Department of Biology and Medical
Parasitology
Pomeranian Medical University
Szczecin, Poland

ISBN 978-3-030-00119-3 ISBN 978-3-030-00121-6 (eBook)
https://doi.org/10.1007/978-3-030-00121-6

Library of Congress Control Number: 2018967754



This Springer imprint is published by the registered company Springer Nature Switzerland AG
The registered company address is: Gewerbestrasse 11, 6330 Cham, Switzerland

*We dedicate this book to all animals that help us recognize the threats to our environment and hopefully prevent a global catastrophe.*

# Preface

This monograph includes a collection of data on the role and concentrations of 17 trace elements occurring in wild warm-blooded vertebrates inhabiting the Northern Hemisphere. These elements include those that are essential or potentially essential for the lives of birds and mammals, as well as those that are commonly regarded as toxic.

Since the mid-twentieth century, we have witnessed rapid development in the field of animal biology, including ecology, biogeography, physiology, genetics, and ecotoxicology, fields of science dealing with the increasing number of environmental issues related to the overwhelming growth of the human population. The consequences of this demographic explosion include profound changes in land use, disruption of the natural geochemical cycles of many elements and their large-scale mobilization from deposits as a result of economic activity, the release of anthropogenic pollutants into the environment, and the spread of some species and decline in the abundance and ranges of other species. These environmental changes have many adverse effects on humans and other vertebrates living in terrestrial environments with different geological structures and varying levels of anthropogenic pressure.

In this context, we endeavored to analyze the available ecotoxicological literature (mainly in English and published between 1960 and 2016), focusing on terrestrial bird and mammal species of different trophic levels to identify the most universal bioindicators of pollution. We also aimed to identify the usefulness of different types of biological samples in biomonitoring of certain elements and to assess the present state of knowledge in this area.

Szczecin, Poland
January 2019

Elżbieta Kalisińska

# Contents

# List of Contributors

**Małgorzata Bąkowska** Chair of Animal Reproduction Biotechnology and Environmental Hygiene, West Pomeranian University of Technology in Szczecin, Szczecin, Poland

**Irena Baranowska-Bosiacka** Department of Biochemistry and Medical Chemistry, Pomeranian Medical University in Szczecin, Szczecin, Poland

**Łukasz J. Binkowski** Institute of Biology, Pedagogical University of Cracow, Kraków, Poland

**Halina Budis** Chair of Health Education Faculty of Physical Education and Health Promotion, Szczecin University, Szczecin, Poland

**Dariusz Chlubek** Department of Biochemistry and Medical Chemistry, Pomeranian Medical University in Szczecin, Szczecin, Poland

**Danuta Czarnomysy-Furowicz** Department of Immunology, Microbiology and Physiological Chemistry, West Pomeranian University of Technology in Szczecin, Szczecin, Poland

**Izabela Gutowska** Department of Biochemistry Human Nutrition, Pomeranian Medical University in Szczecin, Szczecin, Poland

**Elżbieta Kalisińska** Department of Biology and Medical Parasitology, Pomeranian Medical University in Szczecin, Szczecin, Poland

**Jan Korbecki** Department of Biochemistry and Medical Chemistry, Pomeranian Medical University in Szczecin, Szczecin, Poland

**Marta Kołnierzak** Department of Animal Environment Biology, Warsaw University of Life Sciences – SGGW, Warsaw, Poland

**Danuta I. Kosik-Bogacka** Department of Biology and Medical Parasitology, Pomeranian Medical University in Szczecin, Szczecin, Poland

**Tadeusz Kośla** Department of Animal Environment Biology, Warsaw University of Life Sciences – SGGW, Warsaw, Poland

**Karolina Kot** Department of Biology and Medical Parasitology, Pomeranian Medical University in Szczecin, Szczecin, Poland

**Natalia Łanocha-Arendarczyk** Department of Biology and Medical Parasitology, Pomeranian Medical University in Szczecin, Szczecin, Poland

**Iwona Lasocka** Department of Animal Environment Biology, Warsaw University of Life Sciences – SGGW, Warsaw, Poland

**Mariola Marchlewicz** Department of Aesthetic Dermatology, Pomeranian Medical University in Szczecin, Szczecin, Poland

**Andrzej Marciniak** West Pomeranian Research Center in Szczecin, Institute of Technology and Life Sciences in Falenty, Falenty, Poland

**Ewa Nowakowska** Department of Animal Reproduction Biotechnology and Environmental Hygiene, West Pomeranian University of Technology in Szczecin, Szczecin, Poland

**Bogumiła Pilarczyk** Department of Animal Reproduction Biotechnology and Environmental Hygiene, West Pomeranian University of Technology in Szczecin, Szczecin, Poland

**Renata Pilarczyk** Department of Animal Reproduction Biotechnology and Environmental Hygiene, West Pomeranian University of Technology in Szczecin, Szczecin, Poland

**Monika Rać** Department of Biochemistry and Medical Chemistry, Pomeranian Medical University in Szczecin, Szczecin, Poland

**Ewa M. Skibniewska** Department of Animal Environment Biology, Warsaw University of Life Sciences – SGGW, Warsaw, Poland

**Michał Skibniewski** Department of Animal Environment Biology, Warsaw University of Life Sciences – SGGW, Warsaw, Poland

**Lidia Strużyńska** Laboratory of Patoneurochemistry, Department of Neurochemistry, Mossakowski Medical Research Centre, Polish Academy of Sciences, Warsaw, Poland

**Agnieszka Tomza-Marciniak** Department of Animal Reproduction Biotechnology and Environmental Hygiene, West Pomeranian University of Technology in Szczecin, Szczecin, Poland

**Jan Udała** Department of Animal Reproduction Biotechnology and Environmental Hygiene, West Pomeranian University of Technology in Szczecin, Szczecin, Poland

# Part I
# Introduction

# Chapter 1
# Human Population Increase and Changes in Production and Usage of Trace Elements in the Twentieth Century and First Decades of the Twenty-First Century

**Elżbieta Kalisińska**

**Abstract** People currently live in a unique time, the Anthropocene. Since the acceleration of the Industrial Revolution (~1850), humans have become a huge geological force. In 1800, 1 billion people lived in the world, but in 2018 the global human population exceeded 7.6 billion. The beginning of large-scale human impacts during the 1950s was related to the dynamics of global population growth thus far unprecedented in human history. The years 1950–1970 were defined by a quickly expanding chemical industry and the widespread popular belief that so-called progress would result in seemingly endless improvement in the quality of everyday life but that led to destruction and pollution of environment with huge amounts of chemicals (including metals) from industry, agriculture and transport. Anthropogenic metal emission still persists in the world, but its main sources are no longer located in Europe and North America, however, in Asia where half of the global population live. For example, in 2015 aluminium ore mining increased 33 times compared to 1950 and the mining of other economically important metals [iron (Fe), copper (Cu), zinc (Zn)] >6 times. In the case of highly toxic metals such as cadmium (Cd) and lead (Pb), this increase was 4.3 higher, respectively, but there was a 50% decrease in mercury (Hg) production. It is estimated that at least 60 elements (out of 118 naturally occurring on Earth) were mobilized from minerals and introduced into biogeochemical cycles on a larger scale (>50%) as the result of human activity rather than natural causes. Never in Earth's history a single species has dominated the biosphere the way *Homo sapiens* population does now.

E. Kalisińska (✉)
Department of Biology and Medical Parasitology, Pomeranian Medical University in Szczecin, Szczecin, Poland
e-mail: ekalist@pum.edu.pl



E. Kalisińska (ed.), *Mammals and Birds as Bioindicators of Trace Element Contaminations in Terrestrial Environments*,
https://doi.org/10.1007/978-3-030-00121-6_1

# 1 Population Growth in the Twentieth Century

In the twentieth century, especially its later decades, we witnessed the highest growth dynamics of the global population, unprecedented in the recorded history of mankind. Its exponential course began in the late eighteenth century with the advent of the Industrial Revolution. It is estimated that in 1800, 1 billion people lived in the world. During the following 150 years, the number of people on Earth increased more than two and a half times, reaching 2.58 billion in 1951. Even before the end of the twentieth century, the number of people had doubled more than twice, amounting to 6 billion in 1999 (Fig. 1.1). In less than 12 subsequent years, the next billion people appeared, and in 2011 we had a global population of 7 billion (Bongaarts 2009; UNEP 2012).

According to current data, in 2018 the number of people in the world exceeded 7.6 billion and is expected to reach 8 to 9 billion by 2023 and 2037, respectively. Annual natural growth rate had the highest value in the late 1960s, when it was ~2%. The rate of increase has nearly halved since then and will continue to decline in the coming years. It is estimated to reach 1% by 2023. The world population will continue to grow in the twenty-first century, but at a much slower rate in comparison to the twentieth century. It is estimated that by 2055, the global population will reach 10 billion (http://www.worldometers.info). This means that in about 100 years, the human population will have increased almost four times (from 2.58 to 10 billion).

Both in 1950 and 2017, and probably also in 2050, the largest part of the world's population (>50%) inhabited and will inhabit Asia (Fig. 1.2). In 2017, over 4.5 billion people (almost 60% of the global population) lived there, with the greatest number of people in China and India, 1.41 and 1.34 billion, respectively, which amounts together to about 36% of the world's population. The continent with the

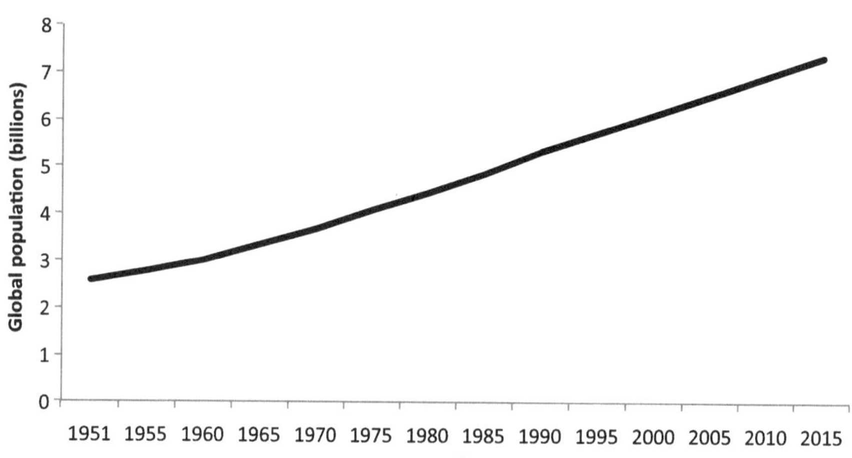


**Fig. 1.1** Increase of global population in years 1951–2015 (Source of data UN WPP 2017)

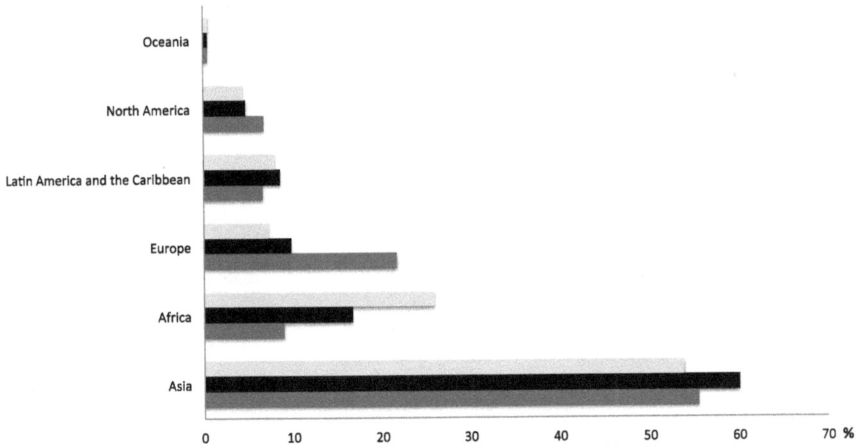


| | Asia | Africa | Europe | Latin America and the Caribbean | North America | Oceania |
|---|---|---|---|---|---|---|
| 2050 | 53.8 | 25.9 | 7.3 | 8 | 4.45 | 0.58 |
| 2017 | 59.97 | 16.64 | 9.83 | 8.56 | 4.78 | 0.54 |
| 1950 | 55.4 | 9 | 21.7 | 6.7 | 6.8 | 0.5 |

**Fig. 1.2** Regional share of the global population in years 1950, 2017 and 2050 with total number of people 2.54, 7.55 and 9.77 billion, respectively (Source: UN WPP 2017; www.worldometrics.info)

second largest number of people is Africa, where almost 17% of the global population live.

In 2017, the group of the ten most populous countries of the world, besides China and India, included three other Asian countries (Indonesia, Pakistan and Bangladesh), two countries in Latin America (Mexico and Brazil), one in North America (USA), one in Africa (Nigeria) and one in Europe (Russian Federation). The biggest growth rate between 2017 and 1950 was observed in Pakistan and Nigeria, as the number of people in these countries increased about five times (Table 1.1).

In 2050, the human population in Asia and Africa will presumably amount to 54% and 26% of the world's population, respectively (Fig. 1.2). In the subsequent few decades, population growth in Asia will slow down, but growth will probably accelerate on the African continent as a result of high fertility as expressed in number of live births per woman. Between 2010 and 2015, only 8% of the world's population lived in countries where fertility was $\geq$5 live births per woman. Of 22 countries with such high levels of fertility in the mentioned period, 20 are found in Africa and 2 in Asia (UN WPP 2017). Ninety-five percent of current global population growth (75–80 million annually) occurs in developing countries and only 5% in developed countries. As most demographers have forecast, in approaching decades, the world's population will become larger, older and more urban than it was in the twentieth century, but with much more variance among different regions and across the world (Cohen 2010; United Nations 2004).

**Table 1.1** Total population in the ten most populous countries in 2017, the number of their inhabitants in 1950, differences between population size and population ratio 2017/1950

| Continent | Country | Population (millions) 2017 | 1950 | Difference between 2017 and 1950 | Increase ratio 2017/1950 |
|---|---|---|---|---|---|
| Asia | China | 1410 | 554 | +856 | 2.55 |
| | India | 1339 | 376 | +963 | 3.56 |
| | Indonesia | 264 | 70 | +194 | 3.77 |
| | Pakistan | 197 | 38 | +159 | 5.18 |
| | Bangladesh | 165 | 39 | +126 | 4.23 |
| North America | USA | 324 | 159 | +165 | 2.04 |
| Latin America | Mexico | 129 | 28 | +101 | 4.61 |
| | Brazil | 211 | 54 | +157 | 3.91 |
| Africa | Nigeria | 191 | 39 | +152 | 4.90 |
| Europe | Russian Federation | 144 | 103 | +41 | 1.40 |

Source of data UN WPP (2017)

## 2 Main Reasons for, and Natural Consequences of Overpopulation

The rapid growth of the global population in the second half of the twentieth century, unprecedented in the history of mankind, was the result of progress in many areas of life, including medicine, hygiene and healthcare, as well as better access to food and increasing material prosperity, which have all created conditions favourable to the extension of the life span. However, at the same time, this demographic increase has caused dramatic environmental changes, especially in ecosystems and resources (Gardner et al. 2015). In the middle of the last century, the countries of North America (the USA and Canada) and Western Europe, including the UK, France, Germany, Sweden, Denmark, the Netherlands and Norway, dominated the world economic map. While the world was recovering from the aftermath of the Second World War, the production of material goods was growing, which was directly related to the increasing demand for mineral and energy resources (mainly coal, oil and natural gas). This development was accompanied by a huge expansion of the chemical industry and a popular belief that it would bring unending improvement to everyday life. At the same time, huge amounts of chemical pollutants from industry, agriculture, transport and increasingly numerous and populous cities—some of which grew into gigantic agglomerations with slums—were introduced into the environment in an uncontrolled manner. As it turned out, pesticides that were commonly used in agriculture for the protection of crops (insecticides, herbicides, fungicides, bactericides) had very toxic effects not only on pathogenic fungi, bacteria, weeds and the insect pests of plants but also on vertebrates, such as birds and mammals (including humans). Dozens of scientific papers have been written on pesticides and other pollutants and their adverse influence on the natural environment. One of the most important environmental science works that drew the attention of the general public to the problem (especially in the USA) was the groundbreaking book *Silent Spring* by Rachel Carson (1962). She wrote: "The most alarming of all man's assaults upon the environment is the contamination of air, earth, water, and sea with dangerous and even lethal materials". Unfortunately, more than 55 years later, the processes of releasing thousands of different chemicals into the environment on a tremendous scale still continue, albeit with altered geographical distribution. Since the late 1970s, mainly in developed countries, certain environmental monitoring systems have gradually been created, and legal restrictions concerning the anthropogenic emissions of pollutants on national and multinational levels were introduced (Gorman and Conway 2005; OECD 2011; Kuklinska et al. 2015). Additionally, such countries have prohibited the use of some of the most dangerous pesticides in agriculture, including DDT (UNEP/FAO UN 1991; Bouwman et al. 2013). Over the last 30 years, the greatest problems with chemical emissions in the Northern Hemisphere have been pushed onto the emerging economies of China and India (Tian et al. 2015; Stokes et al. 2016; Gong et al. 2017). The last analysis carried out by Bernhardt et al. (2017) showed that the use of synthetic chemicals, including pesticides, is increasing just as fast as other agents of global change do or even more rapidly. What is more, the range of substances of concern has dramatically expanded since Carson's days, including not only pesticides other than DDT (dichlorodiphenyltrichloroethane), heavy metals and food contaminants but now even many pharmaceuticals as well as substances which act as endocrine disruptors to vertebrates and invertebrates (Cox 1991; Balmford 2013; Goulson 2014; Hayes and Hansen 2017; Bernhardt et al. 2017). Since 1900, roughly 10 million chemical compounds have been synthesized, with some 150,000 or so put to commercial use—although nobody knows the exact number (Gardner et al. 2015).

Classic examples of the dramatic effects of anthropogenic environmental pollution, mainly in the years 1950–1970, include human and animal mercury poisonings originating from industrial sources (including cases in Minamata Bay, Japan, and in Ontario, Canada) and agricultural sources (Sweden and other developed countries of Western Europe), as well as cadmium and pesticide poisonings, including DDT, in Japan (Borg et al. 1969; Ikeda et al. 2015; Blus 2011; Bouwman et al. 2013; Köhler and Triebskorn 2013; Carvalho 2017; Mosa and Duffin 2017). The history of lead poisoning of people and wildlife has been very well documented in different ways, including various pieces of research focusing on automotive fuels, paints and hunting ammunition containing this metal (Tong et al. 2000; Franson and Pain 2011; Ma 2011; Assi et al. 2016; ATSDR 2017).

The aforementioned substances and many others have contributed directly or indirectly to a decline in the population of some animal species in different parts of the world and thus to the increasing rate of their extinction (Sodhi et al. 2008; Acevedo-Whitehouse and Duffus 2009; Bernanke and Kohller 2009; Mitra et al. 2011; Kohler and Triebskorn 2013; Goulson 2014). Other relevant factors pertaining to global population growth and human activity, which are closely related to the loss of biodiversity and extinction of species, are deforestation, desertification, climate change (including global warming, increased atmospheric methane and dioxide concentrations and enlargement of the hole in the ozone layer over Antarctica), acid rains, habitat loss, overfishing, hunting and the introduction of alien species. According to new research, all these factors pose a threat to life on Earth, raising the risk of extinction by 20–25%, and it is estimated that the current rates of extinction are about 1000 times the background rate of extinction (Nott et al. 1995; McKee et al. 2004; Gaston 2005; Castello 2015; McCauley et al. 2015). It is generally assumed that the rate of extinction in terrestrial and freshwater ecosystems, which are in greater danger of direct anthropopressure and which are better known in this respect, is nine times higher than in the seas (Pimm et al. 2014; Webb and Mindel 2015).

## 3 Changes in Production and Usage of Trace Elements

The dynamic growth of the global population is reflected by a huge increase in demand for food, freshwater and the natural resources necessary for the production of energy and the material goods on which the functioning of various sectors of the economy is based. Many forecasters ask themselves how many people can live on Earth and what is Earth's carrying capacity. The UNEP study (2012) analysed 65 estimates in that respect from the years 1971–2012, according to which the number of human beings that could possibly exist on our planet falls into a wide range from below 2 to over 1000 billion. However, the most frequently (24/65) estimated number ranged between $\leq 8$ and $\leq 16$ billion. There are many indications that in this century, between 2055 and 2070, the world's population will reach 10 billion, thus becoming ten times larger than it was in 1800 (Bongaarts 2009; http://www.worldometers.info). However, the UNEP study (2012) shows that predictions concerning the growth of human population tend to be uncertain, and the further into the future they are made, the more uncertain they become. Unquestionably, since the 1950s there have been enormous technological advances in various areas of life, but as numerous studies indicate, the increase in demand for and exploitation of many resources is faster than the growth of the global population because everyone wants a good quality of life, including the residents of the most populous countries in Asia and Africa (Steffen et al. 2011; Balatsky et al. 2015; Meinert et al. 2016). As stated in a report by Ewing et al. (2010), "if everyone lived the lifestyle of the average American, we would need five planet Earths". Future trends in demography, biosphere exploitation and environmental pollution depend largely on today's policies (UNEP 2012).

Releasing a multitude of elements from different minerals and introducing them into geochemical circulation, landscape transformations, soil erosion, climate warming and changes in water relations in many regions of the world as well as the gigantic amounts of pollutants emitted as the result of economic activity, degenerations of ecosystems and extinction of various species are happening on a scale and at a pace that has never occurred during the 4.5 billion years of Earth's history. Since the acceleration of the Industrial Revolution around 1850, over a period of less than two centuries, which by the geological time clock is merely a fraction of a second, people have become a huge geological force (van der Pluijm 2014). The rapid changes on our planet caused by people have prompted scientists to introduce the term "Anthropocene" into the timescale of Earth's history. This informal term refers to the new epoch, which follows the Holocene (or postglacial epoch) and which began about 11,700 years ago. However, geologists still debate when the exact beginning of the Anthropocene would be (Crutzen and Stoermer 2000; Monastersky 2015). Some scientists identify its onset with the arrival/acceleration of the Industrial Revolution, and this idea has been popularized by environmentalists, politicians and the media (Clemencon 2012; Syvitski 2012; van der Pluijm 2014; Corlett 2015; Monastersky 2015; Olsson et al. 2017).

Out of 118 elements naturally occurring on Earth, only a small number are considered essential to the life of organisms. All living creatures have four basic elements (the big four) in their organisms, carbon (C), hydrogen (H), oxygen (O) and nitrogen (N), which amount to more than 95% of their body masses. Aside from the big four, other major elements include phosphorus, sulphur, sodium, chlorine, potassium, calcium and magnesium. These make up the remaining 3.5% of the body mass of living things. The other elements essential for life, which amount to about 0.5% of the body mass of living organisms, are said to belong to the group of microelements. Individual eukaryotic and prokaryotic taxonomic groups may vary considerably between themselves, mainly with regard to their demand for microelements. Microelements performing important physiological functions in all plants include chlorine (Cl), boron (B), iron (Fe), manganese (Mn), copper (Cu), zinc (Zn), nickel (Ni) and molybdenum (Mo), while the number of all elements essential for life is 17 (Soetan et al. 2010; White and Brown 2010). It is estimated that the bodies of vertebrates, including humans, are composed of about 21–25 elements, while the group of microelements include B, cobalt (Co), Cu, Fe, iodine (I), Mn, Mo, selenium (Se) and Zn. In addition, several further elements such as arsenic (As), fluorine (F), Ni, silicon (Si) and vanadium (V) are believed to have beneficial effects on human health (Nielsen 1984, 1998; WHO 1996; Chellan and Sadler 2015). However, some microelements may become harmful at high levels of exposure or may cause malnutrition in case of deficiency. At the same time, it should be noted that some non-essential elements penetrate from the environment into various organisms, of which human beings are the best known, but also including domesticated animals and arable crops. Their physiological role is not always known. Some are considered neutral; others have proven toxic effects, such as lead (Pb), cadmium (Cd) and mercury (Hg). For many decades some of the non-essential elements, such as As, Cd, Hg and Pb, have attracted attention worldwide because of their toxicity to living organisms.

Numerous interactions among elements in the inanimate environment, and acid rains containing elevated levels of S and N oxides, facilitate increased mobilization of heavy metals in soil and consequently their uptake by plants, which leads to increased supply and absorption of various metals and non-metals by herbivores, then omnivores and predators. There might also be a new environmental menace related to nanomaterials or particles with at least one dimension of 100 nm or less.

Some of them are natural nanoparticles (NPs), but ecotoxicologists specializing in nanoecotoxicology raise concern over the possible adverse influence of nanomaterials on living systems due to increasing diversity within and amounts of engineered NPs. These engineered materials such as carbonaceous NPs, metal oxides, zero-valent metals, semiconductor materials and nanopolymers are introduced into the environment both unintentionally and intentionally (Viswanath and Kim 2016). For a few decades now, many metals and their oxides have been used for the manufacture of NPs, including Cd, Cu, Fe, Mn, Zn, aluminium (Al), titanium (Ti), silver (Ag) and gold (Au) due to the ever-broadening use of these elements in consumer products, chemical and medical equipment, information technology and energy production, among other things. Present knowledge on the behaviour of metals containing NPs and other such materials in the natural environment and their possible effect on biota is extremely limited, but many scientists think that NPs are a class of newly emerging environmental pollutants (Peralta-Videa et al. 2011; Thiery et al. 2012; Ray et al. 2009; Ding et al. 2015; Viswanath and Kim 2016; Giese et al. 2018). Perhaps nanomaterials, along with other new anthropogenically manufactured or used elements such as synthetic chemicals and radionuclides, as well as the man-made enrichment of biogeochemical cycles with dozens of metals and non-metals released from mineral deposits, will become hallmarks of the Anthropocene (Whitmee et al. 2015; Brondizio et al. 2016).

The manufacture of NPs and their dissemination by humans in nature span only the last two or three decades, but these processes have a tendency towards rapid increase (Salata 2004; Inshakova and Inshakov 2017). On the other hand, the acquisition of metal ores and other minerals on a gigantic scale, including those used in the production of energy or for various economic purposes, has been taking place for an incomparably longer time (at least extending to the last two centuries). Metals have long been used by humans since prehistoric times, but their application on a mass scale is a relatively recent phenomenon. In the first part of the twentieth century, world metal production rose from 30 million tons in 1900 to 198 million tons in 1950. After reaching 740 million tons in the 1970s, the output levelled off for the following 20 years. But then, driven mainly by economic expansion in China, another phase of rapid growth occurred. The bulk of these figures are connected with steel production, which expanded at least 56 times since 1900 (Gardner et al. 2015). In 1950 and 2015, steel production reached 134 and 1640 million tons, respectively, which means an over 12-fold increase in this period (Norwood et al. 1951; US GS 2016). As a result of very intense economic activity, including metal acquisition and processing, numerous elements have been introduced into the biogeochemical cycles in quantities often far greater than those resulting from natural biotic and abiotic factors or processes, such as weathering of rocks, erosion, volcanic eruptions and fires. The most attention has been devoted to this problem with regard to biogenic elements (C, N, P, S) and their influence on climate changes, engineered metals (e.g. Ag, Al, Cr, Cu, Fe, Ni, Pb and Zn, among which the major industrial metals are Al, Cu, Fe and Zn in the human-built environment) and toxic elements such as lead (Pb), cadmium (Cd), mercury (Hg) and arsenic (As), which are of interest because of the health hazards they pose to humans and ecosystems (Nriagu and Pacyna 1988; Mackenzie and Chris 1993;

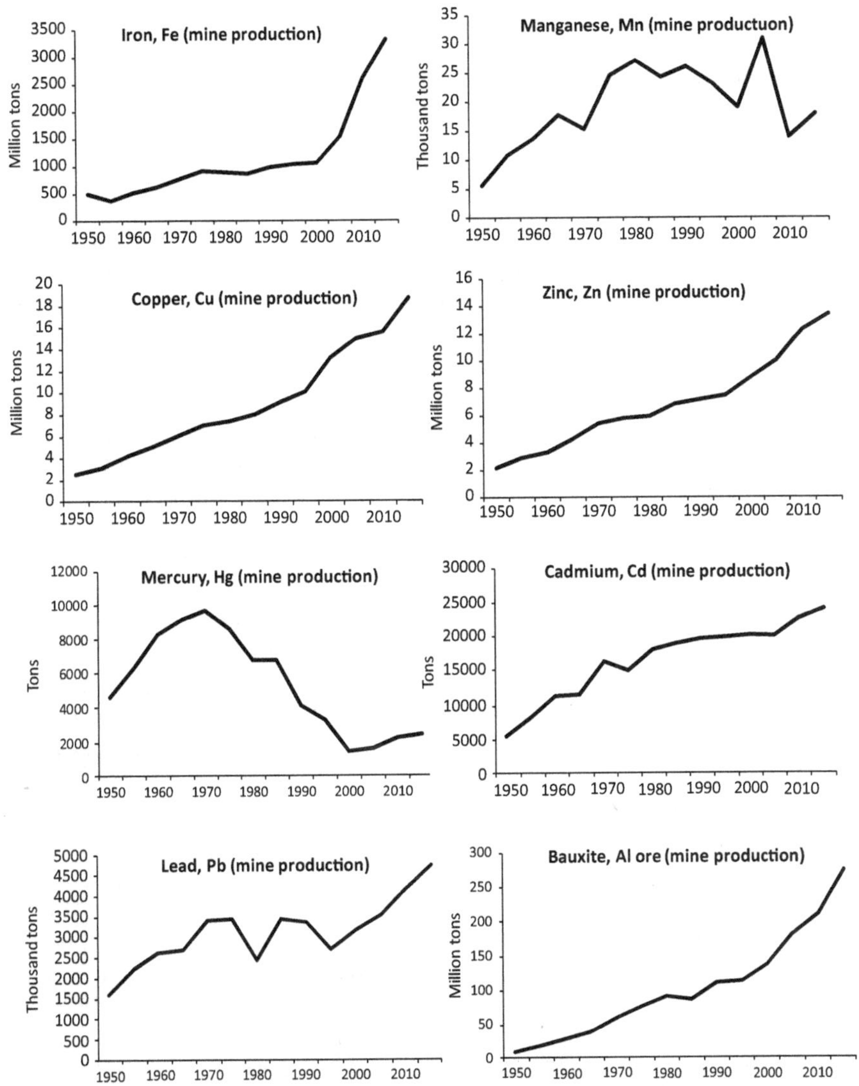


**Fig. 1.3** Selected metal mine production in years 1950–2015 (Source of data: Bureau of Mines, Minerals Yearbook: 1951, 1956, 1961, 1966, 1971, 1981, 1986, 1991, 1996, 2001, 2006, https://minerals.usgs.gov/minerals/pubs/usbmmyb.html; US GS 2011, 2016. Mineral commodity summaries. US Geological Survey, p 202, https://doi.org/10.3133/70140094)

Candelone et al. 1995; Wang et al. 2010; Klee and Graedel 2004; Rauch 2009; Rauch and Pacyna 2009; Sen and Peucker-Ehrenbrink 2012; Galloway et al. 2014; Gardner et al. 2015; Nishijo et al. 2017; Obrist et al. 2018).

Figure 1.3 depicts selected metals (Fe, Mn, Zn, Cu, Hg, Cd, Pb, Al) with varying dynamics in their ore mining in the years 1950–2015, usually with an upward trend.

The exception to this is Hg (characterized by very high toxicity and atmospheric mobility), which when carried by air masses can travel thousands of kilometres and cover distances between continents. Owing to mercury toxicity and concerns for the environment and human health, and as the result of the long-term efforts of some international bodies, the world's nations have adopted the Minamata Convention on Mercury to protect humans and nature from mercury pollution. The Convention came into force on 16 August 2017. In many countries with large mercury ore deposits, including those belonging to the European Union and the USA, extraction of Hg had already been discontinued. The exception is China, where most Hg is currently mined (US GS 2016; UN Environment 2017). In 2015, about 50% less Hg was extracted globally when compared to 1950–2340 and 4605 tons, respectively (Fig. 1.3). It should be noted, however, that mercury had previously been intensively produced, among other things, for the purposes of extracting precious metals (silver and gold) using an amalgamation method that utilizes mercury's ability to form alloys with them, from which it is later evaporated. Particularly large amounts of Hg were used in the period of the gold and silver rushes in North America at the turn of the nineteenth and twentieth centuries, as is reflected by deposits found on this continent containing elevated levels of the toxic metal (Hylander and Meili 2003; Strode et al. 2009; Horowitz et al. 2014).

During the last 65 years, the largest increase in production of the aforementioned metals has occurred with aluminium ore (bauxite)—over 33 times, from 8.2 to 274 million tons. In 2015, the world smelter production of aluminium (the third most abundant element after O and silicon, Si, and the most abundant metal in the Earth's crust) reached over 58 million tons, caused by a huge demand for this metal due to its versatile applications (US GS 2016). During the discussed period, iron ore production increased >6 times (from 498 to 3.320 million tons), copper production (from 2.5 to 18.7 million tons) and zinc (from 2.2 to 13.4 million tons). Furthermore, manganese ore production increased 3.2 times (from 5.6 to 18,000 tons, with a maximum value of 31,000 tons in 2005) and cadmium 4.3 times (from 5.666 to 24.200 tons). In the case of lead, its production increased ~2.9 times (from 1.640 to 4.710 thousand tons), but during the discussed period, there were two distinct decreases, in 1980 and 1995. They were at least partly related to the withdrawal of lead additives to gasoline and paints in well-developed countries. On the other hand, a persistent and clear upward trend continues after 1995, mainly due to the growing demand for lead for battery manufacturing, including those used in vehicles (Davidson et al. 2016; US GS 2016; Mohr et al. 2018). Some experts believe that despite the increase in production of various metals over the past 50 years, their reserves have remained largely unchanged. Additionally, in the case of most metals, recycling currently satisfies 10–20% of demand, and new technologies play an important role in the more economical management of metals (Bloodworth and Gunn 2012; Arndt et al. 2017; Mohr et al. 2018).

These examples of the increased mining of metal ores (except for Hg) in the last 65 years are closely related to the growth of the human population and the extremely intense anthropogenic mobilization of those and many other elements found in the Earth's crust (Halada et al. 2008; van der Voet et al. 2013; Meinert et al. 2016). Their mobilization is the result not only of mining and smelting but also processing (including processing metals from recycling), use in various economic sectors, combustion of fossil fuels and biomass, deforestation and improper agricultural uses of soil conducive to their erosion. These processes are accompanied by the emission of metallic and non-metallic pollutants into the air, water and soil. Various elements are partly subject to redeposition processes on land and ocean surfaces and remineralization in soils and sediments, but they are also partly remobilized, which is well known and broadly described in the case of a small group of elements, including those mentioned above. The most comprehensive current studies in this field include Klee and Graedel (2004) and Sen and Peucker-Ehrenbrink (2012), which analysed 91 and 77 elements, respectively. These reports estimated, among other things, anthropogenic mobilization (AM) and natural mobilization (NM), the ratio between them (AM/NM), total mobilization (TM = AM + NM) and percentage share of AM in TM (AM/TM). According to Klee and Graedel (2004), the anthropogenic mobilization of almost 3/4 of the elements they analysed exceeds natural mobilization, and the AM/TM ratio exceeds 50%. However, Sen and Peucker-Ehrenbrink (2012) suppose that the value >50% of the ratio may concern up to 80% of the elements (62/77), while in anthropogenic contributions, soil erosion and eolian dust (results of unsustainable farming practices and deforestation in large scale) are considered.

In Table 1.2, based on the research by Klee and Graedel (2004), 17 elements are listed, which are discussed in more detail in this book. In the case of Cu, Cr, Fe, Hg, Pb and Sn, it can be assumed that 75–99% of their amounts present in the natural environment were mobilized as the result of human activity. This includes Hg (AM/TM = 95%) and Pb (AM/TM = 84%), i.e. two metals with remarkably toxic effects, as well as Cr (AM/TM = 99%) which is mobilized 140 times more due to human activity than to natural processes. Hazards to both human and animal health and threats to entire ecosystems are caused not only by the increasing amount of metals mobilized from minerals but also by their transformations in the environment affecting the oxidation state and the occurrence of various inorganic and organic compounds with different biochemical properties. For example, Cr(III) is a micronutrient, while Cr(VI) is a carcinogenic and mutagenic mammalian agent. Inorganic Hg forms are less bioavailable than organic methyl mercury (Adriano 2001; Gall et al. 2015; Mikulewicz et al. 2017). Anthropogenic mobilizations of silver (Ag), arsenic (As), fluorine (F), iodine (I) and nickel (Ni) also exceed natural mobilizations. The values of their AM/TM ratio range between 51 and 75%. In the case of six elements (Al, Cd, Mn, Mo, Se, Zn), the values of their AM/TM ratios fall below 50% (range 17–47%).

Since the 1950s humanity has been developing new technologies to meet its rapidly growing needs, which have effected not only quantitative but also qualitative changes in the demand for various raw materials and the methods of their use. The search for new deposits of metal ores and improvements in acquisition of various elements from poorer minerals is ongoing (Halada et al. 2008; Balatsky et al. 2015; Arndt et al. 2017). The world is changing at an unusually fast pace under the enormous pressure of the incessantly growing human population. Many raw materials are being depleted, including fossil fuels and the ores of some metals, natural

**Table 1.2** Calculation of mobilization rates and dominance of selected elements

| Element | Anthropogenic mobilization, AM (Tg/year) | Natural mobilization, NM (Tg/year) | Total mobilization, TM = AM + NM (Tg/year) | Anthropogenic vs. natural ratio (AM/NM) | Anthropogenic as % of total (AM/TM × 100%) |
|---|---|---|---|---|---|
| Ag | 0.02 | 0.01 | 0.03 | 1.4 | 58 |
| Al | 80 | 228 | 309 | 0.4 | 26 |
| As | 0.08 | 0.05 | 0.1 | 1.7 | 63 |
| Cd | 0.03 | 0.1 | 0.16 | 0.2 | 17 |
| Cu | 13 | 2.3 | 16 | 5.9 | 85 |
| Cr | 15 | 0.1 | 15 | 139 | 99 |
| F | 3.0 | 1.4 | 4.4 | 2.2 | 69 |
| Fe | 762 | 86 | 848 | 8.9 | 90 |
| Hg | 0.06 | 0.003 | 0.07 | 18.5 | 95 |
| I | 0.05 | 0.02 | 0.07 | 1.9 | 66 |
| Mn | 8.8 | 32 | 40 | 0.3 | 22 |
| Mo | 0.2 | 0.6 | 0.8 | 0.3 | 24 |
| Ni | 1.6 | 0.7 | 2.3 | 2.3 | 69 |
| Pb | 3.3 | 0.6 | 3.9 | 5.2 | 84 |
| Se | 0.01 | 0.05 | 0.05 | 0.2 | 17 |
| Sn | 0.3 | 0.07 | 0.3 | 3.5 | 78 |
| Zn | 9.4 | 10 | 20 | 0.9 | 47 |

Tg, tetragram = 1,000,000 tons
Source of data: Klee and Graedel (2004)

ecosystems are shrinking and biodiversity is decreasing, while environmental pollution increases, which results in adverse climate changes. According to geological research, five great extinctions of species have occurred in the history of the Earth, the causes of which are still being investigated. Probably one of the most important ones was climate warming and increasing emissions of volcanic dusts and gases containing many toxic substances, including mercury (Bond and Grasby 2016; Racki et al. 2018). The sixth extinction that we are currently witnessing is accompanied by climate warming and anthropogenic emissions of Hg and many other heavy metals, as well as many synthetic pollutants.

In the face of the numerous deep and unfavourable changes taking place on Earth as the result of human activity in such a relatively short period of time and on such a large scale, efforts have been made, at least since the 1970s, to stop them. Public awareness in this field increased greatly with the publication of a report by the United Nations World Commission on Environment and Development. This report, entitled *Our Common Future* (Brundtland 1987), contained the definition of the term "sustainable development" (development that meets the needs of the present without compromising the ability of future generations to meet their own needs). The term became very popular in later years (e.g. in economy, politics, sociology and environmental sciences), but its definition was imprecise and concerned an unspecified period of time, measured by the length of a human lifetime (Bartlett 1998). Referring to the first two laws of sustainability, the author emphasized that population growth cannot be sustained in any society and that the larger the population, the more difficult it will be for the society to achieve sustainability. According to subsequent scientific studies and analyses, reality has turned out to be much more complicated and destructive for Earth than anyone suspected while preparing Brundtland's report and later documents for the United Nations Conference on Environment Development (also known as the Rio de Janeiro Earth Summit) in 1992 or the United Nations Conference on Sustainable Development in 2012 held in Rio again and commonly called Rio + 20 or Rio Earth Summit 2012. The analysis of demographic and economic trends between the first and second Rio Earth Summits shows that the global population increased during this period by 30%, but the production of cement, plastics and steel by 170%, 130% and 100%, respectively. In the two aforementioned decades, one of the few beneficial changes that have taken place in anthropogenic environmental impact has been the decreasing use of ozone-depleting substances by 93% (Clemencon 2012; Gardner et al. 2015; www.worldometers.info). The incessant demographic increase of the global population has rapidly eroded Earth's biosphere (de Sherbinin et al. 2007; Bradshaw and Brook 2014; Balatsky et al. 2015; Fagerberg and Srholec 2017). Unfortunately, most people in the world live only in the shell of their own personal and local problems, focusing on satisfying their own needs and dreaming about "American style" prosperity. Vast masses of people, especially those living in developing countries, but also the majority of political and economic decision-makers at various levels seem not to comprehend the severity of the situation indicated by only too abundant evidence and arguments presented by scientists pertaining to increasing environmental degradation and resource shortages occurring at a rate that can no longer be sustained.

The first manifesto concerning this problem, called *World Scientists' Warning to Humanity*, was penned in 1992, and the second was published 25 years later, in 2017. Many examples of the diverse and effective steps humanity can take to transition to sustainability were presented in the first manifesto. One of the most important ones should be to further reduce fertility rates by ensuring that women and men have access to education and voluntary family-planning services, since the environmental problems of our planet are closely related to progressing overpopulation. Unfortunately, we have not heeded the recommendations presented in the list of very important steps to be taken, so well-articulated by the world's leading scientists who first warned humanity 25 years ago. Time is running out, and soon it will be too late to shift our course away from its present disastrous trajectory (Ripple et al. 2017).

## References

Acevedo-Whitehouse K, Duffus ALJ (2009) Effects of environmental change on wildlife health. Philos Trans R Soc B 364:3429–3438

Adriano DC (2001) Trace elements in terrestrial environments: biogeochemistry, bioavailability, and risk of metals. Springer, New York

Arndt NT, Fontboté L, Hedenquist JW, Kesler SE, Thompson JFH, Wood DG (2017) Future global mineral resources. Geochem Perspect 6:1–171

Assi MA, Hezmee MNM, Haron AW, Sabri MYM, Rajion MA (2016) The detrimental effects of lead on human and animal health. Vet World 9:660–671

ATSDR (2017) Case studies in environmental medicine. Lead toxicity. Course WB2832, https://www.atsdr.cdc.gov/csem/csem.asp?csem=34&po=8

Balatsky AV, Balatsky GI, Borysov SS (2015) Resource demand growth and sustainability due to increased world consumption. Sustainability 7:3430–3440

Balmford A (2013) Pollution, politics, and vultures. Science 339:653–654

Bartlett AA (1998) Reflections on sustainability, population growth, and the environment. Renew Resour J 15:6–23

Bernanke J, Kohller HG (2009) The impact of environmental chemicals on wildlife vertebrates. Rev Environ Contam Toxicol 198:1–47

Bernhardt ES, Rosi EJ, Gessner MO (2017) Synthetic chemicals as agents of global change. Front Ecol Environ 15:84–90

Bloodworth A, Gunn G (2012) The future of the global minerals and metals sector: issues and challenges out to 2050. Geosciences 15:90–97

Blus LJ (2011) DDT, DDD and DDE in birds. In: Beyer WN, Meador JP (eds) Environmental contaminants in biota: interpreting tissue concentrations. CRC Press, Taylor & Francis, Boca Raton, pp 425–447

Bond DPG, Grasby SE (2016) On the causes of mass extinctions. Palaeogeogr Palaeoclimatol Palaeoecol 478:3–29

Bongaarts J (2009) Human population growth and the demographic transition. Philos Trans R Soc B 364:2985–2990

Borg K, Wanntrop H, Erne K, Hanko E (1969) Alkyl mercury poisoning in terrestrial Swedish wildlife. Viltrevy 6:301–379

Bouwman H, Bornman R, van den Berg H, Kylin H (2013) DDT: fifty years since *Silent Spring*. In: Late lessons from early warnings: science, precaution, innovation, Chapter 11. Environment and

Health Environmental Scenarios. EEA Report No 1/2013, Environmental European Agency, Luxembourg, pp 240–259
Bradshaw CJA, Brook BW (2014) Human population reduction is not a quick fix for environmental problems. PNAS 111:16610–16615
Brondizio ES, O'Brien K, Bai X, Biermann F, Steffen W, Berkhout F et al (2016) Re-conceptualizing the Anthropocene: a call for collaboration. Glob Environ Chang 39:318–327
Brundtland GH (1987) Our common future. World Commission on Environment and Development, Oxford University Press, Oxford
Candelone JP, Hong S, Pellone C, Boutron C (1995) Post-industrial revolution changes in large-scale atmospheric pollution of the Northern Hemisphere by heavy metals as documented in central Greenland snow and ice. J Geophys Res 100(D8):16605–16616
Carson R (1962) Silent spring. Houghton Mifflin, Boston
Carvalho FP (2017) Pesticides, environment, and food safety. Food Energy Secur 6:48–60
Castello MJ (2015) Biodiversity: the known, unknown, and rates of extinction. Curr Biol 25:R362–R383
Chellan P, Sadler PJ (2015) The elements of life and medicines. Philos Trans A Math Phys Eng Sci 373:20140182
Clemencon R (2012) Welcome to the anthropocene: Rio+20 and the meaning of sustainable development. J Environ Dev 21:311–338
Cohen JE (2010) Beyond population: everyone counts in development. CGD Working Paper 220, Washington, DC Center for Global Development, http://www.cgdev.org/content/publications/detail/1424318
Corlett LT (2015) The Anthropocene concept in ecology and conservation. Trends Ecol Evol 30:36–41
Cox C (1991) Pesticides and birds: from DDT to today's poisons. J Pestic Reform 11:2–6
Crutzen PJ, Stoermer EF (2000) The "Anthropocene". Glob Change Newsl 41:17–18
Davidson AJ, Binks SP, Gediga J (2016) Lead industry life cycle studies: environmental impact and life cycle assessment of lead battery and architectural sheet production. Int J Life Cycle Assess 21:1624–1636
de Sherbinin A, Carr D, Cassels S, Jiang L (2007) Population and environment. Annu Rev Environ Resour 32:345–373
Ding L, Liu Z, Aggrey MO, Li C, Chen J, Tong L (2015) Nanotoxicity: the toxicity research progress of metal and metal-containing nanoparticles. Mini Rev Med Chem 15:529–542
Ewing B, Reed A, Galli A, Kitzes J, Wackernagel M (2010) Calculation methodology for the national footprint accounts, 2010 edition. Global Footprint Network, Oakland
Fagerberg J, Srholec M (2017) Capabilities, economic development, sustainability. Camb J Econ 41:905–926
Franson JC, Pain D (2011) Lead in birds. In: Beyer WN, Meador JP (eds) Environmental contaminants in biota: interpreting tissue concentrations. CRC Press, Boca Raton, pp 563–593
Gall JE, Boyd RS, Rajakaruna N (2015) Transfer of heavy metals through terrestrial food webs: a review. Environ Monit Assess 187:201
Galloway JN, Schlesinger WH, Clark CM, Grimm NB, Jackson RB, Law BE et al (2014) Biogeochemical cycles. In: Melillo JM, Richmond TC, Yohe GW (eds) Climate Change Impacts in the United States: The Third National Climate Assessment, US Global Change Research Program, pp 350–368
Gardner G, Prugh T, Renner M (2015) State of the world 2015: confronting hidden threats to sustainability. Worldwatch Institute, Washington
Gaston KJ (2005) Biodiversity and extinction: species and people. Prog Phys Geogr 29:239–247
Giese B, Klaessig F, Park B, Kaegi R, Steinfeldt M, Wigger H et al (2018) Risks, release and concentrations of engineered nanomaterial in the environment. Sci Rep 8:1565
Gong X, Yang S, Zhang M (2017) Not only health: environmental pollution disasters and political trust. Sustainability 9:575

Gorman HS, Conway EM (2005) Monitoring the environment: taking a historical perspective. Environ Monit Assess 106:1–10
Goulson D (2014) Pesticides linked to bird declines. Nature 511:295–296. https://doi.org/10.1038/nature13642
Halada K, Shimada M, Ijima K (2008) Forecasting of the consumption of metals up to 2050. Mater Trans 49:402–410
Hayes TB, Hansen M (2017) From silent spring to silent night: agrochemicals and the Anthropocene. Elem Sci Anth 5:57
Horowitz HM, Jacob DJ, Amos HM, Streets DG, Sunderland EM (2014) Historical mercury releases from commercial products: global environmental implications. Environ Sci Technol 48:10242–10250
Hylander LD, Meili M (2003) 500 years of mercury production: global annual inventory by region until 2000 and associated emissions. Sci Total Environ 304:13–27
Ikeda M, Watanabe T, Nakatsuka H, Moriguchi J, Sakuragi S, Ohashi F, Shimbo S (2015) Cadmium exposure in general populations in Japan: a review. Food Saf 3:118–135
Inshakova E, Inshakov O (2017) World market for nanomaterials: structure and trends. MATEC Web of Conferences 129:02013
Klee RJ, Graedel TE (2004) Elemental cycles: a status report on human or natural dominance. Annu Rev Environ Resour 29:69–107
Köhler HR, Triebskorn R (2013) Wildlife ecotoxicology of pesticides: can we track effects to the population level and beyond? Science 341:759–765
Kuklinska K, Wolska L, Namiesnik J (2015) Air quality policy in the U.S. and the EU – a review. Atmos Pollut Res 6:129–137
Ma WC (2011) Lead in mammals. In: Beyer WN, Meador JP (eds) Environmental contaminants in biota: interpreting tissue concentrations. CRC Press, Boca Raton, pp 595–607
Mackenzie FT, Chris S (1993) C, N, P, S biogeochemical cycles and global change. In: Wollast R, Mackenzie FT, Chou L (eds) Interactions of C, N, P and S biogeochemical cycles and global change. Springer, New York, pp 1–62
McCauley DJ, Pinsky ML, Palumbi SR, Estes JA, Joyce FH, Warner RR (2015) Marine defaunation: animal loss in the global ocean. Science 347:1255641
McKee JK, Sciulli PW, Fooce CD, Waite TA (2004) Forecasting biodiversity threats due to human population growth. Biol Conserv 115:61–164
Meinert LD, Gilpin R, Robinson GR, Nassar NT (2016) Mineral resources: reserves, peak production and the future. Resources 5:14. https://doi.org/10.3390/resources5010014
Mikulewicz M, Chojnacka K, Kawala B, Gredes T (2017) Trace elements in living systems: from beneficial to toxic effects. Biomed Res Int 2017:8297814
Mitra A, Chatterjee C, Mandal FB (2011) Synthetic chemical pesticides and their effects on birds. Res J Environ Toxicol 5:81–96
Mohr S, Giurco D, Retamal M, Mason L, Mudd G (2018) Global projection of lead-zinc supply from known resources. Resources 7:17
Monastersky R (2015) The human age. Nature 519:145–147
Mosa A, Duffin J (2017) The interwoven history of mercury poisoning in Ontario and Japan. CMAJ 189:E213–E215
Nielsen FH (1984) Fluoride, vanadium, nickel, arsenic, and silicon in total parenteral nutrition. Bull NY Acad Med 60:177–195
Nielsen FH (1998) Ultratrace elements in nutrition: current knowledge and speculation. J Trace Elem Exp Med 11:251–274
Nishijo M, Nakagawa H, Suwazono Y, Nogawa K, Kido T (2017) Causes of death in patients with Itai-itai disease suffering from severe chronic cadmium poisoning: a nested case–control analysis of a follow-up study in Japan. BMJ Open 7:e015694
Norwood BM, Forbes JM, Harris JCO (1951) Iron and steel. In: Bureau of mines, minerals yearbook, pp 696–714
Nott MP, Rogers E, Pimm S (1995) Modern extinctions inthe kilo-death range. Curr Biol 5:14–17

Nriagu JO, Pacyna J (1988) Quantitative assessment of worldwide contamination of air, water and soil by trace metals. Nature 333:134–139

Obrist D, Kirk JL, Zhang L, Sunderland EM, Jiskra M, Selin NE (2018) A review of global environmental mercury processes in response to human and natural perturbations: changes of emissions, climate, and land use. Ambio 47:116–140

OECD (2011) Celebrating 40 years of the OECD environment policy committee (1971–2011). OECD, http://www.oecd.org/env/48943696.pdf

Olsson P, Moore ML, Westley FR, McCarthy DDP (2017) The concept of the Anthropocene as a game-changer: a new context for social innovation and transformations to sustainability. Ecol Soc 22:31

Peralta-Videa JR, Zhao L, Lopez-Moreno ML, de la Rosa G, Hong J, Gardea-Torresdey JL (2011) Nanomaterials and the environment: a review for the biennium 2008–2010. J Hazard Mater 186:1–15

Pimm SL, Jenkins CN, Abell R, Brooks TM, Gittleman JL, Joppa LN et al (2014) The biodiversity of species and their rates of extinction, distribution, and protection. Science 344:1246752

Racki G, Rakocinski M, Marynowski L, Wignall P (2018) Mercury enrichments and the Frasnian-Famennian biotic crisis: a volcanic trigger proved? Geology 46(6):543–546. https://doi.org/10.1130/G40233.1

Rauch JN (2009) Global mapping of Al, Cu, Fe, and Zn in-use stocks and in-ground resources. PNAS 106:18920–18925

Rauch JN, Pacyna JM (2009) Earth's global Ag, Al, Cr, Cu, Fe, Ni, Pb, and Zn cycles. Global Biogeochem Cycles 23:GB2001

Ray PC, Yu H, Fu PP (2009) Toxicity and environmental risks of nanomaterials: challenges and future needs. J Environ Sci Health C Environ Carcinog Ecotoxicol Rev 27:1–35

Ripple WJ, Wolf C, Newsome TM, Galetti M, Alamgir M, Crist E et al (2017) World scientists' warning to humanity: a second notice. BioScience 67:1026–1028

Salata OV (2004) Applications of nanoparticles in biology and medicine. J Nanobiotechnol 2:3

Sen IS, Peucker-Ehrenbrink B (2012) Anthropogenic disturbance of element cycles at the Earth's surface. Environ Sci Technol 46:8601–8609

Sodhi NS, Bickford D, Diesmos AC, Lee TM, Koh LP, Brook BW et al (2008) Measuring the meltdown: drivers of global amphibian extinction and decline. PLoS One 3:e1636

Soetan K, Olaiya CO, Oyewole OE (2010) The importance of mineral elements for humans, domestic animals and plants: a review. Afr J Food Sci 4:200–222

Steffen W, Persson A, Deutsch L, Zalasiewicz J, Williams M, Richardson K et al (2011) The anthropocene: from global change to planetary stewardship. Ambio 40:739–761

Stokes LC, Giang A, Selin NE (2016) Splitting the south: China and India's divergence in international environmental negotiations. Glob Environ Polit 16:12–31

Strode S, Lyatt Jaegle L, Selin NE (2009) Impact of mercury emissions from historic gold and silver mining: global modeling. Atmos Environ 43:2012–2017

Syvitski JPM (2012) Anthropocene: an epoch of our making. Glob Chang 78:12–15

Thiéry A, De Jong L, Issartel J, Moreau X, Saez G, Barthélémy P et al (2012) Effects of metallic and metal oxide nanoparticles in aquatic and terrestrial food chains: biomarkers responses in invertebrates and bacteria. Int J Nanotechnol 9:181–203

Tian HZ, Zhu CY, Gao JJ, Cheng K, Hao JM, Wang K et al (2015) Quantitative assessment of atmospheric emissions of toxic heavy metals from anthropogenic sources in China: historical trend, spatial distribution, uncertainties, and control policies. Atmos Chem Phys 15:10127–10147

Tong S, von Schirnding YE, Prapamontol T (2000) Environmental lead exposure: a public health problem of global dimensions. Bull WHO 78:1068–1077

UCS (1992) World Scientists' Warning to Humanity. Union of Concerned Scientists, https://www.ucsusa.org

UN Environment (2017) Global mercury supply, trade and demand. UN Environment – Economy Division Chemicals and Health Branch, Geneva, p 96. www.unep.org/chemicalsandwaste/resources/publications
UN WPP (2017) World Population Prospects: The 2017 Revision, Key Findings and Advance Tables. United Nations, Department of Economic and Social Affairs, Population Division, Working Paper No. ESA/P/WP/248, p 53
UNEP (2012) One planet, how many people? A review of Earth's carrying capacity. A discussion paper for the year of RIO+20
UNEP/FAO UN (1991) Decision guidance documents. DDT. United Nations Environment Programme & Food and Agriculture Organization of the United Nations, Rome – Geneva, http://www.pic.int/Portals/5/DGDs/DGD_DDT_EN.pdf
United Nations (2004) World population to 2300. UN Department of Economic and Social Affairs, Population Division, p 240
US GS (2011) Mineral commodity summaries 2011: US Geological Survey, US Department of the Interior
US GS (2016) Iron and steel. In: Mineral commodity summaries 2016. US Geological Survey, US Department of the Interior
van der Pluijm B (2014) Hello Anthropocene, goodbye holocene. Earth's Future 2:566–568
van der Voet E, Salminen R, Eckelman M, Norgate T, Mudd G, Hisschier R et al (2013) Environmental challenges of anthropogenic metals flows and cycles. United Nations Environment Programme, p 235
Viswanath B, Kim S (2016) Influence of nanotoxicity on human health and environment: the alternative strategies. In: de Voogt P (ed) Rev Environ Contam Toxicol 242:61–104
Wang YP, Law RM, Pak B (2010) A global model of carbon, nitrogen and phosphorus cycles for the terrestrial biosphere. Biogeosciences 7:2261–2282
Webb TJ, Mindel BL (2015) Global patterns of extinction risk in marine and non-marine systems. Curr Biol 25:506–511 [Probably a 9-fold lower marine extinction rate of species is currently observed in comparison to non-marine systems. On average between 20% and 25% of species worldwide are threatened with extinction]
White PJ, Brown PH (2010) Plant nutrition for sustainable development and global health. Ann Bot 105:1073–1080
Whitmee S, Haines A, Beyrer C, Boltz F, Capon AG, de Souza Dias BF et al (2015) Safeguarding human health in the Anthropocene epoch: report of The Rockefeller Foundation-Lancet Commission on planetary health. Lancet 386:1973–2028
WHO (1996) Trace elements in human nutrition and health. World Health Organization, Geneva

# Chapter 2
# Endothermic Animals as Biomonitors of Terrestrial Environments

**Elżbieta Kalisińska**

**Abstract** Since the late 1980s, wildlife toxicology has grown considerably as an important field of laboratory and field research. It focuses on the effects of various chemicals on the reproduction, health, and well-being of wildlife, including essential and nonessential elements. Deficiency of essential elements (e.g., copper, manganese, nickel, zinc, selenium) can lead to adverse effects in endothermic vertebrates, while their excess may result in significant intoxication or even death. However, the greatest concern is the contamination with highly toxic nonessential elements such as mercury, lead, cadmium, and arsenic.

Human activity results in the introduction of large amounts of essential and nonessential trace elements into biogeochemical cycles. Particularly exposed to excessive levels of trace elements are top avian and mammalian predators at the end point of biological pathways along which contaminants may accumulate in increasing concentrations. Determinations of trace elements in samples from selected species serving as biomonitors can be used to indirectly assess the condition of terrestrial ecosystems, including herbivorous, omnivorous, and predatory wildlife. Biomonitors are usually native species common in the area (involving hunted animals) but also invasive species (in Europe American mink and raccoon from North America; in the USA and Canada wild boar and common starling from Europe). Biomonitoring using terrestrial birds and mammals can be local, regional, or continental and is well developed in many countries of the Northern Hemisphere, especially in North America and Europe.

E. Kalisińska (✉)
Department of Biology and Medical Parasitology, Pomeranian Medical University in Szczecin, Szczecin, Poland
e-mail: ekalist@pum.edu.pl



E. Kalisińska (ed.), *Mammals and Birds as Bioindicators of Trace Element Contaminations in Terrestrial Environments*,
https://doi.org/10.1007/978-3-030-00121-6_2

# 1 Introduction

Since the 1950s, the dynamic growth of the world's population has been accompanied by growing levels of environmental pollution, resulting in a multifaceted destruction of natural biological systems, as observed both locally and globally. This has been accompanied by an increase in interest in environmental phenomena and developments in ecology, including ecotoxicology. Research has shown that measurements of the levels of various substances in air, water, and soil from chemical monitoring of the environment are still not sufficient to properly assess the health hazards to humans, animals, plants, and entire ecosystems. Currently, the use of living or deceased organisms in biotesting, bioindication, and biomonitoring is an established method of determining inorganic and organic contaminants and pollutions (Burger 2006a; Bealey et al. 2008; Holt and Miller 2011; Markert 2013).

The terms "bioindicator," "biomonitor," "bioaccumulator," and "biomarker" have all been used in varying ways to describe different approaches and techniques for studying biological responses to pollution of the air and other environmental components. In ecological and environmental sciences, the terms "biomonitor" and "bioindicator" have been and still are used interchangeably, and the terminology in this area can be ambiguous. However, since the early 1990s, we may observe a certain distinction in the use of these terms (Burger 2006a, b; Wilkomirski 2013; Sidding et al. 2016). Biomonitoring can include both a qualitative (bioindicator) and quantitative (biomonitor) approach in pollution control. For example, chemical analysis of biomonitors (an organism or its part or a community of organisms) contains information on the quantitative aspects of quality of the environment. A biomonitor is also a bioindicator, except that it quantifies the impact or eventual outcome on an organism or ecosystem and their health (O'Brien et al. 1993; Markert et al. 2003; Burger 2006a; Bealey et al. 2008). Large-scale biomonitoring uses plant and animal bioaccumulators, or organisms that accumulate various chemicals (including contaminants) in the tissues. Bioaccumulation is result of the biological sequestering of many substances often at a higher concentrations than that at which they occur in the surrounding environment or/and in food of animals.

# 2 Trace Elements

Some elements present in inorganic/organic forms in organisms are essential elements and others nonessential. In biochemistry, an essential trace element (or micronutrient) is a dietary mineral that is needed in very minute quantities (expressed in micrograms or milligrams) for the proper growth, development, and physiology of the organism. In humans the requirement per day is below 100 mg, with a deficiency leading to disorders that may even prove fatal.

In endothermic vertebrates, the biochemical functions of essential trace elements include enzyme activity, transport of oxygen (iron and copper), organization and

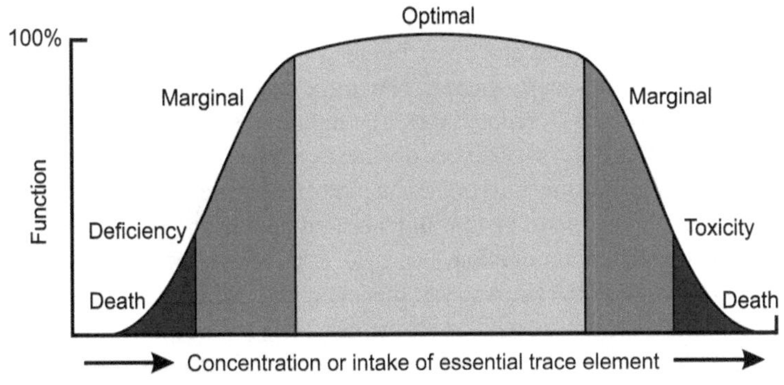


**Fig. 2.1** Dependence of biologic function on the tissue concentration of essential trace elements (according to Aras and Ataman 2006)

structure of macromolecules, vitamin activity (cobalt and vitamin $B_{12}$), or hormonal activity, e.g., iodine and thyroid hormones (Taylor 1996). All essential elements may even be toxic in animals and humans if ingested at sufficiently high levels and for a long enough period (Fig. 2.1). This aspect has been well recognized in humans, domesticated and laboratory animals, yet very poorly in wildlife (WHO 1973; Wada 2004; NRC 2005; Aras and Atman 2006; Lopez-Alonso 2012; Yatoo et al. 2013; Prashanth et al. 2015; Bhattacharya et al. 2016).

According to the National Research Council (NRC 2005), heavy metals (accepting as a criterion a density of at least 5.0 g $cm^{-3}$) such as cobalt (Co), chromium (Cr), copper (Cu), iron (Fe), manganese (Mn), molybdenum (Mo), and zinc (Zn) and two nonmetal elements (iodine (I) and selenium (Se)) are considered essential trace elements for higher animals. NRC (2005) classified nickel (Ni) as a possible essential element, while in the case of fluorine (F), which is nonmetallic, opinions are divided. Some researchers believe that F is an essential element for animals and humans while others consider the available evidence for indispensability to be insufficient. Elements essential to domestic, laboratory, or wild mammals and birds may not be essential to humans and vice versa.

In the document by the WHO (2002), the following trace elements are described as essential for human health: Cu, Zn, Fe, Cr, Mo, Se, Co, and I (the list includes Cr but its status as an essential element is controversial, where CrIII is beneficial for animals and humans but CrVI is a human carcinogen). The next smaller group is composed of Si (silicon), Mn, Ni, B (boron), and V (vanadium) with those elements classified as probably essential elements for humans. There are some differences between the NRC (2005) and WHO (2002) reports concerning the essentiality of elements for animals and humans, and the discussion on essentiality of some of those elements is still open (Aras and Ataman 2006; Bhattacharya et al. 2016; Maret 2016).

The term "trace element" is also used in analytical chemistry and geochemistry. In analytical chemistry it is an element whose average concentration is less than 100 mg $kg^{-1}$ (<100 ppm) but in geochemistry it is less than 1000 mg $kg^{-1}$

(<1000 ppm) or 0.1% of a rock's composition. Elements from mineral deposits are activated as a result of natural processes, but their contribution to the biogeochemical cycles is very much driven by human economic activity, especially over the last 100–150 years (Klee and Graedel 2004). The natural distribution and concentration of elements in the Earth's crust are very diverse as a result of the geological structure, but they are subject to strong anthropogenic modification having a significant impact on the mineral composition of plants, animals, and people and consequently on their condition, health, and reproduction (Adriano 2001; Yaroshevsky 2006; Steinnes 2009; Kabata-Pendias 2011). Among the elements, which can occur in living creatures, special attention is paid to all the essential and some of the nonessential trace elements, the latter having no biological function. For a long time, the greatest concerns have been triggered by heavy metals which are highly toxic to endothermic vertebrates, such as cadmium (Cd), lead (Pb), mercury (Hg), and arsenic (As), the latter being metalloids (Nriagu and Pacyna 1988; Adriano 2001; Anke et al. 2001; Jarup 2003; Hubner et al. 2010; Tchounwou et al. 2012). Additionally, for several decades, there has been an increasing interest on other metallic xenobiotics such as silver (Ag) and aluminum (Al) due to their high neurotoxicity and increasing distribution in the environment, including man-made Ag and Al nanoparticles (Ray et al. 2010; Stensberg et al. 2011; Jaishankar et al. 2014; Karmakar et al. 2014). Table 2.1 summarizes the concentrations and densities of selected trace elements found in the Earth's crust, which in varying amounts accumulate in wild-living endothermic vertebrates (Selinus and Finkelman 2011). Some of them are classified as essential elements and some as nonessential elements, all of which are discussed in more detail in the respective chapters of this book. In the case of wildlife, the ranges of mean concentrations of elements are shown for the liver, because this organ plays a key role in trace element regulation, bioaccumulation, and detoxification (Vikøren et al. 2005; Horai et al. 2006; Neuschwander-Tetri 2007). In addition, the liver is one of the most extensively used biological materials in ecotoxicological studies for quantifying trends in medium- to long-term contaminant exposure, with most data in this field existing from hepatic tissue (Burger et al. 2000; Gamberg et al. 2005; Braune and Malone 2006; Taggart et al. 2006; Vikøren et al. 2011; Gall et al. 2015; Espin et al. 2016; Kitowski et al. 2017).

In the group of essential elements, mean concentrations in the liver may in some cases reach one (Cr, Cu, Mo, Zn) or three (Se) orders of magnitude higher than their level in the Earth's crust. Liver concentrations of nonessential and highly toxic elements such as Cd, Hg, Ag may also be three orders of magnitude higher than in the crust. In the case of As and Sn, average hepatic concentrations may exceed those in the crust by two or one order of magnitude, respectively.

Eukaryotic organisms, including vertebrates, have evolutionarily developed mechanisms that enable them to maintain a proper level of various essential trace elements and homeostasis (Zhang and Gladyshev 2010). Terrestrial vertebrates via physiological and anatomical means have regulated and/or stored essential elements, including heavy metals up to certain exposure levels such that metals may not be present in their bodies in a concentration, form, or place that can result in a toxic effect. In such regulation, the gastrointestinal tract and the liver play crucial roles in

**Table 2.1** Density and concentration of selected elements in the upper continental crust (according to Yaroshevsky 2006), their biochemical status (NRC 2005), and range of mean hepatic levels in wild endothermic animals

| Element | Density ($g\ cm^{-3}$) | Crust concentration | Biochemical status and range of mean hepatic concentration (ppm dw) | | | No. of chapters in this book |
|---|---|---|---|---|---|---|
| | | | Essential | Probably essential | Non-essential | |
| Ag, silver | 10.49 | 0.07 ppm | | | <DL–44.0 | 18 |
| Al, aluminum | 2.70 | 8.05% | | | 0.20–25.35 | 12 |
| As, arsenic | 5.73 | 1.70 ppm | | | 0.05–122.6 | 13 |
| Cd, cadmium | 8.65 | 0.13 ppm | | | 0.05–163.2 | 14 |
| Cr, chromium | 7.14 | 83.0 ppm | 0.05–150.0 | | | 3 |
| Cu, copper | 8.92 | 47.0 ppm | 10–790 | | | 4 |
| F, fluorine | 1.70[a] | 660 ppm | | | <2.6–270[b] | 15 |
| Fe, iron | 7.87 | 4.65% | <100–4920 | | | 6 |
| Hg, mercury | 13.53 | 0.08 ppm | | | <DL–35 | 17 |
| I, iodine | 4.94 | 0.40 ppm | 0.17–0.35[c] | | | 5 |
| Mn, manganese | 7.47 | 1000 ppm | 2.20–38.3 | | | 7 |
| Mo, molybdenum | 10.28 | 1.10 ppm | 0.9–13.0 | | | 8 |
| Ni, nickel | 8.91 | 58.0 ppm | | 0.01–13.0 | | 9 |
| Pb, lead | 11.34 | 16.0 ppm | | | 0.30–48.0 | 16 |
| Se, selenium | 4.82 | 0.05 ppm | <0.20–23.0 | | | 10 |
| Sn, tin | 7.31 | 2.50 ppm | | | <0.01–15.0 | 19 |
| Zn, zinc | 7.14 | 83.0 ppm | 21.0–297.4 | | | 11 |

*ppm* parts per million or $mg\ kg^{-1}$, *dw* dry weight, *DL* detection limit

[a]Density of F in $g\ L^{-1}$

[b]In literature no data on wild animals was found, just fluoride concentrations in the livers of control laboratory rat and fatal human cases because 99% of fluorine is retained in bones and teeth (Inkielewicz and Krechniak 2003; NRC 2005; Martinez et al. 2007)

[c]Data available for farm animals only

the uptake and transport of cations (e.g., Cu, Fe, Zn). The anionic group such as Mo and Se is more water-soluble and is less reactive with nitrogen, sulfur, phosphorus, and oxygen, as well as hydroxide groups, than are cations. They are absorbed very efficiently through the intestine. In general, total body burden is regulated by renal excretion (WHO 1996; Rutherford and Bird 2004; US EPA 2007; Lopez-Alonso 2012; Sakulsak 2012; Kim and Oh 2013). Toxic elements strongly affect some essential element metabolisms because they compete for binders for these elements in the biological system. Concentrations of various essential and nonessential elements in birds and mammals depend on many factors and processes, including their forms, oxidation state, and the amount in habitats; biotransformation, bioavailability, diet, and position in the food chains of endothermic vertebrates; absorption (in which the intestinal route is the most important); and the duration of exposure (Chapman 1996; Adriano 2001; Martelli et al. 2006; Diaz-Bone and van de Wiele 2010; García-Barrera et al. 2012; Bhargava and Bhargava 2013). Heavy metals (both essential and nonessential) and metalloids (such as Se, As) in wildlife are the most often analyzed pollutants (Burger 2006b; Jaishankar et al. 2014; Stankovic et al. 2014; Gall et al. 2015; Espin et al. 2016), wherein pollution is defined as contamination that does or can result in adverse biological effects to resident communities. All pollutants are contaminants (substances which are present at places where they should not be or at concentrations above background), but not all contaminants are pollutants (Chapman 2007). Unlike plants and lichens, domestic and wild animals do not usually show qualitative morphological and/or physiological changes as a consequence of chronic bioabsorption of trace elements and the undesirable effects caused by them, which would allow these animals to be considered as bioindicators. There are only a few examples in this field from areas where a natural excess of these elements is noticed. These include loss of hair and malformations of hooves as a result of excessive selenium in food sources and dental fluorosis as a result of a high uptake of fluoride dissolved in groundwater (James and Shupe 1984; Al-Dissi et al. 2011; Choubisa et al. 2011). Wildlife is very often used as a biomonitor where they chronically bioaccumulate trace elements and other substances, but the reaction of the animals to them are generally invisible. These substances qualitatively and quantitatively can be assayed in laboratories using highly specialized and sensitive equipment, from samples of the appropriate biota (Markert et al. 2003).

## 3 Terrestrial Endothermic Vertebrates as Biomonitors

Since the 1970s there has been a steady and dynamic growth in research and implementation of biomonitoring programs that use organisms from various taxonomic groups as biomonitors of environmental pollution. Of the terrestrial endothermic vertebrates, mainly wild animals but sometimes also breeding birds and mammals (including furbearers) are chosen (Wren 1984; O'Brien et al. 1993; López Alonso et al. 2002; Golden and Rattner 2003; Tataruch and Kierdorf 2003; Ji et al.

2006; Wolfe et al. 2007; Rabinowitz et al. 2009; Reis et al. 2010; Rajaganapathy et al. 2011; Kalisinska et al. 2012a). The consequence of this has been an increase in the number of reports in this field concerning wildlife. Of particular interest are persistent organic pollutants (including organochlorine pesticides and polychlorinated biphenyls (PCBs)), but much attention is also devoted to trace elements (Golden and Rattner 2003; Markert et al. 2003, 2008; Stolen et al. 2005; Burger 2006b; Hollamby et al. 2006; Holt and Miller 2011). Warm-blooded biomonitors can be used for information on:

- Essential and nonessential element concentrations and relations between them in selected species (especially in rare and threatened birds and mammals, common species, including game animals, which are used by humans as food and potentially valuable source of minerals but may also contained toxic contaminants)
- Concentrations and bioavailability of essential elements in an area of interest important for the detection of their deficiency or excess and which may be referenced in proper steps in the management and health protection both in animals and humans (e.g., level of selenium is significant in protection against mercury toxicity)
- Bioindicators can be used as information about various temporal and spatial changes occurring in a specific area (including those from anthropogenic and natural sources such as atmospheric deposition, floods), especially in the case of xenobiotic metals and metalloids
- Differences in trace element concentrations among animals from the same area (or from control and contaminated sites) and the various trophic levels
- Ecotoxicological situation of selected species which are widely distributed in various provinces, states, countries, and even continents

Depending on the purpose of the research or the biomonitoring program being implemented, one or several of the above points may be taken into account, but there may also be others not mentioned above (Talmage and Walton 1991; Stolen et al. 2005; Burger 2006a; Smith et al. 2007; Zhang and Ma 2011; Garcia-Fernandez 2014; Espin et al. 2016; Herzke et al. 2017). Species that are targeted as candidates of bioindicators of trace elements should have desirable characteristics including:

- Are sensitive and indicative of change
- Broad distribution with accompanying data
- Easily measured and readily observable
- Well-known ecology and life history
- Suitable for lab studies
- Significant to humans
- Economical/cost-effective
- Well-developed and usable with existing data
- Common enough not to impact populations

Lists of characteristics may differ from one another to a certain point, and a potential or suitable terrestrial candidate may fulfill only some of the desired features (O'Brien et al. 1993; Hollamby et al. 2006; Espin et al. 2016; Herzke et al. 2017). It seems that the list of avian species is much longer than that of mammalian species. This is due not only to the greater number of bird species found in the world, which is almost twice as much as mammals (9993 and 5416 species, respectively), but also from a much larger and more active group of people professionally and amateurly researching and observing birds (ornithologists, volunteers, and bird-watchers) compared to the analogous "mammalian" group (Jones and Safi 2011; Jetz et al. 2012). The highest biodiversity of birds and mammals is recorded in tropical regions. Mammalian and avian fauna in Europe in comparison to other parts of the Northern Hemisphere is poorly diversified, with the avifauna of eastern Asia about 50% richer than Europe and North America, and Western North America the richest region with 14% and 44% more species than eastern Asia and Europe, respectively (Monkkonen and Viro 1997). However, wildlife researchers and observers mainly operate in Europe and North America, which are dominated by animal species of temperate and boreal biomes, with a much better knowledge of their biology. For example, in Europe about 270 mammalian species and 400 avian species are noticed, and in continental North America (USA and Canada) over 710 and 540 species, respectively (Leveque and Mounolou 2003; Arita et al. 2005; Thuiller et al. 2014; Sauer et al. 2017). For biomonitoring goals, much more numerous and more diverse avian samples and information on them than mammalian have been gathered. It concerns terrestrial wildlife too (Frank 1986; Ma et al. 1991; Furness and Greenwood 1993; Golden and Rattner 2003; Rattner et al. 2005; Burger 2006b; Smith et al. 2007; Schmeller et al. 2012; Carneiro et al. 2016; Sauer et al. 2017).

## *3.1 Mammals as Biomonitors*

In Europe, North America, and Asia (mainly in Korea and Japan), samples in which trace elements are determined usually come from several or a dozen selected species of wildlife found on those continents. Among inland mammals there are mainly representatives of the following animal groups: even-toed ungulates (ordo Artiodactyla), lagomorphs (ordo Lagomorpha including hares and rabbits), carnivores (ordo Carnivora), bats (ordo Chiroptera), and Micromammalia, which comprises both rodents (ordo Rodentia) and insectivores (ordo Insectivora). Many researchers prefer micromammals because of their frequent occurrence in the environment, small individual areas, relatively easy acquisition for research, and the possibility of comparison and/or verification of laboratory rodent species results. In addition, their small size makes it possible to assess trace element content in the whole body and an assessment of their transmission to predatory animals (Wren 1986; Ma et al. 1991; Talmage and Walton 1991; Shore and Douben 1994; Kramarova et al. 2005; Sánchez-Chardi et al. 2007; Wijnhoven et al. 2007; Mendez-Rodriguez and Alvarez-Castaneda 2014; Gall et al. 2015). In addition, micromammals are an important part of the diet of avian and mammalian predators and participate in the transmission of trace elements between the links of terrestrial food chains (Gall et al. 2015; Knopper et al. 2006; Herzke et al. 2017). However, the transformation of trace elements in these small mammals is poorly correlated with that occurring in humans and medium-sized long-lived mammals because micromammals have a much higher metabolic rate, usually a short life (1–2 years), and the samples taken, e.g., kidneys or brain, have very low mass, which may cause some analytical problems, including the risk of contamination of the research material (Speakman 2005; Wijnhoven et al. 2007).

Trace elements in the environment generally occur in low concentrations (including highly toxic metals), but their impact on long-lived organisms, including many animals and humans, lasts many years. In the indirect evaluation of their chronic impact on mammals, medium-sized carnivores have been used successfully such as canids (family Canidae: red fox (*Vulpes vulpes*), Arctic fox (*V. lagopus*), golden jackal (*Canis aureus*), and raccoon dog (*Nyctereutes procyonoides*)), mustelids (family Mustelidae: river otter (*Lontra canadensis*), Eurasian otter (*Lutra lutra*), American mink (*Neovison vison*) (previously *Mustela vison*), voloine (*Gulo gulo*), European badger (*Meles meles*), and martens among others), and raccoon (*Procyon lotor*) belonging to family Procyonidae (Wren 1984, Van den Brink and Ma 1998; Lord et al. 2002; Hoekstra et al. 2003; Millan et al. 2008; Heltai and Markov 2012; Kalisinska et al. 2016; Markov et al. 2016; Herzke et al. 2017). They are positioned on the top of the food pyramid, and their feed consists of field and forest rodents, hares, birds, seeds, fruits, or fish in various amounts in semiaquatic species (otters, American mink, raccoon). Many medium-sized carnivores are widely distributed in forest, agricultural, and urban landscapes of the Northern Hemisphere, with some species introduced into areas beyond their natural occurrence (Gehrt et al. 2011; Lesmeister et al. 2015; Poessel et al. 2017). For example, native North American raccoon and American mink are common as alien species in many European countries, while the raccoon dog present in Eastern and Central Europe originated from Asia (Genovesi et al. 2009). Fish-eating wildlife is particularly exposed to mercury biomethylated in water and sediments, and methylmercury product undergoing biomagnification in food chains. For this reason Hg achieves its highest concentrations in fish and piscivorous birds and mammals from the ends of food chains. In inland ecosystems fish-eating carnivores are preferred in studies on mercury contamination. Many reports concerning Hg (and sometimes other heavy metal levels) in American minks, river otters, and raccoons inhabiting North America have been published over the years (e.g., Wobeser and Swift 1976; Wren et al. 1980; Wren 1986; Lord et al. 2002; Wolfe et al. 2007; Sleeman et al. 2010; Basu 2012), but increasing numbers of European studies using American minks and raccoons have also been observed (Norheim et al. 1984; Kalisinska et al. 2012a, 2016, 2017; Brzezinski et al. 2014; Lanocha et al. 2014; Ljungvall et al. 2017).

In contrast to mesocarnivores, publications on trace element concentrations in large predatory Northern Hemisphere mammals, such as cats, are rare (e.g., Eurasian lynx (*Lynx lynx*), North American cougar (*Puma concolor*), bears, and wolves) due to their usually small population sizes, dispersion, and very large anthropogenic limitations of natural ranges, making it difficult to obtain biological samples from them and perform spatiotemporal comparisons (Gamberg and Braune 1999; Shore et al. 2001; Newman et al. 2004; Millan et al. 2008; Celechovska et al. 2006; Noel et al. 2014; Lazarus et al. 2017).

In ecotoxicology, herbivorous game mammals (especially deer; red deer (*Cervus elaphus*), roe deer (*Capreolus capreolus*), mule deer (*Odocoileus hemionus*), white-tailed deer (*O. virginianus*) among others), moose/elk (*Alces alces*), reindeer/caribou (*Rangifer tarandus*), and hares play an important role. Determination of trace elements in these species allows not only identification of the adverse effects connected with excess or deficiency of micronutrients in the animals themselves and on their populations, identification of the contamination of the food chains by nonessential elements, and estimation of human health risks (Adriano 2001; O'Hara et al. 2003; Tataruch and Kierdorf 2003; Vikøren et al. 2005; Mysłek and Kalisińska 2006; Pedersen and Lierhagen 2006; Kursa et al. 2010; Al-Dissi et al. 2011; Ertl et al. 2016). Venison (mainly muscle and to a lesser extent the liver and other offal) is consumed as an important source of protein and micronutrients, but when it contains elevated amounts of toxic elements (e.g., Pb incorporated in tissues), this may contribute to consumer intoxication (Strmiskova and Strmiska 1992; Borch-Iohnsen et al. 1996; Frank et al. 2000; Wolfe et al. 2010; Roug et al. 2015; Skibniewski et al. 2015; Ertl et al. 2016) and a threat to wild mammalian and avian raptors and scavengers, including threatened species (Rogers et al. 2012; Haig et al. 2014; Behmke et al. 2015; Arnemo et al. 2016; Herring et al. 2016).

Omnivorous animals occupy an intermediate trophic position between herbivorous and carnivorous mammals. In Eurasia, one of the most widespread hunted species in this group is wild boar *Sus scrofa*, the progenitor of the domestic pig, and is very often used in European ecotoxicological studies (Santiago et al. 1998; Kursa et al. 2010; Rudy 2010; Amici et al. 2012; Danieli et al. 2012; Dlugaszek and Kopczynski 2013; Gasparik et al. 2017). In North America, the wild boar (feral hog) is classified as an invasive rapidly spreading species and is now abundant in the south and southwest of the USA (Snow et al. 2017; McClure et al. 2018). Although it is a hunted animal in the USA and its meat is often consumed by people, its tissues are rarely tested for the presence of trace elements. Therefore, this type of data is very seldom used for indirect assessment of environmental pollution and consumer health exposure in North America (Oldenkamp 2016; Oldenkamp et al. 2017; Smith et al. 2018). The trophic chain position of the raccoon in North America and Europe is similar to that of the wild boar. In an effort to protect native fauna, the populations of these (and other) species are being deliberately reduced outside their natural ranges, so their tissues can be easily obtained for ecotoxicological studies and intercontinental comparisons. Selenium concentration comparisons in this aspect in omnivorous wild-living mammals seem particularly interesting. Selenium is an element with a very uneven distribution in the earth's crust. Much of Central and Northern Europe's soils are Se-deficient, while North American soils are generally rich in this microelement; in some areas its levels are even excessive. A comparison of Se concentrations in wild boar muscles from Europe (Czech Republic) and the USA (Georgia) indicates that Se levels in the European population are an order of magnitude smaller than in the USA, at 0.10 mg kg$^{-1}$ vs 1.0 mg kg$^{-1}$ dw (Kursa et al. 2010; Oldenkamp 2016). Considering that Se counteracts the absorption of Hg from the diet, areas with an elevated amount of Hg and food poor in Se (e.g., fish) would exhibit increased Hg intoxication of animals compared to individuals of the same species from areas of comparable Hg concentration but more abundant in Se. In relation to raccoon and American mink from Poland, such a suggestion was put forward by Kalisinska et al. (2017) after comparisons of data on Se and Hg in the muscles of these species in Europe (NW Poland) and North America.

## 3.2 Birds as Biomonitors

Avifauna, especially inland birds, is the longest (over 100 years) and the most intensively methodically observed group of animals in Europe and North America. In contrast, in Asia large-scale observations were initiated as late as the 1970s–1980s (Bibby 2003; Li and Mundkur 2006; Keck 2015). Various bird monitoring programs in Europe and North America, from local to pancontinental, have been introduced for at least 50 years, and some of them include pollution testing (Lambert et al. 2009; Schmeller et al. 2012; Gomez-Ramirez et al. 2014; Ahrestani et al. 2017; Sauer et al. 2017). There are many examples in the history of ecotoxicology where birds have been used as sentinels of environmental and human health. Canaries used to be taken to mines to indicate dangerous concentrations of methane. A dramatic reduction in the populations of birds of prey showed the dangers associated with the widespread use of pesticides in agriculture, including DDT (dichlorodiphenyltrichloroethane), organochlorine substances, and alkyl mercury compounds. The use of the latter, highly neurotoxic and undergoing biomagnification in the trophic chains, resulted in the considerable exposure of piscivorous wildlife and humans to mercury (Scheuhammer 2008; Rabinowitz et al. 2009; Rattner 2009; Basu 2012; Holt et al. 2012; Espin et al. 2016).

Yet another and very spectacular example is the impact of lead contained in hunting ammunition on the health and fitness of individual birds and its effects at the population level. Waterfowl such as ducks and geese (also some landfowls) are a unique group in this respect, because they swallow small pebbles as gastroliths, which are retained in the gizzard and used to grind food. However, the birds do not distinguish pebbles from spent lead shot pellets. Incidental mortality from waterfowl hunting reached population-level effects when over two million ducks and geese (~2% of all waterfowl) in North America were poisoned annually by ingestion of spent lead shot deposited on the grounds and in sediments (Bellrose 1951). Waterfowl, in addition to shot pellets, also swallow leaded fishing gear used in recreational fishing, which eventually also results in the intoxication of animals and people. In addition, waterfowls and other game animals may retain hunting ammunition in their bodies, which can then be swallowed by predators and scavengers. Thanks to numerous field observations of professionals, bird watchers, volunteers, and ecotoxicological research, the use of DDT and pesticides containing mercury was eventually banned in many developed countries (Smith et al. 2007; Espin et al. 2016; Movalli et al. 2017). In the USA, the use of lead pellets in waterfowl hunting was discontinued, as in a few European Union countries. The scientific arguments and the strong voice of the public resulted in a change of policy in the USA and Canada which used the prevalence of lead poisoning among birds as the basis of policy and law introduced to reduce lead use at the continental level, including leaded petrol (Thomas and Guitart 2010; Golden et al. 2016). However, the problem of metallic lead poisoning of rare, endangered birds and the so-called flagship species remains one of the most important in wildlife toxicology, because lead pellets scattered in the environment are still swallowed by waterfowl and landfowl, and lead bullets used in large-game hunts contaminate viscera (offal) left by hunters in the field (Pain et al. 2009; Haig et al. 2014; Espin et al. 2016; Herring et al. 2016). Tranel and Kimmel (2009), based on data from Minnesota (USA), estimated that among terrestrial vertebrates such as reptiles, mammals, and birds, lead ammunition had the greatest effect on birds (about 95%), mostly water birds (38%), raptors, and scavengers (24%). In this respect, the situation may be similar in other parts of the world where hunters use lead ammunition (Pain et al. 2009; Saito 2009; Nadjafzadeh et al. 2013; Golden et al. 2016). Another source of intoxication of birds and humans with lead are remnants of paint containing this metal and leaded gasoline (Nriagu 1990; Cai and Calisi 2016). Therefore, birds are also used in the biomonitoring of cities, e.g., urban pigeons (Ohi et al. 1981; Dauwe et al. 2005; Deng et al. 2007; Roux and Marra 2007; Behmke et al. 2015; Cai and Calisi 2016; Pollack et al. 2017).

In addition to a large number of studies on lead in birds, there is also a considerable body of research on mercury, especially in North America (Rattner et al. 2000, 2005). In inland ecosystems, exposure to mercury is the highest among piscivorous species, and in North America key research in this field includes common loon (*Gavia immer*), bald eagle (*Haliaeetus leucocephalus*), osprey (*Pandion haliaetus*), mergansers, and grebes (DesGranges et al. 1998; Scheuhammer et al. 1998; Stout and Trust 2002; Mierzykowski et al. 2011, 2013; Rutkiewicz et al. 2011; Shore et al. 2011; Depew et al. 2012; Schoch et al. 2014). There are also many studies on other aquatic birds, especially game waterfowls (Gerstenberger 2004; Rothschild and Duffy 2005; Braune and Malone 2006). For a long time, it was thought that mercury is only marginally accumulated in terrestrial songbirds. This view changed after the publication of Cristol et al. (2008) which showed that in areas historically anthropogenically contaminated with mercury it is transferred from the river (the South River, Virginia, USA) and riverside areas to arthropods (spiders and insects) and then songbirds feeding on them. This discovery inspired broader studies on songbirds as sentinels of mercury in terrestrial habitats (Jackson et al. 2015). The flagship species in European studies on mercury contamination is the white-tailed eagle (*Haliaeetus albicilla*) (Norheim and Frøslie 1978; Falandysz et al. 2001; Kenntner et al. 2001; Kalisinska et al. 2016; Kitowski et al. 2017), but observational studies show this species is not at risk of mercury intoxication, as opposed to lead. However, elevated amounts of this metal were found in some of the common merganser (*Mergus merganser*) wintering on the southern coast of the Baltic Sea (Kalisinska et al. 2010).

In ecotoxicological studies, elements are rarely determined in many types of samples from wild birds to characterize their distribution in the body. Such exceptions include two papers describing the distribution of Hg and trace elements in piscivorous great cormorants (*Phalacrocorax carbo*) (Nam et al. 2005; Misztal-Szkudlińska et al. 2018) and one report concerning Hg in young osprey (DesGranges et al. 1998). The muscles of an adult cormorant had the largest amount of Cu (>65%), a significant part of Hg and Cr (about 35%), as well as about 30% of Se and Co accumulated in the body (Nam et al. 2005). In osprey nestlings, about 85% of absorbed Hg gets to feathers during their growth, and from the remaining a dozen or so percent, half of them accumulate in muscles (DesGranges et al. 1998). The quoted works show that the highest amount of Hg in soft tissues is found in the muscles of birds, but its distribution is strongly influenced by the intense transfer of Hg to feathers during their growth.

The usefulness of various bird tissues to monitor the abundance of the environment with elements essential to life and its contamination with toxic metals is constantly discussed. Although the samples most frequently selected in biomonitoring include liver and kidneys, it is important to study their concentration in the muscles and target tissues because of the transfer of various elements up the terrestrial food chains. Interpretation of the obtained concentrations of elements in avian samples requires their reference to threshold values, as in the book by Beyer and Meador (2011) for Cd, Hg, Pb, and Se. However, most trace elements have not been researched in such a thorough fashion for very large and species-diverse clusters of birds. Due to this lack of data, certain reference may come from values calculated for unanalyzed tissues based on the known concentration in the examined tissues through the use of appropriate equations (when the concentrations between these tissues correlate with each other) (Mochizuki et al. 2011; Ackerman et al. 2016; Evers 2018).

### 3.3 *Tissues of Terrestrial Vertebrates Used in Biomonitoring*

In wildlife toxicology, various types of biological samples may be collected from live animals captured then released (mainly feathers, hair/fur, blood; less frequently fragments of claws or oil from the uropygial gland) or from dead birds and mammals (most of all internal tissues such as liver, kidney, muscle, bone, and brain, but also external tissues). Studies on environmental contaminants, including toxic trace elements, often use avian eggs, with one egg usually taken from individual broods, assaying contaminants in the eggshell, whole egg content, or white and yolk separately (Leonzio and Massi 1989; Burger and Gochfeld 2003; Hashmi et al. 2015; Ackerman et al. 2016; Orlowski et al. 2016; Movalli et al. 2017; Pollack et al. 2017). In addition, researchers often use feathers (e.g., in nests or nearby), hair, mammalian scat, cervid antlers, and avian pellets. Those biological materials are taken mainly from endangered, threatened, or sensitive species; such noninvasive sampling methods are recommended as valuable tools to monitor wildlife and minimally affect free-ranging animals. So-called "postlethal" animal samples are obtained from those already killed by hunters, trappers, museum collectors, or vehicles or found in the field (Kierdorf and Kierdorf 2003; Pokorny 2006; Pauli et al. 2010; Movalli et al. 2017; Trapp and Flaherty 2017).

Different trace elements are deposited in different wildlife tissues at different rates and amounts. The liver, kidney, muscle, and bone from internal tissues are major locations where the largest part of the absorbed essential trace elements are deposited, but concentrations in these tissues are not necessarily representative of the entire body burden, and it can be difficult to detect trace element deficiencies within critical organs (Taylor 1996; Demesko et al. 2018). Essential and nonessential trace elements in different tissues and organs may be subject to temporary or long-term accumulation in various body parts, biotransformation (including methylation and demethylation), and removal mainly with feces and urine, and to a small extent also with saliva, sweat, tears and respiration (Nollet et al. 2008; Lopez-Alonso 2012; Jan et al. 2015; Prashanth et al. 2015). Additional methods of metal and metalloid excretion in birds are eggs and feathers and in mammals the fur (Burger et al. 1993; Burger 1994; Leonzio et al. 2009; Rendón-Lugo et al. 2017). The organ or tissue in which trace metal/metalloid toxicity occurs may differ from the organ or tissue(s) where the element bioaccumulates, which may be connected with its kinetics. Target organs (where the toxic effects are produced) may differ between species of endothermic vertebrates, mainly owing to differences in absorption, distribution, and excretion (US EPA 2007). Table 2.2 presents the main target organs/tissues of nonessential elements and internal body parts where the elements achieve typically highest levels in terrestrial endothermic animals.

Among the internal tissues of wildlife, a number of essential and nonessential elements are predominantly measured in the liver and kidney; however, fluoride and lead are mainly investigated in the bones (Mateo et al. 2003; Demesko et al. 2018). For the past two to three decades, nondestructive samples (hair, feathers, and blood) have been preferred, which are often taken from birds and bats (Russo and Jones 2015; Pauli et al. 2010; Wada et al. 2010; Langner et al. 2012; Lodenius and Solonen 2013; Stankovic et al. 2014; Gall et al. 2015; Flache et al. 2015; Ackerman et al. 2016). Sampling of live animals does not reduce the population, which is important in the case of their small numbers, especially with regard to protected species, and such action is usually socially acceptable. It is estimated that plumage and mammalian pelts contain the largest part of methylmercury (MeHg) accumulated in the body. Therefore these tissues are frequently used in the detection of mercury exposure in wildlife, but many other heavy metals are also investigated in these keratin structures. Feathers (similar to hair) are metabolically inert after their formation, so for those avian species with well-known molt schedules, the analyses of specific individual feathers provide unique chemical information of a very discrete time. For many bird species, the molt schedules are poorly recognized, and metal concentrations in feathers are highly variable within an individual bird. Therefore, proper interpretation of chemical results is very difficult or impossible. For these reasons some researchers state that feathers and hair have a low priority as preferred tissues for sampling in ecotoxicological studies (Furness and Greenwood 1993; Leonzio et al. 2009; Ackerman et al. 2016; Rendón-Lugo et al. 2017).

**Table 2.2** Main target organ or tissue as well as internal body parts of terrestrial endothermic animals where nonessential trace elements are accumulated following chronic oral chronic

| Element | Target organ or tissue | Organ or tissue with typically highest concentration | References |
|---|---|---|---|
| Ag, silver | Probably brain and liver | Bones | Connors et al. (1972), Horai et al. (2006), and Kuo et al. (2000) |
| Al, aluminum | Brain and bone | Brain, liver | Llacuna et al. (1995), Krewski et al. (2007), Al-Ganzoury and El-Shaer (2008), and Lucia et al. (2010) |
| As, arsenic | Liver | Liver | Liu and Waalkes (2008), Sanchez-Virosta et al. (2015), and Mandal (2017) |
| Cd, cadmium | Kidneys | Kidneys, liver | Martelli et al. (2006) and Wayland and Scheuhammer (2011) |
| F, fluorine | Skeleton and kidneys | Bones | Bird et al. (1992), Tsunoda et al. (2005), and Kurdi (2016) |
| Hg, mercury | Kidneys for inorganic Hg; brain for organic Hg | Kidney, liver | Evers et al. (2005), Clarkson and Magos (2006), Bridges and Zalups (2010), and Sleeman et al. (2010) |
| Pb, lead | Nervous system, mainly brain | Bones | Silbergeld et al. (1993), Nemsadze et al. (2009), Franson and Pain (2011), Flora et al. (2012), and Kalisinska et al. (2016) |
| Sn, tin | Probably bones and liver | Bones, liver | Kannan and Falandysz (1997), Harding et al. (1998), Nath (2000), ATSDR (2005), and Mizukawa et al. (2009) |

Generally, metal levels in blood samples reflect short-term exposure (immediate dietary intake), the liver and kidney reflect longer terms, while the bones reflect the longest because their mineral remodeling occurs very slowly (Stankovic et al. 2014; Gall et al. 2015; Espin et al. 2016). Cadmium is bioaccumulated in bird and mammal kidneys almost over the entire lifetime, and a strong correlation between nephric Cd level and animal age is observed (Wayland and Scheuhammer 2011; Rendón-Lugo et al. 2017). Many trace elements achieve their highest concentrations in the liver and kidneys, (Table 2.2), but together these organs constitute no more than 4%–6% of the animal's body weight. The muscles (40%–50% of body weight) are most significant in the transfer of trace elements between animals from different trophic levels, depending on the type of consumer (Kalisinska et al. 2017). This is especially important in the case of Hg. The level of intestinal absorption of Hg in terrestrial vertebrates depends on its chemical form, and in animal muscle about 90% Hg is present as MeHg, which is almost completely absorbed from the digestive tract. Hg in the liver and kidney is mostly inorganic Hg with low intestinal absorption (<10%). In the kidneys and liver, the percentage of MeHg in total mercury (THg) can be small (especially when THg reaches high concentrations), which is why these organs play a small role in the transfer of Hg between animals. Unfortunately, few papers provide information about the absolute and relative weight of tissues and organs as well as the percentage composition of the consumer's diet, so it is difficult to estimate the amount of transfer of trace elements between different trophic levels. Among terrestrial birds and mammals, Hg concentration increases from herbivores to omnivores and carnivores, but in the case of other trace elements, this type of regularity is not always clearly determined (Tete et al. 2013; Stankovic et al. 2014; Kalisinska et al. 2017).

The most numerous group of ecotoxicological studies concerns a small group of trace elements ($<10$). They are dominated by toxic elements (Cd, Hg, Pb, As), usually analyzed in 1–3 types of biological samples. Publications in which several or dozens of elements were determined in samples obtained from terrestrial birds and mammals are much less numerous, but this has been made possible due to technical progress in chemical analysis (e.g., Harding et al. 1998; Falandysz et al. 2001; Horai et al. 2006; Deng et al. 2007; Dailey et al. 2008; Zimmerman et al. 2008; Ertl et al. 2016; Lazarus et al. 2017). In literature, data concerning trace elements in soft and hard tissues tend to be presented as mean wet/fresh or dry weight. In scientific studies, the diversity of samples and the multiplicity of the elements determined are subject to various comparisons and discussions. Then it is necessary to present concentrations of elements not only in the same units (mainly expressed as mg $kg^{-1}$, which is analogous to μg $g^{-1}$ or ppm) but also selecting dry or wet weight. Conversion of wet weight to dry weight (or vice versa) requires knowledge of the percentage of water in the samples, but such information is seldom presented in the reports. Furthermore, samples are dried at temperatures ranging from 50 °C to 105 °C (not always to constant weight), depending on the methodology and the analytical requirements. Therefore, various comparisons use the average percentage of water in vertebrate tissues (Ackerman et al. 2016; Zukal et al. 2015). For the purposes of this book, the average water content in the four most commonly analyzed tissues of birds and mammals was calculated using data from seven and ten species, respectively (data for birds were taken from Honda et al. 1985; Cosson et al. 1988: Kalisinska et al. 2010, 2014; Binkowski et al. 2013; for mammals from Weiner 1973; Reinoso et al. 1997; Blus and Henny 1990; Gamberg et al. 2005; Rudy 2010; Sleeman et al. 2010; Kalisinska et al. 2012a, b; Lanocha et al. 2014). Table 2.3 shows the average percentage of water in the tissues of birds and mammals and also proposed coefficients for wet to dry mass conversion.

When collecting samples from wild mammals and birds, it is advisable to obtain and record important information about them, including species, sex, age, location (latitude and longitude), and season/year. This kind of data is needed for intra- and

**Table 2.3** Mean moisture content in tissues of endothermic vertebrates and proposed conversion factors (CF) for normalization of wet weight assay results from tissue samples to dry weight

| Parameter | Liver | Kidney | Muscle | Brain |
|---|---|---|---|---|
| *Mammals* | | | | |
| Moisture in tissues (%) | 70.9 | 75.5 | 74.6 | 77.0 |
| CF | 3.0 | 2.5 | 2.5 | 2.0 |
| *Birds* | | | | |
| Moisture in tissues (%) | 70.2 | 74.3 | 71.4 | 79.9 |
| CF | 3.0 | 2.5 | 3.0 | 2.0 |

interspecies analyses of differences in the concentration of trace elements and may reveal time-spatial changes on a regional, continental, and even global scale (Tataruch and Kierdorf 2003; Hollamby et al. 2006; Burger 2007; Traas and van Leeuwen 2007; Zukal et al. 2015; Gochfeld 2017).

Depending on the assumed objective of research in wildlife toxicology, samples used in analysis may come from one or more species representing the same or different trophic categories (e.g., herbivores, omnivores, predators). Particularly important are studies analyzing the concentration of selected highly toxic elements (Hg, Pb, Cd) due to the range of research and the very large number of samples (sometimes exceeding 1000). For example, a study on Hg levels in North American birds analyzed blood samples of 102 songbird species from terrestrial habitats (Jackson et al. 2015). The review by Ackerman et al. (2016) compiled literature data on Hg in approximately 27,000 samples (eggs, blood, liver, muscle, and feathers) from 225 species of birds from various systematic groups found in western North America. They concluded that avian Hg concentrations were greatest in ocean and salt marsh habitats and lowest in terrestrial habitats. Their analysis identified multiple hotspots contaminated by the metal in the western part of North America. Finally, Jackson et al. (2016) studied Hg in the blood of 20 avian piscivorous species (including a few target species: bald eagle, osprey, common loon) and those species turned out to be much more exposed to Hg than non-piscivorous species including songbirds.

Biomonitoring of heavy metals in Europe uses bird species on a smaller scale. The leading role is played by diurnal and nocturnal avian raptors, mostly tested for lead and to a lesser extent mercury, the two most preferred metals in such studies (Gomez-Ramirez et al. 2014; Espin et al. 2016). Unlike birds, it is difficult to find extensive studies on toxic metals in North American and European mammals that would allow intra- and intercontinental comparisons (Tranel and Kimmel 2009; Yates et al. 2014). Such publications can only be found for mercury in otters. Mercury in white-tailed eagle and Eurasian otter has been of great interest in Europe for years and sporadically in Asia. In relation to these two species, their North American counterparts are the bald eagle and river otter, which have also been extensively studied. Below we present an example of intercontinental comparisons concerning hepatic mercury concentrations in these species (Fig. 2.2). Median hepatic Hg concentrations in both otter species were similar, but Hg levels in the bald eagle were higher than in the white-tailed eagle (Mann-Whitney $U$ test, $p < 0.05$).

Also other piscivorous wildlife species are used in Hg biomonitoring, including both native and alien species occurring in Europe and North America, with well-known biology and reactions to Hg (Table 2.4). However, the volume of European research is much smaller than in North America (e.g., because of lower Hg contamination), and it is difficult to perform comprehensive intercontinental comparisons. For example, there are many American and Canadian papers on Hg in species such as American mink or raccoon (native mammals from North America introduced in Europe), but in Europe the research has been scarce so far. Birds such as common loon or common merganser are native to both continents, but the volume of research

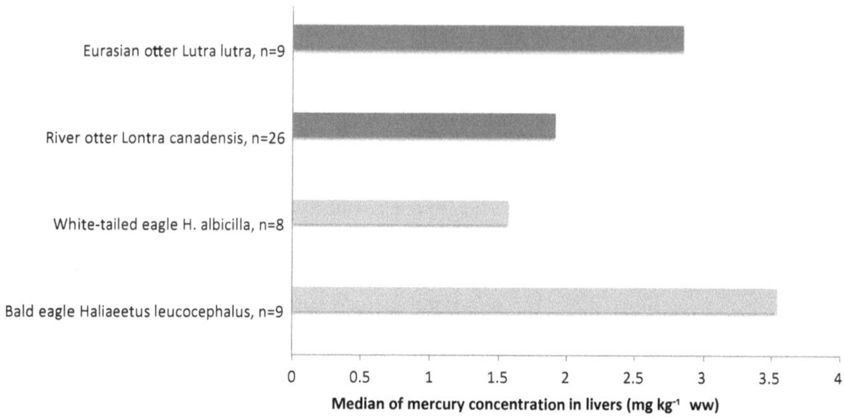


**Fig. 2.2** Hepatic mercury concentrations in counterpart piscivorous species of Europe (Eurasian otter and white-tailed eagle) and North America (river otter and bald eagle). Data sources of Eurasian otter, Madsen et al. (1999), Kruuk et al. (1997), Gutleb et al. (1998), Lemarchand et al. (2010), Walker et al. (2010, 2011), Lodenius et al. (2014); river otter: Wren et al. (1980), Sheffy and Amant (1982), Kucera (1983), Halbrook et al. (1994), Evans et al. (2000), Facemire et al. (1995), Mierle et al. (2000), Fortin et al. (2001), Yates et al. (2005), Grove and Henny (2008), Klenavic et al. (2008), Strom (2008), Sellers (2010), Stansley et al. (2010), Mayack (2012), Keeyask Hydropower Limited Partnership (2012), Dornbos et al. (2013); white-tailed eagle, Norheim and Frøslie (1978), Holt et al. (1979), Falandysz et al. (2001), Kenntner et al. (2001), Kalisinska et al. (2014), Krone et al. (2004, 2006), Kitowski et al. (2017); bald eagle, Evans (1993), Wood et al. (1996), Stout and Trust (2002), Weech et al. (2003), Evers et al. (2005), Mierzykowski et al. (2011, 2013), Rutkiewicz et al. (2011)

**Table 2.4** Candidates of mercury bioindicator species from terrestrial mammals and birds in North America and Europe

| North America | Distribution category and remarks | Europe | Distribution category and remarks |
|---|---|---|---|
| Mammals | | Mammals | |
| American mink *Neovison vison* | Native | American mink *Neovison vison* | Alien, common in Europe |
| Raccoon *Procyon lotor* | Native | Raccoon *Procyon lotor* | Alien common in Europe |
| River otter *Lontra canadensis* | Native | Eurasian otter *Lutra lutra* | Native, counterpart species to river otter |
| Birds | | Birds | |
| Common loon *Gavia immer* | Native | Common loon *Gavia immer* | Native, mainly in Scandinavia |
| Common merganser *Mergus merganser* | Native | Common merganser *Mergus merganser* | Native |
| Bald eagle *Haliaeetus leucocephalus* | Native | White-tailed eagle *Haliaeetus albicilla* | Native, counterpart species to bald eagle |

in North America is also much greater than in Europe (especially with regard to common loon). With time, when the number of European studies on Hg in their bodies will become sufficiently large (especially American mink and raccoon in Europe), it will be possible to deepen intercontinental comparative studies.

Biomonitoring potential is one of the few acceptable effects of introducing alien game animals. It is associated with good knowledge of the biology of most of these species (e.g., American mink, raccoon, wild boar, red fox), social approval for acquiring material for research from specimens during culling of their populations. Nevertheless, in various European countries and some parts of North America (rarely in Asia), biomonitoring programs for various contaminants in terrestrial ecosystems, including trace elements, are created mainly on the basis of selected native species of birds and mammals. An interesting European example is the Norwegian program "Environmental pollutants in the terrestrial and urban environment," now having been conducted for several years and based mainly on the research on the following animals: earthworms, brown rat (*Rattus norvegicus*), red fox, fieldfare (*Turdus pilaris*), Eurasian sparrowhawk (*Accipiter nisus*), and tawny owl (*Strix aluco*) (Herzke et al. 2017).

## 4 Conclusions

The collection and analysis of a sufficiently large number of diverse data on trace elements determined in many species of wildlife allow, among other things, to select candidate species as biomonitors accumulating specific elements in their tissues (e.g., piscivorous species for mercury biomonitoring) and identify existing threats from toxic substances for endangered species, localization of hotspots, and levels of human exposure to trace elements. In order to carry out comparisons in this respect on a large scale, i.e., covering the large terrestrial areas of the Northern Hemisphere, it would be necessary to focus on widespread and numerous species representing different trophic levels.

## References

Ackerman JT, Eagles-Smith CA, Herzog M, Hartman C, Peterson S, Evers DC et al (2016) Avian mercury exposure and toxicological risk across western North America: a synthesis. Sci Total Environ 568:749–769

Adriano DC (2001) Trace elements in the terrestrial environments: biogeochemistry, bioavailability, and risks of metals. Springer, New York

Ahrestani FS, Saracco JF, Sauer JR, Pardieck KL, Royle JA (2017) An integrated population model for bird monitoring in North America. Ecol Appl 27:916–924

Al-Dissi AN, Blakley BR, Woodbury MR (2011) Selenium toxicosis in a white-tailed deer herd. Can Vet J 52:70–73

Al-Ganzoury HH, El-Shaer ME (2008) Aluminum residues in meat and edible tissues of some ruminant and its relation to public health in Sharkia Governorate. SCVMJ 13:361–366

Amici A, Danieli PP, Russo C, Primi R, Ronchi B (2012) Concentrations of some toxic and trace elements in wild boar (*Sus scrofa*) organs and tissues in different areas of the Province of Viterbo, Central Italy. Ital J Anim Sci 11:354–362

Anke M, Muller M, Anke S, Gurtler H, Muller R, Schafer U, Angelow L (2001) The biological and toxicological importance of aluminium in the environment and food chain of animals and humans. In: Ermidou-Pollet S, Pollet S (eds) 3rd international symposium on trace elements in human: new perspectives, Greece

Aras NK, Ataman OY (2006) Trace element analysis of food and diet. RSC Publishing, Cambridge, 344 pp

Arita HT, Rodriguez P, Ella Vazquez-Dominguez E (2005) Continental and regional ranges of North American mammals: Rapoport's rule in real and null worlds. J Biogeogr 32:961–971

Arnemo JM, Andersen O, Stokke S, Thomas VG, Krone O, Pain DJ et al (2016) Health and environmental risks from lead-based ammunition/science versus socio-politics. EcoHealth 13:618–622

ATSDR (2005) Toxicological profile for tin and tin compounds. US Department of Health and Human Services Public Health Service, Agency for Toxic Substances and Disease Registry, 425 pp

Basu N (2012) Piscivorous mammalian wildlife as sentinels of methylmercury exposure and neurotoxicity in humans. In: Ceccatelli S, Aschner M (eds) Methylmercury and neurotoxicity. Current topics in neurotoxicity, vol 2. Springer, Boston, MA, pp 357–370

Bealey WJ, Long S, Spurgeon DJ, Leith I, Cape JN (2008) Review and implementation study of biomonitoring for assessment of air quality outcomes. Environment Agency, Bristol UK, Science report SC030175/SR2, p 181

Behmke S, Fallon J, Duerr AE, Lehner A, Buchweitz J, Katzner T (2015) Chronic lead exposure is epidemic in obligate scavenger populations in eastern North America. Environ Int 79:51–55

Bellrose FC (1951) Effects of ingested lead shot upon waterfowl populations. North Am Wildl Conf Trans 16:125–133

Beyer WN, Meador JP (eds) (2011) Environmental contaminants in biota. CRC Press, Boca Raton

Bhargava S, Bhargava S (2013) Ecological consequences of the acid rain. IOSR JAC 5:19–24

Bhattacharya PT, Misra SR, Hussain M (2016) Nutritional aspects of essential trace elements in oral health and disease: an extensive review. Scientifica, Article ID 5464373

Bibby CJ (2003) Fifty years of bird study. Bird Study 50(3):194–210

Binkowski ŁJ, Sawicka-Kapusta K, Szarek J, Strzyżewska E, Felsmann M (2013) Histopathology of liver and kidneys of wild living mallards *Anas platyrhynchos* and coots *Fulica atra* with considerable concentrations of lead and cadmium. Sci Total Environ 450–451:326–333

Bird DM, Carriere D, Lacombe D (1992) The effect of dietary sodium fluoride on internal organs, breast muscle, and bones in captive American kestrels (*Falco sparverius*). Arch Environ Contam Toxicol 22:242–246

Blus LJ, Henny CJ (1990) Lead and cadmium concentrations in mink from northern Idaho. Northwest Sci 64:2019–2223

Borch-Iohnsen B, Nilssen KJ, Norheim G (1996) Influence of season and diet on liver and kidney content of essential elements and heavy metals in Svalbard reindeer. Biol Trace Elem Res 51: 235–247

Braune BM, Malone BJ (2006) Mercury and selenium in livers of waterfowl harvested in northern - Canada. Arch Environ Contam Toxicol 50:284–289

Bridges CC, Zalups RK (2010) Transport of inorganic mercury and methylmercury in target tissues and organs. J Toxicol Environ Health, Part B 13:385–410

Brzezinski M, Zalewski A, Niemczynowicz A, Jarzyna I, Suska-Maławska M (2014) The use of chemical markers for the identification of farm escapees in feral mink populations. Ecotoxicology 23:767–778

Burger J (1994) Heavy metals in avian eggshells: Another excretion method. J Toxicol Environ Health 41:207–220

Burger J (2006a) Bioindicators: types, development, and use in ecological assessment and research. Environ Bioindic 1:22–39

Burger J (2006b) Bioindicators: a review of their use in the environmental literature 1970–2005. Bioindicators 1:136–144

Burger J (2007) A framework and methods for incorporating gender-related issues in wildlife risk assessment: gender-related differences in metal levels and other contaminants as a case study. Environ Res 104:153–162

Burger J, Gochfeld M (2003) Spatial and temporal patterns in metal levels in eggs of common terns (*Sterna hirundo*) in New Jersey. Sci Total Environ 311:91–100

Burger J, Seyboldt S, Morganstein N, Clark K (1993) Heavy metals and selenium in feathers of three shorebird species from Delaware Bay. Environ Monit Assess 28:189–198

Burger J, Lord CG, Yurkow EJ, McGrath L (2000) Metals and metallothionein in the liver of raccoons: utility for environmental assessment and monitoring. J Toxicol Environ Health Part A 60:243–261

Cai F, Calisi RM (2016) Seasons and neighborhoods of high lead toxicity in New York City: the feral pigeon as a bioindicator. Chemosphere 161:274–279

Carneiro M, Colaço B, Colaço J, Faustino-Rocha AI, Colaço A, Lavin S, Oliveira PA (2016) Biomonitoring of metals and metalloids with raptors from Portugal and Spain: a review. Environ Rev 24(1):63–83

Celechovska O, Literak I, Ondrus S, Pospisil Z (2006) Heavy metals in brown bears from the central European Carpathians. Acta Vet Brno 75:501–506

Chapman PM (1996) Hazard identification, hazard classification and risk assessment for metals and metal compounds in the aquatic environment. ICME, The International Council on Metals and the Environment, Ottawa, 36 pp

Chapman PM (2007) Determining when contamination is pollution—weight of evidence determinations for sediments and effluents. Environ Int 33:492–501

Choubisa SL, Mishra GV, Sheikh Z, Bhardwaj B, Mali P, Jaroli VJ (2011) Osteo-dental fluorosis in domestic horses and donkeys in Rajasthan, India. Adv Pharmacol Toxicol 12:29–37

Clarkson TW, Magos L (2006) The toxicology of mercury and its chemical compounds. Clin Rev Toxicol 36:609–662

Connors PG, Anderlini VC, Risebrough RW, Martin JH, Schroeiber RW, Anderson DW (1972) Heavy metal concentrations in brown pelicans from Florida and California. Cal-Neva Wildl 1972:56–64

Cosson RP, Amiard JC, Amisard-Triquet C (1988) Trace elements in little egrets and flamingos of Camargue, France. Ecotoxicol Environ Saf 15:107–116

Cristol DA, Brasso RL, Condon AM, Fovargue RE, Friedman SL, Hallinger KK et al (2008) The movement of aquatic mercury through terrestrial food webs. Science 320:335

Dailey RN, Raisbeck MF, Siemion RS, Cornish TE (2008) Liver metal concentrations in greater sage-grouse (*Centrocercus urophasianus*). J Wildl Dis 44:494–498

Danieli PP, Serrani F, Primi R, Ponzetta MP, Ronchi B, Amici A (2012) Cadmium, lead, and chromium in large game: a local-scale exposure assessment for hunters consuming meat and liver of wild boar. Arch Environ Contam Toxicol 63:612–627

Dauwe T, Janssens E, Bervoets L, Blust R, Eens M (2005) Heavy-metal concentrations in female laying great tits (*Parus major*) and their clutches. Arch Environ Contam Toxicol 49:249–256

Demesko J, Markowski J, Słaba M, Hejduk J, Minias P (2018) Age-related patterns in trace element content vary between bone and teeth of the European roe deer (*Capreolus capreolus*). Arch Environ Contam Toxicol 74:330–338

Deng H, Zhang Z, Chang C, Wang Y (2007) Trace metal concentration in great tit (*Parus major*) and greenfinch (*Carduelis sinica*) at the Western Mountains of Beijing, China. Environ Pollut 148:620–626

Depew DC, Basu N, Burgess NM, Campbell LM, Evers DC, Grasman KA et al (2012) Derivation of screening benchmarks for dietary methylmercury exposure for the common loon (*Gavia immer*): rationale for use in ecological risk assessment. Environ Toxicol Chem 31:2399–2407

DesGranges JL, Rodrigue J, Tardif B, Laperle M (1998) Mercury accumulation and biomagnification in ospreys (*Pandion haliaetus*) in the James Bay and Hudson Bay Regions of Quebec. Arch Environ Contam Toxicol 35:330–341

Diaz-Bone RA, van de Wiele T (2010) Biotransformation of metal(loid)s by intestinal microorganisms. Pure Appl Chem 82:409–427

Dlugaszek M, Kopczynski K (2013) Elemental composition of muscle tissue of wild animals from central region of Poland. Int J Environ Res 7:973–978

Dornbos P, Strom S, Basu N (2013) Mercury exposure and neurochemical biomarkers in multiple brain regions of Wisconsin river otters (*Lontra canadensis*). Ecotoxicology 22: 469–475

Ertl K, Kitzer R, Goessler W (2016) Elemental composition of game meat from Austria. Food Addit Contam Part B 9:120–126

Espin S, Garcia-Fernandez AJ, Herzke D, Shore RF, van Hattum B, Martinez-Lopez E et al (2016) Tracking pan-continental trends in environmental contamination using sentinel raptors—what types of samples should we use? Ecotoxicology 25:777–801

Evans ED (1993) Mercury and other metals in bald eagle feathers and other tissues from Michigan, Nearby Areas of Minnesota, Wisconsin, Ohio, Ontario and Alaska 1985-1989. Wildlife Division Report No. 3200, Michigan Dep Nat Res, Lansing, 57 pp

Evans RD, Addison EM, Villeneuve JY, MacDonald KS, Joachim DG (2000) Distribution of inorganic and methylmercury among tissues in mink (*Mustela vison*) and otter (*Lutra canadensis*). Environ Res 84:133–139

Evers D (2018) The effects of methylmercury on wildlife: a comprehensive review and approach for interpretation. In: DellaSala DA, Goldstein MI (eds) The Encyclopedia of the Anthropocene, vol 5. Elsevier, Oxford, pp 181–194

Evers DC, Burgess NM, Champoux L, Hoskins B, Major A, Goodale WM et al (2005) Patterns and interpretation of mercury exposure in freshwater avian communities in northeastern North America. Ecotoxicology 14:193–221

Facemire C, Augspurger T, Bateman D, Brim M, Conzelmann P, Delchamps S, Douglas E, Inmon L, Looney K, Lopez F, Masson G, Morrison D, Morse N, Robison A (1995) Impacts of mercury contamination in the southeastern United States. Water Air Soil Pollut 80:923–926

Falandysz J, Ichihashi H, Szymczyk K, Yamasaki S, Mizera T (2001) Metallic elements and metal poisoning among white-tailed sea eagles from the Baltic South Coast. Marine Pollut Bull 42: 1190–1119

Flache L, Becker NI, Kierdorf U, Czamecki S, Düring RA, Encarnacao JA (2015) Hair samples as monitoring units for assessing metal exposure of bats: a new tool for risk assessment. Mamm Biol 80:178–181

Flora G, Gupta D, Tiwari A (2012) Toxicity of lead: a review with recent updates. Interdiscip Toxicol 5:47–58

Fortin C, Beauchamp G, Dansereau M, Larivière N, Bélanger D (2001) Spatial variation in mercury concentrations in wild mink and river otter carcasses from the James Bay territory, Quebec, Canada. Arch Environ Contam Toxicol 40:121–127

Frank A (1986) In search of biomonitors for cadmium: cadmium content of wild Swedish fauna during 1973-1976. Sci Total Environ 57:57–65

Frank A, Danielsson R, Jones B (2000) The 'mysterious' disease in Swedish moose. Concentrations of trace elements in liver and kidneys and clinical chemistry. Comparison with experimental molybdenosis and copper deficiency in the goat. Sci Total Environ 249:107–122

Franson JC, Pain DJ (2011) Lead in birds. In: Beyer WN, Meador JP (eds) Environmental contaminants in biota: interpreting tissue concentrations. CRC Press, Taylor & Francis, Boca Raton, FL, pp 563–593

Furness RW, Greenwood JJD (1993) Birds as monitors of environmental change. Chapman & Hall Press, London

Gall JE, Boyd RS, Rajakaruna N (2015) Transfer of heavy metals through terrestrial food webs: a review. Environ Monit Assess 187:201

Gamberg M, Braune BM (1999) Contaminant residue levels in arctic wolves (*Canis lupus*) from the Yukon Territory, Canada. Sci Total Environ 243–244:329–338

Gamberg M, Boila G, Stern G, Roach P (2005) Cadmium, mercury and selenium concentrations in mink (*Mustela vison*) from Yukon, Canada. Sci Total Environ 351–352:523–529

García-Barrera T, Gómez-Ariza JL, González-Fernández M, Moreno F, García-Sevillano MA, Gómez-Jacinto V (2012) Biological responses related to agonistic, antagonistic and synergistic interactions of chemical species. Anal Bioanal Chem 403:2237–2225

Garcia-Fernandez AJ (2014) Ecotoxicology, avian. In: Wexler P (ed) Encyclopedia of toxicology, vol 2, 3rd edn. Elsevier, Academic Press, San Diego, CA, pp 289–294

Gasparik J, Binkowski LJ, Jahnatek A, Smehyl P, Dobias M, Lukac N et al (2017) Levels of metals in kidney, liver, and muscle tissue and their influence on the fitness for the consumption of wild boar from western Slovakia. Biol Trace Elem Res 177:258–266

Gehrt SD, Riley SPD, Cypher BL (eds) (2011) Urban carnivores: ecology, conflict, and conservation. Johns Hopkins University Press, Baltimore

Genovesi P, Bacher S, Kobelt M, Pascal M, Scalera R (2009) Alien mammals of Europe. In: DAISIE (ed) Handbook of alien species in Europe. Springer, Dordrecht, Netherlands, pp 119–129

Gerstenberger SL (2004) Mercury concentrations in migratory waterfowl harvested from Southern Nevada Wildlife Management areas, USA. Environ Toxicol 19:35–44

Gochfeld M (2017) Sex differences in human and animal toxicology: toxicokinetics. Toxicol Pathol 45:172–189

Golden NH, Rattner BA (2003) Ranking terrestrial vertebrate species for utility in biomonitoring and vulnerability to environmental contaminants. Rev Environ Contam Toxicol 176:67–136

Golden NH, Warner SE, Coffey MJ (2016) A review and assessment of spent lead ammunition and its exposure and effects to scavenging birds in the United States. Rev Environ Contam Toxicol 237:123–191

Gomez-Ramirez P, Shore RF, van den Brink NW, van Hattum B, Bustnes JO, Duke G et al (2014) An overview of existing raptor contaminant monitoring activities in Europe. Environ Int 67: 12–21

Grove RA, Henny CJ (2008) Environmental contaminants in male river otters collected from Oregon and Washington, USA, 1994-1999. Environ Monit Assess 145:49–73

Gutleb AC, Kranz A, Nechay G, Toman A (1998) Heavy metal concentrations in livers and kidneys of the otter (*Lutra lutra*) from central Europe. Bull Environ Contam Toxicol 60:273–279

Haig SM, D'Elia J, Eagles-Smith C, Fair JM, Gervais J, Herring G et al (2014) The persistent problem of lead poisoning in birds from ammunition and fishing tackle. Condor 116:409–428

Halbrook RS, Jenkins JH, Bush PB, Seabolt ND (1994) Sublethal concentrations of mercury in river otters: monitoring environmental contamination. Arch Environ Contam Toxicol 27: 306–310

Harding LE, Harris ML, Elliott JE (1998) Heavy and trace metals in wild mink (*Mustela vison*) and river otter (*Lontra canadensis*) captured on rivers receiving metals discharges. Bull Environ Contam Toxicol 61:600–607

Hashmi MZ, Abbasi NA, Tang X, Malik RN (2015) Egg as a biomonitor of heavy metals in soil. In: Sherametiand I, Varma A (eds) Heavy metal contamination of soils: monitoring and remediation. Springer International, Cham, Switzerland, pp 127–143

Heltai M, Markov G (2012) Red fox (*Vulpes vulpes* Linnaeus, 1758) as biological indicator for environmental pollution in Hungary. Bull Environ Contam Toxicol 89:910–914

Herring G, Eagles-Smith CA, Wagner MT (2016) Ground squirrel shooting and potential lead exposure in breeding avian scavengers. PLoS One 11:e0167926

Herzke D, Nygård T, Heimstad ES (2017) Environmental pollutants in the terrestrial and urban environment 2016. Kjeller, NILU, Norwegian Environment Agency report M-752/2017

Hoekstra PF, Braune BM, Elkin B, Armstrong FAJ, Muir DCG (2003) Concentrations of selected essential and non-essential elements in arctic fox (*Alopex lagopus*) and wolverines (*Gulo gulo*) from the Canadian Arctic. Sci Total Environ 309:81–92

Hollamby S, Afema-Azikuru J, Waigo S, Cameron K, Gandolf AR, Norriset A et al (2006) Suggested guidelines for use of avian species as biomonitors. Environ Monit Assess 118:13–20

Holt EA, Miller SW (2011) Bioindicators: using organisms to measure environmental impacts. Nat Educ Knowl 2:8

Holt G, Frøslie A, Norheim G (1979) Mercury, DDE, and PCB in the avian fauna in Norway 1965-1976. Acta Vet Scand Suppl 70:1–28

Holt YS, Kim YM, Lee KE (2012) Methylmercury exposure and health effect. J Prev Med Public Health 45:353–363

Honda K, Min BY, Tatsukawa R (1985) Heavy metal distribution in organs and tissues of the eastern great white egret *Egretta alba modesta*. Bull Environ Contam Toxicol 35:781–789

Horai S, Minagawa M, Ozaki H, Watanabe I, Takeda Y, Yamada K et al (2006) Accumulation of Hg and other heavy metals in the Javan mongoose (*Herpestes javanicus*) captured on Amamioshima Island, Japan. Chemosphere 65:657–665

Hubner R, Astin KB, Herbert RJH (2010) 'Heavy metal'—time to move on from semantics to pragmatics. J Environ Monit 12:1511–1514

Inkielewicz I, Krechniak J (2003) Fluoride content in soft tissues and urine of rats exposed to sodium fluoride in drinking water. Fluoride 36:263–266

Jackson AK, Evers DC, Adams EM, Cristol DA, Eagles-Smith C, Edmonds ST et al (2015) Songbirds as sentinels of mercury in terrestrial habitats of eastern North America. Ecotoxicology 24:453–467

Jackson A, Evers DC, Eagles-Smith CA, Ackerman JT, Willacker JJ, Elliott JE et al (2016) Mercury risk to avian piscivores across Western United States and Canada. Sci Total Environ 568:685–696

Jaishankar M, Tseten T, Anbalagan N, Mathew BB, Jaishankar BKN et al (2014) Toxicity, mechanism and health effects of some heavy metals. Interdiscip Toxicol 7:60–72

James LF, Shupe JL (1984) Selenium poisoning in livestock. Rangelands 6:64–67

Jan AT, Azam M, Siddiqui K, Ali A, Choi I, Haq QMR (2015) Heavy metals and human health: mechanistic insight into toxicity and counter defense system of antioxidants. Int J Mol Sci 16: 29592–29630

Jarup L (2003) Hazards of heavy metal contamination. Brit Med Bull 68:167–182

Jetz W, Thomas GH, Joy JB, Hartmann K, Moores AO (2012) The global diversity of birds in space and time. Nature 491:444–448

Ji X, Hu W, Cheng J, Yuan T, Xu F, Qu L et al (2006) Oxidative stress on domestic ducks (Shaoxing duck) chronically exposed in a mercury-selenium coexisting mining area in China. Ecotoxicol Environ Saf 64:171–177

Jones KE, Safi K (2011) Ecology and evolution of mammalian biodiversity. Phil Trans R Soc B 366:2451–2461

Kabata-Pendias A (2011) Trace elements in soils and plants. CRC Press, Boca Raton, FL

Kalisinska E, Budis H, Podlasinska J, Łanocha N, Kavetska KM (2010) Body condition and mercury concentration in apparently healthy goosander (*Mergus merganser*) wintering in the Odra estuary, Poland. Ecotoxicology 19:1382–1399

Kalisinska E, Budis H, Łanocha N, Podlasińska J, Baraniewicz E (2012a) Comparison of hepatic and nephric concentrations of mercury between feral and ranch American mink (*Neovison vison*) from NW Poland. Bull Environ Contam Toxicol 88:802–806

Kalisinska E, Lisowski P, Kosik-Bogacka DI (2012b) Red fox *Vulpes vulpes* (L., 1758) as a bio-indicator of mercury contamination in terrestrial ecosystems of north-western Poland. Biol Trace Elem Res 145:172–180

Kalisinska E, Gorecki J, Lanocha N, Okonska A, Melgarejo JB, Budis H et al (2014) Total and methyl mercury in soft tissues of white-tailed eagle (*Haliaeetus albicilla*) and osprey (*Pandion haliaetus*) collected in Poland. Ambio 43:858–870

Kalisinska E, Kosik-Bogacka DI, Lanocha-Arendarczyk N, Budis H, Podlasinska J, Popiolek M et al (2016) Brains of native and alien mesocarnivores in biomonitoring of toxic metals in Europe. PLoS One 11(8):e159935

Kalisinska E, Lanocha-Arendarczyk N, Kosik-Bogacka DI, Budis H, Pilarczyk B, Tomza-Marciniak A et al (2017) Muscle mercury and selenium in fishes and semiaquatic mammals from a selenium-deficient area. Ecotoxicol Environ Saf 136:24–30

Kannan K, Falandysz J (1997) Butyltin residues in sediment, fish, fish-eating birds, harbour porpoise and human tissues from the Polish coast of the Baltic Sea. Mar Pollut Bull 34:203–207

Karmakar A, Zhang Q, Zhang Y (2014) Neurotoxicity of nanoscale materials. J Food Drug Anal 22: 147–160

Keck F (2015) Sentinels for the environment. China Perspect 2:43–52

Keeyask Hydropower Limited Partnership (2012) Keeyask Generation Project. Environmental Impact Statement, 75 pp [http://keeyask.com/wp/wp-content/uploads/2012/07/Section-8-Wildlife-and-Mercury.pdf]

Kenntner N, Tataruch F, Krone O (2001) Heavy metals in soft tissue of white-tailed eagles found dead or moribund in Germany and Austria from 1993 to 2000. Environ Toxicol Chem 20: 1831–1837

Kierdorf U, Kierdorf H (2003) Temporal variation of fluoride concentration in antlers of roe deer (*Capreolus capreolus*) living in an area exposed to emissions from iron and steel industry, 1948–2000. Chemosphere 52:1677–1681

Kim J, Oh JM (2013) Assessment of trace metals in four bird species from Korea. Environ Monit Assess 185:6847–6854

Kitowski I, Jakubas D, Wiącek D, Sujak A (2017) Concentrations of lead and other elements in the liver of the white-tailed eagle (*Haliaeetus albicilla*), a European flagship species, wintering in Eastern Poland. Ambio 46:825–841

Klee RJ, Graedel TE (2004) Elemental cycles: a status report on human or natural dominance. Annu Rev Environ Resour 29:69–107

Klenavic K, Champoux L, O'Brien M, Daoust PY, Evans RD, Evans HE (2008) Mercury concentration in wild mink (*Mustela vison*) and river otters (*Lontra canadensis*) collected from eastern and Atlantic Canada: relationship to age and parasitism. Environ Pollut 156:359–366

Knopper LD, Mineau P, Scheuhammer AM, Bond DE, Mckinnon DT (2006) Carcasses of shot Richardson's ground squirrels may pose lead hazards to scavenging Hawks. J Wildl Manag 70:295–299

Kramarova M, Massanyi P, Jančová A, Toman R, Slamečka J, Tataruch F et al (2005) Concentration of cadmium in the liver and kidneys of some wild and farm animals. Bull Vet Inst Pulawy 49:465–469

Krewski D, Yokel RA, Nieboer E, Borchelt D, Cohen J, Harry J, Kacew S et al (2007) Human health risk assessment for aluminium, aluminium oxide, and aluminium hydroxide. J Toxicol Environ Health Part B Crit Rev 10(Suppl 1):1–269

Krone O, Willie F, Kenntner N, Boertmann D, Tataruch F (2004) Mortality factors, environmental contaminants, and parasites of white-tailed sea eagles from Greenland. Avian Dis 48:417–424

Krone O, Stjernberg T, Kenntner N, Tataruch F, Koivusaari J, Nuuja I (2006) Mortality factors, helminth burden, and contaminant residues in white-tailed sea eagles (*Haliaeetus albicilla*) from Finland. Ambio 35:98–104

Kruuk H, Conroy JWH, Webb A (1997) Concentration of mercury in otters (*Lutra lutra*) in Scotland in relation to rainfall. Environ Pollut 96:13–18

Kucera E (1983) Mink and otter as indicators of mercury in Manitoba waters. Can J Zool 61: 2250–2256

Kuo HW, Kuo SM, Chou C-H, Lee TC (2000) Determination of 14 elements in Taiwanese bones. Sci Total Environ 255:45–54

Kurdi MS (2016) Chronic fluorosis: the disease and its anaesthetic implications. Ind J Anaesth 60:157–162

Kursa J, Herzig I, Trávníček J, Illek J, Kroupová V, Fuksová S (2010) Iodine and selenium contents in skeletal muscles of red deer (*Cervus elaphus*), roe deer (*Capreolus capreolus*) and wild boar (*Sus scrofa*) in the Czech Republic. Acta Vet Brno 79:403–407

Lambert JD, Hodgman TP, Laurent EJ, Brewer GL, Iliff MJ, Dettmers R (2009) The northeast bird monitoring handbook. American Bird Conservancy, The Plains, VA, 32 pp

Langner HW, Greene E, Domenech R, Staats M (2012) Mercury and other mining-related contaminants in ospreys along the Upper Clark Fork River, Montana, USA. Arch Environ Contam Toxicol 62:681–695

Lanocha N, Kalisinska E, Kosik-Bogacka DI, Budis H, Podlasinska J, Jedrzejewska E (2014) Mercury levels in raccoons (*Procyon lotor*) from the Warta Mouth National Park, north-western Poland. Biol Trace Elem Res 159:152–160

Lazarus M, Sekovanić A, Orct T, Reljić S, Kusak J, Jurasović J, Huber D (2017) Apex predatory mammals as bioindicator species in environmental monitoring of elements in Dinaric Alps (Croatia). Environ Sci Pollut Res Int 24:23977–23991

Lemarchand C, Rosoux R, Berny P (2010) Organochlorine pesticides, PCBs, heavy metals and anticoagulant rodenticides in tissues of Eurasian otters (*Lutra lutra*) from upper Loire River catchment (France). Chemosphere 80:1120–1124

Leonzio C, Massi A (1989) Metal biomonitoring in bird eggs: a critical experiment. Bull Environ Contam Toxicol 43:402–406

Leonzio C, Bianchi N, Gustin M, Sorace A, Ancora S (2009) Mercury, lead and copper in feathers and excreta of small passerine species in relation to foraging guilds and age of feathers. Bull Environ Contam Toxicol 83:693–697

Lesmeister DB, Nielsen CK, Schauber EM, Hellgren E (2015) Spatial and temporal structure of a mesocarnivore guild in midwestern North America. Wildl Monogr 191:1–61

Leveque C, Mounolou J (2003) Biodiversity. John Wiley & Sons Ltd, Chichester

Li ZWD, Mundkur T (2006) Monitoring waterbirds in the Asia-Pacific region. In: Boere GC, Galbraith CA, Stroud DA (eds) Waterbirds around the world. The Stationery Office, Edinburgh, UK, pp 339–342

Liu J, Waalkes MP (2008) Liver is a target of arsenic carcinogenesis. Toxicol Sci 105:24–32

Ljungvall K, Magnusson U, Korvela M, Norrby M, Bergquist J, Persson S (2017) Heavy metal concentrations in female wild mink (*Neovison vison*) in Sweden: sources of variation and associations with internal organ weights. Environ Toxicol Chem 36:2030–2035

Llacuna S, Gorizz A, Sanpera C, Nadal J (1995) Metal accumulation in three species of passerine birds (*Emberiza cia*, *Parus major*, and *Turdus merula*) subjected to air pollution from coal-fired power plant. Arch Environ Toxicol 28:298–303

Lodenius M, Solonen T (2013) The use of feathers of birds of prey as indicators of metal pollution. Ecotoxicology 22:1319–1334

Lodenius M, Skaren U, Hellstedt P, Tulisalo E (2014) Mercury in various tissues of three mustelid and other trace metals in liver o European otter from eastern Finland. Environ Monit Assess 186: 325–333

Lopez-Alonso M (2012) Trace minerals and livestock: not too much not too little. ISRN Vet Sci 2012:704825

López Alonso M, Benedito JL, Miranda M, Castillo C, Hernández J, Shore RF (2002) Cattle as biomonitors of soil arsenic, copper, and zinc concentrations in Galicia (NW Spain). Arch Environ Contam Toxicol 43:103–108

Lord CG, Gaines KF, Boring CS, Brisbin IL, Gochfeld M Jr, Burger J (2002) Raccoon (*Procyon lotor*) as a bioindicator of mercury contamination at the U.S. Department of Energy's Savannah River Site. Arch Environ Contam Toxicol 43:356–363

Lucia M, Andre JM, Gontier K, Diot N, Veiga J, Davail S (2010) Trace element concentrations (mercury, cadmium, copper, zinc, lead, aluminum, nickel, arsenic, and selenium) in some aquatic birds of the Southwest Atlantic Coast of France. Arch Environ Contam Toxicol 58: 844–853

Ma WC, Denneman W, Faber J (1991) Hazardous exposure of ground-living small mammals to cadmium and lead in contaminated terrestrial ecosystems. Arch Environ Contam Toxicol 20: 266–270

Madsen AB, Dietz HH, Henriksen P, Clausen B (1999) Survey of Danish free living otters *Lutra lutra*—a consecutive collection and necropsy of dead bodies. IUCN Spec Group Bull 16:65–75

Mandal P (2017) An insight of environmental contamination of arsenic on animal health. Emerg Contam 3:17–22

Maret W (2016) The metals in the biological periodic system of the elements: concepts and conjectures elements. Int J Mol Sci 17:66

Markert B (2013) Bioindication and biomonitoring as innovative biotechniques for controlling heavy metal data of the environment. Proceedings of 15th international conference on heavy metals in the environment, ICHMET, Poland, Gdansk 2010, pp 44–48

Markert B, Breure AM, Zechmeister HG (2003) Definitions, strategies and principles for bioindication/biomonitoring of the environment. In: Markert BA, Breure AM, Zechmeister HG (eds) Bioindicators and biomonitors. Principles, concepts and applications. Elsevier, Amsterdam, pp 3–39

Markert B, Wuenschmann S, Fraenzle S, Wappelhorst O, Weckert V, Breulmann G, Djingova R, Herpin U, Lieth H, Schroder W, Siewers U, Steinnes E, Wolterbeek B, Zechmeister H (2008) On the road from environmental biomonitoring to human health aspects: monitoring atmospheric heavy metal deposition by epiphytic/epigeic plants: present status and future needs. Int J Environ Pollut 32(4):486

Markov G, Kocheva M, Gospodinova M (2016) Assessment of heavy metal accumulation in the golden jackal (*Canis aureus*) as a possible bioindicator in an agricultural environment in Bulgaria. Bull Environ Contam Toxicol 96:458–464

Martelli A, Rousselet E, Dycke C, Bouron A, Moulis JM (2006) Cadmium toxicity in animal cells by interference with essential metals. Biochimie 88:1807–1814

Martinez MA, Ballesteros S, Piga FJ, Sánchez de la Torre C, Cubero CA (2007) The tissue distribution of fluoride in a fatal case of self-poisoning. J Anal Toxicol 31:526–533

Mateo R, Taggart M, Meharg AA (2003) Lead and arsenic in bones of birds of prey from Spain. Environ Pollut 126:107–114

Mayack DT (2012) Hepatic mercury, cadmium, and lead in mink and otter from New York State: monitoring environmental contamination. Environ Monit Assess 184:2497–2516

McClure ML, Burdett CL, Farnsworth ML, Sweeney SJ, Miller RS (2018) A globally-distributed alien invasive species poses risks to United States imperiled species. Sci Rep 8(1):5331

Mendez-Rodriguez LC, Alvarez-Castaneda ST (2014) Influence of trace elements in the epigenetic of mammals. Therya 5:817–829

Mierle G, Addison EM, MacDonald KS, Joachim DG (2000) Mercury levels in tissues of otters from Ontario, Canada: variation with age, sex, and location. Environ Toxicol Chem 19: 3044–3051

Mierzykowski SE, Smith JEM, Todd CS, Kusnierz D, DeSorbo CR (2011) Liver contaminants in bald eagle carcasses from Maine. USFWS. Specific Project Report FY09-MEFO-6-EC. Maine Field Office, Orono, ME, 53 pp

Mierzykowski SE, Todd CS, Pokras MA, Oliveira RD (2013) Lead and mercury levels in livers of bald eagles recovered in New England. USFWS. Specific Project Report FY13-MEFO-2-EC. Maine Field Office, Orono, ME, 26 pp

Millan J, Mateo R, Taggart MA, López-Bao JV, Viota M, Monsalve L et al (2008) Levels of heavy metals and metalloids in critically endangered Iberian lynx and other wild carnivores from southern Spain. Sci Total Environ 399:193–201

Misztal-Szkudlińska M, Kalisińska E, Szefer P, Konieczka P, Namieśnik J (2018) Mercury concentration and the absolute and relative sizes of the internal organs in cormorants *Phalacrocorax carbo* (L. 1758) from the breeding colony by the Vistula Lagoon (Poland). Ecotoxicol Environ Saf 154:118–126

Mizukawa H, Takahashi S, Nakayama K et al (2009) Contamination and accumulation feature of organotin compounds in common cormorants (Phalacrocorax carbo) from Lake Biwa, Japan. In: Obayashi Y, Isobe T, Subramanian A, Suzuki S, Tanabe S (eds) Interdisciplinary studies on environmental chemistry – environmental research in Asia, Terrapub, pp 153–161

Mochizuki M, Kaitsuka C, Mori M, Hondo R, Ueda F (2011) An innovative approach to biological monitoring using wildlife. In: Ekundayo E (ed) Environmental monitoring, InTech

Monkkonen M, Viro P (1997) Taxonomic diversity of the terrestrial bird and mammal fauna in temperate and boreal biomes of the northern hemisphere. J Biogeogr 24:603–612

Movalli P, Bode P, Dekker R, Fornasari L, van der Mije S, Yosef R (2017) Retrospective biomonitoring of mercury and other elements in museum feathers of common kestrel *Falco tinnunculus* using instrumental neutron activation analysis (INAA). Environ Sci Pollut Res 24: 25986–26005

Mysłek P, Kalisińska E (2006) Contents of selected heavy metals in the liver, kidneys, and abdominal muscle of the brown hare (*Lepus europaeus* Pallas, 1778) in Central Pomerania, Poland. Pol J Vet Sci 9:31–41

Nadjafzadeh M, Hofer H, Krone O (2013) The link between feeding ecology and lead poisoning in white-tailed eagles. J Wildl Manag 77:48–57

Nam DH, Anan Y, Ikemoto T, Okabe Y, Kim EY, Subramanian A et al (2005) Specific accumulation of 20 trace elements in great cormorants (*Phalacrocorax carbo*) from Japan. Environ Pollut 134:503–514

Nath R (2000) Tin. In: Nath R (ed) Health and disease role of micronutrients and trace elements: recent advances in the assessment of micronutrients and trace elements deficiency in humans. APH Publishing, New Delhi, pp 385–389

Nemsadze K, Sanikidze T, Ratiani L, Gabunia L, Sharashenidze T (2009) Mechanisms of lead-induced poisoning. Georgian Med News 172–173:92–96

Neuschwander-Tetri BA (2007) Trace elements and the liver. In: Rodés J, Benhamou JP, Blei A, Reichen J, Rizzetto M, Dufour JF, Friedman SL et al (eds) Textbook of hepatology: from basic science to clinical practice. Blackwell Publishing, Oxford, UK, pp 233–241

Newman J, Zillioux E, Rich E, Liang L, Newman C (2004) Historical and other patterns of monomethyl and inorganic mercury in the Florida panther (Puma concolor coryi). Arch Environ Contam Toxicol 48:75–80

Noel M, Spence J, Harris KA, Robbins CT, Fortin JK, Ross PS et al (2014) Grizzly bear hair reveals toxic exposure to mercury through salmon consumption. Environ Sci Technol 48:7560–7567

Nollet L, Huyghebaert G, Spring P (2008) Effect of different levels of dietary organic (bioplex) trace minerals on live performance of broiler chickens by growth phases. J Appl Poult Res 17: 109–115

Norheim G, Frøslie A (1978) The degree of methylation and organ distribution of mercury in some birds of prey in Norway. Acta Pharmacol Toxicol 43:196–204

Norheim G, Sivertsen T, Brevik EM, Frøslie A (1984) Mercury and selenium in wild mink (*Mustela vision*) from Norway. Nord Vet Med 36:43–48 (in Norveriann Norwegian)

NRC (2005) Mineral Tolerance of Animals. National Research Council, National Academy Press, Washington, DC

Nriagu JO (1990) The rise and fall of leaded gasoline. Sci Total Environ 92:13–28

Nriagu JO, Pacyna J (1988) Quantitative assessment of worldwide contamination of air, water and soil by trace metals. Nature 333:134–139

O'Brien DJ, Kaneene JB, Poppenga RH (1993) The use of mammals as sentinels for human exposure to toxic contaminants in the environment. Environ Health Perspect 99:351–368
O'Hara TM, George JC, Blake J, Burek K, Carroll G, Dau J et al (2003) Investigation of heavy metals in a large mortality event in caribou of Northern Alaska. Arctic 56:125–135
Ohi G, Seki H, Minowa K, Ohsawa M, Mizoguchi I, Sugimori F (1981) Lead pollution in Tokyo – the pigeon reflects its amelioration. Environ Res 26:125–129
Oldenkamp RE (2016) Exposure of game species to trace elements and radiocesium on the Savannah River site in South Carolina. MSc Thesis, University of Georgia, 164 pp
Oldenkamp RE, Bryan AL Jr, Kennamer RA, Leaphart JC, Webster SC, Beasley JC (2017) Trace elements and radiocesium in game species near contaminated sites. J Wildl Manag 81: 1338–1350
Orlowski G, Halupka L, Pokorny P, Klimczuk E, Sztwiertnia H, Dobicki W (2016) Variation in egg size, shell thickness, and metal and calcium content in eggshells and egg contents in relation to laying order and embryonic development in a small passerine bird. Auk 133:470–483
Pain DJ, Fisher IJ, Thomas VG (2009) A global update of lead poisoning in terrestrial birds from ammunition sources. In: Watson RT, Fuller M, Pokras M, Hunt WG (eds) Ingestion of lead from spent ammunition: implications for wildlife and humans. The Peregrine Fund, Boise, ID, pp 1–21
Pauli JN, Whiteman JP, Riley MD, Arthur D. Middleton AD 2010 Defining noninvasive approaches for sampling of vertebrates. Conserv Biol 24:349–352
Pedersen S, Lierhagen S (2006) Heavy metal accumulation in arctic hares (*Lepus arcticus*) in Nunavut, Canada. Sci Total Environ 368:951–955
Poessel SA, Gese EM, Young JK (2017) Environmental factors influencing the occurrence of coyotes and conflicts in urban areas. Lands Urban Plan 157:259–269
Pokorny B (2006) Roe deer (*Capreolus capreolus* L.) antlers as an accumulative and reactive bio-indicator of lead pollution near the largest Slovene thermal power plant. Vet Arhiv 76:131–142
Pollack L, Ondrasek NR, Calisi R (2017) Urban health and ecology: the promise of an avian bio-monitoring tool. Curr Zool 63:205–212
Prashanth L, Kattapagari KK, Chitturi RT, Baddam VR, Prasad LK (2015) A review on role of essential trace elements in health and disease. J NTR Univ Health Sci 4:75–85
Rabinowitz P, Scotch M, Conti L (2009) Human and animal sentinels for shared health risks. Vet Ital 45:23–24
Rajaganapathy V, Xavier F, Sreekumar D, Mandal PK (2011) Heavy metal contamination in soil, water and fodder and their presence in livestock and products: a review. J Environ Sci Technol 4:234–249
Rattner BA (2009) History of wildlife toxicology. Ecotoxicology 18:773–783
Rattner BA, Pearson JL, Golden NH, Cohen JB, Ervin RM, Ottinger MA (2000) Contaminant exposure and effect – terrestrial vertebrates database: trends and data gaps for Atlantic coast estuaries. Environ Monit Assess 63:131–142
Rattner BA, Eisenreich KM, Golden NH, McKernan MA, Hothem RL, Custer TW (2005) Retrospective ecotoxicological data and current information needs for terrestrial vertebrates residing in coastal habitat of the United States. Arch Environ Contam Toxicol 49:257–265
Ray PC, Yu H, Fu PP (2010) Toxicity and environmental risks of nanomaterials: challenges and future needs. J Environ Sci Health Part C Environ Carcinog Ecotoxicol Rev 27:1–35
Reinoso RF, Telfer BA, Rowland M (1997) Tissue water content in rats measured by desiccation. J Pharmacol Toxicol Methods 38:87–92
Reis LSLS, Pardo PL, Camargos AS, Oba E (2010) Mineral element and heavy metal poisoning in animals. J Med Med Sci 1:560–579
Rendón-Lugo AN, Santiago P, Puente-Lee I, León-Paniagua L (2017) Permeability of hair to cadmium, copper and lead in five species of terrestrial mammals and implications in biomonitoring. Environ Monit Assess 189:640

Rogers TA, Bedrosian B, Graham J, Foresman KR (2012) Lead exposure in large carnivores in the greater Yellowstone ecosystem. J Wildl Manag 76:575–582

Rothschild RFN, Duffy LK (2005) Mercury concentrations in muscle, brain and bone of Western Alaskan waterfowl. Sci Total Environ 349:277–283

Roug A, Swift PK, Gerstenberg G, Woods LW, Kreuder-Johnson C, Torres SG et al (2015) Comparison of trace mineral concentrations in tail hair, body hair, blood, and liver of mule deer (*Odocoileus hemionus*) in California. J Vet Diagn Invest 27:295–305

Roux KE, Marra PP (2007) The presence and impact of environmental lead in passerine birds along an urban to rural land use gradient. Arch Environ Contam Toxicol 53:261–268

Rudy M (2010) Chemical composition of wild boar meat and relationship between age and bioaccumulation of heavy metals in muscle and liver tissue. Food Addit Contam Part A Chem Anal Control Expo Risk Assess 27:464–472

Russo D, Jones G (2015) Bats as bioindicators: an introduction. Mammal Biol 80:157–158

Rutherford JC, Bird AJ (2004) Metal-responsive transcription factors that regulate iron, zinc, and copper homeostasis in eukaryotic cells. Eukaryot Cell 3:1–13

Rutkiewicz J, Nam DH, Cooley T, Neumann K, Padilla IB, Route W et al (2011) Mercury exposure and neurochemical impacts in bald eagles across several Great Lakes states. Ecotoxicology 20: 1669–1676

Saito K (2009) Lead poisoning of Steller's sea-eagle (*Haliaeetus pelagicus*) and white-tailed eagle (*Haliaeetus albicilla*) caused by the ingestion of lead bullets and slugs, in Hokkaido Japan. In: Watson RT, Fuller M, Pokras M, Hunt WG (eds) Ingestion of lead from spent ammunition: implications for wildlife and humans. The Peregrine Fund, Boise, ID, pp 1–8

Sakulsak N (2012) Metallothionein: an overview on its metal homeostatic regulation in mammals. Int J Morphol 30:1007–1012

Sánchez-Chardi A, Marques CC, Nadal J, Da Luz Mathias M (2007) Metal bioaccumulation in the greater white-toothed shrew, *Crocidura russula*, inhabiting an abandoned pyrite mine site. Chemosphere 67:121–130

Sanchez-Virosta P, Espin S, Garcia-Fernandez AJ, Eeva T (2015) A review on exposure and effects of arsenic in passerine birds. Sci Total Environ 512-513:506–525

Santiago D, Motas-Guzmán M, Reja A, María-Mojica P, Rodero B, García-Fernández AJ (1998) Lead and cadmium in red deer and wild boar from Sierra Morena Mountains (Andalusia, Spain). Bull Environ Contam Toxicol 61:730–737

Sauer JR, Pardieck KL, Ziolkowski DJ, Smith AC, Hudson MAR, Vicente Rodriguez V et al (2017) The first 50 years of the North American breeding bird survey. Condor 119:576–593

Scheuhammer AM (2008) Recent advances in the toxicology of methylmercury in wildlife. Ecotoxicology 17:67–68

Scheuhammer AM, Atchison CM, Wong AHK, Evers DC (1998) Mercury exposure in breeding common loons (*Gavia immer*) in central Ontario, Canada. Environ Toxicol Chem 17:191–196

Schmeller DS, Henle K, Loyau A, Besnard A, Henry P-Y (2012) Bird-monitoring in Europe—a first overview of practices, motivations and aims. Nat Conserv 2:41–57

Schoch N, Jackson AK, Duron M, Evers DC, Glennon MJ, Driscoll CT, Yu X, Simonin H, Sauer AK (2014) Wildlife criterion value for the common loon (*Gavia immer*) in the Adirondack Park, New York, USA. Waterbirds 37(sp1):76–84

Selinus O, Finkelman RB (2011) Geochemical aspects of medical geology. J Geol Soc Sri Lanka 14:1–9

Sellers P (2010) *A* survey *of* chemical contaminants *in* wild meat harvested *from the* traditional territories of Wabauskang First Nation (Wabauskang), Asubpeeschoseewagong Netum Anishinaabek (Grassy Narrows), and Wabaseemong Independent Nation (Whitedog). First Nations Environmental *Contaminants* Program (National) as Partial fulfillment of Project No. HQ0900055, 6 pp

Sheffy TB, Amant JR (1982) Mercury burdens in furbearers in Wisconsin. J Wildl Manag 46: 1117–1120

Shore RF, Douben PE (1994) Predicting ecotoxicological impacts of environmental contaminants on terrestrial small mammals. Rev Environ Contam Toxicol 134:49–89

Shore RF, Casulli A, Bologov V, Wienburg CL, Afsar A, Toyne P et al (2001) Organochlorine pesticide, polychlorinated biphenyl and heavy metal concentrations in wolves (*Canis lupus* L. 1758) from north-west Russia. Sci Total Environ 280:45–54

Shore RF, Pereira MG, Walker LA, Thompson DR (2011) Mercury in nonmarine birds and mammals. In: Beyer WN, Meador JP (eds) Environmental contaminants in biota. CRC Press, Boca Raton, FL, pp 609–642

Sidding AAH, Ellison AM, Ochs A, Villar-Leeman C, Lau MK (2016) How do ecologists select and use indicator species to monitor ecological change? Ecol Indic 60:223–230

Silbergeld EK, Sauk J, Somerman M, Todd A, McNeill F, Fowler B et al (1993) Lead in bone: storage site, exposure source, and target organ. Neurotoxicology 14:225–236

Skibniewski M, Skibniewska EM, Kosla KM (2015) The molybdenum content in the muscles of red deer (*Cervus elaphus*). Acta Sci Pol Zootechnica 14:175–182

Sleeman JM, Cristol DA, White AE, Evers DC, Gerhold RW, Keel MK (2010) Mercury poisoning in a free-living northern river otter (*Lontra canadensis*). J Wildl Dis 46:1035–1039

Smith PN, Cobb GP, Godard-Codding C, Hoff D, McMurry ST, Rainwater TR et al (2007) Contaminant exposure in terrestrial vertebrates. Environ Pollut 150:41–64

Smith JB, Tuberville TD, Beasley JC (2018) Hunting and game consumption patterns of hunters in South Carolina. J Fish Wildl Manag 9:321–329

Snow NP, Jarzyna MA, VerCauteren KC (2017) Interpreting and predicting the spread of invasive wild pigs. J Appl Ecol 54:2022–2032

Speakman JR (2005) Body size, energy metabolism and lifespan. J Exp Biol 208:1717–1730

Stankovic S, Kalaba P, Stankovic AR (2014) Biota as toxic metal indicators. Environ Chem Lett 12: 63–84

Stansley W, Velinsky D, Thomas R (2010) Mercury and halogenated organic contaminants in river otters (*Lontra canadensis*) in New Jersey, USA. Environ Toxicol Chem 29:2235–2242

Steinnes E (2009) Soils and geomedicine. Environ Geochem Health 31:523–535

Stensberg MC, Wei Q, McLamore ES, Porterfield DM, Wei A, Sepúlveda MS (2011) Toxicological studies on silver nanoparticles: challenges and opportunities in assessment, monitoring and imaging. Nanomedicine (Lond) 6:879–898

Stolen ED, Breininger DR, Frederick PC (2005) Using waterbirds as indicators in estuarine systems: successes and perils. In: Bartone S (ed) Estuarine indicators. CRC Press, Boca Raton, FL, pp 409–422

Stout JH, Trust KA (2002) Elemental and organochlorine residues in bald eagles from Adak Island, Alaska. J Wildl Dis 38:511–517

Strmiskova G, Strmiska F (1992) Contents of mineral substances in venison. Nahrung 36:307–308

Strom SM (2008) Total mercury and methylmercury residues in river otters (*Lutra canadensis*) from Wisconsin. Arch Environ Contam Toxicol 54:546–554

Taggart MA, Figuerola J, Green AJ, Mateo R, Deacon C, Osborn D et al (2006) After the Aznalcollar mine spill: arsenic, zinc, selenium, lead and copper levels in the livers and bones of five waterfowl species. Environ Res 100:349–361

Talmage SS, Walton BT (1991) Small mammals as monitors of environmental contaminants. Rev Environ Contam Toxicol 119:48–143

Tataruch F, Kierdorf H (2003) Mammals as biomonitors. In: Markert BA, Breure AM, Zechmeister HG (eds) Bioindicators and biomonitors. principles, concepts and applications. Elsevier, Amsterdam, pp 737–772

Taylor A (1996) Detection and monitoring of disorders of essential trace elements. Ann Clin Biochem 33:486–510

Tchounwou PB, Yedjou CG, Patlolla AK, Sutton DJ (2012) Heavy metals toxicity and the environment. EXS 101:133–164

Tete N, Fritsch C, Afonso E, Coeurdassier M, Lambert JC, Giraudoux P et al (2013) Can body condition and somatic indices be used to evaluate metal-induced stress in wild small mammals? PLoS One 8(6):e66399

Thomas VG, Guitart R (2010) Limitations of European Union policy and law for regulating use of lead shot and sinkers: comparisons with North American regulation. Env Pol Gov 20:57–72

Thuiller W, Pironon S, Psomas A, Barbet-Massin M, Jiguet F, Lavergne S et al (2014) The European functional tree of bird life in the face of global change. Nat Commun 5:3118

Traas TP, van Leeuwen CJ (2007) Ecotoxicological effects. In: Leeuwen CJ, van Vermeire TG (eds) Risk assessment of chemicals: an introduction. Springer, Dordrecht, The Netherlands, pp 281–355

Tranel MA, Kimmel RO (2009) Impacts of lead ammunition on wildlife, the environment, and human health – a literature review and implications for Minnesota. In: Watson RT, Fuller M, Pokras M, Hunt WG (eds) Ingestion of lead from spent ammunition: implications for wildlife and humans. The Peregrine Fund, Boise, ID, pp 1–20

Trapp SE, Flaherty EA (2017) Noninvasive and cost-effective trapping method for monitoring sensitive mammal populations. Wildl Soc Bull 41:770–775

Tsunoda M, Aizawa Y, Nakano K, Liu Y, Horiuchi T, Itai K, Humio Tsunoda H (2005) Changes in fluoride levels in the liver, kidney, and brain and in neurotransmitters of mice after subacute administration of fluoride. Fluoride 38:284–292

US EPA (2007) Framework for metals risk assessment. US Environmental Protection Agency, EPA 120/R-07/001, 172 pp [https://www.epa.gov/sites/production/files/2013-09/documents/metals-risk-assessment-final.pdf]

Van den Brink NW, Ma WC (1998) Spatial and temporal trends in levels of trace metals and PCBs in the European badger *Meles meles* (L., 1758) in The Netherlands: implications for reproduction. Sci Total Environ 222:107–118

Vikøren T, Bernhoft A, Waaler T, Handeland K (2005) Liver concentrations of copper, cobalt, and selenium in wild Norwegian red deer (*Cervus elaphus*). J Wildl Dis 41:569–579

Vikøren T, Kristoffersen AB, Lierhagen S, Handeland K (2011) A comparative study of hepatic trace element levels in wild moose, roe deer, and reindeer from Norway. J Wildl Dis 47:661–672

Wada O (2004) What are trace elements? Their deficiency and excess states. JMAJ 47:351–358

Wada H, Yates DE, Evers DC, Taylor RJ, Hopkins WA (2010) Tissue mercury concentrations and adrenocortical responses of female big brown bats (Eptesicus fuscus) near a contaminated river. Ecotoxicology 19(7):1277–1284

Walker LA, Lawlor AJ, Chadwick EA, Potter E, Pereira MG, Shore RF (2010) Inorganic elements in the liver of Eurasian otters, Lutra lutra, from England and Wales in 2007 & 2008: a Predatory Bird Monitoring Scheme (PBMS) report. Centre for Ecology & Hydrology, Lancaster, UK, 13 pp

Walker LA, Lawlor AJ, Chadwick EA, Potter E, Pereira MG, Shore RF (2011) Inorganic elements in the livers of Eurasian otters, Lutra lutra, from England and Wales in 2009: a Predatory Bird Monitoring Scheme (PBMS) report. Centre for Ecology & Hydrology, Lancaster, UK, 12 pp

Wayland M, Scheuhammer AM (2011) Cadmium in birds. In: Beyer WN, Meador JP (eds) Environmental contaminants in biota: interpreting tissue concentrations. CRC Press, Boca Raton, FL, pp 645–666

Weech SA, Wilson LK, Langelier KM, Elliott JE (2003) Mercury residues in livers of bald eagles (*Haliaeetus leucocephalus*) found dead or dying in British Columbia, Canada (1987–1994). Arch Environ Contam Toxicol 45:562–569

Weiner J (1973) Dressing percentage, gross body composition and caloric value of the roe-deer. Acta Theriol 18:209–222

WHO (1973) Trace elements in human nutrition. Report of a WHO Expert Committee, WHO Technical Report Series, No. 532, World Health Organization, Geneva, Switzerland

WHO (1996) Trace elements in human nutrition and health. World Health Organization, Switzerland, Geneva

WHO (2002) Principles and methods for the assessment of risk from essential trace elements. Environmental Health Criteria 228. World Health Organization, Switzerland, Geneva. http://www.inchem.org/documents/ehc/ehc/ehc228.htm

Wijnhoven S, Leuven RSEW, van der Velde G, Jungheim G, Koelemij EI, de Vries FT et al (2007) Heavy-metal concentrations in small mammals from a diffusely polluted floodplain: importance of species- and location-specific characteristics. Arch Environ Contam Toxicol 52:603–613

Wilkomirski B (2013) History of bioindication. Monit Srod Przyrod 14:137–142

Wobeser G, Swift M (1976) Mercury poisoning in a wild mink. J Wildl Dis 12:335–340

Wolfe MF, Atkeson T, Bowerman W, Burger K, Evers DC, Murray MW et al (2007) Wildlife indicators. In: Harris R, Krabbenhoft DP, Mason R, Murray MW, Reash R, Saltman T (eds) Ecosystem response to mercury contamination: indicators of change. SETAC, CRC Press, Webster, NY, pp 123–189

Wolfe LL, Conner MM, Bedwell CL, Lukacs PM, Miller MW (2010) Select tissue mineral concentrations and chronic wasting disease status in mule deer from North-central Colorado. J Wildl Dis 46:1029–1034

Wood PB, White JH, Steffer A, Wood JM, Facemire CF, Percival HF (1996) Mercury concentrations in tissues of Florida bald eagle. J Wildl Manag 60:178–185

Wren CD (1984) Distribution of metals in tissues of beaver, raccoon and otter from Ontario, Canada. Sci Total Environ 34:177–184

Wren CD (1986) Mammals as biological monitors of environmental metal levels. I. Mercury. Environ Monit Assess 6:127–144

Wren CD, MacCrimmon H, Frank R, Suda P (1980) Total methylmercury levels in wild mammals from the Precambrian shield area of south central Ontario, Canada. Bull Environ Contam Toxicol 25:100–105

Yaroshevsky AA (2006) Abundances of chemical elements in the Earth's crust. Geochem Int 44: 48–55

Yates DE, Mayack DT, Munney K, Evers DC, Major A, Kaur T, Taylor RJ (2005) Mercury levels in mink (*Mustela vison*) and river otter (*Lontra canadensis*) from northeastern North America. Ecotoxicology 14:263–274

Yates DE, Adams EM, Angelo SE, Evers DC, Schmerfeld J, Moore MS et al (2014) Mercury in bats from the northeastern United States. Ecotoxicology 23:45–55

Yatoo MI, Saxena A, Deepa PM, Habeab BP, Devi S, Jatav RS et al (2013) Role of trace elements in animals: a review. Vet World 6:963–967

Zhang Y, Gladyshev VN (2010) General trends in trace element utilization revealed by comparative genomic analyses of Co, Cu, Mo, Ni, and Se. J Biol Chem 285:3393–3405

Zhang WW, Ma JZ (2011) Waterbirds as bioindicators of wetland heavy metal pollution. Proc Environ Sci 10:2769–2774

Zimmerman TJ, Jenks JA, Leslie DM Jr, Neiger RD (2008) Hepatic minerals of white-tailed and mule deer in the southern black hills, South Dakota. J Wild Dis 44:341–350

Zukal J, Pikula J, Bandouchova H (2015) Bats as bioindicators of heavy metal pollution: history and prospect. Mamm Biol 80:220–227

# Part II
# Selected Trace Elements

# Chapter 3
# Chromium, Cr

**Tadeusz Kośla, Iwona Lasocka, and Marta Kołnierzak**

**Abstract** In 1957 chromium(Cr) was proposed as an essential element, vital for the life and normal development of animals and humans. In the natural environment, the activity of compounds containing chromium in the third and the sixth oxidation states differs substantially, Cr(III) and Cr(VI), respectively. Trivalent Cr is essential for humans and animals; it shapes the proper glucose metabolism through participation in the glucose tolerance factor (GTF) and is involved in metabolism of hydrocarbons, proteins, and lipids. Chromium supplementation brings positive results only at small doses though. Chromium-deficient nutrition impairs glucose tolerance and insulin function, alters protein metabolism, and negatively affects both growth and reproduction.

Hexavalent Cr is a strong oxidant, easily penetrating into living organisms, being reduced to Cr(III) in cells. Industrial production is a source of Cr-containing wastes, which contaminate the water and air and, in consequence, the soil. Emission of chromium to the atmosphere is mainly due to combustion of coal and other fossil fuels but also results from iron and nonferrous metal smelting. Hexavalent Cr, which acts oxidatively, is very toxic. Anthropogenic Cr soil contamination is a result of atmospheric deposition of dust but also industrial wastes discharged into the soil from paint factories, tanneries, sewage treatment plants, and chrome-steel scrap piles. Chromium is toxic to plants and accumulated in the roots and is to a limited extent transferred to overground parts of plants. In the cells of mammals and birds, Cr(VI) is reduced to Cr(III), which produces highly toxic-free radicals. Hexavalent Cr is carcinogenic to homeothermic vertebrates. Reproduction disorders were observed in mammals. In males, exposure to high Cr level deteriorates the quality of semen, leads to testicular disorders, and reduces libido. In females, Cr negatively affects fertility. A high level of Cr in the environment is mutagenic, carcinogenic, and teratogenic to birds; concentration of chromium in avian lungs increases with age, which implies equivalence of diet and air as sources of Cr intoxication. Studies

T. Kośla (✉) · I. Lasocka · M. Kołnierzak
Department of Animal Environment Biology, Warsaw University of Life Sciences – SGGW, Warsaw, Poland
e-mail: tadeusz_kosla@sggw.pl



E. Kalisińska (ed.), *Mammals and Birds as Bioindicators of Trace Element Contaminations in Terrestrial Environments*,
https://doi.org/10.1007/978-3-030-00121-6_3

show that chromium exhibits no biomagnification. On the contrary, with an increase in the trophic level, the concentration of Cr considerably decreases. This is referred to as "biominification." In this chapter are presented the effect of exposure to hexavalent Cr and chromium content in avian and mammalian soft tissues as well as in feathers and hair.

# 1 Introduction

More than half a century after chromium was recognized as an essential nutrient (Mertz and Schwarz 1957), experiments yield results that may undermine the opinion that chromium really is a necessary component for the normal development of warm-blooded organisms (Nielsen 2007; Zaccaroni et al. 2008a; Di Bona et al. 2011).

Varied assimilation and toxicity of trivalent and hexavalent chromium has been a major issue in the discussion on the topic (Anke et al. 2005). On the one hand, chromium is a key metal in many industries, including metallurgy, electroplating, chemical industry, leather tanning, pigment production, and wood preservation (Cotton et al. 1999; Gheju and Balcu 2010; Kabata-Pendias 2011; Pati et al. 2014). On the other hand, the toxic, mutagenic, and carcinogenic properties of chromium are well known too (Bartlett 1991; Irwin 1997; Barałkiewicz and Siepak 1999; Dhal et al. 2013; Singh et al. 2015). Chromium induces an oxidative stress and kills cells, damages DNA, and modifies gene expression (Bagchi et al. 2002; Pechova and Pavlata 2007; Gheju and Balcu 2010; Cantu et al. 2014). Hexavalent chromium is toxic to plants, animals, and humans (Borel and Anderson 1984; Mertz 1993; Cohen et al. 1993; Mancuso 1997; Wise et al. 2002; OSHA 2006; Islam et al. 2007; Wise et al. 2008; Singh et al. 2015). It is the most mobile, reactive, and toxic form of chromium (Bartlett 1991; Tribovillard et al. 2006; Dhal et al. 2013). The main mechanism underlying the toxicity of Cr(VI) is the oxidative stress (Bagchi et al. 2003; Cantu et al. 2014). In endothermic vertebrates, long-lasting exposure to chromium(VI) leads to neurotoxic, dermatoxic, and genotoxic effects and to carcinogenic and mutagenic changes; in humans, it additionally leads to skin allergies, ulcers, and kidney failures (Bagchi et al. 2002; Thacker et al. 2006; Singh et al. 2015).

Most authors have confidence that trivalent chromium is an essential element and emphasize its positive influence if included in the nutrition of both humans and animals (Gheju and Balcu 2010; Weksler-Zangen et al. 2012; Lewicki et al. 2014).

Experiments carried out in the 1990s confirm that chromium is a necessary nutrient for various groups of animals, such as cattle, sheep, pigs, and poultry (Pechova and Pavlata 2007).

In animals, Cr(VI) rapidly converts to Cr(III) (Anke et al. 2005). The environmental mobility of chromium(III) is lower, and it is 1000 times less toxic for living cells as compared with hexavalent Cr (Bagchi et al. 2003; Suwalsky et al. 2008;

Gheju and Balcu 2010; Singh et al. 2015). Trivalent, non-resorbable chromium is used as a marker (Anke et al. 2005).

Chromium was identified as a component of biological tissues in the 1940s. Mertz and Schwarz (1957) observed developmental abnormalities in a population of rats fed with a chromium-deficient diet and hypothesized that the feed lacked the glucose tolerance factor (GTF). This has been confirmed in other studies (Schwarz and Mertz 1959; Mertz 1993; Grela et al. 1997; Pechova and Pavlata 2007; Vincent and Stallings 2007).

GTF has been confirmed to contain an active form of trivalent chromium (Mertz 1993). According to Grela et al. (1997), GTF is a trivalent chromium-nicotinic acid occurring in combination with glutamic acid, glycine, and cysteine. It improves insulin binding to its receptors, with the participation of hydrogen sulfide. Trivalent chromium enhances the action of insulin, both in vitro and in vivo. Many experiments on animals, studies on human diabetic patients, and in vitro trials have shown that Cr(III) improves the efficacy of insulin and enhances the actions of tyrosine kinase and the transmembrane tyrosine phosphatase of insulin receptors. As a result, glucose is more efficiently absorbed and utilized by the cells of the body (Krejpcio et al. 2007). The maximum in vitro activity requires a specific chemical form of GTF and is initially identified as chromium-niacin complex (Mertz 1993; Piva et al. 2003). The importance of chromium in terms of human diabetes has been confirmed in a number of studies (Mertz 1993; Grela et al. 1997; Kurył and Dębski 2001; Krejpcio et al. 2007; Pechova and Pavlata 2007).

The results of experiments demonstrate that chromium is an essential micronutrient in the metabolism of carbohydrates, lipids, and proteins in mammals (NRC 1989; Anderson et al. 1991; Anderson 1993; Mertz 1993; Morris et al. 1995; Mordenti and Piva 1997; Şahin et al. 2001; Pollard et al. 2002; Brown 2003; Lewicki et al. 2014). In addition, chromium is involved in the antioxidant processes, RNA and DNA synthesis, and the immune response and also has an effect on the secretion of hormones and some vitamins (Grela et al. 1997; NRC 1997; Thor et al. 2011).

It has been demonstrated in various livestock animals that a diet low in chromium leads to such deficiency symptoms as reduced feed intake, lower weight gains, reproduction disorders, and a higher lipid content in blood serum (Frank et al. 2000; Anke et al. 2001; Bagchi et al. 2002). Parenterally nourished patients showed symptoms of diabetes (including reduced glucose tolerance, weight loss, and neurological disorders), which were subsequently effectively alleviated with chromium administration (Anderson 1995; Gammelgaard et al. 1999). Insufficient intake of chromium leads to symptoms similar to those observed in diabetes and cardiovascular disorders. Chromium is a nutrient, not a medication; thus it is beneficial only for patients diagnosed for its deficiency (Anderson 1997). In endothermic vertebrates, including humans, such cases are rare (Anderson et al. 1991). Chromium deficiency may mostly occur due to low dietary intakes of chromium, less often due to stress (Grela et al. 1997). Children with protein-energy malnutrition, diabetic patients, as well as the elderly and middle-aged definitely require supplementation with chromium (Borel and Anderson 1984).

Chromium supplementation, regardless of its form, produces beneficial outcomes only at low doses; its excess in the diet interferes with the development of animals (Pollard et al. 2002). Supplements of Cr enhance the binding of insulin and increase the number of insulin receptors and their phosphorylation (Anderson 1998; Brown 2003).

Chronic stress may alter the daily requirement for micronutrients, including chromium. Stress situations in animals result from transport, high temperatures in the premises, rapid growth rate, aging, and diets requiring full activation of insulin; also the species and/or breed of the animal may induce the symptoms of chromium deficiency if its intakes are short. Stress factors affecting chromium metabolism in humans include heavy glycemic load, a diet high in monosaccharides, lactation, infection, or injuries. The response of the human body in such situations consists in elevated chromium urinary excretion. If these stress factors, which lead to reduced renal reabsorption of chromium and its increased excretion with urine, add to a reduced Cr level in the diet, all this may result in the symptoms of chromium deficiency (Anderson 1997). Under increased stress, such as fatigue, injuries, and pregnancy, but also as a result of dietary, metabolic, physical, environmental, or emotional stress, the chromium requirement increases (Burton 1995; Piva et al. 2003).

The role of chromium in cholesterol metabolism remains unexplained (Wang et al. 1989; Press et al. 1990; Abraham et al. 1992; Pechova and Pavlata 2007; Hua et al. 2012). A slight chromium deficiency increases the risk of ischemic heart disease and myocardial infarction (Mertz 1993). Grela et al. (1997) suggest that the condition is an important risk factor in the coronary heart disease, since it is accompanied by high blood pressure, high plasma triglyceride levels, and low plasma high-density lipoproteins (HDL) levels.

The animal daily requirement for chromium is probably 300 μg Cr kg dry weight (dw) of feed and is usually covered by the feed ration. If the demand is met, any chromium supplements will fail to improve the health status of the animal (Anke et al. 2005).

Chromium reduces total body weight but can increase lean body mass (Morris et al. 1995; Bielicka et al. 2005). It has been found that Cr also improves lean body mass in animals, enhances growth rate and feed conversion, and stimulates feed intake and energetic efficiency (NRC 1996). Chromium may accelerate the growth rate of skeletal musculature and cardiac muscle (Morris et al. 1995). Studies on farm animals (Grela et al. 1997; Pollard et al. 2002) have shown that chromium yeast supplementation improves their production parameters (better carcass lean content, lower fat percentage, improved weight gains, more effective feed conversion, higher content of polyunsaturated fatty acids). Organic chromium reduces the mortality of piglets and chicks, reduces disease incidence in bulls, and improves reproduction performance. It increases the immunity of the body and improves the response to vaccination (NRC 1997). It enhances the immune response and reduces blood levels of cortisol, which is an indicator of stress. This form of chromium improves milk yields in dairy cows, and increases equine endurance in exercise, reducing muscle lactate content in horses, which improves muscle efficiency (Grela et al. 1997). Chromium ions stimulate lipid metabolism in broiler chickens (Kurył and Dębski 2001; Krejpcio et al. 2007).

A risk of chromium contamination of the natural environment on a global scale does not seem real. However, due to its increased use and significant local introduction of chromium to the atmosphere, water, and soil, an excessive share of chromium in the biogeochemical cycle and increased risks to the environmental health should be anticipated (Kabata-Pendias and Mukherjee 2007). Hence, it is reasonable to collect and study data on chromium concentrations in a variety of organisms, including wild birds and mammals.

## 2 General Properties

Chromium(Cr) was discovered by Vauquelin in 1798, who isolated the element from crocoite, $PbCrO_4$ (Barceloux 1999). In the periodic table of elements, chromium is a member of Group 6 (previously VIb). The valence of chromium may vary from −2 to +6. Its most common forms found in the environment are trivalent, Cr(III), and hexavalent, Cr(VI), and both have a significant impact on living organisms, including humans (WHO 1988; Miksche and Lewalter 1997; Thacker et al. 2006; Oliveira 2012). Divalent chromium ion is rapidly oxidized to trivalent chromium and therefore does not occur in living organisms (Borel and Anderson 1984; Pechova and Pavlata 2007).

Chromium is a hard, heavy, and brittle metal, steely-gray in color with a bluish tint. Its atomic number is 24, and its atomic weight is 51.996 g $mol^{-1}$. There are more than 20 chromium isotopes (with atomic weight from $Cr^{42}$ to $Cr^{67}$), but only four of them are stable ($Cr^{50}$, $Cr^{52}$, $Cr^{53}$, $Cr^{54}$). The most common in nature chromium isotopes are $Cr^{52}$ (83.76%) and $Cr^{53}$ (9.55%) (Borel and Anderson 1984; Rosman and Taylor 1998; Stoecker 2004; Hammond 2005; Izbicki et al. 2008). The melting point and the boiling point of chromium are, respectively, 1857 °C and 2672 °C. Chromium dissolves in dilute sulfuric and hydrochloric acids and passivates in concentrated nitric acid. It forms a range of complex ions (Borel and Anderson 1984; Greenwood and Earnshaw 1997). In complex compounds, the element occurs in the oxidation states from −2 to +6, of which the substances containing Cr in the +2 and +3 oxidation states are most common (Borel and Anderson 1984).

In the rank representing the abundance of elements in the Earth's crust, chromium occupies the 21st place (McGrath and Smith 1990; Barnhart 1997; Cervantes et al. 2001; Zou et al. 2006; Thacker et al. 2006; Pechova and Pavlata 2007; Ebdon et al. 2001), but it is the seventh most abundant element within the globe's surface layer (McGrath and Smith 1990; Kirklin 1999; Cervantes et al. 2001; Anke et al. 2005). Its average content in the Earth's crust is 100 mg $kg^{-1}$ (Cervantes et al. 2001; Anke et al. 2005; Emsley 2011), but it attains its highest concentrations in ultrabasic igneous rocks, 1600–3400 mg $kg^{-1}$, and the lowest in acidic volcanic rocks, limestones and dolomites, 5–16 mg $kg^{-1}$ (Papp 1994; Kabata-Pendias and Mukherjee 2007). Metallic chromium(0) does not occur in the Earth's crust and is biologically neutral (Pechova and Pavlata 2007).

The environmental effects of the compounds of chromium in the third and sixth oxidation state are fundamentally different, and therefore these two should definitely be considered independently (Bielicka et al. 2005).

Trivalent Cr is the most stable ion in the biological system (Borel and Anderson 1984; Mertz 1993; Irwin 1997; Thacker et al. 2006; Pechova and Pavlata 2007) and does not penetrate cell membranes (Mertz 1993). Trivalent Cr tends to form ligand complexes. Chromium hydroxide [$Cr(OH)_3$], which contains Cr(III), is nearly insoluble in water within the range of pH 6–9, and its migration through the soil to the groundwater is limited (Jeyasingh and Philip 2005; Cantu et al. 2014).

Biologically active chromium(III) form, low-molecular-weight chromium-binding substance (LMWCr) may take part in the insulin-signaling pathway (Hua et al. 2012; Vincent 2015). Studies on laboratory and farm animals indicate that organic and inorganic compounds of Cr are beneficial to carbohydrate and lipid metabolism, improve the growth of animals, and reduce mortality (NRC 1989; Anderson 1993; Grela et al. 1997; Şahin et al. 2001; Pollard et al. 2002; Lewicki et al. 2014).

On the other hand, the dietary guidelines for daily chromium intake in the USA were in 2001 lowered from 50–200 μg (for an adult person) to 35 μg for an adult male and to 25 μg for an adult female (Thor et al. 2011).

Hexavalent chromium is highly soluble and reactive; therefore, the ion easily penetrates into living organisms, causing changes in their functioning (Bartlett 1991; Peterson et al. 2008; Wuana and Okieimen 2011; Dhal et al. 2013). The hexavalent form is a strong oxidant; in acidic soils, with a high content of organic matter, this ion can easily convert to the trivalent, nontoxic form of chromium (Borel and Anderson 1984; Cervantes et al. 2001; Jianlong et al. 2004). The reverse process of Cr(III) converting into Cr(VI) is also possible, particularly in the presence of common minerals containing manganese (Mn) oxides, mainly of Mn (IV) (Bielicka et al. 2005). Cr(VI) chromates, which prevail under pH-neutral conditions, are absorbed by cells via sulfate channels and are subject to nonenzymatic activation involving ubiquitous ascorbate and small thiols (Zhitkovich 2011).

On entering the cell, hexavalent chromium degrades to Cr(III), which has a stronger binding affinity to DNA, RNA, proteins, and lipids (De Flora et al. 1990; McGrath and Smith 1990; Cervantes et al. 2001; Pechova and Pavlata 2007). In this process, however, other transitional products are generated, including long-persisting Cr(IV) and Cr(V) (Shi and Dalal 1989, 1990).

A high concentration of hexavalent Cr may suppress environmental capacity able to reduce the ion to Cr(III) and so may create a serious pollution problem. Moreover, Cr(III) can be oxidized to Cr(VI) as a result of excess oxygen (Cervantes et al. 2001).

## 3 Chromium Minerals, Production, and Uses

Industrial production of chromium is based on chromium-rich ore called chromite or ferrochromite ($FeCr_2O_4$) (Irwin 1997) (Fig. 3.1). Chromite may contain up to 55% $Cr_2O_3$. Most chromite deposits, which are of economic importance, contain at least

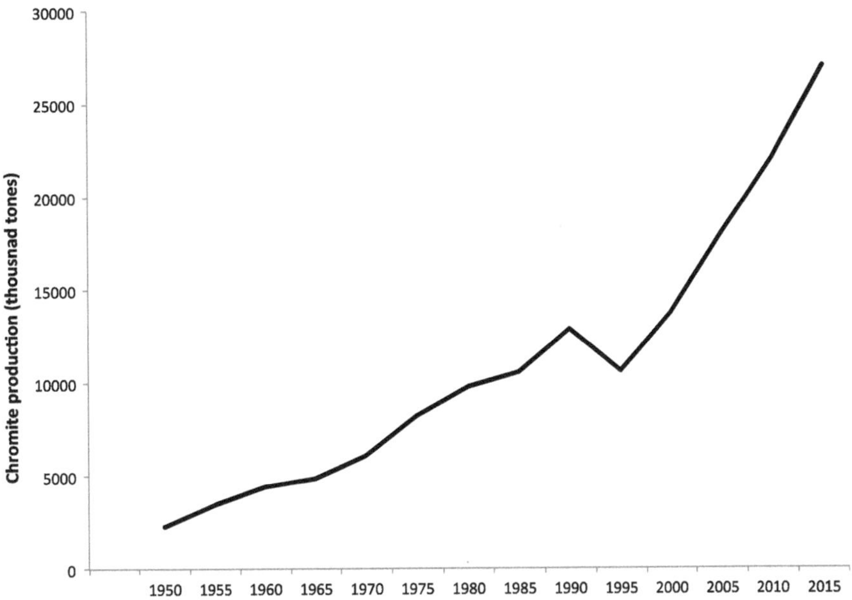


**Fig. 3.1** World chromite production in 1950–2015 (Source of data: US Geological Survey https://minerals.usgs.gov/minerals/pubs/commodity/chromium/index.html#myb)

25% of the mineral (Irwin 1997; Papp and Lipin 2006). Another source of chromium is crocoite ($PbCrO_4$), known as a secondary mineral found in the oxidized zones of lead deposits (Gaines et al. 1997). Chromium was first detected in crocoite by Vauquelin in 1798 (Bequer et al. 2003; Shanker et al. 2005). Crocoite occurs in chromite deposits, coal seams, and quartz veins (Zhangsheng et al. 2001; Bequer et al. 2003). It contains hexavalent form of chromium (Irwin 1997). Another mineral containing chromium is magnesiochromite ($MgCr_2O_4$), belonging to the spinel group, takes part in the process of oxidation and is used in flotational chromite ore beneficiation. Magnesiochromite is an industrially important mineral (Sobieraj and Laskowski 1973; Vert 2016).

The metallurgical, refractory, and chemical industries are the fundamental users of chromium (Jacobs and Testa 2005; ATSDR 2012; Vert 2016). About 60–70% of chromium is used in the production of alloys, including stainless steel, and another 15% is used in chemical industrial processes (US Geological Survey 2010; Papp 2005, 2016). The main applications of Cr involve the production of nonferrous alloys, decorative finishes, metal plating, and green-colored glass (Jacobs and Testa 2005). The aircraft manufacturing industry has been using chromium for aluminum anodizing. Cr is also used as a catalyst in chemical reactions (ATSDR 2012). Oxidizing agents containing Cr are used for qualitative analysis (Jacobs and Testa 2005).

According to the Mineral Commodity Summaries 2015 (US Geological Survey 2015), global reserves of Cr-rich chromite ore are estimated at over 480 million tons,

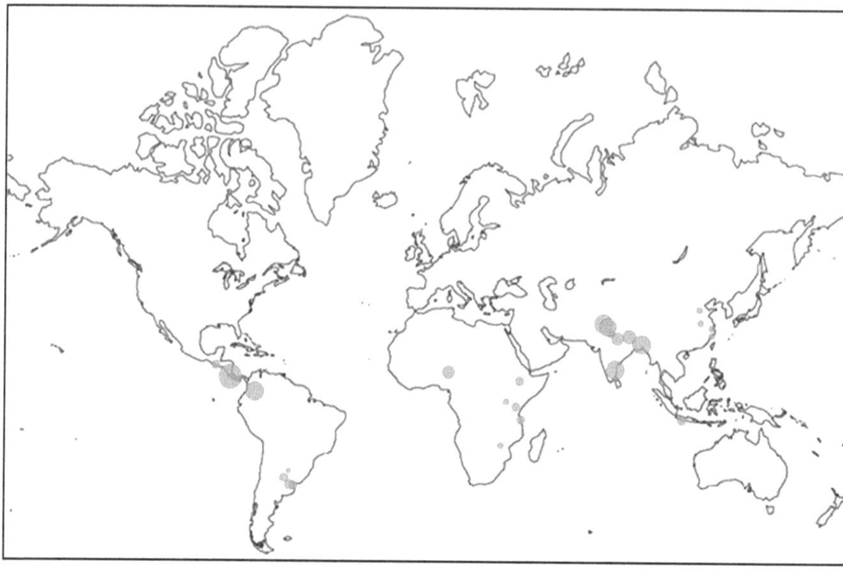

**Fig. 3.2** Chromium pollution of Earth regions from tanneries (Based on https://www.worstpolluted.org/projects_reports/display/88)

most of which is located in Kazakhstan, South Africa, and India (230, 200, and 54 million tons, respectively). South Africa, India, Kazakhstan, and Turkey, which together make up about 80% of the world chromite production (Papp 2011, 2016), have been the leading chromite mining countries for many years.

Industrial wastes carry large amounts of chromium in the form of chromates, dichromates, and other compounds, which threaten the aquatic environments. Both air and water become polluted by industrial plants that combust liquid fuels or wastes. US EPA (2010) has listed chromium as a priority pollutant (Rauch and Pacyna 2009; Friis 2012; Pati et al. 2014).

Dyestuffs and leather tanning agents discharged directly into waste streams, either as liquids or solids, are considered as the main source of chromium pollution (Fig. 3.2).

Both industrial residential sewage treatment plants discharge substantial amounts of Cr. Thus, COPR (chromite ore processing residue) is one of the greatest environmental threats in some regions (Kabata-Pendias and Mukherjee 2007; ATSDR 2012).

# 4 Chromium in Nature

The sources of chromium present in the natural environment are both natural and anthropogenic. Table 3.1 shows the chromium content in the environment and its various components.

**Table 3.1** Global mean chromium concentrations in abiotic reservoirs

| Reservoir | Chromium (mg $kg^{-1}$) |
|---|---|
| Upper continental crust | 92 |
| Oceanic crust | 320 |
| Continental sediment | 74 |
| Oceanic sediment | 79 |
| Loess | 59 |
| Soil | 130 |
| Coal | 16 |
| Crude oil | 0.19 |
| Rivers (particulate) | 100 |
| Rivers (dissolved) | 0.00085 |
| Freshwater lakes (dissolved) | 0.00025 |
| Groundwater (dissolved) | 0.0007 |
| Greenland ice | 0.000038 |
| Antarctic ice | 0.000007 |
| Ocean (dissolved) | 0.00026 |
| Air | 0.015–0.03 mg $m^{-3}$ |
| Air | Average 0.6 ng $m^{-3}$, urban, industrial 1–1100 ng $m^{-3}$ |

Data sources: Kamaludeen et al. (2003), Shanker et al. (2005), Kabata-Pendias and Mukherjee (2007), and Rauch and Pacyna (2009)

An air concentration of chromium in the range 0.01–1 ng $m^{-3}$ is considered normal; however, it may vary greatly depending on the anthropogenic impact (Barałkiewicz and Siepak 1999; Kabata-Pendias 2011; Pati et al. 2014). The average concentration of the metal in the atmosphere ranges from 1 ng $m^{-3}$, in rural areas, to 10 ng $m^{-3}$, in polluted urban sites (Bielicka et al. 2005). If we look at the Northern Hemisphere, chromium concentrations in the air vary as follows (according to Kabata-Pendias 2011): the cleanest air, 0.6–0.8 ng $m^{-3}$, can be found in Greenland, 1–140 in Germany, and 1.3–167 in Japan, and the highest chromium concentrations are found in North America, ranging between 5 and 1000 ng $m^{-3}$. In heavily polluted areas, chromium concentrations in the air may exceed 1000 ng $m^{-3}$. Maximum acceptable yearly average chromium air concentrations are 0.4 for cities and 0.08 for protected areas (Barałkiewicz and Siepak 1999). According to Wise et al. (2008), a chromium level in the range 1–100 ng $m^{-3}$ poses a threat of exposure to hexavalent Cr and afflicts human lungs with 2 μg of chromium per day.

The United States Environmental Protection Agency (EPA 2000) estimated the lifetime cancer risk due to exposure to chromium(VI) to be $1.2 \times 10^{-2}$. According to IARC (1990), neither metallic nor trivalent chromium is classified as a human carcinogen, contrary to the carcinogenic hexavalent form of chromium.

Chromium occurring in the atmosphere is in 60–70% of the anthropogenic origin, including that of iron and steel production and, to a lesser extent, fossil fuel combustion. The remaining 30–40% of chromium in the air occurs naturally (Talebi 2003; Bielicka et al. 2005).

Global emission of chromium to the atmosphere is estimated at about 30,000 tons, most of air pollution from combustion of coal and other fossil fuels, as well as from the smelting of iron and nonferrous metals (Kabata-Pendias 2011). According to the US Agency for Toxic Substances and Disease Registry (ATSDR), chromium VI accounts for approximately one third of the 2700–2900 tons of chromium emitted to the atmosphere annually in the USA (ATSDR 2012). Total atmospheric emissions of chromium in China during the period from 1990 to 2009 were estimated at approximately $1.92 \times 10^5$ tons, at annual growth rates of 8.8% since 1990. Coal combustion was identified as the largest contributor, though a decrease from 60.2% in 1990 to 42.2% in 2009 was noted; nevertheless, a contribution of oil combustion has grown slightly during these years (Cheng et al. 2014). The European Union emission inventory report informs that between 1990 and 2013, Cr emissions in the EU-28 dropped by 74% (EEA 2015).

Hexavalent Cr is very toxic and soluble in water and can migrate down to the groundwater. The pollution of groundwater by Cr(VI) is related to the widespread industrial activity, like the usage of hexavalent chromium in various processes and discharges of Cr-bearing wastes. For the USA, a median value of 10 $\mu g\ L^{-1}$ in fresh water is reported (ATSDR 2012). Chromium concentrations in riverine waters, according to Kabata-Pendias and Mukherjee (2007), range between 0.29 and 11.46 $\mu g\ L^{-1}$. Barałkiewicz and Siepak (1999) report that chromium concentrations in unpolluted lakes are on average 1–2 $\mu g\ L^{-1}$ and in rivers from 5 to 50 $\mu g\ L^{-1}$ (e.g., Germany 2.5–15.5, Russia-Siberia > 10, Italy 0.1–1.2 $\mu g\ L^{-1}$). Also Bielicka et al. (2005) report that the chromium concentration in lakes and rivers is usually between 1 and 10 $\mu g\ L^{-1}$, regardless of the estimated 6.7 million kg of Cr discharged annually into the sea with industrial wastes. A total of 13,400 tons of Cr was discharged to the water in China during the years 1990–2009, mainly by fabricated metal products manufacturers and the leather tanning industry, accounting for 68.0% and 20.0% of the total emission, and representing increases of 15.6% and 10.3% annually, respectively (Cheng et al. 2014).

Over the last decade, there have been many reports on very high levels of Cr(VI) naturally occurring in groundwater. For example, in Cazadero County (California, USA), 12–22 $\mu g\ L^{-1}$ of Cr(VI) was found in spring water from ultramafic rock; in La Spezia (Italy), 5–73 $\mu g\ L^{-1}$ was found in groundwater from ophiolite complex (McNeill and McLean 2012). The US EPA regulates total chromium in drinking water and has set a maximum contaminant level (MCL) of 0.1 $mg\ L^{-1}$ and 100 and 50 $\mu g\ L^{-1}$, respectively, for Cr(III) and Cr(VI) (Das and Singh 2011; Dhal et al. 2013). The World Health Organization (WHO) guideline is 0.05 $mg\ L^{-1}$ for total chromium.

Chromium in soils occurs mainly in its trivalent oxidation state, which is poorly absorbed by vegetation (Bielicka et al. 2005). In strongly acidic soils, trivalent chromium dissolves well, whereas Cr(VI) is well soluble in both acidic and alkaline soils. Chromium(III) mobility is reduced through adsorption by loam and mineral oxides at pH <5 and low solubility at pH > 5 due to formation of $Cr(OH)_3$. Chromium in soil is reduced by a number of organic compounds, primarily by reduced sulfur compounds and Fe(II) compounds (Buerge and Hug 1999). In the presence of organic ligands (humic and fulvic acids), trivalent chromium may undergo complexation, which increases the solubility of trivalent chromium compounds and, therefore, leads to its higher mobility and better accessibility for living organisms (Wyszkowski and Radziemska 2009; Wuana and Okieimen 2011). The most mobile forms of chromium in the soil are $CrO_4^{2-}$ and $HCrO_4^-$ ions (Dhal et al. 2013). Thakur et al. (2007) describe the average content in the soils of the world (Table 3.2).

Barnhart (1997) estimated the average soil Cr concentration in the USA at 40 mg kg$^{-1}$, whereas in the Czech Republic from 4.16 to 7.50 mg kg$^{-1}$ (Mazanec 1996). The concentration range in the soil and regulatory guidelines have been established for some heavy metals, including chromium: 0.05–3950 mg kg$^{-1}$, regulatory limits 100 mg kg$^{-1}$ (Riley et al. 1992; NJDEP 1996; Wuana and Okieimen 2011). Anke et al. (2005) ranked the soils in terms of chromium content as follows (in descending order): Rotliegend, granite, syenite, peat and muck soils, phyllite, shale, sand and Buntsandstein soil, glacial till clays, and sandy alluvial soil.

Canada's chromium in soil guidelines for the protection of environmental and human health is generally lower than soil standards in other countries. On Canadian agricultural, residential, and park soils, the total Cr should be below 64 mg kg$^{-1}$ and Cr(VI) below 0.4 mg kg$^{-1}$. For industrial and commercial lands, the limits are 87 mg kg$^{-1}$ total Cr and 1.4 mg kg$^{-1}$ Cr(VI). The upper limits of typical Ontario chromium concentrations in uncontaminated soils are slightly higher than the

**Table 3.2** Chromium concentrations in soils

| Region | Soil | Concentration (mg Cr kg$^{-1}$) | References |
|---|---|---|---|
| World | Natural | 10–15 | Thakur et al. (2007) |
| World | Serpentine | 634–125,000 | Thakur et al. (2007) |
| World | World soils | 200 (mean value) | Thakur et al. (2007) |
| World | | 100–300 | Thakur et al. (2007) |
| World | | 10–150 | Thakur et al. (2007) |
| USA | Generally | 25–85 | Thakur et al. (2007) |
| *Urban soils* | | | |
| China, Beijing | | 35.60 | Su et al. (2014) |
| China, Shanghai | | 107.90 | Su et al. (2014) |
| China, Changsha | | 121.00 | Su et al. (2014) |
| USA, Cincinnati | | 37.00 | Su et al. (2014) |
| France | | 42.08 | Su et al. (2014) |
| Finland, Turku | | 59.00 | Su et al. (2014) |
| *Arable soils* | | | |
| China, Beijing | | 75.74 | Su et al. (2014) |
| China, Guangzhou | | 64.65 | Su et al. (2014) |
| China, Wuxi | | 58.60 | Su et al. (2014) |
| India | | 1.23–2.19 | Su et al. (2014) |
| USA | | 48.5 | Su et al. (2014) |
| Spain | | 63.48 | Su et al. (2014) |

Canadian guidelines, with 0.66 mg $kg^{-1}$ Cr(VI) and 67–70 mg $kg^{-1}$ total chromium. The Ontario Site Condition Soil Standards have set higher limits than the Canadian guidelines. Limits for soils of all property uses are 160 mg $kg^{-1}$ total chromium and 8–10 mg $kg^{-1}$ Cr(VI). Lower limits for soil within 30 m of a water body are set at 67–70 mg $kg^{-1}$ total chromium and 0.66 mg $kg^{-1}$ Cr(VI) (CCME 1999).

Hexavalent chromium significantly affects biological properties of soil, including the content of nitrogen compounds—has an oxidizing effect and, moreover, penetrates the cell membranes of soil organisms. Due to its high solubility in the soil solution, this species is considered more readily available to plants than Cr(III). It is believed that Cr is not essential for microorganisms or plants and may even be toxic in higher concentrations (Samantaray et al. 1998; Cervantes et al. 2001; Singh et al. 2013). A typical ratio of Cr content in plants to Cr content in soil is estimated at the level 0.0045 (0.45%). It has been demonstrated that Cr in plants, especially in crop plants, occurring in low concentrations in the substrate (0.05–1 mg $L^{-1}$) stimulates plant growth and crop yield. Once Cr is present in the soil solution at a concentration of 1–5 mg $L^{-1}$, it causes adverse changes in the metabolism of plants, leading to growth inhibition and reduction in the chlorophyll synthesis; it also contributes to the development of chlorosis (Cervantes et al. 2001; Dube et al. 2003; Peralta-Videa et al. 2009). Some plant species tend to accumulate the metal in their tissues extensively, without any visible symptoms of poisoning. Those include the crimson-spot rockrose (*Cistus ladanifer*), a shrub native to the western Mediterranean region, which grows on serpentine soils. More than 2600 mg Cr $kg^{-1}$ has been detected in this particular plant (Wuana and Okieimen 2011; Favas et al. 2014). Plants capable of accumulating and tolerating high concentrations of Cr are sometimes found in some areas of high chromium anthropogenic pollution, such as that near a Chinese electroplating facility. The grass *Leersia hexandra* growing in this area had more than 1600 mg Cr $kg^{-1}$ dw (Liu et al. 2011).

In animal bodies, chromium has been estimated in the range from 0.02 to more than 1500 mg $kg^{-1}$ dw, depending on a large number of biotic and abiotic factors (Eisler 1986; Shanker et al. 2005).

Absorption and transport of chromium may be modified by such factors as soil pH, organic matter content, or chelating agents (Han et al. 2004). Geochemical and reductive conditions of the substrate jointly shape chromium solubility, mobility, and bioavailability to organisms (Martello et al. 2007; Peralta-Videa et al. 2009). With the surface runoff, chromium may be transported into water bodies in a soluble form or as a sediment. The soluble and unadsorbed chromium complexes may be washed from the soil to the groundwater. Leeching of Cr(VI) increases with the soil pH. Most of chromium entering natural water bodies binds to various particles and, eventually, is deposited in the sediment (Wuana and Okieimen 2011). The most common anthropogenic chromium soil pollution comes from the atmospheric deposition of dust emitted by electroplating, steel, and automobile manufacturing industries, as well as from any industrial wastes that are introduced to the soil, in particular all kinds of sludge from paint factories, tanneries, sewage treatment plants, or chrome-steel scrap piles. Estimates are that approximately 80–90% of leather is tanned using a chromium-containing substance, of which approx. 40% of Cr is discharged with wastewater as Cr(VI) and Cr(III) (Anke et al. 2005; Dhal et al. 2013). In the USA, Cr(VI) represents approximately 35% of all chromium released from anthropogenic sources (Das and Singh 2011; Dhal et al. 2013).

## 5 Biological Status of Chromium

Chromium(III) is probably an important micronutrient for the glucose metabolism, whereas Cr(VI) is a carcinogenic factor (Peterson et al. 2008).

It has never been considered as essential nutrient in plants; however, positive effects of small Cr doses in terms of growth stimulation were found (Kimborough et al. 1999; Zayed and Terry 2003; Sharma et al. 2003). Also a negative impact of Cr on plants has been reported (Samantaray et al. 1998; Hood 2010; Singh et al. 2013). Toxic effects on plants from chromium exposure are well known, especially in those growing on serpentine soils or in heavily chromium-polluted areas (Anke et al. 2005). Plants can accumulate Cr(III) and Cr(VI) from soil, sediment, water, and atmospheric deposition on leaves. Plant exposure to excess Cr(III) or Cr(VI) can negatively affect plant health and survival. The complex influence of soil chemistry and differences between plant species make it impossible to generalize (Kimborough et al. 1999; Zayed and Terry 2003; Sharma et al. 2003). It has been found that Cr(III) is more toxic against certain plant species as compared to the Cr(VI), unlike in humans (Suwalsky et al. 2008). Higher concentrations of Cr(III) inhibited plant growth or even killed some more susceptible plants (Samantaray et al. 1998; Singh et al. 2013). According to Eisler (1986), chromium is beneficial but not essential to growth in higher plants.

Chromium accumulates mostly in roots, and its transport to shoots, leaves, or fruits of trees and shrubs is limited. Toxic chromium doses hinder plant growth efficiency even by 50% (Samantaray et al. 1998). Toxic effects to plants exposed to excess chromium include reduced growth; decreased chlorophyll production causing yellow leaves, narrow leaves, and small root systems; decreased or complete inhibition of seed germination; delayed growth; decreased seed yield; wilting; and death (Dube et al. 2003; Zayed and Terry 2003; Anke et al. 2005; Mining Watch Canada 2012). Excess chromium damages root membranes and a plant's ability to take up water. It also alters uptake and translocation of essential elements such as nitrogen, iron, potassium, magnesium, manganese, phosphorous, calcium, sulfur, copper, and zinc (Zayed and Terry 2003; Dube et al. 2003; Gardea-Torresdey et al. 2005). It disturbs enzymatic activity and leads to formation of reactive oxygen species, which destroy lipids and DNA in the plant (Singh et al. 2013). Not only is this detrimental to plant health, but a change in plant nutrient balances may also affect wildlife and human health. Sensitivity and effects vary between species, making toxicity predictions difficult without extensive plant studies.

Some researchers suggest that there is a "soil-plant barrier" that protects the terrestrial food chain from excess chromium due to (a) the insolubility of most chromium in soils, (b) >90% of chromium that is taken up is stored in plant roots, and (c) plant toxicity occurring below levels thought to affect plant consumers (Zayed and Terry 2003).

It has been demonstrated that plant bioavailability of Cr(VI) in the soil is about 35% of its total concentration, and *Streptomyces* bacteria are able to transform hexavalent chromium in such a way that 90% of the ion present in soil will be bioavailable to plants (Polti et al. 2009, 2011). Anke et al. (2005) found that parts of plants—such as shoots and bark of various trees and shrubs, shoots of raspberries and blackberries, as well as crop plants—eaten by wild animals in winter contain from 0.70 to 2.80 mg Cr $kg^{-1}$ dw. Chromium toxic concentrations have also been found in tobacco leaves (18–24 mg $kg^{-1}$ dw), maize leaves (4–8 mg $kg^{-1}$ dw), and barley and rice stems (10 and 10–100 mg $kg^{-1}$ dw, respectively). Plant toxicity limit is assumed at the level 10 mg Cr $kg^{-1}$ dw (Anke et al. 2005). Sanyal et al. (2015), who studied aquatic ecosystems of water bodies and rivers, found a high rate of Cr accumulation in the water. The highest concentration of chromium was found in the roots of aquatic plants (877.5 mg $kg^{-1}$). Studies have shown Cr(VI) toxicity to algae and aquatic plants exposed to as little as 1–10 μg $L^{-1}$ (Mining Watch Canada 2012). According to Marchese et al. (2008), Cr(III) induces toxic effects in algae at the level 320 μg $L^{-1}$. Most studies have found Cr(VI) effects occur at lower concentrations and with greater severity. Chromium toxicity to plants has been observed at exposure as low as 160 μg $L^{-1}$ of Cr(VI) (Adema and Henzen 1989) and 104 μg $L^{-1}$ of Cr(III) (Pawlisz et al. 1997) when grown in soilless solution and 1.8 mg $kg^{-1}$ (added as Cr(VI)) when grown in soil (Adema and Henzen 1989).

Anthropogenic chromium pollution of soil in the areas located in the vicinity of cement factories and phosphate-fertilizer plants can significantly increase chromium content in plants, as shown by the example of vegetables as well as legumes and herbs that are eaten on pasture by grazing animals. Chromium content in vegetables collected in an area near a cement factory was from 0.76 to 1.63 mg Cr $kg^{-1}$ dw as compared to non-contaminated areas: 0.34–1.03 mg Cr $kg^{-1}$ dw (Anke et al. 2005).

Fruit and grain eaten by humans and animals usually contain from 0.15 to 0.65 mg Cr $kg^{-1}$ dw and represent a good source of chromium (Anke et al. 2005). There is evidence that plants can convert Cr(VI) to Cr(III) in their roots (Zayed et al. 1998; Howe et al. 2003; Zayed and Terry 2003) though there are differences in observed rates and extent of reduction to Cr(III). Some studies support the concept that plants can quickly convert all Cr(VI) to Cr(III) (Lytle et al. 1998; Zayed et al. 1998; Wang et al. 2011), while other studies have found evidence of Cr(VI) and its intermediates, Cr(V) and Cr(IV), persisting in plants and algae (Micera and Dessi 1988; Liu et al. 1995; Mishra et al. 1995; Aldrich et al. 2003; Howe et al. 2003). As Cr(VI) intermediates are implicated in Cr toxic effects (Stearns et al. 1995), their presence in plants could "produce dangerous effects to ecological cycles" (Micera and Dessi 1988). According ATSDR (2008), there is no documented evidence for biomagnification of Cr from soil to plant to animal, though researchers also acknowledge there is a lack of study regarding the risk of chromium in the food chain (Lind et al. 2001; ATSDR 2008; Peralta-Videa et al. 2009).

The binding of Cr(III) by microalgae is higher at pH ranging 3.5–5.5 than the binding of Cr(VI) at pH $\leq$ 2.0. Reduction of Cr(VI) to Cr(III) increases at a temperature 25–55 °C (Wilde and Benemann 1993). Laboratory tests on *Scenedesmus*, *Selenastrum*, and *Chlorella* algae have shown that Cr(III) is more efficiently removed from an aqueous medium (83–99%) than Cr(VI) (18–22%) (Brady et al. 1994).

Both Cr(III) and Cr(VI) are efficiently removed from aqueous solution through adsorption on various nanomaterials. Cantu et al. (2014) have shown that sorption of Cr(VI) by the oxides of the nanomaterials is in fact reduction to Cr(III), which is next adsorbed on the surface of a metal oxide. Besides metal oxides, multiwalled carbon nanotubes are also effective here.

Chromium is actively transported through cell membranes in prokaryotes and eukaryotes. Most cells are impermeable to Cr(III), which is probably due to the insolubility of its compounds in an alkaline medium (Jianlong et al. 2004). The amount of chromium taken up and stored in living tissue and its effects depend on factors such as species, organism size, sex and developmental stage, water characteristics, and presence of other contaminants (Eisler 1986; Mining Watch Canada 2012).

Yeast exposed to Cr revealed various phenomena inside their cellular processes, including oxidation and reduction reactions, interactions with cellular organelles, binding cytosolic molecules, forming DNA proteins, Cr-DNA adducts, breaks in the DNA strand, and DNA-DNA cross-links (Kaszycki et al. 2004). It has been shown that yeast is able to accumulate high concentrations of both forms of Cr in line with their concentration in the medium (Brady et al. 1994). In *Saccharomyces cerevisiae*, the concentration of chromium may reach as high as 30 mg $g^{-1}$ dw (Batic and Raspor 1998), being usually lower in other species and ranging from 0.45 to 10 mg Cr $g^{-1}$ dw (Muter et al. 2001; Kaszycki et al. 2004; Ksheminska et al. 2005). Yeast Cr tolerance depends on the physiological growth phase, biomass density, and exposure time (Kaszycki et al. 2004).

Soil organisms are important in creating soil structure and cycling organic matter and elements in the environment. Relatively small amounts of Cr(VI) (1–10-mg $kg^{-1}$) and Cr(III) (25–100 mg $kg^{-1}$) can negatively impact the number, type, health, and activity of soil organisms and so impact the health of the soil ecosystem (MOE 2011; Viti et al. 2006). Tests conducted under and near a steel alloy plant slag heap in China found that chromium was related to decreased microorganism numbers and enzyme activity (Huang et al. 2008). Ten thousand to fifteen thousand μg $L^{-1}$ of Cr(VI) in water applied to soil was lethal to two species of earthworms (Eisler 1986). Total chromium levels of 671–1400 mg $kg^{-1}$ killed 50% of a tested earthworm population (*Eisenia fetida*), and 1000 mg Cr $kg^{-1}$ significantly reduced earthworm (*Eisenia andrei*) growth and cocoon production (CCME 1999).

Concentrations of chromium were measured in insects such as Coleoptera, Diptera, Hemiptera, and Hymenoptera captured at ultramafic and non-ultramafic sites in the Highlands of Scotland. Hemiptera showed the greatest concentration differences between the ultramafic and non-ultramafic sites in Cr, with greatly increased levels in ultramafic sites. Coleoptera and Hymenoptera had lower levels of Cr at the ultramafic site than at the non-ultramafic sites. Diptera showed no difference in Cr levels from any site (Davison et al. 1999). Studies by Corbi et al.

(2011) confirm that Cr may also bioaccumulate in the Odonata larvae. Their experiment revealed a higher concentration of Cr in the larvae inhabiting the lower reaches of the river, which is related to its higher concentration in the riverine sediments. The studies on invertebrates did not reveal significant differences in Cr concentrations, though it was noted that habitats more heavily polluted with Cr lack stoneflies (Plecoptera), which are otherwise present in cleaner parts of the river. This may be due to a particular sensitivity of these taxa to chromium pollution (Corbi et al. 2011). Canada has set a quality criteria for agricultural irrigation water and livestock water. The Canadian Water Quality Guideline for the Protection of Freshwater Aquatic Life for Cr(VI) at 1 μg $L^{-1}$ is based on toxicity to a sensitive invertebrate (*Ceriodaphnia dubia*) (Mining Watch Canada 2012). Azam et al. (2015) found that the concentration of toxic metals in insects (a libellulid dragonfly, *Crocothemis servilia*; an acridid grasshopper, *Oxya hyla hyla*; and a nymphalid butterfly, *Danaus chrysippus*) is highest in industrial areas and lowest in places most distant from cities. This implies that these insects represent good bioindicators of environmental pollution (Azam et al. 2015). Long-time exposure of insects to toxic metals may inhibit their growth, impair their development, and lead to reproductive disorders and lower proliferation. Industrial effluent discharged in eastern France contained about 25 mg $L^{-1}$ Cr, which significantly reduced the fertility of the pond snail *Lymnaea palustris* (Coeurdassier et al. 2005). Aquatic organisms may be exposed to various forms of chromium present in both sediments and water column. Crustaceans accumulate chromium in their exoskeleton (Marchese et al. 2008; Peralta-Videa et al. 2009). Weegman and Weegman (2007) investigated whether the population decline of the lesser scaup *Aythya affinis* was related to the diet (zebra mussels, fingernail clam *Sphaerium transversum*, chironomid larvae, gastropods, and amphipods). The authors propose that chromium accumulated in the bivalves, and amphipods may have caused the scaup decline.

Chromium III in water seems to be more toxic to fish than Cr(VI). It can decrease fertilization success and deposit onto gills which damages tissue and function and can cause death at relatively low doses. Trivalent Cr has caused toxic effects on fish at 5 μg $L^{-1}$ and on invertebrates at 44 μg $L^{-1}$ (Marchese et al. 2008). The Canadian Water Quality Guideline for the Protection of Freshwater Aquatic Life for Cr(III) is set at 8.9 μg $L^{-1}$ which is based on toxicity to rainbow trout *Oncorhynchus mykiss*. There are currently no tissue residue guidelines for the protection of animals and people who consume fish (Mining Watch Canada 2012). The one field study located that examined wild fish found no difference between the Cr concentration in unexposed fish and bluegill *Lepomis macrochirus* and largemouth bass *Micropterus salmoides* exposed to hexavalent chromium (0.10–0.30 mg $L^{-1}$). Researchers acknowledge there is still little information on chromium uptake and effects in freshwater species (Marchese et al. 2008). Studies have observed Cr(VI) toxicity on fish exposed to as little as 10 μg $L^{-1}$. Chromium(VI) does not deposit on gills but enters the fish and exerts toxic effects on internal organs such as the kidney (Mining Watch Canada 2012). Fish exposed to hexavalent chromium have shown changes in physical and biochemical conditions, increased hatching time, DNA damage, and reduced survival (Mining Watch Canada 2012).

In Chinook salmon *Oncorhynchus tshawytscha*, the kidney is the most vulnerable organ in terms of exposure to Cr(VI), which causes necrosis of the renal tubular epithelium (Farag et al. 2006). Pathological changes caused by Cr were also found in the spleen and blood. Exposure to hexavalent Cr in a teleost *Channa punctata* led to changes on gills, kidney, and liver (Mishra and Mohanty 2009a) and resulted in an increased blood serum cortisol level (Mishra and Mohanty 2009b). The fish caught near petrochemical industry centers contained 0.72 mg Cr $kg^{-1}$ (Copat et al. 2012). Edible fish collected in Karai and Puliyankannu (southern India) had a Cr concentration at a level 0.8–1.66 mg $kg^{-1}$ (Imam Khasim et al. 1989). According to Ciftci et al. (2010), the highest Cr concentration was detected in the liver of *Glossogobius giuris* (679.7 mg $kg^{-1}$) during monsoon season, followed by the gills of *Mystus bleekeri* (190.0 mg $kg^{-1}$) and intestines of *Glossogobius giuris* (123.7 mg $kg^{-1}$) in summer. The liver and gills were found to be the most Cr-contaminated tissues (Ciftci et al. 2010). Sanyal et al. (2015) observed that Cr also accumulates in the piscine gut and kidney. The authors also propose that the muscles of fish escape Cr contamination.

Mozambique tilapia *Oreochromis mossambicus* exposed to trivalent chromium chloride and hexavalent potassium dichromate showed no humoral and cellular response of lymphoid cells. Both forms suppressed the antibody response, though Cr(VI) is more stronger than Cr(III). Supplementation of Cr(VI) and Cr(III) also reduced the weight of the spleen and the percentage of lymphocytes in blood (Arunkumar et al. 2000). It has been demonstrated that trivalent chromium can also cause changes and mutations in DNA (Costa and Klein 2006; Peralta-Videa et al. 2009). Experiments indicate that a high Cr content in the diet affects the growth and survival rates of birds, e.g., the brood of American black ducks *Anas rubripes* (Eisler 2000; Koivula and Eeva 2010).

In some poultry species (broiler chickens, turkeys, laying hens), Cr supplements improved weight gains and egg production while reducing the cholesterol levels in the muscles. On the other hand, a decrease in the concentration of serum glucose, fat, and total cholesterol was observed in laying hens (Şahin et al. 2001; Anke et al. 2005). Puls (1994) adds improved quality of egg contents to the positive effects of poultry feed Cr(III) supplementation.

Information on the doses of Cr in water or food that cause health problems in mammals come mostly from toxicology lab tests done on mice and rats. Effects observed on animals in experimental doses through food, water, or injection include cancers, reproductive harm, behavioral changes, reduced growth, and reduced survival (Nriagu and Kabir 1995). The liver may be the most important depot of biologically active Cr, including GTF, which is secreted into the bloodstream in response to increased amounts of plasma insulin (Outridge and Scheuhammer 1993).

Although in vitro tests reveal mutagenicity of Cr(III), in vivo studies have never indicated any carcinogenic properties of the ion, irrespective of the route of exposure (Juturu and Komorowski 2003). A 90-day period of exposing rats to a dose that was 30,000 times greater than that recommended as dietary supplementation in humans (as Cr picolinate) revealed no adverse health consequences of Cr bioaccumulation (Anderson et al. 1997). Staniek et al. (2010) observed that rats fed with Cr(III) propionate showed no genotoxic effects, based on the comet assay of lymphocytes, in contrast to Cr(VI) in the form of $K_2Cr_2O_7$.

Experiments on rats (both in vitro and in vivo) demonstrate that 80% of blood Cr is bound to transferrin—the protein transporting and regulating iron concentration (Feng et al. 2003). Chromates are accumulated through a system of sulfate uptake and influence their metabolism (Peitzsch et al. 1998; Juhnke et al. 2002; Nies 2004). Just like in the case of the abiotic reactions, cellular reduction of Cr(VI) results in the thermodynamically stable Cr(III) (Zhitkovich 2011). Chromate reduction to Cr(III) forms free radicals, which make the metal very toxic (Nies 2004).

Chromium metabolism in endothermic vertebrates depends on the degree of oxidation and the properties of the compounds it forms. Most of Cr found in the organisms is ingested with food, in which it is present in the form of Cr(III). Chromium absorbed in blood leaves it quickly; therefore, the blood Cr concentration does not reflect the content of Cr in the tissues (Suwalsky et al. 2008). Cr(VI) and Cr (III) differ in the ability to penetrate cell membranes (Miksche and Lewalter 1997). It has been shown that chromium picolinate and other compounds of Cr(III) have a limited ability to penetrate the cell membrane to get access to DNA, in contrast to Cr (VI), which is carcinogenic (Juturu and Komorowski 2003).

Animal studies have used various experimental models, and various forms (picolinate, propionate, chromium-L-methionine) and concentrations of Cr were applied in the diet (Lewicki et al. 2014). Chromium affected the metabolism of glucose and fat, reduced cholesterol levels, reduced the risk of atherosclerosis, and reduced mortality due to stress in cats, monkeys, guinea pigs, rabbits, squirrels, pigs, cows, calves, poultry, and humans (Anderson 1988; Moonsie-Shageer and Mowat 1993; Piva et al. 2003; Arvizu et al. 2011; Lewicki et al. 2009, 2014). In livestock animals, Cr supplements improved reproductive performance, weight gains, and carcass quality (Mooney and Cromwell 1997; Pollard et al. 2002). Kurył et al. (2006) demonstrated a significant increase in glucose transport in erythrocytes and β-oxidation of fatty acids in the lymphocytes of healthy rats fed with a diet containing fructans and chromium(III), with the effects increasing with the level of these components in the diet.

In pigs, an addition of 200 μg of Cr improved glucose tolerance, increased glucose production, and decreased the half-life of glucose. Such results have not been confirmed in lambs (Anke et al. 2005). Lewicki et al. (2009) demonstrated in vitro (murine C2C12 myocytes) that chromium supplementation had a positive effect (more potent stimulation of chromium chloride than picolinate) on the increase in the intensity of β-oxidation. Cattle fed with a Cr-supplemented feed showed an increased ($p > 0.05$) cellular immunity, a reduced blood cortisol levels, and a higher antibody titer (Mallard and Borgs 1997; Pollard et al. 2002).

The most widely accepted theory explaining the mechanism of Cr effect on glucose metabolism involves chromodulin (Vincent 1999; Peterson et al. 2008). One of the more recent concepts of chromium action is its effect on cell membrane fluidity and—in consequence—on the regulation of glucose uptake by cells. This effect is associated with a lower membrane cholesterol level, which is thought to be a glucose transport inhibitor controlled by the insulin receptor (Pattar et al. 2006). It is also possible that the mechanism of Cr action depends on estrogen receptor activation (Song et al. 2004). The molecular mechanism of estrogen influence on insulin secretion is not fully understood. It can be divided into two parts: a fast signal (associated with membrane estrogen receptor activation) and slow signal (associated with activation of nuclear estrogen receptors) (Lewicki et al. 2014).

It has been demonstrated that Cr may enhance insulin activity through a hypothetic mechanism which consist in:

- Formation of chromodulin (LMWCr), which—after binding to tyrosine kinase—is supposed to stabilize the active conformation of the insulin receptor and amplify insulin signal approx. eightfold, until the blood concentration of this hormone decreases; thereafter, the bonds with tyrosine kinase loosen and chromodulin is secreted from cells to blood, which is followed by its urinary excretion (Chen et al. 2006; Peterson et al. 2008).
- Decrease in the concentration of tyrosine phosphatase, which—by dephosphorylation of tyrosine residues—inactivates the proteins of insulin receptor stimulation cascade, so quenching the effect of insulin (Wang et al. 2006; Krzysik and Grajeta 2010).
- Regulation of translocation of GLUT4 (transmembrane glucose transporter in insulin-dependent cells) by Cr(III), which may occur independently from the insulin-signaling pathway proteins and is associated with changes in cell membrane fluidity resulting from excessive loss of membrane cholesterol (Ginsberg 2000).

The mechanism of hypotriglyceridemic action of chromium has not been precisely explained. It has been proposed that the mechanism is necessary to maintain the antilipolytic effect of insulin, which involves activation of lipoprotein lipase resulting in the hydrolysis of triglycerides, competitive inhibition of lipolysis, and consequent reduced use of free fatty acids in the biosynthesis of triglycerides (Ginsberg 2000; Krzysik and Grajeta 2010).

Chromium deficiency has been described in rats, guinea pigs, and squirrel monkeys; signs include reduced growth, decreased life span, elevated serum cholesterol, increased formation of aortic plaques, and signs resembling those of diabetes mellitus (Eisler 1986). Some of the first symptoms of Cr deficiency are diminution of insulin function and glucose metabolism disturbances. Further changes involve altered metabolism of proteins, general weakness, and circulatory system damage (Kabata-Pendias and Mukherjee 2007).

Various animal species fed with a chromium-deficient diet exhibited reduced glucose tolerance. A dietary Cr level reduced to below 0.31 mg $kg^{-1}$ dw applied to male goats resulted in lower feed intake, reduced glucose tolerance, and poor weight gains. Bucks fed on chromium-deficient diet produced semen with impaired sperm motility, had lower libido, and suffered from skin diseases (Anke et al. 2005). Chromium deficiency in rats results in increased serum cholesterol level and accelerates platelet formation; in rats, mice, and guinea pigs, it hampers growth rate and shortens the life span (Anke et al. 2005). Some researchers suggest that certain level of dietary chromium is necessary for the normal growth of the heart

(Morris et al. 1995). Experiments on male Sprague-Dawley rats fed with a diet supplemented with chromium picolinate at 300 and 1500 ppb demonstrate that even a fivefold increase of the recommended daily dose of chromium does not cause change in the mass of the heart, total protein content in the heart, and the number of myofibrils but reduces by 11% the amount of high ATPase myosin isoform—V1 (Morris et al. 1995).

Chromium ingestion by humans may be a result of contamination of foods and beverages during the processing and from their packaging. Acids in fruits (malic and citric acids) may wash chromium out from the stainless steel of a cooling vat in the process of canned fruit production (Krzysik et al. 2008; Peralta-Videa et al. 2009). Fish and shellfish are common seafood products, which are regarded as Cr transporters in the trophic chain (Marchese et al. 2008; Peralta-Videa et al. 2009).

Chromium is present in the tissues of human fetuses and infants. Its content decreases with age in all organs except the lungs, where a small increase in the content of chromium is detectable from 10 years of age, probably due to an increase in the Cr levels inhaled into the lungs. The highest accumulation of Cr levels (0.2–2.0 mg $kg^{-1}$) was found in hair (Bielicka et al. 2005).

On the whole, the levels of chromium in the diets of wild mammals and birds seem to be meeting their daily requirements. Chromium deficiency cases in animals are rare, and those documented most often involve experimental mammals (Anke et al. 2001, 2005).

## 6 Toxicity of Various Chromium Forms in Homeothermic Animals

Chromium is listed on the Superfund Priority List of Hazardous Substances; it has been among of the top 20 contaminants for the last 15 years (Chrysochoou and Johnston 2012; Friis 2012; Dhal et al. 2013). Humans and animals take chrome into the body with food, water, and air. Chromium excess in drinking water may pose a health threat.

According to Puls (1994), hexavalent chromium is toxic for cattle at the doses 3000 mg $kg^{-1}$ (chromium oxide) and 1000 mg $kg^{-1}$ (CrCl). A dose causing chronic toxicity is 30–40 mg $kg^{-1}$ of zinc chromate for 1 month. Puls (1994) gives the following data on Cr content in poultry feed, eggs, and tissues of chickens (Table 3.3).

A content of 5 mg $kg^{-1}$ Cr as potassium chromate fed to 2-week-old chicks caused 50% mortality in 3 weeks (Puls 1994). Consumption of $Cr^{6+}$ by hens reduced the hatchability of chicks (Kirklin 1999). Chromium transferred vertically by hens into the eggs inhibits the growth of bone in the embryos (Hui 2002). The content of Cr in avian muscles ranges from 0.1 to 15 μg $g^{-1}$ dw, in unpolluted areas, and from 1 to 700 mg $kg^{-1}$ dw, in polluted habitats. The authors conclude that birds inhabiting chromium-polluted areas tend to accumulate the metal. The same cannot be said about mammals (Outridge and Scheuhammer 1993).

**Table 3.3** Content of chromium, Cr, in the feed, eggs, and tissues of laying hens

| Cr level (mg $kg^{-1}$) | Diet | Eggs | Liver | Kidney | Muscle |
|---|---|---|---|---|---|
| Deficient | <3.0 | | | | |
| Adequate | 5.0–20 | 0.05–0.15 | 0.05–0.40 | 0.19–0.29 | 0.05–0.10 |
| High | 100 | 1.25 | | 1.43 | 0.10 |
| Toxic (chromate) | 400–5000 | | 13.0–150.0 | >18.6 | 2.5–14.0 |

Puls (1994)

**Table 3.4** Content of chromium, Cr, in the tissues of rabbit

| Diet | Liver | Kidney |
|---|---|---|
| Normal | 0.3–1.0 | 0.42–1.58 |
| *High/toxic* | | |
| Trivalent | 6.0–50 | 17.0–30.7 |
| Hexavalent | 10–50 | 3.3–11.2 |

mg $kg^{-1}$ wet weight, Puls 1994

A normal Cr content in equine hair ranges between 1.5 and 3.3 mg $kg^{-1}$. Hair is a better indicator of Cr level in the body than blood serum (Puls 1994). According to Outridge and Scheuhammer (1993), the concentration of Cr in the fur of wild mammals inhabiting uncontaminated areas ranges from 0.1 to 10 mg $kg^{-1}$ dw. However, samples collected in Cr-polluted sites varied in Cr content from 0.3 to 20 mg $kg^{-1}$ dw. The authors observed large differences in the hair of the same species animals collected from a close distance from each other, which is difficult to explain. Cr concentrations in the muscle of marine and terrestrial mammals were similar (Outridge and Scheuhammer 1993). Data by Puls (1994) on the content of Cr in the liver and kidneys of rabbits are shown in Table 3.4.

Only for a few species or groups of mammals and birds, mainly domesticated, have been developed the ranges of normal, high, and toxic levels of Cr in the liver, kidneys, blood, and other biological samples. The Wisconsin Veterinary Diagnostic Laboratory regularly publishes important information on various substances, including Cr (WVDL 2015). Table 3.5 shows the reference values and toxic levels for birds and same mammalian groups.

According to the data summary presented by ATSDR (2012), at acute dermal exposure in rabbits, depending on the mode of exposure to Cr(VI), a LOAEL (lowest-observed-adverse-effect level)—$LD_{50}$—ranged from 30 mg $kg^{-1}$ (24 h) to 426 mg $kg^{-1}$, in males, and 553 mg $kg^{-1}$, in females (2-day exposure). At a single incident of exposure, $LD_{50}$ was 763 mg $kg^{-1}$. With a 4-h exposure in rabbits, a LOAEL occurred at a dose of 42–55 mg $kg^{-1}$ resulting in necrosis, erythema, and edema. This was similar to other experiment with guinea pigs administered with 0.04 mg $kg^{-1}$ once, which caused erythematous reaction. In terms of inhaled Cr(VI), a serious LOAEL in rats was observed after 4-h inhalation at a dose of 29–137 mg $m^{-3}$ ($LC_{50}$). Inhaled doses of Cr(VI) in rats resulted in NOAEL at 0.49 mg $m^{-3}$, causing nasal hemorrhage. In another experiment on rats, the results observed at

**Table 3.5** Normal and toxic ranges of Cr in avian, lapine, ursine, and bovine tissues

| Animal group | Tissue | Normal | Toxic | Unit | Normal values expressed as mg $kg^{-1}$ dw[a] |
|---|---|---|---|---|---|
| Avian | Blood | <0.03 | 3.9–25 | mg $dL^{-1}$ | |
| | Serum | 0.19–0.29 | – | | |
| | Kidney | 0.05–0.1 | 19–170 | mg $kg^{-1}$ ww | 0.25–0.50 |
| | Liver | 0.05–0.40 | 13–150 | mg $kg^{-1}$ ww | 0.17–1.33 |
| Lapine | Serum | <0.10 | 1–6 | | |
| | Kidney | 0.42–1.60 | 3.3–31 | mg $kg^{-1}$ ww | 2.1–8.0 |
| | Liver | 0.30–1.00 | 6–50 | mg $kg^{-1}$ ww | 1.0–3.3 |
| Ursine | Kidney | 0.01–0.52 | – | mg $kg^{-1}$ ww | |
| | Liver | 0.01–0.53 | – | mg $kg^{-1}$ ww | |
| Bovine | Blood | <0.07 | 1–4 | mg $dL^{-1}$ | |
| | Kidney | 0.05–6.20 | 15 | mg $kg^{-1}$ ww | |
| | Liver | 0.04–3.80 | 30 | mg $kg^{-1}$ ww | |

WVDL (2015)

[a]We assumed (according to Kalisinska suggestion) that livers and kidneys of mammals and birds contain 70% and 80% moisture, respectively

0.4 mg $m^{-3}$ were defined as NOAEL in the gastrointestinal, hematological, hepatic, and renal systems.

Laboratory studies on mammals show the following LOAELs for Cr(VI) in drinking water: 100 mg $L^{-1}$ for mice, 70 mg $L^{-1}$ for rats, and 62.7 mg $L^{-1}$ for dogs. LOAELs of Cr(III) in drinking water of rats (28 mg $L^{-1}$) and mice (5 mg $L^{-1}$) are notably lower than those of Cr(VI) (Pawlisz et al. 1997).

Carcinogenicity of Cr(VI) was discovered in the late nineteenth century, when cases of nose tumor in workers handling chromium pigments in Scotland were described (Cohen et al. 1993). Research on lung cancer carried out in the 1930s has led to the recognition of chromium as an occupational disease-causing agent for Cr-exposed workers in Germany. Since then, the mainstream research on Cr has been focused on its toxic effects in humans (Pechova and Pavlata 2007). Hexavalent Cr has been introduced in the environment nearly entirely by human activity, and the literature is dominated by reports on its toxicity studies. Over time, a number of experimental studies have shown that Cr(VI) is also carcinogenic for other homeothermic vertebrates (Mertz 1993; Eisler 2000; Koivula and Eeva 2010; Friis 2012). Currently, Cr(VI) is a well-established carcinogen associated with lung, nasal, and sinus cancer, and the International Agency for Research on Cancer (IARC) has classified Cr(VI) compounds as Group 1 substances carcinogenic to humans (WHO/IPCS 2013).

Chromium compounds may damage the respiratory system and gastrointestinal tract; may cause skin lesions; have carcinogenic, mutagenic, embryotoxic, and teratogenic effects; and affect the postnatal development of neonates (Elbetieha and Al-Hamood 1997; Friis 2012). Women occupationally exposed to chromium may encounter impaired fetal maturation and complications during pregnancy and childbirth (Kanojia et al. 1998). Analogous effects have been observed in experimental mammals exposed to hexavalent chromium compounds (Kanojia et al. 1998; Eisler 2000; Jeyasingh and Philip 2005; Koivula and Eeva 2010; Dhal et al. 2013). A high content of chromium in the environment can have mutagenic, teratogenic, and carcinogenic effects on various avian species (Eisler 2000; Koivula and Eeva 2010).

A $LD_{50}$ based on chromium intake in rats ranges from 50 to 100 mg $kg^{-1}$ for Cr (VI) and from 1900 to 3000 mg $kg^{-1}$ for Cr(III) (De Flora et al. 1990; Dhal et al. 2013).

Chromium exposure of the body reduces the numbers of white cells and, consequently, impairs immunity (Vasylkiv et al. 2010). It has been demonstrated in pregnant female rats exposed to chromium that Cr(VI) level in their embryos and fetuses is ten times higher compared to Cr(III) (Kanojia et al. 1998).

Hexavalent chromium compounds studied in vivo exhibit relatively strong mutagenicity, whereas trivalent Cr may cause chromosomal damage only at very high doses. However, studies on isolated DNA demonstrated that trivalent chromium has a more than tenfold stronger mutagenicity compared to chromium(VI). Hence, a direct mutagenic action is attributed to Cr(III). In conclusion, Cr(VI) easily penetrates into the cell where it is next reduced to chromium III. When the reduction takes place outside the cell (or even outside the nucleus), the mutagenic activity of Cr(III) is lower (De Flora 2000; Sobański et al. 2007).

Clinical trials and animal research reveal that Cr accumulates also in the testes, which causes their damage and a decrease in weight, leads to degeneration of the seminiferous tubules, reduces the volume of semen, and lowers the libido (Wise et al. 2008). Numerous experiments on rodents have shown adverse effects of Cr in relation to testicular and epididymal functions and the semen quality (Ernst and Bonde 1992; Saxena et al. 1990). Presumably, an increase in the production of hydrogen peroxide ($H_2O_2$) during Cr supplementation may result in an increased lipid peroxidation of epithelial cells in the gut, liver, brain, and kidney, as well as in the membranes of sperm cells (Subramanian et al. 2006). Loss of germ cells from the seminiferous epithelium and accumulation of Cr in the lumen of the tubules result from a chronic exposure to Cr (Aruldhas et al. 2005). Male mice exposed to Cr(III) compounds showed a significant decrease in fertility but also damages in the epithelial cells of the intestine, liver, brain, and kidney. The number of implantation sites and the number of viable embryos were significantly reduced in the females fertilized by males exposed to Cr(VI) compounds. The incidence of embryonic resorption and fetal mortality was higher in females fertilized by males exposed to both Cr(III) and Cr(VI) (Elbetieha and Al-Hamood 1997). The body weight, the weight of the seminal vesicles, and the prostate gland were significantly reduced in males receiving Cr(III) and Cr(VI), whereas the weight of the testes increased substantially. Females receiving Cr(III) and Cr(VI) exhibited a significant increase in ovarian weight, while the uterine weight was considerably lower under Cr(III) (Elbetieha and Al-Hamood 1997). Exposure to Cr in female mice resulted in a lower number of embryonic implantations and, in consequence, reduced fertility (Kanojia et al. 1998). A high level of Cr in the placenta and increased chromium transport to the fetus directly affect its organs (Kanojia et al. 1998). In conclusion, chromium may have a negative impact on fertility and reproduction (Elbetieha and Al-Hamood 1997).

The most serious global problem of human and environmental health is exposure to carcinogenic metals, including Cr(VI), mainly due to their content in the water. Chronic exposure to Cr(VI) raises the risk of lung cancer, diseases of the gastrointestinal tract and central nervous system, as well as disorders of the reproductive and respiratory systems (Coogan et al. 1991; Koivula and Eeva 2010; Wang et al. 2012b). Chromium(VI) passes through the cell membrane more easily than Cr(III) and binds to the intracellular proteins in various tissues, which explains the higher toxicity of this form (Coogan et al. 1991). The mechanism of carcinogenic activity of Cr(VI) has not been explained, though primary genotoxic effects are most likely to be the case (Wise et al. 2008; Kimura et al. 2010). Elevated Cr(VI) in the body causes a decrease in antioxidant enzymes and an increase in reactive oxygen species in mammals and birds (Koivula and Eeva 2010; Wang et al. 2012b). The carcinogenic potential of chromium VI compounds is well documented in humans and animals. The mechanism of tumor formation in response to exposure to heavy metals consists in an increase in DNA damages, increased production of highly reactive oxygen species, and interference in the process of DNA repair (Kasprzak 1991; Zocche et al. 2010). The lipid bilayer is the main permeability barrier in the membrane, and structural disturbances caused by chromium affect this permeability. Chromium may also impair the function of ion channels, enzymes, and receptors submerged in the lipid layer (Suwalsky et al. 2008). Ingestion of large doses of Cr (VI) may be lethal for people and animals (Zayed and Terry 2003; ATSDR 2012). Intake of Cr(VI) with drinking water is carcinogenic to both sexes of mice and rats (National Toxicology Program 2008; Zhitkovich 2011).

Although much attention is being focused on Cr(VI), information on the toxicity of Cr(III) is also available (Vasylkiv et al. 2010). Chromium(III) may reduce the activity of the immune system and contribute to formation of necroses, but its compounds are not classified as carcinogens, since there is no sufficient evidence to it (Speranza et al. 2007; Suwalsky et al. 2008; ATSDR 2012). Animal and epidemiological studies indicate that the inorganic form of Cr(III) is nontoxic and has no carcinogenic properties (IARC 1990; ATSDR 2000). The lack of toxic effects of Cr(III) compounds results from their poor ability to penetrate into cells, lack of their intracellular accumulation, and good stability of their bonds with ligands, which prevents binding cellular macromolecules (Zhitkovich 2011).

Concentrations of heavy metals, including chromium, were evaluated in the liver of three insectivorous bat species in the Catarinense coal basin, southern Brazil (Table 3.6). The waters, soil, animals, and vegetation over an area of 2000–6000 hectares were directly compromised by heavy metals. Coal mining residues contain large quantities of heavy metals (O'Shea et al. 2001; Sampaio, 2002; Zocche et al. 2010). Exposure to metals or their uptake by aquatic and terrestrial animals may vary and depend on the local status of the environment quality (Zocche et al. 2010). Open-pit mines may fill with water and become lakes with possibly toxic levels of heavy metals and, in consequence, may pose a threat for animals, either through drinking their water or via a contaminated trophic chain (O'Shea et al. 2001;

**Table 3.6** Chromium concentration (mg kg$^{-1}$) in soft tissues of birds and mammals

| Species | Place and years | *n* | Liver | Kidney | Muscle | Brain | Other | References and additional data |
|---|---|---|---|---|---|---|---|---|
| Birds | | | | | | | | |
| Aquatic birds | | | | | | | | |
| *Anas platyrhynchos* Mallard | Russia, SW, 1993–1995 | 4 | | | | | Bone: 2.90 | Lebedeva (1997) |
| *Anas platyrhynchos* | Poland | | | 0 | | | | Bojar and Bojar (2009) ww |
| Mallard | Czestoborowice reservoir, 2002 | 24 | 0.07 | 1.53 | 0.03 | | | |
| | | | 0.05–0.13 | 0.28–3.63 | <DL–0.06 | | | |
| | Przytoczno, | 19 | 0.02 | 0.03 | 0.02 | | | |
| | ref. group 2002 | | 0.02–0.03 | 0.02–0.04 | 0.01–0.03 | | | |
| *Anas platyrhynchos* Mallard | USA, New Jersey, 1980–1981 | 7 | 1.22 ± 0.55 | | | | Salt gland: 165.5 ± 18.5 | Burger and Gochfeld (1985) AM ± SE, ww |
| *Anas platyrhynchos* Mallard | Japan, Izumi coast, 2003 | 13 | 0.26 ± 0.09 (0.182) | 0.15 ± 0.04 (0.1125) | 0.15 ± 0.08 (0.105) | | | Nam et al. (2005b) AM ± SD; dw, (ww) |
| *Anas poecilorhyncha* Spot-billed duck | Japan, Izumi coast, 2003 | 1 | 0.23 (0.161) | 0.59 (0.4425) | 0.056 (0.0392) | | | Nam et al. (2005b) AM ± SD; dw, (ww) |
| *Anas acuta* Pintail | Japan, Izumi coast, 2003 | 2 | 0.27 (0.189) | 0.16 (0.12) | 0.091 (0.064) | | | Nam et al. (2005b) AM ± SD; dw, (ww) |
| *Anas crecca* Common teal | Japan, Izumi coast, 2003 | 2 | 0.45 (0.315) | 0.16 (0.12) | 0.086 (0.060) | | | Nam et al. (2005b) AM ± SD; dw, (ww) |

(continued)

**Table 3.6** (continued)

| Species | Place and years | *n* | Liver | Kidney | Muscle | Brain | Other | References and additional data |
|---|---|---|---|---|---|---|---|---|
| *Aythya affinis*<br>Lesser scaup | Canada, Lower Great Lakes, 1999–2000 | | | | | | | Petrie et al. (2007)<br>GM; dw, (ww) |
| | Fall | 64 | 0.88 (0.616)<br>0.77–1.02 | | | | | |
| | Spring | 56 | ND | | | | | |
| *Aythya marila*<br>Greater scaup | Canada, Lower Great Lakes, 1999–2000 | | | | | | | Petrie et al. (2007)<br>GM; dw, (ww) |
| | Fall | 41 | 1.02 (0.714)<br>0.91–1.28 | | | | | |
| | Spring | 28 | ND | | | | | |
| *Aythya marila*<br>Greater scaup | USA, Connecticut, Branford, 1996–1997 | 9 W<br>11 S | 3.07 (2.149)<br>1.26 (0.882) | 1.80 (1.35)<br>0.98 (0.735) | 2.37 (1.66)<br>1.91 (1.34) | | | Cohen et al. (2000)<br>dw; (ww)<br>W, winter<br>S, spring |
| *Aythya marila*<br>Greater scaup | USA, New Jersey, 1980–1981 | 16 | 1.53 ± 0.44 | | | | Salt gland<br>3.54 ± 1.17 | Burger and Gochfeld (1985)<br>ww; AM ± SE |
| *Anas rubripes*<br>Black duck | USA, New Jersey, 1980–1981 | 14 | 2.05 ± 0.43 | | | | Salt gland<br>18.90 ± 4.81 | Burger and Gochfeld (1985)<br>AM ± SE, ww |
| *Branta canadensis*<br>Canada goose | USA, New Jersey, 2007 | 26 | 0.16 ± 0.04 | | 0.07 ± 0.02 | | | Tsipoura et al. (2011)<br>ww |
| *Cygnus olor*<br>Mute swan | USA, Watson Creek Marsh, 1997 | 13 | 0.24 ± 0.14<br>(0.17) | | | | | Beyer and Day (2004)<br>dw; (ww) |
| *Egretta gularis*<br>Western reef heron | Iran, Hara Biosphere Reserve, 2010 | 15 | 1.05 ± 0.11<br>(0.735)<br>0.37–1.82 | 0.96 ± 0.08<br>(0.72)<br>0.35–1.63 | 0.56 ± 0.04<br>(0.392)<br>0.34–0.04 | | | Mansouri et al. (2012)<br>GM, dw; (ww) |

| | | | | | | | | |
|---|---|---|---|---|---|---|---|---|
| *Larus heuglini*<br>Siberian gull | Iran, Hara Biosphere Reserve, 2010 | 15 | 2.75 ± 0.28<br>(1.925)<br>1.63–4.51 | 2.32 ± 0.25<br>(1.74)<br>1.22–4.35 | 1.02 ± 0.04<br>(0.714)<br>0.78–1.30 | | | Mansouri et al. (2012)<br>GM, dw; (ww) |
| *Larus atricilla*<br>Laughing gull | USA, New Jersey, 1992 | 20M<br>20F | 0.37 ± 0.03<br>0.42 ± 0.06 | 0.21 ± 0.09<br>0.18 ± 0.03 | 0.22 ± 0.02<br>0.28 ± 0.03 | | | Gochfeld et al. (1996)<br>AM ± SE, ww |
| *Haliaeetus leucocephalus*<br>Bald eagle | USA, Great Lakes, 2002–2010 | 46 | 1.28 ± 0.18<br>(0.896) | | | 1.47 ± 0.30<br>(1.18) | | Nam et al. (2012)<br>dw; (ww) |
| *Haliaeetus leucocephalus*<br>Bald eagle | USA, Alaska, Adak Island, 1993–1998 | 26 | 0.73 (0.51)<br><0.61–18.7<br>50% of samples | 0.93 (0.6975)<br><0.61–58.9<br>54% of samples | | | | Stout and Trust (2002)<br>dw; (ww) |
| *Pandion haliaetus*<br>Osprey | USA, 1975–1982 | 21 | 17 > DL<br>0.33<br>0.09–1.70 | | | | | Wiemeyer et al. (1987)<br>AM, ww |
| *Haliaeetus albicilla*<br>White-tailed eagle | Poland, 1991–1995 | | $n = 3$<br>0.09 ± 0.10<br>(0.063) | $n = 6$<br>0.06 ± 0.08<br>(0.04) | | | | Falandysz et al. (2000)<br>AM ± SD; dw, (ww) |
| *Phalacrocorax carbo*<br>Great cormorant | Spain, Murcia, 2009–2010 | 8<br>juv | 0.41 ± 0.24<br>(0.287)<br>0.13–0.78 | 0.49 ± 0.42<br>(0.37)<br>0.05–1.27 | 0.48 ± 0.31<br>(0.34)<br>0.18–1.13 | | Bone:<br>0.21 ± 0.15<br>0.03–0.40 | Navarro et al. (2010)<br>AM ± SD; dw, (ww) |
| *Phalacrocorax carbo*<br>Great cormorant | Japan, 2003 | | | | | | | Nam et al. (2005a)<br>AM ± SD; dw, (ww) |
| | Lake Biwa | 11 | 0.37 ± 0.26[a] | 0.47 ± 0.66[a] | 0.13 ± 0.04 | | | |
| | | | (0.259) | (0.35) | (0.09) | | | |
| | | | 0.18–1.1 | 0.10–2.3 | 0.10–0.21 | | | |
| | Mie | 19 | 0.22 ± 0.10<br>(0.154)<br>0.13–0.53 | 0.17 ± 0.09<br>(0.1275)<br>0.06–0.37 | 0.14 ± 0.03<br>(0.098)<br>0.10–0.18 | | | |

(continued)

Table 3.6 (continued)

| Species | Place and years | *n* | Liver | Kidney | Muscle | Brain | Other | References and additional data |
|---|---|---|---|---|---|---|---|---|
| *Pelecanus occidentalis* Brown pelican | USA, Florida, 1969–1972 | 5 | 0.92 (0.644) 0.80–1.20 | 4.01 (3.07) 2.65–4.71 | | | Bone: 15.40 8.82–22.89 | Connors et al. (1972) dw, (ww) |
| *Pelecanus occidentalis* Brown pelican | USA, California, 1969–1971 | 3 | 1.37 (0.959) 0.70–1.80 | 3.57 (2.68) 1.45–6.27 | | | Bone: 6.03 4.65–8.68 | Connors et al. (1972) dw, (ww) |
| *Pelecanus erythrorhynchos* White pelican | USA, California, 1969–1971 | 1 | 0.70 (0.49) | 1.19 (0.89) | | | Bone: 3.83 | Connors et al. (1972) dw, (ww) |
| Passerine | | | | | | | | |
| *Parus major* Great tit | Belgium, Antwerp, 2000 | 10F | 0.034 (0.024) 0.007–0.051 | 0.049 (0.037) 0.027–0.17 | 0.014 (0.01) 0.003–0.050 | 0.020 (0.016) 0.005–0.052 | Lung: 0.014 1 nd–0.077 | Dauwe et al. (2005) dw; median; (ww) |
| *Parus major* Great tit | China, Western Mountains of Beijing, 2004 | 25 | 1.86 ± 0.33 (1.302) | 6.26 ± 1.08 (4.69) | 1.14 ± 1.20 (0.798) | 1.65 ± 0.26 (1.32) | 2.22 ± 0.13 Lung: 8.90 ± 1.99 | Deng et al. (2007) dw; (ww) |
| *Carduelis sinica* Greenfinch | China, Western Mountains of Beijing, 2004 | 20 | 0.96 ± 0.34 (0.672) | 2.61 ± 0.47 (1.96) | 1.07 ± 0.54 (0.749) | 1.12 ± 0.28 (0.90) | 1.54 ± 0.12 Lung: 1.10 ± 0.11 | Deng et al. (2007) dw; (ww) |
| *Pycnonotus xanthopygos* White-spectacled bulbul | Turkey, 2002–2003 | 21M | 0.13 ± 0.03 | 6.22 ± 1.22 | 1.19 ± 0.21 | Lung M: | Bone M: | Aslan et al. (2006) dw; (ww) |
| | | | (0.091) | (4.66) | (0.83) | 2.07 ± 0.44 | 2.09 ± 0.17 | |
| | | 21F | 0.21 ± 0.04 (0.147) | 6.58 ± 1.33 (4.93) | 1.25 ± 0.22 (0.875) | Lung F: 2.15 ± 0.45 | Bone F: 2.16 ± 0.17 | |
| *Sayornis nigricans* Black phoebe | USA, Texas, 1997 | 8 | | | | | Carcass: 1.69 1.0–3.6 | Mora et al. (2002) dw; AM |

| | | | | | | | | |
|---|---|---|---|---|---|---|---|---|
| *Petrochelidon pyrrhonota*<br>Cliff swallow | USA, Texas, 1997 | 8 | | | | | Carcass:<br>1.58<br>1.3–2.0 | Mora et al. (2002)<br>dw; AM |
| Other | | | | | | | | |
| *Centrocercus urophasianus*<br>Greater sage-grouse | USA, Wyoming and Montana, 2003–2006 | 70 | 0.50 ± 0.02<br>0.17–0.80 | | | | | Dailey et al. (2008)<br>ww |
| Mammals | | | | | | | | |
| Ungulata | | | | | | | | |
| *Cervus elaphus*<br>Red deer | Slovakia | 22 | 0.138<br>0.030–0.368 | 0.139<br>0.010–4.130 | 0.246<br>0.040–0.584 | | | Gasparik et al. (2004)<br>ww, median |
| *Cervus elaphus*<br>Red deer | Croatia, Sedico | | | | | | Bone (teeth)<br>5.489 ± 0.218<br>4.214–6.338 | Zaccaroni et al. (2008a)<br>AM ± SE, dw |
| *Capreolus capreolus*<br>Roe deer | Croatia, Sedico | | | | | | Bone (teeth)<br>4.595 ± 0.255<br>0.954–6.678 | Zaccaroni et al. (2008a)<br>AM ± SE, dw |
| *Capreolus capreolus*<br>Roe deer | Poland, 2009 | 13 | | | 0.61 ± 0.75<br>0.04–2.35 | | | Długaszek and Kopczyński (2013)<br>AM ± SD, ww |
| *Capreolus capreolus*<br>Roe deer | Poland, 2009 | 14 | 0.04 ± 0.01<br>0.03–0.07 | | | | | Długaszek and Kopczynski (2011)<br>AM ± SD, ww |
| *Odocoileus virginianus*<br>White-tailed deer | USA, Texas, 2005 | 20 | <0.5–1.0<br>20% of samples | | <0.5–1.5<br>50% of samples | | | Bruckwicki et al. (2006)<br>dw |

(continued)

**Table 3.6** (continued)

| Species | Place and years | *n* | Liver | Kidney | Muscle | Brain | Other | References and additional data |
|---|---|---|---|---|---|---|---|---|
| *Alces alces* Moose | Yukon, Canada, 1994–2001 | | $n = 56$ 0.52 ± 0.24 | $n = 384$ 0.22 ± 0.15 | $n = 37$ 0.26 ± 0.09 | | | Gamberg et al. (2005), ww |
| *Porcupine caribou* Caribou | Canadian Arctic, 1994–2003 | 331 | | 0.28 ± 1.18 | | | | Gamberg et al. (2005), ww |
| *Rangifer tarandus* Reindeer (semi-domesticated) | Northern Norway, 2004–2005 | 29 | 0.028 <0.028–0.095 | | 0.038 0.028–1.56 | | | Hassan et al. (2013), Med, ww |
| *Rangifer tarandus* Reindeer | Norwegian-Russian border areas, 1990–1991 | | | | | | | Sivertsen et al. (1995), Med, ww |
| | Jarfjord | 31 | 0.07 | | | | | |
| | | | <0.01–0.09 | | | | | |
| | Pasvik | 30 | <0.01 | | | | | |
| | | | <0.01–0.02 | | | | | |
| | Finnmark | 40 | <0.01 | | | | | |
| | | | <0.01–0.03 | | | | | |
| *Alces alces* Moose | Norwegian-Russian border areas, 1990–1991 | | | | | | | Sivertsen et al. (1995), Med, ww |
| | Jarfjord | 31 | <0.01 | | | | | |
| | | | <0.01–0.04 | | | | | |
| | Pasvik | 30 | <0.01 | | | | | |
| | | | <0.01–0.09 | | | | | |
| | Finnmark | 40 | <0.01 | | | | | |
| | | | <0.01–0.06 | | | | | |

| | | | | | | | | |
|---|---|---|---|---|---|---|---|---|
| *Sus scrofa*<br>Wild boar | Poland, 2009 | 11 | | | 0.12 ± 0.06<br>0.04–0.27 | | | Dlugaszek and Kopczynski (2013)<br>AM ± SD, ww |
| *Sus scrofa*<br>Wild boar | Poland, 2009 | 10 | 0.14 ± 0.21<br>0.03–0.72 | | | | | Dlugaszek and Kopczynski (2011)<br>AM ± SD, ww |
| *Sus scrofa*<br>Wild boar | Slovak Republic, 1998–1999<br>Polluted area | 15 | 0.15 ± 0.12<br>0.02–0.36 | 0.19 ± 0.11<br>0.09–0.48 | 0.29 ± 0.42<br>0.10–1.80 | | | Piskorova et al. (2003)<br>AM ± SD, ww |
| *Sus scrofa*<br>Wild boar | Italy, Lazio Region, 2005–2006 | 54 | 0.15 ± 0.11<br><DL–0.63 | | 0.13 ± 0.08<br>0.07–0.69 | | | Danieli et al. (2012)<br>ww |
| *Sus scrofa*<br>Wild boar | Italy, Lazio Region | | $n = 75$<br>0.141 ± 0.110<br>0.003–0.626 | $n = 65$<br>0.097 ± 0.075<br>0.035–0.590 | $n = 78$<br>0.139 ± 0.082<br>0.069–0.692 | | | Amici et al. (2012)<br>ww |
| *Equus ferus caballus*<br>Horse | Germany | 21 | 0.116<br>0.022–0.427 | Cortex<br>0.085<br>0.016–0.051<br>Medulla<br>0.091<br>0.028–0.348 | | | | Paßlack et al. (2014)<br>ww |
| *Bos taurus*<br>Cattle | India, 1991<br>Polluted area | | 10.95<br>0.54–21.37 | 1.09<br>0.65–1.52 | | | | Dogra et al. (1996)<br>In ash |
| Carnivora | | | | | | | | |
| *Neovison vison* | Poland, 2009–2011 | | | | | | | Brzezinski et al. (2014)<br>dw, AM ± SD; (ww) |
| American mink | Drawa N. Park | | | | | | | |
| | | 26 | 0.25 ± 0.50 | 0.62 ± 0.34 | | | | |
| | | | (0.175) | (0.465) | | | | |
| | Narew N. Park | | <DL–2.34 | 0.25–1.65 | | | | |
| | | 26 | 0.48 ± 0.67 | 0.79 ± 0.47 | | | | |
| | | | (0.336) | (0.593) | | | | |
| | | | <DL–3.17 | 0.26–2.39 | | | | |

(continued)

**Table 3.6** (continued)

| Species | Place and years | *n* | Liver | Kidney | Muscle | Brain | Other | References and additional data |
|---|---|---|---|---|---|---|---|---|
| *Neovison vison* American mink | Canada, British Columbia, 1990–1996 | | | | | | | Harding et al. (1998) AM ± SE, dw; (ww) |
| | Kootenay River | 4 | | 0.98 ± 0.34 (0.735) | | | | |
| | L. Fraser River | 8 | 1.16 ± 0.28 (0.812) | 1.68 ± 0.33 (1.26) | | | | |
| *Neovison vison* American mink | USA, South Carolina, 1987–1988 | 55 | 0.156<br>0.01–5.08 | | | | | Carmichael and Baker (1989)<br>GM, ww |
| *Neovison vison* American mink | USA, Illinois, 1984–1989 | 29 | 12/29 (41%)<br>1.72<br>0.60–2.63 | 15/29 (52%)<br>2.16<br>0.66–3.29 | 12/29 (41%)<br>1.80<br>0.66–3.10 | | | Halbrook et al. (1996) AM, ww |
| *Lontra canadensis* River otter | Canada, British Columbia, 1990–1996 | | | | | | | Harding et al. (1998) AM ± SD, dw; (ww) |
| | Kootenay River | 12 | 1.44 ± 0.30 (1.008) | | | | | |
| | L. Fraser River | 3 | 1.40 ± 0.12 (0.98) | | | | | |
| | U Fraser River | 6 | 1.42 ± 0.22 (0.994) | | | | | |
| *Lontra canadensis* River otter | USA, Illinois, 1984–1989 | 8 | 1/8 (12%)<br>0.52 | | 8/8 (100%)<br>2.20 ± 4.61<br>0.21–11.6 | | | Halbrook et al. (1996) AM ± SD, ww |
| *Lutra lutra* Eurasian otter | England and Wales, 2009 | 50 | 0.219<br>(0.153)<br>0.155–0.306 | | | | | Walker et al. (2011) Med., dw; (ww) |

| | | | | | | | | |
|---|---|---|---|---|---|---|---|---|
| *Lutra lutra* Eurasian otter | England and Wales, 2007 | 48 | 0.164 (0.115) | | | | | Walker et al. (2010) Med., dw; (ww) |
| | | | 0.097–0.266 | | | | | |
| | | | 0.302 (0.211) | | | | | |
| | 2008 | 57 | 0.214–0.370 | | | | | |
| *Lutra lutra* Eurasian otter | France | 21 | 0.63 (0.441) 0.07–4.90 | | | | | Ruiz-Olmo et al. (2000) GM, dw; (ww) |
| *Gulo luscus* Wolverine | Canada, British Columbia, 1997 | 11 | 0.50 ± 0.16 (0.35) | | | | | Harding (2004) AM ± SD, dw; (ww) |
| *Martes americana* Marten | Canada, British Columbia, 1997 | 5 | <0.40 | | | | | Harding (2004) AM ± SD, dw |
| *Procyon lotor* Raccoon | USA, South Carolina, SRS 1996–1997 | | | | | | | Burger et al. (2002) SRS, Savannah River Site AM ± SE, ww |
| | On-site area | 46 | 0.36 ± 0.02 | 0.31 ± 0.02 | | | | |
| | | | 0.19–1.08 | 0.14–1.00 | | | | |
| | Off-site area | 25 | 0.32 ± 0.02 | 0.25 ± 0.01 | | | | |
| *Vulpes vulpes* Red fox | Slovak Republic, 1998–1999 | 18 | 0.26 ± 0.19 0.12–0.58 | 0.29 ± 0.16 0.14–0.52 | 0.33 ± 0.20 0.18–0.62 | | | Piskorova et al. (2003) AM ± SD, ww |
| | Polluted area | | | | | | | |
| *Herpestes javanicus* Javan mongoose | Japan, Amami Oshima Islands, 2004–2005 | | $n = 53$ 0.039 ± 0.027 0.006–0.135 | $n = 47$ 0.033 ± 0.021 0.06–0.158 | | | Brain: $n = 10$ 0.047 ± 0.055 0.14–0.197 | Horai et al. (2007) AM ± SD, ww |
| *Vulpes vulpes* Red fox | Poland, 2009 | 14 | | | 0.34 ± 0.23 0.07–0.65 | | | Dlugaszek and Kopczynski (2012) AM ± SD, ww |
| Other species | | | | | | | | |
| *Lepus europaeus* Brown hare | Poland, 2009 | 11 | 0.04 ± 0.03 0.01–0.12 | | | | | Dlugaszek and Kopczynski (2011) AM ± SD, ww |

(continued)

**Table 3.6** (continued)

| Species | Place and years | $n$ | Liver | Kidney | Muscle | Brain | Other | References and additional data |
|---|---|---|---|---|---|---|---|---|
| *Lepus europaeus* Brown hare | Poland, 2009 | 11 | | | 0.09 ± 0.13<br>0.02–0.44 | | | Dlugaszek and Kopczynski (2013)<br>AM ± SD, ww |
| *Lepus europaeus* Brown hare | Germany, E part | 25 | 0.807 ± 0.185 | | | | Bone (rib):<br>7.652 ± 1.343 | Anke et al. (2005) |
| *Rattus rattus* Wild rat | Portugal | | | | | | Spleen: TG ($n = 4$)<br>1.231 ± 0.332 | Pereira et al. (2006)<br>AM ± SE, dw; (ww) |
| | TG, Tapada | 4 | 0.513 ± 0.221 | 0.345 ± 0.051 | | | | |
| | Grande | | (0.359) | (0.259) | | | | |
| | M4, Sulfur mill | 5 | 0.117 ± 0.029 | 0.072 ± 0.012 | | | | |
| | | | (0.082) | (0.054) | | | | |
| *Mus spretus* Algerian mouse | Portugal | | | | | | Spleen: TG ($n = 7$)<br>14.921 ± 6.127<br>M4 ($n = 5$)<br>8.357 ± 2.961 | Pereira et al. (2006)<br>AM ± SE, dw; (ww) |
| | TG, Tapada | 7 | 0.361 ± 0.068 | 1.441 ± 0.343 | | | | |
| | Grande | | (0.253) | (1.01) | | | | |
| | M4, Sulfur mill | 8 | 0.651 ± 0.238<br>(0.456) | 0.985 ± 0.361<br>(0.739) | | | | |
| *Crocidura russula* White-toothed shrew | Spain, Donana Reserve, 1999 | | | | | | | Sánchez-Chardi et al. (2009)<br>dw, AM ± SE; (ww) |
| | Polluted site | | $n = 19$ | $n = 19$ | | | | |
| | | | 1.74 ± 0.60 | 1.91 ± 0.19 | | | | |
| | | | (1.218) | (1.433) | | | | |
| | Reference site | | $n = 10$ | $n = 9$ | | | | |
| | | | 3.00 ± 0.48 | 2.07 ± 0.27 | | | | |
| | | | (2.1) | (1.553) | | | | |

*ww* wet weight, *dw* dry weight, *AM* arithmetic mean, *SD* standard deviation, *SE* standard error, *GM* geometric mean, *ND* not detected, *DL* detection limit, *M* male, *F* female

[a]Significant differences between Biwa and Mie

Zocche et al. 2010). Liver Cr content in *Molossus molossus* bats in the mining area was highest (5.7 mg $kg^{-1}$ dw), while in the other two species, the values were below the limit of detection. There were also no significant differences in the comet assay results in this species in relation to the control. The authors speculate that the accumulation of metals in bats is probably a reflection of the degradation of habitats within the mining area, the availability of food exposed to contamination, as well as the foraging behavior, the use of habitat, physiological differences, and, as indicated by Walker et al. (2007), interspecific variability in adaptation to the environment.

## 7 Toxicokinetics and Effects of Chromium in Wildlife

The distribution of chromium in the body of a homeothermic vertebrate is determined by several factors: the form of Cr; its concentration in the water, air, and food; as well as the route it penetrates the body. Biological membranes are less permeable to Cr(III) compared to Cr(VI) (Upreti et al. 2004; ATSDR 2008; EFSA 2009). Form Cr(VI) does not react with macromolecules like DNA, RNA, proteins, and lipids, but it is metabolically reduced to Cr(III) and the reductional intermediate Cr(V) inside the cell, where they are capable of coordinate and covalent interactions with macromolecules (Upreti et al. 2004).

Hexavalent chromium in the body is quickly reduced by glutathione, ascorbic acid, or cysteine to kinetically much more stable Cr(III), which is bound intracellularly (Miksche and Lewalter 1997; Gagelli et al. 2002; Suwalsky et al. 2008). The cellular metabolism of Cr(VI) may result in both oxidative and non-oxidative DNA damage (Sugden and Stearns 2000; Salnikow and Zhitkovich 2008). Chromium (VI) displays no ability of damaging DNA directly; instead, it must be reduced to gain its genotoxic property. Intracellular reduction is an activation process, which produces oxidation-reduction intermediates Cr(V/IV) and stable Cr(III), forming mutagenic Cr-DNA adducts. Cellular Cr(VI) reduction in vivo is primarily driven by ascorbate (Asc), whereas glutathione (GSH) is the most obvious reducing agent in cultured cells due to the lack of Asc (Zhitkovich 2011). Extracellular reduction of Cr(VI) is a detoxification process, which produces nontoxic Cr(III), poorly permeable through cell membranes (Zhitkovich 2011). Studies on the reductive activity in tissue homogenates and biological fluids have shown that Asc is the key biological reducing agent of Cr(VI), accounting for 80–95% of its metabolism (Suzuki and Fukuda 1990; Standeven and Wetterhahn 1992). Jointly, the activity of Asc, GSH, and Cys is responsible for more than 95% of in vivo Cr(VI) reduction. Asc and GSH concentrations in tissues are usually not very different, and the dominant role of Asc is due to the very high degree of Cr(VI) reduction (Zhitkovich 2011).

Tandon et al. (1979) showed in the study on rats the greater concentration of Cr in the nuclear fraction of liver cell than in the mitochondrial fraction, but they also revealed that it was dose dependent. The increase in the dose of chromium from 1 mg $kg^{-1}$ to 2 or 3 mg $kg^{-1}$ has increased uptake of the metal by the mitochondrial fraction. Feng et al. (1999) using $^{50}Cr(III)$ confirmed that the nucleic fraction has the highest Cr concentration in the liver cell of both normal and diabetic rats. They concluded also that diabetic rats retain more Cr in the mitochondrial and lysosomal fractions of the liver and have lower Cr concentration in the subcellular fractions of the pancreas, testes, and kidney, which can indicate Cr participating in the glucose or lipid metabolism to compensate the low level of insulin in the body of diabetic rats.

On average in mammals, depending on the species, absorption of Cr(III) remains in the range from 0.4% to 5.0% (Dowling et al. 1989; Krejpcio 2001; Pechova and Pavlata 2007; Wang et al. 2012a; Kirman et al. 2013). In an organic form, Cr is absorbed better than the inorganic form, which promotes Cr concentration in the tissues (Ohh and Lee 2005). The highest tissue accumulation of Cr was demonstrated after administration of Cr in the form of nanoparticles (Zha et al. 2007; Wang et al. 2009; Lewicki et al. 2014).

Chromium absorption is low, ranging between 0.4 and 2.0% for inorganic compounds, while the availability of organic Cr is more than ten times higher (Cefalu and Hu 2004; Pechova and Pavlata 2007). Although Anderson et al. (1996) showed in rats fed diets supplemented with Cr-chloride, chromium an apparent absorption rate of up to 0.9 %, while that for organic Cr sources, such as Cr-nicotinate, Cr-picolinate, Cr complex of dinicotinic acid-diglycine-cysteine-glutamic acid: 1.3, 1.1 and 0.6 %, respectively. Absorption of Cr depends on its diet content, chemical form, and other dietary components. Higher phytate, calcium, manganese, titanium, zinc, vanadium, and iron can inhibit Cr absorption (EFSA 2009). A study on rats revealed that Cr(VI) was reduced in gastric juice, whereas chromium picolinate complex was stable (Gammelgaard et al. 1999).

After intestinal absorption, Cr(III) is released to blood and bound by proteins involved in iron metabolism, mainly transferrin, as shown by in vitro and in vivo studies on rats (Feng et al. 2003). In this form, Cr is transported to cells, and the cells' transmembrane transfer efficiency depends on insulin concentration (Clodfelder and Vincent 2005). The common transport mechanism of chromium and iron makes these metals compete for transferrin-binding sites. Rats intraperitoneally administered with Cr ions showed a decrease of Fe bioavailability in the body, and initial symptom characteristics to anemia were observed (Ani and Moshtaghie 1992; Lewicki et al. 2014).

Absorbed chromium distributes to nearly all tissues, with the highest concentrations noted in the kidney and liver, while the bone may contribute to long-term retention kinetics of chromium (Outridge and Scheuhammer 1993; ATSDR 2008). Mouse and human pharmacokinetic data support that even low, environmentally relevant doses of Cr(VI) are likely to escape reduction in the stomach, due to the ability of absorption and gastric emptying to successfully compete with reduction (Collins et al. 2010; Casalegno et al. 2015).

According to Nam et al. (2005a), chromium in the great cormorant (*Phalacrocorax carbo*) is not evenly distributed among the particular parts of the body. The highest percentage of chromium in its body (see Fig. 3.3) was found in the muscle which is ~37%, followed by the feather 20%, skin 16%, and bone 10%. Much lower amount was located in the liver (5%). The highest concentration of Cr (mean ± SD, mg $kg^{-1}$ dw) was recorded in the lung (0.67 ± 0.63) and spleen

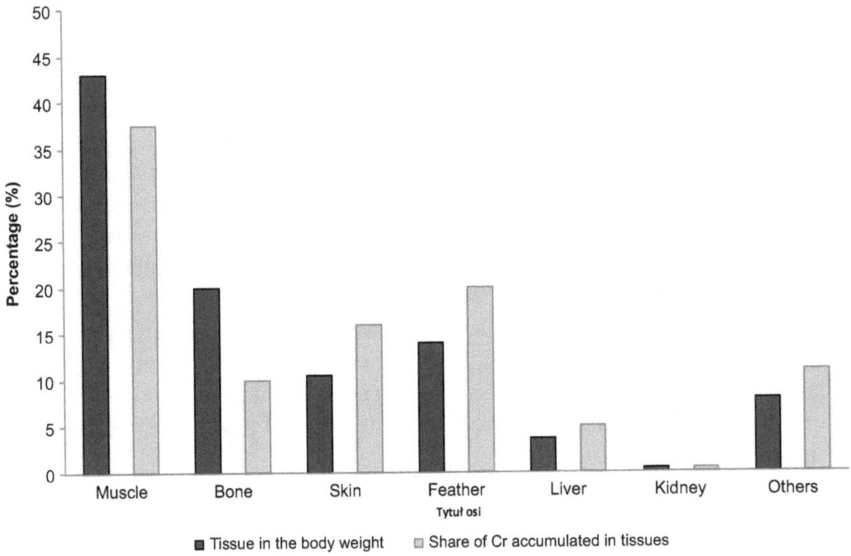


**Fig. 3.3** Percentage share of tissues in the body of the great cormorant and percentage of Cr bioaccumulated in various tissues (source of data Nam et al. 2005a)

(0.51 ± 0.03), whereas the lowest in the bone (0.07 ± 0.01), uropygial gland (0.11 ± 0.02), eyeball (0.12 ± 0.01), pancreas (0.13 ± 0.01), and gonads (0.11–0.14 mg $kg^{-1}$ dw).

ATSDR (2008) informs that toxicokinetic data for chromium in humans, dogs, rats, mice, rabbits, and hamsters, generally correlate well among species however, exposures to chromium(VI) resulted in different organ distribution patterns between rats and mice. Cr accumulates to a greater extent in the blood of rats *vs* mice after short-term exposure by injection; after oral exposure, liver concentrations of chromium were three to four times higher in mice than rats, whereas kidney concentrations were about 50% lower (Kargacin et al. 1993).

Exposure of wild animals to chromium present in the air is poorly known. Outridge and Scheuhammer (1993) suggested that Cr concentrations in fur or feathers can be extremely variable even among individuals within the same habitat, and, at best, it might be used to indicate relative levels of airborne Cr contamination. People working in the environment with chromium-contaminated air exhibit its elevated levels in the lungs. In unexposed people, this level usually remains within 0.07–1 mg $kg^{-1}$ ww; however, chromium concentrations in lung cancer-affected workers are usually higher, from 0.5 to 192 mg $kg^{-1}$ ww (Mancuso 1997; Wise et al. 2008). Birds of urban areas and places with heavily polluted air can accumulate large amounts of Cr in the lungs. For example, the resident great tit *Parus major*, which inhabits a park in the capital of China being confronted with extremely heavily polluted air, had chromium at a level of 8.9 mg $kg^{-1}$ dw (Deng et al. 2007). On the other hand, a great tit living in the capital of Belgium, a European country concerned with the air quality and the environment protection, had lung Cr concentrations lower by two orders of magnitude (0.014 mg $kg^{-1}$ dw) (Dauwe et al. 2005). Compared to these great tits, another passerine species, the white-spectacled bulbul (*Pycnonotus xanthopygos*) inhabiting the Turkish Mediterranean coast, including Antalya, where chromite ore is mined, showed Cr concentrations of 2.1 mg $kg^{-1}$ dw (USGS 2015; Aslan et al. 2006). This is four times lower compared to that measured in the great tit in China and even several hundred times higher than in the same species in Belgium. In birds, especially small passerines (characterized by high metabolism rates and very active respiratory function), the air contaminated with heavy metals, including inhaled Cr, probably promotes accumulation in the lungs. A small number of data in this area do not allow drawing more general conclusions though. Nevertheless, high concentrations of Cr observed in avian lungs in polluted areas, similar to those observed in the lungs of people occupationally exposed to dust contaminated with the metal, may indicate equivalence of food and air as sources of Cr intoxication in passerines.

Deng et al. (2007) examined the concentrations of Cr in tissues of great tits and greenfinches (*Carduelis sinica*) collected at Badachu Park in the Western Mountains of Beijing (China) and concluded that the highest concentrations of Cr were found in their kidneys (6.26 ± 1.08 and 2.61 ± 0.47 mg $kg^{-1}$ dw, respectively). They also noted higher Cr level in tissues of most body parts in great tits which are primarily insectivorous than in greenfinches which feed mainly on seeds and fruits, for example, 1.86 ± 0.33 and 0.96 ± 0.34 for the liver, 4.66 ± 0.95 and 0.69 ± 0.06 for the heart, and 1.65 ± 0.26 and 1.12 ± 0.28 for the brain.

For most wild birds and mammals, their food is the main source of chromium. Animals feeding on aquatic organisms may be dangerously exposed to Cr present in fish, shellfish, molluscs, and plants. Chromium accumulates in the tissues of aquatic animals and thus can have toxic effects also on the consumers of fish (Govind and Madhuri 2014).

On being absorbed in the intestine and/or lungs, Cr is bounded by the blood, and chromium transport depends on its valence. Chromium(VI) easily penetrates to red blood cells and, after reduction to Cr(III), binds to hemoglobin. Chromium(III) is not capable of passing through cell membranes; however, to some extent it may penetrate into erythrocytes (Suwalsky et al. 2008). More than 99% of Cr(III) absorbed in the blood appears in plasma (EFSA 2009). Absorbed chromium can circulate in a free state, may bind to transferrin or plasma proteins, or circulate as GTF complex (Ducros 1992; Piva et al. 2003). According to results of in vitro and in vivo studies on rats receiving Cr-chloride, approximately 90% of Cr(III) in the serum was associated with β-globulin fraction and 80% of all Cr was bounded to transferrin (EFSA 2009). Apart from chromates, all chromium compounds are quickly removed from blood (Stern and Mansdorf 1998). Soluble chromates show similar kinetics of distribution and excretion, despite significant differences in their solubility (Collins et al. 2010). The rapid uptake of Cr(VI) from the gastrointestinal tract appears to result from the transport of anionic chromate and dichromate complexes across cell membranes by the $SO_4^{2-}$ and $PO_4^{3-}$ anion transport system, and Cr(III) crosses cell membranes only by passive diffusion (Collins et al. 2010; Casalegno et al. 2015).

Mertz et al. (1965) proposed body retention of chromium (after administration of $^{51}CrCl_3$ to rats) in terms of three compartment models with three half-lives of 0.5, 5.9, and 83.4 days. Subsequent studies reported similar results in man and rat (EFSA 2009). It has been demonstrated that within 3 h, hens absorb approximately 11.8% of trivalent $^{51}Cr$ administered into the crop. The absorption rate of $^{51}Cr$ was highest within 6 h after administration, representing about 15% of the dose. In rats, orally administered with $^{51}Cr$, the absorption ranged from 14 to 17% (Anke et al. 2005). The respiratory tract is the primary target organ for inhaled chromium, including its water-insoluble forms (Gad 1989).

Chromium absorbed by blood in the lungs or intestines is transported to various tissues and organs and deposited there in varying degrees. Most chromium (>95%) present in the food eaten and water drunk remains unabsorbed, and endothermic vertebrates excrete it with feces; on the other hand, a major part of the absorbed chromium is removed with urine (Gad 1989; Outridge and Scheuhammer 1993; Gammelgaard et al. 1999; ATSDR 2012; Kler et al. 2014). According to Ducros (1992), mammals are able to get rid of about 80% of absorbed chromium with urine and the rest with bile and sweat. Urinary chromium excretion process begins within an hour after oral ingestion and reaches its peak intensity within 12–24 h. Fecal chromium excretion starts within 6–12 h after ruminal administration, reaching its maximum after 24 h and lasting for more than 4 days. Biliary and urinary excretion of absorbed chromium takes more than 4 days (Anke et al. 2005).

It is suggested that, apart from the kidney, aquatic birds may use the salt glands to remove Cr that penetrates the bloodstream (Burger and Gochfeld 1985). The kidney, which is able to accumulate high levels of Cr, is probably the target organ for systemically absorbed chromates (Gad 1989; De Flora 2000). However, a high level of Cr accumulation is sometimes observed not only in the kidney and liver but also in other soft parts of a mammal and avian body (spleen, pancreas, brain, bone marrow) and also in highly mineralized keratin (hair, feathers) and bone growths (Jenkins 1979; Eisler 1986; Burger et al. 1993; Outridge and Scheuhammer 1993; Lebedeva 1997; Piva et al. 2003; Pereira et al. 2006; Deng et al. 2007; Manjula et al. 2015). According to Anke et al. (2005), the largest part of Cr in mammals is accumulated in the skeleton (45%). For example, in the mouflon, *Ovis aries musimon*, wild boar *Sus scrofa*, roe deer *Capreolus capreolus*, and European hare *Lepus europaeus*, living in Thuringia, Germany, the bone Cr concentrations ranged between 4.5 and 7.7 mg kg dw. Some percentage of Cr is found in the skin and hair of mammals (16%), muscle (13%), and more than 5% in the blood, kidney, and liver (7.4%, 5.7%, 5.3%, respectively). Monitoring of Cr content in erythrocytes can be a marker of chromium(VI) exposure (IARC 1990, 1999; Sobański et al. 2007).

## 8 Bioaccumulation of Chromium

Huffman and Allaway (1973) propose that Cr(III) is beneficial for plants in low concentrations. A level of 0.05 to 1 mg Cr $L^{-1}$ was found to promote growth and increase yield, yet it is not considered essential to plants (Oliveira 2012). Zou et al.

(2006) indicate that high concentrations of Cr have been found harmful to plants and—if increasing—adversely affect several biological parameters. Plants tend to absorb Cr(VI) and reduce it to less toxic Cr(III) in roots (Mei et al. 2002), but there is also another suggestion that dissolved Cr(VI) is taken up by plants without reduction (Oliveira 2012). The subcellular localization of Cr as found by electron energy loss spectroscopy and electron spectroscopic imaging suggested that Cr is accumulated mainly in the cell wall and vacuoles (Zou et al. 2006). Cr content in the plants grown on uncontaminated soils is in the range 0.02–1.0 mg kg$^{-1}$, whereas those from contaminated or serpentine-derived soils can accumulate Cr within the range 10–-190 mg kg$^{-1}$, depending upon the soil concentration and the plant species (Hood 2010). Oliveira (2012) indicate that under normal conditions, concentration of Cr in plants is less than 1 mg kg$^{-1}$. Hood (2010) showed distribution of accumulation in different parts of plant, root > leaves or shoot > grain, which indicates restricted translocation of Cr. The high accumulation of Cr in root is likely due to its immobilization in the vacuoles of root cells. Chromium compounds used in dyeing and tanning processes are often found in the soil and groundwater in industrial areas, which need environmental cleanup and phytoremediation (Zayed and Terry 2003; Lytle et al. 1998). Chromium enters the food chain with the plant material (Zou et al. 2006), but there is no indication of biomagnification of chromium along the terrestrial food chain (WHO/IPCS 2013).

Chromium is also not expected to biomagnify in the aquatic food chain (WHO/IPCS 2013). Metal bioconcentration processes depend on the fish species and their trophic level and also on the kind of food, the kind of absorption carried out by the organism, or the phase in which the metal or particulate is dissolved (Voigt et al. 2015). Aquatic animals are more sensitive to metals than are aquatic plants, for example, 62 μg kg$^{-1}$ inhibits growth in algae and 16 μg kg$^{-1}$ inhibits growth in Chinook salmon (Solomon 2008).

Chromium is present in all animal tissues in a concentration ranging from a few to several tens of micrograms per kg, rarely exceeding 100 μg kg$^{-1}$ (NRC 2005). The highest concentration of this element was found in the liver, kidney, and spleen, slightly lower in the heart, muscle, pancreas, lungs, bones, and brain (Feng et al. 2007; Uyanik et al. 2005). It has been demonstrated that some tissues, such as bone, testes, and epididymides, are capable of long-time Cr storing, as compared with the heart, pancreas, and brain (Lewicki et al. 2014).

In ecotoxicological studies, concentration of Cr in endothermic vertebrates is determined mainly in the liver and kidneys and less often in the muscles, lungs, brain, bones, and other samples (Tables 3.6 and 3.7). In the liver of wild herbivorous mammals, Cr concentration normally remains in the range 0.022–0.427 mg kg$^{-1}$ ww (Table 3.6), and the values are within those specified for cattle by the WVDL (2015). In European ungulates, such as roe deer and red deer inhabiting areas either totally uncontaminated or slightly contaminated with heavy metals, hepatic Cr concentrations do not exceed an average of 0.52 mg kg$^{-1}$ ww, like in the moose living in the Arctic area of Canada (Table 3.6). The range of the median concentrations of Cr in the liver of roe deer, hare, and wild boar from central Poland ranged between 0.03 and 0.07 mg kg$^{-1}$ ww (Długaszek and Kopczyński 2011). In turn, the average

**Table 3.7** Chromium in feathers, bone, and mammalian hair

| Species | Place and years | *n* | Mean (mg kg$^{-1}$ dw) | References |
|---|---|---|---|---|
| Avian feathers | | | | |
| Aquatic birds | | | | |
| *Anas platyrhynchos* Mallard | Iran, Fereydoonkenar International Wetland, 2012 | 10 | 2.92 ± 0.83 | Aghdasi et al. (2013) |
| *Ardea alba* Great egret | USA, New Jersey, Barnegat Bay, 1989–2011 | 271 | 0.447 ± 0.376 (own calculation) | Burger (2013) |
| *Aythya ferina* Pochard | Iran, Fereydoonkenar International Wetland, 2012 | 10 | 3.05 ± 0.87 | Aghdasi et al. (2013) |
| *Somateria mollissima* Common eider | USA, Aleutian Islands, 2007 | 30 | 1.78 ± 0.11 | Burger et al. (2009) |
| *Branta canadensis* Canada goose | USA, New Jersey, 2007<br>USA, New Jersey, Mill Creek, 2007 | 26<br>5 | 1.360 ± 0.241<br>2.740 ± 0.545 | Tsipoura et al. (2011) |
| *Cygnus olor* Mute swan | Hungary, Balaton Lake | 17 | 82%<DL<br>1.02 ± 0.09 | Grúz et al. (2015) |
| *Bubulcus ibis* Cattle egret 4–6-day chicks | Pakistan, Rawal Lake Reservoir<br>Chenab River<br>Ravi River, 2007 | 10<br>10<br>10 | 6.6 ± 2.6<br>7.1 ± 2.2<br>5.38 ± 1.0 | Malik and Zeb (2009) |
| *Phalacrocorax carbo* Great cormorant | Spain, Murcia, 2009–2010 | 8 juv | 4.88 ± 1.83<br>2.55–8.59 | Navarro et al. (2010) |
| *Phalacrocorax carbo* Great cormorant | Serbia, 2010 | 21 | <DL | Skoric et al. (2012) |
| *Phalacrocorax carbo* Great cormorant | Japan, 2003 | | | Nam et al. (2005a) |
| | Lake Biwa | 11 | 0.32 ± 0.35 | |
| | | | 0.12–1.3 | |
| | Mie | 19 | 0.23 ± 0.08 | |
| | | | 0.16–0.45 | |
| *Calidris canutus* Red knot | USA, NJ, Delaware Bay | | | Burger et al. (2015) |
| | 1991–1992 | 16 | 0.291 ± 0.042 | |
| | 2011–2012 | 30 | 0.578 ± 0.083 | |
| *Calidris alba* Sanderling | 1991–1992 | 12 | 0.764 ± 0.260 | Burger et al. (2015) |
| | 2011–2012 | 20 | 0.463 ± 0.062 | |
| *Calidris pusilla* Semipalmated sandpiper | 1991–1992 | 12 | 1.149 ± 0.294 | Burger et al. (2015) |
| | 2011–2012 | 30 | 0.523 ± 0.064 | |
| *Calidris canutus* Red knot | USA, NJ, Delaware Bay, 1991 | 16 | 24.10 ± 2.58 | Burger et al. (1993) |

(continued)

**Table 3.7** (continued)

| Species | Place and years | *n* | Mean (mg kg$^{-1}$ dw) | References |
|---|---|---|---|---|
| *Calidris alba* Sanderling | USA, NJ, Delaware Bay, 1991 | 13 | 16.52 ± 2.46 | Burger et al. 1993 |
| *Calidris pusilla* Semipalmated sandpiper | USA, NJ, Delaware Bay, 1991 | 11 | 26.29 ± 4.37 | Burger et al. (1993) |
| *Larus atricilla* Laughing gull | USA, New Jersey, 1992 | 20M<br>20F | 0.76 ± 0.15<br>0.68 ± 0.12 | Gochfeld et al. (1996) |
| *Ciconia ciconia* White stork | Poland, 2005 | 32 | 0.0031 ± 0.0003<br>0.0012–0.0084 | Orłowski et al. (2006) |
| *Nycticorax nycticorax* Black-crowned night heron nestlings | USA, Chesapeake and Delaware Bays, 1998 | | | Golden et al. (2003) |
| | Pea Patch Island | 12 | 2.49 | |
| | Baltimore Harbor | 12 | 3.28 | |
| | Holland Island | 9 | 3.17 | |
| Passerine | | | | |
| *Corvus splendens* House crow | India, Tiruchirappalli, 2013 | | | Manjula et al. (2015) |
| | Urban area | | 76.47 ± 3.88 | |
| | Rural area | | 60.13 ± 8.80 | |
| *Agelaius phoeniceus* Red-winged blackbird | USA, New Jersey, Meadowlands District, 2006 | 29 | 0.607 ± 0.0532 | Tsipoura et al. (2008) |
| *Cistothorus palustris* Marsh wren | USA, New Jersey, Meadowlands District, 2006 | 15 | 1.040 ± 0.109 | Tsipoura et al. (2008) |
| *Tachycineta bicolor* Tree swallow | USA, New Jersey, Meadowlands District, 2006 | 5 | 0.659 ± 0.219 | Tsipoura et al. (2008) |
| Other birds | | | | |
| *Centropus phasianinus* Pheasant coucal | India, Tiruchirappalli, 2013 | | | Manjula et al. (2015) |
| | Urban area | | 36.96 ± 5.72 | |
| | Rural area | | 53.12 ± 5.39 | |
| *Agapornis roseicollis* Rosy-faced lovebird (pet bird) | India, Tiruchirappalli, 2013 | | | Manjula et al. (2015) |
| | Urban area | | 6.11 ± 1.01 | |
| | Rural area | | 6.00 ± 3.20 | |
| All species ($n$ = 11)—free-living birds | India, Tiruchirappalli, 2013 | | | Manjula et al. (2015) |
| | Urban area | | 55.48 | |
| | Rural area | | 51.43 | |
| *Falco biarmicus jugger* Laggar falcon | Pakistan, 1996 | | | Movalli (2000) |
| | Bahawalnagar | 8 | 1.72 | |
| | Bahawalpur | 8 | 1.67 | |
| | Karachi | 11 | 2.31 | |

(continued)

**Table 3.7** (continued)

| Species | Place and years | *n* | Mean (mg $kg^{-1}$ dw) | References |
|---|---|---|---|---|
| Avian bone | | | | |
| Waterbirds | | | | |
| *Anas platyrhynchos* Mallard | Russia, SW, 1993–1995 | 4 | 2.90 | Lebedeva (1997) |
| *Aythya fuligula* Tufted duck | Russia, SW, 1993–1995 | 2 | 1.5 | Lebedeva (1997) |
| *Ardea cinerea* *Grey heron* | Russia, SW, 1993–1995 | 2 | 0.99 | Lebedeva (1997) |
| *Phalacrocorax carbo* Great cormorant | Russia, SW, 1993–1995 | 1 | 0.31 | Lebedeva (1997) |
| Falconiformes and owls | | | | |
| *Buteo buteo* Common buzzard | Russia, SW, 1993–1995 | 2 | 7.59 | Lebedeva (1997) |
| *Falco tinnunculus* Common kestrel | Russia, SW, 1993–1995 | 1 | 14.70 | Lebedeva (1997) |
| Passerine | | | | |
| *Parus major* Great tit | Belgium, Antwerp, 2000 | 10F | 0.26 0.23–0.43 | Dauwe et al. (2005) dw, Med |
| *Parus major* Great tit | Russia, SW, 1993–1995 | 4 | 0.38 | Lebedeva (1997) |
| *Passer montanus* Tree sparrow | Russia, SW, 1993–1995 | 1 | 14.70 | Lebedeva (1997) |
| *Passer domesticus* House sparrow | Russia, SW, 1993–1995 | 4 | 0.41 | Lebedeva (1997) |
| Mammalian hair | | | | |
| *Homo sapiens* Man | USA | | Normal range: 0.0–4.0 Reported range: 0.0–6.43 | Jenkins (1979) |
| *Homo sapiens* Man, children 6–9 years | Spain, 2001 | 117 | 0.64 0.33–1.00 | Pena-Fernández et al. (2014) |
| Ungulates | | | | |
| *Equus ferus caballus* Horse | Poland | 20 | 0.324 ± 0.221 0.029–1.037 | Topczewska (2012) |
| *Equus ferus caballus* Horse | Japan | 24 | 0.22 ± 0.16 | Asano et al. (2002) |
| *Equus ferus caballus* Horse | | | 1.5–3.3 | Puls (1994) |
| Rangifer tarandus Reindeer | Norway, 1988 Spitsbergen | 11 | 1.03 | Drbal et al. (1992) |

(continued)

**Table 3.7** (continued)

| Species | Place and years | *n* | Mean (mg kg$^{-1}$ dw) | References |
|---|---|---|---|---|
| *Bos taurus* Cattle | Poland | 33 | 0.075 ± 0.050 | Gabryszuk et al. (2010) |
| *Capreolus capreolus* Roe deer | Poland, 2009 | 13 | 0.20 ± 0.23<br>Med 0.10 | Długaszek and Kopczyński (2014) |
| Carnivores | | | | |
| *Felis bengalensis* Leopard cat | India, Assam and Meghalaya | 10–15 | 72.3 ± 7 | Dey et al. (1999) AM ± SD |
| *Panthera pardus* Leopard | India, Assam and Meghalaya | 10–15 | <DL | Dey et al. (1999) AM ± SD |
| *Viverra zibetha* Civet cat | India, Assam and Meghalaya | 10–15 | <DL | Dey et al. (1999) AM ± SD |
| *Vulpes vulpes* Red fox | Poland, 2009 | 14 | 0.20 ± 0.23<br>Med 0.10 | Długaszek and Kopczyński (2014) |
| *Vulpes vulpes* Red fox (farmed) | Poland, rural region and urbanized region | 8<br>8 | 0.268 ± 0.137<br>0.145 ± 0.033 | Filistowicz et al. (2012) |
| *Vulpes lagopus* Arctic fox (farmed) | Poland, rural region and urbanized region | 8<br>8 | 0.193 ± 0.079<br>0.147 ± 0.032 | Filistowicz et al. (2012) |
| *Vulpes vulpes* Red fox (farmed) Fire morph Silver | Poland | | | Cholewa et al. (2014) |
| | | 15F | 0.431 ± 0.020 | |
| | | 11F | 4.251 ± 0.063 | |
| *Canis familiaris* Dog | Italy, Campania | | | Zaccaroni et al. (2014) |
| | A, triangle of death | 30 | 0.14 ± 0.03 | |
| | B, urban | 30 | 0.16 ± 0.03 | |
| | C, rural | 30 | 0.38 ± 0.66 | |
| *Canis familiaris* Dog | Korea, urban areas | 204 | 2.41 ± 0.28 | Park et al. (2005) |
| Other species | | | | |
| *Lepus europaeus* European hare | Poland, 2009 | 11 | 0.27 ± 0.45<br>Med 0.14 | Długaszek and Kopczyński (2014) |
| *Erinaceus europaeus* Hedgehogs | Belgium, 2002 | | | D'Havé et al. (2006) |
| | Hoboken, metal-polluted area | 12 | 1.57 ± 0.33 | |
| | Zoersel, control area | 10 | 1.46 ± 0.09 | |
| *Erinaceus europaeus* Hedgehogs | Belgium, Flanders, 2002/2003 | 43 | 5.4 ± 0.7 | D'Havé et al. (2006) |
| *Rattus rattus* Wild rat | Portugal, 2000 | | | Pereira et al. (2006) TG, the mine pit |
| | TG, Tapada | 4 | 1.309 ± 0.197 | |
| | Grande | | | |
| | M4, Sulfur mill | 5 | 0.232 ± 0.066 | |

(continued)

**Table 3.7** (continued)

| Species | Place and years | *n* | Mean (mg kg$^{-1}$ dw) | References |
|---|---|---|---|---|
| *Mus spretus* Algerian mouse | Portugal, 2000 | | | Pereira et al. (2006) TG, the mine pit |
| | TG, Tapada | 7 | 1.673 ± 0.618 | |
| | Grande | | | |
| | M4, Sulfur mill | 8 | 0.579 ± 0.165 | |
| *Spermophilus beecheyi* Ground squirrel | USA, California, 2001–2002 | 181 | 0.45 ± 0.57 | Hubbart (2012) |
| *Rattus norvegicus* Rat (lab) | Germany, | | | Ambeskovic et al. (2013) |
| | Young, 6 months | 7 | 0.599 | |
| | Adult, 18 months | 7 | 0.612 | |
| *Callithrix jacchus* Common marmoset | Germany, Gottingen, German Primate Center | | | Ambeskovic et al. (2013) |
| | Young, 1–2 years | 8 | 1.269 | |
| | Middle-aged, 4 years of age, 6–8 years | 8 | 1.523 | |
| | | 8 | 1.454 | |
| *Macaca mulatta* Rhesus | China SW, animal farm | 28 | 11.33 ± 4.26 | Lee et al. (2012) |
| *Apodemus sylvaticus* Wood mice | Italy, Modena, 2004 | 26 | ~1 | Marcheselli et al. (2010) |
| *Petaurista magnificus* Flying squirrel | India, Assam and Meghalaya | 10–15 | <DL | Dey et al. (1999) |
| *Didelphis virginiana* Opossum | USA, Costa Rica | 12M<br>12F | 1.57 ± 0.19<br>3.56 ± 0.64 | Burger et al. (1994) |
| *Zapus princeps* Western jumping mouse | USA, Wyoming | 3 | 23–45 | Jenkins (1979) |
| *Microtus montanus* Mountain vole | USA, Wyoming | 16 | 4.7–180 | Jenkins (1979) |
| *Antilocapra americana* Pronghorn | USA, Idaho<br>Wyoming | 30<br>7 | 1.9–640<br>0.3–130 | Jenkins (1979) |
| *Canis latrans* Coyote | USA, Wyoming | 15/19 | 0.7–12.0 | Jenkins (1979) |
| *Cervus canadensis* Elk | USA, Idaho | 15 | 1.9–570 | Jenkins (1979) |
| *Odocoileus hemionus* Mule deer | USA, Idaho | 9/11 | 13–630 | Jenkins (1979) |

Feather Cr threshold, 2.8 mg kg$^{-1}$ dw
*DL* detection limit, *F* female, *M* male

hepatic content of Cr in animals caught in the eastern states of Germany was much higher, in the range 0.495 (0.347) in red deer, 0.581 (0.407) in wild boar, 0.628 (0.44) in roe deer, 0.728 (0.51) in mouflon, and 0.807 (0.565) mg $kg^{-1}$ dw (ww) in hare (Anke et al. 2005). Amici et al. (2012) evaluated the content of Cr in the organs of wild boar living in the area of Central Italy. The concentrations of Cr in the liver (Table 3.6) are comparable to those reported by Piskorová et al. (2003). Similar Cr content in the muscle (Table 3.6) was measured by Długaszek and Kopczyński (2013). The renal Cr concentration was different compared to data published by Piskorová et al. (2003). Długaszek and Kopczyński (2013) determined the content of Cr also in the muscles of roe deer and hare (Table 3.6). The results showed strong variability; in roe deer muscle, these values ranged from 0.04 to 2.35 mg $kg^{-1}$ ww. The concentration of Cr in the liver of the reindeer ranged from 0.01 to 0.09 mg $kg^{-1}$ ww (Table 3.6). One reindeer only had a muscle Cr concentration of 1.56 mg $kg^{-1}$ ww (Hassan et al. 2013). Daily adequate intakes of 25 μg for adult female and 35 μg for male were established by the Food and Nutrition Board at the Institute of Medicine of the National Academies (Hassan et al. 2013).

Among the carnivores Cr concentrations were analyzed in tissues of piscivorous otters (European otter *Lutra lutra* and North American river otter *Lontra canadensis*) as well as in tissues of the American mink *Neovison vison*, which is native to North America and is an invasive species in Eurasia. Moreover, Cr was analyzed in Europe in the red fox and in North America in raccoon *Procyon lotor* and wolverine *Gulo gulo* (Table 3.6). Mason and Stephenson (2001) compared the concentration of Cr in the liver of European otters from Denmark, Great Britain, and Ireland. No chromium was detected in the livers of a part of the animals, although in some others the concentration exceeded 4 mg $kg^{-1}$ dw. The highest average (geometric mean) concentration of the metal was measured in the liver of otter living in Britain (0.270 mg $kg^{-1}$ dw), whereas Cr levels in the specimens from Denmark and Ireland were an order of magnitude lower. The relatively high level of Cr in the otters from Great Britain may be a consequence of high water concentrations of chromium, leached from the Cr-rich bedrock.

Both feral and ranch American mink from Poland and the wild mink from Illinois (USA) revealed higher Cr concentrations in the kidney compared to the liver (Table 3.6) (Halbrook et al. 1996; Brzezinski et al. 2014). In raccoon tissues (heart, kidney, muscle, spleen, and liver) from Steel Creek, which was within the watershed of a former reactor effluent stream, the highest Cr concentration was noted in muscle (0.95 ± 0.13 μg $kg^{-1}$ ww) and spleen (0.98 ± 0.18), which differed significantly compared to territory outside of the Department of Energy's Savannah River Site (SRS-South Carolina, USA) (0.24 ± 0.02 μg $kg^{-1}$ ww). Higher ($p <$ 0.001) Cr concentration was found also in the kidney of raccoons from SRS (0.31 ± 0.02) in relation to out-site of SRS (0.25 ± 0.01 μg $kg^{-1}$ ww) (Burger et al. 2002). The mean concentration of Cr in the wild canids was 99 times higher than in the fox *Vulpes vulpes* from a contaminated site in Italy (Alleva et al. 2006). Foxes from Slovakia also showed low levels of Cr content (Piskorová et al. 2003) (see details in Table 3.6).

Analysis of Cr content in various species of herbivorous mice and voles have shown that, despite the differences in diets of these two groups, there were no significant differences in the Cr content in their bodies, which ranged from 1.064 to 1.779 mg kg$^{-1}$ dw. Living in the same area, insectivorous shrews (*Sorex* sp.) had twice that level of Cr which is 3.447 mg kg$^{-1}$ dw (Anke et al. 2005).

The issue of Cr concentration in the organs of livestock animals and their diet chromium supplementation has been widely discussed and investigated. Pigs' diet supplementation with 0.2 mg Cr kg$^{-1}$ resulted in deposition of the element in the liver and kidneys, without a significant impact on the muscles. Cr supplementation of cattle diet remained without detectable changes (Spears et al. 2004). In chickens, Cr was administered in various forms (yeast, chromium picolinate, or chloride), which accumulated in the liver, kidneys, and muscles, without affecting the Cr content in eggs (Dębski et al. 2001; Uyanik et al. 2005; Lewicki et al. 2014). It was found that application of $CrCl_3$ in amounts 25, 100, and 200 μg kg$^{-1}$ increased the accumulation of the metal in the liver and kidney, but did not alter its levels in eggs and meat (Anderson et al. 1989). Laying hens fed with high doses (500, 1000, 2000 mg kg$^{-1}$) of inorganic $CrCl_3$ for 75 days transported a very small part (0.0026%) of the daily intake to the edible parts of the egg (Piva et al. 2003).

Table 3.6 presents the data divided by aquatic birds, passerines, and others. The average hepatic Cr content in these groups ranged between 0.1 and 2.75, in aquatic birds, and between 0.13 and 1.86 mg kg$^{-1}$ dw, in passerines. According to WVDL (2015), the normal range of Cr in the liver of an avian is 0.05–0.40 mg kg$^{-1}$ ww. The great tit inhabiting the Western Mountains of Beijing in China had 50 times more Cr in the liver than birds of the same species in Belgium (Table 3.6), not high enough though to be referred to as toxic, which—according to WVDL (2015)—is 13–150 mg kg$^{-1}$ ww. Renal content of Cr in the great tit was 6.26 mg kg$^{-1}$ dw, whereas the level considered as toxic is in the range 19–170 mg kg$^{-1}$ ww (WVDL 2015). The data presented in Table 3.6 concerning aquatic birds does not show that plant or animal diet raises the level of Cr in avian organs. Ducks of the genera *Anas* and *Aythya* are characterized by variable hepatic level of Cr as the species of *Haliaeetus* sp., *Pandion* sp. and *Larus* sp., or *Parus* and *Carduelis* sp.

Table 3.7 presents data on the concentration of Cr in the feathers, hair, and bone of various animals. Chromium content in the feathers of white stork in Poland was several thousands lower than those measured in the feathers of seagulls and terns (5.5–12.87 mg kg$^{-1}$) in North America (Burger 1996; Orłowski et al. 2006). Burger (2013) studied Cr in the feather of young great egrets *Ardea alba* from Barnegat Bay (New Jersey, USA) and observed that these varied in the range 0.09–1.47 mg kg$^{-1}$ dw in 1989–2011 (Table 3.7), and no significant reductions in Cr levels were observed over the years. Burger et al. (2014) investigated Cr levels in various tissues of the semipalmated sandpiper *Calidris pusilla* during its migratory stopover in Delaware Bay (New Jersey, USA). The highest Cr level was found in feathers (Table 3.7). Tsipoura et al. (2008) measured feather Cr contents in three passerine species feeding in the highly urbanized area of Meadowlands of New Jersey. It ranged between 0.6 and 1.0 mg kg$^{-1}$ dw and was nearly 80 times lower as compared to another passerine species inhabiting a polluted area in India (Table 3.7). Golden et al. (2003) determined the Cr content in feathers of *Nycticorax nycticorax* chicks from Chesapeake and Delaware Bays (USA). They suggested that wading birds may be exposed to industrial and agricultural contaminants; hence, it is necessary to monitor the environmental pollution (Table 3.7). D'Havé et al. (2006) measured Cr concentration (all values in mg $kg^{-1}$ dw) in the hair (5.4), spikes (4.3), liver (3.9), kidneys (3.4), and muscle (4.6) in European hedgehog *Erinaceus europaeus*. It should be noted that the concentration of Cr in the hair and spikes was higher than in the liver and kidney.

## 9 Ecological Effects of Chromium

Chromium belongs to the group of heavy metals that raise serious concerns in terms of increasing environmental pollution resulting from human activity. Few researchers have focused on the analysis of Cr levels in organisms representing various trophic levels, to bring new knowledge on the possible biomagnification of this metal. Inland ecosystems, both aquatic and terrestrial, have been studied in this respect by, e.g., Outridge and Scheuhammer (1993), Zaccaroni et al. (2003), Alhashemi et al. (2011), Sample et al. (2014), and Zojaji et al. (2014). Sample et al. (2014) conducted soil-screening studies in order to find out the reference values, which could be used to evaluate the toxicity of soil in relation to wild animals. Ecological soil screening is a simplified method assuming that the soil contaminant concentration is at the level of toxicity reference value (TRV), without adverse effects. The analyses involved six mammalian and avian species: the meadow vole (*Microtus pennsylvanicus*), northern short-tailed shrew (*Blarina brevicauda*), long-tailed weasel (*Mustela frenata*), American woodcock (*Scolopax minor*), and red-tailed hawk (*Buteo jamaicensis*). The analyses showed that food and soil have the strongest effect on wildlife exposure to chromium. In the soil-plant system, most chromium is retained in roots, and only a small portion of this is transported to the upper parts of the plant (the stem and leaves). Therefore, the authors assume that the risk of Cr intoxication from plants eaten by herbivorous mammals is low.

For example, Cr was last in the sequence of eight metals studied by Ma (1982) in earthworms, with a geometric mean of 0.06. According to other authors (van Gestel et al. 1993), BAF (bioaccumulation factor) for earthworms depend on the concentration in the soil. Beyer et al. (1990), however, did not find a correlation between Cr concentration in the soil and the biota. These authors considered it unlikely for chromium to accumulate in the trophic chain. According to Outridge and Scheuhammer (1993), there is no evidence of chromium biomagnification. On the contrary, with an increase in the trophic level, the concentration of Cr considerably decreases, which is referred to as "biominification." The literature cited by Outridge and Scheuhammer (1993) brings many examples of decreasing Cr concentrations at a higher trophic level. Biominification exists in the wildlife (mammals), both within land and marine trophic chains. In some organs of predators (seals, porpoises, and dolphins), the level of Cr was lower than in the fish they preyed on. A similar situation was found in birds: seagulls had a lower concentration of Cr in their tissues compared to organisms they ate, such as clams, snails, crabs, or finfish. An example of biominification in a terrestrial trophic chain is domestic cattle fed with a *Festuca* sp. grass on a pasture contaminated with chromium-containing wastes. Also in the case of Cr-rich wastewater discharged to a forest, mammals living there, such as voles, mice, or even red deer, demonstrated the effect of biominification. The authors explained this with the fact of low absorption of Cr in the alimentary tract (Outridge and Scheuhammer 1993). Zukal et al. (2015) proposed Cr as one of the heavy metals of concern in terms of wildlife conservation. They reviewed the literature for the use of bats as bioindicators of environmental pollution and pointed out that a major challenge is to create standardized programs to monitor the concentration of elements in the tissues of bats of different species and coming from different countries and continents, using modern analytical techniques.

According to Appendix H of the Technical Support Document for Exposure Assessment and Stochastic Analysis (OEHHA 2000), the bioconcentration factor (BCF) for Cr in fish has been established in the range 1–3.4. The arithmetic mean of these values, which is 2, is recommended as the default BCF for Cr. Bioconcentration of Cr in piscine tissues may lead to impaired respiratory functions and osmotic regulation due to damages in the gill epithelium. Chromium levels measured in fish (0.02–0.75 mg $kg^{-1}$) in most samples exceeded the limits established by the WHO (2000) and FEPA for fish as seafood (0.15 mg $kg^{-1}$). Consumption of fish with such a high content of Cr may lead to health hazards.

For invertebrates, Outridge and Scheuhammer (1993) recommend the BAF 0.6, whereas for small mammalian species, after Beyer et al. (1990), the BAF 0.2. According to Alhashemi et al. (2011), the trend in Cr allocation in plants was roots > stem > leaves. The translocation factor (TF) from stem to leaves was higher than from roots to stem. Chromium bioaccumulation was studied in relation to diet in the white-spectacled bulbul *Pycnonotus xanthopygos* in various tissues (kidney, rib, muscle, lung, liver, heart) and eggs (yolk and shell). Significant differences were found in Cr bioaccumulation; the levels in kidneys and ribs were higher compared to those in other tissues (Aslan et al. 2006). The results of studies show that an excess beyond the reducing capacity of the environment leads to bioaccumulation of Cr(IV) in the environment and to an increase in its pollution (Vajpayee et al. 1999; Jianlong et al. 2004). Zojaji et al. (2014), who irrigated an experimental plantation of maize (*Zea mays*) using wastewater with a high content of Cr (12 μg $kg^{-1}$), demonstrated a high level of Cr accumulation in the soil, roots, stems, and leaves of the plants, with the highest levels found in roots. The results presented by Imam Khasim et al. (1989) indicate a dangerous level of Cr bioaccumulation in tissues of animals representing various levels of the trophic chain, as well as in agricultural produce, due to penetration of Cr from soil and water.

## 10 Bioindicators and Biomarkers of Chromium in Ecotoxicological Studies

Typical Cr levels, considered as physiological, have been established only for a small group of birds and mammals (see Table 3.6, WVDL 2015). Data on Cr concentrations in tissues and hair of livestock animals have been collected by Puls (1994). For man, Yoo et al. (2000) determined in Korean human organs the following Cr concentrations: liver 0.21, kidneys 1.6, heart 0.56, lungs 0.80, spleen 3.4, brain 0.42, bone 0.27, blood 0.24, hair 0.57, and nail 0.54 mg $kg^{-1}$ ww.

The most useful samples in ecotoxicological studies determining Cr concentration in animals tissues seem to be liver, kidneys, (Harding et al. 1998; Piskorova et al. 2003; Pereira et al. 2006), bone and lungs (Dauwe et al. 2005; Horai et al. 2007; Aslan et al. 2006), and in birds also salt glands (Burger and Gochfeld 1985). However, existing empirical records do not suffice to allow drawing definitive conclusions. Further research is needed to collect new data and to monitor Cr concentrations in the environment.

Hair and feathers are promising materials, which can be sampled from living animals (Burger et al. 2015; Manjula et al. 2015). Cr analysis in birds, however, based on feather samples has its limitations. Orlowski et al. (2006), basing on their own research and that by other authors, have shown that feathers are a good indicator of the concentration of Cr in the environment; however, feathers of similar length and collected from corresponding parts of the body, preferably the breast, should be used for analyses. Of the ungulates, wild boars and deer (game animals, widespread across the world) can be used to indirectly assess the impact, which heavy metals, including Cr, have on the environment. Otters are at the top of the trophic chain in aquatic environments and appear to be a useful bioindicator of contaminant metals, including Cr (Harding et al. 1998; Mason and Stephenson 2001; Walker et al. 2010). Mink may also provide data on changes in the environment in terms of chromium levels (Harding et al. 1998; Brzezinski et al. 2014). Also the hair and spikes of the European hedgehog seem to represent suitable biological material to assess environmental pollution with heavy metals, including Cr, since:

- The hedgehog lives in suburban areas and, due to a relatively small foraging territory (1-10-30 hectares) and a low migration index, reflects well the level of local pollution.
- Its long average life span (Reeve 1994) may suggest long-term exposure to contaminants, which may lead to chronic toxic effects (D'Havé et al. 2006).

From among birds, many authors propose passerines, the most numerous avian order, widely distributed, as good bioindicators of Cr pollution (Deng et al. 2007; Tsipoura et al. 2008; Aslan et al. 2006).

Domesticated animals can also be used for environmental pollution assessment, provided they are fed exclusively with feeds from the areas close to their habitat. Studies on livestock animals also show how the diet affects Cr absorption and accumulation (Spears et al. 2004; Dębski et al. 2001).

It is essential to simultaneously obtain information on the geochemical background of the analyzed area and possible anthropogenic sources of Cr and to establish reference values for the content of chromium in various tissues, in order to read the results correctly.

## 11 Conclusions

- Chromium is an essential trace element for both animals and humans.
- In the natural environment, the effects of hexavalent and trivalent chromium compounds differ fundamentally.
- Trivalent Cr is essential for humans and animals; it underlies the normal glucose metabolism by participating in the glucose tolerance factor (GTF) and plays a role in the glucose-signaling pathway.
- Chromium activates insulin activity.
- In homeothermic vertebrates chromium takes part in the metabolism of hydrocarbons, proteins, and lipids.
- In animals chromium deficiency leads to reduced glucose tolerance, impaired insulin function, changes in protein metabolism, general weakness, and damage to the cardiovascular system.
- Chromium deficiency in mammals leads to growth disorders and low sperm motility in the semen.
- Hexavalent chromium is highly toxic for both plants and animals.
- Anthropogenic Cr pollution of air, water, and soil is a result of combustion of coal and other fossil fuels, smelting of iron and nonferrous metals, and discharges of wastes from paint factories and tanneries, from wastewater treatment plants, and from scrap piles of chrome steel.
- In animal cells, hexavalent Cr is reduced to trivalent Cr, which produces free radicals.
- Hexavalent Cr is carcinogenic, mutagenic, and teratogenic in endothermic animals.
- Excess of Cr, like its deficiency, in males deteriorates sperm quality, damages testicles, and reduces libido.
- It was found that a higher level in the trophic chain does not cause biomagnification of chromium; on the contrary, the higher the trophic level, the lower the tissue Cr content, which is referred to as "biominification."

## References

Abraham AS, Brooks BA, Eylath U (1992) The effects of chromium supplementation on serum glucose and lipids in patients with and without non-insulin dependent diabetes. Metabolism 41:768–771

Adema DMM, Henzen L (1989) A comparison of plant toxicities of some industrial chemicals in soil culture and soilless culture. Ecotoxicol Environ Saf 18:19–229

Aghdasi S, Mansoori J, Mashayekhi HA (2013) To study the amount of cadmium and chromium in the feathers of emigrant birds in Fereydoonkenar International Wetland. Int J Agri Crop Sci 5: 1730–1734

Aldrich MV, Gardea-Torresdey JL, Peralta-Videa JR, Parsons JG (2003) Uptake and reduction of Cr(VI) to Cr(III) by mesquite (*Prosopis* spp.): chromate-plant interaction in hydroponics and solid media studied using XAS. Environ Sci Technol 37:1859–1864

Alhashemi AS, Karbassi AR, Kiabi BH, Monavari SM, Nabavi SM, Sekhavatjou MS (2011) Bioaccumulation of trace elements in trophic levels of wetland plants and waterfowl birds. Biol Trace Elem Res 142:500–516

Alleva E, Francia N, Pandolfi M, De Marinis AM, Chiarotti F, Santucci D (2006) Organochlorine and heavy-metal contaminants in wild mammals and birds of Urbino-Pesaro Province, Italy: an analytic overview for potential bioindicators. Arch Environ Contam Toxicol 51:123–134

Ambeskovic M, Fuchs E, Beaumier P, Gerken M, Metz GA (2013) Hair trace elementary profiles in aging rodents and primates: links to altered cell homeodynamics and disease. Biogerontology 14:557–567

Amici A, Danieli PP, Russo C, Primi R, Ronchi B (2012) Concentrations of some toxic and trace elements in wild boar (*Sus scrofa*) organs and tissues in different areas of the Province of Viterbo, Central Italy. Ital J Anim Sci 11(4):e65

Anderson RA (1988) Chromium. In: Smith K (ed) Trace minerals in food. Marcel Dekker, New York, pp 231–247

Anderson RA (1993) Recent advances in clinical and biochemical effect sod chromium deficiency. Prog Clin Biol Res 380:221–234

Anderson RA (1995) Chromium and parenteral nutrition. Nutrition 11:83–86

Anderson RA (1997) Chromium as an essential nutrient for humans. Regul Toxicol Pharmacol 26: 35–41

Anderson RA (1998) Chromium, glucose intolerance and diabetes. J Am Coll Nutr 17:548–555

Anderson RA, Bryden NA, Polansky MM, Richards MP (1989) Chromium supplementation of turkeys: effect on tissue chromium. J Agric Food Chem 37:131–134

Anderson RA, Polansky MM, Bryden NA, Canary JJ (1991) Supplemental-chromium effects on glucose, insulin, glucagon and urinary chromium losses in subjects consuming controlled low-chromium diets. Am J Clin Nutr 54:909–916

Anderson RA, Bryden NA, Polansky MM, Gautschi K (1996) Dietary chromium effects on tissue chromium concentrations and chromium absorption in rats. J Trace Elem Exp Med 9:11–25

Anderson RA, Bryden NA, Polansky MM (1997) Lack of toxicity of chromium chloride and chromium picolinate in rats. J Am Coll Nutr 16:273–279

Ani M, Moshtaghie AA (1992) The effect of chromium on parameters related to iron metabolism. Biol Trace Elem Res 32:57–64

Anke M, Jaritz M, Holzinger S, Arnhold W, Müller R, Angelow L, Choppe C (2001) Nutrients, macro-, trace-, and ultratrace elements in the food chain of mouflons and their mineral status. In: Nahlik A, Uloth W (eds), Proceedings of the third international symposium on Mouflon, Sopron, pp 262–280

Anke M, Schäfer U, Jaritz M, Seifert M, Müller R (2005) Chromium in the food chain – essentiality, metalopharmaceutical effects and toxicity. Proceedings of the 5th international symposium on trace elements in human: new perspectives, Athens/Greece, pp 554–581 (www.med.uoa.gr)

Aruldhas MM, Subramanian S, Sehkar P, Venkatesh G, Chandrakasan G, Govindarajulu P et al (2005) Chronic chromium exposure induced changes in testicular histoarchitechture are associated with oxidative stress: study in a non-human primate (*Macaca radiata* Geoffrey). Hum Reprod 20:801–2813

Arunkumar RI, Rajasekaran P, Michael RD (2000) Differential effect of chromium compounds on the immune response of the African mouth breeder *Oreochromis mossambicus* (Peters). Fish Shellfish Immunol 10:667–676

Arvizu RR, Domínguez IA, Rubio MS, Bórquez JL, Pinos-Rodríguez JM, González M, Jaramillo G (2011) Effects of genotype, level of supplementation, and organic chromium on growth performance, carcass and meat traits grazing lamb. Meat Sci 88:404–408

Asano R, Suzuki K, Otsuka M, Sakurai H (2002) Concentrations of toxic metals and essential minerals in the mane hair of healthy racing horses and their relation to age. J Vet Med Sci 64:607–610

Aslan A, Yavuz M, Kaska Y, Erdoğan A, Kiziroğlu I (2006) Preliminary study on feeding ecology and heavy metal accumulation of white-spectacled bulbul (*Pycnonotus xanthopygos*), Antalya - Turkey. Fresenius Environ Bull 15:1174–1181

ATSDR (Agency for Toxic Substances and Disease Registry) (2000) Toxicological Profile for Chromium, U.S. Department of Health and Human Services, Washington, DC

ATSDR (Agency for Toxic Substances and Disease Registry) (2008) Draft Toxicological Profile for Chromium: for public comment. U.S. Department of Health and Human Services. http://www.atsdr.cdc.gov/toxprofiles/tp7.pdf. Accessed 13 Jan 2012

ATSDR (Agency for Toxic Substances and Disease Registry) (2012) Toxicological profile for chromium. Atlanta, GA, U.S. Department of Health and Human Services, Public Health Service, Agency for Toxic Substances and Disease Registry (http://www.atsdr.cdc.gov/toxprofiles/tp7.pdf)

Azam I, Afsheen S, Zia A, Javed M, Saeed R, Sarwar MK et al (2015) Evaluating insects as bioindicators of heavy metal contamination and accumulation near industrial area of Gujrat, Pakistan. BioMed Res Int. Article ID 942751. doi:https://doi.org/10.1155/2015/942751

Bagchi D, Stohs SJ, Downs BW, Bagchi M, Preuss HG (2002) Cytotoxicity and oxidative mechanisms of different forms of chromium. Toxicology 180:5–22

Bagchi D, Downs BW, Preuss HG (2003) Reply to the editor (cytotoxicity and oxidative mechanisms of different forms of chromium). Toxicology 186:175–177

Barałkiewicz D, Siepak J (1999) Chromium, nickel and cobalt in environmental samples and existing legal norms. Pol J Environ Stud 8(4):201–208

Barceloux DG (1999) Chromium. Clin Toxicol 37:173–194

Barnhart J (1997) Occurrences, uses, and properties of chromium. Regul Toxicol Pharm 26:S3–S7

Bartlett RJ (1991) Chromium cycling in soils and water: links, gaps and methods. Environ Health Perspect 92:17–24

Batic M, Raspor P (1998) Uptake and bioaccumulation of Cr (III) in yeast *Saccharomyces cerevisiae*. Food Technol Biotechnol 36:291–297

Bequer T, Quantin C, Sicot M, Boudot JP (2003) Chromium availability in ultramafic soil from New Caledonia. Sci Total Environ 301:251–261

Beyer WN, Day D (2004) Role of manganese oxides in the exposure of mute swans (*Cygnus olor*) to Pb and other elements in the Chesapeake Bay, USA. Environ Pollut 129:229–235

Beyer WN, Miller G, Simmers JW (1990) Trace elements in soil and biota in confined disposal facilities for dredged material. Environ Pollut 65:19–32

Bielicka A, Bojanowska I, Wiśniewski A (2005) Two faces of chromium-pollutant and bioelement. Pol J Environ Stud 14:5–10

Bojar H, Bojar I (2009) Monitoring of contamination of the Lublin region wetlands using mallards (*Anas platyrhynchos*) as a vector of the contamination by various conditionally toxic elements. Ann Anim Sci 9:195–204

Borel JS, Anderson RA (1984) Chromium. In: Frieden E (ed) Biochemistry of the essential ultratrace elements. Plenum Press, New York and London, pp 175–199

Brady D, Letebele B, Duncan JR, Rose PD (1994) Bioaccumulation of metals by *Scenedesmus, Selenastrum* and *Chlorella* algae. Water SA 20:213–218

Brown M (2003) Harnessing chromium in the fight against diabetes. Drug Discov Today 8:962–963

Bruckwicki P, Giggleman C, Lewis J (2006) An investigation of contaminant levels in white-tailed deer (*Odocoileus virginianus*) collected from Caddo Lake National Wildlife Refuge, Harrison County, Texas 2005 Project ID Nos.: DEC No. 200520002; FFS No. 2N53

Brzezinski M, Zalewski A, Niemczynowicz A, Jarzyna I, Suska-Malawska M (2014) The use of chemical markers for the identification of farm escapees in feral mink populations. Ecotoxicology 23:767–778

Buerge IB, Hug SJ (1999) Influence of mineral surfaces on chromium (VI) reduction by iron (II). Environ Sci Technol 33:4285–4291

Burger J (1993) Metals in avian feathers: bioindicators of environmental pollution. Rev Environ Toxicol 5:203–311

Burger J (1996) Heavy metals and selenium levels in feathers of Franklin's Gulls in interior North America. Auk 113:399–407

Burger J (2013) Temporal trends (1989-2011) in levels of mercury and other heavy metals in feathers of fledgling great egrets nesting in Barnegat Bay, NJ. Environ Res 122:11–17

Burger J, Gochfeld M (1985) Comparisons of nine heavy metals in salt gland and liver of greater scaup (*Aythya marila*), black duck (*Anas rubripes*) and mallard (*A. platyrhynchos*). Comp Biochem Physiol 81C:287–292

Burger J, Seyboldt S, Morganstein N, Clark K (1993) Heavy metals and selenium in feathers of three shorebird species from Delaware Bay. Environ Monit Assess 28:189–198

Burger J, Marquez M, Gochfeld M (1994) Heavy metals in the hair of opossum from Palo Verde, Costa Rica. Arch Environ Contam Toxicol 27:472–476

Burger J, Gaines KF, Lord CG, Brisbin IL, Shukla S, Gochfeld M (2002) Metal levels in raccoon tissues: differences on and off the Department of Energy's Savannah River site in South Carolina. Environ Monit Assess 74:67–84

Burger J, Gochfeld M, Jeitner C, Burke S, Volz CD, Snigaroff R et al (2009) Mercury and other metals in eggs and feathers of glaucous-winged gulls (*Larus glaucescens*) in the Aleutians. Environ Monit Assess 152:179–194

Burger J, Gochfeld M, Niles L, Dey A, Jeitner C, Pittfield T, Tsipoura N (2014) Metals in tissues of migrant semipalmated sandpipers (*Calidris pusilla*) form Delaware Bay, New Jersey. Environ Res 133:362–370

Burger J, Tsipoura N, Niles LJ, Gochfeld M, Dey A, Mizrahi D (2015) Mercury, lead, cadmium, arsenic, chromium and selenium in feathers of shorebirds during migrating through Delaware Bay, New Jersey: comparing the 1990s and 2011/2012. Toxics 3:63–74

Burton JL (1995) Supplemental chromium: its benefits to the bovine immune system. Anim Feed Sci Tech 53:117–133

Cantu Y, Remes A, Reyna A, Martinez D, Villarreal J, Ramos H et al (2014) Thermodynamics, kinetics and activation energy studies of the sorption of chromium (III) and chromium (VI) to a $Mn_3O_4$ nanomaterial. Chem Eng J 254:374–383

Carmichael DB, Baker OE (1989) Pesticide, PCB and heavy metal residues in South Carolina mink. Proc Annu Conf Southeast Assoc Fish Wildl Agencies 43:444–451

Casalegno C, Schifanella O, Zennaro E, Marroncelli S, Briant R (2015) Collate literature data on toxicity of chromium (Cr) and nickel (Ni) in experimental animals and humans. EFSA Supp Publ 12:1–287. www.efsa.europa.eu/publications

CCME (Canadian Council of Ministers of the Environment) (1999) Canadian Water Quality Guidelines for the Protection of Aquatic Life Factsheet. http://ceqg-rcqe.ccme.ca/; Canadian soil quality guidelines for the protection of environmental and human health: chromium (total 1997) (VI 1999). http://ceqgrcqe.ccme.ca/

Cefalu WT, Hu FB (2004) Role of chromium in human health and in diabetes. Diabetes Care 27: 2741–2751

Cervantes C, Campos-García J, Devars S, Gutiérrez-Corona F, Loza-Tavera H, Torres-Guzmán JC, Moreno-Sánchez R (2001) Interactions of chromium with microorganisms and plants. FEMS Microbiol Rev 25:335–347

Chen G, Liu P, Pattar GR, Tackett L, Bhonagiri P, Strawbridge AB, Elmendorf JS (2006) Chromium activates glucose transporter 4 trafficking and enhances insulin-stimulated glucose transport in 3T3-L1 adipocytes via a cholesterol-dependent mechanism. Mol Endocrinol 20: 857–870

Cheng H, Zhou T, Li Q, Lu L, Lin C (2014) Anthropogenic chromium emissions in China from 1990 to 2009. PLoS One 9(2):e87753

Cholewa R, Mleczek M, Kołodziejczyk D, Socha S (2014) Concentration of minerals in the coat of red fox (*Vulpes vulpes* L.) of various color morphs and their crossbreds vs. hair strength properties. Acta Sci Pol Zootech 13:35–44

Chrysochoou M, Johnston CP (2012) Reduction of chromium (VI) in saturated zone sediments by calcium polysulfide and nanoscale zerovalent iron derived from green tea extract. In: Hryciw RD, Athanasopoulos-Zekkos A, Yesiller N (eds) GeoCongress 2012: state of the art and practice in geotechnical engineering, pp 3959–3967

Ciftci N, Cicik B, Erdem C, Ay O, Gunalp C (2010) Accumulation of chromium in liver, gill and muscle tissue of *Oreochromis niloticus*. J Anim Vet Adv 9:1958–1960

Clodfelder BJ, Vincent JB (2005) The time-dependent transport of chromium in adult rats from the bloodstream to the urine. J Biol Inorg Chem 10:383–393

Coeurdassier M, de Vaufleury A, Crini N, Scheifler R, Badot PM (2005) Assessment of whole effluent toxicity on aquatic snails: bioaccumulation of Cr, Zn, and Fe, and individual effects in bioassays. Environ Toxicol Chem 24:198–204

Cohen MD, Kargacin B, Klein CB, Costa M (1993) Mechanisms of chromium carcinogenicity and toxicity. Crit Rev Toxicol 23:255–281

Cohen JB, Barclay JS, Major AR, Fisher JP (2000) Wintering greater scaup as biomonitors of metal contamination in federal wildlife refuges in the Long Island Region. Arch Environ Contam Toxicol 38:83–92

Collins BJ, Stout MD, Levine KE, Kissling GE, Melnick RL, Fennell TR et al (2010) Exposure to hexavalent chromium resulted in significantly higher tissue chromium burden compared with trivalent chromium following similar oral doses to male F344/N rats and female B6C3F1 mice. Toxicol Sci 118:368–379

Connors PG, Anderlini VC, Risebaough RW, Martin JH, Schreiber RW, Anuerson DW (1972) Heavy metal concentrations in brown pelicans from Florida and California. Trans Wildl Sot, Cal-Neva Wildl, pp 56–64

Coogan TP, Squibb KS, Motz J (1991) Distribution of chromium within cells of blood. Toxicol Appl Pharmacol 108:157–166

Copat C, Bella F, Castaing M, Fallico R, Sciacca S, Ferrante M (2012) Heavy metals concentration in fish from Sicily (Mediterranean Sea) and evaluation of possible health risks to consumers. Bull Environ Contam Toxicol 88:78–83

Corbi JJ, dos Santos FA, Zerlin R, dos Santos A, Froehlich CG, Trivinho-Strixino S (2011) Assessment of chromium contamination in the Monte Alegre stream: a case study. Braz Arch Biol Technol 54:613–620

Costa M, Klein CB (2006) Toxicity and carcinogenicity of chromium compounds in humans. Crit Rev Toxicol 36:155–163

Cotton FA, Wilkinson G, Murillo CA, Bochmann M (1999) Advanced inorganic chemistry, 6th edn. John Wiley & Sons, Inc, Hoboken, NJ, p 1356

D'Havé H, Scheirs J, Mubiana VK, Verhagen R, Blust R, De Coen W (2006) Non-destructive pollution exposure assessment in the European hedgehog (*Erinaceus europaeus*): II. Hair and spines as indicators of endogenous metal and as concentrations. Environ Pollut 142:438–448

Dailey RN, Raisbeck MF, Siemon RS, Cornish TE (2008) Liver metal concentrations in greater sage-grouse (*Centrocercus urophasianus*). J Wildl Dis 44:494–498

Danieli PP, Serrani F, Primi R, Ponzetta MP, Ronchi B, Amici A (2012) Cadmium, lead, and chromium in large game: a local-scale exposure assessment for hunters consuming meat and liver of wild boar. Arch Environ Contam Toxicol 63:612–627

Das AP, Singh S (2011) Occupational health assessment of chromite toxicity among Indian miners. Indian J Occup Environ Med 15:6–13

Dauwe T, Janssens E, Bervoets L, Blust R, Eens M (2005) Heavy-metal concentrations in female laying great tits (*Parus major*) and their clutches. Arch Environ Contam Toxicol 49:249–256

Davison G, Lambie CL, James WM, Skene ME, Skene KR (1999) Metal content in insects associated with ultramafic and non-ultramafic sites in the Scottish Highlands. Ecol Entomol 24:396–401

De Flora S (2000) Threshold mechanisms and site specificity in chromium(VI) carcinogenesis. Carcinogenesis 21:533–541
De Flora S, Bagnasco M, Serra D, Zanacchi P (1990) Genotoxicity of chromium compounds: a review. Mutat Res 238:99–172
Dębski B, Zalewski W, Gralak AM (2001) Effect of different dietary chromium sources on broiler glycogen and glucose level. In: Anke M, Muller R, Schafer U (eds) Mineralstoffe, Mengen-, Spuren-, and Ultraspurenelemente in der Prävention. Wiss Verlag, Stuttgart, pp 302–308
Deng H, Zhang Z, Chang C, Yong W (2007) Trace metal concentration in great tit (*Parus major*) and greenfinch (*Carduelis sinica*) at the Western Mountains of Beijing, China. Environ Pollut 148:620–626
Dey S, Stafford R, Deb Roy MK, Bhattacharjee CR, Khating DT, Bhattacharjee PC et al (1999) Metal toxicity and trace element deficiency in some wild animal species from north-east India, as revealed by cellular, bio-inorganic and behavioural studies. Curr Sci 77:276–280
Dhal B, Thatoi HN, Das NN, Pandey BD (2013) Chemical and microbial remediation of hexavalent chromium from contaminated soil and mining metallurgical solid waste: a review. J Hazard Mater 250–251:272–291
Di Bona KR, Love S, Rhodes NR, McAdory D, Sinha SH, Kern N et al (2011) Chromium is not an essential trace element for mammals: effects of a "low-chromium" diet. J Biol Inorg Chem 16: 381–390
Długaszek M, Kopczynski K (2011) Comparative analysis of liver mineral status of wildlife. Probl Hig Epidemiol 9:859–863
Długaszek M, Kopczyński K (2012) Application of atomic absorption spectrometry in environmental monitoring based on comparative analysis of element contents in red fox tissues. Proc Electrotech Inst 255:19–28
Długaszek M, Kopczyński K (2013) Elemental composition of muscle tissue of wild animals from central region of Poland. Int J Environ Res 7:973–978
Długaszek M, Kopczyński K (2014) Correlations between elements in the fur of wild animals. Bull Environ Contam Toxicol 93:25–30
Dogra RKS, Murthy RC, Srivastava AK, Gaur JS, Shukla LJ, Varmani BML (1996) Cattle mortality in the Thane district, India: a study of cause/effect relationships. Arch Environ Contam Toxicol 30(2):292–297
Dowling HJ, Offenbacher EG, Pi-Sunyer FX (1989) Absorption of inorganic trivalent chromium from the vascular perfused rat small intestine. J Nutr 119:1138–1145
Drbal K, Elster J, Komarek J (1992) Heavy metals in water, ice and biological material from Spitsbergen, Svalbard. Polar Res 11:99–101
Dube BK, Tewari K, Chatterjee J, Chatterjee C (2003) Excess chromium alters uptake and translocation of certain nutrients in citrullus. Chemosphere 53:1147–1153
Ducros V (1992) Chromium metabolism, a literature review. Biol Trace Elem Res 32:65–77
Ebdon L, Pitts L, Cornelis R, Crews H, Donard OFX, Quevauviller P (eds) (2001) Trace element speciation for environment, food, and health. Royal Society of Chemistry, Cambridge, UK
EEA (2015) European Union emission inventory report 1990–2013 under the UNECE Convention on Long-range Transboundary Air Pollution (LRTAP). European Environment Agency (EEA), Publications Office of the European Union, Luxembourg, 69 pp
EFSA (2009) Safety and efficacy of chromium methionine as feed additive for all species. EFSA J 1043:9–69
Eisler R (1986) Chromium hazards of fish, wildlife, and invertebrates: a synoptic review. U.S. Fish and Wildlife Service. Biological report No. 6 (1.6), 85 pp
Eisler R (2000) Chromium. Handbook of chemical risk assessment: health hazards to humans, plants, and animals. Lewis Publishers, Boca Raton, FL, pp 45–92
Elbetieha A, Al-Hamood MH (1997) Long-term exposure of male and female mice to trivalent and hexavalent chromium compounds: effect on fertility. Toxicology 116:39–47
Emsley J (2011) Nature's building blocks. Oxford, Oxford University Press

EPA (2000) Method 3060A: Alkaline digestion for hexavalent chromium. SW-846: test methods for evaluating solid waste, physical/chemical methods, US EPA, Washington, DC

EPA (2010) Toxicological review of hexavalent chromium. External Review draft. CAS No. 18540-29-9

Ernst E, Bonde JP (1992) Sex hormones and epididymal sperm parameters in rats following subchronic treatment with hexavalent chromium. Hum Exp Toxicol 11:255–258

Falandysz J, Ichihashi H, Mizera T, Yamasaki S (2000) Mineral composition of selected tissues and organs of white-tailed eagle. Rocz Panstw Zakl Hig 51:1–5 (in Polish)

Farag AM, May T, Marty GD, Easton M, Harper DD, Little EE et al (2006) The effect of chronic chromium exposure on the health of Chinook salmon (*Oncorhynchus tshawytscha*). Aquat Toxicol 76:246–257

Favas PJC, Pratas J, Varun M, D'Souza R, Paul MS (2014) Phytoremediation of soils contaminated with metals and metalloids at mining areas: potential of native flora. In: Hernandez-Soriano MC (ed) Environmental risk assessment of soil contamination. InTech Europe, Rijeka, Croatia

Feng W, Ding W, Qian Q, Chai Z (1999) Comparison of the chromium distribution in organs and subcellular fractions of normal and diabetic rats by using enriched stable isotope Cr-50 tracer technique. Biol Trace Elem Res 71:121–129

Feng W, Li B, Liu J, Chai Z, Zhang P, Gao Y (2003) Study of chromium-containing proteins in subcellular fractions of rat liver by enriched stable isotopic tracer technique and gel filtration chromatography. Anal Bioanal Chem 375:363–368

Feng XH, Zhai LM, Tan WF, Liu F, He JZ (2007) Adsorption and redox reactions of heavy metals on synthesized mn oxide minerals. Environ Pollut 147:366–373

Filistowicz A, Przysiecki P, Nowicki S, Filistowicza A, Durkalec M (2012) Contents of copper, chromium, nickel, lead, and zinc in hair and skin of farm foxes. Pol J Environ Stud 21:865–869

Frank A, Anke M, Danielsson R (2000) Experimental copper and chromium deficiency and additional molybdenum supplementation in goats. I. Feed consumption and weight development. Sci Total Environ 249:133–142

Friis RH (2012) Essentials of environmental health, 2nd edn. Jones & Bartlett Learning, LLC Publishers, Washington, DC

Gabryszuk M, Sloniewski K, Metera E, Sakowski T (2010) Content of mineral elements in milk and hair of cows from organic farms. J Elementol 15:259–267

Gad SC (1989) Acute and chronic systemic chromium toxicity. Sci Total Environ 86:149–157

Gagelli E, Berti F, D'Amelio N, Gagelli N, Valensin G, Bovalini L et al (2002) Metabolic pathways of carcinogenic chromium. Environ Health Perspect 110(Suppl. 5):733–738

Gaines RV, Skinner CW, Foord EE, Mason B, Rosenzweig AR (1997) Dana's new mineralogy, 8th edn. Wiley, New York, 1819 pp

Gamberg M, Braune B, Davey E, Elkin B, Hoekstra PF, Kennedy D et al (2005) Spatial and temporal trends of contaminants in terrestrial biota from the Canadian Arctic. Sci Total Environ 351–352:148–164

Gammelgaard B, Jensen K, Steffansen B (1999) In vitro metabolism and permeation studies on rat jejunum: organic chromium compared to inorganic chromium. J Trace Elements Med Biol 13:82–88

Gardea-Torresdey JL, de la Rosa G, Peralta-Videa JR, Montes M, Cruz-Jimenez G, Cano-Aguilera I (2005) Differential uptake and transport of trivalent and hexavalent chromium by tumbleweed (*Salsola kali*). Arch Environ Contam Toxicol 48:225–232

Gasparik J, Massányi P, Slamecka J, Fabis M, Jurcik R (2004) Concentration of selected metals in liver, kidney, and muscle of the red deer (*Cervus elaphus*). J Environ Sci Health Part A 39: 2105–2111

Gheju M, Balcu I (2010) Hexavalent chromium reduction with scrap iron in continuous-flow system. Part 2: effect of scrap iron shape and size. J Hazard Mater 182:484–493

Ginsberg HN (2000) Insulin resistance and cardiovascular disease. J Clin Invest 106:453–458

Gochfeld M, Belant JL, Shukla T, Benson T, Burger J (1996) Heavy metals in laughing gulls: gender, age and tissue differences. Environ Toxicol Chem 15:2275–2283

Golden NH, Rattner BA, McGowan PC, Parsons KC, Ottinger MA (2003) Concentrations of metals in feathers and blood of nestling black-crowned night-herons (*Nycticorax nycticorax*) in Chesapeake and Delaware Bays. Bull Environ Contam Toxicol 7:385–393

Govind P, Madhuri S (2014) Heavy metals causing toxicity in animals and fishes. Res J Animal Vet Fishery Sci 2:17–23

Greenwood NN, Earnshaw A (1997) Chemistry of the elements, 2nd edn. Butterworth-Heinemann. ISBN 0-08-037941-9

Grela ER, Studziński T, Rabos A (1997) The role of chromium in human and animals nutrition. Med Wet 53:312–315 (in Polish)

Grúz A, Szemerédy G, Kormos E, Budai P, Majoros S, Tompai E, Lehel J (2015) Monitoring of heavy metal burden in mute swan (*Cygnus olor*). Environ Sci Pollut Res 22:15903–15909

Halbrook RS, Woolf A, Hubert GF Jr, Ross S, Braselton WE (1996) Contaminant concentrations in Illinois mink and otter. Ecotoxicology 5:103–114

Hammond CR (2005) The elements. In: Linde DR (ed) CRC Handbook of chemistry and physics. CRC Press, Boca Raton, FL, pp 4/1–4/34

Han FX, Sridhar BBM, Monts DL, Su Y (2004) Phytoavailability and toxicity of trivalent and hexavalent chromium to *Brassica juncea*. New Phytol 162:489–499

Harding LE (2004) Environmental contaminants in wild martens (*Martes americana*) and wolverines (*Gulo luscus*). Bull Environ Contam Toxicol 73:98–105

Harding L, Harris ML, Elliott JE (1998) Heavy and trace metals in wild mink (*Mustela vison*) and river otter (*Lutra canadensis*) captured on rivers receiving metals discharges. Bull Environ Contam Toxicol 61:600–607

Hassan AA, Rylander C, Sandanger TM, Brustad M (2013) Copper, cobalt and chromium in meat, liver, tallow and bone marrow from semi-domesticated reindeer (*Rangifer tarandus tarandus* L.) in Northern Norway. Food Publ Health 3:154–160

Hood P (2010) Trace elements in soils. John Wiley & Sons (part 19.5)

Horai S, Watanabe I, Takada H, Iwamizu Y, Hayashi T, Tanabe S et al (2007) Trace element accumulations in 13 avian species collected from the Kanto area, Japan. Sci Total Environ 373:512–525

Howe JA, Loeppert RH, Derose VJ, Hunter DB, Bertsch PM (2003) Localization and speciation of chromium in subterranean clover using XRF, XANES, and EPR spectroscopy. Environ Sci Technol 37:4091–4097

Hua Y, Clark S, Ren J, Sreejayan N (2012) Molecular mechanisms of chromium in alleviating insulin resistance. J Nutr Biochem 23:313–319

Huang S-H, Peng B, Yang Z-H, Chai L-Y, Zhou L-C (2008) Chromium accumulation, microorganism population and enzyme activities in soils around chromium-containing slag heap of steel alloy factory. Trans Nonferrous Met Soc China 19:241–248

Hubbart JA (2012) Elemental hair analysis of the California ground squirrel (*Otospermophilus beecheyi*): an investigation of age class, gender, seasons and habitats. J Biol Life Sci 3:20–34

Huffman EWD, Allaway WH (1973) Chromium in plants: distribution in tissues, organelles, and extracts and availability of bean leaf Cr to animals. J Agric Food Chem 21:982–984

Hui CA (2002) Concentrations of chromium, manganese, and lead in air and in avian eggs. Environ Pollut 120:201–206

IARC (1990) Chromium, nickel and welding. IARC, International Agency for Research on Cancer, Monographs on the Evaluation of Carcinogenic Risks to Humans. WHO IARC, Lyon, France, 49:49–256

IARC (1999) Re-evaluation of some organic chemicals, hydrazine and hydrogen peroxide. IARC, International Agency for Research on Cancer, Monographs on the Evaluation of Carcinogenic Risks to Humans. IARC Scientific Publications, Lyon, France, 71:1–3

Imam Khasim D, Nanda Kumar NV, Hussain RC (1989) Environmental contamination of chromium in agricultural and animal products near a chromate industry. Bull Environ Contam Toxicol 43:742–746

Irwin RJ (1997) Environmental contaminants encyclopedia chromium VI. National Park Service Water Resource Division, Fort Collins, CO
Islam E, Yang X, He Z, Mahmood Q (2007) Assessing potential dietary toxicity of heavy metals in selected vegetables and food crops. J Zhejiang Univ Sci B 8:1–13
Izbicki JA, Ball JW, Bullen TD, Suttley SJ (2008) Chromium, chromium isotopes and selected trace elements, western Mojave desert, USA. Appl Geochem 23:1325–1352
Jacobs JA, Testa SM (2005) Overview of chromium (VI) in the environment: background and history. In: Guertin J, Jacobs A, Avakian CP (eds) Chromium (VI) handbook. CRC Press, Boca Raton, FL, pp 1–21
Jenkins DW (1979) Toxic trace metals in mammalian hair and nails. EPA-600/4-79-049
Jeyasingh J, Philip L (2005) Bioremediation of chromium contaminated soil: optimization of operating parameters under laboratory conditions. J Hazard Mater 118:113–120
Jianlong W, Zeyu M, Xuan Z (2004) Response of *Saccharomyces cerevisiae* to chromium stress. Process Biochem 39:1231–1235
Juhnke S, Peitsch N, Hübener N, Grosse C, Nies DH (2002) New genes involved in chromate resistance in *Ralstonia metallidurans* strain CH34. Arch Microbiol 179:15–25
Juturu V, Komorowski JR (2003) Chromium compounds: cytotoxicity and carcinogenesis. Letter to the Editor. Toxicology 186:171–173
Kabata-Pendias A (2011) Trace elements in soils and plants, 4th edn. CRC Press, Boca Taton, FL, pp 181–190
Kabata-Pendias A, Mukherjee AB (2007) Trace elements from soil to human. Springer, Berlin, Heidelberg, pp 173–183
Kamaludeen SP, Arunkumar KR, Avudainayagam S, Ramasamy K (2003) Bioremediation of chromium contaminated environments. Indian J Exp Biol 41:972–985
Kanojia RK, Junaid M, Murthy RC (1998) Embryo and fetotoxicity of hexavalent chromium: a long-term study. Toxicol Lett 95:165–172
Kargacin B, Squibb KS, Cosentino S, Zhitkovich A, Costa M (1993) Comparison of the uptake and distribution of chromate in rats and mice. Biol Trace Elem Res 36:307–318
Kasprzak KS (1991) The role of oxidative damage in metal carcinogenicity. Chem Res Toxicol 4: 604–615
Kaszycki P, Fedorovych D, Ksheminska H, Babyak L, Wójcik D, Koloczek H (2004) Chromium accumulation by living yeast at various environmental conditions. Microbiol Res 159:11–17
Kimborough DE, Cohen Y, Winer AM, Creelman L, Mabuni C (1999) A critical assessment of chromium in the environment. Crit Rev Environ Sci Technol 29(1):1–46
Kimura T, Okumura F, Onodera A, Nakanishi T, Itoh N, Isobe M (2010) Chromium (VI) inhibits mouse metallothionein-I gene transcription by modifying transactivation potential of the co-activator p300. Toxicol Lett 196S: S37–S351 (S301)
Kirklin DR (1999) Properties of materials and systems of importance to environmental fates and remediation. III. Review of previous thermodynamic property values for chromium and some of its compounds. J Phys Chem Ref Data 28:1675–1704
Kirman CR, Aylward LL, Suh M, Harris MA, Thompson CM, Haws LC et al (2013) Physiologically based pharmacokinetic model for humans orally exposed to chromium. Chem Biol Interact 204:13–27
Kler TK, Vashishat N, Kumar M (2014) Heavy metal contamination in excreta of avian species from Ludhiana district of Punjab. Int J Adv Res 2:873–879
Koivula MJ, Eeva T (2010) Metal-related oxidative stress in birds. Environ Pollut 158:2359–2370
Krejpcio Z (2001) Essentiality of chromium for human nutrition and health. Pol J Environ Stud 10:399–404
Krejpcio Z, Kurył T, Dębski B, Wójciak RW (2007) Effith fructans and chromium(III) on blood glucose and insulin and beta-oxidation in lymphocytes of type 1 diabetes rats. Med Wet 63(Supp):1494–1496 (in Polish)
Krzysik M, Grajeta H (2010) The role of chromium in etiopathogenesis of selected diseases. Bromatol Chem Toksykol 43:428–435 (in Polish)

Krzysik M, Grajeta H, Prescha A (2008) Chromium content in selected convenience and fast foods in Poland. Food Chem 107:208–212

Ksheminska H, Fedorovych D, Babyak L, Yanovych D, Kaszycki P, Koloczek H (2005) Chromium (III) and (VI) tolerance and bioaccumulation in yeast: a survey of cellular chromium content in selected strains of representative genera. Process Biochem 40:1565–1572

Kurył T, Dębski B (2001) Is metabolism of fatty acids in broiler chicken affected by chromium? Ateroskleroza 5:132–133

Kuryl T, Krejpcio Z, Wójciak RW, Lipko M, Debski B, Staniek H (2006) Chromium(III) propionate and dietary fructans supplementation stimulate erytrocyte glucose uptake and beta-oxidation in lymphocytes of rats. Biol Trace Elem Res 114:237–248

Lebedeva NV (1997) Accumulation of heavy metals by birds in the southwest of Russia. Russ J Ecol 28:41–46

Lee JI, Jung WY, Lee G, Kim MS, Kim YS, Park CG et al (2012) Heavy metal concentrations in hair of newly imported China-origin rhesus macaques (*Macaca mulatta*). Lab Anim Res 28: 151–154

Lewicki S, Rattman D, Kurył T, Snochowski M, Dębski B (2009) The effect of chromium (III) on fatty acid metabolism and insulin path related gene expression in mouse myocytes cells line C2C12. Zywn Nauk Technol 4:183–194 (in Polish)

Lewicki S, Zdanowski R, Krzyżowska M, Lewicka A, Dębski B, Niemcewicz M et al (2014) The role of chromium III in the organism and its possible use in diabetes and obesity treatment. Ann Agric Environ Med 21:331–335

Lind BB, Fallman AM, Larsson LB (2001) Environmental impact of ferrochrome slag in road construction. Water Manag 21:255–264

Liu KJ, Jiang J, Shi X, Gabrys H, Walczak T, Swatz HM (1995) Low frequency EPR study of chromium (V) formation from chromium (VI) in living plants. Biochem Biophys Res Commun 206(3):829–834

Liu J, Duan C, Zhang X, Zhu Y, Lu X (2011) Potential of *Leersia hexandra* Swartz for phytoextraction of Cr from soil. J Hazard Mater 188:85–91

Lytle CM, Lytle FW, Yang N, Qian J-H, Hansen D, Zayed A et al (1998) Reduction of Cr(VI) to Cr (III) by wetland plants: potential for in situ heavy metal detoxification. Environ Sci Technol 32:3087–3093

Ma WC (1982) The influence of soil properties and worm-related factors on the concentration of heavy metals in earthworms. Pedobiologia 24:109–119

Malik RN, Zeb N (2009) Assessment of environmental contamination using feathers of *Bubulcus ibis* L., as a biomonitor of heavy metal pollution, Pakistan. Ecotoxicolology 18:522–536

Mallard BA, Borgs P (1997) Effects of supplemental trivalent chromium on hormone and immune responses of cattle. In: Lyons TP, Jacques KA (eds) Proceedings of the 13th alltech annual symposium on the biotechnology in the feed industry. Nottingham University Press, Loughborough, UK

Mancuso TF (1997) Chromium as an industrian carcinogen. Part II. Chromium in human tissues. Am J Ind Med 2:140–147

Manjula M, Mohanraj R, Devi MP (2015) Biomonitoring of heavy metals in feathers of eleven common bird species in urban and rural environments of Tiruchirappalli, India. Environ Monit Assess 187:267

Mansouri B, Pourkhabbaz A, Babaei H, Houshyari E (2012) Heavy metal contamination in feathers of Western Reef Heron (*Egretta gularis*) and Siberian gull (*Larus heuglini*) from Hara biosphere reserve of Southern Iran. Environ Monit Ass 184:6139–6145

Marchese M, Gagneten AM, Parma MJ, Pave PJ (2008) Accumulation and elimination of chromium by freshwater species exposed to spiked sediments. Arch Environ Contam Toxicol 55: 603–609

Marcheselli M, Sala L, Mauri M (2010) Bioaccumulation of PGEs and other traffic-related metals in populations of the small mammal *Apodemus sylvaticus*. Chemosphere 80:1247–1254

Martello L, Fuchsman P, Sorensen M, Magar V, Wenning RJ (2007) Chromium geochemistry and bioaccumulation in sediments from the lower Hackensack River, New Jersey. Arch Environ Contam Toxicol 53:337–350

Mason CF, Stephenson A (2001) Metals in tissues of European otters (*Lutra lutra*) from Denmark, Great Britain and Ireland. Chemosphere 44:351–353

Mazanec O (1996) The content of risk elements in farm land throughout the Czech Republik. In: Microelements. Ceska spolecnost chemicka. Prague, pp 56–59 (in Czech)

McGrath SP, Smith S (1990) Chromium and nickel. In: Alloway BJ (ed) Heavy metals in soils. Wiley, New York, pp 125–150

McNeill L, McLean J (2012) State of the science of hexavalent chromium in drinking water, Water Research Foundation

Mei B, Puryear JD, Newton RJ (2002) Assessment of Cr tolerance and accumulation in selected plant species. Plant Soil 247:223–231

Mertz W (1993) Chromium in human nutrition: a review. J Nutr 123:626–633

Mertz W, Schwarz K (1957) Relation of glucose tolerance factor to impaired glucose tolerance in stock diets. Am J Physiol 1196:614–618

Mertz W, Roginski EE, Reba RC (1965) Biological activity and fate of trace quantities of intravenous chromium (III) in the rat. Am J Physiol 209:489–494

Micera G, Dessi A (1988) Chromium adsorption by plant roots and formation of long-lived Cr (V) species: an ecological hazard? J Inorg Biochem 34:157–166

Miksche LW, Lewalter J (1997) Health surveillance and biological effect monitoring for chromium-exposed workers. Regul Toxicol Pharm 26:S94–S99

Mining Watch Canada (2012) Potential toxic effects of chromium, chromite mining and ferrochrome production: a literature review. MiningWatch Canada, Mines Alerte, 43 pp (http://www.miningwatch.ca/sites/ www.miningwatch.ca/files/Chromite%20Review.pdf). Accessed 6 Dec 2015

Mishra AK, Mohanty B (2009a) Chronic exposure to sublethal hexavalent chromium affects organ histopathology and serum cortisol profile of a teleost, *Channa punctatus* (Bloch). Sci Total Environ 407:5031–5038

Mishra AK, Mohanty B (2009b) Effect of hexavalent chromium exposure on the pituitary – interrenal axis of a teleost, *Channa punctatus* (Bloch). Chemosphere 76:982–988

Mishra S, Singh V, Srivastava S, Srivatava R, Dass S, Satsangi GP et al (1995) Studies on uptake of trivalent and hexavalent chromium by maize (*Zea mays*). Food Chem Toxicol 33:393–397

MOE: Ministry of the Environment, Government of Ontario (2011) Ontario air standards for hexavalent chromium compounds and chromium and chromium compounds (metallic, trivalent and divalent)

Mooney KW, Cromwell GL (1997) Efficacy of chromium picolinate and chromium chloride as potential carcass modifiers in swine. J Anim Sci 75:2661–2671

Moonsie-Shageer S, Mowat DN (1993) Effect of level of supplemental chromium on performance, serum constituents and immune status of stressed feeder cakves. J Anim Sci 71:232–238

Mora MA, Skiles R, McKinney B, Paredes M, Buckler D, Papoulias D et al (2002) Environmental contaminants in prey and tissues of the peregrine falcon in the Big Bend Region, Texas, USA. Environ Pollut 116:169–176

Mordenti A, Piva G (1997) Chromium in animal nutrition and possible effects on human health. In: Canali S, Tittarelli F, Sequi P (eds) Chromium environmental issues. Franco Angeli s.r.l., pp 131–151

Morris GS, Guidry KA, Hegsted M, Hasten DL (1995) Effects of dietary chromium supplementation on cardiac mass, metabolic enzymes and contractile proteins. Nutr Res 15:1045–1052

Movalli PA (2000) Heavy metal and other residues in feathers of Laggar Falcon (Falco biarmicus juggar) from six districts of Pakistan. Environ Pollut 109:267–275

Muter O, Patmalnieks A, Rapoport A (2001) Interactions of the yeast *Candida utilis* and Cr (VI): metal reduction and its distribution in the cell and medium. Process Biochem 36:963–970

Nam D-H, Anan Y, Ikemoto T, Okabe Y, Kim E-Y, Subramanian A et al (2005a) Specific accumulation of 20 trace elements in great cormorants *(Phalacrocorax carbo)* from Japan. Environ Pollut 134:503–514

Nam D-H, Anan Y, Ikemoto T, Tanabe S (2005b) Multielemental accumulation and its intracellular distribution in tissues of some aquatic birds. Mar Pollut Bull 50:1347–1362

Nam D-H, Rutkiewicz J, Basu N (2012) Multiple metals exposure and neurotoxic risk in bald eagles (*Haliaeetus leucocephalus*) from two Great Lakes states. Environ Toxicol Chem 31:623–631

National Toxicology Program (2008) NTP Toxicology and Carcinogenesis Studies of Sodium Dichromate Dihydrate (CAS No. 7789-12-0) in F344/N Rats and B6C3F1Mice (Drinking Water Studies). Natl Toxicol Program Tech Rep Ser 546:1–192

Navarro J, Oro D, Bertolero A, Genovart M, Delgado A, Forer MG (2010) Age and sexual differences in the exploitation of two anthropogenic food resources for an opportunistic seabird. Mar Biol 157:2453–2459

Nielsen FH (2007) Summary: the clinical and nutritional importance of chromium – still debated after 50 years of research. The Nutritional Biochemistry of Chromium (III) Elsevier, pp 265–276

Nies DH (2004) Essential and toxic effects of elements on microorganisms. In: Merian E, Anke M, Ihnat M, Stoeppler M (eds) Elements and their compounds in the environment. Wiley-VCH Verlag GmbH&Co.KGaA, Weinheim, pp 257–276

NJDEP (1996) New Jersey Department of Environmental Protection. Soil Cleanup Criteria. Proposed Cleanup Standards for Contaminated Sites. NJAC 7:26D

NRC (1989) Recommended dietary allowance, National Research Council, 10th edn. National Academy of Sciences, National Academy Press, Washington, DC

NRC (1996) Nutrient requirements of beef cattle. National Research Council, 7th edn. National Academy of Sciences, National Academy Press, Washington, DC

NRC (1997) The role of chromium in animal nutrition. National Research Council. In: National Academy of Sciences. National Academy Press, Washington, DC

NRC (2005) Mineral tolerance of animals, 2nd rev edn. National Academic Press, Washington, DC

Nriagu JO, Kabir A (1995) Chromium in the Canadian environment. Environ Rev 3:121–144

O'Shea TJ, Everette AL, Ellison LE (2001) Cyclodiene insecticides, DDE, DDT, arsenic and mercury contamination of big brown batts (*Eptesicus fuscus*) foraging at a Colorado Superfund site. Arch Environ Contam Tox 40:112–120

OEHHA (The Office of Environmental Health Hazard Assessment) (2000) Technical support document for exposure assessment and stochastic analysis. Appendix H: Fish Bioconcentration Factors (BCF)

Oh SJ, Lee JY (2005) Dietary chromium-methionine chelate supplementation and animal performance. Asian-Aust J Anim Sci 18:898–907

Oliveira H (2012) Chromium as an environmental pollutant: Insights on induced plant toxicity. J Bot 2012:375843

Orłowski G, Polechoński R, Dobicki W, Dolata PT, Bednarska M (2006) Heavy metal concentrations in feathers of white storks *Ciconia ciconia* nesting in Central Poland: methodological implications for further ecotoxicological studies. In: Tryjanowski P, Sparks TH, Jerzak L (eds) White Stork study in Poland: biology, ecology and conservation. Bogucki Wydawnictwo Naukowe, Poznań, pp 51–61

OSHA – Occupational Safety and Health Administration (2006) Occupational exposure to hexavalent chromium, Final rule. Fed Regist 71:10099–10385

Outridge PM, Scheuhammer AM (1993) Bioaccumulation and toxicology of chromium: implications for wildlife. Rev Environ Contam Toxicol 130:31–77

Papp JF (1994) Chromium life cycle study. U.S. Department of Interior, Bureau of Mines, Information Circular 9411, Washington, DC

Papp JF (2005) Minerals yearbook. Chromium [advance release] USGS Science for a changing world, pp 17.1–17.24

Papp JF (2011) 2011 Minerals yearbook. Chromium [advance release] USGS Science for a changing world, pp 17.1–17.24

Papp JF (2016) USGS 2013 Minerals Yearbook, US Departament of the Interior, US Geological Survey
Papp JF, Lipin BR (2006) Chromite. In: Kogel JE, Trivedi NC, Barker JM, Krukowski ST (eds) Industrial minerals and rocks—Commodities, markets, and uses, 7th edn. Society for Mining, Metallurgy, and Exploration, Inc., Littleton, CO, pp 309–333
Park SH, Lee MH, Kim SK (2005) Studies on Cd, Pb, Hg and Cr values in dog hairs from urban Korea. Asian-Austr J Anim Sci 18:1135–1140
Paßlack N, Mainzer B, Lahrssen-Wiederholt M, Schafft H, Palavinskas R, Breithaupt A et al (2014) Concentrations of strontium, barium, cadmium, copper, zinc, manganese, chromium, antimony, selenium and lead in the equine liver and kidneys. SpringerPlus 3:343
Pati A, Chaudhary R, Subramani S (2014) A review on management of chrome-tanned leather shavings: a holistic paradigm to combat the environmental issues. Environ Sci Pollut Res 21: 11266–11284
Pattar GR, Tackett L, Liu P, Elmendorf JS (2006) Chromium picolinate positively influences the glucose transporter system via affecting cholesterol homeostasis in adipocytes cultured under hyperglycemic diabetic conditions. Mutat Res 610:93–100
Pawlisz AV, Kent RA, Schneider UA, Jefferson C (1997) Canadian water quality guidelines for chromium. Environ Toxicol Water Quality 12:185–193
Pechova A, Pavlata L (2007) Chromium as an essential nutrient: a review. Vet Med Czech 52:1–18
Peitzsch N, Eberz G, Nies DH (1998) *Alcaligenes eutrophus* as a bacterial chromate sensor. Appl Environ Microbiol 64:453–458
Pena-Fernández A, González-Munoz MJ, Lobo-Bedmar MC (2014) "Reference values" of trace elements in the hair of a sample group of Spanish children (aged 6–9 years) – are urban top soils a source of contamination? Environ Toxicol Pharmacol 38:141–152
Peralta-Videa JR, Lopez ML, Narayan M, Saupe G, Gardea-Torresdey J (2009) The biochemistry of environmental heavy metal uptake by plants: implications for the food chain. Int J Biochem Cell Biol 41:1665–1677
Pereira R, Pereira ML, Ribeiro R, Gonçalves F (2006) Tissues and hair residues and histopathology in wild rats (*Rattus rattus* L.) and Algerian mice (*Mus spretus* Lataste) from an abandoned mine area (Southeast Portugal). Environ Pollut 139:561–575
Peterson RL, Banker KJ, Garcia TY, Works CF (2008) Isolation of a novel chromium (III) binding protein from bovine liver tissue after chromium (VI) exposure. J Inorg Biochem 102:833–841
Petrie SA, Badzinski SS, Drouillard KG (2007) Contaminants in lesser and greater scaup staging on the lower Great Lakes. Arch Environ Contam Toxicol 52:580–589
Piskorova L, Vasilkova Z, Krupicer I (2003) Heavy metal residues in tissues of wild boar (*Sus scrofa*) and red fox (*Vulpes vulpes*) in the Central Zemplin region of the Slovak Republic. Czech J Anim Sci 48:134–138
Piva A, Meola E, Gatta PP, Biagi G, Castellani G, Mordenti AL et al (2003) The effect of dietary supplementation with trivalent chromium on production performance of laying hens and the chromium content in the yolk. Anim Feed Sci Technol 106:149–163
Pollard GV, Richardson CR, Karnezos TP (2002) Effects of supplemental organic chromium on growth, feed efficiency and carcass characteristics of feedlot steers. Anim Feed Sci Technol 98:121–128
Polti MA, García RO, Amoroso MJ, Abate CM (2009) Bioremediation of chromium (VI) contaminated soil by *Streptomyces* sp. MC1. J Basic Microb 49:285–292
Polti MA, Atjián MC, Amoroso MJ, Abate CM (2011) Soil chromium bioremediation: synergic activity of actinobacteria and plants. Int Biodeterior Biodegrad 65:1175–1181
Press RI, Geller J, Evans GW (1990) The effect of chromium picolinate on serum cholesterol and apolipoprotein fractions in human subjects. West J Med 152:41–45
Puls R (1994) Mineral levels in animal health. Diagnostic data, 2nd edn. Sherpa International, Clearbrook, BC
Rauch JN and Pacyna JM (2009) Earth's global Ag, Al, Cr, Cu, Fe, Ni, Pb, and Zn cycles, Global Biogeochem Cycles, 23, GB2001 doi:https://doi.org/10.1029/2008GB003376, 1–16
Reeve RN (1994) Environmental analysis. Wiley, Chichester

Riley RG, Zachara JM, Wobber FJ (1992) Chemical contaminants on DOE lands and selection of contaminated mixtures for subsurface science research. US-DOE, Energy Resource Subsurface Science Program, Washington, DC
Rosman KJR, Taylor PDP (1998) Isotopic compositions of the elements 1997. Pure Appl Chem 70: 217–235
Ruiz-Olmo J, Lafontaine L, Prigioni C, López-Martín JM, Santos-Reis M (2000) Pollution and its effects on otter populations in south-western Europe. In: Conroy JWH, Yoxon P, Gutleb AC (eds) Proceedings of the first otter toxicology conference. J Internat Otter Survival Fund 1: 63–82
Şahin K, Küçük O, Şahin N, Ozbey O (2001) Effects of dietary chromium picolinate supplementation on egg production, egg quality and serum concentrations of insulin, corticosterone and some metabolites of Japanese quails. Nutr Res 21:1315–1321
Salnikow K, Zhitkovich A (2008) Genetic and epigenetic mechanisms in metal carcinogenesis and cocarcinogenesis: nickel, arsenic and chromium. Chem Res Toxicol 21:28–44
Samantaray S, Ranjan Rout G, Das P (1998) Role of chromium on plant growth and metabolism. Acta Physiol Plant 20:201–2012
Sampaio CH (2002) Beneficiamento de carvões. In: Teixeira EC (ed) Meio Ambiente e Carvão: Impactos da Exploração e Utilização. Fepam, Porto Alegre, 497 pp
Sample BE, Faitbrother A, Kaiser A, Law S, Adams B (2014) Sensitivity of ecological soil-screening levels for metals to exposure model parameterization and toxicity reference values. Environ Toxicol Chem 33:2386–2398
Sánchez-Chardi A, Oliveira-Ribeiro C, Nadal J (2009) Metals in liver and kidneys and the effects of chronic exposure to pyrite mine pollution in the shrew Crocidura russula inhabiting the protected wetland of Doñana. Chemosphere 76:387–394
Sanyal T, Kaviraj A, Saha S (2015) Deposition of chromium in aquatic ecosystem from effluents of handloom textile industries in Ranaghat–Fulia region of West Bengal, India. J Adv Res 6: 995–1002
Saxena DK, Murthy RC, Lal B, Srivastava RS, Chandra SV (1990) Effect of hexavalent chromium on testicular maturation in the rat. Reprod Toxicol 4:223–228
Schwarz K, Mertz Z (1959) Chromium (III) and glucose tolerance factor. Arch Biochem Biophys 85:292–295
Shanker AK, Cervantes C, Loza-Tavera H, Avudainayaagam S (2005) Chromium toxicity in plants. Environ Int 31:739–753
Sharma DC, Sharma CP, Tripathi RD (2003) Phytotoxic lesions of chromium in maize. Chemosphere 51:63–68
Shi X, Dalal NS (1989) Chromium(V) and hydroxyl radical formation during the glutathione reductase-catalyzed reduction of chromium(VI). Biochem Biophys Res Commun 163:627–634
Shi X, Dalal NS (1990) On the hydroxyl radical formation in the reaction between hydrogen peroxide and biologically generated chromium (V) species. Arch Biochem Biophys 277:342–350
Singh HP, Mahajan P, Kaur S, Batish DR, Kohli RK (2013) Chromium toxicity and tolerance in plants. Environ Chem Lett 11:229–254
Singh N, Uppal H, Chawla S, Singh S, Tripathy S (2015) An efficient and fast process for the removal of trivalent and hexavalent chromium from contaminated water using zinc peroxide nanomaterial. Pharm Anal Acta 6:412
Sivertsen T, Daae HL, Godal A, Sand G (1995) Ruminant uptake of nickel and other elements from industrial air pollution in the Norwegian-Russian border area. Environ Pollut 90:75–81
Skoric S, Visnjić-Jeftic Z, Jaric I, Djikanovic V, Mickovic B, Nikcevic M et al (2012) Accumulation of 20 elements in great cormorant (*Phalacrocorax carbo*) and its main prey, common carp (*Cyprinus carpio*) and Prussian carp (*Carassius gibelio*). Ecotoxicol Environ Saf 80:244–251
Sobański L, Sprzęczka-Niedolaz M, Łebek G (2007) The role of chromium in human nutrition. Bromat Chem Toksykol – XL 2:113–119 (in Polish)

Sobieraj S, Laskowski J (1973) Flotation of chromite: 1-early research and recent trends; 2-flotation of chromite and surface properties of spinel minerals. Trans Inst Min Metall Sect C 82: C207–C213

Solomon F (2008) Impacts of metals on aquatic ecosystems and human health. http://www.infomine.com/publications/docs/mining.com/Apr2008c.pdf

Song RX, Barnes CJ, Zhang Z, Bao Y, Kumar R, Santen RJ (2004) The role of Shc and insulin-like growth factor 1 receptor in mediating the translocation of estrogen receptor alpha to the plasma membrane. Proc Natl Acad Sci USA 101:2076–2081

Spears JW, Kegley EB, Mullis LA (2004) Bioavailability of copper from tribasic copper chloride and copper sulfate in growing cattle. Anim Feed Sci Technol 116:1–13

Speranza A, Ferri P, Battistelli M, Falcieri E, Crinelli R, Scoccianti V (2007) Both trivalent and hexavalent chromium strongly alter in vitro germination and ultrastructure of kiwifruit pollen. Chemosphere 66:1165–1174

Standeven AM, Wetterhahn KE (1992) Ascorbate is the principal reductant of chromium(VI) in rat lung ultrafiltrates and cytosols, and mediates chromium-DNA binding in vitro. Carcinogenesis 13:1319–1324

Staniek H, Kostrzewska-Poczekaj M, Arndt M, Szyfter K (2010) Genotoxicity assessment of chromium (III) propionate complex in the rat model using the comet assay. Food Chem Toxicol 48:89–92

Stearns DM, Wise JP Sr, Patierno SR, Wetterhahn KE (1995) Chromium (III) picolinate produces chromosomal damage in Chinese hamster ovary cells. J Fed Am Soc Exp Biol 9:1643–1649

Stern MB, Mansdorf SZ (1998) Applications and computational elements of industrial hygiene. CRC Press, Taylor & Francis Group, Boca Raton, FL

Stoecker B (2004) Chromium. In: Merian E, Anke M, Ihnat M, Stoeppler M (eds) Elements and their compounds in the environment, 2nd edn. Wiley-VCH Verlag GmbH & Co. KGaA, Weinheim, pp 709–729

Stout JH, Trust KA (2002) Elemental and organochlorine residues in bald eagles from Adak Island, Alaska. Wildl Dis 38:511–517

Su C, Jiang LQ, Zhang WJ (2014) A review on heavy metal contamination in the soil worldwide. Environ Skep Crit 3:24–38

Subramanian S, Rajendiran G, Sekhar P, Gowri C, Govindarajulu P, Aruldhas MM (2006) Reproductive toxicity of chromium in adult bonnet monkeys (*Macaca radiata* Geoffrey). Reversible oxidative stress in the semen. Toxicol Appl Pharm 215:237–249

Sugden KD, Stearns DM (2000) The role of chromium(V) in the mechanism of chromate-induced oxidative DNA damage and cancer. J Environ Pathol Toxicol Oncol 19:215–230

Suwalsky M, Castro R, Villena F, Sotomayor CP (2008) Cr (III) exerts stronger structural effects than Cr (VI) on the human erythrocyte membrane and molecular models. J Inorg Biochem 102:842–849

Suzuki Y, Fukuda K (1990) Reduction of hexavalent chromium by ascorbic acid and glutathione with special reference to the rat lung. Arch Toxicol 64:169–176

Talebi SM (2003) Determination of total and hexavalent chromium concentrations in the atmosphere of the city of Isfahan. Environ Res 92:54–56

Tandon SK, Behari JR, Kachru DN (1979) Distribution of chromium in poisoned rats. Toxicology 13:29–34

Thacker U, Parikh R, Shouche Y, Madamwar D (2006) Hexavalent chromium reduction by *Providencia* sp. Process Biochem 41:1332–1337

Thakur R, Sharma GD, Dwivedi BS, Khatik SK (2007) Chromium: as a pollutant. J Ind Pollut Control 23:197–203

Thor MY, Harnack L, King D, Jasthi B, Pettit J (2011) Evaluation of the comprehensiveness and reliability of the chromium composition of foods in the literature. J Food Comp Anal 24: 1147–1152

Topczewska J (2012) Effects of seasons on the concentration of selected trace elements in horse hair. J Cent Eur Agric 13:671–680

Tribovillard N, Algeo TJ, Lyons T, Riboulleau A (2006) Trace metals as paleoredox and paleo-productivity proxies: an update. Chem Geol 232:12–32
Tsipoura N, Burger J, Feltes R, Yacabucci J, Mizrahi D, Jeitner C et al (2008) Metal concentrations in three species of passerine birds breeding in the Hackensack Meadowlands of New Jersey. Environ Res 107:218–228
Tsipoura N, Burger J, Newhouse M, Jeitner C, Gochfeld M, Mizrahi D (2011) Lead, mercury, cadmium, chromium, and arsenic levels in eggs, feathers, and tissues of Canada geese of the New Jersey Meadowlands. Environ Res 111:775–784
Upreti RK, Shrivastava R, Chaturvedi UC (2004) Gut microflora and toxic metals. Chromium as a model. Indian J Med Res 119:49–59
USGS (2010) Chromium. In: 2007 Minerals Yearbook. U.S. Department of the Interior, U.S. Geological Survey, pp 17.1–17.28
USGS (2015) Chromium. In: Mineral commodity summaries 2015. U.S. Geological Survey, pp 42–43
Uyanik F, Eren M, Guclu BK, Sahin N (2005) Effects of dietary chromium supplementation on performance, carcass traits, serum metabolites and tissue chromium levels of Japanese quails. Biol Trace Elem Res 103:187–197
Vajpayee P, Sharma SC, Tripathi RD, Rai UN, Yunus M (1999) Bioaccumulation of chromium and toxicity to photosynthetic pigments, nitrate reductase activity and protein content of *Nelumbo nucifera* Gaertn. Chemosphere 39:2159–2169
van Gestel CAM, Van Breemen EMD, Baerselman R (1993) Accumulation and elimination od cadmium, chromium and zinc and effects on growth and reproduction in *Eisenia andrei* (Oligochaeta, Annelida). Sci Total Environ 134:585–597
Vasylkiv OY, Kubrak OI, Storey KB, Lushchak VI (2010) Cytotoxicity of chromium ions may be connected with induction of oxidative stress. Chemosphere 80:1044–1049
Vert T (2016) Refractory material selection for steelmaking. The American Ceramic Society and John Wiley & Sons, Hoboken, NJ
Vincent JB (1999) Mechanisms of chromium action: low-molecular-weight chromium-binding substance. J Am Coll Nutr 18:6–12
Vincent JB (2015) Is the pharmacological mode of action of chromium(III) as a second messenger? Biol Trace Elem Res 166:7–12
Vincent JB, Stallings D (2007) Chapter 1 – Introduction: a history of chromium studies (1955–1995). The nutritional biochemistry of chromium. Elsevier, Amsterdam, pp 1–40
Viti C, Mini A, Ranalli G, Lustrato G, Giovannetti L (2006) Response of microbial communities to different doses of chromate in soil microcosms. Appl Soil Ecol 34:125–139
Voigt CL, da Silva CP, Doria HB, Randi MAF, de Oliveira Ribeiro CA, de Campos SX (2015) Bioconcentration and bioaccumulation of metal in freshwater neotropical fish Geophagus brasiliensis. Environ Sci Pollut Res 22:8242–8252
Walker LA, Simpson VR, Rockett L, Wienbeurg CL, Shore RF (2007) Heavy metal contamination in bats Britain. Environ Pollut 148:483–490
Walker LA, Lawlor AJ, Chadwick EA, Potter E, Pereira MG, Shore RF (2010) Inorganic elements in the livers of Eurasian otters, Lutra lutra, from England and Wales in 2007 & 2008: a Predatory Bird Monitoring Scheme (PBMS) report, pp 13, Centre for Ecology and Hydrology
Walker LA, Lawlor AJ, Chadwick EA, Potter E, Pereira MG, Shore RF (2011) Inorganic elements in the Livers of Eurasian Otters, *Lutra lutra*, from England and Wales in 2009 – a Predatory Bird Monitoring Scheme (PBMS) Report. Centre for Ecology & Hydrology, Lancaster, UK (http://nora.nerc.ac.uk/14176/1/PBMS_Metals_ Otters_2009.pdf)
Wang MM, Fox EA, Stoecker BJ, Menendez CE, Chan SB (1989) Serum cholesterol of adults supplemented with brewers's yeast or chromium chloride. Nutr Res 9:989–998
Wang ZQ, Zhang XH, Russell JC, Hulver M, Cefalu WT (2006) Chromium picolinate enhances skeletal muscle cellular insulin signaling in vivo on obese, insulin-resistant JCR: LA-cp rats. J Nutr 136:415–420

Wang MQ, He YD, Lindemann MD, Jiang ZG (2009) Efficacy of Cr (III) supplementation on growth, carcass composition, blood metabolites, and endocrine parameters in finishing pigs. Asian-Aust J Anim Sci 22:1414–1419

Wang ZX, Chen JQ, Chai LY, Yang ZH, Huang SH, Zheng Y (2011) Environmental impact and site-specific human health risks of chromium in the vicinity of a ferro-alloy manufactory, China. J Hazard Mater 190:980–985

Wang M-Q, Li H, He Y-D, Wang C, Tao W-J, Du Y-J (2012a) Efficacy of dietary chromium (III) supplementation on tissue chromium deposition in finishing pigs. Biol Trace Elem Res 148: 316–321

Wang X, Mandal AK, Saito H, Pulliam JF, Lee EY, Ke ZJ et al (2012b) Arsenic and chromium in drinking water promote tumorigenesis in a mouse colitis-associated colorectal cancer model and the potential mechanism is ROS-mediated Wnt/β-catenin signaling pathway. Toxicol Appl Pharm 262:11–21

Weegman MD, Weegman MM (2007) Chromium and selenium in invertebrate prey of lesser scaup. J Wildl Manag 71:778–782

Weksler-Zangen S, Mizrahi T, Raz I, Mirsky N (2012) Glucose tolerance factor extracted from yeast: oral insulin-mimetic and insulin-potentiating agent: in vivo and in vitro studies. Br J Nutr 108:875–882

WHO (1988) Chromium. Environmental Health Criteria, WHO (World Health Organization), Geneva, Switzerland, 61 pp

WHO (2000) World health report 2000: health systems: improving performance. World Health Organization, Geneva

WHO/IPCS (2013) World Health Organization/International Programme on Chemical Safety. Inorganic chromium(VI) compounds. Concise International Chemical Assessment Document 78. http://www.who.int/ipcs/publications/cicad/cicad_78.pdf

Wiemeyer SN, Schmeling SK, Anderson A (1987) Environmental pollutant and necropsy data for ospreys from the Eastern United States, 1975-1982. J Wildl Dis 23:279–291

Wilde E, Benemann JR (1993) Bioremoval of heavy metals by the use of microalgae. Biotechnol Adv 11:781–812

Wise JP, Wise SS, Little JE (2002) The cytotoxicity and genotoxicity of particulate and soluble hexavalent chromium in human lung cells. Mutat Res 517:221–229

Wise JP, Wise SS, Kraus S, Shaffiey F, Grau M, Chen TL et al (2008) Hexavalent chromium is cytotoxic and genotoxic to the North Atlantic right whale (*Eubalaena glacialis*) lung and testes fibroblasts. Mutat Res 650:30–38

Wuana RA, Okieimen FE (2011) Heavy metals in contaminated soils: a review of sources, chemistry, risks and best available strategies for remediation. International Scholarly Research Network ISRN Ecology ID 402647, pp 1–20

WVDL (2015) www.wvdl.wisc.edu/wp-content/uploads/2013/06/WVDL.Info_. Toxicology_ Normal_ Ranges.pdf (Accessed 28 Apr 2015)

Wyszkowski M, Radziemska M (2009) Chromium contamination and the content of nitrogen compounds in soil. Ochr Środ Zasob Natural 40:88–95

Yoo Y, Lee S, Yang J, In S, Kim K, Kim S et al (2000) Distribution of heavy metals in normal Korean tissues. Probl Foren Sci 43:283–289

Zaccaroni A, Amorena M, Naso B, Castellani G, Lucisano A, Stracciari GL (2003) Cadmium, chromium and lead contamination of *Athene noctua*, the little owl, of Bologna and Parma, Italy. Chemosphere 52:1251–1258

Zaccaroni A, Andreani G, Ferrante MC, Carpene E, Isani G, Lucisano A (2008a) Metal concentrations in the liver and kidney of raptor species from the Calabria region, Italy. Acta Vet Beograd 58:315–324

Zaccaroni A, Scaravelli D, De Battisti R, Zanella A, Gelli D (2008b) Toxicological survey of free ranging population of roe deer (*Capreolus capreolus*) and red deer (*Cervus elaphus*) by teeth examination. Nat Croat Zagreb 17:273–281

Zaccaroni A, Corteggio A, Altamura G, Silvi M, Di Vaia R, Formigaro C et al (2014) Elements levels in dogs from "triangle of death" and different areas of Campania region (Italy). Chemosphere 108:62–69

Zayed A, Terry N (2003) Chromium in the environment: factors affecting biological remediation. Plant Soil 249:139–156

Zayed A, Lytle CM, Qian J-H, Terry N (1998) Chromium accumulation, translocation and chemical speciation in vegetable crops. Planta 206:293–299

Zha LY, Xu ZR, Wang MQ, Gu LY (2007) Effects of chromium nanoparticle dosage on growth, body composition, serum hormones and tissue chromium in Sprague-Dawley rats. J Zhejiang Univ Sci B 8(5):323–330

Zhangsheng L, Moore TA, Weaver SD, Finkelman RB (2001) Crocoite: an usual mode of occurrence for lead in cool. Int J Coal Geol 45:289–293

Zhitkovich A (2011) Chromium in drinking waters: sources, metabolism and cancer risks. Chem Res Toxicol 24:1617–1629

Zocche JJ, Leffa DD, Damiani AP, Carvalho F, Mendonca RA, dos Santos CEI et al (2010) Heavy metals and DNA damage in blood cells of insectivore bats in coal mining areas of Catarinense coal basin, Brazil. Environ Res 110:684–691

Zojaji F, Hassani AH, Sayadi MH (2014) Bioaccumulation of chromium by *Zea mays* in wastewater-irrigated soil: an experimental study. Proc Int Acad Ecol Environ Sci 4:62–67

Zou J, Wang M, Jiang W, Liu D (2006) Chromium accumulation and its effects on other mineral elements in *Amaranthus viridis* L. Acta Biol Cracov Bot 48:7–12

Zukal J, Pikula J, Bandouchova H (2015) Bats as bioindicators of heavy metal pollution: history and prospect. Mamm Biol 80:220–227

# Chapter 4
# Copper, Cu

**Natalia Łanocha-Arendarczyk and Danuta I. Kosik-Bogacka**

**Abstract** Copper (Cu) is an essential micronutrient for human and animal organisms, playing a structural and enzymatic role in endothermic animals. The profile of Cu emission to the environment has changed significantly in the past decades, with Asia responsible for 50% of global anthropogenic copper emission. This review of reports on the presence of Cu in different avian and mammalian species, usually evaluated in soft tissues including the liver and kidney, shows that it is mainly influenced by the type of diet and anthropogenic environmental contamination with Cu fungicides and with the high natural soil Cu levels resulting from volcanic activity. The good bioindicators of environmental Cu concentration are birds associated with wetland areas, e.g., herbivores such as the mute swan, ducks from the genera *Anas* and *Aythya*, terrestrial passerines (house sparrow, great tit, and blue tit), and birds of prey including the common buzzard, white-tailed eagle, and bald eagle. In terrestrial mammals, a measurable response to pollution with Cu is exhibited by canids common in natural and seminatural habitats, e.g., the Arctic fox, red fox, raccoon dog, American mink, otters, and ungulates, such as the wild boar and red deer. In Europe, Cu levels may now be tested in increasingly popular alien species, e.g., mink, raccoon, and raccoon dog, which allows wider intercontinental comparative studies. Biomarkers for identification of copper status are still being defined, and still the best solution to evaluate the exposure is to measure Cu concentrations in wildlife and the environment.

## 1 Introduction

Emissions of copper into the environment are from similar volumes of both natural sources (mainly from dust carried by winds, forest fires, and volcanic particles) and anthropogenic sources (e.g., storage of ash, fungicides, Cu nanomaterials, mining,

N. Łanocha-Arendarczyk (✉) · D. I. Kosik-Bogacka
Department of Biology and Medical Parasitology, Pomeranian Medical University in Szczecin, Szczecin, Poland
e-mail: nlanocha@pum.edu.pl



E. Kalisińska (ed.), *Mammals and Birds as Bioindicators of Trace Element Contaminations in Terrestrial Environments*,
https://doi.org/10.1007/978-3-030-00121-6_4

and smelting) (AMAP 2002; Kabata-Pendias and Szteke 2015). About 90% of world copper reserves are sulfide ores, ~9% copper oxides, and the remaining 1% pure copper. The twentieth century saw an unprecedented growth in demand and production of this element, and although annual global production of mined copper has stabilized at about 10 million tonnes, consumption is still very high. World reserves of economically viable copper deposits are currently estimated at 300 million tonnes (USGS 2016).

Copper is an essential trace element involved in the formation and metabolism of bone tissue but most importantly in oxidation-reduction processes where it acts as a coenzyme (Angelova et al. 2011). The persistent presence of Cu compounds in the environment results in exposure to its toxic effects to many aqueous and terrestrial organisms (including invertebrates, birds, and mammals) (Eisler 1998; Hernandez et al. 1999; Millan et al. 2008). The negative effects to wild animals from the various Cu compounds in the environment have been well documented in multifaceted field studies in North America, Canada (mainly trans-Arctic animals), and Europe (mainly herbivores, carnivores, and omnivores) (Clausen and Wolstrup 1978; Beyer et al. 1998; Eisler 1998; Rush et al. 2008; Bilandžić et al. 2012).

Birds and mammals with the highest accumulation of Cu include both herbivores and carnivores (Reglero et al. 2008; Schummer et al. 2011; Bilandžić et al. 2012). Wetland birds are reported to accumulate much higher levels of Cu than terrestrial birds, with many ecotoxicological studies showing their potential usefulness in biomonitoring of inland water ecosystems in North America and Europe (Kalisińska et al. 2004; Taggart et al. 2009; Schummer et al. 2011; Komosa et al. 2012). Importantly, there is a lack of biomonitoring studies on Cu levels in Europe using alien species originating from North America and southeastern Asia, e.g., mink, raccoon, or raccoon dog, namely, animals which in the ecotoxicological studies carried out in North America have given a measurable response to Cu contamination in aqueous environments (Brzezinski et al. 2014; Souza et al. 2013).

Available literature lacks data on background Cu levels for avian and mammalian species. Moreover, there are no threshold levels of copper toxicity for terrestrial animals (birds and mammals), with the proposed physiological Cu ranges in organs and tissues often very different (Eisler 1998; AMAP 2013). Reports also relatively rarely identify a threat of Cu intoxication (many wild animals demonstrate high Cu tolerance). Nonetheless, Cu intoxication has been reported in swans (*Cygnus olor*), geese (*Branta canadensis*), and also red foxes (*Vulpes vulpes*) from highly polluted areas (Millan et al. 2008). In addition to acute Cu poisoning, there are also reports of Cu deficiencies in various wild cervid populations, e.g., in white-tailed deer (*Odocoileus virginianus*), red deer (*Cervus elaphus*), and moose (*Alces alces*) (Skibniewski et al. 2016; Handeland et al. 2017).

Given the important role of Cu in vertebrate organisms, and the reported northern hemisphere cases of health risks associated with a deficiency or excess of Cu in the consumed diet, it seems justified to perform ecotoxicological studies (including comparative research) of this important micronutrient in the tissues of wild birds and mammals in Europe, especially alien species studied in North America.

## 2 General Properties

Copper (Cu) is a metallic element in Group 11 of the periodic table (previously group IB) (atomic number 29, atomic weight 63.546) that occurs naturally as a free metal or more commonly associated with other elements in compounds comprising a variety of minerals (Kabata-Pendias 2011). Almost all copper found in nature exists as one of two stable isotopes, $^{63}Cu$ (69.09%) and $^{65}Cu$ (30.91%). Its specific gravity is 8.96 at 20 °C, melting point at 1083 °C, and boiling point at 2927 °C (Rudnick and Gao 2003). Copper has very high electrical and thermal conductivity and is resistant to corrosion. Moreover, it is a lightweight and malleable metal (ICSG 2014). Most Cu compounds occur in Cu (I) and Cu (II) valence states, with oxidation states from 0 to +4 (Cotton and Wilkinson 1980). Copper is naturally present in the Earth's crust (lithosphere) at a concentration of about 50–70 mg $kg^{-1}$ (0.0068%); the highest concentrations are found in volcanic and basic igneous rock, with the lowest in lime and sandstone. Copper levels in major rock types can be placed in the following decreasing order (mg $kg^{-1}$): basic igneous (100) > shales and clay (40) > acid igneous (10). The concentration of copper in the core of the Earth is estimated to be 125 ppm, while in the bulk of the mantle and crust, it is estimated at 30 ppm (Rudnick and Gao 2003). Copper is generally present in low levels in coal (17 mg $kg^{-1}$), in oil at 0.2–1 mg $kg^{-1}$, and gasoline at <3 mg $kg^{-1}$ (average values) (Rauch and Pacyna 2009; Georgopoulos et al. 2001).

## 3 Copper Minerals, Production, and Uses

Cu occurs very rarely in nature as native copper, being found mostly in the form of sulfide minerals, particularly chalcopyrite ($CuFeS_2$), chalcocite ($Cu_2S$), bornite ($Cu_5FeS_4$), and tetrahedrite $(CuFe)_{12}Sb_4S_{13}$ (British Geological Survey 2013; Kabata-Pendias and Szteke 2015). As a result of chemical weathering of these primary copper sulfide minerals, secondary minerals may be formed, including the oxide mineral cuprite ($Cu_2O$), the carbonate minerals malachite ($Cu_2(CO_3)(OH)_2$) and azurite ($Cu_3(CO_3)_2(OH)_2$), and the sulfate minerals brochantite ($Cu_4SO_4(OH)_6$) and antlerite ($Cu_3SO_4(OH)_4$) (CCME 1999). There are many copper alloys, e.g., brass (Cu + Zn), bronze (Cu + Sn), cupronickel (Cu + Ni), and Ti-Cu alloys, which exhibit strong antibacterial ability (Zhang et al. 2016). Copper is widely used in a variety of industries, e.g., in pipes, sheets, strips, and cables in electrical engineering. It is an important component of brass and bronze. Copper sheets are used as roofing and in the chemical industry, e.g., in the production of coolers, chemical instruments, and heat exchangers. Copper is a component of dental amalgams. Cu is also the primary component of electrical and data communication wiring such as the windings of stators and generator rotors and most electronic circuit boards. It is also still used to produce copper coins, ammunition, and as a catalyst in chemical processes (Bharti et al. 2010; ICSG 2014). Nanocolloidal copper is a powerful disinfectant and fungicide. The Bordeaux mixture ($Ca(OH)_2 + CuSO_4$) is used in medicine, cosmetics, and food industries. Cu compounds are also used as animal feed additives (Brun et al. 2001; Kabata-Pendias and Szteke 2015). Copper oxide nanoparticles (CuO NPs) are used as additives in lubricants, polymers, plastics, metallic coatings, inks, photocatalysts, gas sensors, and an antimicrobial agent (Midander et al. 2009; Hou et al. 2017). It is possible that a new composite containing Cu and graphene (allotropic variety of carbon) will contribute to the development of the electricity industry with a variety of applications in the electronics, machinery, automotive and food industries, as well as in the construction industry and medicine (Nam and Lee 2016). In the USA, in 2015, Cu has been used in building construction (43%), in electric and electronic products (19%), transportation equipment (19%), consumer and general products (12%), industrial machinery and equipment (7%), and for wiring, plumbing, and waterproofing (USGS 2016). Between 1992 and 2003, there was a more than twofold increase in the use of Cu in the construction industry, from 21% to 46%. Moreover, in the twenty-first century, a downward trend was observed in copper use in electrical and electronic products by ~50% compared to the 1990s (USGS 2014). Copper is one of the few materials than can be recycled. Globally, it is estimated that 2/3 of the 550 million tonnes of copper produced since 1900 is still in productive use, of which 70% has been used for electrical applications and 30% for non-electrical. Around 55% of this usage is in buildings, 15% in infrastructure, 10% in manufacturing, 10% in transport, and 10% in equipment manufacture (Glöser et al. 2013; ECI 2014). Countries with the largest natural resources of Cu range downward from Chile > Peru > Australia > Mexico > USA > - China > Indonesia > Russia > Poland, at 150, 90, 80, 38, 35, 30, and 26 Mt, respectively (ISGS 2011). About two thirds of global resources are located in the western hemisphere, with the largest natural deposits of copper in the world in the Atacama desert in the northern part of Chile, with an estimated 200 million tonnes (Kabata-Pendias and Szteke 2015). From 1960 to 2013, copper consumption has shown some interesting trends. From less than 750,000 tonnes in 1960, copper mine production in Latin America surged to over 7.5 million tonnes in 2013, representing 42% of the global total. Asia has also exhibited a significant growth, probably caused by the rapid industrialization of the continent (USGS 2014). China's copper consumption has expanded markedly over the past decade. Between 2001 and 2011, China's copper usage increased by 5.1 million tonnes (million mt), or by 215% (ISGS 2011).

## 4 Copper in Nature (Soil, Water, and Air)

The average Cu content in different types of soil in the world ranges from 13 to 30 mg $kg^{-1}$ dw and usually does not exceed 50 mg $kg^{-1}$ dw (Kovačič et al. 2013; Kabata-Pendias and Szteke 2015). Copper concentrations from four remote locations in Greenland soils were <12–37 mg $kg^{-1}$ dw, probably representing background levels (AMAP 1998). Copper levels in various types of soils can be placed in the following decreasing order: heavy clay soils > cambisols > histosols > rendzina > arenosols. The lowest Cu concentrations (70 mg $kg^{-1}$ dw) were found in arenosols and calcisols and the highest (140 mg $kg^{-1}$ dw) in heavy clay soils, from which copper can be easily extracted (Kabata-Pendias 2011; Kabata-Pendias and Szteke 2015). The mean concentrations of Cu in soils in North America (USA) and in agricultural soils in China range from 5–70 to 5.8–66.1 mg $kg^{-1}$ dw, respectively (Kabata-Pendias 2011; Li et al. 2014). Copper-deficient soils do occur across large areas of Europe (including Poland, northeastern Germany, Sweden, Central France, Spain), while potentially toxic levels are rarely reached (in Italy, Croatia, Greece) (Reimann et al. 2013). A large difference is observed in the concentration of Cu between the soils from northern and southern Europe. In northern Europe, young soils have 2–3 times lower Cu concentrations than the older and more weathered southern European soils (Reimann et al. 2013). Mean levels of Cu < 5 mg $kg^{-1}$ dw have been observed in some European countries, e.g., the UK, Croatia, Finland, Greece, Norway, Sweden, and Belgium (Kovačič et al. 2013). However, levels higher than 100 mg $kg^{-1}$ dw also occur in Italy (probably because of the volcanic origin of these soils), Norway, Spain, and the UK (Mantovi et al. 2003; Heijerick et al. 2006). Cu contamination in the surface layers of soil, mostly in the area of mines and smelters (especially copper smelters), can exceed several thousand mg $kg^{-1}$ dw (e.g., in Canada, Japan, and Bulgaria) (Kabata-Pendias and Szteke 2015). Some soils close to the Russian Severonickel copper smelter (e.g., Kola Peninsula, one of the most polluted ecocatastrophe areas of the former Soviet Union) have Cu concentrations 50–80 times higher than the background level (AMAP/UNEP 2013). Significant sources of soil Cu contamination are copper fungicides, fertilizers, and waste used to fertilize soils. In some regions of Europe, e.g., in Switzerland, soil Cu may exceed 6000 mg $kg^{-1}$ dw on military training grounds (Kabata-Pendias and Szteke 2015).

Copper content in waterway sediment, especially in polluted water reservoirs and rivers, can be as high as 2000 mg $kg^{-1}$ dw (Kabata-Pendias 2011; Kabata-Pendias and Szteke 2015). Cu levels in surface waters range from 0.5 to 1000 μg $l^{-1}$, with a median of 10 μg $l^{-1}$. Seawater contains from <1 to 5 μg $l^{-1}$ (Martin et al. 1993; EPA 2007; AMAP/UNEP 2013). In uncontaminated natural waters, Cu compounds are relatively rare, with water from wetlands or peat containing trace amounts of Cu, up to about 0.01 mg $dm^{-3}$. Small amounts of Cu may leak into the groundwater; in areas of temperate climate, these waters contain on average 5.0 μg $l^{-1}$ (Zhang et al. 2011). Cu concentrations in groundwater can be classified as permissible (20 μg $l^{-1}$), requiring monitoring (50 μg $l^{-1}$) and requiring treatment (500 μg $l^{-1}$) (Dutch Pollutant Standards 2011). The world average concentration of Cu in rivers has been estimated as 23.6 nM Cu dissolved load and 100 ppm Cu suspended particulate load (Martin et al. 1993). Dissolved ambient Cu concentrations in European surface waters typically range from 0.5 in Denmark to 4.7 μg $l^{-1}$ in Ireland (EPA 2007).

The concentrations of copper in the air depend on the proximity of the site to major particulate sources such as smelters, power plants, and incinerators (IPCS 1998; ATSDR 2004). In the air, Cu occurs most frequently in the form of oxides which combine with particulates and are readily soluble in rainwater, especially acidified rainwater (containing sulfur compounds) (Kabata-Pendias and Szteke 2015). Over 90% of copper in the atmosphere falls to the surface of the soil and plants via precipitation (e.g., fog, clouds, and rain) as a result of global cycling (AMAP/UNEP 2013). Copper in the atmosphere occurs in very different concentrations, from 0.03 ng $m^{-3}$ over the South Pole to 5000 ng $m^{-3}$ in industrial areas of Germany. The average amount of natural Cu in the air is 4 ng $m^{-3}$ (Kabata-Pendias and Szteke 2015) and in nonindustrial and urban locations from 2.5–4.0 to 30–200 ng $m^{-3}$, respectively (IPCS 1998). The average annual value of Cu in precipitation has been evaluated at 34.0 g $ha^{-1}$ $year^{-1}$, with the least in Finland (5) and the most in Austria (100) (Nicholson et al. 2003).

## 4.1 Geogenic and Anthropogenic Sources of Copper

Cu has entered the environment via both natural and anthropogenic sources, with both emissions appearing comparable (Table 4.1). The average annual global emissions of Cu (thousand tones) from various natural sources include dust carried by winds (0.9–15), forest fires (0.2–7.5), volcanic particles (0.9–10), biogenic transformation (0.1–6.4), and spraying seawater (0.2–6.9) (AMAP 2002; ATSDR 2004; Kabata-Pendias and Szteke 2015). The mean natural emission rates for copper are $2.6 \times 10^6$ kg $year^{-1}$ for Canada, $5.0 \times 10^7$ kg $year^{-1}$ for North America, and $2.0 \times 10^9$ kg $year^{-1}$ globally (Richardson et al. 2001; AMAP/UNEP 2013). Emissions of Cu into the atmosphere are mainly from nonferrous metal smelting and

**Table 4.1** Global emissions of copper from natural and anthropogenic sources ($\times 10^6$ kg $year^{-1}$)

| | | |
|---|---|---|
| *Natural sources* | | |
| Wind-borne particulates | 8.0 | 0.9–15 |
| Marine-sprayed sea salt and surface organic microlayers | 4.0 | 0.25–7.7 |
| Volcanoes | 9.4 | 0.9–18 |
| Forest fires | 3.8 | 0.1–7.5 |
| Biogenic-continental particulates and volatiles | 2.9 | 0.11–5.6 |
| Total emission | 28 | 2.3–54 |
| *Anthropogenic sources* | | |
| Coal combustion | 5.15 | 2.3–8.0 |
| Oil combustion | 1.86 | 0.42–3.3 |
| Pyrometallurgical | 23.5 | 15–32 |
| Secondary nonferrous metal production | 0.115 | 0.06–0.17 |
| Steel and iron manufacturing | 1.47 | 0.14–2.8 |
| Refuse incineration | 1.5 | 1.0–2.0 |
| Phosphate fertilizers | 0.415 | 0.14–0.69 |
| Wood combustion | 0.9 | 0.60–1.2 |
| Total emission | 35 | 20–50 |

Nriagu and Pacyna (1988), AMAP (2002), and ATSDR (2004)

refining (~70%), with concentrations well in excess of 1 μg $m^{-3}$ frequently recorded near Cu smelters (AMAP 2002), as well as from burning coal for power generation and from municipal waste incinerators. Asia and South America were responsible for 50% and 21%, respectively, of global anthropogenic copper emissions (AMAP 2002; AMAP/UNEP 2008). In 2010, 3.3 kt of copper was emitted to the atmosphere in Europe (AMEC 2014). Between 1990 and 2012, Cu emissions in European countries dropped by 1.3%, and between 2011 and 2012, emissions decreased by a further 1.9%, mainly from reduced emissions in Germany, Italy, and Romania (EEA 2014).

# 5 Copper in Living Organisms

## *5.1 Plants*

Copper is an essential nutrient for plant growth and development, as a component of many enzymes involved in photosynthesis. It has protective properties against plant diseases and affects the metabolism of nitrogen compounds and DNA and RNA synthesis (Mahmood and Islam 2006; Nagajyoti et al. 2010; Szatnik-Kloc 2014). Copper may become phytotoxic and cause metabolic disorders at high concentrations and so lead to a potential threat to animal health through the food chain (Chang and Page 2000; Xu et al. 2013). Cu accumulation in leaves is between 15 and 25 mg $kg^{-1}$ dw (Hladun et al. 2015), with Cu content in plants in contaminated regions usually elevated and exceeding concentrations considered toxic (Kabata-Pendias 2011). A critical Cu deficiency level in vegetative plant parts is generally 3–5 mg $kg^{-1}$ dw (Robson and Reuther 1981). Cu levels in plants are influenced by soil pH; Cu mobility in soils increases only at pH $< 5.0$. It has also been demonstrated that soil liming reduces the accumulation of Cu (Kabata-Pendias 2011; Kabata-Pendias and Szteke 2015). Nitrogen and phosphorus fertilization adversely affects Cu uptake due to elevated levels of phosphorus, nitrogen, or zinc in the soil. Cu content in plants varies depending on the parts: fruit (3–8 mg $kg^{-1}$ dw), nuts (0.2–24 mg $kg^{-1}$ dw), grasses (2–10 mg $kg^{-1}$ dw), and clover (7–15 mg $kg^{-1}$ dw). Plants growing on Cu-polluted sites tend to accumulate increased amounts of this metal, especially near industrial areas and in soils treated with Cu-bearing herbicides (Eisler 1998). Some plants from the *mint* family (e.g., *Aeollanthus biformifolius*) are useful indicators of Cu levels in soils, with samples growing on mineralized land containing 13,700 mg Cu $kg^{-1}$ dw (Kabata-Pendias 2011).

## *5.2 Biological Effects, Metabolism, and Toxicity of Cu in Homeothermic Animals*

Copper is an essential trace element involved in the formation and metabolism of bone tissue. However, its primary function is its participation in oxidation-reduction reactions as a coenzyme, a regulator of iron metabolism and transport, as well as collagen metabolism (Honda et al. 1997; Angelova et al. 2011; AMAP/UNEP 2013; Cornu et al. 2017). Cu metabolism disorders can lead to hereditary diseases, e.g., Menkes disease (steely hair syndrome) and Wilson's disease (Roberts and Schilsky, 2008; Ojha and Prasad 2016). These diseases are characterized by deposition of Cu in the internal organs, leading to their damage. In the blood, Cu occurs in complexes with histidine, threonine, and glutamic acid. In complexes with amino acids (e.g., Cu-lysine) and albumin, Cu is transported to the liver, kidney, intestines, and other tissues. From the ingested food (mainly from plants), about 50%–70% of Cu is absorbed in the digestive tract, particularly in the small intestine, with small quantities being absorbed in the stomach, and then transported via albumins, then hepatocytes and hepatic ceruloplasmin (Cp) (IPCS 1998). Cu absorbed into the intestinal endothelium may be sequestered by metallothionein (MT), or it may migrate to the portal circulation. Most Cu is transported across the mucous membrane involving copper translocation, while intestinal metallothionein may participate in the detoxification of excess copper (Kabata-Pendias and Szteke 2015). In the bodies of some animals, e.g., in sheep, a much smaller proportion of copper in the liver is bound to MT, as sheep have a limited ability to increase metallothionein synthesis in response to increased liver copper. In mammals, the liver is the main organ responsible for the accumulation of Cu (Osredkar and Sustar 2011). Outside the liver, this element is associated mainly with enzymatic proteins: SOD (superoxide dismutase), tyrosinase, cytochrome c oxidase, and lysyl oxidase (IPCS 1998; WHO 2004). Copper is present in all tissues at levels ranging from 0.7 to 7.8 mg $kg^{-1}$ body weight and varies between animal species.

Antagonistic interactions occur between Cu and Zn, Mo, Cd; a high level of zinc in the diet reduces the absorption and bioavailability of Cu. Copper deficiency in animals results in anemia, loss of appetite, damage to internal organs, bone deformities, reduced reproductive capacity, changes in coat, myocardial fibrosis, and chronic debilitating diarrhea. Copper deficiency may be primary, i.e., associated with a low Cu intake, or secondary, when Cu absorption or metabolism is adversely affected due to high levels of molybdenum, iron, sulfur, or zinc in the diet (Vengust et al. 2015). Among domesticated species, signs of Cu deficiency are observed mainly in cattle, with sheep showing a high sensitivity to even low concentrations of Cu in the diet (Puls 1994). Copper deficiency has been reported in various wild cervid populations (O'Hara et al. 2001; Handeland et al. 2017). Subclinical Cu deficiency may produce marginal signs of poor health and result in morbidity and mortality from other opportunistic factors, such as secondary infections, predation, and weather events (O'Hara et al. 2001). In addition, white-tailed deer (*Odocoileus virginianus*) from Texas, USA, with mean liver Cu levels of 16.7 mg $kg^{-1}$ ww, had stunted and twisted antlers (King et al. 1984). In moose, Frank (1998) regarded a Cu concentration of >5 mg $kg^{-1}$ ww in the liver as an indication of severe Cu deficiency. In addition, in red foxes, Cu deficiencies are usually observed in offspring; puppies are born weak, poorly developed, and with high mortality rates. Cu deficiency-related anemia results in a pale skin and mucous membranes, discoloration, and loss of elasticity and silkiness of hair (Frank 1998).

Excess copper is harmful to wildlife, but it occurs rarely. Mammals and birds are at least 100 times more resistant to Cu than other organisms. An excessive dietary intake of Cu by 20- to 50-fold over normal levels may, however, have serious adverse effects on birds and mammals. No data are available on the toxicity of copper to avian wildlife (Eisler 1998). Research on the adverse effects of higher concentrations concern only domestic waterfowl fed with extremely high doses of copper (diet containing up to 500 mg Cu per 1 kg of feedstuff). The more common effects of high copper concentrations in the environment on birds can be seen through the tropic chain where the prey of some avian species is eliminated by pollution (which limits the nutrition sources) (Eisler 1998). In mammals, the mechanism of copper toxicity is complex. Copper can increase cell permeability in erythrocytes leading to lysis and inhibition of intracellular enzymes. Thus, copper poisoning can lead to oxidative stress in erythrocytes and to accelerated loss of intracellular glutathione. In addition, copper ions can cause mitochondrial swelling and inhibit oxygen consumption, which leads to cell degeneration (EPA 2007). Scientific literature describes cases of acute and chronic toxicity in laboratory and farm animals, although there is little data on contamination with Cu compounds in wildlife.

Acute responses to copper vary with the species and copper compound. Ferrets, sheep, dogs and cats are more sensitive to copper than rodents, pigs, and poultry (WHO 2004). Acute poisoning from the ingestion of excessive Cu can cause temporary gastrointestinal distress with symptoms such as nausea, vomiting, and abdominal pain. Liver toxicity was seen in doses high enough to result in death. High levels of exposure to copper can cause the destruction of red blood cells, possibly resulting in anemia. An oral $LD_{50}$ (median lethal dose) of 300 mg cupric sulfate/kg in rats has been reported (Siegel and Sisler 1977). Acute copper poisoning can occur at copper intakes of 20–100 mg $kg^{-1}$ in sheep and young calves and 200–800 mg $kg^{-1}$ in adult cattle. The chronic LOAEL (*lowest-observed-adverse-effect level*) in bobwhite quail (*Colinus virginianus*) was 289 mg $kg^{-1}$ dw Cu as metallic copper. An acute oral $LD_{50}$ for this bird exposed to copper sulfate was also reported as 616 mg $kg^{-1}$ dw (ATSDR 2004). Luo et al. (2005) fed 450 mg $kg^{-1}$ dw of copper sulfate to male chicks for 21 days and noted reduced feeding and less weight gain in the exposed birds. Liver Cu residue is a commonly used and reliable index of exposure in cases of acute Cu toxicosis. Reported liver Cu levels in wild adult mallard ducks ranged from 35 to 585 mg $kg^{-1}$ dw (Eisler 1998). Mute swans diagnosed as having Cu poisoning in the north American Mamaroneck Harbor had a mean Cu concentration in the liver of ~3900 mg $kg^{-1}$ dw. Significantly, some mute swans tolerated liver Cu residues at levels up to 1000 mg $kg^{-1}$ dw. In wild birds with symptoms of inanition, anemia, and generalized weakness, concentrations of Cu in the liver and kidney were 3000 and 50 mg $kg^{-1}$ dw, respectively.

Mammals have efficient mechanisms to regulate Cu stores in the body such that they are generally protected from excess dietary copper levels. In chronic poisonings, sheep and cattle are the most affected species, with copper exerting a hepatotoxic effect (Oruc et al. 2009). Fuentealba and Aburto (2003) established a normal copper content in the liver at between 10 and 50 mg $kg^{-1}$ ww. Buck and

Sharma (1969) observed sheep liver Cu concentrations ranging from 160 to 550 mg $kg^{-1}$ ww in animals diagnosed with chronic copper poisoning and kidney concentrations between 12 and 220 mg $kg^{-1}$ ww. Chronic copper intoxication in calves could only be seen at levels higher than 200 mg $kg^{-1}$ ww (Croubels et al. 2001), while Hadrich (1996) reported liver Cu content between 500 and 800 mg $kg^{-1}$ dw in cases of poisoning. Grobler and Swan (1999) observed liver Cu concentrations above 150 mg $kg^{-1}$ (for chronic copper poisoning in domestic sheep) in impala (*Aepyceros melampus*) within a high risk zone (near a copper smelter of a mine). Gilbert et al. (1996) fed copper sulfate at concentrations of 78 and 1437 mg $kg^{-1}$ bw to laying hens for 2 weeks. At the highest concentration, hens produced fewer eggs, consumed less feed, and developed ulcers in the gizzard and oral cavity (Gilbert et al. 1996; Boone et al. 2012). Other studies on chickens that were fed copper sulfate reported oral lesions proportional to the dose of Cu and conflicting effects on feeding rate and weight gain (Boone et al. 2012). Scientific literature lacks data on toxic thresholds of Cu indicating subclinical or clinical poisoning in wild birds and mammals (AMAP/UNEP 2013). There are reports describing a wildlife toxicity reference value (TRV) and a no effect hazard concentration (NEHC) for the American bald eagle (*Haliaeetus leucocephalus*), at 4.05 and 40.5 mg $kg^{-1}$ dw, respectively. The same parameters were established for ranch mink at 5.60 and 41.2 mg $kg^{-1}$ dw, respectively (EPA 2007).

Cu compounds can have a mutagenic effect on the cells of mammals. Copper is currently categorized by the EPA as a Group D carcinogen (inadequate evidence to classify) and has not yet been reviewed for placement into any of the new cancer classification categories (WHO 2004). Cu may affect cancer growth and cell proliferation and stimulate blood vessel formation (Boone et al. 2012). Developmental effects have been observed in only a few studies giving animals high doses of copper, such as delayed growth and development, delayed bone formation, and reduced litter size and body weights (ATSDR 2004).

## 5.3 *Copper in Biological Samples from Birds and Mammals*

Ecotoxicological studies on Cu in birds usually determine levels in the feathers, eggs, blood, the liver and kidneys and much less frequently in the muscles, brain, and bones (Kalisińska et al. 2004, 2007; Schummer et al. 2011; Binkowski and Meissner 2013). In mammals, the concentration of Cu is commonly determined in parenchymal organs (liver and kidney), which are essential for detoxification. There is a lack of publications on Cu in the brain, muscle, or bone fragments, with the latter preferred in assessing long-term changes in the accumulation of elements, including Cu (Kalisińska et al. 2004, 2006; Lanocha et al. 2012). In mammals, apart from blood levels that reflect short-term exposure, Cu is increasingly often determined in the hair, which consists of keratin with cysteine sulfhydryl groups capable of binding metals. In birds, Cu is also determined in the feces and urine, which have been suggested as useful nondestructive indicators of Cu contamination in the diet and environment (Berglund et al. 2015).

Cu concentrations in biological material from birds usually can be placed in the following descending order: liver > kidney > feathers > muscle > brain > blood > eggs > bone (Ek et al. 2004; Lucia et al. 2010). The muscles and feathers typically contain the highest levels of total Cu (up to 53% and 34%, respectively); sometimes in birds associated with water ecosystems, e.g., the great cormorant (*Phalacrocorax carbo*), muscle tissues had the highest total Cu levels at ~70% (Honda et al. 1986; Nam et al. 2005), ~11% in the liver, 7% in feathers, and 2% in bone tissues (Nam et al. 2005). In birds, the concentrations of Cu in feathers are higher than in the eggs and blood due to the affinity of Cu to keratin (the principal structural protein in building a feather, rich in sulfhydryl groups (SH)), although >20% of the total content of Cu is incorporated into newly formed feathers (Honda et al. 1986; Ek et al. 2004; Costa et al. 2013). As shown in environmental monitoring, Cu concentrations in bird feathers occur over a wide range, from ~0.5 to 88 mg $kg^{-1}$ dw, with generally higher concentrations detected in birds living in contaminated environments (Dauwe et al. 2002). St Clair et al. (2015) have suggested that a toxicity threshold has not been established for Cu in feathers. Lester and van Riper (2014) stated in southwestern song sparrows (*Melospiza melodia fallax*) that the concentration of Cu in the feathers, ranging from ~10 to ~15 mg $kg^{-1}$ dw, did exceed background levels.

In terrestrial mammals, Cu concentrations in biological materials can usually be arranged in the following descending order: liver > kidney > heart > brain > lung > muscle > bone, with variations in mineral content not only from organ to organ but also between animal species (Hanusova et al. 2007). Han et al. (2002) described the distribution of Cu in the European otter (*Lutra lutra*), where 30% of total Cu found was in the liver, 30% in hair, 20% in kidneys, and 10% each in muscle and bone tissues. It is estimated that 30% of total Cu in mammals is in the hair, as the process of normal hair and wool pigmentation requires Cu (Brzezinski et al. 2014).

### 5.3.1 Bioaccumulation of Copper in Avian Liver, Kidney, and Muscle

Although most of the data on Cu concentrations in birds concern the liver, scientific literature does not present values indicative of elevated concentrations of Cu against the geochemical background. The no-effect level of Cu in bird livers has been stated at <60 mg $kg^{-1}$ dw, with the toxicity threshold >540 mg $kg^{-1}$ dw (U.S. Department of the Interior 1998). It has been found that a concentration ~50 mg $kg^{-1}$ dw in the kidney may cause a nephrotoxic effect in the mute swan (Frank and Borg 1979). However, in scientific literature we have only found information on a so-called typical range for Cu in the liver for wild birds, e.g., the wild Canadian goose and mute swan: ranging from 6 to 30 mg $kg^{-1}$ ww (~20–100 mg $kg^{-1}$ dw) and 120 to 360 mg $kg^{-1}$ dw, respectively (Puls 1994; Isanhart et al. 2011).

The concentrations of Cu in internal tissues represent the available levels of this metal in the diet, which in turn can reflect the degree of contamination of the ecosystem (Carneiro et al. 2016). Taking into account the trophic groups of birds associated with northern hemisphere terrestrial ecosystems, the concentrations of Cu in the parenchymal organs (liver and kidney) in each group trophic can be arranged in the following descending order: herbivores > omnivores > piscivores > birds of prey (Fig. 4.1). In omnivorous birds, the mean concentrations of Cu in the liver are not as high as in herbivores (although in some cases they can reach >200 mg kg$^{-1}$ dw) (Hernandez et al. 1999; Taggart et al. 2006). Omnivorous avian species have high average concentrations of Cu in the kidneys (~45 mg kg$^{-1}$ dw), although these levels may differ significantly between studies (Frank and Borg 1979; Schummer et al. 2011; Komosa et al. 2012). Jackson et al. (1979) found a concentration of Cu in the kidneys between ~11 and 29 mg kg$^{-1}$ dw in dunlin (*Calidris alpina*) that could be associated with toxicity in other avian species, where Cu levels in the kidneys can remain stable even with toxic doses in the diet (St Clair et al. 2015).

Studies on Cu in wild birds from the northern hemisphere published after 1980 usually concern wetland birds and rarely discuss the concentration of Cu in the kidney. Only a few species of terrestrial birds have average Cu concentrations in the kidney above the avian nephrotoxic level, ~60 mg kg$^{-1}$ dw, including Eurasian wigeon (*Anas penelope*), mute swan, mallard (*Anas platyrhynchos*), spectacled eider

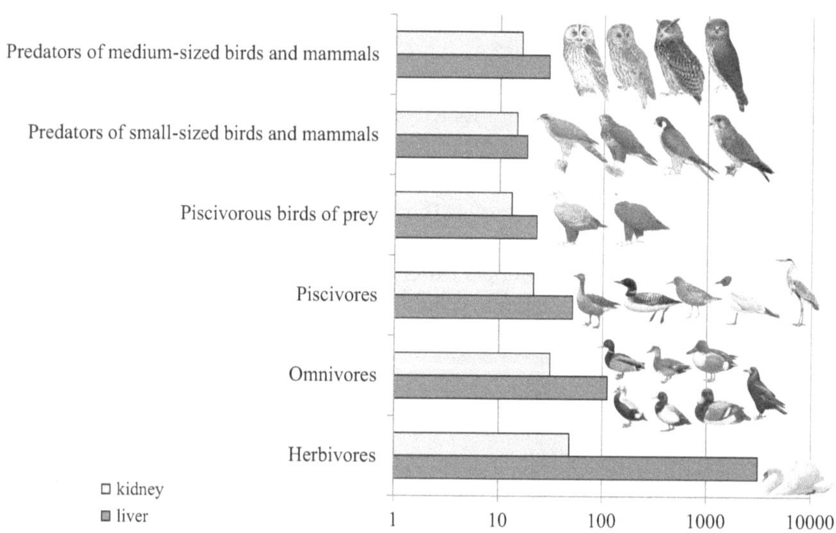


**Fig. 4.1** The concentration of Cu (mg kg$^{-1}$ dw) in the livers and kidneys of birds with different diets; logarithmic scale based on herbivores (Beyer et al. 1998; Frank and Borg 1979; Schummer et al. 2011), omnivores (Hernandez et al. 2017; Kalisińska et al. 2004; Kim and Oh 2012; Lucia et al. 2008; Orlowski et al. 2007; Taggart et al. 2006; Trust et al. 2000), *piscivores* (Kim and Oh 2015; Lucia et al. 2008; Lucia et al. 2010; Mateo and Guitart 2003; Orlowski et al. 2007; Schummer et al. 2011), *fish eaters* (Falandysz et al. 2001; Falandysz et al. 1988; Falandysz and Szefer 1983; Kalisińska et al. 2006; Kim and Oh 2012; Mierzykowski and Todd 2012; Stout and Trust 2002), small-sized bird and mammal eaters (Ek et al. 2004; Hontelez et al. 1992; Jager et al. 1996; Kalisińska et al. 2008; Komosa et al. 2012; Licata et al. 2012; Naccari et al. 2009), and medium-sized bird and mammal eaters (Kalisińska et al. 2009a; Komosa et al. 2012)

(*Somateria fischeri*), and brown hawk-owl (*Ninox scutulata*) (Clausen and Wolstrup 1978; Trust et al. 2000; Lucia et al. 2008; Kim and Oh 2012). One of the highest mean kidney Cu concentrations was found by Kim and Oh (2012) in the Korean brown hawk-owl, at ~67 mg $kg^{-1}$ dw, and the authors suggest that as this bird is a summer visitor, the pollutant levels may reflect both Korean and wintering site levels. Furthermore, in the American spectacled eider, the concentrations of Cu in the kidney and liver exceeded levels that can cause nephro- and hepatotoxic effects in birds: means were ~67.5 and ~559 mg $kg^{-1}$ dw, respectively (Trust et al. 2000).

The highest concentrations of Cu in the liver, indicative of poisoning, have been reported in herbivorous birds related to freshwater ecosystems, at times reaching 5000 mg $kg^{-1}$ dw in North American and 3820 mg $kg^{-1}$ dw in European specimens (Frank and Borg 1979; Beyer et al. 1998). High levels of hepatic copper observed in the mute swan can be explained by the fact that the swans consume daily up to 35%–43% of their body mass in aquatic vegetation (Komosa et al. 2012). Copper is an essential micronutrient for all higher plants and is easily absorbed by aquatic vegetation (Xue et al. 2010). Such a high consumption of water plant biomass may contribute to the high levels of Cu in the livers of the mute swans. Schummer et al. (2011) found concentrations of Cu in the liver >2000 mg $kg^{-1}$ dw in swans residing in the lower Great Lakes in Canada (Fig. 4.1). Toxicological thresholds for Cu in mute swans are unknown, and mortality rates noted in studies from Sweden, New York, and northeastern America may have resulted from contaminants other than Cu or a lethal combination of contaminants (Frank and Borg 1979; Schummer et al. 2011). In the livers of grey heron (*Ardea cinerea*) and intermediate egret (*Mesophoyx intermedia*), Horai et al. (2007) found concentrations of Cu as high as 4970 and 2420 mg $kg^{-1}$ dw, respectively.

Many scientific publications document the usefulness of ducks (genera *Anas* and *Aythya*) in biomonitoring studies, and it has been reported that wetland birds can accumulate much larger amounts of Cu than terrestrial birds, with the largest concentrations recorded in contaminated areas (Hernandez et al. 1999; Horai et al. 2007; Schummer et al. 2011). Hernandez et al. (1999) analyzed the concentration of Cu in omnivorous wetland birds in a polluted area of the Doñana National Park (NDP) in Spain and showed that liver Cu levels were lower before the ecological catastrophe in 1998, when waste water heavily contaminated with heavy metals (including Cu) leaked from a damaged mine tank into groundwaters. Concentrations of Cu in the liver of birds surveyed before the disaster were below 35 mg $kg^{-1}$ dw, and after 1998 ranged from ~52 to 1300 mg $kg^{-1}$ dw, and at almost 480 mg $kg^{-1}$ dw in the common pochard (Hernandez et al. 1999).

The accumulation of Cu in the liver may differ significantly between bird species. Within the *Anatidae*, the mute swan can accumulate more Cu in the liver than other species even in the absence of environmental contamination (Beyer et al. 1998). Parslow et al. (1982) detected the highest and the lowest Cu concentrations in common pochard and common snipe (*Gallinago gallinago*): 603 and 33 mg $kg^{-1}$ dw, respectively. Moreover, comparisons between species living in the same areas indicate that *Aythya* and *Netta* genera, closely related but with different feeding habits, may also accumulate more Cu in the liver than other waterfowl species (Figuerola and Green 2000). In El Hondo, Spain, a wetland by a river heavily polluted by agricultural pollution, samples analyzed from marbled teal (*Marmaronetta angustirostris*) and white-headed ducks (*Oxyura leucocephala*) showed that ~40% of the animals had liver Cu levels above 100 mg $kg^{-1}$ ww (~350 mg $kg^{-1}$ dw) (Taggart et al. 2009), these being above the level indicating acute Cu poisoning in a study of Canadian geese (Henderson and Winterfield 1975). The authors suggest that such a large concentration of Cu in the liver of the birds, indicating intoxication, probably depended on several polluting factors, e.g., geological, agricultural (including plant Cu fungicide), or industrial sources near El Hondo, or key invertebrate food bioaccumulating Cu at unusual levels (Taggart et al. 2009). The interpretation of biomonitoring results obtained from terrestrial birds and waterfowl is thus very difficult, and Pillatzki et al. (2011) suggest that many factors can influence the interpretation of hepatic Cu levels, including various interspecies sensitivities, differences in hepatic accumulation by species and age, interaction of copper with other elements, and seasonal variation (Eisler 1998; Pillatzki et al. 2011).

Birds of prey are particularly susceptible to the effects of heavy metals as they occupy the uppermost positions in the food pyramid, and their diet can accumulate a significant amount of particularly toxic trace elements (Jager et al. 1996; Ek et al. 2004; Kalisińska et al. 2008; Zaccaroni et al. 2008; Naccari et al. 2009; Licata et al. 2012). Based on an analysis of scientific papers, it was found that the average concentration of Cu in the liver and kidney in wild birds is ~17 and 14 mg $kg^{-1}$ dw, respectively (Fig. 4.1) (Hontelez et al. 1992; Zaccaroni et al. 2008; Komosa et al. 2012). In the liver and kidney of the common buzzard (*Buteo buteo*) and common kestrel (*Falco tinnunculus*), birds with a similar diet sourced from Poland and the Czech Republic, Kalisińska et al. (2009a) determined Cu levels and noted significant differences in kidney Cu levels between these two species of birds: Cu was significantly lower in the common buzzard than in the common kestrel (~16.0 and 10.0 mg $kg^{-1}$ dw, respectively). In the livers of both species, the concentration of Cu was below 17.2 mg $kg^{-1}$ dw.

Ecotoxicological analysis on Cu in the common buzzard in the Netherlands was conducted by Jager et al. (1996) and Hontelez et al. (1992), with concentrations of Cu in the livers in both studies similar, ranging from 13 to 16 mg $kg^{-1}$ dw. In the case of birds from the south of Europe, e.g., Sicily in Italy, we noticed larger concentrations of Cu in the liver, at >39 mg $kg^{-1}$ dw, i.e., more than two times higher than in the buzzards from the Netherlands, Poland, and Czech Republic, which may be associated with the higher natural level of Cu in the soils in Italy and a higher affinity of this element to the volcanic rocks (Naccari et al. 2009; Licata et al. 2012).

In European literature we found several publications on Cu in owls (Fig. 4.1). In different areas of the Calabria region in Italy, Zaccaroni et al. (2008) observed that Cu concentrations were generally higher in the liver of the marsh harrier (*Circus spilonotus*) and tawny owl (*Strix aluco*), at 14.6 and 49.4 mg $kg^{-1}$ dw, respectively, while in the marsh harrier and barn owl (*Tyto alba*), the kidney Cu concentration did not exceed 17.5 mg $kg^{-1}$ dw, suggesting that the values of Cu in the parenchymal organs of the analyzed birds of prey are within the range for physiological values and do not show the presence of any deficiency or excess. Komosa et al. (2012) in the long-eared owl (*Asio otus*) from eastern Poland and Kim and Oh (2012) in eagle owls (*Bubo bubo*) and brown hawk-owls from eastern Asia detected Cu levels similar to those observed in birds of prey from Italy, with concentrations of Cu in the liver not exceeding 17 mg kg$^{-1}$ dw. Data on Cu in wild birds obtained from the northern hemisphere are in many cases difficult to interpret, especially given the small number of examined specimens. Cu concentrations in the liver and kidney of various birds of prey are generally lowest in piscivorous species and highest in medium-size birds and carnivorous species (Fig. 4.1). In comparison of piscivorous bird species, e.g., European white-tailed eagle (*Haliaeetus albicilla*) and North American bald eagle sometimes show substantial differences. The highest concentration of Cu in the muscle tissues of white-tailed eagle was found by Falandysz et al. (1988) at ~30 mg kg$^{-1}$ dw (range, 15–320 mg kg$^{-1}$ dw) and in key parenchymal organs ~20 mg kg$^{-1}$ dw. In white-tailed eagle specimens collected in northwestern Poland in 1991–1995, Falandysz et al. (2001) observed hepatic and nephric Cu levels ranging from ~6–36 to ~7–40 mg kg$^{-1}$ dw, respectively. Six years later in the same Polish region, Kalisińska et al. (2006) observed even lower concentrations of Cu in the liver and kidney in white-tailed eagles, with the average concentration of Cu below 8 mg kg$^{-1}$ dw in either organ. Much higher concentrations of Cu in the liver and the kidney have been reported in bald eagle in North America; the average levels in both organs ~33 mg kg$^{-1}$ dw range between 9 and 395 mg kg$^{-1}$ dw (Stout and Trust 2002; Mierzykowski and Todd 2012). In Cu-contaminated sites, high Cu levels in kidneys and muscles have been reported in omnivorous and herbivorous bird species such as the common pochard and Eurasian coot (*Fulica atra*), at ~48 and 44 mg kg$^{-1}$ dw, respectively (Gomez et al. 2004). Lower concentrations in these organs have been detected in crab- and fish-eating predators such as the black-headed gull (*Chroicocephalus ridibundus*) at <12 and <17 mg kg$^{-1}$ dw, respectively (Hernandez et al. 1999; Gomez et al. 2004; Orlowski et al. 2007). Muscle tissues of birds are tested only occasionally for Cu, even though it is a muscle (representing 30%–40% of the weight of the bird) that can be the source of this metal for predatory species. Available scientific literature presents a certain tendency where the highest Cu muscle levels are recorded in piscivorous birds from freshwater ecosystems, such as the grey plover (*Pluvialis squatarola*), greylag goose, and red knot, 51, 43, and ~30 mg kg$^{-1}$ dw, respectively. Lower concentrations of Cu have been found in the muscle of birds of prey and omnivorous birds (Lucia et al. 2008, 2010; Kalisińska et al. 2004, 2008; Bojar and Bojar 2009; Licata et al. 2012). In birds of prey, e.g., white-tailed eagle and Eurasian buzzard, the average concentration of Cu in the muscle is similar at below 32 mg kg$^{-1}$ dw (Falandysz et al. 1988; Licata et al. 2012). The biggest difference in this respect was noted in the case of a highly emaciated female peregrine falcon, in which the concentration of Cu in the muscle was more than two times higher (~67 mg kg$^{-1}$ dw) (Kalisińska et al. 2008). The lowest muscle Cu levels are observed in birds with varied diets, including mallard and tufted ducks (*Aythya fuligula*) from Polish territories (~1.6–6 and ~6 mg kg$^{-1}$ dw, respectively) (Szymczyk and Zalewski 2003; Kalisińska et al. 2004).

Binkowski et al. (2013) suggested that Cu levels in bird tissues may significantly differ between waterfowl, where the concentration of Cu in the pectoral muscle (~40 mg kg$^{-1}$ dw) of coot from the Zator area (fishing farms), southern Poland, was almost 7 times higher than in mallard from northwestern Poland as determined by Kalisińska et al. (2004). In some water ponds, some Cu compounds are used as algaecides to control the growth of phytoplankton and filamentous algae and to control certain fish diseases. Their possible transfer into the food chain may cause a significant increase in the concentration of Cu in the muscles of birds.

Passerines such as the house sparrow (*Passer domesticus*) and great and blue tit (*Parus major* and *Cyanistes caeruleus*) play an important role in biomonitoring in terrestrial ecosystems and have been studied in Finland, Belgium, Serbia, Portugal, Turkey, and China. In these birds, Cu concentrations are most often analyzed in the feathers and eggs and much less frequently in the liver, kidneys, and muscles (Dauwe et al. 2002; Nam et al. 2005; Gong et al. 2012; Costa et al. 2013). In Finland, Kekkonen et al. (2012) found that concentrations of Cu in the liver of house sparrows sampled in the 1980s were higher in urban than in rural areas, at 4.03 and 3.7 mg kg$^{-1}$ dw, respectively. More than 30 years later, Millaku et al. (2015) studied the same species from a polluted area in Serbia and observed an order of magnitude higher concentration of Cu in the liver (20.2 mg kg$^{-1}$ dw) and kidney (28.4 mg kg$^{-1}$ dw) and considered the species a good bioindicator of the terrestrial environment. The authors of many studies suggest that the accumulation of Cu in the respective tissues depends on the degree of exposure (Kekkonen et al. 2012; Millaku et al. 2015). In China, Gong et al. (2012) showed that hepatic Cu in the tree sparrow (*Passer montanus*) differed depending on the area of foraging, mining area (MA), urban district (UD), or reference site (RS), where hepatic Cu could be arranged in the following descending order: UD > MA > RS (~25.6, 22.7, and 12.2 mg kg$^{-1}$ dw, respectively). In addition to anthropogenic pollution, higher Cu concentrations in the livers in these insectivorous birds may be associated with the local use of copper-based agricultural fungicides and diet. Some terrestrial songbirds fed mainly on spiders, which contain high levels of Cu in their hemocyanin (Eeva et al. 2005, Belskii and Belskaya 2013). The concentration of Cu in the muscles of passerine species sometimes varies between contaminated sites, e.g., in China and Turkey, respectively; muscle Cu levels were <7 and 25 mg kg$^{-1}$ dw (Albayrak and Mor 2011; Gong et al. 2012).

### 5.3.2 Relationship Between Cu Levels and Age and Sex of Terrestrial Birds

Studies on the relationship between the age of wild birds and the concentration of Cu in their organs do not allow firm conclusions to be drawn. Sometimes, liver Cu levels show a significant correlation with age and sex, e.g., in coots and mallards, while there are discrepancies in others. Taggart et al. (2006) in Spain did not note an increase in Cu concentration with age as reported in the greylag goose by Mateo and Guitart (2003), in the barn owl by Esselink et al. (1995), in the bald eagle by Stout and Trust (2002), in the white-tailed eagle by Kalisińska et al. (2006), or in passerine bird species by Berglund et al. (2015). In the black duck (*Anas rubripes*), Gochfeld and Burger (1987) observed significantly higher liver Cu concentration in males than in females. Similarly, Schummer et al. (2011) showed differences in liver Cu between females and males, suggesting that the differences reflected the different feeding areas of the males and females. Moreover, Barjaktarovic et al. (2002) and Kalisińska et al. (2004) observed that for scoter (*Melanitta* sp.) from Canada and Mallard from Poland, sex was not a factor correlating to kidney Cu.

The aforementioned test results indicate that copper impurities are present in the environment for a long time, during which both typically aqueous and terrestrial birds are exposed. There is only fragmentary data on Cu in the organs and tissues of typically terrestrial species in Asia, including China. More comprehensive studies are recommended in this regard, especially in light of the rapid industrialization in Asia in recent decades and the greatest share of this continent in global anthropogenic atmospheric Cu emission.

## 5.4 *Bioaccumulation of Copper in Mammalian Liver, Kidney, and Muscle Tissues*

Cu concentrations in the organs crucial for detoxification in wild mammal taxa from the northern hemisphere can be arranged in the following ascending order: piscivorous < omnivorous < carnivorous < herbivorous (Arnhold et al. 2002). Herbivores accumulate the highest amount of Cu and are the subject of intense and multifaceted research in the USA, Canada, and Europe, despite the fact that Cu deficiencies are often reported in herbivorous and omnivorous mammals (Skibniewski et al. 2016). There are also many signs that predatory mammals, acting as the final link in the trophic chain, tend to accumulate Cu in tissues and organs. It seems that this observation is strongly pronounced in Europe, especially in southern parts, e.g., in Spain, where the environment has been exposed to a lot of anthropogenic deposition of Cu (Millan et al. 2008). In mammals associated with freshwater ecosystems, Cu levels are not so high; in this group the highest concentrations have been found in piscivores. Copper poisoning in terrestrial mammals appears to be rare, thanks to their efficient mechanisms to regulate Cu stores, so they are generally protected from excess dietary Cu levels.

### 5.4.1 Piscivorous Mammals (Fish-Eating Mammals)

Piscivorous mammals play an important role in biomonitoring, e.g., American mink (*Neovison vison*) and European otter (*Lutra lutra*), which have been studied in the USA, Canada, and various European countries (Fig. 4.2). In the 1980s, Stejskal et al. (1989) determined reference ranges for Cu concentrations in the liver and kidney of

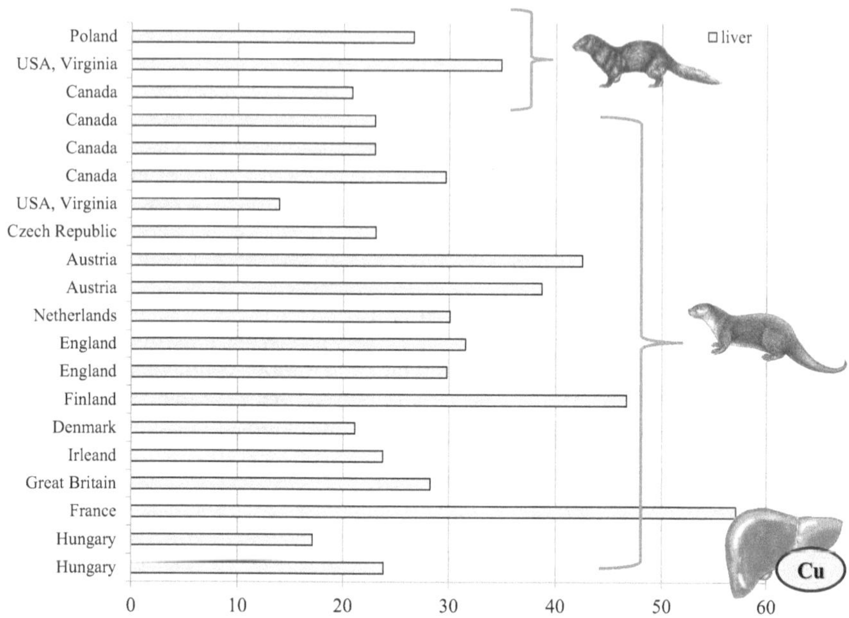


**Fig. 4.2** The concentration of Cu (mg kg$^{-1}$ dw) in livers of American mink and otter from Europe and North America based on data by American minks (Brzezinski et al. 2014; Harding et al. 1998; Ogle et al. 1985; Wren 1984) and otters (Anderson-Bledsoe and Scanlon 1983; Broekhuizen 1987; Gutleb 1992; Gutleb et al. 1998; Lanszki et al. 2009; Lemarchand et al. 2010; Lodenius et al. 2014; Mason and Stephenson 2001)

mink at ~15–250 and ~10–33 mg kg$^{-1}$ dw, respectively. A review of scientific literature shows that the concentrations of Cu in the liver and kidney of North American mink are usually slightly lower than the corresponding specimens in the Central Europe, not exceeding 35 mg kg$^{-1}$ dw (Ogle et al. 1985; Harding et al. 1998; Brzezinski et al. 2014). Sometimes individual specimens of these piscivorous mammals have liver Cu > 190 mg kg$^{-1}$ dw, albeit this value is considered normal for these mink (Ogle et al. 1985; Stejskal et al. 1989). We have found only one European work, Brzezinski et al. (2014), on the concentrations of Cu in feral and ranch minks, observing that liver Cu concentrations in ranch mink were almost two times higher than in feral mink from Polish National Parks, i.e., 42.8 vs. 22.4–28.1 mg kg$^{-1}$ dw, respectively. The authors suggest that this is associated with the high Cu level in feed additives given to ranch mink to maintain normal hair pigmentation (Brzezinski et al. 2014).

Ecotoxicological studies conducted in European countries, including Denmark, Czech Republic, France, Hungary, Finland, Ireland, and Austria, have shown that the concentrations of Cu in the liver of the Eurasian otter, a top predator in aquatic food chains, range from ~23 to 57 mg kg$^{-1}$ dw, while kidney levels do not exceed 20 mg kg$^{-1}$ dw (Gutleb et al. 1998; Mason and Stephenson 2001; Lanszki et al. 2009; Lemarchand et al. 2010; Walker et al. 2010, 2011; Lodenius et al. 2014)

(Fig. 4.2). Kang et al. (2015) showed that hepatic copper concentrations in Eurasian otter (*Lutra lutra*) from South Korea did not exceed 30 mg $kg^{-1}$ dw, similar to levels reported in England, Hungary, and Austria (Gutleb et al. 1998; Lanszki et al. 2009; Walker et al. 2010, 2011), but lower than those reported in Finland and France where hepatic copper levels were >46 mg $kg^{-1}$ dw (Lemarchand et al. 2010; Lodenius et al. 2014) (Fig. 4.2). In North American river otter (*Lontra canadensis*), concentrations of Cu in the liver and kidney appear to be consistent with the results obtained in individuals from Europe, at <35 and 6 mg $kg^{-1}$ dw (Fig. 4.2). There is very little data on the concentration of Cu in muscle tissues in piscivorous species. Lodenius et al. (2014) showed that the concentration of Cu in the muscle of otters from Finland was below 6.5 mg $kg^{-1}$ dw, an order of magnitude lower than those reported for the liver and kidney.

### 5.4.2 Carnivores and Omnivores (Mustelidae)

In scientific literature we found no data on toxic thresholds for Cu in wild predatory mammals. It is known that in the domestic dog (*Canis lupus familiaris*), natural Cu concentrations in the liver are from 200 to 400 mg $kg^{-1}$ dw (Skibniewska et al. 2012). Meanwhile, Puls (1994) had shown that a concentration of 20 mg $kg^{-1}$ ww (~67 mg $kg^{-1}$ dw) could cause potential nephrotoxic effects. In the kidney of wild carnivores and omnivores, Cu concentrations typically range from 11 to 77 mg $kg^{-1}$ dw (Fig. 4.3) and in the muscle from 2 to ~41 mg $kg^{-1}$ dw (Millan et al. 2008; Bilandžić et al. 2012). Most of the ecotoxicological works mentioning concentrations of copper in the tissues and organs of carnivores and omnivores concern European countries, while the few papers mentioning North American predatory mammals were in relation to trans-Arctic mammals such as polar bears (*Ursus maritimus*) and polar foxes (*Alopex lagopus*) (Hoekstra et al. 2003; Rush et al. 2008; Routti et al. 2011).

Mean Cu concentrations in the liver of wild carnivores vary widely among species. In mammalian predators from the northern hemisphere, the average concentration of Cu in the liver ranged from ~5.2 to 230 mg $kg^{-1}$ dw, with only sporadically levels >400 mg $kg^{-1}$ dw reported, e.g., indicating subclinical hepatitis in European red foxes from Spain (between 205 and 950 mg $kg^{-1}$ dw) (Millan et al. 2008). Recent ingestion of contaminated water or food is likely to be responsible for these single cases of increased Cu. The highest obtained Cu concentrations in terrestrial top carnivores and omnivores measured in the liver can be placed in the following descending order: Arctic wolf (*Lupus canis*) > polar bear > stone marten and pine marten (*Martes foina* and *Martens martens*) > Eurasian badger (*Meles meles*) > wolverine (*Gulo gulo*) > brown bear (*Ursus arctos*) > red fox > Iberian lynx (*Lynx pardinus*) > raccoon dog (*Nyctereutes procyonoides*) (Fig. 4.3) (Medvedev 1999; Gamberg and Braune 1999; Kannan et al. 2007; Millan et al. 2008; Bilandžić et al. 2010; Routti et al. 2011; Bilandžić et al. 2012). The highest concentrations of Cu in the liver of the Canadian Arctic wolf ranged from ~157 to 230 mg $kg^{-1}$ dw (Gamberg and Braune 1999). Much lower concentrations of Cu

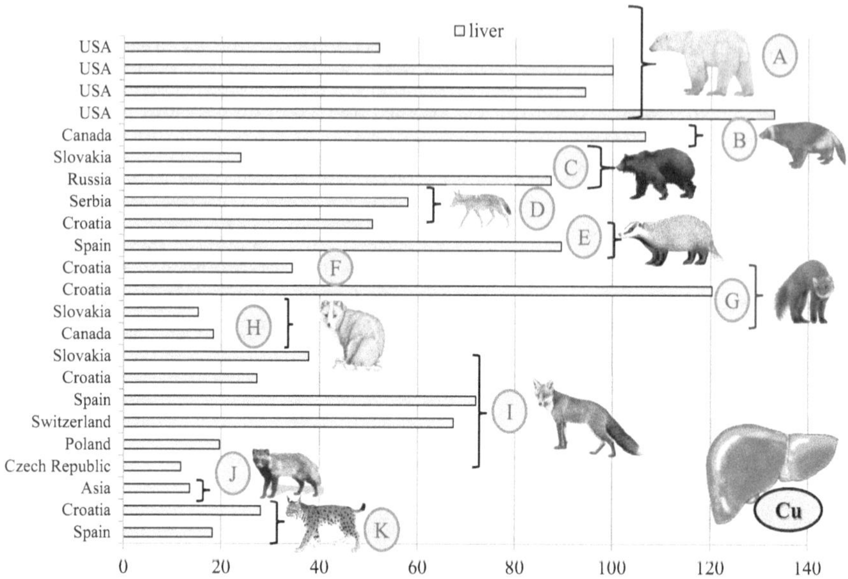


**Fig. 4.3** The concentration of Cu (mg kg$^{-1}$ dw, after conversion from wet weight to dry weight; we assumed that the liver contain 70% of water) in livers of carnivores and omnivores from North America, Europe, and Asia: A. polar bear (Kannan et al. 2007; Routti et al. 2011; Rush et al. 2008; Woshner et al. 2001), B. wolverine (Hoekstra et al. 2003), C. brown bear (Čelechovská et al. 2006; Medvedev 1999), D. golden jackal (Ćirović et al. 2015), E. Eurasian badger (Bilandžić et al. 2012; Millan et al. 2008), F. European pine marten (Bilandžić et al. 2012), G. stone marten (Bilandžić et al. 2010), H. polar fox (Hanusova et al. 2007; Hoekstra et al. 2003), I. red fox (Bilandžić et al. 2010; Dip et al. 2001; Hanusova et al. 2007; Jankovská et al. 2010; Millan et al. 2008), J. raccoon dog (Hou et al. 2017), K. Eurasian lynx and Iberian lynx (Bilandžić et al. 2012; Millan et al. 2008)

were found in individuals of the European gray wolf from Croatia, an order of magnitude smaller at <22 mg kg$^{-1}$ dw (Bilandžić et al. 2012). Hoffmann et al. (2010) suggested that the differences in metal levels (including Cu) between organs are consistent with the variability observed in other studies on metal levels in Arctic and sub-Arctic wolf prey, including the moose (*Alces alces*) and caribou (*Rangifer tarandus*) (Aastrup et al. 2000). Only in some predatory mammals did the average concentration of Cu in the liver exceed 100 mg kg$^{-1}$ dw, including the wolverine, stone marten, and polar bear (Fig. 4.3). Occasionally, in some omnivores from the Mustelidae family, e.g., a stone marten from suburban area in Croatia, concentrations of Cu in the liver exceeded ~120 mg g$^{-1}$ dw (~36.1 mg kg$^{-1}$ ww) albeit not approaching the toxic level for domestic dog (Bilandžić et al. 2010). In polar bears the Cu concentrations were comparable to those reported for other marine mammals, including sea otters (*Enhydra lutris*) from coastal California (Kannan et al. 2007).

In Europe, ecotoxicological studies most commonly use the red fox, which meets a number of the established criteria for bioindication (Dip et al. 2001; Naccari et al. 2013; Binkowski et al. 2016), and studies involving these mammals have been carried out in various countries, e.g., Czech Republic, Slovakia, Switzerland,

Spain, Italy, Croatia, Hungary, and Poland. It has been shown that the concentrations of Cu in the liver, kidney, and muscles of this species ranged from ~12 to 90, 6 to 60, and 9 to 27 mg $kg^{-1}$ dw, respectively (Millan et al. 2008; Bilandžić et al. 2010; Naccari et al. 2013; Binkowski et al. 2016). In North American polar foxes, the average concentration of Cu in the liver did not exceed ~20 mg $kg^{-1}$ dw (Hoekstra et al. 2003). Only in one case of a gray wolf from Canada did the average concentration of Cu in the kidney exceed ~67 mg $kg^{-1}$ dw, a level deemed likely to cause nephrotoxic effects in mammals (Puls 1994; Gamberg and Braune 1999). A concentration of Cu in the kidney >25 mg $kg^{-1}$ dw (~9 mg $kg^{-1}$ ww) was recorded in a Croatian stone marten, higher than the levels observed in the polar and brown bears, red foxes, and badgers from the northern hemisphere (Fig. 4.4).

Little is known about muscle Cu levels in carnivores and omnivores. Concentrations of ~12 mg $kg^{-1}$ dw have been found in Mustelidae and the European badger (Millan et al. 2008; Bilandžić et al. 2010, 2012). In addition, muscle Cu concentrations <9 mg $kg^{-1}$ dw have been reported in canines, such as raccoon dogs, red foxes, polar foxes, the Egyptian mongoose (*Herpestes ichneumon*), and common

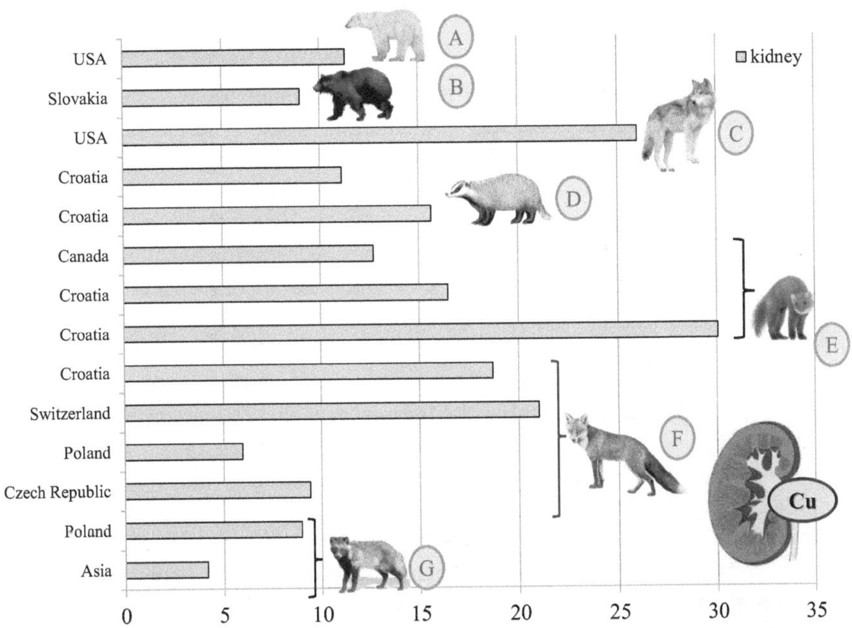


**Fig. 4.4** The concentration of Cu (mg $kg^{-1}$ dw, after conversion from wet weight to dry weight; we assumed that the kidney contain 75% of water) in the kidney of carnivores and omnivores from Europe, North America, and Asia. Based on data: A. polar bear (Woshner et al. 2001); B. brown bear (Čelechovská et al. 2006); C. grey wolf (Hoffmann et al. 2010); D. Eurasian badger (Bilandžić et al. 2012); E. American marten (Harding et al. 1998), pine marten (Bilandžić et al. 2012), stone marten (Bilandžić et al. 2010); F. red fox (Bilandžić et al. 2010; Dip et al. 2001; Jankovská et al. 2010; Kopczewski and Kopczewska 1996); G. raccoon dog (Hou et al. 2012; Mertin et al. 2006)

genet (*Genetta genetta*) from Spain (Mertin et al. 2006; Hanusova et al. 2007; Millan et al. 2008; Bilandžić et al. 2010). The lowest concentration of Cu in the muscle, at <2 mg kg$^{-1}$ dw, has been detected, for example, in felines in southern Europe, including Iberian and Eurasian lynxes (Millan et al. 2008; Bilandžić et al. 2012). However, the greatest concentration of Cu in the muscles, >41 mg kg$^{-1}$ dw, was observed by Horai et al. (2006) in the Javan mongoose (*Herpestes javanicus*) in Japan. Differences in diet are likely to provide the main explanation for resultant differences in Cu levels in the tissue of wild mammals. The mongoose family prefers to eat lizards, snakes, small mammals and birds, as well as invertebrates, mainly spiders associated with a soil habitat (Kalisińska et al. 2009b). Similarly, badgers feed mainly on earthworms and insects, where the known application of agricultural chemicals can lead to higher accumulation of Cu. The lynx showed the lowest levels of Cu in muscles than other carnivores; the diet of this mammal comprises 85%–90% rabbits. Canidae and Mustelidae have more varied feeding habits, including small birds, reptiles, and eggs (Millan et al. 2008).

### 5.4.3 Omnivorous Mammals (Raccoon and Wild Boar)

The raccoon (*Procyon lotor*) is a predator with a diet similar to the wild boar (*Sus scrofa*). Raccoons come from North America and are an invasive species in Europe (Lanocha et al. 2014). In European biomonitoring studies, the extent of environmental pollution with trace elements is assessed in the omnivorous wild boar; however, we have not found any publications about Cu concentration in raccoons from Europe (Amici et al. 2012; Roslewska et al. 2016). Comparing the North American raccoon to the European wild boar, the concentration of Cu in the raccoon liver is greater than in wild boar, at <47 and <26 mg kg$^{-1}$ dw, respectively (Fig. 4.5) (Burger et al. 2000, 2002; Levengood 2001; Zaccaroni et al. 2003; Gasparik et al. 2012; Souza et al. 2013; Hernandez et al. 2017). Sometimes concentrations of Cu in the liver are found an order of magnitude higher, e.g., in wild boar from Italy, at ~154 mg kg$^{-1}$ dw (Fig. 4.5) (Amici et al. 2012). In many areas of Italy, soils are volcanic. Cu that occurs in the soil and vegetables is tightly linked to the volcanic origin, and the high hepatic Cu in the wild boar may likely be due to such high dietary exposure (Amici et al. 2012).

In the case of the kidney, an opposite trend can be observed, with the concentration of Cu in the kidney of the North American raccoon not exceeding 20 mg kg$^{-1}$ dw and in the European wild boar ~30 mg kg$^{-1}$ dw (Fig. 4.5) (Wren 1984; Burger et al. 2000, 2002; Levengood 2001; Długaszek and Kopczynski 2011; Gasparik et al. 2012; Souza et al. 2013). Concentrations of hepatic Cu detected in raccoon from polluted areas in North America (including those exposed to coal ash dust) ranged from ~26 to ~47 mg kg$^{-1}$ dw, and the kidney Cu did not exceed ~17 mg kg$^{-1}$ dw, well below the reference values for the liver and kidneys in unexposed piscivorous minks (<250 and <33 mg kg$^{-1}$ dw, respectively) (Stejskal et al.1989; Burger et al. 2002; Souza et al. 2013). Little is known about the concentrations of Cu in the muscles of North American raccoon, with studies sometimes significantly different

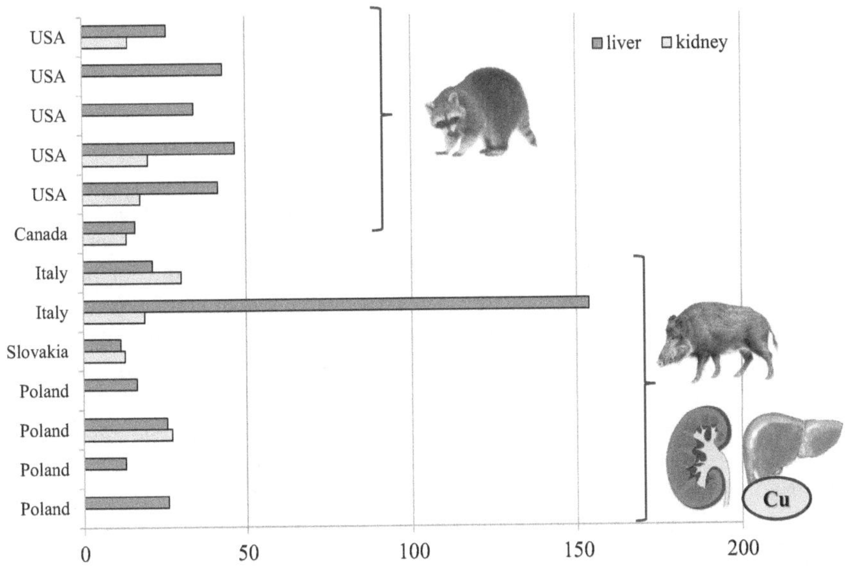


**Fig. 4.5** The concentration of Cu (mg kg$^{-1}$ dw, after conversion from wet weight to dry weight; we assumed that the kidneys contain 80% of water and the liver as well as muscles 70% of water) in the kidney and liver of omnivores from North hemisphere. Based on data: raccoon, North America (Burger et al. 2002; Hernandez et al. 2017; Levengood 2001; Souza et al. 2013); wild boar, Europe (Amici et al. 2012; Bakowska 2014; Burger et al. 2002; Długaszek and Kopczynski 2011; Gasparik et al. 2012; Krynski et al. 1991; Mazurek et al. 1991; Wren 1984; Zaccaroni et al. 2008)

from each other, but in contaminated sites, levels range from >6 to <38 mg kg$^{-1}$ dw (Burger et al. 2002; Souza et al. 2013).

Publications after 1990 on Cu in wild boar have concerned only European countries. Pollock (2005) proposed a biochemical criteria for determining the concentration of Cu in the liver, which indicated the following levels: deficiency (<35 mg kg$^{-1}$ dw), marginal (35–87.5 mg kg$^{-1}$ dw), and optimal (>87.5 mg kg$^{-1}$ dw). The concentration of Cu in the liver of European wild boar ranged from ~3.4 mg kg$^{-1}$ ww (11.4 mg kg$^{-1}$ dw) in Slovakia to ~25.5 mg kg$^{-1}$ dw in Poland (Gasparik et al. 2012; Bakowska 2014). Given the above, we can conclude that wild boars from Poland and Slovakia were deficient in Cu. The kidneys of these mammals had similar values of Cu in the liver, at 3.8 mg kg$^{-1}$ ww (12.7 mg kg$^{-1}$ dw) to ~27.0 mg kg$^{-1}$ dw (Gasparik et al. 2012; Bakowska 2014). The concentration of Cu in European wild boar muscle, especially in the areas of Cu-rich soil, appears to be greater than has been reported for the North American raccoon, ranging between ~1.6 mg kg$^{-1}$ ww (5.5 mg kg$^{-1}$ dw) in Hungary, ~7.5 mg kg$^{-1}$ ww (25 mg kg$^{-1}$ dw) in Poland, and ~12.2 mg kg$^{-1}$ ww (40.7 mg kg$^{-1}$ dw) in Italy (Skobrak et al. 2010; Amici et al. 2012; Roslewska et al. 2016). The greatest levels (~90 mg kg$^{-1}$ dw) were recorded in the muscle of wild boar in Italy, although those high values do not appear to cause particular concern when compared with exposure in humans (Gupta and Gupta 1998; Amici et al. 2012).

### 5.4.4 Herbivorous Mammals

Differences in the concentrations of the hepatic, kidney, and muscle Cu in between examined wild mammals likely result from the varied diets. Herbivorous animals take in fiber from vegetable food, a component consisting of poly- and oligosaccharides, which increase the absorption and retention of Cu and other elements. In addition, the absorption and metabolism of Cu are significantly affected by elements such as sulfur (S) and molybdenum (Mo). The presence of S and Mo contribute to Cu deficiency, with lower than normal concentrations of these elements in the diet of animals conducive to the bioaccumulation of Cu. The antagonistic relationship between Cu, Mo, and S in ruminants lowers Cu bioavailability and has a negative impact on Cu absorption (Kabata-Pendias and Szteke 2015). In red deer and roe deer (*Capreolus capreolus*) as well as other herbivorous species including moose and European hare (*Lepus europaeus*), the highest concentrations of Cu are recorded in the liver (Gasparik et al. 2004; Reglero et al. 2008; Jarzyńska and Falandysz 2011; Skibniewski et al. 2016). In a study by Ivan (1993), the normal level of Cu in the liver of ruminants ranges from 100 to 400 mg kg$^{-1}$ dw, and a value <25 mg kg$^{-1}$ dw indicates a possible Cu deficiency. It seems that the capacity of the liver to accumulate Cu differs between ruminants. Lazarus et al. (2008) showed that hepatic copper levels in Croatian deer ranged from ~1.6 to 54 mg kg$^{-1}$ ww (~5.3–180 mg kg$^{-1}$ dw). Concentrations of Cu in the liver were comparable with levels reported for red deer in Spain (51.7 mg kg$^{-1}$ dw), Slovak Republic (44.3 mg kg$^{-1}$ dw), Poland (59 mg kg$^{-1}$ dw), and Norway (86.7 mg kg$^{-1}$ dw) (Gasparik et al. 2004; Vikøren et al. 2005; Reglero et al. 2008; Jarzyńska and Falandysz 2011). At the same time, in some specimens from Poland and Germany (areas with copper-deficient soils), the concentration of Cu in the liver was ~3.0 mg kg$^{-1}$ ww (10 mg kg$^{-1}$ dw), which may suggest a deficiency of this element; these low levels imply that red deer have the ability to adapt to low nutritional levels of Cu (Hecht 1996; Skibniewski et al. 2015). In addition, a possible Cu deficiency has been reported in free-ranging moose from Alaska, Northwest Minnesota, and Sweden. Ecotoxicological studies on herbivores were carried out by Vikøren et al. (2011) in Norway, who observed that moose had a significantly higher Cu concentration in the liver (222 mg kg$^{-1}$ dw) than roe deer (112 mg kg$^{-1}$ dw) and reindeer (105 mg kg$^{-1}$ dw). Moreover, the Cu status of moose and roe deer in Norway are among the highest reported in Europe. Cu content in the liver of Alaskan elks reported by Gamberg et al. (2005) was definitely higher than that observed in Poland: ~43.5 vs. 23.1 mg kg$^{-1}$ ww, respectively (145.1 and 76.9 mg kg$^{-1}$ dw) (Skibniewski et al. 2016). Importantly, red deer from European countries did not seem to accumulate and retain Cu amounts that would cause a harmful effect. Much higher mean concentrations of Cu in the liver have been recorded in some North American white-tailed deer, at ~27.4–122 mg kg$^{-1}$ ww (91.3–406.7 mg kg$^{-1}$ dw); still, these concentrations did not exceed ~150 mg kg$^{-1}$

ww (500 mg kg$^{-1}$ dw), i.e., a level associated with Cu poisoning (Sleeman et al. 2010). In the kidneys of Canadian porcupine caribou (*Rangifer tarandus granti*), red deer from Croatia and Slovakia, and in moose from Canada and Russia, concentrations of Cu did not exceed ~19.0 mg kg$^{-1}$ dw (Gasparik et al. 2004; Gamberg et al. 2005; Lazarus et al. 2005, 2008). Data obtained from various studies may sometimes differ, i.e., in the case of porcupine caribou from Canada and red deer from Poland, with the concentrations of Cu in the kidney in both studies >66.0 mg kg$^{-1}$ dw (Jarzyńska and Falandysz 2011). The concentration of Cu in the muscles of herbivores usually does not exceed ~12 mg kg$^{-1}$ dw, e.g., in red deer from Poland (~10 mg kg$^{-1}$ dw), Croatia (~12 mg kg$^{-1}$ dw), and Slovakia (~8 mg kg$^{-1}$ dw), in roe deer from Poland (~11 mg kg$^{-1}$ dw), and moose from Russia (~13 mg kg$^{-1}$ dw) (Medvedev 1999; Karpinski 1999; Gasparik et al. 2004; Lazarus et al. 2008; Skibniewski et al. 2015).

### 5.4.5 Copper Levels in Relation to Age and Sex of Mammals

In addition to diet, trophic level, and environmental pollution, the level of Cu bioaccumulation depends on biological factors including age and sex. The results of research on the relationship between Cu content and the age and gender of wild mammals are in many cases markedly varied and difficult to interpret. Among predatory mammals, the relationship between the concentration of Cu in the liver and kidney compared to age has been analyzed in the mink, otter, red fox, polar fox, gray wolf, and polar bear, and no significant relationship has been found between these levels (Ogle et al. 1985; Dip et al. 2001; Kannan et al. 2007; Hoffmann et al. 2010; Kang et al. 2015; Binkowski et al. 2016). In contrast to the abovementioned examples, Skobrak et al. (2010) observed significant age-related differences in the concentration of Cu in the muscle of wild boar from Hungary. In ungulates, Vikøren et al. (2011) observed that Cu concentration increased significantly with age for moose, but no age relationship was found for reindeer. In predatory mammals, e.g., mink from North America, otter, red fox, polar fox, polar bear, wolverine, golden jackal, and gray wolf, no significant differences were found between females and males and the concentration of Cu (Ogle et al. 1985; Hoekstra et al. 2003; Kannan et al. 2007; Rush et al. 2008; Hoffmann et al. 2010; Ćirović et al. 2015; Kang et al. 2015; Binkowski et al. 2016). In contrast to the cited papers, Brzezinski et al. (2014) detected a statistically significant difference in the concentration of Cu in the kidney between male and female mink, similar to Mertin et al. (2006), who observed such a relationship in raccoon dog muscles. In addition, differences in the concentrations of Cu between sexes have been observed in ungulates, for example, in wild boar, but in the case of red deer from Norway, such gender-related differences have not been observed (Roslewska et al. 2016).

## 6 Biomarkers of Cu in Ecotoxicological Studies

Good biomarkers of Cu levels in humans and domestic animals seem to be found in serum, urine, hair, and the liver. Little is known about wild animals, e.g., due to the difficulty in obtaining biological samples (ATSDR 2004). Some researchers suggest that liver analysis is the gold standard for evaluating Cu reserves in ungulates, but liver samples can only be collected from dead animals. For live ruminants, serum ceruloplasmin can be used as a nonlethal biomarker of liver Cu in species such as muskoxen, sheep, and cattle, but it must be validated by age, sex, and species (Barboza and Blake 2001). Burger et al. (2000) indicated that sometimes the induction of metallothionein levels in raccoon tissues may be a more rapid and cost-effective method for screening metals, including Cu.

## 7 Conclusions

1. Proper interpretation of results requires knowledge of physiological concentrations of Cu in the tissues and organs, values reflecting the geochemical background, and consideration of the specificity of the animal species and diet.
2. Characteristics of useful bioindicators of terrestrial environmental pollution with Cu can be found among birds (birds associated with wetland areas, e.g., herbivores such as the mute swan, ducks from the genera *Anas* and *Aythya*, including the mallard, shoveler, pochard) and terrestrial passerines (house sparrow, great tit, and blue tit). Birds of prey are important for biomonitoring studies, as are rare species threatened with extinction and protected by law. Good bioindicators include the common buzzard, Eurasian kestrel, Eurasian eagle-owl, tawny owl, northern goshawk, peregrine falcon, white-tailed eagle, and bald eagle. In ecotoxicological studies on terrestrial mammals, a measurable response to pollution with Cu is exhibited by canids common in natural and seminatural habitats, e.g., the Arctic fox, red fox, raccoon, American mink, otters, and ungulates, such as the wild boar and red deer. It should be noted that in Europe, there is now the possibility of testing the concentration of Cu in increasingly popular alien species, e.g., mink, raccoon, and raccoon dog, which allows wider intercontinental comparative studies. In addition to information on common terrestrial species, there is a widely felt need for collection of comparative ecotoxicological data on Cu concentrations in other less abundant species (including mongoose, marten, and badger).
3. In biomonitoring studies on wild birds, it is advisable to perform noninvasive sampling of eggs and feathers and of hair from mammals. Internal tissues and organs (mainly liver, kidney, muscles, and bones for analyses of long-term impact) may be collected from dead specimens (hunting and/or roadkills). The deliberate killing of animals for research should be avoided for ethical reasons. In

field studies on wild mammals, Cu is analyzed in the liver, kidneys, muscles and much less frequently in the brain, bones, and hair.

4. The ecotoxicological effects of nano-Cu may be expected to be closely linked with aquatic environments and affect species inhabiting water ecosystems, mainly piscivores. Moreover, in the immediate future, the continued and increased reliance on coal combustion for electricity and heat production will result in an increased diffuse contamination of soils and waters with various elements, including Cu.
5. No efficient and good biomarkers for Cu exposure have been found, and still the best solution to evaluate the exposure is to measure Cu concentrations in wildlife and the environment.

## References


Aastrup P, Riget F, Dietz R, Asmund G (2000) Lead, zinc, cadmium, mercury, selenium and copper in Greenland caribou and reindeer (*Rangifer tarandus*). Sci Total Environ 245:149–159

Albayrak T, Mor F (2011) Comparative tissue distribution of heavy metals in house sparrow (*Passer domesticus*, Aves) in polluted and reference sites in Turkey. Bull Environ Contam Toxicol 87:457–462

AMAP (1998) Assessment report: arctic pollution issues, 859 p

AMAP Assessment (2002) Persistent organic pollutants in the Arctic. AMAP, Oslo, Norway, 112 pp

AMAP/UNEP (2008) Technical background report to the global atmospheric mercury assessment. Arctic Monitoring and Assessment Programme/UNEP Chemicals Branch, 159 pp

AMAP/UNEP (2013) Technical background report for the global mercury assessment 2013. Arctic Monitoring and Assessment Programme, Oslo, Norway/UNEP Chemicals Branch, Geneva, Switzerland, 263 pp

AMEC (2014) Environment & Infrastructure UK Limited in partnership with Bio Intelligence Service, Milieu, IEEP and REC. Doc Reg No. 32790-01 FR 13298i5, 293 pp

Amici A, Danieli PP, Russo C, Primi R, Ronchi B (2012) Concentrations of some toxic and trace elements in wild boar (*Sus scrofa*) organs and tissues in different areas of the Province of Viterbo, Central Italy. Ital J Anim Sci 11(e65):354–361

Anderson-Bledsoe KL, Scanlon PF (1983) Heavy metal concentrations in tissues of Virginia river otter. Bull Environ Contam Toxicol 30:442–447

Angelova M, Asenova S, Nedkova V, Koleva-Kolarov R (2011) Copper in the human organism. Trakia J Sci 9:88–98

Arnhold W, Anke M, Goebel S (2002) The copper, zinc and manganese status in opossum and gray fox. Z Jagdwiss 48:77–86

ATSDR (2004) Toxicological profile for copper. Agency for Toxic Substances and Disease Registry, U.S department of Health and Human Services, Public Health Services, Atlanta, GA

Bakowska M (2014) The concentration Analysis of toxic metals (Pb, Cd) and bioelements (Cu, Zn) in the liver and kidneys of free-living animals (roe deer, red deer, wild boar) from the area of Poland (In Polish), Doctoral dissertation

Barboza PS, Blake JE (2001) Ceruloplasmin as an indicator of copper reserves in wild ruminants at high latitudes. J Wildl Dis 37:324–331

Barjaktarovic L, Elliott JE, Scheuhammer AM (2002) Metal and metallothionein concentrations in Scoter (*Melanitta* spp.) from the Pacific northwest of Canada, 1989-1994. Arch Environ Contam Toxicol 43:486–491

Belskii E, Belskaya E (2013) Diet composition as a cause of different contaminant exposure in two sympatric passerines in the Middle Urals, Russia. Ecotoxicol Environ Saf 97:67–72
Berglund ÅM, Rainio MJ, Eeva T (2015) Temporal trends in metal pollution: using bird excrement as indicator. PLoS One 10:e0117071
Beyer WN, Franson JC, Locke LN, Stroud RK, Sileo L (1998) Retrospective study of the diagnostic criteria in a lead-poisoning survey of waterfowl. Arch Environ Contam Toxicol 35:506–512
Bharti R, Wadhwani KK, Tikku AP, Chandra A (2010) Dental amalgam: an update. J Conserv Dent 13:204–208
Bilandžić N, Deždek D, Sedak M, Dokić M, Solomun B, Verenina I et al (2010) Concentrations of trace elements in tissues of red fox (*Vulpes vulpes*) and stone marten (*Martes foina*) from suburban and rural areas in Croatia. Bull Environ Contam Toxicol 85:486–491
Bilandžić N, Deždek D, Sedak M, Dokić M, Simic B, Rudan N et al (2012) Trace elements in tissues of wild carnivores and omnivores in Croatia. Bull Environ Contam Toxicol 88:94–99
Binkowski LJ, Meissner W (2013) Levels of metals in blood samples from Mallards (*Anas platyrhynchos*) from urban areas in Poland. Environ Pollut 178:336–342
Binkowski ŁJ, Stawarz RM, Zakrzewski M (2013) Concentrations of cadmium, copper and zinc in tissues of mallard and coot from southern Poland. J Environ Sci Health Part B 48:410–415
Binkowski ŁJ, Merta D, Przystupińska A, Sołtysiak Z, Pacoń J, Stawarz R (2016) Levels of metals in kidney, liver and muscle tissue and their relation to the occurrence of parasites in the red fox in the Lower Silesian Forest in Europe. Chemosphere 149:161–167
Bojar H, Bojar I (2009) Monitoring of contamination of the Lublin region wetlands using mallards (*Anas platyrhynchos*) as a vector of the contamination by various conditionally toxic elements. Ann Anim Sci 9:195–204
Boone C, Jervais G, Luukinen B, Buhl K, Stone D (2012) Copper sulfate technical fact sheet. National Pesticide Information Center, Oregon State University Extension Services, Corvallis, OR
British Geological Survey (2013) World mineral production 2007-2011. British Geological Survey, Nottingham, 78 pp
Broekhuizen S (1987) First data on contamination of otters in the Netherlands. IUCN Otter Spec Group Bull 2:27–32
Brun LA, Maillet J, Hinsinger P, Pepin M (2001) Evaluation of copper availability to plants in copper-contaminated vineyard soils. Environ Pollut 11:293–302
Brzezinski M, Zalewski A, Niemczynowicz A, Jarzyna I, Suska-Maławska M (2014) The use of chemical markers for the identification of farm escapees in feral mink populations. Ecotoxicology 23:767–778
Buck WB, Sharma RM (1969) Copper toxicity in sheep. Iowa State Univ Vet 31:1
Burger J, Lord CG, Yurkow EJ, McGrath L, Gaines KF, Brisbin IL Jr et al (2000) Metals and metallothionein in the liver of raccoons: utility for environmental assessment and monitoring. J Toxicol Environ Health Part A 60:243-261
Burger J, Gaines KF, Lord CG, Brisbin IL Jr, Shukla S, Gochfeld M (2002) Metal levels in raccoon tissues: differences on and off the Department of Energy's Savannah River Site in South Carolina. Environ Monit Assess 74:67-84
Carneiro M, Colaço B, Colaço J, Faustino-Rocha AI, Colaço A, Lavin S et al (2016) Biomonitoring of metals and metalloids with raptors from Portugal and Spain: a review. Environ Rev 24:63–83
CCME (1999) Canadian Council of Ministers of the Environment. Canadian soil quality guidelines for the protection of environmental and human health: Copper 1999. In: Canadian environmental quality guidelines, Canadian Council of Ministers of the Environment, Winnipeg
Čelechovská O, Literák I, Ondruš S, Pospíšil Z (2006) Heavy metals in brown bears from the central European Carpathians. Acta Vet Brno 75:501–506
Chang AC, Page AL (2000) Trace elements slowly accumulating, depleting in soils. Calif Agric 54:49–55

Ćirović D, Gizejewska A, Jovanović V, Penezić A, Milenković M, Vujošević M et al (2015) Concentration of selected trace elements in the golden jackal (*Canis aureus* L., 1758) population from Serbia. Acta Zool Bulg 67:409–414

Clausen B, Wolstrup C (1978) Copper load in mute swans (*Cygnus olor*) found in Denmark. Nord Vet Med 30:260–266

Cornu JY, Huguenot D, Jézéquel K, Lollier M, Lebeau T (2017) Bioremediation of copper-contaminated soils by bacteria. World J Microbiol Biotechnol 33:26

Costa RA, Eeva T, Eira C, Vaqueiro J, Vingada JV (2013) Assessing heavy metal pollution using great tits (*Parus major*): feathers and excrements from nestlings and adults. Environ Monit Assess 185:5339–5344

Cotton FA, Wilkinson G (1980) Copper. Advanced inorganic chemistry. John Wiley & Sons, New York, pp 798–821

Croubels S, Baert K, Torck T, Deprez P, De Backer P (2001) Chronic copper intoxication in veal calves. Vlaam Diergeneeskund Tijdschr 70:142–146

Dauwe T, Bervoets L, Blust R, Eens M (2002) Tissue levels of lead in experimentally exposed zebra finches (*Taeniopygia guttata*) with particular attention on the use of feathers as biomonitors. Arch Environ Contam Toxicol 42:88–92

Dip R, Stieger C, Deplazes P, Hegglin D, Müller U, Dafflon O et al (2001) Comparison of heavy metal concentrations in tissues of red foxes from adjacent urban, suburban, and rural areas. Arch Environ Contam Toxicol 40:551–556

Długaszek M, Kopczynski K (2011) Comparative analysis of liver mineral status of wildlife. Probl Hig Epidemiol 9:859–863

Dutch Pollutant Standards (2011) Soils and ground water criteria used in the Netherlands for contaminated land

ECI (2014) European Copper Institute, Annual Report 2014, 20 pp

EEA (2014) European Union emission inventory report 1990–2012 under the UNECE Convention on Long-range Transboundary Air Pollution (LRTAP), 130 pp

Eeva T, Ryömä M, Riihimäki J (2005) Pollution-related changes in diets of two insectivorous passerines. Oecologia 145:629–639

Eisler R (1998) Copper hazards to fish, wildlife, and invertebrates: a synoptic review. In: Biological Report No. 33, US Fish and Wildlife Service, Washington, DC

Ek KH, Morrison GM, Lindberg P, Rauch S (2004) Comparative distribution of metals in birds in Sweden using ICP-MS and laser ablation ICP-MS. Arch Environ Contam Toxicol 47:259–269

EPA (2007) United States Environmental Protection Agency, Aquatic Life Ambient Freshwater Quality Criteria-Copper 2007 Revision- EPA 822-R-07-001

Esselink H, van der Geld FM, Jager LP, Posthuma-Trumpie GA, Zoun PE, Baars AJ (1995) Biomonitoring heavy metals using the barn owl (*Tyto alba guttata*): sources of variation especially relating to body condition. Arch Environ Contam Toxicol 28:471–486

Falandysz J, Szefer P (1983) Metals and organochlorines in a specimen of white-tailed eagle. Environ Conser 10:256–258

Falandysz J, Jakuczun B, Mizear T (1988) Metals and organochlorines in four female white-tailed eagles. Mar Pollut Bull 19:521–526

Falandysz J, Ichihashi H, Szymczyk K, Yamasaki S, Mizera T (2001) Metallic elements and metal poisoning among white-tailed sea eagles from the Baltic South Coast. Mar Pollut Bull 42:1190–1193

Figuerola J, Green AJ (2000) The evolution of sexual dimorphism in relation to mating patterns, cavity nesting, insularity and sympatry in the Anseriformes. Funct Ecol 14:701–710

Frank A (1998) 'Mysterious' moose disease in Sweden. Similarities to copper deficiency and/or molybdenosis in cattle and sheep. Biochemical background of clinical signs and organ lesions. Sci Total Environ 209:17–26

Frank A, Borg K (1979) Heavy metals in tissues of the mute swan (*Cygnus olor*). Acta Vet Scand 20:447–465

Fuentealba IC, Aburto EM (2003) Animal models of copper-associated liver disease. Comp Hepatol 2:5

Gamberg M, Braune BM (1999) Contaminant residue levels in arctic wolves (*Canis lupus*) from the Yukon Territory, Canada. Sci Total Environ 243–244:329–338

Gamberg M, Braune B, Davey E, Elkin B, Hoekstra PF, Kennedy D et al (2005) Spatial and temporal trends of contaminants in terrestrial biota from the Canadian Arctic. Sci Total Environ 351–352:148–164

Gasparik J, Massányi P, Slamecka J, Fabis M, Jurcik R (2004) Concentration of selected metals in liver, kidney, and muscle of the red deer (*Cervus elaphus*). J Environ Sci Health Part A Tox Hazard Subst Environ Eng 39:2105–2111

Gasparik J, Dobias M, Capcarova M, Smehyl P, Slamecka J, Bujko J et al (2012) Concentration of cadmium, mercury, zinc, copper and cobalt in the tissues of wild boar (*Sus scrofa*) hunted in the western Slovakia. J Environ Sci Health Part A 47:1212–1216

Georgopoulos G, Roy A, Yonone-Lioy MJ, Opiekun RE, Lioy PJ (2001) Environmental copper: its dynamics and human exposure issues. J Toxicol Environ Health Part B 4:341–394

Gilbert RW, Sander JE, Brown TP (1996) Copper sulfate toxicosis in commercial laying hens. Avian Dis 40:236–239

Glöser S, Soulier M, Tercero Espinoza LA (2013) Dynamic analysis of global copper flows. Global stocks, postconsumer material flows, recycling indicators, and uncertainty evaluation. Environ Sci Technol 18:6564–6572

Gochfeld M, Burger J (1987) Heavy metal concentrations in the liver of three duck species: influence of species and sex. Environ Pollut 45:1–15

Gomez G, Baos R, Gomara B, Jimenez B, Benito V, Montoro R et al (2004) Influence of a mine tailing accident near Donana National Park (Spain) on heavy metals and arsenic accumulation in 14 species of waterfowl (1998 to 2000). Arch Environ Contam Toxicol 47:521–529

Gong Q, Jin Z, Zou H (2012) Concentrations of copper, zinc and manganese in Tree Sparrow (*Passer montanus*) at Jixi, Heilongjiang Province, China. Can J For Res 23:319

Grobler DG, Swan GE (1999) Copper poisoning in the Kruger National Park: field investigation in wild ruminants. Onderstepoort J Vet Res 66:157–168

Gupta UC, Gupta SC (1998) Trace element toxicity relationships to crop production and livestock and human health: implication for management. Commun Soil Sci Plan 29:1491–1522

Gutleb AC (1992) The Otter in Austria: A Review on the Current State of Research. IUCN Otter Spec Group Bull 7:4–9

Gutleb AC, Kranz A, Nechay G, Toman A (1998) Heavy metal concentrations in livers and kidneys of the otter (*Lutra lutra*) from central Europe. Bull Environ Contam Toxicol 60:273–279

Hadrich J (1996) High amounts of copper in calf livers. Recent data and estimation of potential health hazards. Dtsch Lebensmitt Rundsch 92:103–113

Han SY, Son SW, Ando M, Sasaki H (2002) Heavy metals and PCBs in Eurasian otters (Lutra lutra) in South Korea. Proceedings VIIth International Otter Colloquium, 103–109

Handeland K, Viljugrein H, Lierhagen S, Opland M, Tarpai A, Vikøren T (2017) Low copper levels associated with low Carcass Weight in Wild Red Deer (*Cervus elaphus*) in Norway. J Wildl Dis 53:176–180

Hanusova E, Mertin D, Suvegova K, Szeleszczuk O (2007) Comparison of content of mineral elements in selected organs in carnivorous fur animals. Trace Elem Electrolytes 24:12–18

Harding L, Harris M, Elliott J (1998) Heavy and trace metals in wild mink (*Mustela vison*) and river otter (*Lontra canadensis*) captured on rivers receiving metals discharges. Bull Environ Contam Toxicol 61:600–607

Hecht H (1996) Kupfer in Muskeln und Lebern von Kalbern und anderen Saugetieren. Fleischuntersuchung 76:492–494

Heijerick DG, Van Sprang PA, Van Hyfte AD (2006) Ambient copper concentrations in agricultural and natural European soils: an overview. Environ Toxicol Chem 25:858–864

Henderson BM, Winterfield RW (1975) Acute copper toxicosis in the Canada goose. Avian Dis 19:385–387

Hernandez LM, Gomara B, Fernandez M, Jimenez B, Gonzalez MJ, Baos R et al (1999) Accumulation of heavy metals and as in wetland birds in the area around Donana National Park affected by the Aznalcollar toxic spill. Sci Total Environ 242:293–308

Hernandez F, Oldenkamp RE, Webster S, Beasley JC, Farina LL, Wisely SM (2017) Raccoons (*Procyon lotor*) as sentinels of trace element contamination and physiological effects of exposure to coal fly ash. Archiv Environ Contam Toxicol 72:235–246

Hinck JE, Schmitt CJ, Chojnacki K, Tillitt DE (2009) Environmental contaminants in freshwater fish and their risk to piscivorous wildlife based on a National Monitoring Program. Environ Monit Assess 152:469–494

Hladun KR, Parker DR, Trumble JT (2015) Cadmium, copper, and lead accumulation and bioconcentration in the vegetative and reproductive organs of *Raphanus sativus*: implications for plant performance and pollination. J Chem Ecol 41:386–395

Hoekstra PF, Braune BM, Elkin B, Armstrong FAJ, Muir DCG (2003) Concentrations of selected essential and non-essential elements in arctic fox (*Alopex lagopus*) and wolverines (*Gulo gulo*) from the Canadian Arctic. Sci Total Environ 309:81–92

Hoffmann SR, Blunck SA, Petersen KN, Jones EM, Koval JC, Misek R et al (2010) Cadmium, copper, iron, and zinc concentrations in kidneys of grey wolves, *Canis lupus*, from Alaska, Idaho, Montana (USA) and the Northwest Territories (Canada). Bull Environ Contam Toxicol 85:481–485

Honda K, Min BY, Tatsukawa R (1986) Distribution of heavy metals and their age-related changes in the eastern great white egret, *Egretta alba modesta*, in Korea. Arch Environ Contam Toxicol 15:185–197

Honda R, Tsuritani I, Ishizaki M, Yamada Y (1997) Zinc and copper levels in ribs of cadmium-exposed persons with special reference to osteomalacia. Environ Res 75:41–48

Hontelez LCMP, Van den Dungen HM, Baars AJ (1992) Lead and cadmium in birds in the Netherlands: a preliminary survey. Arch Environ Contam Toxicol 23:453

Horai S, Minagawa M, Ozaki H, Watanabe I, Takeda Y, Yamada K et al (2006) Accumulation of Hg and other heavy metals in the Javan mongoose (*Herpestes javanicus*) captured on Amami oshima Islands in Japan. Chemosphere 65:657–665

Horai S, Watanabe I, Takada H, Iwamizu Y, Hayashi T, Tanabe S et al (2007) Trace element accumulations in 13 avian species collected from the Kanto area, Japan. Sci Total Environ 373:512–525

Hou Z, Shen M, Chai H, Ma J, Hua Y (2012) Iron, copper, zinc and selenium concentrations and their interaction in organs of white nose disease of raccoon dogs. Asian J Anim Vet Adv 7:1110–1119

Hou J, Wang X, Hayat T, Wang X (2017) Ecotoxicological effects and mechanism of CuO nanoparticles to individual organisms. Environ Pollut 221:209–217

ICSG (2014) International Copper Study Group, The World Copper FactBook, 59 pp

IPCS (1998) International Programme on Chemical Safety, Environmental Health Criteria 200. Copper. World Health Organization, Geneva

Isanhart JP, Wu H, Pandher K, MacRae RK, Cox SB, Hooper MJ (2011) Behavioral, clinical, and pathological characterization of acid metalliferous water toxicity in mallards. Arch Environ Contam Toxicol 61:653–667

ISGS (2011) International Copper Study Group "Release of ICSG 2011 Statistical Yearbook," press release, 7 Sep 2011

Ivan M (1993) How dietary copper affects ruminants. Communications Branch (red) Agriculture Canada, Ottawa, ON

Jackson N, Stevenson MH, Kirkpatrick GM (1979) Effects of the protracted feeding of copper sulphate-supplemented diets to laying, domestic fowl on egg production and on specific tissues, with special reference to mineral content. Br J Nutr 42:253–266

Jager FV, Rijnierse AJ, Esselink H, Baars AJ (1996) Biomonitoring with the Buzzard *Buteo buteo* in the Netherlands: heavy metals and sources of variation. J Ornithol 137:295–318

Jankovská I, Miholová D, Bejček V, Vadlejch J, Šulc M, Száková J et al (2010) Influence of parasitism on trace element contents in tissues of red fox (*Vulpes vulpes*) and its parasites *Mesocestoides* spp. (Cestoda) and *Toxascaris leonina* (Nematoda). Arch Environ Contam Toxicol 58:469–477

Jarzyńska G, Falandysz J (2011) Selenium and 17 other largely essential and toxic metals in muscle and organ meats of red deer (*Cervus elaphus*)-consequences to human health. Environ Int 37:882–888

Kabata-Pendias A (2011) Trace elements in soils and plants. CRC Press/Taylor & Francis Group, Boca Raton, FL, 548 pp

Kabata-Pendias A, Szteke B (2015) Trace elements in abiotic and biotic environments. CRC Press, Boca Raton, FL, 468 pp

Kalisińska E, Salicki W, Mysek P, Kavetska KM, Jackowski A (2004) Using the Mallard to biomonitor heavy metal contamination of wetlands in northwestern Poland. Sci Total Environ 320:145–161

Kalisińska E, Salicki W, Jackowski A (2006) Six trace metals in white-tailed eagle from northwestern Poland. Pol J Environ Stud 15:727–737

Kalisińska E, Salicki W, Kavetska KM, Ligocki M (2007) Trace metal concentrations are higher in cartilage than in bones of scaup and pochard wintering in Poland. Sci Total Environ 388:90–103

Kalisińska E, Lisowski P, Czernomysy-Furowicz D, Kavetska KM (2008) Serratospiculiasis, mycosis, and haemosiderosis in wild peregrine falcon from Poland. A case report. Bull Vet Inst Pulawy 52:75–79

Kalisińska E, Lanocha N, Budis H, Wilk A, Kavetska K, Krolaczyk K (2009a) Essential trace elements in the liver and kidneys of the common Buzzard *Buleo buteo* and the kestrel *Falco tinnunculus*. In: Wiącek J, Polak M, Kucharczyk M, Grzywaczewski G, Jerzak L (eds) Birds – Environment – Threats – Protection: selected aspects of the ecology of birds. Lub Tow Ornitol, Lublin, pp 215–226 (in Polish)

Kalisińska E, Lisowski P, Salicki W, Kucharska T, Kavetska K (2009b) Mercury in wild terrestrial carnivorous mammals from north-western Poland and unusual fish diet of red fox. Acta Theriol 54:345–356

Kang S, Kang JH, Kim S, Lee SH, Lee S, Yu HJ et al (2015) Trace element analysis of three tissues from Eurasian otters (*Lutra lutra*) in South Korea. Ecotoxicology 24:1064–1072

Kannan K, Agusa T, Evans TJ, Tanabe S (2007) Trace element concentrations in livers of polar bears from two populations in Northern and Western Alaska. Arch Environ Contam Toxicol 53:473–482

Karpinski M (1999) Concentration of selected macro and microelements in the tissues of red deer (*Cervus elaphus*) and roe deer (*Capreolus capreolus*) from the central-eastern region of Poland II. Ann UMSC 40:311–316

Kekkonen J, Hanski IK, Väisänen RA, Brommer JE (2012) Levels of heavy metals in House sparrows (*Passer domesticus*) from urban and rural habitats of southern Finland. Ornis Fennica 89:91–98

Kim J, Oh JM (2012) Biological monitoring of heavy metal contaminations using owls. J Environ Monit 14:1091–1097

Kim J, Oh JM (2015) Comparison of trace element concentrations in grey heron and black-crowned night heron chicks. Environ Monit Assess 187:4124

King KA, Leleux J, Mulhern BM (1984) Molybdenum and copper levels in white-tailed deer near uranium mines in Texas. J Wildl Manag 48:267–270

Komosa A, Kitkowski I, Komosa Z (2012) Essential trace (Zn, Cu, Mn) and toxic (Cd, Pb, Cr) elements in the liver of birds from eastern Poland. Acta Vet 62:579–589

Kopczewski A, Kopczewska T (1996) Iron, copper and manganese content in the liver and kidneys of foxes. Med Wet 52:524–525

Kovačič G, Lešnik RM, Vršič S (2013) An overview of the copper situation and usage in viticulture. Bulg J Agric Sci 19:50–55

Krynski A, Zarski TP, Rokicki E (1991) Free living animals as bioindicators of environmental pollution. 7th International Congress on Animal Hygiene, vol 1, Leipzig, pp 325–330
Lanocha N, Kalisińska E, Kosik-Bogacka DI, Budis H, Noga-Deren K (2012) Trace metals and micronutrients in bone tissues of the red fox *Vulpes vulpes* (L., 1758). Acta Theriol 57:233–244
Lanocha N, Kalisińska E, Kosik-Bogacka DI, Budis H, Podlasinska J, Jedrzejewska E (2014) Mercury levels in raccoons (*Procyon lotor*) from the Warta Mouth National Park, north-western Poland. Biol Trace Elem Res 159:152–160
Lanszki J, Orosz E, Sugar L (2009) Metal levels in tissues of Eurasian otters (*Lutra lutra*) from Hungary: variation with sex, age, condition and location. Chemosphere 74:741–743
Lazarus M, Vicković I, Sostarić B, Blanusai M (2005) Heavy metal levels in tissues of red deer (*Cervus elaphus*) from Eastern Croatia. Arh Hig Rada Toksikol 56:233–240
Lazarus M, Orct T, Blanuŝ a M, Vicković I, Ŝoŝtarić B (2008) Toxic and essential metal concentrations in four tissues of red deer (*Cervus elaphus*) from Baranja, Croatia. Food Addit Contam 25:270–283
Lee HCH, Liu WW, Chai SP, Mohamed AR, Lai CCH, Khe CHS et al (2016) Synthesis of single-layer graphene: a review of recent development. Proc Chem 19:916–921
Lemarchand C, Rosoux R, Berny P (2010) Organochlorine pesticides, PCBs, heavy metals and anticoagulant rodenticides in tissues of Eurasian otters (*Lutra lutra*) from upper Loire River catchment (France). Chemosphere 80:1120–1124
Lester MB, van Riper C (2014) The distribution and extent of heavy metal accumulation in song sparrows along Arizona's upper Santa Cruz River. Environ Monit Assess 186:4779–4791
Levengood JM (2001) Concentrations of selected elements in Illinois raccoons. Trans Ill State Acad Sci 94:89–99
Li Z, Zongwei M, Tsering JV, Zengwei Y, Lei H (2014) A review of soil heavy metal pollution from mines in China: pollution and Health risk assessment. Sci Total Environ 468:843–853
Licata P, Naccari F, Dugo G, Fotia V, Lo Turco V, Potorti AG et al (2012) Organochlorine pesticides and polychlorinated biphenyls in common buzzard (*Buteo buteo*) from Sicily (Italy). Environ Monit Assess 184:2881–2892
Lodenius M, Skaren U, Hellstedt P, Tulisalo E (2014) Mercury in various tissues of three mustelid and other trace metals in liver o European otter from Eastern Finland. Environ Monit Assess 186:325–333
Lucia M, André J-M, Bernadet M-D, Gontier K, Gérard G, Davail S (2008) Concentrations of metals (zinc, copper, cadmium, and mercury) in three domestic ducks in France: Pekin, muscovy, and mule ducks. J Agric Food Chem 56:282–288
Lucia M, André J-M, Gontier K, Diot N, Veiga J, Davail S (2010) Trace element concentrations (mercury, cadmium, copper, zinc, lead, aluminium, nickel, arsenic, and selenium) in some aquatic birds of the southwest Atlantic coast of France. Arch Environ Contam Toxicol 58:844–853
Luo XG, Ji F Lin YX, Steward FA, Lu L, Liu B et al (2005) Effects of dietary supplementation with copper sulfate or tribasic copper chloride on broiler performance, relative copper bioavailability, and oxidation stability of vitamin E in feed. Poult Sci 84:888–893
Mahmood T, Islam KR (2006) Response of rice seedlings to copper toxicity and acidity. J Plant Nutr 29:943–957
Mantovi P, Bonazzi G, Maestri E, Marmiroli N (2003) Accumulation of copper and zinc from liquid manure in agricultural soils and crop plants. Plant Soil 250:249–257
Martin JM, Guan DM, Elbazpoulichet F, Thomas AJ, Gordeev V (1993) Preliminary assessment of the distributions of some trace elements (As, Cd, Cu, Fe, Ni, Pb, and Zn) in a pristine aquatic environment-The Lena River estuary (Russia). Mar Chem 43:185–199
Mason CF, Stephenson A (2001) Metals in tissues of European otters (Lutra lutra) from Denmark, Great Britain and Ireland. Chemosphere 44:351–353
Mateo R, Guitart R (2003) Heavy metals in livers of waterbirds from Spain. Arch Environ Contam Toxicol 44:398–404

Mazurek J, Rokicki E, Krynski A, Zarski TP, Gorska M (1991) Comparative investigations of the contamination of the liver of domestic pig and wild boar with chosen heavy metals in the region of Legnica-Glogow Copper District and eastern Podlasie. Ann Warsaw Agric Univ SGGW Anim Sci 26:15–19

Medvedev N (1999) Levels of heavy metals in Karelian wildlife, 1989–91. Environ Monit Assess 56:177–193

Mertin D, Szeleszczuk O, Süvegova K, Niedbala P, Hanusova E (2006) Content of microelements in the selected organs of raccoon dog (*Nyctereutes procynoides*). Ecol Chem Eng 13:85–90

Midander K, Cronholm P, Karlsson HL, Elihn K, Moller L, Leygraf C et al (2009) Surface characteristics, copper release, and toxicity of nano- and micrometer-sized copper and copper (II) oxide particles: a cross-disciplinary study. Small 115:389–399

Mierzykowski SE, Todd CS (2012) Environmental contaminants in across bill bald eagle recovered in Maine. USFWS Special Project Report. FY12-MEFO-3-EC.Maine Fiel Office. Orono, ME, 28 pp

Millaku L, Imeri R, Trebicka A (2015) Bioaccumulation of heavy metals in tissues of house sparrow (*Passer domesticus*). Res J Environ Toxicol 9:107–112

Millan J, Mateo R, Taggart MA, López-Bao JV, Viota M, Monsalve L et al (2008) Levels of heavy metals and metalloids in critically endangered Iberian lynx and other wild carnivores from southern Spain. Sci Total Environ 399:193–201

Naccari C, Cristani M, Cimino F, Arcoraci T, Trombetta D (2009) Common buzzards (*Buteo buteo*) bioindicators of heavy metals pollution in Sicily (Italy). Environ Int 35:594–598

Naccari C, Giangrosso G, Macaluso A, Billone E, Cicero A, D'Ascenzi C et al (2013) Red foxes (*Vulpes vulpes*) bioindicator of lead and copper pollution in Sicily (Italy). Ecotoxicol Environ Saf 90:41–45

Nagajyoti PC, Lee KD, Sreekanth TVM (2010) Heavy metals, occurrence and toxicity for plants: a review. Environ Chem Lett 8:199–216

Nam VB, Lee D (2016) Copper nanowires and their applications for flexible, transparent conducting films: a review. Nanomaterials 6:47

Nam D, Anan Y, Ikemoto T, Tanabe S (2005) Multielemental accumulation and its intracellular distribution in tissues of some aquatic birds. Mar Pollut Bull 50:1347–1362

Nicholson FA, Smith SR, Alloway BJ, Carlton-Smith C, Chambers BJ (2003) An inventory of heavy metals inputs to agricultural soils in England and Wales. Sci Total Environ 311:205–219

Nriagu JO, Pacyna JM (1988) Quantitative assessment of worldwide contamination of air, water and soils by trace metals. Nature 333:134–139

O'Hara TM, Carroll G, Barboza P, Mueller K, Blake J, Woshner V, Willetto C (2001) Mineral and heavy metal status as related to a mortality event and poor recruitment in a moose population in Alaska. J Wildl Dis 37:509–522

Ogle MC, Scanlon PF, Kirkpatrick RL, Gwynn JV (1985) Heavy metal concentrations in tissues of mink in Virginia. Bull Environ Contam Toxicol 35:29–37

Ojha R, Prasad AN (2016) Menkes disease: what a multidisciplinary approach can do. J Multidiscip Healthc 17:371–385

Orlowski G, Polechonski R, Dobicki W, Zawada Z (2007) Heavy metal concentrations in the tissues of the black-headed gull Larus ridibundus L. nesting in the dam reservoir in south-western Poland. Pol J Ecol 55:783–793

Oruc HH, Cengiz M, Beskaya A (2009) Chronic copper toxicosis in sheep following the use of copper sulfate as a fungicide on fruit trees. J Vet Diagn Investig 21:540–543

Osredkar J, Sustar N (2011) Copper and zinc, biological role and significance of copper/zinc. J Clinic Toxicol S3:001

Parslow JLF, Thomas GJ, Williams TD (1982) Heavy metals in the livers of waterfowl from the Ouse Washes, England. Environ Pollut 29:317–327

Pillatzki AE, Neiger RD, Chipps SR, Higgins KF, Thiex N, Afton AD (2011) Hepatic element concentrations of lesser scaup (*Aythya affinis*) during spring migration in the upper Midwest. Arch Environ Contam Toxicol 61:144–150

Pollock B (2005) Trace elements status of white-tailed red deer (*Odocoileus virginianus*) and moose (*Alces alces*) in Nova Scotia. Wildlife Damage Management, Internet Center for Canadian Cooperative Wildlife Health Centre: Newsletters & Publications, University of Nebraska–Lincoln

Puls R (1994) Mineral levels in animal health: diagnostic data, 2nd edn. Sherpa International, Canada, BC

Rauch JN, Pacyna JM (2009) Earth's global Ag, Al, Cr, Cu, Fe, Ni, Pb, and Zn cycles. Global Biogeochem Cycles 23:GB2001

Reglero MM, Monsalve-González L, Taggart MA, Mateo R (2008) Transfer of metals to plants and red deer in an old lead mining area in Spain. Sci Total Environ 15:287–297

Reimann C, Birke M, Demetriades A, Filzmoser P, O'Connor P (eds) (2013) Chemistry of Europe's agricultural soils. Geologisches Jahrbuch (Reihe B), Schweizerbarth, Hannover

Richardson GMR, Garrett I, Mitchell I, Mah-Paulson M, Hackbarth T (2001) Critical review on natural global and regional emissions of six trace metals to the atmosphere, report Risklogic Scientific Services, Ottawa, ON, Canada

Roberts EA, Schilsky ML (2008) Diagnosis and treatment of Wilson disease: an update. Hepatology 47:2089–2111

Robson AD, Reuther DJ (1981) Diagnosis of copper deficiency and toxicity. In: Loneragan JF, Robson AD, Graham RD (eds) Copper in soils and plants. Academic Press, Orlando, pp 287–312

Roslewska A, Stanek M, Janicki B, Cygan-Szczegielniak D, Stasiak K, Buzała M (2016) Effect of sex on the content of elements in meat from wild boars (*Sus scrofa* L.) originating from the Province of Podkarpacie (south-eastern Poland). J Elem 21:823–832

Routti H, Letcher RJ, Born EW, Branigan M, Dietz R, Evans TJ et al (2011) and temporal trends of selected trace elements in liver tissue from polar bears (*Ursus maritimus*) from Alaska, Canada and Greenland. J Environ Monit 13:2260–2267

Rudnick RL, Gao S (2003) Composition of the continental crust. In: Treatise on Geochemistry, Elsevier, 659 pp

Rush SA, Borgå K, Dietz R, Born EW, Sonne C, Evans T et al (2008) Geographic distribution of selected elements in the livers of polar bears from Greenland, Canada and the United States. Environ Pollut 153:618–626

Schummer ML, Petrie SA, Badzinski SS, Deming M, Chen Y-W, Belzile N (2011) Elemental contaminants in livers of mute swans on Lakes Erie and St. Clair. Arch Environ Contam Toxicol 61:677–687

Siegel MR, Sisler HD (1977) Antifungal compounds, vol 1. Marcel Dekker, New York, 507 pp

Skibniewska EM, Skibniewski M, Kosla T (2012) Dependence between Cu concentration in the liver, kidneys and skeletal muscles of canine females. Cent Eur J Biol 7:817–824

Skibniewski M, Skibniewska EM, Kośla T (2015) The content of selected metals in muscles of the red deer (*Cervus elaphus*) from Poland. Environ Sci Pollut Res Int 22:8425–8431

Skibniewski M, Skibniewska EM, Kośla T, Olbrych K (2016) The content of copper and molybdenum in the liver, kidneys, and skeletal muscles of Elk (*Alces alces*) from North-Eastern Poland. Biol Trace Elem Res 169:204–210

Skobrak EB, Javor A, Gunde LJ, Bodnar K (2010) Analyses of macro- and microelements of wild boar meat in three different regions of Hungary. Lucrari Stiintifice, Seria Agronomie 53:22–25

Sleeman JM, Magura K, Howell J, Rohm J, Murphy LA (2010) Hepatic mineral values of white-tailed deer (*Odocoileus virginianus*) from Virginia. J Wildl Dis 46:525–531

Souza MJ, Donnell R, Ramsay E (2013) Metal accumulation and health effects in raccoons (*Procyon lotor*) associated with coal fly ash exposure. Arch Environ Contam Toxicol 64:529–536

St Clair CT, Baird P, Ydenberg R, Elner R, Bendell LI (2015) Trace elements in Pacific Dunlin (*Calidris alpina pacifica*): patterns of accumulation and concentrations in kidneys and feathers. Ecotoxicology 24:29–44

Stejskal S, Aulerich RJ, Slanker MR, Braselton WE, Lehning EJ, Napolitano AC (1989) Element concentrations in livers and kidneys of ranch mink. J Vet Diagn Invest 1:343–348
Stout JH, Trust KA (2002) Elemental and organochlorine residues in bald eagles from Adak Island, Alaska. J Wildl Dis 38:511–517
Szatnik-Kloc A (2014) Application of adsorption methods to determine the effect of pH and Cu-stress on the changes in the surface properties of the roots. Int Agrophys 28:511–520
Szymczyk K, Zalewski K (2003) Copper, zinc, lead and cadmium content in liver and muscles of Mallards (*Anas platyrhynchos*) and other hunting fowl species in Warmia and Mazury in 1999-2000. Pol J Environ Stud 3:381–386
Taggart MA, Figuerola J, Green AJ, Mateo R, Deacon C, Osborn D et al (2006) After the Aznalcollar mine spill: arsenic, zinc, selenium, lead and copper levels in the livers and bones of five waterfowl species. Environ Res 100:349–361
Taggart MA, Green AJ, Mateo R, Svanberg F, Hillström L, Meharg AA (2009) Metal levels in the bones and livers of globally threatened marbled teal and white-headed duck from El Hondo. Spain. Ecotoxicol Environ Saf 72:1–9
Trust KA, Rummel KT, Scheuhammer AM, Brisbin IL Jr, Hooper MJ (2000) Contaminant exposure and biomarker responses in spectacled eiders (*Somateria fischeri*) from St. Lawrence Island, Alaska. Arch Environ Contam Toxicol 38:107–113
US Department of the Interior (1998) Guidelines for interpretation of the biological effects of selected constituents in biota, water, and sediment. National Irrigation Water Quality Program Information Report No 3, 198 pp
USGS (2014) US Geological Survey, Mineral commodity summaries 2014: US Geological Survey, 196 pp
USGS (2016) US Geological Survey, Mineral commodity summaries 2016: US Geological Survey, 205 pp
Vengust G, Svara T, Gombac M, Zele D (2015) Enzootic ataxia associated with copper deficiency in a farmed red deer: a case report. Vet Med 60:522–526
Vikøren T, Bernhoft A, Waaler T, Handeland K (2005) Liver concentrations of copper, cobalt, and selenium in wild Norwegian red deer (*Cervus elaphus*). J Wildl Dis 41:569–579
Vikøren T, Kristoffersen AB, Lierhagen S, Handeland K (2011) A comparative study of hepatic trace element levels in wild moose, roe deer, and reindeer from Norway. J Wildl Dis 47:661–672
Walker LA, Lawlor AJ, Chadwick EA, Potter E, Pereira MG, Shore RF (2010) Inorganic elements in the liver of Eurasian otters, *Lutra lutra*, from England and Wales in 2007 & 2008: a Predatory Bird Monitoring Scheme (PBMS) report. Centre for Ecology & Hydrology, Lancaster, UK, 13 pp. http://nora.nerc.ac.uk/10822/2010
Walker LA, Lawlor AJ, Chadwick EA, Potter E, Pereira MG, Shore RF (2011) Inorganic elements in the livers of Eurasian otters, *Lutra lutra*, from England and Wales in 2009: a Predatory Bird Monitoring Scheme (PBMS) report. Centre for Ecology & Hydrology, Lancaster, UK
WHO (2004) Guidelines for drinking-water quality, 3rd edn. World Health Organization, Geneva
Woshner VM, O'Hara TM, Bratton GR, Suydam RS, Beasley VR (2001) Concentrations of selected essential and nonessential elements in ringed seals and polar bears of arctic Alaska. J Wildl Dis 37:711–721
Wren CD (1984) Distribution of metals in tissues of Beaver, Raccoon and Otter from Ontario, Canada. Sci Total Environ 34:177–184
Xu Y, Yu W, Ma Q, Zhou H (2013) Accumulation of copper and zinc in soil and plant within ten-year application of different pig manure rates. Plant Soil Environ 11(492):499
Xue PY, Li GX, Liu WJ, Yan CZ (2010) Copper uptake and translocation in a submerged aquatic plant *Hydrilla verticillata* (L.f.) Royle. Chemosphere 81:1098–1103
Zaccaroni A, Andreani G, Zucchini M, Merendi F, Simoni P (2003) Heavy metals in wild boar (*Sus scrofa*) and related lesions. Hystrix Ital J Mammal, supp 107
Zaccaroni A, Andreani G, Errante MC, Carpenè E, Isani G. Lucisano A (2008) Metal concentrations in the liver and kidney of raptor species from the Calabria region, Italy Acta Vet (Beograd) 4:315–324

Zhang H, Cui B, Zhang K (2011) Heavy metal distribution of natural and reclaimed tidal riparian wetlands in south estuary, China. J Environ Sci 23:1937–1946

Zhang E, Wang X, Chen M, Hou B (2016) Effect of the existing form of Cu element on the mechanical properties, bio-corrosion and antibacterial properties of Ti-Cu alloys for biomedical application. Mater Sci Eng C Mater Biol Appl 69:1210–1221

# Chapter 5
# Iodine, I

**Bogumiła Pilarczyk, Agnieszka Tomza-Marciniak, Renata Pilarczyk, Andrzej Marciniak, Małgorzata Bąkowska, and Jan Udała**

**Abstract** Iodine is not a common element in nature, so many regions of the world suffer rather from a deficiency than an overabundance. The northern part of North America, most of South America, Africa, Asia (Himalaya), and Europe are classified as endemic regions with low iodine content.

Iodine plays a crucial role in the thyroid for hormone synthesis, which in turn regulates the regularity of most of metabolic pathways in mammalian cells. An iodine deficiency in animals causes malformation, growth retardation, decreased fertility, increased perinatal mortality, and lowered animal performance and productivity, e.g., reduced growth of wool, egg, and milk production. Moreover it has been shown that in roe deer, iodine-containing hormones play a role in the complex neurohormonal regulation of antler development. Antler weight increases with increasing iodine concentration.

The content of iodine in wild range animals is subject to evaluation only occasionally. For most wild-ranging species of mammals and birds, there is no information about iodine requirements. As more than 90% of dietary iodine is expelled by the kidneys, urine is found to be a good study material to assess recent iodine intake.

B. Pilarczyk (✉) · A. Tomza-Marciniak · A. Marciniak · M. Bąkowska · J. Udała
Department of Animal Reproduction Biotechnology and Environmental Hygiene, West Pomeranian University of Technology in Szczecin, Szczecin, Poland
e-mail: bogumila.pilarczyk@zut.edu.pl

R. Pilarczyk
Laboratory of Biostatistics, West Pomeranian University of Technology in Szczecin, Szczecin, Poland



E. Kalisińska (ed.), *Mammals and Birds as Bioindicators of Trace Element Contaminations in Terrestrial Environments*,
https://doi.org/10.1007/978-3-030-00121-6_5

# 1 Introduction

Iodine (I, in Latin iodum) is an element that was discovered in 1811 by Bernard Courtois (Baumann 1896). Its name comes from the characteristic violet color of its fumes (in Greek iodes). It is an uncommon element in nature, present in only trace amounts.

In the animal organism, iodine plays several important biological functions. As a main component of the thyroid hormones, triiodothyronine ($T_3$) and thyroxine ($T_4$), every single molecule of $T_3$ and $T_4$ contains three atoms of iodine. According to the World Health Organization (WHO), this element is considered to be one of the factors with a direct effect on the health of the human population. In men, thyroid swelling (endemic goiter) and impaired intellectual development are the main symptoms of iodine deficiency. Iodine-deficiency disorders (IDD) include the increased risk of spontaneous miscarriages, stillbirths and deformations, increased mortality in infants in the perinatal period, irreversible brain damage in the fetus and infants, as well as thyroid tumors. The evident consequences of iodine deficiency are a decreased rate of metabolism, which then causes body weight gain, feeling cold, overstrain, and depressive disorders. Deficiencies of selenium, iron, and vitamin A exacerbate the effects of an iodine deficiency (American Thyroid Association 2003). In animals, a deficiency of iodine causes malformations, growth retardation, decreased fertility, increased perinatal mortality, and reduced animal performance and productivity, e.g., reduced growth of wool, egg, and milk production. Calves born to iodine-deficient dams may be stillborn, with goiter and areas of alopecia and subcutaneous edema (Potter et al. 1980, 1981). In roe deer the iodine-containing hormones produced by the thyroid gland play a role in the complex neurohormonal regulation of antler development. Antler weight increases with increasing iodine concentration (Lehoczki et al. 2011).

The thyroid gland in birds is located in the thoracic inlet. In this group of vertebrates, the clinical symptom of goiter is the effect of the pressure placed on organs adjacent to the gland. A loud wheezing respiration with the neck extended may occur if there is pressure against the trachea (Tollefson 1982).

It is estimated that about 1 million years ago, the distribution of iodine across the Earth was rather uniform, but subsequent Ice Ages caused a disturbance in that balance. Changes in the shape of the Earth's surface lead to iodine impoverishment in some regions and to enrichment in another (Merke 1965). This process is still in progress due to constant movements in the tectonic plates, earthquakes, and volcanic eruptions. The formation of iodine-deficient regions is observed particularly in areas with heavy precipitation (mountain areas) and repeated floods.

Globally, it is estimated that about 2 billion individuals have an insufficient iodine intake, with South Asia and sub-Saharan Africa particularly affected (Zimmermann 2009). In Europe, around 60% of the population consumes too little iodine (Delange et al. 2001). The rate of iodine deficiency is lowest in North and South America, at 10.1%. Populations settling in mountainous areas are most exposed to lack of iodine. As reported by Lee (2002), endemic low iodine areas include the northern part of

North America (the Great Lakes region, northwestern Pacific coast, surroundings of Mississippi, and Quebec province in Canada), most of South America, Africa, Asia (Himalaya), and Europe (Alps, Balkan Peninsula, Scandinavia, Carpathian and Sudety mountains, Germany particularly surrounding Munich, and Scotland).

## 2 General Properties

Iodine is a chemical element with ordinal 53 and atomic weight 126.9045, not common in the Earth's crust. Iodine is slightly soluble in water (Przewłocki and Śliwowski 2007).

Chemically separated iodine at room temperature forms gray-black crystals with a metallic gloss and characteristic smell (boiling point +184.35 °C). It sublimes easily, forming violet fumes. Its density is 4940 kg $m^{-3}$. Iodine is located in main group VII of the periodic table within the group of halogens (Przewłocki and Śliwowski 2007). The main valence of halogens is −1, but in special conditions (e.g., in photolysis), halogens may have a positive valence: +1, +3, +5, and +7. There are 36 isotopes of iodine, and 14 of these yield significant radiation. The only naturally occurring isotopes of iodine are the stable isotope $^{127}I$ and the radioactive isotope $^{129}I$.

Binary compounds with other elements called halides form salts belonging to fluorides, chlorides, bromides, and iodides. Also, compounds of halogens with nonmetallic elements are known, as well as halogenated organic compounds with carbohydrates and carboxyl acids.

Iodine is a widespread trace element in the hydrosphere, lithosphere, atmosphere, and biosphere, although it usually only occurs in low concentrations. Iodine occurs in the Earth's crust at an average level of 0.45 mg $kg^{-1}$ (Muramatsu and Wedepohl 1998; Wong 1991). The concentration of iodine in bedrock varies between 0.5 and 380 mg $kg^{-1}$, depending on whether the rock is igneous or sedimentary (ATSDR 2004).

## 3 Iodine Minerals, Production, and Uses

The primary source of iodine is rocks rich in this element, such as oil shales, which release iodine into the seas and oceans following erosion, making these and the creatures within (e.g., algae, shellfish, and fish) the most iodine abundant (Manz 1992). The concentration of iodine in seawater is about 50 μg $L^{-1}$ and is 10 times higher than in freshwater (Manz 1992). The average iodine content in deep water carbonate rocks is about 0.03 ppm (or 30 ppb). Most of the iodine (>70%) in the Earth's surfaces exists in the oceans, at 45–60 μg $L^{-1}$ (Wong 1991; Muramatsu and Wedepohl 1998). The continental crust, the oceanic crust (including seawater), and the remainder of the Earth's crust contains 119, 777, and 300 ppb of iodine, respectively. Nearly 70% of total iodine is estimated to exist in ocean sediments (Muramatsu and Wedepohl 1998).

In the seas, iodine is usually present as $IO^{-}$ and in a much smaller amount as iodides ($I^{-}$). Under solar light both forms decompose to elementary iodine, which evaporates into the atmosphere and is transferred back into land via winds and precipitation. The concentration of iodine in river water ranges between 0.1 and 18 μg $L^{-1}$, in rainwater from 0.1 to 15 μg $L^{-1}$, and in underground water 1 μg $L^{-1}$. The concentration of iodine in river water from urban areas often increases due to the discharge of municipal waste into refining installations (Kabata-Pendias and Pendias 2000).

With the increase in distance inland from the coast, the content of iodine in the air decreases (usually it is about 0.7 μg $m^{-3}$) (Hetzel and Marberly 1986). The content of iodine in the air depends on the vicinity of salt waters, wind direction (the lowest concentration is noted in continental winds), and environmental pollution (compounds of sulfur, mercury, lead).

Iodine (as iodide) is present in soils. The content may fluctuate widely within regions and between regions as a result of a number of factors (e.g., differences that occurred during geological formation, impact of glaciation, flooding, and soil erosion) (Rohner et al. 2014).

Areas with a low iodine content in the soil include Asia (including parts of China, India, Bangladesh, the Himalayan hillsides, Indonesia), Africa (mountain regions of Morocco and Algeria; large part of Western and Central Africa: Nigeria, Cameroon, the Central African Republic, Democratic Republic of Congo, and some regions of Eastern Africa like Uganda, Ethiopia), Europe (regions of the Alps and Pyrenees, inland areas of England and Wales, Greece, and the Netherlands), South America (including the Andes and inland Brazil), North America (Midwestern United States), Southern Australia, and the New Guinea Highlands (WHO 1999; Kabata-Pendias and Pendias 2000; Eastman and Zimmermann 2009; Zimmermann 2010). Additionally, postglacial areas, high mountains (Alps, Andes, Himalayas), are a great distance from the sea coast and floodplains (surroundings of the Ganges river in Southeastern Asia). One exception is Japan, as most Japanese soils (excluding paddy soils) are rich in iodine, at 14.2 mg $kg^{-1}$ (Yuita and Kihou 2005), which is an effect of the mild climate, the oceans that surround this insular country, and the volcanic nature of the geology (Yuita 1994; Johnson 2003). By comparison, the mean content of iodine in soils in Great Britain is 8 mg $kg^{-1}$ (Johnson 2003), in Germany and Austria 2.2 mg $kg^{-1}$ (Schnell and Aumann 1999; Gerzabek et al. 1999), in Russia 3.8 mg $kg^{-1}$, and in India 3.65–12.59 mg $kg^{-1}$ (Ghose et al. 2003). The primary source of iodine in the Earth's crust is volcanic rock: granite, basalt, diorites, and tonalites. The mean content of iodine in such rocks is from 4 to 9 ppb (Johnson 2003) (Table 5.1).

Iodine occurs as iodides and iodates in seawater, most mineral waters, and in some marine algae. Moreover it is an additive (sodium iodide) in Chilean saltpeter (nitratine). For industrial needs, large amounts of iodine are obtained from mineral deposits found in Chile (as a coproduct of surface mineral deposits used to produce nitrate fertilizers) and from brines in the USA (Colorado, Nevada, New Mexico).

**Table 5.1** Concentrations of iodine in various rocks

| Material | Concentration (mg $kg^{-1}$) | References |
|---|---|---|
| Igneous rocks | 5–200 | Fuge and Johnson (1986) |
| Granite | 0.25 | Fuge and Ander (1998) |
| All other intrusives | 0.22 | Fuge and Ander (1998) |
| Basalts | 0.22 | Fuge and Ander (1998) |
| All other volcanics | 0.24 | Fuge and Ander (1998) |
| Volcanic glasses | 0.52 | Fuge and Ander (1998) |
| *Sedimentary rocks* | | |
| Shales | 2.3 | Fuge and Johnson (1986) |
| Sandstones | 0.8 | Fuge and Johnson (1986) |
| Limestones | 2.7 | Fuge and Ander (1998) |
| Organic-rich shales | 16.7 | Fuge and Ander (1998) |
| Carbonates | 2.7 | Fuge and Johnson (1986) |

However, this element is mostly obtained from marine algae (sometimes containing up to 14% of iodine in dry matter) (Szymańska and Bruchajzer 2010). World iodine production in 2012 was about 28,400 Mt (USGS 2013). For years, the main iodine producers in the world have been Chile (~66%) and Japan (~33%), where in 2012 17,000 and 9400 Mt, respectively, were obtained (USGS 2013). In recent years (2000–2013) no significant changes in the production of this element have been noted (Table 5.2).

Iodine and its compounds are used in photography (for the production of light-colored material), the printing industry, laboratory diagnostics, dye production (erythrosine, cyanine – Bengali red), pharmacology as a disinfecting agent (alcohol solutions of iodine and sodium iodide), and also for iodination of salt and disinfection of water (Backer and Hollowell 2000).

The highest occupational exposures are found in employees of chemical industries (e.g., during the synthesis of dyes), pharmacy, printing industry, and those using iodine-containing disinfection agents (in medicine, veterinary, and water purification), as well as in agriculture and the food industry (food and fodder supplementation) (Szymańska and Bruchajzer 2010).

The release of iodine into the environment occurs from both natural sources and human activity. Sources of iodine from human activities include effluents from municipal plants and the combustion of waste and fossil fuels (ATSDR 2004). After the experimental nuclear tests during World War II, and as a result of leakages from nuclear power plants, the environment has been burdened with anthropogenic-origin iodine as a product of uranium and plutonium fission (Likhtarev et al. 1993; Moore and Groszko 1999; Rao and Fehn 1999; Johnson 2003). Radioactive iodines from fuel reprocessing plants enter the environment primarily in a gaseous state and are incorporated into the food chain by deposition onto vegetation or via inhalation. Iodine-129 can also be incorporated into food chains via deposition onto soil surfaces and subsequent uptake by plant roots (Price et al. 1981; Johnson 2003).

**Table 5.2** Production of iodine 2000–2013 (metric tons)

| Country | Year | | | | | | | | | | | | | |
|---|---|---|---|---|---|---|---|---|---|---|---|---|---|---|
| | 2000 | 2001 | 2002 | 2003 | 2004 | 2005 | 2006 | 2007 | 2008 | 2009 | 2010 | 2011 | 2012 | 2013 |
| Chile | 10,474 | 11,355 | 11,648 | 13,916 | 14,931 | 15,346 | 16,494 | 15,473 | 15,503 | 17,399 | 15,793 | 16,000 | 17,494 | 20,656 |
| Japan | 6157 | 6643 | 6548 | 6524 | 7264 | 8095 | 8724 | 9282 | 9500 | 8232 | 9216 | 9277 | 9315 | 9400 |
| USA | 1470 | 1290 | 1420 | 1090 | 1130 | 1570 | 1220 | 1200 | 1200 | 1250 | 1250 | 1270 | 1270 | 1270 |
| China | 500 | 500 | 500 | 500 | 550 | 550 | 560 | 570 | 570 | 580 | 590 | 590 | 600 | 600 |
| Turkmenistan | 200 | 200 | 200 | 200 | 250 | 270 | 270 | 270 | 500 | 270 | 270 | 270 | 270 | 270 |
| Azerbaijan | 2 | – | – | – | – | – | – | 21 | 116 | 149 | 191 | 200 | 240 | 249 |
| Russia | 24 | 46 | 58 | 58 | 105 | 105 | 105 | 105 | 105 | 105 | 105 | 105 | 105 | 105 |
| Indonesia | 75 | 75 | 75 | 75 | 75 | 75 | 75 | 75 | 75 | 75 | 75 | 75 | 75 | 75 |
| World | 18,902 | 20,109 | 20,449 | 22,363 | 24,305 | 26,011 | 27,448 | 269,96 | 27,569 | 28,060 | 27,490 | 27,787 | 29,369 | 32,625 |

Source: British Geological Survey (2006) World Mineral Production 2000–2004; British Geological Survey (2010) World Mineral Production 2004–2008; British Geological Survey (2015) World Mineral Production 2009–2013

## 4 Iodine in Nature: Geogenic and Anthropogenic Sources (Global Iodine Budgets)

Once released into the atmosphere, iodine may be present in different physicochemical forms: elementary iodine (I and $I_2$) and organic iodine (methyl iodide, $CH_3I$; hypoiodous acid, HOI). It is estimated that global emissions of $CH_3I$ types are as follows: oceanic 130–1300 Gg year$^{-1}$ (Rasmussen et al. 1982; Nightingale 1991; Reifenhauser and Heumann 1992; Campos et al. 1996; Moore and Groszko 1999), terrestrial biomass burning <10 Gg year$^{-1}$ (Andreae et al. 1996), plant-soil systems (rice paddies) 20–71 Gg year$^{-1}$ (Muramatsu and Yoshida 1995; Redeker et al. 2000), peatland ecosystems 1.4 Gg year$^{-1}$ (Dimmer et al. 2000), and wetlands 7.3 Gg year$^{-1}$ (Dimmer et al. 2000).

## 5 Biological Status of Iodine

The content of iodine in surface and groundwaters, from where it is absorbed by plants, depends on the content in precipitation and the soil. The content of iodine in precipitation is higher than in the air. Iodine can penetrate into plants by direct deposition on overground parts of plants or from the soil by the root system (soil-accumulated iodine). The concentration of iodine in plants depends on their absorptive potential and the content in the soil. Plants show a low ability to bind iodine from the soil and air, which—with a simultaneous deficiency in selenium—would exacerbate the problem. Iodine absorbed by a plant is in 60% accumulated in leaves, with the remainder in the stalks and roots. Iodine ions penetrate through trichomes and together with plant enzymes are transported to the rhizosphere. Enzyme-rich substances secreted by trichomes activate particular morphological types of soil microorganisms (Strzetelski 2005). The content of iodine in land plants is on average about 1 mg kg$^{-1}$ dw (Hetzel and Marberly 1986). The plants richest in iodine are those residing in alluvial riverside soils and on clays and sands that have been deposited by water flow (Hetzel and Marberly 1986).

In homoeothermic vertebrates, the rate of absorption of iodine via the alimentary tract varies and depends on both the form of iodine and the species of mammal (Akiba and Matsumoto 1976; Thrall and Bull 1990; Johnson 2003; ATSDR 2004). It is estimated that in dogs, it is about 80–92% (Alexander et al. 1967; Nath et al. 1992; Flachowsky 2007; Ghazvinian et al. 2012). In cattle, between 70% and 90% of dietary iodine is absorbed directly via the bovine rumen, reticulum, and omasum. In turn in humans, gastrointestinal absorption of iodine (taken as water-soluble iodide salts) is almost 100% (ATSDR 2004). Fisher et al. (1965) say that fecal excretion of $^{131}I$ was <1% of the dose in seven euthyroid adult subjects who received a single tracer dose of $^{131}I$ and that daily urinary iodine excretion was approximately 80–90% of the estimated daily intake, which unambiguously indicates the near-complete absorption of the ingested radioiodine. Similar results were obtained in an acute ingestion study of nine healthy subjects (Ramsden et al. 1967—quoted from ATSDR 2004). Urinary radioiodine accounted for 97% (±5, SD) of a single ingested tracer dose of radioiodine ($^{131}I$ or $^{132}I$).

Thrall and Bull (1990), in studies on the effect of a chemical form on the uptake and distribution of radioiodine in fed and fasted rats, showed that the initial distribution of $^{125}I$ to the thyroid depended sharply on the chemical form, being greater when iodide rather than iodine was administered, irrespectively of whether the animals were fed or fasted. The authors have also found that after oral administration of $^{125}I_2$ or $^{125}I^-$, the concentration of $^{125}I$ in the blood reached a maximum after 2 h and was comparable between the group of feed animals supplemented with $^{125}I^-$ and the fasting group supplemented either with $^{125}I_2$ or $^{125}I^-$.

Available scientific data indicate that there are large differences in the transfer of iodine from feed to animal tissues and products such as eggs or milk. For example, the transfer of iodine from feed to pork and beef was only 0.3% and <1%, whereas in milk and eggs, it was 30–40% and 10–20%, respectively (Richter 1995; Yalcin et al. 2004).

Only 10% of iodine demand in humans comes from drinking water (Ziemlański et al. 2001). Inorganic iodine compounds (e.g., potassium iodide) are absorbed very quickly and almost entirely, while the absorption of organic compounds is limited (Bobek 1998; Nath et al. 1992; Ziemlański et al. 2001). Other vectors of iodine penetration are the pulmonary system and the skin. These tracts for most mammals and birds are less relevant (excluding the littoral areas, where the air is rich in this element). After being absorbed by an organism, inorganic iodine is transformed into organic compounds, which are vital for the functioning of the thyroid, which produces tetraiodothyronine ($T_4$) and triiodothyronine ($T_3$). The thyroid gland is the critical target organ in vertebrates, including humans, following excessive repeated oral intake of stable iodine. As said by Nath et al. (1992), an organism accumulates iodine mostly in the thyroid gland (70–80%). The other stores are the salivary glands, skeletal muscles, mammary glands, gastric mucosa, and in females the ovaries. Moreover iodine also accumulates in the epidermis and hair follicles in rats and probably in other mammals as well (Brown–Grant 1961; Ziemlański et al. 2001). Absorbed iodine is excreted primarily in the urine and feces but can also be excreted in breast milk, exhaled air, sweat, and tears. Urinary excretion normally accounts for >97% of the elimination of absorbed iodine, while fecal excretion accounts for approximately 1–2% (ATSDR 2004).

## *5.1 Toxicity of Various Iodine Forms in Homoeothermic Animals*

A large excess of iodine can be harmful to the thyroid, inhibiting the process of synthesis and release of thyroid hormones, the Wolff-Chaikoff effect (WHO 2004). The mechanism of iodine toxicity is related to its strongly caustic activity, affecting protein degradation and cellular dilapidation, while the basic symptom of iodine toxicity in low doses or small concentrations is the irritation of mucosal membranes. In countries where the consumption of seafood is high, for instance, in Japan, iodine intoxication is more likely (Kostogrys et al. 1999).

In available literature no reports have been found that consider the mutagenic effects of nonradioactive iodine. No mutagenic activity by iodine was shown in mouse L5178Y lymphocytes or colonies of Balb/c mice 3T3 cells (iodine concentration $0.1 \div 10\ \mu g\ mL^{-1}$) (Merkle and Zeller 1970; Kessler et al. 1980) nor in a study performed with *Saccharomyces cerevisiae* yeast (Mehta and Borstel 1982). Still, genotoxic activity has been proven in the case of $^{131}I$ (Joseph et al. 2009). Also, Wlodkowski et al. (1975) noted that Povidone-iodine, used as a disinfectant, is capable of specifically altering DNA and the induction of base-substitution mutations, which potentially may lead to the development of cancers.

Although no unambiguous data about the carcinogenic activity of iodine is available in present literature, the results of some studies suggest that an elevated iodide intake may be a risk factor for thyroid cancer in certain populations, particularly in populations in iodine-deficient regions with endemic goiter (Zimmermann and Galetti 2015).

## 5.2 Toxicokinetics and Effects of Iodine

As reported by Szymańska and Bruchajzer (2010) and RTECS (2006), based on medial lethal doses for laboratory animals, iodine is unranked in acute toxicity ($DL_{50}$ for rats after an intragastric administration is 14,000 mg $kg^{-1}$ body weight, bw). No changes were found in hamsters (*Cavia porcellus*) and rats after a single exposure (inhalation) to iodine at 5 mg $m^{-3}$. To a small extent, exposure of rats to 8.6 mg $m^{-3}$ and dogs to 10 mg $m^{-3}$ of iodine produced an effect in the pulmonary system (Casarett 1975). At 73–100 mg $m^{-3}$ of iodine, an irritating effect was observed, with an adverse effect on lung function. At an inhaled concentration of 4.7 mg $m^{-3}$ for 3–4 months in rats, some changes in the structure of mucosal membranes of the mouth, eyes, and tooth enamel were observed. As reported by Biegishev (1975), the exposure of hamsters and rats to iodine at 0.5 or 3.1 mg $m^{-3}$ (at an increased temperature) resulted in temporal disorders of smell, slight kidney damage, and decreases in body weight and oxygen uptake.

An oral $LD_{50}$ of 3320 mg I $kg^{-1}$ bw has been reported in rats for both sodium iodide and for potassium iodide, and an oral $LD_{100}$ of 1425 mg I $kg^{-1}$ bw has been reported for mice (Clayton and Clayton 1981; EFSA 2006). According to Clayton and Clayton (1981), an acute oral $LD_{50}$ value for potassium iodide in rats was 3320 mg I $kg^{-1}$ bw, and the lowest oral lethal dose in mice was 1425 mg I $kg^{-1}$ bw. Amounts at 200–500 mg $kg^{-1}$ bw can cause death in experimental animals (SCOGS 1975).

Humans seem to be less sensitive to iodine than rodents, concerning thyroid disturbances, even though the basic hypothalamic-pituitary-thyroid axis functions in a similar way in animals and humans. Animal data are therefore of limited value in relation to human toxicity (Nielsen et al. 2014).

A toxic level in total diet for ruminants is 50 ppm (excluding goats, 8 ppm); pigs, 800 ppm; horses, 4.8 ppm; and poultry, 625 ppm (NRC).

## 5.3 Bioaccumulation of Iodine

The collocation of iodine in the environment is very varied, as shown in Fig. 5.1. The content of iodine in water plants is different and mostly depends on the type of water (fresh, sea). Freshwater algae contain $10^{-5}$% by weight of iodine, whereas marine algae contain $10^{-3}$% by weight. In freshwater fish, iodine concentrations in tissues ranged from 0.003 to 0.81 mg kg$^{-1}$ ww, while in marine fish, they ranged between 0.023 and 0.11 ppm ww (ATSDR 2004). The average iodine concentration in terrestrial plants was 0.42 mg kg$^{-1}$ dw (Moiseyev et al. 1984).

The content of iodine in vertebrate tissues depends mostly on the species of the animal and the environment in which it remains. Kaňa et al. (2015) found a highest content of iodine in the skeletal muscles of sea fish, especially in Atlantic cod (*Gadus morhua*), 1.484 mg kg$^{-1}$ ww, and a lower concentration in blue shark (*Prionace glauca*), at 0.0506 mg kg$^{-1}$ ww. According to Kaňa et al. (2015), the levels of iodine in the tissues of farm animals are much lower than in fish, where in pigs in the thyroid gland, the level of iodine was 0.513 mg kg$^{-1}$ ww and in the kidney, liver,

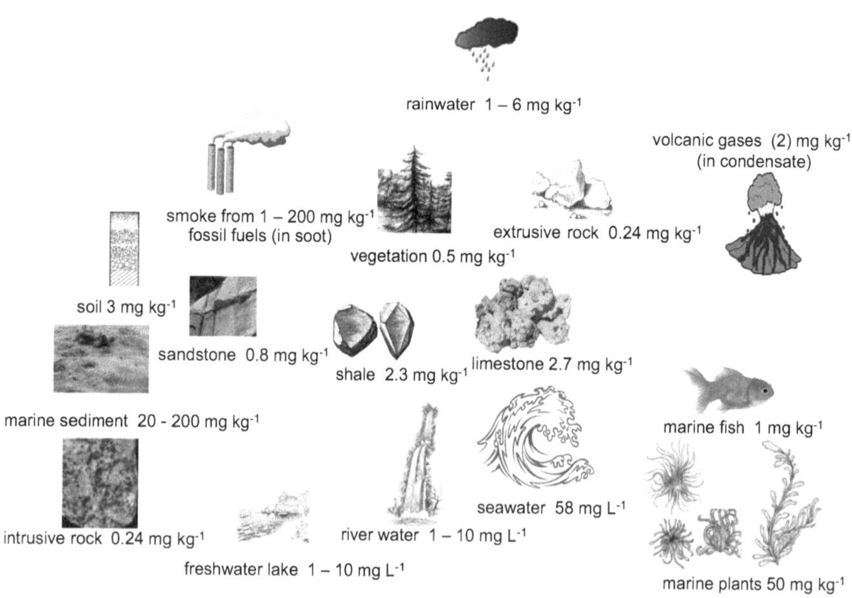


**Fig. 5.1** Collocation of iodine in the environment (based on Johnson 1980)

and muscle 0.103, 0.051, and 0.034 mg $kg^{-1}$ ww, respectively. In chickens, relatively more iodine is present in the gut (0.123 mg $kg^{-1}$ ww), with the levels in the liver and muscle (0.068 and 0.029 mg $kg^{-1}$ ww) similar to that observed in pigs. A higher content of iodine was found in turkey livers than in chickens and pigs (0.105 mg $kg^{-1}$ ww).

In plants growing in iodine-deficient soils, concentrations may be as low as 10 mg $kg^{-1}$ dw, compared with approximately 1 mg $kg^{-1}$ in iodine-sufficient soils (Zimmermann 2009), with average iodine concentrations in fleshy fungi at 6.2 mg $kg^{-1}$ dw, ferns 5.7 mg $kg^{-1}$ dw, coniferous trees 3.9 mg $kg^{-1}$ dw, monocotyledons 5.9 mg $kg^{-1}$ dw, dicotyledonous trees 2.7 mg $kg^{-1}$ dw, and dicotyledonous vegetables 6.9 mg $kg^{-1}$ dw (Shacklette and Cuthbert 1967).

Studies in goats and sheep have revealed that in temperate climate (Iran), seasons affected the content of iodine in animal organisms. The highest concentration of iodine was noted in autumn with the lowest in spring (Gharahveysi et al. 2012). Also Topczewska (2012) showed significant differences between seasons in the concentrations of iodine in horse hair in Poland. The author observed the highest level of iodine in hair in winter and the lowest in spring.

## 5.4 *Ecological Effects of Iodine*

The content of iodine in free ranging animals is measured only occasionally. For the most wild mammals and birds, no iodine requirements have been suggested.

Due to the suitability of meat for consumption from different animals, including game, muscle tissue is usually considered to be an appropriate tissue for studies in both food quality and ecotoxicology. However, literature data does not indicate the suitability of muscles in the evaluation of environmental levels of iodine (Rambeck et al. 1997; Schone et al. 1986; Schone 1999). Admittedly, German scientists introduced additional iodine supplementation (potassium iodide) to the diet of porkers and observed an increased level in muscle tissues and internal organs (liver, kidneys) of animals, yet the results were not satisfactory (Schone et al. 1986; Rambeck et al. 1997; Schone 1999). The skeletal muscles of farm and game animals living without supplemental sources of iodine contain 0.10–0.30 mg $kg^{-1}$ ww (Anke 2007). Most likely studies on the concentration of this element in the thyroid or other organs that are known to be sensitive to a deficiency of iodine would provide a better view on the relationship between the presence of iodine in the environment and its content in homoeothermic animals, yet the availability of data in this field is very scarce.

The iodine-containing hormones produced by the thyroid gland play a role in the complex neurohormonal regulation of antler development (Lehoczki et al. 2011). This study revealed that the level of iodine in the environment has a determinant effect on roe deer (*Capreolus capreolus*) antler weight and that iodine deficiency (ID) can be an additional limiting factor constraining optimal deer performance for antler weight. Bubenik et al. (1987) studied white-tailed deer (*Odocoileus virginianus*) and showed that the T3 utilization rate in the growing antler correlated with the intensity of antler growth. Iodine levels may so influence roe deer performance (antler development) through the mechanisms of thyroid function. As reported by Watkins et al. (1983), the iodine requirements of white-tailed deer can be met by feed (dry matter) containing 0.26 ppm of iodine. This recommended iodine content in feed is required to cover the average iodine needs of domestic ruminants (generally 0.1–0.25 mg $kg^{-1}$ of consumed fodder and 0.5–0.6 ppm for lactating animals) (Schöne and Rajendram 2009).

Problems with iodine deficiency may also occur in birds, especially those feeding on seeds, as observed in individuals of different species kept in captivity or nesting in human neighborhoods. This element is particularly important for the budgerigar (*Melopsittacus undulatus*) and pigeons, which appear to be particularly susceptible to thyroid problems. In these species, iodine deficiency may lead to goiter (enlargement of the thyroid glands), while such an affliction is not common in other groups of birds (Tollefson 1982).

## 5.5 *Bioindicators and Biomarkers of Iodine in Ecotoxicological Studies*

In humans a reliable method of evaluating iodine supply is population studies in which iodine urinal excretion, frequency of goiter occurrence, and concentrations of TSH (thyroid-stimulating hormone) and thyroglobulins are considered.

Because more than 90% of dietary iodine eventually appears in the urine, the amount of iodine excreted with the urine is a very good indicator of recent iodine intake (Nicola et al. 2009; WHO 2007). In humans, based on this, the degree of iodine deficiency can be defined for further development of prophylactic programs.

It is assumed that the content of iodine in the urine translates into the supply of this element in recent days, the concentration of thyroglobulins for weeks and months, and the presence of goiter for years (Zimmermann 2008; Erdman et al. 2012).

Detection of an enlarged thyroid gland and a ratio of thyroid weight to body weight greater than 0.4 g $kg^{-1}$ provides an indication of goiter in lambs and kids. Iodine concentrations in doe milk less than 0.8 μmol $L^{-1}$ indicate an inadequate iodine intake (Caple et al. 1983). A concentration of thyroxine in blood serum of neonatal lambs lower than 50 μmol $L^{-1}$ indicate hypothyroidism (Caple and Nugent 1982).

Typical biomarkers of hyperthyroidism are an elevation in the circulating levels of $T_4$ and/or $T_3$ above their normal ranges, which is always accompanied by a depression of TSH below the normal range. The clinical manifestation of abnormally elevated circulating levels of $T_4$ and/or $T_3$ is often referred to as thyrotoxicosis or Graves' disease or Basedows' disease. Serum thyroglobulin represents a very sensitive index of the state of thyroid hyperstimulation in humans (Karasek and

Lewiński 2003). In domestic ruminants, an assessment of iodine status is done by measurement of serum I, protein-bound I, thyroxine (T4), or the presence of goiter (McDowell 1992). Also, to diagnose iodine deficiencies, female milk can be used. As reported by Caple et al. (1983), iodine concentrations in cow milk less than 0.2 μmol $L^{-1}$ indicate inadequate iodine intake. Unfortunately, in the wild, obtaining urine or milk is a difficult problem, and practically such studies are no longer performed.

In wild ruminants, including cervids, the content of iodine depends on the concentration of this element in their plant food (Watkins and Ullrey 1983; Whitehead 1984; Anke et al. 1995). However, Ceacero et al. (2009) suggest that deer were able to modulate mineral intake to meet requirements and that, on the other hand, relative ID can occur if goitrogenic compounds were present in the food (Whitehead 2000; Tripathi and Mishra 2007). In the case of evaluation of anthropogenic iodine in the environment, probably the thyroid glands of cervids may turn out to be good bioindicators of this element, as was shown in the studies by Price et al. (1981). The authors established the current levels of $^{129}I$ in the environs of the Hanford Site (USA) prior to the proposed restart of fuel reprocessing at the PUREX plant (nuclear production complex on the Columbia River in the US state of Washington). The results of this study clearly demonstrated the longevity of $^{129}I$ in the biosphere, following an earlier gaseous release from the nuclear facility. Price et al. (1981) demonstrated that the thyroid glands of deer living within 160 km (Wooten Game Range) of Hanford had elevated levels of $^{129}I$ when compared to distant Pacific Northwest locations (Centralia or Bend). Levels of $^{129}I$ in deer thyroid from Bend or Centralia (15 fCi $g^{-1}$ wet weight) were about five times higher than those reported in the Central United States, with Hanford samples about 2700 times higher. The average concentration of $^{129}I$ in deer thyroids collected at Hanford in 1978 was similar to samples collected 14 years earlier (Price et al. 1981).

It also cannot be excluded that bird eggs may turn out to be good indicators of environmental iodine levels, as was shown in farm birds where they can easily accumulate this element. After weeks of feeding hens with iodine-supplemented fodder, an increased content was observed in the eggs, reaching a maximal value during the eighth week (Ryś et al. 1995).

## 6 Conclusions

Literature data on iodine in wildlife is very scarce, while problems with health or the environmental effects of a possible iodine deficiency or surplus are not yet sufficiently well understood. In land habitats we usually observe the problem of a deficiency rather than an overabundance. For this reason, regarding the possible indirect effect of this element on the animal reproductive potential, we can suppose that iodine clearly may affect the size and condition of populations in some species of animals and can also play a role in shaping the structure of particular ecosystems, especially in isolated groups of animals in mountain areas.

## References


Alexander WD, Harden RM, Harrison MT, Shimmins J (1967) Some aspects of the absorption and concentration of iodide by the alimentary tract in man. Proc Nutr Soc 26:62–66

Akiba Y, Matsumoto T (1976) Antithyroid activity of goitrin in chicks. Poult Sci 55(2):562–567

American Thyroid Association (2003) ATA Hypothyroidism Booklet, p 3

Andreae MO, Atlas E, Harris GW, Helas G, de Kock A, Koppmann R, Maenhaut W, Manø S, Pollock WH, Rudolph J, Scharff DE, Schebeske G, Welling M (1996) Methyl halide emissions from savanna fires in southern Africa. J Geophys Res 101:603–623

Anke M (2007) Iod. In: Dunkelberg H, Gebel T, Hartwig A (Hrsg) Handbuch der Lebensmitteltoxikologie Belastungen, Wirkungen, Lebensmittelsicherheit, Hygiene (In German). Wiley, Weinheim, pp 2317–2379

Anke M, Groppel B, Müller M, Scholz E, Krämer K (1995) The iodine supply of humans depending on site, food offer and water supply. Fresenius J Anal Chem 352:97–10112

ATSDR (2004) Toxicological Profile for iodine. U.S. Department of Health & Human Services, Public Health Service, Agency for Toxic Substances and Disease Registry

Backer H, Hollowell J (2000) Use of iodine for water disinfection: iodine toxicity and maximum recommended dose. Environ Health Perspect 108:679–684

Baumann F (1896) Ueber das normale Vorkommen von Jod im Thierkorper. (I. Mittheilung). Z Physiol Chem 21:319–330

Biegishev A (1975) Data on the toxicity characteristics of iodine vapours under conditions of increased air temperature (experimental data). Gig Tr Prof Zabol 6:38–41

Bobek S (1998) Iodine prophylaxis in animals. Med Weter 54:80–86

British Geological Survey (2006) World mineral production 2000–2004. British Geological Survey, Keyworth

British Geological Survey (2010) World mineral production 2004–2008. British Geological Survey, Keyworth

British Geological Survey (2015) World mineral production 2009–2013. British Geological Survey, Keyworth

Brown–Grant K (1961) Extrathyroidal iodide concentrating mechanism. Physiol Rev 41:189–213

Bubenik GA, Sempéré AJ, Hamr J (1987) Developing antler, a model for endocrine regulation of bone growth. Concentration gradient of $T_3$, $T_4$, and alkaline phosphatase in the antler, jugular, and the saphenous veins. Calcif Tissue Int 41:38–43

Campos MLAM, Nightingale PD, Jickells TD (1996) A comparison of methyl iodide emissions from seawater and wet depositional fluxes of iodine over the southern North Sea. Tellus Ser B 48:106–114

Caple IW, Nugent GF (1982) Relationships between plasma thyroxine concentrations and the responses of newborn lambs to hypothermia. Proc Aust Soc Anim Prod 14:657

Caple LW, Azuolas KK, Nugent GF, Corbett HR (1983) Goitre, iodine deficiency and hypothyroidism in goats. In: Trace element review papers, 1982. Agricultural Services Library, Department of Agriculture, Victoria

Casarett LJ (1975) Toxicology of the respiratory system. In: Toxicology: the basic science of poisons. Macmillan, New York, pp 201–224

Ceacero F, Landete-Castillejos T, García A, Estévez JA, Martinez A, Calatayud A, Gaspar-López E, Gallego L (2009) Free-choice mineral consumption in Iberian red deer (*Cervus elaphus hispanicus*) response to diet deficiencies. Livest Sci 122:345–348

Clayton GD, Clayton FE (1981) In: Clayton GD, Clayton FE (eds) Patty's industrial toxicology and hygiene, vol 2B, 3rd edn, p 2975

Delange F, de Benoist B, Pretell E, Dunn JT (2001) Iodine deficiency in the world: where do we stand at the turn of the century? Thyroid 11:437–447

Dimmer CH, Simmonds PG, Nickless G, Bassford MR (2000) Biogenic fluxes of halomethanes from Irish peatland ecosystems. Atmos Environ 35:321–330

Eastman CJ, Zimmermann MB (2009) The iodine deficiency disorders. In: Thyroid disease manager. Available from http://www.thyroidmanager.org/chapter/the-iodine-deficiency-disorders/
EFSA (2006) Opinion of the Scientific Committee on Food on the Tolerable Upper Intake Level of iodine (expressed on 26 September 2002). In: Tolerable Upper Intake Levels for vitamins and minerals. Scientific Committee on Food. Scientific Panel on Dietetic Products, Nutrition and Allergies. European Food Safety Authority. February 2006
Erdman JW, MacDonald IA, Zeisel SH (2012) Present knowledge in nutrition. Wiley, Hoboken
Fisher DA, Oddie TH, Epperson D (1965) Effect of increased dietary iodide on thyroid accumulation and secretion in euthyroid Arkansas subjects. J Clin Endocrinol 25:1580–1590
Flachowsky G (2007) Iodine in animal nutrition and iodine transfer from feed into food of animal origin. Lohman Inf 42:47–59
Fuge R, Ander EL (1998) Geochemical barriers and the distribution of iodine in the secondary environment: implications for radio-iodine. In: Nicholson K (ed) Energy and the environment: geochemistry of fossil, nuclear and renewable resources. MacGregor, Washington, pp 163–170
Fuge R, Johnson CC (1986) The geochemistry of iodine – a review. Environ Geochem Health 8:31–54
Gerzabek MH, Muramatsu Y, Strebl F, Yoshida S (1999) Iodine and bromine contents of some Austrian soils and relations to soil characteristics. J Plant Nutr Soil Sci 162:415–419
Ghazvinian K, Gharahveysi S, Taghipur T, Jamshidi R, Ali Salar N (2012) The study of iodine levels in the blood serum of the Iranian sheep and goat. Euro J Exp Bio 2(3):814–816
Ghose NC, Das K, Saha D (2003) Distribution of iodine in soil water system in the Gandak basin. J Geol Soc India 62:91–98
Hetzel BS, Marberly GF (1986) Iodine. In: Trace elements in human and animal nutrition, vol 5. Academic Press, London, pp 139–208
Johnson CC (1980) The geochemistry of iodine and a preliminary investigation into its potential use as a pathfinder element in geochemical exploration. PhD thesis, University College of Wales, Aberystwyth
Johnson CC (2003) The geochemistry of iodine and its application to environmental strategies for reducing the risks from iodine deficiency disorders (IDD). British Geological Survey Commissioned Report, CR/03/057N, p 54
Joseph LJ, Bhartiya US, Raut YS, Kand P, Hawaldar RW, Nair N (2009) Micronuclei frequency in peripheral blood lymphocytes of thyroid cancer patients after radioiodine therapy and its relationship with metastasis. Mutat Res 675:35–40
Kabatas-Pendias A, Pendias H (2000) Trace elements in soils and plants, 3rd edn. CRC Press, Boca Raton
Kaňa A, Hrubá L, Vosmanská M, Mestek O (2015) Analysis of iodine and its species in animal tissues. Chem Speciat Bioavailab 27:81–91
Karasek M, Lewiński A (2003) Etiopathogenesis of Grave' disease. Neuroendocrinol Lett 24:161–166
Kessler FK, Laskin DL, Borzelleca JF (1980) Assessment of somatogenotoxicity of povidone-iodine using two in vitro assays. J Environ Pathol Toxicol 4:327–335
Kostogrys R, Sikora E, Pisulewski P (1999) Problems of iodine deficiency in the human body – a review of research. Żyw Człow Metab 26:330–342
Lee SL (2002) Iodine deficiency. J Med 3:1–3
Lehoczki R, Erdélyi K, Sonkoly K, Szemethy L, Csányi S (2011) Iodine distribution in the environment as a limiting factor for roe deer antler development. Biol Trace Elem Res 139:168–176
Likhtarev IA, Shandala NK, Goulko GM, Kairo IA (1993) Exposure doses to thyroid of the Ukrainian population after the Chernobyl accident. Health Phys 64:594–599
Manz F (1992) Jodmangel: Grunde, Folgen und Vorbeugemassnahmen. Prev Vet Med 15:111–117
McDowell LR (1992) Minerals in animal and human nutrition. Academic Press, New York
Mehta RD, von Borstel RC (1982) Effect of growth phase and different solvents on the genetic activity and cell toxicity of diethylstilbestrol in *Saccharomyces cerevisiae*. Environ Mutagen 4:417

Merke F (1965) Die Eiszeit als primordiale Ursache des endemischen Kropfes. Schweiz Med Wochenschr 95:1183–1192
Merkle J, Zeller H (1970) Absence of povidone-iodine induced mutagenicity in mice and hamster. J Pharm Sci 68:100–102
Moiseyev IT, Tikhomirov FA, Perevezentsev VM, Rerikh LA (1984) Role of soil properties, interspecific plant differences, and other factors affecting the accumulation of radioactive iodine in crops. Soviet Soil Sci 16:60–66
Moore RM, Groszko W (1999) Methyl iodide distribution in the ocean and fluxes to the atmosphere. J Geophys Res 104:163–171
Muramatsu Y, Wedepohl KH (1998) The distribution of iodine in the Earth's crust. Chem Geol 147:201–216
Muramatsu Y, Yoshida S (1995) Volatilization of CH3I from the soil-plant system. Atmos Environ 29:21–25
Nath SK, Moinier B, Thuillier F, Rongier M, Desjeux JF (1992) Urinary excretion of iodide and fluoride from supplemented food grade salt. Int J Vitam Nutr Res 62:66–72
Nicola JP, Basquin C, Portulano C, Reyna-Neyra A, Paroder M, Carrasco N (2009) The Na/I-symporter mediates active iodide uptake in the intestine. Am J Physiol Cell Physiol 296:C654–C662
Nielsen E, Greve K, Larsen J, Meyer O, Krogholm K, Hansen M (2014) Iodine, inorganic and soluble salts. The Danish Environmental Protection Agency, Copenhagen
Nightingale PD (1991) Low molecular weight halocarbons in seawater. PhD thesis, University of East Anglia, Norwich
Potter BJ, Jones GB, Buckley RA, Belling GB, McIntosh GH, Hetzel BS (1980) Production of severe iodine deficiency in sheep using a prepared low-iodine diet. Aust J Biol Sci 33:53–619
Potter BJ, McIntosh GH, Hetzel BS (1981) The effect of iodine deficiency on fetal brain development in the sheep. In: Hetzel BS, Smith RM (eds) Fetal brain disorders: recent approaches to the problem of mental deficiency. Elsevier/North Holland Biomedical, Amsterdam, pp 119–148
Price KR, Cadwell LL, Schreckhise RG, Brauer FP (1981) Iodine-129 in Forage and Deer on the Hanford Site and Other Pacific Northwest Locations, PNL-3357. Pacific Northwest Laboratory, p 17
Przewłocki K, Ślizowski K (2007) Iodine occurrence in the natural environment. Gospod Surowcami Min 23:17–26
Rambeck WA, Kaufmann S, Feng J, Hollowich W, Arnold R (1997) Improving the human iodine supply by iodination of swine feed. Tierärztl Prax 25:312–315
Ramsden SA, John PK, Kronast B, Benesch R (1967) Evidence for a thermonuclear reaction in a ⍰-pinch plasma from the scattering of a ruby laser beam. Phys Rev Lett 19(12):688–689
Rao U, Fehn U (1999) Sources and reservoirs of anthropogenic iodine 129 in Western New York. Geochim Cosmichim Acta 63:1927–1938
Rasmussen RA, Khalil MAK, Gunawardena R, Hoyt SD (1982) Atmospheric methyl iodide. J Geophys Res 87:3086–3090
Redeker KR, Wang NY, Low J, McMillan A, Tyler SC, Cicerone R (2000) Emissions of methyl halides from a CA rice field. Science 290:966–968
Reifenhauser W, Heumann KG (1992) Determinations of methyl iodide in the Antarctic atmosphere and the South Polar Sea. Atmos Environ Part A 26:2905–2912
Richter G (1995) Beziehung zwischen dem Jodgehalt im Futter und im Ei. Proc Soc Nutr Physiol 4:86
Rohner F, Zimmermann M, Jooste P, Pandav C, Caldwell K, Raghavan R, Raiten DJ (2014) Biomarkers of nutrition for development – iodine review. J Nutr 144:1322S–1342S
RTECS (2006) Registry of toxic effects of chemical substances. National Institutes for Occupational Safety and Health, Cincinnati
Ryś R, Wir–Kornaś E, Pyska H, Kuchta M (1995) The effect of different types and levels of iodine additives in feed on iodine deposition in eggs. Roczn Nauk Zootech 23:187–190
Schnell D, Aumann DC (1999) The origin of iodine in soil: iodine in soils of Germany. Chem Erde 59:69–76

Schone F (1999) Iodine deficiency, iodine requirement and iodine excess of farm animals – experiments on growing pigs. Berl Münch Tierärztl Wochenschr 112:64–70
Schöne F, Rajendram R (2009) Iodine in farm animals. In: Preedy VR, Burrow GN, Watson RR (eds) Comprehensive handbook of iodine: nutritional, biochemical pathological and therapeutic aspects. Academic Press, Burlington, pp 151–170
Schone F, Ludke H, Jahreis G, Hennig A (1986) Effects of iodine, cooper and zinc supplements to rations with a high quota of rapeseed extract meal on the growth and thyroid function of fattening swine. The effects on weight and histomorphometric findings in the thyroid and serum T3 and T4 concentrations. Arch Tierernahr 36:371–380
SCOGS (1975) Evaluation of the health aspects of KI, KIO3, Ca(IO3)2 as food ingredients. Select Committee on GRAS Substances, Report No. 39, Life Sciences Research Office, FASEB, Bethesda, MD
Shacklette HT, Cuthbert ME (1967) Iodine content of plant groups as influenced by variation in rock and soil type. Geol Soc Am Spec Pap 90:31–46
Strzetelski P (2005) Presence and migration of iodine in the soil-plant system. Post Nauk Roln 6:85–100
Szymańska J, Bruchajzer E (2010) Iodine. PiMOŚP 65:61–83
Thrall KD, Bull RJ (1990) Differences in the distribution of iodine and iodide in the Sprague-Dawley rat. Fundam Appl Toxicol 15:75–81
Tollefson IC (1982) Nutrition. In: Petrak ML (ed) Diseases of cage and aviary birds, 2nd edn. Lea & Febiger, Philadelphia, pp 220–249
Topczewska J (2012) Effects of seasons on the concentration of selected trace elements in horse hair. JCEA 13:671–680
Tripathi MK, Mishra AS (2007) Glucosinolates in animal nutrition: a review. Anim Feed Sci Technol 132:1–27
USGS (2013) Iodine. In: U.S. Geological Survey, Mineral Commodity Summaries, pp 76–77
Watkins BE, Ullrey DE (1983) Iodine concentration in plants used by white-tailed deer in Michigan. J Wildl Manag 47:1220–1226
Watkins BE, Ullrey DE, Nachreiner RF, Schmitt SM (1983) Effect of supplemental iodine and season on thyroid activity of white-tailed deer. J Wildl Manag 47:45–58
Whitehead DC (1984) The distribution and transformations of iodine in the environment. Environ Int 10:321–339
Whitehead DC (2000) Nutrient elements in grassland: soil–plant–animal relationships. CABI, Wallingford, New York
Wlodkowski TJ, Speck WT, Rosenkranz HS (1975) Genetics of povidone-iodine. J Pharm Sci 64:1235–1237
Wong GTF (1991) The marine geochemistry of iodine. Rev Aquat Sci 45:45–73
World Health Organization (1999) United Nations Children's Fund, International Council for Control of Iodine Deficiency Disorders. Progress towards the elimination of iodine deficiency disorders (IDD), Geneva
World Health Organization (2004) Vitamin and mineral requirements in human nutrition, 2nd edn. World Health Organization and Ford
World Health Organization (2007) United Nations Children's Fund, International Council for the Control of Iodine Deficiency Disorders Assessment of iodine deficiency disorders and monitoring their elimination. 3rd edn. Geneva
Yalçin S, Kahraman Z, Yalcin S, Yalcin SS, Dedeoglu HE (2004) Effects of supplementary iodine on the performance and egg traits of laying hens. Br Poult Sci 45:499–503
Yuita K (1994) Overview and dynamics of iodine and bromine in the environment. 1. Dynamics of iodine and bromine in soil-plant system. Jarq-Jpn Agric Res Q 28:90–99
Yuita K, Kihou N (2005) Behavior of iodine in a forest plot, an upland field and a paddy field in the upland areas of Tsukuba, Japan. Vertical distribution of iodine in the soil horizons and layers to a depth of 50 m. Soil Sci Plant Nutr 51:455–467

Ziemlański Ś, Bułhak–Jachymczyk B, Niedźwiecka–Kącik D, Panczewska–Kresowska B, Wartanowicz M (2001) Standards of human nutrition. PZWL, Warszawa
Zimmermann MB (2008) Methods to assess iron and iodine status. Br J Nutr 3:S2–S9
Zimmermann M (2009) Iodine deficiency. Endocr Rev 30:376–408
Zimmermann MB (2010) Symposium on 'geographical and geological influences on nutrition': iodine deficiency in industrialised countries. Proc Nutr Soc 69:133–143
Zimmermann M, Galetti V (2015) Iodine intake as a risk factor for thyroid cancer: a comprehensive review of animal and human studies. Thyroid Res 18:8

# Chapter 6
# Iron, Fe

**Danuta Kosik-Bogacka, Natalia Łanocha-Arendarczyk, Elżbieta Kalisińska, Karolina Kot, Danuta Czernomysy-Furowicz, Bogumiła Pilarczyk, and Agnieszka Tomza-Marciniak**

**Abstract** Iron (Fe) is an essential micronutrient for human and animal organisms, playing an important role in the metabolic process in endothermic animals. This metal is one of the most abundant elements in the Earth. In this review the recent studies on Fe existence in environment and terrestrial ecosystems and their organisms are presented. The reports show that Fe concentration in tissues of different avian and mammalian species can be influenced by the type of diet and biological parameters. Owing to the wide range of Fe concentration in animal tissues of different taxonomic groups and between individuals of the same species, it is hard to indicate the bioindicator of environmental Fe concentration.

## 1 Introduction

Iron (Fe) is the second most abundant metal in the Earth's crust, and it is known to be biologically essential element of every living organism (Soetan et al. 2010; Kabata-Pendias and Szteke 2012). Despite its common occurrence in water, soils, and plants, iron bioavailability is very low due to the fact that in contact with oxygen Fe forms insoluble oxides (Abbaspour et al. 2014; Kabata-Pendias and Szteke 2012). Iron is essential in transport of oxygen in the body, and it is involved in many enzymatic reactions. Its deficient or excessive amount in the body can cause disease symptoms

D. Kosik-Bogacka (✉) · N. Łanocha-Arendarczyk · E. Kalisińska · K. Kot
Department of Biology and Medical Parasitology, Pomeranian Medical University in Szczecin, Szczecin, Poland
e-mail: kodan@pum.edu.pl

D. Czernomysy-Furowicz
Department of Immunology, Microbiology and Physiological Chemistry, West Pomeranian University of Technology in Szczecin, Szczecin, Poland

B. Pilarczyk · A. Tomza-Marciniak
Chair of Animal Reproduction, Biotechnology and Environmental Hygiene, West Pomeranian University of Technology in Szczecin, Szczecin, Poland



E. Kalisińska (ed.), *Mammals and Birds as Bioindicators of Trace Element Contaminations in Terrestrial Environments*,
https://doi.org/10.1007/978-3-030-00121-6_6

(Soetan et al. 2010; MacKenzie et al. 2008). Iron deficiency is manifested by a decrease in hemoglobin in the blood, which leads to anemia. Excess accumulation of Fe in the body can lead to iron overload disease (Soetan et al. 2010; Abbaspour et al. 2014; Powell et al. 2016).

The research on the Fe concentration, usually evaluated in the liver, are conducted both in domesticated and wild avian and mammalian species. The ranges of Fe concentration in mammals and birds of different taxonomic groups and between individuals of the same species from North America and Europe vary greatly, and the scientists from such studies deter from interpretation. Nonetheless, Fe deficiency and iron storage disease have been reported in farm and wild animals (Underwood 1977; Dierenfeld et al. 1994; Andrews 1999; Cork 2000; Clauss and Paglia 2012).

Given the important role of Fe in human and animal organisms and the reported cases of health risks associated with a deficiency or excess of Fe, it seems worthwhile to monitor the Fe concentration in the environment and tissues of wild birds and mammals.

## 2 General Properties

Iron (Latin: ferrum, Fe) is a metallic chemical element belonging to group 8 in the periodic table. Its atomic number is 26, atomic weight 55.847, specific gravity 7.874, melting point 2795 °F (1535 °C), and boiling point 4982 °F (2750 °C).

Pure iron is a glossy, pliable, and ductile metal, which undergoes oxidation (passivation). One of the most characteristic features of iron is its ferromagnetism or ability to become strongly and permanently magnetized in a magnetic field. In its pure form, iron is relatively soft and slightly magnetic, but its magnetism increases after hardening. There are four stable allotropic iron species, alpha iron ($\alpha$), beta iron ($\beta$), gamma iron ($\gamma$), and delta iron ($\delta$), and one high-pressure form: epsilon iron ($\varepsilon$). Iron allotropes differ from one another in their durability, crystalline lattice structure, and magnetism (Boehler 2000). Iron has many isotopes, of which only four occur naturally in the environment. These are durable isotopes with atomic weights of 54 (5.8%), 56 (91.7%), 57 (2.2%), and 58 (0.3%). The other isotopes of iron are artificial and radioactive (Audi et al. 2004).

Iron creates compounds at the second and third stages of oxidation. In an abiotic environment with a pH of 4.0–4.5, Fe occurs at the second stage of oxidation, while at pH $> 4.5$, the element is oxidized to $Fe^{3+}$. In aluminosilicate minerals, iron easily substitutes magnesium and aluminum. Its circulation in the environment is closely linked to the cycling of O, S, and C. Iron is one of the most common elements on Earth. Its content in the Earth's crust is about 5%. The highest levels of Fe are found in ultra-alkaline and alkaline igneous rocks ($<$10%). In contrast, in acidic igneous rocks, iron content does not exceed 3%. In sedimentary rocks, the Fe content is very different (0.4–4.8%), while iron-rich sedimentary rocks can contain up to 15% Fe. The smallest amount of Fe is found in limestones and dolomites (Kabata-Pendias and Pendias 2001).

## 3 Ferrous Minerals and the Production and Uses of Iron

A high concentration of iron is found in metamorphic limonite deposits. Fe is also found in at least 25 minerals. The most important iron ores in terms of industrial significance are magnetite (72% Fe), hematite (70% Fe), goethite (63% Fe), and siderite (48% Fe). As an element, iron is easily oxidized, reduced, or hydrolyzed (depending on the prevailing environmental conditions) and is quickly released from minerals such as aluminosilicates, silicates, and carbonates. Only a small amount of Fe is contained in minerals which are resistant to weathering processes. Iron is the least expensive and the most widely used metal. Iron must be processed before use, in order to improve its qualities. More than 90% of mined iron ore is used for the production of steel and metal alloys. Global production of pig iron in 2014 and 2015 was 1170 and 1180 million tons, respectively. The production of raw steel in 2014 and 2015 was at a similar level, amounting to 1650 million tons (USGS 2016).

Iron ores are distributed widely around the earth. Currently, it is mainly ores containing a high level of pure iron which are mined. The level of iron ore mining around the world has risen from 2283 million tons to 3328 million tons over the years 2009–2015. Today, China is the largest iron ore producer in the world (1381 million tons in 2015), having increased its production fourfold in the last 10 years. A substantial growth in iron ore production in China occurred between 2009 and 2014, from 880 million tons to 1514 million tons, while in 2015 production fell to 1381 million tons. Major mining areas in China include the provinces of Liaoning (especially around the city of Anshan), Hebei (NE China), Inner Mongolia (N China), and Sichuan on Hainan Island (S China). Australia is the second largest producer of iron ore. Its mining output has increased from 394 million tons in 2009 to 816.7 million tons. Its main iron ore mining areas include Hamersley (NW - Australia) and the area northwest of Adelaide. Australia exports 90% of its mined iron ore. In Asia, India is an important producer of iron ore (over 155 million tons in 2015). Iron ore deposits in India are very rich and of high quality. The main deposits of magnetite and hematite occur in the central states of the country (Maharashtra, Goa, Andhra Pradesh, Odisha, and Bihar). North America has a much smaller share of iron ore production and has seen a 3% decline due to lower production in the USA (42.5 million tons). Brazil and Australia account for half of the world's exports of iron ore. In Europe, the most important iron ore deposits are located in northern Sweden (in the Kiruna and Gallivare basins). Japan is the largest importer of iron ore in the world (O'Brien 2009; Indian Bureau of Mine 2014; British Geological Survey 2015).

Iron is rarely used in its chemically pure form due to its poor mechanical properties, with industry using alloys produced by alloying metals and nonmetals. Steel is an alloy of iron and processed carbon (about 0.8%). In addition, steel contains about 2% of an admixture of chromium and nickel. It is estimated that steel is four times stronger than pure iron and, due to its properties, is used for manufacturing springs and parts of machines which are exposed to heavy loads. Another type of steel is stainless steel, which is made up of iron, 0.2% carbon and

13% chromium, and is used for manufacturing equipment for the food industry, as well as measuring devices and surgical instruments. Cast iron is another example of an iron alloy, with a carbon content of up to 6%. This alloy can be cast very effectively and, due to its properties, is used for casting complex shapes such as engine hulls and lathes. Between 2009 and 2015, world production of crude steel increased by an average of 2–3% annually. Asian countries, whose share of global steel production is about 70%, contributed to this increase the most significantly. In contrast, European Union (EU) countries' share of world steel production, which at present is about 10%, has been steadily decreasing. The apparent consumption of steel in the EU amounted to 187 million tons in 2008 and 149 million tons in 2015. The largest users of steel in the EU are the construction (35%) and automotive (18%) industries. The apparent use of steel products globally increased by over 70% between 1997 and 2007. Between 2009 and 2015, this accounted for an increase in steel production from 1232 million to 1620 million tons. The countries with the highest apparent per capita use of steel are South Korea (1057 kg per head), Taiwan (793 kg), the Czech Republic (547 kg), Japan (516 kg), and China (515 kg) (British Geological Survey 2015).

## 4 Iron in Nature

In the soil, iron occurs in the form of silicates, sulfates, and carbonates, as well as oxides and hydroxides. The latter may form crystalline or amorphous structures. Iron oxides include hematite ($\alpha Fe_2O_3$), maghemite ($\gamma Fe_2O_3$), magnetite ($Fe_3O_4$), ferrihydrite ($Fe_2O_3 \cdot H_2OFe_2O_3 \cdot H_2O$), goethite ($\alpha FeOOH$), lepidocrocite ($\alpha FeOOH$), ilmenite ($FeTiO_3$), and pyrite ($FeS_2$) (Cornell and Schwertmann 2003). All Fe (II) compounds are mobile, and their solubility increases proportionally to the degree of soil acidity. In addition, they exhibit low stability, as opposed to Fe (III) compounds, which are durable. Fe content in soils can vary to a considerable degree, ranging between 0.1% and 5% (Kabata-Pendias and Pendias 2001). Cornell and Schwertmann (2003) found that Fe concentration in arable soils varied between 20 and 40 g $kg^{-1}$ dry weight (dw). Szymański et al. (2017) observed that Fe content was significantly lower in topsoil and eluvial horizons (5.22 g $kg^{-1}$ dw) than in the lower horizons (7.87 g $kg^{-1}$ dw), indicating translocation of Fe with clay minerals further down the soil profile. The natural causes of Fe in bodies of water are the processes of weathering and soil migration. Iron in the form of fine and colloidal hydroxides is subjected to sorption on the surface of solid particles that form a suspension and are transported over long distances, while Fe in minerals which are hardly subject to weathering (e.g., silicates, magnetite) is rapidly sedimented. In addition, the presence of Fe in bodies of water results from urban and industrial effluents (Park et al. 2006; Stopić et al. 2007).

Fe content in freshwaters usually ranges from 1 to 28 μmol $L^{-1}$ and in marine waters from <0.2 to 0.8 nmol $L^{-1}$ (Johnson et al. 1997; Kritzberg and Ekström 2012). Because Fe occurs predominantly in surface water in the form of insoluble trivalent compounds that precipitate as a sediment, the concentration of Fe in soluble form is usually small. Schönfelder et al. (2002) reported that Fe concentration in freshwaters (69 lakes and 15 rivers) ranged from 11 to 318 μg $L^{-1}$, with an average of 59.9 μg $L^{-1}$. However, it has been noted that during floods the concentration of this form of Fe in water is three times greater (Baixing et al. 2016).

In river bottom sediments, Fe content is generally >10%. However, Kanbar et al. (2017) demonstrated that if a river runs through a location affected by mining or metallurgical activities, iron content can reach 29%. In lake bottom sediments, the concentration of Fe is 8–47 g $kg^{-1}$ (Czaplicka et al. 2016; Rajkowska and Protasowicki 2011); the organic fraction of the sediments contains about ten times more Fe than the mineral fraction (Rajkowska and Protasowicki 2011). Estuary sediments can contain an amount of Fe 5 up to 25 times greater than the levels of background metals (Carral et al. 1995). In river or lake bottom sediments, as with soils, Fe concentration increases with depth (Kanbar et al. 2017).

The source of Fe in atmospheric air is crustal and industrial dust. Depending on the degree of industrialization, iron content may range from 50 to 14,000 ng $m^{-3}$ (Reimann and Caritat 1998). Iron in atmospheric dust is mainly found in the largest fractions: as shown in a study carried out by Rogula-Kozłowska et al. (2015), in dust fraction $PM_{0.03–0.17}$, no Fe was found to be present, whereas in fractions $PM_{0.26–0.4}$ and $PM_{0.4–0.65}$, Fe contents of 0.4 and 7.5 ng $m^{-3}$, respectively, were found. It is estimated that the annual fall of Fe on urban areas is between 16.8 and 43.2 kg $ha^{-1}$ (Kabata-Pendias and Mukherjee 2007).

## 5 Biological Functions of Iron

Iron is essential for the growth and development of all living organisms. It plays a significant role in many different biological processes, including oxygen transport and storage, oxidative phosphorylation, regulation of cell growth and differentiation, detoxification, cellular respiration, the activity of many enzymes, and the promotion of apoptosis (MacKenzie et al. 2008). Iron participates in DNA and RNA synthesis, gene regulation, and catalysis of many metabolic redox reactions (Thelander et al. 1983; Coffey and Ganz 2017). It is also responsible for the generation of reactive oxygen species (ROS), which induce oxidative stress and lipid and DNA damage (Zhang 2014). It participates in erythropoiesis, formation of leukocytes, and immune reactions, influencing the humoral and cellular immunity of the body (Beard 2001; Zhang 2014).

In endothermic animals Fe is necessary for cytochrome c production and is the core of molecules such as hemoglobin and myoglobin (Li and Ginzburg 2010), compounds that in mammals contain about 70% and 5–10% of Fe in the body, respectively (Albretsen 2006). Iron is also found in iron-binding proteins, including ferritin, hemosiderin, transferrin, ovotransferrin, and lactoferrin (Ganz 2013). Ovotransferrin is found in bird egg white, while lactoferrin in the colostrum and milk of mammals and in the secretions of glandular cells and neutrophils (Adlerova et al. 2008). In mammals, ferritin, hemosiderin, and transferrin in the liver, spleen, and bone marrow contain up to 25% of Fe (as $Fe^{3+}$) (Kohgo et al. 2008).

Ferritin is the main cytoplasmic protein regulating the labile iron pool. Its serum level is a sensitive indicator of Fe status in cattle (Puls 1994). Produced in the liver, and also in Sertoli cells, and oligodendroglia, this alpha globulin has a very high affinity to $Fe^{3+}$ (Espinosa de los Monteros et al. 1990; Lécureuil et al. 2004). It occurs in the blood, lymph, cerebrospinal fluid, bile, amniotic fluid, and milk. It transports Fe from the duodenum and mononuclear phagocyte system to cells, especially the bone marrow and rapidly dividing cells. Transferrin transports Fe to the inside of cells via the transferrin receptor (TfR). The iron transport to a cell is based on endocytosis of the transferrin complex saturated with iron together with TfR (Luck and Mason 2012). In cytoplasm, trivalent iron (III) is reduced to the divalent form (II). Formed in a vesicle, the complex of apotransferrin and receptor is transported onto the cell surface, and apotransferrin dissociates from TfR (Bakker and Boyer 1986; Brock 1989).

Iron metabolism is a semi-closed system without natural mechanisms that could remove any excess of iron. Iron homeostasis in mammals includes cells for transport to other cells and organs (Pantopoulos et al. 2012). Systemic Fe regulatory processes occur in the following order: (1) intestinal iron absorption, (2) cell transport (into cells), (3) ion transport across cell membranes, (4) participation in metabolic process, (5) iron recycling from erythrocytes, and (6) Fe storage (Duck and Connor 2016). Gastrointestinal iron absorption occurs primarily in duodenum and constitutes 5–10% Fe from the diet. The amount of iron absorbed from the gastrointestinal tracts is equivalent to the amount of iron excreted from the body. Among other things, the absorption of Fe depends on (1) age, (2) Fe status and state of health of the animal, (3) conditions within the gastrointestinal tract, (4) the amount and chemical form of the Fe ingested, (5) the diet, and (6) season of the year. Demand for Fe is lower in adult animals than in young animals, which need it for maintaining a correct concentration of hemoglobin and rate of tissue growth (Underwood 1977).

Rapidly proliferating cells have a much higher demand for iron than cells with a small proliferative potential. High iron concentrations in food may result in the development of iron-dependent microorganisms (including *Yersinia* bacteria, such as *Yersinia pseudotuberculosis*, causing cecal abscesses in birds (Stovel 1963; Czernomysy-Furowicz and Furowicz 1999)). On the other hand, Fe serum levels have been observed to be reduced down to 40% during infectious diseases with a fever (Puls 1994).

Low serum Fe levels and low saturation of transferrin with iron result in a release of stored iron and transport to the bone marrow. High Fe concentrations in serum and high saturation of transferrin with iron cause an increase in iron storage, e.g., in hepatocytes (Williams et al. 1982). In physiological conditions, high transferrin levels in serum accompany elevated iron concentrations in the blood. During pregnancy the concentration of free transferrin, e.g., in mares, increases due to the elevated demand for iron in the developing fetus. Serum Fe concentration decreases in the 8th–9th months of pregnancy until the 1st month of lactation (Puls 1994).

Transferrin does not undergo filtration in the glomeruli, which protects a healthy body against a loss of iron (Czernomysy-Furowicz 2007).

Iron accumulates in various tissues and organs of endothermic vertebrates, becoming toxic above 100 mg $kg^{-1}$ body weight (bw) and causing anemia below 1 mg $kg^{-1}$ bw. High Fe concentration may indicate inflammation and rapid immune response (Ganz and Nemeth 2015). Excessive Fe accumulation in the liver can lead to high hepatic load, with hepatic Fe reaching thousands of mg $kg^{-1}$ dry weight (dw), a pathological phenomenon known as hemosiderosis (Cork 2000). In cases where Fe concentration exceeds the binding ability of apoferritin, it undergoes additional accumulation as insoluble hemosiderin.

In humans, Fe deficiency is associated with malnutrition and parasitic and neoplastic diseases and accompanies chronic inflammatory diseases, including intestinal and gastrointestinal bleeding (Abbaspour et al. 2014). This leads to severe anemia, immune and cognitive disorders, and growth impairment.

The symptoms of Fe deficiency are observed quite often in farm animals, which is associated with the low level of assimilable forms of Fe in feed or with problems with Fe absorption. Deficiency symptoms manifest themselves in generic symptoms of anemia, which are tiredness, lassitude and general feelings of a lack of energy, poor immunity, and retarded growth. Iron is much more common than other trace elements, and iron excess can lead to poor absorption of copper and manganese (Underwood 1977). In a study by Puls (1994), the normal level of Fe in the liver of cattle ranged from 45 to 300 mg $kg^{-1}$ wet weight (ww), and a value <30 mg $kg^{-1}$ ww indicates a possible Fe deficiency. Moreover, Fe deficiency in the liver of domestic dogs (*Canis lupus familiaris*) was observed at concentrations less than 25 mg $kg^{-1}$ ww (Puls 1994). In young pigs (*Sus scrofa f. domestica*), Fe deficiency was found if the diet was not supplemented with Fe. Hepatic deficient Fe concentrations in adult and day 3 weaning pigs usually ranged from 30–40 and 10–15 mg $kg^{-1}$ ww, respectively (Puls 1994).

Free Fe ions ($Fe^{2+}$) are toxic to cells and tissues (Goyer 1996). The exact biochemical basis for Fe toxicosis is not well known, although it is suggested that free radicals form when free iron is present (Bacon and Britton 1990). Excess Fe in the cell leads to the formation of hydrogen peroxide in the cytosol during iron-catalyzed dissociation, followed by a diffusion of hydrogen peroxide through the nuclear membrane, DNA strand breaks, and purine purification (Yamazaki and Piette 1990). In order to protect the cell against Fe toxicity, Fe is incorporated into ferritin present in each cell (Piperno 1998).

Acute and subacute Fe poisonings in animals are mainly found in dogs, foals, piglets, and occasionally in other species. The causes of acute poisoning in animals may include the use of preparations used to destroy moss on lawns. No toxic dose has been established to date, with an oral dose of 20–60 mg $kg^{-1}$ bw considered to cause mild to moderate poisoning and 100–200 mg $kg^{-1}$ bw lethal for all animal species (Albretsen 2006).

Excess of Fe induces damage in the liver, heart, pancreas, thyroid, and the central nervous system, due to the overproduction of ROS in the presence of excess iron. This is followed by cell death, fibrosis, and carcinogenesis. Iron toxicosis causes coagulation disturbances that are related to thrombocytopenia, hypoprothrombinemia, and impaired clotting factor synthesis (Greentree and Hall 1995; Hillman 1995). In addition to iron deposition in the liver and heart, pancreatic beta cells are another important target of iron toxicity, which causes glucose intolerance and diabetes mellitus (Kohgo et al. 2008).

Iron toxicosis in mammals manifests clinically in four stages. The first stage (6 h after the overdose) is manifest by gastrointestinal symptoms, including vomiting, diarrhea, and gastrointestinal bleeding (Osweiler et al. 1985; Greentree and Hall 1995; Hillman 1995; Liebelt 1998). Then 6–24 h after the overdose, a latent period is observed (Greentree and Hall 1995). The third stage of iron toxicosis occurs about 12–96 h after the initial clinical signs: it is characterized by lethargy, a recurrence of gastrointestinal signs, metabolic acidosis, shock, hypotension, tachycardia, cardiovascular collapse, coagulation deficits, hepatic necrosis, and possibly death (Greentree and Hall 1995; Hillman 1995; Liebelt 1998). The last stage, which may occur 2–6 weeks after the iron overdose, is when the animals that survived gastrointestinal ulceration are healing.

## 5.1 *Iron Storage Disease (ISD)*

The pathogenesis of ISD (iron storage disease) is poorly understood. It is characterized by the accumulation of Fe in the liver until heart failure, hypoalbuminemia, and death (Cork 2000). To date, ISD has been described in wild birds, including toucans (Rhamphastidae), mynahs (Sturnidae), birds-of-paradise (Paradisaeidae), curassows (Cracidae), quetzals (*Pharomachrus* species), tanagers, and hornbills, and in *Turdus* species (Crissey et al. 2000; Dierenfeld et al. 1994; Turner 1994; Pavone et al. 2014). Excess iron stores in the body are classified as hemochromatosis or hemosiderosis (Pietrangelo 2010; Powell et al. 2016). Hemochromatosis is characterized by the accumulation of ferritin and hemosiderin in hepatocytes, Kupffer cells, bile duct epithelium, heart, pancreas, joints, and fibroblasts in the dermis. In humans, hereditary hemochromatosis is autosomal recessive and characterized by normal iron-driven erythropoiesis and toxic accumulation of iron in parenchymal cells of vital organs that can be caused by mutations in any gene that limits iron entry into the blood (Pietrangelo 2010). Patients with hemochromatosis may develop liver failure and may die of cardiac arrhythmias; this condition usually develops in men after 40 years of age (Nichols and Bacon 1989).

Hemosiderosis is characterized by storage of increased amounts of iron in Kupffer cells of the liver with no apparent hepatocellular damage. It can be caused by high concentrations of iron in the diet, transfusions, or thalassemias, which are diseases with defective hemoglobin metabolism (Halliday and Searle 1996). Hemosiderosis has been reported in a wide range of species in zoo and private collections (Randell et al. 1981; Gosselin and Kramer 1983; Spelman et al. 1989; Crawshaw et al. 1995; Lowenstine and Munson 1999; Paglia and Dennis 1999). In wild mammals and birds, hemosiderosis is very poorly investigated. In mammals, hemosiderosis in the liver may be associated with inflammations in the intestine due to bacterial and parasitic infections (Andrews 1999). This pathological Fe storage has been described in lemurs, hyraxes, and some carnivores, including procyonids and pinnipeds (Clauss and Paglia 2012).

In birds, hemosiderosis is characterized by the accumulation of iron in body tissues, especially the liver. Based on a retrospective study of 180 necropsy cases representing 40 different species of birds, Cork et al. (1995) showed that hepatic hemosiderosis is a common histological finding in most avian species. Although not necessarily associated with overt liver disease, it is often associated with concurrent malignant and infectious diseases. The authors observed that the presence of excess stainable iron in the liver is probably a reflection of an altered iron metabolism associated with an increased turnover of tissue iron and may be caused by conditions such as starvation or trauma. In the liver of Falconiformes, Kalisinska et al. (2008) observed Fe liver concentrations over 6000 mg $kg^{-1}$ dw, where usually in this group of birds, the physiological level ranges from ~430 to 2300 mg $kg^{-1}$ dw (Jager et al. 1996; Kalisinska et al. 2006). In some cases ISD or hemosiderosis in birds is associated with elevated levels of lead (Pb) in tissues (Lewis et al. 2001). Hemosiderosis has also been described in a trematode infection of the liver and spleen of the mute swan (*Cygnus olor*) (van Bolhuis et al. 2004). Fruit-eating and insect-eating birds have a predisposition to develop hemosiderosis (Dierenfeld et al. 1994; Cork et al. 1995).

## 6 Iron Bioaccumulation in Mammals and Birds

Wild animals are affected by a variety of environmental factors, which, to a large extent, determine the process of accumulation of heavy metals, both toxic ones and those essential to life. Analysis of heavy metal concentrations in animal tissues and organs usually provides indirect information about the quality of the environment. However, Fe concentration in animal tissues generally does not reflect its content in the environment. This probably results from effective homeostatic regulation within mammals and birds, the low food chain transfer of this element, and the forms in which iron is found in the environment, which determine its bioavailability (Wren et al. 1988; Cork et al. 1995; Miret et al. 2002; Soetan et al. 2010). The percentage of Fe absorbed from food (its bioavailability) varies but can range from 1% to 50%. Intestinal absorption of the metal can be affected by the form of ingested Fe and other food compounds including ascorbic acid, phytates, and proteins (Finch and Cook 1984; Miret et al. 2002; Hurrell and Egli 2010). Moreover, biological parameters such as age, physiological stage, genetic factors, and diet have a substantial influence on the content of Fe in endothermic vertebrates (Borch-Iohnsen et al. 1991; Puls 1994; Borch-Iohnsen and Thorstensen 2009; Olias et al. 2011). Animals with an active life, which run a lot and are in good shape, for example, horses and hares, contain more myoglobin, which is an iron-rich protein, and so the Fe concentration in their tissues is higher than in weak organisms of the same species (Mendel and

Wiechetek 2006; Myslek and Kalisinska 2006). Excessive storage of Fe (iron storage disease) in mammals and birds, especially in the liver, can occur as a result of genetically determined diseases, seasonal changes in their diet, and/or periodic starvation (Borch-Iohnsen et al. 1991; Dierenfeld et al. 1994; Clauss et al. 2002; Borch-Iohnsen and Thorstensen 2009; Clauss and Paglia 2012). At the same time, it should be noted that iron stimulates pathogenic bacteria to produce toxins. That is why, in response to inflammation, animal organisms retain iron in the cells of the reticuloendothelium system, intestinal epithelium, and other parts of the body, which may contribute to its increased concentration in different tissues (Puls 1994; Czernomysy-Furowicz and Furowicz 1999; Filipczyk et al. 2010; Pantopoulos et al. 2012).

## 6.1 Iron Concentration in Mammals

The average Fe content in the fresh tissue of adult mammals is about 0.005% (Hanusová et al. 2007). The main areas in which Fe is accumulated in mammals are the liver and spleen and, to a lesser degree, the kidneys, skeletal muscles, and bone marrow. In wild mammals, iron concentration is most frequently tested in their edible parts, such as the muscles, liver, and kidneys of ungulate game animals, and more rarely of omnivores and predators (Clauss and Paglia 2012). However, few studies offer information on physiological concentrations of Fe in wild herbivorous, omnivorous, or carnivorous mammals. Herbivores accumulate the lowest amount of Fe. This is probably because the Fe form present in plants is non-heme (inorganic) and hard to assimilate. The hemolytic form of readily assimilable iron occurs in foods of animal origin, which may explain the higher content of this element in piscivorous and carnivorous animals than in herbivores (Soetan et al. 2010).

Ranges of Fe concentration obtained during ecotoxicological studies from vertebrates of different taxonomic groups and between individuals of the same species vary greatly, making it difficult to interpret the results, which, in turn, deters many scientists from such studies, and so they are rarely undertaken.

### 6.1.1 Bioaccumulation of Fe in Herbivorous Mammals

In the diet of herbivorous mammals, iron interferes with many other elements. Cobalt (Co), manganese (Mn), and nickel (Ni) inhibit the uptake and transport of Fe, whereas zinc (Zn), phosphorus (P), potassium (K), sulfur (S), and calcium (Ca) decrease its bioavailability and disturb its metabolism (Kabata-Pendias and Szteke 2012; Sears 2013). Iron deficiency is not normally observed in ruminants (Doyle and Spaulding 1978). Mean hepatic Fe concentrations in large and medium wild herbivorous mammals range from 100 to 2800 mg $kg^{-1}$ dw (Table 6.1). According to Puls (1994), a typical Fe concentration in the liver of domesticated ruminants changes between 100 and 200 mg $kg^{-1}$ ww (or 330–670 mg $kg^{-1}$ dw).

**Table 6.1** The iron concentration in selected tissues of herbivorous mammals

| Species | Location | Fe concentration (mg kg$^{-1}$ dw) | | | References |
|---|---|---|---|---|---|
| | | Liver | Kidney | Muscle | |
| *Ruminants* | | | | | |
| Red deer *Cervus elaphus* | Poland, central part | 147.8[a] | 345.0[a] | 109.2[a] | Michalska and Zmudzki (1992) |
| | Poland, S part | 213 | 105 | 91 | Swiergosz et al. (1993) |
| | Poland, NW part | 102.5 | 66.98 | | Wieczorek-Dabrowska et al. (2013) |
| | Poland, NW part | 126.7[a] | 330[a] | 101.7[a] | Falandysz (1994) |
| | Sweden | | | 110[a] | Wilkund et al. (2014) |
| | Croatia | 228.3[a] | 505[a] | 247.7[a] | Lazarus et al. (2008) |
| | Croatia | | 605[a] | | Lazarus et al. (2005) |
| Fallow deer *Dama dama* | Slovenia | 470[a] | | | Vengušt and Vengušt (2004) |
| White-tailed deer *Odocoileus virginianus* | USA, Georgia | 320[a] | | | Lewis et al. (2001) |
| | USA, Virginia | S: 620[a] | | | Sleeman et al. (2010) |
| | | H: 485.7[a] | | | |
| | USA, Dakota | 537.8[a] | | | Zimmerman et al. (2008) |
| Mule deer *Odocoileus hemionus* | USA, California | 600[a] | | | Roug et al. (2015) |
| | USA, South Dakota | 514.6[a] | | | Zimmerman et al. (2008) |
| Roe deer *Capreolus capreolus* | Poland, central part | 684.0[a] | | | Dlugaszek and Kopczynski (2011) |
| | Poland, NW part | | | 127[a] | Dlugaszek and Kopczynski (2013) |
| | Poland, NW part | 240.2[a] | 304.0[a] | 117.3[a] | Michalska and Zmudzki (1992) |
| | Poland, NW part | 1007[a] | | | Chudzinska-Popek and Majdecka (2010) |
| | Poland, NW part | 201.4 | 106.3 | | Wieczorek-Dabrowska et al. (2013) |
| | Poland, northern part | 150[a] | 422.5[a] | 91[a] | Falandysz (1994) |
| Elk/moose *Alces alces* | Russia, NW part, Karelia | 578.1[a] | 161.4[a] | 170.5[a] | Medvedev (1999) |
| | Poland, NW part | | | 90[a] | Falandysz (1994) |
| | Canada, NW part | 693 | 171 | 232 | McDonald et al. (2005) |
| | USA, Minnesota | 848.5 | | | Custer et al. (2004) |
| | Canada, Ontario | 368.4 | 236.1 | 145.7 | Parker and Hamr (2001) |
| Caribou *Rangifer tarandus* | USA, Alaska, northern part | 2797[a] | 892.5[a] | 316.7[a] | O'Hara et al. (2003) |
| Reindeer *Rangifer tarandus* | Sweden | 467[a] | | 98[a] | Odsjö et al. (2007) |
| | Sweden | | | 113[a] | Wilkund et al. (2014) |
| | Russia, NW part, Karelia | | | 179.9[a] | Medvedev (1999) |

(continued)

**Table 6.1** (continued)

| Species | Location | Fe concentration ($mg\ kg^{-1}$ dw) Liver | Kidney | Muscle | References |
|---|---|---|---|---|---|
| *Lagomorphs* | | | | | |
| European hare *Lepus europaeus* | Poland, central part | 758[a] | | | Dlugaszek and Kopczynski (2011) |
| | Poland, central part | | | 95.3[a] | Dlugaszek and Kopczynski (2013) |
| | Poland, NW part | 730[a] | 600.0[a] | 138.7[a] | Myslek and Kalisinska (2006) |
| | Slovakia | 163.3 | 159.0 | | Czajkowska et al. (2011) |

We assumed that the kidneys contain 80% of water, liver and muscles 70% of water
*H* healthy, *S* sick with diarrhea
[a]Values converted from wet weight to the dry weight, dw

Medvedev (1999) showed that hepatic levels of Fe in the elk *Alces alces* (known as the moose in North America), which is a cervid species from the ruminant group, inhabiting the northern parts of Russia (Karelia) ranged from 115 to 273 mg kg$^{-1}$ ww (385–910 mg kg$^{-1}$ dw), and these were considered normal physiological levels. These values are similar to those determined by Puls (1994). Compared to the studies by Puls (1994) and Medvedev (1999), the values of Fe concentration proposed as physiological in the cervid liver by the Wisconsin Veterinary Diagnostic Laboratory (WVDL 2015) are lower (40–90 mg kg$^{-1}$ ww or 130–300 mg kg$^{-1}$ dw). The data on hepatic Fe in cervid species (elk/moose *Alces alces*, red deer *Cervus elaphus*, roe deer *Capreolus capreolus*, mule deer *Odocoileus virginianus*, fallow deer *Dama dama*, and reindeer/caribou *Rangifer tarandus*) as presented in Table 6.1 show that the mean values of Fe concentration are never lower than 100 mg Fe kg$^{-1}$ dw. They usually range from 100 to 600 mg kg$^{-1}$ dw, while the range is significantly higher in red deer from Poland and Croatia (Central and Southern Europe, respectively) (100–230 mg kg$^{-1}$ dw) and similar to that defined by WVDL (2015). A significant Fe concentration in the liver, exceeding 600 mg kg$^{-1}$ dw, was observed in moose and caribou from North America (O'Hara et al. 2003; Custer et al. 2004; McDonald et al. 2005) and in roe deer from Poland (Chudzinska-Popek and Majdecka 2010; Dlugaszek and Kopczynski 2011) as well as in the European hare *Lepus europaeus* from Poland (Myslek and Kalisinska 2006).

Although iron toxicity is rare in ruminants, it does occur (Doyle and Spaulding 1978; Clauss and Paglia 2012). In reindeer from Svalbard (Norway), whose diet varies during the year with respect to Fe content and who starve during winter, a loss of body mass is observed, including the liver (about 40% and 65%, respectively). In this species so-called seasonal liver siderosis was described. In late winter, the liver of reindeer foraging on iron-rich plants or plants with a normal iron level can contain 2910 and 1650 mg kg$^{-1}$ ww (or 9700 and 5500 mg kg$^{-1}$ dw), respectively (Borch-Iohnsen and Nilssen 1987). Another example of cervids with iron overload affecting the liver is the red deer from Germany. Iron storage disease was diagnosed in a few

2-year-old males of the same herd, and on top of that, cachexia and weight loss were observed. Hepatic Fe concentrations were very high in these animals (1108–-2275 mg $kg^{-1}$ ww or 3690–7580 mg $kg^{-1}$ dw). A possible genetic basis connected with the cervid hemochromatosis gene (HFE) was identified and compared between diseased and healthy red deer, but the results did not reveal any differences in HFE sequences (Olias et al. 2011). The authors found significantly increased Fe concentration in the water used by the animals and suggested that polymorphisms in other non-HFE genes involved in iron metabolism may have led to a higher sensitivity to iron in some members of the investigated herd. Probably an emaciation of diseased red deer and the loss of liver weight connected with this resulted in a simultaneous increase of iron concentrations in hepatic cells. The work by Borch-Iohnsen and Nilssen (1987) and Olias et al. (2011) indicates that in comparative studies related to hepatic Fe concentration in wild animals, greater emphasis should be paid to the condition of the animals and to seasonal changes in the quality of their food.

Ecotoxicological studies in Central Europe related to hepatic concentration of Fe in the European hare have shown that specimens from Poland contained over four times more of this metal than hares from neighboring Slovakia (~740 and 160 mg $kg^{-1}$ dw, respectively). Also in roe deer from various parts of Poland, average Fe concentration in the liver ranged widely – from 150 to >1000 mg $kg^{-1}$ dw. Thus, even in populations of the same species of mammal living at relatively short distances from one another, hepatic Fe concentrations can reach significantly different values (Table 6.1). This may be related to the effects of other heavy metals present in the environment and/or to the condition of the tested animals (Dlugaszek and Kopczynki 2011; Wieczorek-Dabrowska et al. 2013).

Iron is more rarely assayed in the kidneys than the liver of wildlife. Average Fe concentration in the kidneys of herbivorous mammals changes within the range of 70–1500 mg $kg^{-1}$ dw and is smaller when compared to the liver range. On the whole, hepatic Fe level is higher than the nephric one (Table 6.1). According to Puls (1994), adequate nephric Fe level in ruminants ranged from 30 to 150 mg $kg^{-1}$ ww (150–750 mg $kg^{-1}$ dw), but in cervids this range is probably narrower: 45–90 mg $kg^{-1}$ ww or 225–450 mg $kg^{-1}$ dw (WVDL 2015). However, as indicated in Table 6.1, mean nephric Fe concentration in cervids usually remains within the range of 100–500 mg $kg^{-1}$ dw, in which the lower threshold is lower than that in the studies cited here. The levels of Fe in the kidneys of herbivores are rarely higher than 600 mg kg $^{1}$ dw, for example, in caribou from Alaska, USA, and in the European hare from Poland (O'Hara et al. 2003; Myslek and Kalisinska 2006).

Mean Fe concentrations in the muscle of wild herbivores range from 90 to 320 mg $kg^{-1}$ dw and usually remain within the range of 90–150 mg $kg^{-1}$ dw (Table 6.1). Doyle and Spaulding (1978) as well as Puls (1994) describe a somewhat narrower range for the meat of domesticated ruminants: 64–130 and 45–54 mg Fe $kg^{-1}$ dw, respectively. Generally speaking, Fe content in the flesh of domesticated animals is lower than in wild ruminants (Wilkund et al. 2014).

On the basis of the analyzed studies, average concentrations of Fe in the liver, kidneys, and muscles of large- and medium-sized herbivorous mammals generally remain within the range of 100–600, 100–500, and 90–150 mg $kg^{-1}$ dw, respectively.

### 6.1.2 Bioaccumulation of Fe in Omnivorous Mammals

From the ecological point of view, the raccoon *Procyon lotor* (native to North America) and wild boar *Sus scrofa* (indigenous to Eurasia) can be ranked as omnivorous mammals. However, they belong to separate taxonomic orders (carnivore Carnivora and even-toed ungulates Artiodactyla, respectively). Both species are widespread as invasive species outside the area of their natural distribution (Genovesi et al. 2009; Snow et al. 2017). The raccoon is an opportunistic carnivore that relies on both aquatic and terrestrial habitats for foraging and may accumulate heavy metals present in either or both habitats (Souza et al. 2013). The wild boar is a typical terrestrial species, and its diet is dominated by plant material (Ballari and Barrios-García 2014). Assaying the content of elements in the tissues and organs of raccoons and wild boars is one way of evaluating their nutritional status and the partial environmental exposure of these animals, which is important for potential game consumers (Souza et al. 2013; Dlugaszek and Kopczynski 2011). Although, as an invasive species, the wild boar inhabits North America, and the raccoon Europe, there are no studies on Fe in these animals living outside their natural range. So far, no physiological Fe concentrations have been established either for the wild boar or the raccoon. In the case of wild boar, it can be assumed that they are close to the adequate values given for porcine and swine by Puls (1994) and WVDL (2015). Both of these reports of hepatic concentrations were given as normal levels of 100–200 mg Fe kg$^{-1}$ ww (330–670 mg Fe kg$^{-1}$ dw). In the livers of wild boar from various parts of Poland, mean Fe concentration ranges between 100 and 870 mg kg$^{-1}$ dw, but does not usually exceed 300 mg kg$^{-1}$ dw (Michalska and Zmudzki 1992; Swiergosz et al. 1993; Dlugaszek and Kopczynski 2011). In Italy the hepatic Fe level of the species was found to be ~420 mg kg$^{-1}$ dw (Zaccaroni et al. 2003). These data indicate that mean Fe concentration in the liver of wild boar oscillated around the lower values given for swine. Michalska and Zmudzki (1992) noticed that in good nourished wild boar, hepatic Fe concentration was significantly higher in autumn than in spring (290 and 240 mg kg$^{-1}$ dw, respectively). This observation also applies to the kidneys and muscles of wild boar obtained in autumn. In spring their nephric Fe and muscular concentrations were about 560 and 420 and 163 and 76 mg kg$^{-1}$ dw, respectively. In comparison with data from Michalska and Zmudzki (1992) concerning Fe levels in the kidneys of wild boar from South Poland, Swiergosz et al. (1993) found that concentration of this metal in animals from the central part of that country was twice as low (55 mg kg$^{-1}$ dw). There are significantly more reports on muscle Fe than offal Fe of the species because its meat has the highest value for consumers. Iron content in meat depends on the amount of fat and blood (Strazdina et al. 2013). Therefore, mean muscle Fe concentration in European wild boar can vary from 60 (in Romania) to 170 (in Poland) mg kg$^{-1}$ dw (Crăciunescu et al. 2014; Roslewska et al. 2016), while the values usually fall within the range of 80–120 mg kg$^{-1}$ dw (Swiergosz et al. 1993; Dlugaszek and Kopczynski 2013; Strazdina et al. 2013).

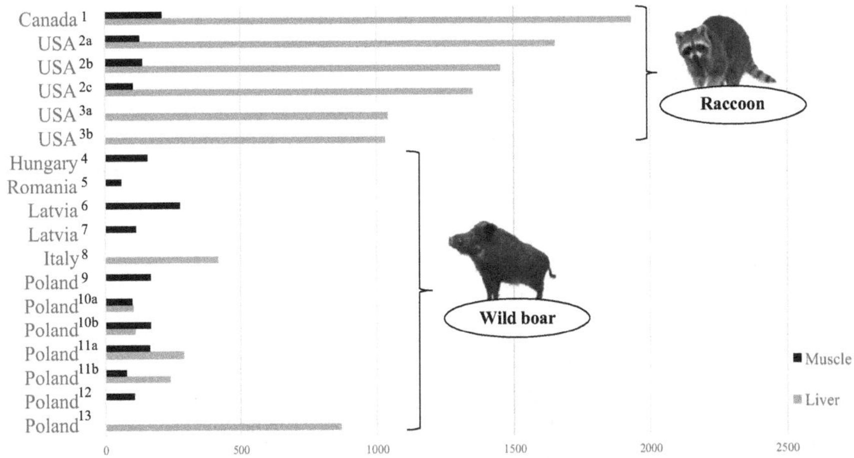


**Fig. 6.1** Hepatic and muscle iron concentrations (mg kg$^{-1}$ dw) in raccoons from North America and wild boar from Europe (raccoon: 1. Wren 1984; 2. Souza et al. 2013: 2a polluted area in 2009, 2b polluted area in 2010, 2c unpolluted area; 3. Lewis et al. 2001: 3a studied group, 3b control group), wild boar (4. Skobrák et al. 2011; 5. Crăciunescu et al. 2014; 6. Strazdina et al. 2014; 7. Strazdina et al. 2013; 8. Zaccaroni et al. 2003; 9. Roslewska et al. 2016; 10. Swiergosz et al. 1993: 10a Niepolomicka Forest, 10b Beskid Sadecki; 11. Michalska and Zmudzki 1992: 11a in autumn, 11b in spring; 12. Dlugaszek and Kopczynki 2013; 13. Dlugaszek and Kopczynki 2011)

In North America raccoon is classified as a game species, and its meat is used for human consumption and for bioindicative purposes yet more rarely than that of wild boar in Europe (Fig. 6.1). Raccoon is solely opportunistic in its feeding habits. Fruits, insects, acorns, and crayfish are normally the main elements of its diet (Wilhide et al. 1992). Average concentration of Fe in the liver and kidneys of raccoon ranges from 1030 to 1940 and 458 to 1970 mg kg$^{-1}$ dw, respectively (Wren 1984; Lewis et al. 2001; Souza et al. 2013).

Souza et al. (2013) determined Fe concentration in the liver, kidney, and muscle of raccoons originating from areas unexposed and exposed to coal fly ash. Only muscle Fe concentration was significantly higher in raccoons from the polluted areas (~130 mg kg$^{-1}$ dw) than the unpolluted areas (~100 mg kg$^{-1}$ dw). When comparing Fe concentration in the livers of two omnivorous species from Europe and North America, it may be noticed that wild boar livers contain 2–3 times less Fe than those of raccoons, which may result in a higher percentage of animal material in the diet of raccoons than in the diet of wild boar, for whom plants contribute to even 90% of their food (Ballari and Barrios-García 2014).

In the case of muscle Fe, such clear interspecies differences are not observed (Fig. 6.1). What is more, as far as Fe concentrations in raccoon are concerned, there is clear and strong differentiation between tissues (liver > kidney > muscle), but no such trend has been observed in wild boar. Sometimes the concentration of Fe in the kidney was higher than in the liver (Michalska and Zmudzki 1992), or Fe concentrations in the liver and muscle were similar (~100 mg kg$^{-1}$ dw) and lower in the kidneys by half (Swiergosz et al. 1993).

### 6.1.3 Bioaccumulation of Fe in Carnivorous Mammals

With respect to their diet, carnivores dwelling on land can be divided into piscivores, which are associated with aquatic habitats (including river otter *Lontra canadensis* and American mink *Neovison vison*), typical terrestrial predators (mainly canids such as gray wolf *Canis lupus*, red fox *Vulpes vulpes*, arctic fox *V. lagopus*, and raccoon dog *Nyctereutes procyonoides*), and opportunistic predators, which feed on small- and medium-sized endothermic animals and, to a lesser degree, on fruit and other plant material. This last group includes, for example, representatives of the genus *Martes*. The species most commonly used in ecotoxicological studies of Fe include American mink, otters, and canids, but the number of such studies is not large. Most data on normal metal concentrations found in the literature pertain to American mink and dog, because they are domestic or partially domestic mammals (Stejskal et al. 1989; Puls 1994). American mink, as a fur-bearing species, is farmed in North America, Europe, and on other continents, but the animals sometimes escape from farms and live as feral (Bowman et al. 2017). Stejskal et al. (1989) determined mean Fe concentrations in the liver and kidneys of the adults of natural dark ranch mink to be as high as 392 and 206 mg kg$^{-1}$ ww (or 1310 and 1030 mg kg$^{-1}$ dw, respectively). In wild minks from Canada (British Columbia and Ontario), the mean values of hepatic Fe concentrations were similar to those described by Stejskal et al. (1989) (Fig. 6.2). However, nephric Fe levels in mink from Ontario (Wren et al. 1988) and British Columbia (Harding et al. 1998) were lower than those reported by Stejskal et al. (1989) by about 17% and 43%,

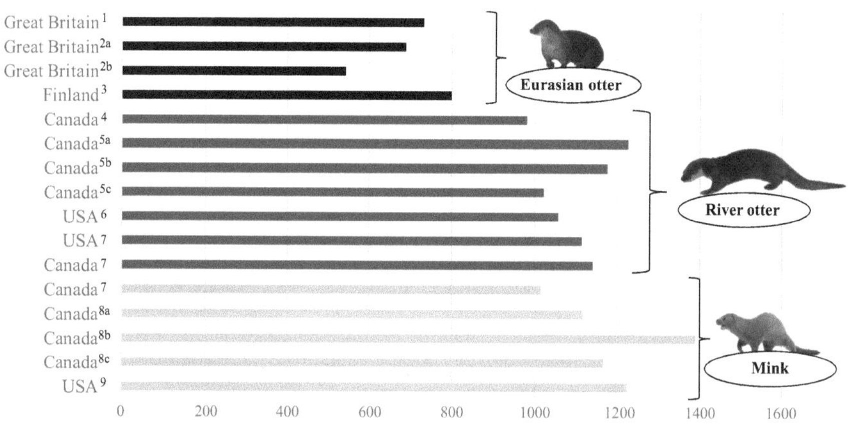


**Fig. 6.2** Hepatic iron concentration (mg kg$^{-1}$ dw) in piscivorous mammals (from Europe – Eurasian otter: 1. Walker et al. 2011; 2. Walker et al. 2010: 2a year 2007, 2b year 2008; 3. Lodenius et al. 2014; from North America: river otter; 4. Wren 1984; 5. Wren et al. 1988: 5a Muskoka, 5b Sudbury, 5c Turkey Lake; 6. Grove and Henny 2008; 7. Harding et al. 1998; wild American mink: 7. Harding et al. 1998; 8. Wren et al. 1988: 8a Muskoka, 8b Sudbury, 8c Turkey Lake; ranch mink: 9. Stejskal et al. 1989)

respectively. Although the concentrations of different metals were investigated in feral American mink living in Europe, they did not concern iron (Brzezinski et al. 2014; Ljungvall et al. 2017).

As indicated in Fig. 6.2 in river otters from Canada and the USA, hepatic Fe concentrations change within a small range (980–1230 mg $kg^{-1}$ dw). Only in young specimens can it be significantly lower (~830 mg $kg^{-1}$ dw) than in adults, as indicated by Grove and Henry (2008). Usually mean Fe concentrations in the liver of the river otter are similar to those of wild and ranch mink. Similarly to the wild mink, kidney Fe levels in river otters are generally lower than liver Fe levels. River otters had mean values of nephric Fe concentration within the range of 130–195 mg $kg^{-1}$ ww or 650–970 mg $kg^{-1}$ dw (Wren 1984; Wren et al. 1988). In Eurasian otters from Great Britain and Finland, only hepatic Fe concentrations were measured, and their average values were within the range of 540–800 mg $kg^{-1}$ dw (Fig. 6.2). It appears therefore that in comparison with North American otters, mean Fe concentration in the liver was lower by a few hundred mg $kg^{-1}$ dw. These differences in hepatic Fe concentrations between river and Eurasian otters may result from the biological parameters of the studied animals and be affected by local environmental factors, including the presence of other heavy metals in the otters' food (Wren et al. 1988; Grove and Henny 2008; Walker et al. 2010, 2011; Lodenius et al. 2014).

Mean values of Fe concentration in the kidneys of wild piscivores from North America ranged from ~460 to 960 mg $kg^{-1}$ dw and were lower than those in their livers (Wren 1984; Wren et al. 1988; Harding et al. 1998). It seems that nephric Fe concentration in wild mink is lower in comparison with ranch animals: 456–880 versus 1030 mg $kg^{-1}$ dw (Wren et al. 1988; Stejskal et al. 1989; Harding et al. 1998). In other North American mustelid species, like marten (*Martes americana*), Fe levels in the kidney were similar to those in wild mink and reached 777 mg $kg^{-1}$ dw (Harding 2004).

In ecotoxicological studies of heavy metals, different canid species are utilized. Natural hepatic Fe concentration in the domestic dog is 400–1200 mg $kg^{-1}$ dw (Puls 1994). According to WVDL (2015), normal, deficient, and toxic hepatic Fe concentration in canines ranges from 100 to 300 mg $kg^{-1}$ ww (330–1000 mg $kg^{-1}$ dw), 20 to 60mg $kg^{-1}$ ww (70–200 mg $kg^{-1}$ dw), and above 500 mg $kg^{-1}$ ww (>1700 mg $kg^{-1}$ dw), respectively. Normal nephric levels remain within the range 66–150 mg $kg^{-1}$ ww or 330–750 mg $kg^{-1}$ dw (WVDL 2015). Schultheiss et al. (2002) determined liver Fe concentration and analyzed pathological symptoms in the organs of dogs. They found hepatic Fe levels between 177 and 7680 mg $kg^{-1}$ dw. Dogs with concentrations >2400 mg $kg^{-1}$ dw had severe inflammation and fibrosis, but they did not appear to have hemochromatosis. Some pathological changes (mainly mild periportal inflammation) also occurred in some dogs whose Fe concentration in the liver ranged between 1200 and 2400 mg $kg^{-1}$ dw. In the progenitor of the dog, namely, the wolf (in individuals from four parts of North America), nephric Fe concentrations were assayed, and the mean values ranged between 102 and 172 mg $kg^{-1}$ ww, but usually did not exceed 120 mg $kg^{-1}$ ww or 600 mg $kg^{-1}$ dw (Hoffman et al. 2010). It is only in wolves from Alaska, USA, that increased Fe levels were observed in the kidneys when compared to the normal values determined for dogs (WVDL 2015). In wild Arctic foxes in North America (Canada) and Europe (Norway) and in red fox in Poland, the mean Fe concentration in the liver remains in the range between 110 and 344 mg $kg^{-1}$ ww (370–1147 mg $kg^{-1}$ dw) and generally does not deviate from the values determined for the house dog (Prestrud et al. 1994; Hoekstra et al. 2003; Binkowski et al. 2016).

However, in different canid species (silver fox, Arctic fox, and raccoon dog) kept at fur farms, mean hepatic concentration of Fe generally ranged between 1300 and 1500 mg $kg^{-1}$ dw, which may indicate the occurrence of some pathological changes (Hanusová et al. 2007; WVDL 2015). Hepatic, nephric, and muscle Fe concentrations were investigated in farm raccoon dogs in Europe and Asia. In a European study, Hanusová et al. (2007) detected in the livers of males and females about 1300 and 1470 mg Fe $kg^{-1}$ dw, respectively, but in an Asian analysis, the hepatic Fe concentration was at least 3.5 times lower in the control raccoon dog group in comparison with the cited report (Hou et al. 2012). For that reason using data obtained from farm animals to interpret Fe concentration in the livers of wild animals may have a rather limited significance. It is also worth noting that Fe concentration in the kidneys of farm raccoon dogs from Europe and Asia, as well as in wild red fox from Poland, was similar and ranged from 180 to 270 mg $kg^{-1}$ dw (Hanusová et al. 2007; Hou et al. 2012; Binkowski et al. 2016).

Iron concentration in the muscles of carnivores is rarely studied. In wild animals, the highest value of Fe concentration was found in the river otter, an animal which swims and dives a lot throughout its life (158 mg $kg^{-1}$ ww or 527 mg $kg^{-1}$ dw), and values were much lower in red fox and brown bear: 103 and 218 mg $kg^{-1}$ dw, respectively (Wren 1984; Medvedev 1999; Binkowski et al. 2016). In farm canids Fe concentrations in the muscle were within the range of 180–390 mg $kg^{-1}$ dw (Hanusová et al. 2007; Hou et al. 2012).

### 6.1.4 Bioaccumulation of Fe in Micromammals

Fe concentration in the livers of micromammals followed a descending order: insectivorous shrew *Crocidura russula* > plant-eating rodents (bank vole *Clethrionomys glareolus*, Algerian mice *Mus spretus*, wood mice *Apodemus sylvaticus*, and yellow-necked mice *Apodemus flavicollis*). Mean hepatic Fe concentrations of insectivores and other micromammals are in the ranges ~730–3300 and 100–1080 mg $kg^{-1}$ dw, respectively (Damek-Poprawa and Sawicka-Kapusta 2003; Sanchez-Chardi et al. 2007, 2009a, b; Marques et al. 2008). On the whole, Fe levels in the liver and kidneys of rodents are within 400–650 and 300–500 mg $kg^{-1}$ dw (Damek-Poprawa and Sawicka-Kapusta 2003; Topolska et al. 2004; Martiniaková et al. 2010). Such a wide range of Fe concentration in the organs of micromammals can be most likely explained by differences in metabolism rates, altered feeding patterns, seasonal food availability, habitat suitability, and connectivity, as well as life-stage-related food (Martiniaková et al. 2010).

Most data related to medium and large wild mammals concern Fe concentration in the liver. As far as Fe concentration is concerned, they can be arranged in the following order of magnitude: piscivorous > carnivorous > omnivorous > herbivorous. These data indicate that Fe concentrations in the liver, kidneys, and muscles of mammals vary not only among different tissues, are dependent on the species and may differ depending on the age of an animal and its trophic level. Sleeman et al. (2010), Grove and Henny (2008), and Puls (1994) observed significantly higher Fe concentration in the liver of juvenile white-tailed deer, river otters, and dogs than in adults. Moreover, Lazarus et al. (2008) found that Fe concentration was higher in the kidneys and muscles of juvenile red deer than in adults of the species. However, although the relationship between Fe levels in the liver, kidneys, and muscles and the age of an animal have been analyzed in roe deer, fallow deer, reindeer, raccoon dog, red fox, gray wolf, shrew, and otter, no significant relationship has been found (Prestrud et al. 1994; Medvedev 1999; Hoekstra et al. 2003; Vengušt and Vengušt 2004; Hanusová et al. 2007; Sanchez-Chardi et al. 2007; Dlugaszek and Kopczynski 2011, 2013; Walker et al. 2011; Binkowski et al. 2016).

Generally, there were no significant differences in Fe concentration in the tissues of many wild mammalian species related to their sex (Prestrud et al. 1994; Medvedev 1999; Hoekstra et al. 2003; Zaccaroni et al. 2003; Skobrak et al. 2011; Sleeman et al. 2010; Walker et al. 2011; Binkowski et al. 2016; Roslewska et al. 2016). In contrast to the papers cited above, Vengušt and Vengušt (2004) found significantly higher Fe concentration in the liver of female fallow deer when compared to that of males.

## 6.2 Bioaccumulation of Fe in Birds

Concentrations of trace elements, including Fe, in birds depend on a number of factors, including their position in the trophic chain, feeding and dietary habits, health condition, physiological state, and age, as well as the susceptibility of a given species to absorption of different elements (Kim et al. 1996; Kalisinska et al. 2009). Iron is accumulated in differing concentrations in the tissues and organs of birds. It becomes toxic above 100 mg kg body weight (bw), while at 1 mg kg bw, it results in anemia. High concentrations of Fe in the liver may indicate inflammatory processes in the body and immune response (Cook et al. 1974). Many factors affect the pathological deposition of Fe in birds' livers. It has been suggested, however, that parenchymal siderosis can hardly be explained by liver weight loss (Borch-Iohnsen et al. 1991; Dierenfeld et al. 1994; Sheppard and Dierenfeld 2002). It may well be related to diseases that the birds had previously suffered from and/or be a result of inflammatory processes in the body (Kalisinska et al. 2008). Saiz et al. (1990) suggested that excess Fe is stored in nontoxic form as ferritin molecules, mainly in the liver. It is therefore not unusual to find high hepatic Fe concentrations, which may reflect physiological variations in Fe levels related to the egg-laying process. High levels of Fe in the liver can be related to lead and zinc poisoning (Carpene et al. 1995).

Studies of Fe concentration in the organs of birds are not very common. In birds, Fe concentration is commonly determined in the liver, kidneys, muscles, feathers, blood, brain, and bones. Even though Fe concentrations in wild birds are most often studied in the liver, the scientific literature has not proposed normal Fe values for different avian species. The Wisconsin Veterinary Diagnostic Laboratory determined normal, deficient, high, and toxic hepatic Fe concentrations in poultry, which ranged from 60–300mg kg$^{-1}$ ww (200–1000 mg kg$^{-1}$ dw), 30–35mg kg$^{-1}$ ww (100–120 mg kg$^{-1}$ dw), 300–2000 mg kg$^{-1}$ ww (1000–6600 mg kg$^{-1}$ dw), and above 8000 mg kg$^{-1}$ ww (>26,000 mg kg$^{-1}$ dw), respectively. Normal concentration in the kidneys of poultry ranges from 45 to 100 mg kg$^{-1}$ ww (225–500 mg kg$^{-1}$ dw), while the toxic level begins above 200 mg kg$^{-1}$ ww (>1000 mg kg$^{-1}$ dw) (WVDL 2015).

### 6.2.1 Bioaccumulation of Fe in Avian Liver, Kidney, Muscles, and Feathers

Among the various bird species, Fe concentration in different organs and tissues is most commonly analyzed in ducks, including genera *Anas* and *Aythya*. Because of their wide geographical ranges, long life, the fact that the species and the sex of those birds are easily recognizable, as well as the large sizes of their populations, this group of birds is considered a good biomonitor (Onderscheka et al. 1985; Furness and Greewood 1993; Kalisinska et al. 2004).

In adult ducks the highest iron concentrations were found in liver < kidneys < muscle (Kozulin and Pavluschick 1993; Kalisinska et al. 2004). Physiological concentration not exceeding 1000 mg kg$^{-1}$ dw (WVDL 2015) was determined in mallards (*Anas platyrhynchos*) from Iran (Sinka-Karimi et al. 2015) and Korea (Kim and Oh 2012b), in pochards (*Aythya ferina*) from Iran (Sinka-Karimi et al. 2015), and spot-billed ducks (*Anas poecilorhyncha*) from Korea (Kim and Oh 2012a). Iron concentration exceeding 1000 mg kg$^{-1}$ dw (ranging between about 1200 and 2800 mg kg$^{-1}$ dw), indicating high Fe accumulation in the liver (WVDL 2015), was observed in common eider (*Somateria mollissima*) from Finland (Franson et al. 2000), mallards from central Poland (Kalisinska et al. 2004), and lesser scaup (*Aythya affinis*) and redhead (*Aythya americana*) from North America (Michot et al. 1994; Custer et al. 2003; Pillatzki et al. 2011). The highest average Fe concentration in the liver of Anseriformes was observed in great scaup (*Aythya marila*) from Alaska and in mallard from a polluted area in Russia, ranging from around ~4500 to 5900 mg kg$^{-1}$ dw, respectively (Kozulin and Pavluschick 1993; Badzinski et al. 2009). However, these concentrations did not exceed 26,000 mg kg$^{-1}$ dw, which would indicate toxic Fe concentration (WVDL 2015). In some cases, including the mynah bird (*Gracula religiosa*), which is one of the species most commonly reported to be susceptible to iron overload, hepatic Fe concentration ranged from ~6700 to 23,000 mg kg$^{-1}$ dw (Mete et al. 2003).

In the kidneys, normal Fe concentration not exceeding 500 mg kg$^{-1}$ dw was assayed in mallard and pochard from Iran (Sinka-Karimi et al. 2015). Higher Fe concentration was observed in the kidneys of mallard from Poland and Russia and ranged around 800 mg kg$^{-1}$ dw (Kozulin and Pavluschick 1993; Kalisinska et al.

2004). Taking into consideration their habitat, nephric Fe concentrations in ducks from Asia were at physiological levels, but they reached high levels in the same species of birds from Europe (WVDL 2015).

Based on the data available in the literature, it was found that Fe concentration in the muscle of Anseriformes ranged from ~140 to ~370 mg $kg^{-1}$ dw, which reflected physiological iron levels in the tissues (Proske et al. 1993; Kalisinska et al. 2004; Sinka-Karimi et al. 2015).

Passerines play an important role in the biomonitoring of terrestrial ecosystems. Some of the studied species include the house sparrow (*Passer domesticus*), the great tit (*Parus major*), and the blue tit (*Cyanistes caeruleus*), which have been studied in China, Finland, Poland, and Spain. However, there are only a few studies reporting on Fe concentrations in the organs of those birds, and they usually pertain to hepatic concentrations of this element. In the livers of great tit from Europe, Fe concentration ranged around ~1500 mg $kg^{-1}$ dw (Ingervo et al. 1995; Llacuna et al. 1995; Sawicka-Kapusta et al. 1986). In house sparrow (*Passer domesticus*), hepatic Fe concentration did not exceed 570 mg $kg^{-1}$ dw (Kekkonen et al. 2012). Twofold higher Fe levels were observed in the liver of the tree sparrow (*Passer montanus*) from industrial areas in China. In its muscles Fe level ranged from ~170 to ~270 mg $kg^{-1}$ dw (Chao et al. 2003).

High Fe concentrations of several, sometimes even a few tens of thousands of mg $kg^{-1}$ dw, were found in dead birds of prey, which had died in various circumstances and been found in the field. In the liver of these birds, which originated from Europe and Asia, Fe concentrations ranged from ~1030 to ~2500 mg $kg^{-1}$ dw (Hontelez et al. 1992; Jager et al. 1996; Falandysz et al. 1988; Kalisinska et al. 2009; Kim and Oh 2015; Kitowski et al. 2017a, b). A significantly higher Fe concentration was reported, among others, in a severely emaciated female peregrine falcon (*Falco peregrinus*) from Poland, and amounted to over 6000 mg $kg^{-1}$ dw (Kalisinska et al. 2008).

Even higher concentrations were reported by Kitowski et al. (2017a) in a few specimen of buzzard (*Buteo buteo*). In the liver of these birds, Fe concentration exceeded 18,000 mg $kg^{-1}$ dw, which could indicate Fe hyper-accumulation. A high Fe level in the liver could be associated with serious bacterial and helminthological infections as well as cadmium (Cd), lead (Pb), or zinc (Zn) poisoning (Lewis et al. 2001).

Nephric Fe concentration in birds of prey ranged from ~530 to ~820 mg $kg^{-1}$ dw (Jager et al. 1996; Falandysz et al. 1988; Kalisinska et al. 2006, 2008, 2009). Higher levels of nephric Fe of ~1200 mg $kg^{-1}$ dw were reported by Hontelez et al. (1992) and Falandysz et al. (2000) in buzzard from the Netherlands and white-tailed eagle from Poland (*Haliaeetus albicilla*), respectively. Mean concentration of Fe in the muscles of white-tailed eagle did not exceed 400 mg $kg^{-1}$ dw (Falandysz et al. 1988, 2000; Kalisinska et al. 2006; Mihaljev et al. 2012).

In birds associated with aquatic ecosystems, including the gray heron (*Ardea cinerea*) and black-crowned night heron (*Nycticorax nycticorax*), Fe concentrations were mainly assayed in the liver. The average Fe concentration in the liver of those birds varied from ~600 to ~820 mg $kg^{-1}$ dw (Kim and Oh 2015). Higher concentrations were, however, found in the gray heron from Serbia (Mihaljev et al. 2012) and in the bald eagle (*Haliaeetus leucocephalus*) from Alaska (Stout and Trust 2002). Such wide ranges of Fe concentration in a specimen of the same species and the same trophic group make it difficult to interpret the results. The average nephric Fe concentrations in the intermediate egret (*Egretta intermedia*) and in little egret (*Egretta garzetta*) were similar and ranged between 350 and 450 mg $kg^{-1}$ dw (Kim et al. 2010; Kim and Oh 2015). Much lower Fe concentrations in the kidney were noted in the western reef egret (*Egretta gularis schistacea*), and they did not exceed 40 mg $kg^{-1}$ dw (Mansouri et al. 2012).

Ecotoxicological studies researching Fe concentration in the tissues and organs of herbivorous birds were carried out in Korea, Russia, and the USA. Fe concentration in the liver of these birds depended, among other factors, on their habitat. In the mute swan (*Cygnus olor*) from Lake Erie, USA, hepatic Fe ranged from ~600 to ~12,000 mg $kg^{-1}$ dw, and in birds of the same species from St. Clair, USA, it ranged between ~700 and 6000 mg $kg^{-1}$ dw (Schummer et al. 2011). The concentration of Fe in the liver of the southern subpopulation of the lesser snow goose (*Anser caerulescens*) ranged between ~400 and ~4600 mg $kg^{-1}$ dw and, in the liver of the northern subpopulation, from ~1200 to ~4200 mg $kg^{-1}$ dw (Hui et al. 1998). The authors suggested that geese from the northern subpopulation feed in pastures and on coastal marshes and migrate along the coast, while those from the southern subpopulation feed predominantly in rice fields and migrate over farmland. Increased concentrations of various elements, including the organic forms of Hg, in rice paddy fields may possibly lead to changes in liver function, thus increasing, inter alia, hepatic Fe levels (Strickman and Mitchell 2017).

Rose and Parker (1982) determined that grouse's feathers act as a natural indicator of iron levels present in the environment – not, however, through endogenous incorporation of iron into growing plumage but through exogenous adsorption of iron onto plumage surfaces during the plumage year. There is little research pertaining to Fe concentration in bird feathers. There are just a few studies concerning birds associated with land and wetland ecosystems, including the cattle egret (*Bubulcus ibis*) and the black-crowned night heron (Abdullah et al. 2015; Kim and Koo 2007; Malik and Zeb 2009; Manjula et al. 2015; Ullah et al. 2014). The average Fe concentration in the feathers of these birds ranged from ~100 to ~340 mg $kg^{-1}$ dw. The highest Fe concentrations were found in birds from highly urbanized areas (Abdullah et al. 2015; Manjula et al. 2015). High Fe concentrations in feathers may reflect diet and the mobilization quantities stored during the period of feather growth (Dauwe et al. 2000; Rattner et al. 2008). Fe concentration in the feathers of birds of prey, including ospreys (*Pandion haliaetus*) originating from the heavily polluted coastal waters of Chesapeake Bay (USA), differed significantly in the years 2000 and 2001 and amounted to ~170 mg $kg^{-1}$ dw and ~80 mg $kg^{-1}$ dw, respectively (Rattner et al. 2008). The authors suggest that Fe concentrations in feathers may be affected, among other factors, by extensive metal-working and petroleum refinery activities in this area.

Generally, there were no significant differences in Fe concentration in the tissues of various birds with respect to sex and age (Custer et al. 2003; Jager et al. 1996;

Janiga et al. 1990; Kekkonen et al. 2012; Malinga et al. 2010; Mansouri et al. 2012; Michot et al. 1994; Stout and Trust 2002). However, Sinka-Karni et al. (2015) observed that sex affected Fe levels in mallards, as females had a higher Fe hepatic concentration than males. This could result from differences in the production of metalloprotein, which plays a fundamental role in the transport, storage, and excretion of metals. Ingervo et al. (1995) observed that Fe concentrations in the livers of female great and blue tits were lower than in the males. Schummer et al. (2011) observed significantly higher hepatic Fe concentration in female mute swans than in males. Borch-Iohnsen et al. (1991) attributed high iron content in the livers of female birds partly to catabolism of lean body tissue during incubation. Moreover, Proske et al. (1993) studied ducklings hatched in captivity, aged 1 day to 22 weeks, and found that the liver iron content increased between the ages of 1 and 16 weeks. Mallards from the two areas in northwestern Poland studied by Kalisinska et al. (2004) showed positive and significant correlations between muscle iron content and age. Greater Fe content in the muscles of adult ducks most likely results from physiological differences in muscle structure. Adult ducks protect young birds during flight and thus carry out more work.

Generally, hepatic Fe concentration in birds can be ranked in the following order: carnivores > herbivores > omnivores > piscivores (Fig. 6.3). Differences in Fe

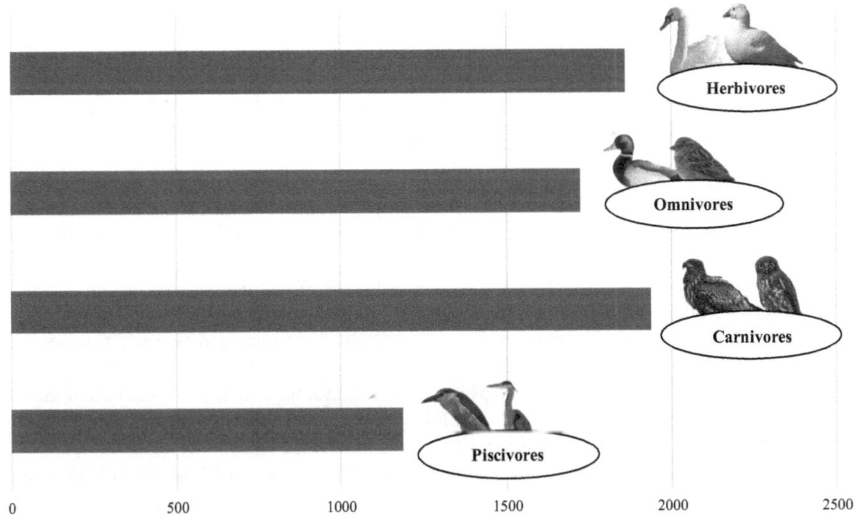


**Fig. 6.3** The concentration of Fe (mg kg$^{-1}$ dw) in the livers of birds with different diets based on herbivores (Hui et al. 1998; Schummer et al. 2011; Kim and Oh 2013b), omnivores (Sawicka-Kapusta et al. 1986; Michot et al. 1994; Ingervo et al. 1995; Franson et al. 2000; Chao et al. 2003; Custer et al. 2003; Kalisinska et al. 2004; Badzinski et al. 2009; Pillatzki et al. 2011; Kekkonen et al. 2012; Kim and Oh 2012b; Sinka-Karimi et al. 2015), carnivores (Falandysz et al. 1988, 2000; Hontelez et al. 1992; Esselink et al. 1995; Jager et al. 1996; Kalisinska et al. 2006, 2008, 2009; Kim et al. 2008; Kim and Oh 2012a, 2015; Mihaljev et al. 2012), piscivorous (Carpene et al. 1995; Stout and Trust 2002; Kim et al. 2010; Mansouri et al. 2012; Mihaljev et al. 2012; Kim and Oh 2013a, 2015)

content in bird tissues may be due to different physiological demands for this element at different times of life and the intensity of enzyme reactions in the cells. Despite various studies focusing on iron content in wild birds, our understanding of the concentration of this element in the parenchyma is only fragmentary, and further studies are needed. Such research should take into account various factors affecting Fe concentration in bird organs and tissues, including age, sex, diet, health condition, seasonal changes, and migrations. A multivariate analysis of this kind would certainly facilitate the interpretation of results.

## References


Abbaspour N, Hurrell R, Kelishadi R (2014) Review on iron and its importance for human health. J Res Med Sci 19:164–174

Abdullah M, Fasola M, Muhammad A, Malik SA, Bostan N, Bokhari H et al (2015) Avian feathers as a non-destructive bio-monitoring tool of trace metals signatures: a case study from severely contaminated areas. Chemosphere 119:553–561

Adlerova L, Bartoskova A, Faldyna M (2008) Lactoferrin: a review. Vet Med 53:457–468

Albretsen J (2006) The toxicity of iron, an essential element. Vet Med 101:82–90

Andrews NC (1999) Disorders of iron metabolism. New Engl J Med 341:1886–1995

Audi G, Bersillon O, Blachot J, Wapstra AH (2004) The NUBASE evaluation of nuclear and decay properties. 2004. <in2p3-00020241>, http://hal.in2p3.fr/in2p3-00020241

Bacon BR, Britton RS (1990) The pathology of hepatic iron overload: a free radical-mediated process? Hepatology 11:127–137

Badzinski SS, Flint PL, Gorman KB, Petrie SA (2009) Relationships between hepatic trace element concentrations, reproductive status, and body condition of female greater scaup. Environ Pollut 157:1886–1893

Baixing Y, Guan J, Shesterkin V, Zhu V (2016) Variations of dissolved iron in the Amur River during and extreme flood event in 2013. Chin Geogr Sci 26:679–686

Bakker GR, Boyer RF (1986) Iron incorporation into apoferritin. The role of apoferritin as a ferroxidase. J Biol Chem 261:13182–13185

Ballari SA, Barrios-García MN (2014) A review of wild boar *Sus scrofa* diet and factors affecting food selection in native and introduced ranges. Mammal Rev 44:124–134

Beard JL (2001) Iron biology in immune function, muscle metabolism and neuronal functioning. J Nutr 131:568S–580S

Binkowski LJ, Merta D, Przystupinska A, Soltysiak Z, Pacon J, Stawarz R (2016) Levels of metals in kidney, liver and muscle tissue and their relation to the occurrence of parasites in the red fox in the Lower Silesian Forest in Europe. Chemosphere 149:161–167

Boehler R (2000) High-pressure experiments and the phase diagram of lower mantle and core materials. Rev Geophys 38:221–245

Borch-Iohnsen B, Nilssen KJ (1987) Seasonal iron overload in Svalbard reindeer liver. J Nutr 117:2072–2078

Borch-Iohnsen B, Thorstensen K (2009) Iron distribution in the liver and duodenum during seasonal iron overload in Svalbard reindeer. J Comp Pathol 141:27–40

Borch-Iohnsen B, Holm H, Jørgensen A, Norheim G (1991) Seasonal siderosis in female eider nesting in Svalbard. J Comp Pathol 104:7–15

Bowman J, Beauclerc K, Farid AH, Fenton H, Klütsch CFC, Schulte-Hostedde AI (2017) Hybridization of domestic mink with wild American mink (*Neovison vison*) in eastern Canada. Can J Zool 95:443–451

British Geological Survey (2015) World Mineral Production 2011–2015

Brock JH (1989) Iron-binding proteins. Acta Paediatr Scand Suppl 361:31–43
Brzezinski M, Zalewski A, Niemczynowicz A, Jarzyna I, Suska-Malawska M (2014) The use of chemical markers for the identification of farm escapees in feral mink populations. Ecotoxicology 23:767–778
Carpene E, Serra R, Isani G (1995) Heavy metals in some species of waterfowl of northern Italy. J Wildl Dis 31:49–56
Carral E, Puente X, Villares R, Carballeira A (1995) Background heavy metal levels in estuarine sediments and organisms in Galicia (northwest Spain) as determined by modal analysis. Sci Total Environ 172:175–188
Chao P, Guangmei Z, Zhenwang Z, Chengyi Z (2003) Metal contamination in tree sparrows in different locations of Beijing. Bull Environ Contam Toxicol 71:142–147
Chudzinska-Popek M, Majdecka T (2010) Problems of mineral metabolism in roe deer (*Capreolus capreolus* L) – preliminary study. Proc ECOpole 4:325–328 [in Polish]
Clauss M, Paglia DE (2012) Iron storage disorders in captive wild mammals: the comparative evidence. J Zoo Wildl Med 43:6–18
Clauss M, Lechner-Doll M, Hänichen T, Hatt JM (2002) Excessive iron storage in captive mammalian herbivores—a hypothesis for its evolutionary etiopathology. Proc Eur Assoc Zoo Wildl Vet 4:123–131
Coffey R, Ganz T (2017) Iron homeostasis – an anthropocentric perspective. J Biol Chem 292 (31):12727–12734. https://doi.org/10.1074/jbc.R117.781823
Cook JA, Marconi EA, Diluzio NR (1974) Lead, cadmium, endotoxin interaction: effect on morality and hepatic function. Toxicol Appl Pharmacol 28:292–302
Cork SC (2000) Iron storage diseases in birds. Avian Pathol 29:7–12
Cork SC, Alley MR, Stockdale PH (1995) A quantitative assessment of haemosiderosis in wild and captive birds using image analysis. Avian Pathol 24:239–254
Cornell RM, Schwertmann U (2003) The iron oxides, 2nd edn. Wiley-VCH, Weinheim
Crăciunescu A, Moatar M, Stanciu SM (2014) Study of hunting spread in winter-spring, calculation of necessity of food and comparison with the potential natural food in the hunting fund no. 3 Borlova. Anim Sci Biotechnol 47:22–24
Crawshaw G, Oyarzun S, Valdes E, Rose K (1995) Hemochromatosis (iron storage disease) in fruit bats. Proc AZA Nutr Advis Group 1:136–147
Crissey SD, Ward AM, Block SE, Maslanka MT (2000) Hepatic iron accumulation time in European starlings (*Sturnus vulgaris*) fed two levels of iron. J Zoo Wildl Med 31:491–496
Custer CM, Custer TW, Anteau MJ, Afton AD, Wooten DE (2003) Trace elements in lesser scaup (*Aythya affinis*) from the Mississippi flyway. Ecotoxicology 12:47–54
Custer TW, Cox E, Gray B (2004) Trace elements in moose (*Alces alces*) found dead in Northwestern Minnesota, USA. Sci Total Environ 330:81–87
Czajkowska M, Chrobaczynska M, Kuczkowska-Kuźniar A, Gal A (2011) Accumulation of zinc, iron and copper in the liver and kidney of the european hare (*Lepus Europaeus*). Episteme 1:27–32
Czaplicka A, Bazan S, Szarek-Gwiazda E, Slusarczyk Z (2016) Spatial distribution of manganese and iron in sediments of the Czorsztyn reservoir. Environ Prot Eng 42:179–188
Czernomysy-Furowicz D (2007) Effect of anti-*Yersinia pseudotuberculosis* vaccine and *Propionibacterium acnes* immunostimulator on transferrin and iron concentration in mare and foal sera. Med Weter 63:467–470
Czernomysy-Furowicz D, Furowicz AJ (1999) Nutritional infection triggered by *Yersinia enterocolitica* and *Yersinia pseudotuberculosis*. In: Boroń-Kaczmarska A, Furowicz AJ (eds) Carried zoonoses road nutritional, Wyd Lek PZWL, Poland [in Polish]
Damek-Poprawa M, Sawicka-Kapusta K (2003) Damage to the liver, kidney, and testis with reference to burden of heavy metals in yellow-necked mice from areas around steelworks and zinc smelters in Poland. Toxicology 186:1–10
Dauwe T, Bervoets L, Blust R, Pinxten R, Eens M (2000) Can excrement and feathers of nestling songbirds be used as biomonitors for heavy metals pollution? Arch Environ Contam Toxicol 39:541–546

Dierenfeld ES, Pini MT, Sheppard CD (1994) Hemosiderosis and dietary iron in birds. J Nutr 124:2685–2686

Dlugaszek M, Kopczynski K (2011) Comparative analysis of liver mineral status of wildlife. Probl Hig Epidemiol 92:859–863 [in Polish]

Dlugaszek M, Kopczynski K (2013) Elemental composition of muscle tissue of wild animals from central region of Poland. Int J Environ Res 7:973–978

Doyle JJ, Spaulding JE (1978) Toxic and essential trace elements in meat-a review. J Anim Sci 47:398–419

Duck KA, Connor JR (2016) Iron uptake and transport across physiological barriers. Biometals 29:573–591

Espinosa de los Monteros A, Kumar S, Scully S, Cole R, de Vellis J (1990) Transferrin gene expression and secretion by rat brain cells in vitro. J Neurosci Res 25:576–580

Esselink H, van der Geld FM, Jager LP, Posthuma-Trumpie GA, Zoun PE, Baars AJ (1995) Biomonitoring heavy metals using the barn owl (*Tyto alba guttata*): sources of variation especially relating to body condition. Arch Environ Contam Toxicol 28:471–486

Falandysz J (1994) Some toxic and trace metals in big game hunted in the northern part of Poland in 1987-1991. Sci Total Environ 141:59–73

Falandysz J, Jakuczun B, Mizera T (1988) Metals and organochlorines in four female white-tailed eagles. Mar Pollut Bull 19:521–526

Falandysz J, Ichihashi H, Mizera T, Yamasaki SI (2000) Mineral composition of selected tissues and organs of white-tailed sea eagle. Roczn PZH 51:1–5

Filipczyk L, Krol P, Wystrychowski A (2010) Hepcidin – a hepatic hormone that controls iron homeostasis. Forum Nefrol 3:233–242 [in Polish]

Finch CA, Cook JD (1984) Iron deficiency. Am J Clin Nutr 39:471–477

Franson JC, Hollmen T, Poppenga RH, Hario M, Kilpi M (2000) Metals and trace elements in tissues of common eiders (*Somateria mollissima*) from the Finnish archipelago. Ornis Fennica 77:57–63

Furness RW, Greenwood JJD (1993) Birds as monitors of environmental change. Chapman & Hall, London

Ganz T (2013) Systemic iron homeostasis. Physiol Rev 93:1721–1741

Ganz T, Nemeth E (2015) Iron homeostasis in host defence and inflammation. Nat Rev Immunol 15:500–510

Genovesi P, Scalera R, Solarz W, Roy D (2009) Towards an early warning and information system for invasive alien species threatening biodiversity in Europe EEA Technical report no 5/2010

Gosselin SJ, Kramer LW (1983) Pathophysiology of excessive iron storage in mynah birds. J Am Vet Med Assoc 32:1238–1240

Goyer RA (1996) Toxic effects of metals. In: Klaassen CD (ed) Casarett and Doull's toxicology: the basic science of poisons, 5th edn. McGraw-Hill, New York, pp 715–716

Greentree WF, Hall JO (1995) Iron toxicosis. In: Bonagura JD (ed) Kirk's current therapy XII. McGraw-Hill, New York

Grove RA, Henny CJ (2008) Environmental contaminants in male river otters from Oregon and Washington, USA, 1994-1999. Environ Monit Assess 145:49–73

Halliday JW, Searle J (1996) Hepatic iron deposition in human disease and animal models. Biometals 2:205–209

Hanusová E, Mertin D, Süvegová K, Szeleszczuk O (2007) Comparison of content of mineral elements in selected organs in carnivorous fur animals. Trace Elem Electrolytes 24:12–18

Harding LE (2004) Environmental contaminants in wild martens (*Martes americana*) and wolverines (*Gulo luscus*). Bull Environ Contam Toxicol 73:98–105

Harding LE, Harris ML, Elliott JE (1998) Heavy and trace metals in wild mink (*Mustela vison*) and river otter (*Lontra canadensis*) captured on rivers receiving metals discharges. Bull Environ Contam Toxicol 61:600–607

Hillman RS (1995) Hematopoietic agents: growth factors, minerals, and vitamins. In: Hardman JG, Limbird LE, Molinoff PB et al (eds) Goodman and Gilman's the pharmacological basis of therapeutics, 9th edn. McGraw-Hill, New York, pp 1311–1340

Hoekstra PF, Braune BM, Elkin B, Armstrong FA, Muir DC (2003) Concentrations of selected essential and non-essential elements in arctic fox (*Alopex lagopus*) and wolverines (*Gulo gulo*) from the Canadian Arctic. Sci Total Environ 309:81–92

Hoffmann SR, Blunck SA, Petersen KN, Jones EM, Koval JC, Misek R et al (2010) Cadmium, copper, iron, and zinc concentrations in kidneys of grey wolves, *Canis lupus*, from Alaska, Idaho, Montana (USA) and the Northwest Territories (Canada). Bull Environ Contam Toxicol 85:481–485

Hontelez LCMP, van den Dungen HM, Baars AJ (1992) Lead and cadmium in birds in the Netherlands: a preliminary survey. Arch Environ Contam Toxicol 23:453–456

Hou Z, Shen M, Chai H, Ma J, Hua Y (2012) Iron, copper, zinc and selenium concentrations and their interaction in organs of white nose disease of raccoon dogs. Asian J Anim Vet Adv 7:1110–1119

Hui A, Takekawa JY, Baranyuk VV, Litvin KV (1998) Trace element concentrations in two subpopulations of lesser snow geese from Wrangel Island, Russia. Arch Environ Contam Toxicol 34:197–203

Hurrell R, Egli I (2010) Iron bioavailability and dietary reference values. Am J Clin Nutr 91:1461–1467

Indian Bureau of Mine (2014) Indian minerals yearbook 2012. Part III: mineral reviews, 51st edn. Indira Bhavan, Civil Lines, Nagpur

Ingervo S, Strandberg C, Nuorteva P (1995) Trace metals in the liver of Finish *Parus* species. Ornis Fennica 72:127–131

Jager LP, Rijnierse FV, Esselink H, Baars AJ (1996) Biomonitoring with the buzzard *Buteo buteo* in the Netherlands: heavy metals and sources of variation. J Field Ornithol 137:295–318

Janiga M, Mankovska B, Bobal'ova M, Durcova G (1990) Significance of concentrations of lead, cadmium and iron in the plumage of the feral pigeon. Arch Environ Contam Toxicol 19:892–897

Johnson KS, Gordon RM, Coale KH (1997) What controls dissolved iron concentrations in the world ocean? Mar Chem 57:137–161

Kabata-Pendias A, Mukherjee AB (2007) Trace elements from soil to human. Springer, Heidelberg, 550 pp

Kabata-Pendias A, Pendias H (2001) Trace elements in soils and plants, 3rd edn. CRC Press, Boca Raton

Kabata-Pendias A, Szteke B (2012) Trace elements in geo- and biosphere. Publishing House IUNG-PIB, Pulawy [in Polish]

Kalisinska E, Salicki W, Myslek P, Kavetska KM, Jackowski A (2004) Using the Mallard to biomonitor heavy metal contamination of wetlands in north-western Poland. Sci Total Environ 320:145–161

Kalisinska E, Salicki W, Jackowski A (2006) Six trace metals in white-tailed eagle from north-western Poland. Pol J Environ Stud 15:727–737

Kalisinska E, Lisowski P, Czernomysy-Furowicz D, Kavetska KM (2008) Serratospiculiasis, mycosis, and haemosiderosis in wild peregrine falcon from Poland. a case report. Bull Vet Inst Pulawy 52:75–79

Kalisinska E, Lanocha N, Budis H, Wilk A, Kavetska K, Królaczyk K (2009) Essential trace elements in the liver and kidneys of the common buzzard *Buteo buteo* and the Kestrel *Falco tinnunculus*. In: Więcek J, Polak M, Kucharczyk M, Grzywaczewski G, Jerzak L (eds) Ptaki-Środowisko-Zagrożenia-Ochrona. Wybrane aspekty ekologii ptaków. LTO, Lublin [in Polish]

Kanbar JH, Montargès-Pelletiera E, Lossond B, Bihannica I, Gleya R, Bauera A et al (2017) Iron mineralogy as a fingerprint of former steelmaking activities in river sediments. Sci Total Environ 599–600:540–553

Kekkonen J, Hanski IK, Väisänen RA, Brommer JE (2012) Levels of heavy metals in House Sparrows (*Passer domesticus*) from urban and rural habitats of southern Finland. Ornis Fennica 89:91–98

Kim J, Koo TH (2007) The use of feathers to monitor heavy metal contamination in herons, Korea. Arch Environ Contam Toxicol 53:435–541

Kim J, Oh JM (2012a) Biological monitoring of heavy metal contaminations using owls. J Environ Monit 14:1091–1097
Kim J, Oh JM (2012b) Metal levels in livers of waterfowl from Korea. Ecotoxicol Environ Saf 78:162–169
Kim J, Oh JM (2013a) Assessment of trace metals in four bird species from Korea. Environ Monit Assess 185:6847–6854
Kim J, Oh JM (2013b) Tissue distribution of metals in white-fronted geese and spot-billed ducks from Korea. Bull Environ Contam Toxicol 91:18–22
Kim J, Oh JM (2015) Tissue distribution of heavy metals in heron and egret chicks from Pyeongtaek, Korea. Arch Environ Contam Toxicol 68:283–291
Kim EY, Ichihashi H, Saeki K, Atrashkevich G, Tanabe S, Tatsukawa R (1996) Metal accumulation in tissues of seabirds from Chaun, northeast Siberia, Russia. Environ Pollut 92:247–252
Kim J, Lee H, Koo TH (2008) Heavy-metal concentrations in three owl species from Korea. Ecotoxicology 17:21–28
Kim J, Koo TH, Oh JM (2010) Monitoring of heavy metal contamination using tissues of two ardeids chicks, Korea. Bull Environ Contam Toxicol 84:754–758
Kitowski I, Jakubas D, Wiacek D, Sujak A, Pitucha G (2017a) Trace element concentrations in livers of common buzzards *Buteo buteo* from eastern Poland. Environ Monit Assess 189:421
Kitowski I, Jakubas D, Wiącek D, Sujak A (2017b) Concentrations of lead and other elements in the liver of the white-tailed eagle (*Haliaeetus albicilla*), a European flagship species, wintering in Eastern Poland. Ambio 46:825–841. https://doi.org/10.1007/s13280-017-0929-3
Kohgo Y, Ikuta K, Ohtake T, Torimoto Y, Kato J (2008) Body iron metabolism and pathophysiology of iron overload. Int J Hematol 88:7–15
Kozulin A, Pavluschick T (1993) Content of heavy metals in tissues of mallards *Anas platyrhynchos* wintering in polluted and unpolluted habitats. Acta Ornit 28:55–61
Kritzberg ES, Ekström SM (2012) Increasing iron concentrations in surface waters – a factor behind brownification? Biogeosciences 9:1465–1478
Lazarus M, Vicković I, Sostarić B, Blanusai M (2005) Heavy metal levels in tissues of red deer (*Cervus elaphus*) from Eastern Croatia. Arh Hig Rada Toksikol 56:233–240
Lazarus M, Orct T, Blanusa M, Vickovic I, Sostarić B (2008) Toxic and essential metal concentrations in four tissues of red deer (*Cervus elaphus*) from Baranja, Croatia. Food Addit Contam Part A Chem Anal Control Expo Risk Assess 25:270–283
Lécureuil C, Saleh MC, Fontaine I, Baron B, Zakin MM, Guillou F (2004) Transgenic mice as a model to study the regulation of human transferrin expression in Sertoli cells. Hum Reprod 19:1300–1307
Lewis LA, Poppenga RJ, Davidson WR, Fischer JR, Morgan KA (2001) Lead toxicosis and trace element levels in wild birds and mammals at a firearms training facility. Arch Environ Contam Toxicol 41:208–214
Li H, Ginzburg YZ (2010) Crosstalk between iron metabolism and erythropoiesis. Adv Hematol 2010:605435
Liebelt EL (1998) Iron. In: Haddad LM, Shannon MW, Winchester JF (eds) Clinical management of poisoning and drug overdose, 3rd edn. WB Saunders, Philadelphia, pp 757–766
Ljungvall K, Magnusson U, Korvela M, Norrby M, Bergquist J, Persson S (2017) Heavy metal concentrations in female wild mink (*Neovison vison*) in Sweden: sources of variation and associations with internal organ weights. Environ Toxicol 9999:1–6
Llacuna S, Gorriz A, Sanpera NJ (1995) Metal accumulation in three species of passerine birds (*Emberiza cia, Parus major, and Turdus merula)* subjected to air pollution from a coal-fires power plant. Arch Environ Contam Toxicol 28:298–303
Lodenius M, Skarén U, Hellstedt P, Tulisalo E (2014) Mercury in various tissues of three mustelid species and other trace metals in liver of European otter from Eastern Finland. Environ Monit Assess 186:325–333
Lowenstine LJ, Munson L (1999) In: Fowler ME, Miller RE (eds) Iron overload in the animal kingdom, zoo and wild animal medicine, current therapy, 4th edn. WB Saunders, Philadelphia, pp 260–268

Luck AN, Mason AB (2012) Transferrin-mediated cellular iron delivery. Curr Top Membr 69:3–35
MacKenzie EL, Iwasaki K, Tsuji Y (2008) Intracellular iron transport and storage: from molecular mechanisms to health implications. Antioxid Redox Signal 10:997–1030
Malik RN, Zeb N (2009) Assessment of environmental contamination using feathers of *Bubulcus ibis* L., as a biomonitor of heavy metal pollution, Pakistan. Ecotoxicology 18:522–536
Malinga M, Szefer P, Gabrielsen GW (2010) Age, sex and spatial dependent variations in heavy metals levels in the Glaucous Gulls (*Larus hyperboreus*) from the Bjørnøya and Jan Mayen, Arctic. Environ Monit Assess 169:407–416
Manjula M, Mohanraj R, Devi MP (2015) Biomonitoring of heavy metals in feathers of eleven common bird species in urban and rural environments of Tiruchirappalli, India. Environ Monit Assess 187:267
Mansouri B, Pourkhabbaz A, Babaei H, Hoshyari E, Khodaparast SH, Mirzajani A (2012) Assessment of trace-metal concentrations in Western Reef heron (*Egretta gularis*) and Siberian gull (*Larus heuglini*) from southern Iran. Arch Environ Contam Toxicol 63:280–287
Marques CC, Gabriel SI, Pinheiro T, Viegas-Crespo AM, da Luz Mathias M, Bebianno MJ (2008) Metallothienin levels in Algerian mice (*Mus spretus*) exposed to elemental pollution: an ecophysiological approach. Chemosphere 71:1340–1347
Martiniaková M, Omelka R, Grosskopf B, Jančová A (2010) Yellow-necked mice (*Apodemus flavicollis*) and bank voles (*Myodes glareolus*) as zoomonitors of environmental contamination at a polluted area in Slovakia. Acta Vet Scand 5:52–58
McDonald C, Elkin B, Gunn A (2005) Analysis of the elemental composition of tissues and faecal ash in a moose (*Alces alces*) exposed to tailings at the abandoned colomac gold mine, NWT. Department of Resources, Wildlife and Economic Development Government of Northwest Territories, Yellowknife, NT
Medvedev N (1999) Levels of heavy metals in Karelian wildlife, 1989-91. Environ Monit Assess 56:177–193
Mendel M, Wiechetek M (2006) Iron poisoning in animals. Med Weter 62:1357–1361
Mete A, Hendriks HG, Klaren PH, Dorrestein GM, van Dijk JE, Marx JJ (2003) Iron metabolism in mynah birds (*Gracula religiosa*) resembles human hereditary haemochromatosis. Avian Pathol 32:625–632
Michalska K, Zmudzki J (1992) Metals content in the tissues of boars, roe deers and red deers in the Wielkopolska region. Med Weter 48:160–162 [in Polish]
Michot TC, Garvin MC, Weidner EH (1994) Survey for blood parasites in redheads (*Aythya americana*) wintering at the Chandeleur Islands, Louisiana. J Wildl Dis 31:90–92
Mihaljev Ž, Živkov–Baloš M, Kapetanov M, Jakšić S (2012) Content of microelements in wild birds in Vojvodina. International symposium on hunting. Modern aspects of sustainable management of game population. Zemum-Belgrade, Serbia
Miret S, Simpson RJ, McKie AT (2002) Physiology and molecular biology of dietary iron absorption. Annu Rev Nutr 23:283–301
Myslek P, Kalisinska E (2006) Contents of selected heavy metals in the liver, kidneys and abdominal muscle of the brown hare (*Lepus europaeus* Pallas, 1778) in Central Pomerania, Poland. Pol J Vet Sci 9:31–41
Nichols GM, Bacon BR (1989) Hereditary hemochromatosis: pathogenesis and clinical features of a common disease. Am J Gastroenterol 84:851–862
O'Brien R (2009) Australia's iron ore product quality. Geoscience Australia, Onshore Minerals and Energy Division, 9 pp
O'Hara TM, George JC, Blake J, Burek K, Carrolll G, Dau J et al (2003) Investigation of heavy metals in a large mortality event in caribou of Northern Alaska. Arctic 56:125–135
Odsjö T, Räikkönen J, Bignert A (2007) Time trends of metals in liver and muscle of reindeer (*Rangifer tarandus*) from northern and central Lapland, Sweden, 1983–2005. Swedish Museum of Natural History, Stockholm, 33 pp
Olias P, Weiss AT, Gruber AD, Klopfleisch R (2011) Iron storage disease in red deer (*Cervus elaphus elaphus*) is not associated with mutations in the HFE gene. J Comp Pathol 145:207–213

Onderscheka K, Tataruch F, Steineck T (1985) Das freilebende tierals indikator fur die umwelt. Allg Forstztg 96:80–83

Osweiler GD, Carson TL, Buck WB, Van Gelder GA (1985) Iron. In: Clinical and diagnostic veterinary toxicology, 3rd edn. Kendall/Hunt Publishing, Dubuque, pp 104–106

Paglia DE, Dennis P (1999) Role of chronic iron overload in multiple disorders of captive black rhinoceros. Proc AAZVAM:163–171

Pantopoulos K, Porwal SK, Tartakoff A, Devireddy L (2012) Mechanisms of mammalian iron homeostasis. Biochemistry 51:5705–5724

Park C, Muller CD, Abu-Orf MM, Novak JT (2006) The effect of wastewater cations on activated sludge characteristics: effects of aluminum and iron in floc. Water Environ Res 78:31–40

Parker GH, Hamr J (2001) Metal levels in body tissues, forage and fecal pellets of elk (*Cervus elaphus*) living near the ore smelters at Sudbury, Ontario. Environ Pollut 113:347–355

Pavone S, Salamida S, Pecorelli I, Rossi E, Manuali E (2014) Deadly outbreak of iron storage disease (ISD) in Italian birds of the family Turdidae. J Vet Med Sci 76:1209–1212

Pietrangelo A (2010) Hereditary hemochromatosis: pathogenesis, diagnosis, and treatment. Gastroenterology 139:393–408

Pillatzki AE, Neiger RD, Chipps SR, Higgins KF, Thiex N, Afton AD (2011) Hepatic element concentrations of lesser scaup (*Aythya affinis*) during spring migration in the upper Midwest. Arch Environ Contam Toxicol 61:144–150

Piperno A (1998) Classification and diagnosis of iron overload. Haematologica 83:447–455

Powell LW, Seckington RC, Deugnier Y (2016) Haemochromatosis. Lancet 388:706–716

Prestrud P, Norheim G, Sivertsen T, Daae HL (1994) Levels of toxic and essential elements in artic fox in Svalbard. Polar Biol 14:155–159

Proske U, Kolb E, Klemm R, Salomon FV (1993) Der Gehalt an Eisen, Kupfer und Zink in 7 Geweben von Hybridenten (Moschusente x Pekingente) und von Stockenten im Verlaufe des Wachstums in 12 verschiedenen Altersgruppen. Arch Geflügelk 57:113–120

Puls R (1994) Mineral levels in animal health: diagnostic data, 2nd edn. Sherpa International, Clearbrook, BC

Rajkowska M, Protasowicki M (2011) Distribution of selected metals in bottom sediments of lakes Insko and Wisola (Poland). Ecol Chem Eng A 18:805–812

Randell MG, Patnaik AK, Gould WJ (1981) Hepatopathy associated with excessive iron storage in mynah birds. J Am Vet Med Assoc 179:1214–1217

Rattner BA, Golden NH, Toschik PC, McGowan PC, Custer TW (2008) Concentrations of metals in blood and feathers of nestling ospreys (*Pandion haliaetus*) in Chesapeake and Delaware Bays. Arch Environ Contam Toxicol 54:114–122

Reimann C, de Caritat P (1998) Chemical elements in the environment: factsheets for the geochemist and environmental scientist. Springer, Berlin Heidelberg

Rogula-Kozłowska W, Majewski G, Czechowski PO (2015) The size distribution and origin of elements bound to ambient particles: a case study of a Polish urban area. Environ Monit Assess 187:240

Rose GR, Parker GH (1982) Effects of smelter emissions on metal levels in the plumage of Ruffed Grouse near Sudbury, Ontario, Canada. Can J Zool 60:2659–2667

Roslewska A, Stanek M, Janicki B, Cygan-Szczegielniak D, Stasiak K, Buzała M (2016) Effect of sex on the content of elements in meat from wild boars (*Sus scrofa* L.) originating from the Province of Podkarpacie (south-eastern Poland). J Elem 21:823–832

Roug A, Swift PK, Gerstenberg G, Woods LW, Kreuder-Johnson C, Torres SG et al (2015) Comparison of trace mineral concentrations in tail hair, body hair, blood, and liver of mule deer (*Odocoileus hemionus*) in California. J Vet Diagn Investig 27:295–305

Saiz MP, Martí MT, Mitjavila MT, Planas J (1990) Sexual and age variations of organ iron content in Shaver chickens. Br Poult Sci 31:339–349

Sanchez-Chardi A, Marques CC, Nadal J, da Luz Mathias M (2007) Metal bioaccumulation in the greater white-toothed shrew, *Crocidura russula*, inhabiting an abandoned pyrite mine site. Chemosphere 67:121–130

Sanchez-Chardi A, Peñarroja-Matutano C, Borrás M, Nadal J (2009a) Bioaccumulation of metals and effects of a landfill in small mammals part III: structural alterations. Environ Res 109:960–967

Sanchez-Chardi A, Ribeiro CA, Nadal J (2009b) Metals in liver and kidneys and the effects of chronic exposure to pyrite mine pollution in the shrew *Crocidura russula* inhabiting the protected wetland of Doñana. Chemosphere 76:387–394

Sawicka-Kapusta K, Kozłowski J, Sokołowska T (1986) Heavy metals in tits from polluted forests in southern Poland. Environ Pollut 42:297–310

Schönfelder I, Gelbrecht J, Schönfelder J, Steinberg CEW (2002) Relationships between littoral diatoms and their chemical environment in northeastern German lakes and rivers. J Phycol 38:66–82

Schultheiss PC, Bedwell CL, Hamar DW, Fettman MJ (2002) Canine liver iron, copper, and zinc concentrations and association with histologic lesions. J Vet Diagn Investig 14:396–402

Schummer ML, Petrie SA, Badzinski SS, Deming M, Chen YW, Belzile N (2011) Elemental contaminants in livers of mute swans on lakes Erie and St. Clair. Arch Environ Contam Toxicol 61:677–687

Sears ME (2013) Chelation: harnessing and enhancing heavy metal detoxification – a review. Sci World J 2013:219840

Sheppard C, Dierenfeld E (2002) Iron storage disease in birds: speculation on etiology and implications for captive husbandry. J Avian Med Surg 16:192–197

Sinka-Karimi MH, Pourkhabbaz AR, Hassanpour M, Levengood JM (2015) Study on metal concentrations in tissues of Mallard and Pochard from two major wintering sites in Southeastern Caspian Sea, Iran. Bull Environ Contam Toxicol 95:292–297

Skobrák EB, Bodnár K, Jónás EM, Gundel J, Jávor A (2011) The comparison analysis of the main chemical composition parameters of wild boar meat and pork. Sci Pap Anim Sci Biotechnol 44(1):105–112

Sleeman JM, Magura K, Howell J, Rohm J, Murphy LA (2010) Hepatic mineral values of white-tailed deer (*Odocoileus virginianus*) from Virginia. J Wildl Dis 46:525–531

Snow NP, Lavelle MJ, Halseth JM, Blass CR, Foster JA, Vercauteren KC (2017) Strength testing of raccoons and invasive wild pigs for a species-specific bait station. Wildl Soc Bull 41:264–270

Soetan KO, Olaiya CO, Oyewole OE (2010) The importance of mineral elements for humans, domestic animals and plants: a review. Afr J Food Sci 45:200–222

Souza MJ, Ramsay EC, Donnell RL (2013) Metal accumulation and health effects in raccoons (*Procyon lotor*) associated with coal fly ash exposure. Arch Environ Contam Toxicol 64:529–536

Spelman LH, Osborn KG, Anderson MP (1989) Pathogenesis of hemosiderosis in lemurs: role of dietary iron, tannin, and ascorbic acid. Zoo Biol 8:239–251

Stejskal SM, Aulerich RJ, Slanker MR, Braselton WE, Lehning EJ, Napolitano AC (1989) Element concentrations in livers and kidneys of ranch mink. J Vet Diagn Investig 1:343–348

Stopić S, Pavlović J, Friedrich B (2007) Treatment of highly contaminated waste waters in a continuous cascade line reactor. World Metall – Erzmetall 60:155–162

Stout JH, Trust KA (2002) Elemental and organochlorine residues in bald eagles from Adak Island, Alaska. J Wildl Dis 38:511–517

Stovel PL (1963) Epizootical factors in 3 outbreaks of *Pseudotuberculosis* in British Columbia canaries. MSc thesis, University of British Columbia

Strazdina V, Jemeljanovs A, Sterna V, Ikauniece D (2013) Nutrition value of deer, wild boar and beaver meat hunted in Latvia. In: 2nd international conference on nutrition and food sciences, vol 53, pp 71–76

Strazdina V, Jemeljanovs A, Sterna V, Ikauniece D (2014) Nutritional characteristics of wild boar meat hunter in Latvia. Proc Foodbalt 1:32–36

Strickman RJ, Mitchell CP (2017) Accumulation and translocation of methylmercury and inorganic mercury in *Oryza sativa*: an enriched isotope tracer study. Sci Total Environ 574:1415–1423

Swiergosz R, Perzanowski K, Makosz U, Biłek I (1993) The incidence of heavy metals and other toxic elements in big game tissues. Sci Total Environ 134:225–231
Szymański W, Skiba M, Błachowski A (2017) Influence of redox processes on clay mineral transformation in Retisols in the Carpathian Foothills in Poland. Is a ferrolysis process present? J Soils Sediments 17:453–470
Thelander L, Gräslund A, Thelander M (1983) Continual presence of oxygen and iron required for mammalian ribonucleotide reduction: possible regulation mechanism. Biochem Biophys Res Commun 110:859–865
Topolska K, Sawicka-Kapusta K, Cieslik E (2004) The effect of contamination of the Krakow Region on Heavy Metals Content in the Organs of Bank Voles (*Clethrionomys glareolus*, Schreber, 1780). Pol J Environ Stud 13:103–109
Turner R (1994) Iron storage disease (hemochromatosis) in the curassow. In: Proceeding of Annual Conference Association Avian Veterinarians, Teaneck, NJ, pp 265–267
Ullah K, Hashmi MZ, Malik RN (2014) Heavy-metal levels in feathers of cattle egret and their surrounding environment: a case of the Punjab Province, Pakistan. Arch Environ Contam Toxicol 66:139–153
Underwood EJ (1977) Trace elements in human and animal nutrition. Academic Press, New York
USGS (2016) Iron. Mineral Commodity Summaries 2016, U.S. Geological Survey, p 202, https://doi.org/10.3133/70140094
van Bolhuis GH, Rijks JM, Dorrestein GM, Rudolfova J, van Dijk M, Kuiken T (2004) Obliterative endophlebitis in mute swans (*Cygnus olor*) caused by *Trichobilharzia* sp. (*Digenea*: *Schistosomatidae*) infection. Vet Pathol 41:658–665
Vengušt G, Vengušt A (2004) Some minerals as well as trace and toxic elements in livers of fallow deer (*Dama dama*) in Slovenia. Eur J Wildl Res 50:59–61
Walker LA, Lawlor AJ, Chadwick EA, Potter E, Pereira MG, Shore RF (2010) Inorganic elements in the livers of Eurasian otters, *Lutra lutra*, from England and Wales in 2007 and 2008: a Predatory Bird Monitoring Scheme (PBMS) report, Centre for Ecology and Hydrology, p 13
Walker LA, Lawlor AJ, Chadwick EA, Potter E, Pereira MG, Shore RF (2011) Inorganic elements in the livers of Eurasian otters, *Lutra lutra*, from England and Wales in 2009 – a Predatory Bird Monitoring Scheme (PBMS) report. Centre for Ecology and Hydrology
Wieczorek-Dabrowska M, Tomza-Marciniak A, Pilarczyk B, Balicka-Ramisz A (2013) Roe and red deer as bioindicators of heavy metals contamination in north-western Poland. Chem Ecol 29:100–110
Wiklund E, Farouk M, Finstad G (2014) Venison: meat from red deer (*Cervus elaphus*) and reindeer (*Rangifer tarandus tarandus*). Anim Front 4:55–61
Wilhide JD, McDaniel VR, Tumlison R (1992) Age specific analysis of food habits for Arkansas raccoons (*Procyon lotor*). Ark Acad Sci 46:112–113
Williams A, Hoy T, Pugh A, Jacobs A (1982) Pyridoxal complexes as potential chelating agents for oral therapy in transfusional iron overload. J Pharm Pharmacol 34:730–732
Wren CD (1984) Distribution of metals in tissues of beaver, raccoon and otter from Ontario. Can Sci Total Environ 34:177–184
Wren CD, Fischer KL, Stokes PM (1988) Levels of lead, cadmium and other elements in mink and otter from Ontario, Canada. Environ Pollut 52:193–202
WVDL (2015) www.wvdl.wisc.edu/wp-content/uploads/2013/06/WVDL.Info_Toxicology_Normal_Ranges.pdf (Accessed 28 Apr 2015)
Yamazaki I, Piette LH (1990) ESR spin-trapping studies on the reaction of $Fe^{2+}$ ions with $H_2O_2$-reactive species in oxygen toxicity in biology. J Biol Chem 265:13589–13594
Zaccaroni A, Andreani G, Zucchini M, Merendi F, Simoni P (2003) Heavy metals in wild boar (*Sus scrofa*) in related lesions. Hystrix It J Mammal 2003:14 (Congress Italian Teriologia)
Zhang C (2014) Essential functions of iron-requiring proteins in DNA replication, repair and cell cycle control. Protein Cell 5:750–760
Zimmerman TJ, Jenks JA, Leslie DM Jr, Neiger RD (2008) Hepatic minerals of white-tailed and mule deer in the southern Black Hills, South Dakota. J Wildl Dis 44:341–350

# Chapter 7
# Manganese, Mn

**Elżbieta Kalisińska and Halina Budis**

**Abstract** Manganese is considered an important essential trace element, and there is extensive literature concerning its accumulation in ecosystems. This metal is continuously released into the biosphere by volcanoes and the natural weathering of rocks, but also by numerous anthropogenic activities such as mining, fossil fuel combustion and industrial and urban waste. In this review, recent studies on manganese presence in soil, freshwater and terrestrial ecosystems and their organisms are presented. Metal speciation, natural and anthropogenic sources and level of bioaccumulation in biota, as well as abiotic and biotic factors affecting their bioavailability are reviewed, in addition to the use of bioindicator organisms for the biomonitoring of this metal in the environment and the related toxicity mechanisms and ecological effects of manganese pollution.

## 1 Introduction

Manganese (Mn) is an essential metal for humans, animals and plants. It is one of the most commonly used metals, especially in metallurgic and chemical industries; hence, its elevated levels in the environment cause excessive Mn exposure to terrestrial animals. Manganese plays an important role in the biochemical reactions of several enzymes, but its excessive amount in the body can cause both acute and chronic diseases (ATSDR 2012; O'Neal and Zheng 2015; Prashanth et al. 2015). Generally, Mn is found in low levels in all types of animal diet. Ingestion is the principal route by which animals are exposed to this metal, although toxicologically

E. Kalisińska (✉)
Department of Biology and Medical Parasitology, Pomeranian Medical University in Szczecin, Szczecin, Poland
e-mail: ekalist@pum.edu.pl

H. Budis
Chair of Health Education Faculty of Physical Education and Health Promotion, Szczecin University, Szczecin, Poland



E. Kalisińska (ed.), *Mammals and Birds as Bioindicators of Trace Element Contaminations in Terrestrial Environments*,
https://doi.org/10.1007/978-3-030-00121-6_7

significant inhalation exposures also occur in humans and laboratory animals as it has been demonstrated (ATSDR 2012; Zeman et al. 2015; Wang et al. 2015).

The gastrointestinal and hepatobiliary systems play crucial role in regulating and maintaining Mn tissue levels within a relatively narrow physiologic range (Aschner and Aschner 2005; Foster et al. 2015; Zeman et al. 2015). It was observed that, in mammals, a moderate increase of Mn levels in the diet causes a compensatory decrease of Mn absorption in the gut and an increased Mn concentration in the liver. In such a condition, an elevated biliary excretion of Mn helps to maintain normal Mn level in the brain and other tissues (Aschner and Aschner 2005; Foster et al. 2015). Excessive exposure to Mn and/or results in disturbances in liver functioning can overwhelm normal homeostatic controls, which, in turn, results in elevated or toxic Mn levels in different tissues, especially in the central nervous system (O'Neal and Zheng 2015). The research on the role of Mn and its normal, deficient and excessive concentrations predominantly focuses on studying the tissues of domesticated and laboratory animals. Much fewer reports describe wild mammals and birds in which Mn uptake from the environment occurs mainly through food and water. Taking into consideration that the amount of anthropogenic Mn in nature has been steadily increasing, including increasing number of nanoparticles containing Mn that are ever more often released into aquatic and terrestrial ecosystems, it would seem worthwhile to monitor its concentration not only in air, water and soil but also in biotas (Karmakar et al. 2014; Pinsino et al. 2012).

In such biomonitoring, careful attention should be paid selecting species of animals representing different trophic groups and using appropriate types of samples. Systematic research conducted in different areas would allow for a better indirect assessment of Mn pollution in the environment and potential hazards to human health as well as for a direct assessment of health status of the analysed animals.

## 2 General Properties

Manganese (Lat. manganum, Mn) is a chemical element; it has an atomic number of 25 and an atomic weight of 54.94. It belongs to group 7 in the periodic table, next to iron (Fe). These metals have similar chemical properties and often coexist in different minerals and ores. The melting and boiling points of manganese are 1244 °C and 1962 °C, respectively, and its density is 7470 kg $m^{-3}$. Manganese compounds occur at oxidation states I, II, III, IV, VI and VII. The higher the level of oxidation, the greater the acidic character of manganese. At oxidation state II, Mn forms stable $Mn^{2+}$ cations and, at oxidation state VII, $MnO^{4-}$ permanganate anions. Manganese is a silvery metal, harder than iron, but more brittle. It reacts to diluted acids and water with hydrogen separation. This element is ubiquitous in the earth's crust and accounts for about 0.095% of its composition (Lide 2005; Yaroshevsky 2006). In nature, Mn most commonly occurs in the form of oxides, carbonates and silicates. Minerals in which Mn is most abundant include pyrolusite ($MnO_2$), of which it constitutes 60–63% of its composition. Other important minerals containing Mn include rhodochrosite ($MnCO_3$), manganite ($Mn_2O_2{\bullet}H_2O$), hausmannite ($Mn_3O_4$), braunite ($3Mn_2O_3{\bullet}MnSiO_3$) and rhodonite ($MnSiO_3$). Minerals rich in Mn usually also contain other heavy metals: iron, cobalt, zinc or nickel (Adriano 2001).

## 3 Production and Uses of Manganese

The most valuable manganese ores are those containing more than 20% of this metal, but in some regions of the world, deposits with less than 5% of Mn are also mined (USGS 2017a). USGS (2017a, b) estimated world manganese ore production in 2015 to be 17,500 thousand metric tons (Mt). The largest producers of this metal are South Africa, Australia and China, which account for 34%, 17% and 14% of world production, respectively. More than 90% of land-based manganese resources belong to six countries: South Africa, Ukraine, Brazil, Australia, India and China (USGS 2017b).

Manganese is mainly used in ferroalloy production (about 90% of total Mn) and as an oxidizer in other metallurgical processes. In addition, it is used in the chemical, textile, ceramic and electrochemical industries (welding electrodes) and in the manufacturing of dyes, plant protection agents and fertilizers (Howe et al. 2004; USGS 2016). Manganese is an indispensable ingredient in the production of steel because it removes sulphur from molten pig iron, causes the deoxidation of steel and acts as a fusing agent. It also helps to increase the hardness and elasticity of steel, as well as its resistance to stress.

Manganese oxide is commonly used as a cathode in zinc manganese batteries. In the USA, such alkaline cells are commonly used in households, mainly in wireless electrical appliances and torches (Aschner et al. 2006). In the course of the development of the automotive industry, so-called leaded and highly toxic fuels have gradually been replaced by unleaded fuels, and a compound containing Mn—methylcyclopentadienyl manganese tricarbonyl, or MMT—was used as an antiknock agent to improve the work of car engines. This fuel additive was developed in the 1950s to increase the octane level of gasoline. As MMT is combusted in gasoline engines, manganese oxide, the main product of combustion, is released into the atmosphere. In urban environments with heavy traffic, the concentration of manganese in the air is many times greater than in suburban areas (Loranger and Zayed 1997; Zayed et al. 1999; Vezer et al. 2005). Over the course of time, due to the negative impact of MMT on human health and the environment, the use of this substance has been restricted or prohibited in many countries. Recent data suggest that MMT is rarely used in developed countries, including Canada and the USA, nowadays (ATSDR 2012).

## 4 Manganese in Nature: Geogenic and Anthropogenic Sources

Manganese is a trace element present in rocks, soils, waters, air and food (Reimer 1999). In soils, this metal is found in mineral and organic compounds, and its concentration is estimated at 500–900 mg $kg^{-1}$ dry weight, dw (WHO 1999, 2004). In the air, manganese is a constituent of dust, the transport of which depends on the size and density of the particles forming it and the speed and direction of the wind. Weathering rocks and soil are the main sources of atmospheric Mn. Other important sources include ocean evaporation, forest fires, plant vegetation and volcanic activity (Schroeder et al. 1987). At the end of the 1980s, it was estimated that 2/3 of Mn present in the atmosphere came from natural sources (Stokes et al. 1988). However, since the beginning of the 1990s, there has been a dramatic increase in the content of this metal in the environment, especially in the air, as the result of human economic activity. In 1995, emissions of Mn compounds into the atmosphere were 30% higher than in 1983, most of which came from the industrialized countries of Asia, following those in the North America and Europe (Pacyna and Pacyna 2001).

Nearly 80% of industrial emissions of Mn are related to the production of iron and steel (EPA 2003). Concentration of this metal near industrial centres ranges from 0.22 to 0.30 μg Mn $m^{-3}$ (WHO 2004), while in areas remote from them, it does not exceed 0.01 μg Mn $m^{-3}$ (Sweet et al. 1993).

The dust, containing Mn compounds, which migrates to the atmosphere in the process of manganese ore mining, metal melting and fuel production, is particularly dangerous to human health as it can cause various respiratory and nervous system diseases (Williams-Johnson 1999). Neurological symptoms occur in people exposed to Mn levels of about 2000 μg $m^{-3}$.

Manganese migrates into the water from soil, dead plant parts and industrial pollutants dumped into surface waters where it can undergo oxidation or adsorption to sediment particles, depending on factors such as pH, presence of anions and oxidation potential (ATSDR 2000). Groundwater with poor oxygen content, or entirely depleted of it, often shows elevated concentrations of dissolved manganese. Bivalent manganese ($Mn^{2+}$) is prevalent in most waters with pH values between 4 and 7. Manganese can occur at higher oxidation states if the pH value of water is higher, but its higher valency may also be due to microbial oxidation. Generally, mean concentrations of Mn are 16 and 5 μg Mn $L^{-1}$ for surface and groundwater, respectively. Natural Mn concentrations in seawater range from 0.4 to 10 μg Mn $L^{-1}$ . In some groundwaters, lakes and reservoirs, Mn concentrations can reach up to 1300 μg $L^{-1}$ in neutral conditions and 9600 μg $L^{-1}$ in acidic environments (ATSDR 2012). The presence of Mn in groundwater is often detected as the result of its ubiquity in soils and rocks. Similarly, Mn is also found in surface waters, but its concentrations are generally too low to cause any adverse health effects. Manganese is commonly found in sediments, plant and animal tissues, and the process of manganese accumulation in aquatic ecosystems is much more intense than in terrestrial ones (US EPA 2003; WHO 2011). Concentrations of Mn found in some water organisms (phytoplankton, algae, molluscs and some fish) suggest only slight bioaccumulation, while in higher organisms, bioaccumulation does not occur (ATSDR 2000; US EPA 2003; Niemiec and Wisniowska-Kielan 2015). Biomagnification of this element in the food chain appears to be of little importance (ATSDR 2000). Nriagu and Pacyna (1988) estimated that in 1983, between 109,000 and 414,000 tons of Mn of anthropogenic origin found its way into the water environment, mainly from municipal sewage and landfills. At the same time, it should be noted that potassium permanganate ($KMnO_4$), which has strong oxidizing, bactericidal and fungicidal properties, is commonly used in many countries for water purification. It is also used to reduce and eliminate unpleasant odours and as air freshener, especially in paint factories and processing plants (ATSDR 2012).

Manganese is a naturally occurring component of almost all soils, and its concentration (ATSDR 2000) ranges from <2 to 7000 mg $kg^{-1}$ (or ppm) dw and an average of 550 ppm dw (Shacklette and Boerngen 1984; ATSDR 2000). Mn accumulation takes place in the substrate rather than on the surface of the soil (ATSDR 2000). It is estimated that 60–90% of Mn in the soil is bonded with a fraction of sand (WHO 1981).

## 5 The Biological Role of Manganese

Manganese is an essential element for plants because it plays an important role in many processes, including chlorophyll production. It works as an activating factor in more than 35 different plant enzymes. The uptake and transfer of manganese in plants occurs in the form of $Mn^{2+}$. Younger plant organs are richer in Mn than older ones (Mousavi et al. 2011). A wide range of Mn concentrations were found in plants (10–600 mg $kg^{-1}$ dw mean is about 50 mg $kg^{-1}$ dw). However, high Mn levels may be toxic for plants and induce iron deficiency (Lohry 2007). Manganese is also essential for the proper development and functioning of invertebrates and endothermic vertebrates. It is included in many enzymes and is involved in reproduction and growth processes, the metabolism of carbohydrates and fats, the functioning of the immune system and the processes of cartilage and bone formation (Erikson and Aschner 2003; Erikson et al. 2005; Baden and Eriksson 2006). In freshwater crustaceans from pristine areas, the highest Mn concentrations (~100 or more mg $kg^{-1}$ dw) are found in midgut glands, exoskeletons and gills. Mn concentration in the muscles is usually several and rarely several dozen, mg Mn $kg^{-1}$ dw (Baden and Eriksson 2006). Benthic organisms are especially sensitive to the accumulation of Mn, which is much higher in water sediments. In these organisms, average concentration of this metal is about 25 mg $kg^{-1}$ dw (Niemiec and Wisniowska-Kielan 2015). In fish, depending on the species, the highest Mn levels are observed in the skin and gills (6–8 and 2–9 mg $kg^{-1}$ ww, respectively) and are significantly lower in the liver

(~0.8 mg $kg^{-1}$ ww) and muscles (~0.2 mg $kg^{-1}$ ww). In benthic fish, such as bream *Abramis brama*, a much higher Mn concentration was found in the skin and gills than in the analogous parts of predatory pike *Esox lucius* (Rajkowska and Protasowicki 2013).

In mammals, manganese is found in every kind of tissue, mainly in the form of metalloenzymes, sometimes in combination with pyruvate carboxylase, superoxide dismutase and glycosyltransferases. Its concentration ranges from 0.3 to 2.9 mg $kg^{-1}$ ww, while tissues rich in mitochondria and pigments (including retina and dark skin) tend to have high Mn concentrations (Prashanth et al. 2015). Metalloenzymes containing Mn play an important role in the metabolism of amino acids, cholesterol and carbohydrates (Reynolds et al. 1998). Manganese is also involved in the processes of ossification and synthesis of cartilage and mucopolysaccharides, i.e. in the processes of formation of chondrocytic tissues, which is especially intensive in young organisms (Schroeder et al. 1966; Hurley and Keen 1987; Levander 1988; Freeland-Graves and Llanes 1994; Wedler 1994; Reynolds et al. 1998; Smrcka 2005).

The uptake of Mn in mammals and birds occurs through the digestive and respiratory tracts. Manganese absorption through the airways is poorly understood. It is known, however, that grains of Mn dust (especially in the form of oxides: $MnO_2$ i $Mn_3O_4$) with a diameter of less than a few tenths of a micrometre are absorbed into the bloodstream from the lungs, which, in cases of chronic exposure, may trigger respiratory disease. Mn molecules of small enough size enter the bloodstream, where they are found in compounds with transferrin and α-macroglobulin (ATSDR 2008). The main source of Mn in animals is food. The highest concentrations of this metal (expressed dw) are found in nuts (up to 47 mg $kg^{-1}$), legumes (up to 7 mg $kg^{-1}$) and cereal grains (up to 41 mg $kg^{-1}$) and the smallest in meat, fish and eggs (0.10–3.99 mg $kg^{-1}$) and fruit (0.20–10.38 mg $kg^{-1}$) (WHO 2011). In fields and grasslands treated with fertilizers containing Mn, leaves and roots may reach hundreds or even thousands of mg $kg^{-1}$ dw. Such high Mn concentrations in plants may be toxic for them and also become a cause of excessive Mn accumulation in herbivorous animals (Millaleo et al. 2010; Reis et al. 2010).

Only 3–7% of absorbed Mn from food remains in the body. It is most effectively deposited in the liver, pancreas, kidneys and bones, in which over 40% of all Mn in the body is found (O'Neal and Zheng 2015). Mn is transformed in the liver from its $Mn^{2+}$ form into $Mn^{3+}$, which is associated with transferrins and transported to cells. In this transformation Mn can be replaced by other metals, including magnesium (Scheuhammer and Cherian 1983). Manganese absorption from the gastrointestinal tract occurs mainly through the mucous membrane, in duodenum and ileum. This process is influenced by many factors, including the presence and concentration of other metals in food and the body (iron, calcium), the age of the organism and environmental exposure (CICAD 1999).

Manganese is metabolized in the liver. Inorganic and organic Mn compounds are mainly excreted in faeces (40–70%). Small amounts are also expelled in urine, perspiration and milk (Newland 1999).

## 5.1 Manganese Deficiency and Excess

Manganese deficiency in various organisms may cause growth disorders, skeletal deformities and glucose intolerance problems (Erikson et al. 2005). Deformities in the bone structure caused by this deficiency are the result of enzymatic defects in the synthesis of glycosaminoglycans, which are part of the sugar part of glycoproteins forming the organic framework of connective tissue, including cartilage and bone. Cartilage tissue makes up a large share of the skeleton in the early postnatal period.

Therefore, most abnormalities in the skeletal structure are related to morphological changes in the epiphyseal cartilage and epiphyseal plate, within which the growth of long bones occurs (Mills et al. 1976; Watts 1990). Manganese deficiency has been demonstrated in several animal species, including laboratory animals, swine, cattle and poultry. Symptoms of Mn deficiency in animals include skeletal deformities, inhibited growth and neurological disorders, such as problems with motor coordination (Gehrke 1997; Takeda 2003; Spears 2011). As a result of experimental insufficient Mn supply in the diet of laboratory and farm animals, their bones became shorter and more fragile than normal (Ellis et al. 1947; Spears 2011). Moreover, it was found that the complete absence of Mn in food causes dysfunction in milk production in mammals, and the atrophy of certain organs, as well as a decrease in bone density, and infertility in cattle (ATSDR 2012). In birds, Mn deficiency results in *perosis* (or slipped tendon), skeletal deformities, impaired growth and eggshell formation and, in young chicks, cartilage dystrophy (Underwood 1981; Soetan et al. 2010).

Not only Mn deficiency but also its excess can be harmful to animals because it can cause metabolic disturbances in other elements, such as iron. Manganese introduced into mammalian organisms in too large quantities acts as a neurotoxin and osteotoxin and disrupts the physiology of many other tissues and organs (Mergler and Baldwin 1997). Manganese penetrates through the blood-brain barrier, and therefore with large Mn accumulation in the body, symptoms of brain damage appear, especially extrapyramidal symptoms, including those related to the corpus striatum and globus pallidus. Muscle stiffness and dystonia, motor slowdown, spasmodic movements and tremors may also occur (Davis 1999; Newland 1999; O'Neal and Zheng 2015).

In humans, exposure to excessive amounts of Mn results in clinical signs and symptoms resembling Parkinson's disease (O'Neal and Zheng 2015). With smaller Mn exposure, mammals experience neurobehavioral disorders, especially in cognitive functioning (Levy and Nassetta 2003; Knauer et al. 2017). Schneider et al. (2006) observed the influence of Mn on the behaviour of monkeys. Long-tailed macaque (*Macaca fascicularis*), which suffered from chronic exposure to this metal, exhibited abnormal behaviours, including uncontrolled reactions, decreased motor activity and problems with executing movements.

The adverse effects of Mn excess on the bones and the symptoms accompanying it have been relatively well documented in laboratory animals. Komura and Sakamoto (1992) reported damage to the central nervous system and growth inhibition in mice. It has also been shown that an excess of Mn causes a decrease in proteoglycan levels, inhibiting DNA and collagen synthesis in chondrocytes (Litchfield et al. 1998). It was observed that toxic levels of Mn in mice caused poor rib development in their foetuses, with reduction or complete lack of ossification in segments of the sternum, as well as the parietal and occipital bones (Sanchez et al. 1993; Torrente et al. 2000). In addition, Doyle and Kapron (2002) demonstrated the adverse effect of excessive Mn concentration on the calcification process of growth plate in murine chondrocyte cell cultures. These researchers believe that excessive levels of Mn in the mother's body, which occurs in the embryo at the chondrogenesis stage, might result in skeletal deformities of the foetus.

Little information can be found in the literature regarding the effects of Mn overload on the skeletal and nervous systems of wild endothermic vertebrates. The existing data are scarce and pertain mostly to birds and few species of mammals, especially herbivores. Even less information in that respect can be found on predators.

## 5.2 *Bioindicators and Biomarkers of Manganese in Ecotoxicological Studies*

Evaluation of the quality of the natural environment can be carried out based on indicator organisms, i.e. bioindicators. Vertebrates, especially birds and mammals, have been increasingly used as bioindicators. Based on analysis of the mineral composition of various tissues and organs of endothermic vertebrates, attempts are made to indirectly assess the state of the environment.

Different taxonomic and trophic groups of endothermic vertebrates, originating most commonly from Europe and North America, underwent comparative analysis. For most wild mammals and birds, normal, marginal and toxic ranges of manganese concentration in different tissues were not defined. The normal values assumed by Puls (1994) and Wisconsin Veterinary Diagnostic Laboratory (WVDL 2015) for some species and/or taxonomic groups (bovine, cervid, lapine, porcine, canine, poultry, birds) may serve as a certain frame of reference in relation to wild animals. The data available in the literature, mainly pertaining to domesticated animals, are shown in Tables 7.1 and 7.2. In ecotoxicology, trace elements are primarily assayed in the liver and kidneys, which are responsible for detoxification of the organism and at the same time are places of accumulation of high concentrations of various substances. Considerable attention is paid to the muscles of game animals (because of their usefulness as food), and data on other parts of the body (including the brain, bone, lung, hair and feathers) are significantly fewer. Tables 7.1 and 7.2 show various data pertaining to Mn concentrations in biological materials obtained from mammals and birds associated with land as well as inland aquatic ecosystems.

**Table 7.1** Manganese mean concentrations in selected tissues of mammals

| Species or animal group | Location | Mn concentration (mg kg$^{-1}$ dw) | | | References |
|---|---|---|---|---|---|
| | | Liver | Kidney | Muscle | |
| Domestic and ranch animals | | | | | |
| Cattle *Bos taurus* | Canada | | | | Puls (1994) |
| | Adequate, ww | 2.5–6.0 | 1.2–2.0 | | |
| | dw | 8.3–20[a] | 6.0–10.0 | 2.0–3.8 | |
| | Marginal, ww | 1.5–3.0 | 0.93–1.2 | | |
| | dw | 5.0–10[a] | 04.6–6.0[a] | | |
| Swine *Sus scrofa* | Canada | | | | Puls (1994) |
| | Adequate, ww | 2.0–4.0 | 1.3–2.0 | | |
| | | 6.7–13.3[a] | 6.5–10[a] | | |
| | Marginal, ww | 2.8–3.1 | 0.75–1.13 | | |
| | | 9.3–10.3[a] | 3.75–5.65[a] | | |
| Dog *Canis lupus* | Canada | | | | Puls (1994) |
| | Adequate, ww | 3.0–5.0 | 1.2–1.8 | | |
| | | 10.0–16.7[a] | 6.0–9.0[a] | | |
| Raccoon dog *Nyctereutes procyonoides* | Poland, ranch | F: 5.89<br>M: 3.62 | F: 2.46<br>M: 2.05 | F: 1.07<br>M: 1.04 | Mertin et al. (2006) |
| Bovine | Normal, ww | 2.5–6.0 | 1.2–2.0 | | WVDL (2015) |
| | dw | 8.3–20.0[a] | 6.0–10.0[a] | | |
| Cervid | Normal, ww | 2.5–8.0 | 1.0–4.0 | | |
| | dw | 8.3–26.7[a] | 5.0–20.0[a] | | |
| Lapine | Normal, ww | 1.0–2.0 | 2.0–3.0 | | |
| | dw | 3.3–6.7[a] | 10.0–15.0[a] | | |
| Porcine | Normal, ww | 2.3–4.0 | 1.3–2.0 | | |
| | dw | 7.7–13.3[a] | 6.5–10.0[a] | | |
| Canine | Normal, ww | 3.0–5.0 | 1.2–1.8 | | |
| | dw | 10–16.7[a] | 6.0–9.0 | | |
| American mink *Neovison vison* | Ranch (dark) | 5.97[a] | 4.15[a] | | Stejskal et al. (1989) |
| Wild animals | | | | | |
| Herbivorous ungulates | | | | | |
| Red deer *Cervus elaphus* | Poland, central part | | | | Michalska and Żmudzki (1992) |
| | Autumn | 10.4[a] | 10.0[a] | 1.30[a] | |
| | Spring | 7.4[a] | 6.90[a] | 1.07[a] | |
| | Poland, N part (1987) | 14.3[a] | 27.5[a] | 2.4[a] | Falandysz (1994) |
| | Poland, NE part | 12.0 | 6.6 | 2.3 | Jarzynska and Falandysz (2011) |
| | Poland, S part | 19.8 | 4.2 | 28.0 | Świergosz et al. (1993) |
| | Romania | | | 2.04 | Craciunescu et al. (2014) |

(continued)

**Table 7.1** (continued)

| Species or animal group | Location | Mn concentration (mg kg$^{-1}$ dw) | | | References |
|---|---|---|---|---|---|
| | | Liver | Kidney | Muscle | |
| | Austria | | | 0.50 | Ertl et al. (2016) |
| | Scotland, NW part | | | | French et al. (2017) |
| | M | 9.29 | | | |
| | F | 14.6 | | | |
| | Slovakia, W part | 11.6[a] | 6.7[a] | 6.8[a] | Gasparik et al. (2004) |
| Roe deer *Capreolus capreolus* | Austria | | | 0.81[a] | Ertl et al. (2016) |
| | Poland, central part | | | | Michalska and Żmudzki (1992) |
| | Autumn | 10.9[a] | 14.5[a] | 1.50[a] | |
| | Spring | 7.6[a] | 5.7[a] | 0.90[a] | |
| | Poland, central part | 7.5[a] | | 1.2[a] | Dlugaszek and Kopczynski et al. (2013) |
| | Poland, N part (1987) | 16.7[a] | 27.5[a] | 2.8[a] | Falandysz (1994) |
| | Poland, N part | | | | Chudzinska-Popek and Majdecka (2010) |
| | Male | 22.3[a] | | | |
| | Female | 19.0[a] | | | |
| Reindeer *Rangifer tarandus* | Norway, Svalbard | 8.7[a] | 6.5[a] | | Borch-Iohnsen et al. (1996) |
| | Sweden, Lapland | 8.5[a] | | 0.49[a] | Odsjo et al. (2007) |
| Moose (elk) *Alces alces* | USA, Minnesota | | | | Custer et al. (2004) |
| | Agriculture and prairie | 7.9 | | | |
| | Bog and forest | 8.0 | | | |
| | Canada, Nova Scotia | 9.2 | 13.0 | | Pollock and Roger (2007) |
| | Canada, Nova Scotia | 10.1[a] | 14.0[a] | | Frank et al. (2004) |
| | Sweden, NW part | | | | Frank et al. (2000) |
| | Healthy | 14.4[a] | 21.9[a] | | |
| | Affected | 13.6[a] | 15.8[a] | | |
| White-tailed deer *Odocoileus virginianus* | Canada, Nova Scotia | 11.6 | | | Pollock and Roger (2007) |
| | USA, Texas | 13.7 | | 0.67 | Bruckwicki (2006) |
| | USA, South Dakota | 13.2[a] | | | Zimmerman et al. (2008) |
| Mule deer *Odocoileus hemionus* | USA, South Dakota | 12.4[a] | | | Zimmerman et al. (2008) |
| | USA, | | | | Wolfe et al. (2010) |
| | Colorado | | | | |
| | Infected | 8.5 | | | |
| | Uninfected | 9.5 | | | |
| | USA, California | 11.7 | | | Roug et al. (2015) |

(continued)

**Table 7.1** (continued)

| Species or animal group | Location | Mn concentration (mg $kg^{-1}$ dw) Liver | Kidney | Muscle | References |
|---|---|---|---|---|---|
| Japanese serow *Capricornis crispus* | Japan | 14.4[a] | 27.6[a] | 0.87[a] | Honda et al. (1987) |
| Lagomorphs | | | | | |
| European hare *Lepus europaeus* | Austria | | | 0.69[a] | Ertl et al. (2016) |
| | Slovakia | | | 1.5[a] | Strmiskova and Strmiska (1992) |
| | Poland, N part | 11.7[a] | 10.5[a] | 7.16[a] | Myslek and Kalisinska (2006) |
| | Poland, central part | 6.0[a] | | 0.33[a] | Dlugaszek and Kopczynski et al. (2013) |
| Rodents | | | | | |
| Canadian beaver *Castor canadensis* | Canada, Ontario | 12.3[a] | 17.0[a] | 2.0[a] | Wren (1984) |
| Omnivores | | | | | |
| Wild boar *Sus scrofa* | Poland, central part | | | | Michalska and Zmudzki (1992) |
| | Autumn | 10.5[a] | 11.7[a] | 1.6[a] | |
| | Spring | 7.4[a] | 6.9[a] | 0.97[a] | |
| | Poland, S part | 2.2 | 1.0 | 3.4 | Świergosz et al. (1993) |
| | Poland, N part, 1987 | 4.7[a] | 5.0[a] | 0.93[a] | Falandysz (1994) |
| | Poland, central part | 4.2[a] | | 1.0[a] | Dlugaszek and Kopczynski et al. (2013) |
| | Austria | | | 0.33[a] | Ertl et al. (2016) |
| | Slovakia | | | 2.1[a] | Strmiskova and Strmiska (1992) |
| | Romania | | | 0.28 | Craciunescu et al. (2014) |
| Carnivores | | | | | |
| Gray fox *Urocyon cinereoargenteus* | USA, California (Zoo) | 8.4 | 4.7 | | Arnhold et al. (2002) |
| Arctic fox *Vulpes lagopus* | Norway, Svalbard | 8.0[a] | | | Prestrud et al. (1994) |
| | Canada, Nunavut | 12.8[a] | | | Hoekstra et al. (2003) |
| Red fox *Vulpes vulpes* | Czech Republic | 6.45 | 3.84 | | Jankovska et al. (2010) |
| | Romania | 15.9 | | | Farkas et al. (2017) |

(continued)

**Table 7.1** (continued)

| Species or animal group | Location | Mn concentration ($mg\ kg^{-1}$ dw) | | | References |
|---|---|---|---|---|---|
| | | Liver | Kidney | Muscle | |
| Golden jackal *Canis aureus* | Romania | 13.8 | | | Farkas et al. (2017) |
| Marten *Martes americana* | Canada, Northwest Territories | | 2.16 | | Poole et al. (1998) |
| | Canada, British Columbia | | 4.32 | | Harding (2004) |
| Wolverine *Gulo gulo* | Canada, British Columbia | 9.43 | | | Harding (2004) |
| | Canada, Nunavut | 9.7[a] | | | Hoekstra et al. (2003) |
| Javan mongoose *Herpestes javanicus* | Japan, Amami Oshima Island | 14.6[a] | 4.0[a] | | Horai et al. (2006) |
| Raccoon *Procyon lotor* | Canada, Ontario | 10.3[a] | 6.5[a] | 1.3[a] | Wren (1984) |
| | USA, South Carolina | 17.0[a] | 7.5[a] | 1.2[a] | Burger et al. (2002) |
| | USA, Tennessee | | | | Souza et al. (2013) |
| | Polluted area | | | | |
| | 2009 | 9.3[a] | 6.0[a] | 0.62[a] | |
| | 2010 | 9.5[a] | 5.5[a] | 0.60[a] | |
| | Unpolluted area | 7.3[a] | 6.2[a] | 0.65[a] | |
| River otter *Lontra canadensis* | Canada, Ontario | 9.7[a] | 3.0[a] | 1.7[a] | Wren (1984) |
| | Canada, Ontario, Sudbury | 9.3[a] | 3.5[a] | | Wren et al. (1988) |
| | Canada, British Columbia, Fraser River | 11.5 | | | Harding (2004) |
| | USA, Illinois | 10.1[a] | | 1.3[a] | Halbrook et al. (1996) |
| | USA, SW Washington | 10.7 | | | Grove and Henny (2008) |
| Eurasian otter *Lutra lutra* | Finland | 11.4[a] | | | Skaren (1992) |
| | Finland | 10.7[a] | | | Lodenius et al. (2014) |
| | Denmark<br>Great Britain<br>Ireland | 3.53<br>7.37<br>4.91 | | | Mason and Stephenson (2001) |
| | England and Wales | | | | Walker et al. (2010) |
| | 2007 | 8.30 | | | |
| | 2008 | 8.98 | | | |
| | South Korea | 9.18 | 2.92 | | Kang et al. (2015) |

(continued)

**Table 7.1** (continued)

| Species or animal group | Location | Mn concentration (mg kg$^{-1}$ dw) | | | References |
|---|---|---|---|---|---|
| | | Liver | Kidney | Muscle | |
| American mink *Neovison vison* | Canada, Ontario, Sudbury | 14.1[a] | 5.3[a] | | Wren et al. (1988) |
| | Canada, NT, Inuvik | | 2.88 | | Poole et al. (1998) |
| | Canada, British Columbia, Fraser River | 8.98 | 3.56 | | Harding (2004) |
| | USA, Illinois | 21.8[a] | 16.2[a] | 4.9[a] | Halbrook et al. (1996) |
| | Poland | | | | Brzezinski et al. (2014) |
| | Feral: WMNP | 6.87 | 2.90 | | |
| | Feral: NNP | 7.13 | 3.34 | | |
| | Ranch | 5.02 | 2.83 | | |
| Insectivores | | | | | |
| White-tooted shrew *Crocidura russula* | Spain | | | | Sanchez-Chardi et al. (2009) |
| | Polluted site | 38.27 | 20.02 | | |
| | Reference site | 36.78 | 17.81 | | |
| European hedgehog *Erinaceus europaeus* | Finland | 6.33 | 2.38 | | Rautio et al. (2010) |

*im* immature, *ad* adult *M* male, *F* female, *WMNP* Warta Mouth National Park, *NNP* Narew National Park

[a]Values converted from wet weight to dry weight according to the author's data, or we assume that the kidneys consist of 80% water and the liver and muscle 70%

### 5.2.1 Mammals

There are not many studies in which Mn concentration be assayed in different tissues and organs of specimen belonging to the same species. One such study is a report by MacDonald et al. (2005) on one moose *Alces alces* from the area of an abandoned gold mine (Colomac Mine, Northwest Territories, Canada). The highest Mn concentration (expressed in dw) was found in various parts of its stomach and intestines (from 177 to 948 and from 101 to 154 mg kg$^{-1}$ dw, respectively). Manganese concentration ranging between 11.9 and 13.6 mg Mn kg$^{-1}$ dw was detected in the cranial part of the lung, kidney and liver, while skeletal muscle contained 2.0 mg Mn kg$^{-1}$. External tissues, such as skin and antlers, contained 9.8 and 18.6 mg Mn kg$^{-1}$ dw, respectively. The data on Mn concentrations that can be found in the available literature mostly relate to herbivores (mainly ruminants) and carnivorous mammals.

**Table 7.2** Manganese mean concentrations in selected avian tissues

| Species | Location | Mn concentration (mg kg$^{-1}$ dw) | | | References |
|---|---|---|---|---|---|
| | | Liver | Kidney | Muscle | |
| Domesticated animals | | | | | |
| Poultry | Deficient | | | | Puls (1994) |
| | ww | 1.5–4.0 | | | |
| | dw | 5–13.3[a] | | | |
| | Adequate ww | | | | |
| | dw | 2.0–4.0 | 1.5–2.5 | | |
| | High | 6.7–13.3[a] | 6.0–10[a] | | |
| | ww | | | | |
| | dw | 4.0–6.0 | 2.5–6.0 | | |
| | | 13.3–20.0[a] | 10–24[a] | | |
| Wild birds | | | | | |
| Galliformes | | | | | |
| Pheasant *Phasianus colchicus* | Slovakia | | | 0.53[a] | Strmiskova and Strmiska (1992) |
| | Austria | | | 0.70[a] | Ertl et al. (2016) |
| | Italy | | | <1.1[a] | Roselli et al. (2016) |
| Quail *Coturnix coturnix* | Romania | | | 3.5[a] | Roselli et al. (2016) |
| Columbiformes | | | | | |
| Wood pigeon *Columba palumbus* | Italy | | | 1.2[a] | Roselli et al. (2016) |
| Turtle dove *Streptopelia turtur* | Italy | | | 1.8[a] | Roselli et al. (2016) |
| Feral pigeon *Columba livia* | Japan, Kanto | 7.12 | 20.4 | 1.32 | Horai et al. (2007) |
| Mourning dove *Zenaida macroura* | USA, South Carolina | | | | Burger et al. (1997) |
| | 1992 | 15.40[a] | | 2.18[a] | |
| | 1993 | 18.64[a] | | 1.83[a] | |
| Anseriformes | | | | | |
| Swans and geese | | | | | |
| Mute swan *Cygnus olor* | Poland, E part | 7.92 | | | Komosa et al. (2012) |
| | Canada, Ontario | F: 10.59<br>M: 7.85 | | | Schummer et al. (2011) |
| Whooper swan *Cygnus cygnus* | Japan | 8.6[a] | 10.0[a] | 1.43[a] | Honda et al. (1990) |
| Bewick's swan *Cygnus columbianus* | Japan | 15.4[a] | 12.2[a] | 1.40[a] | Honda et al. (1990) |

(continued)

**Table 7.2** (continued)

| Species | Location | Mn concentration (mg $kg^{-1}$ dw) | | | References |
|---|---|---|---|---|---|
| | | Liver | Kidney | Muscle | |
| White-fronted goose *Anser albifrons* | Korea | 28.1 | 40.6 | 3.29 | Kim and Oh (2013) |
| Lesser snow goose *Anser caerulescens* | Russia | SP: 10.3<br>NP: 9.9 | | | Hui et al. (1998) |
| Ducks | | | | | |
| Lesser scaup *Aythya affinis* | USA, Mississippi flyway | 17.9 | | | Custer et al. (2003) |
| | USA | F: 21.1[a]<br>M: 21.2[a] | | | Pillatzki et al. (2011) |
| Spot-billed duck *Anas poecilorhyncha* | Korea | 17.7 | 12.5 | 9.46 | Kim and Oh (2013) |
| Mallard *Anas platyrhynchos* | Japan | 13.7 | 10.6 | 1.79 | Nam et al. (2005b) |
| | USA, New Jersey | 10.5[a] | | | Burger and Gochfeld (1985) |
| | Poland, NW part, Szczecin | ad: 21.8[a]<br>im: 17.0[a] | ad: 33.3[a]<br>im: 46.8[a] | ad: 2.19[a]<br>im: 2.00[a] | Kalisinska et al. (2004) |
| Black duck *Anas rubripes* | USA, New Jersey | 6.2[a] | | | Burger and Gochfeld (1985) |
| Eurasian wigeon *Anas penelope* | Spain, NE part | 8.7 | | | Mateo and Guitard (2003) |
| Common teal *Anas crecca* | | 8.6 | | | |
| Gadwall *Anas strepera* | | 12.0 | | | |
| Northern pintail *Anas acuta* | | 9.0 | | | |
| Northern shoveler *Anas clypeata* | | 13.0 | | | |
| Red-crested pochard *Netta rufina* | | 11.0 | | | |
| Common pochard *Aythya ferina* | | 8.1 | | | |
| Redhead *Aythya americana* | USA, Texas and Louisiana | 7.27 | | | Michot et al. (1994) |
| Gruiformes | | | | | |
| Moorhen *Gallinula chloropus* | Japan, Kanto | 9.67 | 13.1 | 1.10 | Horai et al. (2007) |

(continued)

**Table 7.2** (continued)

| Species | Location | Mn concentration (mg kg$^{-1}$ dw) | | | References |
|---|---|---|---|---|---|
| | | Liver | Kidney | Muscle | |
| Falconiformes | | | | | |
| Bald eagle *Haliaeetus leucocephalus* | USA, Alaska | 9.94 | 5.57 | | Stout and Trust (2002) |
| | USA, Maine | 11.35 | | | Mierzykowski et al. (2011) |
| | USA, Michigan, Minnesota | 14.5 | | | Nam et al. (2012) |
| White-tailed eagle *Haliaeetus albicilla* | Poland, NW part | 12.13[a] | 7.64[a] | 2.06[a] | Kalisinska et al. (2006) |
| | Poland, NW part | 8.5 | 4.6 | | Falandysz et al. (2001) |
| Common buzzard *Buteo buteo* | Poland, E part | 5.30 | | | Komosa et al. (2012) |
| | Italy, Sicily | 9.0 | 1.41 | 0.90 | Licata et al. (2010) |
| | Netherlands | 11.35 | 5.52 | | Jager et al. (1996) |
| Black kite *Milvus migrans* | Japan, Kanto | 9.39 | 4.02 | 1.32 | Horai et al. (2007) |
| Northern goshawk *Accipiter gentilis* | Japan, Kanto | 13.4 | 5.74 | 2.07 | Horai et al. (2007) |
| Strigiformes | | | | | |
| Barn owl *Tyto alba guttata* | Netherlands | 9.8 | 6.7 | | Esselink et al. (1995) |
| Ural owl *Strix uralensis* | Japan, Kanto | 12.4 | 9.19 | 0.88 | Horai et al. (2007) |
| Brown hawk-owl *Ninox scutulata* | Japan, Kanto | 12.0 | 5.27 | 1.59 | Horai et al. (2007) |
| Passeriformes | | | | | |
| Greenfinch *Chloris chloris* | China | 4.28 | 6.08 | 1.87 | Deng et al. (2007) |
| Great tit *Parus major* | China | 5.14 | 6.58 | 1.82 | Deng et al. (2007) |
| | Belgium | 2.71 | 4.85 | 0.94 | Dauwe et al. (2005) |
| | Spain | P: 5.98<br>NP: 8.72 | P: 8.26<br>NP: 8.19 | P: 2.10<br>NP: 2.21 | Llacuna et al. (1995) |
| Rock bunting *Emberiza cia* | Spain | P: 7.90<br>NP: 6.18 | P: 10.34<br>NP: 9.91 | P: 1.56<br>NP: 1.97 | Llacuna et al. (1995) |
| Blackbird *Turdus merula* | Spain | P: 4.84<br>NP: 3.90 | P: 11.97<br>NP: 8.56 | P: 1.59<br>NP: 1.31 | Llacuna et al. (1995) |
| Rook *Corvus frugilegus* | Poland, E part | 2.53 | | | Komosa et al. (2012) |
| Jungle crows *Corvus macrorhynchos* | Japan, Kanto | 4.23 | 8.35 | 1.36 | Horai et al. (2007) |

(continued)

**Table 7.2** (continued)

| Species | Location | Mn concentration (mg kg$^{-1}$ dw) | | | References |
|---|---|---|---|---|---|
| | | Liver | Kidney | Muscle | |
| Carrion crow *Corvus corone* | Japan, Kanto | 4.07 | 9.39 | 1.32 | Horai et al. (2007) |
| Pelecaniformes | | | | | |
| Great cormorant *Phalacrocorax carbo* | Spain, Murcia | 9.41 | 4.67 | 1.79 | Navarro et al. (2010) |
| | Serbia | 9.38 | | | Skoric et al. (2012) |
| | Japan | 19.0 | 11.0 | 1.95 | Nam et al. (2005a) |
| | Japan | 16.6 | 8.64 | 1.98 | Nam et al. (2005b) |
| | Japan, Kanto | 15.7 | 7.74 | 1.57 | Horai et al. (2007) |
| Great white egret *Egretta alba* | Japan | | | 2.29 | Honda et al. (1985) |
| Great white egret *Egretta alba* | Japan, Kanto | 9.85 | 5.46 | 1.83 | Horai et al. (2007) |
| Intermediate egret *Egretta intermedia* | Japan, Kanto | 12.0 | 8.0 | 1.52 | Horai et al. (2007) |
| Grey heron *Ardea cinerea* | Poland, E part | 6.43 | | | Komosa et al. (2012) |
| | Poland, NE part | 5.3[a] | | | Babinska et al. (2008) |
| | Japan | 13.4 | 8.43 | 1.71 | Horai et al. (2007) |

*ad* adult, *im* immature, *M* male, *F* female, *SP* southern population, *NP* northern population
[a]Value converted from wet weight to the dry weight according to author's data, or we assumed that the kidney contains 75% of water, liver and muscle 70% of water

### 5.2.1.1 Hepatic Manganese

In ruminants such as bovines and cervids, normal Mn hepatic concentrations remain within the following ranges: 2.5–6.0 and 2.5–8.0 mg kg$^{-1}$ ww or 8.3–20.0 and 8.3–26.7 mg kg$^{-1}$ dw, respectively (Puls 1994; WVDL 2015). The overwhelming majority of the data related to mean Mn levels in the livers of wild cervids (expressed in dw), shown in Table 7.1, stays within the aforementioned range or is slightly lower than 8 mg kg$^{-1}$ dw. Average Mn concentrations most commonly range between 8.5 and 14.5 mg kg$^{-1}$ dw. Only in two studies were hepatic Mn levels in red deer *Cervus elaphus* and roe deer *Capreolus capreolus* (from N and S Poland, respectively) ≥19 mg Mn kg$^{-1}$ dw (Falandysz 1994; Chudzinska-Popek and Majdecka 2010). Significant differences were noted between Mn hepatic concentrations in autumn and spring of both these species, living in Poland, while the spring value was lower by ~3 mg kg$^{-1}$ dw than the autumn value (Michalska and Zmudzki 1992). Zimmerman et al. (2008) also detected inter-seasonal differences in Mn concentration in the same organs of white-tailed deer females *Odocoileus virginianus* from South Dakota, USA. What is more, different diseases can contribute to lowering Mn concentration in the liver of the moose *Alces alces* from Sweden, as demonstrated by Frank et al. (2000).

In the omnivorous ungulate wild boar *Sus scrofa*, average hepatic Mn levels change from 2.2 to 10.5 mg kg$^{-1}$ dw (Table 7.1). For farm swine, Mn concentrations ranging from 9.3 to 10.3 mg kg$^{-1}$ dw (or 2.8–3.1 mg kg$^{-1}$ ww) indicate marginal levels of this element (Puls 1994). Probably, the natural demand for Mn is lower in the house pig's ancestor than in intensively fattened farm animals.

Mean Mn concentration in the liver has a somewhat wider range in carnivores compared to ruminants: from 3.5 (in the Eurasian otter from Denmark) to 21.8 mg kg$^{-1}$ dw (in the American mink from Illinois, USA). Usually, in the livers of wild and feral American mink and otters from North America and Europe, mean Mn levels stay in the range between 7 and 14 mg kg$^{-1}$ dw. The cited values are higher for wild/feral animals than for ranch minks, 5–6 mg kg$^{-1}$ dw (Table 7.1), as had been previously observed by Halbrook et al. (1996) in their research on Illinois minks. A similar tendency can be observed in canids. For the domesticated dog *Canis lupus*, the analogous range falls between 10 and 16.7 mg kg$^{-1}$ dw (Puls 1994), but it is slightly wider in wild canids: 6.4–15.9 mg kg$^{-1}$ dw (Table 7.1). In comparison with herbivorous, omnivorous and carnivorous species, the highest mean Mn concentrations in the liver were found in the small insectivorous white-tailed shrew *Crocidura russula* from Spain. In the livers of animals from polluted and reference sites, similar concentrations (~38 mg Mn kg$^{-1}$ dw) were detected (Sanchez-Chardi et al. 2009). This might be related to fast metabolism in that species. With the exception of the aforementioned species, hepatic mean levels of Mn in other wild terrestrial mammals usually range between 6 and 16 mg kg$^{-1}$ dw.

#### 5.2.1.2 Nephric Manganese

The highest mean Mn concentration in the kidneys has been observed in three cervid species: Japanese serow *Capricornis crispus* from Japan and red deer and roe deer from Poland (27.5–27.6 mg kg$^{-1}$ dw), while it was lowest in wild boar (1.0 mg kg$^{-1}$ dw) from a southern part of Poland polluted by heavy metals (Table 7.1). For cervids and swines, normal levels of Mn in the kidneys are as follows: 5–20 and 6.5–-10 mg kg$^{-1}$ dw, respectively (WVDL 2015). In wild cervids and other herbivorous species (European hare *Lepus europaeus*, Canadian beaver *Castor canadensis*), nephric Mn levels usually range between 6 and 17 mg kg$^{-1}$ dw. In wild boar from Poland, Mn levels varied between 1 and 12 mg kg$^{-1}$ dw, whereas in different carnivorous species, Mn nephric levels were clearly lower than in herbivores and did not exceed 6 mg kg$^{-1}$ dw. The exception here is the raccoon *Procyon lotor*, whose mean Mn concentration values in the kidneys ranged from 5.5 to 7.5 mg Mn kg$^{-1}$ dw (Table 7.1), but its diet contains substantial amounts of plant material. Normal Mn concentration in the kidneys of wild animals is lower than in the liver, but in some cases, the opposite is true (Table 7.1). For example, higher nephric than hepatic values were observed in moose from Canada and Sweden (Frank et al. 2000, 2004; Pollock and Roger 2007). Moreover, in moose from NW Sweden, nephric Mn levels were about 28% higher in healthy individuals than in diseased animals (Frank et al. 2000).

Similarly to the liver, variations in nephric Mn concentrations are observed both between herbivorous species and individuals of the same species occurring in the same area. There was a spectacular case of a moose affected by a disease with an unknown aetiology and healthy animals from NW Sweden. In the affected moose, nephric Mn level was almost 30% lower than in healthy individuals (Table 7.1). Additionally, concentration of molybdenum was 36% higher in the livers of the diseased animals, and the concentration of copper was about 60% lower than in animals from the control group. The changes in these and other trace element concentrations in the kidney and liver, as well as some clinical biochemical parameters observed in these animals, corresponded to molybdenosis and secondary copper deficiency in domesticated ruminants (Frank et al. 2000). These studies indicate the existence of complex and still poorly understood relationships between concentrations of various elements in mammalian detoxification organs.

#### 5.2.1.3 Manganese in Skeletal Muscle and Other Tissues

Manganese concentration in the muscles is mainly assayed in game animals, which are obtained for food. They are mostly cervids, wild boars, hares and raccoons (mainly in North America). In these mammals, Mn concentration in the skeletal muscles ranges from <0.30 to >7.0 mg $kg^{-1}$ dw, but sporadically can exceed 25 mg Mn $kg^{-1}$ dw as was the case in red deer from a southern area of Poland polluted with heavy metals. Generally, in wild herbivores and boars, the values of mean Mn muscle levels change around 1 mg $kg^{-1}$ dw (Table 7.1). Analogical values for herbivorous cattle and sheep, but not for omnivorous swine (<0.5 mg Mn $kg^{-1}$ dw), were described by Doyle and Spaulding (1978). Puls (1994) found that the adequate Mn levels in sheep and cattle muscle tissues changed in ranges 0.24–0.40 and 2.0–3.8 mg $kg^{-1}$ dw, respectively, which clearly diverges from the value of 1 mg Mn $kg^{-1}$ dw usually observed in game mammals. It appears that Mn concentration in the muscles of animals whose diet predominantly includes plants with various concentrations of this metal can differ to a considerable degree. Seasonal changeability observed in Mn concentrations in the muscle (also in liver and kidneys) in red deer, roe deer and wild boar seems to support this point of view (Michalska and Zmudzki 1992). These researchers demonstrated that in the autumn, i.e. after the abundance of readily available young plants, the muscles of these animals contained more Mn than in the early spring, preceded by a long and difficult winter season. In temperate climates, during winter, especially when there is a lot of snow, vegetation is more difficult to access, and the older parts surviving above ground contain significantly less Mn than young shoots and leaves (Millaleo et al. 2010; Mousavi et al. 2011).

The data on Mn concentrations in the muscle of piscivorous mammals are scarce (Table 7.1). It appears that the average Mn concentration exceeds 1 mg Mn $kg^{-1}$ dw, but its range does not vary much (1.2–4.9 mg $kg^{-1}$ dw) compared to the muscle of omnivorous boar/swine and herbivorous ruminants (0.3–28 mg Mn $kg^{-1}$ dw). This may be the result of a fish diet (which has a lower and more constant Mn content than a plant-based diet) and the high degree of physical activity of piscivorous animals.

Manganese crosses the blood-brain barrier and is accumulated in varying concentrations in different parts and structures of mammalian brains. In mammals, one of the most important target tissues in Mn toxicity is the brain (Zheng et al. 2010). Increased brain Mn concentration is a critical step in the pathogenesis of manganese-induced neurotoxicity (Aschner et al. 2006, 2009). Accumulation of Mn occurs mainly in the basal ganglia and leads to a syndrome called manganism, whose symptoms, including cognitive dysfunction and motor impairment, resemble Parkinson's disease. In this structure, which is a group of subcortical nuclei, and in the cortex, intra- and interspecies differences are observed. For example, in primates such as humans, rhesus macaque *Macaca mulatta* and marmoset *Callithrix jacchus*, mean Mn concentrations were, in the basal ganglia, 2.0–2.5, 0.34–0.48 and 0.21 mg kg$^{-1}$ ww, respectively, and in the cortex as follows: 1.0–1.2, 0.25 and 0.20 mg kg$^{-1}$ ww (Dorman et al. 2006; Ramos et al. 2014; Knauer et al. 2017). In fact, studies designed to explain the effects of Mn concentration on the central nervous system functioning due to manganese-induced neurotoxicity as seen in humans, primarily occupationally and/or environmentally exposed to elevated concentrations of this metal in air, water and food, have been carried out only on humans, laboratory rodents and primates (Takeda et al. 1998; Dorman et al. 2006; Erikson et al. 2005; Fitsanakis et al. 2008; Ramos et al. 2014; Knauer et al. 2017).

Only two studies in which brain Mn levels in wild mammals were assayed were found in the available literature. They concerned a small marsupial living in North America—the northern opossum *Didelphis virginiana* and raccoon. The brains of immature and adult opossums contained 3.8 and 2.2 mg Mn kg$^{-1}$ dw, respectively (Arnhold et al. 2002). This observation is analogous to the human brain because children accumulate higher levels of Mn than adults (O'Neal and Zheng 2015). In the whole brains of raccoons, mean values ranged from 0.33 to 0.37 mg Mn kg$^{-1}$ ww, depending on the year and site of collection, but such differences were insignificant (Souza et al. 2013). Assuming that the brain of a medium-sized mammal consists of 77% water (Kalisinska et al. 2016), Mn levels in the brain of the adult opossums were ~0.51 mg kg$^{-1}$ ww. Similarly, to the aforementioned raccoons and opossums, the brain Mn levels (in the cortex/cerebrum) assayed in dogs (aged 1–7 years old), cats and lambs were 0.33, 0.41 and 0.44–0.61 mg kg$^{-1}$ ww, respectively (Michalska et al. 1991; Soltysiak et al. 1997; Bakirdere et al. 2011).

Hesketh et al. (2007) and White et al. (2010) compared Mn concentrations in different parts of the central nervous system of ruminants suffering from animal prion diseases such as bovine spongiform encephalopathy (BSE), scrapie of sheep and chronic wasting disease (CWD) of cervids. In sheep and cattle from control groups, Mn concentrations in the frontal cortex were 0.26 and 0.33 mg kg$^{-1}$ ww, respectively (Hesketh et al. 2007). Those values were similar to Mn concentration in the parietal/optic lobes of healthy Rocky Mountain elks (*Cervus elaphus nelsoni*): 1.32 mg kg$^{-1}$ dw (White et al. 2010) or 0.26 mg kg$^{-1}$ ww. Since the brain of large ruminants consists of ~80% water (Jung et al. 2015), considering the aforementioned values, it can be assumed that Mn concentration in the cerebrum and/or cortex parts of mammals differs across various species, but it changes only slightly, i.e. from 0.20 to 0.65 mg $kg^{-1}$. It should be stressed that substantial differences in Mn concentration in the central nervous system occur naturally, and in the case of some diseases (BSE, scrapie), a significant increase in the concentration of this metal has been documented, compared to controls, in the brain stem, spinal cord and cerebellum, but not in the frontal cortex (Hesketh et al. 2007).

Analysis of the research presented here reveals that in farm mammals, and especially in wild animals, the concentration of Mn in the brain is very rarely assayed. This is the case for several reasons. Firstly, the brain is enclosed in the skull, which makes obtaining samples of nervous tissue a more difficult task. Secondly, the distribution of Mn in the brain is heterogeneous, which in the case of samples taken from different parts of the brain may complicate the performance of comparative ecotoxicological studies. Finally, the value of the brain of pigs, cattle, sheep, cervids and wild boar as food is negligible, and so the mineral composition of their brains is only sporadically analysed.

However, given the local environmental Mn pollution found in dust and food, especially in food originating from plants, its increased penetration into the mammalian brain cannot be excluded, and to determine this would require extensive further field studies.

About 100 years ago, Mn was recognized as an essential nutrient for proper bone formation (Erikson et al. 2005). It is believed that the mammalian skeletal contains up to 40–43% of all Mn accumulated in an organism (O'Neal et al. 2014). The bone is one of the major organs for long-term storage of Mn in the body, and Mn half-life there is much longer than in soft tissues such as liver and kidney or body fluids. In adult mammals, the main function of the skeleton is bone remodelling, which involves both bone formation and resorption. Metals incorporated into the skeleton are subject to bone turnover and remodelling processes that occur at approximately 10% per year in adult individuals. For these reasons, bone tissue is considered to be a suitable long-term bioindicator of environmental metal exposure (Martiniakova et al. 2012; O'Neal et al. 2014). O'Neal et al. (2014) and O'Neal and Zheng (2015) suggested that the bone reflects the Mn body burden much better than body fluids or keratinized tissues such as nails and hair. Additionally, bone Mn level correlates with Mn concentrations in some brain structures. All of these features indicate that bone tissue is a useful biomarker in epidemiological and ecotoxicological studies of Mn.

Despite the many advantages of the bone as a long-term bioindicator in the environmental investigation of the Mn body burden, surprisingly little data have been published on its concentrations in the bones of wildlife, with the exception of cervid antlers, which are produced annually by males (Arnhold et al. 2002; Dobrowolska 2002; McDonald et al. 2005; Sanchez-Chardi and Lopez-Fuster 2009; Budis et al. 2013; Ceacero et al. 2015; Cappelli et al. 2017). Recently Buddhachat et al. (2016) conducted an elemental analysis of the humerus bone of 14 species representing carnivores, ungulates, primates, elephants and dolphins, which originated from the collection of Chiang Mai University in Thailand. They have shown that Mn occurred in all the studied bones and found interspecies differences. Distribution of many elements (including Mn) was determined using a handheld X-ray fluorescence device. Each element was expressed as a percentage obtained from the area under the peak of each element divided by the total area for all the elements recorded in the scan. Elemental values represented a relative amount (elemental fingerprint), but not the actual concentrations of each element in a tested sample. In humerus bone, the highest Mn percentage values were detected in the Asian elephant *Elephas maximus* and dog (0.068% and 0.059%) and moderate values in the lion *Panthera leo* and Assam macaque *Macaca assamensis* (0.035% and 0.032%). In other analysed animals, the values of the analysed parameter did not exceed 0.015% (ranging between 0.008 and 0.015%). An important feature of this report is the comparative analysis of the same bone (humerus bone), because Mn concentrations can vary greatly between different bones from the same individual (e.g. rib, carpal bone, thigh bone and lower jaw). This has been demonstrated in experimental studies, among others, on sheep and other domestic ruminants (Hidiroglou 1980). Buddhachat et al. (2016) observed this kind of difference between the antlers and frontal bones of cervids (0.028% vs. 0.006%). It was also observed that Mn concentrations vary in different fragments of the same bone (layers of cartilage, compact bone and spongy bone coming from the femur head of the hip joint). Significantly higher Mn concentrations were found in samples of canid cartilage and cartilage/compact bone than in spongy bone (Budis et al. 2013, 2015). Moreover, experimental research on Mn in rat bones showed that the shortest half-lives of this metal were related to weight-bearing bones such as the femur and tibia, while the longest half-lives were related to parietal bones that form part of the skull (O'Neal et al. 2014).

In wild mammals, relatively high Mn levels in large bones were found in insectivorous white-toothed shrews from Spain. The animals from the Ebro Delta, which is contaminated by heavy metals, differed significantly from those originating from the reference area (Medas Islands). Their mean Mn bone levels were 7.80 vs. 4.58 mg kg$^{-1}$ dw. Adult individuals from the contaminated site were characterized by higher Mn bone concentrations in comparison with juveniles and seniles: 10.81, 7.45 and 7.02 mg kg$^{-1}$ dw, respectively. However, in insectivores from the Medes Islands, the values in analogous age groups were similar and ranged between 4.06 and 4.89 mg kg$^{-1}$ dw (Sanchez-Chardi and Lopez-Fuster 2009). In the ribs of immature and adult northern opossums (from San Diego, USA), which are scavenger omnivores, values of 11.0 and 3.1 mg Mn kg$^{-1}$ dw were found (Arnhold et al. 2002). In comparison with the white-toothed shrew and the northern opossum, Mn concentration a few times lower (~1.0 mg kg$^{-1}$ dw) was found in the femur head of the red fox, whereas in the samples from cartilage and compact bone, it was significantly higher than in spongy bone: 1.14 vs. 0.92 mg kg$^{-1}$ dw (Budis et al. 2013). In two other canids, dog and ranch raccoon dog (studied by Budis et al. 2015 and Mertin et al. 2006, respectively), mean bone Mn levels were as follows: ~0.60 mg kg$^{-1}$ dw in dog and 2.60 and 1.40 mg kg$^{-1}$ dw in male and female raccoon dogs. The cited values were, respectively, lower and higher than in the wild red fox, but all of them remained in a rather narrow range (0.6–2.6 mg Mn kg$^{-1}$ dw). These differences may be related to the different composition of the diets of canids, their biology and types of bone. A much wider range of mean Mn concentrations

(1.0–10 mg $kg^{-1}$ dw) was found in the bones of domestic ruminants in a number of nutritional experiments in which different doses of Mn were administered to animals of various ages (Hidiroglou 1979, 1980; Bellof et al. 2007). An even greater range of Mn concentrations (0.30–18.0 mg $kg^{-1}$ dw) was observed in the antlers of cervids from Eurasia and North America (Dobrowolska 2002; McDonald et al. 2005; Ceacero et al. 2015; Capelli et al. 2015, 2017). This is probably due to interspecies differences, Mn content in the diet and/or sample collection sites: for example, Mn concentration observed in the outer rim of the antlers was 200 times higher than in the adjacent bone tissue (Turyanskaya et al. 2016). It can be surmised from the above data that Mn concentrations in mammalian skeletons vary slightly, from a few decimals to 10 mg $kg^{-1}$ dw. The highest Mn concentration in the bones occurs in small insectivorous mammals with very high metabolisms, followed by herbivores, and is the lowest in carnivores. In all these animal groups, the concentration of Mn in bones was significantly influenced by the contents of their diet. However, it should be kept in mind that Mn concentrations may vary within the same organism, in different bones, which have various structures and functions. Therefore, in ecotoxicological comparative studies, not only should the same bones be chosen but also the same parts, such as the head of a femur, in the case of large- and medium-sized mammals, or the whole femur of micromammals. In addition, ecological retrospective studies of the bones and antlers of museum specimens can provide valuable insights into the dynamics of Mn levels in the environment over the past decades and centuries.

### 5.2.2 Birds

Of nine different soft tissues and bones in which Dauwe et al. (2005) assayed Mn concentration in the great tit *Parus major*, the highest mean values were found in the intestine, bone and kidney (10.7, 5.08, 4.85 mg $kg^{-1}$ dw) followed by the ovary, stomach and liver (4.14, 3.86 and 2.71 mg $kg^{-1}$ dw). Based on the research by Nam et al. (2005a), it can be assumed that most of the Mn accumulated in avian organisms is found in the feathers, bones, muscles and liver (28.8, 25.1, 20.5 and 16.7%, respectively). Manganese contained in the skin and kidneys amounts to 2.8% and 1.4% and in all other body parts to 4.7%. Manganese penetrates into feathers only during their growth and later on is permanently bound in them and does not participate in biological changes. It has been shown that concentration of Mn bound in plumage is affected by its colour. Feathers containing large quantities of melanin, including black, brown and greenish ones, have a higher concentration of Mn than white feathers, which are devoid of this dye (Niecke et al. 1999; Burger and Gochfeld 2009). For that reason, feathers are rarely used in ecotoxicological studies pertaining to Mn, though about 30% of Mn in the organism is found in plumage. In that respect, the liver is considered to be a better biomarker; however, the kidneys, muscles and bones are also important for various reasons. Although the brain is an important organ and in endothermic vertebrates excess Mn there leads to severe dysfunction of the whole body, Mn concentrations in avian nerve tissue are rarely examined.

#### 5.2.2.1 Manganese in Avian Liver, Kidney and Muscle

In wild birds, mean hepatic Mn concentrations range from 2.5 to 28.1 mg kg$^{-1}$ dw. Taking into account the adequate values of hepatic Mn in poultry suggested by Puls (1994) (2.0–4.0 mg kg$^{-1}$ ww or 7–14 mg kg$^{-1}$ dw), it can be noted that 56.2% out of 64 hepatic Mn concentrations in wild birds (shown in Table 7.2) remain between 7 and 14 mg kg$^{-1}$ dw. Only in 21.9% of cases did Mn concentrations fall below 7 or above 14 mg Mn kg$^{-1}$ dw. The lowest Mn concentrations (<5 mg kg$^{-1}$ dw) were observed only in passerines from Europe and Asia (Llacuna et al. 1995; Deng et al. 2007; Dauwe et al. 2005; Horai et al. 2007; Komosa et al. 2012), which may indicate naturally lower hepatic Mn levels in this bird group. A tendency to accumulate higher hepatic Mn levels (>15 mg kg$^{-1}$ dw) is observed mainly in herbivorous birds (mourning dove *Zenaida macroura*, Bewick's swan *Cygnus columbianus* and white-fronted goose *Anser albifrons*) and in some omnivorous ducks whose diet contains large quantities of plant material (lesser scaup *Aythya affinis*, mallard *Anas platyrhynchos* and spot-billed duck *A. poecilorhyncha*), particularly in areas treated with Mn for agricultural and/or industrial purposes (Table 7.2). The mean Mn concentration in the kidneys of wild birds ranges between 1.4 and 46.8 mg kg$^{-1}$ dw, which is wider than the respective hepatic range. This metal is more rarely assayed in avian kidneys than in livers. According to Puls (1994), adequate nephric Mn concentration ranges between 1.5 and 2.5 mg kg$^{-1}$ ww (or 6.0–10 mg kg$^{-1}$ dw). Of the 37 cases summarized in Table 7.2, almost 43.3% lie within the range of adequate values, while 29.7% and 27.0% remain, respectively, above or below the limit values set for poultry. Much higher Mn concentration (>20 mg Mn kg$^{-1}$ dw) was found in the kidneys of the feral pigeon *Columba livia*, the white-fronted goose and the mallard whose diets are mainly composed of plants, which are generally much richer in Mn than animal tissues.

Nephric Mn levels were a little above or below 5 mg kg$^{-1}$ dw in some diurnal and nocturnal predators (black kite *Milvus migrans*, common buzzard *Buteo buteo*, white-tailed eagle *Haliaeetus albicilla*, bald eagle *H. leucocephalus*, northern goshawk *Accipiter gentilis*, brown hawk-owl *Ninox scutulata*), in piscivores (great cormorant *Phalacrocorax carbo* and great white egret *Egretta alba*) and sporadically in passerines (Table 7.2). On the whole, hepatic Mn concentrations in birds are higher than their nephric values (22 out of 37 cases, ~60%, presented in Table 7.2), but sometimes the opposite is true, especially among passerines.

Avian skeletal muscles and bones are the major internal body pools of Mn. Unlike the liver and kidneys, mean concentrations of Mn in the muscles of wild birds change within a very narrow range, from 0.5 to 3.5 mg kg$^{-1}$ dw (Table 7.2). Only exceptionally does Mn muscle concentration reach a higher level (~9.5 mg kg$^{-1}$ dw), such as in the spot-billed duck *Anas poecilorhyncha*, which prefers plant food (Kim and Oh 2013). With regard to potential consumers Mn content, along with that of other metals, is most often studied in the muscles of game birds, such as pigeons, galliforms, geese and ducks. Their meat generally contain low concentration of Mn, most commonly ranging between 0.7 and 2.0 mg kg$^{-1}$ dw (Table 7.2).

### 5.2.2.2 Manganese in Avian Brain and Bones

In the brains of wild birds, average Mn concentrations remain in a very narrow range, between 0.22 and 0.44 mg kg$^{-1}$ dw or 1.10–2.20 mg kg$^{-1}$ ww. The highest average concentrations of this metal (>1.80 mg kg$^{-1}$ dw) were found in herbivorous and omnivorous anseriforms (whooper swan *Cygnus cygnus*, white-fronted goose, bean goose *Anser fabalis* and mallard) as well as in some piscivorous birds such as the bald eagle and grey heron *Ardea cinerea* (Honda et al. 1990; Kalisinska 2000; Kalisinska et al. 2004; Horai et al. 2007; Nam et al. 2012). In other piscivorous species, the average brain Mn concentration is ~1.7 mg kg$^{-1}$ dw, including the white-tailed eagle, great white egret *Egretta alba* and intermediate egret *E. intermedia*. In birds foraging on land, such as the feral pigeon *Columba livia*, blackbird *Turdus merula* and black kite *Milvus migrans*, Mn concentration in the brain usually does not exceed 1.60 mg kg$^{-1}$ dw, while in some passerines (carrion crow *Corvus corone*, jungle crow *C. macrorhynchos* and great tit) markedly lower values, ranging from 1.1 to 1.2 mg kg$^{-1}$ dw, were observed (Llacuna et al. 1995; Dauwe et al. 2005; Horai et al. 2007). There are no rules for this however. For example, large amounts of Mn penetrated from food and air into the organisms of the great tit and greenfinch *Chloris chloris* from the polluted city of Beijing, China, and Mn concentrations in their brains were 1.91 and 1.68 mg kg$^{-1}$ dw, respectively (Deng et al. 2007). Kalisinska (2000), investigating migrating geese, who, during their flights, stop in large numbers at the floodplains of Warta and Odra River (W Poland), has also shown that Mn concentration in their brains is affected by the environment. Immature and adult bean geese flying from their breeding grounds (located in NE Europe and even beyond the Ural) have lower Mn concentrations in the brain (~0.30 mg kg$^{-1}$ ww or 1.50 mg kg$^{-1}$ dw) when they stop in the floodplains than geese returning from wintering grounds located in W Europe (~0.40 mg kg$^{-1}$ ww or 2.00 mg kg$^{-1}$ dw). Their western European feeding grounds are fields and permanent grasslands, which are fed with mineral fertilizers containing Mn. This element penetrates from their food to different parts of the body of the geese, including the brain. Thus, in the cases of the great tit, greenfinch and geese, their brains proved to be good bioindicator of the Mn environmental burden (Kalisinska 2000; Deng et al. 2007).

Compared to Mn levels in the avian brain, the concentrations of this metal in the bones of wild birds vary considerably (from ~1 to >20 mg kg$^{-1}$ dw). The highest Mn concentrations in the bones are characteristic of herbivorous geese and swans, as well as omnivorous mallard, in which it exceeds 13 mg kg$^{-1}$ dw (Honda et al. 1990; Kalisinska et al. 2004; Kim and Oh 2013). In passerines (black bird, great tit and rock bunting *Emberiza cia*) and piscivorous birds (great white egret and great cormorant), the level of the bone Mn ranges from 4.5 to 7.0 mg kg$^{-1}$ dw but sometimes is much lower (1.15 mg $kg^{-1}$ dw), as in the great cormorant from Serbia (Llacuna et al. 1995; Dauwe et al. 2005; Scoric et al. 2012). It appears that the lowest concentrations of this metal (2–3 mg Mn $kg^{-1}$ dw) occur in nocturnal and diurnal predatory birds such as the common buzzard and barn owl *Tyto alba guttata* (Jager et al. 1996; Esselink et al. 1995). Furthermore, it should be noted that Mn concentrations in different bird bones (e.g. the femur and sternum) from individuals of the same species may differ in values, as was noted by Deng et al. (2007).

## 6 Conclusions

Manganese content in the diet of wildlife significantly influences its concentration in their tissues. The available data indicate that the highest concentrations of this metal are recorded in herbivorous mammals and birds and the lowest in predatory species of both groups. Manganese reaches the highest concentrations in various parts of the digestive tract (including the intestines), but is most commonly assayed in the liver and kidney. Many authors point out that bone and nervous tissues are good bioindicative material for ecotoxicological studies of Mn. However, in comparative studies, attention should be paid to whether samples come from analogous places, as Mn is not evenly distributed in the brain and bones of skeleton.

## References

Adriano DC (2001) Trace elements in the terrestrial environments: biogeochemistry, bioavailability, and risks of metals. Springer, New York

Arnhold W, Anke M, Goebel S (2002) The copper, zinc and manganese status in opossum and gray fox. Z Jagdwiss 48:77–86

Aschner JL, Aschner M (2005) Nutritional aspects of manganese homeostasis. Mol Asp Med 26:353–362

Aschner M, Lukey B, Tremblay A (2006) The manganese health research program (MHRP): status report and future research needs and directions. Neurotoxicology 27:733–736

Aschner M, Erikson KM, Hernandez EH, Tjalkens R (2009) Manganese and its role in Parkinson's disease: from transport to neuropathology. NeuroMolecular Med 11:252–266

ATSDR (2000) Toxicological profile for manganese. U.S. Department of Health and Human Services, Public Health Service, Agency for Toxic Substances and Disease Registry, Atlanta, GA

ATSDR (2012) Toxicological profile for manganese. U.S. Department of Health and Human Services, Public Health Service, Agency for Toxic Substances and Disease Registry, Atlanta, GA

Babińska I, Szarek J, Binkowski ŁJ, Skibniewska K, Wojtacka J, Markiewicz E et al (2008) Grey herons (*Ardea cinerea* L.) as a tool for monitoring the environment for metal concentrations in the vicinity of a pesticide tomb in the Iławskie lake district. Fresenius Environ Bull 17:98–102

Baden SP, Eriksson SP (2006) Role, routes and effects of manganese in crustaceans. Oceanogr Mar Biol 44:61–83

Bakirdere S, Kizilkan N, Yaman M (2011) Determination of zinc, copper, iron, and manganese in different regions of lamb brain. Biol Trace Elem Res 142:492–499
Bellof G, Most E, Pallauf J (2007) Concentration of copper, iron, manganese and zinc in muscle, fat and bone tissue of lambs of the breed German Merino Landsheep in the course of the growing period and different feeding intensities. J Anim Physiol Anim Nutr (Berl) 91:100–108
Borch-Iohnsen B, Nilssen KJ, Norheim G (1996) Influence of season and diet on liver and kidney content of essential elements and heavy metals in Svalbard reindeer. Biol Trace Elem Res 51:235–247
Bruckwicki P (2006) An investigation of contaminant levels in white-tailed deer (*Odocoileus virginianus*) collected from Caddo Lake National Wildlife Refuge, Harrison County, Texas 2005. https://www.fws.gov/southwest/es/ArlingtonTexas/pdf/CLNWRDeerStudy2006.pdf
Brzeziński M, Zalewski A, Niemczynowicz A, Jarzyna I, Suska-Malawska M (2014) The use of chemical markers for the identification of farm escapees in feral mink populations. Ecotoxicology 23:767–778
Buddhachat K, Klinhom S, Siengdee P, Brown JL, Nomsiri R, Kaewmong P et al (2016) Elemental analysis of bone, teeth, horn and antler in different animal species using non-invasive handheld X-ray fluorescence. PLoS One 11(5):e0155458
Budis H, Kalisinska E, Lanocha N, Kosik-Bogacka D (2013) The concentration of manganese, iron and strontium in bone of red fox *Vulpes vulpes* (L. 1758). Biol Trace Elem Res 155:361–369
Budis H, Kalisinska E, Lanocha N, Kosik-Bogacka DI (2015) Concentrations of manganese, iron and strontium in bones of the domestic dog (*Canis lupus familiaris*). Turk J Vet Anim Sci 39:279–286
Burger J, Gochfeld M (1985) Comparisons of nine heavy metals in salt gland and liver of greater scaup (*Aythya marila*), black duck (*Anas rubripes*) and mallard (*A. platyrhynchos*). Comp Biochem Physiol C 81:287–292
Burger J, Gochfeld M (2009) Comparison of arsenic, cadmium, chromium, lead, manganese, mercury and selenium in feathers in bald eagle (*Haliaeetus leucocephalus*), and comparison with common eider (*Somateria mollissima*), glaucous-winged gull (*Larus glaucescens*), pigeon guillemot (*Cepphus columba*), and tufted puffin (*Fratercula cirrhata*) from the Aleutian Chain of Alaska. Environ Monit Assess 152:357–367
Burger J, Kennamer RA, Brisbin IL Jr, Gochfeld M (1997) Metal levels in mourning doves from South Carolina: potential hazards to doves and hunters. Environ Res 75:173–186
Burger J, Gaines KF, Lord CG, Brisbin JR, Shukla S, Gochfeld M (2002) Metal levels in raccoon tissues: differences on and off the Department of Energy's Savannah River Site in South Carolina. Environ Monit Assess 74:67–84
Cappelli J, Garcia A, Ceacero F, Gomez S, Luna S, Gallego L et al (2015) Manganese supplementation in deer under balanced diet increases impact energy and contents in minerals of antler bone tissue. PLoS One 10(7):e0132738
Cappelli J, Atzori AS, Ceacero F, Landete-Castillejos T, Cannas A, Gallego L et al (2017) Morphology, chemical composition, mechanical properties and structure in antler of Sardinian red deer (*Cervus elaphus corsicanus*). Hystrix 28: https://doi.org/10.4404/hystrix-28.1-12270
Ceacero F, Pluhacek J, Landete-Castillejos T, Garcia AJ, Gallego L (2015) Inter-specific differences in the structure and mechanics but not the chemical composition of antlers in three deer species. Ann Zool Fenn 52:368–376
Chudzińska-Popek M, Majdecka T (2010) Problems of mineral metabolism in roe deer (*Capreolus capreolus* L) – preliminary study. Proc ECOpole 4:325–328
Crăciunescu A, Stanciu S, Mihaela Moatăr M (2014) Research on meat quality of some wild animal species for human consumption. J Agroaliment Process Technol 20:80–86
Custer CM, Custer TW, Anteau MJ, Afton AD, Wooten DE (2003) Trace elements in lesser scaup (*Aythya affinis*) from the Mississippi flyway. Ecotoxicology 12:47–54
Custer TW, Cox E, Gray B (2004) Trace elements in moose (*Alces alces*) found dead in Northwestern Minnesota, USA. Sci Total Environ 330:81–87

Dauwe T, Janssens E, Bervoets L, Blust R, Eens M (2005) Heavy-metal concentrations in female laying great tits (*Parus major*) and their clutches. Arch Environ Contam Toxicol 49:249–256

Davis JM (1999) Inhalation health risk of manganese: an EPA perspective. Neurotoxicology 20:511–518

Deng H, Zhang Z, Chang C, Wang Y (2007) Trace metal concentration in great tit (*Parus major*) and greenfinch (*Carduelis sinica*) at the Western Mountains of Beijing, China. Environ Pollut 148:620–626

Dlugaszek M, Kopczynski K (2013) Elemental composition of muscle tissue of wild animals from central region of Poland. Int J Environ Res 7:973–978

Dobrowolska A (2002) Chemical composition of the red deer (*Cervus elaphus*) antlers, with a particular reference to the toxic metal contents. Z Jagdwiss 48:148–155

Dorman DC, Struve MF, Marshal MW, Parkinson CU, Arden RJ, Wong BA (2006) Tissue manganese concentrations in young male rhesus monkeys following subchronic manganese sulfate inhalation. Toxicol Sci 92:201–210

Doyle D, Kapron CM (2002) Inhibition of cell differentiation by manganese chloride in micromass cultures of mouse embryonic limb bud cells. Toxicol In Vitro 16:101–106

Doyle JJ, Spaulding JE (1978) Toxic and essential trace elements in meat: a review. J Anim Sci 47:398–419

Ellis GH, Smith S, Ande E, Gates M (1947) Further studies of manganese deficiency in the rabbit. J Nutr 34:21–31

Erikson KM, Aschner M (2003) Manganese neurotoxicity and glutamate-GABA interaction. Neurochem Int 43:475–480

Erikson KM, Syversen T, Aschner T, Aschner JL, Aschner M (2005) Interactions between excessive manganese exposure and dietary iron-deficiency in neurodegeneration. Environ Toxicol Pharmacol 19:415–421

Ertl K, Kitzer R, Goessler W (2016) Elemental composition of game meat from Austria. Food Addit Contam Part B 9:120–126

Esselink H, van der Geld FM, Jager LP, Posthuma-Trupie GA, Zoun PEF, Baars AJ (1995) Biomonitoring heavy metals using the barn owl (*Tyto alba guttata*): sources of variation especially relating to body condition. Arch Environ Contam Toxicol 28:471–486

Falandysz J (1994) Some toxic and trace metals in big game hunted in the northern part of Poland in 1987-1991. Sci Total Environ 141:59–73

Falandysz J, Ichihashi H, Szymczyk K, Yamasaki S, Mizera T (2001) Metallic elements and metal poisoning among white-tailed sea eagles from the Baltic South Coast. Mar Pollut Bull 42:1190–1193

Farkas A, Bidló A, Bolodár-Varga B, Jánoska F (2017) Accumulation of metals in liver tissues of sympatric golden jackal (*Canis aureus*) and red fox (*Vulpes vulpes*) in the southern part of Romania. Bull Environ Contam Toxicol 98:513–520

Fitsanakis VA, Zhang N, Anderson JG, Erikson KM, Avison MJ, Gore JC et al (2008) Measuring brain manganese and iron accumulation in rats following 14 weeks of low-dose manganese treatment using atomic absorption spectroscopy and magnetic resonance imaging. Toxicol Sci 103:116–124

Foster ML, Bartnikas TB, Johnson LC, Herrera C, Pettiglio MA, Keene AM et al (2015) Pharmacokinetic evaluation of the equivalency of gavage, dietary, and drinking water exposure to manganese in F344 rats. Toxicol Sci 145:244–251

Frank A, Danielsson R, Jones B (2000) The 'mysterious' disease in Swedish moose. Concentrations of trace elements in liver and kidneys and clinical chemistry. Comparison with experimental molybdenosis and copper deficiency in the goat. Sci Total Environ 249:107–122

Frank A, Mc Partlin J, Danielsson R (2004) Nova Scotia moose mystery – a moose sickness related to cobalt and vitamin B12 deficiency. Sci Total Environ 318:89–100

Freeland-Graves JH, Llanes C (1994) Models to study manganese deficiency. In: Klimis-Tavantzis DJ (ed) Manganese in health and disease. CRC Press, Boca Raton, pp 59–86

French AS, Shaw D, Gibb SW, Taggart MA (2017) Geochemical landscapes as drivers of trace and toxic element profiles in wild red deer (*Cervus elaphus*). Sci Total Environ 601–602:1606–1618
Gasparik J, Massányi P, Slamecka J, Fabis M, Jurcik R (2004) Concentration of selected metals in liver, kidney, and muscle of the red deer (*Cervus elaphus*). J Environ Sci Health A Tox Hazard Subst Environ Eng 39:2105–2111
Gehrke M (1997) Copper and manganese in the pathogenesis of diseases of the osseous system in animals. Med Wet 53:644–646 [in Polish]
Grove RA, Henny CJ (2008) Environmental contaminants in male river otters collected from Oregon and Washington, USA, 1994-1999. Environ Monit Assess 145:49–73
Halbrook RS, Woolf A, Hubert GF Jr, Ross S, Braselton WE (1996) Contaminant concentrations in Illinois mink and otter. Ecotoxicology 5:103–114
Harding LE (2004) Environmental contaminants in wild martens (*Martes americana*) and wolverines (*Gulo gulo*). Bull Environ Contam Toxicol 73:98–105
Hesketh S, Sassoon J, Knight R, Brown DR (2007) Elevated manganese levels in blood and CNS in human prion disease. Mol Cell Neurosci 37:590–598
Hidiroglou M (1979) Manganese in ruminant nutrition. Can J Anim Sci 59:217–236
Hidiroglou M (1980) Zinc, copper and manganese deficiencies and the ruminant skeleton: a review. Can J Anim Sci 60:579–590
Hoekstra PF, Braune BM, Elkin B, Armstrong FA, Muir DC (2003) Concentrations of selected essential and non-essential elements in arctic fox (*Alopex lagopus*) and wolverines (*Gulo gulo*) from the Canadian Arctic. Sci Total Environ 309:81–92
Honda K, Min BY, Tatsukawa R (1985) Heavy metal distribution in organs and tissues of the eastern great white egret *Egretta alba modesta*. Bull Environ Contam Toxicol 35:781–789
Honda K, Ichihashi H, Tatsukawa R (1987) Tissue distribution of heavy metals and their variations with age, sex, and habitat in Japanese serows (*Capricornis crispus*). Arch Environ Contam Toxicol 16:551–561
Honda K, Lee DP, Tatsukawa R (1990) Lead poisoning in swans in Japan. Environ Pollut 65:209–218
Horai S, Minagawa M, Ozaki H, Watanabe I, Takeda Y, Yamada K et al (2006) Accumulation of Hg and other heavy metals in the Javan mongoose (*Herpestes javanicus*) captured on Amami oshima Island in Japan. Chemosphere 65:657–665
Horai S, Watanabe I, Takada H, Iwamizu Y, Hayashi T, Tanabe S, Kuno K (2007) Trace element accumulations in 13 avian species collected from the Kanto area, Japan. Sci Total Environ 373:512–525
Howe PD, Malcolm HM, Dobson S (2004) Manganese and its compounds: environmental aspects. WHO, CICAD 63, Geneva
Hui A, Takekawa JY, Baranyuk VV, Litvin KV (1998) Trace element concentrations in two subpopulations of Lesser Snow Geese from Wrangel Island, Russia. Arch Environ Contam Toxicol 34(2):197–203
Hurley LS, Keen CL (1987) Manganese. In: Mertz W (ed) Trace elements in human and animal nutrition, 5th edn. Academic Press, San Diego, pp 185–223
Jager LP, Rijnierse FVJ, Esselink H, Baars AJ (1996) Biomonitoring with the buzzard *Buteo buteo* in the Netherlands: heavy metals and sources of variation. J Ornithol 137:295–318
Jankovská I, Miholová D, Bejcek V, Vadlejch J, Sulc M, Száková J et al (2010) Influence of parasitism on trace element contents in tissues of red fox (*Vulpes vulpes*) and its parasites *Mesocestoides* spp. (Cestoda) and *Toxascaris leonina* (Nematoda). Arch Environ Contam Toxicol 58:469–477
Jarzynska G, Falandysz J (2011) Selenium and 17 other largely essential and toxic metals in muscle and organ meats of red deer (*Cervus elaphus*) – consequences to human health. Environ Int 37:882–888
Jung JM, Kim KH, Kwon EE, Kim HW (2015) Analysis of the lipid profiles in a section of bovine brain via non-catalytic rapid methylation. Analyst 140:6210

Kalisinska E (2000) Lead and other heavy metals in the brain of geese hunted in the vicinity of Slonsk, Poland. Biol Bull Poznań 37:273–286

Kalisinska E, Salicki W, Myslek P, Kavetska KM, Jackowski A (2004) Using the mallard to biomonitor heavy metal contamination of wetlands in north-western Poland. Sci Total Environ 320:145–161

Kalisinska E, Salicki W, Jackowski A (2006) Six trace metals in white-tailed eagle from north-western Poland. Pol J Environ Stud 15:727–737

Kalisinska E, Lanocha-Arendarczyk N, Kosik-Bogacka D, Budis H, Podlasinska J, Popiolek M et al (2016) Brains of native and alien mesocarnivores in biomonitoring of toxic metals in Europe. PLoS One 11:e0159935

Kang S, Kang JH, Kim S, Lee SH, Lee S, Yu HJ et al (2015) Trace element analysis of three tissues from Eurasian otters (*Lutra lutra*) in South Korea. Ecotoxicology 24:1064–10672

Karmakar A, Zhang Q, Zhang Y (2014) Neurotoxicity of nanoscale materials. J Food Drug Anal 22:147–160

Kim J, Oh JM (2013) Tissue distribution of metals in white-fronted geese and spot-billed ducks from Korea. Bull Environ Contam Toxicol 91:18–22

Knauer B, Majka P, Watkins KJ, Taylor AWR, Malamanova D, Paul B et al (2017) Whole-brain metallomic analysis of the common marmoset (*Callithrix jacchus*). Metallomics 9:411

Komosa A, Kitowski I, Komosa Z (2012) Essential trace (Zn, Cu, Mn) and toxic (Cd, Pb, Cr) elements in the liver of birds from eastern Poland. Acta Vet (Beograd) 62:579–589

Komura J, Sakamoto M (1992) Effects of manganese forms on biogenic amines in the brain and behavioral alternation in mouse: long-term oral administration of several manganese compounds. Environ Res 57:34–44

Levander OA (1988) Selenium, chromium and manganese: (C) manganese. In: Shils ME, Young VR (eds) Modern nutrition in health and disease, 7th edn. Lea & Febiger, Philadelphia, p 274

Levy BS, Nassetta WJ (2003) Neurologic effects of manganese in humans: a review. Int J Occup Environ Health 9:153–163

Licata P, Naccari F, Turco VL, Rando R, Bella GD, Dugo G (2010) Levels of Cd (II), Mn (II), Pb (II), Cu (II), and Zn (II) in common buzzard (*Buteo buteo*) from Sicily (Italy) by derivative stripping potentiometry. Int J Ecol 2010:541948

Lide DR (ed) (2005) CRC handbook of chemistry and physics. CRC Press, Boca Raton, Internet Version 2005, http://www.hbcpnetbase.com

Litchfield TM, Ishikawa Y, Wu LNY, Wuthier RE, Sauer GR (1998) Effects of metal ions on calcifying growth plate cartilage chondrocytes. Calcif Tissue Int 62:341–349

Llacuna S, Gorriz A, Sanpera C, Nadal J (1995) Metal accumulation in three species of passerine birds (*Emberiza cia*, *Parus major*, and *Turdus merula*) subjected to air pollution from a coal-fired power plant. Arch Environ Contam Toxicol 28:298–303

Lodenius M, Skaren U, Hellstedt P, Tulisalo E (2014) Mercury in various tissues of three mustelid and other trace metals in liver o European otter from Eastern Finland. Environ Monit Assess 186:325–333

Lohry R (2007) Micronutrients: functions, sources and application methods. Indiana CCA Conference Proceedings, 15 pp

Loranger S, Zayed J (1997) Environmental contamination and human exposure to airborne total and respirable manganese in Montreal. J Manage Air Waste Assoc 47:938–989

MacDonald C, Elkin B, Gunn A (2005) Analysis of the elemental composition of tissues and faecal ash in a moose (*Alces alces*) exposed to tailings at the abandoned colomac gold mine, NWT. Department of Resources, Wildlife and Economic Development Government of Northwest Territories, Yellowknife, NT

Martiniakova M, Omelka R, Stawarz R, Formicki G (2012) Accumulation of lead, cadmium, nickel, iron, copper, and zinc in bones of small mammals from polluted areas in Slovakia. Pol J Environ Stud 21:153–158

Mason CF, Stephenson A (2001) Metals in tissues of European otters (*Lutra lutra*) from Denmark, Great Britain and Ireland. Chemosphere 44:351–353

Mateo R, Guitart R (2003) Heavy metals in livers of waterbirds from Spain. Arch Environ Contam Toxicol 44:398–404
Mergler D, Baldwin M (1997) Early manifestations of manganese neurotoxicity in humans: an update. Environ Res 78:92–100
Mertin D, Szeleszczuk O, Suvegova K, Niedbala P, Hanusova E (2006) Content of microelements in the selected organ s of raccoon dog (*Nyctereutes procyonoides*). Chem Inz Ekol 13:85–90
Michalska K, Zmudzki J (1992) Metals content in the tissues of boars, roe deers and red deers in the Wielkopolska region. Med Weter 48:160–162 [in Polish]
Michalska Z, Soltysiak Z, Milian A (1991) The level of heavy metals in the brain of dog. Med Weter 47:410–411 [in Polish]
Michot TC, Custer TW, Nalut AJ, Mitchell CA (1994) Environmental contaminants in redheads wintering in coastal Louisiana and Texas. Arch Environ Contam Toxicol 26:425–434
Mierzykowski SE, Smith JEM, Todd CS, Kusnierz D, DeSorbo CR (2011) Liver contaminants in bald eagle carcasses from Maine. USFWS. Special Project Report FY09-MEFO-6-EC. Maine Field Office. Orono, ME, p 53
Millaleo R, Reyes-Diaz M, Ivanov AG, Mora ML, Alberdi M (2010) Manganese as essential and toxic element for plants: transport, accumulation and resistance mechanisms. J Soil Sci Plant Nutr 10:476–494
Mills CF, Dalgarno AC, Wenham G (1976) Biochemical and pathological changes in tissues of Friesian cattle during the experimental induction of copper deficiency. Br J Nutr 35:309–331
Mousavi SR, Shahsavari M, Rezaei M (2011) A general overview on manganese (Mn) importance for crops production. Aust J Basic Appl Sci 5:1799–1803
Myslek P, Kalisinska E (2006) Contents of selected heavy metals in the liver, kidneys and abdominal muscle of the brown hare (*Lepus europaeus* Pallas, 1778) in Central Pomerania, Poland. Pol J Vet Sci 9:31–41
Nam DH, Anan Y, Ikemoto T, Okabe Y, Kim EY, Subramanian A, Saeki K, Tanabe S (2005a) Specific accumulation of 20 trace elements in great cormorants (*Phalacrocorax carbo*) from Japan. Environ Pollut 134:503–514
Nam DH, Anan Y, Ikemoto T, Tanabe S (2005b) Multielemental accumulation and its intracellular distribution in tissues of some aquatic birds. Mar Pollut Bull 50:1347–1362
Nam DH, Rutkiewicz J, Basu N (2012) Multiple metals exposure and neurotoxic risk in bald eagles (*Haliaeetus leucocephalus*) from two Great Lakes states. Environ Toxicol Chem 31:623–631
Navarro G, Jerez S, Farinós P, Robledano F, Motas M (2010) Assessment of inorganic element exposure (Cr, Mn, Ni, Cu, Zn, As, Se, Cd and Pb) in great cormorants (*Phalacrocorax carbo sinensis*) from the Mar Menor coastal lagoon. An Vet (Murcia) 26:97–110
Newland CM (1999) Animal models of manganese's neurotoxicity. Neurotoxicology 20:415–432
Niecke M, Heid M, Kruger A (1999) Correlations between melanin pigmentation and element concentration in feathers of white-tailed eagles (*Haliaeetus albicilla*). J Ornithol 140:355–362
Niemiec M, Wisniowska-Kielian B (2015) Manganese accumulation in selected links of food chain of aquatic ecosystems. J Elem 20:945–956
Nriagu JO, Pacyna J (1988) Quantitative assessment of worldwide contamination of air, water and soils by trace elements. Nature 333:134–139
O'Neal SL, Zheng W (2015) Manganese toxicity upon overexposure: a decade in review. Curr Environ Health Rep 2:315–328
O'Neal SL, Hong L, Fu X, Jiang W, Jones A, Nie LH et al (2014) Manganese accumulation in bone following chronic exposure in rats: steady-state concentration and half-life in bone. Toxicol Lett 229:93–100
Odsjo T, Raikkonen J, Bignert A (2007) Time trends of metals in liver and muscle of reindeer (*Rangifer tarandus*) from northern and central Lapland, Sweden, 1983-2003. Swedish Museum of Natural History, 33 pp
Pacyna JM, Pacyna EG (2001) An assessment of global and regional emissions of trace metals to the atmosphere from anthropogenic sources worldwide. Environ Rev 9:269–298
Pillatzki AE, Neiger RD, Chipps SR, Higgins KF, Thiex N, Afton AD (2011) Hepatic element concentrations of lesser scaup (*Aythya affinis*) during spring migration in the upper Midwest. Arch Environ Contam Toxicol 61:144–150

Pinsino A, Matranga V, Roccheri MC (2012) Manganese: a new emerging contaminant in the environment. In: Srivastava J (ed) Environmental contamination. InTech Open Access, London, pp 17–36
Pollock B, Roger E (2007) Trace elements status of moose and white-tailed deer in Nova Scotia. Alces 43:61–77
Poole KG, Elkin BT, Bethke RW (1998) Organochlorine and heavy metal contaminants in wild mink in Western Northwest Territories, Canada. Arch Environ Contam Toxicol 34:406–413
Prashanth L, Kattapagari KK, Chitturi RT, Baddam VRR, Prasad LK (2015) A review on role of essential trace elements in health and disease. J Dr NTR Univ Health Sci 4:75–86
Prestrud P, Norheim G, Sivertsen T, Daae HL (1994) Levels of toxic essential elements in arctic fox in Svalbard. Polar Biol 14:155–159
Puls R (1994) Mineral levels in animal health: diagnostic data, 2nd edn. Sherpa International, Clearbrook, BC
Rajkowska M, Protasowicki M (2013) Distribution of metals (Fe, Mn, Zn, Cu) in fish tissues in two lakes of different trophy in Northwestern Poland. Environ Monit Assess 185:3493–3502
Ramos P, Santos A, Pinto NR, Mendes R, Magalhaes T, Almeida A (2014) Anatomical region differences and age-related changes in copper, zinc, and manganese levels in the human brain. Biol Trace Elem Res 161:190–201
Rautio A, Kunnasranta M, Valtonen A, Ikonen M, Hyvärinen H, Holopainen IJ et al (2010) Sex, age and tissue specific accumulation of eight metals, arsenic, and selenium in the European hedgehog (*Erinaceus europaeus*). Arch Environ Contam Toxicol 59:642–651
Reimer PS (1999) Environmental effects of manganese and proposed freshwater guidelines to protect aquatic life in British Columbia. MSc thesis, Vancouver, BC, University of British Columbia
Reis LS, Pardo PE, Camargos AS, Oba E (2010) Mineral element and heavy metal poisoning in animals. J Med Med Sci 1:560–579
Reynolds N, Blumhson A, Baxter JP, Houston G, Pennington CR (1998) Manganese requirements and toxicity in patients on home parenteral nutrition. Clin Nutr 17:227–230
Roselli C, Desideri D, Meli MA, Fagiolino I, Feduzi L (2016) Essential and toxic elements in meat of wild birds. J Toxicol Environ Health A 79:1008–1014
Roug A, Swift PK, Gerstenberg G, Woods LW, Kreuder-Johnson C, Torres SG et al (2015) Comparison of trace mineral concentrations in tail hair, body hair, blood, and liver of mule deer (*Odocoileus hemionus*) in California. J Vet Diagn Investig 27:295–305
Sanchez D, Domingo JL, Llobet JM, Keen CL (1993) Maternal and developmental toxicity of manganese in the mouse. Toxicol Lett 69:45–52
Sanchez-Chardi A, Lopez-Fuster MJ (2009) Metal and metalloid accumulation in shrews (Soricomorpha, Mammalia) from two protected Mediterranean coastal sites. Environ Pollut 157:1243–1248
Sánchez-Chardi A, Ribeiro CA, Nadal J (2009) Metals in liver and kidneys and the effects of chronic exposure to pyrite mine pollution in the shrew *Crocidura russula* inhabiting the protected wetland of Doñana. Chemosphere 76:387–394
Scheuhammer AM, Cherian MG (1983) The influence of manganese on the distribution of essential trace elements. II. The tissue distribution of manganese, magnesium, zinc, iron and copper in rats after chronic manganese exposure. J Toxicol Environ Health 12:361
Schneider JS, Decamp E, Koser AJ, Fritz S, Gonczi H, Syversen T, Guilarte TR (2006) Effects of chronic manganese exposure on cognitive and motor functioning in non-human primates. Brain Res 1118:222–231
Schroeder HA, Balassa JJ, Tipton IH (1966) Essential trace metals in man: manganese: a study in homeostasis. J Chronic Dis 19:545–571
Schroeder WH, Dobson M, Kane DM (1987) Toxic trace elements associated with airborne particulate matter: a review. J Air Pollut Control Assoc 37:1267–1285
Schummer ML, Petrie SA, Badzinski SS, Deming M, Chen YW, Belzile N (2011) Elemental contaminants in livers of mute swans on lakes Erie and St. Clair. Arch Environ Contam Toxicol 61:677–687

Shacklette HT, Boerngen JG (1984) Element concentrations in soils and other surficial materials of the conterminous United States: U.S. Geological Survey Professional Paper 1270:105
Skaren U (1992) Analysis of one hundred otters killed by accidents in Central Finland IUCN Otter Spec. Group Bull 7:9–12
Skoric S, Visnjić-Jeftic Z, Jaric I, Djikanovic V, Mickovic B, Nikcevic M et al (2012) Accumulation of 20 elements in great cormorant (*Phalacrocorax carbo*) and its main prey, common carp (*Cyprinus carpio*) and Prussian carp (*Carassius gibelio*). Ecotoxicol Environ Saf 80:244–251
Smrcka V (2005) Trace elements in bone tissue. Charles University in Prague, The Karolinum Press, Prague
Soetan KO, Olaiya CO, Oyewole OE (2010) The importance of mineral elements for humans, domestic animals and plants: a review. Afr J Food Sci 4:200–222
Soltysiak Z, Michalska Z, Milian A (1997) The content of heavy metals in internal organs of cats. Med Weter 53:337–340 [In Polish]
Souza MJ, Ramsay EC, Donnell RL (2013) Metal accumulation and health effects in raccoons (*Procyon lotor*) associated with coal fly ash exposure. Arch Environ Contam Toxicol 64:529–536
Spears JW (2011) Importance of manganese in cattle and poultry. Salt Institute Newsletter 1st Quarterly, pp 1–6
Stejskal S, Aulerich RJ, Slanker MR, Braselton WE, Lehning EJ, Napolitano AC (1989) Element concentrations in livers and kidneys of ranch mink. J Vet Diagn Investig 1:343–348
Stokes PM, Campbell PGC, Schroeder WH, Trick C, France RL, Puckett KJ et al (1988) Manganese in the Canadian environment. National Research Council, Ottawa
Stout JH, Trust KA (2002) Elemental and organochlorine residues in bald eagles from Adak Island, Alaska. J Wildl Dis 38:511–517
Strmiskova G, Strmiska F (1992) Contents of mineral substances in venison. Nahrung 36:307–308
Sweet CW, Vermette SJ, Landsberger S (1993) Sources of toxic trace elements in urban air in Illinois. Environ Sci Technol 27:2502–2510
Swiergosz R, Perzanowski K, Makosz U, Bilek I (1993) The incidence of heavy metals and other toxic elements in big game tissues. Sci Total Environ 1:225–231
Takeda A (2003) Manganese action in brain function. Brain Res Brain Res Rev 41:79–87
Takeda A, Kodama Y, Ishiwatari S, Okada S (1998) Manganese transport in the neural circuit of rat CNS. Brain Res Bull 45:149–152
Torrente M, Albina ML, Colomina MT, Corbella J, Domingo JL (2000) Interactions in developmental toxicology: effects of combined administration of manganese and hydrocortisone. Trace Elem Electrolytes 17:173–179
Turyanskaya A, Rauwolf M, Roschger A, Prost J, Pemmer B, Simon R et al (2016) Manganese distribution in bone tissue by SR-μXRF. Bone Abstr 5:P312
Underwood EJ (1981) The mineral nutrition of livestock, 2nd edn. Commonwealth Agricultural Bureaux, Slough
US EPA (2003) Health effects support document for manganese. United States Environmental Protection Agency, Washington, DC
USGS (2017a) 2014 Minerals yearbook – manganese. United States Geological Survey 2017. United States
USGS (2017b) Mineral commodity summaries 2017. U.S. Geological Survey, Reston, VA
Vezer T, Papp A, Hoyk Z, Varga C, Naray M, Nagymajtenyi L (2005) Behavioral and neurotoxical effect of subchronic manganese exposure in rats. Environ Toxicol Pharmacol 19:797–810
Walker LA, Lawlor AJ, Chadwick EA, Potter E, Pereira MG, Shore RF (2010) Inorganic elements in the livers of Eurasian otters, *Lutra lutra*, from England and Wales in 2007&2008—a Predatory Bird Monitoring Scheme (PBMS) report. Centre for Ecology and Hydrology, Lancaster
Wang F, Zou Y, Shen Y, Zhong Y, Lv Y, Huang D, Chen K, Li Q, Qing L, Xia B, Cheng S, Ma S, Yang X, Miao X (2015) Synergistic impaired effect between smoking and Manganese dust exposure on pulmonary ventilation function in Guangxi Manganese-exposed workers healthy cohort (GXMEWHC). PLoS One 10(2):e0116558
Watts DL (1990) The nutritional relationships of manganese. J Orthomol Med 5:219–222

Wedler FC (1994) Biochemical and nutritional role of manganese: an overview. In: Klimis-Tavantzis DJ (ed) Manganese in health and disease. CRC Press, Boca Raton, FL, pp 1–36
White SN, O'Rourke KI, Gidlewski T, VerCauteren KC, Mousel MR, Phillips GE, Spraker TR (2010) Increased risk of chronic wasting disease in Rocky Mountain elk associated with decreased magnesium and increased manganese in brain tissue. Can J Vet Res 74:50–53
WHO (1981) Environmental Health Criteria, No 17. Manganese. World Health Organization, Geneva
WHO (1999) Manganese and its compounds. Concise International Chemical Assessment Document 12. World Health Organization, Geneva
WHO (2004) Manganese and its compounds: environmental aspects. Concise International Chemical Assessment Document 63. World Health Organisation, Geneva, pp 1–70
WHO (2011) Manganese in drinking-water background document for development of WHO Guidelines for drinking-water quality. World Health Organisation WHO/SDE/WSH/03.04/104/Rev/1
Williams-Johnson M (1999) Manganese and its compounds: foreword. IPCC Concide International Chemical Assessment Documents 12:1–42
Wolfe LL, Conner MM, Bedwell CL, Lukacs PM, Miller MW (2010) Select tissue mineral concentrations and chronic wasting disease status in mule deer from North-central Colorado. J Wildl Dis 46:1029–1034
Wren CD (1984) Distribution of metals in tissues of beaver, raccoon and otter from Ontario, Canada. Sci Total Environ 34:177–184
Wren CD, Fischer KL, Stokes PM (1988) Levels of lead, cadmium and other elements in mink and otter from Ontario, Canada. Environ Pollut 52:193–202
WVDL (2015) www.wvdl.wisc.edu/wp-content/uploads/2013/06/WVDL.Info_Toxicology_Normal_Ranges.pdf. Accessed 28 Apr 2015
Yaroshevsky AA (2006) Abundances of chemical elements in the Earth's crust. Geochem Int 44:48–55
Zayed J, Hong B, Esperance G (1999) Characterization of manganese-containing particles collected from the exhaust emissions of automobiles funning with MMT additive. Environ Sci Technol 33:3341–3346
Zeman T, Buchtova M, Docekal B, Misek I, Navratil J, Mikuska P et al (2015) Organ weight changes in mice after long-term inhalation exposure to manganese oxides nanoparticles. J Phys Conf Ser 617:012018
Zheng W, Fu SX, Dydak U, Cowan DM (2010) Biomarkers of manganese intoxication. Neurotoxicology 32:1–8
Zimmerman TJ, Jenks JA, Leslie DM Jr, Neiger RD (2008) Hepatic minerals of white-tailed and mule deer in the southern black hills, South Dakota. J Wildl Dis 44:341–350

# Chapter 8
# Molybdenum, Mo

**Tadeusz Kośla, Michał Skibniewski, Ewa M. Skibniewska, Iwona Lasocka, and Marta Kołnierzak**

**Abstract** Molybdenum in plants, animals, and humans occurs in small quantities; however, being an essential element, the metal is a component of a range of enzymes. In animals and humans, these are xanthine oxidase, aldehyde oxidase, and sulfite oxidase. Goats that had been for many generations fed a semisynthetic, Mo low diet exhibited impaired growth, lower fertility, poor fetal survival, and a higher incidence of miscarriages. The importance of molybdenum consists in its toxicity, which affects primarily ruminants, cattle in particular. Molybdenum penetrates animal bodies through oral intake or inhalation. From the gastrointestinal tract, the element is transported to the kidneys, liver, and bones. In monogastric animals, molybdenum is absorbed through the wall of the stomach. In ruminants, its absorption takes place in the intestinal wall. Molybdenum poisoning is accompanied by a range of symptoms, many of which result from secondary copper deficiency. Typical signs of acute, uncomplicated molybdenosis include impaired bone formation, which may be a result of altered phosphorus metabolism. These changes lead to deformation of bones and joints, spontaneous pathological fractures of long bone metaphyses, and exostosis. A decrease can be observed in the activity of alkaline phosphatase and the content of proteoglycans in articular cartilage. Reproductive disorders affecting individuals of either sex also appear. Renal excretion rate of Mo in laboratory animals is rather quick, taking about 2 weeks. So far no biomagnification and bioaccumulation of molybdenum in animal bodies have been confirmed.

Monogastric animals are less susceptible to Mo toxicity, and the most tolerant to elevated doses seem to be horses. Sheep and cattle, on the other hand, exhibit the lowest level of resistance to Mo toxic effects. Regardless of the species, acute

T. Kośla (✉) · E. M. Skibniewska · I. Lasocka · M. Kołnierzak
Department of Animal Environment Biology, Warsaw University of Life Sciences – SGGW, Warsaw, Poland
e-mail: tadeusz_kosla@sggw.pl

M. Skibniewski
Department of Morphological Sciences, Warsaw University of Life Sciences – SGGW, Warsaw, Poland



E. Kalisińska (ed.), *Mammals and Birds as Bioindicators of Trace Element Contaminations in Terrestrial Environments*,
https://doi.org/10.1007/978-3-030-00121-6_8

molybdenum poisoning leads to growth hindrance, reduced body weight, and increased mortality.

Anthropogenic environmental Mo contamination is related to its mining, production of steels for special purposes, and industrial waste discharge. When it comes to fossil fuels, the element accumulates and builds up in coals. Molybdenum contents in soils vary greatly. The highest amounts of molybdenum are found in soils formed from granite and clays. In plants growing on muschelkalk, keuper, and alluvial soils, on the other hand, Mo content is low.

Data on the way molybdenum affects free-living mammals are scarce. Some well-documented sources point to the "mysterious" moose disease in Sweden, caused by copper deficiency and molybdenosis. The latter disease evokes symptoms resembling copper deficiency.

Research is carried out mainly on domesticated avian species in which toxic molybdenum effects were observed at the dietary concentration of its absorbable forms exceeding 200 mg $kg^{-1}$. Among undomesticated birds, analyses involved bobwhite quail (*Colinus virginianus*) treated with molybdenum disulfide (insoluble) and a soluble form of sodium molybdate dihydrate. Toxicity and bioavailability of both Mo forms were compared. These results are more realistic in showing the susceptibility and the risk birds are exposed to when it comes to environmental molybdenum. Human impact associated with molybdenum environmental contamination involves mainly combustion of fossil fuels, mining, and metallurgy. Molybdenum does not belong to metals of special significance in terms of environmental toxicology. The metal has not been classified as a major toxic element by the EPA. In Central Europe, plants growing on the soils formed from granite, gneiss, rotliegend, and shale, as well as on muck and peat soils, may represent a health hazard for free-living ruminants.

## 1 Introduction

As a metal, molybdenum was known in the ancient times (Anke 2004). In the 16-km layer of the Earth's crust, the content of molybdenum is approx. 1.4 mg $kg^{-1}$, making it the 40th most abundant element in nature (Anke et al. 2000). Molybdenum is an essential element for the life and normal growth of plants, animals, and humans (Anke et al. 2000). The biological role of molybdenum was discovered in 1830 by Boussingault, who demonstrated that clover and pea increase their nitrogen contents using *Acetobacter* sp. bacteria, which have Mo in their enzymes. In 1940, after many years of studies, Bortels proved that molybdenum is essential for higher plants (Anke 2004).

As a heavy metal, molybdenum has drawn attention of animal scientists; in 1938 a group of researchers led by Ferguson found that the element was responsible for teart disease in cattle in Mo-rich areas of England (Anke 1986).

In both animals and humans, molybdenum is recognized as an important constituent of three enzymes: xanthine oxidase, aldehyde oxidase, and sulfite oxidase

(Cohen et al. 1971; Feldman and Weiner 1972; Anke et al. 2000). Although Mo is a component of a range of enzymes, no symptoms of molybdenum primary deficiency have been so far observed in humans and animals (Kośla et al. 1989). Only high doses of Mo-antagonistic elements (tungsten) result in sulfite oxidase activation, which leads to Mo deficiency symptoms (Johnson 1997).

Anke et al. (1978, 1985a, b), who administered a molybdenum-poor (<24 μg $kg^{-1}$ dw of feed) semisynthetic dose to several generations of goats, observed depressed growth, reduced fertilization rates, and low fetal survival. A significant increase in the prevalence of miscarriage was also noted. Those experiments demonstrated the essentiality of Mo for animals, without using antagonistic substances (Kośla et al. 1989).

The importance of molybdenum consists in its toxicity (Anke et al. 2000). In combination with Cu and S, molybdenum causes secondary copper deficiency, which is the case primarily in ruminants (Anke et al. 2000). In this group, molybdenum toxicity (molybdenosis) affects mainly cattle (Anke 2004). In horses, a high, species-specific hepatic content of Mo is significantly higher compared to other domestic and wild animals (Kośla et al. 1989).

## 2 General Properties

Molybdenum occurs in the Earth's crust at a level of 1–2 ppm, only in the form of compounds with other elements (Sebenik et al. 2002; Anbar 2004; Kabata-Pendias and Mukherjee 2007). Its concentrations are highest in granite rocks and loam sediments ("argillaceous sediments") (Kabata-Pendias and Mukherjee 2007). Molybdenum is the 24th most abundant element in the oceans, with an average of 10 parts per billion (Emsley 2001; Considine 2005). Although the most important source is molybdenite, $MoS_2$, molybdenum can be found in other minerals, i.e., molybdite, $MoO_3$; wulfenite, $PbMoO_4$; and powellite, $CaMoO_4$ (Kabata-Pendias and Mukherjee 2007). In 1778 Carl Wilhelm Scheele separated molybdenum from graphite and lead and derived the oxide from molybdenite (Mitchell 2003). Three years later Peter Jacob Hjelm extracted the metal by reducing its oxide using carbon.

Molybdenum is a transient metal located in Group VI of the periodic table. It is a silvery white, literary lead-like metal, and very hard, with a Mohs hardness of 5.5. It occurs in several oxidation states, from −2 to 6, with Mo(6) being the most common species in most natural waters, including seawater, in the form of molybdate ion, $MoO_4^{2-}$ (Reid 2012). The boiling point of molybdenum is 4639 °C; its melting point, 2617 °C, is one among the highest of all the elements (Reid 2012). There are seven stable isotopes of molybdenum, 92, 94, 95, 96, 97, 98, and 100. The most common is $^{98}Mo$ (24.13%), whereas $^{99}Mo$ has found its application in the nuclear sector (Anbar 2004). Molybdenum does not visibly react with oxygen or water at room temperature; however, at higher temperatures (600 °C), the metal oxidizes quickly and forms molybdenum(6) oxide, $MoO_3$. In strongly alkaline water, $MoO_3$ is transformed into molybdates ($MoO_4^{2-}$). Molybdenum oxidation states four (e.g.,

$MoS_2$) and six are the most stable species. The element is prone to bioaccumulation and has accumulated in coals, which—if combusted—release dispersing molybdenum that finally reaches the soil and sediments. Volatile ashes may contain 5–33 mg $kg^{-1}$ of the element (Pasieczna 2012). Under anoxic conditions, concretions containing up to 2000 mg $kg^{-1}$ of molybdenum can form in marine sediments (Anbar 2004; Arnold et al. 2004).

Molybdenum binds into complexes with many inorganic and organic ligands, thereby playing a significant role in the biochemical process. It is involved in redox reactions and is a constituent of many enzymes, which are essential in aldehyde oxidation (aldehyde oxidase), purine and pyrimidine metabolism (xanthine oxidase), or sulfite detoxification (sulfite oxidase) (Skibniewski et al. 2015). No biomagnification and bioaccumulation of Mo in animals have been observed so far (Ikemoto et al. 2008; Reid 2012).

## 3 Molybdenum Minerals, Production, and Uses

Molybdenum is mostly extracted from its ore, molybdenite, but also produced as a side product in copper smelting (Kabata-Pendias and Mukherjee 2007).

Molybdenum is primarily used as an alloy hardening component, in heat- and wear-resistant steel production (Kabata-Pendias and Mukherjee 2007). Ferromolybdenum and molybdic oxide are the main molybdenum forms used in combination with chromium, manganese, nickel, niobium, tungsten, etc. (Polyak 2016b). It is also used in the aerospace, in arms and electronics industries, in catalyst manufacturing, and in the production of dyes, lubricants, and elements of incandescent lamps; its complexes are also utilized as catalysts in oil-processing industry, especially for sulfur removal from petroleum products (NRC 2004; Kabata-Pendias and Mukherjee 2007). Molybdenum disulfide is an excellent high-temperature-resistant lubricant and is applied in heat-resistant metal coatings. Its role in the so-called green technologies is increasingly more and more important, and it has found applications in the production of biofuels and ethanol, as well as in the manufacturing of solar panels and wind turbines (Polyak 2013). The International Molybdenum Association (IMOA 2014) reports that molybdenum compounds will soon play an important role in the production of stable and environmentally safe dyes.

China, the USA, and Chile are the world's leaders in molybdenum ore mining (Fig. 8.1). Other producing countries (2015 production) include Armenia (7300), Russia (4800), Iran (4000), Mongolia (2000), Turkey (1400), and Uzbekistan (520) (Polyak 2016a).

An upward trend in molybdenum mining is particularly apparent in China, where the production has increased 2.5 times since 2005; this is due to China's galloping economic growth and an improved competitiveness of steel exports (Magyar 2007, 2009; Polyak 2016a). The whole identified Mo resources in the USA reach about 5.4

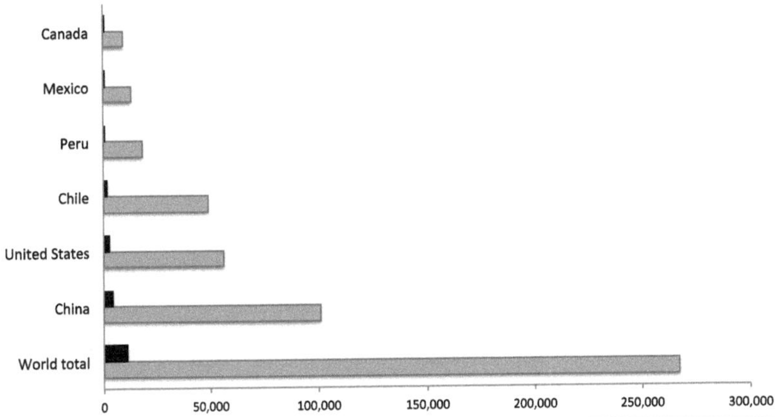


| | World total | China | United States | Chile | Peru | Mexico | Canada |
|---|---|---|---|---|---|---|---|
| ■ Reserves in thousand tons | 11,000 | 4,300 | 2,700 | 1,800 | 450 | 130 | 260 |
| ▨ Mine production in tons | 267,000 | 101,000 | 56,300 | 49,000 | 18,100 | 13,000 | 9,300 |

**Fig. 8.1** World in total and six of the most important countries in molybdenum mining production (in tons) in 2015 and the estimated reserves (in 1000 tons) (Polyak 2016a)

million tons, which adds to about 14 million tons located in the other parts of the world (Polyak 2016a).

Anthropogenic molybdenum pollution results from its mining processes, production of heat- and corrosion-resistant steel alloys, and industrial wastewater discharge including those from tanneries (Reid 2012).

# 4 Molybdenum in Nature

Molybdenum is widely distributed in nature. It is found in such minerals as molybdenite ($MoS_2$—the major ore of molybdenum), wulfenite, ferrimolybdate, jordisite, and powellite (Leichtfried 1990). Environmental release of molybdenum compounds from industrial activities affects the air (stack emissions), water (liquid effluents), or earth in the form of solid wastes (sludge) (Leichtfried 1990).

The content of molybdenum in rocks depends on their type (Table 8.1). Lithosphere molybdenum is mainly associated with acid igneous rocks or argillaceous formations (Kabata-Pendias 2011; Pasieczna 2012). In respect to rock type, caustobioliths and copper and oil shales are the richest rocks in Mo, followed by argillaceous sediments and clays. The lowest molybdenum content is found in limestones, sandstones, and graywackes (Thornton 1981; Wedepohl 2004; He et al. 1998, 2005; Kabata-Pendias and Mukherjee 2007). The continental crust contains Mo in the range 1–1.5 mg $kg^{-1}$, whereas the oceanic basalt crust contains 0.8 mg $kg^{-1}$ (Budaveri 1996; Wedephol 2004).

The content of molybdenum in soils is presented in Table 8.2. The average soil concentration of molybdenum may range between 1.0 and 2.0 mg $kg^{-1}$ (Kubota

**Table 8.1** Concentrations of molybdenum in rocks (mg $kg^{-1}$)

| Rock | Content | References |
|---|---|---|
| Continental crust | 1–1.5 | Budaveri (1996) |
| | 1.1 | Wedepohl (2004) |
| Oceanic basalt crust | 0.8 | Wedepohl (2004) |
| Caustobioliths | >2% | Enzmann (1972) and Pasieczna (2012) |
| Copper and oil shales | >200 | Enzmann (1972) and Pasieczna (2012) |
| Argillaceous sedimentary rocks | 2.5–3.0 | Thornton (1981), Reimann et al. (1998), He et al. (1998, 2005), and Pasieczna (2012) |
| Clays | 2–2.5 | Kabata-Pendias and Mukherjee (2007) |
| Granites | 1.0–6.0 | Thornton (1981), Wedepohl (2004), and He et al. (1998, 2005) |
| Shales | 1.3–2.5 | Thornton (1981), Wedepohl (2004), and He et al. (1998, 2005) |
| Gneisses | 1.5 | Wedepohl (2004) |
| Basalts, gabbroic rocks | 0.9–7.0 | Thornton (1981), Wedepohl (2004), and He et al. (1998, 2005) |
| Igneous rocks | 0.3–2.0 | Reimann et al. (1998) and Pasieczna (2012) |
| Graywackes | 0.7 | Wedepohl (2004) |
| Sandstones | 0.2–0.8 | Kabata-Pendias and Mukherjee (2007) |
| Limestones | 0.2–0.4 | Thornton (1981), Wedepohl (2004), Kabata-Pendias and Mukherjee (2007), and He et al. (1998, 2005) |

**Table 8.2** Concentrations of molybdenum in soils (mg $kg^{-1}$ dry wt)

| Soil type | Content | Reference |
|---|---|---|
| Soils of the world | 0.1–7 | Kabata-Pendias and Mukherjee (2007) |
| On average | 1.8 | Kabata-Pendias and Mukherjee (2007) |
| | 1.2 | Koljonen (1992) |
| | 1.0–2.0 | Kubota (1977), Adriano (1986) and USGS (2015) |
| European topsoil | <0.1–17.2 | de Vos and Tarvainen (2006) |
| Median | 0.62 | de Vos and Tarvainen (2006) |
| Soils used for agricultural purposes | 0.6–72 | Reimann et al. (2003) |
| Light sandys | 0.1–3.7 | Kabata-Pendias and Mukherjee (2007) |
| Medium clays | 0.4–6.4 | Kabata-Pendias and Mukherjee (2007) |
| Heavy clays | 0.7–7.2 | Kabata-Pendias and Mukherjee (2007) |
| Calcareous soils | 0.3–7.4 | Kabata-Pendias and Mukherjee (2007) |
| Organic soils | 0.3–3.2 | Kabata-Pendias and Mukherjee (2007) |
| Soils in the USA | 0.1–4.0 | Friberg et al. (1975) and Chappel et al. (1979) |
| On average | 1.2 | Friberg et al. (1975) and Chappel et al. (1979) |

1977; Adriano 1986; Koljonen 1992; Kabata-Pendias and Mukherjee 2007; USGS 2015). Soils containing more washable particles also contain more molybdenum. Molybdenum content in the soils of Saxony averages 0.30–6.0 mg $kg^{-1}$ (Barth et al.

1996), 0.2–48.6 mg kg$^{-1}$ in Slovakia (Eurlik and Šefeik 1999), 0.24–3.13 mg kg$^{-1}$ in Lithuania (Kadūnas et al. 1999; Gregorauskienė and Kadūnas 2000), and from <1 to up to 20 mg kg$^{-1}$ in the Kola Peninsula (Reimann et al. 1998).

Soils in arid and semiarid areas, especially ferrasoils, usually contain relatively high amounts of molybdenum. Reclamation of such soils involves reduction of Mo phytoavailability (Kabata-Pendias and Mukherjee 2007). Eisler (1989) states that the levels of Mo are higher in the vicinity of molybdenum mines and combined heat and power plants. The soil may contain up to 35 mg kg$^{-1}$ Mo in the areas around molybdenum processing plants in the USA and up to 38 mg kg$^{-1}$ in Chile (Kabata-Pendias and Mukherjee 2007). Molybdenum is different from most trace metals in a way that its compounds are poorly soluble at a low soil pH (<4.5) and highly soluble in alkaline soils (pH > 6.5), which is reflected in Mo uptake by plants growing on different soil types (Kabata-Pendias 2011).

Molybdenum is likely to form compounds with organic matter and some metals (Kabata-Pendias and Mukherjee 2007). Some soils require supplemental enrichment with Mo. These are soils with low organic matter content, exposed to severe erosion or strongly weathered, low in total molybdenum, as well as sandy and acidic soils (pH < 6.3) (Hornick et al. 1977). In the USA, Mo concentrations in soils increase moving from east to west (Kubota 1977; Adriano 1986).

Water and wind are the main factors of molybdenum spreading across the surface of the Earth (Wedepohl 2004). The median of Mo concentration in the oceanic waters was calculated at the level 10 μg L$^{-1}$ (Reimann and de Caritat 1998; Wedephol 2004; Nozaki 2005). Rainwater Mo content may vary depending on the pollution from 0.01 to 2.6 μg L$^{-1}$, between 0.04 and 1.3 μg L$^{-1}$ in river water (Gaillardet et al. 2004), or 1 μg L$^{-1}$ according to Wedephol (2004). In the USA, molybdenum concentrations are noted at levels ranging 1.2–4.1 μg L$^{-1}$ in rivers, <1 μg L$^{-1}$ in groundwaters, and 5–57 mg kg$^{-1}$ dw in fluvial deposits (Friberg et al. 1975; Chappell et al. 1979). Surface water concentration of this element ranges from 0.4 μg L$^{-1}$ in unpolluted rivers of North America to 100,000 μg L$^{-1}$ in mining wastewater. In the groundwaters of the USA, Mo concentration is usually lower than 1 μg L$^{-1}$; however, high levels reaching 50,000 μg L$^{-1}$ were measured near uranium mills in Colorado (Eisler 1989). Natural molybdenum concentrations in surface waters seldom exceed 20 μg L$^{-1}$; higher levels are probably due to industrial pollution (Eisler 1989).

The average molybdenum content in the atmosphere is below 0.2 ng m$^{-3}$, ranging from <0.5 ng m$^{-3}$ in the isles of Svalbard and 0.2 ng m$^{-3}$ in Greenland to 1–10 ng m$^{-3}$ in urban areas (Kabata-Pendias and Mukherjee 2007).

If we look at the concentration of Mo in fossil fuels (Table 8.3), it appears that the element accumulates and builds up in coals. If combusted, these fuels release molybdenum, which spreads around and finally gets into soils and sediments. Fly ash contains Mo at a level between 5 and 33 mg kg$^{-1}$ (Bhattacharyya et al. 2009; Pandey et al. 2009; Pasieczna 2012).

**Table 8.3** Molybdenum content in fossil energy materials (mg $kg^{-1}$)

| Fuel | Content | Reference |
|---|---|---|
| Coal | 3–6 | Kabata-Pendias and Mukherjee (2007) |
| On average | 4 | Wedephol (2004) |
| Lignite | 2.8 | Wedephol (2004) |
| Fly ash | 5–33 | Bhattacharyya et al. (2009), Pandey et al. (2009) and Pasieczna (2012) |
| On average | 14.6 | Kabata-Pendias and Mukherjee (2007) |
| Petroleum | 10 mg $L^{-1}$ | Wedephol (2004) |
| Gasoline | 10 mg $L^{-1}$ | Kabata-Pendias and Mukherjee (2007) |

## 5 Biological Status of Molybdenum

In human and animal bodies, molybdenum occurs in low concentrations; still, Mo is an essential element, a component of several enzymes (Stiefel 2002; Williams and da Silva 2002; Kabata-Pendias 2011; Reid 2012), which catalyze redox reactions. In plants, on the other hand, molybdenum causes reduction of molecular nitrogen and nitrates (Sellman 1993; Kim et al. 1993; Burris and Roberts 1993; Reid 2012), which is important in plant growth processes. Despite its key role, Mo is present in plant tissues in very small amounts, 0.5 mg $kg^{-1}$ dry matter on average (Spears 1992; Turnlund et al. 1995). According to Kabata-Pendias (2011), normal concentrations in plants may vary, ranging from 0.1 to 1.5 mg $kg^{-1}$ dw. In areas where intensive industrial activities take place, herbaceous plants contain Mo in amounts of up to 231 mg $kg^{-1}$ dw. Molybdenum is poorly absorbed from acid soils; hence it is recommended to provide the plants with additional amounts of this element through soil liming or by applying Mo supplementation on pastures (Hornick et al. 1977). Its deficiency signs are pale patches on leaves, withering of buds, and impaired leaf development; particularly susceptible are brassicas (Pasieczna 2012).

Molybdenum concentration in the range 0.1–0.5 mg $kg^{-1}$ suffices for the plant metabolism, whereas levels of 10–50 mg $kg^{-1}$ are toxic to most plants. Molybdenum deficiencies are likely in acid soils (pH $< 5.5$), in those where a low Mo content is coupled with a high level of iron oxide, or in peat soils (Kabata-Pendias and Mukherjee 2007; Kabata-Pendias 2011).

Average Mo content in grain is 0.45 mg $kg^{-1}$ in rye, 0.5 in barley, 0.55 in oats, and 1.0 in wheat (always mg $kg^{-1}$ dw). There is 1.2–1.8 in peas, 0.9–1.6 in beans, 0.1–0.25 in potatoes, and 0.07 in apples on average (always mg $kg^{-1}$). In grasses growing over Mo-contaminated areas, the concentrations of the metal range from 1.5 mg $kg^{-1}$, in Great Britain, to 50 mg $kg^{-1}$, in Russia; in legumes, the levels range from 5.2 mg $kg^{-1}$, in Canada, to 26.6 mg $kg^{-1}$, in Poland (Kabata-Pendias and Mukherjee 2007). Excess soil molybdenum supply will increase its content in plants, which may lead to molybdenosis in ruminants (Underwood 1977; Kabata-Pendias and Mukherjee 2007). In soils, molybdenum is likely to form complex anionic

**Table 8.4** Effect of soils formed from various soils on relative molybdenum (Mo) content in plants

| Geological origin of the soil | Relative Mo content in plants |
|---|---|
| Granite | 100 |
| Clay | 88 |
| Phyllite—a fine-grain shale | 88 |
| Gneiss—a metamorphic rock | 78 |
| Buntsandstein | 78 |
| Peat, muck soils | 71 |
| Rotliegend | 71 |
| Loess | 64 |
| Diluvial sands | 64 |
| Shale soils (Devonian, Siluran, Culm Supergroup) | 51 |
| Alluvial soils | 47 |
| Keuper | 47 |
| Muschelkalk | 38 |

Anke et al. (2000)

compounds, which are mobile in neutral and alkaline substrates. In acid substrates, on the other hand, Mo binds to a hydroxide of aluminum, iron, or manganese or to organic matter and argillaceous minerals, which limits its phytoavailability (Pasieczna 2012). Legumes require more molybdenum compared to other plants, since symbiotic bacteria of their root nodules use it for binding atmospheric nitrogen (Sequi 1973; Ivanova 1973; Regius and Anke 1989; Schnabel and Bunke 1989; Kabata-Pendias 2011).

Plants sometimes exhibit explicit symptoms of Mo deficiency, such as classic whiptail in cauliflower or yellow patches on citrus plants; however, most often Mo deficiency symptoms are invisible, and nitrogen deficiency symptoms may appear instead (Hornick et al. 1977; Lăcătuşu and Borza 1989).

The highest levels of molybdenum available for plants, despite a low pH, are in soils formed from granite rock (Table 8.4). Also vegetation growing on soils created from clays, phyllite, and gneiss contain much molybdenum. Muschelkalk, keuper, and alluvial soils, on the other hand, are low in terms of molybdenum levels in plants (Anke et al. 2000). Plants growing on these soils are likely to suffer Mo deficiencies (Holzinger et al. 1998).

Although animal molybdenum deficiencies are rare, broiler chickens are susceptible, and slow growth rates and poor plumage quality may be symptoms of such (Kabata-Pendias and Mukherjee 2007). Higher concentrations of molybdenum are said to stimulate growth; hence some groups of animals (chickens, rainbow trout) obtain Mo supplementation to enhance growth (Reid 2012).

Anke et al. (1983, 1985b) measured molybdenum in calves, lambs, and newborn children and found that Mo levels in all studied tissues were significantly lower during the period of maternal milk feeding, as compared with those during further, post-suckling nutrition period. Similar pattern was observed in various ungulate species. Studies on herbivorous mammals in the San Diego Zoo (California, USA)

**Table 8.5** The content (mg kg$^{-1}$ dw) of Mo in the liver in different species of wild ruminants

| Name of species | | Adult | Newborn |
|---|---|---|---|
| Species (n;n) | Latin name | Mean ± SD | Mean ± SD |
| Sable antelope (4;3) | *Hippotragus niger* | 1.51 ± 0.96 | 0.28 ± 0.14 |
| Gemsbok (15;15) | *Oryx gazella* | 1.78 ± 0.37 | 0.98 ± 0.40 |
| Pampas deer (5;2) | *Ozotoceros bezoarticus* | 1.89 ± 0.63 | 1.29 ± 0.14 |
| Chital or cheetal (8;14) | *Axis axis* | 1.95 ± 0.85 | 1.12 ± 3.39 |
| Gaur or Indian bison (3;5) | *Bos gaurus* | 2.09 ± 0.88 | 1.13 ± 0.27 |
| Hog deer (7;6) | *Axis porcinus* | 2.18 ± 0.93 | 1.23 ± 0.38 |
| Sika deer (8;6) | *Cervus nippon* | 2.21 ± 0.98 | 1.59 ± 0.25 |
| Impala (4;2) | *Aepyceros melampus* | 2.29 ± 0.78 | 0.41 ± 0.02 |
| Markhor (6;4) | *Capra falconeri* | 2.37 ± 0.91 | 1.69 ± 0.35 |
| European bison (4;5) | *Bison bonasus* | 2.59 ± 0.85 | 1.28 ± 0.39 |
| Nilgai (4;5) | *Boselaphus tragocamelus* | 2.59 ± 0.24 | 0.85±0.52 |
| Kudu (4;3) | *Tragelaphus imberbis* | 2.66 ± 0.10 | 1.09 ± 0.19 |
| Barasingha or swamp deer (6;8) | *Rucervus duvaucelii syn. Cervus duvaucelii* | 2.83 ± 0.80 | 0.98 ± 0.12 |
| Blackbuck/Indian antelope (4;17) | *Antilope cervicapra* | 3.03 ± 0.47 | 0.63 ± 0.17 |
| Mufflon or mouflon (12;12) | *Ovis musimon* | 3.27 ± 0.39 | 1.30 ± 0.63 |
| Grant's gazelle (6;3) | *Nanger granti* | 3.81 ± 0.89 | 0.94 ± 0.53 |
| Alpine ibex (12;17) | *Capra ibex* | 3.88 ± 0.85 | 1.86 ± 0.44 |
| Thomson's gazelle (7;3) | *Eudorcas thomsonii* | 3.89 ± 1.01 | 0.70 ± 0.15 |
| Dama gazelle (5;6) | *Nanger dama* | 4.31 ± 0.85 | 0.69 ± 0.34 |
| *Gazella subgutturosa* (7;8) | *Gazella subgutturosa* | 4.64 ± 2.88 | 1.83 ± 1.21 |
| Mountain gazelle (8;6) | *Gazella gazella* | 4.67 ± 2.11 | 0.84 ± 0.35 |
| Soemmerring's gazelle (5;5) | *Nanger soemmerringii* | 5.15 ± 1.07 | 1.31 ± 0.18 |

Anke et al. (2000)

revealed that hepatic Mo content in neonates was significantly lower compared to adult animals (Table 8.5); nonsignificant differences were found only in 3 out of 22 species studied. Also, molybdenum hepatic levels in adult animals differ significantly ($p \leq 0.001$) between species; between the liver of Soemmerring's gazelle (*Nanger soemmerringii*) (5.15 mg kg$^{-1}$), in which Mo content is highest, and the sable antelope (*Hippotragus niger*) (1.5 mg kg$^{-1}$), with the lowest Mo hepatic content, the difference was 342%. A similar pattern can be seen in their neonates; the difference between extreme species reached 664% (Anke et al. 2000).

Molybdenum is present in all tissues in humans and animals. The lowest level in a human body occurs in blood, 3.4–14.9 ng L$^{-1}$, and the highest in the kidneys and liver, 0.4 mg kg$^{-1}$ (Li 2000). In animals, molybdenum is involved in oxidation (hydroxylation) of xanthine to uric acid and other purines and aldehydes (Schindelin et al. 1996; Reid 2012). Molybdenum in animals also affects protein synthesis and metabolism of a range of elements, i.e., phosphorus, sulfur, potassium, iron, copper, zinc, and iodine (Reid 2012).

Molybdenum is also an essential element for humans. The molybdenum cofactor is necessary for the functioning of at least three enzymes: (1) sulfite oxidase (catalyzes sulfite-to-sulfate oxidation and is necessary for sulfur amino acid metabolism), which occurs in the liver (Mills and Davis 1987; Anke and Risch 1989; Hille et al. 2011; Kabata-Pendias 2011); (2) xanthine oxidase (catalyzes purine and pyridine hydroxylation, including conversion of hypoxanthine to xanthine and xanthine to uric acid); and (3) aldehyde oxidase (oxidizes purines, pyrimidines, and pteridines and is involved in nicotinic acid metabolism), which are both present in the liver and intestine. Xanthine oxidase has been isolated from cow's milk. Molybdenum in milk occurs mainly as a component of this enzyme (Mills and Davis 1987).

The main sources of Mo in the human diet are vegetables, legumes, and offal (Combs 2005). Excess dietary molybdenum reflects in its elevated levels in serum, urine, or hair (Kabata-Pendias and Mukherjee 2007). Low intake of Mo reduces the concentrations of uric acid in urine and serum and leads to excessive excretion of xanthine. Molybdenum deficiency in humans leads to neurological symptoms and premature death (Spears 1992; Van Gennip et al. 1994; Turnlund et al. 1995).

## 6 Toxicity of Various Molybdenum Forms in Homeothermic Animals

Symptoms of either acute or chronic molybdenum toxicity depend on its chemical form and the species of the exposed organism. Generally insoluble Mo compounds are characterized by a lower toxicity compared to its soluble forms. These include primarily $MoS_2$, $MoO_2$, and the metallic form of the element (Stokinger 1981; EFSA 2006). Those of higher toxicity—apart from such water-soluble compounds as calcium molybdate and molybdenum trioxide—include thiomolybdates and oxythiomolybdates, as well as molybdenum forms present in plants. Their bioavailability for laboratory animals and ruminants ranged between 75 and 97% of ingested quantity (EFSA 2006). If we consider molybdenum toxicity in terms of the species, a regularity is that monogastric animals are less susceptible to its toxic effects, as compared to ruminants (Davis and Mertz 1987; Mills and Davis 1987; Blood and Radostits 1989; Frank et al. 2004a, b; Hall 2012).

Excess molybdenum is a commonplace phenomenon, especially in cattle, affecting the growth and health of the animals. It may lead to osteoporosis and bone deformities, since it interferes with Ca and F metabolism in the bone (Anke and Groppel 1985; Kabata-Pendias and Mukherjee 2007). Copper deficiency caused by Mo elevated levels is often the case in ruminants (Jones 2005). Some pastures may have an extremely high concentration of Mo (as a rule, this is the vegetation associated with alkaline soils), which may result in Mo toxicity in sheep and cattle (Hornick et al. 1977). The recommended molybdenum concentration in soils used for agricultural purposes has been established at the level 50 mg $kg^{-1}$ dry weight, dw (Hornick et al. 1977).

Most reports on animal Mo research carried out so far concern laboratory mammals and domesticated species; those on free-living animals are sparse. Underwood (1977) claims that animals of different groups vary in relation to dietary molybdenum dose tolerance. This is particularly dependent on copper and inorganic sulfates contents in the diet. In the group of domesticated animals, horses are the most tolerant species to elevated molybdenum intake, followed by pigs, rats, rabbits, and guinea pigs. Cattle and sheep, on the other hand, are definitely the most susceptible species to Mo toxicity (Underwood 1977; Anke and Groppel 1985; Rajagopalan 1988; Tallkvist and Oskarsson 2015). In cattle, a concentration of 20 mg Mo $kg^{-1}$ dw of feed results in severe diarrhea and body weight loss (teart); this level causes no effect in horses and very little in sheep (Underwood 1977).

In pigs, administration of 1 g Mo $kg^{-1}$ dw of feed lasting for several months produced no symptoms of toxic effect. With acute Mo toxicity, all animal species exhibit growth inhibition, body weight loss, and increased mortality (Underwood 1977). Laboratory animals affected by molybdenosis, however, do not suffer diarrhea, which is the case in cattle. A strong protective activity of copper and inorganic sulfates has been found (Underwood 1977). The $LD_{50}$ for rodents in response to oral administration of molybdenum trioxide was 188 mg $kg^{-1}$ body weight, which corresponds to 125 mg Mo $kg^{-1}$ body weight; for ammonium molybdate, on the other hand, the value was 680 mg $kg^{-1}$ body weight, i.e., 370 mg Mo $kg^{-1}$ body weight. The $LD_{100}$ after oral administration of ammonium molybdate to guinea pigs, rabbits, and cats were, respectively, 2200, 1870, and 2400 mg $kg^{-1}$ body weight, which corresponds to 1200, 1020, and 1310 mg Mo $kg^{-1}$ body weight (Venugopal and Luckey 1978). Reproducible lethal dose for mice, guinea pigs, and rabbits ranges between 60 and 330 mg Mo $kg^{-1}$ body weight (Mills and Davis 1987). The NOAEL for rats is 40 mg Mo $kg^{-1}$ body weight per day, whereas for rabbits 23 mg Mo $kg^{-1}$ body weight per day (Bompart et al. 1990; Vyskocil and Viau 1999).

Studies on the effects of molybdenum on ruminants revealed that symptoms of acute molybdenosis in cattle appear at the concentration of 7400 mg $kg^{-1}$ of feed, in the form of sodium molybdate. This dose caused focal necrosis in the kidneys and liver of the animals (Swan et al. 1998). As in the case of acute toxicity, chronic toxicity also depends on the animal species. An oral lethal dose of the chronic exposure in rats, mice, rabbits, and guinea pigs remains in the range from 60 to 333 mg Mo $kg^{-1}$ of body weight per day. Ruminants are much less resistant to molybdenum, as similar outcomes are observed at only 3 mg Mo $kg^{-1}$ body weight per day (Hall 2012).

Molybdenum toxicity symptoms include diarrhea (cattle only), anorexia, achromotrichia, and disturbation of neurological conditions and may lead to premature death (Anke 2004). Contamination with molybdenum caused in cattle severe diarrhea, which occurred during the period of grazing. This was accompanied by secondary copper deficiency (Anke 1986). Sulfur levels also played a role here. With participation of sulfates, inorganic Mo compounds were converted to thiomolybdates (e.g., $MoS^{2-}$) by microorganisms in the rumen. Thiomolybdates immediately react with various proteins, thereby forming Cu-binding complexes (Mills et al. 1978; Anke 1986). High concentrations of Mo in the diet of cattle, sheep, horses, rabbits, and rats damage the animal skeleton (Anke and Groppel 1985).

Under natural conditions, toxicity of molybdenum starts immediately after ingestion. Some compounds of the metal may also penetrate the body through inhalation or cause skin irritation. These include molybdenum trioxide. As in the case of the gastrointestinal tract, absorption in the airways also depends on the solubility of the substances. It has been found that soluble species are absorbed in the lungs, whereas insoluble compounds are not (Stokinger 1981; Friberg and Lener 1986). Chan et al. (1998) observed that rats and mice exposed to this compound in an amount of 10–100 mg $m^{-3}$ for 6 h a day over the period of 5 days developed adenomas in their respiratory system, as well as chronic inflammation of the lining of the respiratory system and degeneration of the cartilage structure of the bronchial tree. In rabbits treated intratracheally with molybdenum trioxide in particulate form, in an amount of 70–80 mg $kg^{-1}$ body weight for a period of 9 months, induced pneumoconiosis accompanied by effusion into the lumen of pulmonary alveoli and bronchi (Friberg and Lener 1986). Molybdenum compounds inhaled into the airways also have system-wide effects. Lukashev et al. (1971) observed that rabbits and rats exposed to molybdenum trioxide in particulate form administered in the dose changing from 210 to 10 mg $m^{-3}$ in 25-min intervals, for 4 h per day, over the period of 3.5 months, resulted in renal tubular atrophy. Compounds causing skin irritation include $Na_2MoO_4$, which induces dermal reaction within 24 h following the contact. The changes disappear within 72 h from the cessation of irritation (EFSA 2006).

Data concerning the effects of molybdenum on free-living mammals are sparse. Good documented sources deal only with a cervid species inhabiting Scandinavia. Particular focus has been on copper and molybdenum contents in the tissues of the moose living in Sweden in connection with the "mysterious" moose disease, which affected the population in this country in the mid-1980s. It was then a disorder of unknown etiology (Frank et al. 2000). The clinical symptoms and organ changes observed postmortem resembled those caused by copper deficiencies and molybdenosis in cattle and sheep (Frank 1998, Frank et al. 2000). Several years of investigations revealed that in the period 1982–1994, copper hepatic levels in unaffected individuals dropped twofold, whereas molybdenum concentrations in the same period and in the same animals increased 20–40% (Frank et al. 1994, 2000; Frank and Galgan 1997; Frank 1998). Based on microbiological and anatomopathological examinations and chemical composition analyses of parenchymal organs in the moose, it was eventually concluded that the most probable cause of the disease is copper deficiency and molybdenosis (Frank 1998; Frank et al. 2000). Frank (2003) also reported another case of molybdenosis in a moose in Sweden, which was caused by imbalance in the proportion between Cu and Mo in the liver, and this affected the metabolism of sulfur.

There is little information on molybdenum toxicity in free-living birds. As compared with mammals, reports on avian species are sparse and deal mainly with domestic birds. Toxic levels of dietary Mo in birds have been found at its absorbable form concentration of 200 mg $kg^{-1}$. The most apparent symptom was growth inhibition. Molybdenum applied to turkey chicks at a dose of 300 mg $kg^{-1}$ feed caused a considerable growth inhibition (Underwood 1977). At a Mo level of 500 mg $kg^{-1}$ in feed, reproduction disorders were observed, whereas a feed concentration of 6000 mg $kg^{-1}$ applied over 4 weeks resulted in 33% mortality. An increase of Mo dose to 8000 mg $kg^{-1}$ feed over the same period of time killed 61% of chicks. Those that survived weighed 16% less than the control group birds (Friberg et al. 1975; Eisler 1989).

Stafford et al. (2016) analyzed toxicity of molybdenum disulfide ($MoS_2$, a dominant form found in molybdenite ore) in relation to bobwhite quail (*Colinus virginianus*). The chemical form and bioavailability of Mo are important in terms of its toxicity. The trial on birds involved a soluble Mo form, sodium molybdate dihydrate (SMD, $Na_2MoO_4{\cdot}2H_2O$), whereas $MoS_2$ is generally insoluble, poorly available under special environmental conditions. The observations included survival and health status (body weight and feed intake) of 9-day-old bobwhite quails exposed to soluble Mo (SMD) and Mo ore ($MoS_2$) in the diet for 30 days. The two forms of Mo were compared in terms of toxicity and bioavailability (also tissue penetration). Histopathological examinations and analyses of the serum, kidneys, liver, and bone were carried out. Copper, linked with Mo in terms of toxicity, was also determined in the diet and the tissues. There were no deaths or changes in groups of any form of Mo.

Analyses of tissues revealed increased levels of Mo in the serum, kidneys, liver, and bone during the experiment with SMD. After the period of exposure, a reduction in Mo was observed in these tissues. For the soluble form, the no-observed-adverse-effect concentration (NOAEC) was 1200 mg Mo as SMD/kg feed (134 mg SMD/kg body weight per day). There was no adverse exposure effect of $MoS_2$ at the maximum dose 5000 mg $MoS_2$ $kg^{-1}$ feed (545 mg $MoS_2$ $kg^{-1}$ body weight per day). This reveals that the effect of $MoS_2$, a more common and less bioavailable form of Mo, is similar to that of SMD. The NOAEC of $MoS_2$ is 5000 mg (545 mg $MoS_2$ $kg^{-1}$ body weight per day). These results are more realistic in showing the susceptibility and the risk birds are exposed to when it comes to environmental molybdenum (Stafford et al. 2016).

## 7 Toxicokinetics and Effects of Molybdenum in Wildlife

There are two main routes by which the animal body may be exposed to molybdenum, namely, a dietary and a respiratory pathway. The latter has been relatively poorly described and reports refer mostly to laboratory animals (Tallkvist and Oskarsson 2015). Aerial exposure of guinea pigs to 285 mg Mo $m^{-3}$ in the form of molybdenum disulfide powder did not result in elevated absorption of the metal via airways (Fairhall et al. 1945). The ingested dose of molybdenum is predominantly responsible for its uptake. Animal studies show that a single dose of hexavalent molybdenum compounds will be efficiently absorbed in the gastrointestinal tract. Experimental data reveal that from 40% to 90% of ingested metal is absorbed in the gut (Friberg and Lener 1986; Turnlund et al. 1995; Anke 2004). The metal is next transported to the kidneys, liver, and bone; this applies to both short- and long-term exposure. It should be stressed, however, that molybdenum absorption levels differ greatly between monogastric organisms and ruminants (Miller et al. 1972; Nielsen 1996; Frank et al. 2000; Hall 2012). In the former group of animals, molybdenum is absorbed through the walls of the stomach, but also in the other parts of the gastrointestinal tract. In ruminants, on the other hand, absorption takes place via the intestinal walls, whereas the walls of the multi-chambered ruminant stomach create a barrier preventing molybdenum penetration. Molybdenum is absorbed by active transport, which is also used for sulfate intestinal absorption; hence, sulfates present in the lumen of the gastrointestinal tract can significantly reduce the absorption of molybdenum (Mason and Cardin 1977). Sulfates or sulfites present in ruminal digesta lead to formation of di-, tri-, and tetramolybdates, which are strong copper-binding ligands. These capture copper by forming insoluble Cu-thiomolybdate complexes (Price et al. 1987; Gooneratne et al. 1989; Smith and White 1997; Gu et al. 2015; Skibniewski et al. 2016). This type of binding is much stronger than that with metallothionein, which is the main copper-binding tissue protein (Allen and Gawthorne 1987; Frank et al. 2000). The complexes are inhibitory in relation to the activity of copper enzymes resulting from their strong affinity to copper ions (Humphries et al. 1983). Price et al. (1987) observed that about 30% of ruminal fluid molybdenum occurred in the form of di-, tri-, and tetrathiomolybdates combined with the solid phase—thus being undetected in the liquid phase. Tetrathiomolybdates are thought to be bound to the solid phase of the digesta and, in consequence, cannot be absorbed through the walls of the rumen. Probably, large quantities of molybdenum, reaching toxicity levels, may limit the ability of $SO_4$-reducing ruminal bacteria to live in an environment with high sulfur content, which in consequence leads to an increased production of $H_2S$ (Kessler et al. 2012). The main function of molybdenum is to take part in redox reactions in cells, since the element may assume a valency of +4, +5, or +6 in the catalytic reactions of the enzymes it is bound to. Molybdenum-containing enzymes have been identified in all living organisms. In mammals, they are involved in the metabolism of purines, pyrimidines, pteridines, and aldehydes, as well as in the oxidation of sulfites. Most prominent molybdenum enzymes include aldehyde oxidase, sulfite oxidase, and nitrate reductase (Hille et al. 2011). Generally, reactions catalyzed by Mo enzymes consist in oxygen atom transfer from water to or from the substrate as a result of a redox reaction, in which molybdenum changes its valency from IV to VI (Mason 1986; Kisker et al. 1997; Hille 2002; Mendel and Bittner 2006).

Molybdenum toxicity is linked with a wide spectrum of symptoms, of which many arise from molybdenum-caused secondary copper deficiency. Typical symptoms of acute uncomplicated molybdenosis include impaired bone formation, which may be a result of impaired phosphorus metabolism. This leads to bone and joint deformities and spontaneous pathological metaphyseal fractures of the long bones and may cause exostosis. Molybdenosis leads to a decrease in alkaline phosphatase activity and a lower content of proteoglycans in the articular cartilage. Reproductive disorders affecting both sexes may also be the case. Females suffer inhibition of the estrous cycles, while interstitial testicular degeneration may afflict males, probably by affecting sex hormone receptors. Laboratory animals exhibit growth inhibition as well as morphological and functional changes in the liver, kidney, and spleen. Long-term exposure also results in anemia and diarrhea (Tallkvist and Oskarsson 2015).

Molybdenum causes disorders of the copper metabolic functions. This is due to the fact that the elements are antagonistic to each other, which is due to the electron configuration of the atoms in their ionic forms. The effect of copper deficiency caused by excess molybdenum is apparent mainly in ruminants; however, some studies suggest that excessive levels of molybdenum disrupt copper metabolism also in monogastric animals (Halverson et al. 1960; Mills and Davis 1987). In ruminants, the nutritional requirement of the ruminal microbiota is an extremely important issue. The microbial demand for copper, which is lower than that of the tissues of the host organism, is 1.57 μmol $kg^{-1}$ dw of feed. As far as molybdenum is concerned, this value is much higher, from 104.2 to 2.85 mmol $kg^{-1}$ dw of feed (Dziekan et al. 2007). The copper-molybdenum antagonism is based primarily on molybdenum reactions with sulfides formed through bacterial reduction of sulfates in the lumen of the gastrointestinal tract. Another mechanism of copper content reduction involves copper reactions with thiomolybdates. In monogastric animals and in humans, considerable amounts of sulfates may form in the colon. As a consequence, active sulfur compounds are inhibited. Molybdenum is removed from the body with urine and, to a lesser extent, with bile (Vyskocil and Viau 1999; NRC 2006). Molybdenum urinary excretion in laboratory animals is relatively quick, as its total removal in guinea pigs, rats, goats, and pigs takes about 2 weeks. In goats, small amounts of molybdenum, 2% and 0.2% of molybdenum uptake, respectively, have also been observed to be removed with milk and lost hair (Anke et al. 1983).

There is little information on molybdenum toxicity in free-living birds, and the research carried out so far has been focusing on domesticated avian species. In the latter group of birds, molybdenum hepatic concentrations, ranging from 22 to 36 mg $kg^{-1}$ dw, positively correlated with the toxicity symptoms (Puls 1988). Experiments revealed that the level of the element in the avian liver depended not only on molybdenum content in the diet but also on the dietary concentrations of sulfates and copper. The problem of the interaction between these elements in terms of their absorption, excretion, and kinetics belongs to the most important issues of veterinary toxicology, and to date the complex nature of these processes has not been fully explained. The toxicity threshold for avian embryos is 23–33 mg Mo $kg^{-1}$ of egg. Accordingly, in natural conditions molybdenum toxicity to avian embryos is thought to be extremely rare (Eisler 1989).

As is the case of birds, data on molybdenum toxicity in relation to undomesticated mammals have been reported sparsely. The only species well described in terms of this issue is the moose (*Alces alces*). In the mid-1980s, moose population inhabiting the southwestern part of Sweden started showing symptoms of a previously unknown disease. The number of dead or culled animals reached about 150 individuals a year, which represented about 3% of the total regional population (Frank

1998). Investigations revealed that the most possible cause of the "mysterious" moose disease was copper deficiency and molybdenosis (Frank 1998; Frank et al. 2000, 2004a). The molybdenosis hypothesis of the moose illness was later supported by unexpected manifestation of type 2 diabetes mellitus in those of the afflicted animals in which protein glycation occurred due to prolonged hyperglycemia (Frank et al. 2002). The clinical signs of the disease, as well as anatomopathological changes, were similar to those of copper deficiency and molybdenosis in cattle and sheep (Gooneratne et al. 1998; Frank 1998; Frank et al. 2000). The symptoms included diarrhea, anorexia, emaciation, achromotrichia, alopecia, apathy, abnormal behavior, and motor dysfunction. The anatomopathological changes included mucosal edema, hyperemia, and petechiae of the mucosa in the gastrointestinal tract, spleen and liver hemosiderosis, dilation of the heart, lung alveolar emphysema, neuropathies, as well as uni- or bilateral corneal opacity (Frank et al. 2000, 2004a, b; Frank 1998). Reports on molybdenum-exposed mule deer reveal that the species has a similar tolerance to molybdenum as pigs and horses. Animals fed a dose of 2500 mg Mo $kg^{-1}$ body weight for 25 days exhibited only reduced feed intake and moderate diarrhea. At a dose of 5000–7000 mg Mo $kg^{-1}$ body weight administered for 3–15 days, the symptoms aggravated; however, relief of symptoms was observed nearly immediately on the restoration of the normal diet (Chappell et al. 1979; Eisler 1989).

# 8 Bioaccumulation of Molybdenum

Molybdenum is a constituent of numerous enzymes found in all living organisms. This gets into the body as a molybdate anion; however, its biochemical marker has not been identified yet. In order to attain biological activity, molybdenum must be complexed by a pterin compound, this way forming molybdenum cofactor (Mendel and Bittner 2006). Its structure is unique and probably results from the necessity to control and sustain the oxidation-reduction properties of molybdenum. The role of this cofactor consists in maintaining the catalyst, i.e., molybdenum, in the active center, which in consequence enables electron exchange between the pterin ring and molybdenum atom (Kisker et al. 1997).

In both humans and animals, the tissue content of molybdenum and the processes of its removal mostly depend on the dose ingested orally; however, this level does not fully reflect molybdenum status in the body due to numerous interactions with other metals and substances present in the tissue (Kisker et al. 1997; Mendel and Bittner 2006).

Molybdenum is distributed to all the tissues of the living system, although its highest concentrations occur in the liver, kidney, and bone. In the light of its role and importance for metabolic processes, it must be noted that the quantities of this essential element present in the system are relatively low. After absorption, molybdenum is transported with erythrocytes and in the form of molybdate ions (Allway et al. 1968; Versieck et al. 1981). In the red blood cells, molybdenum binds to their membrane proteins, particularly with spectrin, whereas Mo present in the plasma binds with $\alpha_2$-macroglobulin (Bibr et al. 1983). Either taken in or produced on an ongoing basis, thiomolybdates may bind copper, forming copper-thiomolybdate complexes which—remaining in the bloodstream—cannot be used by the cells of the living system (Hall 2012). It has been generally accepted that molybdenum is accumulated in parenchymal organs, primarily in the kidney and liver (Rousseaux et al. 2002).

Studies on molybdenum distribution in the tissues of laboratory animals show that—within 1–24 h following a single intravenous injection of Mo-99 isotope—the highest levels occurred in the kidney, liver, and pancreas. With a lapse of time, molybdenum concentrations decreased in the kidneys and pancreas, whereas the hepatic level of the element did not change significantly over the entire period of the experiment (Rosoff and Spencer 1973). A similar pattern of molybdenum tissue distribution was found in cattle and goats subjected to long-term exposure, with the highest values observed in the kidneys (Anke et al. 2007; Tallkvist and Oskarsson 2015). An experiment on rats that were orally or subcutaneously administered a dose of 25–40 micrograms molybdenum per kilogram of body weight revealed that within 14 days Mo concentration reached a high level in the kidneys, liver, and skin. The levels in these tissues of the rats depended on the dose (Rosoff and Spencer 1973). Pott et al. (1999) report that two compartments may be present in the liver, of which one is responsible for free molybdenum circulation, whereas the other is characteristic for its rapid distribution. Molybdenum also binds to dermal collagen. It is also accumulated in the bone tissue and hyaline cartilage; hence it may be found in the cancellous bone of epiphyses and the epiphyseal plates. The metal may also penetrate dentition, the highest concentrations being found in the dentin and much lower in the enamels (Lener 1978). Pott et al. (1999) claim that plasma, muscle, and kidney levels are the best measure of molybdenum bioavailability.

## 9 Ecological Effects of Molybdenum

Molybdenum is a relatively rare element, finding its application mostly in smelting steel used in many industries and medicine. Anthropogenic activities causing molybdenum contamination of the environment involve mainly fossil fuel combustion, mining, and metallurgical industry. In a general outline, molybdenum chemistry is complex and not fully understood (Eisler 1989). The metal was discovered about 200 years ago and found a wide application in metallurgy, superseding the traditionally used toxic metals, such as hexavalent chromium (Heijerick et al. 2012b; Shields 2013; Wang et al. 2016). In nature, molybdenum does not occur in its pure form, but in compounds with sulfur, oxygen, tungsten, lead, uranium, iron, magnesium, cobalt, vanadium, bismuth, or calcium (Eisler 1989).

If molybdenum concentration in fodder plants exceeds 5 or 10 mg $kg^{-1}$ dw, it may cause molybdenosis in ruminants. This may be the case in plants growing on undrained soils rich in organic components (He et al. 2005). In terms of environmental toxicology, the most important molybdenum form is molybdate anion, $MoO_4^{2-}$, which is of particular significance to animals inhabiting aquatic ecosystems, including freshwater systems. It was found to form under similar environmental conditions from various molybdenum-containing substances (De Schamphelaere et al. 2010; Heijerick et al. 2012a, b). Currently, the most common molybdenum-containing chemicals include ammonium molybdate, potassium molybdate, sodium molybdate, molybdenum disulfide, molybdic acid, and molybdenum trioxide, the latter being produced on the largest scale (Wang et al. 2016). Although environmental levels of molybdenum are relatively low, human impact may lead to a considerable increase, which in aquatic ecosystems may reach toxic concentrations. According to Shan et al. (2012), Mo concentrations in industrial wastes range from 4 to 145 mg $L^{-1}$. It must be stressed, however, that various forms of molybdenum have a different toxicity. In rainbow trout, the median lethal concentration ($LC_{50}$) for sodium molybdate was found to be 800 mg $L^{-1}$, whereas for ammonium molybdate 420 mg $L^{-1}$ (Sigma-Aldrich 2015a, b, c). Similar results were reported by Wang et al. (2016), who found that acute toxicity of various molybdenum compounds against *Daphnia magna* after 48-h exposure increased in the following order: sodium molybdate, molybdenum trioxide, and ammonium molybdate. Bioconcentration of molybdenum on each level of the trophic chain, however, was found to be low. In aquatic organisms living in uncontaminated waters, average molybdenum concentration remains at a level below 1 mg $kg^{-1}$ dw, whereas in strongly polluted areas, it does not exceed 10 mg $kg^{-1}$ dw (Ikemoto et al. 2008; Regoli et al. 2012). Considering fish inhabiting polluted waters, the highest levels of molybdenum are measured in the liver, gills, and kidney. Contrary to these organs, Mo levels in the muscle tissue are much lower (Regoli et al. 2012). Given the research on various species inhabiting terrestrial ecosystems, molybdenum is not a metal of significant importance for environmental toxicology; hence, no legal regulations have been established so far on molybdenum dietary exposure for fish and terrestrial wild animals. Molybdenum has not been classified as an element of major toxicity by the EPA (US EPA 1992; US Department of Interior 1998).

## 10 Bioindicators and Biomarkers of Molybdenum in Ecotoxicological Studies

Environmental molybdenum contamination may be an outcome of using fertilizers containing this element but also results from mining and metallurgical industry emissions (Buekers et al. 2010). It may accumulate in animal tissues and is removed primarily with feces and urine. Herbivorous animals, both small (like the muskrat) and big (e.g., the moose), are more susceptible to increased Mo levels in the tissues than carnivorous and omnivorous animals, since the element accumulates in aquatic and terrestrial plants, reaching concentrations exceeding 1000 mg $kg^{-1}$ dw (Fitzgerald et al. 2007). Mochizuki et al. (2002) report that—depending on the way ducks feed (Table 8.6)—their molybdenum hepatic levels increase in the following order: carnivorous > omnivorous > herbivorous ones. Also, the authors point out that molybdenum hepatic and renal levels are higher in dabbling ducks (including spot-billed duck (*Anas poecilorhyncha*), pintail (*A. acuta*), wigeon (*A. penelope*)) compared to diving duck species such as scaup (*Aythya marila*) and tufted duck (*A. fuligula*) (Table 8.6).

Both in birds and mammals, molybdenum is mainly taken in with food and water, absorbed in the gastrointestinal tract. The highest Mo contents are usually found in the liver, kidney, and spleen (Table 8.6). Such all species of ducks, as well as muskrat and moose, may accurately reflect their natural habitat status in terms of the levels of molybdenum, which accumulates in plants, is present in the water column, and occurs in the benthic deposits, where these animals forage for food (Mochizuki et al. 2002; Frank 1998; Custer et al. 2004). Diet difference and its influence on Mo bioaccumulation in small terrestrial animals were also observed. Anke et al. (2007) investigated carcasses of two groups of micromammals (rodents and shrews) from habitats on Triassic (Muschelkalk) soils in Germany. In seven species rodent group (mainly herbivorous), Mo concentrations in their carcasses were similar and mean values ranged from 0.35 to 0.65 mg $kg^{-1}$ dw. However, insectivorous common shrew (*Sorex araneus*) and pigmy shrew (*Sorex minutus*) accumulate significantly more Mo in their body (2.40 and 1.40 mg $kg^{-1}$ dw, respectively) than rodents.

Elevated levels of Mo in ecosystems may result from the presence of large industrial objects or intensive agricultural production in the area but may also be a consequence of natural changes, as is the case of copper deficiency in the soils of northern Saskatchewan and Manitoba and parts of Alaska (North America). Frank (1998) demonstrated that in the livers of moose living in the region affected by acid rains (southwest Sweden), the content of molybdenum increased by 40% (Table 8.6).

Analyses presented by Anke et al. (2000) reveal that molybdenum contents in the liver and kidneys of red deer differ significantly ($p \leq 0.001$; Table 8.6) depending on the region (East Germany—Erzgebirge, Germany). Wild animals are susceptible to molybdenum tissue accumulations due to high concentrations of molybdate, copper deficiencies, and excess dietary sulfur, which is crucial in the case of ruminants. Mule deer and Alaskan moose tolerate high doses of molybdenum, compared with cattle. Mouflons (*Ovis musimon*) respond with immediate diarrhea if a Mo level in alfalfa reaches 300 mg $kg^{-1}$ dw (Anke 1986).

In Central Europe, plants growing on soils formed from granite, rotliegend, or shale, as well as on peat and muck soils, may cause Mo contamination in free-living ruminants (Anke et al. 2000).

**Table 8.6** Molybdenum in soft tissues of birds and mammals in mg $kg^{-1}$

| Species | Place and years | *n* dw or ww | Liver | Kidney | Muscle | Brain | Lung | References |
|---|---|---|---|---|---|---|---|---|
| Birds | | | | | | | | |
| *Ducks* | | | | | | | | |
| Spot-billed duck *Anas poecilorhyncha* | Japan, 1993–1995 | 19 dw | 9.27 ± 3.59 0.02–53.84 | 5.76 ± 1.78 0.02–32.34 | | | | Mochizuki et al. (2002) |
| Pintail *Anas acuta* | | 11 dw | 3.30 ± 0.56 0.04–6.97 | 9.17 ± 3.21 0.06–33.03 | | | | Mochizuki et al. (2002) |
| Wigeon *Anas penelope* | | 15 dw | 13.04 ± 5.80 0.08–91.78 | 8.58 ± 2.46 0.07–35.47 | | | | Mochizuki et al. (2002) |
| Scaup *Aythya marila* | | 6 dw | 3.35 ± 1.52 0.66–10.75 | 3.11 ± 1.61 0.70–10.81 | | | | Mochizuki et al. (2002) |
| Tufted duck *Aythya fuligula* | | 6 dw | 1.76 ± 0.98 0.02–6.32 | 2.63 ± 0.65 1.00–5.67 | | | | Mochizuki et al. (2002) |
| *Other aquatic birds* | | | | | | | | |
| Grey heron *Aldea cinerea* | Japan, Kanto, 2001–2002 | 31 dw | 3.18 ± 0.72 2.04–5.15 | 2.64 ± 0.58 1.82–4.69 | 0.06 ± 0.03 0.02–0.16 | *n* = 10 0.11 ± 0.13 0.03–0.47 | 0.19 ± 0.13 0.07–0.80 | Horai et al. (2007) |
| Great white egret *Egretta alba* | | 22 dw | 1.55 ± 0.27 1.03–2.09 | 2.17 ± 0.389 1.28–2.84 | 0.05 ± 0.02 0.02–0.09 | *n* = 18 0.15 ± 0.27 0.024–1.20 | 0.20 ± 0.16 0.07–0.76 | Horai et al. (2007) |
| Intermediate egret *Egretta intermedia* | | 17 dw | 1.69 ± 0.20 1.35–2.10 | 2.63 ± 0.436 1.94–3.52 | 0.08 ± 0.02 0.04–0.13 | 0.16 ± 0.19 0.05–0.69 | 0.31 ± 0.35 0.08–1.57 | Horai et al. (2007) |

(continued)

**Table 8.6** (continued)

| Species | Place and years | $n$<br>dw or ww | Liver | Kidney | Muscle | Brain | Lung | References |
|---|---|---|---|---|---|---|---|---|
| Great cormorant<br>*Phalacrocorax carbo* | | 4<br>dw | 1.69 ± 0.38<br>1.27–2.17 | 1.57 ± 0.207<br>1.37–1.86 | 0.05 ± 0.02<br>0.03–0.07 | $n = 1$<br>0.18 | 0.10 ± 0.04<br>0.06–0.16 | Horai et al. (2007) |
| Moorhen<br>*Gallinula chloropus* | | 3<br>dw | 5.87 ± 6.09<br>1.79–12.9 | 1.80 ± 0.154<br>1.64–1.94 | 0.11 ± 0.04<br>0.08–0.16 | 0.11 ± 0.01<br>0.10–0.13 | 0.30 ± 0.01<br>0.29–0.31 | Horai et al. (2007) |
| *Predatory birds* | | | | | | | | |
| Northern gos-hawk<br>*Accipiter gentiles* | | 12<br>dw | 2.13 ± 0.64<br>1.01–3.53 | 2.95 ± 1.03<br>1.75–5.10 | 0.09 ± 0.04<br>0.04–0.21 | | 0.12 ± 0.07<br>0.06–0.30 | Horai et al. (2007) |
| Japanese sparrowhawk<br>*Accipiter gularis* | | 5<br>dw | 1.64 ± 0.46<br>1.18–2.22 | 1.88 ± 0.41<br>1.47–2.47 | 0.08 ± 0.03<br>0.049–0.12 | | 0.12 ± 0.05<br>0.07–0.20 | Horai et al. (2007) |
| Black kite<br>*Milvus migrans* | | 14<br>dw | 1.61 ± 0.48<br>0.85–2.53 | 1.83 ± 0.689<br>0.570–2.83 | 0.07 ± 0.02<br>0.04–0.11 | $n = 8$<br>0.10 ± 0.04<br>0.06–0.19 | 0.18 ± 0.14<br>0.01–0.45 | Horai et al. (2007) |
| Brown hawk-owl<br>*Ninox scutulata* | | 7<br>dw | 1.19 ± 0.31<br>0.76–1.66 | 1.61 ± 0.40<br>1.12–2.23 | 0.09 ± 0.04<br>0.05–0.19 | | 0.13 ± 0.03<br>0.08–0.17 | Horai et al. (2007) |
| Ural owl<br>*Strix uralensis* | | 12<br>dw | 1.20 ± 0.42<br>0.38–1.84 | 1.41 ± 0.985<br>0.190–3.10 | 0.07 ± 0.02<br>0.04–0.12 | | 0.21 ± 0.11<br>0.07–0.39 | Horai et al. (2007) |
| *Others* | | | | | | | | |
| Feral pigeon<br>*Columba livia* | | 11<br>dw | 6.30 ± 1.98<br>3.41–8.94 | 8.54 ± 1.92<br>4.5–11.1 | 0.12 ± 0.06<br>0.07–0.26 | $n = 10$<br>0.48 ± 0.23<br>0.26–0.98 | 0.33 ± 0.27<br>0.12–1.00 | Horai et al. (2007) |

| | | | | | | | | |
|---|---|---|---|---|---|---|---|---|
| Jungle crow<br>*Corvus macrorhynchos* | | 14<br>dw | 2.17 ± 0.52<br>1.43–3.03 | 1.33 ± 0.27<br>0.93–1.98 | *n* = 12<br>0.13 ± 0.04<br>0.08–0.21 | *n* = 8<br>0.64 ± 0.25<br>0.425–1.20 | 0.39 ± 0.32<br>0.14–1.39 | Horai et al. (2007) |
| Carrion crow<br>*Corvus corone* | | 5<br>dw | 2.32 ± 0.53<br>1.52–2.95 | 1.52 ± 0.54<br>0.986–2.21 | *n* = 4<br>0.03 ± 0.03<br><DL-0.08 | *n* = 4<br>0.13 ± 0.08<br>0.07–0.25 | 0.27 ± 0.14<br>0.15–0.51 | Horai et al. (2007) |
| Mammals | | | | | | | | |
| *Ungulates* | | | | | | | | |
| Red deer<br>*Cervus elaphus* | Poland, 2008–2009 | 27<br>ww | | | 0.05 median<br>0.04–0.07 | | | Skibniewski et al. (2015) |
| Red deer<br>*Cervus elaphus* | Germany, E part | 35<br>22<br>dw | 1.2 ± 0.62<br>2.1 ± 0.80 | 0.85 ± 0.46<br>1.7 ± 0.76 | | | | Anke et al. (2000) |
| Elk<br>*Alces alces* | Poland, NE part, 2010 | 35<br>ww | 0.92 ± 0.46 | 0.42 ± 0.62 | 0.05 ± 0.03 | | | Skibniewski et al. (2016) |
| Moose<br>*Alces alces* | Sweden, SW part, 1982 reference area | ww | | | | | | Frank (1998) |
| | | 124 | 0.80–0.90<br>0.24–0.27 | | | | | |
| | 1988 | 26 | 1.20–1.30<br>0.36–0.39 | | | | | |
| | 1992 | 84 | 1.0–1.1<br>0.30–0.33 | | | | | |
| | 1996 | 67 | 1.10–1.20<br>0.33–0.36 | | | | | |

(continued)

**Table 8.6** (continued)

| Species | Place and years | $n$ dw or ww | Liver | Kidney | Muscle | Brain | Lung | References |
|---|---|---|---|---|---|---|---|---|
| Wild boar *Sus scrofa* | Germany | dw | $n = 95$ 3.38 ± 1.40 | $n = 10$ 2.08 ± 1.10 | | | | Anke et al. (2007) |
| Fallow deer *Dama dama* | Germany | dw | $n = 99$ 1.95 ± 1.17 | $n = 100$ 1.62 ± 0.92 | | | | Anke et al. (2007) |
| Red deer *Cervus elaphus* | Germany | dw | $n = 35$ 1.19 ± 0.62 | $n = 25$ 0.85 ± 0.46 | | | | Anke et al. (2007) |
| Roe deer *Capreolus capreolus* | Germany | dw | $n = 16$ 0.62 ± 0.37 | $n = 3$ 0.21 ± 0.14 | | | | Anke et al. (2007) |
| Horse *Equus ferus caballus* | Germany | dw | $n = 152$ 6.30 ± 5.80 | $n = 133$ 1.44 ± 0.71 | | | | Anke et al. (2007) |
| Porcupine caribou *Rangifer tarandus granti* | Canada, Yukon, 1996 | 25 dw | | 2.07 ± 0.53 | | | | Gamberg (1997) |
| Moose *Alces alces* | Canada, Yukon, 1996 | 63 dw | | 1.39 ± 0.62 | | | | Gamberg (1997) |
| Moose *Alces alces* | USA, Minnesota, 1998–1999, bog and forest agricultural areas | dw | | | | | | Custer et al. (2004) |
| | | 32 | 2.3 | | | | | |
| | | 47 | 3.4 | | | | | |
| White-tailed deer *Odocoileus virginianus* | USA, Texas, 2005 | 20 dw | 20% of samples <2 | | 50% of samples <2 | | | Bruckwicki et al. (2006) |

| | | | | | | | | |
|---|---|---|---|---|---|---|---|---|
| *Carnivores* | | | | | | | | |
| *Lutra lutra*<br>Eurasian otter | England and Wales,<br>2007–2008 | dw | | | | | | Walker et al.<br>(2010) |
| | | 48 | 1.24 | | | | | |
| | | | 1.04–1.43 | | | | | |
| | | 57 | 1.2 | | | | | |
| | | | 0.95–1.33 | | | | | |
| American mink<br>*Neovison vison* | Poland, W part,<br>2009–2011 | dw<br>30 | 1.16 ± 1.01 | 0.87 ± 1.461 | | | | Brzezinski et al.<br>(2014) |

*dw* dry weight or *ww* wet weight, *DL* detection limit

## 11 Conclusion

No symptoms of primary molybdenum deficiency have ever been observed in humans or animals. The importance of molybdenum consists in its toxicity. Molybdenosis, or molybdenum toxicity, primarily affects ruminants, mainly cattle. Monogastric animals are less susceptible to Mo toxic effects. No biomagnification and bioaccumulation of Mo in animals have been confirmed so far. Anthropogenic environmental Mo contamination is linked with Mo ore mining, production of steels for special purposes, and industrial waste discharge. Excess molybdenum, especially in cattle, is a commonplace phenomenon in the world. Data on the effects of molybdenum in free-living mammals are sparse. Well-documented sources deal only with a cervid inhabiting Scandinavia. Moose in Sweden suffer from a disease most probably caused by copper deficiency and molybdenosis. The symptoms include diarrhea, anorexia, emaciation, achromotrichia, alopecia, apathy, as well as behavioral and motor disorders. Necropsy findings included mucosal edema, atrophied lymphoid tissues of the mucous membranes of the alimentary tract, splenic and hepatic hemosiderosis, hypertrophic cardiomyopathy, alveolar emphysema, neuropathy, and uni- or bilateral corneal opacity. Hepatic molybdenum concentrations have reportedly increased by 40% in the moose inhabiting areas affected by acid rain.

There is very little information on molybdenum toxicity in free-living birds. Research was carried out on bobwhite quail (*Colinus virginianus*) treated with molybdenum disulfide (insoluble) and soluble sodium molybdate dihydrate. Toxicity and bioavailability of both forms of Mo were compared. For the soluble form, the NOAEC (no-observed-adverse-effect concentration) was 1200 mg Mo $kg^{-1}$ feed (134 mg $kg^{-1}$ body weight per day). For $MoS_2$, the NOAEC was found to be 5000 mg (545 mg $kg^{-1}$ body weight per day). These results reflect how avian species are exposed to environmental Mo. Molybdenum does not belong to metals of special significance in relation to environmental toxicology; hence no dietary exposure regulations for fish and free-living terrestrial animals have been established. The metal has not been classified by the EPA as an element of major toxicity. In Central Europe, plants growing on the soils formed from granite, gneiss, rotliegend, and shales, as well as on muck and peat soils, may cause Mo contamination in free-living ruminants.

## References

Adriano DC (1986) Trace elements in the terrestrial environment. Springer, New York, p 533

Allen JD, Gawthorne JM (1987) Effect of molybdenum treatments on the distribution of Cu and metallothionein in tissue extracts from rats and sheep. J Inorg Biochem 31:161–170

Allway WH, Kubota J, Losee F, Roth M (1968) Selenium, molybdenum, and vanadium in human blood. Arch Environ Health 16:342–349

Anbar AD (2004) Molybdenum stable isotopes: observations, interpretations and directions. Rev Mineral Geochem 55:429–454
Anke M (1986) Schädliche Wirkungen lebensnodwendiger anorganischer Futterbestandteile. Jahrestagung, Wissenschaftliche Gesellschaft für Veterinärmedizin der DDR, Karl-Marx-Stadt, pp 45–66
Anke M (2004) Molybdenum. In: Merian E et al (eds) Elements and their compounds in the environment, 2nd edn. Wiley-VCH, Weinheim, pp 1007–1037
Anke M, Groppel B (1985) Possibilities to reduce the toxic effect of sulphur, cadmium, molybdenum, zinc and nickel in useful animals. In: Legutko S (ed) Protection of the natural environment and development of animal production, Kraków, pp 41–63
Anke M, Risch M (1989) Importance of molybdenum in animal and man. In: Anke M et al (eds) 6th international trace element symposium, Molybdenum, vanadium, vol 1. University of Leipzig and Jena, Germany, pp 303–321
Anke M, Grün M, Partschefeld M, Groppel B (1978) Molybdenum deficiency in ruminants. In: Kirchgessner M (ed) Trace element metabolism in man and animals (TEMA–3). Technical University Munich, Freising-Weihenstephan, Germany, pp 230–232
Anke M, Groppel B, Kronemann H, Grün M, Szentmihalyi S (1983) Die Versorgung von Pflanze, Tier und Mensch mit Molybdän. In: Anke M et al (eds) Mengen- und Spurenelemente. University of Leipzig, Arbeitstagung, pp 22–36
Anke M, Groppel B, Grün M (1985a) Essentiality, toxicity, requirement and supply of molybdenum in human and animals. In: Mills CF, Bremner I, Chester JK (eds) Trace element in man and animals. Commonwealth Agricultural Bureaux, Farnham Royal, Slough, pp 154–157
Anke M, Groppel B, Kronemann H, Grün M (1985b) Molybdenum supply and status in animals and human beings. Nutr Res (Suppl I):180–186
Anke M, Arnhold W, Müller R, Schäfer U, Dorn W, Gunstheimer G et al (2000) Der Kupfer- und Molybdänstatus des ostdeutschen Rotwildes im vergleich zu anderen Wildwiederkäuerarten. Beitrage zur Jagd-und Wildforschung 25:77–87
Anke M, Seifert M, Holzinger S, Müller R, Schäfer U (2007) The biological and toxicological importance of molybdenum in the environment and in the nutrition of plants, animals and man. Part 2. Molybdenum in animals and man. Acta Biol Hung 58:325–333
Arnold GL, Anbar AD, Barling J, Lyons TW (2004) Molybdenum isotope evidence for widespread anoxia in Mid-Proterozoic oceans. Science 304:87–90
Barth N, Palchen W, Rank G, Heilmann H (1996) Bodenatlas des Freistaates Sachsen. Teil 1. Freistaat Sachsen. Landesamt fur Umwelt und Geologie, Dresden
Bhattacharyya S, Donahoe RJ, Patel D (2009) Experimental study of chemical treatment of coal fly ash to reduce the mobility of priority trace elements. Fuel 88:1173–1184
Bibr B, Marik T, Ksetikova M, Lener J (1983) Binding of Mo-99 to red-cell membrane-proteins is not affected by reducing agent. Br J Haematol 53:172–173
Blood DC, Radostits OM (1989) Veterinary medicine: a textbook of the diseases of cattle, sheep, pigs, goats and horses, 7th edn. Bailliere-Tindall, London
Bompart G, Pécher C, Prévot D, Girolami JP (1990) Mild renal failure induced by subchronic exposure to molybdenum: urinary kallikrein excretion as a marker for distal tubular effect. Toxicol Lett 52:293–300
Bruckwicki P, Giggleman C, Lewis J (2006) An investigation of contaminant levels in white-tailed deer (*Odocoileus virginianus*) collected from Caddo Lake National Wildlife Refuge, Harrison County, Texas 2005 Project ID Nos: DEC No 200520002; FFS No 2N53
Brzeziński M, Zalewski A, Niemczynowicz A, Jarzyna I, Suska-Malawska M (2014) The use of chemical markers for the identification of farm escapees in feral mink populations. Ecotoxicology 23:767–778
Budaveri S (1996) The Merck index: an encyclopedia of chemicals, drugs, and biological, 12th edn. Merck & Co, Whitehouse Station, New Jersey

Buekers J, Mertens J, Smolders E (2010) Toxicity of the molybdate anion in soil is partially explained by effects of the accompanying cation or by soil pH. Environ Toxicol Chem 29:1274–1278

Burris RH, Roberts GP (1993) Biological nitrogen fixation. Annu Rev Nutr 13:317–335

Chan PC, Herbert RA, Roycroft JH, Haseman JK, Grumbein SL, Miller RA et al (1998) Lung tumor induction by inhalation exposure to molybdenum trioxide in rats and mice. Toxicol Sci 45:58–65

Chappell WR, Meglen RR, Moure-Eraso R, Solomons CC, Tsongas TA, Walravens PA (1979) Human health effects of molybdenum in drinking water. Cincinnati, OH, United States Environmental Protection Agency (EPA-600A-79-006)

Cohen HJ, Fridovich I, Rajagopalan KV (1971) Hepatic sulfite oxidase: a functional role for molybdenum. J Biol Chem 246:374–382

Combs GF Jr (2005) Geological impacts on nutrition. In: Selinus O et al (eds) Essentials of medical geology, impacts of the natural environment on public health. Elsevier, Amsterdam, pp 162–177

Considine GD (ed) (2005) Molybdenum. Van Nostrand's encyclopedia of chemistry. Wiley-Interscience, New York, pp 1038–1040

Custer TW, Cox E, Gray B (2004) Trace elements in moose (*Alces alces*) found dead in north western Minnesota, USA. Sci Total Environ 330:81–87

Davis GK, Mertz W (1987) Copper. In: Mertz W (ed) Trace elements in human and animal nutrition, vol I. Academic Press, San Diego, pp 301–364

De Schamphelaere KAC, Stubblefield W, Rodriguez P, Vleminckx K, Janssen CR (2010) The chronic toxicity of molybdate to freshwater organisms. I. Generating reliable effects data. Sci Total Environ 408:5362–5371

De Vos W, Tarvainen T (2006) Geochemical atlas of Europe. Part 2, Geological Survey of Finland, Espoo

Dziekan P, Kleczkowski M, Kluciński W, Jakubowski T, Dembele K, Sikora J (2007) Influence of fodder irrigated with utility refuses from potato industry on copper and malondialdehyde concentration in the blood of cows. Med Weter 63:1111–1114

EFSA (2006) European food safety authority. Tolerable upper intake levels for vitamins and minerals, Scientific Committee and Food. Scientific Panel on Dietetic Products. Nutrition and allergies, pp 77–86 [http://www.efsa.eu.int]

Eisler R (1989) Molybdenum hazards to fish, wildlife, and invertebrates: a synoptic review. US Fish and Wildlife Service, Biol Rep 85 (1.19)

Emsley J (2001) Nature's building blocks. Oxford University Press, Oxford, pp 262–266

Enzmann RD (1972) Molybdenum: element and geochemistry. In: The encyclopedia of geochemistry and environmental sciences. Van Nostrand Reinhold Co, New York

Eurlik J, Šefeik P (1999) Geochemical atlas of the Slovak Republic: soils. Ministry of the Environment of the Slovak Republic

Fairhall LT, Dunn RC, Sharpless NE, Pritchard EA (1945) The toxicity of molybdenum. US Public Health Bull 293:36

Feldman RI, Weiner H (1972) Horse liver aldehyde dehydrogenase. I. Purification and characterizations. J Biol Chem 247:260–266

Fitzgerald D, Nicholson R, Regoli L (2007) Environmental management criteria for selenium and molybdenum: a review relevant to the mining industry. MEND REPORT 10.1.1. The Mining Association of Canada (MAC). http://mend-nedem.org/wp-content/uploads/ 2013/01/10.1.1.pdf

Frank A (1998) 'Mysterious' moose disease in Sweden. Similarities to copper deficiency and/or molybdenosis in cattle and sheep. Biochemical background of clinical signs and organ lesions. Sci Total Environ 209:17–26

Frank A (2003) Molybdenosis leading to type 2 diabetes mellitus in Swedish moose. In: Skinner HCW, Berger AR (eds) Geology and health-closing the gap. Oxford University Press, Oxford, pp 79–81

Frank A, Galagan V (1997) The moose (*Alces alces* L.), a fast and sensitive monitor of environmental changes. In: Subramanian KS, Iyengar GV (eds) Environmental biomonitoring. ACS Symposium Series 654:57–64

Frank A, Galagan V, Petersson LR (1994) Secondary copper deficiency, chromium deficiency and trace element imbalance in the moose (*Alces alces* L.): effect of anthropogenic activity. Ambio 23:315–317

Frank A, Danielsson R, Jones B (2000) The 'mysterious' disease in Swedish moose. Concentrations of trace elements in liver and kidneys and clinical chemistry. Comparison with experimental molybdenosis and copper deficiency in the goat. Sci Total Environ 249:107–122

Frank A, Wibom R, Danielsson R (2002) Myocardial cytochrome c oxidase activity in Swedish moose (*Alces alces* L.) affected by molibdenosis. Sci Total Environ 290:121–129

Frank A, Danielsson R, Selinus O (2004a) Comparison of two monitoring systems for Cu and Mo in the Swedish environment. Sci Total Environ 330:131–143

Frank A, Mc Partlin J, Danielsson R (2004b) Nova Scotia moose mystery – a moose sickness related to cobalt-end vitamin $B_{12}$ deficiency. Sci Total Environ 318:89–100

Friberg L, Lener J (1986) Molybdenum. In: Friberg L et al (eds) Handbook on the toxicology of metals, 2nd edn. Elsevier, North-Holland Biomedical Press, New York, pp 446–461

Friberg L, Boston P, Nordberg G, Piscator M, Robert KH (1975) Molybdenum a toxicological appraisal. US Environmental Protection Agency Report 600/1–75-004, p 142

Gaillardet J, Viers J, Dupré B (2004) Trace elements in river waters. Treat Geochem 5:225–227

Gamberg M (1997) Contaminants in Yukon Moose and Caribou – 1996. Report for Yukon Contaminants Committee and Department of Indian and Northern Affairs Northern Contaminants Program Whitehorse, Yukon. http://northerncontaminants.ca/old-ncp-site/done/reports/1g-contamMoosCarReports/reports/1996%20Report.pdf

Gooneratne SR, Buckley WT, Christensen DA (1989) Review of copper deficiency and metabolism in ruminants. Can J Anim Sci 69:819–845

Gooneratne SR, Chaplin RK, Trent AM, Christensen DA (1998) Effect of tetrathiomolybdate administration on the excretion of copper, zinc, iron and molybdenum in sheep bile. Br Vet J 145:62–72

Gregorauskiene V, Kadûnas V (2000) Chemical composition of soil land lake sediments – an indicator of geological processes in Lithuania. Geol Q 44:347–354

Gu X, Chen R, Hu G, Zhuang Y, Luo J, Zhang C et al (2015) Cell apoptosis of caprine spleen induced by toxicity of cadmium with different levels of molybdenum. Environ Toxicol Pharmacol 40:49–56

Hall JO (2012) Molybdenum. In: Gupta RC (ed) Veterinary toxicology. Elsevier, Amsterdam

Halverson AW, Phifer JH, Monty KJ (1960) A mechanism for the copper-molybdenum interrelationship. J Nutr 71:95–100

He ZL, Zhou QX, Xie ZM (1998) Soil-chemical balances of pollution and beneficial elements. China Environmental Science Press, Beijing, pp 1–34

He ZL, Yang XE, Stoffellab PJ (2005) Trace elements in agro ecosystems and impacts on the environment. J Trace Elem Med Biol 19:125–140

Heijerick DG, Regoli L, Carey S (2012a) The toxicity of molybdate to freshwater and marine organisms. II. Effects assessment of molybdate in the aquatic environment under REACH. Sci Total Environ 435–436:179–187

Heijerick DG, Regoli L, Stubblefield W (2012b) The chronic toxicity of molybdate to marine organisms. I. Generating reliable effects data. Sci Total Environ 430:260–269

Hille R (2002) Molybdenum and tungsten in biology. Trends Biochem Sci 27:360–367

Hille R, Nishinoa T, Bittner F (2011) Molybdenum enzymes in higher organisms. Coord Chem Rev 255:1179–1205

Holzinger S, Anke M, Röhrig B, Gonzalez D (1998) Molybdenum intake of adults in Germany and Mexico. Analyst 123:447–450

Horai S, Watanabe I, Takada H, Iwamizu Y, Hayashi T, Tanabe S et al (2007) Trace element accumulations in 13 avian species collected from the Kanto area, Japan. Sci Total Environ 373:512–525
Hornick SB, Baker DE, Guss SB (1977) Molybdenum in the environment, 2nd edn. Marcel Dekker, New York
Humphries WR, Phillippo M, Young BW, Bremer I (1983) The influence of dietary iron and molybdenum on copper metabolism in calves. Br J Nutr 49:77–86
Ikemoto T, Tu NPC, Okuda N, Iwata A, Omori K, Tanabe S et al (2008) Biomagnification of trace elements in the aquatic food Web in the Mekong Delta, South Vietnam, Rusing stable carbon and nitro gen isotope analysis. Arch Environ Contam Toxicol 54:504–515
IMOA (2014) Industrial and Environmental Exposure of Humans – www.imoa.info/HSE/environ mental-data/human-health/industrial-exposure.php [Access of 10 Oct 2016]
Ivanova NN (1973) Molybdenum in plant nutrition. Agrochimica 17:96
Johnson JL (1997) Molybdenum. In: O'Dell BL, Sunde RA (eds) Handbook of nutritionally essential mineral elements. Marcel Dekker, New York, pp 413–438
Jones B (2005) Animals and medical geology. In: Selinus O, Alloway BJ, Centeno JA, Finkelman LB, Fuge R, Lindh U et al (eds) Essentials of medical geology. Elsevier, Amsterdam, pp 513–526
Kabata-Pendias A (2011) Trace elements in soils and plants, 4th edn. Taylor and Francis Group, Boca Raton, pp 190–198
Kabata-Pendias A, Mukherjee AB (2007) Trace elements from soil to human. Springer, Berlin
Kadûnas V, Budavièius R, Gregorauskiené V, Katinas V, Kliaugiene E, Radzevièius A et al (1999) Geochemical atlas of Lithuania. Geological Survey of Lithuania, Geology Institute, Vilnius
Kessler KL, Olson KC, Wright CL, Austin KJ, Johnson PS, Cammack KM (2012) Effects of supplemental molybdenum on animal performance, liver copper concentrations, ruminal hydrogen sulfide concentrations, and the appearance of sulfur and molybdenum toxicity in steers receiving fiber-based diets. J Anim Sci 90:5005–5012
Kim J, Woo D, Rees DC (1993) X-ray crystal structure of the nitrogenase molybdenum-iron protein from *Clostridium pasteurianum* at 3.0-A resolution. Biochemistry 32:7104–7115
Kisker C, Schindelin H, Rees DC (1997) Molybdenum-cofactor-containing enzymes: structure and mechanism. Annu Rev Biochem 66:233–267
Koljonen T (1992) The geochemical atlas of Finland. Part 2: Till. Geological Survey of Finland, Espoo
Kośla T, Anke M, Lösch E (1989) Molybdänstatus-bedarf und-versorgung des Pferdes. In: Anke M et al (eds) 6th international trace element symposium, Molybdenum, vanadium, vol 1. University of Leipzig and University of Jena, Germany, pp 337–345
Kubota J (1977) Molybdenum status of U.S. soils and plants. In: Chappell WR, Petersen KK (eds) Molybdenum in the environment, The geochemistry, cycling, and industrial uses of molybdenum, vol 2. Marcel Dekker, New York, pp 555–581
Lăcătuşu R, Borza I (1989) Induced molybdenum deficiency in sunflower. In: Anke M et al (eds) 6th international trace element symposium, Molybdenum, vanadium, vol 1. University of Leipzig and University of Jena, Germany, pp 295–302
Leichtfried G (1990) Ullmann's encyclopedia of industrial chemistry, vol A16, 5th edn. Wiley-VCH, Germany, p 668
Lener J (1978) Molybdenum as an environmental factor from the viewpoint of some of its effects. DSc thesis, Charles University, Prague
Li Y-H (ed) (2000) A compendium of geochemistry: from solar nebula to the human brain. Princeton University Press, Princeton
Lukasev AA, Siskova NK, Knys VS (1971) Pathomorphological studies of animal organs in chronic molybdenum poisoning. Tr Inst Kraev Patol Kaz SSR 22:202–205
Magyar MJ (2007) Molybdenum. U.S. Geological Survey, Mineral Commodity Summaries, pp 110–111

Magyar MJ (2009) Molybdenum. U.S. Geological Survey, Mineral Commodity Summaries, pp 108–109
Mason J (1986) Thiomolybdates: mediators of molybdenum toxicity and enzyme inhibitors. Toxicology 42:99–109
Mason J, Cardin CJ (1977) The competition of molybdate and sulphate ions for a transport system in the ovine small intestine. Res Vet Sci 22:313–315
Mendel RR, Bittner F (2006) Cell biology of molybdenum. Biochim Biophys Acta 1763:621–635
Miller JK, Moss BR, Bell MC, Sneed NN (1972) Comparison of $^{99}$Mo metabolism in young cattle and swine. J Anim Sci 34:846–850
Mills CF, Davis GK (1987) In: Mertz W (ed) Trace elements in human and animal nutrition, vol 1, 5th edn. Academic Press, San Diego, pp 429–463
Mills CF, Bremner I, El-Gallad TT, Dalgarno AC, Young BW (1978) Mechanisms of the molybdenum/sulphur antagonism of copper utilisation by ruminants. In: Kirchgessner M (ed) Trace element metabolism in man and animals, vol 3. Technical University of Munich, Germany, pp 50–158
Mitchell PCH (2003) Molybdenum. Chem Eng News 81:108
Mochizuki M, Sasaki R, Yamashita Y, Akinaga M, Anan N, Sasaki S et al (2002) The distribution of molybdenum in the tissue of wild duck. Environ Monit Assess 77:155–161
Nielsen FH (1996) Other trace elements. In: Ziegler EE, Filer LJ (eds) Present knowledge in nutrition, 7th edn. International Life Science Institute Press, Washington, DC, pp 353–377
Nozaki Y (2005) A fresh look at element distribution in the North Pacific, AGU. http://www.agu.org/eos.elec/97025e.html
NRC (2004) National Research Council. Canadian minerals yearbook: molybdenum. National Resources Canada. Available online: http://www.nrcan-rncan.gc.ca/mms-smm/busi-indu/cmy-amc/content/2004/40.pdf
NRC (2006) National Research Council. Molybdenum. In: Mineral tolerance of animals, 2nd edn. National Academies Press, Washington, DC, pp 262–275
Pandey VC, Abhilash PC, Upadhyay RN, Tewari DD (2009) Application of fly ash on the growth performance and translocation of toxic heavy metals within *Cajanus cajan* L. Implication for safe utilization of fly ash for agricultural production. J Hazard Mater 166:255–259
Pasieczna A (2012) Molybdenum and tin in soils of Poland. Biul Państw Inst Geol 450:75–82
Polyak DE (2013) Molybdenum. U.S. Geological Survey, Mineral Commodity Summaries, pp 106–107
Polyak DE (2016a) Molybdenum. U.S. Geological Survey, Mineral Commodity Summaries, pp 112–113
Polyak DE (2016b) Molybdenum. 2014 Minerals Yearbook. U.S. Geological Survey, pp 50.1–50.12
Pott EB, Henry PR, Zanetti MA, Rao PV, Hinderberger EJ Jr, Ammerman CB (1999) Effects of high dietary molybdenum concentration and duration of feeding time on molybdenum and copper metabolism in Steep. Anim Feed Sci Technol 79:93–105
Price J, Will AM, Paschaleris G, Chesters JK (1987) Identification of thiomolybdates in digesta and plasma from sheep after administration of Mo$^{99}$-labeled compounds into the rumen. Br J Nutr 58:127–138
Puls R (1988) Mineral levels in animal health: diagnostic data, 2nd edn. Clearbrook, British Columbia, p 240
Rajagopalan KV (1988) Molybdenum: an essential trace element in human nutrition. Annu Rev Nutr 8:401–427
Regius A, Anke M (1989) The Cu-, Mo- and protein content of different sorts of alfalfa on soils of similar and different origin. In: Anke M et al (eds) 6th international trace element symposium, Molybdenum, vanadium, vol 1. University of Leipzig and University of Jena, Germany, pp 276–285

Regoli L, Van Tilborg W, Heijerick D, Stubblefield W, Carey S (2012) The bioconcentration and bioaccumulation factors for molybdenum in the aquatic environment from natural environmental concentrations up to the toxicity boundary. Sci Total Environ 435–436:96–106
Reid SD (2012) Molybdenum and chromium. Homeost Toxicol Essent Met 31A(8):375–415
Reimann C, de Caritat P (1998) Chemical elements in the environment. Springer-Verlag, Berlin Heidelberg
Reimann C, Äyräs S, Chekusin V, Bogatyrev I, Boyd R, Caritat P et al (1998) Environmental geochemical atlas of the central Barents region. Geological Survey of Norway, Trondheim
Reimann C, Siewers U, Tarvainen T, Bitukova L, Eriksson J, Gilucis A et al (2003) Agricultural soils in Northern Europe: a geochemical atlas. Bundesanstalt für Geowissenschaften und Rohstoffe und Staatliche Geologische Dienste in der Bundesrepublik Deutschland, Hannover
Rosoff B, Spencer H (1973) The distribution and excretion of molybdenum-99 in mice. Health Phys 25:173–175
Rousseaux CG, Haschek WM, Wallig MA (2002) Handbook of toxicologic pathology, 2nd edn. Academic Press, San Diego
Schindelin H, Kisker C, Hilton J, Rajagopalan KV, Rees DC (1996) Crystal structure of DMSO reductase: redox-linked changes in molybdopterin coordination. Science 272:1615–1621
Schnabel R, Bunke C (1989) On the distribution of molybdenum in soil – plant system. In: Anke M et al (eds) 6th international trace element symposium, Molybdenum, vanadium, vol 1. University of Leipzig and Jena, Germany, pp 192–198
Sebenik RF, Burkin AR, Dorfler RR, Laferty JM, Leichtfried G, Meyer-Grünow H et al (2002) Molybdenum and molybdenum compounds. In: Ullmann's encyclopedia of industrial chemistry. Wiley-VCH, Weinheim
Sellmann D (1993) X-ray structure analysis of FeMo nitrogenase – is the problem of $N_2$ fixation solved? Angew Chem Int Ed 32:64–67
Sequi P (1973) Molybdenum in earth. Agrochimica 17:119–140
Shan WJ, Fang DW, Shuang Y, Kong YX, Zhao ZY, Xing ZQ et al (2012) Equilibrium, kinetics and thermodynamics on the recovery of Rhenium(VII) and Molybdenum(VI) from industrial wastewater by chemically modified waste paper gel. J Chem Eng Data 57:290–297
Shields JA Jr (2013) Applications of molybdenum metal and its alloys. International Molybdenum Association, London
Sigma-Aldrich (2015a) Material Safety Data Sheet of Ammonium Molybdate. http://www.sigmaaldrich.com/taiwan.html
Sigma-Aldrich (2015b) Material Safety Data Sheet of Molybdenum(VI) Oxide. http://www.sigmaaldrich.com/taiwan.html
Sigma-Aldrich (2015c) Material Safety Data Sheet of Sodium Molybdate. http://www.sigmaaldrich.com/taiwan.html
Skibniewski M, Skibniewska E, Kośla T, Kołnierzak M (2015) The molybdenum content in the muscles of red deer (*Cervus elaphus*). Acta Sci Pol Zootech 14:175–183
Skibniewski M, Skibniewska EM, Kośla T, Olbrych K (2016) The content of copper and molybdenum in the liver, kidneys, and skeletal muscles of elk (*Alces alces*) from North-Eastern Poland. Biol Trace Elem Res 169:204–210
Smith GM, White CL (1997) A molybdenum–sulfur–cadmium interaction in sheep. Aust J Agric Res 48:147–154
Spears JW (1992) Reevaluation of the metabolic essentiality of the minerals. Asian Australas J Anim Sci 12:1002–1008
Stafford JM, Lambert CE, Zyskowski JA, Engfehr CL, Fletcher OJ, Clark SL et al (2016) Dietary toxicity of soluble and insoluble molybdenum to northern bobwhite quail (*Colinus virginianus*). Ecotoxicology 25:291–301
Stiefel EI (2002) The biogeochemistry of molybdenum and tungsten. In: Sigel A, Sigel H (eds) Metal ions in biological system, Molybdenum and tungsten: their roles in biological processes, vol 39. Marcel Dekker, New York, pp 1–29

Stokinger HE (1981) Molybdenum. In: Clayton GD, Clayton FE (eds) Patty's industrial hygiene and toxicology, Toxicology, vol 2A, 3rd edn. John Wiley & Sons, New York, pp 1807–1820
Swan DA, Creeper JH, White CL, Ridings M, Smith GM, Costa ND (1998) Molybdenum poisoning in feedlot cattle. Aust Vet J 76:345–349
Tallkvist J, Oskarsson A (2015) Molybdenum. In: Nordberg GF, Fowler BA, Nordberg M, Friberg L (eds) Handbook on the toxicology of metals, 4th edn. Elsevier, Amsterdam, pp 731–741
Thornton I (1981) Geochemical aspects of the distribution and forms of heavy metals in soils. In: Lepp NW (ed) Effect of heavy metal pollution on plants: metals in the environment. Applied Science Publishers, London, pp 1–34
Turnlund JR, Keyes WR, Peiffer GL, Chiang G (1995) Molybdenum absorption, excretion, and retention studied with stable isotopes in young men during depletion and repletion. Am J Clin Nutr 61:1102–1109
Underwood EJ (1977) Trace elements in human and animal nutrition, Molybdenum, 4th edn. Academic Press, New York, pp 109–131
United States Department of the Interior (1998) National Irrigation Water Quality Program Information Report no. 3. Guidelines for interpretation of the biological effect of selected constituents in biota, water and sediment molybdenum, p 115
US EPA (1992) Guidelines for exposure assessment. US Environmental Protection Agency, Risk Assessment Forum, Washington, DC, EPA/600/Z-92/001, 1992
USGS (2015) U.S. geological survey. Mineral resources on-line spatial data. http://mrdata.usgs.gov/soilgeochemistery/#/detail/element/42
Van Gennip AH, Abeling NG, Stroomer AEM, Overmars H, Bakker HD (1994) The detection of molybdenum cofactor deficiency: clinical symptomatology and urinary metabolic profile. J Inherit Metab Dis 17:142–145
Venugopal B, Luckey TD (1978) Molybdenum. In: Metal toxicity in mammals. 2. Chemical toxicity of metals and metalloids. Plenum Press, New York, pp 253–257
Versieck J, Hoste J, Vanballengerghe L, Barbier F, Cornelis R, Waelput I (1981) Serum molybdenum in diseases of the liver and biliary system. J Lab Clin Med 97:535–544
Vyskocil A, Viau C (1999) Assessment of molybdenum toxicity in humans. J Appl Toxicol 19:185–192
Walker LA, Lawlor AJ, Chadwick EA, Potter E, Pereira MG, Shore RF (2010) Inorganic elements in the livers of Eurasian otters, *Lutra lutra*, from England and Wales in 2007 & 2008: A Predatory Bird Monitoring Scheme (PBMS) report. Centre for Ecology and Hydrology, Lancaster
Wang CW, Liang C, Yeh HJ (2016) Aquatic acute toxicity assessments of molybdenum (VI) to *Daphnia magna*. Chemosphere 147:82–87
Wedepohl KH (2004) The composition of Earth's upper crust, natural cycles of elements, natural resources. In: Merian E et al (eds) Elements and their compounds in the environment, 2nd edn. Wiley-VCH, Weinheim, pp 3–16
Williams RJP, da Silva JRF (2002) The involvement of molybdenum in life. Biochem Biophys Res Commun 292:293–229

# Chapter 9
# Nickel, Ni

**Łukasz J. Binkowski**

**Abstract** Nickel (Ni) is a transition metal whose average concentration in the Earth's crust is not high. Very high concentrations are thought to be found in the Earth's inner core and have been discovered in soils based on serpentine rocks. The most common forms of Ni are oxides, and of the five stable isotopes, $^{58}$Ni is the most abundant. The present occurrence of Ni in the environment is mainly connected with industrial use, especially in metallurgy. Because of its density, Ni is considered a heavy metal. It is an essential element for plants and microorganisms, but its physical role in vertebrates is still not fully understood. On the other hand, its toxic, genotoxic, and carcinogenic properties are known. Nickel concentrations considered normal in birds and mammals fall into a range of 0.05–0.5 for kidneys and 0.04–0.3 mg kg$^{-1}$ for liver, but other materials, such as feathers and fur, usually accumulate it in higher amounts. Studies on the evaluation of possible biomarkers of Ni exposure have been carried out mainly on humans, but no clear and measurable relationship has been found so far. Some initial findings linked Ni exposure with a decrease in δ-aminolevulinic acid dehydratase activity, but the most useful methods of its detection continue to be based on the measurement of concentrations in select tissues or materials.

## 1 Introduction

Nickel (chemical symbol Ni) belongs to the group of transition metals. According to the most common definition of heavy metals (based on density), Ni can be considered one of these elements (Duffus 2002). Its concentration in the Earth's crust is not high, 0.008% (IARC 1990). However, significant Ni content is thought to occur in the Earth's inner core (around 5.5%), where it likely is found as part of the alloy known as FeNi (also called NiFe) (McDonough and Sun 1995).

Ł. J. Binkowski (✉)
Institute of Biology, Pedagogical University of Cracow, Kraków, Poland
e-mail: lukasz.binkowski@up.krakow.pl



E. Kalisińska (ed.), *Mammals and Birds as Bioindicators of Trace Element Contaminations in Terrestrial Environments*,
https://doi.org/10.1007/978-3-030-00121-6_9

The main use of Ni is in industry; thus, the main source of this metal in the environment is metallurgy. Nickel is mainly used in the production of stainless steel and iron magnetic alloys known as alnico (an acronym formed from the symbols for aluminum, nickel, and cobalt).

The essentiality of Ni in plants and microorganisms is generally known. Its role in animals, especially vertebrates, remains incompletely understood. Numerous studies have shown its toxicity, genotoxicity, effect on reproduction, and even carcinogenic properties (Nordberg et al. 2007; Kong et al. 2014; Wang et al. 2016). Additionally, the ecotoxicological status of Ni nanoparticles (whose use has increased significantly) has not been fully elucidated, but their toxicity for animals and whole ecosystems is suspected (Ray et al. 2009; Magaye et al. 2012).

The main organizations that study Ni reserves, as well as market trends for the metal, are the International Nickel Study Group (INSG 2016) and the United States Geological Survey (USGS 2016).

## 2 General Properties

Nickel (Latin *Niccolum*) lies between cobalt and copper in the tenth group of the periodic table with atomic number 28. Ni's atomic mass and density are 58.7 and 8.90 g $cm^{-3}$, respectively. This metal, which is silver-white at room temperature, occurs in a solid state; it is hard but malleable; it reaches its melting point at a high temperature of 1455 °C (Hammond 2004). The most common Ni form is the oxide, so the metallic form is rarely observed. Properties of the metallic form are quite similar to those of cobalt (Sienko and Plane 1979). Nickel has 5 naturally occurring stable isotopes (58, 60, 61, 62, and 64) and 19 unstable ones (Hammond 2004). In nature, nickel exists for the most part in the form of isotope $^{58}Ni$ (68.3%). The element is often encountered in the 0, +1, +3, and +4 oxidation states in compounds, but states −2 and −1 are also known (Earnshaw and Greenwood 1997).

## 3 Nickel Minerals, Production and Uses

Nickel resembles iron and cobalt in that it occurs in rocks; its largest deposits are found in alkaline (up to 160 mg $kg^{-1}$) and ultraalkaline (up to 2000 mg $kg^{-1}$) rocks. Acidic rocks contain substantially lower concentrations (Kabata-Pendias 2011). In rocks, Ni occurs most often in the form of oxides, sulfides, and silicates. Over 100 minerals contain a significant substratum of Ni; the most common of these are garnierite, millerite, nickeline, nickel galena, and pentlandite (Cempel and Nikel 2006; NPI 2015). Nickel is also abundant in iron meteorites in taenite and kamacite alloys (Rasmussen et al. 1988).

Identified land resources that contain at least 1% of Ni are estimated to hold at least 130 million tons of Ni. The largest part (ca. 60%) is bound in laterites, with the rest in sulfide deposits. The world total available Ni reserves is estimated at 80 million metric tons, of which the largest parts lie in Australia (24 million tons), New Calcedonia (12 million tons), and Brazil (8.7 million tons) (USGS 2012a).

The global market demand for steel (38.13 million tons in 2013) (Statista 2015a) entails an increase in global Ni production (Fig. 9.1). Approximately 20 countries mine Ni ore, and around 25 countries smelt it (INSG 2016). In the fourth quarter of the twentieth century, the largest Ni mining operations were located in the USSR and Canada (USGS 1980, 1995). Since then different leaders in production have emerged. In 2013 the leading producers were the Philippines and Indonesia, where annual Ni production was estimated at 440,000 tons. Filipino production almost doubled over 2011 levels. Russian production has remained stable in recent years and has been hovering around 260,000 tons (Statista 2015b).

Because of Ni's high resistance to oxidation and corrosion, it is widely used around the globe in industry (including nickel plating on steel) and daily life (e.g., in coin production). The industry that uses Ni most intensively is metallurgy for the production of stainless steel, estimated at 65% of global use (INSG 2016). A rather recent new use has been observed in the hunting industry. Owing to regulations banning lead shot use in waterfowl hunting, a few so-called nontoxic forms of ammunition have been produced. One is produced from an alloy of iron, tungsten, and nickel (Brewer et al. 2003; Binkowski and Sawicka-Kapusta 2015).

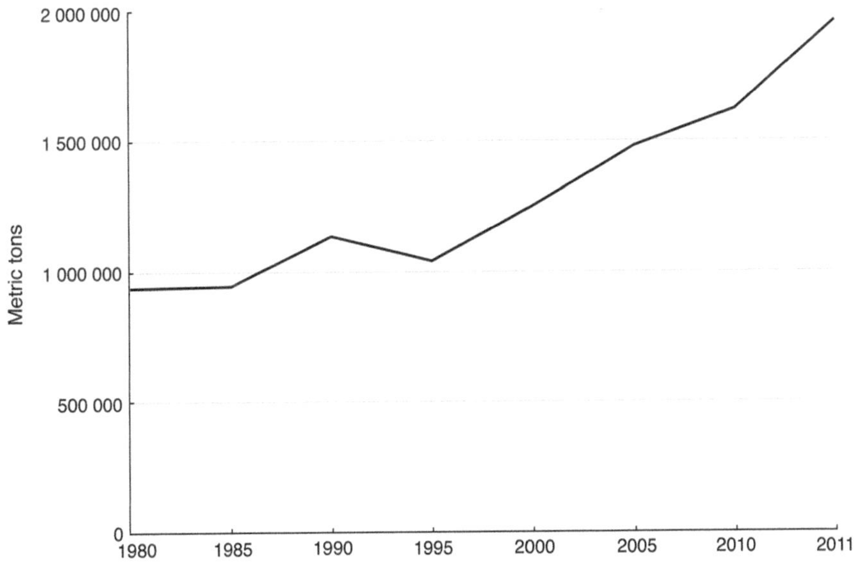


**Fig. 9.1** Dynamic of Ni mining production between 1980 and 2011 (USGS 1980, 1985, 1990, 1995, 2000, 2007, 2012b, 2013)

The general Ni usage is divided into two branches. The so-called first use is the use of metal produced indirectly from ores mainly in the production of stainless steel, alloyed steels, high-nickel alloys, castings, and chemicals. The second branch is the use of recycled metal from scrap, which has enjoyed very wide usage (INSG 2016).

# 4 Nickel in Nature: Geogenic and Anthropogenic Sources

Apart from ores, which consist of minerals rich in oxides, sulfides, and silicates of nickel, the element occurs widely in carbon deposits (around 15 mg $kg^{-1}$) and in petroleum (the range 20–100 mg $kg^{-1)}$ (Fig. 9.2) (Kabata-Pendias 2011).

Nickel background levels are hard to assess currently since numerous pollutant sources are distributed around the world. The range of normal values in the air over the continents is 0.1–1 ng $m^{-3}$ (Livett 1992; Kabata-Pendias and Pendias 1999). The concentrations observed over Spitsbergen in the 1980s were up to 0.95 ng $m^{-3}$ (Maenhaut et al. 1989). Concentrations in natural waters varied between fresh waters (2–10 μg $L^{-1}$) and marine waters (0.2–0.7 μg $L^{-1}$) (Rojas et al. 1999). Nickel concentrations in nonpolluted soils significantly depend on the soil type. Usually the entire range fits between 10 (organic soils) and 34 mg $kg^{-1}$ (rendzinas) (Kabata-Pendias 2011), but soils based on serpentine rocks reach significantly higher concentrations—in some cases even up to 2% of the soil composition (Shallari et al. 1998; Brooks 1999; Marsh and Anderson 2011; Altinozlu et al. 2012).

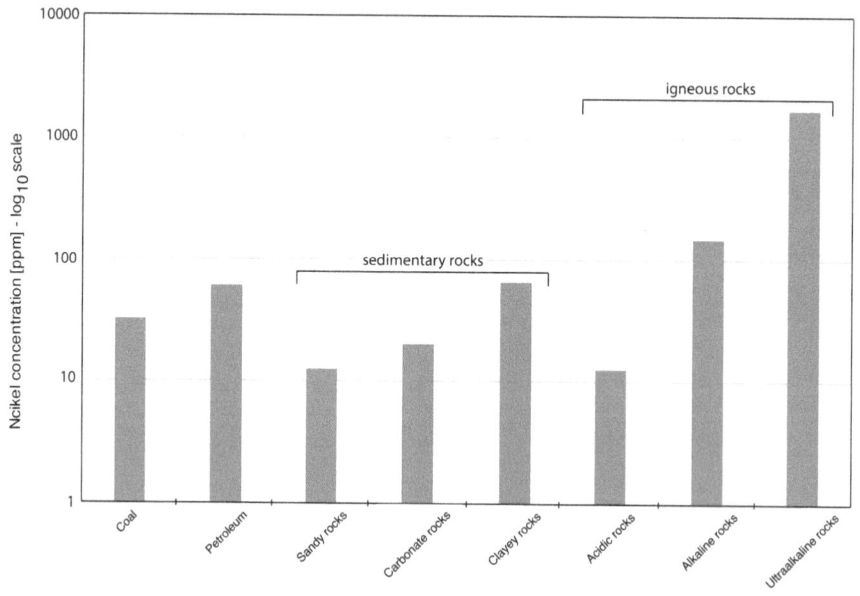


**Fig. 9.2** Nickel concentrations in biolites and rocks (Kabata-Pendias and Pendias 1999; Adriano 2001; Chmielnicka 2002; Kabata-Pendias 2011)

In anthropogenically changed environments, concentrations in air usually fit the range 1–150 ng $m^{-3}$ (Kabata-Pendias and Pendias 1999). In European cities, the average values are 9–60 ng $m^{-3}$, but in strictly industrialized areas, concentrations reach 110–180 ng $m^{-3}$ or more (Bennett 1994; WHO 2000; Cempel and Nikel 2006). Concentrations in potable water are generally low (around 10 μg $L^{-1}$), but levels in waters of most European rivers reach 75 μg $L^{-1}$ or more (Kabata-Pendias and Pendias 1999; Cempel and Nikel 2006; EEA 2009). Nickel concentrations in soils and deposits from industrial regions significantly depend on the industry type and intensity, for example, in some areas, following the emergence of a particular industry, a 25-fold increase (up to 3073 mg $kg^{-1}$ dw) in Ni concentrations in superficial sediments has been observed (Dauvalter 2003).

## 5 Biological Status of Nickel

The essentiality of nickel in microorganisms and plants is generally known (Muyssen et al. 2004; Sydor and Zamble 2013). Some studies have pointed out the side effects of low Ni intake in birds and mammals. Nevertheless, the study of physiological role of Ni in animals remains incomplete (Nielsen 1974, 2000). Nickel activates numerous enzymes, regulates hormonal activity, and stabilizes the structure of ribonucleic acid. Some observations have led researchers to conclude that the metal is under homeostatic regulation in mammals but not in birds (Gamberg 1998). Research on Ni deficiency has revealed pigmentation changes, thicker legs, dermatitis, and ultrastructural changes in liver (Sunderman et al. 1972; Nielsen 1974). The physiological influence of Ni in rats and chickens is connected with lipid metabolism in liver and erythrocytes and in serum lipoproteins (Nielsen et al. 1975; Stangl and Kirchgessner 1996, 1997), as well as with Ni flux in everted intestinal sacs (Stangl et al. 1998). Nickel deficiency causes iron deficiency in rats owing to the impairment of iron absorption by the intestines (Kirchgessner and Schnegg 1976). Nickel deficiency may also lower reproductive performance in rats as a result of diminishing sperm movement and quantity (Yokoi et al. 2003). Supplementing pigs with nickel and cobalt can ameliorate vitamin B-12 deficiency and hyperhomocysteinemia (Stangl et al. 2000).

On the basis of various sources of information about Ni essentiality in animals, its recommended daily allowance for animals and humans was estimated at 0.5 mg $kg^{-1}$ body weight (Nordberg et al. 2007). Other sources suggest 25–35 μg $day^{-1}$. Such amounts should be easily supplied by conventional diets (which usually provide 100–150 μg Ni $day^{-1}$). Diets rich in chocolate, nuts, grains, peas, and beans may even provide 900 μg $day^{-1}$ (Pennington and Jones 1987; Anke et al. 1993; Nielsen 2000).

## 5.1 Toxicity of Various Forms of Nickel in Homoeothermic Animals

Most studies on Ni toxicity relate to plants, microorganisms, invertebrates, and fish (Chen et al. 2009; Parsons et al. 2010; Macomber and Hausinger 2011). Research on mammals is scant, and the number of studies on birds is even smaller.

Exposure to Ni varies between animals and humans. In animals, the main route of exposure is the gastrointestinal tract. In contrast, absorption through the skin from various products, such as nickel-plated tools, medical instruments, jewelry, and household consumer products, constitutes the major pathway of human exposure (Basketter et al. 2003). The harmful effects of Ni in organisms depend significantly on the properties of the compound, the time and strength of exposure, and individual variables (Adriano 2001). The most toxic Ni compound for animals is nickel carbonyl, known generally as one of the most toxic inhaled poisons. According to the International Agency for Research on Cancer (IARC), metallic Ni and its alloys are classified as possibly carcinogenic to humans (group 2B) and nickel compounds as carcinogenic to humans (group 1) (IARC 2016). Carcinogenic Ni activity is mostly connected with its inhaled compounds. Such problems are diagnosed in the nose and lungs of workers in nickel-connected industries (Adriano 2001; Denkhaus and Salnikow 2002). Experiments on intratracheal instillation of Ni compound also showed histopathological and bronchoalveolar changes in rat lungs (Bajpai et al. 1999). There have even been occurrences of deaths from adult respiratory distress syndrome following occupational Ni exposure (Sandström et al. 1989; Rendall et al. 1994). In the second example cited, a man spraying Ni did not use protective equipment. After several days of exposure, his urinary Ni concentrations reached 700 μg $L^{-1}$. The estimation of air concentrations during work was 382.1 mg $m^{-3}$ (with 64.6% of the particles smaller than 1.4 μm) and the total amount of Ni inhaled ca. 1 g (Rendall et al. 1994).

After entering the body, Ni induces synthesis of metalothioneine. Next to genotoxicity, developmental toxicity, hematotoxicity, immunotoxicity, neurotoxicity, hepatoxicity, and reproductive toxicity noted in birds and mammals, Ni induces an oxidative stress and crosses the placental barrier (Hoffman 1979; Domingo 1994; Chen and Lin 1998; Mathur and Shanker 2001; Das et al. 2008; Thomas et al. 2009; Saini et al. 2014; Casalegno et al. 2015). Induction is probably connected with further cancerogenesis, but still the direct correlation is not fully understood (Salnikow et al. 1994; Das et al. 2008). Following oral administration in rodents of higher doses of Ni compounds (chloride or sulfate), deleterious effects on organisms, such as histological lesions, liver and body mass decrease, liver cell apoptosis, necrosis, enzyme-level changes, and others, have been observed (Obone et al. 1999; Pari and Amudha 2011; Gathwan et al. 2012). Administration of Ni in food can also disturb reproduction efficiency through its effects on the level of sex hormones such as testosterone (Pandey and Singh 2001; Mathur et al. 2010; Forgacs et al. 2012). Apart from hormonal changes, Ni's influence on reproduction has been additionally observed in the histological level of testis structure in rats (Toman et al. 2012). The connection between increased Ni concentrations ($NiCl_2$ per os) and decreased egg production and quality has also been reported in the literature (Arpasova et al. 2007). Moreover, it has also been found that birds exposed to nickel per os revealed lower concentrations of magnesium and triglycerides in blood serum, as well as activity of alanine aminotransferase. Other biochemical parameters did not appear to be disturbed (Capcarova et al. 2008; Kolesarova et al. 2008). The lowest-observed-effect concentration (LOEC) noted by oral administration in chickens and mallards was respectively equal to 300 and 700 mg $kg^{-1}$ in diet (DeForest et al. 2012).

Injections of Ni compounds also revealed deleterious effects. Pancreatic, hepatic, and osteogenic toxicity and tumor formation at injection points have been observed in rats (Novelli et al. 1998).

## 5.2 *Toxicokinetics and Effects of Nickel in Wildlife*

Most studies concerned with Ni concentrations in animals also address other metals. Nickel alone and its concentrations in vertebrates are rarely studied, so suspicions regarding Ni toxicity in live wild animals are usually explained as Ni's partial influence together with other metals (e.g., Sánchez-Chardi et al. 2008). Laboratory experiments have revealed that the general scheme of increasing concentrations in mammals exposed to Ni is heart = liver < spleen < lung = brain < testes < kidneys (Obone et al. 1999). In birds, the scheme seems to be different (Table 9.1). The highest concentrations are usually encountered in external matter such as fur or feathers, especially in polluted areas. Among internal tissues, usually bones accumulate the highest concentrations (Outridge and Scheuhammer 1993). Some studies have also revealed high Ni concentrations in bone marrow, even six times higher than in liver (Hassan et al. 2012).

Nickel inhalation is connected with higher bioaccumulation than other routes (Reichrtova et al. 1988). Accumulation from food and water is weak, on the level of 5%. This is supported by studies that revealed significant concentrations in ingesta and low concentrations in internal tissues (Hui et al. 1998; Hui 1998; Kabata-Pendias and Pendias 1999). However, in cases of significant exposure, accumulation is observed in bones and several soft tissues (including heart muscle, lungs, skin, fur, and feathers) (Kabata-Pendias and Pendias 1999; ATSDR 2005; Nordberg et al. 2007). Oral absorption of soluble Ni forms is higher than insoluble ones. Soluble Ni compounds are usually accumulated in liver and kidneys and insoluble forms mainly in lungs and pancreas (Casalegno et al. 2015).

**Table 9.1** Nickel concentrations (mg kg$^{-1}$ dry weight; mainly means) in chosen tissues and materials collected from birds in the Northern Hemisphere

| Species | Area | Kidneys | Liver | Muscles | References |
|---|---|---|---|---|---|
| *Birds* | | | | | |
| American coot *Fulica americana* | USA, San Francisco Bay area | | <0.05–1.61 | | Hui (1998) |
| Blue-winged teal *Anas discors* | USA, Southern Texas | | 0.95 | | Warren et al. (1990) |
| Canvasback *Aythya valisineria* | USA, San Francisco Bay area | | Up to 3.5 | | Miles and Ohlendorf (1993) |
| Eider *Somateria mollissima* | Norway, Taura | 2 | 1 | 2 | Lande (1977) |
| Gray plover *Pluvialis squatarola* | France, Southwest Atlantic coast | 0.8 | 0.4 | 0.7 | Lucia et al. (2010) |
| Greylag goose *Anser anser* | France, Southwest Atlantic coast | 0.8 | 0.7 | 1.8 | Lucia et al. (2010) |
| Mallard *Anas platyrhynchos* | Belarus, Swisloch River (above Minsk) | 1.7 | 1.0 | 1.3 | Kozulin and Pavluschick (1993) |
| Mallard *Anas platyrhynchos* | Belarus, Swisloch River (below Minsk) | 3.7 | 4.3 | 1.1 | Kozulin and Pavluschick (1993) |
| Red knot *Calidris canutus* | France, Southwest Atlantic coast | 0.3 | 0.4 | 0.3 | Lucia et al. (2010) |
| Peregrine falcon *Falco peregrinus* | Poland, northwest | 1.15 | 0.11 | 0.07 | Kalisińska et al. (2008) |
| White-tailed eagle *Haliaeetus albicilla* | Poland, north and northwest | 6.5 | 13 | | Falandysz et al. (2000) |
| Great tit *Parus major* | Belgium, Antwerp | 0.024 | 0.014 | 0.09 | Dauwe et al. (2005) |
| Great tit *Parus major* | Finland, various regions | | 0.3 | | Ingervo et al. (1995) |
| Blue tit *Parus caerulescens* | Finland, various regions | | 0.3 | | Ingervo et al. (1995) |

(continued)

**Table 9.1** (continued)

| Species | Area | Kidneys | Liver | Muscles | References |
|---|---|---|---|---|---|
| *Mammals* | | | | | |
| White-tailed deer *Odocoileus virginianus* | USA, Texas | | <0.5–3.0 | <0.5–2.1 | Bruckwicki (2006) |
| Wild boar *Sus scrofa* | Poland, central and central-east | | 0.47 | | Długaszek and Kopczyński (2011) |
| American mink *Neovison vison* | Poland, Drawa National Park | 0.29 | 0.27 | | Brzeziński et al. (2014) |
| American mink *Neovison vison* | Poland, Narew National Park | 1.17 | 0.16 | | Brzeziński et al. (2014) |
| Marten *Martes americana* | Canada, British Columbia | 1.20 | | | Harding (2004) |
| Volwerine *Gulo luscus* | Canada, British Columbia | | 2.00 | | Harding (2004) |
| Eurasian otter *Lutra lutra* | Ireland and Great Britain | | 0.035–0.54 | | Mason and Stephenson (2001) |
| Eurasian otter *Lutra lutra* | England and Wales, various regions | | 0.10 | | Walker et al. (2011) |
| Eurasian otter *Lutra lutra* | France, various regions | | 1.51 | | Ruiz-Olmo et al. (2000) |
| Wild rat *Rattus rattus* | Portugal, Tapada Grande | 0.06 | 0.06 | | Pereira et al. (2006) |
| Algerian mouse *Mus spretus* | Portugal, Tapada Grande | 0.36 | 0.19 | | Pereira et al. (2006) |
| European hare *Lepus europaeus* | Poland, central and central-east | | 0.41 | | Długaszek and Kopczyński (2011) |

Concentrations in tissues in mg kg$^{-1}$ dw but where needed recalculated from ww according to Binkowski (2012)

## 5.3 Bioaccumulation of Nickel

Nickel has not been commonly studied in wild living birds and mammals as cadmium or lead. Comprehensive reviews of Ni concentrations in wildlife have been conducted by Outridge and Scheuhammer (1993) and Eisler (1998), but the presented data mainly focus on birds and mammals of North America. The information gathered concerns mainly liver and kidney concentrations; other biological samples have been significantly less studied. Most studies were carried out on tissues taken during necropsy.

Some evidence suggested that in the same areas birds accumulate higher Ni concentrations than mammals (Outridge and Scheuhammer 1993). The tissues of wild birds and mammals from pristine environments generally contain up to 5 mg $kg^{-1}$, whereas in Ni-polluted environments, concentrations reached 10 in mammals and even 80 mg $kg^{-1}$ dw in birds. Concerning liver and kidneys, the higher concentrations occurred in kidneys and generally do not exceed 3 mg $kg^{-1}$ dw (Outridge and Scheuhammer 1993). This statement can be supported by several other studies, but average values in kidneys of birds and mammals were established in the range 0.06–0.12 and 0.05–0.5 mg $kg^{-1}$, respectively (Table 9.1). The average values noted for liver are 0.04–0.1 mg $kg^{-1}$ in birds and 0.1–0.3 mg $kg^{-1}$ in mammals (WVDL 2015).

In vivo studies of birds are scarce, but mostly such projects reveal the current state of exposure. In the Northern Hemisphere, only Ni concentrations in mallard (*Anas platyrhynchos*) blood collected in Poland are known, where small but significant variation between concentrations in birds from industrialized and country areas was observed. There, the maximum value was 3.71 and the minimum was <0.04 mg $kg^{-1}$ ww (Binkowski and Meissner 2013). A comparison can be done only with the results of studies from southern Africa where concentrations found in the blood of red knobbed coot (*Fulica cristata*), sacred ibis (*Threskiornis aethiopicus*), and reed cormorant (*Microcarbo africanus*) were significantly higher (van Eeden and Schoonbee 1996). Thus, the means were 4.76, 3.65, and 5.12 mg $kg^{-1}$ ww. The likely reason for such differences is the quality of food and the environment because in birds exposed to higher Ni concentrations, the Ni showed up in, among other places, blood accumulation (Eastin and O'Shea 1981).

Apart from internal bird tissues, Ni is also found in eggs. Some studies revealed increased concentrations in egg yolk and egg white, and some has been found in eggshells of birds inhabiting polluted areas (Darolova et al. 1989; Orłowski et al. 2014). Values noted in eggshells of rook (*Corvus frugilegus*) in Poland fell within the range 1.15–4.07 mg $kg^{-1}$ dw. Seabirds occupy a different position in the trophic net. Nickel studies of this group remain very scarce, but the available data suggest that Ni levels in seabirds are not high (Barbieri et al. 2010; Jerez et al. 2013).

Comprehensive data on Ni concentrations in human tissues (mainly blood) can be found in the literature (Tomei et al. 2004; Stridsklev et al. 2004; Nordberg et al. 2007; Ikeda et al. 2011; Silva et al. 2013; Caciari et al. 2013; Khlifi et al. 2013). Among other mammals, rodents have been widely studied. This group represents popular and efficient Ni bioaccumulators, very tolerant to its deleterious effects and easily adaptive to long-term exposure (Marques et al. 2007). Research on bank voles (*Myodes glareolus*) in Slovakia revealed that Ni in bones reaches rather low concentrations (up to 9.52 mg $kg^{-1}$ dw). Concentrations in soft tissues are low, often below the detection limit of the methods used (Cloutier et al. 1986). An increase in Ni concentrations in soft tissues may be observed in connection with in the aspect of environmental pollution, but generally, even statistically increased values are low (Marques et al. 2007; Sánchez-Chardi et al. 2007). Research studies from Spain showed that concentrations in the liver and kidneys of the shrew (*Crocidura russula*) in a reference pristine area were 0.67 mg $kg^{-1}$ dw and 5.40 mg $kg^{-1}$, which were ca. three times lower than at a mining site—respectively 1.48 and 15.28 mg $kg^{-1}$ dw (Sánchez-Chardi et al. 2007). Studies on roe deer (*Capreolus capreolus*) antlers in Poland revealed that exposure over the course of 40 years in the middle of the twentieth century did not increase, and no relationship between concentrations in samples and environmental pollution was noted (Sawicka-Kapusta 1979). Renal Ni concentrations in wild Yukon moose (*Alces alces*) and caribou (*Rangifer tarandus*) revealed concentrations comparable to those in domestic cattle, with a maximum mean up to 0.78 mg $kg^{-1}$ dw (Gamberg 1998).

Nickel concentrations may be correlated with chromium in liver, as well as in hair (Pereira et al. 2006). The explanation of this observation is not clear, but it may be due to a mutual source of the metals in the environment. Concentrations in soft tissues of mammals can show a significant negative correlation with age (Smith and Rongstad 1982; Sánchez-Chardi et al. 2007). Also, in some studies, a sex dependency of Ni concentrations was observed. Furthermore, observations were inconsistent with the generally suspected mechanism of the reduction in concentrations in females during reproduction (Sánchez-Chardi et al. 2007). Seasonal fluctuation in general is not observed (Cloutier et al. 1986).

## *5.4 Ecological Effects of Nickel*

Apart from indirect contact with environmental pollution, some herbivorous birds and mammals can be potentially exposed through the trophic net. Generally, most plants are sensitive to higher Ni levels in soil, which results in a distribution of plants of optimum concentration areas. In such places concentrations in plants are low, so exposure of herbivores is also low. However, in places of significantly elevated concentrations in soil (lying on ultramafic bedrock) some plant species (ca. 400 species of various families around the world) evolved mechanisms to tolerate and safely accumulate Ni (van der Ent et al. 2015). In such plants, accumulation can reach on average 1000 mg $kg^{-1}$ in the foliage (particular specimens can even accumulate 20,000 mg $kg^{-1}$ ww) (Reeves et al. 1996; Robinson et al. 1997; Li et al. 2003; Altinozlu et al. 2012; van der Ent et al. 2013). In such cases the exposure of animals can be extremely high.

Debate about metal biomagnification has been ongoing for dozens of years. Many authors have questioned its occurrence in terrestrial environments, but generally it is believed that it can occur only in specific food chains (Laskowski 1991; van Straalen and Ernst 1991). Few research studies on Ni biomagnification have been carried out. However, data available for aquatic and terrestrial ecosystems do not confirm the process in birds and mammals (Gamberg 1998; Phipps et al. 2002; EPA 2011; Iamiceli et al. 2015). The groups that might be exposed the most through the trophic net are those that forage on aquatic animals, fish, and invertebrates, but only in specific aquatic environments, and bivalves, gastropods, and barnacles may show signs of Ni biomagnification (Gamberg 1998; Cardwell et al. 2013).

### 5.5 Bioindicators and Biomarkers of Nickel in Ecotoxicological Studies

The most advanced works on the potential biomarkers of Ni exposure have been carried out among people, especially industrial workers (Nordberg et al. 2007). No clearly measurable relationships with physiological parameters and enzyme activities have been identified. There were some suspicions that connection between the exposure and genotoxic effects may be useful, but after the evaluation it seems to be not accurate (Kiilunen et al. 1997; Burgaz et al. 2002). However, recent studies have demonstrated a connection between Ni concentrations in organism and the activity of δ-aminolevulinic acid dehydratase (ALA-d), which is a heme biopathway enzyme. The activity of ALA-d is widely used as a biomarker of lead poisoning that causes a significant drop (even up to 99% in blood) of activity (Binkowski and Sawicka-Kapusta 2015). Animals injected with Ni salts also revealed a similar negative correlation, but the observed decrease in activity was not very significant, up to 38% in liver and 53% in blood (Sulinskiene et al. 2014).

Still the most useful and reliable practice is measuring the concentrations in tissues, especially in liver, kidneys, and bones. In the case of humans, some thresholds in urine or serum have been established (Nordberg et al. 2007). Among animals, no certain values have been given, but some studies have presented common ranges of concentrations (Sect. 5.3).

## 6 Conclusions

- Nickel is a transition metal that is toxic for birds and mammals at higher concentrations. However, there is also evidence that it is essential in birds and mammals.
- The main source of Ni in the environment is industry (mainly metallurgy), but some areas have naturally very high concentrations in so-called serpentine soils.
- Nickel bioaccumulation in tissues and materials of birds and mammals depends on environmental factors, but in most cases, in pristine areas, values do not exceed 5 mg $kg^{-1}$ dw. In Ni-polluted regions, birds seem to be more efficient accumulators since concentrations in their soft tissue may reach 80 mg $kg^{-1}$, while in mammals this figure is 10 mg $kg^{-1}$ dw.
- No efficient and reliable biomarkers of Ni exposure have been found, and the best method of evaluating exposure is still to measure Ni concentrations in wildlife and the environment.

## References

Adriano DC (2001) Nickel. In: Trace elements in terrestrial environments, pp 677–705
Altinozlu H, Karagoz A, Polat T, Unver I (2012) Nickel hyperaccumulation by natural plants in Turkish serpentine soils. Turk J Bot 36:269–280
Anke M, Angelow L, Muller M, Glei M (1993) Dietary trace element intake and excretion of man. In: Anke M, Meissner D, Mills CF (eds) Trace elements in man and animals – TEMA-8. Verlag Media Touristik, Gersdorf, pp 180–188
Arpasova H, Capcarova M, Kalafova A, Lukac N, Kovacik J, Formicki G, Massanyi P (2007) Nickel induced alteration of hen body weight, egg production and egg quality after an experimental peroral administration. J Environ Sci Health Part B 42:913–918
ATSDR (2005) Toxicological profile for nickel. U.S. Department of Health and Human Services, Public Health Service, Atlanta
Bajpai R, Waseem M, Khanna AK, Kaw JL (1999) Comparative pulmonary toxicity of cadmium and nickel: histopathological and bronchoalveolar lavage analysis. Indian J Exp Biol 37:541–545
Barbieri E, Passos Ede A, Filippini A, dos Santos IS, Garcia CAB (2010) Assessment of trace metal concentration in feathers of seabird (*Larus dominicanus*) sampled in the Florianópolis, SC, Brazilian coast. Environ Monit Assess 169:631–638
Basketter DA, Angelini G, Ingber A, Kern PS, Menné T (2003) Nickel, chromium and cobalt in consumer products: revisiting safe levels in the new millennium. Contact Dermatitis 49:1–7
Bennett BJ (1994) Environmental nickel pathways to man. In: Sunderman FWJ (ed) Nickel in the human environment. IARC, Lyon, pp 487–495
Binkowski ŁJ (2012) The effect of material preparation on the dry weight used in trace elements determination in biological samples. Fresenius Environ Bull 21:1956–1960
Binkowski ŁJ, Meissner W (2013) Levels of metals in blood samples from Mallards (*Anas platyrhynchos*) from urban areas in Poland. Environ Pollut 178:336–342
Binkowski ŁJ, Sawicka-Kapusta K (2015) Lead poisoning and its in vivo biomarkers in Mallard and Coot from hunting activity areas. Chemosphere 127:101–108
Brewer L, Fairbrother A, Clark J, Amick D (2003) Acute toxicity of lead, steel, and an iron-tungsten-nickel shot to mallard ducks (*Anas platyrhynchos*). J Wildl Dis 39:638–648
Brooks RR (1999) Serpentine and its vegetation: a multidisciplinary approach. Dioscorides Press, Portland
Bruckwicki P (2006) An investigation of the contaminant levels in white-tailed deer (*Odocoileus virginianus*) collected from Caddo Lake National Wildlife Refuge, Harrison County, Texas 2005. US Fish and Wildlife Service
Brzeziński M, Zalewski A, Niemczynowicz A, Jarzyna I, Suska-Malawska M (2014) The use of chemical markers for the identification of farm escapees in feral mink populations. Ecotoxicology 23:767–778
Burgaz S, Demircigil GÇ, Yılmazer M, Ertaş N, Kemaloğlu Y, Burgaz Y (2002) Assessment of cytogenetic damage in lymphocytes and in exfoliated nasal cells of dental laboratory technicians exposed to chromium, cobalt, and nickel. Mutat Res Toxicol Environ Mutagen 521:47–56
Caciari T, Rosati MV, Di Giorgio V, Casale T, Pimpinella B, Scala B, Giubilati R, Capozzella A, Tomei G, Tomei F (2013) Urinary nickel and prolactin in workers exposed to urban stressors. Environ Sci Process Impacts 15(11):1–8
Capcarova M, Kolesarova A, Arpasova H, Massanyi P, Lukac N, Kovacik J, Kalafova A, Schneidgenova M (2008) Blood biochemical dynamics and correlations in laying hens after experimental nickel administration. Int J Poult Sci 7:538–547
Cardwell RD, DeForest DK, Brix KV, Adams WJ (2013) Do Cd, Cu, Ni, Pb, and Zn biomagnify in aquatic ecosystems? Rev Environ Contam Toxicol 226:101–122
Casalegno C, Schifanella O, Zennaro E, Marroncelli S, Chemservice S (2015) Collate literature data on toxicity of chromium (Cr) and nickel (Ni) in experimental animals and humans. CehService Support Publ EN 478:1–287

Cempel M, Nikel G (2006) Nickel: a review of its sources and environmental toxicology. Pol J Environ Stud 15:375–382
Chen CY, Lin TH (1998) Nickel toxicity to human term placenta: in vitro study on lipid peroxidation. J Toxicol Environ Health Part A 54:37–47
Chen C, Huang D, Liu J (2009) Functions and toxicity of nickel in plants: recent advances and future prospects. Clean: Soil, Air, Water 37:304–313
Chmielnicka J (2002) Metals and Metalloids (in Polish, Metale i metaloidy). In: Toksykologia. Wydawnictwo Lekarskie PZWL, Warszawa, pp 484–490
Cloutier NR, Clulow FV, Lim TP, Davé NK (1986) Metal (Cu, Ni, Fe, Co, Zn, Pb) and Ra-226 levels in tissues of meadow voles *Microtus pennsylvanicus* living on nickel and uranium mine tailings in Ontario, Canada: site, sex, age and season effects with calculation of average skeletal radiation dose. Environ Pollut 41:295–314
Darolova A, Reichtrova E, Pavelka J (1989) Bioaccumulation of metals from nickel works waste in the gull (*Larus ridibundus* L., 1766). Biologia (Bratisl) 44:567–573
Das KK, Das SN, Dhundasi SA (2008) Nickel, its adverse health effects & oxidative stress. Indian J Med Res 128:412–425
Dauvalter V (2003) Impact of mining and refining on the distribution and accumulation of nickel and other heavy metals in sediments of subarctic Lake Kuetsjärvi, Murmansk Region, Russia. J Environ Monit 5:210–215
Dauwe T, Janssens E, Bervoets L, Blust R, Eens M (2005) Heavy-metal concentrations in female laying great tits (*Parus major*) and their clutches. Arch Environ Contam Toxicol 49:249–256
DeForest DK, Schlekat CE, Brix KV, Fairbrother A (2012) Secondary poisoning risk assessment of terrestrial birds and mammals exposed to nickel. Integr Environ Assess Manag 8:107–119
Denkhaus E, Salnikow K (2002) Nickel essentiality, toxicity, and carcinogenicity. Crit Rev Oncol Hematol 42:35–56
Długaszek M, Kopczyński K (2011) Comparative analysis of liver mineral status of wildlife. Probl Hig i Epidemilogii 92:859–863
Domingo JL (1994) Metal-induced developmental toxicity in mammals: a review. J Toxicol Environ Health 42:123–141
Duffus JH (2002) "Heavy metals" – a meaningless term? Pure Appl Chem 74:793–807
Earnshaw AA, Greenwood N (1997) Chemistry of the elements. Elsevier, New York
Eastin WC, O'Shea TJ (1981) Effects of dietary nickel on mallards. J Toxicol Environ Health 7: 883–892
EEA (2009) European Environment Agency website. http://www.eea.europa.eu/data-and-maps/figures/concentration-of-heavy-metals-in-the-rivers-rhine-and-elbe. Accessed 19 July 2016
Eisler R (1998) Nickel hazards to fish, wildlife, and invertebrates: a synoptic review. USGS, Laurel
EPA (2011) Bioacummulation summary – Nickel
Falandysz J, Ichihashi H, Mizera T, Yamasaki S-I (2000) Skład mineralny wybranych tkanek i narządów bielika (Mineral composition of selected tissues and organs of white-tailed sea eagle). Rocz Państwowego Zakładu Hig 51:1–5 (in Polish)
Forgacs Z, Massanyi P, Lukac N, Somosy Z (2012) Reproductive toxicology of nickel – review. J Environ Sci Health Part A 47:1249–1260
Gamberg M (1998) Contaminants in Yukon Moose and Caribou. p 16
Gathwan KH, Al-Karkhi IHT, Jaffar AL-Mulla EA (2012) Hepatic toxicity of nickel chloride in mice. Res Chem Intermed 39:2537–2542
Hammond CR (2004) The elements. In: Lide DR (ed) CRC handbook of chemistry and physics, 86th edn. CRC Press, Boca Raton, pp 1–34
Harding LE (2004) Environmental contaminants in wild martens (*Martes americana*) and wolverines (*Gulo luscus*). Bull Environ Contam Toxicol 73:98–105
Hassan AA, Rylander C, Brustad M, Sandanger TM (2012) Level of selected toxic elements in meat, liver, tallow and bone marrow of young semi-domesticated reindeer (*Rangifer tarandus tarandus* L.) from Northern Norway. Int J Circumpolar Health 71:1–7

Hoffman DJ (1979) Embryotoxic effects of crude oil containing nickel and vanadium in mallards. Bull Environ Contam Toxicol 23:203–206
Hui C (1998) Elemental contaminants in the livers and ingesta of four subpopulations of the American coot (*Fulica americana*): an herbivorous winter migrant in San Francisco Bay. Environ Pollut 101:321–329
Hui A, Takekawa JY, Baranyuk V, Litvin KV (1998) Trace element concentrations in two subpopulations of lesser snow geese from Wrangel Island, Russia. Arch Environ Contam Toxicol 34:197–203
Iamiceli AL, Ubaldi A, Lucchetti D, Brambilla G, Abate V, De Felip E, De Filippis SP, Dellatte E, De Luca S, Ferri F, Fochi I, Fulgenzi AR, Iacovella N, Moret I, Piazza R, Roncarati A, Melotti P, Fanelli R, Fattore E, di Domenico A, Miniero R (2015) Metals in Mediterranean aquatic species. Mar Pollut Bull 94:278–283
IARC (1990) Chromium, nickel and welding. IARC, Lyon
IARC (2016) IARC monographs on the evaluation of cancerogenic risk to humans. http://monographs.iarc.fr/ENG/Classification/index.php. Accessed 27 July 2016
Ikeda M, Ohashi F, Fukui Y, Sakuragi S, Moriguchi J (2011) Cadmium, chromium, lead, manganese and nickel concentrations in blood of women in non-polluted areas in Japan, as determined by inductively coupled plasma-sector field-mass spectrometry. Int Arch Occup Environ Health 84:139–150
Ingervo S, Stranberg C, Nuorteva P (1995) Trace metals in the livers of Finnish Parus species. Ornis Fenn 72:127–131
INSG (2016) International nickel study group website. http://www.insg.org. Accessed 15 July 2016
Jerez S, Motas M, Benzal J, Diaz J, Barbosa A (2013) Monitoring trace elements in Antarctic penguin chicks from South Shetland Islands, Antarctica. Mar Pollut Bull 69:67–75
Kabata-Pendias A (2011) Trace elements in soils and plants, 4th edn. CRC Press, Boca Raton
Kabata-Pendias A, Pendias H (1999) Biogeochemistry of trace elements (in Polish, Biogeochemia pierwiastków śladowych), 2nd edn. Wydawnictwo Naukowe PWN, Warszawa
Kalisińska E, Lisowski P, Czernomysy-Furowicz D, Kavetska KM (2008) Serratospiculiasis, mycosis, and haemosiderosis in wild peregrine falcon from Poland: a case report. Bull Vet Inst Pulawy 52:75–79
Khlifi R, Olmedo P, Gil F, Feki-Tounsi M, Chakroun A, Rebai A, Hamza-Chaffai A (2013) Blood nickel and chromium levels in association with smoking and occupational exposure among head and neck cancer patients in Tunisia. Environ Sci Pollut Res Int 20:8282–8294
Kiilunen M, Utela J, Rantanen T, Norppa H, Tossavainen A, Koponen M, Paakkulainen H, Aitio A (1997) Exposure to soluble nickel in electrolytic nickel refining. Ann Occup Hyg 41:167–188
Kirchgessner M, Schnegg A (1976) Malate dehydrogenase and glucose-6-phosphate dehydrogenase activity in livers of Ni-deficient rats. Bioinorg Chem 6:155–161
Kolesarova A, Capcarova M, Arpasova H, Kalafova A, Massanyi P, Lukac N, Kovacik J, Schneidgenova M (2008) Nickel-induced blood biochemistry alterations in hens after an experimental peroral administration. J Environ Sci Health Part B 43:625–632
Kong L, Tang M, Zhang T, Wang D, Hu K, Lu W, Wei C, Liang G, Pu Y (2014) Nickel nanoparticles exposure and reproductive toxicity in healthy adult rats. Int J Mol Sci 15: 21253–21269
Kozulin A, Pavluschick T (1993) Content of heavy metals in tissues of mallards *Anas platyrhynchos* wintering in polluted and unpolluted habitats. Acta Ornithol 28:55–61
Lande E (1977) Heavy metal pollution in trondheimsfjorden, Norway, and the recorded effects on the fauna and flora. Environ Pollut 12:187–198
Laskowski R (1991) Are the top carnivores endangered by heavy metal biomagnification? Oikos 60:387–390
Li Y-M, Chaney R, Brewer E, Roseberg R, Angle JS, Baker A, Reeves R, Nelkin J (2003) Development of a technology for commercial phytoextraction of nickel: economic and technical considerations. Plant Soil 249:107–115
Livett EA (1992) Heavy metal pollution of the atmosphere. In: Nierenberg WA (ed) Encyclopedia of earth system science. Academic Press, San Diego, pp 507–518

Lucia M, André JM, Gontier K, Diot N, Veiga J, Davail S (2010) Trace element concentrations (mercury, cadmium, copper, zinc, lead, aluminium, nickel, arsenic, and selenium) in some aquatic birds of the southwest atlantic coast of France. Arch Environ Contam Toxicol 58: 844–853

Macomber L, Hausinger RP (2011) Mechanisms of nickel toxicity in microorganisms. Metallomics 3:1153

Maenhaut W, Cornille P, Pacyna JM, Vitols V (1989) Trace element composition and origin of the atmospheric aerosol in the Norwegian arctic. Atmos Environ 23:2551–2569

Magaye R, Zhao J, Bowman L, Ding M (2012) Genotoxicity and carcinogenicity of cobalt-, nickel- and copper-based nanoparticles (Review). Exp Ther Med 4:551–561

Marques CC, Sánchez-Chardi A, Gabriel SI, Nadal J, Viegas-Crespo AM, da Luz Mathias M (2007) How does the greater white-toothed shrew, *Crocidura russula*, responds to long-term heavy metal contamination? A case study. Sci Total Environ 376:128–133

Marsh E, Anderson E (2011) Ni-Co laterites – a deposit model. U.S. Geological Survey

Mason CF, Stephenson A (2001) Metals in tissues of European otters (*Lutra lutra*) from Denmark, Great Britain and Ireland. Chemosphere 44:351–353

Mathur AK, Shanker R (2001) Dermal toxicity of linear alkylbenzene sulphonate and nickel in guinea pigs. Cutan Ocul Toxicol 20:23–27

Mathur N, Pandey G, Jain GC (2010) Male reproductive toxicity of some selected metals: a review. J Biol Sci 10:396–404

McDonough WF, Sun SS (1995) The composition of the Earth. Chem Geol 120:223–253

Miles AK, Ohlendorf HM (1993) Environmental contaminants in canvasbacks wintering on San Francisco Bay, California. Calif Fish Game 79:28–38

Muyssen BT, Brix KV, DeForest DK, Janssen CR (2004) Nickel essentiality and homeostasis in aquatic organisms. Environ Rev 12:113–131

Nielsen FH (1974) Essentiality and function of nickel. In: Hoekstra WG, Suttie JW, Ganther HE, Mertz W (eds) Trace element metabolism in animals – 2. University Park Press, Baltimore, pp 381–395

Nielsen FH (2000) Importance of making dietary recommendations for elements designated as nutritionally beneficial, pharmacologically beneficial, or conditioinally essential. J Trace Elem Exp Med 13:113–129

Nielsen FH, Myron DR, Givand SH, Ollerich DA (1975) Nickel deficiency and nickel-rhodium interaction in chicks. J Nutr 105:1607–1619

Nordberg GF, Fowler BA, Nordberg M, Friberg LT (2007) Handbook on the toxicology of metals. Elsevier, London

Novelli ELB, Novelli Filho JLVB, Rodrigues NL, Ribas BO, Barbosa LL (1998) Long-term toxicity following acute administration of nickel. Toxicol Subst Mech 17:175–185

NPI (2015) Nickel and compounds. National Pollutant Invectory website. http://www.npi.gov.au/resource/nickel-compounds. Accessed 28 Feb 2015

Obone É, Chakrabarti SK, Bai C, Anwar Malick M, Lamontagne L, Subramanian KS (1999) Toxicity and bioaccumulation of nickel sulfate in Sprague-Dawley rats following 13 weeks of subchronic exposure. J Toxicol Environ Health Part A 57:379–401

Orłowski G, Kasprzykowski Z, Dobicki W, Pokorny P, Wuczyński A, Polechoński R, Mazgajski TD (2014) Residues of chromium, nickel, cadmium and lead in Rook Corvus frugilegus eggshells from urban and rural areas of Poland. Sci Total Environ 490:1057–1064

Outridge PM, Scheuhammer AM (1993) Bioaccumulation and toxicology of nickel: implications for wild mammals and birds. Environ Rev 1:172–197

Pandey R, Singh SP (2001) Seminal toxicity of nickel sulfate in mice. Biol Trace Elem Res 82: 211–215

Pari L, Amudha K (2011) Hepatoprotective role of naringin on nickel-induced toxicity in male Wistar rats. Eur J Pharmacol 650:364–370

Parsons JG, Lopez ML, Gonzalez CM, Peralta-Videa JR, Gardea-Torresdey JL (2010) Toxicity and biotransformation of uncoated and coated nickel hydroxide nanoparticles on mesquite plants. Environ Toxicol Chem 29:1146–1154

Pennington JA, Jones JW (1987) Molybdenum, nickel, cobalt, vanadium, and strontium in total diets. J Am Diet Assoc 87:1644–1650

Pereira R, Pereira ML, Ribeiro R, Gonçalves F (2006) Tissues and hair residues and histopathology in wild rats (*Rattus rattus* L.) and Algerian mice (*Mus spretus Lataste*) from an abandoned mine area (Southeast Portugal). Environ Pollut 139:561–575

Phipps T, Tank SL, Wirtz J, Brewer L, Coyner A, Ortego LS, Fairbrother A (2002) Essentiality of nickel and homeostatic mechanisms for its regulation in terrestrial organisms. Environ Rev 10: 209–261

Rasmussen KL, Malvin DJ, Wasson JT (1988) Trace element partitioning between taenite and kamacite; relationship to the cooling rates of iron meteorites. Meteorit Planet Sci 23:107–112

Ray PC, Hongtao Y, Fu PP (2009) Toxicity and environmental risks of nanomaterials: challenges and future needs. J Environ Sci Health Part C 27:1–35

Reeves RD, Baker AJM, Borhidi A, Berazain R (1996) Nickel-accumulating plants from the ancient serpentine soils of Cuba. New Phytol 133:217–224

Reichrtova E, Takac L, Sulicova L, Foltinova J (1988) Biological monitoring of airborne metal particles originated from nickel refinery dump. In: Abbou R (ed) Hazardous waste: detection, control, treatment. Elsevier, Amsterdam, pp 931–936

Rendall REG, Phillips JI, Renton KA (1994) Death following exposure to fine particulate nickel from a metal arc process. Ann Occup Hyg 38:921–930

Robinson BH, Chiarucci A, Brooks RR, Petit D, Kirkman JH, Gregg PEH, De Dominicis V (1997) The nickel hyperaccumulator plant *Alyssum bertolonii* as a potential agent for phytoremediation and phytomining of nickel. J Geochem Explor 59:75–86

Rojas E, Herrera LA, Poirier LA, Ostrosky-Wegman P (1999) Are metals dietary carcinogens? Mutat Res Genet Toxicol Environ Mutagen 443:157–181

Ruiz-Olmo J, Lafontaine L, Prignioni C, Lopez-Martin J, Santos-Reis M (2000) Pollution and its effects on otter populations in south-western Europe. In: Conroy JWH, Yoxon P, Gutleb AC (eds) Proceedings of the first otter toxicology conference. International Otter Survival Fund, Isle of Skye, pp 1–20

Saini S, Nair N, Saini MR (2014) Prenatal exposure to nickel on pregnant Swiss albino mice and fetal development. Toxicol Environ Chem 96:650–659

Salnikow K, Gao M, Voitkun V, Huang X, Costa M (1994) Altered oxidative stress responses in nickel-resistant mammalian cells. Cancer Res 54:6407–6412

Sánchez-Chardi A, Marques CC, Nadal J, da Luz Mathias M (2007) Metal bioaccumulation in the greater white-toothed shrew, *Crocidura russula*, inhabiting an abandoned pyrite mine site. Chemosphere 67:121–130

Sánchez-Chardi A, Marques CC, Gabriel SI, Capela-Silva F, Cabrita AS, López-Fuster MJ, Nadal J, Mathias ML (2008) Haematology, genotoxicity, enzymatic activity and histopathology as biomarkers of metal pollution in the shrew *Crocidura russula*. Environ Pollut 156:1332–1339

Sandström AIM, Wall SGI, Taube A (1989) Cancer incidence and mortality among Swedish smelter workers. Br J Ind Med 46:82–89

Sawicka-Kapusta K (1979) Roe deer antlers as bioindicators of environmental pollution in southern Poland. Environ Pollut 19:283–293

Shallari S, Schwartz C, Hasko A, Morel JL (1998) Heavy metals in soils and plants of serpentine and industrial sites of Albania. Sci Total Environ 209:133–142

Sienko MJ, Plane RA (1979) Chemistry: principles and applications. McGraw Hill, New York

Silva N, Senanayake H, Waduge V (2013) Elevated levels of whole blood nickel in a group of Sri Lankan women with endometriosis: a case control study. BMC Res Notes 6:13

Smith GJ, Rongstad OJ (1982) Small mammal heavy metal concentrations from mined and control sites. Environ Pollut 28:121–134

Stangl GI, Kirchgessner M (1996) Nickel deficiency alters liver lipid metabolism in rats. J Nutr 126: 2466–2473
Stangl GI, Kirchgessner M (1997) Effect of nickel deficiency on fatty acid composition of total lipids and individual phospholipids in brain and erythrocytes of rats. Nutr Res 17:137–147
Stangl GI, Eidelsburger U, Kirchgessner M (1998) Nickel deficiency alters nickel flux in rat everted intestinal sacs. Biol Trace Elem Res 61:253–262
Stangl GI, Roth-Maier DA, Kirchgessner M (2000) Vitamin B-12 deficiency and hyperhomocysteinemia are partly ameliorated by cobalt and nickel supplementation in pigs. J Nutr 130: 3038–3044
Statista (2015a) Statistics and facts about nickel. In: Statistics Portal. http://www.statista.com/topics/1572/nickel/. Accessed 28 Feb 2015
Statista (2015b) Major countries in worldwide nickel mine production from 2010 to 2013. In: Statistics Portal. http://www.statista.com/statistics/264642/nickel-mine-production-by-country/. Accessed 28 Feb 2015
Stridsklev IC, Schaller K-H, Langård S (2004) Monitoring of chromium and nickel in biological fluids of stainless steel welders using the flux-cored-wire (FCW) welding method. Int Arch Occup Environ Health 77:587–591
Sulinskiene J, Baranauskiene D, Naginiene R, Ivanov L (2014) Protective effect of zinc ions against lead and nickel induced inhibition of δ-aminolevulinic acid dehydratase activity in mice liver. Trace Elem Electrolytes 32:91–96
Sunderman FW, Nomoto SJ, Morang R, Nechay MW, Burke CN, Nielsen SW (1972) Nickel deprivation in chicks. J Nutr 102:259–267
Sydor AM, Zamble DB (2013) Nickel metallomics: general themes guiding nickel homeostasis. Met Ions Life Sci 12:375–416
Thomas VG, Roberts MJ, Harrison PTC (2009) Assessment of the environmental toxicity and carcinogenicity of tungsten-based shot. Ecotoxicol Environ Saf 72:1031–1037
Toman R, Massanyi P, Adamkovicova M, Lukac N, Cabaj M, Martiniaková M (2012) Quantitative histological analysis of the mouse testis after the long-term administration of nickel in feed. J Environ Sci Health Part A 47:1272–1279
Tomei F, Rosati MV, Ciarrocca M, Marchetti MR, Baccolo TP, Anzelmo V, Tomao E (2004) Urban pollution and nickel concentration in serum. Int J Environ Health Res 14:65–74
USGS (1980) 1980 Minerals yearbook nickel. US Geological Survey
USGS (1985) 1985 Minerals yearbook nickel. US Geological Survey
USGS (1990) 1990 Minerals yearbook nickel. US Geological Survey
USGS (1995) 1995 Minerals yearbook nickel. US Geological Survey
USGS (2000) 2000 Minerals yearbook nickel. US Geological Survey
USGS (2007) 2005 Minerals yearbook nickel. US Geological Survey
USGS (2012a) Mineral Commodity Summaries: Nickel
USGS (2012b) 2010 Minerals yearbook nickel. US Geological Survey
USGS (2013) 2011 Minerals yearbook nickel. US Geological Survey
USGS (2016) U.S. Geological Survey website
van der Ent A, Baker AJM, Reeves RD, Pollard AJ, Schat H (2013) Hyperaccumulators of metal and metalloid trace elements: facts and fiction. Plant Soil 362:319–334
van der Ent A, Erskine P, Sumail S (2015) Ecology of nickel hyperaccumulator plants from ultramafic soils in Sabah (Malaysia). Chemoecology 25:243–259
van Eeden PH, Schoonbee HJ (1996) Metal concentrations in liver, kidney, bone and blood of three species of birds from a metal-polluted wetland. Water SA 22:351–358
van Straalen NM, Ernst WHO (1991) Metal biomagnification may endanger species in critical pathways. Oikos 62:255–256
Walker LA, Lawlor AJ, Chadwick EA, Potter E, Pereira MG, Shore RF (2011) Inorganic elements in the livers of the Eurasian otter, Lutra lutra, from England and Wales in 2009 – a Predatory Bird Monitoring Scheme report

Wang Y, Hao Z, Zhang L, Liang C (2016) Nanomaterials: friend or foe to male fertility? World J Urol:1–3

Warren RJ, Wallace BM, Bush PB (1990) Trace elements in migrating blue-winged teal: seasonal-, sex- and age-class variations. Environ Toxicol Chem 9:521–528

WHO (2000) Nickel. In: Air quality guidelines, 2nd edn. WHO Regional Office for Europe, Copenhagen

WVDL (2015) Normal range values for WVDL toxicology. https://www.yumpu.com/en/document/view/52919318/normal-range-values-for-wvdl-toxicology. Accessed 28 April 2015

Yokoi K, Uthus EO, Nielsen FH (2003) Nickel deficiency diminishes sperm quantity and movement in rats. Biol Trace Elem Res 93:141–154

# Chapter 10
# Selenium, Se

**Bogumiła Pilarczyk, Agnieszka Tomza-Marciniak, Renata Pilarczyk, Andrzej Marciniak, Małgorzata Bąkowska, and Ewa Nowakowska**

**Abstract** Selenium (Se), in small amounts, is necessary for the proper development and health of humans and animals. This element plays a structural and enzymatic role in the organism. Soils in many regions of Europe are Se deficient in comparison to the USA, Canada, and China. Anthropogenic activity contributes to the introduction of selenium into the environment, as well as to mobilization from subsoils from various land exploitation processes. From the beginning of the twenty-first century, China has emitted the most Se into the atmosphere from anthropogenic sources and, significantly, shows dynamic growth in this matter. The measurement of Se levels in animal tissues is the most commonly used biomarker in evaluation of exposure to this element. Wild ruminants from Cervidae like moose, red deer, roe deer, mule deer, white-tailed deer, and elk, as well as some representatives of the Bovidae family (pronghorn and bighorn sheep), are considered as good bioindicators of environmental selenium levels. From the carnivorous animals, common selenium bioindicators are a fox, mink, and raccoon. The content of Se in the organism is usually evaluated in the liver and kidneys.

B. Pilarczyk (✉) · A. Tomza-Marciniak · A. Marciniak · M. Bąkowska · E. Nowakowska
Department of Animal Reproduction Biotechnology and Environmental Hygiene, West Pomeranian University of Technology in Szczecin, Szczecin, Poland
e-mail: bogumila.pilarczyk@zut.edu.pl

R. Pilarczyk
Laboratory of Biostatistics, West Pomeranian University of Technology in Szczecin, Szczecin, Poland



E. Kalisińska (ed.), *Mammals and Birds as Bioindicators of Trace Element Contaminations in Terrestrial Environments*,
https://doi.org/10.1007/978-3-030-00121-6_10

# 1 Introduction

Selenium (Se) was discovered in 1817 by Swedish chemist J. J. Berzelius. For many years it was considered to be a dangerous element to human beings and animals as it had been identified by its toxic features. Only in 1957 did Schwarz and Foltz in their studies on rats demonstrated that Se had prevented necrosis of the liver, which indicated the important biological role of selenium (Brown and Arthur 2001). It was proven then that this element in small concentrations is crucial for the proper development and health of humans and animals.

The Earth's crust is the most significant source of selenium, although the distribution and concentration of this element vary greatly between regions. For this reason some regions of the world are characterized by excessive levels of Se, and some remain Se deficient. The amount of selenium available to humans and animals depends on its content in soils, the chemical forms in which selenium occurs, the pH and redox potential of the soil, as well as the presence of other organic compounds, climatic factors, and the activity of soil microorganisms. Anthropogenic activity contributes to the introduction of Se into the environment, as well as to mobilization from subsoils in various land exploitation processes (Rayman 2000; Fordyce 2013).

Most Se in vertebrate tissues is integrated into selenoproteins, with only a small amount in a free form. Selenium plays a structural and enzymatic role in the organism. The most important selenoproteins include glutathione peroxidase, iodothyronine deiodinase, P and W selenoproteins, as well as thioredoxin reductase (Flohe et al. 2000; Rayman 2000; Gladyshev 2001). Selenium participates in thyroid hormone metabolism as part of the important enzymes engaged in the functions of this organ, including thyroxine 5-deiodinase which catalyzes the deiodination of thyroxine and its conversion into an active form of triiodothyronine. A deficiency of Se in humans and animals causes disturbances in bone mineralization and tooth growth (Moreno-Reyes et al. 2001; Fordyce 2013; Flueck 2015). In humans, dilated cardiomyopathy (Keshan disease) and osteochondropathy (Kashin-Beck disease) are the most well-known examples of Se deficiency (Kim et al. 2001; Zhang et al. 2010).

Selenium assists in the functions of the immune system and stimulates the production of antibodies, where a deficiency may contribute to a suppression of the immune response to bacterial, viral, and fungal infections by inhibiting the synthesis of prostaglandin and immunoglobulin, as well as by decreasing the activity of T lymphocytes, NK cells, and macrophages, which are engaged in the immunological mechanisms of homeothermic vertebrates (Johnson et al. 2000; Rayman 2000; Arthur et al. 2003; Hartikainen 2005).

A surplus of Se in the organism causes some symptoms of intoxication (selenosis endemica). Typical disorders caused by excessive levels of Se include liver damage, depression, emaciation, nervous system dysfunction, tooth gnashing, salivation, hair loss or rough hair coat, improper growth or necrosis of hooves, stiffness and irritation of limbs, dyspnea, hearth muscle atrophy, and anemia. An excessive intake of Se leads to the inhibition of cellular proliferation, DNA replication, and protein synthesis and supports oxidative stress (Kim and Mahan 2001; Dodig and Cepelak 2004; Tinggi 2005; Zachary and McGavin 2014).

Soils across a large part of Europe are relatively selenium deficient in comparison to the USA, Canada, and China. In recent years, Se content in northern European ungulates has fallen. This deficiency is likely the effect of a decrease of selenium levels in the environment, caused by the introduction of new technologies in agriculture, which adversely influence the assimilation of selenium into plants. The limited transport of Se from soils to plants, and later to animals, manifests through the diminishing content of this element in these organisms (Winkel et al. 2015).

In wild animals, clinical and subclinical cases of Se deficiency remain practically unnoticed. The situation however is different in farm animals, especially in grazing livestock, in which the symptoms of selenium deficiency have been described frequently. In contrast, in free-ranging animals, even in highly Se-deficient areas, by ranging across much larger areas, they can benefit from a more diversified diet. Moreover, they have probably developed more effective mechanisms of Se management (Wichtel 1998; Spears 2011; Flueck et al. 2012). Pathological changes caused by a Se and iodine deficiency have been found in one wild and endangered species of deer in a Se- and iodine-deficient area in Chile (Wichtel 1998; Spears 2011; Flueck 2015).

Due to the important role of Se in vertebrate organisms, as well as the health threats caused by its deficiency or surplus in other regions of the world, ecotoxicological research (including comparative studies) and biomonitoring of this important microelement are indicated, especially in wild-ranging birds and mammals.

## 2 General Properties

Selenium belongs to the group of trace elements, with the average concentration of Se in the Earth's crust estimated at 0.05 mg $kg^{-1}$ (Taylor and McLennan 1995). The Se concentration in the Earth's crust is ~0.09 mg Se $kg^{-1}$ (NRC 1983). It is located in the group VI of the periodic table. The atomic mass of Se is 78.96 (Table 10.1). Together with sulfur, oxygen, tellurium, and polonium, it builds the oxygen family, known also as chalcogens.

Selenium has been discovered in slime from the production of sulfuric acid from pyrites. This element comprises two amorphous and four crystalline forms. In nature, six stable isotopes of Se are present: $^{80}$Se (79.61%), $^{78}$Se (23.77%), $^{76}$Se (9.37%), $^{82}$Se (8.73%), $^{77}$Se (7.63%), and $^{74}$Se (0.89%) (Taylor and McLennan 1995). Selenium may occur in the following oxidation states: −2 (selenide), 0 (elementary Se), +4 (selenite), and +6 (selenate). It composes selenides with metals, hydrogen selenide $H_2Se$ with hydrogen (selan), and oxides with oxygen ($SeO_2$ and $SeO_3$), which are the selenic (IV) and selenic (VI) acid anhydrides $H_2SeO_3$ and $H_2SeO_4$, respectively (Broadley et al. 2006; Kabata-Pendias and Szteke 2015).

**Table 10.1** Basic physicochemical properties of selenium

| Feature | |
|---|---|
| Atomic number | 34 |
| Atomic mass | 78.96 |
| Density (g $cm^{-3}$) | 4.26–4.79 |
| Melting point (°C) | 217 |
| Boiling point (°C) | 685.4 |
| Electronegativity | 2.55 (Pauling) |
| Electronegativity (eV) | 2.4 |
| Oxidation states | −2, 0, +4, +6 |
| Mass of stable isotopes | 74, 76, 77, 78, 80, 82 |
| Electron configuration | $1s^2\ 2s^2\ p^6\ 3s^2\ p^6\ d^{10}\ 4s^2\ p^4$ |

Newland (1982)

## 3 Selenium Minerals, Production, and Uses

Many factors influence the presence and circulation of Se in nature. It is not usually distributed evenly—there are some areas abundant in Se (large areas of North and South America, part of China) as well as Se-deficient areas (a large part of Europe, including some regions of Poland, a few Chinese provinces, New Zealand). A significant diversity occurs in China, where the southwestern and northeastern parts are characterized by extremely low Se levels and the central provinces represent the Se-rich soils, with a risk of intoxication in grazing animals (Xu and Jiang 1985; Amouroux et al. 2001; Blazina et al. 2014).

Particularly high contents of Se in the soil are present in Wyoming and North and South Dakota in the USA (Reilly 1996), Enshi County in China, and in parts of Ireland, Colombia, and Venezuela (Combs 2001). As noted by Haug et al. (2007), many European soils show relatively high concentrations of Se from the naturally high deposition of this element from the sea (e.g., Ireland, England, Scotland, and the Netherlands).

Selenium is a natural and widespread component of the Earth's crust, despite its disparate distribution. It is contained in a variety of rocks, minerals, and soils and moreover occurs in volcanic material. Selenium does not appear individually, but 40 minerals are known to be mostly formed with this element and 37 others where Se is a minor component, mainly as sulfides (NRC 1983; Butterman and Brown 2004). The average Se concentration is much higher in sedimentary rocks, especially shales and coal, than in igneous rocks (Krauskopf 1982). Selenium is found in nature in pyrites of copper and iron and sulfide ores of copper, lead, nickel, gold, or silver, at variable levels between 0.1 and 2 mg Se $kg^{-1}$ dry weight dw (ppm). The most important selenium minerals are berzelianite ($Cu_2Se$), naumannite ($Ag_2Se$), and chalcomenite ($CuSeO_3{\cdot}2H_2O$). High content of Se is accompanied by deposits of sulfur and sulfides. However, its ratio to sulfur in these deposits is only 1:6000. Minerals particularly rich in selenium are sulfides (up to 120 mg Se $kg^{-1}$ dw) and volcanic soils, where Se content can reach 200 $kg^{-1}$ dw (Smakowski et al. 2011).

**Table 10.2** Concentrations of selenium in rocks and other natural sources

| Material | Concentration (mg kg$^{-1}$ dw) | Range (mg kg$^{-1}$ dw) | References |
|---|---|---|---|
| Earth's crust | | 0.05–0.09 | Lakin (1972) and Frost and Ingvoldstad (1975) |
| Igneous rocks | | | |
| Ultrabasic (dunit, peridotite, pyroxenite) | | 0.02–0.05 | Kabata-Pendias and Pendias (2000) |
| Basic (basalt, gabbro) | | 0.01–0.05 | Kabata-Pendias and Pendias (2000) |
| Intermediate (syenite) | | 0.02–0.05 | Kabata-Pendias and Pendias (2000) |
| Intermediate to acidic (rhyolite, trachyte, dacite) | | 0.02–0.05 | Kabata-Pendias and Pendias (2000) |
| Acidic (granite, gneiss) | | 0.01–0.05 | Kabata-Pendias and Pendias (2000) |
| Basic volcanic rocks | 0.155 | | Koljonen (1973) |
| Acid volcanic rocks | 0.116 | | Koljonen (1973) |
| Volcanic tuff (China, South Qinling Mountain) | 32 | | Kunli et al. 2004 |
| Black carbonaceous slate (China, South Qinling Mountain) | 22 | | Kunli et al. (2004) |
| Sandstones | <0.05 | <0.01–0.05 | Turekian and Wedepohl (1961) and Ebens and Shacklette (1982) |
| Carbonates | 0.08 | | Turekian and Wedepohl (1961) |
| Ocean sediments | | 0.34–4.8 | de Goeij et al. (1974) |
| Chondrites | 8.0 | | Ebens and Shacklette (1982) |
| Shales (general) | 0.06 | 0.05–0.06 | Green (1959) and Ebens and Shacklette (1982) |
| Limestones | 0.08 | | Ebens and Shacklette (1982) |
| Phosphate rocks | | 1–300 | Mayland et al. (1989) |

Parent materials with the highest concentration of Se are black shales and phosphate rocks, with about 600 and 1–300 mg Se kg$^{-1}$ dw, respectively (Presser 1994). Black shales of the Selwyn Basin (Yukon) contain a relatively high amount of selenium, up to 1 mg kg$^{-1}$ dw (Gamberg et al. 2005b). Much higher concentrations were found in cretaceous sedimentary rock formations in central Saskatchewan, especially in White-Speckled Shales from the Lower Cretaceous: up to 11.7 mg kg$^{-1}$ dw (Dunn 1990). Magmatic rocks (granites and basalts) do not contain much Se, mostly <0.05 mg kg$^{-1}$ dw. Some sedimentary rocks with the exception of sandstone (0.02–0.05 kg$^{-1}$ dw) contain much higher selenium concentrations (shales 0.6 mg kg$^{-1}$ dw; argillaceous sediments 0.4–0.6 mg kg$^{-1}$ dw, limestones and dolomites 0.03–0.10 mg kg$^{-1}$ dw) (Kabata-Pendias and Pendias 2000) (Table 10.2). Selenium is also present in fossil fuels in concentrations ranging from 0.046 to 10.65 kg$^{-1}$ dw in coal (3.0 mg $kg^{-1}$ dw average) and from 0.006 to 2.2 mg $kg^{-1}$ in oil (0.6 mg $kg^{-1}$ average) (Marier and Jaworski 1983). Natural background concentrations of Se in air in nonvolcanic areas are within the range 0.01–1.0 ng $m^{-3}$ (Frankenberger and Karlson 1994).

Selenium is obtained almost exclusively as a coproduct of copper processing (anodic slimes produced by copper refining contain roughly 5–25% selenium), as well as furnace dust from copper ironworks. It is estimated that in the known deposits of copper ore, world Se resources amount to about 90,000 tons (NRC 1983; Plant et al. 2003). The authors suggested that the amount of selenium in undeveloped and potential deposits may even be 2.5 times higher.

Although a relatively high concentration of Se is present in coal (1.5 ppm on average, 80–90 times more than in copper ores), the technical potential of it is not being recovered from coal. Small amounts of Se are also present in lead, nickel, platinum, gold, and zinc ores. The Wolverine deposit of zinc–silver (Zn–Ag) ore in Canada is an exception, where the proportion of Se is very high. Wolverine is the richest of Se deposits and has a very high concentration compared to all known deposits. Other similar massive volcanogenic sulfide deposits characterized by high selenium content include those in the Mattagami District, Quebec, and Skellefte District in Sweden (both of which are nearing depletion). The mining and smelting of copper–nickel ores at Sudbury, Ontario, Canada, alone release about 2 tons of Se into the environment daily, representing the greatest single source of Se emitter in the world (NRC 1983; Plant et al. 2003). The production of Se recovered from secondary sources (electrical equipment) in 2000 reached about 20% of total supply; however, in recent years this has almost entirely disappeared (Smakowski et al. 2011). Over the next years, the recovery of selenium from secondary sources may increase due to the recycling of solar batteries (with Cu-In-Ga-Se-CIGS).

Selenium is widely spread in nature, and in most rocks and soils, the content ranges from 0.1 to 2.0 ppm (Fishbein 1983). In soils, the concentration usually does not exceed 2 mg $kg^{-1}$ dw (with the exception of loamy soils: 2.3–4.2 mg $kg^{-1}$ dw). The mean concentration of Se in the soil is usually between 0.1 and 0.7 mg $kg^{-1}$ dw. In clay soils it is usually higher (from 0.8 to 2 mg $kg^{-1}$ dw) and is highest in tropical soils: from 2 to 4.5 mg $kg^{-1}$ dw. Volcanic soils and granite are lacking in Se. These soils are found, for example, in the mountainous countries of Northern Europe, such as Finland, Sweden, and Scotland (Table 10.3; Fig. 10.1). The low content of Se in Scandinavian soils is due to the commingling of postglacial material by water and wind as well as by leeching of maternal rocks in geological processes (Winkel et al. 2015). So-called selenium provinces can be found in the world. The content of Se in such soils may reach 1200 mg $kg^{-1}$ dw in provinces located in Japan, China, Canada, and the USA, among others (Winkel et al. 2015) (Figs. 10.2 and 10.3). Selenium in soils is positively correlated with the amount of atmospheric precipitation and negatively with the distance to the sea as the content of Se in the precipitation decreases (Wang et al. 1994; Blazina et al. 2014).

The presence of Se in natural waters is mostly the result of seleniferous rock erosion (Callahan 1979). The concentration of Se in such waters depends on the geochemical surrounding, the elution from rocks, and possible contamination (Table 10.4). In natural waters, selenium is present in various forms ($SeO_3^{2-}$, $HSeO_4^-$, $HSeO_3^-$, $H_2SeO_3$). As reported by Kabatas-Pendias and Pendias (2000), the highest

**Table 10.3** Concentrations of selenium in various soils

| Material | Concentration (mg kg$^{-1}$ dw) | Range (mg kg$^{-1}$ dw) | References |
|---|---|---|---|
| Worldwide | 0.4 | 0.1–2 | Berrow and Ure (1989) |
| *Asia* | | | |
| China (general) | | 0.02–3.81 | Tan (1989) |
| India | 3.63 | 0.25–4.55 | Dhillon and Dhillon (2003) |
| Taiwan | | 0.03–0.65 | Wang and Chen (2003) |
| Japan | | 0.11–2.72 | Kang et al. (1990) |
| *Europe* | | | |
| Denmark | | 0.14–0.52 | Hamdy and Gissel-Nielsen (1976) |
| Finland | | 0.05–1.24 | Aro and Alfthang (1998) |
| Norway | | 0.043–2.73 | Wu and Lag (1988) |
| France | | 0.10–0.70 | INRA (2008) |
| Germany | 0.123 | 0.02–0.42 | Hartfield and Bahners (1988) |
| Poland (PL) | | 0.04–0.64 | Piotrowska (1984) |
| Kujawy and Pomorze, PL | 0.138 | 0.035–0.332 | Borowska et al. (2007) |
| Wielkopolska, PL | 0.19 | 0.00–0.57 | Tomza-Marciniak et al. (2010) |
| Lithuania | | 0.144–0.228 | Antanaitis et al. (2008) |
| Slovakia | | 0.2–0.33 | Ducsay et al. (2007) |
| Romania | | 0.143–0.237 | Lăcătuşu et al. (2010) |
| Serbia | | 0.12–0.44 | Jakovljeviã et al. (1995) |
| Great Britain | | 0.1–4 | Broadley et al. (2006) |
| Great Britain | 0.5 | 0.01–4.66 | Thornton (1983) |
| Scotland, Glasgow | 1.0 | 0.1–6.60 | Fordyce et al. (2009) |
| Northern Ireland | 0.80 | 0.1–7.8 | GSNI (in preparation) |
| *North America* | | | |
| Canada | | 0.41–2.09 | Levesque (1977) |
| USA (general) | | 0.10–4.30 | Jacobs (1989) |
| USA, New Mexico | 0.23 | 0.039–1.4 | Shacklette and Boemgen (1984) |

levels of selenium are present in ground and underground water that flushes through formations rich in this element, for example, over pyrite shales, and through landfill areas that contain dusts with a high Se content from coal combustion. It is estimated that only about 15% of selenium occurs in a dissolved form, while large amounts are deposited as suspensions and sediments (Kabata-Pendias and Szteke 2015). The content of Se in rain and snow is usually similar to fresh water and ranged between 0.03 and 0.3 μg L$^{-3}$ (Ĉuvardić 2003) but may change according to the part of the world.

In atmospheric air, the concentration of Se ranges from 0.008 to 5 ng m$^{-3}$ (Wang et al. 1994). The lowest concentration of selenium was noted over the South Pole 0.004 ng m$^{-3}$, with the highest air concentration over urban and industrial areas (up to tens of ng m$^{-3}$). Moreover, considerable concentrations of Se (1 ng m$^{-3}$) have

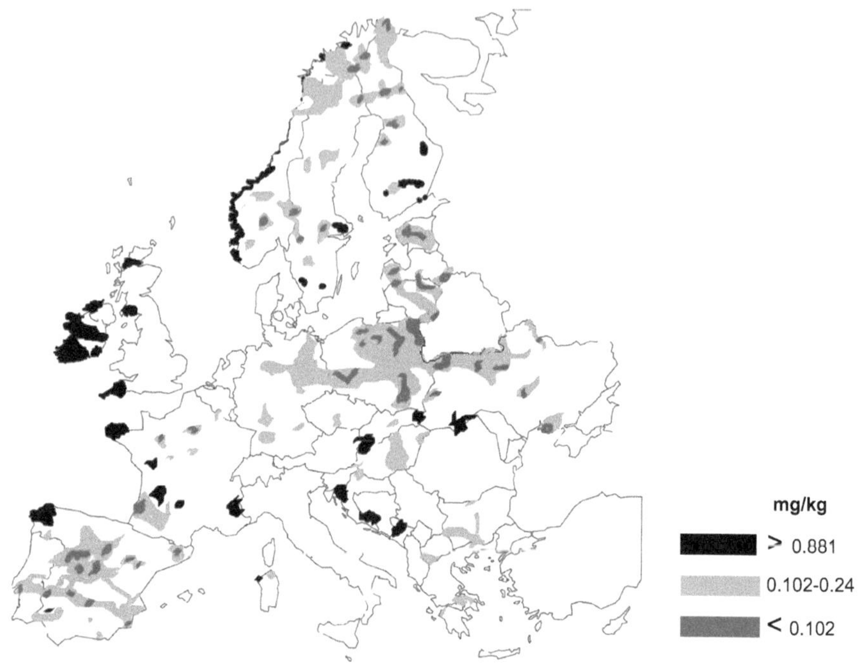


**Fig. 10.1** The content of selenium in the soils of Europe (based on the map by Reimann et al. 2013)

been found over dumps and swamps (Beavington et al. 2004; Wen and Carignan 2007).

Selenium is emitted into the atmosphere mostly by the combustion of minerals (including coal and oil processing products) in volatile forms (elementary selenium, selenium dioxide, dimethyl selenide). These compounds may persist in the atmosphere for up to 45 days (Wang et al. 1994; Blazina et al. 2014).

## 4 Selenium in Nature: Geogenic and Anthropogenic Sources

As an effect of volcanic rock erosion, Se is transferred to the atmosphere and then to oceans, seas, and lakes as well as soils. Selenium is also transferred from the eroding rocks directly to underground and groundwaters, and via the soil-forming processes, also to soils. The process of rock erosion releases to the atmosphere from 100,000 to 200000 tons of this element every year (Andren and Klein 1975). Selenium is assimilated from the soil by plants and through them moves to humans and animals, mostly by ingestion and to a smaller extent by inhalation from the fresh atmospheric air. It returns to soil by the dry and wet deposition from decaying plants and animals as well as from human and animal excrements (Fig. 10.4).

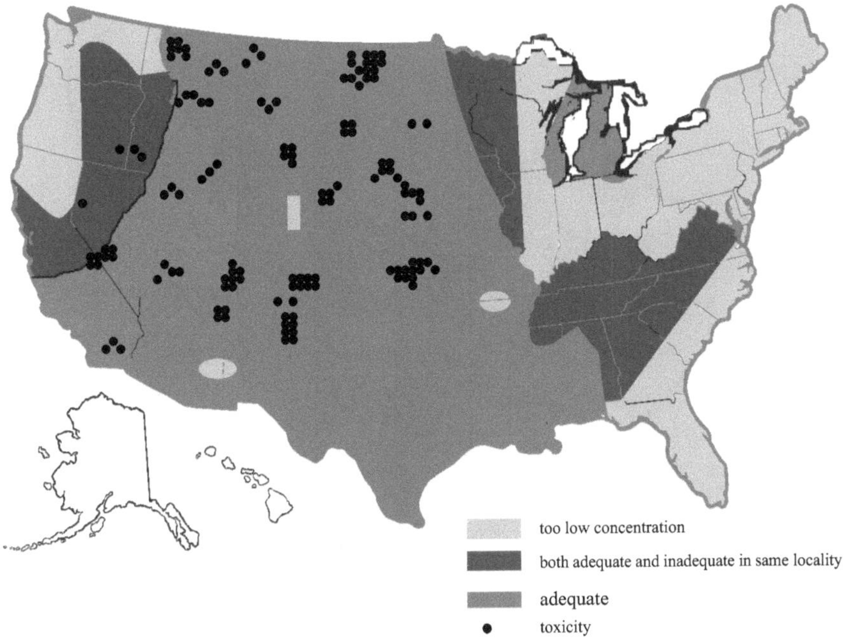


**Fig. 10.2** The content of selenium in the soils of USA, based on the map http://www.swampyacresfarm.com/SeleniumSupplements.html

In Europe, a toxic soil selenium level was found in a few places, including Wales and Ireland (Fleming 1962). These seleniferous soils are typically low lying, poorly drained, and of high pH and organic matter status. The soils have been influenced to a large degree by percolating waters from Se-rich rocks where black shales are the predominant facies. Seleniferous soils may contain very much selenium, even up to 200 mg $kg^{-1}$. Soils that contain more than 5 mg $kg^{-1}$ are considered as toxic (McGrath et al. 2008). Toxic concentrations of Se are found in some parts of China, Hawaii, Mexico, Columbia, and western part of the USA (McNeal and Balistrieri 1989; Blazina et al. 2014). However, to a global scale, the deficiency of selenium is a more severe problem than its abundance.

The atmosphere is an important temporal reservoir of Se in the global store of this element, while the main sources of emission are the oceanic and continental biospheres (Table 10.5). It is estimated that natural and anthropogenic sources emit into the atmosphere 60% and 40%, respectively, and in recent decades, the total annual emission of Se in a global scale was evaluated at 13–19 thousand tons (Mosher and Duce 1987; Wen and Carignan 2007; Winkel et al. 2015).

Selenium evaporates from the surface of the seas and oceans as a dimethyl, which is then incorporated in processes carried out by microbes and higher plants. Selenium is also emitted from volcanoes, such as the European Mt Etna (Mosher and Duce 1987).

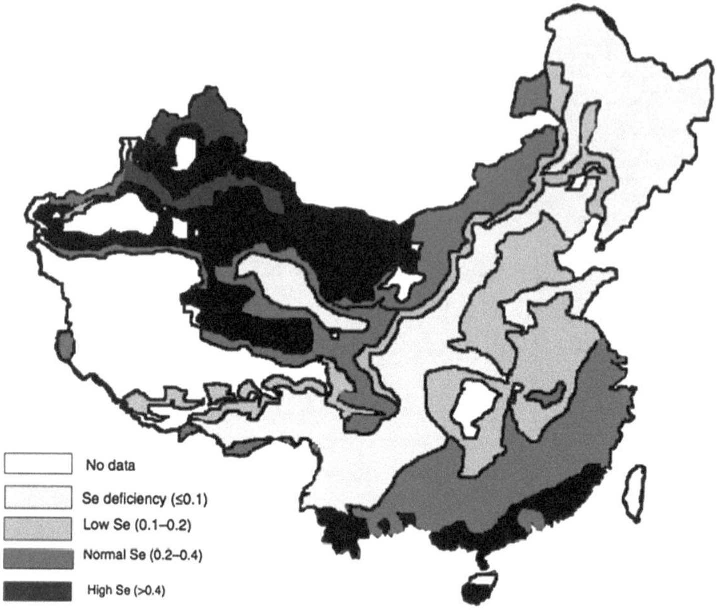


**Fig. 10.3** Distribution of total selenium in the soils of China, based on the map by Blazina et al. (2014)

The contamination of the environment by Se is caused by fossil fuels combustion (oil, hard coal) and emissions from industrial plants (Pezzarossa et al. 2007). In the past, Se was used in pesticide production, but due to its stability in soils and subsequent contamination of food crops, its use has been limited (ATSDR 2003). An increased concentration of Se in land plants, lichens, and mosses located at a distance from anthropogenic sources of emission suggests that precipitation is an important vector of contamination (Wen and Carignan 2007).

Selenium is also present in coal as well as in coal dust and ash (Tian et al. 2011). In the industrialized areas, the processes of coal combustion have contributed to a large release of Se into the atmosphere; however, the simultaneous emission of $SO_2$ caused a decrease in its assimilation. Studies in coal power plants have demonstrated that as a result of coal combustion, 10–60% of the Se associated with coal was released to the atmosphere (Klusek et al. 1983; Conzemius et al. 1984). Modern technologies in coal combustion limit the emission of contaminants such as Se, for example, wet flue gas desulfurization (WFGD) in power plants (Tian et al. 2010). The washing of coal before combustion is an effective way to reduce ash and $SO_2$ emission. It can reduce sulfur pyrite content by 40%, along with trace elements concentrations (You and Xu 2010). The biggest amount of Se emission from combusted coal is in China, which is connected with the rapid growth of the economy in that country. The use of coal provides more than 75% of the country's total energy sources (You and Xu 2010). From 1980 to 2007, ~145% increase in Se emissions took place in China (Fig. 10.5) (Tian et al. 2010). The World Coal

**Table 10.4** The content of selenium in waters

| Water type | Content (μg $L^{-1}$) | References |
|---|---|---|
| Rain water | | |
| Worldwide | 0.03–1.7 | Reimann et al. (1998) |
| | 0.04–1.4 | Plant et al. (2004) |
| Polar ice | 0.02 | Frankenberger and Karlson (1994) |
| Rivers | | |
| Worldwide | 0.07 (0.02–0.5) | Gaillardet et al. (2003) |
| USA, Gunnison | 10 | Jacobs (1989) |
| Germany | 0.015 | Ebens and Shacklette (1982) |
| Jordan, River Jordan | 0.25 | Nishri et al. (1999) |
| Brazil, Amazon River | 0.021 | Ebens and Shacklette (1982) and Jacobs (1989) |
| Japan | 0.03–0.09 | Nriagu and Wong (1983) |
| USA | 0.14 | Robberecht and Grieken (1982) |
| China | 0.04–5 | Wang and Gao (2001) |
| Seas and oceans | | |
| Worldwide | 0.2 (0.1–0.35) | Reimann et al. (1998) |
| Seawater | | |
| Worldwide | 0.009–0.045 | Ebens and Shacklette (1982) |
| Dead Sea | 0.09–0.45 | Whittle et al. (1977) |
| | 0.8 | Robberecht and Grieken (1982) |
| Atlantic Ocean | 0.075–0.096 | Schutz and Turekian (1965) |
| Pacific Ocean | 0.104 | Schutz and Turekian (1965) |
| Freshwater | 0.2 (0.02–10) | Reimann et al. (1998) |
| Drinking water | | |
| Worldwide | 0.12–0.44 | Robberecht and Grieken (1982) |
| Germany | 0.02–0.03 | Veber et al. (1994) |
| Slovenia | 0.2 | Veber et al. (1994) |
| New Mexico | 5 | Martin (1975) |
| USA, New York | 0.12–0.44 | Cutter (1989) |
| Belgium | 0.13–0.14 | Ediger (1975) |
| Australia | <1 | Mesman and Thomas (1975) |
| Groundwaters | | |
| Argentina | 48–67 | Robberecht and Grieken (1982) |
| Australia | 0.008–0.33 | Robberecht and Grieken (1982) |
| Israel | 0.9–27 | Robberecht and Grieken (1982) |
| Poland | 0.25–1.80 | Siepak et al. (2003) |
| Slovakia | 0.5–45 | Rapant et al. (1996) |
| Norway | 0.01–4.82 | Reimann et al. (1998) |
| Italy | 0.002–1.94 | Dall'Aglio et al. (1978) |
| USA | <1–480 | Engberg (1973) |
| Lake waters | | |
| Poland | 0.15–0.35 | Siepak et al. (2003) |
| USA, Lake Michigan | 0.8–10 | Jacobs (1989) |

(continued)

**Table 10.4** (continued)

| Water type | Content (μg $L^{-1}$) | References |
|---|---|---|
| Sweden | 0.04–0.21 | Nriagu and Wong (1983) |
| Australia, Lake | 0.3–5.0 | Peters et al. (1999) |
| Macquarie | <0.5 | Maier et al. (1979) |
| Germany | <0.1 | Speyer (1980) |
| Canada | 0.16–0.25 | Kucukbay and Demir (2001) |
| Turkey, Malta Poland | 0.12–0.45 | Niedzielski (2006) |

Institution estimates that global coal production in 2030 will reach 7000 Mt, of which about a half will be made in China (Wen and Carignan 2007). Also, the production of energy in many other countries is based on coal and contributes to a lesser extent to selenium contamination of the atmosphere, relative to the content of selenium in the coal as well as to the appropriate filters in the power plant equipment preventing dust and exhaust gases (Table 10.6). For example, about 50% of electricity is generated from coal combustion in the USA, 70% in India, Australia, Greece, and the Czech Republic, while in Poland and South Africa, it is nearly 95%. The content of Se in coal ranges from 0.13 to 4.30 mg $kg^{-1}$, and in the ash from 0.5 to 15.5 mg $kg^{-1}$ (Table 10.6; Bielowicz 2013).

As reported by Wen and Carignan (2007), the northern hemisphere is more burdened with anthropogenic activity, so 70–80% of total Se emission is located there. From the beginning of the industrial era, Se emissions from anthropogenic sources have increased. Anthropogenic sources mainly comprise combustion (coal, oil, wood, biomass), nonferrous metal smelting, manufacturing, and utilization of agriculture products (Mosher and Duce 1987). In the USA, more than 90% of Se production comes from the anodic slime release by the electrochemical refining of copper (Nriagu and Pacyna 1988). The emission of Se to the atmosphere in European Union countries (EU) changed to a limited extent, showing a slight decreasing tendency. A comparison of 1990 and 2011 has shown a decrease in annual Se emission from 267 to 234 tons, which is 12% (EEA 2013) (Figs. 10.6 and 10.7). In 2011, the highest percentage shares in industrial selenium emissions were from Spain, Portugal, and the UK. As noted by Dodig and Cepelak (2004), in the USA about 1500 tons of annual emissions of Se come from fuel combustion. The same amount is emitted by municipal landfills and industrial plants.

It seems that since the beginning of the twenty-first century, the highest amount of Se emitted into the atmosphere from anthropogenic sources was by China, which additionally shows a constant and dynamic increase in this matter. Between 2000 and 2012, this emission increased by 44.9%. In 2000 and 2015, the emissions of Se into the environment by the UK were 2113 and 3062 tons per year, respectively, which was at least 10 times higher than the rest of the EU.

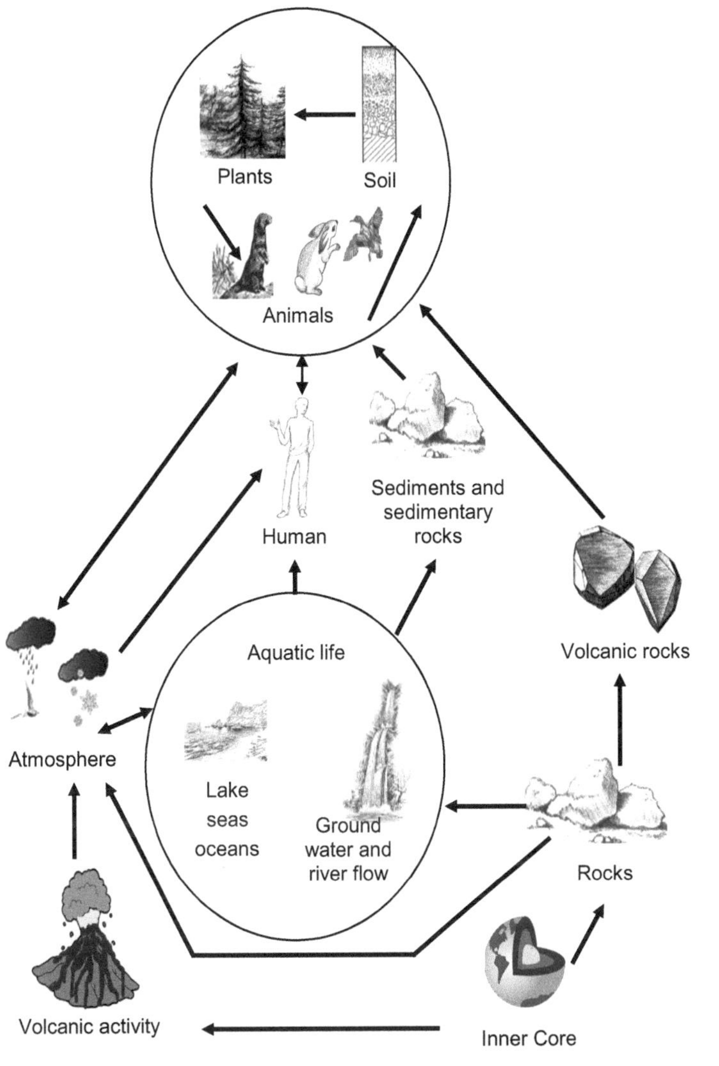


**Fig. 10.4** The circulation of selenium in nature (based on the diagram by Diplock 1985)

**Table 10.5** The estimated global flux of atmospheric Se from natural sources ($10^9$ g $year^{-1}$)

| Source | Amount |
|---|---|
| Crustal weathering | 0.003–0.035 |
| Sea salt | Up to 1.1 |
| Wild forest fires | Up to 0.52 |
| Marine biosphere | 0.4–9 |
| Volcanoes | 0.1–1.8 |
| Continent biosphere | 0.15–5.25 |

Sources: Mosher and Duce (1987), Nriagu and Pacyna (1988), Nriagu and Wong (1983)

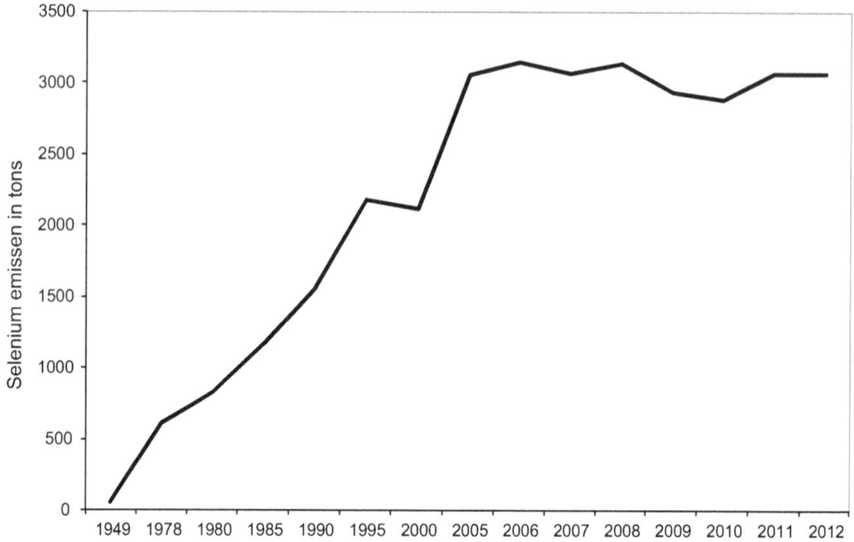


**Fig. 10.5** Emission of Se from the main anthropogenic sources in China (Data source: Tian et al. (2015)

**Table 10.6** Concentrations of selenium in coals and other natural fuel sources

| Country and material | Concentration (mg $kg^{-1}$) | Range (mg $kg^{-1}$) | References |
|---|---|---|---|
| USA, coals | 3.36 | 0.46–10.65 | Kuhn et al. (1980) |
| Hard coal | 1.6 | | Yudovich and Ketris (2006) |
| Brown coal | 1.0 | | Yudovich and Ketris (2006) |
| Brown coal | 1.0 | | Ketris and Yudovich (2009) |
| Kentucky coals | 4.3 | 2.9–5.5 | Eble and Hower (1997) |
| Poland, coals | 0.13 | 0.11–0.49 | Symanowicz et al. (2013) |
| Brown coal | 3.08 | | Goldsztejn (2007) |
| Brown coal | <3 | | Bielowicz (2013) |
| Germany, coals | 2.9 | | Sabbioni et al. (1983) |
| Indonesia, coals | 0.52 | | ACARP (2006) |
| China, coals | 1.5 | 0.36–12.10 | Tian et al. (2010) |
| | 3.91 | | Wang et al. (2010) |
| China, fossil fuels | | 1–10 | Harr (1978) |
| China, petroleum | | 500–1650 | USPHS (1996) |

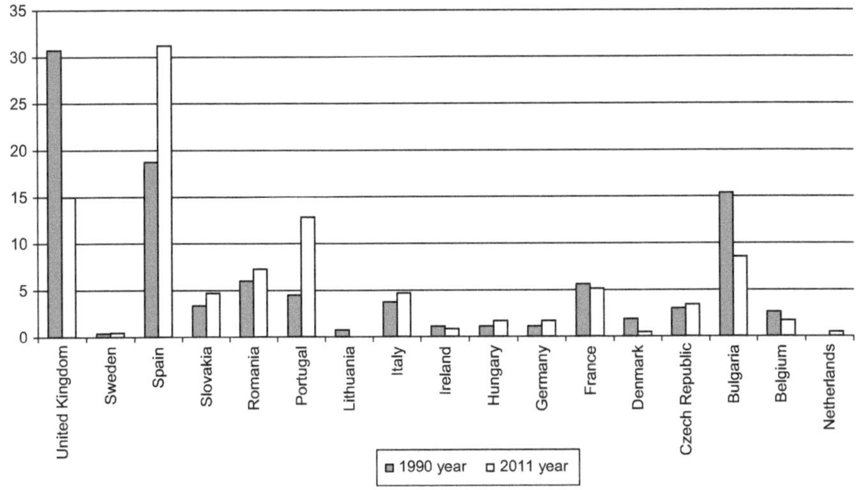


**Fig. 10.6** The percentage share of EU countries in the anthropogenic emission of selenium (Data source: EEA 2013)

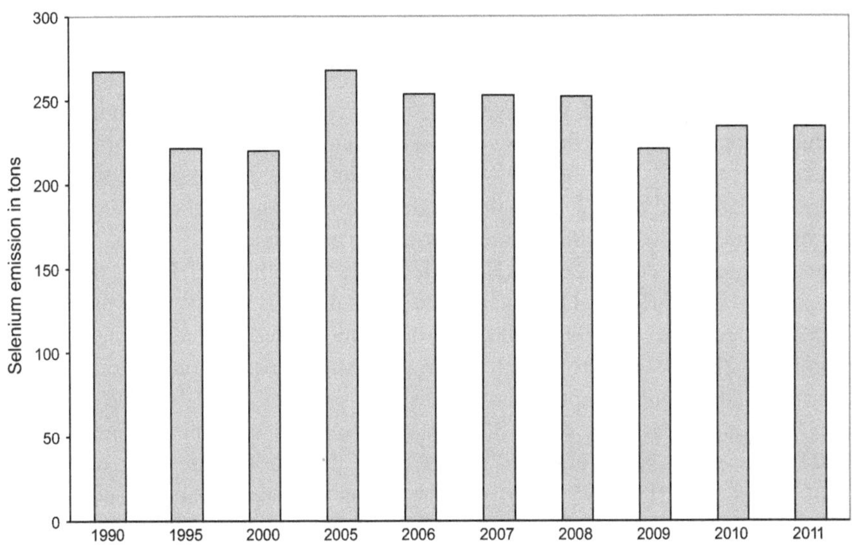


**Fig. 10.7** Anthropogenic selenium emissions in the European Union (Data source: EEA 2013)

# 5 Production and Uses

Four types of Se are offered in the world trade: (1) high-quality Se, purity above 99.99%; (2) pigment selenium, purity above 99.8% Se; (3) standard Se, purity from 99.0 to 99.9% Se; and (4) trade Se, purity above 97.0% Se (George 2004). It is hard

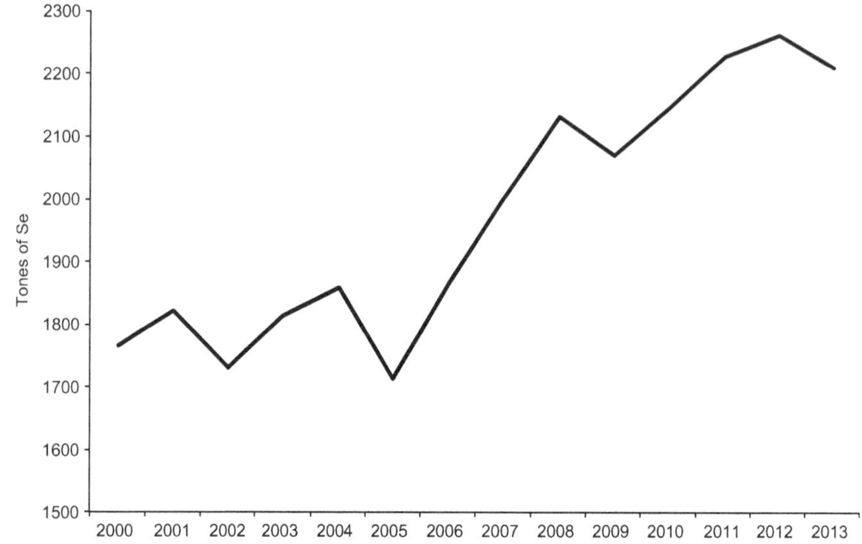


**Fig. 10.8** Global production of selenium (data sources: British Geological Survey 2015; World Mineral Production 2009–13)

to say how the demand for Se will look in future years, yet it is expected that demand will increase due to the development of some new technologies (including nanotechnologies). In industry, Se is used as a semiconductor in rectifiers, photocells, and anticorrosive shells (e.g., on steel), as an addition to some copper alloys, and in coloring glass. Red Se at 2–3% will dye glass ruby red (signal glass). According to the ordinance no. 1272/2008 from the European Parliament and EU Council, Se has been classified as Acute Tox 3 (H331, toxic after swallowing; H301, toxic after inhalation), STOT RE 2 (H373, may cause organ damage by prolonged or repeated exposure), Aquatic Acute 1 (H400, very toxic for water organisms), and Aquatic Chronic 1 (H410, very toxic for water organisms, causing prolonged effects). Selenium compounds in pollution are usually selenites ($SeO_3^{2-}$) and selenates ($SeO_4^{2-}$), and to a lesser extent its organic forms, as selenomethionine (George 2004; Pilon-Smits and LeDuc 2009). In 1987, the world production of Se was 1400 tones (USPHS 1996). The present production of Se cannot be precisely evaluated as there is no comprehensive data available, but a clear increasing tendency is visible (Fig. 10.8). In 2014 it surely exceeded 2275 tones (U.S. Geological Survey 2015).

In recent years, Japan and Germany have been considered as the largest Se producers (34% and 30% of world production, respectively) (Table 10.7), with the major consumers of selenium in order of importance: Europe, the USA, and Pacific Asia, in particular China. The scale of selenium demand in China correlates with the use of manganese in the steel industry.

**Table 10.7** Production of selenium metal (tones)

| Country | 2000 | 2001 | 2002 | 2003 | 2004 | 2005 | 2006 | 2007 | 2008 | 2009 | 2010 | 2011 | 2012 | 2013 | 2014 | 2015 |
|---|---|---|---|---|---|---|---|---|---|---|---|---|---|---|---|---|
| Belgium | 200 | 200 | 200 | 200 | 200 | 200 | 200 | 200 | 200 | 200 | 200 | 200 | 200 | 200 | 200 | 200 |
| Canada | 335 | 238 | 175 | 253 | 277 | 107 | 117 | 141 | 191 | 173 | 97 | 35 | 144 | 159 | 154 | 154 |
| Chile | 47 | 84 | – | – | – | – | – | – | – | – | – | – | – | – | | |
| China | 65 | 65 | 65 | 65 | 65 | 65 | 65 | 65 | 65 | 65 | 65 | 65 | 65 | 65 | 65 | 65 |
| Finland | 37 | 42 | 37 | 52 | 65 | 66 | 70 | 52 | 65 | 59 | 73 | 86 | 93 | 101 | 94 | 93 |
| Germany | 100 | 100 | 100 | 100 | 100 | 120 | 120 | 120 | 250 | 230 | 250 | 250 | 250 | 250 | 700[a] | 700[a] |
| India | 10 | 10 | 6 | 2 | – | 8 | – | – | – | – | – | – | – | – | – | – |
| Japan | 612 | 735 | 752 | 734 | 599 | 625 | 730 | 806 | 754 | 709 | 804 | 809 | 820 | 739 | 782 | 772 |
| Kazakhstan | 40 | 40 | 40 | 40 | 100 | 60 | 70 | 155 | 130 | 120 | 130 | 130 | 130 | 130 | 130 | 130 |
| Mexico | – | – | – | – | – | – | – | – | – | – | 62 | 95 | 95 | 132 | 120 | 107 |
| Peru | 21 | 16 | 21 | 24 | 76 | 70 | 75 | 59 | 60 | 61 | 59 | 54 | 42 | 40 | 49 | 50 |
| Philippines | 38 | 43 | 40 | 40 | 48 | 68 | 65 | 65 | 65 | 65 | 65 | 65 | 70 | 70 | 70 | 70 |
| Poland | 65 | 65 | 68 | 78 | 83 | 82 | 87 | 85 | 82 | 80 | 79 | 85 | 90 | 80 | 90 | 87 |
| Russia | 41 | 48 | 60 | 81 | 85 | 100 | 110 | 110 | 110 | 160 | 170 | 265 | 172 | 155 | 131 | 143 |
| Serbia and Montenegro | 21 | 14 | 17 | 7 | – | – | – | – | – | – | – | – | – | – | 16 | 16 |
| Sweden | 104 | 88 | 118 | 107 | 131 | 122 | 135 | 126 | 139 | 129 | 72 | 70 | 70 | 70 | 20 | 20 |
| Uzbekistan | 20 | 20 | 20 | 20 | 20 | 20 | 20 | 20 | 20 | 20 | 20 | 20 | 20 | 20 | 20 | 20 |
| Zambia | 10 | 13 | 11 | 10 | 10 | – | – | – | – | – | – | – | – | – | – | – |

Sources: British Geological Survey (2015, 2017), World Mineral Production 2009–2013; British Geological Survey (2006), World Mineral Production 2000–2004; British Geological Survey (2010), World Mineral Production 2004–2008

[a]Includes selenium produced from imported material

According to analysts, the greatest respective increases in selenium demand will be connected with the progress in photovoltaic technologies, especially the development of thin-layer solar batteries in the latest Cu-In-Ga-Se (CIGS) technology, as well as in the boom in demand for glass in architecture (Andersson 2005). In many countries, especially in the USA, CIGS and CdTe battery recycling facilities are well developed, from which the demand for selenium and tellurium for thin-layer cells will be partially covered (George 2004).

The structure of selenium use in 2004 was dominated by the glass industry (35%) while in 2009, by metallurgic use (40%), the glass industry dropping to 25% (this decrease is explained by the crisis in that real estate market). The share of other directions of Se use is as follows: agriculture 10%, chemical pharmaceutical and pigment industries 10%, electronics 10%, and others 5%.

Considerable amounts of selenium are used in the glass and ceramic industries for red mosaic dying and in agriculture as an additive for selenium-enriched fertilizers in Se-deficient soils, for instance, in China, New Zealand, and Australia, and as a fodder supplement for farmed animals. Cadmium sulfoselenide until recently was used as a dye to manufacture colored artistic glass and intensive red lamps in traffic lights. In the metallurgy industry, selenium is used as an additive to improving the casting, forming, and machining properties of iron, steel, copper, and lead alloys, of which the last is used in car batteries (George 2004).

In the pharmaceutical industry, selenium is used to produce anti-dandruff shampoos and anti-inflammatory and antifungal drugs for dermatology. The use of selenium in the electronics industry is still high due to the constant demand for devices that use photoelectricity and DC converters (Canadian Council of Ministers of the Environment).

## 6 Biological Status of Selenium

The concentration of selenium in soils and plants varies depending on the geographic location (including the distance to sea), type of maternal rock, intensity of flushing and washing of rock particles into water reservoirs, climatic conditions (temperature, amount of precipitation), type of soil and its pH and salinity, processes of sorption by iron oxides and loamy minerals in the soil, content of organic matter, level of environmental contamination (presence of antagonistic elements as Pb, Cd), the chemical form of Se, and finally the specific composition of soil microorganisms and plants (Pezzarossa et al. 2007; Kabata-Pendias and Mukherjee 2007; Kabata-Pendias and Szteke 2015; Winkel et al. 2015). In soils rich in iron and organic matter, as well as in salified soils, a higher content of selenium is found, while lower levels pertain to acidic soils. The soils of the central region of North America are formed from cretaceous shales, having favorable pH characteristics that increase Se availability for uptake (Mayland 1994; Kabata-Pendias and Mukherjee 2007).

In selenites, Se may occur as Se (IV) or Se (VI). Selenites (IV) may be absorbed to a higher extent by iron oxides/hydroxides than selenates (VI) (Barrow and Whelan

1989). The intensity of this process decreases with a rise in pH. Iron oxy-hydroxides (hematite, getite) are characterized by a high potential to absorb Se, which causes an inhibition in their transport and retention in soil (Rovira et al. 2008).

In soil, selenium is present in nonvolatile forms, as selenides, seleno-amino acids, and dimethyloselenium—as an ion or as a volatile methylated (DMSe) and dimethylated (DMDSe) form (Cooke and Bruland 1987). Due to the activity of microorganisms, inorganic selenium compounds after methylation change into volatile compounds that may be transferred into the atmosphere from the surface of both water and soil (Zayed et al. 1998; Azaizeh et al. 2003). In anaerobic conditions, microorganisms convert the elementary selenium into hydrogen selenide (Barrow and Whelan 1989). The process of methylation depends on the content and source of carbon, on soil temperature and type, as well as on the redox conditions. Selenium may also be released into the atmosphere by higher-order plants (Laser 2004).

The binding of Se (IV) to solid soil particles occurs by ligand exchange pathway, using the hydroxyl group that is present on the surface of colloid soil particles and hydrated metal oxides. This mechanism is very dependent on the pH. Se (VI) is different, as the connection to other solid components in the soil, water molecules are used between the external layer and the adsorbed ion (Neal 1995).

The content of selenium in plants depends on the abundance of Se in the soil, its chemical form, and the species of plants, which assimilate selenium mostly as selenite Se (IV) and selenate Se (VI), and some species—as selenides. A significant difference in the interaction of selenites (IV) and selenates (VI) has been found which concerns the different mechanism of transport of both chemical selenium forms through cellular membranes. Inorganic Se (VI) ions easily penetrate the root, but are transported by the apoplastic way—without entering the root cells, but rather in the water stream in an unchanged chemically ionic form. In turn, inorganic Se (IV) ions, after penetrating the root, are able to enter the cellular plasmalemma (symplastic transport) and are biotransformed into organic compounds. Selenium is stored by plants as different organic compounds, including amino acids, such as selenomethionine, selenocysteine, Se-methylselenocysteine, selenocystathionine, and selenohomocysteine. Translocation of selenium from the roots to the shoots depends primarily on the chemical form. Selenates (VI) are more easily transported than selenites (IV). This is caused by the quicker transformation of selenites (IV) to forms that are retained in the roots. Therefore the transfer of selenium to the above ground parts of a plant is limited (Terry et al. 2000; Li et al. 2008; Michalska-Kacymirow et al. 2014).

Most crop plants contain less than 25 mg Se $kg^{-1}$ dw, and in only some of them does the concentration exceed 100 mg Se $kg^{-1}$ dw. However, a low content of Se in plants is not always requisite with a low selenium content in the soil, as the concentration of Se in plants depends on many factors: plant species, type of soil and its Se content, climatic and vegetative conditions, fertilizers, chemical form of Se and its bioavailability for plants, pH of the soil, organic matter content, and the presence of competitive anions (Bell et al. 1992; Ellis and Salt 2003; Fig. 10.9).

Due to the varying ability of plants to assimilate and accumulate Se, they can be divided into three groups (Bell et al. 1992; Ellis and Salt 2003): selenophilic plants

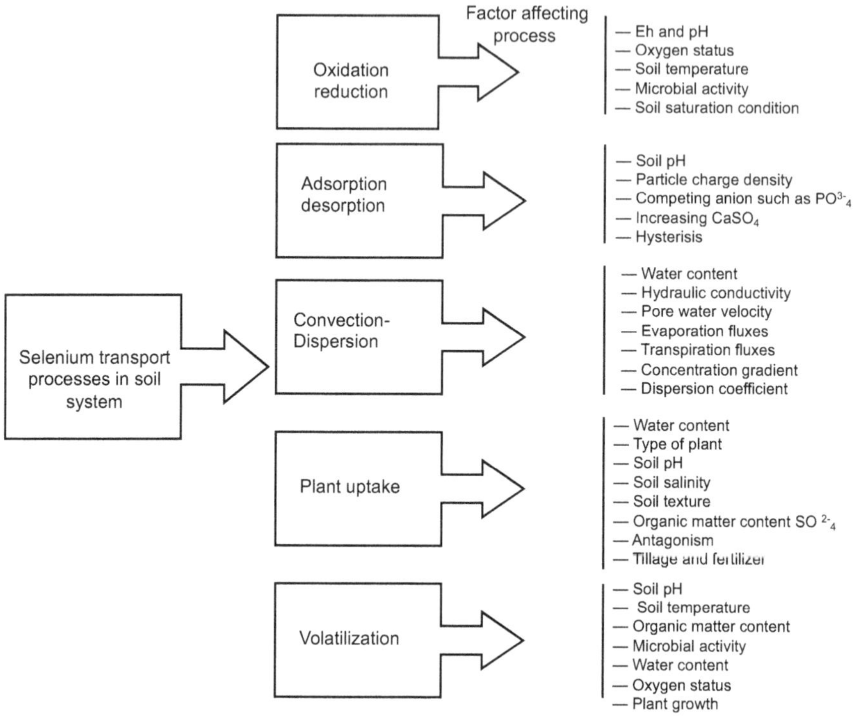


**Fig. 10.9** The transport processes and the related factors, based on the diagram by Mirbagheri (2004)

(selenium accumulators), which can store from a thousand to a few thousands mg of Se $kg^{-1}$ dw in their overground parts and are toxic to humans and animals; plants that moderately accumulate selenium (secondary selenium accumulators); and plants that assimilate low amounts of selenium and are characterized by a lower tolerance to this element and do not develop well in selenium-abundant soils.

The concentration of Se in the soil decreases with depth, and in the lowest layers is low, so plants with a deep root system do not have wide access to this element (Hartfiel and Bahners 1987). The location of Se in different parts of a plant depends on the species, development stage, and physiological status. Plants classified as Se accumulators store this element mostly in the leaves in early stages of development and in seeds during the reproductive stage. Then, a decrease of selenium in the leaves is observed. In mature grain plants, the content of Se is at a similar level in the caryopses and roots, with a lesser level in the stalks and leaves (Terry et al. 2000).

The concentration of Se in the animal organism depends on the content of this element in food and on its digestibility. Food is the main route of Se intake for vertebrates and invertebrates. Animals can (depending on the species, the form of selenium and other factors) intake from 44% to 95% of the selenium contained in the food (Opresko 1993). For most of the Se forms (selenite, selenate, selenomethionine), the level of assimilation is very high (from 80% to over 90%). Selenium can also enter organisms through inhalation, while some Se compounds, such as selenyl chloride, are easily absorbed by the skin. The assimilation of Se is strictly dependent on the chemical form of Se. Vertebrates assimilate Se better in organic forms: selenomethionine (SeMet) and selenocysteine (SeCys). The absorption of organic Se forms is via the active intestinal transport, while the inorganic forms are absorbed by the passive transport (diffusion in the presence of sodium ions) in the small intestine (Fordyce 2013; Sampaio da Silva et al. 2013). Absorbed Se is bound by erythrocytes and blood plasma albumins and globulins and is transported to the tissues. Inorganic Se forms are converted into organic selenocomplexes in the muscles, hemoglobin, and blood plasma. Selenates (VI) in the liver, spleen, blood, and plasma are reduced by enzymes to selenites (IV) or selenium. Selenites (IV) may build complexes with proteins and in this way are more easily bound to glutathione peroxidase than selenates (VI). Selenites (IV) penetrate the blood-placenta barrier and enter the fetus. Selenites (IV) show a higher affinity to tissues than selenates (VI). The excretion of selenium from the organism happens in the urine ($TMeSe^+$), sweat, and as DMeSe with exhaled air. Much lower amounts of this element are excreted with feces (Fishbein 1991; Fordyce 2013). In homeothermic animals, 85%–100% of dietary Se in plants is absorbed, compared to 20%–50% of Se present in meat and fish (Fordyce 2013). The main vector of Se intake in fish is not the water but the food chain (Luoma et al. 1992).

Selenium shows a tendency to bioaccumulate in tissues and causes toxicological effects. Bioaccumulation of Se in lower trophic level invertebrates (zooplankton, clams) is important in evaluating the effect of Se on the higher trophic levels of predators (fish and birds that feed on invertebrates). Absorption of Se dissolved in the water is not as significant as the assimilation from the diet (Luoma et al. 1992; Lemly 1993). Direct uptake of particulate selenium by invertebrates via filter feeding or deposit feeding is the primary route for Se to enter the food chain (Luoma and Presser 2000). Microorganisms connected with the detritus of sediments may be important in the incorporation of Se into the aquatic food chain (Hamilton and Buhl 2003; Sampaio da Silva et al. 2013). It has been shown that differences in the structure of the food chain in lentic and lotic habitats affect the level of Se in the aquatic food chain. Organisms from the higher trophic levels in lentic habitats collect higher concentrations of selenium than organisms in lotic environments (Orr et al. 2006; Sampaio da Silva et al. 2013). The cycle of Se in dentric sediment is where Se is transformed into selenomethionine and is subsequently transferred to superior trophic levels (Sampaio da Silva et al. 2013).

One of the many factors that affects selenium assimilation by animals is their taxonomic affiliation, with the biggest such differences described between single-stomached and ruminant animals (which absorb selenium less effectively). It has been concluded that in monogastric animals, about 80%–90% of Se is absorbed in the digestive tract, while in ruminants it is much less—about 30%–35% (Spears 2003; Lyons et al. 2007; Fordyce 2013). This dependence is caused by the processes of bacterial reduction in the rumen, which leads to a conversion of mineral selenium links into nondigestible forms, the consequence of which is a low biological availability of the Se. An additional factor that reduces the absorption of Se in ruminants is the low pH in the rumen. Selenium assimilation also depends on the concentration of calcium in particular ingredients of the diet, which also relates to sulfur (Cristaldi et al. 2005; Mynhardt et al. 2006).

## *6.1 Toxicity of Various Selenium Forms in Homeothermic Animals*

In the thirteenth century, Marco Polo had already described symptoms typical for selenosis, such as hair loss in horses and humans, salivation, tooth loss, and hoof crumbling, in his China travel diary. Selenite, selenate, and selenomethionine are among the most acutely toxic selenium compounds (Högberg and Alexander 1986). The particular chemical forms of selenium are categorized by toxicity. A lethal dose of Se (IV) for rats is 3.2 mg $kg^{-1}$ body weight (bw), while for dimethyl selenide—1600 mg $kg^{-1}$ b.w. Inorganic selenium compounds such as selenite (IV) and selenate (VI) are the most toxic forms of this element. In nature, selenium most often occurs as selenomethionine and selenocysteine, so in a form connected to amino acids. The least toxic forms of selenium are volatile methyl compounds of Se that are the metabolites of the detoxification process (Orr et al. 2006).

A surplus of Se in food causes symptoms of intoxication in animals (selenosis endemica), while in humans, hair loss, nail cracking, skin changes, and polyneuritis have been observed. The toxic activity of Se in animals may not only occur in places environmentally abundant in this element but also in the areas polluted by metallurgical and mining industry wastes (Wayland and Crosley 2006). Cases of selenosis have been recorded in those regions of the world where the natural Se content in the environment is very high (e.g., some regions of China, northwestern USA). Selenophylic plants (Se accumulators) located in soils abundant in Se are almost always responsible for cases of acute or chronic intoxication with selenium in animals, as the above ground parts of such plants may contain thousands of mg of Se to 1 kg dw converted (Ellis and Salt 2003). Cases of an acute intoxication with selenium have been observed in animals after consuming plants containing 400–800 mg Se $kg^{-1}$ dw, yet as noted by Mayland (1994), a content of Se in fodder above 3 mg $kg^{-1}$ dw may already be toxic for some species of mammals. Raisbeck (2000) claims that a content of Se higher than 0.5 mg $kg^{-1}$ dw in cattle fodder may cause appetite loss, hair loss, inflammation, horn and hoof fragility, hobbling, or even death in these ruminants. The medial lethal doses of Se for adult cattle and for lambs ($LD_{50}$) are 0.501 and 0.455 mg $kg^{-1}$ bw, respectively (Grace 1994; Tinggi 2005). A toxic concentration of Se in the blood is >2 mg $L^{-1}$ in cattle and >0.6–0.7 mg $L^{-1}$ in sheep (Levander 1986).

The afflictions caused by an excessive supply of Se in animals occur in two clinical forms: chronic (called the alkali disease) and acute (called blind circling). The symptoms of the chronic form include liver damage (metabolic discrepancies and lack of appetite), depression, emaciation, hair loss, roughness of hair, improper development or necrosis of hooves, limb stiffness, heart atrophy, and anemia (NRC 1980; Clayton and Clayton 1994; Raisbeck 2000; Tinggi 2005). An acute form of intoxication manifests in disturbances in the nervous system, including increased excitability or, conversely, stupor. In some cases, animals blindly push forward, grate their teeth, and salivate. Moreover, sometimes animals suffer colic, diarrhea, and impairment of vision. The final phase includes dyspnea and limb paralysis. Death is caused by respiratory failure (Raisbeck 2000; Kim and Mahan 2001; Dodig and Cepelak 2004).

An excessive concentration of Se in vertebrates causes an increased substitution of sulfur by Se in cysteine, named the "sulfuric amino acid," which is very common in keratin (which composes the hair, feathers, and other horny formations of vertebrates), as well as in other sulfur-containing molecules. Moreover, an excess of Se leads to inhibitions of cellular proliferation, replication of DNA, and protein synthesis while supporting an increase in oxidative stress, intensified lipid peroxidation, and metal complexing, which then accumulate in brain cells (Raisbeck 2000; Kim and Mahan 2001; Dodig and Cepelak 2004; McKittrick et al. 2012; Zachary and McGavin 2014).

Studies performed before 1987 in animals and humans, by the International Agency for Research on Cancer (IARC), have not shown any carcinogenic activity of selenium nor of its compounds, so these substances have been classified within group 3 "Not classifiable as to its carcinogenicity to humans" (IARC 2015). However, since that time it has been shown that selenium sulfide is likely a human carcinogen. Selenium sulfide is the only Se compound that causes tumors in laboratory rats and mice (ADSTR 2003).

It is commonly believed that selenium has a preventive activity against some tumors (of the large intestine, lungs, larynx, prostate gland, stomach, and esophagus) when Se supplementation is used in animals with a deficiency of this element. In the case of animals with a marginal concentration of Se, any additional supplementation with selenium did not cause any benefits and may even cause an opposite effect (Goodman et al. 2001; Duffield-Lillico et al. 2002; Grau et al. 2006; Lippman et al. 2009).

Data concerning the genotoxicity of Se compounds are unambiguous. It has been observed that inorganic Se compounds may act twofold: genotoxically and antigenotoxically. Antigenotoxic effects usually occur in lower levels of exposure than genotoxic activity (ATSDR 2003).

## 6.2 The Role of Selenium in the Organism

The concentration of Se in an organism is the main factor regulating the activity of glutathione peroxidase and other selenoproteins that participate in metabolic pathways. In the processes of cellular protection against the effects of reactive oxygen forms, a group of five glutathione selenoperoxidases, three thioredoxin reductases, and a selenoprotein P are engaged (Arthur 2000; Flohe et al. 2000; Gladyshev 2001; Behne and Kyriakopoulos 2001). Selenium, as an antioxidant, limits the harmful effects of the peroxidation of lipids, DNA, and RNA, protecting cells from deformation and genetic damage. Moreover it causes the apoptosis of tumor cells (Griffin 1982; Behne et al. 1996). Epidemiological studies have shown a relationship between a deficiency of selenium and the growth of some types of tumors. Selenium has an inhibiting effect on tumor cell proliferation by enhancing the expression of suppressor gene (p53), as well as on regulating the expression of apoptosis suppressor gene (Bcl-2) (Jackson and Combs 2008; Björkhem-Bergman et al. 2012).

In males, Se accumulates in the testes and is excreted from the organism with semen (Hansen and Deguchi 1996). A deficiency of this element negatively affects the development, motility, and number of sperm cells in an ejaculate, as well as the volume of seminal fluid. A low concentration of Se increases the susceptibility of sperm to free oxygen radicals, which disturb the biochemical processes in the acrosome (Kantola et al. 1988). In males, the large level of Se deposited in the testes is used for the protection of sperm mitochondria from oxidation processes (Roveri et al. 2001). Sperm glutathione peroxidase protects the DNA of the sperm against oxidation damage and participates in the condensation of chromatin (Pfeifer et al. 2001). Some authors also suggest that Se may play a role in the biosynthesis of testosterone (Bedwal et al. 1993).

Selenium is a deactivator of toxic heavy metals (including mercury, cadmium, and lead), as well as organic compounds, which are released in infections, injuries, and stress (Rayman 2000; Kalisińska et al. 2014). In mammalian organisms, selenites of cadmium, mercury, lead, silver, and tallium are easily formed, which enables the removal of these metals from the organism. Selenium is also a factor that decreases the toxicity of methylmercury (Ralston and Raymond 2010). Selenium is also an element that neutralizes the activity of carcinogenic aflatoxins (Navarro-Alarcon et al. 1998; Maehira et al. 2002).

The activity of Se in the cardiovascular system includes protecting LDLs from oxidants, interaction with alpha-tocopherol, and heavy metal detoxication, influencing the metabolism of homocysteine and tissue sensitivity to insulin (Marcus 1993). The protection of LDL cholesterol fractions against oxidative modification is via the antioxidative properties of glutathione peroxidase and other selenoenzymes present in blood plasma, including selenoprotein P (Köhrle et al. 2000).

Administration of Se reduces rheumatic pain and increases mobility in inflamed joint, as well as induces antiviral (e.g., inhibits the progression of HIV virus) and antidepressive activity (Stone et al. 2010). Moreover, it seems that Se is relevant for the proper functioning of the brain, where in the brain tissues of patients diagnosed with Alzheimer's and Parkinson's disease, a lower concentration of selenium was found in comparison to the tissues of healthy people (Atroshi et al. 2007).

### 6.2.1 Deficiency of Selenium in Animals

Deficiency of selenium in homeothermic vertebrates causes a decrease in the concentration of glutathione, glutathione peroxidase, and the enzymes responsible for thyroid hormone metabolism, including thioredoxin reductase, which causes a decrease in thyroid activity (Arthur et al. 1993; Fischer et al. 2008). A deficiency in selenium can also be seen in oxidative damage to thyroid tissues and in a decreased activity of T3 in the metabolism of an organism (Corvilain et al. 1993). In farm animals, the most well-known disease connected with selenium deficiency is white muscle disease (WMD), which manifests with limb stiffness, mioglobinuria, and difficulties in food uptake and swallowing, which often leads to aspiration pneumonia (McCann and Ames 2011).

### 6.2.2 High Levels of Selenium in Animals

High concentrations of Se in males lead to a decrease in the number of sperm cells as well as an increase in the number of improperly shaped sperm. Other typical symptoms are changes in sperm motility indicators (progression or forward velocity), testicular hypertrophy, changes in the female reproductive cycle in rats, and changes in the menstrual cycle (anovulation, short luteal and follicular phases) in monkeys (Cukierski et al. 1989). In pregnant women who drank water with a chronically increased dose of Se, a slightly increased number of spontaneous abortions were observed, although recent studies have not proven that a surplus of Se could cause any birth defects in humans or other mammals. In people exposed to Se dust and selenium compounds in workplace air, dizziness, fatigue, and irritation of the mucous membranes, and in extremal cases also accumulation of fluid in lungs (pulmonary edema) and severe bronchitis, were observed (ATSDR 2003).

### 6.2.3 Toxicokinetics and the Effects of Selenium in Wildlife

According to the amount of consumed Se, clear differences in its distribution are observed. At a Se deficiency, firstly it is incorporated into specific proteins (such as selenoprotein P, 5′-thyronin deiodinase). Next, Se is delivered to non-specific proteins. The brain, endocrine glands, and reproductive organs are the first to accumulate selenium before the liver, heart, and skeletal muscles. For this reason, in a Se deficiency, the discrepancies refer to these tissues and organs first (Behne et al. 1996). The highest amounts of selenium accumulate in the cortical part of the kidneys, pancreas, pituitary, and liver, but a lot of selenium is also transferred to hair, feathers, and horns (Daun and Akesson 2004). The organs in which the highest concentrations of Se are found are the kidneys, as they produce glutathione peroxidase (GSH-Px) for blood plasma (Zachara et al. 2006).

In a proper supply of selenium in an organism, its concentration in the liver and kidneys is higher than in the other organs. It is estimated that in humans, about 30% of the total Se pool is located in the liver, about 15% in the kidneys, 30% in the muscles, and 10% in blood plasma (WHO/FAO 2002). However, in mammals, significant differences in these values are observed (Whanger 1996).

The concentration of selenium in bird tissues depends on age, health, diet, presence of metals, and many other factors. The highest concentrations of Se are observed in the tissues of piscivorous birds and the lowest in herbivorous birds. Middle levels are noted in omnivorous birds (Mora and Anderson 1995). In birds, a deficiency of Se contributes to a decrease in the reproductive potential and immunocompetence, which results in an increased susceptibility to numerous infections. However, too high a concentration of Se is the cause of many illnesses that sometimes lead to death, which also correlates with the size of the population. The most well-known case of wildfowl selenium intoxication happened in Kesterson Reservoir (located on the Kesterson National Wildlife Refuge) in California in the 1980s (Ohlendorf 1989, 2002). Reproductive disorders and high mortality rates of the birds were observed. An excess of Se in the diet of female birds during the period just before egg-laying can result in the transfer of harmful levels of Se to the eggs or other tissues. This incident has shown a close relationship between the transfer of Se from an aquatic habitat rich in this element to the vertebrate organisms that feed in such habitats. In aquatic ecosystems, inorganic Se is rapidly assimilated by primary producers (bacteria, fungi, algae, and plants) and is transformed into organic selenium compounds (Ohlendorf 1989, 2002).

In nature, Se occurs in two different chemical forms that differ in toxicity to birds. The four common oxidation states are selenide (−2), elementary Se (0), selenite (+4), and selenate (+6). Elemental Se is virtually insoluble in water and presents little risk to the birds. Both selenite and selenate are toxic to the birds, but organic selenides pose the greatest hazard. Among the organic selenides, selenomethionine has been shown to be highly toxic to birds and to be the form most likely to harm wild birds, as it results in high bioaccumulation of Se in their eggs (Golubkina and Papazyan 2006; Ohlendorf and Heinz 2011a, b). The negative consequences of an excess of Se in birds have been described by many authors (Heinz 1996; Eisler 2000; Hoffman 2002; Ohlendorf 2003). In Se-rich areas, this element may accumulate in plants and invertebrates, which may then become part of birds' diet and cause intoxication through bioaccumulation.

Literature offers various values as the highest Se level that does not cause any toxic effects in birds (no-observed-adverse-effect level, NOAEL) (Table 10.8). Lam et al. (2005) suppose that an estimated value of NOAEL for medium concentrations of Se in eggs is from 0.9 to 1.45 mg $kg^{-1}$ dw, whereas Latshaw et al. (2004) did not note any effect of 2.05 mg $kg^{-1}$ dose on the development of pheasant embryos. Other authors propose much higher NOAEL values for selenium (Table 10.8). In studies on mallard *Anas platyrhynchos*, an addition of 3.5 mg $kg^{-1}$ of selenium (as seleno-DL-methionine) to fodder did not affect reproductive performance with the lowest Se level that caused a negative effect was 7 mg $kg^{-1}$ (Stanley et al. 1996).

**Table 10.8** Values of NOAEL (no-observed-adverse-effect level) and LOAEL (lowest-observed-adverse-effect level) for selenium

| Species | Selenium form | Endpoint | NOAEL mg kg day$^{-1}$ | LOAEL mg kg day$^{-1}$ | References |
|---|---|---|---|---|---|
| *Birds* | | | | | |
| Mallard *Anas platyrhynchos* | Sodium selenite | Reproduction | 0.5 | 1.0 | Heinz et al. (1987) |
| | Selenomethionine | Reproduction | 0.4 | 0.8 | Heinz et al. (1989) |
| Screech owl *Otus asio* | Selenomethionine | Reproduction | 0.44 | 1.5 | Wiemeyer and Hoffman (1996) and Sample et al. (1996) |
| Black-crowned night heron *Nycticorax nycticorax* | Selenomethionine | Reproduction | 1.8 | | Smith et al. (1988) |
| Osprey[a] *Pandion haliaetus* | Selenomethionine | Reproduction | 0.44 | 1.5 | Sample et al. (1996) |
| Great blue heron[a] *Ardea herodias* | Selenomethionine | Reproduction | 0.40 | 0.8 | Sample et al. (1996) |
| Wild turkey[a] *Meleagris gallopavo* | Selenomethionine | Reproduction | 0.40 | 0.8 | Sample et al. (1996) |
| Belted kingfisher[a] *Megaceryle alcyon* | Selenomethionine | Reproduction | 1.8 | | Sample et al. (1996) |
| Great blue heron | Selenomethionine | Reproduction | 1.8 | | Sample et al. (1996) |
| *Mammals* | | | | | |
| Lab mouse *Mus musculus* | Selenate | Reproduction | 0.076 | 0.76 | Schroeder and Mitchner (1971) |
| | Sodium selenite | Reproduction | 0.46 | 4.6 | Chernoff and Kavlock (1982) |
| Lab rat *Rattus norvegicus* | Potassium selenate | Reproduction | 0.20 | 0.33 | Rosenfeld and Beath (1954) |
| Long-tailed macaques *Macaca fascicularis* | L-Selenomethionine | Reproduction | 0.025 | | Tarantal et al. (1991) |

[a]Estimated wildlife

Hoffman (2002) in his field studies on two species of wading birds has demonstrated that even a high mean concentration of Se in the water, reaching 190 μg $L^{-1}$, did not translate into discrepancies in their reproduction. Biochemical indicators in the liver as well as the weight of chick livers did not indicate a toxic activity of selenium, yet the different sensitivity of the examined species to a particular concentration of Se was observed. These studies suggest that the lowest adverse effect level (LOAEL) of selenium, which is already dangerous for health, may vary across the different species of birds. This relationship results from the differences in earlier life history, habitat interactions, and other factors (Table 10.8).

## 6.3 *Bioaccumulation of Selenium in Wildlife*

In environmental studies concerning biomonitoring, the measurement of selected elements is mostly performed using internal organs, body fluids, and the hair/feathers of animals typical for the particular habitats (Kalisińska and Salicki 2010; Jarzyńska and Falandysz 2011). The concentrations of Se in animal tissues show a tendency to reflect Se levels in the food consumed from the different habitats, particularly when the diet includes natural ingredients (Heinz et al. 1989; Stowesand et al. 1990). Ecotoxicologists for years have been evaluating and indicating the most useful animal species and types of biological samples which could be used in bioindication of environmental Se levels.

### 6.3.1 Selenium in the Soft Tissues of Endothermic Vertebrates

Tables 10.9, 10.10, 10.11, and 10.12 show various data concerning the concentrations of Se in biological materials from homeothermic vertebrates related to land ecosystems and inland waters. A proper interpretation of the data requires knowledge of the physiological concentrations of Se in the tissues and organs, values of selenium reflecting the geological background, and the effect of the species specificity. For most of the wild animal species, no data is available in this topic, although it has been developed for some groups of mammals and birds.

In European ruminants, the concentrations of selenium were most often analyzed in roe deer *Capreolus capreolus* and red deer *Cervus elaphus*, while in North America in mule deer *Odocoileus hemionus*, white-tailed deer *O. virginianus*, and elk *Alces alces*. The majority of studies concerned the liver, kidneys, and muscles. European ruminants, in comparison to American, were characterized by much lower concentrations of Se in the examined organs. Studies on Se content in cervids have been performed in Europe, including countries deficient in Se such as Poland and Norway, as well as in North America where the environment contains much higher amounts of this element. The highest level of selenium was found in North American herbivorous white-tailed deers and omnivorous raccoons *Procyon lotor* (Tables 10.9 and 10.10). In Europe the highest content of Se was noted in red deer livers in Poland

Table 10.9 Selenium concentrations in soft tissues and blood of ruminants and other herbivorous mammals

| Species | Localization | Concentration | References |
|---|---|---|---|
| *Liver (mg kg$^{-1}$ dry wt)* | | | |
| Red deer *Cervus elaphus* | Norway | 0.39[a] | Vikøren et al. (2005) |
| | | 0.71[a] | Frøslie et al. (1984) |
| | Poland | 0.34[a] | Pilarczyk et al. (2011c) |
| | | 0.23[a] | Jarzyńska and Falandysz (2011) |
| | | 0.36 | Pilarczyk et al. (2009) |
| | Croatia | 0.86[a] | Lazarus et al. (2008) |
| | Spain | 0.31 | Berzas Nevado et al. (2012) |
| | | 0.23 (M) | Reglero et al. (2009) |
| | | 0.16 | Reglero et al. (2009) |
| Elk (or wapiti) *Cervus canadensis* | USA | 0.25[a] | Fielder (1986) |
| Moose *Alces alces* | Sweden | 0.89[a] | Galgan and Frank (1995) |
| | Norway | 1.68[a] | Vikøren et al. (2011) |
| | | 1.50[a] | Frøslie et al. (1984) |
| | | 0.46–4.28[a] | Ytrehus et al. (1999) |
| | USA, Minnesota | 1.07–2.43[a] | Custer et al. (2004) |
| | Canada, Yukon | 5.71[a] | Gamberg et al. (2005a) |
| Roe deer *Capreolus capreolus* | Poland | 0.18–0.35[a] | Pilarczyk et al. (2011c) |
| | | 0.62 | Pilarczyk et al. (2009) |
| | | 0.57[a] | Pilarczyk et al. (2009) |
| | | 0.71 | Pilarczyk et al. (2008) |
| | | 0.31[a] | Nowakowska et al. (2014) |
| | Germany | 0.96[a] | Humann-Ziehank et al. (2008) |
| | Norway | 0.61[a] | Vikøren et al. (2011) |
| Reindeer *Rangifer tarandus* | Norway | 0.75[a] | Vikøren et al. (2011) |
| | | 1.79[a] | Frøslie et al. (1984) |
| | Greenland | 0.90[a] | Aastrup et al. (2000) |
| Mule deer *Odocoileus hemionus* | USA, Washington | 0.43[a] | Fielder (1986) |
| | USA | 2.29[a] | Zimmerman et al. (2008) |
| White-tailed deer *Odocoileus virginianus* | USA | 0.86[a] | Brady et al. (1978) |
| | | 0.68[a] | McDowell et al. (1995) |
| | | 2.89[a] | Zimmerman et al. (2008) |
| | | 0.29[a] | Sleeman et al. (2009, 2010) |
| Pronghorn *Antilocapra americana* | USA | 0.40[a] | Dunbar et al. (1999) |
| | | 0.43–0.99[a] | Stoszek et al. (1980) |
| Mountain goat *Oreamnos americanus* | USA | 0.07[a] | Fielder (1986) |
| Bighorn sheep *Ovis canadensis* | USA | 0.57[a] | Cox (2006) |
| European hare *Lepus europaeus* | Croatia | 0.51(P)[a] | Linšak et al. (2014) |
| | | 1.10[a] | Linšak et al. (2014) |
| | Poland | 0.27[a] | Drozd et al. (2015) |
| | | 0.84[a] | Dębski et al. (2005) |

(continued)

**Table 10.9** (continued)

| Species | Localization | Concentration | References |
|---|---|---|---|
| *Kidneys (mg kg$^{-1}$ dry wt)* | | | |
| European hare | Croatia | 2.71 (P)[a] | Linšak et al. (2014) |
| | | 2.80[a] | Linšak et al. (2014) |
| Red deer | Spain | 2.60 | Berzas Nevado et al. (2012) |
| Roe deer | Poland | 2.18[a] | Nowakowska et al. (2014) |
| | | 3.09 | Pilarczyk et al. (2008) |
| | | 2.99 | Pilarczyk et al. (2009) |
| | | 2.72 | Pilarczyk et al. (2009) |
| *Skeletal muscle (mg kg$^{-1}$ dry wt)* | | | |
| Red deer | Poland | 0.15[a] | Jarzyńska and Falandysz (2011) |
| | Croatia | 0.19[a] | Lazarus et al. (2008) |
| | Spain | 0.12 | Berzas Nevado et al. (2012) |
| | Czech Republic | 0.058[a] | Kursa et al. (2010) |
| | | 0.13[a] | Kursa et al. (2010) |
| White-tailed deer | USA | 0.14[a] | Ullrey et al. (1981) |
| | | 0.25[a] | Brady et al. (1978) |
| Reindeer | Greenland | 0.37[a] | Aastrup et al. (2000) |
| European hare | Croatia | 0.43 (P)[a] | Linšak et al. (2014) |
| | | 0.40[a] | Linšak et al. (2014) |

*P* polluted area, *M* mine
[a]Values were converted from wet weight into dry weight

(Pilarczyk et al. 2009) but slightly less than in the studies by Vikøren et al. (2005) in west Norway (Table 10.9).

Bioaccumulation of trace elements depends on many factors such as sex, age, and trophic level in a food chain. The results of studies on the dependency between selenium level and sex of the animals were ambiguous in the analyzed species. In an examination of roe deer livers from Poland, no significant differences were found between males and females in Se concentration (Pilarczyk et al. 2011a). This lack of differences between the sexes was also observed in other species, including ungulates (wild boar *Sus scrofa*) and carnivores, for example, in the domestic dog *Canis lupus f. domestica* and polar bear *Ursus maritimus* (Nicpoń et al. 2005; Rush et al. 2008; Pilarczyk et al. 2010b). In contrast to those works, Millán et al. (2008) showed significant differences in the concentrations of Se in the liver between males and females in the red fox *Vulpes vulpes* and Egyptian mongoose *Herpestes ichneumon* from Spain, as well as Vikøren et al. (2005), who analyzed hepatic concentrations of Se in mature Norwegian red deer.

In the case of free-living birds, not many such studies on this topic have been performed. Pilarczyk et al. (2012) in their analysis of marine ducks (velvet scoter *Melanitta fusca*, common scoter *M. nigara*, and long-tailed duck *Clangula hyemalis*), wintering in the south coast of the Baltic Sea, did not find any significant differences between selenium concentrations in the liver, kidneys, lungs, or heart

Table 10.10 The content of selenium in selected tissues of omnivores

| Species | Localization | Concentration | References |
|---|---|---|---|
| *Liver (mg kg$^{-1}$ dry wt)* | | | |
| Wild boar *Sus scrofa* | Spain | 0.84 | Berzas Nevado et al. (2012) |
| | | 0.45 (M) | Reglero et al. (2009) |
| | | 0.59 | Reglero et al. (2009) |
| | Poland | 0.68[a] | Pilarczyk et al. (2010b) |
| | | 0.86[a] | Jankowiak et al. (2015) |
| Raccoon *Procyon lotor* | USA, Illinois | 7.06[a] | Levengood and Hubert (2001) |
| | Canada | 9.24[a] | Wren (1984) |
| | USA, East, Central Michigan | 7.56[a] | Herbert and Peterle (1990) |
| | USA, New York | 3.17[a] | Valentine et al. (1988) |
| | USA, California (Kesterson) | 14.52–34.65[a] | Clark et al. (1989) |
| | USA, Volta | 1.69 | Clark et al. (1989) |
| *Kidneys (mg kg$^{-1}$ dry wt)* | | | |
| Raccoon | USA, Illinois | 7.28[a] | Levengood and Hubert (2001) |
| | Canada | 7.60[a] | Wren (1984) |
| | USA, New York | 12.32[a] | Valentine et al. (1988) |
| Wild boar | Poland | 5.45[a] | Pilarczyk et al. (2010b) |

*P* polluted area, *M* mine
[a]Values were converted from wet weight into dry weight

between males and females of the mentioned species. An analogous conclusion was made by Conover and Vest (2009) in comparison studies on the livers of males and females of the black-necked grebe *Podiceps nigricollis* from Great Salt Lake (Utah, USA).

Recent studies on the relationships between the age of wild mammals and Se concentrations in their organs were also ambiguous. For example, McDowell et al. (1995) stated that age made a significant difference in the content of selenium in the kidneys of white-tailed red deer, with the animals younger than 12 months characterized by the lowest concentrations of selenium. Also Vikøren et al. (2005) found the lowest hepatic Se concentrations in the youngest group of Norwegian red deer. A similar tendency was observed by Pilarczyk et al. (2010b) in studies on Polish wild boar—the individuals under 1 year old were characterized by higher concentrations of selenium than the older animals. The lower concentration of selenium in the livers of the youngest ungulates was most probably caused by their faster metabolism, which contributed to the formation of larger amounts of free radicals and neutralized by Se released by the liver. In a representative of predators (Eurasian otter, *Lutra lutra*), a significant difference in Se concentration was found between young and adult individuals in the kidneys—higher values were typical for adults, yet hepatic concentrations of this element were similar (Kang et al. 2015). In contrast to the above, Pilarczyk et al. (2011a) and Jankowiak et al. (2015), in studies on ungulates (roe deer and wild boar) from Poland, where a deficiency of selenium in the environment is noted, did not find any dependence between the age of the animals

**Table 10.11** The concentration of selenium in selected tissues of carnivores

| Species | Localization | Concentration | References |
|---|---|---|---|
| *Liver (mg kg$^{-1}$ dry wt)* | | | |
| Red fox *Vulpes vulpes* | Poland | 0.88 | Pilarczyk et al. (2011) |
| | USA, Central New York | 1.90 | Valentine et al. (1988) |
| Eurasian otters *Lutra lutra* | South Korea | 1.90 | Kang et al. (2015) |
| | England and Wales | 6.15 | Walker et al. (2010) |
| | England and Wales | 6.92 | Walker et al. (2011) |
| American mink *Neovison vison* | Canada, Yukon | 4.62[a] | Gamberg et al. (2005a) |
| | Canada, British Columbia | 2.24 | Harding et al. (1998) |
| | Norway | 1.50 | Norheim et al. (1984) |
| | Poland | 2.40 | Brzezinski et al. (2014) |
| River otter *Lontra canadensis* | Canada, British Columbia | 6.13 | Harding et al. (1998) |
| Wolverine *Gulo luscus* | Canada, British Columbia | 6.28 | Harding (2004) |
| *Kidneys (mg kg$^{-1}$ dry wt)* | | | |
| Red fox | Poland | 2.36 | Pilarczyk et al. (2011b) |
| | USA, Central New York | 5.50 | Valentine et al. (1988) |
| Eurasian otter | South Korea | 3.40 | Kang et al. (2015) |
| American mink | Canada, Yukon | 8.28[a] | Gamberg et al. (2005a) |
| | Canada, British Columbia | 4.00 | Harding et al. (1998) |
| | Poland | 4.92 | Brzezinski et al. (2014) |
| American marten *Martes americana* | Canada, British Columbia | <4.00 | Harding (2004) |

[a]Values were converted from wet weight into dry weight

and the concentration of selenium in their organs. Between the predators, a connection between Se in the liver and age was analyzed in the Canadian American mink *Neovison vison*, and no significant relationship was found (Gamberg et al. 2005a).

Stussy et al. (2000), basing on the results of studies on female elk aged from 1 to 23 years, denied the usefulness of hepatic Se measurements as a good indicator of this element's status in the organism. The authors did not note any relationship between liver Se concentration and the age of the animals, as well as no dependence between Se content and the condition and progression of pregnancy or lactation. The authors claim that the total pool of selenium in the liver did not fully reflect the level of the bioactive forms of this element available for the biosynthesis of glutathione peroxidase and other selenium-dependent enzymes. Moreover they indicated that a low concentration of Se in the blood may occur with either a low or high content of selenium in the liver.

Some studies have indicated that one of the relevant factors determining the concentration of Se in wild ungulates may be the season. Such studies were performed in Poland in roe deer and red deer showing selenium deficiencies during winter and spring in all of the sampled deer (Pilarczyk et al. 2008, 2009). An optimal concentration of hepatic Se was found only during autumn and in about 30% of the

**Table 10.12** The concentration of selenium in selected soft tissues of birds

| Species | Localization | Concentration | References |
|---|---|---|---|
| *Liver (mg kg$^{-1}$ dry wt)* | | | |
| Common merganser *Mergus merganser* | Poland | 3.08 | Kalisińska et al. (2014b) |
| | Canada, Ontario | 9.7 | Scheuhammer et al. (1998) |
| Bald eagle *Haliaeetus leucocephalus* | USA, Alaska islands | 10.2 | Stout and Trust (2002) |
| Black-tailed godwit *Limosa limosa* | France | 16 | Lucia et al. (2012) |
| Great cormorant *Phalacrocorax carbo* | Japan | 7.3 | Nam et al. (2005) |
| Mallard *Anas platyrhynchos* | Japan | 3.4 | Nam et al. (2005) |
| Spot-billed duck *Anas poecilorhyncha* | Japan | 2.3 | Nam et al. (2005) |
| Pintail *Anas acuta* | Japan | 23 | Nam et al. (2005) |
| Common teal *Anas crecca* | Japan | 2.3 | Nam et al. (2005) |
| Common loon *Gavia immer* | Canada, Ontario | 15 | Scheuhammer et al. (1998) |
| Eurasian coot *Fulica atra* | Spain | 1.19 (P)[a] | Taggart et al. (2006) |
| Mallard | Spain | 1.32 (P)[a] | Taggart et al. (2006) |
| Gadwall *Anas strepera* | Spain | 1.25 (P)[a] | Taggart et al. (2006) |
| Common pochard *Aythya ferina* | Spain | 1.25 (P)[a] | Taggart et al. (2006) |
| Red-crested pochard *Netta rufina* | Spain | 1.12 (P)[a] | Taggart et al. (2006) |
| Western swamphen *Porphyrio porphyrio* | Spain | 0.59 (P)[a] | Taggart et al. (2006) |
| Bufflehead *Bucephala albeola* | Canada | 10.35 | Braune and Malone (2006) |
| | USA | 5.30 | Michot et al. (1998) |
| | USA | 32.1 | Custer and Custer (2000) |
| Common goldeneye *Bucephala clangula* | Canada | 7.3–13 | Braune and Malone (2006) |
| | USA | 16.0–36.2 | Custer and Custer (2000) |
| Canvasback *Aythya valisineria* | USA | 20 | Custer and Custer (2000) |
| Lesser scaup *Aythya affinis* | USA | 21.7 | Custer and Custer (2000) |
| Black-capped chickadee *Poecile atricapillus* | USA, Alaska | 4.2 | Handel and Hemert (2015) |
| Great tit *Parus major* | China, Beijing | 4.17 | Deng et al. (2007) |

(continued)

**Table 10.12** (continued)

| Species | Localization | Concentration | References |
|---|---|---|---|
| Greenfinch *Chloris chloris* | China, Beijing | 3.46 | Deng et al. (2007) |
| Cliff swallow *Hirundo pyrrhonota* | USA, Arizona | 3.86 | Estrada and Maughan (2000) |
| Red-winged blackbird *Agelaius phoeniceus* | USA, Arizona | 12.30 | Estrada and Maughan (2000) |
| Western kingbird *Tyrannus verticalis* | USA, Arizona | 6.86 | Estrada and Maughan (2000) |
| *Kidney (mg kg$^{-1}$ dry wt)* | | | |
| Common merganser | Poland | 2.32 | Kalisińska et al. (2014) |
| | Canada, Ontario | 8.50 | Scheuhammer et al. (1998) |
| Black-tailed godwit | France | 11.0 | Lucia et al. (2012) |
| Bald eagle | USA, Alaska islands | 13.1 | Stout and Trust (2002) |
| Great cormorant | Japan, Izumi | 9.5 | Nam et al. (2005) |
| Mallard | Japan, Izumi | 4.9 | Nam et al. (2005) |
| Spot-billed duck | Japan, Izumi | 4.3 | Nam et al. (2005) |
| Pintail | Japan, Izumi | 16.0 | Nam et al. (2005) |
| Common teal | Japan, Izumi | 4.1 | Nam et al. (2005) |
| Common loon | Canada, Ontario | 15.0 | Scheuhammer et al. (1998) |
| Great tit | China, Beijing | 5.47 | Deng et al. (2007) |
| Greenfinch | China, Beijing | 5.5 | Deng et al. (2007) |
| *Muscle (mg kg$^{-1}$ dry wt)* | | | |
| Common merganser | Poland | 0.54 | Kalisińska et al. (2014) |
| | Canada, Ontario | 1.80 | Scheuhammer et al. (1998) |
| Black-tailed godwit | France | 3.20 | Lucia et al. (2012) |
| Bald eagle | Aleutian Islands, Alaska | 12.87[a] | Burger et al. (2012) |
| Great cormorant | Japan, Izumi | 1.80 | Nam et al. (2005) |
| Mallard | Japan, Izumi | 1.10 | Nam et al. (2005) |
| Spot-billed duck | Japan, Izumi | 2.30 | Nam et al. (2005) |
| Pintail | Japan, Izumi | 3.00 | Nam et al. (2005) |
| Common teal | Japan, Izumi | 1.10 | Nam et al. (2005) |
| Common loon | Canada, Ontario | 2.80 | Scheuhammer et al. (1998) |
| Great tit | China, Beijing | 1.47 | Deng et al. (2007) |
| Greenfinch | China, Beijing | 2.91 | Deng et al. (2007) |

*P* polluted area, *M* mine
[a]Values were converted from wet weight into dry weight

examined red deers (Pilarczyk et al. 2011a). The lower level of selenium in animals that manifested strongly in particular seasons is probably the effect of the limited period of availability of selenium-rich foods such as mushrooms. In roe deer, in each of the analyzed organs (liver, kidneys, lungs, heart), the highest concentrations of selenium were noted in spring and the lowest in autumn and winter (Pilarczyk et al.

2011a). This seems to have a biological explanation. The autumn period is a time of increased food intake and intensive metabolic changes related to the energetic reserve storage in adipocytes for winter.

According to Pollock (2005), the biochemical criteria used to diagnose the presence of a selenium deficiency based on hepatic selenium levels in game animals are as follows: below mg kg$^{-1}$ dw, deficiency; 0.6–0.88 mg kg$^{-1}$ dw, marginal level; and above 0.88 mg kg$^{-1}$ dw, optimal level. In studies by McDowell et al. (1995) in white-tailed deer in South Florida (USA), a low concentration of selenium was found in 13% of liver samples (<0.25 mg kg$^{-1}$ dw) and in 36% of kidney samples (<3.0 mg kg$^{-1}$ dw). This situation was caused by a limited availability of Se-rich food at that time. The situation is different for the hare. Drozd et al. (2015) found a higher Se level in hare livers in winter, during which a diversified diet seems to be quite unavailable.

In wild boar, Pilarczyk et al. (2010b) noted the highest hepatic concentration of Se in spring and the lowest in autumn. The differences were statistically significant. In the kidneys, the highest concentration was noted in summer and was significantly higher in comparison to the spring and autumn periods. This situation shows that wild boar consumes, apart from forest and meadow plants, also field crops (corn, potatoes, cereals) as well as animal feed. For this reason, wild boar feed is more diversified during the whole year than in roe and red deers (Pilarczyk et al. 2010b).

One of the most interesting research problems is the analysis of relationships between Se concentrations in the different tissues and organs. In wild land mammals and birds, the most commonly examined organs are the liver and kidneys and less frequently the muscles, lungs, heart, and gonads. The presence of a correlation between selenium content in the particular organs and tissues of free-ranging animals has not been unambiguous and not always statistically confirmed. For example, a positive correlation between Se levels in the liver and kidneys has been found in ungulates (wild boar and roe deer) from Poland (Pilarczyk et al. 2010b; Nowakowska et al. 2014), yet no such correlation was found in foxes from the same country (Pilarczyk et al. 2011a). Moreover, in the roe deer and fox, a positive and statistically significant relationship between selenium levels in two other pairs of organs, liver-lungs and liver-heart, has been found (Pilarczyk et al. 2011a; Nowakowska et al. 2014).

In juvenile ducks from the Kesterson Reservoir in California, a relationship between Se content in the muscles and the liver was observed ($r = 0.67$) (Ohlendorf et al. 1990).

McDowell et al. (1995) suppose that concentrations of selenium in the kidneys of red deer lower than 3.0 mg kg$^{-1}$ dw indicate a deficiency. Using the criteria for roe deer from Poland, Pilarczyk et al. (2009) also showed a deficiency of this element in the kidneys, mostly during the summer period (87.5%) and less in autumn (40.9%).

In the herbivorous European hare, *Lepus europaeus* in Croatia and Poland, a deficiency of this element was shown in the both countries, with the mean concentration between 0.08 and 0.33 mg kg$^{-1}$ ww and the lowest concentration in the individuals from Poland (Dębski et al. 2005; Drozd et al. 2015; Linšak et al. 2014) (Table 10.9).

In predator mammals, depending upon the species, diet, and place of origin, differences in hepatic Se levels are also present (Table 10.11).

According to Puls (1994), in canids when the Se concentration in the liver does not exceed 0.3 mg $kg^{-1}$ ww, a deficiency of this element can be declared, while optimal values are placed between 0.50 and 1.50 mg $kg^{-1}$ ww. In the Arctic fox *Vulpes lagopus* from the Canadian Arctic and Svalbard (Norwegian archipelago in the Arctic Ocean), about 0.8 and 2.7 mg Se $kg^{-1}$ ww, respectively, was observed in the livers, which is classified as an optimal value for the first population and visibly raised for the second (Prestrud et al. 1994; Hoekstra et al. 2003). In comparison to the Arctic fox, in red foxes from central and southern Europe, the hepatic Se concentrations turned out to be lower. In individuals from Poland, where there is a deficiency in environmental selenium supply, the concentration was extremely low (<0.3 mg $kg^{-1}$ ww), and in the red fox from Spain, it was optimal, reaching slightly more than 0.50 mg $kg^{-1}$ ww (1.72 mg $kg^{-1}$ dw) (Millán et al. 2008; Pilarczyk et al. 2011a).

In ecotoxicological studies on Se, birds play a significant role, including the species connected with freshwater habitats (Ohlendorf et al. 1988; Albers et al. 1996; Wu 2004; Paveglio and Kilbridge 2007). The natural level of selenium in the liver of birds ranges from 4 to 10 mg $kg^{-1}$ dw (Ohlendorf 1989). Aquatic birds with a mixed diet (plants, insects, small crustaceans, and fish), such as the black-headed gull (*Chroicocephalus ridibundus*), consume more Se in an organic form (selenomethionines) and smaller amounts of Se as selenocysteines (Spallholz and Hoffman 2002).

The liver reacts quickly to Se intoxication in food (Heinz et al. 1990). Such a situation occurred in adult birds (coots, stilts, and ducks) from Kesterson Reservoir. Ohlendorf et al. (1990) demonstrated the difference in Se content according to the place and time of exposure. In the livers of water birds classified as selenium intoxicated, hunted directly in Kesterson Reservoir or in its neighborhood, but without any symptoms of intoxication, the content of this element was several times higher than normal ranging between 26 to 86 and 38 to 85 mg $kg^{-1}$ dw (Ohlendorf et al. 1988; Wu 2004). This may be evidence of the adaptation of birds to high levels of this element in the environment and in the diet. Experimental studies on mallard *Anas platyrhynchos* indicated that with Se concentrations higher than 10 mg $kg^{-1}$ dw, disturbances in the reproductive process may be expected, and above 33 mg $kg^{-1}$ dw can adversely affect health and survival (Heinz et al. 1989; Heinz 1996). A comparative analysis of selenium concentrations in water bird livers from California (USA), in three species of ducks, mallard, northern shoveler *Anas clypeata*, and northern pintail *A. acuta,* has shown that hepatic Se background levels, respectively, reached 4.1, 8.1, and 5.5 mg $kg^{-1}$ dw. All these species of ducks collect mixed food with different proportions of invertebrates and plants (both land and aquatic), while in another species from that area—American coot *Fulica americana*—which feeds mostly on algae, the background level turned out to be the lowest at 3.2 mg $kg^{-1}$ dw (Paveglio and Kilbridge 2007). Mentioned authors suppose that values of hepatic Se background level may be diverse due to the diet.

For some time, interest in heavy metal and Se contamination in urban and suburban environments increased, requiring an appropriate selection of species to study with regard to their food preferences. In China (Western Mountains of Beijing), a study was performed on two settled passerine species (insectivorous great tit *Parus major*), seed- and fruit-eating greenfinch *Chloris chloris*, in which 11 elements including Se were measured in 10 different parts of the body (Table 10.12). In both species the highest (and similar) concentrations of Se were found in the kidneys and liver (about 5.5 and 4 mg $kg^{-1}$ dw, respectively), with statistically significant differences in levels found only in the muscles and feathers. Clearly more selenium in the muscles and feathers was observed in the greenfinch in comparison to the insectivorous great tit (Deng et al. 2007). Because the concentration of Se in the pectoral muscle of the greenfinch exceeded 3 mg $kg^{-1}$ dw (reflecting the background level), we may conclude that in Beijing and the surroundings of the Chinese capital city, an elevated level of selenium in the environment exists.

Ohlendorf and Heinz (2011a, b) have proposed the thresholds for bird tissues. For the liver and muscles of freshwater species, <10 and 1–3 mg Se $kg^{-1}$ dw were established as concentrations that reflect the background level. Kalisińska et al. (2014) found neither hepatic nor muscle concentrations of Se exceeded background levels in common merganser *Mergus merganser* form Poland. Puls (1994) reported that an adequate level of Se in the muscle of poultry is 0.49–4.9 mg $kg^{-1}$ dw. Between Polish common mergansers examined by Kalisińska et al. (2014), muscle values $\leq$0.5 mg $kg^{-1}$ dw were noted in 37% of the analyzed birds, and the mean concentration (0.54 mg $kg^{-1}$ dw) was close to the lower value of the adequate range. A background Se level in the avian kidney has not been clearly defined, but an adequate value for poultry ranges from 2.2 to 5.2 mg $kg^{-1}$ dw (Puls 1994; Ohlendorf and Heinz 2011a, b). Kalisińska et al. (2014) found the Se concentration in the kidneys $\leq$2.2 mg $kg^{-1}$ dw in 42% of the studied birds. Mean concentrations of hepatic and nephric Se in the piscivorous ducks were 3.1 and 2.3 mg $kg^{-1}$ dw, respectively, and were 70–130% and 60–180% lower than analogous values for the nearby sea ducks (*Melanitta fusca, M. nigra*, and *Clangula hyemalis*) feeding on zoobenthos and wintering in the Pomeranian Bay (Pilarczyk et al. 2012). These birds nest in areas considered as selenium deficient (Scandinavia). In studies from other European countries (Leonzio et al. 1986; Lucia et al. 2012) not classified as Se deficient, mean concentrations of Se were several times higher than in the study by Kalisińska et al. (2014). In the liver, kidney, and muscles, the selenium levels were between the following respective ranges: 10–16, 7–18, and 3.2–6.7 mg $kg^{-1}$ dw. In comparison to two piscivorous avian species from North America (Scheuhammer et al. 1998), the Polish common mergansers were characterized by much lower Se levels in all three types of tissues. In Canadian common merganser and common loon (*Gavia immer*), Se concentrations in the liver were 9.7 and 15 mg $kg^{-1}$ d.w, in the kidneys 8.5 and 15 mg $kg^{-1}$ d.w, and in the muscles 1.8 and 2.8 mg $kg^{-1}$ dw, respectively. These values are higher than those found in Polish common merganser by 3–6 times, depending on the tissue and compared species.

In contrast to the liver, background Se levels in bird kidneys have not been clearly set, similarly to the proportion between the concentration of selenium in the liver and kidneys expressed as the liver/kidney ratio, although usually a significant correlation between Se levels in these organs is indicated. In birds from areas with an optimal Se content, nephric Se concentrations are usually slightly higher, similar to found in contaminated areas (Ohlendorf et al. 1988, 1990; Agusa et al. 2005; Deng et al. 2007). However in the piscivorous common merganser from eastern Canada, Scheuhammer et al. (1998) found higher concentrations of Se in the liver than in the kidneys, at 9.7 and 8.5 mg $kg^{-1}$ dw, respectively (the L/K factor value: 1.14). In the same species located in European selenium-deficient areas, the content of this element in the liver and kidneys was three times lower than in those from Canada, and also the hepatic concentration was higher than the nephric: 3.08 and 2.32 mg $kg^{-1}$ dw, respectively (Kalisińska et al. 2014). Apart from the liver and kidneys, Se is also measured in avian muscle. Background Se concentrations in birds muscles usually range from 1 to 3 mg $kg^{-1}$ dw. It was found that there is a clear relationship between the concentration of Se in feed and its content in muscles.

Because changes in Se content in bird muscles progress slowly, they are not considered a good indicator of Se levels, either in the organism or in the environment (Ohlendorf and Heinz 2011b).

In contrast to water birds, in land birds from the passeriformes group, Se is much less frequently measured, but there are some reports on this topic from the USA and China (Estrada and Maughan 2000; Deng et al. 2007), for example, related to the red-winged blackbird *Agelaius phoeniceus* and western kingbird *Tyrannus verticalis* (mostly insectivorous species).

Although the concentration of Se in the avian liver depends on the amount contained in food, this organ is still not treated as a good estimator of the pathological condition of a bird (Hoffman et al. 1991). When birds consume Se-rich food, the level of this element quickly passes to the eggs, liver, and blood and slowly to the muscles. A similar situation occurs when birds change from a Se-rich diet to a diet with lower selenium content or when they migrate from a high-Se area to a Se-normal area (Ohlendorf et al. 2008). The study by Albers et al. (1996) has also shown that the concentration of selenium in the liver is 2–3 times higher than in the food, while in the kidneys, it is about 1.8 times higher.

In the study by Santolo et al. (1999), selenium bioconcentration factors for diet-blood and diet-eggs in the American kestrel *Falco sparverius* are 1.0 and 2.2, respectively.

Between many tissues, blood is in a small extent used in studies of Se content in birds because collection in free-ranging animals is quite complicated, while in ecotoxicological studies in this matter, a more important role is played by eggs.

#### 6.3.1.1 Selenium in Avian Eggs

Because eggs, as one of bird's stages of life, are particularly sensitive to selenium, they are often used in biomonitoring studies mostly concerning water habitats and wetlands (Ohlendorf et al. 2008). Selenium levels of 3 mg $kg^{-1}$ dw in the bird egg are considered to be the threshold for concern for teratogenesis (Lemly 1993). Mean background concentrations of Se in eggs in freshwater and terrestrial birds have been established at <3 mg $kg^{-1}$ dw (usually from 1.5 to 2.5 mg $kg^{-1}$). Generally in various species, Se concentrations in avian eggs range from roughly equivalent to about three or four times the concentrations in the diet of the female during the laying period (Ohlendorf 2003). It has to be underlined that in different species of birds, the concentrations of Se in eggs that would affect a decrease in the reproductive parameters of birds may strongly differ, as some species are able to regulate the amount of selenium transferred to the eggs or they are characterized by an increased tolerance to this element during embryonic development (Harding 2008; Ohlendorf et al. 2008). It seems that the species that show a high level of tolerance to Se in eggs include red-winged blackbirds from North America, with a threshold for reduced egg hatchability estimated at 22 mg $kg^{-1}$ dw, while in other bird species, it ranges from 5 to 16 mg $kg^{-1}$ dw (Harding 2008; Ohlendorf et al. 2008).

Like the concentration of Se in different internal tissues of the birds, a significant impact on the amount of Se in the eggs is exerted by the diet of females. Moreover, it has been recently shown that the concentration of this element in eggs correlates with Se concentrations in the mother's liver (Ackerman et al. 2016). In passerine birds in the USA (from the border of Arizona and California, where the habitat is rich in Se), quite high concentrations of Se were found in the eggs of five bird species, with higher levels found in the insectivorous species in comparison to the grain eaters (Estrada and Maughan 2000; Table 10.13). In comparison to the number of reports concerning the presence of Se in the eggs of birds that live in Se-rich and selenium-excessive habitats, not many analogous studies in medium and poor level habitats are available, which impedes making proper comparisons of the results obtained in this research field (Guitart et al. 2003; Ohlendorf and Heinz 2011a, b).

#### 6.3.1.2 Selenium in Hair and Feathers

The content of selenium in animal hair may be successfully used to diagnose both a deficiency and a surplus of this element in the organism. The concentration of Se in human hair turned out to be proportional to the concentration of this element in the blood (Yang et al. 1989). Selenium in the hair or feathers is incorporated instead of sulfur into sulfuric amino acids (Wichert et al. 2002). Animal hair is a valuable research material due to its stable level of selenium (Dunnet and Lees 2003), the simple and noninvasive collection (no stress caused by blood sampling), and ease in storage (Wichert et al. 2002). In the case of free-ranging animals, hair for selenium measurements is used occasionally due to problems with obtaining material intravitally.

Background concentrations of Se in feathers range from 1 to 4 mg $kg^{-1}$ dw but are usually <2 mg $kg^{-1}$ dw (Ohlendorf and Heinz 2011b). Chemical analysis of feathers may deliver useful information about the exposure to pollutants consumed with food. However it has to be mentioned that different types of feathers from one

**Table 10.13** Concentration of selenium in avian eggs

| Species | Localization | Concentration (mg kg$^{-1}$ dry wt) | References |
|---|---|---|---|
| Yellow-breasted chat *Icteria virens* | USA, Arizona | 3.1 | Mora (2003) |
| Yellow warbler *Setophaga petechia* | USA, Arizona | 2.8 | Mora (2003) |
| Bell's vireo *Vireo bellii* | USA, Arizona | 1.93 | Mora (2003) |
| Willow flycatcher *Empidonax traillii* | USA, Arizona | 3.43 | Mora (2003) |
| Common yellowthroat *Geothlypis trichas* | USA, Arizona | 4.95 | Mora (2003) |
| Black-throated gray warbler *Setophaga nigrescens* | USA, Arizona | 3.8 | Mora (2003) |
| Summer tanager *Piranga rubra* | USA, Arizona | 2.4 | Mora (2003) |
| Vermilion flycatcher *Pyrocephalus obscurus* | USA, Arizona | 3.2 | Mora (2003) |
| Song sparrow *Melospiza melodia* | USA, Arizona | 2.77 | Mora (2003) |
| Brown-headed cowbird *Molothrus ater* | USA, Arizona | 2.3 | Mora (2003) |
| Lesser goldfinch *Spinus psaltria* | USA, Arizona | 2.1 | Mora (2003) |
| Red-winged blackbirds *Agelaius phoeniceus* | British Columbia | 0.89–9.9 | Harding (2008) |
| | USA, Arizona | 4.54 | Estrada and Maughan (2000) |
| Cliff swallow *Hirundo pyrrhonota* | USA, Arizona | 2.56 | Estrada and Maughan (2000) |
| Verdin *Auriparus flaviceps* | USA, Arizona | 2.9 | Estrada and Maughan (2000) |
| Western kingbird *Tyrannus verticalis* | USA, Arizona | 5.99 | Estrada and Maughan (2000) |

bird may be characterized by different selenium concentrations. Also in a feather itself, differences in levels of Se are possible: the highest concentration is usually found at the tip and the lowest at the base. For this reason feathers for analysis should be chosen very precisely. It is considered that feathers are not a good indicator of the current exposure of birds to Se, as the accumulation of this element takes place during the feather growth period, and not constantly (Burger 1993; Eisler 2000). The content of Se in feathers depends on the species and origin of birds (Table 10.14).

**Table 10.14** Concentration of selenium in bird feathers

| Species | Location | Concentration ($mg\ kg^{-1}$ dry wt) | References |
|---|---|---|---|
| Golden eagle *Aquila chrysaetos* | USA | 0.89 | Harmata and Restani (2013) |
| Black-tailed godwit *Limosa limosa* | France | 1.90 | Lucia et al. (2012) |
| Black skimmer *Rynchops niger* | USA | 1.22 | Burger and Hochfeld (1992) |
| Song sparrow *Melospiza melodia* | USA, Arizona | 1.20 | Lester and Riper (2014) |
| Great tit *Parus major* | China, Beijing | 2.17 | Deng et al. (2007) |
| | Portugal | 0.93 | Costa et al. (2013) |
| Greenfinch *Chloris chloris* | China, Beijing | 1.24 | Deng et al. (2007) |

## *6.4 Ecological Effects of Selenium*

Selenium belongs to the group of trace elements, which are crucial for the proper functioning of terrestrial vertebrate organisms. In nature and biotas, Se may be present in inorganic and organic forms with diverse bioavailability. Vertebrates are characterized by different demands for this microelement in global areas of low, proper, and excessive amounts of selenium. A review study concerning the "soil to small mammal" selenium bioconcentration factor derived final mean and median factors of 0.35 and 0.16, respectively (Sample et al. 1998). In global ecotoxicological literature, many reports concern the effects of too high selenium levels in aquatic and soil habitats, as well as the interactions with organisms from different trophic levels.

The Se concentration in water, at which toxicity is observed, ranges widely and varies between organisms like algae, invertebrates, and vertebrates: 0.01–80, 0.07–200, and 0.09–82 mg Se $L^{-1}$, respectively (Maier et al. 1987). Environmental intoxication also concerns organisms that live on land. Relevant studies have indicated that Se mobilization is caused by earthworks from changing land use (Ohlendorf et al. 1988; Albers et al. 1996). In California National Park, 20% of water bird offspring developed abnormalities along with increased chick morbidity due to the flow of excessively polluted water from a nearby reservoir (Lenz and Piet 2009). Coal and crude oil combustion are a considerable source of Se contamination, as well as copper and lead ore smelting, and some specific branches of industry (fat processing, pharmacy, dyes etc.)

Clinical symptoms of Se deficiency in free-ranging animals are noted occasionally; however, it is suspected that some subclinical symptoms may be noted by woodsmen, veterinary doctors, and hunters. In European wild ungulates, including roe deer, red deer, elk (*Alces alces* in Eurasia), and fallow deer *Dama dama*, WDM may occur to a level, which impairs the ability to escape quickly from predators. This disease occurs mostly between the 3rd and 6th weeks of life, where in such young individuals damage to the heart muscle may also be caused by the deficiency (Flueck et al. 2012).

In females with a deficiency in Se, milk production may decrease which in turn results in early weaning of offspring and added risk from predators. A deficiency of selenium may also manifest as a decrease in female fertility and a higher number of weak offspring. In males, it may reduce reproductive potential, which would result in discrepancies in the gender and age structure of the population (Flueck and Smith-Flueck 2008; Flueck et al. 2012). Moreover in males, due to a deficiency in this microelement, diarrhea, weight loss, weak growth, and antler deformations may take place. It was also shown that a deficiency of Se in free-ranging animals causes disturbances in bone mineralization and osteoblast activity, decrease of bone density, arthritis, improper teeth formation, premature tooth loss, as well as periodontal diseases. These various consequences of a selenium deficiency may together contribute to a decrease in the wild ruminant population size in a selected area (Hnilicka et al. 2004; Flueck and Smith-Flueck 2008; Flueck et al. 2012).

In other groups of free-ranging mammals (e.g., in representatives of omnivores, wild boar; carnivores, raccoon dog *Nyctereutes procyonoides*, red fox, American mink), a prolonged deficiency of Se influences the condition of the animals, litter size, embryo and fetal morbidity, placenta retention, decreased fetal growth rate, and delay in puberty onset (Flueck and Smith-Flueck 2008; Flueck et al. 2012). One of the reasons of a reduced population in some free-ranging animals may also be a decrease in their immunity caused by a lack of Se (Flueck et al. 2012). In wild boar, as in pigs, a deficiency of this element may be a cause of MHD (mulberry heart disease), as well as hepatosis diaetetica, while in sows it may cause premature fetal death due to impaired development, postparturitional placenta retention, and genital tract inflammations (Radostits et al. 2000). In Canidae (wolf, dog, foxes), a deficiency in selenium may be (as in dogs) a reason behind liver alimentary myopathy, muscle degeneration, tumors, subcutaneous endemas, shortness of breath, myocardial necrosis and kidney failure (Kuchan and Milner 1991; Green et al. 2001; Evans et al. 2004; Pilarczyk et al. 2010a, 2013).

In wild fowl, as in domestic fowl, a deficiency of Se may cause exudative diathesis and skeletal, stomachic and cardiac muscle degeneration (necrotic centers), increased morbidity, decreased fertility, weak feathering fibrosis and atrophy of the pancreas, as well as decreased digestive enzyme synthesis (Paton et al. 2002; Pappas et al. 2005).

Al-Dissi et al. (2011) in red deer, at liver Se concentrations reaching 2.7–8.97 mg $kg^{-1}$ ww, found symptoms of intoxication characterized by myocardial necrosis, muscle mineralization, and heart fibroplasia, with impaired hoof development and necrosis. The skin over the antler pedicles had areas of erosion and ulceration extending to the bony cranium. Within the meninges and Virchow Robin's spaces, an inflammatory infiltrate composed of neutrophils, lymphocytes, plasma cells, and macrophages was present. The pyloric area of the abomasum had isolated areas of erosion and ulceration.

## 6.5 *Bioindicators and Biomarkers of Selenium in Ecotoxicological Studies*

Selenium is present in every tissue of mammals and accumulates mostly in the liver, kidneys, hair, and nails. About 30% of the total Se pool is located in the liver, 15% in the kidneys, 30% in muscles, and 10% in blood plasma (Navarro-Alarcón and López-Martínez 2000). The content of selenium in body fluids is usually low. Erythrocytes contain more Se than whole blood in total, wherein the content of Se in serum and blood plasma is the same. The ratio of Se in red cells to its content in blood plasma is 3:1 (Sager 1993).

In the evaluation of Se status, it is important to obtain data about the geochemical background and anthropological sources of Se located within the examined areas, and to develop so-called typical (physiological) values for the most important organs and tissues in ecotoxicological studies, in order to interpret the obtained results in the proper way. The present knowledge database in this area is lacking (Table 10.15). The difference between a necessary (physiological) and toxic dose of selenium is very little. Many authors claim that the content of selenium in blood plasma or in full blood does not reflect the real level of this element in the organism (Burk and Hill 1994; Breedlove et al. 2006). It is said that an interrupted conversion of thyroxine to triiodothyronine, glutathione peroxidase activity, and the ratio between liver and kidney Se content in homeothermic mammals are good markers of Se levels (Oh et al. 1976).

Some Se-biomarkers, such as the selenoproteins and particularly GPX3 and SEPP1, provide information about the functioning of the organism directly and are

**Table 10.15** Levels of Se in animal organisms (mg kg$^{-1}$ ww), according to Puls (1994)

| Tissue | Animal group | Deficient | Marginal | Normal | High | Toxic |
|---|---|---|---|---|---|---|
| Kidney | Avian | 0.1–0.4 | 0.4–0.5 | 0.5–1.2 | 1.5–5.2 | – |
| | Bovine | 0.18–0.4 | 0.4–1 | 1–1.5 | 2–2.5 | 2.5–8 |
| | Canine | – | – | 1–1.5 | – | – |
| | Lapine | < 0.4 | 0.6–0.8 | 1–2 | – | 12 |
| | Ovine | 0.05–0.6 | 0.7–1.1 | 0.9–3 | 4–6 | 6–15 |
| | Porcine | 0.4–0.77 | 0.7–1.1 | 1.5–2.9 | 3–18 | 3.8–90 |
| Liver | Avian | 0.05–0.25 | 0.25–0.35 | 0.35–1 | 2–6 | 4–23 |
| | Bovine | 0.02–0.17 | 0.12–0.25 | 0.25–0.5 | 0.75–1.25 | 1.25–47 |
| | Canine | 0.1–0.3 | 0.3–0.5 | 0.5–1.5 | – | – |
| | Caprine | 0.01–0.1 | 0.1–0.2 | 0.25–1.2 | – | – |
| | Cervid | – | – | 0.25–1.4 | – | – |
| | Lapine | <0.4 | 0.6–1 | 1–2 | – | 7 |
| | Ovine | 0.01–0.1 | 0.15–0.25 | 0.25–1.5 | 2–10 | 15–30 |
| | Porcine | 0.03–0.1 | 0.12–0.25 | 0.4–1.2 | 1.5–12 | 3–120 |
| Muscle | Bovine | 0.01–0.05 | 0.05–0.07 | 0.07–0.15 | 0.25–0.5 | 0.08–1.5 |
| | Cervid | 0.05–0.26 | – | 0.25–0.49 | – | – |
| | Ovine | – | 0.9–1.2 | 1–1.3 | 1.1–1.6 | – |

of value in identifying a nutritional Se deficiency and in tracking the responses to Se-treatment in deficient individuals (Combs 2015).

Pollock (2005) stated that the concentration of selenium in liver is a better indicator of its status in the organism in comparison to the kidneys, as liver is the main organ responsible for selenium homeostasis in an organism. Selenium is stored in hepatocytes where the synthesis and distribution of not only Se-GSH-Px and other enzymatic selenoproteins take place. In a selenium deficiency, the organism is emptied of Se reserves, and the reserves located in the liver are mobilized first (Georgieva 2005). However Hoffman et al. (1991) indicated that liver does not reflect the selenium level in an organism in a precise way due to the fact that the concentration of Se in the liver depends on the absorbed dose of selenium, the relatively quick process of Se concentration balancing in the liver (7–8 days), as well as short half-life period of Se in this organ (18.7 days). For these reasons, the liver should be treated as a medium-term indicator of Se levels in an organism. Oh et al. (1976) claim that the relative proportions between Se content in the kidneys and liver are very important. In the case of Se-poor fodder, a higher concentration of this element is found in the kidneys than in the liver. Inversely, in selenium-rich fodder, the hepatic Se concentration is higher than the kidneys.

For toxic Se levels in a selected area, some authors consider bird eggs (as well as fish muscle) as good indicators. Selenium concentration in the eggs reflects the unfavorable changes in the environment to a higher extent, as selenium is very easily transferred to eggs (Focardi et al. 1988), which makes them good indicators of this element (Ohlendorf et al. 2011) as selenomethionine readily accumulates in the protein of egg albumen. Reproductive impairment is considered to be the most sensitive indicator of selenium toxicity in birds (Ohlendorf 2003; Seiler et al. 2003). As reported by Ohlendorf et al. (2011) eggs are treated as good Se indicators, while the feathers are considered weak indicators (Ohlendorf and Heinz 2011a, b). Bird embryos are very sensitive to the toxic effects of Se, including the teratogenic activity on embryonic development, which in turn translates into a decreased number of chicks hatching (Janz et al. 2010).

Information about the environmental Se status may be obtained through studying samples of inanimate (soil, water, air) and active (tissues, body fluids) parts of the habitat. The most used biomarkers of selenium exposure in animal organisms are tissue samples. Wild-ranging animals, due to a full integrity with the environment, are a very good indicator of Se levels in ecosystems (Pilarczyk et al. 2010b).

For some time, interest in heavy metal and Se contamination in urban and suburban environments has been increasing, entailing a proper selection of species regarding their nutritional preferences. In homeothermic vertebrates (farm animals), selenium status is usually based on analyses of fresh blood samples. Unfortunately in free-ranging animals, there is almost no possibility to obtain fresh blood, so the level of Se is usually measured in the liver and kidneys and less often in the muscles, blood, brain, feathers (hair), eggs, and bones of hunted or dead animals (Tables 10.9, 10.10, 10.11, 10.12, 10.13, and 10.14).

Despite the good documentation on the usefulness of ducks for Se biomonitoring in aquatic habitats, not many reports concerning the concentration of Se in their organisms are available outside of North America. Because of the wide spread of the species included into this group of birds (for instance, species belonging to *Anas*, *Aythya*, and *Mergus* genera) and usually the large number of individuals, it is possible to perform studies of a general biological and ecotoxicological character to a much higher extent and to do different comparisons of results between birds from areas rich and poor in selenium. In comparison to water birds, not many studies on samples obtained from typically land birds, for example, Passeriformes, are available, even though they are considered as potentially good indicators of selenium levels in the environment (Den et al. 2007; Costa et al. 2013).

In China (in the Western Mountains of Beijing), a study was conducted on two settled passerine species (the insectivorous great tit *Parus major* and the herbivorous greenfinch *Chloris chloris*), in which concentrations of 11 elements (including Se) were measured in 10 of their body parts (Table 10.12). In both species the highest and similar concentrations of selenium were found in the kidneys and liver (about 5.5 and 4 mg $kg^{-1}$ dw, respectively), and significant differences in Se level were noted only in the muscles and feathers. Clearly more Se was present in the muscles and less in the feathers of the greenfinch (Tables 10.12 and 10.14) than the insectivorous great tit (Deng et al. 2007). As the concentration of Se in the pectoral muscle of the greenfinch exceeded 3 mg $kg^{-1}$ dw (reflecting the background level), we may conclude that in Beijing and its surroundings, the level of Se in the environment is elevated (Table 10.12). To ease identification and systematic classification, the longevity (to evaluate the effects of the environment over time) and mobility (to monitor the level of pollution over a wider area) of birds are great advantages as bioindicators. If a species lives in a tight association with a specific area, it can reflect the pollution in that environment more specifically (air, food, water) (Ohlendorf and Heinz 2011a, b).

Typical land birds, such as tiny Passeriformes, can successfully be used to monitor the environment. They are considered good biomonitors of Se content as they occur commonly, are ubiquitous and abundant, and are eager to live in birdhouses, making it relatively easy to obtain study material from living individuals (blood, feathers, feces, and eggs). Moreover, the size of the population may be easily monitored (Costa et al. 2013).

As reported by Hobson et al. (2000), the use of free-ranging bird chicks in biomonitoring has two main advantages. Firstly, it avoids the consequences of bioaccumulation related to aging, as the exposure time is short and similar in all individuals from the same nest or colony. Secondly, all the chicks are fed by the parents with food collected from the surroundings and rich in local nutrients and contaminants. According to Ohlendorf and Heine (2011a, b), feathers are not a good indicator of selenium content.

In many works on selenium status in organisms of homeothermic mammals, studies mostly focused on evaluating the level of Se in the liver and kidneys, as well as often defining the mutual relationships in the obtained results (Pilarczyk et al. 2010a, b, 2011a; Flueck et al. 2012). However, as it has been shown, a proper evaluation of Se status should be based on the analysis of Se content in several different tissues, like the liver and kidneys as well as the muscles and/or lungs.

Moreover in such an evaluation, age, sex, and season should be taken into consideration (Garcia et al. 2000). Unfortunately such a multifaceted analysis is very rarely possible to conduct in free-ranging animals, as it is very difficult to obtain enough research material from the same year and the same area (region) to perform a reliable statistical interpretation of the results, representing different age and sex groups and in a particular season.

Wild ruminants from cervid group like the moose/elk, red deer, roe deer, mule deer, white-tailed deer, as well those of the Bovidae family, such as pronghorn and bighorn sheep, are considered good bioindicators of selenium levels in the environment. From the carnivores such as a role may be played by the fox, mink, and raccoon.

The concentration of Se in males and females living in the same area may differ. No unambiguous results of studies are available that would describe the dependence between sex and the content of selenium in the body. For this reason future experiments should regard both males and females settling in the same area (Nicpoń et al. 2005; Pilarczyk et al. 2010a; Rush et al. 2008; Pilarczyk et al. 2012). Also, data on the dependence of age and Se content are not yet clear enough. (Pilarczyk et al. 2010b, 2011a; Jankowiak et al. 2015; Vikøren et al. 2005; McDowell et al. 1995). Therefore in environmental studies, we should choose animals with regard to the particular age groups. In most of the present publications about selenium levels in animal organs, no such data is provided. With a low number of examined individuals, the results may be distorted due to any large interindividual variability. The evaluation of environmental levels of Se, as well as its status in homeothermic animals, should be performed on the basis of a population.

## 7 Conclusions

Wild ruminants belonging to the *Cervidae* family, like the moose, red deer, roe deer, mule deer, white-tailed deer, and elk, as well those in the *Bovidae* family, like pronghorn and bighorn sheep, are considered to be good bioindicators of selenium levels in the environment. From the carnivores, such a role may be played by the fox, mink, and raccoon.

In birds, many species of ducks (e.g., *Anas*, *Aythya*, and *Mergus* genera) and Passeriformes (e.g., greenfinch, great tit, song sparrow) may be used as selenium biomonitors. Typical land birds, such as tiny passerines, may successfully be used to monitor the environment.

From the European ruminants, the concentration of selenium has most often been analyzed in roe deer and red deer, while in North America in mule deer, white-tailed deer, and elk, using mostly samples from the liver, kidneys, and muscles. European ruminants, in comparison to North American, are characterized by much lower concentrations of selenium in analyzed organs.

The status of selenium in free-ranging homeothermic organisms should be evaluated mostly in the liver and kidneys. In ecotoxicological studies on Se, the liver and kidneys should primarily be used and to a lesser extent the muscles, eggs and feathers. It is necessary to assess the mutual relationships between these concentrations. Also age, sex and season should be taken into account. In an evaluation of selenium status, it is particularly important to obtain data about the geochemical background and anthropogenic sources of selenium located in the examined area, to ensure a proper interpretation of the results.

## References


Aastrup P, Riget F, Dietz R, Asmund G (2000) Lead, zinc, cadmium, mercury, selenium and copper in Greenland caribou and reindeer (*Rangifer tarandus*). Sci Total Environ 245:149–159

ACARP (2006) Trace elements in coal. Report Dale L. Csiro Energy Technology

Ackerman JT, Eagles-Smith CA, Herzog MP, Hartman CA (2016) Maternal transfer of contaminants in birds: Mercury and selenium concentrations in parents and their eggs. Environ Pollut 210:145–154

ADSTR (2003) Public Health Statement: selenium, cas# 7782-4 [http://www.atsdr.cdc.gov/ToxProfiles/tp92-c1-b.pdf]

Agusa T, Matsumoto T, Ikemoto T, Anan Y, Kubota R, Yasunaga G et al (2005) Body distribution of trace elements in black-tailed gulls from Rishiri Island, Japan: age-dependent accumulation and transfer to feathers and eggs. Environ Toxicol Chem 24:2107–2120

Albers PH, Green DE, Sanderson CJ (1996) Diagnostic criteria for selenium toxicosis in aquatic birds: dietary exposure, tissue concentrations, and macroscopic effects. J Wildl Dis 32:468–485

Al-Dissi AN, Blakley BR, Woodbury MR (2011) Selenium toxicosis in a white-tailed deer herd. Can Vet J 52:70–73

Amouroux D, Liss PS, Tessier E, Hamren-Larsson M, Donard O (2001) Role of oceans as biogenic sources of selenium. Earth Planet Sci Lett 189:277–283

Andersson E (2005) Hazardous substances in electrical and electronic equipment (EEE) – Expanding the scope of the RoHS directive. Document prepared for the Swedish Chemicals Inspectorate at the Department of Risk Reduction in Sundybberg as part of a course for the Department of Applied Environmental Science at Göteborg University, Sweden. Available at http://forum.europa.eu.int/ Public/irc/env/weee_2008/library?l=/characteristics/hazardous_substances/_EN_1.0_&a=d

Andren A, Klein D (1975) Selenium in coal-fired steam plant emissions. Environ Sci Technol 9: 856–858

Antanaitis A, Lubyte J, Antanaitis S, Staugaitis G, Viskelis P (2008) Selenium concentration dependence on soil properties. J Food Agric Environ 6:163–167

Aro A, Alfthang G (1998) Effects of selenium supplementation fertilizers on human nutrition and selenium status. In: Frankenberger WT Jr, Engberg RA (eds) Environmental chemistry of selenium. Marcel Dekker, New York, pp 81–97

Arthur JR (2000) The glutathione peroxidases. Cell Mol Life Sci 57:1825–1835

Arthur J, Nicol F, Beckeit G (1993) Selenium deficiency, thyroid hormone metabolism, and thyroid hormone deiodinases. Am J Clin Nutr 57:236–239

Arthur JR, McKenzie RC, Beckett GJ (2003) Selenium in the immune system. J Nutr 133: 1457S–1459S

Atroshi F, Antila E, Westermarck T (2007) The role of selenium in epilepsy and other neurological disorders. Epileptologia 15:211–224

ATSDR (2003) Toxicological profiles for selenium. Agency for Toxic Substances and Disease Registry, U.S department of Health and Human Services, Public Health Services, Atlanta, GA. http://www.atsdr.cdc.gov/toxpro2.html

Azaizeh HA, Salhani N, Sebesvari Z, Emons H (2003) The potential of rhizosphere microbes from constructed wetland to biomethylate selenium. J Environ Qual 32:55–62
Barrow NJ, Whelan BR (1989) Testing a mechanistic model. VIII. The effect of time and temperature of incubation on the sorption and subsequent desorption selenite and selenate by a soil. J Soil Sci 40:29–37
Beavington F, Cawse PA, Wakenshaw A (2004) Comparative studies of atmospheric trace elements: improvements in air quality near a copper smelter. Sci Total Environ 332:39–49
Bedwal RS, Nair N, Sharma NP, Mathur RS (1993) Selenium – its biological perspectives. Med Hypotheses 41:150–159
Behne D, Kyriakopoulos A (2001) Mammalian selenium-containing proteins. Annu Rev Nutr 21: 453–473
Behne D, Kyriakopoulos A, Weiss-Nowak C, Kalckloesch M, Westphal C, Gessner H (1996) Newly found selenium-containing proteins in the tissues of the rat. Biol Trace Elem Res 55: 99–100
Bell PF, Parker DR, Page AL (1992) Contrasting selenate sulfate interactions in selenium accumulating and nonaccumulating plant species. Soil Sci Soc Am J 56:1818–1824
Berrow ML, Ure AM (1989) Geologic material and soil. In: Inat M (ed) Occurrence and distribution of selenium. CRC Press, Boca Raton, FL, pp 213–242
Berzas Nevado JJ, Rosa C, Rodríguez Martín-Doimeadios RC, Mateo R, Rodríguez Fariñas N, Rodríguez-Estival J et al (2012) Mercury exposure and mechanism of response in large game using the Almand mercury mining area (Spain) as a case study. Environ Res 112:58–66
Bielowicz B (2013) Occurrence of selected harmful elements in Polish lignite. Gospod Sur Min 29: 47–59 [in Polish]
Björkhem-Bergman L, Ekström L, Eriksson LC (2012) Exploring anticarcinogenic agents in a rat hepatocarcinogenesis model–focus on selenium and statins. In Vivo 26:527–536
Blazina T, Sun Y, Voegelin A, Lenz M, Berg M, Winkel L (2014) Terrestrial selenium distribution in China is potentially linked to monsoonal climate. Nat Commun 5:4717
Borowska K, Koper J, Tykwińska T (2007) Zawartość selenu w wybranych typach gleb mineralnych regionu Kujaw i Pomorza na tle aktywności oksydoreduktaz. Ochr Śr Zasobów Nat 31: 8–23
Brady PS, Brady LJ, Whetter PA, Ullrey DE, Fay LD (1978) The effect of dietary selenium and vitamin E on biochemical parameters and survival of young among white-tailed deer (*Odocoileus virginianus*). J Nutr 108:1439–1448
Braune BM, Malone BJ (2006) Mercury and selenium in livers of waterfowl harvested in northern - Canada. Arch Environ Contim Toxicol 50:284–289
Breedlove HA, Smith AM, Burk RF, Hill KE, Shapiro CL (2006) Serum selenium measurements in women with early-stage breast cancer with and without chemotherapy-induced ovarian failure. Breast Cancer Res Treat 97:225–230
British Geological Survey (2006) Word mineral statistics 2000–2004. British Geological Survey, Keyworth
British Geological Survey (2010) Word mineral statistics 2004–2008. British Geological Survey, Keyworth
British Geological Survey (2015) World mineral production 2009–2013. British Geological Survey, Keyworth and Nottingham
British Geological Survey (2017) Word mineral statistics 2012–2016. British Geological Survey, Keyworth
Broadley MR, White PJ, Bryson RJ, Meacham MC, Bowen HC, Johnson SE, Hawkesford MJ, McGrath SP, Zhao FJ, Breward N, Harriman M, Tucker M (2006) Biofortification of UK food crops with selenium. Proc Nutr Soc 65:169–181
Brown KM, Arthur JR (2001) Selenium, selenoproteins and human health: a review. Public Health Nutr 4:593–599

Brzezinski M, Zalewski A, Niemczynowicz A, Jarzyna I, Suska-Malawska M (2014) The use of chemical markers for the identification of farm escapees in feral mink populations. Ecotoxicology 23:767–778

Burger J (1993) Metals in avian feathers: bioindicators of environmental pollution. Rev Environ Toxicol 5:203–311

Burger J, Hochfeld M (1992) Heavy metal and selenium concentrations in black skimmers (*Rynchops niger*): gender differences. Arch Environ Contam Toxicol 23:431–434

Burger J, Gochfeld M, Jeitner Ch, Donio M, Pittfield T (2012) Interspecific and intraspecific variation in selenium:mercury molar ratios in saltwater fish from the Aleutians: potential protection on mercury toxicity by selenium. Sci Total Environ 431:46-56

Burk RF, Hill KE (1994) Selenoprotein P. A selenium-rich extracellular glycoprotein. J Nutr 124:1891–1897

Butterman WC, Brown RD (2004) Selenium. Mineral Commodity Profiles. Raport U.S. Department of the Interior U.S. Geological Survey, pp 1–20

Callahan MA (1979) Water-related fate of 129 priority pollutants. EPA 440/4-79-029a and 029b. US Environmental Protection Agency Office of Water Planning and Standards, Washington DC

Chernoff N, Kavlock RJ (1982) An in vivo teratology screen utilizing pregnant mice. J Toxicol Environ Health 10:541–550

Clark DR Jr, Ogasawara PA, Smith GJ, Ohlendorf HM (1989) Selenium accumulation by raccoons exposed to irrigation drainwater at Kesterson National Wildlife Refuge, California, 1986. Arch Environ Contam Toxicol 18:787–794

Clayton GD, Clayton FE (1994) Patty's Industrial Hygiene and Toxicology. Part A. JohnWiley & Sons, Toronto

Combs GF (2001) Selenium in global food systems. Br J Nutr 85:517–547

Combs GF (2015) Biomarkers of selenium status. Nutrients 7:2209–2236

Conover MR, Vest JL (2009) Concentrations of selenium and mercury in eared grebes (*Podiceps nigricollis*) from Utah's Great Salt Lake, USA. Environ Toxicol Chem 28:1319–1323

Conzemius RJ, Welcomer TD, Svec HJ (1984) Elemental partitioning in ash depositories and material balances for a coal burning facility by spark source mass spectrometry. Environ Sci Technol 18:12–18

Cooke TD, Bruland KW (1987) Aquatic chemistry of selenium: evidence of biomethylation. Environ Sci Technol 21:1214–1219

Corvilain B, Contempr B, Longomb A, Govens P, Gervy-Decoster C, Lamy F et al (1993) Selenium and the thyroid: how the relationship was established. Am J Clin Nutr 57:244–248

Costa RA, Eeva T, Eira C, Vaqueiro J, Vingada JV (2013) Assessing heavy metal pollution using Great Tits (*Parus major*): feathers and excrements from nestlings and adults. Environ Monit Assess 185:5339–5344

Cox MK (2006) Effects of mineral supplements on California bighorn sheep in northern Nevada. Bienn Symp North Wild Sheep Goat Council 15:107–120

Cristaldi LA, McDowell LR, Buergelt CD, Davis PA, Wilkinson NS, Martin FG (2005) Tolerance of inorganic selenium in wether sheep. Small Rumin Res 56:205–213

Cukierski MJ, Willhite CC, Lasley BL, Hendrie TA, Book SA, Cox DN, Hendrickx AG (1989) 30-Day oral toxicity study of L-selenomethionine in female long-tailed macaques (*Macaca fascicularis*). Fundam Appl Toxicol 13:26–39

Custer CM, Custer TW (2000) Organochlorine and trace element contamination in wintering and migrating diving ducks in the southern Great Lakes, USA, since the zebra mussel invasion. Environ Toxicol Chem 19:2821–2829

Custer TW, Cox E, Gray B (2004) Trace elements on moose (*Alces alces*) found dead in Northwestern Minnesota, USA. Sci Total Environ 330:81–87

Cutter GA (1989) Freshwater systems. In: Ihnat M (ed) Occurrence and distribution of selenium. CRC Press, Boca Raton, FL, pp 243–262

Ĉuvardić MS (2003) Selenium in soil. Proc Nat Sci 1004:23–37

Dall'Aglio M, Ghiara E, Proietti W (1978) New data on the hydrogeochemistry of selenium. Rend Soc Ital Mineral Petrol 34:591–604

Daun C, Akesson B (2004) Glutathione peroxidase activity and content of total and soluble selenium in five bovine and porcine organs used in meat production. Meat Sci 66:801–807

De Goeij JJM, Guinn VP, Young DR, Mearns AJ (1974) Neutron activation analysis trace-element, studies of dover sole liver and marine sediments. Comparative Studies of Food and Environmental Contamination. International Atomic Energy Agency Vienna, pp 189–200

Dębski B, Kryński A, Skrzymowska K (2005) Selenium concentration in musk rat, hare, cow tissues and in cow's milk, as an indicator of its status in local ecosystem. ISAH 2:442–445

Deng H, Hang Z, Chang C, Wang Y (2007) Trace metal concentration in great tit (Parus major) and greenfinch (Carduelis sinica) at the Western Mountains of Beijing, China. Environ Pollut 148: 620–626

Dhillon KS, Dhillon SK (2003) Distribution and management of seleniferous soils. Adv Agron 79: 19–85

Diplock AT (1985) Selenium and health. SC Rev Limited, London, p 81

Dodig S, Cepelak I (2004) The facts and controversies about selenium. Acta Pharm 54:261–276

Drozd R, Pilarczyk R, Pilarczyk B, Drozd A, Tomza-Marciniak A, Bombik T et al (2015) Activity of selected antioxidant enzymes, selenium content and fatty acid composition in the liver of the Brown Hare (*Lepus europaeus* L.) in relation to the season of the year. Biol Trace Elem Res 168:421–428

Ducsay L, Ložek O, Varga L, Lošak T (2007) Effects of winter wheat supplementation with selenium. Ecol Chem Eng 14(3–4):289–294

Duffield-Lillico AJ, Reid ME, Turnbull BW, Combs GF Jr, Slate EH, Fischbach LA et al (2002) Baseline characteristics and the effect of selenium supplementation on cancer incidence in a randomized clinical trial: a summary report of the Nutritional Prevention of Cancer Trial. Cancer Epidemiol Biomark Prev 11:630–639

Dunbar MR, Velarde R, Gregg MA, Bray M (1999) Health evaluation of a pronghorn antelope population in Oregon. J Wildl Dis 35:496–510

Dunn CE (1990) Lithogeochemical study of the Cretaceous in central Saskatchewan – preliminary report; in Summary of Investigations 1990, Saskatchewan Geological Survey, Saskatchewan Energy and Mines, Miscellaneous Report 90–94

Dunnet M, Lees P (2003) Trace element, toxin and drug elimination in hair with particular reference to the horse. Res Vet Sci 75:89–101

Ebens RJ, Shacklette HT (1982) Geochemistry of some rocks, mine spoils, stream sediments, soils, plants, and waters in the Western Energy Region of the Conterminous United States. U.S. Geological Survey Professional Paper 1237. U.S. Government Printing Office, Washington, DC, p 173

Eble C, Hower J (1997) Coal quality trends and distribution of potentially hazardous trace elements in Eastern Kentucky coals. Fuel 76:711–715

Ediger RD (1975) Atomic absorption analysis with the graphite furnace using matrix modification. Atom Absorption Newsl 14:127–130

EEA (2013) Technical report No 10/2013. European Union emission inventory report 1990–2011 under the UNECE Convention on Long-range Transboundary Air Pollution (LRTAP). Luxembourg, Publications Office of the European Union

Eisler R (2000) Selenium. In: Handbook of chemical risk assessment: health hazards to humans, plants, and animals. CRC Press, Boca Raton, FL, pp 1649–1705

Ellis DR, Salt DE (2003) Plants, selenium and human health. Curr Opin Plant Biol 3:273–279

Engberg RA (1973) Selenium in Nebraska's groundwater and streams. Nebraska Water-Supply Paper Number 35. University of Nebraska Conservation and Survey Division, Lincoln

Estrada KD, Maughan E (2000) Reproductive impacts of elevated selenium levels. A final report prepared for the U.S. Fish and Wildlife Service, Ecological Services Office, Environmental Contaminants Division http://ecos.fws.gov/ServCatFiles/reference/holding/21868?accessType= DOWNLOAD

Evans J, Levesque D, Shelton GD (2004) Canine inflammatory myopathies: a clinicopathologic review of 200 cases. J Vet Intern Med 18:679–691

Fielder PC (1986) Implications of selenium levels in Washington mountain goats, mule deer, and Rocky Mountain elk. Northwest Sci 60:15–20

Fischer J, Bosse A, Pallauf J (2008) Effect of selenium deficiency on the antioxidative status and muscle damage in growing turkeys. Arch Anim Nutr 62:485–497

Fishbein L (1983) Environmental selenium and its significance. Fundam Appl Toxicol 3:411–419

Fishbein L (1991) Selenium. In: Merian E (ed) Metals and their compounds in the environment. Verlag Chemie, Weinheim, New York, pp 1153–1190

Fleming GA (1962) Selenium in Irish soils and plant. Soil Sci 94:28–35

Flohe L, Andreesen JR, Brigelius-Flohe R, Maiorino M, Ursini F (2000) Selenium, the element of the moon, in life on earth. IUBMB Life 49:411–420

Flueck WT (2015) Osteopathology and selenium deficiency co-occurring in a population of endangered Patagonian huemul (*Hippocamelus bisulcus*). BMC Res Notes 8:330

Flueck WT, Smith-Flueck JM (2008) Age-independent osteopathology in skeletons of a South American cervid, the Patagonian huemul (*Hippocamelus bisulcus*). J Wildl Dis 44:636–648

Flueck WT, Smith-Flueck JM, Mionczynski J, Mincher BJ (2012) The implications of selenium deficiency for wild herbivore conservation: a review. Eur J Wildl Res 58:561–780

Focardi S, Fossi C, Lambertini M, Leonzio C, Massi A (1988) Long-term monitoring of pollutants in eggs of Yellow-legged Herring Gull from Capraia Island (*Tuscan Archipelago*). Environ Monit Assess 10:43–50

Fordyce F (2013) Selenium deficiency and toxicity in environment. In: Selinus O, Alloway B, Centeno JA, Finkelman RB, Fuge R, Lindh U, Smedley P (eds) Essentials of medical geology. Impacts of the natural environment on public health. Elsevier, Academic Press, Boston, San Diego, London, pp 373–415

Fordyce FM, Nice SE, Lister TR, Ódochartaigh BÉ, Cooper R, Allen M et al (2009) Urban Soil Geochemistry of Glasgow. British Geological Survey Open Report, OR/08/002, British Geological Survey, Edinburgh

Frankenberger WT Jr, Karlson U (1994) Microbiological volatilization of selenium from soil and sediments. In: Frankenberger WT Jr, Benson S (eds) Selenium in the environment. Marcel Dekker, New York, pp 369–387

Frøslie A, Norheim G, Rambaek JB, Steinnes E (1984) Levels of trace elements in liver from Norwegian moose, reindeer and red deer in relation to atmospheric deposition. Acta Vet Scand 25:333–345

Frost DV, Ingvoldstad D (1975) Ecological aspects of selenium and tellurium in human and animal health. Chem Scr 8A:96–107

Gaillardet J, Viers J, Dupre B (2003) Trace elements in river waters. In: Holland HD, Turekian KK (eds) Treatise on geochemistry, vol 5. Elservier, Oxford, pp 225–227

Galgan V, Frank A (1995) Survey of bioavailable selenium in Sweden with the moose (*Alces alces*) as monitoring animal. Sci Total Environ 172:37–45

Gamberg M, Boila G, Stern G, Roach P (2005a) Cadmium, mercury and selenium concentrations in mink (*Mustela vison*) from Yukon, Canada. Sci Total Environ 351–352:523–529

Gamberg M, Palmer M, Poach P (2005b) Temporal and geographic trends in trace element concentrations in moose from Yukon. Sci Total Environ 351(352):530–538

Garcia MJ, Alegria A, Barberà R, Farré R, Lagarda MJ (2000) Selenium, copper, and zinc indices of nutritional status. Influence of sex and season on reference values. Biol Trace Elem Res 73: 77–83

George MW (2004) Minerals yearbook 2004: Selenium and Tellurium. U.S. Geological Survey. Available at http://minerals.usgs.gov/minerals/pubs/commodity/selenium/selenmyb04.pdf

Georgieva NV (2005) Oxidative stress as a factor of disrupted ecological oxidative balance in biological systems: a review. Bul J Vet Med 8:1–11

Gladyshev VN (2001) Selenium in biology and human health: controversies and perspectives. In: Hatfield DL (ed) Selenium: Its molecular biology and role in human health. Kluwer, Boston, pp 313–317

Goldsztejn P (2007) Koncentracje wybranych pierwiastków w węglu brunatnym ze złoża ościsłowo w rejonie Konina. Prace Naukowe Instytutu Górnictwa Politechniki Wrocławskiej. Studia i Materiały 118:17–24

Golubkina NA, Papazyan TT (2006) Selenium distribution in eggs of avian species. Comp Biochem Physiol B Biochem Mol Biol 145:384–388

Goodman GE, Schaffer S, Bankson DD, Hughes MP, Omenn GS (2001) Carotene and retinol efficacy trial co-investigators. Predictors of selenium in cigarette smokers and lack of association with lung and prostate cancer risk. Cancer Epidemiol 10:1069–1076

Grace ND (1994) Selenium. In: Grace ND (ed) Managing trace element deficiencies. New Zealand Pastoral Agricultural Research Institute, Simon Print, Palmerston North, New Zeland

Grau MV, Rees JR, Baron JA (2006) Chemoprevention in gastrointestinal cancers: current status. Basic Clin Pharmacol Toxicol 98:281–287

Green J (1959) Geochemical table of the elements for 1959. Geol Soc Am Bull 70:1127–1184

Green SL, Bouley DM, Pinter MJ, Cork LC, Vatassery GT (2001) Canine motor neuron disease: clinicopathologic features and selected indicators of oxidative stress. J Vet Intern Med 15: 112–119

Griffin AC (1982) The chemopreventive role of selenium in carcinogenesis. In: Arnott MS, van Eys J, Wang YM (eds) Molecular interrelations of nutrition and cancer. Raven Press, New York, NY, pp 401–408

GSNI (in preparation) The Tellus Geochemical Atlas of Northern Ireland. Geological Survey of Northern Ireland, Belfast

Guitart R, Mateo R, Sanpera C, Hernández-Matías A, Ruiz X (2003) Mercury and selenium levels in eggs of common terns (*Sterna hirundo*) from two breeding colonies in the Ebro Delta, Spain. Bull Environ Contam Toxicol 70:71–77

Hamdy AA, Gissel-Nielsen G (1976) Relationships between soil factors and selenium content of Danish. Soil and Plants. Risø Report No 349, Denmark

Hamilton SJ, Buhl KJ (2003) Selenium and other trace elements in water sediment aquatic plants aquatic invertebrates and fish from streams in southeastern Idaho near phosphate mining operations. Final Report as Part of the USGS Western US Phosphate Project

Handel CM, Hemert C (2015) Environmental contaminants and chromosomal damage associated with beak deformities in a resident North American Passerine. Environ Toxicol Chem 34: 314–327

Hansen JC, Deguchi Y (1996) Selenium and fertility in animals and man: a review. Acta Vet Scand 37:19–30

Harding LE (2004) Environmental contaminants in wild martens (*Martes americana*) and volvorines (*Gulo luscus*). Bull Environ Contam Toxicol 73:98–105

Harding LE (2008) Non-linear uptake and hormesis effects of selenium in red-winged blackbirds (*Agelaius phoeniceus*). Sci Total Environ 389:350–366

Harding L, Harris M, Elliott J (1998) Heavy and trace metals in wild mink (*Mustela vison*) and river otter (*Lontra canadensis*) captured on rivers receiving metals discharges. Bull Environ Contam Toxicol 61:600–607

Harmata AR, Restani M (2013) Lead, mercury, selenium, and other trace elements in tissues of golden eagles from southwestern Montana, USA. J Wildl Dis 49(1):114–124

Harr JR (1978) Biological effects of selenium. In: Oehme FW (ed) Toxicity of heavy metals in the environment. Marcel Dekker, New York, pp 393–426

Hartfiel W, Bahners N (1987) Selenmangel in der Bundesrepublik Deutschland. Vita Min Spur 2: 125–131

Hartfield W, Bahners N (1988) Selenium deficiency in the Federal Republic of Germany. Biol Trace Elem Res 15:1–12

Hartikainen H (2005) Biogeochemistry of selenium and its impact on food chain quality and human health. J Trace Elem Med Biol 18:309–318
Haug A, Graham RD, Christophersen OA, Lyons GH (2007) How to use the world's scarce selenium resources efficiently to increase the selenium concentration in ford. Microb Ecol Health D 19:209–228
Heinz GH (1996) Selenium in birds. In: Beyer WN, Heinz GH, Redmon-Norwood AW (eds) Environmental contaminants in wildlife: Interpreting environmental contaminants in animal tissues. Lewis Publishers, Boca Raton, FL, pp 447–458
Heinz GH, Hoffman DJ, Krynitsky AJ, Weller DMG (1987) Reproduction in mallards fed selenium. Environ Toxicol Chem 6:423–433
Heinz GH, Hoffman DJ, Gold LG (1989) Impaired reproduction of mallards fed an organic form of selenium. J Wildl Manag 53:418–428
Heinz GH, Pendleton GW, Krynitsky AJ, Gold LG (1990) Selenium accumulation and elimination in mallards. Arch Environ Contam Toxicol 19:374–379
Herbert GB, Peterle TJ (1990) Heavy metal and organochlorine compound concentrations in tissues of raccoons from east-central Michigan. Bull Environ Contam Toxicol 44:331–338
Hnilicka PA, Mionczynski J, Mincher BJ, States J, Hinschberger M, Oberlie S, Thompson C, Yates B, Siemer DD (2004) Bighorn sheep lamb survival, trace minerals, rainfall, and air pollution: are there any connections? Biennial Symp North Wild Sheep Goat Council 13:69–94
Hobson KA, Sirois J, Gloutney ML (2000) Tracing nutrient allocation to reproduction with stable isotopes: a preliminary investigation using colonial waterbirds of Great Slave Lake. Auk 117:760e774
Hoekstra P, Braune BM, Elkin B, Armstrong FA, Muir DC (2003) Concentrations of selected essential and non-essential elements in arctic fox (*Alopex lagopus*) and wolverines (*Gulo gulo*) from the Canadian Arctic. Sci Total Environ 309:81–92
Hoffman DJ (2002) Role of selenium and toxicity and oxidative stress in aquatic birds. Aquat Toxicol 57:11–26
Hoffman DJ, Heinz GH, Le Captain LJ, Bunck CM, Green DE (1991) Subchronic hepatotoxicity of selenomethionine ingestion in mallard ducks. J Toxicol Environ Health 32:449–464
Högberg J, Alexander J (1986) Selenium. In: Friberg L, Nordberg GF, Vouk VB (eds) Handbook on the toxicology of metals. Elsevier, Amsterdam, pp 482–520
Humann-Ziehank E, Ganter M, Hennig-Pauka I, Binder A (2008) Trace mineral status and liver and blood parameters in sheep without mineral supply compared to local roe deer (*Capreolus capreolus*) populations. Small Rumin Res 75:185–191
IARC (2015) IARC monographs on the evaluation of carcinogenic risks to humans, vol 1–112
INRA (2008) Information sur les éléments traces dans les sols en France. Institut National de la Recherche Agronomique. http://etm.orleans.inra.fr/webetmidf.html
Jackson MI, Combs GF Jr (2008) Selenium and anticarcinogenesis: underlying mechanisms. Curr Opin Clin Nutr Metab Care 11:718–726
Jacobs LW (ed) (1989) Selenium in agriculture and the environment. Soil Science Society of America Special Publication no. 23, Madison, WI
Jakovljeviã M, Stevanoviã D, Blagojeviã S, Kostiã N, Martinoviã LJ (1995) The content of selenium in the soils of northern Pomoravlje, Symposium on Selenium. Serbian Academy of Sciences and Arts, pp 43–49
Jankowiak D, Pilarczyk R, Drozd R, Pilarczyk B, Tomza-Marciniak A, Wysocka G et al (2015) Activity of antioxidant enzymes in the liver of wild boars (*Sus scrofa*) from a selenium-deficient area depending on sex, age, and season of the year. Turk J Biol 39:129–138
Janz D, DeForest D, Brooks M, Chapman P, Gilron G, Hoff D et al (2010) Selenium toxicity to aquatic organisms. In: Chapman PM, Adams WJ, Brooks ML, DelosCG LSN, Maher WA, Ohlendorf HM, Presser TS, Shaw DP (eds) Ecological assessment of selenium in the aquatic environment. SETAC, Pensacola, FL, pp 141–231

Jarzyńska G, Falandysz J (2011) Selenium and 17 other largely essential and toxic metals in muscle and organ meats of red deer (*Cervus elaphus*)—consequences to human health. Environ Int 37:882–888

Johnson VJ, Tsunoda M, Sharma RP (2000) Increased production of proinflammatory cytokines by murine macrophages following oral exposure to sodium selenite but not to seleno-L-methionine. Arch Environ Contam Toxicol 39:243–250

Kabata-Pendias A, Mukherjee AB (2007) Trace elements from soil to human. Springer, Berlin/Heidelberg/New York

Kabata-Pendias A, Szteke B (2015) Trace elements in abiotic and biotic environments. CRC Press, Boca Raton, FL

Kabatas-Pendias A, Pendias H (2000) Trace elements in soils and plants. CRC Press, Boca Raton, FL

Kalisińska E, Salicki W (2010) Lead and cadmium levels in muscle, liver and kidney of scaup *Aythya marila* from Szczecin Lagoon, Poland. Pol J Environ Stud 19:1213–1222

Kalisińska E, Górecki J, Okońska A, Pilarczyk B, Tomza-Marciniak A, Budis H et al (2014) Hepatic and nephric mercury and selenium concentrations in common mergansers *Mergus merganser* from Baltic region, Europe. Environ Toxicol Chem 33:421–430

Kang Y, Yamada H, Kyuma K, Hattori T (1990) Selenium content and distribution of various Japanese soils. Soil Sci Plant Nutr 36:475–482

Kang S, Kang JH, Kim S, Lee SH, Lee S, Yu HJ et al (2015) Trace element analysis of three tissues from Eurasian otters (*Lutra lutra*) in South Korea. Ecotoxicology 24:1064–1072

Kantola M, Saaranen M, Vanha-Perttula T (1988) Selenium and glutathione peroxidase in seminal plasma of men and bulls. J Reprod Fert 83:785–794

Ketris MP, Yudovich YE (2009) Estimations of clarkes for carbonaceous biolithes: world averages for trace element contents in black shales and coals. Int J Coal Geol 78:135–148

Kim YY, Mahan DC (2001) Comparative effects of high dietary levels of organic and inorganic selenium on selenium toxicity of growing-finishing pigs. J Anim Sci 79:942–948

Kim H, Lee TH, Hwang YS, Bang MA, Kim KH, Suh JM, Chung HK, Yu DY, Lee KK, Kwon OY, Ro HK, Shong M (2001) Methimazole as an antioxidant and immunomodulator in thyroid cells: mechanisms involving interferon-signaling and $H_2O_2$ scavenging. Mol Pharmacol 60:972–980

Klusek CS, Miller KM, Heit M (1983) Trace element and radionuclide mass balances at a coal-fired electric generating station. Environ Int 9:139–144

Köhrle J, Brigelius-Flohe R, Bock A, Gartner R, Meyer O, Flohe L (2000) Selenium in biology: facts and medical perspectives. Biol Chem 38:849–864

Koljonen T (1973) Selenium in certain igneous rocks. Bull Geol Soc Finland 45:9–22

Krauskopf KB (1982) Introduction to geochemistry, 2nd edn. McGraw-Hill Book Company., Appendix 3, Auckland, Singapore, p 546

Kuchan MJ, Milner JA (1991) Influence of supplemental glutathione on selenite-mediated growth inhibition of canine mammary cells. Cancer Lett 57:181–186

Kucukbay F, Demir M (2001) Selenium speciation in Karakaya Dam Lake's water (Malatya-Turkey). Turk J Chem 25:341–347

Kuhn JK, Fiene FL, Cahill RA, Gluskoter HJ, Shimp NF (1980) Abundance of trace and minor elements in organic and mineral fractions of coal. Environmental Geology Notes 88. Illinois State Geological Survey Division, Urbana, IL, p 67

Kunli L, Lirong X, Jian'an T, Druhu W, Lianhua X (2004) Selenium source in the selenosis area of the Daba region, South Qinling Mountain, China. Environ Geol 45:426–432

Kursa J, Herzig I, Trávníček J, Illek J, Kroupová V, Fuksová S (2010) Iodine and selenium contents in skeletal muscles of red deer (*Cervus elaphus*), roe deer (*Capreolus capreolus*) and wild boar (*Sus scrofa*) in the Czech Republic. Acta Vet Brno 79:403–407

Lăcătuşu R, Lungu M, Aldea M, Lăcătuşu A, Stroe V, Lazăr R et al (2010) Selenium in the rock-soil system from south-eastern part of Romania. Present Environ Sustain Dev 4:145–158

Lakin HW (1972) Selenium accumulation in soils and its absorption by plants and animals. Geol Soc Am Bull 83:181–189

Lam JCW, Tanabe S, Lam MHW, Lam PK (2005) Risk to breeding success of waterbirds by contaminants in Hong Kong: evidence form trace elements in eggs. Environ Pollut 135(3): 481–490

Laser H (2004) Pflanzenbauliche Ansätze zur Selen-Versorgung von Mutterkühen und Fleischrindern in Weidesystemen. Habilitationsschrift. Universität Gießen, p 213

Latshaw JD, Morishita TY, Sarver CF, Thilsted J (2004) Selenium toxicity in breeding ring-necked pheasants (Phasianus colchicus). J Avian Dis 48:935–939

Lazarus M, Orct T, Blanuŝa M, Vicković I, Ŝoŝtarić B (2008) Toxic and essential metal concentrations in four tissues of red deer (*Cervus elaphus*) from Baranja, Croatia. Food Addit Contam 25:270–283

Lemly AD (1993) Guidelines for evaluating selenium data from aquatic monitoring and assessment studies. Environ Monit Assess 28:83–100

Lenz M, Lens Piet NL (2009) The essential toxin: the changing perception of selenium in environmental science. Sci Total Environ 407:3620–3633

Leonzio C, Fossi C, Focardi S (1986) Heavy metals and selenium variation in a migratory bird wintering in a mercury-polluted lagoon. Bull Environ Contam Toxicol 37:219–225

Lester MB, van Riper C III (2014) The distribution and extent of heavy metal accumulation in song sparrows along Arizona's upper Santa Cruz River. Environ Monit Assess 186:4779–4791

Levander OA (1986) Selenium. In: Mertz W (ed) Trace elements in human and animal nutrition. Academic Press, London, pp 139–197

Levengood JM, Hubert GF, Jr (2001) Concentrations of selected elements in Illinois raccoons. Trans Ill State Acad Sci 94:89-99

Levesque M (1977) Some aspects of selenium relationships in Eastern Canadian soils and plants. Can J Soil Sci 54:205–214

Li HF, McGrath SP, Zhao FJ (2008) Selenium uptake, translocation and speciation in wheat supplied with selenate or selenite. New Phytol 178:92–102

Linšak DT, Linšak Z, Špirić Z, Srebočan E, Glad M, Cenov A, Jakovac H, Milin C (2014) Influence of cadmium on metallothionein expression and products of lipid peroxidation in the organs of hares ( Lepus europaeus Pallas). J Appl Toxicol 34:289–295

Lippman SM, Klein EA, Goodman PJ, Lucia MS, Thompson IM, Ford LG et al (2009) Effect of selenium and vitamin E on risk of prostate cancer and other cancers. The selenium and vitamin E cancer Prevention Trial (SELECT). JAMA 301:39–51

Lucia M, Bocher P, Cosson RP, Churlaud C, Robin F, Bustamante P (2012) Insight on trace element detoxification in the Black-tailed Godwit (*Limosa limosa*) through genetic, enzymatic and metallothionein analyses. Sci Total Environ 423:73–83

Luoma SN, Presser TS (2000) Forecasting selenium discharges to the San Francisco Bay-Delta Estuary: ecological effects of a proposed San Luis Drain Extension, USGS report 00-416

Luoma SN, Johns C, Fisher NS, Steinberg NA, Oremland RS, Reinfelder JR (1992) Determination of selenium bioavailability to a benthic bivalve from particulate and solute pathways. Environ Sci Technol 26:485–491

Lyons MP, Papazyan TT, Sural PF (2007) Selenium in food chain and animal nutrition: lessons from nature. Asian-Aust J Anim Sci 20:1135–1155

Maehira G, Luyo GA, Oshiro M (2002) Alterations of serum selenium concentrations in the acute phase of pathological conditions. Clin Chim Acta 316:137–146

Maier D, Sinemus HW, Wiedeking E (1979) AAS-Bestimmung geliister Spurenelemente im Bodenseewasser des Uberlinger Sees. Fresenius Z Anal Chem 296:114–124

Maier KJ, Foe C, Ogle RS, Williams MJ, Knight AW, Kiffney P, Melton LA (1987) The dynamics of selenium in aquatic ecosystems. Trace Subst Environ Health 21:361–408

Marcus RW (1993) Myopathy and cardiomyopathy associated with selenium deficiency: case report, literature review, and hypothesis. Md Med J 42:669–674

Marier JR, Jaworski JF (1983) Interactions of Selenium. National Research Council Canada, Associate Committee on Scientific Criteria for Environmental Quality, Subcommittee on Heavy Metals and Certain Other Elements. NRCC No. 20643

Martin TD (1975) Determining selenium in wastewater, sediment, and sludge by flameless atomic absorption. Atomic Absorption Newsl 14:109–116

Mayland HF (1994) Selenium in plants and animal nutrition. In: Frankenberg WT Jr, Benson S (eds) Selenium in the environment. Marcel Dekker Inc, New York, pp 29–45

Mayland HF, James LF, Panter KE, Sonderegger JL (1989) Selenium in Seleniferous Environments. In: Jacobs LW (ed) Selenium in Agriculture and the Environment. Soil Science Society of America Special Publication no. 23, Madison, WI, pp 15–50

McCann JC, Ames BN (2011) Adaptive dysfunction of selenoproteins from the perspective of the triage theory: why modest selenium deficiency may increase risk of diseases of aging. FASEB J 25:1793–1814

McDowell L, Forrester D, Linda S, Wright SD, Wilkinson NS (1995) Selenium status of white-tailed deer in southern Florida. J Wildl Dis 31:205–211

McGrath D, Fleming GA, Culleton N (2008) Trace elements and heavy metals in Irish soils. Teagasc, Wexford, Ireland, p 210

McKittrick J, Chen PY, Bodde SG, Yang W, Novitskaya EE, Meyers MA (2012) The structure, functions, and mechanical properties of keratin. JOM 64:449–468

McNeal JM, Balistrieri LS (1989) Geochemistry and occurrence of selenium: an overview. In: Jacobs LW (ed) Selenium in agriculture and the environment. Soil Science Society of America Special Publication no. 23, Madison, WI, pp 1–14

Mesman BB, Thomas TC (1975) A study of two atomic absorption methods for the determination of sub-microgram amounts of arsenic and selenium. Anal Lett 8:449–459

Michalska-Kacymirow M, Kurek E, Smolis A, Wierzbicka M, Bulska E (2014) Biological and chemical investigation of *Allium cepa* L. response to selenium inorganic compounds. Anal Bioanal Chem 406:3717–3722

Michot TC, Benson WH, O'Neil JM (1998, September) Trace element concentrations from greater scaup and other diving ducks wintering in Apalachee Bay, Florida, seagrass beds. Scaup Workshop, Jamestown, ND, 9–10

Millán JR, Mateo M, Taggart MA, López-Bao JV, Viota M, Monsalve L et al (2008) Levels of heavy metals and metalloids in critically endangered Iberian lynx and other wild carnivores from Southern Spain. Sci Total Environ 399:193–201

Mirbagheri SA (2004) Modeling contaminant transport in soil column and ground water pollution control. Int J Environ Sci Technol 1:141–150

Mora MA (2003) Heavy metals and metalloids in egg contents and eggshells of passerine birds from Arizona. Environ Pollut 125:393–400

Mora MA, Anderson DW (1995) Selenium, boron, and heavy metals in birds from the Mexicali Valley, Baja California, Mexico. Bull Environ Contamin Toxicol 54:198–206

Moreno-Reyes R, Egrise D, Neve J, Pasteels JL, Schoutens A (2001) Selenium deficiency-induced growth retardation is associated with an impaired bone metabolism and osteopenia. J Bone Miner Res 16:1556–1563

Mosher BW, Duce RA (1987) A global atmospheric selenium budget. J Geophys Res 92: 13289–13298

Mynhardt H, van Ryssen JBJ, Coertze RJ (2006) The effect of the heat processing of soybean seed on the metabolism of its selenium in lambs. Anim Feed Sci Technol 128:122–134

Nam D, Anan Y, Ikemoto T, Tanabe S (2005) Multielemental accumulation and its intracellular distribution in tissues of some aquatic birds. Mar Pollut Bull 50:1347–1362

National Research Council (NRC) (1980) Selenium. In: Mineral Tolerance of Domestic Animals. National Academy of Sciences (NAS), Washington, DC, pp 392–415

National Research Council (NRC) (1983) Selenium in Nutrition. Agricultural Board, Committee on Animal Nutrition, National Academy Press, Washington, DC

Navarro-Alarcón M, López-Martínez MC (2000) Essentiality of selenium in the human body: relationship with different diseases. Sci Total Environ 249:347–371

Navarro-Alarcon L, Lopez de la Serrana H, Perez-Valero V, López-Martínez C (1998) Serum selenium levels in indicators of body status in cancer patients and their relationship with other nutritional and biochemical markers. Sci Total Environ 212:195–202
Neal RH (1995) Selenium. In: Alloway BJ (ed) Heavy metals in soils. Blackie Academic and Professional, London, pp 260–283
Newland LW (1982) Handbook of environmental chemistry. Springer, New York, pp 45–57
Nicpoń J, Balicka-Ramisz A, Jankowski M, Kubiak K, Pilarczyk B, Ramisz A (2005) Selenium level of dog sera from Lower Silesia. Med Weter 61:170–172
Niedzielski P (2006) Microtrace metalloids speciation in lasek water samples (Poland). Environ Monit Assess 118:231–246
Nishri A, Brenner IB, Hall GEM, Taylor HE (1999) Temporal variations in dissolved selenium in Lake Kinneret (Israel). Aquat Sci 61:215–233
Norheim G, Sivertsen T, Brevik EM, Frøslie A (1984) Mercury and selenium in wild mink (*Mustela vison*) from Norway. Nord Vet Med 36:43–48
Nowakowska E, Pilarczyk B, Pilarczyk R, Tomza-Marciniak A, Bąkowska M (2014) Selenium content in selected organs of roe deer (*Capreolus capreolus*) as a criterion to evaluate environmental abundance of this element in Poland. Int J Environ Res 8:569–576
Nriagu JO, Pacyna JM (1988) Quantitative assessment of worldwide contamination of air, water and soils by trace metals. Nature 333:134–139
Nriagu JO, Wong HK (1983) Selenium pollution of lakes near the smelters at Sudbury, Ontario. Nature 301:55–57
Oh S, Sunde R, Pope A, Hoekstra W (1976) Glutathione peroxidase response to selenium intake in lambs fed a torulabased, artificial milk. J Anim Sci 42:977–983
Ohlendorf HM (1989) Bioaccumulation and effects of selenium in wildlife. In: Jacobs LW (ed) Selenium in agriculture and the environment. American Society of Agronomy and Soil Science Society of America, Madison, pp 133–177
Ohlendorf HM (2002) The birds of Kesterson reservoir: a historical perspective. Aquat Toxicol 57:1–10
Ohlendorf HM (2003) Ecotoxicology of selenium. In: Hoffman DJ, Rattner BA, Burton GA, Cairns J (eds) Handbook of ecotoxicology. Lewis Publishers, Boca Raton, FL, pp 465–501
Ohlendorf H, Heinz G (2011a) Selenium in birds. In: Beyer WN, Meador JP (eds) Environmental contaminants in biota: interpreting tissue concentrations, 2nd edn. CRC Press, Boca Raton, FL, p 768
Ohlendorf HM, Heinz GH (2011b) Selenium in birds. In: Beyer WN, Meador JP (eds) Environmental contaminants in biota: interpreting tissue concentrations, 2nd edn. CRC Press/Taylor and Francis Group, Boca Raton, FL, pp 669–701
Ohlendorf HM, Kilness AW, Simmons JL, Stroud RK, Hoffman DJ, Moore JF (1988) Selenium toxicosis in wild aquatic birds. J Toxicol Environ Health 24:67–92
Ohlendorf HM, Hothem RL, Bunck CM, Marois KC (1990) Bioaccumulation of selenium in birds at Kesterson Reservoir, California. Arch Environ Contam Toxicol 19:495–507
Ohlendorf HM, Covington S, Byron E, Arenal C (2008) Approach for conducting site-specific assessments of selenium bioaccumulation in aquatic systems. North America Metals Council – Selenium Working Group, Washington, DC
Ohlendorf HM, Covington SM, Byron ER, Arenal CA (2011) Conducting site-specific assessments of selenium bioaccumulation in aquatic systems. Integr Environ Assess Manag 7:314–324
Opresko DM (1993) Toxicity summary for selenium and selenium compounds. Chemical hazard evaluation group, Biomedical environmental information analysis section, Health and Safety Research Division, Oak Ridge National Laboratory, Tennessee, prepared for Oak Ridge Reservation Environmental Restoration Program. Managed by Martin Marietta Energy Systems, Inc, for the U.S. Department of Energy under Contract No. DE-AC05-84OR21400
Orr PL, Guiguer KR, Russel CK (2006) Food chain transfer of selenium in lentic and lotic habitats of a western Canadian watershed. Ecotoxicol Environ Safe 63:175–188

Pappas AC, Karadas F, Surai PF, Speake BK (2005) The selenium intake of the female chicken influences the selenium status of her progeny. Comp Biochem Physiol B 142:465–474

Paton ND, Cantor AH, Pescatore AJ, Ford MJ, Smith CA (2002) The effect of dietary selenium source and level on the uptake of selenium by developing chick embryos. Poult Sci 81: 1548–1554

Paveglio FL, Kilbridge KM (2007) Selenium in aquatic birds from central California. J Wildl Manag 71:2550–2555

Peters GM, Maher WA, Krikowa F, Roach AC, Jeswani HK, Barford JP et al (1999) Selenium in sediments, pore waters and benthic infauna of Lake Macquarie, New South Wales, Australia. Mar Environ Res 47:491–508

Pezzarossa B, Petruzzelli G, Malorgio F, Ferri T (2007) Absorption of selenium by Lactuca sativa as affected by carboxymethylcellulose. Chemosphere 67:322–329

Pfeifer H, Conrad M, Roethlein D (2001) Identification of a specific sperm nuclei selenoenzyme necessary for protamine thiol cross-linking during sperm maturation. FASEB 15:1236–1238

Pilarczyk B, Balicka-Ramisz A, Ramisz A, Adamowicz E, Bujak T, Tomza-Marciniak A et al (2008) Selenium concentration in roe deer from the Western Pomerania, Poland. Bull Vet Inst Pulawy 52:631–633

Pilarczyk B, Balicka-Ramisz A, Ramisz A, Adamowicz E, Pilarczyk R, Tomza-Marciniak A et al (2009) Selenium concentration in liver and kidney of free living animals (roe and red deer) from West Pomerania (Poland). Eur J Wildl Res 55:279–283

Pilarczyk B, Balicka-Ramisz A, Ramisz A, Vovk S, Vantukh A, Bąkowska M et al (2010a) Selengehalt im Serum von Hunden und Katzen in Westpommern und der Westukraine. Tierärztl Prax Kleintiere 38:374–378

Pilarczyk B, Hendzel D, Pilarczyk R, Tomza-Marciniak A, Błaszczyk B, Dąbrowska-Wieczorek M et al (2010b) Liver and kidney concentrations of selenium in wild boars (*Sus scrofa*) from northwestern Poland. Eur J Wildl Res 55:797–802

Pilarczyk B, Drozd R, Pilarczyk R, Tomza-Marciniak A, Jankowiak D, Hendzel D et al (2011a) Glutathione peroxidase (GSHPx) activity in the liver of red deer in relation to hepatic selenium concentrations, sex, body weight and season of the year. Biol Trace Elem Res 144:560–569

Pilarczyk B, Pilarczyk R, Tomza-Marciniak A, Hendzel D, Bąkowska M, Stankiewicz T (2011b) Evaluation of selenium status and its distribution in organs of free living foxes (*Vulpes vulpes*) from an Se deficient area. Pol J Vet Sci 14:453–457

Pilarczyk B, Tomza-Marciniak A, Pilarczyk R, Hendzel D, Błaszczyk B, Bąkowska M (2011c) Tissue distribution of selenium and effect of season and age on selenium content in roe deer from northwestern Poland. Biol Trace Elem Res 140:299–307

Pilarczyk B, Tomza-Marciniak A, Pilarczyk R, Kavetska K, Rząd I, Hendzel D et al (2012) Selenium status in sea ducks (*Melanitta fusca*, *Melanitta nigra* and *Clangula hyemalis*) wintering on the southern Baltic coast, Poland. Mar Biol Res 8:1019–1025

Pilarczyk B, Tomza-Marciniak A, Pilarczyk R, Bąkowska M, Gaik M, Wilk M et al (2013) Relationship between serum Se concentration in dogs and incidence of some disease conditions. Cent Eur J Biol 8:527–533

Pilon-Smits EAH, LeDuc DL (2009) Phytoremediation of selenium using transgenic plants. Curr Opin Biotechnol 20:207–212

Piotrowska A (1984) Content of selenium in arable soils in Poland. Rocz Glebozn 35:24–31

Plant JA, Kinninburg DG, Smedley PL, Fordyce FM, Klink BA (2003) Arsenic and selenium. In: Lollar BS, Heinrich D, Karl K (eds) Treatise on geochemistry, Environmental geochemistry. Elsevier, London, pp 17–66

Plant JA, Kinniburgh DG, Smedley PL, Fordyce FM, Klinck BA (2004) Arsenic and selenium. In: Holland HD, Turekian KK (eds) Treatise on geochemistry, Environmental geochemistry, vol 9. Elsevier, London, pp 17–66

Pollock B (2005) Trace elements status of white-tailed red deer (*Odocoileus virginianus*) and moose (*Alces alces*) in Nova Scotia. Wildlife Damage Management, Internet Center for Canadian

Cooperative Wildlife Health Centre: Newsletters & Publications, University of Nebraska-Lincoln, p 45
Presser TS (1994) Geologic origin and pathways of selenium from the California coast ranges to the west-central San Joaquin Valley. In: Frankenberger WT Jr, Benson S (eds) Selenium in the environment. Marcel Dekker, New York, pp 139–155
Prestrud P, Norheim G, Sivertsen T, Daae HL (1994) Levels of toxic and essential elements in arctic fox in Svalbard. Polar Biol 14:155–159
Puls R (1994) Mineral levels in animal health: diagnostic data. Sherpa International, Clearbrook, p 356
Radostits OM, Gay CC, Blood DC, Hinchcliff KW (2000) Veterinary medicine: a textbook of the diseases of cattle, sheep, pigs, goats and horses. Harcourt Publishers Ltd 1877, London
Raisbeck MF (2000) Selenosis. Vet Clin North Am Food Anim Pract 16:465–480
Ralston NVC, Raymond LJ (2010) Dietary selenium's protective effects against methylmercury toxicity. Toxicology 278:112–123
Rapant S, Vrana K, Bodis D (1996) Geochemical Atlas of Slovakia: Part 1. Groundwater. Geological Survey of Slovak Republic
Rayman MP (2000) The importance of selenium to human health. Lancet 356:233–241
Reglero M, Taggart M, Monsalve-González L, Mateo R (2009) Heavy metal exposure in large game from a lead mining area: effects on oxidative stress and fatty acid composition in liver. Environ Pollut 157:1388–1395
Reilly C (1996) Se in food and health. Blackie Academic and Professional, London
Reimann C, Ayras M, Chekushin V, Bogatyev I, Boyd R, Caritat P et al (1998) Environmental geochemical atlas of the Central Barents Region. Geological Survey of Norway, Trondheim
Reimann C, Birke M, Demetriades A, Filzmoser P, O'Connor P (2013) Chemistry of Europe's agricultural soils. Geologisches Jahrbuch (Reihe B)
Robberecht H, Von Grieken R (1982) Selenium in environmental waters: determination, speciation and concentration levels. Talanta 29:823–844
Rosenfeld I, Beath OA (1954) Effect of selenium on reproduction in rats. Proc Soc Exp Biol Med 87:295–297
Roveri A, Orsini F, Flohe L, Miorino M (2001) PHGPx and spermatogenesis. Biofactors 12: 213–222
Rovira M, Giménez J, Martínez M, Martínez-Lladó X, de Pablo J, Martí V et al (2008) Sorption of selenium (IV) and selenium (VI) onto natural iron oxides: goethite and hematite. J Hazard Mater 150:279–284
Rush SA, Borgå K, Dietz R, Born EW, Sonne C, Evans T et al (2008) Geographic distribution of selected elements in the liver of polar bears from Greenland, Canada and the United States. Environ Pollut 153:618–626
Sabbioni E, Goetz L, Springer A, Pietra R (1983) Trace metals from coal-fired power plants: derivation of an average data base for assessment studies of the situation in the European communities. Sci Total Environ 29:213–227
Sager M (1993) Selenium occurrence and ecology. Stud Environ Sci 55:459–473
Sampaio da Silva D, Lucotte M, Paquet S, Brux G, Lemire M (2013) Inverse mercury and selenium concentration patterns between herbivorous and piscivorous fish in the Tapajos River, Brazilian Amazon. Ecotoxicol Environ Saf 97:17–25
Sample BE, Opresko DM, Suter II GW (1996) Toxicological benchmarks for wildlife: revision. Prepared by the Risk Assessment Program Health Sciences Research Division Oak Ridge, Tennessee and Prepared for the U.S. Department of Energy Office of Environmental Management
Sample BE, Beauchamp JJ, Efroymson RA, Suter GW II (1998) Development and Validation of Bioaccumulation Models for Small Mammals. Prepared for the US Department of Energy. Office of Environmental Management. Oak Ridge National Laboratory. ES/ER/TM-219
Santolo GM, Yamamoto JT, Pisenti JM, Wilson BW (1999) Selenium accumulation and effects on reproduction in captive American kestrels fed selenomethionine. J Wildl Manag 63:502–511

Scheuhammer AM, Wong AHK, Bond D (1998) Mercury and selenium accumulation in common loons (*Gavia immer*) and common mergansers (*Mergus merganser*) from eastern Canada. Environ Toxicol Chem 17:197–201

Schroeder HA, Mitchener M (1971) Toxic effects of trace elements on the reproduction of mice and rats. Arch Environ Health 23:102–106

Schutz DF, Turekian KK (1965) The investigation of the geographical and vertical distribution of several trace elements in seawater using neutron activation analysis. Geochim Cosmochim Acta 29:259–313

Seiler RL, Skorupa JP, Naftz DL, Nolan BT (2003) Irrigation-induced contamination of water, sediment, and biota in the western United States-synthesis of data from the National Irrigation Water Quality Program. U.S. Geological Survey Professional Paper 1655, 131 pp

Shacklette HT, Boemgen JG (1984) Element concentrations in soils and other superficial materials of the conterminous United States. U.S. Geological Survey Professional Paper 1270, 105 pp

Siepak M, Niedzielski P, Przybyłek J (2003) Badania rozprzestrzeniania się mikropierwiastków w wodach podziemnych z wykorzystaniem analizy specjacyjnej. Współczesne Problemy Hydrogeologii 11:305–313

Sleeman JM, Manning EJ, Rohm JH, Sims JP, Sanchez S, Gerhold RW et al (2009) Johne's disease in a free-ranging white-tailed deer from Virginia and subsequent surveillance for Mycobacterium avium subspecies paratuberculosis. J Wildl Dis 45:201–206

Sleeman JM, Magura K, Howell J, Rohm J, Murphy LA (2010) Hepatic mineral values of white-tailed deer (*Odocoileus virginianus*) from Virginia. J Wildl Dis 46:525–531

Smakowski T, Neya R, Galosa K (2011) Bilans Gospodarki surowcami mineralnymi Polski i świata 2009. Instytut Gospodarki Surowcami Mineralnymi i Energią. Polskiej Akademii Nauk. Pracownia Polityki Surowcowe Kraków 2011

Smith GJ, Heinz GH, Hoffman DJ, Spann JW, Krynitsky AJ (1988) Reproduction in black-crowned night-herons fed selenium. Lake Reservoir Manag 4:175–180

Spallholz JE, Hoffman DJ (2002) Selenium toxicity: cause and effects on aquatic birds. Aquat Toxicol 57:27–37

Spears JW (2003) Trace mineral bioavailability in ruminants. J Nutr 133:1506S–1509S

Spears JW (2011) Selenium deficiency and its prevention in grazing ruminants. Salt Inst Newsl 4 www.saltinstitute.org

Speyer HR (1980) Mercury and selenium concentrations in fish, sediment and water of two Northwestern Quebec lakes. Bull Environ Contam Toxicol 24:427–432

Stanley TR Jr, Smith GJ, Hoffman DJ, Heinz GH, Rosscoe V (1996) Effects of boron and selenium on mallard reproduction and duckling growth and survival. Environ Toxicol Chem 15:1124-1132

Stone CA, Kawai K, Kupka R, Fawzi W (2010) Role of selenium in HIV infection. Nutr Rev 68:671–681

Stoszek MJ, Willmes H, Jordan NL, Kessler WB (1980) Natural trace mineral deficiency in native pronghorn antelope populations. Proceedings of the 9th Biennial Pronghorn Antelope Workshop, Rio Rico, AZ, pp 71–76

Stout JH, Trust KA (2002) Elemental and organochlorine residues in bald eagles from Adak Island, Alaska. J Wildl Dis 38:511–517

Stowesand GS, Anderson JL, Weinstein LH, Osmloski JF, Gutenmann WH, Lisk DJ (1990) Selenium in tissues of rats fed rutabagas grown on soil covering a cola fly ash landfill. Bull Environ Contam Toxicol 44:681–685

Stussy RJ, Findholt SL, Johnson BK, Noyes JH, Dick BL (2000) Selenium levels and productivity in three Oregon elk herds. Northwest Sci 74:97–101

Symanowicz B, Kalembasa S, Jaremko D, Niedbała M (2013) Polish brown coals waste – potential source of plants nutrients. Ann UMCS Sect E, pp 21–27

Taggart MA, Figuerola J, Greek AJ, Mateo R, Deacon C, Osborn D et al (2006) After the Aznalcollar mine spill: arsenic, zinc, selenium, lead and copper levels in the livers and bones of five waterfowl species. Environ Res 100:349–361

Tan J (ed) (1989) The Atlas of endemic diseases and their environments in the People's Republic of China. Science Press, Beijing

Tarantal AF, Willhite CC, Lasley BL, Murphy CJ, Miller CJ, Cukierski MJ et al (1991) Developmental toxicity of l-selenomethionine in Macaca fascicularis. Fundam Appl Toxicol 16: 147–160

Taylor SR, McLennan SM (1995) The geochemical evolution of the continental crust. Rev Geophys 33:241–265

Terry N, Zayed M, DE Souza MP, Tarun AS (2000) Selenium in higher plants. Annu Rev Plant Physiol Plant Mol Biol 51:401–432

Thornton M (1983) Geochemical aspects of selenium in British soils and implications to animal health. Trace Subst Environ Health 27:391–398

Tian HZ, Wang Y, Xue ZG, Cheng K, Qu YP, Chai FH et al (2010) Trend and characteristics of atmospheric emissions of Hg, As, and Se from coal combustion in China, 1980–2007. Atmos Chem Phys 10:11905–11919

Tian H, Wang Y, Xue Z, Qu Y, Chai F, Hao J (2011) Atmospheric emissions estimation of Hg, As, and Se from coal-fired power plants in China, 2007. Sci Total Environ 409:3078–3081

Tian HZ, Zhu CY, Gao JJ, Cheng K, Hao JM, Wang K et al (2015) Quantitative assessment of atmospheric emissions of toxic heavy metals from anthropogenic sources in China: historical trend, spatial variation distribution, uncertainties and control policies. Atmos Chem Phys 15: 12107–12166

Tinggi U (2005) Selenium toxicity and its adverse health effects. In: Preedy R, Watson RR (eds) Reviews in food and nutrition toxicity. Taylor & Francis, Boca Raton, FL, pp 29–55

Tomza-Marciniak A, Bąkowska M, Pilarczyk B, Semeniuk M, Hendzel D, Udała J et al (2010) Selenium concentration in soil and selected tissues of roe deer (Capreolus capreolus) from Wielkopolska region. Acta Sci Pol Zootech 9:251–260 [in Polish]

Turekian KK, Wedepohl KH (1961) Distribution of the elements in some major units of the Earth's crust. Bull Geol Soc Am 72:175–192

U.S. Geological Survey (2015) Mineral commodity summaries 2015: U.S. Geological Survey. https://doi.org/10.3133/70140094

U.S. Public Health Service (USPHS) (1996) Toxicological profile for selenium (update). U.S. Department of Health and Human Services, PHS, Agency for Toxic Substances and Disease Registry, Atlanta, GA, 324 pp

Ullrey DE, Youatt WG, Whetter PA (1981) Muscle selenium concentrations in Michigan deer. J Wildl Manag 45:534–536

Valentine RL, Bache CA, Gutenmann WH, Lisk DJ (1988) Tissue concentrations of heavy metals and polychlorinated biphenyls in raccoons in central New York. Bull Environ Contam Toxicol 40:711–716

Veber M, Cujes K, Gomiscek S (1994) Determination of selenium and arsenic in mineral waters with hydride generation atomic absorption spectrometry. JAAS 9:285–290

Vikøren T, Bernhoft A, Waaler T, Handelan K (2005) Liver concentrations of copper, cobalt and selenium in wild Norwegian red deer (*Cervus elaphus*). J Wildl Dis 41:569–579

Vikøren T, Kristoffersen AB, Lierhagen S, Handeland K (2011) A comparative study of hepatic trace element levels in wild moose, roe deer, and reindeer from Norway. J Wildl Dis 41:569–579

Walker LA, Lawlor AJ, Chadwick EA, Potter E, Pereira MG, Shore RF (2010) Inorganic elements in the livers of Eurasian otters, Lutra lutra, from England and Wales in 2008 – a Predatory Bird Monitoring Scheme (PBMS) report. Centre for Ecology & Hydrology, Lancaster

Walker LA, Lawlor AJ, Chadwick EA, Potter E, Pereira MG, Shore RF (2011) Inorganic elements in the livers of Eurasian otters, Lutra lutra, from England and Wales in 2009 – a Predatory Bird Monitoring Scheme (PBMS) report. Centre for Ecology &Hydrology, Lancaster

Wang MC, Chen HM (2003) Forms and distribution of selenium at different depths and among particle size fractions of three Taiwan soils. Chemosphere 52:585–593

Wang ZJ, Gao YX (2001) Biogeochemical cycling of selenium in Chinese environments. Appl Geochem 16:1345–1351

Wang D, Alfthan G, Aro A (1994) Determination of total selenium and dissolved selenium species in natural waters by fluorometry. Environ Sci Technol 28:383–387

Wang L, Ju Y, Liu G, Chou C, Zheng L, Qi C (2010) Selenium in Chinese coals: distribution, occurrence, and health impact. Environ Earth Sci 60:1641–1651
Wayland M, Crosley R (2006) Selenium and other trace elements in aquatic insects in coal mine-affected streams in the Rocky Mountains of Alberta, Canada. Arch Environ Contam Toxicol 50: 511–522
Wen H, Carignan J (2007) Reviews on atmospheric selenium: emissions, speciation and fate. Atmos Environ 41:7151–7165
Whanger PW (1996) Selenium metabolism in animals and humans. J Environ Sci 8:328–340
Whittle KJ, Hardy R, Holden AV, Johnston R, Pentreath RJ (1977) Occurrence and fate of organic and inorganic contaminants in marine animals. Ann NY Acad Sci 298:47–79
WHO/FAO (2002) Human vitamin and mineral requirements: report of a joint WHO/FAO Expert Committee, World Health Organization/Food and Agricultural Organisation, Rome pp 235–255
Wichert B, Frank T, Kienzle E (2002) Zinc, copper and selenium intake and status of horses in Bawaria. J Nutr 132:1776S–1777S
Wichtel JJ (1998) A review of selenium deficiency in grazing ruminants. Part 1. New roles for selenium in ruminant metabolism. N Z Vet J 46:47–52
Wiemeyer SN, Hoffman DJ (1996) Reproduction of eastern screech-owls fed selenium. J Wildl Manag 60:332–341
Winkel LHE, Vriens B, Jones GD, Schneider LS, Pilon-Smits E, Banuelos GS (2015) Selenium cycling across soil-plant-atmosphere interfaces: a critical review. Nutrients 7: 4199–4239
Wren CD (1984) Distribution of metals in tissues of beaver, raccoon, and otter from Ontario, Canada. Sci Total Environ 34:177–184
Wu L (2004) Review of 15 years of research on ecotoxicology and remediation of land contaminated by agricultural drainage sediment rich in selenium. Ecotoxicol Environ Saf 57:257–269
Wu L, Lag J (1988) Selenium in Norwegian farmland soils. Acta Agric Scand 38:271–276
Xu GI, Jiang YF (1985) Selenium and the prevalence of Keshan and Kashin-Beck diseases in China. In: Thornton I (ed) Proceedings of the 1st International Symposium on Geochemistry and Health (held at the Royal Society, London, 16–17 April 1985). Science Reviews, Northwood, UK, pp 192–204
Yang G, Yin S, Zhou R, Gu L, Liu Y, Liu X (1989) Studies of safe maximal dietary Se intake in a seleniferous area in China. II. Relation between Se intake and the manifestation of clinical signs and certain biochemical alterations in blood and urine. J Trace Elem Electrolyte Health Dis 3: 123–130
You CF, Xu XC (2010) Coal combustion and its pollution control in China. Energy 35:4467–4472
Ytrehus B, Skagemo H, Stuve G, Sivertsen T, Handeland K, Vikoren T (1999) Osteoporosis, bone mineralization, and status of selected trace elements in two populations of moose calves in Norway. J Wildl Dis 35:204–211
Yudovich YE, Ketris MP (2006) Selenium in coal: a review. Int J Coal Geol 67:112–126
Zachara AB, Gromadzińska J, Wąsowicz W, Zbróg Z (2006) Red blood cell and plasma glutathione peroxidase activities and selenium concentration in patients with chronic kidney disease: a review. Acta Bioch Pol 53(4):663–677
Zachary JF, McGavin MD (eds) (2014) Pathologic basis of veterinary disease, 5th edn. Mosby-Elsevier, St. Louis, MO, 1344 pp
Zayed A, Lytle CM, Terry N (1998) Accumulation and volatilization of different chemical species of selenium by plants. Planta 206:284–292
Zhang B, Yang L, Wang W, Li Y, Li H (2010) Quantification and comparison of soil elements in the Tibetan plateau Kashin-Beck disease area: a case study in Zamtang County, Sichuan Province, China. Biol Trace Elem Res 138:69–78
Zimmerman TJ, Jenks JA, Leslie DM Jr, Neiger RD (2008) Hepatic minerals of white-tailed and mule deer in the southern Black Hills, South Dakota. J Wildl Dis 44:341–350

# Chapter 11
# Zinc, Zn

**Danuta I. Kosik-Bogacka and Natalia Łanocha-Arendarczyk**

**Abstract** Zinc (Zn) is an essential trace element having a wide range of biological roles. The presented data concern Zn concentrations in various organs of terrestrial mammals and birds. In studies on terrestrial mammals, a measurable response to Zn pollution is exhibited by canids common in natural and seminatural habitats, including the red fox, raccoon, American mink, otters, and ungulates, such as the white-tailed deer, reindeer, red deer, and wild boar. Birds are used as bioindicators because they are abundant and widely distributed, have long lifespans, and feed at different trophic levels and are often the top consumers. At the same time, they are more sensitive to Zn contamination than other vertebrates and therefore seem to be better bioindicators. Nestling passerines are potential good biomonitors for Zn pollution because Zn is ingested in a clearly defined time period and originates from a limited parental foraging area. Zinc concentrations in the tissues of mammals and birds depend both on biological factors (e.g., age, physiological condition, animal species, sex, and age) and environmental factors (e.g., supply of Zn in the diet). Long-term bioindication research conducted on the organs and tissues of mammals and birds, as well as noninvasive sampling of eggs and feathers, indicates the usefulness of this type of material for evaluating the state of the environment.

## 1 Introduction

Zinc (Zn) is one of the most common elements in the Earth's crust. It is found in the air, soil, and water and is present in all foods. This essential trace element plays catalytic, structural, and regulatory roles in more than 300 Zn-metalloenzymes identified in biological systems. It plays an important role in the polymeric organization of macromolecules like DNA and RNA, protein synthesis, cell division, and

D. I. Kosik-Bogacka (✉) · N. Łanocha-Arendarczyk
Department of Biology and Medical Parasitology, Pomeranian Medical University in Szczecin, Szczecin, Poland
e-mail: kodan@pum.edu.pl



E. Kalisińska (ed.), *Mammals and Birds as Bioindicators of Trace Element Contaminations in Terrestrial Environments*,
https://doi.org/10.1007/978-3-030-00121-6_11

stability of biological membranes through protection against oxidative injury. Physiologically, Zn is vital for growth and development, sexual maturation and reproduction, dark vision adaptation, olfactory and gustatory activity, insulin storage and release, and for a variety of host immune defenses. Improvement in sperm production and fertility has been achieved following the supplementary feeding of Zn. Crop residues, which form the bulk of rations in India, are deficient in Zn, and deficiency may lead to male reproductive failure in domestic livestock.

For years, ecotoxicologists have tried to assess and identify the most useful animal species and types of biological samples that could be used in the bioindication of environmental levels of Zn. The herbivorous species, which often forage across very large territories, are not suitable for determining the specific source of a toxic hazard. It seems that species such as the red fox, which adapts to different environmental conditions and ranges within small territories of 0.5 $km^2$ and less, would be more suitable for biomonitoring research. Another species that seems appropriate in this regard is the American mink, a predatory mammal in which Zn concentrations in the kidney, liver, and bone are similar to those in other mammalian species.

## 2 General Properties of Zinc

Zinc is an essential trace element appearing in Group II-B of the periodic table, with atomic number 30, atomic weight 65.38, melting point 419.5 °C, and boiling point 908 °C and is a relatively soft metal with a density of 7.13 g $cm^{-3}$ (Adriano 2001). As a fairly active element, Zn dissolves in both acids and alkalis. It has two common oxidation states, Zn (0) and Zn (+2). Zinc forms a variety of compounds such as zinc chloride, zinc oxide, and zinc sulfate (ATSDR 2005). This element has five stable isotopes: $^{64}Zn$, $^{66}Zn$, $^{67}Zn$, $^{68}Zn$, and $^{70}Zn$, with average natural abundances at 48.89, 27.81, 4.11, 18.56, and 0.62% of the total volume, respectively. There are also six radioactive zinc isotopes: $^{62}Zn$, $^{63}Zn$, $^{65}Zn$, $^{69}Zn$, $^{72}Zn$, and $^{73}Zn$, with $^{65}Zn$ ($t_{1/2} = 245$ days) and $^{69}Zn$ ($t_{1/2} = 55$ min) being the most commonly used (Adriano 2001). Zinc has high lability (Stone and Marsalek 1999). In weathering processes, all Zn compounds are readily soluble, especially in acid (Kabata-Pendias and Pendias 1992). According to Goldschmidt's classification, Zn belongs to the chalcophile elements with an affinity to sulfur higher than that of iron, and correspondingly lower relative to oxygen (it mainly creates sulfides) (Goldschmidt 1954). According to the classification by Vernadsky (1945), Zn belongs to cyclical elements.

## 3 Zinc Production and Uses

Global production of Zn in 2014 was 813.372 million metric tonnes (MT), according to the International Lead and Zinc Study Group. The major world producers include Canada, the former Soviet Union, and Japan, which collectively account for about half the production (Elinder 1986). Zinc is used in many industries, including the production of noncorrosive alloys and brass and in galvanizing steel and iron products. This element undergoes oxidation on the surface, thereby protecting the underlying metal from degradation. Galvanized products are widely used in construction materials, automobile parts, and household appliances (Elinder 1986). Zinc oxide (ZnO) is used in the manufacture of paints, white pigments in rubber processing, the coating of photocopy paper, and in cosmetics, pharmaceuticals, floor covering, plastics, printing inks, soap, textiles, and electrical equipment (EPA 1987; PHS 1989). Zinc sulfate ($ZnSO_4$) is used in making luminous dials, X-ray films, TV screens, paints and fluorescent lights, as a cooperative agent in fungicides, and as a protective agent against Zn deficiency in soils (Maita et al. 1981). Zinc chloride ($ZnCl_2$) is used in vulcanization, refining of oils, and dyes and disinfectants. In 2001 more than 9 million MT of Zn were produced for galvanizing (47%), brass and bronze production (19%), Zn-based alloys (14%), chemicals (9%), and other uses (11%) (NRC 2005).

Zinc has been identified in at least 985 of the 1662 hazardous waste sites proposed for inclusion in the EPA National Priorities List (NPL) (HazDat 2005). This element is released into the environment in the form of particulates from Zn smelters, around which soil Zn concentrations range from 10 to 80 mg $kg^{-1}$. Another source of this element in the soil are certain plant protection chemicals and phosphate fertilizers. Zinc contained in industrial and mining waste often reaches groundwater. An important source of environmental pollution by Zn is the combustion of coal and petroleum and their products. Incineration of solid municipal waste may be responsible for about 75% of airborne Zn in urban areas; waste waters also generally contain significant amounts of zinc (Senczuk 2006).

## 4 Zinc in Nature

Zinc is ubiquitous in the environment and is the 24th most abundant element, with an average concentration of 70 mg $kg^{-1}$ in the Earth's crust (Mason and Moore 1982) and in the silicate crust, core, and mantle of the Earth at 55, 0, and 40 mg $g^{-1}$, respectively (McDonough and Sun 1995). This element is present in nature primarily as zinc sulfide (ZnS) and zinc carbonate ($ZnCO_3$) and it is found in the air, soil, and water and is present in all foods (Kabata-Pendias and Pendias 1992). Most rocks of the Earth's crust contain Zn in various concentrations depending on the type of rock (Aubert and Pinta 1977). Zinc is a chalcophile metallic element, easily combining with carbonates and organic compounds. There are approximately 55 mineralized

**Table 11.1** Zinc (Zn) in certain minerals

| Minerals | Zinc percentages |
|---|---|
| Sphalerite (ZnS) | 67.0 |
| Hemimorphite ($Zn_4Si_2O_7(OH)_2H_2O$ | 54.2 |
| Smithsonite ($ZnCO_3$) | 52.0 |
| Hydrozincite ($Zn_5(OH)_6(CO3)_2$) | 56.0 |
| Zincite (ZnO) | 80.3 |
| Willemite ($Zn_2SiO_4$) | 58.5 |
| Franklinite ((Zn, Fe, Mn)(Fe, Mn)$_2$O$_4$) | 15–20 |

Goodwin (1998)

forms of Zn, with the most significant being sphalerite, smithsonite, and hemimorphite (Table 11.1). Zinc is also widely dispersed as a trace element in pyroxene, amphibole, mica, garnet, and magnetite. Significant deposits of Zn ore can be found in China, the USA, Canada, Australia, Kazakhstan, Peru, and Mexico. World reserves of Zn are estimated at about 180 million MT of pure metal (US Geological Survey 2010).

Although Zn occurs naturally in soils, these levels have risen unnaturally due to anthropogenic additions. Mean Zn concentrations in soils were estimated at about 50 mg kg$^{-1}$, ranging between 10 and 300 mg kg$^{-1}$ (Malle 1992). It was later found that in noncontaminated soil environments, the average Zn concentration was 63 mg kg$^{-1}$, while in light sandy, medium loam, heavy loam, calcareous, and organic soils, levels fell in ranges of 31–61, 47–61, 35–75, 50–100, and 57–100 mg kg$^{-1}$, respectively (Kabata-Pendias and Mukherjee 2007; Kabata-Pendias and Szteke 2012). Zinc concentration is highest in lithosols (95 mg kg$^{-1}$) and lowest in oxisols (34 mg kg$^{-1}$). Zinc levels in soils have been found to be very heterogeneous in many parts of the world, especially in areas affected by glacial and periglacial processes, with a wide range of soils developed on drift deposits (Alloway 2008). The element's deficiency in soils and plants is a widespread micronutrient deficiency problem in many countries (Alloway 2008). This problem concerns arid and semiarid regions due to low organic matter and soil moisture, as well as high pH and levels of calcium carbonate ($CaCO_2$) (Cakmak 2008), typically in sampling in Iraq (57% of samples), Turkey (35%), and Pakistan (20%) (Sillanpaa 1982). Zinc is low in the Southeast USA and moderately higher in California, the Southwest, Colorado, and in the Lower Mississippi Valley.

Zinc is present in water in the form of hydrated cations and soluble salts, mainly carbonates and sulfates, as well as compounds with organic colloids (Kabata-Pendias and Szteke 2012). Despite its high mobility, Zn reaches groundwater in relatively low quantities. It has been established that an acceptable Zn level in groundwater is 50 μg L$^{-1}$; at 200 μg L$^{-1}$ it requires control, while at 800 μg L$^{-1}$ water needs to be treated (VROM 2012). In natural surface waters, Zn concentration is usually below 10 μg L$^{-1}$ and in groundwater 10–40 μg L$^{-1}$ (Elinder 1986). The mean Zn concentrations in ambient water and drinking water range from 0.02 to 0.05 mg L$^{-1}$ and from 0.01 to 0.1 mg L$^{-1}$, respectively (ATSDR 2005). The average Zn concentrations measured in rainwater, rivers, and seawater were 1.2–6.6, 1.5–3.3,

**Table 11.2** Global emissions of Zn into the atmosphere: natural and anthropogenic sources

| Sources | Emissions ($\times 10^6$ kg year$^{-1}$) | References |
|---|---|---|
| Natural | | |
| Windborne soil particles | 19 | Nriagu (1989) |
| Sea salt spray | 0.44 | |
| Volcanoes | 9.6 | |
| Wild forest fires | 7.5 | |
| Biogenic | | |
| Continental particulates | 2.6 | |
| Continental volatiles | 2.5 | |
| Marine | 3.0 | |
| Total in 1983 | 45 | |
| Total in 2000 | 301 | Rauch and Pacyna (2009) |
| Anthropogenic | | |
| Stationary fossil fuel combustion | 9.42 | Pacyna and Pacyna (2001) |
| Nonferrous metal production | 40.87 | |
| Iron and steel production | 2.12 | |
| Cement production | 2.67 | |
| Waste disposal | 1.93 | |
| Total in 1983 | 131.88 | Nriagu (1989) |
| Total in 2000 | 99.80 | Rauch and Pacyna (2009) |

and 0.04–5.0 mg L$^{-1}$, respectively (Reimann and de Caritat 1998; Kabata-Pendias and Mukherjee 2007)

Zinc may enter waters from a variety of sources, including mine drainage, industrial and municipal wastes, urban runoff, and soil erosion particles containing Zn (US EPA 1980). This element in water is easily bioaccumulated in phyto- and zooplankton and accumulates in bottom sediments. The background levels in river sediments are 110 mg kg$^{-1}$ in the Vistula River and 115 mg kg$^{-1}$ in the Rhein River, while in polluted parts of these rivers Zn concentrations may exceed 2000 and 14,000 mg kg$^{-1}$, respectively (Kabata-Pendias and Mukherjee 2007).

Mean Zn concentrations in the air at remote and polluted sites are estimated to be 7 and 900 ng m$^{-3}$, respectively. Such Zn occurs in different forms, for example, as free cations and suspended particulate matter (Reimann and de Caritat 1998). Zinc levels in rural air are between 10 and 200 ng m$^{-3}$, while in urban air Zn can reach 16,000 ng m$^{-3}$ (Eriksson et al. 2001). The element is released into the environment from both natural and anthropogenic sources; however, anthropogenic emissions are greater than those from natural sources (Table 11.2).

Zinc is an essential trace element (micronutrient) required by plants in small but critical amounts. It is necessary for growth and plays an important role in several plant metabolic processes. It is a functional, structural, or regulatory cofactor of enzymes, participating in protein synthesis, in carbohydrate, nucleic acid, and lipid metabolism, in the regulation of auxin synthesis, and in pollen formation (Pahlsson 1989). When accumulated at excess levels in plant tissues, Zn causes alterations in vital growth processes and membrane integrity (Doncheva et al. 2001). Zinc is taken up by plants as cations ($Zn^{+2}$, $ZnOH^{+}$), organic chelates, and insoluble Zn compounds (Kabata-Pendias and Szteke 2012). In general, this element from anthropogenic sources is more accessible for plants. Because of the high solubility of its compounds, the bioavailability of Zn for plants is high, which results in a high risk of Zn being introduced into the food chain (Broadley et al. 2007). Mean Zn levels in the aerial part of plants in noncontaminated areas range from 10 to 70 mg $kg^{-1}$ dry weight (dw) (Kucharczak and Moryl 2010). Most Zn can be found in the roots, especially in the rhizosphere. Zinc is most often associated with low-molecular-weight proteins or other soluble organic compounds. Relatively high levels of Zn can be found in cereal grains (18–23 mg $kg^{-1}$), vegetables (23–73 mg $kg^{-1}$), fruit (1.2–2.8 mg $kg^{-1}$), grass (25–47 mg $kg^{-1}$), and clover (24–39 mg $kg^{-1}$) (Kabata-Pendias and Szteke 2012).

## 5 Biological Effects, Metabolism, and Toxicity of Zinc in Endothermic Animals

Zinc plays a structurally and functionally significant role in more than 300 metalloenzymes from all six classes of enzymes (McCall et al. 2000). Zinc ions exist primarily as complexes with proteins that participate in all aspects of intermediary metabolism; they are also neurotransmitters, with cells in the salivary glands, prostate, immune system, and intestine using Zn signaling (Tapeiro and Tew 2003; Herschfinkel et al. 2007). Zinc serves as a cofactor in RNA polymerase and reverse transcriptase and in zinc-finger proteins that are adducts to DNA (Oberleas and Harland 2008). It plays a role in the synthesis, storage, and secretion of insulin and is necessary for the proper functioning and development of teeth and the skeletal system (NRC 2005; Chausmer 1998; Smrcka 2005). Zinc is a component of thymulin, a thymus hormone necessary for the maturation and differentiation of T cells (Dardenne et al. 1982). This element is also a major regulator of lymphocyte apoptosis in vitro and in vivo (Dardenne 2002). Zinc is involved in the control of oxidative stress and could contribute to membrane stabilization, acting at the cytoskeletal level (Dardenne et al. 1982). It is required for normal testicular development (Merck 1986). In birds, Zn is necessary for the growth and development of the skeleton, the formation and maintenance of epithelial tissues, and for egg production (Gordon 1977). In swine, Zn prevents thickening or hyperkeratinization of the epithelial cells of the skin and esophagus (parakeratosis), while in chicks it prevents a similar disease (Soetan et al. 2010).

Zinc homeostasis is primarily maintained via the gastrointestinal system, especially the small intestine, liver, and pancreas, by a process of absorption of exogenous zinc and gastrointestinal secretion and excretion of endogenous Zn (Krebs 2000). In humans and nonruminant animals, the highest absorption rate for Zn was observed in the jejunum, and in ruminants also in the rumen (Georgieskii et al. 1982;

Krebs 2000). Zinc absorption from the gut is dependent on the level in the diet, and generally ranges from 15% to 60% (McDowell 2003). Many dietary factors influence Zn absorption. Protein in the meal, amino acids, and other low-molecular-weight ions are known to have a positive effect on Zn absorption, while phytate forms, iron (Fe), calcium (Ca), and cadmium (Cd), have a negative effect on Zn absorption (Lonnerdal 2000). A decrease in Zn absorption is associated with age, but zinc excretion also decreases with age, and Zn homeostasis is not downregulated with age (Semrad 1999). The element is transported to the liver via portal circulation and then to other tissues, mainly in a complex with albumin (McMahon and Cousins 1998). In blood, 60% of Zn is bound to albumin, 30% to α2-macroglobulin, and 10% to transferrin (Scott and Bradwell 1983). Other plasma proteins that bind to zinc are transferrin, histamine-rich glycoprotein, and metallothionein (NRC 2005). Zinc physiological concentrations in mammalian serum range from 0.8 to 0.9 μg $mL^{-1}$, whereas in organs and tissues it has been found at 10–200 mg $kg^{-1}$ wet weight (ww) (Goyer et al. 1995; Kabata-Pendias and Pendias 1999).

Zinc is excreted mainly in the feces and urine (McDowell 2003). Increased Zn excretion in the urine can be caused by trauma, muscle catabolism, and administration of chelating agents (Hambridge et al. 1986). Additional zinc is lost daily in seminal emissions, menstrual losses, and hair and nail growth.

Zinc is not mutagenic and does not represent a carcinogenic risk to humans (Leonard et al. 1986), and it is relatively nontoxic to birds and animals. Rats, pigs, poultry, sheep, cattle, and humans exhibit a tolerance to high intake of Zn. In contrast, a zinc deficiency in animals causes numerous pathological changes, including skin parakeratosis, growth retardation, gastrointestinal malfunction, testicular and lymphoid tissue atrophy, poor wound healing, general debility, lethargy, poor appetite, and increased susceptibility to infection (Prasad et al. 1979; Dardenne 2002). In addition, long-term Zn deficiency increases the susceptibility to damage induced by oxidative stress; a low Zn level increases the level of lipid peroxidation in mitochondrial membranes. A dietary Zn deficiency significantly reduces red blood cell carbonic anhydrase activity, which may impair respiratory functions (Lukaski 2005). Zinc deficiency makes bones thin and fragile with excessive bone resorption and in male rats caused a delay in the growth and development of testes and prohibited spermatogenesis, while in mice it decreased natural killer cell activity and responses to cutaneous sensitization (Charles et al. 2001; Fernandes et al. 1979; Gilabert et al. 1996). Studies in rats, mice, pigs, and sheep showed that a Zn deficiency increased fetal death due to spontaneous abortions or multiple congenital anomalies (King 2000). Zinc deficiency occurs in cattle and calves and is characterized by reduced growth and feed intake, loss of hair and skin lesions that are most severe on the legs, neck, and head and around the nostrils, with excessive salivation, swollen feet with open, scaly lesions, and impaired reproduction (Spears 1994; Radostits et al. 2007). In pigs, Zn deficiency causes a marked depression of appetite, growth rate, and parakeratosis, while in young birds it results in poor growth, severe dermatitis, especially of the feet, poor feathering, abnormal respiration, and skeletal abnormalities causing leg weakness and ataxia (Soetan et al. 2010). Long bones are shortened and thickened and are sometimes crooked, and the joints are enlarged and rigid (Gordon 1977). The element's deficiency in mothers can cause an increased incidence of congenital malformations in infants (Elinder 1986). Gestational Zn deficiency in mice produced short- and long-term deleterious effects and showed a depressed immune function on offspring. Offspring from marginally Zn-deficient mice reduced lymphoid organ size and immunoglobulin concentrations (Beach et al. 1982).

Zinc toxicity depends on the Zn concentration and duration of exposure, age, sex, species, nutritional status, and composition of the diet. Young animals may be more vulnerable than older animals because they tend to have a higher efficiency of Zn absorption (NRC 2005). The effects of this element's toxicity depend on its chemical forms and presence or absence of other cations in the diet (Abdel-Mageed and Oehme 1990). Perhaps one of the major causes of Zn toxicity is its effect on the metabolism of essential metals. It has been found that the excessive consumption of Zn interferes with the absorption of Cu, probably via the induction of metallothionein, which has a greater affinity to copper (Cu) than to Zn. After binding to metallothionein, Cu is excreted from the body.

High doses of Zn interfere with metabolic processes and accumulate in the kidneys, liver, and gonads. Too high doses of Zn contribute to a gradual reduction in bone mass and calcium ion concentration in bones and blood serum (Charles et al. 2001). The half-life of Zn elimination is 162–500 days, and the content in the body decreases with age. Levengood et al. (1999) found clinical signs of Zn poisoning in mallards, with liver concentrations of 473–1990 mg $kg^{-1}$ dw; similarly, Sileo et al. (2003) diagnosed Zn poisoning in wild waterfowl, with liver concentrations of 280–2900 mg $kg^{-1}$ dw, while Doneley (1992) observed moderate to severe nephrosis in caged and aviary birds with hepatic Zn levels of 320 and 534 mg $kg^{-1}$ dw, respectively. Eisler (1993) concluded that Zn levels in the tissues of birds and mammals are typically <210 mg $kg^{-1}$ dw, and that Zn poisoning usually occurs in birds at liver or kidney concentrations >2100 mg $kg^{-1}$ dw and in mammals when kidney, liver, or pancreas levels exceed 274, 465, or 752 mg $kg^{-1}$ dw. In cattle, adequate Zn levels are 25–100 mg $kg^{-1}$ ww (or 83.33–333.33 mg $kg^{-1}$ dw) in the liver and 18–20 mg $kg^{-1}$ ww (or 72–80 mg $kg^{-1}$ dw) in the kidney, with Zn levels not considered "elevated" (not necessarily toxic) until 300–500 mg $kg^{-1}$ ww (or 1000–1666.66 mg $kg^{-1}$ dw) in the liver and 50–140 mg $kg^{-1}$ ww (or 200–560 mg $kg^{-1}$ dw) in the kidney (Puls 1994).

Zinc poisoning has been described in dogs, cats, ferrets, birds, cattle, sheep, and horses, usually as a result of ingesting galvanized metal objects, certain paints and fertilizers, zinc-containing coins, and skin and sunblock preparations containing zinc oxide (Wentink et al. 1985; Ogden et al. 1988; Lu and Combs 1988; Binnerts 1989; Robinette 1990). In several different species high doses, from 2 to 8 mg $kg^{-1}$, of orally ingested Zn generally resulted in gastrointestinal distress with nausea, vomiting, abdominal cramps, and diarrhea (ATSDR 2005). Acute Zn toxicosis has been described in dogs that had ingested large numbers of pennies, metal nuts from dog kennels, and other metal objects (Hornfeldt and Koepke 1984; Caldwell 1994; Mikszewski et al. 2003). Signs of zinc toxicosis in dogs include anorexia, vomiting, depression, fever (39.9 °C), intravascular hemolysis, pigmenturia, acute pancreatitis, hepatomegaly, and renal disease (Mikszewski et al. 2003; Hammond et al. 2004; Gurnee and Drobatz 2007). In chickens, high dietary Zn halted egg production and lowered body weight (McCormick and Cunningham 1984) and at 500 mg $kg^{-1}$ in the diet caused dysfunctions of pancreatic acinar cells and exocrine pancreatic insufficiency (Lu et al. 1990). In mice, the reported $LCT_{50}$ (product of lethal concentration and time to kill 50% of animals) of zinc chloride was 11,800 mg $min^{-1}$ $m^{-3}$ (Schenker et al. 1981). The $LD_{50}$ for several zinc compounds (ranging from 186 to 623 mg $kg^{-1}$ $day^{-1}$) has been determined for rats and mice (Domingo et al. 1988). In general, mice appear to be more sensitive to the lethal effects of Zn than rats.

Zinc status may be reflected by biochemical markers, i.e., biomarkers that are related to the structural, regulatory, and catalytic roles of Zn (van Riet et al. 2015). Although plasma Zn level is the most used biomarker for Zn status, other markers, such as albumin, metallothionein, and alkaline phosphatase, also play a role in Zn metabolism and homeostasis. Zinc is bound to albumin after absorption, metallothionein is involved in regulating the quantity of absorbed Zn, and alkaline phosphatase is a Zn-dependent enzyme in which Zn ions are present in active center (McDowell 2003; Coleman 1992).

# 6 Animals in Biomonitoring Studies

## 6.1 *Bioaccumulation of Zinc in Wildlife*

Environmental studies, including those concerned with biomonitoring, often use wild animal species as bioindicators due to the processes of bioaccumulation and biomagnification, which are especially pronounced in those species. Wild animals are also suitable bioindicators due to their large geographical distribution, limited feeding range, feeding habits, relatively long life span (sometimes 20–30 years), and easy sampling via regular hunting activities (Duffy et al. 2001). Measurements of trace element concentrations, including Zn, are performed in various organs and tissues, body fluids, hair, or feathers of animals associated with certain habitats (Kalisinska and Salicki 2010; Jarzynska and Falandysz 2011). In wild land mammals and birds, these are most often the liver and kidney and, less frequently, muscle, lung, heart, bone, and intestine.

Tables 11.4, 11.5, 11.6, and 11.7 show summarized data on Zn concentrations in various biological materials from endothermic vertebrates associated with land ecosystems and inland waterways. A correct interpretation of these results requires knowledge of the physiological Zn concentrations in the tissues and organs, Zn levels reflecting the geochemical background, and the specificity of species. For most species of wild animals there are no relevant data in this field, with the exception of domesticated animals. Table 11.3 shows deficient, marginal, optimal, high, and toxic Zn concentrations in the liver and kidney in cattle, dogs, horses,

**Table 11.3** Deficient, marginal, optimal, and high Zn levels (mg $kg^{-1}$) in liver, kidney, and muscle of domestic and wild animals

| Animals | Deficient | Marginal | Adequate | High | Toxic |
|---|---|---|---|---|---|
| | | | Liver | | |
| Cervid[a] | | | 30–110 | | |
| Cattle[b] | <20–40<br><67–133[d] | 25–40<br>83–133[d] | 25–100<br>167–333[d] | 300–500<br>1000–1667[d] | 120–500<br>400–1667[d] |
| Sheep[b] | 20–30<br>67–100[d] | | 30–75<br>100–250[d] | 100–400<br>333–1333[d] | >400<br>>1333[d] |
| Canine or dog[a,b] | <15<br><50[d] | | 30–70<br>100–233[d] | | 370<br>1233[d] |
| Musteline or mink[a,b] | | | 25–100<br>83–333[d] | 200<br>667[d] | |
| Mink[c] | | | 27.4<br>91[d] | | |
| Pig[b] | 9.6–25<br>32–83[d] | 25–35<br>83–117[d] | 40–90<br>133–300[d] | >200<br>>667[d] | 500–3100<br>1667–10333[d] |
| Birds/poultry[a,b] | 18–40<br>60–133[d] | 20–40<br>67–133[d] | 25–40<br>117–133[d] | 90–300<br>300–1000[d] | 200–700<br>667–2333[d] |
| | | | Kidney | | |
| Cervid[a] | | | 19–33 | | |
| Cattle[b] | 16–20<br>80–100[d] | | 18–25<br>90–125[d] | 50–140<br>250–700[d] | 130–480<br>650–2400[d] |
| Sheep[b] | 15–30<br>75–150[d] | | 20–40<br>100–200[d] | 50–1000<br>250–5000[d] | 240–1600<br>1200–8000[d] |
| Canine[a] | <8<br><40[d] | | 16–30<br>80–150[d] | | 300<br>1500[d] |
| *Dog*[b] | | | | | |
| Mustaline or mink[a,b] | | | 18–20<br>90–100[d] | | >100?<br>>500?[d] |
| Mink[c] | | | 19.9<br>99.5[d] | | |
| Pig[b] | 9.6–25<br>48–125[d] | 25–35<br>125–175[d] | 40–90<br>200–450[d] | >200<br>>1000[d] | 500–3100<br>2500–15500[d] |
| Birds/poultry[a,b] | 17–22<br>85–110[d] | | 22–32<br>110–160[d] | 120<br>600[d] | 300–800<br>1500–4000[d] |
| | | | Muscle | | |
| Sheep[b] | | | 75–130 (dw) | | 80–130 (dw) |
| Pig[b] | | | 21–24<br>105–120[d] | | |

We assume that the kidneys consist of 80% water and the liver and muscle 70%
*dw* dry weight, *ww* wet weight
[a]WVDL (2015), [b]Puls (1994), [c]Stejskal et al. (1989)
[d]Values converted from wet weight to dry weight

mink, pigs, and poultry (Puls 1994; WVDL 2015). Importantly, in the liver in waterfowl on the basis of field and laboratory studies, it was found that the background Zn concentration, sublethal effects, and mortality in adults amounted to

**Table 11.4** Zinc concentrations in liver, kidney, and muscle of ruminants and other herbivorous mammals

| Species | Location | Zn concentration | References |
|---|---|---|---|
| *Liver (mg kg$^{-1}$ dw)* | | | |
| Red deer *Cervus elaphus* | Northwest Poland | 100 | Jarzynska and Falandysz (2011) |
| | West–central Poland | 126.67[a] | Michalska and Zmudzki (1992) |
| | Netherlands, Veluwe | 108–124 (age: <0.6 and 1.5–5 years) | Wolkers et al. (1994) |
| | Eastern Croatia | 100.00[a] | Lazarus et al. (2008) |
| | Western Slovakia | 87.45[a] | Gasparik et al. (2003) |
| Roe deer *Capreolus capreolus* | Central Poland | 116.67[a] | Dlugaszek and Kopczynski (2011) |
| | Southern Poland | 93.33[a] | Lech and Gubala (1996) |
| | Northern Slovenia | 108.67[a] | Pokorny and Ribaric-Lasnik (2000) |
| White-tailed deer *Odocoileus virginianus* | Nova Scotia, Canada | 99.7 | Pollock (2005) |
| | Illinois, USA, | 70 | Woolf et al. (1982) |
| | Georgia, USA | 80[a] | Lewis et al. (2001) |
| Mule deer *Odocoileus hemionus* | California, USA | 39 | Roug et al. (2015) |
| Elk *Cervus elaphus* | Ontario, Canada | 73.62 | Parker and Hamr (2001) |
| Moose *Alces alces* | Northwest Russia | 135.53[a] | Medvedev (1999) |
| | Nova Scotia, Canada | 75 | Pollock (2005) |
| Caribou *Rangifer tarandus* | Northwest Russia | 123.33[a] | Bernhoft et al. (2002) |
| | Norway | 103 | Vikoren et al. (2011) |
| | Northern Alaska, USA | 153[a] | O'Hara et al. (2003) |
| | Northwest Canada | 276.33[a] | MacDonald et al. (2002) |
| *Kidney (mg kg$^{-1}$ dw)* | | | |
| Red deer *Cervus elaphus* | Northeast Poland | 130 | Jarzynska and Falandysz (2011) |
| | Northern Poland | 196[a] | Falandysz (1994) |
| | Netherlands, Veluwe | 144–165 (age: <0.6 and 1.5–5 years) | Wolkers et al. (1994) |
| | Southern Spain | 97.6 | Reglero et al. (2008) |
| | Slovenia | 114.4–188.4[a] | Pokorny (2000) |
| | Eastern Croatia | 196.8–202.8[a] | Lazarus et al. (2008) |
| | Western Slovakia | 124.47[a] | Gasparik et al. (2003) |
| Roe deer *Capreolus capreolus* | Northern Poland | 196[a] | Falandysz (1994) |
| | Southern Poland | 140[a] | Lech and Gubala (1996) |
| | Norway | 125 | Vikoren et al. (2011) |
| | Northern Slovenia | 188.4[a] | Pokorny and Ribaric-Lasnik (2000) |

(continued)

**Table 11.4** (continued)

| Species | Location | Zn concentration | References |
|---|---|---|---|
| White-tailed deer *Odocoileus virginianus* | Nova Scotia, Canada | 79.7 | Pollock (2005) |
| | Virginia, USA | 119.33[a] | Sleeman et al. (2010) |
| | Georgia, USA | 116.4[a] | Lewis et al. (2001) |
| Elk *Cervus elaphus* | Ontario, Canada | 164.47 | Parker and Hamr (2001) |
| Moose *Alces alces* | Northwest Russia | 114.92[a] | Medvedev (1999) |
| | Nova Scotia, Canada | 99.7 | Pollock (2005) |
| | Yukon, Canada | 117.56[a] | Gamberg et al. (2005) |
| Caribou *Rangifer tarandus* | Svalbard, Norway | 146.67[a] | Borch-Iohnsen et al. (1996) |
| | Northern Alaska, USA | 127.6[a] | O'Hara et al. (2003) |
| | Northwest Canada | 464[a] | MacDonald et al. (2002) |
| *Muscle (mg kg$^{-1}$ dw)* | | | |
| Red deer *Cervus elaphus* | Southwest and North-east Poland | 93.32–103.12[a] | Skibniewski et al. (2015) |
| | Central–Eastern Poland | 288[a] | Karpinski (1999) |
| | Eastern Croatia | 144.67[a] | Lazarus et al. (2008) |
| | Western Slovakia | 219.04[a] | Gasparik et al. (2003) |
| Roe deer *Capreolus capreolus* | Central Poland | 124[a] | Dlugaszek and Kopczynski (2013) |
| | Northern Poland | 144[a] | Falandysz (1994) |
| | Northern Slovenia | 206.4[a] | Pokorny and Ribaric–Lasnik (2000) |
| Elk *Cervus elaphus* | Ontario, Canada | 48.19 | Parker and Hamr (2001) |
| Moose *Alces alces* | Northwest Russia | 147.36[a] | Medvedev (1999) |
| Caribou *Rangifer tarandus* | Northern Alaska, USA | 134.8[a] | O'Hara et al. (2003) |

[a]Values converted from wet weight to dry weight

28.2–54.5 mg kg$^{-1}$ ww (or 94–82 mg kg$^{-1}$ dw), ≥84.8 mg kg$^{-1}$ ww (or ≥283 mg kg$^{-1}$ dw), and >333 mg kg$^{-1}$ ww (or >1111 mg kg$^{-1}$ dw), respectively (Pillatzki et al. 2011).

**Table 11.5** Zinc concentrations in selected tissues of omnivorous mammals

| Species | Localization | Zn concentration | References |
|---|---|---|---|
| *Liver (mg kg$^{-1}$ dw)* | | | |
| Wild boar *Sus scrofa* | Western Poland | 157.27–171.3[a] | Kucharczak et al. (2003) |
| | Slovakia | 94[a] | Gasparik et al. (2012) |
| | Central Italy | 165.87[a] | Amici et al. (2012) |
| | Southern Spain | 113 | Reglero et al. (2009) |
| Raccoon *Procyon lotor* | Illinois, USA | 150–186.67[a] | Levengood et al. (2002) |
| | Michigan, USA | 148[a] | Herbert and Peterle (1990) |
| | Georgia, USA | 100.33[a] | Lewis et al. (2001) |
| | Tennessee, USA, PA | 132.83[a] | Souza et al. (2013) |
| | Tennessee, USA, UA | 136.33 | |
| | South Carolina, USA, PA | 112.98 | Hernandez et al. (2017) |
| | South Carolina, USA, UA | 106.02 | |
| | Ontario, Canada | 34.4 | Wren (1984) |
| *Kidney (mg kg$^{-1}$ dw)* | | | |
| Wild boar *Sus scrofa* | Southern Poland | 82 | Swiergosz et al. (1993) |
| | Western Poland | 111.56–122.12[a] | Kucharczak et al. (2003) |
| | Slovakia | 83.92[a] | Gasparik et al. (2012) |
| | Central Italy | 129.84[a] | Amici et al. (2012) |
| Rraccoon *Procyon lotor* | Illinois, USA | 92–108[a] | Levengood et al. (2002) |
| | Michigan, USA | 74.4[a] | Herbert and Peterle (1990) |
| | Georgia, USA | 65.6[a] | Lewis et al. (2001) |
| | Tennessee, USA, PA | 79[a] | Souza et al. (2013) |
| | Tennessee, USA, UA | 79.4[a] | |
| | Ontario, Canada | 29.5 | Wren (1984) |
| *Muscle (mg kg$^{-1}$ dw)* | | | |
| Wild boar *Sus scrofa* | Central Poland | 126[a] | Długaszek and Kopczyński (2011) |
| | Western Poland | 42.68–249.16[a] | Kucharczak et al. (2003) |
| | Hungary | 151.48–204.68[a] | Skobrak et al. (2011) |
| | Slovakia | 53.92[a] | Gasparik et al. (2012) |
| | Central Italy | 212.84[a] | Amici et al. (2012) |
| Raccoon *Procyon lotor* | Tennessee, USA, PA | 221.6[a] | Souza et al. (2013) |
| | Tennessee, USA, UA | 254[a] | |

*PA* polluted area, *UA* unpolluted area
[a]Values converted from wet weight to dry weight

**Table 11.6** Zinc concentrations in selected tissues in carnivorous mammals

| Species | Localization | Zn concentration | References |
|---|---|---|---|
| *Liver (mg kg$^{-1}$ dw)* | | | |
| Red fox *Vulpes vulpes* | Southwest Poland | 128.26 | Binkowski et al. (2016) |
| | Northern Czech Republic | 96.17 | Jankovska et al. (2010) |
| | Central Hungary | 156.93 | Heltai and Markov (2012) |
| | North-central Switzerland | 149.67[a] | Dip et al. (2001) |
| | Northwest Spain | 77 | Perez-Lopez et al. (2016) |
| | Southern Spain | 118.7 | Millan et al. (2008) |
| Polar fox *Alopex lagopus* | Svalbard, Norway | 106.67[a] | Prestrud et al. (1994) |
| | Nunavut, Canada | 29 | Hoekstra et al. (2003) |
| California gray fox *Urocyon cinereoargenteus californicus* | California, USA, ZOO | 109 | Arnhold et al. (2002) |
| Arctic wolf *Canis lupus* | Keewatin, Canada | 67.0 | Lamothe (1991) |
| Eurasian otter *Lutra lutra* | Hungary | 99.75 | Lanszki et al. (2009) |
| European otter *Lutra lutra* | Central and Eastern Finland | 133.6[a] | Skaren (1992) |
| | | 120[a] | Lodenius et al. (2014) |
| | England and Wales | 102 | Walker et al. (2011) |
| | Great Britain | 111.67 | Mason and Stephenson (2001) |
| | Ireland | 83.80 | |
| | Denmark | 92.13 | |
| | Netherlands | 131 | Broekhuizen (1987) |
| | Austria | 92.6 | Gutleb et al. (1998) |
| | Hungary | 96.2 | |
| | Czech | 60.7 | |
| River otter *Lontra canadensis* | Illinois, USA | 80.33[a] | Halbrook et al. (1996) |
| | Virginia, USA | 62.63 | Anderson-Bledsoe and Scanlon (1983) |
| | Ontario, Canada | 26.5 | Wren (1984) |
| | British Columbia, Canada | 86 | Harding et al. (1998) |
| American mink *Neovision vision* | Poland | 67.22–135.38 | Brzezinski et al. (2014) |
| | British Columbia, Canada | 95 | Harding et al. (1998) |
| | Virginia, USA | 123.24 | Ogle et al. (1985) |

(continued)

**Table 11.6** (continued)

| Species | Localization | Zn concentration | References |
|---|---|---|---|
| *Kidney (mg kg$^{-1}$ dw)* | | | |
| Red fox *Vulpes vulpes* | Poland, SW part | 58.58 | Binkowski et al. (2016) |
| | Northern Czech Republic | 79.29 | Jankovska et al. (2010) |
| | Central Hungary | 87.16 | Heltai and Markov 2012 |
| | North-central Switzerland | 84.8[a] | Dip et al. (2001) |
| | Northwestern Spain | 17 | Perez-Lopez et al. (2016) |
| California gray fox *Urocyon cinereoargenteus californicus* | San Diego, CA, USA, ZOO | 74 | Arnhold et al. (2002) |
| Grey wolf *Canis lupus* | USA | 28.7–39.1 | Hoffmann et al. (2010) |
| | Northwest Territories, Canada | 28.0 | |
| Arctic wolf *Canis lupus* | Keewatin, Canada | 85.0 | Lamothe (1991) |
| Eurasian otter *Lutra lutra* | England and Wales | 93.5–106.5 | Walker et al. (2011) |
| | Netherlands | 95 | Broekhuizen (1987) |
| | Austria | 138.2 | Gutleb et al. (1998) |
| | Hungary | 55.6 | |
| | Czech | 107.2 | |
| River otter *Lontra canadensis* | Illinois, USA | 82[a] | Halbrook et al. (1996) |
| | Virginia, USA | 78.91 | Anderson-Bledsoe and Scanlon (1983) |
| | Ontario, Canada | 19.6 | Wren (1984) |
| American mink *Neovision vision* | Poland | 76.22–115.40 | Brzeziński et al. (2014) |
| | British Columbia, Canada | 65 | Harding et al. (1998) |
| | Virginia, USA | 93.42 | Ogle et al. (1985) |
| *Muscle (mg kg$^{-1}$ dw)* | | | |
| Red fox *Vulpes vulpes* | Southwest Poland | 87.31 | Binkowski et al. (2016) |
| | Northwest Spain | 77 | Perez-Lopez et al. (2016) |
| | Southern Spain | 118.7 | Millan et al. (2008) |
| River otter *Lutra canadensis* | Illinois, USA | 149.2[a] | Halbrook et al. (1996) |
| | Ontario, Canada | 50.5 | Wren (1984) |

[a]Values converted from wet weight to dry weight

**Table 11.7** Zinc concentrations in different organs of birds

| Species | Localization | Zn concentration | References |
|---|---|---|---|
| *Liver (mg kg$^{-1}$ dw)* | | | |
| White-tailed eagle *Haliaeetus albicilla* | Poland, NW and E parts | 88.67[a] | Kalisinska et al. (2006) |
| | | 170 | Falandysz et al. (2000) |
| | Southwest Poland, Baltic Sea coast | 62 | Falandysz et al. (1988) |
| Bald eagle *Haliaeetus leucocephalus* | Alaska, USA | 127 | Stout and Trust (2002) |
| | Michigan and Minnesota, USA | 237 | Nam et al. (2012) |
| Eurasian eagle owls *Bubo bubo* | South Korea | 212 | Kim and Oh (2016) |
| Osprey *Pandion haliaetus* | Eastern USA | 171.3[a] | Wiemeyer et al. (1987) |
| Black kite *Milvus migrans* | Central India | 84.33[a] | Gupta and Kanaujia (2014) |
| Eurasian buzzard *Buteo buteo* | Northwest Poland, and Eastern Czech Republic | 126.2 | Kalisinska et al. (2009) |
| | Eastern Poland | 121.51 | Komosa et al. (2012) |
| | Sicily, Italy | 137.5 | Naccari et al. (2009) |
| | Southern Italy | 158.33 | Zaccaroni et al. (2011) |
| | Northwestern Spain | 297.4 | Perez-Lopez et al. (2008) |
| | South Korea | 144 | Kim and Oh (2016) |
| Peregrine falcon *Falco peregrinus* | Sweden | 72 | Ek et al. (2004) |
| Common kestrel *Falco tinnunculus* | South Korea | 132 | Kim and Oh (2016) |
| Mallard *Anas platyrhynchos* | Southeast Poland | 144.67[a] | Bojar and Bojar (2009) |
| | Northeast and Southern Poland | 80.97–87.03[a] | Szymczyk and Zalewski (2003) |
| | Spain, DNP, 1998 | 221.33 | Taggart et al. (2006) |
| | Spain, DNP, 1998 and 2000 | 136.0 | Gomez et al. (2004) |
| | South Korea | 103.00 | Kim and Oh (2012) |
| | Southeastern Iran | 62.4[a] | Sinka-Karimi et al. (2015) |
| | Virginia, USA | 161 | DiGulio and Scanlon (1984) |
| Greylag goose *Anser anser* | France | 355.8 | Lucia et al. (2010) |
| | Southwestern Spain | 102–196 | Mateo et al. (2006) |
| Eurasian teal *Anas crecca* | Spain, DNP | 83.91 | Hernandez et al. (1999) |

(continued)

**Table 11.7** (continued)

| Species | Localization | Zn concentration | References |
|---|---|---|---|
| Spot-billed ducks *Anas poecilorhyncha* | South Korea | 131.0 | Kim and Oh (2012) |
| Eurasian wigeons *Anas penelope* | South Korea | 88.4 | Kim and Oh (2012) |
| Greater scaup *Aythya marila* | Alaska, USA | 139 | Badzinski et al. (2009) |
| | Virginia, USA | 117 | Di Giulio and Scanlon (1984) |
| Pochard *Aythya ferina* | Spain, DNP, 1998 | 345.67 | Taggart et al. (2006) |
| | Spain, DNP, 1998 and 2000 | 283 | Gomez et al. (2004) |
| | Southeastern Iran | 93.4[a] | Sinka-Karimi et al. (2015) |
| Redhead duck *Aythya americana* | Louisiana, USA | 122 | Michot and Chadwick (1994) |
| Great tit *Parus major* | Northeastern and Southern Poland | 163 | Sawicka-Kapusta et al. (1986) |
| | Northern Belgium | 36.0 | Dauwe et al. (2005) |
| | Central Norway | 69.60–72.32 | Hogstad (1996) |
| | Northern China | 117.15 | Deng et al. (2007) |
| | Northeastern Spain, UA | 112.80 | Llacuna et al. (1995) |
| | Northeastern Spain, PA | 98.78 | |
| Rock bunting *Emberiza cia* | Northeastern Spain, UA | 71.50 | Llacuna et al. (1995) |
| | Northeastern Spain, PA | 91.40 | |
| Blackbird *Turdus merula* | Northeastern Spain, UA | 51.00 | Llacuna et al. (1995) |
| | Northeastern Spain, PA | 43.00 | |
| *house sparrow Passer domesticus* | Southern Finland, RUA | 18.35 | Kekkonen et al. (2012) |
| | Southern Finland, URA | 21.08 | |
| | Albania, PA | 68.25–94.47 | Millaku et al. (2015) |
| | Albania, UA | 61.48 | |
| | Turkey, UA | 43.18 | Albayrak and Mor (2011) |
| | Turkey, PA | 101.76 | |
| | West Bank, Palestine | 131.4 | Swaileh and Sansur (2006) |
| | Northern Pakistan | 26.16[a] | Mustafa et al. (2015) |
| Italian sparrow *Passer domesticus italiae* | Southern Italy, RUA | 154 | Gragnaniello et al. (2001) |
| | Southern Italy, URA | 204 | |
| Tree sparrow *Passer montanus* | Northern China | 104–137 | Chao et al. (2003) |
| | China | 65.03–82.90 | Gong et al. (2012) |
| Savannah sparrow *Passerculus sandwichensis* | Northwest Alaska, USA | 82.3 | Brumbaugh et al. (2010) |

(continued)

**Table 11.7** (continued)

| Species | Localization | Zn concentration | References |
|---|---|---|---|
| *Kidney (mg kg$^{-1}$ dw)* | | | |
| White-tailed eagle *Haliaeetus albicilla* | Northwestern Poland | 80.16[a] | Kalisinska et al. (2006) |
| | Western Poland | 140 | Falandysz et al. (2000) |
| Bald eagle *Haliaeetus leucocephalus* | Alaska, USA | 96.4 | Stout and Trust (2002) |
| Osprey *Pandion haliaetus* | Eastern USA | 171.3[a] | Wiemeyer et al. (1987) |
| Black kite *Milvus migrans* | Central India | 82[a] | Gupta and Kanaujia (2014) |
| Eurasian buzzard *Buteo buteo* | Northwestern Poland and Eastern Czech Republic | 75.0 | Kalisinska et al. (2009) |
| | Sicily, Italy | 62.25 | Licata et al. (2010) |
| | Southern Italy | 367.6[a] | Zaccaroni et al. (2011) |
| Peregrine falcon *Falco peregrinus* | Sweden | 70 | Ek et al. (2004) |
| Mallard *Anas platyrhynchos* | Southeastern Poland | 88.8[a] | Bojar and Bojar (2009) |
| | Spain, DNP, 1998 and 2000 | 70.4 | Gomez et al. (2004) |
| | Southeastern Iran | 44.88[a] | Sinka-Karimi et al. (2015) |
| | Virginia, USA | 86 | DiGulio and Scanlon (1984) |
| Pochard *Aythya ferina* | Southeastern Iran | 54.44[a] | Sinka-Karimi et al. (2015) |
| Gadwall *Anas strepera* | Spain, DNP | 138.9 | Taggart et al. (2006) |
| | Spain, DNP | 296.2 | Hernandez et al. (1999) |
| Greylag goose *Anser anser* | France | 189.2 | Lucia et al. (2010) |
| Great tit *Parus major* | Northern Belgium | 53.4 | Dauwe et al. (2005) |
| | Northern China | 85.50 | Deng et al. (2007) |
| | Northeastern Spain, UA | 96.43 | Llacuna et al. (1995) |
| | Northeastern Spain, PA | 85.89 | |
| Rock bunting *Emberiza cia* | Northeastern Spain, UA | 93.80 | Llacuna et al. (1995) |
| | Northeastern Spain, PA | 100.22 | |
| Blackbird *Turdus merula* | Northeastern Spain, UA | 113.49 | Llacuna et al. (1995) |
| | Northeastern Spain, PA | 106.31 | |

(continued)

**Table 11.7** (continued)

| Species | Localization | Zn concentration | References |
|---|---|---|---|
| House sparrow *Passer domesticus* | Albania, PA | 61.63–77.76 | Millaku et al. (2015) |
| | Albania, UA | 47.22 | |
| | Turkey, UA | 19.72 | Albayrak and Mor (2011) |
| | Turkey, PA | 31.51 | |
| | Northern Pakistan | 25.28[a] | Mustafa et al. (2015) |
| Italian sparrow *Passer domesticus italiae* | Southern Italy, RUA | 133 | Gragnaniello et al. (2001) |
| | Southern Italy, URA | 162 | |
| *Muscle (mg kg$^{-1}$ dw)* | | | |
| White-tailed eagle *Haliaeetus albicilla* | Northwestern Poland | 86.17[a] | Kalisinska et al. (2006) |
| | Western Poland | 12.00 | Falandysz et al. (2000) |
| Eurasian buzzard *Buteo buteo* | Sicily, Italy | 52.76 | Naccari et al. (2009) |
| | Sicily, Italy | 52.01 | Licata et al. (2010) |
| Mallard *Anas platyrhynchos* | Southeastern Poland | 74.33[a] | Bojar and Bojar (2009) |
| | Northeastern and Southern Poland | 34.77–66[a] | Szymczyk and Zalewski (2003) |
| | Spain, DNP, 1998 and 2000 | 32.3 | Gomez et al. (2004) |
| | Southeastern Iran | 31.53[a] | Sinka-Karimi et al. (2015) |
| Pochard *Aythya ferina* | Spain, DNP, 1998 and 2000 | 80.6 | Gomez et al. (2004) |
| | Southeastern Iran | 33.07[a] | Sinka-Karimi et al. (2015) |
| Greylag gees *Anser anser* | France | 93.8 | Lucia et al. (2010) |
| | Southwestern Spain | 52–111 | Mateo et al. (2006) |
| Great tit *Parus major* | Northern Belgium | 11.9 | Dauwe et al. (2005) |
| | Northern China | 32.15–92.09 | Deng et al. (2007) |
| | Northeastern Spain, UA | 31.57 | Llacuna et al. (1995) |
| | Northeastern Spain, PA | 31.14 | |
| Rock bunting *Emberiza cia* | Northeastern Spain, UA | 29.18 | Llacuna et al. (1995) |
| | Northeastern Spain, PA | 28.24 | |
| Blackbird *Turdus merula* | Northeastern Spain, UA | 26.82 | Llacuna et al. (1995) |
| | Northeastern Spain, PA | 29.95 | |
| House sparrow *Passer domesticus* | Northern Pakistan | 27.5[a] | Mustafa et al. (2015) |
| | Turkey, UA | 19.90 | Albayrak and Mor (2011) |
| | Turkey, PA | 34.03 | |
| | West Bank, Palestine | 61.9 | Swaileh and Sansur (2006) |
| | India, URA | 43.78 | Sundaramahalingam et al. (2016) |
| | India, RUA | 32.4 | |

(continued)

**Table 11.7** (continued)

| Species | Localization | Zn concentration | References |
|---|---|---|---|
| Tree sparrow *Passer montanus* | Northern China | 48.6–75.5 | Chao et al. (2003) |
| | China | 22.50–32.03 | Gong et al. (2012) |
| *Bone (mg kg$^{-1}$ dw)* | | | |
| Mallard *Anas platyrhynchos* | Northwestern Poland | 97.89 (ad), 111.39 (imm) | Kalisinska et al. (2004) |
| | Western Poland | 87.73 (ad), 111.77 (imm) | |
| Pochard *Aythya ferina* | Spain, DNP, 1998 | 154.9 | Taggart et al. (2006) |
| Great tit *Parus major* | Northeastern and Southern Poland | 183–284 | Sawicka-Kapusta et al. (1986) |
| | Northeastern Spain, UA | 248.5 | Llacuna et al. (1995) |
| | Northeastern Spain, PA | 255.9 | |
| Rock bunting *Emberiza cia* | Northeastern Spain, UA | 166.2 | Llacuna et al. (1995) |
| | Northeastern Spain, PA | 209.8 | |
| Blackbird *Turdus merula* | Northeastern Spain, UA | 152.6 | Llacuna et al. (1995) |
| | Northeastern Spain, PA | 149.9 | |
| House sparrow *Passer domesticus* | Albania, PA | 342.8 and 291.5 | Millaku et al. (2015) |
| | Albania, UA | 235.1 | |
| | West Bank, Palestine | 150.4 | Swaileh and Sansur (2006) |
| | India, URA | 25 | Sundaramahalingam et al. (2016) |
| | India, RUA | 27.2 | |
| Tree sparrow *Passer montanus* | Northern China | 207–255 | Chao et al. (2003) |

*DNP* Donana National Park, *PA* polluted area, *UA* unpolluted area, *RUA* rural area, *URA* urban area, *ad* adultus, *imm* immaturus

[a]Values converted from wet weight to dry weight

## 6.2 Mammals

### 6.2.1 Zinc Concentration in Mammalian Tissues

Zinc concentration in mammals can be analyzed on the basis of data on selected animals from different trophic groups. Among European ruminants, Zn levels are most often analyzed in livers, kidneys, and muscles of red deer (*Cervus elaphus*) and roe deer (*Capreolus capreolus*) and in North America in the white-tailed deer (*Odocoileus virgianus*), mule deer (*Odocoileus hemionus*), and caribou (*Rangifer tarandus*). Caribou and reindeer are the same species and occur in the Arctic and subarctic regions of North America and Eurasia. Some studies have also examined Zn concentrations in moose (North America) and elk (Eurasia) (*Alces alces*), the largest extant species in the deer family. Those data indicate that the mean Zn concentration in ruminant livers is in a range of 70–170 mg kg$^{-1}$ dw (Table 11.4). Zinc levels in the liver below 40 mg kg$^{-1}$ dw were found in mule deer from California, USA (Roug et al. 2015), while levels exceeding 270 mg kg$^{-1}$ dw were found in caribou from the Northwest Territories and Nunavut, Canada (MacDonald et al. 2002). In ruminant kidneys, mean Zn concentrations were 80–200 mg kg$^{-1}$ dw (Table 11.4), while in caribou from the Northwest Territories and Nunavut, Zn levels in the kidney were higher, 77–159 mg kg$^{-1}$ ww or 306–636 mg kg$^{-1}$ dw (MacDonald et al. 2002). In ruminant muscle, mean Zn levels were 93–156 mg kg$^{-1}$ dw (Table 11.4). Lower muscle Zn levels were found in elk in the region of Ontario Sudbury ore smelters, Canada (48.19 mg kg$^{-1}$ dw) (Parker and Hamr 2001).

Higher muscle Zn levels were found in European red deer from central-eastern Poland (72.53 mg kg$^{-1}$ ww or 288 mg kg$^{-1}$ dw) and Western Slovakia (54.76 mg kg$^{-1}$ ww or 219.04 mg kg$^{-1}$ dw) and in roe deer from an industrial area, the Koroska Region, with a centuries-old tradition of mining and manufacturing of lead and zinc (51.6 mg kg$^{-1}$ ww or 206.4 mg kg$^{-1}$ dw) (Gasparik et al. 2003; Karpinski 1999; Parker and Hamr 2001; Pokorny and Ribaric-Lasnik 2000).

In the case of omnivorous animals in Europe, Zn concentrations are most frequently analyzed in wild boar (*Sus scrofa*) and in North America in the raccoon (*Procyon lotor*). Average Zn concentrations in these animals can be arranged in the following descending order: muscle > liver > kidney (Table 11.5). Mean hepatic Zn levels usually ranged from 90 to 190 mg kg$^{-1}$ dw, and in the kidney Zn levels were 70–130 mg kg$^{-1}$ dw (Table 11.5). The one significant exception was in raccoon from Ontario, Canada, in which Zn concentrations in the liver and kidney were 34.4 and 29.5 mg kg$^{-1}$ dw, respectively (Wren 1984). In the muscles of omnivorous animals, mean Zn levels were usually in the range 110–250 mg kg$^{-1}$ dw (Table 11.5), while in some wild boar populations from Poland and Slovakia they were much lower, 10.67 mg kg$^{-1}$ ww (or 42.68 mg kg$^{-1}$ dw) and 13.48 mg kg$^{-1}$ ww (or 53.92 mg kg$^{-1}$ dw) (Kucharczak et al. 2003; Gasparik et al. 2012).

Zinc levels in European carnivores have most frequently been analyzed in the red fox (*Vulpes vulpes*) and Eurasian otter (*Lutra lutra*). In North America such research has been mainly carried out on the Arctic wolf (*Canis lupus*) and river otter (*Lontra canadensis*). Research on American mink (*Neovison vison*) has been conducted on both continents. In carnivores, Zn concentrations have usually been studied in connection with the liver, kidney, and muscle, less often bone, hair, and other tissues. Liver Zn levels usually ranged from 70 to 150 mg kg$^{-1}$ dw (Table 11.6). Much lower hepatic Zn levels were found in river otter from Ontario, Canada, and in the polar fox (*Alopex lagopus*) from the Canadian Arctic, at 26.5 and 29 mg kg$^{-1}$ dw, respectively (Hoekstra et al. 2003; Wren 1984). Higher liver Zn levels were found in the Arctic wolf (*Canis lupus*) from Yukon, Canada (122–130 mg kg$^{-1}$ ww or 406.67–433.33 mg kg$^{-1}$ dw) (Gamberg and Braune 1999) and Eurasian otter from Styria, Austria (556.7 mg kg$^{-1}$ dw) (Gutleb 1992). In the kidney, mean Zn levels were 60–120 mg kg$^{-1}$ dw (Table 11.6). Lower Zn concentrations were found in the red fox from Galicia, Spain (17 mg $kg^{-1}$ dw) (Perez-Lopez et al. 2016), gray wolf (*Canis lupus*) from different regions in Canada and the USA (25.8–39.1 mg $kg^{-1}$ dw) (Hoffmann et al. 2010), and river otter from Ontario (19.6 mg $kg^{-1}$ dw) (Wren 1984), and higher Zn concentrations were found in the Eurasian otter from some parts of Austria (327 mg $kg^{-1}$ dw) (Gutleb 1992). In muscles in carnivores, mean Zn concentrations were 50–150 mg $kg^{-1}$ dw (Table 11.6).

### 6.2.2 Factors Affecting Zinc Concentrations in Mammal Tissues

#### 6.2.2.1 Sex of Mammals and Zinc Concentrations

The results of studies on the relationship between Zn concentrations and the sex of wild mammals are not clear. In moose (*Alces alces*) and roe deer from Norway, Vikoren et al. (2011) found significant sex-related differences in Zn concentrations in the liver, with females having lower Zn concentrations than males. Hyvarinen et al. (2003) found that in Eurasian otters from Finland, Zn concentrations in the kidney were higher in males (135.9 mg $kg^{-1}$ ww or 453 mg $kg^{-1}$ dw) than in females (101 mg $kg^{-1}$ ww or 336.6 mg $kg^{-1}$ dw). Similarly, in raccoon dogs from farms in Poland, Hanusova et al. (2007) found significant differences, with kidney Zn concentrations in females higher than in males at 80.05 and 60.14 mg $kg^{-1}$ dw, respectively, and in the liver at 107.04 and 75.83 mg $kg^{-1}$ dw, respectively. Similarly, Suvegova et al. (1993) in silver foxes and Hanusova et al. (2007) in polar foxes from farms found higher Zn concentrations in the livers of females. Hanusova et al. (2007) studied the polar fox (*Alopex lagopus*) and the silver variety of farm fox (*Vulpes vulpes*) from the Research Institute for Animal Production in Nitra, Slovakia, and found that Zn levels in liver and muscle of the polar fox were greater in females (82.27 and 108.27 mg $kg^{-1}$ dw, respectively) than in males (69.5 and 98 mg $kg^{-1}$ dw, respectively). In contrast, Bernhoft et al. (2002) found no difference in hepatic Zn concentrations between male and female reindeer from Rybatsjij Ostrov, northwestern Russia. Similarly, Medvedev (1999) found no sex-related differences in Zn concentrations in various organs of the moose, wild boar, or brown bear (*Ursus arctos*) from Russia. Roslewska et al. (2016) and Dlugaszek and Kopczynski (2013) found no differences in Zn concentrations in the muscles of female and male wild boars from Poland. Similarly, Perez-Lopez et al. (2016) found no sex-related differences in Zn levels in the liver and kidney of red fox from Spain and Ogle et al. (1985) in American mink from the state of Virginia, USA.

#### 6.2.2.2 Age of Mammals and Zinc Concentrations

Existing studies on the relationship between the age of wild mammals and Zn concentrations in their organs do not allow firm conclusions to be drawn. For example, Lazarus et al. (2005, 2008) found a higher Zn level in the kidney of the oldest red deer than in the younger animals. Similarly, Holterman et al. (1984) found that red deer 6–11 years of age had approximately twice as high kidney cortex Zn levels than younger animals. Lazarus et al. (2005) also found an age-related difference in Zn concentrations in the kidney cortex of red deer. Custer et al. (2004) found that 7-year-old moose had higher Zn concentrations in the liver than 1-year-old moose. Parker and Hamr (2001) found much higher Zn concentrations in the liver and bone of elk fetuses (*Cervus elaphus*) (467 and 227 mg $kg^{-1}$ dw, respectively) than in calves (73.62 and 128.77 mg $kg^{-1}$ dw) and adults (74.77 and 102.02 mg $kg^{-1}$ dw), but the authors found no such regularities in the muscle. It has been found that Zn bioaccumulation usually changes with the age of carnivores, being higher in old individuals (Hyvarinen et al. 2003). Cybulski et al. (2009) found higher Zn concentrations in serum, liver, and kidney in older silver fox in comparison to younger animals. An age-related difference has not been shown in the livers of some ruminants, such as reindeer (Bernhoft et al. 2002). Medvedev (1999) found no age-related Zn concentration differences in various organs of the moose (*Alces alces*). This relationship is also not found in different organs of various carnivore species, including the red fox (Dip et al. 2001), Arctic fox (Prestrud et al. 1994), American mink (Ogle et al. 1985; Stejskal et al. 1989; Brzezinski et al. 2014), Eurasian otter (Hyvarinen et al. 2003; Kang et al. 2015), and raccoon (Herbert and Peterle 1990).

#### 6.2.2.3 Environments and Zinc Concentration

For some time we have seen a growing interest in the heavy metal contamination of polluted and unpolluted environments, which requires a proper selection of species in terms of their dietary preferences. Some researchers believe that small ruminants may be good biomonitors of soil Zn levels. Sileo and Beyer (1985) found that mean hepatic and renal Zn concentrations were significantly higher in white-tailed deer caught near zinc smelters at Palmerton, Pennsylvania, USA, than in deer collected farther away from smelters. In white-tailed deer in the vicinity of zinc smelters ($<8$ km), Zn levels in the liver and kidney were 310 and 256 mg $kg^{-1}$ dw, respectively, while in animals hunted in areas located $>100$ km from smelters, the levels were much lower (145 and 132 mg $kg^{-1}$ dw, respectively). Bernhoft et al. (1999) found that reindeer from Rybatsjij Ostrov, Russia, an area with nickel smelters emitting large amounts of trace elements, Zn concentrations in the liver were 1.3–1.8 times higher than was found in reindeer from western Finmark, Norway (Sivertsen et al. 1995; Lovberg and Sivertsen 1997). Also, Lopez-Alonso et al. (2002) found a positive association between Zn concentrations in the soil and levels in the livers of calves. Sileo and Beyer (1985) and Reglero et al. (2008) found no significant differences in Zn level in the liver of red deer living close to lead mines compared to those in control areas in Spain. Dip et al. (2001) found no differences in Zn concentrations in the livers and kidneys of red foxes from urban, suburban, and rural areas.

#### 6.2.2.4 Season and Zinc Concentrations

Some studies have shown that the season can be an important factor determining Zn concentrations in wild ruminants. In some hoofed mammals living in a temperate climate such as roe deer and wild boar, the detected muscle, liver, and kidney Zn levels were greater in autumn than in spring (Michalska and Żmudzki 1992). A likely explanation of this is that in autumn, the animals have already had a few months of eating wholesome food, and at the same time it is also a period increased food intake and intense metabolism related to the storing up of energy reserves for winter. However, the organs of red deer living in the same area exhibited an inverse relationship with Zn levels. Borch-Iohnsen et al. (1996), studying reindeer inhabiting the Norwegian archipelago of Svalbard, found that liver Zn concentrations in summer were 2.5 times higher than in winter, and kidney Zn concentrations in July/August were two times higher than in April. These animals feed mainly on grasses and sedge like horsetail, willowherb, and shoots of deciduous trees (Venalainen 2007). Summer food is high herbaceous plants, while in late summer and early autumn reindeer eat mushrooms, and the winter diet mainly consists of lichen (Elkin and Bethke 1995). Moreover, in coastal reindeer, high hepatic Zn concentrations were found in April, which could be due to the low dietary Fe levels, which would improve Zn bioavailability (Borch-Iohnsen et al. 1996). In addition, in winter, those animals feed on marine algae, which have Zn content higher than normal winter foraging plants. Aastrup et al. (2000) found that Zn concentrations in caribou livers were higher in late winter after a season of feeding on long-lived lichens. It was further found that the Zn concentration increased in plant biomass following burn-off of the meadow (Ohr and Bragg 1985). The regrowing plant formations provided herbivorous mammals with significant amounts of Zn, which resulted in a greater accumulation of this element in the liver and other organs. Zimmerman et al. (2008) showed that in white-tailed deer, hepatic Zn concentration was higher in animals from the southern Black Hills in South Dakota, USA, following burn-off (39.09 mg $kg^{-1}$ ww or 130.3 mg $kg^{-1}$ dw) than in tilled areas (35.69 mg $kg^{-1}$ ww or 118.97 mg $kg^{-1}$ dw). A similar relationship was also found in mule deer, in which liver Zn concentrations were 43.53 mg $kg^{-1}$ ww (or 144.33 mg $kg^{-1}$ dw) in burned-off areas and 40.56 mg $kg^{-1}$ ww (or 135.2 mg $kg^{-1}$ dw) in tilled habitats (Zimmerman et al. 2008). The results of all these studies indicate that research on Zn in wild ruminants should consider seasonal changes in diet.

### 6.2.3 Correlations in Zinc Concentration in Different Tissues

One of the problems researched in ecotoxicology is the relationship between Zn levels in various tissues and organs. However, the occurrence of such correlations in free-living animals is not always clear and may not always be confirmed statistically. Millan et al. (2008) found correlations between muscle and liver Zn concentrations in a few carnivorous species: red fox, Iberian lynx (*Lynx pardinus*), Egyptian mongoose (*Herpestes ichneumon*), common genet (*Genetta genetta*), and Eurasian badger (*Meles meles*) ($r = 0.388$, $p = 0.021$). On the other hand, Medvedev (1999) found no relationship between Zn concentrations in the various organs of the moose, wild boar, and brown bear from Russia, similar to Harding et al. (1998), who analyzed the relationship between the concentrations of Zn in the liver, kidney, and spleen of American mink in Canada.

### 6.2.4 Zinc Versus Other Metals in Mammalian Tissues

Zinc interacts with other trace elements, including Cd and Cu. Cadmium has a slight influence on Zn metabolism, owing to its ability to induce production of metallothionein, as well as a competition with Zn for binding metallothionein (Goyer et al. 1995). It was found that high dietary Cu reduces Zn levels in the liver (Puls 1994). All of these metals and mercury (Hg) have a high affinity to small-molecule proteins from the metallothionein group and may induce their formation and compete for binding sites in those proteins (Henkel and Krebs 2004). Gasparik et al. (2012) found a positive correlation between two metal pairs, Zn–Cu and Zn–Hg, in the muscle of wild boar. In other studies, exposure to Cd led to increased Zn levels in the liver and kidney of rats and cattle (Ashby et al. 1981; Lopez-Alonso et al. 2002). In rats, biliary Zn concentrations greatly increased and Cu decreased following Cd administration (Ashby et al. 1981). In contrast, no correlation between Zn and Cd was found in muscle, liver, or kidney tissues or in cattle in Galicia, Northwest Spain, an area characterized by low environmental concentrations of heavy metals (Lopez-Alonso et al. 2004). Given the aforementioned evidence, it seems that research on Zn concentrations in various organs of mammals should also examine the concentrations of other trace elements, mainly Cd and Cu.

For years, ecotoxicologists have tried to assess and identify the most useful animal species and types of biological samples that could be used in the bioindication of environmental levels of Zn. It seems that data for herbivorous species, which often forage across very large territories, are not suitable for determining the specific source of a toxic hazard. Apparently species such as the red fox, which adapts to different environmental conditions and ranges within small territories of 0.5 $km^2$ and less, would be more suitable for biomonitoring research. Another species that seems appropriate in this regard is the American mink, a predatory mammal in which Zn concentrations in the kidney, liver, and bone are similar to other mammalian species. Finally, studies of this type must also take into account age, gender, and environmental pollution.

## 6.3 Birds

### 6.3.1 Zinc Concentration in Bird Tissues

Various biomonitoring studies on birds have analyzed Zn concentrations in various organs, including the liver, kidney, muscle, and bone. In general, Zn concentrations in birds are highest in the liver, kidney, bone, and feathers and lowest in muscle, blood, feces, and eggs (Eisler 1984; Ek et al. 2004). Birds of prey, piscivores, and insectivores have similar Zn concentration ranges (bone > liver > kidney) (Zaccaroni et al. 2011). Kalisinska et al. (2004) found the highest concentration of Zn in mallards in bone tissues (87.73–111.77 mg $kg^{-1}$ dw) and the liver (34.88–49.78 mg $kg^{-1}$ or 116.27–165.93 mg $kg^{-1}$ dw) and the lowest in the kidney (19.91–24.65 mg $kg^{-1}$ or 79.64–98.6 mg $kg^{-1}$ dw), muscle (11.99–12.96 mg $kg^{-1}$ or 39.97–43.2 mg $kg^{-1}$ dw), and brain (10.63–12.04 mg $kg^{-1}$ or 53.15–60.20 mg $kg^{-1}$ dw).

Zinc concentrations in the liver, kidney, and muscle can be considered indicative of Zn chronic exposure (Naccari et al. 2009). Because the determination of Zn concentrations in internal organs requires that the subject animals be dead, researchers increasingly prefer intravital sampling, for example, blood, excrement, feather, and egg samples (Burger 1993; Dauwe et al. 2000). Birds can excrete Zn through feces or by deposition in the uropygial salt glands (Burger and Gochfeld 1985) and feathers (Burger 1993). Females can also eliminate this metal by deposition in the eggs. Zinc concentrations in feathers may increase with age due to exogenous contamination with atmospheric deposition, so feathers can reflect contamination of the local habitat (Dmowski and Golimowski 1993). This method of research is often used for birds of prey, including the white-tailed eagle (*Haliaeetus albicilla*), bald eagle (*Haliaeetus leucocephalus*), osprey (*Pandion haliaetus*), and Eurasian buzzard (*Buteo buteo*), for water birds, including the mallard (*Anas platyrhynchos*), greylag goose (*Anser anser*), and pochard (*Aythya ferina*), and for passerines such as the great tit (*Parus major*), house sparrow (*Passer domesticus*), and tree sparrow (*Passer montanus*) (Table 11.7).

In Anseriformes, bone Zn levels range from 90 to 120 mg $kg^{-1}$ dw (Table 11.7). However, much higher Zn concentrations in bone have been found in pochard from Donana Natural Park (DNP) (154.9 mg $kg^{-1}$ dw) (Taggart et al. 2006). In passerines, mean bone Zn levels were greater than the average level for Anseriformes at 150–250 mg $kg^{-1}$ dw. Even higher Zn concentrations have been described in the great tit from a polluted zone in Spain (255.90 mg $kg^{-1}$ dw) (Llacuna et al. 1995) and from Niepołomice Forest, Poland (284 mg $kg^{-1}$ dw) (Sawicka-Kapusta et al. 1986) and in a tree sparrow (*Passer montanus*) in Beijing, China (255 mg $kg^{-1}$ dw) (Chao et al. 2003). An even greater Zn concentration has been described in the tibia of house sparrows in two polluted areas in Albania (291.50 and 342.80 mg $kg^{-1}$ dw) (Millaku et al. 2015).

Typical Zn concentrations in avian livers are 25–40 mg $kg^{-1}$ ww (or 83–133 mg $kg^{-1}$ dw) (WVDL 2015). Usually, mean hepatic Zn concentrations in birds of prey range from 60 to 170 mg $kg^{-1}$ dw (Table 11.7). However, in some species, especially those inhabiting areas with significant environmental pollution with heavy metals, levels can exceed 220 mg $kg^{-1}$ dw. Almost 240 mg Zn $kg^{-1}$ dw was detected in the liver of a bald eagle in the area of the Great Lakes in North America (Nam et al. 2012). An even greater Zn level, nearly 300 mg $kg^{-1}$ dw, was found in a Eurasian sparrowhawk from Galicia, Spain (Perez-Lopez et al. 2008). This could be due to the high Zn concentrations in soils in Galicia, ranging from 25 to 400 mg $kg^{-1}$ dw (Perez-Lopez et al. 2008).

In wild waterfowl, average hepatic Zn levels generally occur over a wider range (80–220 mg $kg^{-1}$ dw) than in birds of prey (Table 11.7). Sometimes the levels are much lower, for example 62.4 mg $kg^{-1}$ dw in mallard from the Gomish International Wetland, Iran (Sinka-Karimi et al. 2015). A few cases of waterfowl research showed very high liver Zn, in excess of 280, and even 2900 mg $kg^{-1}$ dw (Sileo et al. 2003). Experimental studies have shown that in mallards with clinical signs of Zn poisoning, liver Zn was 473–1990 mg $kg^{-1}$ dw (Levengood et al. 1999). Sileo et al. (2003) found Zn poisoning in three Canada geese (*Branta canadensis*) and mallard from the Tri-State Mining District (TSMD), Oklahoma, Kansas, and Missouri, USA, contaminated with Pb, Cd, Zn, and from mining, milling, and smelting. Zinc poisoning was diagnosed based on mild to severe degenerative abnormalities of the exocrine secretions of the pancreas; the liver Zn concentrations in the geese and mallard were 1000–2900 mg $kg^{-1}$ and 280 mg $kg^{-1}$ dw, respectively (Sileo et al. 2003). In addition, Beyer et al. (2005) noted that Zn concentrations in liver in three mallards, also from the TSMD, ranged from 770 to 1100 mg $kg^{-1}$ dw. The mallard with the highest concentration of Zn had caseous typhlitis (inflammation of the caeca). Typhlitis wad has also been observed in mallards experimentally poisoned with Zn shot (Levengood et al. 1999).

The literature also describes Zn poisoning in birds kept in zoos and parks in cages with nets made out of galvanized wire or in birds residing in bodies of water into which people threw coins made of alloys containing Zn. Fragments of such wires and coins are a significant source of poisoning of birds held in captivity. Doneley (1992) observed moderate to severe nephrosis in caged and aviary birds containing hepatic Zn levels of 320 and 534 mg $kg^{-1}$ dw, respectively.

In passerines, mean hepatic Zn levels were lower than the previously mentioned groups of birds, and usually range from 40 to 130 mg $kg^{-1}$ dw (Table 11.7). Lower mean levels were found in house sparrows from Sargodha city and Bhalwal Punjab, Pakistan (7.85 mg $kg^{-1}$ or 26.16 mg $kg^{-1}$ dw) (Mustafa et al. 2015) and from rural and urban areas in Finland (18.35 and 21.08 mg $kg^{-1}$ dw, respectively) (Kekkonen et al. 2012), while higher Zn concentrations in the liver were found in the great tit from Bialowieza and Niepolomice Forests, Poland (163 mg $kg^{-1}$ dw) (Sawicka-Kapusta et al. 1986) and in the Italian sparrow *Passer domesticus Italiae* in rural and urban areas in Italy (154 and 204 mg $kg^{-1}$ dw) (Gragnaniello et al. 2001).

Ecotoxicological reports on Zn concentrations in animals also contain relatively large amounts of data regarding the kidneys. Normal Zn concentrations in avian kidneys are 22–32 mg $kg^{-1}$ ww or 88–128 mg $kg^{-1}$ dw (WVDL 2015). In birds of prey, Anseriformes and passerines, mean nephric Zn is usually from 40 to 190 mg $kg^{-1}$

dw (Table 11.7). Lower mean nephric Zn levels have been found in house sparrows in reference (village Ciglik, Antalya) and polluted areas (Yatagan Thermal Power Plant in Mugla Province) in Turkey, at about 20 and 30 mg $kg^{-1}$ dw, respectively (Albayrak and Mor 2011). These small nephric Zn concentrations in polluted areas could be due to low levels of Zn in the environment. Demirak and Balci (2005) found that the Zn content in the underground water around the Yatagan Thermal Power Plant in Turkey were lower than those noted in European Economic Community and World Health Organization guidelines. Higher nephric Zn levels have been found in the Eurasian buzzard in Italy, at 91.1 mg $kg^{-1}$ ww or 367.6 mg $kg^{-1}$ dw (Zaccaroni et al. 2011). Also, higher Zn levels were found in the kidney of the gadwall (*Anas strepera*) in DNP, Spain, at 296.2 mg $kg^{-1}$ dw (Hernandez et al. 1999) and in mallard in the TSMD, USA, at 290–620 mg $kg^{-1}$ dw (Beyer et al. 2005). Sileo et al. (2003) found that Zn concentrations in the kidney in three Canada geese and a mallard displaying Zn poisoning from the TSMD were 510–970 and 220 mg $kg^{-1}$ dw, respectively. Experimental studies have shown that in Zn-intoxicated birds, microscopic changes in the kidneys include varying degrees of acute tubular necrosis, occasional secondary renal or visceral gout, and moderate interstitial nephritis in addition to nephrosis (Puschner et al. 1999).

In contrast to the liver and kidney, the average Zn concentration in the muscles of these birds is considerably lower and does not exceed 100 mg $kg^{-1}$ dw. The muscles of birds of prey, waterfowl, and passerines contain an average of 50–90, 30–95, and 10–95 mg Zn $kg^{-1}$ dw, respectively (Table 11.7).

Among the different types of samples taken from birds, ecotoxicologists indicate the significant usefulness of feathers because they can be obtained from living individuals. Some researchers believe that the feathers of nonmigratory bird species are suitable bioindicators of regional contamination (Burger 1993). Metal concentrations in feathers reflect the levels in the blood during the period of growth when the feather is connected with blood vessels and metals are incorporated into the keratin structure (Dauwe et al. 2000). The relative high Zn concentration in feathers is caused by the participation of Zn in the keratinization process (Burger 1993). Because birds can excrete Zn into growing feathers, Zn concentrations in feathers can be higher than in other tissues (Janssens et al. 2002). Metal levels in feathers reflect the levels in food during the period of feather growth, including during the growth of young birds, or during the molting period of fully grown birds (Solonen et al. 1999). In Falconiformes and Strigiformes, mean Zn concentrations in feathers ranged from 30 to 175 mg $kg^{-1}$ dw (Table 11.8). Much higher Zn concentrations in feathers have been found in the barn owl (*Tyto alba*) in contaminated and control sites in Belgium, at about 360 mg $kg^{-1}$ dw (Denneman and Douben 1993).

Zinc concentrations in feathers are higher in nocturnal raptors compared to diurnal raptors and bird eaters (Solonen et al. 1999). This is an essential element, and there seems to be no significant food-chain-related differences in birds of prey. However, Zn levels in feathers in passerines vary greatly, from 50 to 280 mg $kg^{-1}$ dw. Zinc concentrations much higher than this upper limit were found in great tit in the Bialowieza Forest, Poland (~360 mg $kg^{-1}$ dw) (Sawicka-Kapusta et al. 1986) and blue tit (*Parus caeruleus*) in a reference site in Belgium (~400 mg $kg^{-1}$ dw) (Eens et al. 1999).

Zinc is involved in the formation of certain colored feathers. Pigments such as eumelanin have a great capacity for binding Zn (Niecke et al. 1999). The black-brown pigmentation of feathers often results from the presence of melanin and a high concentration of Fe, Ca, Cu, and Zn (Scanlon et al. 1980; Goede 1985). Darker individuals have higher zinc levels in their feathers compared with paler ones when kept in standardized conditions.

## 6.3.2 Factors Affecting Zinc Concentrations in Bird Tissues

### 6.3.2.1 Age of Birds and Zinc Concentrations

The results of research are not unambiguous on the relationship between Zn content and age of birds. Taggart et al. (2006) found a significant difference in bone Zn concentrations between different age groups of birds. Kalisinska et al. (2004) studied two populations of mallards from an area less polluted with heavy metals in the city of Szczecin (including Zn) and a more polluted area of Slonsk (Poland) and found an age-related difference in Zn concentrations between both mallard populations. Zinc concentrations in the kidney of adults in Szczecin (22.87 mg $kg^{-1}$ ww or 91.48 mg $kg^{-1}$ dw) were higher than in juveniles (19.91 mg $kg^{-1}$ ww or 79.64 mg $kg^{-1}$ dw) ($p \leq 0.05$), while immature mallards from Slonsk had higher Zn levels in bone ($p \leq 0.01$) and liver tissues ($p \leq 0.05$) (111.77 mg $kg^{-1}$ ww or 166.82 mg $kg^{-1}$ dw and 49.78 mg $kg^{-1}$ ww or 165.93 mg $kg^{-1}$ dw, respectively) than adults (87.73 mg $kg^{-1}$ ww or 130.94 mg $kg^{-1}$ dw and 43.32 mg $kg^{-1}$ ww or 144.4 mg $kg^{-1}$ dw, respectively). Similarly, in piscivorous osprey from the Eastern USA, Wiemeyer et al. (1980) found higher Zn concentrations in the liver of immature birds (223.33 mg $kg^{-1}$ dw) than in adults (126.67 mg $kg^{-1}$ dw). Their later studies, however, showed that Zn levels in the same osprey were similar in both age groups (Wiemeyer et al. 1987). Swaileh and Sansur (2006) found that Zn concentration in the liver in adult house sparrows were about 1.5 times higher than in 1-month-old juveniles. In contrast, Licata et al. (2010) found no effect of age on Zn concentrations in Eurasian buzzard. Similarly, Hogstad (1996) found no statistically significant differences in Zn concentration in the liver between adults (72.70 mg $kg^{-1}$ dw) and juveniles (67.06 mg $kg^{-1}$ dw) in five passerine species. Changes in Zn concentration can be due to the fact that this metal is involved in many processes occurring in the egg and during feather development (Morera et al. 1997).

### 6.3.2.2 Sex of Birds and Zinc Concentrations

Gender-related differences in metal concentrations, including Zn, might be expected if males and females eat different food, different sized foods, and different proportions of various types of food (Burger 1995). In addition, some gender-related

**Table 11.8** Zinc concentrations in feathers of birds

| Species | Localization | Zn (mg $kg^{-1}$ dw) | References |
|---|---|---|---|
| Eurasian buzzard *Buteo buteo* | Sicily, Italy | 60.1 | Naccari et al. (2009) |
| | Southern Finland | 140 | Solonen et al. (1999) |
| Peregrine falcon *Falco peregrinus* | Sweden | 47.0 | Ek et al. (2004) |
| | Alaska | 141.9–149.3 | Parrish et al. (1983) |
| Laggar falcon *Falco biarmicus jagger* | Pakistan | 110 | Movalli (2000) |
| Eurasian sparrowhawk *Accipiter nisus* | Southern Finland | 130 | Solonen et al. (1999) |
| | Belgium | 35 | Dauwe et al. (2003) |
| | Sweden | 41 | Ek et al. (2004) |
| Northern goshawk *Accipiter gentilis* | Southern Finland | 130 | Solonen et al. (1999 |
| Barn owl *Tyto alba* | Belgium | 62 | Dauwe et al. (2003) |
| | Netherlands, PA | 363 | Denneman and Douben (1993) |
| | Netherlands, UA | 360 | |
| Little owl *Athene noctua* | Belgium | 31 | Dauwe et al. (2003) |
| Tawny owl *Strix aluco* | Southern Finland | 120 | Solonen et al. (1999) |
| Laggar falcon *Falco biarmicus jugger* | Pakistan | 107.40 | Movalli (2000) |
| Osprey *Pandion haliaetus* | Central California, USA | 173 | Cahill et al. (1998) |
| | Southern Finland | 110 | Solonen et al. (1999) |
| Mallard *Anas platyrhynchos* | Central California, USA | 170 | Cahill et al. (1998) |
| Great tit *Parus major* | Northeastern and Southern Poland | 173–357 | Sawicka-Kapusta et al. (1986) |
| | Belgium, PA | 172.66 | Eens et al. (1999) |
| | Belgium, UA | 178.56 | |
| | Belgium, PA | 97.9 | Dauwe et al. 2000) |
| | Belgium, UA | 127.2 | |
| | Finland, PA | 132.4 | Eeva et al. (2009 |
| | Russia, UA | 126.5 | |
| | Portugal, PA | 111.0 | Costa et al. (2013) |
| | Portugal, UA | 112.9 | |
| | Northern China | 276.60 | Deng et al. (2007) |
| | Northeastern Spain, UA | 185.18 | Llacuna et al. (1995) |
| | Northeastern Spain, PA | 166.60 | |
| Blue tit *Parus caeruleus* | Belgium, PA | 252.64 | Eens et al. (1999) |
| | Belgium, UA | 403.70 | |
| | Belgium, PA | 317.4 | Dauwe et al. (2005) |
| | Belgium, UA | 311.0 | |

(continued)

**Table 11.8** (continued)

| Species | Localization | Zn (mg kg$^{-1}$ dw) | References |
|---|---|---|---|
| Rock bunting *Emberiza cia* | Northeastern Spain, UA | 177.60 196.1 | Llacuna et al. (1995) |
| | Northeastern Spain, PA | 196.10 | |
| Blackbird *Turdus merula* | Northeastern Spain, UA | 131.90 | Llacuna et al. (1995) |
| | Northeastern Spain, PA | 155.40 | |
| House sparow *Passer domesticus* | West Bank, Palestine | 54.9 | Swaileh and Sansur (2006) |
| | India, URA | 48.99 | Sundaramahalingam et al. (2016) |
| | India, RUA | 31.5 | |
| Tree sparrow *Passer montanus* | China | 83.40–126.97 | Gong et al. (2012) |
| Song sparrow *Melospiza melodia fallax* | Arizona, USA | 195.1–206.3 | Lester and van Riper (2014) |

*PA* polluted area, *UA* unpolluted area

differences may be associated with differences in the metabolic profiles of metals involved and the activity of sex hormones, the intake or uptake of metals and nutritional requirements, or interactions between elements (Vahter et al. 2007). The sexes may differ in their production of various metalloproteins ,which play fundamental roles in the transport, storage, and excretion of metals (Gochfeld and Burger 1987). Moreover, differences in Zn levels may be due to physiological differences and the ability of female birds to excrete Zn into eggs (Naccari et al. 2009).

Moreover, because some species such as the house sparrow are considered to be sedentary, females are more prone to dispersal from native areas than males, making them less indicative of the area of environmental contamination (Skjelseth et al. 2007). Eeva et al. (2009) suggest that these differences are also due to the higher reproductive effort that makes females more susceptible to the negative health effects of pollution stress. Taggart et al. (2006) found that in female waterfowl, the liver Zn concentration was higher (87.7 mg kg$^{-1}$ dw) than in males (68.8 mg kg$^{-1}$ dw) in DPN, Spain. Danczak et al. (1997), in their research on mallard from the Slonsk Reserve (Poland), showed that the mean Zn concentration in the liver was higher in females (53.9 mg kg$^{-1}$ ww or 179.67 mg kg$^{-1}$ dw) than in males (43.6 mg kg$^{-1}$ ww or 145.33 mg kg$^{-1}$ dw), while both Swaileh and Sansur (2006) and Albayrak and Mor (2011) found no statistical difference in Zn concentrations between the livers of male and female house sparrows.

Despite the ambiguity concerning the relation between Zn content and sex of birds, it seems that this factor should be taken into consideration when analyzing the results of research.

### 6.3.2.3 Diet of Birds and Zinc Concentrations

Differences in Zn concentration between species and areas of habitation are the result of different diets, feeding strategies, and metabolic rates (Costa et al. 2013), as well as differences in the physiology of bird species (Deng et al. 2007). Animals usually regulate Zn effectively, and consequently hepatic Zn concentrations do not vary in proportion to dietary variability (Sileo et al. 2003). However, homeostatic mechanisms do fail at extremely high concentrations, with significant differences observed between species. In experimental studies on chickens, hepatic Zn levels increased more than 10-fold when the dietary concentration increased to about 2200 mg kg$^{-1}$, but Zn levels remained constant as the dietary concentration increased from 37 to about 110 mg kg$^{-1}$ ww (Stahl et al. 1989). In domestic mallard liver Zn concentration increased from 54 to 401 mg kg$^{-1}$ ww (or 180 to 336.67 mg kg$^{-1}$ dw) as dietary Zn concentration increased from the control concentration to 3000 mg kg$^{-1}$ (Gasaway and Buss 1972). Mute swans (*Cygnus olor*) fed a contaminated and suboptimal diet accumulated three times higher Zn concentrations in the liver than those fed a commercial waterfowl maintenance diet with the same Zn and other metal levels (Day et al. 2003).

Zinc concentrations in tissues may be related to the amount of plant food in the diet. Parslow (1982) studied the concentration of heavy metals in the liver of 16 different bird species from Ouse Washes, England. They found the highest Zn concentrations in the liver of the Eurasian wigeon (*Anas penelope*) and the lowest concentrations in the shoveler (*Anas clypeata*) and tufted duck (*Aythya fuligula*). They surmised that this could be because wheat (31.2 mg Zn kg$^{-1}$ ww) and barley (31.6 mg Zn kg$^{-1}$ ww) seeds contain more zinc than potatoes (7.7 mg Zn kg$^{-1}$ ww). Similarly, Gochfeld and Burger (1987) observed that birds eating other animals accumulate more metals than birds feeding on vegetation. Zaccaroni et al. (2011) studied Zn levels in groups of species with regard to different feeding habits, with higher Zn concentrations in the liver of piscivorous birds (95.1 mg kg$^{-1}$ ww or 317 mg kg$^{-1}$ dw) than insectivorous birds (82 mg kg$^{-1}$ ww or 274 mg kg$^{-1}$ dw). Bone Zn concentration was higher in birds of prey (217 mg kg$^{-1}$ ww or 324 mg kg$^{-1}$ dw) than in fish-eating birds (194 mg kg$^{-1}$ ww or 290 mg kg$^{-1}$ dw).

Lucia et al. (2008) found that in overfed birds Zn concentrations in the kidney (138.1 mg kg$^{-1}$ dw) were higher than in muscle (73.9 mg kg$^{-1}$ dw) and liver (38.4 mg kg$^{-1}$ dw), and in nonoverfed birds liver Zn levels (271.9 mg kg$^{-1}$ dw) were greater than in kidney (95.5 mg kg$^{-1}$ dw) and muscle (67.4 mg kg$^{-1}$ dw). In feathers of both the nonoverfed and overfed birds Zn concentrations were 102.8 and 164.9 mg kg$^{-1}$ dw, respectively. This suggests that the homogeneity in Zn levels in avian species from different habitats is related to the role of zinc in biological systems. Methallothionein can counteract the effects of exposure to high Zn levels, shifting its metabolism to bone accumulation typical of chronic exposure, which is most evident in piscivorous birds whose Zn levels are low in the kidney (47.5 mg kg$^{-1}$ ww or 190 mg kg$^{-1}$ dw) and much higher in bone (194 mg kg$^{-1}$ ww or 290 mg kg$^{-1}$ dw) (Zaccaroni et al. 2011). Therefore, bioindicative studies should take into account food chain differences between bird species.

#### 6.3.2.4 Habitats and Zinc Concentration

In recent years, many studies have investigated Zn concentrations in the organs of birds from polluted and unpolluted sites. Millaku et al. (2015) found that Zn concentrations were higher in the liver and kidney of the house sparrow from polluted areas (94.47 and 61.63 mg kg$^{-1}$ dw, respectively) in comparison to a reference site (61.48 and 47.22 mg kg$^{-1}$ dw, respectively) in Albania. Beyer et al. (2005) found that Zn poisoning in the TSMD USA, seems to be hazardous primarily to waterfowl. They also found that in waterfowl, including Canada geese, mallard, common pintail (*Anas acuta*), green-winged teal (*A. crecca*), ring-necked duck (*Aythya collaris*), and lesser scaup (*A. affinis*) from the TSMD, Zn levels in the liver and kidney were higher (440 and 210 mg kg$^{-1}$ dw, respectively) than in mallard at a reference site (93 and 80 mg kg$^{-1}$ dw, respectively). In intestinal digesta of the waterfowl from the TSMD, Zn concentration was 1100 mg kg$^{-1}$ dw compared to 130 mg kg$^{-1}$ dw at the reference site. The digesta contained a mixture of plant and invertebrate material with some sediments and soil. Sediments in the TSMD contained an extremely high Zn concentration, for example, in the Spring River, Zn levels in sediments were 22,000–25,000 mg kg$^{-1}$ (Ferrington 1989). Zinc concentrations exceeding 1000 mg kg$^{-1}$ were also found in freshwater vascular plants from several polluted sites (Outridge and Noller 1991). Canada geese and swans may be especially susceptible to metal poisoning because they ingest substantial amounts of sediment when they feed (Beyer et al. 2005). Van der Merwe et al. (2011) studied Canada geese in the TSMD with regard to their mean Zn concentrations in different organs at four mine-waste-contaminated sites and a reference site, where those from the confluence between the Spring River and Short Creek, where mine waste was deposited in the streambed of the Spring River and northwestern Galena, Kansas, mean Zn concentrations in the liver were 83 mg kg$^{-1}$ ww (or 277 mg kg$^{-1}$ dw) and 178 mg kg$^{-1}$ ww (or 595 mg kg$^{-1}$ dw), respectively, and in the kidney 22.8 mg kg$^{-1}$ ww (or 91.3 mg kg$^{-1}$ dw) and 48.7 mg kg$^{-1}$ ww (or 194.6 mg kg$^{-1}$ dw), respectively.

In two birds northwest of Galena, liver Zn levels were above a level indicating Zn poisoning in domestic poultry (>200 mg kg$^{-1}$ ww or >666 mg kg$^{-1}$ dw) (Puls 1994). In April 1998, a holding lagoon containing pyrite-ore-processing waste failed and released acidic metal-rich sludge and water into the Rio Guadiamar flowing through DNP (Galan et al. 2002). Benito et al. (1999) and Hernandez et al. (1999), examining blood, livers, and eggs of birds from the DNP since that accident, found that Zn concentration was elevated in relation to uncontaminated areas. Two years after the spill, Gomez et al. (2004) found increased Zn concentrations in the tissues of waterbirds from DNP, with Zn concentration in the liver 2.92–1084 mg kg$^{-1}$ dw. Taggart et al. (2006) found in waterfowl from the DNP maximum Zn concentration in liver at 220 mg kg$^{-1}$ ww (734 mg kg$^{-1}$ dw). Kozulin and Pavluschick (1993), examining mallards from an unpolluted section of the Svisloch River in Minsk, Belarus, and another section heavily polluted by urban and industrial outflows, found no difference in Zn concentration in the liver (135.1 and 126.1 mg kg$^{-1}$ dw, respectively), kidney (113.4 and 93.6 mg kg$^{-1}$ dw, respectively), and muscle

(50.4 and 41.8 mg kg$^{-1}$ dw, respectively). In contrast, house sparrows studied in Albania had higher Zn levels in the liver and muscle at a polluted site than those from an uncontaminated reference area (Albayrak and Mor 2011).

Kekkonen et al. (2012) found that the livers of house sparrows from an urban area had higher Zn levels than at a rural site. Swaileh and Sansur (2006) found the livers of house sparrows from urban areas (131.4 mg kg$^{-1}$ dw) had significantly higher Zn levels than at a rural site (97 mg kg$^{-1}$ dw). Similarly, Sundaramahalingam et al. (2016) found higher Zn levels in the muscles, bones, and feathers of sparrows from an urban site compared to a rural area. Gragnaniello et al. (2001), studying the liver and kidney of Italian sparrows, found higher Zn concentrations in species collected from an urban area than in those collected from rural localities.

Nestling birds are potential good biomonitors for terrestrial point-source pollution (Burger 1993; Janssens et al. 2002). In areas with significant environmental pollution (especially Cd), Zn deficiency may occur in intensively growing chicks. Zinc concentrations lower than that resulting from adequate nutrition were found in dead nestlings of the rook *Corvus frugilegus* in the liver, kidney, and muscle, with average Zn levels at 3.3, 1.6, and 1.3 mg kg$^{-1}$ dw, respectively (Orlowski et al. 2012).

An excess of Zn in the environment can lead to increased concentrations in the tissues of birds, including adults. Zinc concentrations in various organs or tissues taken from birds from a polluted area were greater than at the reference site in a study by Gomez et al. (2004), who found that, 7 months after a spill accident in DNP, Zn levels in 14 waterbird species were higher than those reported before the spill. In songbirds from a site severely contaminated with Zn from smelting, Zn concentration was only 20% greater than in birds from a reference site, although Zn soil concentration at the contaminated site was >10 times higher than at a reference site (Beyer et al. 1985). Hogstad (1996) found no significant differences in liver Zn concentrations in juveniles of three *Parus* species, the great tit, the marsh tit (*P. palustris*), and the willow tit, and the finches, bullfinch (*Pyrrhula pyrrhula*), and greenfinch (*Carduelis chloris*) from industrial areas (72.50 mg kg$^{-1}$ dw) and forest areas (57.99 mg kg$^{-1}$ dw). This may prove that birds adapt to high levels of this element in the environment and diet.

Interesting results are also provided by the analysis of Zn content in the feathers of birds. Janssens et al. (2001) found that mean Zn concentrations in the tail feathers of great tits near a metallurgical factory were higher than in a reference area (264 vs. 119.5 mg kg$^{-1}$ dw). Similarly, Manjula et al. (2015) found that in the feathers of house sparrows from an urban area, Zn concentrations were higher (98.16 mg kg$^{-1}$ dw) than those from a rural area (75.91 mg kg$^{-1}$ dw) ($p = 0.04$). In contrast, there was no difference between Zn levels in feathers of great tits from polluted and reference sites, but Zn levels in feathers were higher in the blue tit from a reference site than a polluted site (Eens et al. 1999). Based on a field experiment with free-living great tits, Jaspers et al. (2004) stated that Zn concentrations in feathers were probably due to endogenous deposition. Therefore, Zn levels in feathers of adult passerines can be used as monitors of local exposure.

Often studies also used feathers of nestlings, because the metal concentration in such feathers small reflect local pollution levels far better than those from in adults, because in nestlings, exposure occurs over a defined time period and is limited to the parents' foraging area (Furness 1993). However, it appears that in the case of Zn, rather young birds do not meet the criteria for good material bioindication. Janssens et al. (2002) studied 15-day-old great tit nestlings in Belgium, gathering feathers collected from multiple positions at different degrees from a large nonferrous smelter.

In feathers of nestlings collected at a distance of 400–600 m from the smelter, Zn concentration was significantly lower (16.5 mg $kg^{-1}$ dw) than at locations at distances of 0–350 m (31.3 mg $kg^{-1}$ dw), 2500 m (40.2 mg $kg^{-1}$ dw), and 4000 m (43.35 mg $kg^{-1}$ dw). Furthermore, in other studies, feathers of great tit nestlings from reference and polluted sites showed no significant differences in Zn concentrations (Dauwe et al. 2000; Janssens et al. 2002; Costa et al. 2013). These authors suggested that the feathers of great tit nestlings could not be used as biomonitors for Zn because they may not adequately reflect nestlings' body burden.

Eggs play an important role in ecotoxicological studies. Bird eggs have been used as bioindicators because they come from a specific fragment of the population, namely laying females. They are formed only during a specific period, have a consistent composition, and are easily sampled, and the removal of one egg from a nest has only a minor effect on population parameters (Furness 1993). Trace elements essential to embryonic development including Zn are transferred from the tissues of the females and ultimately reflect the extent of pollution in the females' environment (Nyholm 1998). The mean Zn concentrations are higher in egg content than eggshell because Zn is embedded in the quaternary structure of proteins whose concentrations are higher in eggs than in eggshells. It seems that the Zn supply to eggs, which are the nucleus of the future of the body, is subject to regulation developed during the evolution of the species, because Zn levels in eggs from areas heavily polluted with Zn are similar to those from uncontaminated areas. Dauwe et al. (1999) found no differences between Zn concentrations in eggs and eggshells from great and blue tits in polluted (62 and 69 mg $kg^{-1}$ dw, respectively) and reference sites in Belgium (28 and 19 mg $kg^{-1}$ dw, respectively). Dauwe et al. (1999) found that the eggshell could be used as an indicator for heavy metal pollution, especially at contaminated sites. It has also been found that egg white, ovalbumin, and conalbumin bind Zn (Richards and Steel 1987). Metals excreted in eggs reflect both stored body burdens and food choices of females during egg formation (Ek et al. 2004).

The aforementioned data indicate that Zn pollution is reflected in water and marsh birds (which collect food from the aqueous environment and contaminated sediments) and passerines.

#### 6.3.2.5 Season and Zinc Concentrations

Differences in Zn concentrations in avian organs may also be related to season. Parslow (1982) demonstrated that Zn concentrations in the liver from each of the waterfowl species they studied, including mallard shot in November (119.8 mg $kg^{-1}$

dw), were lower than in December and January (approximately 166 mg $kg^{-1}$ dw), which could be related to periodic molting and changes in the composition and quality of food. Gomez et al. (2004) found that in 14 waterfowl species from DNP, liver Zn concentrations tended to decrease slightly during summer. Gong et al. (2012) found that Zn concentrations in the pectoral muscle, heart, liver, and primary feathers of tree sparrows from Heilongjiang province, China, were highest in summer. In contrast, Hogstad (1996) found no statistically confirmed differences between hepatic Zn levels in juveniles and adults of three *Parus* species from Central Norway between January–March and October–December. Despite this inconsistency, biomonitoring should take into account the season of the year during which the material was collected and seasonal differences in the composition of the diet.

### 6.3.3 Zinc Versus Other Metals in Avian Tissues

Zinc is a redox-inactive metal, able to interact with other chemicals and produce altered types of toxicity, accumulation, and metabolism in birds (Koivula and Eeva 2010). In birds, Zn has been found to have antagonistic relations with Pb and Cd (Kaminski 1998). The correlation between Zn and Cd depends on the degree of Cd contamination, the correlation being more significant at a higher Cd burden (Wenzel et al. 1996). Levengood and Skowron (2007) found that Cd was significantly associated with Zn in the livers of sentinel ducks. The relationship between Cd, Zn, and metallothionein in sentinel ducks varied with the location of foraging and diet. Levengood et al. (1999) found that exposure to high Zn concentrations reduced hepatic Zn concentrations and increased renal Cd levels in mallard.

In avian kidneys a positive relationship has been described between Zn and Cd because the increased synthesis of Cd-induced metallothionein at the same time increases the number of binding sites available to Zn (Walsh 1990). Zinc is thought to provide protection against the renal toxicity of Cd (Hutton 1981). It was found that a high concentration of Zn interferes with the absorption of Cd, while a Zn:Cd ratio of about 150:1 probably protects terrestrial food chains from Cd toxicity (Chaney et al. 2001). Kim and Oh (2012) found that in the liver of mallard Zn concentrations were significantly correlated with Pb and Cd. A positive correlation between Zn and Cu has been demonstrated in the avian liver, which is most likely due to the similar metabolism of the metals (Wenzel et al. 1996).

## 7 Conclusions

In studies on terrestrial mammals, a measurable response to Zn pollution is exhibited by animals that are common in natural and seminatural habitats such as the red fox, raccoon, American mink, otter, and ungulates (including the white-tailed deer, reindeer, red deer, and wild boar). Birds are used as bioindicators because they are abundant and widely distributed, have long lifespans, and feed at different trophic levels, often being the top consumers. At the same time, they are more sensitive to Zn contamination than other vertebrates and therefore seem to be better bioindicators. Nestling passerines are potential good biomonitors for Zn pollution as Zn is intake in a clearly defined time period and originates from a limited parental foraging area. Zinc concentrations in the tissues of mammals and birds depend both on biological factors (e.g., age, physiological condition, animal species, sex, and age) and environmental factors (such as the supply of zinc in the diet). Long-term bioindication research conducted on the organs and tissues of mammals and birds, as well as noninvasive sampling of eggs and feathers, indicates the usefulness of this type of material for evaluating the state of the environment.

## References


Aastrup P, Riget F, Dietz R, Asmund G (2000) Lead, zinc, cadmium, mercury, selenium and copper in Greenland caribou and reindeer (*Rangifer tarandus*). Sci Total Environ 245:149–159

Abdel-Mageed AB, Oehme FW (1990) A review of the biochemical roles, toxicity and interactions of zinc, copper and iron: II. Copper. Vet Hum Toxicol 32:230–234

Adriano DC (2001) Trace elements in terrestrial environments: biogeochemistry, bioavailability, and risks of metals, 2nd edn. Springer Verlag, New York

Albayrak T, Mor F (2011) Comparative tissue distribution of heavy metals in house sparrow (*Passer domesticus*, Aves) in polluted and reference sites in Turkey. Bull Environ Contam Toxicol 87:457–462

Alloway BJ (2008) Zinc in soils and crop nutrition. Paris/Brussels IFA/IZA

Amici A, Danieli PP, Russo C, Primi R, Ronchi B (2012) Concentrations of some toxic and trace elements in wild boar (*Sus scrofa*) organs and tissues in different areas of the Province of Viterbo, Central Italy. Ital J Anim Sci 11:354–362

Anderson-Bledsoe KL, Scanlon PF (1983) Heavy metal concentrations in tissues of Virginia river otters. Bull Environ Contam Toxicol 30:442–447

Arnhold W, Anke M, Goebel S (2002) The copper, zinc and manganese status in opossum and gray fox. Z Jagdwiss 48:77–86

Ashby SL, King LJ, Parker D (1981) The effect of cadmium administration on the biliary excretion of Cu and Zn and tissue disposition of metals. Environ Res 26:95–104

ATSDR (2005) Toxicological profile for zinc. United States Government. http://www.atsdr.cdc.gov/toxprofiles/tp60.html. Accessed 7 Jan 2012

Aubert H, Pinta M (1977) Trace elements in soils, Developments in soil science, 7. Elsevier, Amsterdam

Badzinski SS, Flint PL, Gorman KB, Petrie SA (2009) Relationships between hepatic trace element concentrations, reproductive status, and body condition of female greater scaup. Environ Pollut 157:1886–1893

Beach RS, Gershwin ME, Hurley LS (1982) Gestational zinc deprivation in mice: persistence of immunodeficiency for three generations. Science 218:469–471

Benito V, Devesa V, Muñoz O, Suñer MA, Montoro R, Baos R et al (1999) Trace elements in blood collected from birds feeding in the area around Doñana National Park affected by the toxic spill from the Aznacóllar mine. Sci Total Environ 242:309–323

Bernhoft A, Waaler T, Mathiesen SD, Flaoyen A (1999) Trace elements in reindeer from Rybatsjij Ostrov, north western Russia. National Veterinary The Tenth Arctic Ungulate Conference, University of Tromsø, Norway, 9–13 August 1999

Bernhoft A, Waaler T, Mathiesen SD, Flaoyen A (2002) Trace elements in reindeer from Rybatsjij Ostrov, north western Russia. Rangifer 22:67–73

Beyer WN, Pattee OH, Sileo L, Hoffman DJ, Mulhern BM (1985) Metal contamination in wildlife living near two zinc smelters. Environ Pollut Ser A 38:63–86
Beyer WN, Dalgarn J, Dudding S, French JB, Mateo R, Miesner J et al (2005) Zinc and lead poisoning in wild birds in the tri-state mining district (Oklahoma, Kansas, and Missouri). Arch Environ Contam Toxicol 48:108–117
Binkowski ŁJ, Merta D, Przystupińska A, Soltysiak Z, Pacon J, Stawarz R (2016) Levels of metals in kidney, liver and muscle tissue and their relation to the occurrence of parasites in the red fox in the Lower Silesian Forest in Europe. Chemosphere 149:161–167
Binnerts W (1989) Zinc status of cows as deduced from the liver zinc content. Netherlands J Agr Sci 37:107–117
Bojar H, Bojar S (2009) Monitoring of contamination of the Lublin region wetlands using mallards (*Anas platyrhynchos*) as a vector of the contamination by various conditionally toxic elements. Ann Anim Sci 9:195–204
Borch-Iohnsen B, Nilssen KJ, Norheim G (1996) Influence of season and diet on liver and kidney content of essential elements and heavy metals in Svalbard reindeer. Biol Trace Elem Res 51:235–247
Broadley MR, White PJ, Hammond JP, Zelko I, Lux A (2007) Zinc in plants. New Phytol 173:677–702
Broekhuizen S (1987) First data on contamination of otters in the Netherlands. IUCN Otter Spec Group Bull 2:27–32
Brumbaugh WG, Mora MA, May TW, Phalen DN (2010) Metal exposure and effects in voles and small birds near a mining haul road in Cape Krusenstern National Monument, Alaska. Environ Monit Assess 170:73–86
Brzezinski M, Zalewski A, Niemczynowicz A, Jarzyna I, Suska-Malawska M (2014) The use of chemical markers for the identification of farm escapees in feral mink populations. Ecotoxicology 23:767–778
Burger J (1993) Metals in avian feathers: bioindicators of environmental pollution. Rev Environ Contam Toxicol 5:203–311
Burger J (1995) Heavy metal and selenium levels in feathers of herring gulls (*Larus argentatus*): differences due to year, gender, and age at Captree, Long Island. Environ Monit Assess 38:37–50
Burger J, Gochfeld M (1985) Nest site selection by Laughing Gulls: comparison of tropical colonies (Culebra, Puerto Rico) with temperate colonies (New Jersey). Condor 87:364–373
Cahill TM, Anderson DW, Elbert RA, Perley BP, Johnson DR (1998) Elemental profiles in feather samples from a mercury-contaminated lake in central California. Arch Environ Contam Toxicol 35:75–81
Cakmak I (2008) Enrichment of cereal grains with zinc: agronomic or genetic biofortification? Plant Soil 302:1–17
Caldwell DE (1994) Zinc toxicity in the dog. Canine Pract 19:6–7
Chaney RL, Ryan JA, Reeves PG (2001) Strategies in soil protection – missions and visions. Presented at Symposium on Soil protection in the United Europe, Vienna, Austria, 5 September 2001
Chao P, Guangmei Z, Zhengwang Z, Chengyi Z (2003) Metal contamination in tree sparrows in different locations of Beijing. Bull Environ Contam Toxicol 71:142–147
Charles CH, Cronin MJ, Conforti NJ, Dembling WZ, Petrone DM, McGuire JA (2001) Anticalculus efficacy of an antiseptic mouthrinse containing zinc chloride. J Am Dent Assoc 132:94–98
Chausmer AB (1998) Zinc, insulin and diabetes. J Am Coll Nutr 17:109–115
Coleman JE (1992) Zinc proteins: enzymes, storage proteins, transcription factors, and replication proteins. Annu Rev Biochem 61:897–946
Costa RA, Eeva T, Eira C, Vaqueiro J, Vingada JV (2013) Assessing heavy metal pollution using Great Tits (*Parus major*): feathers and excrements from nestlings and adults. Environ Monit Assess 185:5339–5344

Custer T, Cox E, Gray B (2004) Trace elements in moose (Alces alces) found dead in northwestern Minnesota, USA. Sci Total Environ 330:81–87
Cybulski W, Jarosz L, Chałabis-Mazurek A, Jakubczak A, Kostro K, Kursa K (2009) Contents of zinc, copper, chromium and manganese in silver foxes according to their age and mineral supplementation. Pol J Vet Sci 12:339–345
Danczak A, Ligocki M, Kalisińska E (1997) Heavy metals in the organs of Anseriform birds. Pol J Environ Stud 5:39–42
Dardenne M (2002) Zinc and immune function. Eur J Clin Nutr 56:S20–S23
Dardenne M, Pleau JM, Nabarra B, Lefrancier P, Derrien M, Choay M et al (1982) Contribution of zinc and other metals to the biological activity of the serum thymic factor. Proc Natl Acad Sci USA 79:5370–5373
Dauwe T, Bervoets L, Blust R, Pinxten R, Eens M (1999) Are eggshells and egg contents of Great and Blue Tits suitable as indicators of heavy metal pollution? Belg J Zool 129:439–447
Dauwe T, Bervoets L, Blust R, Pinxten R, Eens M (2000) Can excrement and feathers of nestling songbirds be used as biomonitors for heavy metal pollution? Arch Environ Contam Toxicol 39:541–546
Dauwe T, Bervoets L, Pinxten R, Blust R, Eens M (2003) Variation of heavy metals within and among feathers of birds of prey: effects of molt and external contamination. Environ Pollut 124:429–436
Dauwe T, Janssens E, Bervoets L, Blust R, Eens M (2005) Heavy-metal concentrations in female laying great tits (*Parus major*) and their clutches. Arch Environ Contam Toxicol 49:249–256
Day DD, Beyer WN, Hoffman DJ, Morton A, Sileo L, Audet DJ et al (2003) Toxicity of lead-contaminated sediment to mute swans. Arch Environ Contam Toxicol 44:510–522
Demirak A, Balci A, Dalman O. Tufekci M (2005) Chemical investigation of water resources around the Yatagan Thermal Power Plant of Turkey. Water Air Soil Pollut 162:171-181
Deng H, Zhang Z, Chang C, Wang Y (2007) Trace metal concentration in Great Tit (*Parus major*) and Greenfinch (*Carduelis sinica*) at the Western Mountains of Beijing, China. Environ Pollut 148:620–626
Denneman WD, Douben PE (1993) Trace metals in primary feathers of the Barn Owl (*Tyto alba guttatus*) in the Netherlands. Environ Pollut 82:301–310
Di Giulio RT, Scanlon PF (1984) Effects of cadmium and lead ingestion on tissue concentrations of cadmium, lead, copper, and zinc in mallard ducks. Sci Total Environ 39:103–110
Dip R, Stieger C, Deplazes P, Hegglin D, Mueller U, Dafflon O et al (2001) Comparison of heavy metal concentrations in tissues of red foxes from adjacent urban, suburban, and rural areas. Arch Environ Contam Toxicol 40:551–556
Dlugaszek M, Kopczynski K (2011) Comparative analysis of liver mineral status of wildlife. Probl Hig Epidemiol 92:859–863
Dlugaszek M, Kopczynski K (2013) Elemental composition of muscle tissue of wild animals from central region of Poland. Int J Environ Res 7:973–978
Dmowski K, Golimowski J (1993) Feathers of magpie (*Pica pica*) as a bioindicative material for heavy metal pollution assessment. Sci Total Environ 139:251–258
Domingo JL, Llobet JM, Paternain JL, Corbella J (1988) Acute zinc intoxication: comparison of the antidotal efficacy of several chelating agents. Vet Hum Toxicol 30:224–228
Doncheva S, Stoynova Z, Velikova V (2001) Influence of succinate on zinc toxicity of pea plants. J Plant Nutr 24:789–804
Doneley R (1992) Zinc toxicity in caged and aviary diseases – new wire disease. Aust Vet Pract 22:6–11
Duffy JY, Miller CM, Rutschilling GL, Ridder GM, Clegg MS, Keen CL et al (2001) A decrease in intracellular zinc level precedes the detection of early indicators of apoptosis in HL-60 cells. Apoptosis 6:161–172
Eens M, Pinxten R, Verheyen RF, Blust R, Bervoets L (1999) Great and blue tits as indicators of heavy metal contamination in terrestrial ecosystems. Ecotoxicol Environ Saf 44:81–85

Eeva T, Hakkarainen H, Belskii E (2009) Local survival of pied flycatcher males and females in a pollution gradient of a Cu smelter. Environ Pollut 157:1857–1861
Eisler R (1984) Trace metal changes associated with age of marine vertebrates. Biol Trace Elem Res 6:165–180
Eisler R (1993) Zinc hazards to fish, wildlife, and invertebrates: a synoptic review. Contaminant Hazard Reviews, Biological Report 10. United States Fish and Wildlife Service, Laurel, MD
Ek KH, Morrison GM, Lindberg P, Rauch S (2004) Comparative tissue distribution of metals in birds in Sweden using ICP-MS and laser ablation ICP-MS. Arch Environ Contam Toxicol 47:259–269
Elinder CG (1986) Zinc. In: Friberg L, Nordberg GF, Vouk VB (eds) Handbook on the toxicology of metals, 2nd edn. Elsevier Science Publishers, Amsterdam, pp 664–679
Elkin BT, Bethke RW (1995) Environmental contaminants in caribou in the Northwest Territories, Canada. Sci Total Environ 160:307–321
EPA (1987) Ambient aquatic life water quality criteria for zinc. Office of Water, U.S. Environmental Protection Agency, Washington, DC EPA 440/5-87-003
Eriksson M, Lord J, Jacobson S (2001) Wear and contact conditions of brake pads: dynamical in situ studies of pad on glass. Wear 249:272–278
Falandysz J (1994) Some toxic and trace metals in big game hunted in the northern part of Poland in 1987-1991. Sci Total Environ 141:59–73
Falandysz J, Jakuczun B, Mizera T (1988) Metals and organochlorines in four female white-tailed eagles. Mar Pollut Bull 19:521–526
Falandysz J, Strandberg L, Mizera T, Kalisinska E (2000) The contamination of white-tailed sea eagles with organichlorines in Poland. Rocz Panstw Zakl Hig 51:7–13
Fernandes G, Nair M, Onoe K, Tanaka T, Floyd R, Good RA (1979) Impairment of cell-mediated immunity functions by dietary zinc deficiency in mice. Proc Natl Acad Sci USA 76:457–461
Ferrington LC (1989) Occurrence and biological effects of cadmium, lead, manganese and zinc in the Short Creek/Empire Lake aquatic system in Cherokee County, Kansas. Kansas Water Resources Research Institute Report, contribution no. 277. Kansas Water Resources Research Institute, Manhattan, KS
Furness RW (1993) Birds as monitors of pollutants. In: Furness RW, Greenwood JJ (eds) Birds as monitors of environmental change. Chapman & Hall, London, pp 86–143
Galan E, Gonzalez I, Fernandez-Caliani JC (2002) Residual pollution load of soils impacted by the Aznalcóllar (Spain) mining spill after clean-up operations. Sci Total Environ 286:167–179
Gamberg M, Braune BM (1999) Contaminant residue levels in arctic wolves (*Canis lupus*) from the Yukon Territory, Canada. Sci Total Environ 243(244):329–338
Gamberg M, Braune B, Davey E, Elkin B, Hoekstra PF, Kennedy D et al (2005) Spatial and temporal trends of contaminants in terrestrial biota from the Canadian Arctic. Sci Total Environ 351–352:148–164
Gasaway WC, Buss IO (1972) Zinc toxicity in the mallard duck. J Wildl Manag 36:1107–1117
Gasparik J, Massanyi P, Slamecka J, Fabis M, Jurcik R (2003) Concentration of selected metals in liver, kidney and muscle of the red deer (Cervus elaphus). Rizikové faktory potravového reťazca III, Nitra
Gasparik J, Dobias M, CapCarova M, Smehyl P, Slamecka J, Bujko J et al (2012) Concentration of cadmium, mercury, zinc, copper and cobalt in the tissues of wild boar (*Sus scrofa*) hunted in the western Slovakia. J Environ Sci Health Part A Tox Hazard Subst Environ Eng 47:1212–1216
Georgieskii VI, Annenkov BN, Samokhin VT (1982) Mineral nutrition of animals. Butterworths, Boston, MA
Gilabert ER, Ruiz E, Osorio C, Ortega E (1996) Effect of dietary zinc deficiency on reproductive function in male rats: biochemical and morphometric parameters. J Nutr Biochem 7:403–407
Gochfeld M, Burger J (1987) Heavy metal concentrations in the liver of three duck species: influence of species and sex. Environ Pollut 45:1–15
Goede AA (1985) Mercury, selenium, arsenic and zinc in waders from the Dutch Wadden Sea. Environ Pollut 37:287–309

Goldschmidt VM (1954) Geochemistry. Clarendon Press, Oxford

Gomez G, Baos R, Gomara B, Jimenez B, Benito V, Montoro R et al (2004) Influence of a mine tailing accident near Doñana National Park (Spain) on heavy metals and arsenic accumulation in 14 species of waterfowl (1998 to 2000). Arch Environ Contam Toxicol 47:521–529

Gong Q, Jin Z, Zou H (2012) Concentrations of copper, zinc and manganese in Tree Sparrow (*Passer montanus*) at Jixi, Heilongjiang Province, China. J For Res 23:319–322

Goodwin FE (1998) Zinc compounds. In: Kroschwitz J, Howe-Grant M (eds) Krik-Othmer encyclopedia of chemical technology. John Wiley & Sons, New York, pp 840–853

Gordon RF (1977) Poultry diseases. The English Language Book Society and Bailliere Tindall, London

Goyer R, Klaassen CD, Waalkes MP (1995) Metal toxicology. Academic Press, San Diego, pp 35–37

Gragnaniello S, Fulgione D, Milone M, Soppelsa O, Cacace P, Ferrara L (2001) Sparrows as possible heavy-metal biomonitors of polluted environments. Bull Environ Contam Toxicol 66:719–726

Gupta R, Kanaujia A (2014) Metal toxicity in black kites, *Milvus migrans govinda* in Bundelkhand Region of India. World J Biol Med Science 1:76–98

Gurnee CM, Drobatz KJ (2007) Zinc intoxication in dogs: 19 cases (1991-2003). J Am Vet Med Assoc 230:1174–1179

Gutleb AC (1992) The otter in Austria: a review on the current state of research. IUCN Otter Spec Group Bull **7**:4–9

Gutleb AC, Kranz A, Nechay G, Toman A (1998) Heavy metal concentrations in livers and kidneys of the otter (*Lutra lutra*) from Central Europe. Bull Environ Contam Toxicol 60:273–279

Halbrook RS, Woolf A, Hubert GF, Ross R Jr, Braselton WE (1996) Contaminant concentrations in Illinois mink and otter. Ecotoxicology 5:103–114

Hambridge KM, Casey CE, Krebs NF (1986) Zinc. In: Mertz W (ed) Trace elements in human and animal nutrition, vol 2, 5th edn. Academic Press, New York, pp 1–137

Hammond GM, Loewen ME, Blakley BR (2004) Diagnosis and treatment of zinc poisoning in a dog. Vet Hum Toxicol 46:272–275

Hanusova E, Mertin D, Suvegová K, Szeleszczuk O (2007) Comparison of content of mineral elements in selected organs in carnivorous fur animals. Trace Elem Electrolytes 24:12–18

Harding LE, Harris ML, Elliott JE (1998) Heavy and trace metals in wild mink (*Mustela vison*) and river otter (*Lontra canadensis*) captured on rivers receiving metals discharges. Bull Environ Contam Toxicol 61:600–607

HazDat (2005) HazDat database: ATSDR's hazardous substance release and health effects database. Atlanta, GA

Heltai M, Markov G (2012) Red fox (*Vulpes vulpes* Linnaeus, 1758) as biological indicator for environmental pollution in Hungary. Bull Environ Contam Toxicol 89:910–914

Henkel G, Krebs B (2004) Metallothioneins: zinc, cadmium, mercury, and copper thiolates and selenolates mimicking protein active site features—structural aspects and biological implications. Chem Rev 104:801–824

Herbert GB, Peterle TJ (1990) Heavy metaland organochlorinecompound concentrations in tissues of raccoons from east-central Michigan. Bull Environ Contam Toxicol 44:331–338

Hernandez LM, Gomara G, Fernandez M, Jimenez B, Gonzalez MJ, Baos R et al (1999) Accumulation of heavy metals and As in wetland birds in the area around Doñana National Park affected by the Aznalcóllar toxic spill. Sci. Total Environ 242:293–308

Hernandez F, Oldenkamp RE, Webster S, Beasley JC, Farina LL, Wisely SM (2017) Raccoons (Procyon lotor) as sentinels of trace element contamination and physiological effects of exposure to coal fly ash. Arch Environ Contam Toxicol 72:235–246

Herschfinkel M, Silverman WF, Sekler I (2007) The zinc sensing receptor, a link between zinc and cel signaling. Mol Med 13:331–336

Hoekstra PF, Braune BM, Elkin B, Armstrong FA, Muir DC (2003) Concentrations of selected essential and non-essential elements in arctic fox (*Alopex lagopus*) and wolverines (*Gulo gulo*) from the Canadian Arctic. Sci Total Environ 20:81–92
Hoffmann SR, Blunck SA, Petersen KN, Jones EM, Koval JC, Misek R et al (2010) Cadmium, copper, iron, and zinc concentrations in kidneys of grey wolves, *Canis lupus*, from Alaska, Idaho, Montana (USA) and the Northwest Territories (Canada). Bull Environ Contam Toxicol 85:481–485
Hogstad O (1996) Accumulation of cadmium, copper and zinc in the liver of some passerine species wintering in central Norway. Sci Total Environ 183:187–194
Holterman WF, de Voogt P, Peereboom-Stegeman JH (1984) Cadmium/zinc relationships in kidney cortex and metallothionein of horse and red deer: histopathological observations on horse kidneys. Environ Res 35:466–481
Hornfeldt CS, Koepke TE (1984) A case of suspected zinc toxicity in a dog. Vet Hum Toxicol 26:214
Hutton M (1981) Accumulation of heavy metals and selenium in three seabird species from the United Kingdom. Environ Pollut 26:129–145
Hyvarinen H, Tyni P, Nieminen P (2003) Effects of moult, age, and sex on the accumulation of heavy metals in the otter (*Lutra lutra*) in Finland. Bull Environ Contam Toxicol 70:278–284
Jankovska I, Miholova D, Bejcek V, Vadlejch J, Sulc M, Szakova J et al (2010) Influence of parasitism on trace element contents in tissues of red fox (*Vulpes vulpes*) and its parasites *Mesocestoides* spp. (Cestoda) and *Toxascaris leonina* (Nematoda). Arch Environ Contam Toxicol 8:469–477
Janssens E, Dauwe T, Bervoets L, Eens M (2001) Heavy metals and selenium in feathers of great tits (Parus major) along a pollution gradient. Environ Toxicol Chem 20:2815–2820
Janssens E, Dauwe T, Bervoets L, Eens M (2002) Inter- and intraclutch variability in heavy metals in feathers of great tit nestlings (*Parus major*) along a pollution gradient. Arch Environ Contam Toxicol 43:323–329
Jarzynska G, Falandysz J (2011) Selenium and other 17 largely essential and toxic metals in muscle and organ meats of red deer (*Cervus elaphus*)—consequences to human health. Environ Int 37:882–888
Jaspers V, Dauwe T, Pinxten R, Bervoets L, Blust R, Eens M (2004) The importance of exogenous contamination on heavy metal levels in bird feathers. A field experiment with free-living great tits, Parus major. J Environ Monit 6:356–360
Kabata-Pendias A, Mukherjee AB (2007) Trace elements from soil to human. Springer, Berlin
Kabata-Pendias A, Pendias H (1992) Trace elements in soils and plants, 2nd edn. CRC Press, Boca Raton, FL, p 365
Kabata-Pendias A, Pendias H (1999) Biogeochemistry of trace elements. PWN, Warsaw
Kabata-Pendias A, Szteke B (2012) Trace elements in geo- and biosphere. Institute of Soil Science and Plant Cultivation, Pulawy, Poland
Kalisinska E, Salicki W (2010) Lead and cadmium levels in muscle, liver and kidney of scaup *Aythya marila* from Szczecin Lagoon, Poland. Pol J Environ Stud 19:1213–1222
Kalisinska E, Salicki W, Myslek P, Kavetska KM, Jackowski A (2004) Using the mallard to biomonitor heavy metal contamination of wetlands in north-western Poland. Sci Total Environ 320:145–161
Kalisinska E, Salicki W, Jackowski A (2006) Six trace metals in white-tailed eagle from north-western Poland. Pol J Environ Stud 15:727–737
Kalisinska E, Budis H, Wilk A, Lanocha N, Jackowski A (2009) Lead and cadmium in kidney and liver of nocturnal and diurnal raptors. Ochr Sr Zasobow Nat 41:102–110
Kaminski P (1998) The impact of calcium and heavy metals upon the nest development of the tree sparrow (*Passer montanus*). Wyd. UMK, Toruń
Kang S, Kang JH, Kim S, Lee SH, Lee S, Yu HJ et al (2015) Trace element analysis of three tissues from Eurasian otters (*Lutra lutra*) in South Korea. Ecotoxicology 24:1064–1072

Karpinski M (1999) Concentrations of selected macro- and microelements in the tissues of red deer (*Cervus elaphus*) and roe deer (*Capreolus capreolus*) from the central-eastern region of Poland. II. Annales UMCS sec EE Zootechnica 17:311–316

Kekkonen J, Hanski IK, Vaisanen RA, Brommer JE (2012) Levels of heavy metals in House Sparrows (Passer domesticus) from urban and rural habitats of southern Finland. Ornis Fennica 89:91–98

Kim J, Oh JM (2012) Metal levels in livers of waterfowl from Korea. Ecotoxicol Environ Saf 78:162–169

Kim J, Oh JM (2016) Assessment of trace element concentrations in birds of prey in Korea. Arch Environ Contam Toxicol 71:26–34

King JC (2000) Determinants of maternal zinc status during pregnancy. Am J Clin Nutr 71:1334S–1343S

Koivula MJ, Eeva T (2010) Metal-related oxidative stress in birds. Environ Pollut 158:2359–2370

Komosa A, Kitowski I, Komosa Z (2012) Essential trace (Zn, Cu, Mn) and toxic (Cd, Pb, Cr) elements in the liver of birds from Eastern Poland. Acta Vet (Beograd) 62:579–589

Kozulin A, Pavluschick T (1993) Content of heavy metals in tissues of mallards *Anas platyrhynchos* wintering in polluted and unpolluted habitats. Acta Ornithol 28:55–61

Krebs NF (2000) Overview of zinc absorption and excretion in the human gastrointestinal tract. J Nutr 130:1374S–1377S

Kucharczak E, Moryl A (2010) Contents of metals in cultivated plants in Zgorzelec-Bogatynia region parts 2. Arsenic, chromium, zinc, copper. Ochr Środ i Zasob Natur 43:7–17

Kucharczak E, Jopek Z, Moryl A (2003) Influence of environment on content of selected metals (Pb, Cd, Zn, Cu) in tissues of roes and wild pigs. Acta Sci Pol Med Vet 2:37–47

Lamothe AR (1991) Winter food habits and foodchain transfer of metals in wolves, (Canis lupus) of the Keewatin District, Northwest Territories. M.Sc. Thesis, Laurentian University

Lanszki J, Orosz E, Sugar L (2009) Metal levels in tissues of Eurasian otters (*Lutra lutra*) from Hungary: variation with sex, age, condition and location. Chemosphere 74:741–743

Lazarus M, Vickovic I, Sostaric B, Blanusai M (2005) Heavy metal levels in tissues of red deer (*Cervus elaphus*) from Eastern Croatia. Arh Hig Rada Toksikol 56:233–240

Lazarus M, Orct T, Blanusa M, Vickovic I, Sostarić B (2008) Toxic and essential metal concentrations in four tissues of red deer (*Cervus elaphus*) from Baranja, Croatia. Food Addit Contam Part A Chem Anal Control Expo Risk Assess 25:270–283

Lech T, Gubala W (1996) Heavy metals in the liver and kidneys of roe-deer from the region of Cracow, 1996. Bromatol Chem Toksykol 31:287–290

Leonard A, Gerber GB, Leonard F (1986) Mutagenicity, carcinogenicity and teratogenicity of zinc. Mutat Res 168:343–353

Lester MB, van Riper C 3rd (2014) The distribution and extent of heavy metal accumulation in song sparrows along Arizona's upper Santa Cruz River. Environ Monit Assess 186:4779–4791

Levengood JM, Skowron LM (2007) Coaccumulation of cadmium and zinc in tissues of sentinel mallards (*Anas platyrhynchos*) using a former dredge-disposal impoundment. Arch Environ Contam Toxicol 53:281–286

Levengood JM, Sanderson GC, Anderson WL, Foley GL, Skowron LM, Brown PW et al (1999) Acute toxicity of ingested zinc shot to game-farm mallards. Bull Ill Nat Hist Surv 36:1–36

Levengood JM, Lichtensteiger CA, Amdor BA (2002) Exposure to selected elements and health of Raccoons from Lake DePue, Illinois. Center for Wildlife Ecology, Illinois Natural History Survey

Lewis LA, Poppenga RJ, Davidson WR, Fischer JR, Morgan KA (2001) Lead toxicosis and trace element levels in wild birds and mammals at a firearms training facility. Arch Environ Contam Toxicol 41:208–214

Licata P, Naccari F, Lo Turco V, Rando R, Di Bella G, Dugo G (2010) Levels of Cd (II), Mn (II), Pb (II), Cu (II), and Zn (II) in common buzzard (*Buteo buteo*) from Sicily (Italy) by derivative stripping potentiometry. Int J Ecol. https://doi.org/10.1155/2010/541948

Llacuna S, Gorriz A, Sanpera C, Nadal J (1995) Metal accumulation in three species of passerine birds (*Emberiza cia, Parus major, and Turdus merula*) subjected to air pollution from a coal-fired power plant. Arch Environ Contam Toxicol 28:298–303
Lodenius M, Skaren U, Hellstedt P, Tulisalo E (2014) Mercury in various tissues of three mustelid species and other trace metals in liver of European otter from Eastern Finland. Environ Monit Assess 186:325–333
Lonnerdal B (2000) Dietary factors influencing zinc absorption. J Nutr 130:1378S–1383S
Lopez-Alonso M, Benedito JL, Miranda M, Castillo C, Hernández J, Shore RF (2002) Cattle as biomonitors of soil arsenic, copper, and zinc concentrations in Galicia (NW Spain). Arch Environ Contam Toxicol 43:103–108
Lopez-Alonso M, Prieto Montana F, Miranda M, Castillo C, Hernandez J, Luis Benedito J (2004) Interactions between toxic (As, Cd, Hg and Pb) and nutritional essential (Ca, Co, Cr, Cu, Fe, Mn, Mo, Ni, Se, Zn) elements in the tissues of cattle from NW Spain. Biometals 17:389–397
Lovberg KL, Sivertsen T (1997) Uptake of elements from industrial air pollution in South Varanger reindeer – a follow up study. Research Report for Directorate for Nature Management. DN-rapport 1997
Lu J, Combs GF Jr (1988) Effect of excess dietary zinc on pancreatic exocrine function in the chick. J Nutr 118:681–689
Lu J, Combs GF Jr, Fleet JC (1990) Time-course studies of pancreatic exocrine damage induced by excess dietary zinc in the chick. J Nutr 120:389–397
Lucia M, Andre JM, Bernadet MD, Gontier K, Gerard G, Davail S (2008) Concentrations of metals (zinc, copper, cadmium, and mercury) in three domestic ducks in France: Pekin, Muscovy, and Mule ducks. J Agric Food Chem 56:281–288
Lucia M, Andre JM, Gontier K, Diot N, Veiga J, Davail S (2010) Trace elements concentrations (mercury, cadmium, copper, zinc, lead, aluminium, nickel, arsenic, and selenium) in some aquatic birds of the Southwest Atlantic coast of France. Arch Environ Contam Toxicol 58:844–853
Lukaski HC (2005) Low dietary zinc decreases erythrocyte carbonic anhydrase activities and impairs cardiorespiratory function in men during exercise. Am J Clin Nutr 81:1045–1051
MacDonald CR, Elkin BT, Roach P, Gamberg M, Palmer M (2002) Inorganic elements in caribou in the Yukon, NWT, and Nunavut from 1992 to 2000: spatial and temporal trends and the effect of modifying factor. Unpublished manuscript prepared for the Northern Contaminants Program, Ottawa, ON, 32 pp
Maita K, Hirano M, Mitsumori K, Takahashi K, Shirasu Y (1981) Subacute toxicity with zinc sulfate in mice and rats. J Pest Sci 6:327–336
Malle KG (1992) Zink in der Umwelt. Acta Hydrochim Hydrobiol 20:196–204
Manjula M, Mohanraj R, Devi MP (2015) Biomonitoring of heavy metals in feathers of eleven common bird species in urban and rural environments of Tiruchirappalli, India. Environ Monit Assess 187:267
Mason B, Moore CB (1982) Principles of geochemistry. John Wiley & Sons, New York
Mason CF, Stephenson A (2001) Metals in tissues of European otters (*Lutra lutra*) from Denmark, Great Britain and Ireland. Chemosphere 44:351–353
Mateo R, Taggart MA, Green AJ, Cristofol C, Ramis A, Lefranc H et al (2006) Altered porphyrin excretion and histopathology of greylag geese (*Anser anser*) exposed to soil contaminated with lead and arsenic in the Guadalquivir Marshes, southwestern Spain. Environ Toxicol Chem 25:203–212
McCall KA, Huang C, Fierke CA (2000) Function and mechanism of zinc metalloenzymes. J Nutr 130:1437S–1446S
McCormick CC, Cunningham DL (1984) High dietary zinc and fasting as methods of forced resting: a performance comparison. Poult Sci 63:1201–1206
McDonough WF, Sun SS (1995) The composition of the Earth. Chem Geol 120:223–253
McDowell LR (2003) Minerals in animal and human nutrition, 2nd edn. Elsevier Science, Amsterdam, Netherlands, pp 357–396

McMahon RJ, Cousins RJ (1998) Mammalian zinc transporters. J Nutr 128:667–670
Medvedev N (1999) Levels of heavy metals in Karelian wildlife, 1989-91. Environ Monitor Assess 56:177–193
Merck VM (1986) The Merck veterinary manual, 6th edn. A handbook of diagnosis, therapy and disease prevention and control for the veterinarian. Merck and Co., Inc., Rahway, NJ
Mertin D, Szeleszczuk O, Suvegova K, Niedbala P, Hanusova E (2006) Content of microelements in the selected organ s of raccoon dog (*Nyctereutes procyonoides*). Ecol Chem Eng 13:85–90
Michalska K, Zmudzki J (1992) Metal concentrations in wild pig, roe and deer tissues of the Wielkopolska region. Med Wet 48:160–162
Michot TC, Chadwick PC (1994) Winter biomass and nutrient values of three seagrass species as potential foods for redheads (*Aythya americana* Eyton) in Chandeleur Sound, Louisiana. Wetlands 14:276–283
Mikszewski JS, Saunders HM, Hess RS (2003) Zinc-associated acute pancreatitis in a dog. J Small Anim Pract 44:177–180
Millaku L, Imeri R, Trebnicka A (2015) Bioaccumulation of heavy metals in tissues of house sparrow (Passer domesticus). Res J Environ Toxicol 9:107–112
Millan J, Mateo R, Taggart MA, Lopez-Bao JV, Viota M, Monsalve L et al (2008) Levels of heavy metals and metalloids critically endangered Iberian lynx and other wild carnivores from Southern Spain. Sci Total Environ 399:193–201
Morera M, Sanpera C, Crespo S, Jover L, Ruiz X (1997) Inter- and intraclutch variability in heavy metals and selenium levels in Audouin's gull eggs from the Ebro Delta, Spain. Arch Environ Contam Toxicol 33:71–75
Movalli PA (2000) Heavy metal and other residues in feathers of laggar falcon *Falco biarmicus* jugger from six districts of Pakistan. Environ Pollut 109:267–275
Mustafa I, Ghani A, Arif N, Asif S, Khan MR, Waqas A et al (2015) Comparative metal profiles in different organs of house sparrow (*Passer domesticus*) and black kite (*Milvus migrans*) in Sargodha District, Punjab, Pakistan. Pak J Zool 47:1103–1108
Naccari C, Cristani M, Cimino F, Arcoraci T, Trombetta D (2009) Common buzzards (*Buteo buteo*) bio-indicators of heavy metals pollution in Sicily (Italy). Environ Int 35:594–598
Nam DH, Rutkiewicz J, Basu N (2012) Multiple metals exposure and neurotoxic risk in bald eagles (*Haliaeetus leucocephalus*) from two Great Lakes states. Environ Toxicol Chem 31:623–631
National Research Council (NRC) (2005) Mineral tolerance of animals, 2nd Rev edn. National Academic Press, Washington, DC
Niecke M, Heid M, Kruger A (1999) Correlations between melanin pigmentation and element concentration in feathers of white-tailed eagles (Haliaeetus albicilla). J Ornithol 140:355–362
Nriagu JO (1989) A global assessment of natural sources of atmospheric trace metals. Nature 338:47–49
Nyholm NE (1998) Influence of heavy metal exposure during different phases of the ontogenyon the development of Pied Flycatcher, Ficedula hypoleuca, in natural populations. Arch Environ Contam Toxicol 35:632–637
O'Hara TM, George JC, Blake J, Burek K, Carroll G, Dau J et al (2003) Investigation of heavy metals in a large mortality event in caribou of northern Alaska. Arctic 56:125–135
Oberleas D, Harland BF (2008) Treatment of zinc deficiency without zinc fortification. J Zhejiang Univ Sci B 9:192–196
Ogden L, Edwards WC, Nail NA (1988) Zinc intoxication in a dog from the ingestion of copper-clad zinc pennies. Vet Hum Toxicol 30:577–578
Ogle MC, Scanlon PF, Kirkpatrick RL, Gwynn JV (1985) Heavy metal concentrations in tissues of mink in Virginia. Bull Environ Contam Toxicol 35:29–37
Ohr KM, Bragg TB (1985) Effect of fire on nutrient and energy concentration of five prairie grass species. Prairie Nat 17:113–126
Orlowski G, Kaminski P, Kasprzykowski Z, Zawada Z, Koim-Puchowska B, Szady-Grad M et al (2012) Essential and nonessential elements in nestling rooks *Corvus frugilegus* from eastern

Poland with a special emphasis on their high cadmium contamination. Arch Environ Contam Toxicol 63:601–611
Outridge PM, Noller BN (1991) Accumulation of toxic trace elements by freshwater vascular plants. Rev Environ Contam Toxicol 121:1–63
Pacyna JM, Pacyna E (2001) An assessment of global and regional emissions of trace metals to the atmosphere from anthropogenic sources worldwide. Environ Rev 9:269–298
Pahlsson AM (1989) Toxicity of heavy metals (Zn, Cu, Cd, Pb) to vascular plants: a literature review. Water Air Soil Pollut 47:287–319
Parker GH, Hamr J (2001) Metal levels in body tissues, forage and fecal pellets of elk (*Cervus elaphus*) living near the ore smelters at Sudbury, Ontario. Environ Pollut 113:347–355
Parrish JR, Rogers DT Jr, Ward FP (1983) Identification of natal locales of Peregrine Falcons (*Falco peregrinus*) by trace element analysis of feathers. Auk 100:560–567
Parslow J (1982) Heavy metals in the livers of waterfowl from the Ouse Washes, England. Environ Pollut 29:317–327
Perez-Lopez M, Hermoso de Mendoza M, Lopez Beceiro A, Soler Rodríguez F (2008) Heavy metal (Cd, Pb, Zn) and metalloid (As) content in raptor species from Galicia (NW Spain). Ecotoxicol Environ Saf 70:154–162
Perez-Lopez M, Rodríguez FS, Hernandez-Moreno D, Rigueira L, Luis Eusebio Fidalgo LE, Lopez Beceiro A (2016) Bioaccumulation of cadmium, lead and zinc in liver and kidney of red fox (*Vulpes vulpes*) from NW Spain: influence of gender and age. Toxicol Environ Chem 98:109–117
Pillatzki AE, Neiger RD, Chipps SR, Higgins KF, Thiex N, Afton AD (2011) Hepatic element concentrations of lesser scaup (*Aythya affinis*) during spring migration in the upper Midwest. Arch Environ Contam Toxicol 61:144–150
Pokorny B (2000) Roe deer *Capreolus capreolus* as an accumulative bioindicator of heavy metals in Slovenia. Roe deer *Capreolus capreolus* as an accumulative bioindicator of heavy metals in Slovenia. Web Ecol 1:54–62
Pokorny B, Ribaric-Lasnik C (2000) Lead, cadmium, and zinc in tissues of roe deer (*Capreolus capreolus*) near the lead smelter in the Koroska region (northern Slovenia). Bull Environ Contam Toxicol 64:20–26
Pollock B (2005) Trace elements status of white-tailed deer (*Odocoileus virginianus*) and moose (*Alces alces*) in Nova Scotia. Canadian Cooperative Wildlife Health Centre: Newsletters & Publications. Paper 45
Prasad AS, Rabbani P, Warth JA (1979) Effect of zinc on hyperammonemia in sickle cell anemia subjects. Am J Hematol 7:323–327
Prestrud P, Norheim G, Sivertsen T, Daae HL (1994) Levels of toxic and essential elements in arctic fox in Svalbard. Polar Biol 14:155–159
Puls R (1994) Mineral levels in animal health: diagnostic data. Sherpa International, Clearbrook, Canada
Puschner B, St Leger J, Galey FD (1999) Normal and toxic zinc concentrations in serum/plasma and liver of psittacines with respect to genus differences. J Vet Diagn Invest 11:522–527
Radostits OM, Gay CC, Blood DC, Hinchliff KW (2007) Veterinary medicine. In: Saunders WB (ed) A textbook of the diseases of cattle, sheep, goats and horses, 10th edn. WB Saunders Co., pp 1730–1733
Rauch JN, Pacyna M (2009) Earth's global Ag, Al, Cr, Cu, Fe, Ni, Pb, and Zn cycles. Global Geochemical Cycles 23:1–16
Reglero MM, Monsalve-Gonzalez L, Taggart MA, Mateo R (2008) Transfer of metals to plants and red deer in an old lead mining area in Spain. Sci Total Environ 406:287–297
Reglero MM, Taggart MA, Monsalve-Gonzalez L, Mateo R (2009) Heavy metal exposure in large game from a lead mining area: effects on oxidative stress and fatty acid composition in liver. Environ Pollut 157:1388–1395
Reimann C, de Caritat P (1998) Chemical elements in the environment—factsheets for the geochemist and environmental scientist. Germany Springer-Verlag, Berlin

Richards MP, Steel NC (1987) Trace element metabolism in thedeveloping avian embryo: a review. J Exp Zool Supp 1:39–51
Robinette CL (1990) Toxicology of selected pesticides, drugs, and chemicals. Zinc. Vet Clin North Am Small Anim Pract 20:539–544
Roslewska A, Stanek M, Janicki B, Cygan-Szczegielniak D, Stasiak K, Buzala M (2016) Effect of sex on the content of elements in meat from wild boars (*Sus scrofa* L.) originating from the province of podkarpacie (south-eastern Poland). J Elem 21:823–832
Roug A, Swift PK, Gerstenberg G, Woods LW, Kreuder-Johnson C, Torres SG et al (2015) Comparison of trace mineral concentrations in tail hair, body hair, blood, and liver of mule deer (*Odocoileus hemionus*) in California. J Vet Diagn Invest 27:295–305
Sawicka-Kapusta K, Kozlowski J, Sokolowska T (1986) Heavy metal in tits from polluted forests in Southern Poland. Environ Pollut A 42:297e310
Scanlon PF, Oderwald RG, Dietrick TJ, Coggin JL (1980) Heavy metal concentrations in feathers of ruffed grouse shot by Virginia hunters. Bull Environ Contam Toxicol 25:947–949
Schenker MB, Speizer FE, Taylor JO (1981) Acute upper respiratory symptoms resulting from exposure to zinc chloride aerosol. Environ Res 25:317–324
Scott BJ, Bradwell AR (1983) Identification of the serum binding proteins for iron, zinc, cadmium, nickel, and calcium. Clin Chem 29:629–633
Semrad CE (1999) Zinc and intestinal function. Curr Gastroenterol Rep 1:398–403
Senczuk W (2006) Modern toxicology. PZWL, Warsaw
Sileo L, Beyer WN (1985) Heavy metals in white-tailed deer living near a zinc smelter in Pennsylvania. J Wildl Dis 21:289–296
Sileo L, Nelson Beyer W, Mateo R (2003) Pancreatitis in wild zinc-poisoned waterfowl. Avian Pathol 32:655–660
Sillanpaa M (1982) Micronutrients and the nutrient status of soils: a global study. FAO Soils Bulletin No. 48, FAO, Rome
Sinka-Karimi MH, Pourkhabbaz AR, Hassanpour M, Levengood JM (2015) Study on metal concentrations in tissues of mallard and pochard from two major wintering sites in southeastern Caspian Sea, Iran. Bull Environ Contam Toxicol 95:292–297
Sivertsen T, Daae HL, Godal A, Sand G (1995) Ruminant uptake of nickel and other elements from industrial air pollution in the Norwegian-Russian border area. Environ Pollut 90:75–81
Skaren U (1992) Analysis of one hundred otters killed by accidents in central Finland. IUCN Otter Spec Group Bull 7:9–12
Skibniewski M, Skibniewska EM, Kosla T (2015) The content of selected metals in muscles of the red deer (*Cervus elaphus*) from Poland. Environ Sci Pollut Res Int 22:8425–8431
Skjelseth S, Ringsby TH, Tufto J, Jensen H, Saether BE (2007) Dispersal of introduced house sparrows *Passer domesticus*: an experiment. Proc Biol Sci 274:1763–1771
Skobrak EB, Bodnar K, Jonas EM, Gundel J, Javor A (2011) The comparison analysis of the main chemical composition parameters of wild boar meat and pork. Anim Sci Biotechnol 44:105A
Sleeman JM, Magura K, Howell J, Rohm J, Murphy LA (2010) Hepatic mineral values of white-tailed deer (*Odocoileus virginianus*) from Virginia. J Wildl Dis 46:525–531
Smrcka V (2005) Trace elements in bone tissue. Charles University in Prague. The Karolinum Press, Prague
Soetan KO, Olaiya CO, Oyewole OE (2010) The importance of mineral elements for humans, domestic animals and plants: a review. Afr J Food Sci *4*:200–222
Solonen T, Lodenius M, Tulisalo E (1999) Metal levels of feathers in birds of various food chains in southern Finland. Ornis Fennica 76:25–32
Souza MJ, Ramsay EC, Donnell RL (2013) Metal accumulation and health effects in raccoons (*Procyon lotor*) associated with coal fly ash exposure. Arch Environ Contam Toxicol 64:529–536
Spears JW (1994) Minerals in forages. In: Fahey GC (ed) Forage quality, evaluation, and utilization. American Society of Agronomy, Inc., Madison, WI, p 281

Stahl JL, Greger JL, Cook ME (1989) Zinc, copper and iron utilisation by chicks fed various concentrations of zinc. Br Poult Sci 30:123–134

Stejskal SM, Aulerich RJ, Slanker MR, Braselton WE, Lehning EJ, Napolitano AC (1989) Element concentrations in livers and kidneys of ranch mink. J Vet Diagn Invest 1:343–348

Stone M, Marsalek J (1999) Trace metal composition and spetiation in street sediment. Water Air Soil Pollut 87:149–169

Stout JH, Trust KA (2002) Elemental and organochlorine residues in bald eagles from Adak Island, Alaska. J Wildl Dis 38:511–517

Sundaramahalingam B, Baskaran S, Pandiarajan J (2016) An opportunistic evaluation of heavy metal accumulation in house sparrow (*Passer domesticus*). Res Rev Res J Biol 4:38–41

Suvegova K, Mertin D, Sviatko E, Oravcova E (1993) Content of some mineral elements in chosen organs of silver foxes (*Vulper vulper*). Scientifur 17:257–262

Swaileh KM, Sansur R (2006) Monitoring urban heavy metal pollution using the House Sparrow (*Passer domesticus*). J Environ Monit 8:209–213

Swiergosz R, Perzanowski K, Makosz U, Bilek I (1993) The incidence of heavy metals and other toxic elements in big game tissues. Sci Total Environ Suppl Pt 1:225–231

Szymczyk K, Zalewski K (2003) Copper, zinc, lead and cadmium content in liver and muscles of Mallards (*Anas platyrhynchos*) and other hunting fowl species in Warmia and Mazury in 1999-2000. Pol J Environ Stud 12:381–386

Taggart MA, Figuerola J, Green AJ, Mateo R, Deacon C, Osborn D et al (2006) After the Aznalcóllar mine spill: arsenic, zinc, selenium, lead and copper levels in the livers and bones of five waterfowl species. Environ Res 100:349–361

Tapeiro H, Tew KD (2003) Trace elements in human physiology and pathology: zinc and metallothioneins. Biomed Pharmacother 57:399–411

US EPA (1980) Ambient water quality criteria for zinc. Prepared by the Office of Water Regulations and Standards, Washington, DC. EPA 440/5-80-079

US Geological Survey (2010) The preliminary determination of epicenters (PDE) bulletin: U.S. Geological Survey Earthquake Hazards Program, http://earthquake.usgs.gov/research/data/pde.php. Accessed 1 Mar 2012

US Public Health Service (PHS) (1989) Toxicological profile for zinc. U.S. Public Health Service, Agency for Toxic Substances and Disease Registry, Atlanta, GA, p 121

Vahter M, Akesson A, Liden C, Ceccatelli S, Berglund M (2007) Gender differences in the disposition and toxicity of metals. Environ Res 104:85–95

van der Merwe D, Carpenter JW, Nietfeld JC, Miesner JF (2011) Adverse health effects in Canada geese (*Branta canadensis*) associated with waste from zinc and lead mines in the Tri-State Mining District (Kansas, Oklahoma, and Missouri, USA). J Wildl Dis 47:650–660

van Riet MM, Millet S, Nalon E, Langendries KC, Cools A, Ampe B et al (2015) Fluctuation of potential zinc status biomarkers throughout a reproductive cycle of primiparous and multiparous sows. Br J Nutr 114:544–552

Venalainen ER (2007) The levels of heavy metals in moose, reindeer and hares in Finland – results of twenty years monitoring. Academic Dissertation, Faculty of Natural and Environmental Sciences, University of Kuopio, Finland

Vernadsky WI (1945) The biosphere and the noösphere. Am Sci 33:1–12

Vikoren T, Kristoffersen AB, Lierhagen S, Handeland K (2011) A comparative study of hepatic trace element levels in wild moose, roe deer, and reindeer from Norway. J Wildl Dis 47:661–672

VROM (Dutch Ministry of Housing, Spatial Planning and the Environment) (2012) Soil Remediation Circular 2009. Staatscourant 3 April 2012, Nr. 6563. Ministry of Housing, Spatial Planning and the Environment, The Hague

Walker LA, Lawlor AJ, Chadwick EA, Potter E, Pereira MG, Shore RF (2011) Inorganic elements in the livers of eurasian otters, *Lutra lutra*, from England and Wales in 2009 – a Predatory Bird Monitoring Scheme (PBMS) Report. Centre for Ecology & Hydrology, Lancaster, UK. http://nora.nerc.ac.uk/14176/1/PBMS_Metals_Otters_2009.pdf. Accessed 25 Feb 2016

Walsh PM (1990) The use of seabirds as monitors of heavy metals in the marine environment. In: Furness RW, Rainbow PS (eds) Heavy metals in the marine environment. CRC Press, New York, p 256
Wentink GH, Spierenburg TJ, de Graaf GJ, van Exsel AC (1985) A case of chronic zinc poisoning in calves fed with zinc-contaminated roughage. Vet Q 7:153–157
Wenzel C, Adelung D, Theede H (1996) Distribution and age-related changes of trace elements in kittiwake Rissa tridactyla nestlings from an isolated colony in the German Bight, North Sea. Sci Total Environ 193:13–26
Wiemeyer SN, Lamont TG, Locke LN (1980) Residues of environmental pollutants and necropsy data for eastern United States ospreys, 1964-1973. Estuaries 3:155–167
Wiemeyer SN, Schmeling SK, Anderson A (1987) Environmental pollutant and necropsy data for ospreys from the eastern United States, 1975-1982. J Wildl Dis 23:279–291
Wolkers H, Wensing T, Groot Bruinderink GW (1994) Heavy metal contamination in organs of red deer (*Cervus elaphus*) and wild boar (*Sus scrofa*) and the effect on some trace elements. Sci Total Environ 144:191–199
Woolf A, Smith JR, Small L (1982) Metals in livers of white-tailed deer in Illinois. Bull Environ Contam Toxicol 28:189–194
Wren CD (1984) Distribution of metals in tissues of beaver, raccoon, and otter from Ontario, Canada. Sci Total Environ 34:177–184
WVDL (2015) Normal range values for WVDL toxicology. www.wvdl.wisc.edu/wp-content/uploads/2013/06/WVDL.Info_. Toxicology_Normal_Ranges.pdf. Accessed 28 Apr 2015
Zaccaroni A, Niccoli C, Andreani G, Scaravelli D, Ferrante MC, Lucisano A et al (2011) Trace metal concentration in wild avian species from Campania, Italy. Cent Eur J Chem 9:86–93
Zimmerman TJ, Jenks JA, Leslie DM, Neiger RD (2008) Hepatic minerals of white-tailed and mule deer in the southern Black Hills, South Dakota. J Wildl Dis 44:341–350

# Chapter 12
# Aluminum, Al

**Ewa Skibniewska and Michał Skibniewski**

**Abstract** Aluminum is the third most abundant element in nature, after oxygen and silicon. Its content in the Earth's crust has been estimated at a level of 8%. In spite of this, the element has never been engaging in the metabolic processes of the evolving living organisms. Aluminum reaches the body of an animal mostly ingested with food. Crossing the intestinal barrier, the metal gets to the bloodstream and so is transported to various tissues using the iron-transport routes. Of the total aluminum uptake, the majority is deposited in the bone (60%) and lungs (25%), whereas much lower amounts accumulate in the muscles (10%) and the liver (3%). Cerebral accumulation of the total uptake is about 1%. Besides blood, the metal is also found in all the other body fluids of a homeothermic organism, e.g., cerebrospinal fluid, lymph, semen, sweat, or urine. Studies on aluminum toxicity involving various taxonomic groups enable concluding that the mechanisms are similar across the taxa and consist mainly in evoking oxidative stress in cells. At the cellular level, aluminum reacts with cell membranes, cytoskeletal structures, and nucleic acids. In terrestrial vertebrates, aluminum impact results in altered enzymatic activity in the central nervous system and other organs and systems of the body. The metal affects the bone tissue metabolism, impairs the function of the excretory system and liver, and also has a negative effect on erythropoiesis. Human activity observed over the last centuries has led to a rapid growth in the production of aluminum obtained from the natural sources and, as a result, to its inclusion into the trophic chains of various ecosystems. In consequence, since 1970, aluminum has been treated as a xenobiotic accumulating in living organisms, whose bioavailability is continuously increasing.

E. Skibniewska (✉)
Department of Animal Environment Biology, Warsaw University of Life Sciences – SGGW, Warsaw, Poland
e-mail: ewa_skibniewska@sggw.pl

M. Skibniewski
Department of Morphological Sciences, Warsaw University of Life Sciences – SGGW, Warsaw, Poland



E. Kalisińska (ed.), *Mammals and Birds as Bioindicators of Trace Element Contaminations in Terrestrial Environments*,
https://doi.org/10.1007/978-3-030-00121-6_12

# 1 Introduction

Aluminum (Al), which belongs to the group of light metals, is the most abundant metallic element in the Earth's crust. Naturally found in the soil, rocks, and numerous minerals, Al does not occur in its pure metallic form (de la Fuente et al. 2007; Exley 2009). Since the dawn of history, aluminum-containing minerals have been used in a variety of industries. They were used already in ancient Greece and Rome, mainly in the manufacturing of dyes, in tanning, in water treatment, and as a component of drugs to treat ulcers (Steinegger et al. 1990; Kawahara and Kato-Negishi 2011).

Although the element is ubiquitous in nature, it is considered as a xenobiotic or a foreign substance, which has no role to play in any of biochemical or physiological functions of the organism. As various forms of life were evolving on Earth, many metals were gradually incorporated in the metabolic cycles and became essential nutrients in the biochemical processes running in the cell. Aluminum has remained an element that has no structural or catalytic function in microorganisms, fungi, plants, or animals. Despite this, the metal is biologically reactive and—in excess—is harmful to living organisms (Exley 2003, 2009).

Due to its specific properties, including the electrochemical characteristics, over the last 100 years of industrialization, Al has become an extremely important element and its importance continues to grow. Aluminum has become a ubiquitous component of the human environment, so the times we live now are sometimes referred to as the age of aluminum (Steinegger et al. 1990; Ranau et al. 2001; Exley 2009). In terms of environmental impact, the metal has a major part in the destruction of many ecosystems due to its role in acid rains. Therefore, ecotoxicological studies on the impact of aluminum on aquatic and terrestrial environments have been carried out for decades (Martin 1994). Due to its technological properties, Al is used in many industries. Not only is the metal used in civil engineering, aviation industry, and machine construction, but it is also indispensable in metallurgy and the generally understood chemical industry. Its physical and chemical properties, such as high resistance to moisture, gases, organic solvents, and light, make Al an almost perfect material for containers intended to store various substances. As a result, the metal finds wide application in the manufacturing of packages used in the food processing and cosmetics industry. Its compounds are used as food stabilizers and—as it has been since ancient times—to produce dyes. Moreover, Al compounds are used in medicine to reduce stomach acidity and as antiseptics and astringents (Peterson et al. 1993; Ranau et al. 2001; Codex Alimentarius Commission 2010; Hirata-Koizumi et al. 2011; Sheasby and Pinner 2011). The nanomaterial industry has been developing rapidly over the recent years, and its products find increasingly more applications. Both metallic Al and Al oxide nanoparticles are used in the industry. Therefore, effects of different forms of Al on the environment, including the impact of the nanoparticles, should be investigated (Krewski et al. 2007; Chen et al. 2008).

When it comes to toxic properties, many issues have been described so far in relation to living organisms, especially mammals. One of the most remarkable is Al ability to penetrate into the structures of the central nervous system (Yokel 2000; Cannon and Greenamyre 2011). Rapid advances are observed in the studies on the Al pro-oxidative function (Exley et al. 1996; Berthon 2002; Exley 2004). Despite intensive research though, there are many unanswered questions as to the effects of aluminum on cellular processes in humans and animals, especially with regard to the central nervous system. Also, there is much to explain and understand in terms of molecular mechanisms of toxicity, especially concerning Al introduced to the environment in the form of nanoparticles (Kearns 2004; Chen et al. 2008; Chaturvedi et al. 2012). A separate but equally important sphere of studies on aluminum is the ecotoxicological research on different groups of organisms, including vertebrates that inhabit inland ecosystems (Rosseland et al. 1990; Barabasz et al. 2002).

## 2 General Properties of Aluminum

Aluminum (Al, Latin *aluminum*) is a major component of the Earth's crust, as it represents 7.91% of the lithosphere. After oxygen (49.9%) and silicon (26.9%), Al is the third most abundant element in the lithosphere by mass (ATSDR 2008; Kumar and Gill 2009; Stahl et al. 2011). If we consider its oxide, Al is thought to closely follow silicon dioxide, representing 15% of the entire lithosphere mass. Its common occurrence in nature is reflected in the name *sial*, formerly referring to the composition of the outermost layer of the Earth, which consists mainly of silicon and Al. The term *sial* is an acronym that combines the symbols of these elements, Si and Al. Currently the upper layer of the crust is more commonly referred to as the granite layer (Degens 1989; Shenglin et al. 1996). Aluminum occurs mainly in its oxidation state III and, extremely rarely, I or II. In its pure form, it is a silvery-white metal, both malleable and ductile. In contact with the air, the light metal undergoes a slow process of oxidation and passivation, its surface covering with an oxide layer. As a result of its high chemical reactivity, the element does not naturally occur in its metallic form, but in combination with other elements. The minerals include oxides but also silicates, sulfates, and phosphates (ATSDR 2008; Priest 2004; Lide 2005; Kumar and Gill 2009). Aluminum is tasteless and does not corrode. It represents group 13 in the periodic table of elements. Due to its low density, 2.7 g cm $^{3}$, aluminum belongs to the group of light metals.

Although being so abundant in the minerals already known since ancient times, aluminum in its metallic form has long remained unknown. The first step to the discovery of metallic aluminum was the work by Johann Heinrich Pott, a German chemist, who extracted aluminum oxide (or alumina, $Al_2O_3$) in 1746, from naturally occurring alum, a mineral in common use at the time. This discovery caused expectations that alumina could have been an oxide of an unknown metal, which—in a relatively pure state—was eventually extracted by H.C. Ørsted and F. Wöhler in 1825. Some physical and chemical properties of metallic aluminum remained unknown for another 20 years. It was not until 1886 that large-scale production of metallic aluminum became feasible; at that time, the method was developed allowing extracting aluminum from molten mixture of cryolite and bauxite by electrolysis (Steinegger et al. 1990; Priest 2004; Kawahara and Kato-Negishi 2011). Carl Josef Bayer invented the process of extracting aluminum oxide from bauxite. His method, patented in 1888, has been fundamental to the industrial production of aluminum to this day (Plunkert 2000).

## 3 Aluminum in Nature: Geogenic Sources of the Metal

If substances that are capable of forming complex compounds are present in the substrate, aluminum may occur in the form of the toxic ion, $Al(H_2O_6)^{3+}$, as a monomeric or polymeric complex with $SO_4^{2-}$ or $Cl^-$, as well as in the form of other species or complexes with organic ligands. In the natural environment, only a small fraction of aluminum compounds are soluble; therefore it is assumed that the level of its toxic species does not increase significantly as a result of exposure to environmental factors (Klöppel et al. 1997). Although the lithosphere contains enormous amounts of aluminum, its geogenic release is relatively low, especially as compared to quantities released from anthropogenic sources. The geogenic release has been estimated at approx. 26% (Lantzy and MacKenzie 1979; Klee and Graedel 2004; ATSDR 2008; Rauch and Pacyna 2009). Both geogenic and anthropogenic origin aluminum may be found in the geochemical cycle. Natural sources of aluminum include both primary and secondary minerals, as well as colloids, in which previously released aluminum returns to the lithosphere through sedimentation. Its retention is extremely efficient, since as little as estimated 0.001% of released aluminum ions never reach the crust again (Exley 2003). Primary minerals with a high content of aluminum are formed as a result of the weathering of the bedrock, which was formed millions of years ago from solidified magma. Aluminum contents in basalt and granite rocks average 87.6 and 77.0 g $kg^{-1}$, respectively (Steinegger et al. 1990). Their disintegration underlies the formation of secondary minerals, such as kaolinite, hydroxy aluminosilicates, and aluminum hydroxide (Berg and Banwart 2000).

Aluminum is also mined from other resources, which are thought to be practically inexhaustible, i.e., anorthite (calcium aluminosilicate), alunite (hydrated aluminum potassium sulfate), post-coal wastes, oil shale, sillimanite, and kyanite (disthene). Aluminum compounds appear in the minerals as a result of rock erosion and leaching. Aluminum's most common natural forms are hydroxides, such as bauxites, kaolins, aluminosilicates, and clays. Bauxite is the basic aluminum ore, mostly composed of gibbsite (hydrargillite). Apart from this, boehmite, diaspore, and corundum represent minerals containing aluminum ions (Kabata-Pendias and Mukherjee 2007). Bauxites consist of up to 55% of $Al_2O_3 \cdot H_2O$, and their deposits—although located mainly in the tropical regions—occur in other parts of the world as well. The world's resources of bauxite are estimated between 55 and 75 billion tons. According to the U.S. Geological Survey for 2016, most bauxites occur in Africa (32%), followed by Oceania (23%), South America and Caribbeans (21%), Asia (18%), and the rest of the world (6%).

Crude oil ash content of aluminum may range widely, from 2000 to 20,000 ppm. The differences result from varied properties of the rock formations found around the oil fields but also may be a result of contamination while drilling, pumping, or preparing oil for transport to refineries.

Aluminum fraction that takes part in the geochemical cycle usually involves monomeric compounds, as polymers are strongly bound to colloids present in the substrate. Released Al may precipitate, forming this way the before-mentioned secondary minerals, of which gibbsite is most common (Steinegger et al. 1990). Cation aluminum species, referred to as mobile Al, are responsible for basic soil properties. Their presence in the soil also increases its sorption capacity (Widłak 2013). Depending on the bedrock and the pH of the soil solution, Al soil concentrations change in a fairly wide range, from 1% to 30%, mostly not exceeding 4% (US EPA 2003; Kabata Pendias and Mukherjee 2007). Solubility of aluminum compounds in the aqueous soil environment is closely related to its pH (Fig. 12.1).

Bedrock weathering processes also release aluminum to aquatic ecosystems, where the element is mostly bound with colloid matrix, representing more than 50% of its content. Soluble fractions in an aqueous medium comprise only 23%, whereas the rest of aluminum remains in its molecular form (Gundersen and Steinnes 2003; Rengel 2004). Particularly important is the mobility of aluminum from water to sediments and back, especially at a changing pH of the environment. Hence, a proper evaluation of a possible threat posed by toxic aluminum species to aquatic

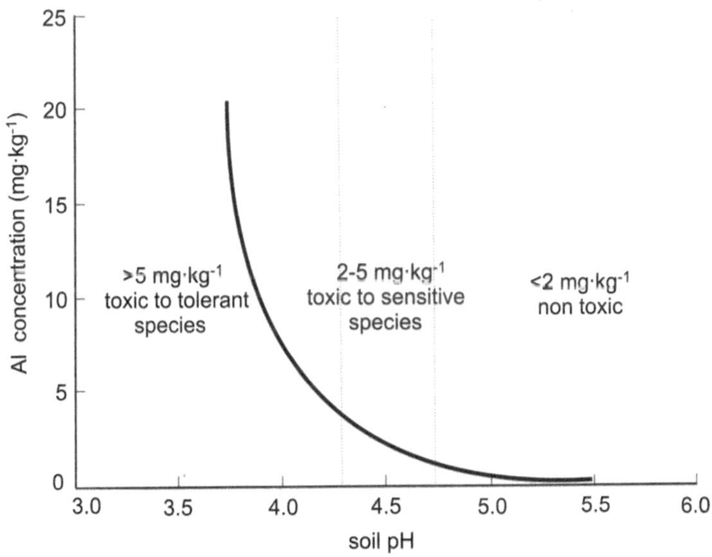


**Fig. 12.1** Dependence between soil pH, aluminum concentration, and its toxicity to plants (data source: http://soilquality.org.au/factsheets/soil-acidity-qld)

organisms requires speciation analysis, which should be carried out not only in the water column but also in the sediments. Gardner and Comber (2003) established a threshold value, pH 6.8, at which the concentration of reactive aluminum is lowest. Below this value, $Al(OH)^{2+}$ and $Al^{3+}$ species are dominant; above this pH, the fraction of $Al(OH)^{4-}$ increases. At pH 8 or higher, nearly 100% of dissolved aluminum is represented by the $Al(OH)^{4-}$ ion.

In the surface waters of North America, levels of aluminum range from 0.012 to 2.25 mg $L^{-1}$ (Jones and Bennett 1986). Miller et al. (1984) report that there is much more aluminum in surface waters compared to groundwaters. It was evidenced by their results that detectable quantities of aluminum were observed only in 9% of groundwaters, whereas as much as 78% of surface water contained this metal as well. Aluminum concentrations in European flowing waters vary greatly, depending on the geological formation, topography, and climate. The values range from 0.1 to 812 μg $L^{-1}$, with an average 17.7 μg $L^{-1}$. The lowest levels of the metal (below 6.5 μg $L^{-1}$) occur in the north of France and northeastern Germany, as well as along the continent strip stretching from southeastern France to Greece. The highest values, exceeding 339 μg $L^{-1}$, are found in the waters of Scandinavia (Salminen et al. 2005). Potable water treated with the most commonly used agents, i.e., aluminum chloride or sulfate, contains mostly the monomeric form of aluminum (Schintu et al. 2000). Going through this process, drinking water may acquire considerable amounts of aluminum, reaching 70 μg $L^{-1}$ on average; its bioavailability, however, is relatively low (Yokel and McNamara 2001). Increased aluminum concentrations in surface waters, resulting from its acidification and exceeding the threshold of toxicity, negatively affect fish and other aquatic organisms (Roux et al. 1996; Bi 2001; Herrmann 2001). A similar effect is observable in acidified soils which contain a high level of mobile aluminum fractions. These are toxic to various taxa of soil organisms (Maňkovská and Steinnes 1995; Markich et al. 2002). Apart from the noxious effects of mobile aluminum species present in terrestrial ecosystem components, one may also observe their negative impact on vegetation. This is thought to be the key factor reducing plant production efficiency on acid soils globally (Foy 1988).

Acidification of the environment, as well as the resulting acidic precipitation containing mostly sulfuric acid and nitric acid, plays an important role in aluminum retention changes. Acid rains are a consequence of emission of sulfur dioxide, nitrogen oxides, and ammonia to the atmosphere. These acids decrease the pH and increase aluminum concentration in surface waters (Bi 2001). This leads to poorer biodiversity in aquatic ecosystems and increases soil acidity (Driscoll et al. 2003; Jönsson et al. 2003).

A fraction of aluminum released from the crust enters the atmosphere. Weathering of rocks is the natural process behind the aluminum's presence in the air. As a result of this process, but also due to the industrial activity of man, aluminum-containing dusts, including particulate matter, are released to the atmosphere (Varrica et al. 2000; ATSDR 2008). Aluminum is the most important metallic component of particulate matter in the environment, where it occurs in the relatively stable form of $Al_2O_3$ ultrafine particle (Reff et al. 2009).

# 4 Aluminum Production and Uses: Anthropogenic Sources of Aluminum

Although Al is present in a number of minerals, of which many have been known since ancient times, its metallic form long remained unknown. The first step in obtaining metallic aluminum was a discovery by the German chemist, Johann Heinrich Pott, who in 1746 derived *alumina* (aluminum oxide, $Al_2O_3$) from then commonly used alum. Originally, alum (aluminum potassium sulfate dodecahydrate) was widely used as astringent, for treating minor bleedings. The discovery by J.H. Pott raised speculations that the new substance was an oxide of an unknown metal, whose relatively pure form was finally obtained by H.C. Ørsted and F. Wöhler in 1825. Some physical and chemical properties had not been described until 20 years later. It was not until 1886 that large-scale production of metallic aluminum was started, as the method was developed of electrolytic extraction of aluminum from molten mixture of cryolite and bauxite (Steinegger et al. 1990; Priest 2004; Kawahara and Kato-Negishi 2011).

Pure, silvery-white metal is obtained in a combination of chemical and electrolytic processes. Aluminum oxide is the main precursor of the metallic form. More than 95% of aluminum production is obtained from bauxites, following the Bayer process. This technology, developed in 1888 by Carl Josef Bayer, consists in heating bauxite ore to high temperatures at high pressure, in caustic soda, to obtain sodium aluminate solution, separated from insoluble residue (Plunkert 2000). In the next step, sodium aluminate is filtered and crystals of aluminum hydroxide are precipitated. Combined with the Hall-Héroult process, the Bayer process has found its application in industrial-scale aluminum extraction (Grojtheim et al. 1982; Steinegger et al. 1990; Lewis 2001).

Aluminum smelting consumes a considerable amount of energy and is responsible for both vast emission of greenhouse gases and extensive environmental acidification (Sheasby and Pinner 2011; Liu et al. 2013). Besides the previously mentioned Bayer process, which is still the basic technology of aluminum production, there exist a limited number of other technologies using different processes. It should be stressed, however, that irrespective of the smelting technology, bauxites remain the main source of the metal, as 95% of world aluminum production is obtained from this ore. Constituent minerals of bauxites are mostly polymorphic aluminum compounds, referred to as gibbsite [$Al(OH)_3$, containing 65% of $Al_2O_3$], boehmite [$AlO(OH)$], and diaspore ($HAlO_2$). Bauxite deposits are easily accessible and contain 30–65% of aluminum oxide (Steinegger et al. 1990; Sheasby and Pinner 2011; Reimann et al. 2014). Efforts are being made to use kaolin and other clays of high aluminum content. Since the early twentieth century, the primary aluminum production has been showing a strong, rising trend. According to some forecasts, it is estimated that the total world production of aluminum will have reached 60 million tons by 2020. It should also be emphasized that the process consumes about 2% of global energy production. Including bauxite ore preprocessing, one kilogram of pure aluminum needs about 23–24 kWh of energy. Aluminum production-related

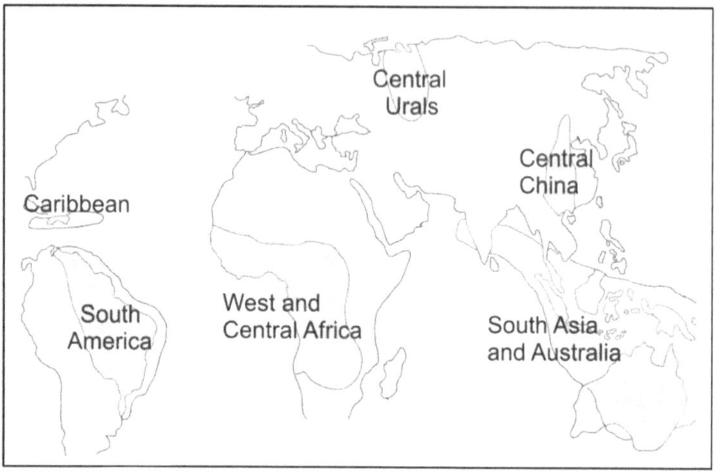


**Fig. 12.2** Distribution of major world bauxite reserves. On the base of the data published by U.S. Geological Survey (2016)

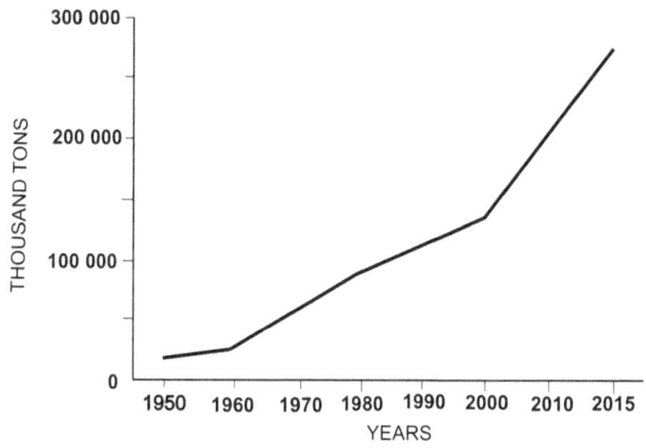


**Fig. 12.3** World bauxite mining in the years 1950–2015. On the base of the data published by U.S. Geological Survey (2016)

industries significantly contribute to the global economy and comprise the basis of national economies in more than 30 countries. Globally, 45 million tons of semi-fabricated aluminum products are made annually, including 14 million tons of recycled metal. The sector involved in the production and processing of aluminum is a direct employer for about 1 million people (Das and Yin 2007; Das 2011). Figure 12.2 illustrates the distribution of world bauxite deposits (www.aluminumleader.com).

The global bauxite mine production in 1900 was 88 thousand tons, produced by France, the USA and the UK (Plunkert 2000). In 2015, the production was 3000 times higher, 274 million tons (Fig. 12.3), with the world's total resources of bauxites estimated at 55–75 billion tons (U.S. Geological Survey 2016).

**Table 12.1** Top countries in and primary smelter aluminum production

| Production | Thousand metric tons |
|---|---|
| *Bauxite mine production in 2015* | |
| Suriname | 2200 |
| Kazakhstan | 5200 |
| Russia | 6600 |
| Jamaica | 10,700 |
| Guinea | 17,700 |
| India | 19,200 |
| Malaysia | 21,200 |
| Brazil | 35,000 |
| China | 60,000 |
| Australia | 80,000 |
| World | 274,000 |
| *Primary aluminum production in 2015* | |
| Iceland | 820 |
| Bahrain | 960 |
| Norway | 1320 |
| USA | 1600 |
| Australia | 1650 |
| United Arab Emirates | 2340 |
| India | 2350 |
| Canada | 2900 |
| Russia | 3500 |
| China | 32,000 |
| World | 58,300 |

Data source: U.S. Geological Survey (2016)

In Table 12.1 are presented the countries with the highest mine production of bauxites, exceeding 2 million tons, and with the highest aluminum primary production. Globally, nearly 40% of bauxite in 2015 was produced by four Asian countries (China, Malaysia, India, and Kazakhstan), and the two largest and most populated countries in this group, i.e., China and India, produced nearly 59% of aluminum extracted from bauxites. The global alumina (aluminum oxide) and bauxite production in 2015 increased by 9% and 12%, respectively, as compared to 2014. Particularly dynamic was the increase in China, where nearly 50% of the global aluminum production takes place, i.e., 66 million tons per year. A part of this quantity is obtained from imported ores. In 2015, Australia sold 20 million tons bauxite ore to China. A similar quantity in the same year was exported to China from Malaysia (U.S. Geological Survey 2016).

It is estimated that about 90% of alumina is used in metallic aluminum smelting (www.world-aluminum.org; http://www.alueurope.eu). An important characteristic of aluminum is its resistance to corrosion, as the metal is passivated by a native oxide layer. Aluminum and its alloys are used mainly in transport and civil engineering, both for support structure construction and in finishing elements (Priest 2004; Menzie et al. 2010).

Natural aluminum minerals, in particular bentonite and zeolite, are used in water purification, sugar refining, and beer production. Large quantities of aluminum hydroxide are used in the production of water treatment agents, refractory products, fuel additives, ceramics, abrasive materials, roofing materials, glass, electrical insulators, catalysts, medications, and cosmetics. Aluminum finds its applications in the chemical industry, mining (production of explosives), and food packaging industry (cans, foil). Aluminum oxides are used as food additives (Priest 2004; Krewski et al. 2007; Kawahara and Kato-Negishi 2011; www.world-aluminum.org; http://www.alueurope.eu).

Since aluminum smelting from bauxite ore is a highly energy-consuming process, it is important to reclaim the metal from waste materials. Recycling aluminum from scrap saves 95% of the energy that would otherwise be consumed to smelt this metal from its virgin source. Typical recycled aluminum materials include machining scrap, used beverage cans (UBC), foils, industrial scrap, shaves, and old metal. Aluminum recycling reduces air pollution by 95% and water pollution by 97%, as compared with its virgin source production. Aluminum scrap represents a significant resource of the metal and is used to manufacture a standard value product in the form of clean aluminum or its alloys. Moreover, recycling may be a highly profitable business, as the total production costs of secondary aluminum are 60% lower compared to aluminum production from bauxite ore (Fundacja RECAL—www.recal.pl. 12 June 2008).

Over the past few decades, we have been witnessing a significant increase in the demand for aluminum, which is applied in numerous conventional sectors of economy. Moreover, with new emerging technologies, the ways the metal and its compounds are used are changing rapidly (Menzie et al. 2010; Nappi 2013; Chaturvedi et al. 2012). The properties of nano- and micron-sized aluminum particles, as well as aluminum oxide nanoparticles, have drawn much attention lately, particularly in terms of their application as a catalyst (Kearns 2004). They are used as fuel additives or in ready-made mixtures of fuel and aluminum-based oxidants. They find application as combustion catalysts in propellants and pyrotechnic chemicals. In the thermite reaction, alumina rapidly reduces metal oxides, which is accompanied by one of the highest industrially attainable temperatures, about 3000 °C—or even 3800 °C, if temperature-boosting materials are added. Also, an intensive emission of light is released. Nanoscale aluminum particles have also found application as high-speed fuel combustion agents in jet engines (Piercey and Klapötke 2010; Gan and Qiao 2011; Sahu et al. 2014; Venkatesan 2015). Trimethylaluminum (TMA) is added to aircraft fuel. During a flight of a jet-propulsion aircraft, TMA is degraded to carbon dioxide, carbon monoxide, and aluminum oxide. The latter occurs in the form of nanoparticles, which remain suspended in the atmosphere for up to 18 months, to finally fall with precipitation and reach both aquatic and terrestrial ecosystems.

A separate and very poorly understood issue is the problem of how aluminum nanoparticles behave in the biotic and abiotic environments. It is presumed that they might be toxic to plants and other organisms (Chen et al. 2008; Li et al. 2009; Burklew et al. 2012; Chifiriuc et al. 2016). Bearing in mind a constantly increasing proportion of the air transportation, we may presume that the amount of aluminum oxide particulate dust affecting the natural environment will be increasing too. As it was mentioned earlier, the mining and processing of aluminum have a negative impact on ecosystems in many parts of the world. Not only does the primary aluminum production contribute to the emission of greenhouse gases but also entails the necessity of production wastes disposal. The most common waste is the red mud (or red sludge), a side product in the Bayer process. Estimates range between 70 and 120 million tons of red mud produced per year (Mayes et al. 2011a, b). It is usually composed of oxides and hydroxides of iron (hence its color), aluminum and titanium, contains radionuclides, $^{266}$Ra and $^{230}$Th, as well as a range of heavy metals, such as As, Cr, Co, Cd, Ni, and V (Mayes et al. 2011a, b; Ruyters et al. 2011; Burke et al. 2013). The strongly alkaline sludge is stored in open holding ponds. On October 4, 2010, a broken dike of a red sludge pond in Hungary caused an ecological disaster, in which approximately 1 million cubic meters of released sludge contaminated an area of about 40 square kilometers (Burke et al. 2013).

## 5 Biological Status of Aluminum

Although aluminum is a ubiquitous element, there is no evidence of any biological function for the metabolism of human or animal body (Exley 2003, 2009). Its complex electrochemical nature makes aluminum able to migrate from a solid phase to an aqueous phase, which eventually defines the role of the metal in the natural environment (Kabata-Pendias and Mukherjee 2007). The element penetrates into any form of life on Earth and, in consequence, passes from one level of the trophic chain to another. This effect results from the fact that the biosphere is formed of overlapping biotic cycles of microorganisms, plants, invertebrates, and vertebrates, with aluminum possibly penetrating each of them (Watanabe and Osaki 2002; Exley 2003).

Human activity over the last centuries has led to a rapid growth in the production of aluminum derived from its virgin sources, which results in an extra amount of the metal available for the trophic chains of various ecosystems. As a consequence, since 1970, aluminum has been regarded as a harmful element, accumulating in the tissues of living forms, constantly increasing in bioavailability (Gromysz-Kałkowska and Szubartowska 1999; Anke et al. 2001; Berthon 2002; Olariu et al. 2004; Kucharczak et al. 2005; Druga et al. 2005; Trif et al. 2005).

Microbial interactions with aluminum present an interesting issue. Illmer et al. (1995) found that the level of aluminum is the main limiting factor for the biomass of soil microorganisms. Studies on how aluminum affects microorganisms, however, show significant limitations, since it is hard to sort out the effects of the culture medium pH from those possibly resulting from the level of aluminum. This is a direct consequence of the fact that the availability of aluminum in the medium depends on its pH but also significantly affects its acidity (Illmer and Mutschlechner 2004). To date, little attention has been paid to the effects of aluminum on microbial cells, as can be compared with the research effort spent on plants and animals. The complex chemistry of aluminum, which is capable of polymerization and may form complexes with phosphates and medium-acidifying organic acids, often makes result interpretation complicated (Garcidueñas-Piña and Cervantes 1996). Guida et al. (1991), who studied aluminum toxicity in *Escherichia coli*, observed that growth inhibition of bacterial colonies was markedly dependent on pH. Aluminum concentrations below 100 μmol, or even 50 μmol, were found to have an inhibitive effect toward *Bradyrhizobium* spp. (Wood et al. 1988; Vargas and Graham 1988). Appanna et al. (1994) report that *Pseudomonas fluorescens* is tolerant to 50 mmol of aluminum if it occurs in the medium as aluminum citrate complex. The citrate residue, which is the only carbon donor in this aluminum complex, is utilized by the bacteria (Appanna and St. Pierre 1994). Plant nodulation is a process linked with the presence of aluminum in the soil. Johnson and Wood (1990) found in *Rhizobium* cells that aluminum binds to bacterial DNA, which disrupts its synthesis in susceptible strains and has no effect for the strains characterized by a high tolerance to aluminum. Richardson et al. (1988) observed that even a small amount of aluminum such as 7.5 μmol may restrain the expression of genes responsible for the processes of nodulation. Data reported by Brady et al. (1993) demonstrate that aluminum substrate concentrations below 5 μmol may inhibit soybean nodulation. Husaini and Rai (1992) demonstrated that aluminum is toxic to *Cyanobacterium*, and the toxicity increases with acidity of the medium, within pH decreasing from 7.5 to 4.5. The authors claim that with decreasing pH values, aluminum present in the medium affects the ATP level. The studies revealed that aluminum inhibits ATP synthesis at a high pH, whereas in an acidic medium, the metal binds to ATP making it biologically unavailable.

Aluminum belongs to metals that are important to plant growth. Its concentrations in plants depend on the species, variety, and developmental status. Some plants are more resistant to aluminum and tend to accumulate large amounts of the metal. Generally, plants growing on acidic soils contain large quantities of mobile, available fractions of aluminum (Domingos et al. 2003; Rengel 2004). The sensitivity to aluminum has been found to be regulated genetically. The tolerance level varies not only between species but also between varieties of the same plant. For instance, some cultivars of barley (*Hordeum* sp.) were demonstrated to have a varying susceptibility to soil aluminum (Steinegger et al. 1990). Plants tolerating high concentrations of aluminum in the soil may belong to one of two groups: one group is represented by species that do not accumulate the element, and the other comprises those being capable of storing considerable quantities of aluminum in their tissues. It has been found that the aluminum level in the leaves of plants belonging to the latter group may reach up to 37 g $kg^{-1}$ dry weight (dw), which corresponds to the content of potassium and nitrogen, the essential nutrients needed for a normal growth and development of the plant (Masunaga et al. 1998). The median value of aluminum content in Chinese tea shrubs (*Camellia sinensis*) is 2969 mg $kg^{-1}$ dw (Houba and Uittienbogaard 1994). Equally high aluminum

**Table 12.2** Mean aluminum contents in selected plants in mg kg$^{-1}$ dry weight

| Species | Al |
|---|---|
| Wheat (*Triticum*) | 31 |
| Barley (*Hordeum*) | 38 |
| Oat (*Avena*) | 47 |
| Onion (*Allium cepa*) | 63 |
| Potatoes–tubers (*Solanum tuberosum*) | 76 |
| Mushrooms (*Fungi*) | 25–130 |
| Grass (*Poaceae*) | 7–3410 |

Kabata-Pendias and Mukherjee (2007)

concentrations have been measured in grasses and legumes, in which these may exceed 3000 mg kg$^{-1}$ dw (Kabata-Pendias and Mukherjee 2007) (Table 12.2).

Aluminum toxic effects involve disrupted intake and transport of nutrients, lower biomass growth, as well as changes in the genotype of the plant. First symptoms of aluminum toxicity include inhibited root growth (Zhang et al. 1998; Zhang and Rengel 1999; Matsumoto 2000; Ahn et al. 2001; Barcelo and Poschenrieder 2002; Ma et al. 2002; Yamamoto et al. 2003). This response results mainly from disrupted metabolism in the growth zone of the root. The disorder may be a consequence of oxidative stress, disrupted calcium homeostasis, inhibition of ATPase activity, and changes in the cytoskeleton. Excess aluminum in plant cells damages polysaccharide-degrading enzymes, slows down the process of cellular respiration, and hampers the transport and absorption of water and macronutrients, including phosphorus, calcium, and magnesium (Ahn et al. 2002, 2004; Schwarzerova et al. 2002; Ma et al. 2003; Rengel and Zhang 2003; Rengel 2004).

Aluminum-caused oxidative stress is manifested in lipid peroxidation in cell membranes (Yamamoto et al. 2003). It also affects calmodulin, a modulator protein in eukaryotic cells, and changes the structure of chromatin in plant cells (Haug and Shi 1991). Plants have a specific protection system against the adverse effects of xenobiotics, which includes vacuoles, relatively large organelles that are able to accumulate aluminum ions and, in consequence, reduce their impact on the cell metabolism (Rengel 1997, 2004). Pathological changes in plants caused by aluminum are not limited to roots. They are also evident in the leaves; however, accurate diagnosis is difficult as the symptoms mimic calcium and phosphorus deficiency (Steinegger et al. 1990). Compared to animal studies, the research on aluminum uptake by plant cells requires a different approach, as the wall present in plant cells may bind large quantities of various forms of the metal (Haug and Shi 1991; Rengel 2004). Aluminum is thought to penetrate into the cells though endocytosis. Aluminum-dependent reactions with the components of intracellular signaling pathways affect the metabolism of plant cells (Haug and Shi 1991).

Freshwater invertebrates, which are food organisms for many species, play an important role in mediating aluminum to the bodies of terrestrial organisms. Results of analyses indicate that aquatic invertebrates may contain considerable amounts of aluminum, which is mostly deposited in the outer layers of their bodies and does not penetrate into the deeper structures and organs (Krantzberg 1989). The bioconcentration factor (BCF) for aluminum in the commonly found cladoceran, *Daphnia magna*, ranges from 10,000 to 0, with the value decreasing with an increase in water acidity (Havas 1985; Frick and Hermann 1990). A similar relationship was also found in freshwater crayfish, *Trichoptera* caddisflies, *Unionidae* mussels, and *Chironomidae* flies (Malley et al. 1988; Otto and Svensson 1983; Servos et al. 1985; Young and Harvey 1991). Toxic aluminum afflicts the gills of freshwater bivalves and crayfish, as the metal stimulates excessive production of mucus, which prevents an efficient gaseous exchange and disrupts the osmotic regulation (Exley et al. 1996; Woodburn et al. 2011). Crayfish inhabiting the bodies of water with low aluminum concentrations do not accumulate the metal in their internal organs, and only a small fraction of the aluminum uptake reaches the hepatopancreas (Alexopoulos et al. 2003). Crayfish caught in strongly aluminum-contaminated waters, or those kept in crayfish farms and fed freshwater snails obtained from polluted waters, may accumulate considerable amounts of aluminum in the hepatopancreas, which proves that the metal gets to their bodies by ingestion (Walton et al. 2010).

Aluminum bioconcentration factor in fishes ranges between 400 and 1365 (Roy 1999). As in the case of the gill-breathing invertebrates, the gills are a particularly affected organ in bony fishes too. Aluminum accumulates both on the surface and within the inner parts of the organ (Spry and Weiner 1991). Its presence in the gills leads to acute respiratory disorders, which, in consequence, kill the fish. Aluminum binds with piscine gills in a complex process, comprising three individual mechanisms. One consists in attracting water-dissolved positively charged aluminum species onto the gill parts that have a negative potential. The other mechanism involves polymerization and precipitation of aluminum within the gill due to a higher pH of the gill tissue in relation to the water. Finally, aluminum binds of the mucus covering the gills (Wilkinson and Campbell 1993; Playle and Wood 1989; Teien et al. 2006).

Excess aluminum in the food may pose a threat to the health of both humans and animals. The metal penetrates into the body of a homeothermic vertebrate in three ways: by ingestion, by respiration, and transdermally—ingestion being the most important route. The fraction of aluminum intestinally absorbed from food and water is relatively low. Animal studies show that intestinal uptake of aluminum averages around 1%; however, a strong interspecific variability has been observed, with the absorbed fractions ranging from 0.06% to 27% (Berthon 1996, 2002). It is generally accepted that aluminum ions are not absorbed spontaneously from the digesta, since they are unable to cross the lipoprotein membrane of the duodenum epithelial cells. Only a small fraction of ingested aluminum is absorbed as organic ligand complexes, which naturally occur in the food (Schuping 1996; Ranau et al. 2001).

The levels of aluminum or its various compounds in the air, water, and foodstuffs are regulated by law. The levels are set on the basis of the research carried out by Golub and German (2001), who studied nervous system developmental effects of dietary aluminum lactate ingestion in mice. The substance was administered to test animals as from the first day of gestation until birth, as well as in lactation, i.e., until 35 days of the postnatal life. The NOAEL (no-observable-adverse-effect level) was established at 26 mg aluminum per kg of body weight per day, whereas the LOAEL (lowest-observable-adverse-effect level) at 130 mg per kg of body weight per day. The estimates were used to set the MRL (minimal risk level) for humans, which is 1 mg Al per kg body weight for both intermediate- and chronic-duration daily oral exposure (ATSDR 2008). According to the WHO (2004) recommendations, potable water aluminum level should not exceed 0.1 mg $L^{-1}$, for large water treatment plants, or 0.2 mg $L^{-1}$, for small facilities. The publication by EPA (2006) does not include information on an RfD (reference dose) of aluminum; however, an RfD has been set for aluminum phosphate, a substance commonly used as insecticide, rodenticide, and fumigant. The reference dose of this substance is $4 \times 10^{-4}$ mg $kg^{-1}$ per day. The NOAEL of aluminum phosphate is 0.043 mg $kg^{-1}$ body weight per day (IRIS 2008).

According to the EFSA (2008, 2012), the ADI (acceptable daily intake) of aluminum is 0.14 mg $kg^{-1}$, whereas the TWI (tolerable weekly intake) is 1 mg $kg^{-1}$ of body weight. According to the SCAN (2003), an average concentration of aluminum in pasture soils is below 100 mg $kg^{-1}$ dw. Under adverse environmental conditions, which to a large extent include soil acidification, aluminum concentrations may reach as much as 1000 mg $kg^{-1}$ dw. Animals grazing on such pastures may take up aluminum, which under these conditions represents 1.5% dry weight of the forage. Although there are no compelling evidence that aluminum is essential for the proper metabolism in homeothermic animals, it has been suggested that, in certain concentration ranges, its presence in the tissues may be needed. Hence, Anke et al. (2005), basing on their own research, recommend a level of aluminum intake for animals below 10 mg $kg^{-1}$ dry weight of the feed. In addition, the authors suggest that the possible aluminum requirements of animals are satisfied from its natural presence in feed and water.

The available literature sources lack data on the maximum tolerated levels (MTLs) of aluminum in free-living homeothermic animals inhabiting terrestrial ecosystems. Based on research data, the MTL for rodents was set to 200 mg $kg^{-1}$ dry weight. For farm animals, such as poultry, cattle, sheep, pigs, and horses, the level is five times higher. It should be noted that there is no experimental data for horses and pigs; instead, estimates have been made by interspecific extrapolation (NRC 2005).

## *5.1 Toxicity of Various Forms of Aluminum in Homeothermic Animals*

As long as the life has been evolving on Earth, aluminum was never incorporated into the metabolic processes of the living organisms (Exley 2003, 2009). According to the National Research Council (NRC 2005), aluminum belongs to elements that are not essential in terms of body homeostasis, as no evidence has ever been found that it should be needed for growth, reproduction, or survival of animals. This is probably due to two basic reasons, i.e., low aluminum bioavailability and its adverse chemical properties in terms of biochemistry (Berthon 2002; Anke et al. 2009). At this point, aluminum is thought to have no biological function (Kawahara et al. 2007; Verstraeten et al. 2008). As a xenobiotic, the element is generally considered toxic; however, its effects on living organisms should be analyzed in the evolutionary context (Nicolini et al. 1991; Berthon 1996).

Although aluminum has been classified as a nonessential element by the NRC (2005), many reports suggest that its deficiency symptoms may be induced under controlled laboratory conditions. Chickens fed a synthetic, low aluminum diet showed reduced weight gains, whereas goats exhibited a higher rate of miscarriages, slower growth, weakness and motor impairment of the pelvic limbs, as well as a shorter lifespan (Nielsen 1996; Anke et al. 2005). The literature, however, is predominated by reports on harmful effects of aluminum in relation to living organisms. Research on aluminum toxicity, which has been carried out over recent years on animals representing various taxonomic groups, reveals that aluminum has similar activity mechanisms in all organisms. These predominantly involve oxidation stress leading to cellular death (Strong et al. 1996; Yokel 2000; Crisponi et al. 2012). The interest in aluminum as a toxic metal affecting various organs and systems in the human body increased when dialysis in patients suffering from renal failure became widespread. In such patients, long-term treatment results in accumulation of aluminum in the structures of the central nervous system, which cause neurodegenerative diseases, as well as bone metabolism disturbances leading to osteomalacia. A high concentration of aluminum salts, used in dialysis fluids as phosphate-binding agent, is considered as the main cause of these disorders (Ward et al. 2001; Exley and House 2011; Crisponi et al. 2012).

The gastrointestinal tract is most important route of aluminum reaching the bodies of homeothermic vertebrates, although a negligible amount of this element contained in the intestinal digesta penetrates into the bloodstream. This extremely low level of absorption (usually $<1\%$) results from the fact that the gastrointestinal tract is an effective barrier protecting the body against aluminum penetration (Yokel and McNamara 2001; Yokel 2004; Kośla et al. 2006; EFSA 2010). Absorption of aluminum compounds heavily depends on their water solubility and their concentration in the liquid of the lumen of the intestine (Crisponi et al. 2012). Highly soluble compounds are absorbed more efficiently compared to poorly soluble substances (Berthon 2002). Highly water-soluble aluminum chloride and aluminum nitrate were found more toxic than much less soluble aluminum sulfate, which have been confirmed experimentally on rats (Steinegger et al. 1990). The solubility of aluminum compounds is determined by the content of $Al^{3+}$ ions in the intestinal lumen, whose fate depends on the pH and the presence of neutralizing anions, which are thought to facilitate the penetration of the metal through the intestinal barrier (Berthon 2002). At a pH $\leq$5.0, aluminum in biological fluids exists mostly as a hexahydrate $[Al(H_2O)_6]^{3+}$. As the pH of the medium increases, other ion forms appear, such as $[Al(OH)]^{2+}$ or $[Al(OH)_2]^+$, as well as chemically inert aluminum hydroxide, $Al(OH)_3$. The latter species appear at the neutral pH and—at a further pH increase—form a soluble ionic species $[Al(OH)_4]^-$ (Crisponi et al. 2012). The presence of highly soluble fractions of aluminum ions their transport with the fluids in the paracellular space of enterocytes. Absorption is also possible via routes used for essential nutrients, which may be substituted by or absorbed along with aluminum ions (Berthon 2002). Aluminum adverse effects can be seen in the gastrointestinal tract, where phosphate binding takes place and where, at high doses, aluminum may cause hypophosphatemia. This leads to bone tissue metabolic disorders. Rats fed a diet containing aluminum at a dose of 6–10 mg $kg^{-1}$ body weight (BW) for 4 weeks exhibited growth impairment and rickets. Some experiments on effects of aluminum on the bone tissue involved intraperitoneal injections of aluminum preparations; this allowed bypassing the alimentary tract, where the binding of phosphates takes place, preventing their absorption. After 63 days of the experiment, in which the animals were administered 38–109 mg aluminum per 1 kg BW, its content in the bone reached a level of 163 mg $kg^{-1}$, which also caused osteomalacia (Steinegger et al. 1990). Bone metabolic disorders caused by aluminum occur at extremely high concentrations, which as a rule do not happen in nature. The problem of a possible linkage between aluminum and osteomalacia was first addressed in 1978 (Ward et al. 1978). It was noticed that excessive aluminum supply in dialysis patients leads to qualitative and quantitative changes in the bone tissue resulting in the weakening of the structure and spontaneous fractures. Aluminum is capable of being transferred to the structures responsible for the bone mineralization, in which it acts as inhibitor of the process. Osteoblasts, responsible for bone formation, have transferrin receptors on their surface, which are sites of transferrin-bound aluminum uptake. Aluminum acts as an antiproliferative agent, and osteoblast activity inhibition results in abnormal mechanical properties of the bone and, consequently, in fractures (Van Landeghem et al. 1998; Kasai et al. 1991; Crisponi et al. 2012). Apart from this, aluminum is an inhibitor of bone tissue metabolic processes through acting on parathyroid glands, as aluminum and transferrin complexes bind also to the cells of the glands. Therefore, the presence of aluminum in the parathyroid glands results in hypoparathyroidism (Smans et al. 2000).

Despite the negligible fraction of aluminum absorbed from the gastrointestinal tract, the element is definitely negative in relation to both bone and nervous tissues. As far as nervous tissue is concerned, aluminum toxicity is thought to result primarily from the oxidation stress affecting cells of the central nervous system (CNS). It is also stressed that any quantity of bioavailable aluminum that reaches CNS inflicts a neurotoxic damage and results in neurochemical response disturbances with the symptoms depending on the degree of the damage to particular structures. Numerous experiments on animal models have proven the toxic effects of excessive $Al^{3+}_{(aq)}$ concentrations against the encephalic tissue, which mainly results from aluminum inhibitory activity in relation to many enzymes (Zaida et al. 2007; Exley and House 2011). Sharma et al. (2013) demonstrate that long-term exposure to aluminum results in oxidative damages as a consequence of increased levels of reactive oxygen species (ROS) in cells, oxidation of mitochondrial DNA, and reduced activity, allosterically inhibited by ATP citrate synthase. This reaction chain leads to reduced expression of the mitochondrial genes, which in consequence impairs the functioning of neurons, as the nervous tissue is particularly sensitive to mitochondrial dysfunction (Fiskum 2000). Moreover, aluminum in CNS cells reduced the expression of COX-1, a marker protein of mitochondrial biogenesis. The protein is a component of the electron transport chain encoded for in the mitochondrial genome. Its reduced expression is inevitably linked with a drop in the activity of other enzymes involved in electron transport and leads to disturbances in the functioning of the cellular energy centers (Sharma et al. 2013). In addition, the presence of aluminum ions inhibits the activity of cytochrome oxidase, the final component in the mitochondrial electron transport chain. This leads to a cascade of biochemical reactions involving oxygen utilization disturbances, which result in ATP depletion and increased calcium concentration outside the cell; in consequence, the cell dies (Crisponi et al. 2012). Besides biochemical disturbances, aluminum induces morphology changes in the CNS (Strong et al. 1996). These involve formation of protein aggregates called neurofibrillary tangles, first described by Alzheimer in 1907, as well as amyloids—also referred to as senile plaques. Aluminum plays a role in the deposition of the insoluble proteins, i.e., amyloid beta and hyperphosphorylated tau protein, which form these neurofibrillary tangles (Paik et al. 1997; Uversky et al. 2001; Rengel 2004; Exley 2012). Besides the studies on $Al^{3+}_{(aq)}$ ions, also those on aluminum oxide nanoparticles reveal their potential negative impact on the mammalian brain (Chen et al. 2008; Li et al. 2009).

Aluminum-caused changes in the enzymatic activity do not pertain only to the nervous system but also affect other organs and systems of the body. The metal may interfere with the process of erythropoiesis through an impact on heme biosynthesis. This interference is manifested with changes in the activity of enzymes involved in the formation of heme (including delta-aminolevulinic acid) and reduced activity of uroporphyrinogen decarboxylase and ferrochelatase. High aluminum doses reduce the lifespan of red blood cells and cause a decrease in the activity of ATPase and such enzymes as hexokinase, alkaline phosphatase, choline acetyltransferase, and ferroxidase (Barabasz et al. 2002). At the cellular level, aluminum reacts with cell membranes, cytoskeletal elements, and the genetic material. Experiments have revealed that the metal interacts with both DNA and RNA. Its presence in the nuclei and a well-documented adverse effect on gene expression are facts that confirm interactions between aluminum and the nucleic acids (Exley 2012). Moreover, aluminum may be nephrotoxic; its increased concentration in the kidney will result in changes in the renal function, including impaired renal removal of the metal from the body. Increased aluminum concentrations in the kidney reduce the viability of its cells, destroy the cell membrane integrity, and damage intercellular junctions (Meshitsuka and Inoue 1998; Shirley et al. 2004). These processes do not severely disrupt the renal function in general; they modify, however, the regulatory processes in the cells, which affect the transport of certain substances and their metabolism. Aluminum in renal tubular cells induces oxidation stress, as well as disorders in p-aminohippuric acid transport and phosphorus absorption, which leads to changes in the body water management and problems with sodium retention. In such cases, female sex hormones act protectively in relation to the kidney, whereas male sex hormones contribute to its gradual damage. Aluminum may also have adverse effects on the liver, to which it is transported from the site of absorption via the portal circulation. Accumulation of aluminum in the liver induces oxidative stress accompanied by an increase in the content of thiobarbituric acid reactive substances (TBARS) and a corresponding reduction in the antioxidant defense, which involves reduced glutathione (GSH) and antioxidant enzymes, such as catalase (CAT) and glutathione peroxidase (GSH-Px). Outcomes of these reactions include abnormal secretion of bile (de Carmen-Contini et al. 2011).

In vitro tests revealed that aluminum compounds, as well as aluminum oxide nanoparticles, are both mutagenic and genotoxic. Most short-term mutagenicity test have shown their negative effects (Krewski et al. 2007; Sjörgen et al. 2007; ATSDR 2008; Hashimoto et al. 2016). A test on human peripheral blood lymphocytes apparently showed formation of micronuclei and chromosomal aberrations caused by aluminum ions (ATSDR 2008). It should be stressed, however, that these responses were observed at relatively high aluminum concentrations, which can be attained in laboratory conditions only (Krewski et al. 2007; Sjögren et al. 2007; ATSDR 2008; EFSA 2010). Interference with CNS cellular function results in changes in the activity of neurotransmitters. Experiments on laboratory rats demonstrate that an elevated aluminum concentration may cause a pronounced cholinergic deficit. In vitro studies have shown that aluminum has a biphasic effect on acetylcholinesterase activity, a marker of cholinergic activity (Kumar 1998).

Research has shown that aluminum, as a strong pro-oxidant, is involved in the processes of carcinogenesis (Exley et al. 2007; Skibniewska 2010). It has been demonstrated that breast tumors accumulate aluminum ions, and this applies to both humans and animals (Majewska et al. 1997; Skibniewska 2010). Most probably, this is related with biochemical properties of a given cancer tissue, which are characterized by overexpression of osteopontin, which forms complexes with aluminum ions; these act in two ways: on estrogen receptors and through binding with DNA in the cells of the mammary gland, which results in genomic instability (Banasik et al. 2013; Pereira et al. 2013; Darbre et al. 2011, 2013). Aluminum salts act as a catalyst of Fenton's reaction, which produces free radicals damaging cellular structures. Female patients suffering from breast cancer exhibited a significant accumulation of certain products of oxidation in the microenvironment of the altered breast tissue, corresponding to the elevated levels of aluminum (Mannello et al. 2009, 2010). Besides its effect on the genome, aluminum has the ability to bind to estrogen receptors; hence it is referred to as metalloestrogen. The signs of its activity include an impact on estrogen-dependent gene expression in response to the activity of these hormones (Darbre et al. 2011, 2013). Moreover, aluminum interacts with other elements. With cations such as $Mg^{2+}$ and $Ca^{2+}$, aluminum competes for coordination sites in enzymes, signaling molecules, receptors, transport proteins, membrane channels, nucleic acids, and other ligands (Exley and House 2011). Ward et al. (2001) demonstrate that an increase in tissue aluminum concentrations has a dramatic consequence for iron homeostasis, as it leads to its increase being proportional to the aluminum load of the tissues. Their results suggest that both ions may get into cells via similar transport pathways. Aluminum compounds, which interfere with zinc, copper, calcium, and chromium, change their bioavailability (Priest 2004).

The cellular mechanisms discussed above induce various clinical symptoms as a consequence of aluminum excessive content in the tissues and organs of homeothermic organisms. As is the case in mammals, aluminum is not an essential nutrient for birds either. In avian species, aluminum-caused disorders result from its impact on calcium and phosphorus metabolism (Llacuna et al. 1995). Aluminum binds to phosphorus present in the intestine, which creates insoluble complexes, and, in consequence, reduces the bioavailability of this important nutrient element (Sparling and Lowe 1996). Toxicity symptoms in birds are mainly associated with phosphorus deficiency, which leads to impaired growth, loss of appetite, lower laying rates, loss of plumage, and rickets. The morbidity symptoms are most apparent if the diet is low in phosphorus and if the level of aluminum exceeds dietary phosphorus by 50%. If the diet is enriched with phosphorus, the symptoms of noxious aluminum excess retreat (Scheuhammer 1991a, b). In addition, excess aluminum interferes with the formation of the calcareous layer of eggshells and causes bleeding in the uterine part of the oviduct (Nyholm 1981). Birds are generally vulnerable to aluminum poisoning due to its presence in the diet. As in mammals, low intestinal absorption of aluminum and its effective fecal excretion are the metal's toxicity limiting factors. Therefore, even waterbirds are to a lesser degree exposed to the toxic effects of aluminum as compared to fish, amphibians, or aquatic invertebrates. Bones, particularly in young birds, and female reproductive organs are most vulnerable to aluminum toxic activity. The metal also penetrates into the avian brain, kidneys, liver, and integumentary appendages (Sparling and Lowe 1996). Bortolotti and Barlow (1988) found that aluminum content in the feathers of birds of prey is positively correlated with its concentration in the diet.

## 5.2 *Toxicokinetics and Effects of Aluminum in Wildlife*

Aluminum is present in all the body fluids, including blood, cerebrospinal fluid, lymph, semen, and urine (Exley 2008; Exley and Mold 2015). The metal penetrates to the body mostly ingested with food. Its concentrations in the diet may vary greatly, with an average of 0.2 g $kg^{-1}$ dry weight (dw). Some plants, mainly those growing on acidic soils, are able to accumulate large amounts of aluminum, with tissue concentrations exceeding 1 g $kg^{-1}$ (Steinegger et al. 1990). The levels of aluminum in the diet of homeothermic vertebrates may range from 0.73 to 3656.7 mg $kg^{-1}$ (Sample et al. 1996).

Aluminum represents a particular case, as its bioavailability from the gastrointestinal tract is relatively low in endothermic organisms. On absorption, the metal is rapidly distributed to all tissues and then effectively removed by the excretory system; hence its concentration measured in biological samples does not in many cases reflect the load of the body with the bioavailable forms of the metal (Berthon 1996, 2002). An excellent example that illustrates this effect is the experiment by Jouhanneau et al. (1997), who used $^{26}Al$ isotope on rats. The isotope was detected in the bone within an hour from its oral administration and remained there for another

30 days. Contrary to the bone tissue, which is characterized by a relatively slow rate of ion exchange, the content of aluminum in parenchymal organs at retained homeostasis is more labile.

Absorption of aluminum from the gastrointestinal tract in homeothermic vertebrates strongly depends on the solubility of its particular forms (Harris 1996; Crisponi et al. 2012). Soluble aluminum compounds are absorbed much more efficiently than its insoluble substances (Berthon 1996, 2002). The routes that aluminum gets into the internal environment of the body lead mainly across the intercellular spaces, as well as through the inner space of the cell. Comparing to the former route, transcellular transport is slightly less important, due to the fact that aluminum ions may be accumulated inside cells, which represents a barrier preventing its further uptake (Exley and Mold 2015). The form of aluminum that penetrates into the body determines both the way of its transport with the bloodstream and the site of its deposition (Exley et al. 2007). Although aluminum is absorbed mainly within the intercellular spaces and transcellularly, there are also three other ways aluminum can break through biological barriers. These include active transport, membrane channel system, and endocytosis. Exley and Mold (2015) suggest that there are five main aluminum forms that are capable of getting to various parts of the body via these routes of transport. The most important here is free trivalent cation [$Al^{3+}{}_{(aq)}$], which is capable of crossing biological barriers being transported through the cell membranes and across intercellular spaces. Significant also are aluminum complexes, such as low molecular weight inert soluble complexes, LMW-$Al^{0}{}_{(aq)}$; high molecular weight inert soluble complexes, HMW-$Al^{0}{}_{(aq)}$; low molecular weight soluble non-inert complexes, LMW-$Al(L)^{x+/-}{}_{n(aq)}$; as well as nano- and microparticles, $Al(L)_{n(s)}$. After breaking through the intestinal barrier, aluminum reaches the bloodstream. In animals, its serum concentration was found to have increased immediately after administration (Lote et al. 1995; Glynn et al. 1999a, b, 2001; Swegert et al. 1999; Vanholder et al. 2002). It has been thought until recently that transferrin was the main protein transporting aluminum in the bloodstream, whereas albumins and ferritin seemed to be of little importance for the process (Harris 1996; Ward et al. 2001; Rengel 2004; Krewski et al. 2007; Sjörgen et al. 2007). Aluminum binds to iron-binding sites in the transferrin molecule; however, it does not substitute iron under normal physiological conditions. This is due to the fact that aluminum-transferrin bond strength is much, approximately 100 times, weaker than that of iron. Under normal physiological conditions, about 70% of serum transferrin is not bound with iron and remains "free" for aluminum, which is transported this way (Crisponi et al. 2012). It is accepted now that transferrin is not the only aluminum transporting protein, since the transferrin-$Al^{3+}{}_{(aq)}$ ion binding/dissociation kinetics do not explain its high rate of urinary excretion through the renal glomeruli. A key feature of transferrin is its high molecular weight, which prevents its glomerular filtration and, in consequence, prevents iron loss. Glomerular filtration barrier threshold in a healthy kidney ranges from 18 to 58 kDa, whereas the molecular weight of the transferrin-aluminum complex is considerably higher (Exley et al. 2007). About 10% of all plasma aluminum is bound to low molecular weight ligands, which include citrates, phosphates, and citrate-phosphate complexes.

Citrates are serum second to transferrin aluminum-binding ligands. They have been estimated to bind 7–8% of serum aluminum (Crisponi et al. 2012).

Exley and Mold (2015) presume that rapid urinary aluminum excretion is a result of its paracellular transport through the glomerular endothelium in the form of inert, low molecular weight complexes [LMW-$Al^{0}_{(aq)}$]. This effect has not been fully explained though and requires further research to describe both the specific transport mechanisms and the species of aluminum involved in the process.

Relevant literature lacks well-documented sources dealing with the presence of aluminum in the tissues and organs of wild animals. Most reports are on humans and laboratory animals; fewer deal with livestock animals, including ruminants. Authors, however, claim that there are analogies in the aluminum effects on the cellular metabolism that are common to all living organisms. Estimates are that an average daily aluminum intake by an adult human in the USA ranges from 8 to 9 mg in males and is about 7 mg in females. In Italy and the UK, a daily intake of this metal has been estimated at a level of, respectively, from 2.5 to 2.3–6.3 and 3.4 mg (Gramiccioni et al. 1996; Ysart et al. 2000). According to data published by the WHO (1996), an adult human ingests on average 5 mg aluminum with food and water per day, of which as little as 0.1% remains in the body, which is less than 10 μg. The rest is removed from the system (Priest et al. 1998). Acids present in the food enhance the solubility of the compounds of aluminum, which assumes its ionic form $Al^{3+}$; this species is characterized by a higher bioavailability in the initial part of the gastrointestinal tract, due to its easier transport through the epithelium. This is due to aluminum capability of inert complex formation (Desroches et al. 2000; Dayde et al. 2003). The acidic gastric environment plays the key role by modifying the amount of soluble aluminum species passing to the small intestine, where the absorption takes place (Powell et al. 1999). The alkaline environment of the intestine should prevent the absorption of aluminum; however, acids contained in the diet, such as glutamic, maleic, or tartaric acid, may serve as organic ligands to increase aluminum uptake. For this reason such ligands should be avoided if any aluminum salts are used for therapeutic purposes (Venturini and Berthon 2001). Studies carried out on laboratory rats revealed that absorption of aluminum from drinking water depended not only on the concentration of aluminum in the water but even more on the capability of binding aluminum by the components present in the stomach (Glynn et al. 2001). This information implies that aluminum bioavailability should not be determined based on its concentration in the food and water before ingestion due to ample changes that may occur in the digestive system. Tests using $^{26}Al$ isotope revealed that as little as 0.06–0.10% of ingested aluminum is absorbed (Moore et al. 2000). Experiments allowed determination that 90–95% of aluminum in the gut is found in its solid phase formed by insoluble and precipitating fractions of the digesta and endogenous material (sloughed epithelial cells of the intestine, bacteria colonizing the digestive tract and mucus). Eventually, most of soluble aluminum forms in the small intestine are bound in the mucus, which occurs in two different fractions. The outer layer is insoluble, adhering to the mucosal surface and forming a kind of "carpet" that lines the surface of the villi, while another layer consists of partially degraded mucus found in the lumen. Aluminum ions were detected in both fractions, however much more gathered in the insoluble, outer layer. It is responsible also for binding other metals, with particular affinity to trivalent ions, such as $Fe^{3+}$. It is suggested that intestinal mucus is the main factor limiting aluminum uptake. The relatively low fraction of aluminum passing to the bloodstream—in relation to its high concentration in the mucus lining the intestinal mucosa—may explain this effect (Powell et al. 1999; Berthon 2002). Moreover, some ligands, such as diet citrates, may compete for aluminum with the intestinal mucus. It was found that its absorbability may substantially increase due to the ease with which the resulting complex penetrates the intestinal barrier to reach the systemic circulation and, in consequence, other tissues and organs of the body (Williams 1996; Whitehead et al. 1997; Powell et al. 1999). Experiments show that the presence of such ligands results in doubled retention of aluminum in the liver and brain (Jouhanenau et al. 1997). In both humans and animals, aluminum citrate complex occurs in significant quantities in the blood plasma of healthy individuals and is the main compound which binds aluminum in the cerebrospinal fluid in dialysis patients with acute aluminum toxicosis. It represents also one of the main aluminum forms that penetrate cell membranes of the neurons (Polak et al. 2001; Yokel et al. 2002). It is believed that aluminum citrate may be nontoxic in human and animals cells, but it rather serves as a substance transporting the ions of the metal into the cells, where they interfere with metabolic pathways (Zatta and Zambenedetti 1996; Levesque et al. 2000; Zatta et al. 2002). There are also other ligands, such as polyphenols, phosphates, and silicates, which form insoluble aluminum complexes in the intestinal lumen, thus limiting the uptake of the metal (Powell et al. 1993; Powell and Thomson 1993; Berthon 1996, 2002).

In mammals, the highest concentrations of aluminum uptake are deposited in the nervous tissue (Al-Ganzoury and El-Shaer 2008; Anke et al. 2009), liver, kidneys, and bone (Tang et al. 1999; Al-Ganzoury and El-Shaer 2008; Anke et al. 2009; Teixeira et al. 2013).

Kidneys are most important for the process of aluminum removal from the body (Sutherland and Greger 1998; Berthon 2002). In the cells of the proximal renal tubules, aluminum is absorbed by lysosomes and precipitates in the form of insoluble aluminum phosphate, which is then transferred into the tubular lumen and removed with the urine (Steinegger et al. 1990). Experiments carried out on rodents with the use of $^{26}Al$ isotope revealed that 45–75% of intravenously administered aluminum is removed from the body within 24 h (Crisponi et al. 2012). It has been confirmed that the body of a healthy human in normal conditions is capable of removing nearly the entire dose of absorbed aluminum via the excretory system. Under heavy aluminum load, its renal excretion rate may be insufficient, and the tissue accumulation may induce cell metabolic disorders (Berthon 2002; Ezomo et al. 2009). Even in physiological conditions, some fraction of aluminum that has not been removed from the bloodstream through the kidneys will be accumulated in the tissues. For this reason, the plasma aluminum level only reflects the current situation and cannot be treated as an indicator of long-term exposure (Boyce et al. 1987). Comparing to other polyvalent ions, aluminum forms relatively weak bonds with serum components, which facilitates its transport and tissue deposition (Priest 2004).

Aluminum accumulates in the CNS, bone, liver, kidneys, integumentary appendages, and heart (Lote et al. 1995; Glynn et al. 1999b, 2001; Swegert et al. 1999; Vanholder et al. 2002). Molecular mechanisms inducing aluminum uptake in particular tissues differ significantly depending on their type (Crisponi et al. 2012). Therefore, it is not equally distributed to different tissues of the body. In humans, if alimentary exposure is the case, the highest fraction of total aluminum uptake is deposited in the bone (60%) and lungs (25%), and much lower percentage in the muscle (10%) and the liver (3%). The brain accounts for approx. 1% of this (Krewski et al. 2007). Probably the observed pattern of aluminum distribution among the organs may be associated with differences in transferrin receptor density (ATSDR 2008).

The presence of aluminum in the bone is of particular importance, since the tissue represents a specific reservoir which retains the metal for a long period of time. This is due to a long exchange cycle of bone structural components, which in mature individuals is about 3% per year, in the cortical bone, and 20%, in the cancellous bone. The biological half-life of bone-absorbed aluminum is estimated to last several decades (Priest 1990). Aluminum accumulation in the bone interferes its metabolism, which results in osteomalacia. Although molecular mechanisms responsible for aluminum deposition in the bone tissue have not been fully explained, the process is thought to depend on three basic effects. First, aluminum binds on the surface of the bone in exchange of calcium. Ions bound this way belong to the most mobile fraction, which is subject to reversed exchange by binding to complexes in the body fluids located near the bone surface and, next, with plasma transferrin. Another mechanism of binding aluminum to the bone consists in precipitation of aluminum together with calcium in the formation sites of bone mineral, where hydroxyapatite crystals are synthesized. This leads to deposition of aluminum relatively strongly bound with the bone tissue. The third way of aluminum deposition in the bone structures is in the form of ions bound to organic complexes. These ways of aluminum binding to bone tissue result in its presence in all the bone structures, mostly in the surface layers (Priest 2004).

Elevated levels of aluminum in the kidneys are mainly associated with disorders of their excretory function. Impaired renal function leads to deposition of aluminum, which in excess is toxic (Sanches-Iglesias et al. 2007). The renal medulla has a particularly strong capability of accumulating aluminum (Cacini and Yokel 1988).

Although a relatively low fraction of aluminum uptake finally reaches the nervous tissue, it induces a range of negative changes in the CNS, which leads to pathological conditions. Previous studies on the concentration of aluminum in the CNS did not bring much success in terms of clear indication of its toxicity threshold concentration in animals. In humans, the average aluminum cerebral concentration can be found in the range from 0.10 to 4.5 mg $kg^{-1}$ dw, with an apparent age-related increasing trend (Roider and Drasch 1999). Increased cerebral levels of aluminum are linked with such conditions as Alzheimer's disease (11.5 mg $g^{-1}$ dw), dialysis encephalopathy syndrome (14.1 mg $g^{-1}$ dw), and a variety of other aluminum-dependent encephalopathies (up to 47.4 mg $g^{-1}$ dw) (Exley and House 2011).

Experiments on rats reveal that gray matter concentrations of aluminum are higher (up to 40 mg $kg^{-1}$ dw) compared to those in the white matter (Roider and Drasch 1999; Exley and House 2011). This is probably due to the fact that the metal accumulates mainly in the somata of the neurons, particularly in the nuclei (Galassi et al. 1995; Reusche et al. 2001; Solomon et al. 2001; Shirabe et al. 2002; Exley and House 2011).

## 5.3 *Bioaccumulation of Aluminum in Mammalian and Avian Species*

Determination of aluminum concentrations in biological samples is prone to uncertainty, since its low uptake from the gastrointestinal tract on the one hand and efficient removal with urine on the other result in the fact that aluminum tissue concentrations are generally low. Additional bias may result from possible contamination of the sample, as various forms of aluminum are ubiquitous in the environment (Cannata-Andia and Diaz-Lopez 1990; Hewitt et al. 1990; Steinegger et al. 1990).

The specificity of aluminum biokinetics is that after entering the systemic circulation, the metal is quickly distributed among all organs and tissues (Exley 2008; Exley and Mold 2015). Bearing this in mind, it is important to decide which tissue reflects the true aluminum load of the body. Diagnostic tests on humans and domestic animals, including laboratory animals, most often involve sampling whole venous blood or serum. However, results of numerous analyses imply that this is not a proper material for aluminum measurement in mammalian bodies, due to the dynamics of its binding and removal. Although data on the aluminum tissue concentration in free-living wild animals are difficult to interpret, we may, to some extent, use for this purpose the results of experiments on laboratory animals or results of medical tests on humans. In the majority of patients suffering from dialysis encephalopathy syndrome, blood serum aluminum levels increase significantly in relation to healthy subjects, reaching values higher than 200 μg $L^{-1}$. There are reports, however, that these are much lower in patients with aluminum toxicosis and do not exceed 100 μg $L^{-1}$. It was found that measuring blood serum aluminum in 2- or 3-month intervals may obscure the outcomes of its excessive exposure resulting in its accumulation in body tissues (Cannata-Andia and Diaz-Lopez 1990). Data reported by various authors on blood serum aluminum in people not exposed to its elevated levels in the environment are usually similar and range within 5–7 μg $L^{-1}$ (Mussi et al. 1984; Schaller and Valentin 1984). As compared with human serum aluminum levels, those measured in the serum of cattle managed in Poland on organic and conventional farms are considerably higher, 277 and 1567 μg $L^{-1}$, respectively (Tomza-Marciniak et al. 2011). In the wild Iberian ibex (*Capra pyrenaica*), whole blood aluminum concentrations ranged between 310 and 390 μg $L^{-1}$ and were similar to bovine plasma levels in the cattle from organic farms but much higher than in human plasma (Ráez-Bravo et al. 2016). This information suggests that herbivorous animals ingest much more aluminum with the fodder than do humans who consume diversified diets.

Ecotoxicological studies on wildlife and domesticated ruminants (Table 12.3) usually involve aluminum determination in the parenchymal organs, such as the liver, kidney, skeletal muscles, and bones, and to a lesser extent in integumentary appendages, such as hair and feathers. Aluminum concentrations in the liver and kidneys were generally found not to reflect elevated environmental levels to which healthy individuals are exposed, though their renal function status plays a significant role in this effect (Scheuhammer 1987; Lucia et al. 2010). Most data on tissue content of aluminum has been collected for ruminants; in the bovine liver, the normal concentration range and the threshold aluminum toxicity values have been established. The normal and toxic values of hepatic aluminum concentrations were estimated at <1–5 and 6.3–11 mg $kg^{-1}$ dw, respectively (WVDL 2015). Hepatic aluminum concentrations in various ruminants may range widely, from 0.83 to 104.3 mg $kg^{-1}$ dw, and exceeds 14 mg $kg^{-1}$ dw in most of farm animals (Al-Ganzoury and El-Shaer 2008; Anke et al. 2009; Gamberg et al. 2016). Typical aluminum levels from 1.53 to 3.6 mg $kg^{-1}$ dw were observed in wild cervids in the USA and Poland (Zimmerman et al. 2008; Długaszek and Kopczyński 2011). Hepatic aluminum levels in wild boars, measured by Kucharczak and Moryl (2012), ranged between 3.37 and 10.67 mg $kg^{-1}$ dw.

Aluminum levels in the bone tissue of ruminants range between 0.36 and 73 mg $kg^{-1}$ dw (Tang et al. 1999; Anke et al. 2009), though much higher values are found in the nervous system, from 48 to 301 mg $kg^{-1}$ dw (Al-Ganzoury and El-Shaer 2008). Anke et al. (2009) report a reversed pattern in the same tissues in lagomorphs; there was 56 ± 20 mg $kg^{-1}$ dw in the nervous tissue and 86 ± 30 mg $kg^{-1}$ dw in the bone of the European hare. A high content of aluminum, at a level of 87 mg $kg^{-1}$ dw, has also been found in the bone of the domestic cat, *Felis catus* (Anke et al. 2001, 2009).

In birds, aluminum significantly affects the quality of the eggshell; hence its analysis is often used for environmental evaluation of exposure to various forms of the metal (Drent and Woldendorp 1989; Oelke 1989; Eeva and Lehikoinen 1995; Miljeteig et al. 2012). The highest aluminum levels in free-living birds were measured in their feathers, where it may vary in a wide range, from 2.25 to 328 mg $kg^{-1}$ dw (Lucia et al. 2010). High concentrations, at a level of 74.49 mg $kg^{-1}$ dw, were also found in avian bone tissue (Llacuna et al. 1995).

Anke et al. (2001, 2009) also found high aluminum concentrations in chickens (*Gallus gallus domesticus*), namely, 50.0 and 69.0 mg $kg^{-1}$ dw, in the nervous and bone tissues, respectively. In terms of analysis of tissue aluminum content, especially in highly mineralized biological materials, a comparative analysis of its concentration in relation to age and sex seems reasonable. Research studies carried out so far have been inconclusive. Some reports suggest that aluminum accumulation in animal

**Table 12.3** Aluminum concentrations in wild and domestic ruminants

| Species | Country | Liver | Kidney | Muscle | Brain | Bone | References |
|---|---|---|---|---|---|---|---|
| Cattle<br>*Bos taurus taurus* | | Normal value:<br><1–5 ww or<br><3.3–16.7 dw | | | | | WVDL (2015) |
| | | Toxic value:<br>6.3–11 ww<br>or 21–36.7 dw | | | | | |
| Cattle | USA | | | | | 0.66 dw | Tang et al. (1999) |
| Cattle | Germany | 32 dw | 32 dw | 20 dw | 48 dw | 68 dw | Anke et al. (2009) |
| Cattle | Sweden | 0.068 ± 0.047 ww | 0.063 ± 0.029 ww | 0.05 ± 0.006 ww | | | Jorhem et al. (1989) |
| | | (0.226 ± 0.16 dw) | (0.25 ± 0.12 dw) | (0.2 ± 0.024 dw) | | | |
| Cattle | Brazil | 70.2 dw | 9.63 dw | 9.69 dw | | 4.57 dw | Teixeira et al. (2013) |
| Cattle | Egypt | 14.4 ww | 10.5 ww | 8.4 ww | 33.1 ww | | Al-Ganzoury and El-Shaer (2008) |
| | | (48 dw) | (42 dw) | (33.6 dw) | (165.5 dw) | | |
| Buffalo<br>*Bison bison* | Egypt | 15.6 ww | 11.0 ww | 5.7 ww | 31.0 ww | | Al-Ganzoury and El-Shaer (2008) |
| | | (52 dw) | (44dw) | (22.8 dw) | (155 dw) | | |
| Sheep<br>*Ovis aries* | Germany | 37 dw | 43 dw | | 61 dw | 59 dw | Anke et al. (2009) |
| Sheep | Egypt | 25.7 ww | 20.6 ww | 5.9 ww | 50.6 ww | | Al-Ganzoury and El-Shaer (2008) |
| | | (85.7 dw) | (82.4 dw) | (23.6 dw) | (253 dw) | | |
| Goat<br>*Capra aegagrus hircus* | Egypt | 26.3 ww | 19.5 ww | 7.9 ww | 53.0 ww | | Al-Ganzoury and El-Shaer (2008) |
| | | (87.7 dw) | (78 dw) | (31.6 dw) | (265 dw) | | |
| Goat | USA | | | | | 0.36 dw | Tang et al. (1999) |
| Camel<br>*Camelus* sp. | Egypt | 31.3 ww | 21.2 ww | 8.1 ww | 60.2 ww | | Al-Ganzoury and El-Shaer (2008) |
| | | (104.3 dw) | (84.8 dw) | (32.4 dw) | (301 dw) | | |
| Fallow deer<br>*Dama dama* | Germany | 30 dw | 34 dw | | 56 dw | 73 dw | Anke et al. (2009) |

(continued)

**Table 12.3** (continued)

| Species | Country | Liver | Kidney | Muscle | Brain | Bone | References |
|---|---|---|---|---|---|---|---|
| Roe deer *Capreolus capreolus* | Poland | 1.08 ww (3.6 dw) | | | | | Długaszek and Kopczyński 2011 |
| Roe deer | Poland | | | 0. 58 ww (2.32 dw) | | | Długaszek and Kopczyński (2013) |
| Roe deer | Poland | Herd 1: 0.41 ww (1.37 dw) | Herd 1: 0.96 ww (3.84 dw) | Herd 1: 1.67 ww (6.68 dw) | | | Kucharczak et al. (2005) |
| | | Herd 2: 1.21 ww (4.27 dw) | Herd 2: 1.92 ww (7.68 dw) | Herd 2: 7.63 ww (30.52 dw) | | | |
| Roe deer | Poland | 0.59 ± 0.32 ww (1.97 ± 1.07 dw) | 0.77 ± 0.12 ww (3.08 dw) | | | | Kucharczak and Moryl (2012) |
| Roe deer | Slovakia | 0.74 ww (2.47 dw) | 0.65 ww (dw) | 1.29 ww (5.16 dw) | | | Bíres et al. (1992) |
| Red deer *Cervus elaphus* | Slovakia | 1.18 ww (3.93 dw) | 0.82 ww (dw) | 0.92 ww (3.68 dw) | | | Bíres et al. 1992 |
| White-tailed deer *Odocoileus virginianus* | USA, Texas | <2 dw | | 4.1 dw | | | Bruckwicki et al. (2006) |
| White-tailed deer | USA, South Dakota | 0.46 ww (1.53 dw) | | | | | Zimmerman et al. (2008) |
| Mule deer *Odocoileus hemionus* | USA, South Dakota | 1.41 ww (4.7 dw) | | | | | Zimmerman et al. (2008) |
| Caribou *Rangifer tarandus* | Greenland | Herd 1: 2.2 dw | Herd 1: 5.34 dw | Herd 1: 0.79 dw | | | Gamberg et al. (2016) |
| | | Herd 2: 0.83 dw | Herd 2: 0.87 dw | Herd 2: 0.20 dw | | | |

mg $kg^{-1}$ dry weight, dw; wet weight, ww, was converted to dry weight using the following % moistures: kidney 75%, liver 70%, muscle 75%, brain 80%

tissues increases with age (Yokel and Golub 1997), but others reveal no effect of sex or age on the concentration of aluminum in the bodies of homeothermic vertebrates (Lucia et al. 2010).

## 5.4 *Ecological Effects of Aluminum*

In the natural environment, aluminum occurs mainly as a component of aluminosilicate minerals, and only its minor amounts are present in a soluble form, which are bioavailable for living organisms (May and Nordstrom 1991). These soluble compounds rapidly increase in quantity along with environment acidity, which is a result of unbalanced nitrogen, sulfur, and carbon cycles (Bolan and Hedley 2003; Tang and Rengel 2003). Apart from these factors, environmental acidification is caused by massive uptake of cations, in relation to anions, and the widespread presence of amides- and ammonia-containing fertilizers (Mahler et al. 1985). Over the last decades, acidic precipitations have been the most serious problem, significantly changing aluminum ions mobility in various environments and, in consequence, changing its bioavailability (Graveland 1990; Ormerod and Wade 1990; Alewell 2003). The phenomenon was first described in the nineteenth century, and the expression "acid rain" was first used in 1852 (Graveland 1990). Intensive research on its effect on aluminum mobility in the ecosystems started in the 1950s, with a majority of reports focusing on aquatic organisms. In subsequent years, more attention was also paid to avian species, which—in response to increasing acidity of the environment—were more exposed to toxic metals, experienced negative changes in food availability and quality, and had to face radical changes in their habitat (Peakall 2000). Aluminum has a significant effect on aquatic organisms, not only under environmental acidification. Freshwater bivalves exhibit severe filtration abnormalities in response to both short-term (1 h) and long-term (15 days) exposure to aluminum at a neutral pH (Kadar et al. 2002). The metal was also found to accumulate in freshwater snails, including the great pond snail, *Lymnaea stagnalis*, at pH 7, which led to sublethal changes in their behavior (Truscott et al. 1995; Elangovan et al. 2000). In this species, aluminum induced changes in the electrical activity of nerve cells, which suggests that the metal interferes with the ion channel electrophysiology in cell membranes (Csoti et al. 2001). This information contradicted the previous opinion that the metal is relatively poorly available in aquatic environments with a pH 6–8, due to its insolubility (Driscoll and Schecher 1989). Acidification of surface waters increases bioavailability of aluminum, but also of other toxic metals, including cadmium. This leads to a simultaneous drop in calcium bioavailability, which hits populations of aquatic invertebrates—food organisms of numerous vertebrate species (Peakall 2000). Markich et al. (2002) found that aluminum becomes toxic to benthic invertebrates at concentrations of its inorganic forms between 0.1 and 0.3 mg $L^{-1}$. A high level of soluble, toxic aluminum forms observed in surface waters may also be related to the activity of water treatment plants, which use aluminum-containing chemicals as coagulants

(Steinegger et al. 1990). An elevated concentration of aluminum is toxic to fish, whose increased mortality in consequence reduces populations of piscivorous birds (Ormerod and Wade 1990). The gills in fish were found to be most susceptible to aluminum toxic exposure, which causes ion exchange problems in this piscine organ. Aluminum also causes increased secretion of mucus, which hinders gas exchange and causes gill hyperplasia (Dietrich and Schlatter 1989). Reductions in fish populations, which results from water acidification and acidity-related increased aluminum uptake, affect the composition of vertebrates living around the body of water. Although occupying land habitats, these animals rely on the aquatic ecosystems. The group of homeothermic vertebrates includes numerous avian species, such as sandpipers (*Scolopacidae*), dippers (*Cinclidae*), tyrant flycatchers (*Tyrannidae*), wood-warblers (*Parulidae*), and wagtails (*Motacillidae*). In some circumstances, we can observe a growth in a population of insectivorous birds that do not need to compete for food with insect-eating fish (Ormerod and Wade 1990). In conclusion, increased bioavailability of aluminum presumably negatively affects the birds of the terrestrial ecosystems. Graveland et al. (1994), who studied birds living on strongly acidified soils, observed frequent abnormalities in the shell quality of their eggs. The shells were darker, more porous and thinner, which had a consequence in poorer reproduction.

Adverse changes resulting from ecosystem acidification also affect semiaquatic mammals, which nearly entirely feed on aquatic organisms. In Europe, the Eurasian otter (*Lutra lutra*) and water shrew (*Neomys fodiens*) are examples of such mammals. The impact of toxic aluminum forms killing fish and aquatic invertebrates may lead to food shortages affecting these species (Ormerod and Wade 1990). In terms of other terrestrial species, toxicity of aluminum is thought to be relatively low, as compared to other metals and to animals inhabiting aquatic ecosystems (Ormerod and Wade 1990; Steinegger et al. 1990; Roseland et al. 1990). However, a comparative study on two ruminant species revealed that an acidified habitat contributes to elevated levels of tissue aluminum, which can be observed in individuals of the same species living in a non-acidified area, as shown in Table 12.3 (Anke et al. 2009).

Pursuing pro-environmental policies has fortunately reduced the problem of acid rains in Europe and the USA; however, an increase in the pH of stream waters in many recovering areas is not as quick as expected (Menz and Seip 2004; Lajtha and Jones 2013). Contrary to Europe and North America, environmental acidification is increasing in many parts of Asia, especially in China and Russia, due to the intensive development of industries relying on coal-based energy sources and due to inefficient removal of sulfur and nitrogen oxides from combustion exhaust gases (Lu et al. 2010; Bhargava and Bhargava 2013). These oxides (key factors of acid rains) may move over long distances with masses of air, even between continents. Therefore, on a global scale, environmental acidification will long remain a serious ecological problem, which increases the bioavailability of aluminum and other toxic metals, harmful for both humans and ecosystems (Rosseland et al. 1990; Sparling and Lowe 1996; Bhargava and Bhargava 2013). Moreover, the twenty-first century seems to bring another aluminum-related issue in the form of nanoparticles; their impact on living organisms and whole ecosystems is largely a mystery (Burklew et al. 2012; Zhang et al. 2015; Chifiriuc et al. 2016) (Table 12.4).

**Table 12.4** Influence of acid rain on aluminum concentrations in tissues of free-living and domestic ruminants in Germany

| Tissue | Acidic rain | | Statistical significance of difference |
|---|---|---|---|
| | With | Without | |
| Wild deer: fallow deer (*Dama dama*) and roe deer (*Capreolus capreolus*) | | | |
| Bone (rib) | 79 ± 20 | 73 ± 34 | NS |
| Brain | 68 ± 16 | 56 ± 20 | NS |
| Kidney | 50 ± 12 | 34 ± 20 | $p < 0.01$ |
| Liver | 49 ± 7.1 | 30 ± 18 | $p < 0.01$ |
| Cattle (*Bos taurus taurus*) | | | |
| Bone (rib) | 73 ± 22 | 68 ± 15 | NS |
| Brain | 69 ± 15 | 48 ± 20 | $p < 0.05$ |
| Kidney | 39 ± 24 | 32 ± 13 | NS |
| Liver | 44 ± 16 | 32 ± 13 | $p < 0.05$ |

mg $kg^{-1}$ dry weight; AM ± SD, arithmetic mean ± standard deviation
*NS* nonsignificant
Data source: Anke et al. (2009)

## 5.5 *Bioindicators and Biomarkers for Aluminum in Ecotoxicological Studies*

Selecting terrestrial homeothermic vertebrate species that could be of use as aluminum ecosystem contamination bioindicators is not an easy task. This is partly due to scarce literature sources dealing with aluminum in the wildlife but also due to a small number of organisms in which the level of the element has been established. This is largely associated with a low intestinal uptake of aluminum and its consequent low toxicity to animals but also with a highly varied impact of particular aluminum forms on the metabolic processes in mammals and birds. Aluminum concentrations have been analyzed in various species of these two vertebrate groups; however, ecotoxicologists present different opinions as to whether the animals are of value in terms of environmental biomonitoring of terrestrial ecosystems.

Studies carried out so far indicate that birds may be good bioindicators of aluminum in land ecosystems. Namely, birds—as compared to mammals—are much more intensively exposed to toxic effects of the metal, which they consume with food in much higher amounts. Up till now, mainly herbivorous mammals have been designated as biomonitors of the presence of elevated levels of bioavailable aluminum in terrestrial ecosystems. The most common species include cervids, widely distributed in Eurasia and North America, such as roe deer, fallow deer, white-tailed deer, or mule deer.

Other free-living mammals (Table 12.5), such as fallow deer, reindeer, wild boar, or hare, can also serve as biomonitors (Wren 1986; Bíres et al. 1992; Godal et al. 1995; Kålås et al. 1995; Kucharczak et al. 2005; Kośla et al. 2006; Anke et al. 2001, 2009; Długaszek and Kopczyński 2014). Within small mammals, grey red-backed vole and common shrew are used as bioindicators of aluminum environmental

**Table 12.5** Aluminum concentrations in mammalian tissues

| Species | Localization | Mean | Liver | Kidney | Muscle | Hair | References |
|---|---|---|---|---|---|---|---|
| Domestic mouse *Mus musculus f. dom.* | Poland Control group in lab study | AM ± SD | 0.07 ± 0.02 ww<br>(0.23 ± 0.07dw) | 0.18 ± 0.09 ww<br>(0.72 ± 0.36 dw) | | | Długaszek and Kopczyński (2000) |
| Grey-sided vole *Clethrionomys rufocanus* | Norway Svanvik | Med | ad 1.19 dw | | | | Kålås et al. (1995) |
| Mountain hare *Lepus timidus* | Norway Jarfjord | Med | juv 1.15 dw<br>ad 4.37dw | juv 0.76 dw<br>ad 1.55 dw | | | Kålås et al. (1995) |
| Mountain hare | Reference area | Med | juv 2.35 dw<br>ad 4.45 dw | juv 1.39 dw<br>ad 1.27 dw | | | Kålås et al. (1995) |
| Snowshoe hare *Lepus americanus* | Northwest, Canada | AM ± SE | | 1.80 ± 0.28 dw | | | Poole et al. (1998) |
| European hare *Lepus europaeus* | Germany | AM ± SD | 16 ± 30 dw | 33 ± 13 dw | 16 ± 6 dw | 35 ± 10 dw | Anke et al. (2009) |
| European hare | Poland, Wroclaw | AM ± SD | 1.58 ± 0.62 ww<br>(5.27 ± 2.07 dw) | 0.71 ± 0.21w<br>(2.84 ± 0.84 dw) | | | Kucharczak and Moryl (2012) |
| European hare | Poland | Med | 1.10 ww<br>(3.67 dw) | | | | Długaszek and Kopczyński (2011) |
| European hare | Poland | Med | | | 1.26 ww<br>(7.04 dw) | | Długaszek and Kopczyński (2013) |
| European hare | Poland | Med | | | | 0.34 dw | Długaszek and Kopczyński (2014) |

| | | | | | | | |
|---|---|---|---|---|---|---|---|
| Wild boar *Sus scrofa* | Poland Wroclaw | AM ± SD | 3.20 ± 2.40 ww | 3.10 ± 1.28 ww | 2.18 ± 0.84 ww | 350.0 ± 66.5 dw | Kucharczak et al. (2005) |
| | | | (10.67 ± 8 dw) | (12.4 ± 5.12 dw) | (8.72 ± 3.36 dw) | | |
| | Piaseczno | | 1.01 ± 0.27 ww | 1.44 ± 0.72 ww | 2.74 ± 1.36 ww | 201.1 ± 29.7 dw | |
| | | | (3.37 ± 0.9 dw) | (5.76 ± 2.88 dw) | (10.96 ± 5.44 dw) | | |
| Wild boar | Poland Bogatynia | AM ± SD | 0.83 ± 0.22 ww | 1.19 ± 0.63 ww | | | Kucharczak and Moryl (2012) |
| | | | (2.77 ± 0.73 dw) | (4.76 ± 2.52 dw) | | | |
| Wild boar | Poland | Med | 1.02 ww | | | | Długaszek and Kopczyński (2011) |
| | | | (3.4 dw) | | | | |
| Wild boar | Poland | Med | | | 0.54 ww | | Długaszek and Kopczyński (2013) |
| | | | | | (2.16 dw) | | |
| Wild boar | Poland | Med | | | | 1.53 dw | Długaszek and Kopczyński (2014) |
| Pig *Sus scrofa f. dom.* | Sweden | AM ± SD | 0.028 ± 0.011 ww | 0.024 ± 0.014 ww | 0.032 ± 0.011 ww | | Jorhem et al. (1989) |
| | | | (0.09 ± 0.04 dw) | (0.10 ± 0.06 dw) | (0.12 ± 0.04 dw) | | |
| Pig | Germany | AM ± SD | 18.0 ± 10 dw | 42.0 ± 24dw | 12.0 ± 47 dw | 39.0 ± 25 dw | Anke et al. (2001, 2009) |
| Mink *Mustela vision* | USA, Illinois | AM | 4.52 ww | 6.51 ww | 7.85 ww | | Halbrook et al. (1996) |
| | | | (15.07 dw) | (26.04 dw) | (31.4 dw) | | |
| American mink *Neovison vison* | Northwest, Canada | AM ± SE | | 8.41 ± 0.75 dw | | | Poole et al. (1995) |
| American mink | Northwest, Canada | AM ± SE | | 5.63 ± 0.76 dw | | | Poole et al. (1998) |
| Northern red-backed voles *Myodes rutilus* | Northwest, Canada | AM ± SE | 25.35 ± 3.55 dw | | | | Poole et al. (1998) |
| River otter *Lontra canadensis* | USA, Illinois | AM ± SD | | | 2.46 ± 1.02 ww | | Halbrook et al. (1996) |
| | | | | | (9.84 ± 4.08 dw) | | |

(continued)

**Table 12.5** (continued)

| Species | Localization | Mean | Liver | Kidney | Muscle | Hair | References |
|---|---|---|---|---|---|---|---|
| River otter *Lontra canadensis* | Canada, British Columbia Upper Fraser River | AM ± SE | 3.33 ± 0.99 dw | | | | Harding et al. (1998) |
| Wild mink *Mustela vison* | Canada, British Columbia | AM ± SE | | | | | Harding et al. (1998) |
| | Upper Fraser River | | 3.50 ± 0.47 dw | | | | |
| | Lower Fraser River | | 3.63 ± 0.91 dw | 7.18 ± 1.75 dw | | | |
| Marten *Martes americana* | Canada, British Columbia | AM ± SD | | 11.3 ± 5.5 dw | | | Harding (2004) |
| Marten | Northwest, Canada | AM ± SE | | 3.51 ± 0.47 dw | | | Poole et al. (1995) |
| Wolverine *Gulo luscus* | Canada, British Columbia | AM ± SD | 11.0 ± 11dw | | | | Harding (2004) |
| Dog, Yorkshire terriers *Canis lupus dom.* | Poland | AM ± SD | | | | 93.8 ± 72.81 dw | Kośla and Skibniewska (2010) |
| Cat *Felis catus* | Poland | AM ± SD | | | | Free living 38.0 ± 32 dw | Kośla et al. (2004) |
| | | | | | | Homebred 20.8 ± 20.5 dw | |
| Cat | Germany | AM ± SD | 21.0 ± 16 dw | 24.0 ± 16 dw | | 44.0 ± 22 dw | Anke et al. (2001, 2009) |

mg $kg^{-1}$ dw, dry weight; ww, wet weight
*AM* arithmetic mean, *SE* standard error, *SD* standard deviation, *Med* median, *ad* adult, *juv* juvenile

exposure (Kålås et al. 1995). Avian species (Table 12.6) used as environmental aluminum bioindicators include greylag goose, wood grouse, red knot, grey plover, great tit, yellowhammer, rock bunting, or common blackbird (Kålås et al. 1995; Llacuna et al. 1995; Lucia et al. 2010).

The small amount of data on the concentrations of Al in the wildlife collected so far, as well as the progressive environmental changes, imply that this list remains open and will be gradually updated with new species, perhaps more suitable for this type of biomonitoring.

Parenchymal organs, bones, and skin appendages represent the main biomarkers of bioavailable aluminum forms in the bodies of homeothermic vertebrates (Anke et al. 2001, 2009; Kucharczak et al. 2005). The important point is the degree in which they reflect the short- or long-term exposure. Blood, plasma, or urine of wild animals are rarely used for analysis, primarily because the sampling process is usually cumbersome and, also, because the fluids reflect the short-term exposure only (Kowalczyk et al. 2004; Exley 2013).

Studies on birds confirm that bone is the most susceptible tissue in terms of aluminum accumulation, particularly in young individuals and breeding females. Further tissues mentioned are the brain, liver, and kidney (Sparling and Lowe 1996). Numerous studies suggest that arguably the most appropriate biomarkers of long-term aluminum exposure in birds and mammals are keratin appendages of the integumentary system, mainly hair and feathers. The process of their formation consists in transformation of living epithelial cells into dead, cornified, and stable structures showing no metabolic activity; thus the metal ions remain "trapped" within for a long period of time and the lack of metabolic activity prevents their removal (Tomlinson et al. 2004). The metal contamination contained in the feathers or hair may be of either endo- or exogenous origin. Many metals accumulate in the structures during food and water ingestion, though only during their growth; after the completion of growth, various exogenous contaminants from the air, water, or soil accumulate on the surface of feathers and hair. These are usually removed before analysis. Studies on mammals have usually focused on coat samples (Table 12.7), which proved to be a good marker of the effects of long-term exposure to some essential or toxic metals (Anke and Risch 1979; Yokel 1982; Kośla et al. 2004, 2006; Kośla and Skibniewska 2010; Skibniewska et al. 2011).

Birds' feathers have similar properties. Concentrations of endogenous aluminum in feathers vary greatly, ranging from a few to several hundred mg $kg^{-1}$ dw (Bortolotti and Barlow 1988; Lewis and Furness 1991; Monteiro 1996; Bond and Lavers 2011; Lodenius and Solonen 2013; Borghesi et al. 2016). Moreover, in contrast to blood plasma—where metal concentrations change rapidly in response to changes in the physiological status of the body—keratin structures respond to such changes much slower (allowing analysis within at least several weeks before sampling); hence they seem to be a good biomarker of long-term exposure (Kośla et al. 2011; Stachurska et al. 2011).

**Table 12.6** Aluminum concentrations

| Species | Localization | Mean | Liver | Kidney | Muscle | Feathers | References |
|---|---|---|---|---|---|---|---|
| Western capercaillie *Tetrao urogallus* | Norway Pasvik | Med | juv 1.13 dw<br>ad 0.20 dw | juv 3.10 dw<br>ad 0.86 dw | | | Kålås et al. (1995) |
| Willow ptarmigan *Lagopus lagopus* | Norway Jarfjord Pasvik Reference area | Med | ad 0.82 dw<br>ad 0.38 dw<br>ad 0.20 dw | ad 1.86 dw<br>ad 0.93 dw<br>ad 0.73 dw | | | Kålås et al. (1995) |
| Greylag goose *Anser anser* | France | AM ± SE | 11.8 ± 12.7 dw | 6.1 ± 3.0 dw | 11.6 ± 2.4 dw | 226 ± 515 dw | Lucia et al. (2010) |
| Red knot *Calidris canutus* | France | AM ± SE | 3.2 ± 0.2 dw | 8.9 ± 6.1 dw | 2.5 ± 0.4 dw | 107 ± 50.1 dw | Lucia et al. (2010) |
| Grey plover *Pluvialis squatarola* | France | AM ± SE | 4.2 ± 1.9 dw | 3.4 ± 1.5 dw | 7.7 ± 11.3 dw | 2.5 ± 3.8 dw | Lucia et al. (2010) |
| Great tit *Parus major* | North Spain | AM ± SD | 21.95 ± 35.0 dw | 8.47 ± 6.9 dw | 12.71 ± 8.61 dw | 170.16 ± 68.8 dw | Llacuna et al. (1995) |
| Rock bunting *Emberiza cia* | North Spain | AM ± SD | 9.88 ± 8.2 dw | 7.88 ± 7.68 dw | 9.63 ± 5.15 dw | 328.2 ± 236.7 dw | Llacuna et al. (1995) |
| Blackbird *Turdus merula* | North Spain | AM ± SD | 7.71 ± 7.37 dw | 1.46 ± 1.05 dw | 1.88 ± 2.5 dw | 113.3 ± 69.7 dw | Llacuna et al. (1995) |
| Bald eagle *Haliaeetus leucocephalus* | USA, Maine | AM ± SD | 3.98 ± 3.19 dw | | | | Mierzykowski et al. (2011) |
| Great cormorant *Phalacrocorax carbo* | Serbia | AM ± SD | | | | juv 66.19 ± 79.88 dw<br>ad 65.94 ± 31.77 dw | Skoric et al. (2012) |
| Hen *Gallus gallus dom.* | Germany | AM ± SD | 42.0 ± 22 dw | | 6.40 ± 25 dw | 38.0 ± 14 dw | Anke et al. (2001, 2009) |

mg $kg^{-1}$ dw, dry weight; ww, wet weight in avian tissues
*AM* arithmetic mean, *SE* standard error, *SD* standard deviation, *Med* median, *juv* juvenile, *ad* adult

**Table 12.7** Aluminum concentrations in the hair of ruminants

| Species | Country | Aluminum ($mg\ kg^{-1}$ dw) | References |
|---|---|---|---|
| Roe deer *Capreolus capreolus* | Poland | 0.6 ± 1.1 | Długaszek and Kopczyński (2014) |
| | | 26.33 ± 3.3 | Kucharczak et al. (2005) |
| European bison *Bison bonasus* | Poland | 59.6 ± 63.9 | Kośla et al. (2006) |
| Cattle *Bos taurus taurus* | Germany | 19.0 ± 62 | Anke et al. (2001, 2009) |

# 6 Conclusion

- Aluminum is the third most abundant element present in the Earth's crust.
- The most common oxidation state of the element is +3, very rarely +1 or +2.
- Metallic aluminum oxidizes in the air, undergoing the process of passivation.
- In the natural environment, aluminum is found in various forms, depending on pH, Eh potential, and the presence of inorganic and organic ligands.
- Minerals containing aluminum include bauxite, cryolite, kaolinite, corundum, albite, orthoclase, gibbsite, and many others.
- Soil content of aluminum varies considerably and depends on the bedrock and the type of soil.
- Despite its abundance and ubiquity, the element has no significant biological function in either animal or human bodies.
- Aluminum is commonly found in plants with its concentrations depending on soil pH.
- In animals, the element occurs in trace quantities in the bone, lungs, and soft tissues.
- The element gets into the tissues of homeothermic vertebrates mainly through alimentary route, followed by aerogenic intake and, to a lesser extent, transdermal absorption.
- Human body may be exposed to aluminum through dialysis fluids or some medications, such as gastric acid suppressors, analgesics, or anti-inflammatory drugs.
- Aluminum nanoparticles, a common aircraft fuel additive, have recently added to aluminum contamination of both aquatic and terrestrial ecosystems.
- Aluminum toxicity to animals representing various taxonomic groups involves oxidative stress resulting in increased cell lethality.
- In homeothermic vertebrates, the toxic aluminum effects are manifested mainly within the central nervous system and in bone tissue and kidneys.
- Aluminum ore mining and processing within the metallic form production pose a threat to the natural environment.

# References

Ahn SJ, Sivaguru M, Osawa H, Chung GC, Matsumoto H (2001) Aluminum inhibits the $H^+$-ATPase activity by permanently altering the plasma membrane surface potentials in squash roots. Plant Physiol 126:1381–1390

Ahn SJ, Sivaguru M, Chung GC, Rengel Z, Matsumoto H (2002) Aluminum-induced growth inhibition is associated with impaired efflux and influx of $H^+$ across the plasma membrane in root apices of squash (*Cucurbita pepo*). J Exp Bot 53:1959–1966

Ahn SJ, Rengel Z, Matsumoto H (2004) Aluminum-induced plasma membrane surface potential and $H^+$–ATPase activity in near isogenic wheat lines differing in tolerance to aluminum. New Phytol 162:71–79

Alewell C (2003) Acid input into the soil from acid rain. In: Rengel Z (ed) Handbook of soil acidity. Marcel Dekker, New York, pp 83–115

Alexopoulos E, McCrohan CR, Powell JJ, Jugdaohsingh R, White KN (2003) Bioavailability and toxicity of freshly neutralised aluminum to the freshwater crayfish *Pacifastacus leniusculus*. Arch Environ Contam Toxicol 45:509–514

Al-Ganzoury HH, El-Shaer ME (2008) Aluminum residues in meat and edible tissues of some ruminant and its relation to public health in Sharkia Governorate. SCVMJ 13:361–366

Anke M, Risch M (1979) Haaranalyse und Spurenelementstatus. VEB Gustav Fischer Verlag, Jena, pp 23–26 and 41–43

Anke M, Müller M, Müller R, Schäfer U, Angelow L (2001) The biological and toxicological importance of aluminum in the environment and food chain of animals and humans. In: Ermidou-Pollet S, Pollet S (eds) 3rd international symposium on trace elements in human: new perspectives. Proceedings Book, Athens/Greece, pp 230–247

Anke M, Müller M, Hoppe C (2005) Recept progress in exploring the essentiality of the ultratrace element aluminum to the nutrition of animal and man. Biomed Res Trace Elem 16:183–187

Anke M, Müller M, Müller R, Schäfer U, Zerull J, Schafer U (2009) The biological importance of aluminum in the food chain of animals and man: intake, apparent absorption rate, balance and limiting concentrations. Trace Elem Med 10:1–16

Appanna VD, St Pierre M (1994) Influence of phosphate on aluminum tolerance in *Pseudomonas fluorescens*. FEMS Microbiol Lett 124:327–332

Appanna VD, Kepes M, Rochon P (1994) Aluminum tolerance in *Pseudomonas fluorescens* ATCC 13525: involvement of a gelatinous lipid-rich residue. FEMS Microbiol Lett 119:295–302

ATSDR (2008) Toxicological profile for aluminum. Atlanta, GA, United States Department of Health and Human Services, Public Health Service, Agency for Toxic Substances and Disease Registry

Banasik A, Lankoff A, Piskulak A, Adamowska K, Lisowska H, Wójcik A (2013) Aluminum-induced micronuclei and apoptosis in human peripheral-blood lymphocytes treated during different phases of the cell cycle. Environ Toxicol 20:402–406

Barabasz W, Albińska D, Jaśkowska M, Lipiec J (2002) Ecotoxicology of aluminum. Pol J Environ Stud 11:199–203

Barcelo J, Poschenrieder C (2002) Fast root growth responses, root exudates, and internal detoxification as clues to the mechanisms of aluminum toxicity and resistance: a review. Environ Exp Bot 48:75–92

Berg A, Banwart SA (2000) Carbon dioxide mediated dissolution of Ca-feldspar: implications for silicate weathering. Chem Geol 163:25–42

Berthon G (1996) Chemical speciation studies in relation to aluminum metabolism and toxicity. Coord Chem Rev 149:241–280

Berthon G (2002) Aluminum speciation in relation to aluminum bioavailability, metabolism and toxicity. Coord Chem Rev 228:319–341

Bhargava S, Bhargava S (2013) Ecological consequences of the acid rain. J Appl Chem 5:19–24

Bi S (2001) Speciation of aluminum in the stream waters from the Susquehanna River watershed, Chesapeake Bay. Environ Geol 40:300–304

Bolan NS, Hedley MJ (2003) Role of carbon, nitrogen and sulfur cycles in soil acidification. In: Rengel Z (ed) Handbook of soil acidity. Marcel Dekker, New York, pp 29–56

Bond AL, Lavers JL (2011) Trace element concentrations in feathers of flesh-footed Shearwaters (*Puffinus carneipes*) from across their breeding range. Arch Environ Contam Toxicol 61:318–326

Borghesi F, Migani F, Andreotti A, Baccetti N, Bianchi N, Birke M et al (2016) Metals and trace elements in feathers: a geochemical approach to avoid misinterpretation of analytical responses. Sci Total Environ 544:476–494

Bortolotti OR, Barlow JC (1988) Some sources of variation in the elemental composition of bald eagle feathers. Can J Zool 66:1948–1951

Boyce NW, Holdsworth SR, Thomson NM, Atkins RC (1987) Rapid alterations in plasma aluminum in association with varied oral aluminum intake in dialysis patients. Nephron 45:164

Brady DJ, Edwards DG, Asher CJ, Blamey FPC (1993) Calcium amelioration of aluminum toxicity effects on root hair development in soybean [*Glycine* max (L) Merr]. New Phytol 123:531–538

Bruckwicki P, Giggleman C, Lewis J (2006) An investigation of the contaminant levels in white-tailed deer (*Odocoileus virginianus*) collected from Caddo Lake National Wildlife Refuge, Harrison County, Texas 2005 Project ID Nos: DEC No 200520002; FFS No 2N53 U.S. Fish and Wildlife Service Region 2, Arlington, Texas

Burke JT, Peacock CL, Lockwood LC, Stewart DJ, Mortimer RJG, Ward MB et al (2013) Behavior of aluminum, arsenic, and vanadium during the neutralization of red mud leachate by HCl, gypsum, or seawater. Environ Sci Technol 47:12

Burklew CE, Ashlock J, Winfrey WB, Zhang B (2012) Effects of aluminum oxide nanoparticles on the growth, development, and microRNA expression of tobacco (*Nicotiana tabacum*). PLoS One 7(5):e34783

Bíres J, Kanka P, Jesenská M, Kovářová M (1992) Toxic element exposure to wildlife, fishes and bees in the Žiar and Horn region. In: Bartko P, Kanka P (eds) Proceeding of seminar on environmental problems, the Žiar Basin, University of Veterinary Medicine, Košice, pp 60–71

Cacini W, Yokel RA (1988) Accumulation of aluminum in rabbit renal cortex. Res Commun Chem Pathol Pharmacol 59:93–105

Cannata-Andia JB, Diaz-Lopez JB (1990) The diagnosis of aluminum toxicity. In: Broe ME, Coburn JW (eds) Aluminum and renal failure. Kluwer Academic, Dordrecht, pp 287–308

Cannon JR, Greenamyre JT (2011) The role of environmental exposures in neurodegeneration and neurodegenerative diseases. Toxicol Sci 124:225–250

Chaturvedi S, Dave PN, Shah NK (2012) Applications of nano-catalyst in new era. J Saudi Chem Soc 16:307–325

Chen L, Yokel RA, Henning B, Toborek M (2008) Manufactured aluminum oxide nanoparticles decrease expression of tight junction proteins in brain vasculature. J Neuroimmune Pharmacol 3:286–295

Chifiriuc MC, Ratiu AC, Popa M, Ecovoiu AA (2016) Drosophotoxicology: an emerging research area for assessing nanoparticles interaction with living organisms. Int J Mol Sci 17:36

Codex Alimentarius Commission (2010) Codex General Standard for Food Additives (GSFA, CODEX STAN 192-1995), revised at 2010, pp 71–72. Available http://www.codexalimentarius.net/gsfaonline/CXS_192e.pdf (Accessed 24 Dec 2010)

Crisponi G, Nurchi VM, Bertolasi V, Remelli M, Faa G (2012) Chelating agents for human diseases related to aluminum overload. Coord Chem Rev 256:89–104

Csoti T, Gyori J, Salanki J, Erdelyi L (2001) pH-dependent actions of aluminum on voltage-activated sodium currents in snail neurons. Neurotoxicology 22:109–116

Darbre PD, Pugazhendhi D, Mannello F (2011) Aluminum and human breast diseases. J Inorg Biochem 105:1484–1488

Darbre PD, Mannello F, Exley C (2013) Aluminum and breast cancer: sources of exposure, tissue measurements and mechanisms of toxicological actions on breast biology. J Inorg Biochem 128:257–261

Das S (2011) Aluminum recycling in a carbon constrained world: observations and opportunities. JOM 63:137–140

Das SK, Yin W (2007) The worldwide aluminum economy: the current state of the industry. JOM 59:57–63
Dayde S, Brumas V, Champmartin D, Rubini P, Berthon G (2003) Aluminum speciation studies in biological fluids. Part 9. A quantitative investigation of aluminum(III)-glutamate complex equilibria and their potential implications for aluminum metabolism and toxicity. J Inorg Chem 97:104–117
de la Fluente D, Otero-Huerta E, Morcillo M (2007) Studies of long-term weathering of aluminum in the atmosphere. Corros Sci 49:3134–3148
Degens ET (1989) Sial: relic of the primordial regolith. Naturwissenschaften 76:165–166
del Carmen-Contini M, Millen N, González M, Mahieu S (2011) Melatonin prevents oxidative stress in ovariectomized rats treated with aluminum. Biol Trace Elem Res 144:924–943
Desroches S, Dayde S, Berthon G (2000) Aluminum speciation studies in biological fluids. Part 6. Quantitative investigation of aluminum(III)-tartrate complex equilibria and their potential implications for aluminum metabolism and toxicity. J Inorg Chem 81:301–312
Dietrich D, Schlatter C (1989) Low levels of aluminum causing death of brown trout (*Salmo trutta fario*, L.) in a Swiss alpine lake. Aquat Sci 51:279–295
Długaszek M, Kopczyński K (2000) Effects of various aluminium compounds given orally to mice on Al tissue distribution and tissue concentrations of essential elements. Pharmacol Toxicol 86:135–139
Długaszek M, Kopczyński K (2011) Comparative analysis of liver mineral status of wildlife. Probl Hig Epidemiol 9(4):859–863
Długaszek M, Kopczyński K (2013) Elemental composition of muscle tissue of wild animals from central region of Poland. Int J Environ Res 7:973–978
Długaszek M, Kopczyński K (2014) Correlations between elements in the fur of wild animals. Bull Environ Contam Toxicol 93:25–30
Domingos M, Klumpp A, Rinaldi MCS, Modesto IF, Klumpp G, Delitti WBC (2003) Combined effects of air and soil pollution by fluoride emissions on *Tibouchina pulchra* Cogn., at Cubatao, SE Brazil, and their relations with aluminum. Plant Soil 249:297–308
Drent PJ, Woldendorp JW (1989) Acid rain and eggshells. Nature 339:431
Driscoll CT, Schecher WD (1989) Aqueous chemistry of aluminum. In: Gitelman HJ (ed) Aluminum and health, a critical review. Marcel Dekker, New York, pp 27–65
Driscoll CT, Driscoll KM, Mitchell MJ, Raynal DJ (2003) Effects of acidic deposition on forest and aquatic ecosystems in New York State. Environ Pollut 123:327–336
Druga M, Trif A, Druga M, Brudiu I, Stef D (2005) The consequences of dietary aluminum sulphate intake on some biochemical parameters in broilers. Proceedings of the 5th international symposium on trace elements in human: new perspectives, Athens/Greece, pp 477–480
Eeva T, Lehikoinen E (1995) Egg shell quality, clutch size and hatching success of the great tit (*Parus major*) and the pied flycatcher (*Ficedula hypoleuca*) in an air pollution gradient. Oecologia 102:312–323
EFSA (2008) Safety of aluminum from dietary intake: scientific opinion on food additives, flavourings, processing aids and food contact materials (AFC). EFSA J 754:1–34
EFSA (2010) Selected trace and ultra trace elements: biological role, content in feed and requirements in animal nutrition – elements for risk assessment, pp 1–15
EFSA (2012) Conclusion on the peer review of the pesticide risk assessment of the active substance aluminum ammonium sulfate 1 (approved as aluminum ammonium sulphate). EFSA J 10:2491
Elangovan R, Mccrohan CR, Balance S, Powell JJ, White KN (2000) Localization and fate of aluminum in the digestive gland of freshwater snail *Lymnaea stagnalis*. Tissue Cell 32:79–87
EPA (2006) US Toxicological Profile for Aluminum. Environmental Protection Agency, Washington
Exley C (2003) A biogeochemical cycle for aluminum? J Inorg Biochem 97:1–7
Exley C (2004) The pro-oxidant activity of aluminum. Free Radic Biol Med 36:380–387
Exley C (2008) Aluminum and medicine. In: Merce ALR, Felcman J, Recio MAL (eds) Molecular and supramolecular bioinorganic chemistry. Nova Science, New York, p 45

Exley C (2009) Darwin, natural selection and the biological essentiality of aluminum and silicon. Trends Biochem Sci 34:589–593
Exley C (2012) The coordination chemistry of biology in neurodegenerative disease. Coord Chem Rev 256:2142–2146
Exley C (2013) Human exposure to aluminium. Environ Sci Processes Impacts 15:1807–1816
Exley C, House ER (2011) Aluminum in the human brain. Monatsh Chem 142:357–363
Exley C, Mold MJ (2015) The binding, transport and fate of aluminum in biological cells. J Trace Elem Med Biol 30:90–95
Exley C, Burgess E, Day JP, Jeffery EH, Melethil S, Yokel RA (1996) Aluminum toxicokinetics. J Toxicol Environ Health 48:569–584
Exley C, Charles LM, Barr L, Martin C, Polwart A, Darbre PD (2007) Aluminum in human breast tissue. J Inorg Biochem 101:1344–1346
Ezomo OF, Matsushima F, Meshitsuka S (2009) Up-regulation in the expression of renin gene by the influence of aluminum. J Inorg Biochem 103:1563–1570
Fiskum GG (2000) Mitochondrial participation in ischemic and traumatic neural cell death. J Neurotrauma 17:843–855
Foy CD (1988) Plant adaptation to acid, aluminum-toxic soils. Commun Soil Sci Plant Anal 19:959–987
Frick KG, Herrmann J (1990) Aluminum accumulation in a lotic mayfly at low pH a laboratory study. Ecotoxicol Environ Saf 19:81–88
Fundacja RECAL – www.recal.pl. 12 June 2008
Galassi G, Cappeli G, Crisi G, Botticelli AR, Lursvarghi E, Winkelmann MD et al (1995) Neuronal accumulation of aluminum in dialysis encephalopathy: laser microprobe study. Trace Elem Electrolytes 12:68–73
Gamberg M, Cuyler C, Wang X (2016) Contaminants in two West Greenland caribou populations. Sci Total Environ 554–555:329–336
Gan Y, Qiao L (2011) Combustion characteristics of fuel droplets with addition of nano and micron-sized aluminum particle. Combust Flame 158:354–368
Garcidueñas-Piña R, Cervantes C (1996) Microbial interactions with aluminum. Biometals 9:311–316
Gardner MJ, Comber SDW (2003) Aluminum speciation in effluents and receiving waters. J Environ Monit 5:902–905
Glynn AW, Sparen A, Danielsson LG, Sundstrom B, Jorhem L (1999a) Concentration-dependent absorption of aluminum in rats exposed to labile aluminum in drinking water. J Toxicol Environ Health 56:501–512
Glynn AW, Thuvander A, Sundstrom B, Sparen A, Danielsson LG, Jorhem L (1999b) Does aluminum stimulate the immune system in male rats after oral exposure? Food Addit Contam 16:129–135
Glynn AW, Sparen A, Danielsson LG, Sundstrom B, Jorhem L (2001) The influence of complexing agents on the solubility and absorption of aluminum in rats exposed to aluminum in water. Food Addit Contam 18:515–523
Godal A, Langseth W, Sivertsen T, Lund W (1995) Determination of aluminum in liver from reindeer, moose and sheep by electrothermal atomic absorption spectrometry. Sci Total Environ 168:249–254
Golub MS, Germann SL (2001) Long-term consequences of developmental exposure to aluminum in a suboptimal diet for growth and behavior of Swiss Webster mice. Neurotoxicol Teratol 23:365–372
Gramiccioni L, Ingrao G, Milana MR, Santaroni P, Tomassi G (1996) Aluminum levels in Italian diets and in selected foods from aluminum utensils. Food Addit Contam 13:767–774
Graveland J (1990) Effect of acid precipitation on reproduction in birds. Experientia 46:962–970
Graveland J, Van der Wal R, Van Balen JH, Van Nordwijk AJ (1994) Poor reproduction in forest passerines from decline of snail abundance on acidified soil. Nature 368:446–448
Grojtheim K, Krohn K, Malinovsky M, Matoasovsky K, Thonstad J (1982) Fundamentals of Hall-Héroult process. In: Aluminum electrolysis, 2nd edn. Aluminum-Verlag, Düsseldorf

Gromysz-Kałkowska K, Szubartowska E (1999) Aluminum: its ecological role and toxicity for animals. Med Weter 55:229–233

Guida AL, Saidi Z, Hughes MN, Poole RK (1991) Aluminum toxicity and binding to *Escherichia coli*. Arch Microbiol 156:507–512

Gundersen P, Steinnes E (2003) Influence of pH and TOC concentration on Cu, Zn, Cd, and Al speciation in rivers. Water Res 37:307–318

Halbrook RS, Woolf A, Hubert GF, Rosss S, Braselton WE (1996) Contaminant concentrations in Illinois mink and otter. Ecotoxicology 5:103–114

Harding LE (2004) Environmental contaminants in wild martens (*Martes americana*) and wolverines (*Gulo luscus*). Bull Environ Contam Toxicol 73:98–105

Harding L, Harris M, Elliott J (1998) Heavy and trace metals in wild mink (*Mustela vison*) and river otter (*Lontra canadensis*) captured on rivers receiving metals discharges. Bull Environ Contam Toxicol 61:600–607

Harris WR (1996) Binding and transport of aluminum by serum proteins. Coord Chem Rev 149:347–365

Hashimoto M, Sasaki J, Imazato S (2016) Investigation of the cytotoxicity of aluminum oxide nanoparticles and nanowires and their localization in L929 fibroblasts and RAW264 macrophages. J Biomed Mater Res B Appl Biomater 104:241–252

Haug A, Shi B (1991) Biochemical basis of aluminum tolerance in plant cells. In: Wright RJ, Baligar VC, Murrmann RP (eds) Plant-soil interactions at low pH: developments in plant and soil sciences, vol 45. Springer, Dordrecht, pp 839–850

Havas M (1985) Aluminum bioaccumulation and toxicity to *Daphnia magna* in soft water at low pH. Can J Fish Aquat Sci 42:1741–1748

Herrmann J (2001) Aluminum is harmful to benthic invertebrates in acidified waters, but at what threshold(s)? Water Air Soil Pollut 130:837–842

Hewitt CD, Wills MR, Savory J (1990) Aluminum determination in biological samples. In: de Broe ME, Coburn JW (eds) Aluminum and renal failure: developments in nephrology, vol 26. Springer, Dordrech, pp 57–73

Hirata-Koizumi M, Fujii S, Ono A, Hirose A, Imai T, Ogawa K et al (2011) Evaluation of the reproductive and developmental toxicity of aluminum ammonium sulfate in a two-generation study in rats. Food Chem Toxicol 49:1948–1959

Houba VJG, Uittienbogaard J (1994) Chemical composition of various plant species. Wageningen Agricultural University, The Netherlands, p 226

Husaini Y, Rai LC (1992) pH dependent aluminum toxicity to *Nostoc linckia*: studies on phosphate uptake, alkaline and acid phosphatase activity, ATP content, photosynthesis and carbon fixation. J Plant Physiol 139:703–707

Illmer P, Mutschlecher W (2004) Effect of temperature and pH on the toxicity of aluminum towards two new, soil born species of *Arthrobacter* sp. J Basic Microbiol 44:98–105

Illmer P, Marschall K, Schinner F (1995) Influence of available aluminum on soil micro-organisms. Lett Appl Microbiol 21:393–397

IRIS (2008) Integrated Risk Information System

Johanson AC, Wood M (1990) DNA, a possible site of action of aluminum in *Rhizobium* spp. Appl Environ Microbiol 56:3629–3633

Jones KC, Bennett BG (1986) Exposure of man to environmental aluminum an exposure commitment assessment. Sci Total Environ 52:65–82

Jönsson U, Rosengren U, Thelin G, Nihlgard B (2003) Acidification-induced chemical changes in coniferous forest soils in southern Sweden 1988-1999. Environ Pollut 123:75–83

Jorhem L, Sundström B, Åstrand H, Haegglund G (1989) The levels of zinc, copper, manganese, selenium, chromium, nickel, cobalt, and aluminum in meat, liver and kidney of Swedish pigs and cattle. Z Lebensm Unters Forsch 188:39–44

Jouhanneau P, Raisbeck GM, Yion F, Lacour B, Banide H, Drüeke TB (1997) Gastrointestinal absorption, tissue retention, and urinary excretion of dietary aluminum in rats determined by using $^{26}$Al. Clin Chem 43:1023–1028

Kabata-Pendias A, Mukherjee AB (2007) Trace elements from soil to human. Springer, Berlin

Kadar E, Salanki J, Powell J, White KN, McCrohan CR (2002) Effect of sub-lethal concentrations of aluminum on the filtration activity of the freshwater mussel *Anodonta cygnea* L. at neutral pH. Acta Bot Hungar 53:485–493

Kålås JA, Ringsby TH, Lierhangen S (1995) Metals and selenium in wild animals from Norwegian areas close to Russian nickel smelters. Environ Monit Assess 36:251–270

Kasai K, Hori MT, Goodman WG (1991) Transferrin enhances the antiproliferative effect of aluminum on osteoblast-like cells. Am J Physiol Endocrinol Metab 260:E537–E543

Kawahara M, Kato-Negishi M (2011) Link between aluminum and the pathogenesis of Alzheimer's disease: the integration of the aluminum and amyloid cascade hypotheses. Int J Alzheimers Dis 2011:276393

Kawahara M, Konoha K, Nagata T, Sadakane Y (2007) Aluminum and human health: its intake, bioavailability and neurotoxicity. Biomed Res Trace Elem 18:211–220

Kearns M (2004) Development and applications of ultrafine aluminum powders. Mater Sci Eng A 375–377:120–126

Klee RJ, Graedel TE (2004) Elemental cycles: a status report on human or natural dominance. Annu Rev Environ Resour 29:69–107

Klöppel H, Fliedber A, Kördel W (1997) Behaviour and ecotoxicology of aluminum in soil and water-review of the scientific literature. Chemosphere 35:353–363

Kośla T, Skibniewska EM (2010) The content of aluminum in the hair of Yorkshire terrier dogs from the Warsaw area depending on sex, age and keeping conditions. Trace Elem Electrolytes 27:209–213

Kośla T, Skibniewska EM, Skibniewski M (2004) Contamination with aluminum of cats free living in Warsaw agglomeration and its effect on haematologic and biochemical indexes of blood as compared to cats kept in flats. In: Macro and trace elements, vol 22. Fridrich-Schiller University, Jena, pp 280–285

Kośla T, Skibniewski M, Skibniewska EM, Urbańska-Słomka G (2006) Aluminum status in free living European Bisons from Białowieża Primeval Forest. Pol J Environ Stud 15:374–377

Kośla T, Skibniewska EM, Skibniewski M (2011) The state of bioelements in the hair of free-ranging European bisons from Bialowieża Primeval Forest. Pol J Vet Sci 14:81–86

Kowalczyk E, Kopff A, Kędziora J, Błaszczyk J, Kopff M, Niedworok J, Fijałkowski P (2004) Effect of long-term aluminum chloride intoxication on selected biochemical parameters and oxidative–antioxidative balance in experimental animals. Pol J Environ Stud 13:41–43

Krantzberg G (1989) Metal accumulation by chironomid larvae: the effects of age and body weight on metal body burdens. Hydrobiologia 188/189:497–506

Krewski D, Yokel RA, Nieboer E, Borchelt D, Cohen J, Harry J et al (2007) Human health risk assessment for aluminum, aluminum oxide, and aluminum hydroxide. J Toxicol Environ Health B Crit Rev 10:1–269

Kucharczak E, Moryl A, Szyposzyński K, Jopek Z (2005) Influence of environment on aluminum content in game animals tissues. Med Weter 61:1277–1279

Kucharczak E, Moryl A (2012) Influence of environment on content of arsenic and aluminium in hunting animals parenchymal organs. Environ Protection Natural Res 53:89–96

Kumar S (1998) Biphasic effect of aluminum on cholinergic enzyme of rat brain. Neurosci Lett 248:121–123

Kumar V, Gill KD (2009) Aluminum neurotoxicity: neurobehavioural and oxidative aspects. Arch Toxicol 83:956–978

Lajtha K, Jones J (2013) Trends in cation, nitrogen, sulfate and hydrogen ion concentrations in precipitation in the United States and Europe from 1978 to 2010: a new look at an old problem. Biogeochemistry 116:303–334

Lantzy RJ, MacKenzie FT (1979) Atmospheric trace metals: global cycles and assessment of man's impact. Geochim Cosmochim Acta 43:511–525

Levesque L, Mizzen CA, McLachlan DR, Fraser PE (2000) Ligand specific effects on aluminum incorporation and toxicity in neurons and astrocytes. Brain Res 877:191–202

Lewis RJ (ed) (2001) Hawley's condensed chemical dictionary. Wiley, New York, pp 39–46, 118 and 555

Lewis SA, Furness RW (1991) Mercury accumulation and excretion by laboratory reared black-headed gulls (*Larus ridibundus*) chicks. Arch Environ Contam Toxicol 21:316–320

Li X, Zheng H, Zhang Z, Li M, Huang Z, Schluesener HJ et al (2009) Glia activation induced by peripheral administration of aluminum oxide nanoparticles in rat brains. Nanomedicine: NBM 5:473–479

Lide DR (2005) CRC handbook of chemistry and physics. CRC Press, New York, pp 4–79

Liu G, Bangs CE, Müller DB (2013) Stock dynamics and emission pathways of the global aluminum cycle. Nat Clim Chang 3:338–342

Llacuna S, Gorizz A, Sanpera C, Nadal J (1995) Metal accumulation in three species of passerine birds (*Emberiza cia*, *Parus major*, and *Turdus merula*) subjected to air pollution from coal-fired power plant. Arch Environ Toxicol 28:298–303

Lodenius M, Solonen T (2013) The use of feathers of birds of prey as indicators of metal pollution. Ecotoxicology 22:1319–1334

Lote CJ, Willmott K, Wood JA, Thewles A, Freeman M (1995) Renal excretion of aluminum in the rat: effect of citrate infusion. Hum Exp Toxicol 14:945–948

Lu Z, Streets DG, Zhang Q, Wang S, Carmichael GR, Cheng YF et al (2010) Sulfur dioxide emissions in China and sulfur trends in East Asia since 2000. Atmos Chem Phys 10:6311–6331

Lucia M, Andre JM, Gontier K, Diot N, Veiga J, Davail S (2010) Trace element concentrations (mercury, cadmium, copper, zinc, lead, aluminum, nickel, arsenic, and selenium) in some aquatic birds of the Southwest Atlantic Coast of France. Arch Environ Contam Toxicol 58:844–853

Ma QF, Rengel Z, Kuo J (2002) Aluminum toxicity in rye (*Secale cereale*): root growth and dynamics of cytoplasmic $Ca^{2+}$ in intact root tips. Ann Bot 89:241–244

Ma G, Rengasamy P, Rathjen AJ (2003) Phytotoxicity of aluminum to wheat plants in high-pH solutions. Aust J Exp Agric 43:497–501

Mahler RL, Halvorson AR, Koehler FE (1985) Long-term acidification of farmland in northern Idaho and eastern Washington. Commun Soil Sci Plant Anal 16:83–95

Majewska U, Braziewicz J, Banas D, Kubala-Kukuś A, Gozdz S, Pajek M et al (1997) An elemental correlation study in cancerous breast tissue by total reflection X-ray fluorescence. Biol Trace Elem Res 60:91–100

Malley DF, Heubner JD, Donkersloot K (1988) Effects on ionic composition of blood and tissues of *Anodonta grandis grandis* (Bivalvia) of an addition of aluminum and acid to a lake. Arch Environ Contam Toxicol 17:479–491

Maňkovská B, Steinnes E (1995) Effect of pollutants from an aluminum reduction plant on forest ecosystems. Sci Total Environ 163:11–23

Mannello F, Tonti GA, Medda V (2009) Protein oxidation in breast microenvironment: nipple aspirate fluid collected from breast cancer women contains increased protein carbonyl concentration. Cell Oncol 31:383–392

Mannello F, Tonti GA, Pederzoli A, Simone P, Smaniotto A, Medda V (2010) Detection of superoxide dismutase-1 in nipple aspirate fluids: a reactive oxygen species-regulating enzyme in the breast cancer microenvironment. Clin Breast Cancer 10:238–245

Markich SJ, Warne MSJ, Westbury AM, Roberts CJ (2002) A compilation of data on the toxicity of chemicals to species in Australasia. Part 3: metals. Aus J Ecotoxicol 8:1–137

Martin RB (1994) Aluminum: a neurotoxic product of acid rain. Acc Chem Res 27:204–210

Masunaga T, Kubota D, Hotta M, Wakatsuki T (1998) Nutritional characteristics of mineral elements on leaves of tree species in tropical rain forest, West Sumatra, Indonesia. Soil Sci Plant Nutr 44:315–329

Matsumoto H (2000) Cell biology of aluminum toxicity and tolerance in higher plants. Int Rev Cytol 200:1–46

May HM, Nordstrom DK (1991) Assessing the solubilities and reaction kinetics of aluminous minerals in soils. In: Ulrich B, Sumner ME (eds) Soil acidity. Springer, Berlin, pp 125–148

Mayes WM, Jarvis AP, Burke IT, Walton M, Gruiz K (2011a) Trace and rare earth element dispersal downstream of the Ajka red mud spill, Hungary. In: Rüde RT, Freund A, Wolkersdorfer C (eds) Mine Water: managing the challenges. International Mine Water Association, Wendelstein, pp 29–498

Mayes WM, Jarvis AP, Burke IT, Walton M, Feigl V, Klebercz O et al (2011b) Dispersal and attenuation of trace contaminants downstream of the Ajka bauxite residue (red mud) depository failure, Hungary. Environ Sci Technol 45:5147–5155
Menz FC, Seip HM (2004) Acid rain in Europe and the United States: an update. Environ Sci Pol 7:253–265
Menzie WD, Barry JJ, Bleiwas DI, Bray EL, Goonan TG, Matos G (2010) The global flow of aluminum from 2006 through 2025. US Geological Survey Open-File Report 2010-1256, p 73. http://pubs.usgs.gov/of/2010/1256/
Meshitsuka S, Inoue M (1998) Urinary excretion of aluminum from antacid ingestion and estimation of its apparent biological half-time. Trace Elem Electrolytes 15:132–135
Mierzykowski SE, Smith JEM, Todd CS, Kusnierz D, DeSorbo CR (2011) Liver contaminants in bald eagle carcasses from Maine. USFWS. Special Project Report FY09 MEFO 6 EC. Maine Field Office Orono, ME, p 53
Miljeteig C, Gabrielsen GW, Strøm H, Gavrilo MV, Lie E, Jenssen BM (2012) Eggshell thinning and decreased concentrations of vitamin E are associated with contaminants in eggs of ivory gulls. Sci Total Environ 431:92–99
Miller RG, Kopfler FC, Kelty KC, Stober JA, Ulmer NS (1984) The occurrence of aluminum in drinking water. J Am Water Works Assoc 76:84–91
Monteiro LR (1996) Seabirds as monitors of mercury in the marine environment. Water Air Soil Pollut 80:851–870
Moore PB, Day JP, Taylor GA, Ferrier IN, Fifield LK, Edwardson JA (2000) Absorption of aluminum-26 in Alzheimer's disease, measured using accelerator mass spectrometry. Dement Geriatr Cogn Disord 11:66–69
Mussi I, Calzaferri G, Buratti M, Alessio L (1984) Behaviour of plasma and urinary aluminum levels in occupationally exposed subjects. Int Arch Occup Environ Health 54:155–161
Nappi C (2013) The global aluminum industry 40 years from 1972, World Aluminum, p 27. www.world-aluminum.org
National Research Council of the National Academies-NRC (2005) Mineral tolerance of animals, 2nd edn. The National Academies Press, Washington DC
Nicolini M, Zatta PF, Corain B (eds) (1991) Aluminum in chemistry, biology and medicine. Cortina International, Verona, Raven Press, New York
Nielsen FH (1996) How should dietary guidance be given for mineral elements with beneficial actions or suspected of being essential? J Nutr 126:S2377–S2385
Nyholm NEI (1981) Evidence of involvement of aluminum in causation of defective formation of eggshells and of impaired breeding in wild passerine birds. Environ Res 26:363–371
Oelke H (1989) Effect of the acid rain syndrome on bird populations (Harz Mountains, Lower Saxony, FR Germany). Beitr Naturk Nieders 42:109–128
Olariu L, Chisu I, Tulcan C, Triff A, Druga M (2004) The influence of aluminum intake on some oxidoreductase in broiler chicken. In: Macro and trace elements, vol 22. Fridrich-Schiller University, Jena, pp 313–318
Ormerod SI, Wade KR (1990) The role of acidity in the ecology of Welsh lakes and streams. In: Edwards RW et al (eds) Acid waters in Wales. Kluwer Academic, Dordrecht, pp 93–119
Otto C, Svensson BS (1983) Properties of acid brown water streams in south Sweden. Arch Hydrobiol 99:15–36
Paik SR, Lee JH, Kim DH, Chang CS, Kim J (1997) Aluminum-induced structural alterations of the precursor of the non-A beta component of Alzheimer's disease amyloid. Arch Biochem Biophys 344:325–334
Peakall DB (2000) Avian data bases and their use in environmental assessment. Ecotoxicology 9:239–253
Pereira S, Cavalie I, Camilleri V, Gilbin R, Adam-Guillermin C (2013) Comparative genotoxicity of aluminum and cadmium in embryonic zebrafish cells. Mutat Res Genet Toxical Environ 750:19–26
Peterson CL, Perry DL, Masood H, White JL, Hem SL, Fritsch C et al (1993) Characterization of antacid compounds containing both aluminum and magnesium. II. Codried powders. Pharm Res 10:1005–1007

Piercey DG, Klapötke TM (2010) Nanoscale aluminum-metal oxide (thermite) reactions for application in energetic materials. Cent Eur J Energetic Mater 7:115–129

Playle RC, Wood CM (1989) Water pH and aluminum chemistry in the gill microenvironment of rainbow trout during acid and aluminum exposure. J Comp Physiol B 159:539–550

Plunkert PA (2000) Bauxite and alumina. US Geological Survey Minerals Yearbook 2000

Polak TB, Milacic R, Mitrovic B, Benedik M (2001) Speciation of low molecular weight Al complexes in serum of CAPD patients. J Pharm Biomed Anal 26:189–201

Poole KG, Elkin BT, Bethe RW (1995) Environmental contaminants in wild and in the Northwest Territories, Canada. Sci Total Environ 160/161:473–486

Poole KG, Elkin BT, Bethe RW (1998) Organochlorine and heavy metal in wild mink in Western Northwest Territories, Canada. Arch Environ Contam Toxicol 34:406–413

Powell JJ, Thomson RPH (1993) The chemistry of aluminum in the gastrointestinal lumen and its uptake and absorption. Proc Nutr Soc 52:241–253

Powell JJ, Greenfield SM, Parkes HG, Nicholson JK, Thomson RPH (1993) Gastro-intestinal availability of aluminum from tea. Food Chem Toxicol 31:449–454

Powell JJ, Whitehead MW, Ainley CC, Kendall MD, Nicholson JK, Thomson RPH (1999) Dietary minerals in the gastrointestinal tract: hydroxypolymerisation of aluminum is regulated by luminal mucins. J Inorg Biochem 75:167–180

Priest ND (1990) The distribution and behaviour of metals in the skeleton and body: studies with bone-seeking radionuclides. In: Priest ND, Van de Vyver F (eds) Trace metals and fluoride in bones and teeth. CRC Press, Boca Raton, pp 83–139

Priest ND (2004) The biological behaviour and bioavailability of aluminum in man, with special reference to studies employing aluminum-26 as a tracer: review and study update. J Environ Monit 6:375–403

Priest ND, Talbot RJ, Newton D, Day JP, King SJ, Fifield LK (1998) Uptake by man of aluminum in a public water supply. Hum Exp Toxicol 17:296–301

Ráez-Bravo A, Granados JE, Cano-Manuel FJ, Soriguer RC, Fandos P, Pérez JM et al (2016) Toxic and essential element concentrations in Iberian Ibex (*Capra pyrenaica*) from the Sierra Nevada Natural Park (Spain): reference intervals in whole blood. Bull Environ Contam Toxicol 96:273–280

Ranau R, Oehlenschläger J, Steinhart H (2001) Aluminum levels of fish fillets baked and grilled in aluminium foil. Food Chem 73:1–6

Rauch JN, Pacyna JM (2009) Earth's global Ag, Al, Cr, Cu, Fe, Ni, Pb, and Zn cycles. Glob Biogeochem Cycles 23:1–16

Reff A, Bhave PV, Simon H, Pace TG, Pouliot GA, Mobley JD et al (2009) Emissions inventory of $PM_{2.5}$ trace elements across the United States. Environ Sci Technol 43:5790–5796

Reimann C, Birke M, Demetriades C, Filzmoser P, O'Connor P (2014) Chemistry of Europe's agricultural soils, Part A, Bundesanstalt für Geowissenschaften und Rohstoffe (BGR), Hannover, pp 143–148

Rengel Z (1997) Mechanisms of plant resistance to toxicity of aluminum and heavy metals. In: Basra AS, Basra RK (eds) Mechanisms of environmental stress resistance in plants. Harwood Academic, Amsterdam, pp 241–276

Rengel Z (2004) Aluminum cycling in the soil-plant-animal-human continuum. Biometals 17:669–689

Rengel Z, Zhang WH (2003) Role of dynamics of intracellular calcium in aluminum toxicity syndrome. New Phytol 159:295–314

Reusche E, Pilz P, Oberascher G, Lindner B, Egensperger R, Gloeckner K et al (2001) Subacute fatal aluminum encephalopathy after reconstructive otoneurosurgery: a case report. Hum Pathol 32:1136–1140

Richardson AE, Simpson RJ, Djordjevic MA, Rolfe BJ (1988) Expression of nodulation genes in *Rhizobium leguminosarum* bv. *trifolii* is affected by low pH and by Ca and AI ions. Appl Environ Microbiol 54:2541–2548

Roider G, Drasch G (1999) Concentration of aluminum in human tissues-investigations on an occupationally non-exposed population in Southern Bavaria (Germany). Trace Elem Electrolytes 16:77–86

Rosseland BO, Eldhuset TD, Staurnes M (1990) Environmental effects of aluminum. Environ Geochem Health 12:17–27

Roux DJ, Jooste SHJ, Mackay HM (1996) Substance-specific water quality criteria for the protection of South African freshwater ecosystems: methods for derivation and initial results for some inorganic toxic substances. S Afr J Sci 92:198–206

Roy R (1999) The chemistry, bioaccumulation and toxicity of aluminum in the aquatic environment for the PSL2 assessment of aluminum salts. Report prepared by Fisheries and Oceans Canada for Environment Canada, p 110

Ruyters S, Mertens J, Vassilieva E, Dehandschutter B, Poffijn A, Smolders E (2011) The red 519 mud accident in Ajka (Hungary): plant toxicity and trace metal bioavailability in red mud 520 contaminated soil. Environ Sci Technol 45:616–1622

Sahu RK, Hiremath SS, Manivannan PV, Singaperumal M (2014) An innovative approach for generation of aluminum nanoparticles using micro electrical discharge machining. Proc Mater Sci 5:1205–1213

Salminen R, Batista MJ, Bidovec M, Demetriades A, De Vivo B, De Vos W, Duris M et al (2005) Al, aluminum. In: Geochemical atlas of Europe: background information, methodology and maps. Geological Survey of Finland, pp 53–57. http://weppi.gtk.fi/publ/foregsatlas/text/Al.pdf

Sample BE, Opresko DM, Suter GW II (1996) Toxicological benchmarks for wildlife: 1996 revision. Oak Ridge National Laboratory, ES/ER/TM-86/R3, Oak Ridge, Tennessee, p 217

Sanchez-Iglesias S, Soto-Otero R, Iglesias-Gonzalez J, Barciela-Alonso MC, Bermejo-Barrera P, Mendez-Alvarez E (2007) Analysis of brain regional distribution of aluminum in rats via oral and intraperitoneal administration. J Trace Elem Med Biol 21:31–34

Schaller KH, Valentin H (1984) In: Alessio L, Berlin A, Boni M, Roi R (eds) Biological indicators for the assessment of human exposure to industrial chemicals, Commission of the European Communities, p 24

Scheuhammer AM (1987) The chronic toxicity of aluminum, cadmium, mercury, and lead in birds: a review. Environ Pollut 46:263–295

Scheuhammer AM (1991a) Acidification-related changes in the biogeochemistry and ecotoxicology of mercury, cadmium, lead and aluminum: overview. Environ Pollut 71:87–90

Scheuhammer AM (1991b) Effects of acidification on the availability of toxic metals and calcium to wild birds and mammals. Environ Pollut 71:329–375

Schintu M, Meloni P, Contu A (2000) Aluminum fractions in drinking water from reservoirs. Ecotoxicol Environ Saf 46:29–33

Schuping B (1996) A model describing the complexing effect in the leaching of aluminum from cooking utensils. Environ Pollut 92:85–90

Schwarzerova K, Zelenkova S, Nick P, Opatrny Z (2002) Aluminum induced rapid changes in the microtubular cytoskeleton of tobacco cell lines. Plant Cell Physiol 43:207–216

Scientific Committee on Animal Nutrition (SCAN) (2003) Opinion of the Scientific Committee on Animal Nutrition on Undesirable Substances in Feed

Servos MR, Rooke JB, Mackie GL (1985) Reproduction of selected mollusca in some low alkalinity lakes in south-central Ontario. Can J Zool 63:511–515

Sharma DR, Sunkaria A, Yousuf-Wani W, Sharma RK, Kandimalla RJL, Bal A et al (2013) Aluminum induced oxidative stress results in decreased mitochondrial biogenesis via modulation of PGC-1α expression. Toxicol Appl Pharmacol 273:365–380

Sheasby PG, Pinner R (2011) Introduction: aluminum, its properties, alloys and finishes. In: The surface treatment and finishing of aluminum and its alloys. ASM International, Ohio, pp 1–10

Shenglin P, Dejiang J, Jianging L, Yongjun S, Mu Y, Xu C (1996) New evidences for the formation of granitic rocks by remelting of the sial crust in southern Hunan. J Cent South Univ Technol 3:1–3

Shirabe T, Irie K, Uchida M (2002) Autopsy case of aluminum encephalopathy. Neuropathology 22:206–210

Shirley DG, Walter MF, Walter SJ, Thewles A, Lote CJ (2004) Renal aluminum handling in the rat: a micropuncture assessment. Clin Sci 107:159–165

Sjögren B, Iregren A, Elinder CG, Yokel RA (2007) Aluminum. In: Nordberg GF, Fowler B, Nordberg M, Friberg LT (eds) Handbook on the toxicology of metals. Elsevier, Amsterdam, pp 339–352

Skibniewska EM (2010) Aluminum content in canine mammary glands. Fresenius Environ Bull 19:390–392

Skibniewska EM, Skibniewski M, Kośla T, Urabańska-Słomka G (2011) Hair zinc levels in pet and feral cats (*Felis catus*). J Elem 16:481–488

Skoric S, Visnjić-Jeftic Z, Jaric I, Djikanovic V, Mickovic B, Nikcevic M, Lenhardt M (2012) Accumulation of 20 elements in great cormorant (*Phalacrocorax carbo*) and its main prey, common carp (*Cyprinus carpio*) and Prussian carp (*Carassius gibelio*). Ecotoxicol Environ Saf 80:244–251

Smans KA, D'Hase PC, Van Landeghem GF, Andries LJ, Lamberts LV, Hendy GN et al (2000) Transferrin-mediated uptake of aluminum by human parathyroid cells results in reduced parathyroid hormone secretion. Nephrol Dial Transplant 15:1328–1336

Solomon B, Koppel R, Jossiphov J (2001) Immunostaining in calmodulin and aluminum in Alzheimer's disease-affected brains. Brain Res Bull 55:253–256

Sparling DW, Lowe TP (1996) Environmental hazards of aluminum to plants, invertebrates, fish, and wildlife by Springer-Verlag New York, Inc. Rev Environ Contam Toxicol 145:1–127

Spry DJ, Weiner JG (1991) Metal bioavailability and toxicity to fish in low-alkalinity lakes: a critical review. Environ Pollut 71:243–304

Stachurska A, Wałkuska G, Cebera M, Jaworski Z, Chałabis-Mazurek A (2011) Heavy metal status of Polish Konik horses from stable-pasture and outdoor maintenance systems in the Mazurian environment. J Elem 16:623–633

Stahl T, Taschan H, Brunn H (2011) Aluminum content of selected foods and food products. Environ Sci Eur 23:37–48

Steinegger A, Rickenbacher U, Schlatter C (1990) Aluminum. In: Adams F, Blunden SJ, van Cleuvenbergen R, Evans CJ, Fishbein L, Rickenbacher UJ et al (eds) Anthropogenic compounds. Springer, Berlin, pp 156–180

Strong MJ, Garruto RM, Joshi JG, Mundy WR, Shafer TJ (1996) Can the mechanisms of aluminum neurotoxicity be integrated into a unified scheme? J Toxicol Environ Health 48:599–613

Sutherland JE, Greger IL (1998) Effect of the size of an oral dose of aluminum on the relative importance of biliary v. urinary aluminum excretion in conscious rats. Food Chem Toxicol 36:505–512

Swegert CV, Dave KR, Katyare SS (1999) Effect of aluminum induced Alzheimer like condition on oxidative energy metabolism in rat liver, brain and heart mitochondria. Mech Ageing Dev 112:27–42

Tang C, Rengel Z (2003) Role of plant cation/anion uptake ratio in soil acidification. In: Rengel Z (ed) Handbook of soil acidity. Marcel Dekker, New York, pp 57–81

Tang S, Parsons PJ, Perl D (1999) Longitudinal and lateral variations in the aluminum concentration of selected caprine, bovine, and human bone samples. Biol Trace Elem Res 68:267–279

Teien HC, Kroglund F, Salbu B, Rosseland BO (2006) Gill reactivity of aluminum species following liming. Sci Total Environ 358:206–220

Teixeira AO, Leonel FP, Knoop R, Ferreira VPA, Ribeiro ET, Moreira LM et al (2013) Mineral deposition in tissues of cattle fed with different phosphates and relationships phosphorus: fluorine. Rev Bras Saúde Prod Anim 14:831–847

Tomlinson DJ, Mülling CH, Fakler TM (2004) Formation of keratins in the bovine claw: roles of hormones, minerals, and vitamins in functional claw integrity. J Dairy Sci 87:797–809

Tomza-Marciniak A, Pilarczyk B, Bąkowska M, Pilarczyk R, Wójcik J (2011) Heavy metals and other elements in serum of cattle from organic and conventional farms. Biol Trace Elem Res 143:863–870

Trif A, Druga M, Druga M, Muselin F, Brudiu I, Dumitrescu E (2005) Aluminum transplacental passage – experimental studies in rats. In: Proceedings of the 5th international symposium on trace elements in human: new perspectives, Athens, pp 212–214

Truscott R, McCrohan CR, Bailey SER, White KN (1995) Effect of aluminum and lead on activity in the freshwater snail *Lymnaea stagnalis*. Can J Fish Aquat Sci 52:1623–1629

US EPA (2003) Ecological soil screening level for aluminum Interim Final OSWER Directive 9285.7-60, p 34

US Geological Survey (2016) Mineral commodity summaries 2016. U.S. Geological Survey, p 202. https://doi.org/10.3133/70140094

US GS (2016) Mineral Commodity Summaries. Bauxite and alumina. US Geological Survey, pp 32–33

Uversky VN, Li J, Fink AL (2001) Metal-triggered structural transformations, aggregation, and fibrillation of human alpha-synuclein. A possible molecular NK between Parkinson's disease and heavy metal exposure. J Biol Chem 276:44284–44296

Van Landeghem GF, Haese PCD, Lamberts LV, Djukanovic L, Pejanovic S, Goodman WG et al (1998) Low serum aluminum values in dialysis patients with increased bone aluminum levels. Clin Nephrol 50:69–76

Vanholder R, Cornelis R, Dhondt A, Lameire N (2002) The role of trace elements in uraemic toxicity. Nephrol Dial Transplant 17:2–8

Vargas AAT, Graham PH (1988) *Phaseolus vulgaris* cultivar and *Rhizobium* strain variation in acid-pH tolerance and nodulation under acid conditions. Field Crop Res 19:91–101

Varrica D, Aiuppa A, Dongarra G (2000) Volcanic and anthropogenic contribution to heavy metal content in lichens from Mt. Etna and Vulcano Island (Sicily). Environ Pollut 108:153–162

Venkatesan SP (2015) Influence of aluminum oxide nanoparticle additive on performance and exhaust emissions of diesel engine. Am-Euras J Sci Res 10:88–92

Venturini SM, Berthon G (2001) Aluminum speciation studies in biological fluids. Part 7. A quantitative investigation of aluminum(III)-malate complex equilibria and their potential implications for aluminum metabolism and toxicity. J Inorg Chem 85:143–154

Verstraeten SV, Aimo L, Oteiza PI (2008) Aluminum and lead: molecular mechanisms of brain toxicity. Arch Toxicol 82:789–802

Walton R, McCrohan KR, Livens F, White KN (2010) Trophic transfer of aluminum through an aquatic grazer–omnivore food chain. Aquat Toxicol 99:93–99

Ward MK, Feest TG, Ellis HA, Parkinson IS, Kerr DN (1978) Osteomalacic dialysis osteodystrophy: evidence for a water-borne aetiological agent, probably aluminum. Lancet 1:841–845

Ward RJ, Zhang Y, Crichton RR (2001) Aluminum toxicity and iron homeostasis. J Inorg Chem 87:9–14

Watanabe T, Osaki M (2002) Mechanisms of adaptation to high aluminum condition in native plant species growing in acid soils: a review. Commun Soil Sci Plant Anal 33:1247–1260

Whitehead MW, Farrar G, Christe G, Blair JA, Thomson RPH, Powell JJ (1997) Mechanisms of aluminum absorption in rats. Am J Clin Nutr 65:1446–1452

WHO (1996) Guidelines for drinking-water quality. Health criteria and other supporting information. World Health Organization, Geneva

WHO (2004) Guidelines for drinking-water quality. Recommendation. World Health Organization, Geneva

Widłak M (2013) Evaluation of exchangeable aluminum content and selected soil parameters of Swietokrzyskie region. Proc ECOpole 7:413–420

Wilkinson KJ, Campbell PGC (1993) Aluminum bioconcentration at the gill surface of juvenile atlantic salmon in acidic media. Environ Toxicol Chem 12:2083–2095

Williams RJP (1996) Aluminum and biological systems: an introduction. Coord Chem Rev 149:1–9

Wood M, Cooper JE, Bjourson AJ (1988) Response of *Lotus rhizobia* to acidity and aluminium in liquid culture and in soil. Plant Soil 107:227–231

Woodburn K, Walton R, McCrohan C, White K (2011) Accumulation and toxicity of aluminum-contaminated food in the freshwater crayfish, *Pacifastacus leniusculus*. Aquat Toxicol 105:535–542

Wren CD (1986) Mammals as biological monitors of environmental metal levels. Environ Monit Assess 6:127–144

WVDL (2015) Normal Range Values for WVDL Toxicology. www.wvdl.wisc.edu/wp-content/uploads/2013/06/WVDL.Info_.Toxicology_Normal_Ranges.pdf (Accessed 28 Apr 2015)

Yamamoto Y, Kobayashi Y, Devi SR, Rikiishi S, Matsumoto H, Abe J (2003) Oxidative stress triggered by aluminum in plant roots. Plant Soil 255:239–243

Yokel RA (1982) Hair as an indicator of excessive aluminum exposure. Clin Chem 28:662–665

Yokel RA (2000) The toxicology of aluminum in the brain: a review. Neurotoxicology 21:813–828

Yokel RA (2004) Aluminum. In: Merian E, Anke M, Ihnat M, Stoeppler M (eds) Elements and their components in the environment, 2nd edn. Wiley, Weinheim, pp 635–658

Yokel RA, Golub MS (eds) (1997) Research issues in aluminum toxicity. Taylor & Francis, Washington, DC, p 256

Yokel RA, McNamara PJ (2001) Aluminum toxicokinetics: an updated mini review. Pharmacol Toxicol 88:159–167

Yokel RA, Wilson M, Harris WR, Halestrap AP (2002) Aluminum citrate uptake by immortalized brain endothelial cells: implications for its blood-brain barrier transport. Brain Res 930:101–110

Young LB, Harvey HH (1991) Metal concentrations in chironomids in relation to the geochemical characteristics of surficial sediments. Arch Environ Contam Toxicol 21:202–211

Ysart G, Miller P, Croasdale M, Crews H, Robb P, Baxter M et al (2000) 1997 UK total diet study: dietary exposures to aluminum, arsenic, cadmium, chromium, copper, lead, mercury, nickel, selenium, tin and zinc. Food Addit Contam 17:775–786

Zaida F, Chadrame S, Sedki A, Lekouch N, Bureau F, Arhan P et al (2007) Lead and aluminum levels in infants' hair, diet, and the local environment in the Moroccan city of Marrakech. Sci Total Environ 377:152–158

Zatta P, Zambenedetti P (1996) Aluminum speciation and morphological differentiation in murine neuroblastoma cells. Biol Trace Elem Res 51:77–85

Zatta P, Ibn-Lkhayat-Idrissi M, Zambenedetti P, Kilyen M, Kiss T (2002) In vivo and in vitro effects of aluminum on the activity of mouse brain acetylcholinesterase. Brain Res Bull 59:41–45

Zhang WH, Rengel Z (1999) Aluminum induces an increase in cytoplasmic calcium in intact wheat root apical cells. Aust J Plant Physiol 26:401–409

Zhang WH, Rengel Z, Kuo J (1998) Determination of intracellular $Ca^{2+}$ in cells of intact wheat roots: loading of acetoxymethyl ester of Fluo-3 under low temperature. Plant J 15:147–151

Zhang X, Xu Y, Zhou L, Zhang C, Meng Q, Wu S et al (2015) Sex-dependent depression-like behavior induced by respiratory administration of aluminum oxide nanoparticles. Int J Environ Res Public Health 12:15692–15705

Zimmerman TJ, Jenks JA, Leslie DM Jr, Neiger RD (2008) Hepatic minerals of white-tailed and mule deer in the southern black hills, South Dakota. J Wildl Dis 44:341–350

# Chapter 13
# Arsenic, As

**Łukasz J. Binkowski**

**Abstract** Arsenic is a semimetal that forms a part of more than 200 minerals. In many places of the world concentrations of arsenic in water are high, which is an issue of high importance in connection with human health. It has three allotrope forms; the gray one is the most common. Among numerous arsenic isotopes only $^{75}$As is stable. The element is produced mainly in the form of a trioxide. Arsenic is used in electronic, metallurgy, pesticides, and defoliants. The most common use is in the production of wood preservatives (which, along with fossil fuel combustion, represents the largest anthropogenic arsenic source in the environment). In some parts of the world arsenic compounds are used as a supplement in poultry farming. Recent research also shows its potential use in medicine. Arsenic toxicity depends on its form (organic and inorganic), as well as on its oxidation state, solubility, and species exposed. In the body, the methylation of its inorganic form takes place mainly in the liver. Following exposure to arsenic, it can be found in various tissues, organs and materials, as kidneys, blood, lungs, feathers, hair, and fur, but mainly in the liver. Arsenic bioaccumulation is low, and biomagnification is still questioned in terrestrial ecosystems. Some biomarkers of exposure, apart from concentration measurements (especially in urine, blood, hair, fur, and feathers) may be used. Among internal tissues, the liver is the most commonly studied.

## 1 Introduction

Arsenic (chemical symbol As) is a metalloid (semimetal), but it is commonly included in the list of "heavy metals" based upon its toxicity (IUPAC 1971; Duffus 2002). As a semimetal it presents some properties of metals and nonmetals. It is a component of numerous minerals and reaches a concentration of 2 mg kg$^{-1}$ in rocks (Mandal and Suzuki 2002). Arsenic's main toxicity combines with its inorganic

Ł. J. Binkowski (✉)
Institute of Biology, Pedagogical University of Cracow, Kraków, Poland
e-mail: lukasz.binkowski@up.krakow.pl



E. Kalisińska (ed.), *Mammals and Birds as Bioindicators of Trace Element Contaminations in Terrestrial Environments*,
https://doi.org/10.1007/978-3-030-00121-6_13

forms occurring in groundwater in many places around the world, including more than 70 countries (Ravenscroft et al. 2009). This is the main reason for the global interest in As occurrence, availability, and exposure.

Arsenic, along with its compounds, has been widely used in industry and agriculture. It has found primary application in wood preservatives, insecticides, and poisons (Rahman et al. 2004). It has been also used in medical treatment, pharmaceuticals, and even in supplementation in animals (Jones 2007). Other applications include alloy production, glass processing, and the production of semiconductors, ammunition, batteries, pigments, paper, and metal adhesives. Elevated As concentrations in organisms cause poisoning and stimulate cancer development (IARC 2012). However, numerous studies also reveal that As deficiency in the diet of birds and mammals causes physiological disorders, especially with respect to methionine metabolism (Uthus 2003).

## 2 General Properties

Arsenic (Lat. *Arsenicum*) lies between germanium and selenium in the nitrogen group (pnictogen) of the periodic table. Its atomic number is 33 and atomic weight is 74.9 Da (Haynes 2014). Among the three As allotropes, the most stable is the gray form (also called α-crystalline). A yellow cubic form is less stable and after warming reverts to the basic form. The black β form is also stable. The density of the gray form is 5.73 g $cm^{-3}$, and its melting temperature is 817 °C; sublimation occurs at 616 °C (Norman 1998). Generally, 29 As isotopes have been identified ($^{64}$As–$^{92}$As). However, some scientists include an additional four ($^{60}$As–$^{63}$As). The only isotope considered stable and naturally occurring is $^{75}$As and due to that fact As is often treated a monoisotopic element.

Arsenic occurs in four oxidation states: −3, 0, +3, and +5 (Adriano 2001). In nature, two major groups of As compounds occur, inorganic and organic (Lunde 1977; Andreae 1978). Compounds with the element on +3 (arsenite) and +5 (arsenate) oxidation levels dominate (Andreae 1978; Morita and Edmonds 1992; Rosen 2002). Apart from those, compounds as arsines and methylarsines with As on the −3 level also occur, but they are unstable in the air. Free arsenic As(0) is rarely encountered in nature (Eisler 1988). In terms of As use and application, the most important form is As trioxide ($As_2O_3$).

## 3 Arsenic Minerals, Production, and Uses

Arsenic constitutes a part of more than 200 minerals, of which 60% are treated as mainly As ones (Kabata-Pendias 2011). The most common are arsenopyrite (FeAsS), arsenolite ($As_4O_6$), loellingite ($FeAs_2$), orpiment ($As_2S_3$), and realgar (AsS). Arsenic also occurs in ores of other metals (such as iron, nickel, cobalt, and

**Fig. 13.1** Dynamic of arsenic trioxide ($As_2O_3$) production between 1950 and 2015 (USGS 1950, 1955, 1960, 1965, 1970, 1975, 1980, 1985, 1990, 1995, 2000, 2006, 2011, 2015)

copper) and reaches high concentrations in sulfide deposits: arsenides (27 minerals), sulfides (13 minerals), and sulfosalts (65 minerals) (Adriano 2001; Hammond 2004).

Arsenic is produced mainly in the form of $As_2O_3$ (most reports present production data expressed in terms of this compound's values). Main global As production in 2015 was estimated at 36,000 metric tons (Fig. 13.1). For many years, largest quantities of $As_2O_3$ were produced (expressed in metric tons) in China (25,000), Chile (10,000), Morocco (7500), and Russia (1500) (USGS 2015, 2016). The USA has not produced As since 1985 (USGS 2011). At present, $As_2O_3$ is produced mainly by volatilization during the mining and production of other elements. Probably only China still mines As ores intentionally (Grund et al. 2005). Metallic As is produced in significantly smaller quantities, but detailed data are not available (USGS 2015). Of the total As imported by the USA, no more than 4% is in metallic form, which is usually produced by the reduction of ores or $As_2O_3$ with coal monoxide (Mandal and Suzuki 2002; Solo-Gabriele et al. 2003; USGS 2006).

The metallic form is used in electronics and nonferrous alloys. $As_2O_3$ has been used mainly in agriculture and forestry as an ingredient of pesticides and defoliants. The most common use is in wood preservatives (most often chromated copper arsenate, CCA) (USGS 2006). Since the 1980s among various As pesticides only CCA was still approved for use. This is why, in the 1990s, more than 80% of total As in the USA was used as a wood preservative (Solo-Gabriele et al. 2003). In 2004 the US Environmental Protection Agency (EPA) introduced a ban on CCA use for residences, led to a drastic decrease in CCA consumption (Jones 2007). Arsenic compounds have also been used as feed additives for poultry, which resulted in increased growth rates, improved feed spent, and better pigmentation. However, because of As's toxicity, As compounds were withdrawn from use in the European Union (EU) in 1998. In the USA they are still in use (Nachman et al. 2005, 2013).

Arsenic is known in history as one of the most commonly used poisons for homicidal and suicidal purposes (Mandal and Suzuki 2002). However, previously, around the world but still in many countries today, As and its compounds were widely used in medicine, especially in treatments of syphilis, various parasitic infections, amoebic dysentery, and trypanosomiasis. In the second half of the twentieth century it was also used in stomatology, where $As_2O_3$ was applied to devitalize the dental pulp (Aso and Abiko 1978). Now recent research has shown the efficiency of As compounds in the treatment of relapsed or refractory acute promyelocytic leukemia (Shen et al. 1997; Antman 2001; Firkin 2014).

## 4 Arsenic in Nature: Geogenic and Anthropogenic Sources

Arsenic is the 20th most abundant element in the Earth's crust, with an average concentration of 0.00005% (Mandal and Suzuki 2002). A natural source of As in the environment is volcanic activity (USGS 2011). Its concentrations in rocks vary significantly around the world and in some geographical regions reach high values (Duker et al. 2005). Arsenic's highest concentrations are found in sedimentary rocks, especially clayey ones (Fig. 13.2). In some offshore areas, claystone concentrations run as high as 490 mg $kg^{-1}$.

In river sediments, As concentrations are even higher—up to 4000 mg $kg^{-1}$ (Mandal and Suzuki 2002). The element occurs in almost all soil types and other environmental matrixes, but its major repositories are aquatic systems (Adriano 2001; Smedley and Kinniburgh 2002; Nordstrom 2002; Kabata-Pendias 2011;

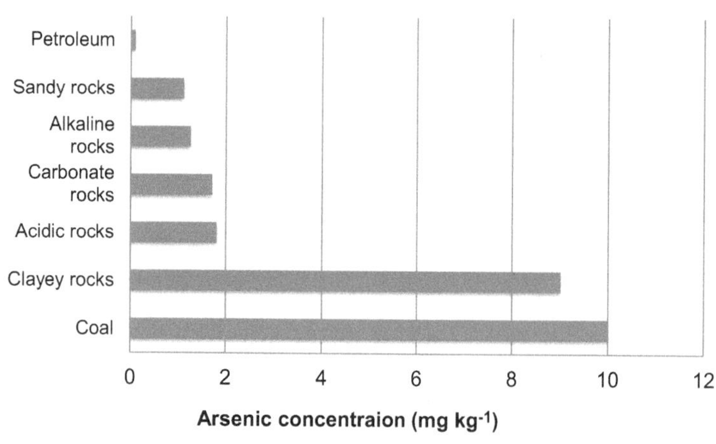


**Fig. 13.2** Average arsenic concentrations in fossil fuels and rocks (data from Kabata-Pendias and Pendias 1999)

Magellan et al. 2014). Arsenic occurrence in soils is strictly connected with the initial material from which soils were formed, and As's form depends on many factors, including oxidation, pH, and microbial activity (Xu et al. 1991). Background As levels in various soils usually do not exceed 10 mg kg$^{-1}$ (Kabata-Pendias 2011). However, owing to the common use of pesticides and other As products, concentrations found in agricultural soils are much higher, even up to 2500 mg kg$^{-1}$ in Japan, the UK, and the USA (Adriano 2001; Kabata-Pendias 2011).

Arsenic is released from soils and rocks into the atmosphere by high-temperature processes and erosion. Later these forms are dispersed with the air on land and in water. However, the most dangerous geogenic exposure to inorganic As (the most toxic form) of humans and animals is through drinking groundwater in a number of places around the world, such as Mexico, the USA, Argentina, Chile, Bangladesh, India, and China (Welch et al. 2000; WHO 2010). The biggest problem is in the Bengal Basin (in Bangladesh and partially in India), where almost 60 million people drink water that contains elevated As levels. One million people have already developed strong symptoms of arsenicosis (Henke 2009).

Among fossil fuels, coal has relatively high As concentrations, within a range of 5–15 mg kg$^{-1}$. Petroleum's concentrations are lower, 0.005–0.14 ppm, with an average value of 0.07 ppm (Fig. 13.2). Concentrations usually found in gasoline fall within a range of 0.02–2 ppm (Kabata-Pendias and Pendias 1999). Fossil fuel combustion and metal smelters are the main anthropogenic As sources next to pesticides and wood-preservative run-off (USGS 2015). In the EU a recent decrease in such activities resulted in a 68% reduction of atmospheric As emissions in the period 1990–2013 (EEA 2016). However, a substantial part of industry still depends on coal combustion, which is linked with As emissions, mainly through particulate matter. Arsenic is observed mainly in the air in the form of arsenites and arsenates. The exceptions are areas where pesticides based on other As forms are sprayed (Davidson et al. 1985). The lowest air As concentration was observed over the South Pole (0.007 ng m$^{-3}$) and Spitsbergen (0.01–1.5 ng m$^{-3}$) (Kabata-Pendias and Pendias 1999). Arsenic concentrations in certain American cities average 2 ng m$^{-3}$ (Chen and Lippmann 2009). Average concentrations in remote areas in the USA were estimated to fall within a range of 1–3 ng m$^{-3}$, in urban areas 20–100 ng m$^{-3}$, and in industrial areas 70–770 ng m$^{-3}$ (ATSDR 2007a; Geiger and Cooper 2010). Arsenic concentrations in Europe are generally low, and the EU As target value in ambient air was established at a level of 6 ng m$^{-3}$ (EU 2005, 2008; Strincone et al. 2013; Guerreiro et al. 2014).

Arsenic compounds are used as feed additives in animal farming, so the possibility of its deposition in manure arises. Simulations show that using manure to enrich agriculture soils in nutrients may lead to pollution of groundwater and the creation of another pathway of exposure, but environmental studies have not confirmed this problem (Nachman et al. 2005; Jones 2007).

## 5 Biological Status of Arsenic

Resistance to and metabolism of As is generally known in bacteria that exert an influence on the global As geocycle (Mukhopadhyay et al. 2002; Stolz et al. 2006). Even As's physiological role is suspected in some types of microbial photosynthesis in biofilms, but it has not been fully proven (Kulp et al. 2008; Schoepp-Cothenet et al. 2009). Arsenic is known as a nonessential and toxic element for plants, but some specimens evolved to metabolize it efficiently (Finnegan and Chen 2012). The main mechanism of its detoxification is the reduction process of arsenate into arsenite, controlled by the arsenate reductase enzyme (Chao et al. 2014). Some fern species are even As hyperaccumulators, but still the reaction of most plants to As compounds makes it possible to use them as ingredients in herbicides and defoliants (NAS 1977; Zhao et al. 2009).

Arsenic essentiality in insects is not known either, and the use of herbicides has demonstrated the sensitivity of insects and other invertebrates to this element (Eisler 1988). The sensitivity of different species may vary significantly, and some of them may play an important role in the retention and cycle of As (Riedel et al. 1989; Schaller et al. 2010).

In birds and mammals, the problem of As essentiality is still disputed. Some observations suggest that inorganic As may be an essential nutrient for goats, chicks, minipigs, and rats (EPA 1998; Adriano 2001). The positive influence of As on animal growth has been long observed and resulted in the use of its compounds in animal breeding as food additives. Studies on birds revealed increased body weight and immune organs of chickens after As supplementation (Ai-zhi and Zhen-yong 2007). However, the main mechanism remained unknown for a long time (Anke 1986). Probably the increased growth of animals bred with the aforementioned feed additives is connected with intestinal health. Organoarsenic additives (the most common being Roxarsone) are very toxic to parasites and significantly decrease their number, which results in a better general condition of animals (Lasky et al. 2004; Jones 2007; FDA 2011). Bearing this in mind, such an influence cannot be treated as a positive function in physiology, but rather as a drug treatment.

### *5.1 Toxicity of Various Arsenic Forms in Homeothermic Animals*

Toxic As's effect is undisputed and significantly depends on its form. Arsenic compounds that are still used in medicine showed adverse effects on the body, including lethal cardiac dysfunctions (Ohnishi et al. 2000; Lin et al. 2005). In spite of the fact that various forms stimulate different levels of toxicity, signs of poisoning are similar (Woolson 1975; NRCC 1978). Generally for all organisms, inorganic As forms are more toxic than organic ones (Tamaki and Frankenberger Jr 1992). However, some observations dispute this statement. The positive relationship between the toxicity of As compounds and solubility in water has been noted (Eisler 1988). The solution showed a toxicity that was as much as ten times higher than that of the undissolved form (Schwartze 1922; Harrisson et al. 1958). Two oxidation states of arsenic are usually discussed in connection with effects on animals: As(III) and As(V). As(III) forms strong bonds with the thiolates of cysteine residues and are regarded as more toxic than As(V)—by as much as 60 times (Rosen 2002; Ventura-Lima et al. 2011). The level of As toxicity also depends on other factors, for example, the species. Comparison of the resistance to As influence between rats and humans revealed that humans are more sensitive than rats (NAS 1977).

The species affects not only the toxicity but also accumulation and distribution of As in the body (Ducoff et al. 1948). The oral $LD_{50}$ of $As_2O_3$ is on the level of 31.5 mg $kg^{-1}$ in mouse and 14.6 mg $kg^{-1}$ in rat. The $LD_{50}$ of As given intraperitoneally is 46.2 mg for mouse and 13.4 for rat. Adequate values of oral intoxication were consecutively 145 mg $kg^{-1}$ and 763 mg $kg^{-1}$, respectively (TOXNET 2015). The acute minimum As lethal dose in humans fell in the 70–200 mg range, or 1 mg $kg^{-1}$ per day (Dart 2004).

In addition to toxicity through ingestion, inhalation of As compounds is also harmful. Lethal cases, diarrhea, respiratory distress, and decreased body weight have been observed in rodents exposed to As pesticide fumes (Stevens et al. 1979). The penetration of organic As in fetuses is negligible, but inorganic As compounds may cross the placental barrier and even cause death of newborns (Lugo et al. 1969).

Medical studies that reveal positive As impacts in leukemia treatment point out also observations regarding the further development of thyroid cancer in patients, probably because of As's carcinogenicity (Firkin 2014). Co-occurrence of lung cancer among people chronically exposed to airborne As compounds has also been observed (Nordberg et al. 2007). The carcinogenic properties of inorganic forms have been confirmed, but the main mechanism is not fully understood (Sakurai 2003). Arsenic and inorganic As compounds have been classified in human carcinogen group 1 based on consistent evidence of associations mainly with lung, skin, and bladder cancers. Arsenobetaine and other organic As compounds have not been classified as carcinogens (IARC 2012). The interactions between As and other elements, such as zinc, selenium, and antimony, are suspected in the etiology of carcinogenicity (Gebel 2000).

Most of the organisms already studied show evolved mechanisms of defense against As toxicity (Rosen 2002; Cullen 2014). Arsenic methylation, which leads to the transfer of inorganic forms into less toxic methylarsenic(V) [MMA(V)], was long treated as a very efficient detoxification process. However, further research showed that methylation may lead to the production of other organic compounds such as methylarsenic(III) (Cullen 2014) (Fig. 13.3). Some methylated organic compounds [such as monomethylarsonate MMA(III)] are more toxic to plants and animals (including humans) than inorganic forms and certain organic forms containing As (V) like dimethyloarsenic [DMA(V)] (Meharg and Hartley-Whitaker 2002; Rahman et al. 2012). Research carried out on human liver cells revealed that the toxicity of various As forms can be presented in the order: MMA(III) > arsenite > arsenate > MMA(V) = DMA(V) (Petrick et al. 2000). MMA and DMA may negatively influence enzymes that work in the energetic cycles in cells. Inorganic As interacts

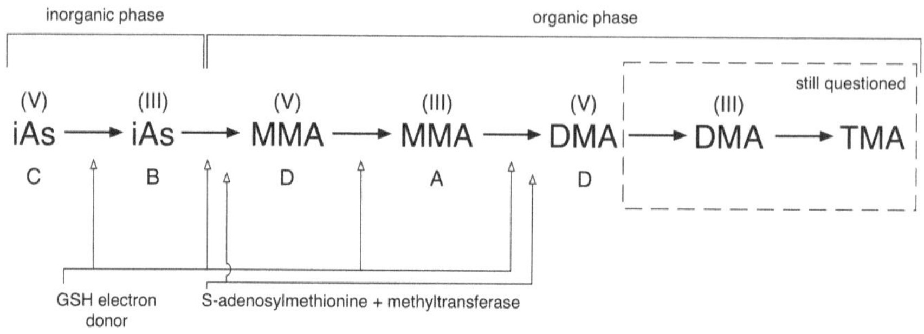


**Fig. 13.3** Methylation of arsenic in liver. Letters (A–D) indicate increasing toxicity of compounds. Roman numbers indicate level of oxidation. *iAs*, inorganic arsenic; *MMA*, monomethylarsenic; *DMA*, dimethyloarsenic; *TMA*, trimethyloarsenic (original scheme)

with sulfhydryl groups; the toxicity of MMA is mainly associated with the thiol groups reaction, and the toxicity of DMA with decreasing oxidative phosphorylation (Khan et al. 2014). Some evidence also suggest that MMA(v) is a carcinogen in rodents. There are also suspicions that in humans, DMA may be methylated further into arsenobetaine, which is characterized by a low toxicity for animals (Kaise et al. 1985; Newcombe et al. 2010).

## 5.2 *Toxicokinetics and Effects of Arsenic in Wildlife*

Both main types of inorganic As—arsenite and arsenate—are well absorbed by ingestion and inhalation. Significantly lower absorption occurs through the skin (ATSDR 2007b). Experiments with $As_2O_3$ in rats showed elevated As concentration in kidneys, liver, lungs, skin, spleen, and blood 24 h following administration. A similar dynamic was observed in humans (Graeme and Pollack 1998). Arsenic accumulates in blood cells, so concentrations in blood and spleen 2 months after subcutaneous implantation remained high. Interestingly, accumulation was not observed in hair and brain (Vallee et al. 1960; Aso and Abiko 1978), but studies on patients during leukemia treatment revealed elevated concentrations in hair, nails, and urine following intravenous infusion of As drugs (Shen et al. 1997).

In higher animals and humans, following various administrations (including oral, fume, and injection exposure), arsenates are partially reduced to arsenites (Vahter and Marafante 1983; Buchet et al. 1998; ATSDR 2007a). This occurs because of the activity of glutathione, which is an electron donor for the reduction (Styblo et al. 2000). Following parental administration of $As_2O_3$ to rabbits, As(III) was the major form of the element detected in blood, lungs, and liver (Vahter and Marafante 1983; Lin et al. 2005). However, the main organ containing As following exposure is the liver (Vahidnia et al. 2007b). Inorganic forms of the element in humans are methylated into MMA and DMA, and partially further to trimethylarsenic (TMA) compounds in liver (Yamauchi and Yamamura 1985; Styblo et al. 2000). The process takes several steps, mainly in the liver, but other organs also showed methylation activity (Fig. 13.3) (Khan et al. 2014). Methylation is catalyzed by methyltransferase, which uses *S*-adenosylmethionine as the methyl group donor (Zakharyan et al. 1995). First, inorganic As(III) is converted into MMA(V). Then MMA(V) is reduced to MMA(III), which is further methylated into DMA(V). Next methylation step (into TMA) is also proceeded by the reduction into DMA(III) (Styblo et al. 2000; Cullen 2014). However, TMA metabolism is still disputed (Kaise et al. 1985; Newcombe et al. 2010) (Fig. 13.3).

The speed of methylation varies among species and is higher, for example, in mice than in rabbits. The excretion of organic forms is faster than that of inorganic forms (Vahter and Marafante 1983). Following absorption, more than 90% of inorganic As is cleared from the blood in 2–4 h and as much as 70% of the intake is excreted in 48 h (Jones 2007).

Arsenic significantly affects the central and peripheral nervous systems. Its effects had already been observed in chronic and acute exposure, but the main mechanisms remain unclear. There are some observations that, not inorganic forms, but organic forms of As are responsible for the impact of the element on nerve cells (Vahidnia et al. 2007a). The time after exposure in single-dose studies varied between 10 days and 3 weeks, but the initial effects may be observed even after a couple of hours (Winship 1984; Vahidnia et al. 2007b). The most characteristic clinical signs of As-induced neuropathy are numbness, paresthesias, and pain (especially in feet soles). They are connected with axonal degeneration and disorganization of the cytoskeletal framework (Vahidnia et al. 2007b).

Studies strictly examining the toxicokinetics and effects of As in birds mainly concern farm species that may be intentionally exposed via feedstuff and unintentionally in some cases through drinking water (Khan et al. 2014). Wild living birds are rarely studied in this respect. Generally, there are numerous observations of lower appetite, weight loss, deterioration of blood parameters, depression, ataxia, dullness, and other neurological disorders among poultry exposed through drinking water or food (Halder et al. 2007; Islam et al. 2009; Sharaf et al. 2013; Khan et al. 2014). Weight loss and even death were observed in mallard (*Anas platyrhynchos*) ducklings (Hoffman et al. 1992). The negative influence of As on the heart (ecchymotic hemorrhages), liver (congestion and hemorrhages), spleen (regression and hemorrhages), kidneys (swelling), and intestinal mucosa (congestion) were already observed in broiler chicks after administration of sodium arsenite (Kalavathi et al. 2011). Studies on bird histopathology showed atrophy of bursa of Fabricius stimulated by a mixture of chemicals including As (also cadmium, lead, benzene, and trichloroethylene), as well as liver lesions (Hoffman et al. 1992; Vodela et al. 1997). Separate analysis of sodium arsenite revealed the substantial negative impact on among others cardiac muscle (disruption of bundles), kidneys (infiltration of mononuclear cells), and spleen (depletion of lymphocytes). The toxic effects of As in these chicks were partially counteracted by supplementation with ascorbic acid and vitamin E (Kalavathi et al. 2011). Increased concentrations of plasma calcium and decreases in plasma glutathione activity are also linked with exposure to As in birds. All toxic effects are more common in birds with limited access to food (Hoffman et al. 1992).

## 5.3 Bioaccumulation of Arsenic

There is a discrepancy in the scientific literature regarding the occurrence and efficiency of As bioaccumulation. Human studies suggest that the element does not bioaccumulate on a large scale, even over time (Jones 2007). Similar observations were made among aquatic organisms, in which the bioconcentration factor (except for algae) is relatively low (Eisler 1988). However, phytoplankton that bioaccumulate As compounds are a major food source for animals of higher levels, so they might be exposed to the element. For this reason, some species of fish are being used in biomonitoring of arsenicals (Rahman et al. 2012).

In terrestrial trophic chains, the situation is different. Generally, inorganic As forms dominate in soil. Soil microbiota may change them into organic ones, methylate and demethylate them, and carry out these processes in opposite ways (Turpeinen et al. 1999). They may get into plants or invertebrates, exposing the animals at higher levels of the trophic pyramid, including birds and mammals, that metabolize the compounds into organic species (Tamaki and Frankenberger Jr 1992; Vahter 2000; Meharg and Hartley-Whitaker 2002).

Despite the fact that As is a known xenobiotic and carcinogen, few studies have been conducted on its concentration and influence in birds. This knowledge gap is especially significant in passerines among which around 75% of studies were conducted only on the great tit (*Parus major*) and pied flycatcher (*Ficedula hypoleuca*). The most often internal tissues have been studied (32.5%), followed by feces (27.5%) and blood (15%) (Sánchez-Virosta et al. 2015). The values considered normal were estimated on levels (mg kg$^{-1}$ dw) 0.01–0.25 for liver and 0.01–0.2 for kidneys (WVDL 2015). Concentrations in internal tissues exceeding 10 mg kg$^{-1}$ wet weight (ww) (~41.6 mg kg$^{-1}$ dry weight, dw, recalculated according to Binkowski 2012) are treated as symptoms of As poisoning (Goede 1985). Such high concentrations are not very common, and in most cases accumulation does not exceed background levels—for liver on average 1.5 mg kg$^{-1}$ dw at pristine sites and 5.8 mg kg$^{-1}$ dw at polluted sites (Berglund et al. 2012; Sánchez-Virosta et al. 2015). However, particular specimens may reach significantly higher concentrations, as high as around 13 mg kg$^{-1}$ dw (pied flycatcher) (Berglund et al. 2012). It is worth mentioning that such high concentrations are harmful to animals since significantly lower concentrations were the cause of disturbances in their biochemistry and growth (Sánchez-Virosta et al. 2015). Among other bird groups, studies on dunlins (*Calidris alpina*) across Europe (including the Netherlands, Norway, and Sweden) revealed that in many cases concentrations were lower than the detection limit. The highest mean concentration was noted for juvenile dunlin and reached 6.2 mg kg$^{-1}$ dw, but the mean value for all studied specimens was lower than 3 mg kg$^{-1}$ dw. Additionally, birds from pristine areas (Scandinavia vs. Western Europe) accumulated significantly lower amounts of the element (Goede et al. 1989). Arsenic concentrations found in the liver of dabbling ducks are similar. A study carried out on common species, including the mallard, blue-winged teal (*Anas discors*), and shoveler (*Anas clypeata*), revealed a mean value of 4.76 mg kg$^{-1}$ dw

(Pereda-Solis et al. 2012). Higher As concentrations were suspected among birds exposed to spent lead shot that contained detectable As levels (Hall and Fisher 1985). Studies verifying this hypothesis in nature did not show high As concentrations in lead-poisoned birds. However, the conclusions are not very obvious because a strong correlation was observed between lead and arsenic in exposed birds (Mateo et al. 2003), so this issue requires further investigation.

Concentrations in feathers are generally lower than in internal organs, and they are used in in vivo biomonitoring. Molting stage and age of the feather are very important here (Burger 1993). The interesting question is whether the contamination of feathers comes mainly from internal distribution or external deposition. A strong argument for the second route is the correlation of As concentrations in preen glands and in feathers (Goede and De Bruin 1984; Goede 1985; Goede et al. 1989). Concentrations of As in feathers of various species of Anseriformes are similar, but a slightly higher one was noted locally for diving duck (pochard *Aythya ferina*), where the mean reached 0.50 μg $g^{-1}$ dw (Karimi et al. 2016). These observations rank Anseriformes in the middle of the common range of species from different trophic levels (e.g., raven, condor, and red knot), whose mean was 0.96 μg $g^{-1}$ dw (Burger 1993). Karimi et al. (2016) also noted an interesting positive but weak correlation between lead and arsenic concentrations in primary feathers of Anseriformes. A similar but stronger relationship was observed in great tit feathers (Janssens et al. 2001). Some studies have also been conducted on As concentrations in bones of birds. The reason for this is that arsenate is structurally very similar to phosphate, which builds bones, so the possibility of phosphate substitution by arsenate may occur (Adriano 2001; WHO 2001; Mateo et al. 2003).

Arsenic is not as widely studied an element in mammals as, for example, lead or cadmium, but because of the potential harm to residential wildlife, studies are being carried out (Saunders et al. 2011). However, among such studies, laboratory acute toxicity issues dominate, and As chronic toxicity studies among wild living animals remain scarce (Drouhot et al. 2014). Normal values, for example in deer, do not exceed 0.5 mg $kg^{-1}$ dw in liver and kidneys. Normal concentrations in other mammals are even lower (WVDL 2015). An interesting procedure for monitoring As levels in shrews was developed by Moriarty et al. (2012). It entails using for analysis the entire torso of an individual. This study revealed that shrews are efficient at processing and excreting As. Animals from heavily contaminated areas may accumulate as much as twice the As body burden as animals from uncontaminated sites. However, not only environmental contamination but also other factors, such as habitat, diet preferences, and animal mobility, play a significant role in exposure and accumulation. Arsenic concentrations in the stomach contents of various small rodent species observed in southern France fell within a very wide range, from below the detection limit to 1669 mg $kg^{-1}$ dw (but in most cases the upper limit did not exceed 50 mg $kg^{-1}$ dw). These values did not correspond clearly to concentrations in soil from different emission zones, and animals from cleaner zones sometimes had higher concentrations in their stomach content. Internal concentrations (mg $kg^{-1}$ dw) fell within a range of 0.05–90.4 for liver, 0.24–50.9 for kidneys, and 0.31–37.7 for lungs (Drouhot et al. 2014). Studies on As accumulation in small mammals led to varying conclusions (Erry et al. 2000). Some revealed that in polluted areas As is accumulated by animals, some revealed no accumulation, and others only in some organs (Sharma and Shupe 1977; Smith and Rongstad 1982; Ismail and Roberts 1992; Peles and Barrett 1997). Studies on large animals, like cattle, showed that particular tissues, such as blood, kidneys, liver, and muscles, accumulate As at statistically different rates of efficiency. What is more, these animals accumulate As only in areas of higher soil concentrations. Maximum values (mg $kg^{-1}$ dw) may reach 122.6 in liver, 135.6 in kidneys, and 8.55 in muscles [values recalculated from ww according to Binkowski (2012)] (López Alonso et al. 2002).

## 5.4 Ecological Effects of Arsenic

The main processes under dispute in terms of the ecological aspects of a given element or compound are bioaccumulation and biomagnification. They are usually separately evaluated for aquatic and terrestrial environments, but generally they are more efficient in aquatic ones. In aquatic environments (both marine and freshwater), inorganic As species dominate, but they are methylated into organic species by aquatic organisms. Because bioaccumulation of total As in fish reaches as high as 22.1, the exposure of predators through fish is likely (Kar et al. 2011). These may include aquatic birds and mammals. In some areas, to limit exposure, As bioremediation with algae is proposed (Magellan et al. 2014).

In the case of terrestrial ecosystems, As bioaccumulation may be observed, as mentioned earlier (Sect. 5.3). However, studies on mammals also reveal that the bioaccumulation factor is lower than 1 (0.69), which means that As bioaccumulation does not occur in these animals (Erry et al. 2000). In both types of environment, a further step in bioaccumulation—As biomagnification—has been widely questioned (Woolson 1975; NRCC 1978; Eisler 1988; Jones 2007). However, the lack of biomagnification does not mean that As does not affect the ecosystem as a whole. It does affect the ecosystem in areas of polluted water or massive amounts of poultry feces deposition, where animals are supplemented with As compounds (Eisler 1988).

## 5.5 Bioindicators and Biomarkers of Arsenic in Ecotoxicological Studies

Metabolomic studies on As bioindicators and biomarkers are mainly done on rodents. Only a few studies have been carried out on humans. It is worth emphasizing that As, after cadmium, is the most frequently studied element in these aspects (García-Sevillano et al. 2015). Despite the fact that the main mechanisms of its toxicity remain unclear, its connection with enzymatic inhibition and oxidative stress is widely observed, which may be employed in biomarker studies. However, there is still a need for As-sensitive and appropriate biomarkers in environmental studies (Marchiset-Ferlay et al. 2012).

It is known that As affects certain enzymes in heme biosynthesis, such as aminolevulinate synthase, porphobilingoen deaminase, and heme oxygenase (Garcia-Vargas and Hernandez-Zavala 1996). The exposure to a mixture of elements (arsenic, cadmium, lead) causes perturbations in lipid and amino acid metabolism in blood serum (Dudka et al. 2014). Additionally, the connection between As exposure and blood porphyrins and their urinary excretion has been observed (Garcia-Vargas and Hernandez-Zavala 1996; Marchiset-Ferlay et al. 2012). All these relationships may be implied to some extent as biomarkers, but their usefulness, especially in animal studies, awaits confirmation. In the range of genotoxicity, DNA damage, chromosomal aberrations, sister chromatic exchange and micronuclei formation are linked with As exposure. Based on these effects, genotoxicity biomarkers may be used (Liou et al. 1999; Chen et al. 2005; Marchiset-Ferlay et al. 2012).

Blood and urine are the most frequently used biomarkers of As exposure. Arsenic is purged from the blood, so the concentration in blood reveals only present and recent exposures (Andrade et al. 2015). A better biomarker is probably concentration in urine, because it reflects chronic exposure. Other very good biomarkers of chronic exposure are hair and nails (thus feathers and claws in animals), but the concentrations here cannot be recalculated as the dose ingested (Marchiset-Ferlay et al. 2012). Arsenic exposure may also be evaluated on the basis of its concentration and distribution in internal tissues. Probably the most commonly studied and useful tissue in this connection is liver tissue.

## 6 Conclusions

- Arsenic is a semimetal that is toxic for birds and mammals at elevated concentrations. Inorganic forms of arsenic possess carcinogenic properties. Additionally, the occurrence of arsenic in water in many parts of the world is an issue of high importance because of the high incidence of arsenicosis in people. Despite its toxicity, As has been used in medicine and as a supplement for farm poultry in some parts of the world (e.g., USA).
- Arsenic forms a part of more than 200 minerals, and its natural sources are rocks, soils, and volcanic activity. In industry, arsenic is used and produced mainly in the form of trioxide. The most prevalent uses of As are in the production of wood preservatives (such as chromated copper arsenate), pesticides, nonferrous alloys, and electronics.
- Bioaccumulation of As in birds and mammals is not high, and biomagnification is still disputed. Concentrations of up to 0.25 mg $kg^{-1}$ dw in bird liver and kidneys are treated as normal. In mammals such values do not exceed 0.5 mg $kg^{-1}$ dw.
- A potential relationship between exposure to arsenic and enzyme activities has been observed, but the precise biomarker needs to be found. Exposure may be

evaluated on the basis of internal concentrations in hair, fur, nails, claws, urine, and organs, preferably in liver.

## References


Adriano DC (2001) Arsenic. In: Trace elements in the terrestrial environments: biogeochemistry, bioavailability, and risks of metals. Springer, New York, pp 219–261

Ai-zhi C, Zhen-yong W (2007) Effect of different supplemented arsenic preparation on growth of body weight and main immune organs in chickens. J Domest Anim Ecol 1:63–65

Andrade VM, Mateus ML, Batoreu MC, Aschner M, Marreilha dos Santos AP (2015) Lead, arsenic, and manganese metal mixture exposures: focus on biomarkers of effect. Biol Trace Elem Res 166:13–23

Andreae MO (1978) Distribution and speciation of arsenic in natural waters and some marine algae. Deep Sea Res 25:391–402

Anke M (1986) Arsenic. In: Mertz W (ed) Trace elements in human and animal nutrition, vol 2. Academic Press, New York, pp 347–372

Antman KH (2001) Introduction: the history of arsenic trioxide in cancer therapy. Oncologist 6 (Suppl 2):1–2

Aso T, Abiko Y (1978) Tissue distribution of arsenic after subcutaneous implantation of arsenic trioxide pellet in rats. J Toxicol Sci 3:109–116

ATSDR (2007a) ToxGuide for Arsenic. CAS # 7440-38-2. U.S Public Health Service, Agency for Toxic Substances and Disease Registry, Atlanta

ATSDR (2007b) Toxicological Profile for Arsenic. Agency for Toxic Substances and Disease, Atlanta

Berglund ÅMM, Rainio MJ, Eeva T (2012) Decreased metal accumulation in passerines as a result of reduced emissions. Environ Toxicol Chem 31:1317–1323

Binkowski ŁJ (2012) The effect of material preparation on the dry weight used in trace elements determination in biological samples. Fresenius Environ Bull 21:1956–1960

Buchet JP, Apostoli P, Lison D (1998) Arsenobetaine is not a major metabolite of arsine gas in the rat. Arch Toxicol 72:706–710

Burger J (1993) Metals in avian feathers: bioindicators of environmental pollution. Rev Environ Toxicol 5:203–311

Chao D-Y, Chen Y, Chen J, Shi S, Chen Z, Wang C, Danku JM, Zhao F-J, Salt DE (2014) Genome-wide association mapping identifies a new arsenate reductase enzyme critical for limiting arsenic accumulation in plants. PLoS Biol 12:e1002009

Chen LC, Lippmann M (2009) Effects of metals within ambient air particulate matter (PM) on human health. Inhal Toxicol 21:1–31

Chen CJ, Hsu LI, Wang CH, Shih WL, Hsu YH, Tseng MP et al (2005) Biomarkers of exposure, effect, and susceptibility of arsenic-induced health hazards in Taiwan. Toxicol Appl Pharmacol 206:198–206

Cullen WR (2014) Chemical mechanism of arsenic biomethylation. Chem Res Toxicol 27:457–461

Dart RC (2004) Medical toxicology. Lippincott Williams & Wilkins, Philadelphia

Davidson CI, Wiersma GB, Brown KW, Goold WD, Mathison TP, Reilly MT (1985) Airborne trace elements in Great Smoky Mountains, Olympic, and Glacier National Parks. Environ Sci Technol 19:27–35

Drouhot S, Raoul F, Crini N, Tougard C, Prudent AS, Druart C, Rieffel D, Lambert JC, Tête N, Giraudoux P, Scheifler R (2014) Responses of wild small mammals to arsenic pollution at a partially remediated mining site in Southern France. Sci Total Environ 470–471:1012–1022

Ducoff HS, Neal WB, Straube R, Jacobson L, Brues A (1948) Biological studies with arsenic; excretion and tissue localization. Proc Soc Exp Biol Med 69:548–554

Dudka I, Kossowska B, Senhadri H, Latajka R, Hajek J, Andrzejak R, Antonowicz-Juchniewicz J, Gancarz R (2014) Metabonomic analysis of serum of workers occupationally exposed to arsenic, cadmium and lead for biomarker research: a preliminary study. Environ Int 68:71–81

Duffus JH (2002) “Heavy metals” – a meaningless term? Pure Appl Chem 74:793–807

Duker AA, Carranza EJM, Hale M (2005) Arsenic geochemistry and health. Environ Int 31:631–641

EEA (2016) European Union emission inventory report 1990–2014 under the UNECE Convention on Long-range Transboundary Air Pollution (LRTAP). Europaen Environment Agency

Eisler R (1988) Arsenic hazards to fish, wildlife, and invertabrates: a synoptic review. Patuxent Wildlife Research Center, Laurel

EPA (1998) Integrated risk information system – arsenic, inorganic (CASRN 7440-38-2). Europaen Environment Agency

Erry BV, MacNair MR, Meharg AA, Shore RF (2000) Arsenic contamination in wood mice (*Apodemus sylvaticus*) and bank voles (*Clethrionomys glareolus*) on abandoned mine sites in southwest Britain. Environ Pollut 110:179–187

EU (2005) Directive 2004/107/EC of the European Parliament and of the Council of 15/12/2004 relating to arsenic, cadmium, mercury, nickel and polycyclic aromatic hydrocarbons in ambient air. Off J Eur Union 23:3–16

EU (2008) Directive 2008/50/EC of the European Parliament and of the Council of 21 May 2008 on ambient air quality and cleaner air for Europe. Off J Eur Communities 152:1–43

FDA (2011) Questions and answers regarding 3-nitro (roxarsone). In: U.S. Food Drug Administration website http://www.fda.gov/AnimalVeterinary/SafetyHealth/ProductSafetyInformation/ucm258313.htm. Accessed 30 Sep 2015

Finnegan PM, Chen W (2012) Arsenic toxicity: the effects on plant metabolism. Front Physiol 3:1–18

Firkin F (2014) Carcinogenic risk of retained arsenic after successful treatment of acute promyelocytic leukemia with arsenic trioxide: a cause for concern? Leuk Lymphoma 55:977–978

García-Sevillano MÁ, García-Barrera T, Gómez-Ariza JL (2015) Environmental metabolomics: biological markers for metal toxicity. Electrophoresis 36:2348–2365

Garcia-Vargas GG, Hernandez-Zavala A (1996) Urinary porphyrins and heme biosynthetic enzyme activities measured by HPLC in arsenic toxicity. Biomed Chromatogr 10:278–284

Gebel T (2000) Confounding variables in the environmental toxicology of arsenic. Toxicology 144:155–162

Geiger A, Cooper J (2010) Overview of airborne metal regulations, exposure limits, health effects and contemporary research. Cooper Environmental Services, Portland

Goede AA (1985) Mercury, selenium, arsenic and zinc in waders from the Dutch Wadden Sea. Environ Pollut 37:287–309

Goede AA, De Bruin M (1984) The use of bird feather parts as a monitor for metal pollution. Environ Pollut 8:281–298

Goede AA, Nygard T, de Bruin M, Steinnes E (1989) Selenium, mercury, arsenic and cadmium in the lifecycle of the dunlin, *Calidris alpina*, a migrant wader. Sci Total Environ 78:205–218

Graeme KA, Pollack CV (1998) Heavy metal toxicity, part I: arsenic and mercury. J Emerg Med 16:45–56

Grund SC, Hanusch K, Wolf HU (2005) Arsenic and arsenic compounds. In: Ullmann’s encyclopedia of industrial chemistry. Wiley, Weinheim, pp 31–34

Guerreiro CBB, Foltescu V, De Leeuw F (2014) Air quality status and trends in Europe. Atmos Environ 98:376–384. https://doi.org/10.1016/j.atmosenv.2014.09.017

Halder G, Mondal S, Paul SK, Roy B, Samanta G (2007) Chronic arsenic toxicity with and without excess supplementation of methionine on the performance and metabolizability of nutrients in layer chicken. Asian J Anim Sci 1:18–25

Hall SL, Fisher FM (1985) Lead concentrations in tissues of marsh birds: relationship of feeding habits and grit preference to spent shot ingestion. Bull Environ Contam Toxicol 35:1–8

Hammond CR (2004) The elements. In: Lide DR (ed) CRC Handbook of chemistry and physics, 86th edn. CRC Press, Boca Raton, pp 1–34
Harrisson JW, Packman EW, Abbott DD (1958) Acute oral toxicity and chemical and physical properties of arsenic trioxides. AMA Arch Ind Health 17:118–123
Haynes WM (2014) Handbook of chemistry & physics, 95th edn. CRC Press, Boca Raton
Henke KR (2009) Arsenic: environmental chemistry, health threats and waste treatment. John Wiley & Sons, Chichester
Hoffman DJ, Sanderson CJ, LeCaptain LJ, Cromartie E, Pendleton GW (1992) Interactive effects of arsenate, selenium, and dietary protein on survival, growth, and physiology in mallard ducklings. Arch Environ Contam Toxicol 22:55–62
IARC (2012) IARC Working Group on the Evaluation of Carcinogenic Risks to Humans. A review of human carcinogens. Part C: arsenic, metals, fibres, and dusts. International Agency for Research on Cancer, Lyon
Islam MS, Awal MA, Mostofa M, Begum F, Khair A, Myenuddin M (2009) Effect of spirulina on biochemical parameters and reduction of tissue arsenic concentration in arsenic induced toxicities in ducks. Int J Poult Sci 8:69–74
Ismail A, Roberts RD (1992) Arsenic in small mammals. Environ Technol 13:1091–1095
IUPAC (1971) Nomenclature of inorganic chemistry. International Union of Pure and Applied Chemistry, London
Janssens E, Dauwe T, Bervoets L, Eens M (2001) Heavy metals and selenium in feathers of great tits (*Parus major*) along a pollution gradient. Environ Toxicol Chem 20:2815–2820
Jones FT (2007) A broad view of arsenic. Poult Sci 86:2–14
Kabata-Pendias A (2011) Trace elements in soils and plants, 4th edn. CRC Press, Boca Raton
Kabata-Pendias A, Pendias H (1999) Biogeochemistry of trace elements (in Polish, Biogeochemia pierwiastków śladowych), 2nd edn. Wydawnictwo Naukowe PWN, Warszawa
Kaise T, Watanabe S, Itoh K (1985) The acute toxicity of arsenobetaine. Chemosphere 14:1327–1332
Kalavathi S, Kumar AA, Reddy AG, Srilatha C, Reddy AR (2011) Sodium arsenite toxicity in broiler chicks and its amelioration: haemato-biochemical and pathological studies. Indian J Vet Pathol 35:171–176
Kar S, Maity JP, Jean J-S, Liu C-C, Liu C-W, Bundschuh J, Lu H-Y (2011) Health risks for human intake of aquacultural fish: arsenic bioaccumulation and contamination. J Environ Sci Health Part A 46:1266–1273
Karimi M-HS, Hassanpour M, Pourkhabbaz A-R, Błaszczyk M, Paluch J, Binkowski ŁJ (2016) Trace element concentrations in feathers of five *Anseriformes* in the south of the Caspian Sea. Iran Environ Monit Assess 188:1–7
Khan A, Hussain HI, Sattar A, Khan MZ, Abbas RZ (2014) Toxico-pathological aspects of arsenic in birds and mammals: a review. Int J Agric Biol 16:1213–1224
Kulp TR, Hoeft SE, Asao M, Madigan MT, Hollibaugh JT, Fisher JC et al (2008) Arsenic(III) fuels anoxygenic photosynthesis in hot spring biofilms from Mono Lake, California. Science 321:967–970
Lasky T, Sun W, Kadry A, Hoffman MK (2004) Mean total arsenic concentrations in chicken 1989-2000 and estimated exposures for consumers of chicken. Environ Health Perspect 112:18–21
Lin C-J, Wu M-H, Hsueh Y-M, Sun SS-M, Cheng A-L (2005) Tissue distribution of arsenic species in rabbits after single and multiple parenteral administration of arsenic trioxide: tissue accumulation and the reversibility after washout are tissue-selective. Cancer Chemother Pharmacol 55:170–178
Liou S, Lung J, Chen Y, Yang T, Hsieh L, Chen C (1999) Increased chromosome-type chromosome aberration frequencies as biomarkers of cancer risk in a blackfoot endemic area increased chromosome-type chromosome aberration frequencies as biomarkers of cancer risk in a blackfoot endemic area. Cancer Res 59:1481–1484

López Alonso M, Benedito JL, Miranda M, Castillo C, Hernández J, Shore RF (2002) Cattle as biomonitors of soil arsenic, copper, and zinc concentrations in Galicia (NW Spain). Arch Environ Contam Toxicol 43:103–108

Lugo G, Cassady G, Palmisano P, Birmingham A (1969) Acute maternal arsenic. Am J Dis Child 117:328

Lunde G (1977) Occurrence and transformation of arsenic in the marine environment. Environ Health Perspect 19:47–52

Magellan K, Barral-Fraga L, Rovira M, Srean P, Urrea G, García-Berthou E, Guasch H (2014) Behavioural and physical effects of arsenic exposure in fish are aggravated by aquatic algae. Aquat Toxicol 156:116–124

Mandal BK, Suzuki KT (2002) Arsenic round the world: a review. Talanta 58:201–235

Marchiset-Ferlay N, Savanovitch C, Sauvant-Rochat MP (2012) What is the best biomarker to assess arsenic exposure via drinking water? Environ Int 39:150–171

Mateo R, Taggart MA, Meharg AA (2003) Lead and arsenic in bones of birds of prey from Spain. Environ Pollut 126:107–114

Meharg AA, Hartley-Whitaker J (2002) Arsenic uptake and metabolism in arsenic resistant and nonresistant plant species. New Phytol 154:29–43

Moriarty MM, Koch I, Reimer KJ (2012) Arsenic speciation, distribution, and bioaccessibility in shrews and their food. Arch Environ Contam Toxicol 62:529–538

Morita M, Edmonds JS (1992) Determination of arsenic species in biological and environmental samples (Technical Report). Pure Appl Chem 64:575–590

Mukhopadhyay R, Rosen BP, Phung LT, Silver S (2002) Microbial arsenic: from geocycles to genes and enzymes. FEMS Microbiol Rev 26:311–325

Nachman KE, Graham JP, Price LB, Silbergeld EK (2005) Arsenic: a roadblock to potential animal waste management solutions. Environ Health Perspect 113:1123–1124

Nachman KE, Baron PA, Raber G, Francesconi KA, Love DC (2013) Arsenic levels in chicken. Environ Health Perspect 121:A267

NAS (1977) Biologic effects of arsenic on plants and animals. In: Arsenic: medical and biological effects of environmental pollutants. National Academy of Sciences, Washington, pp 117–172

Newcombe C, Raab A, Williams PN, Deacon C, Haris PI, Meharg AA, Feldmann J (2010) Accumulation or production of arsenobetaine in humans? J Environ Monit 12:832–837

Nordberg GF, Fowler BA, Nordberg M, Friberg LT (2007) Handbook on the toxicology of metals. Elsevier, London

Nordstrom DK (2002) Public health: worldwide occurrences of arsenic in ground water. Science 296:2143–2145

Norman NC (1998) Chemistry of arsenic, antimony and bismuth. Thomson Science, London

NRCC (1978) Effects of arsenic in the Canadian environment. National Research Council Canada, Ottawa

Ohnishi K, Yoshida H, Shigeno K, Nakamura S, Fujisawa S, Naito K, Shinjo K, Fujita Y, Matsui H, Takeshita A, Sugiyama S, Satoh H, Terada H, Ohno R (2000) Prolongation of the QT interval and ventricular tachycardia in patients treated with arsenic trioxide for acute promyelocytic leukemia. Ann Intern Med 133:881–885

Peles JD, Barrett GW (1997) Assessment of metal uptake and genetic damage in small mammals inhabiting a Fly Ash Basin. Bull Environ Contam Toxicol 59:279–284

Pereda-Solis ME, Martinez-Guerrero JH, Toca-Ramirez JA (2012) Detection of zinc, lead, cadmium and arsenic in dabbling ducks from Durango, Mexico. Asian J Anim Vet Adv 7:761–766

Petrick JS, Ayala-Fierro F, Cullen WR, Carter DE, Vasken Aposhian H (2000) Monomethylarsonous acid (MMA(III)) is more toxic than arsenite in Chang human hepatocytes. Toxicol Appl Pharmacol 163:203–207

Rahman FA, Allan DL, Sadowsky MJ (2004) Arsenic availability from Chromated Copper Arsenate (CCA)–treated wood. J Environ Qual 33:173–180

Rahman MA, Hasegawa H, Lim RP (2012) Bioaccumulation, biotransformation and trophic transfer of arsenic in the aquatic food chain. Environ Res 116:118–135

Ravenscroft P, Brammer H, Richards K (2009) Arsenic pollution: a global synthesis. Willey-Blackwell, Chichester
Riedel GF, Sanders JG, Osman RW (1989) The role of three species of benthic invertebrates in the transport of arsenic from contaminated estuarine sediment. J Exp Mar Bio Ecol 134:143–155
Rosen BP (2002) Biochemistry of arsenic detoxification. FEBS Lett 529:86–92
Sakurai T (2003) Biomethylation of arsenic is essentially detoxicating event. J Health Sci 49:171–178
Sánchez-Virosta P, Espín S, García-Fernández AJ, Eeva T (2015) A review on exposure and effects of arsenic in passerine birds. Sci Total Environ 512–513:506–525
Saunders JR, Hough C, Knopper LD, Koch I, Reimer KJ (2011) Arsenic transformations in terrestrial small mammal food chains from contaminated sites in Canada. J Environ Monit 13:1784–1792
Schaller J, Weiske A, Mkandawire M, Dudel EG (2010) Invertebrates control metals and arsenic sequestration as ecosystem engineers. Chemosphere 79:169–173
Schoepp-Cothenet B, Duval S, Santini JM, Nitschke W (2009) Comment on "Arsenic(III) fuels anoxygenic photosynthesis in hot spring biofilms from Mono Lake, California". Science 323:583
Schwartze EW (1922) The so-called habituation to "arsenic:" variation in the toxicity of arsenious oxide. J Pharmacol Exp Ther 20:181–203
Sharaf R, Khan A, Khan MZ, Hussain I, Abbas RZ, Gul ST et al (2013) Arsenic induced toxicity in broiler chicks and its amelioration with ascorbic acid: clinical, hematological and pathological study. Pak Vet J 33:277–281
Sharma RP, Shupe JL (1977) Lead, cadmium, and arsenic residues in animal tissues in relation to those in their surrounding habitat. Sci Total Environ 7:53–62
Shen ZX, Chen GQ, Ni JH, Li XS, Xiong SM, Qiu QY et al (1997) Use of arsenic trioxide (As2O3) in the treatment of acute promyelocytic leukemia (APL) II. Clinical efficacy and pharmacokinetics in relapsed patients. Blood 89:3354–3360
Smedley P, Kinniburgh D (2002) A review of the source, behaviour and distribution of arsenic in natural waters. Appl Geochem 17:517–568
Smith GJ, Rongstad OJ (1982) Small mammal heavy metal concentrations from mined and control sites. Environ Pollut 28:121–134
Solo-Gabriele H, Sakura-Lemessy DM, Townsend T, Du-bey B, Jambeck J (2003) Quantities of arsenic within the state of Florida. Report #03–06. Florida Center for Solid and Hazardous Waste Management, Gainesville
Stevens JT, DiPasquale LC, Farmer JD (1979) The acute inhalation toxicology of the technical grade organoarsenical herbicides, cacodylic acid and disodium methanearsonic acid; a route comparison. Bull Environ Contam Toxicol 21:304–311
Stolz JF, Basu P, Santini JM, Oremland RS (2006) Arsenic and selenium in microbial metabolism. Annu Rev Microbiol 60:107–130
Strincone M, Fino A, Cattani G, Catrambone M, Pirrone N (2013) Emissions, air concentrations and atmospheric depositions of arsenic, cadmium, lead and nickel in Italy in the last two decades: a review of recent trends in relation to policy strategies adopted locally, regionally and globally. E3S Web Conf 1:38003
Styblo M, Del Razo LM, Vega L, Germolec DR, LeCluyse EL, Hamilton GA et al (2000) Comparative toxicity of trivalent and pentavalent inorganic and methylated arsenicals in rat and human cells. Arch Toxicol 74:289–299
Tamaki S, Frankenberger WT Jr (1992) Environmental biochemistry of arsenic. In: Ware GW (ed) Reviews of environmental contamination and toxicology. Springer, New York, pp 79–110
TOXNET (2015) Toxicology Data Network. NIH U.S. National Library of Medicine. http://toxnet.nlm.nih.gov/. Accessed 4 Oct 2015
Turpeinen R, Pantsar-Kallio M, Häggblom M, Kairesalo T (1999) Influence of microbes on the mobilization, toxicity and biomethylation of arsenic in soil. Sci Total Environ 236:173–180
USGS (1950) 1950 Minerals yearbook arsenic. U.S. Geological Survey

USGS (1955) 1955 Minerals yearbook arsenic. U.S. Geological Survey
USGS (1960) 1960 Minerals yearbook arsenic. U.S. Geological Survey
USGS (1965) 1965 Minerals yearbook minor metals. U.S. Geological Survey
USGS (1970) 1970 Minerals yearbook minor metals. U.S. Geological Survey
USGS (1975) 1975 Minerals yearbook minor metals. U.S. Geological Survey
USGS (1980) 1980 Minerals yearbook minor metals. U.S. Geological Survey
USGS (1985) 1985 Minerals yearbook other metals. U.S. Geological Survey
USGS (1990) 1990 Minerals yearbook arsenic. U.S. Geological Survey
USGS (1995) 1995 Minerals yearbook arsenic. U.S. Geological Survey
USGS (2000) 2000 Minerals yearbook arsenic. U.S. Geological Survey
USGS (2006) 2005 Minerals yearbook arsenic. U.S. Geological Survey
USGS (2011) 2010 Minerals yearbook arsenic. U.S. Geological Survey
USGS (2015) 2013 Minerals yearbook arsenic. U.S. Geological Survey
USGS (2016) Mineral commodity summary arsenic. U.S. Geological Survey
Uthus AO (2003) Arsenic essentially: a role affecting methionine metabolism. J Trace Elem Exp Med:345–355
Vahidnia A, van der Straaten RJHM, Romijn F, van Pelt J, van der Voet GB, de Wolff FA (2007a) Arsenic metabolites affect expression of the neurofilament and tau genes: an in-vitro study into the mechanism of arsenic neurotoxicity. Toxicol In Vitro 21:1104–1112
Vahidnia A, van der Voet GB, de Wolff FA (2007b) Arsenic neurotoxicity – a review. Hum Exp Toxicol 26:823–832
Vahter M (2000) Genetic polymorphism in the biotransformation of inorganic arsenic and its role in toxicity. Toxicol Lett 112–113:209–217
Vahter M, Marafante E (1983) Intracellular interaction and metabolic fate of arsenite and arsenate in mice and rabbits. Chem Biol Interact 47:29–44
Vallee BL, Ulmer DD, Wacker WEC (1960) Arsenic toxicology and biochemistry. Arch Ind Health 21:132–151
Ventura-Lima J, Bogo MR, Monserrat JM (2011) Arsenic toxicity in mammals and aquatic animals: a comparative biochemical approach. Ecotoxicol Environ Saf 74:211–218
Vodela JK, Renden JA, Lenz SD, McElhenney WH, Kemppainen BW (1997) Drinking water contaminants (arsenic, cadmium, lead, benzene, and trichloroethylene). 1. Interaction of contaminants with nutritional status on general performance and immune function in broiler chickens. Poult Sci 76:1474–1492
Welch AH, Westjohn DB, Helsel DR, Wanty RB (2000) Arsenic in ground water of the United States: occurrence and geochemistry. Ground Water 38:589–604
WHO (2001) Environmental health criteria 224: arsenic and arsenic compunds. World Health Organization, Geneva
WHO (2010) Exposure to arsenic: a major public health concern. World Health Organization, Geneva
Winship KA (1984) Toxicity of inorganic arsenic salts. Adverse Drug React Acute Poisoning Rev 3:129–160
Woolson EA (1975) Arsenical pesticides. ACS Symp Ser 7:126–136
WVDL (2015) Normal range values for WVDL toxicology. https://www.yumpu.com/en/document/view/52919318/normal-range-values-for-wvdl-toxicology. Accessed 28 April 2015
Xu H, Allard B, Grimvall A (1991) Effects of acidification and natural organic materials on the mobility of arsenic in the environment. Water Air Soil Pollut 57–58:269–278
Yamauchi H, Yamamura Y (1985) Metabolism and excretion of orally administrated arsenic trioxide in the hamster. Toxicology 34:113–121
Zakharyan R, Wu Y, Bogdan GM, Aposhian HV (1995) Enzymatic methylation of arsenic compounds: assay, partial purification, and properties of arsenite methyltransferase and monomethylarsonic acid methyltransferase of rabbit liver. Chem Res Toxicol 8:1029–1038
Zhao FJ, Ma JF, Meharg AA, McGrath SP (2009) Arsenic uptake and metabolism in plants. New Phytol 181:777–794

# Chapter 14
# Cadmium, Cd

**Agnieszka Tomza-Marciniak, Bogumiła Pilarczyk, Andrzej Marciniak, Jan Udała, Małgorzata Bąkowska, and Renata Pilarczyk**

**Abstract** Cadmium (Cd) is a nonessential and toxic element to animals and plants. In recent years, clear changes in the use of this element have been evident, which has resulted in reduced amounts of Cd being released into the environment. Despite the emissions of Cd gradually decreasing since the 1990s, its presence in different components of the environment is still a severe ecological and health problem. Cadmium, due to its short biological half-life period, acts as a cumulative poison. It shows an explicit tendency to bioaccumulate and even at low levels of exposure can accumulate in animal tissues, reaching values significant as toxic interactions. The highest concentrations of Cd are found in the tissues of free-ranging animals that live in the impact range of Cd emitters. Internal concentrations may even be up to dozens of times higher than in animals from nonpolluted areas. In the tissues of game animals, sometimes levels are found so high that they are rejected for human consumption. Such contamination in the environment causes not only numerous disorders within the organism itself but may also result in changes in the whole ecosystem, especially the negative effects of cadmium in reproductive disorders and increased infant mortality in animals.

## 1 Introduction

Cadmium (Cd) is a toxic element that occurs naturally in the earth's crust. Its average concentration in the lithosphere ranges between 0.10 and 0.20 mg kg$^{-1}$ (or part per million, ppm). Cadmium is emitted from numerous natural and anthropogenic

A. Tomza-Marciniak (✉) · B. Pilarczyk · A. Marciniak · J. Udała · M. Bąkowska
Department of Animal Reproduction Biotechnology and Environmental Hygiene, West Pomeranian University of Technology in Szczecin, Szczecin, Poland
e-mail: agnieszka.tomza-marciniak@zut.edu.pl

R. Pilarczyk
Laboratory of Biostatistics, West Pomeranian University of Technology in Szczecin, Szczecin, Poland



E. Kalisińska (ed.), *Mammals and Birds as Bioindicators of Trace Element Contaminations in Terrestrial Environments*,
https://doi.org/10.1007/978-3-030-00121-6_14

sources into the atmosphere as well as into aquatic and terrestrial ecosystems. It is thought that Cd is emitted mainly by human industrial activities (mining, metal production, fuel combustion, various industrial processes, as well as the use of phosphate fertilizers in agriculture), yet natural emissions may even be 30 times higher than anthropogenic sources (UNEP 2010). Air mass movements play a particular role in the global distribution of Cd. Depending on the atmospheric conditions, dust-suspended Cd may travel up to 1300 km in 3 days, so that even areas located at large distances from the main sources of cadmium emissions are still endangered by its influence. The contamination of soils, water, and air with cadmium is a severe problem as this element does not decompose in the environment and may remain for years, being tied into the net of trophic dependencies. For this reason, as it becomes available to plants, it then follows to animals and humans. Cadmium is toxic to microorganisms, plants, and animals (ATSDR 2012). In birds and mammals, it has been showed that the absorption and accumulation of Cd depend on age and, in a lower level, on gender. Usually old individuals accumulate more Cd in their tissues in comparison to the young, and females accumulate more than males (Scheirs et al. 2006; Mayack 2012). In the body, Cd accumulates mostly in the liver and kidneys, the target organs for the toxic influence of this element. In these organs, Cd induces the synthesis of metallothioneins (MT) that bind Cd ions Cd (II) into CdMT complexes (Klassen et al. 1999). On the basis of the mutual proportions of Cd levels between the kidneys and the liver, we can suppose about the nature of the exposure. A higher concentration of Cd in the liver compared to the kidneys suggests exposure to high doses of Cd, while a higher level in the kidneys indicates chronical exposure to low doses of Cd.

Environmental exposure does not usually cause acute intoxication. Definitely long-term exposure causes intoxication more often. The harmful effects of Cd are more pronounced by disorders in liver and kidney function. Cadmium shows carcinogenic and teratogenic activity and may also induce oxidative stress in cells. This metal has been considered as an endocrine modulator that acts in estrogenic and antiandrogenic ways. By disrupting the hypothalamus-pituitary-gonad axis, Cd affects reproductive potential by interfering with the regulation of the reproductive system. In birds, dietary exposure to Cd resulted in reduced egg production and eggshell thickness. Cd is also a neurotoxin that in wild animals may affect sexual behavior. Moreover, Cd acts negatively on bone metabolism, causing a decrease in bone formation and an increase in bone resorption. Long-term exposure to low environmental levels of Cd results in a decrease in mineral bone density, which is the effect of calcium (Ca) release from bones, as well as disorders in collagen production (ATSDR 2012).

In general, in the tissues of wild animals, different levels of cadmium have been observed. For example, in terrestrial birds, it varies over a wide range in the liver, from undetectable to about 18 mg $kg^{-1}$ dry weight (dw), and in the kidneys from undetectable to almost 60 mg $kg^{-1}$ dw. In the liver and kidneys of terrestrial mammals from different regions of the world, it differs in an even larger scale—from below the limits of detection (<DL) to as high as 75 mg $kg^{-1}$ dw in the liver and from <DL to about 390 mg $kg^{-1}$ dw in the kidneys. Animals from contaminated

regions (both rural and industrial) are characterized by a few to a dozen times higher concentrations of Cd in comparison to uncontaminated reference regions. For this reason, it is believed that wild animals reflect the degree of environmental pollution in which they live. Thus, research on such groups provides useful information on the state of the environment, the scale of exposure, and the possible toxicological threat in a selected area.

## 2 General Properties

Cadmium (Cd, Latin *cadmium*) is a chemical element in the group of transitional metals placed in the group 12 of the periodic table (IIB). It was discovered in 1817 by German chemist Friedrich Strohmeyer, and the name comes from the word *kadmeia*, the Greek name of the zinc ore calamine in which this element was found. Cadmium is a silver-white metal with a blue gloss belonging to the zinc family. Its molecular weight is 8.642 g $cm^{-3}$, which classifies it as a heavy metal. It develops a hexagonal structure. Vapor pressures at 400 °C and 500 °C are 0.2 kPa (1.4 mmHg) and 2.1 kPa (16 mmHg), respectively. The melting point of Cd is 321 °C and boiling point is 767 °C. Compounds of cadmium are characterized by their differing solubility in water. Some of them, e.g., cadmium acetate, cadmium chloride, or cadmium sulfate, are well soluble, while cadmium oxide and cadmium sulfide are almost insoluble. In nature, cadmium does not exist in a free form and creates many compounds in which it is always in a $2^+$ oxidative state. It has six natural, stable isotopes, including $^{106}Cd$, $^{108}Cd$, $^{110}Cd$, $^{111}Cd$, $^{112}Cd$, and $^{114}Cd$, as well as two radioactive isotopes: $^{113}Cd$ (double beta decay, half-life period of $7.7 \times 10^{15}$ years) and $^{116}Cd$ (two-neutrino double beta decay, half-life period of $3.0 \times 10^{19}$ years). There are also a few isotopes than do not occur naturally in the environment. The most common natural isotopes of Cd are $^{110}Cd$, $^{111}Cd$, $^{112}Cd$, and $^{114}Cd$ and constitute 12%, 13%, 24%, and 29% of natural cadmium, respectively (Peterson 2001).

Cadmium is a natural compound of the Earth's crust and is about 0.00005% of its total weight. In general, it is combined with zinc and lead ores. This element is strongly dispersed in the Earth's crust, and its average concentration in the lithosphere ranges between 0.15 and 0.20 mg $kg^{-1}$. This element is located mostly in alkaline igneous rocks, less frequently in acidic (>0.2 mg $kg^{-1}$) and sedimentary rocks (>0.3 mg $kg^{-1}$). Higher concentrations of Cd are observed in phosphate sediments, which may contain even up to 15 mg $kg^{-1}$. Still, in the phosphorites of oceanic islands, levels of Cd may reach even 100 mg $kg^{-1}$ (Adriano 2001).

## 3 Cadmium Minerals, Production, and Uses

The geochemical properties of cadmium are similar to zinc (Zn), but it is more mobile and susceptible to bind with sulfur (S). Cadmium rarely builds its own minerals, e.g., greenockite (CdS), cadmoselite (CdSe), and octavite ($CdCO_3$). Usually Cd occurs as an addition to sulfides of zinc, mercury, lead, and copper, most often in sphalerite (ZnS) in which its content may reach 0.1–5% and 0.3% on average. It also composes wurtzite (ZnS) and galena (PbS). Despite the fact that in none of these Cd-containing minerals does this metal occur in a concentration sufficient to extract it as an ore, the main source of Cd is zinc ores (Butterman and Plachy 2002; De Vos et al. 2006). World resources of Cd have been estimated at about 6 million metric tonnes (based on zinc resources), and world reserves of Cd were estimated at 500,000 tonnes (USGS 2014).

Cadmium is mobilized in the processes of aeration and is bound by loam materials, iron hydroxides, and organic matter, which may explain its susceptibility to accumulate in biolites (Kabata-Pendias and Pendias 1999). The typical levels of Cd in selected biolites are shown in Table 14.1. Mostly the content of Cd in hard coal is higher than in brown coal.

Cadmium, like other elements, may come from natural or anthropogenic sources. Parental rock is the most significant natural source, and more specifically the products of its aeration or weathering, for example, by flowing water. The concentration of cadmium in soils is very varied and usually reaches 0.01–2 mg $kg^{-1}$ dw and most often depends on the content of this element in the paternal rock and the granulometric composition of the soil. The highest concentrations of Cd (even up to 11 mg $kg^{-1}$) are found in soils from very small or small crumb rock, while the lowest are from igneous rock. Soils from metamorphic rock usually contain about 0.1–1 mg $kg^{-1}$ Cd. In general, heavy soils are characterized by even several times the concentration of cadmium than light soils (Alloway 1995).

Soils in uncontaminated regions contain less than 1 mg Cd $kg^{-1}$, while in polluted areas the level of Cd ranges between ten to hundreds of mg $kg^{-1}$ (Table 14.2).

**Table 14.1** The content of cadmium (mg $kg^{-1}$) in selected biolites

| Fossil fuel | Country | Mean | Range | References |
|---|---|---|---|---|
| Hard coal | Bulgaria | 8.60 | ND | Kortenski and Sotirov (2002) |
| | Poland | 0.20 | <0.20–7.70 | Bojakowska (2009) |
| | Spain | 0.30 | ND | Querol et al. (1996) |
| | USA | 0.47 | 0.07–0.14 | Finkelman (1993) |
| Brown coal | Bulgaria | 3.30 | ND | Kortenski and Sotirov (2002) |
| | Poland | 2.00 | <0.2–2.00 | Bojakowska (2009) |
| | USA | ND | 0.04–0.07 | Stricker et al. (2007) |
| Peat | Poland | 0.20 | <0.20–1.80 | Bojakowska (2009) |
| | Russia | 0.56 | 0.20–1.33 | Stepanova et al. (2015) |

*ND* no data, *DL* detection limit

Table 14.2 The concentration of cadmium in various soils

| Country | Concentration (mg $kg^{-1}$ dry wt) | Remarks | References |
|---|---|---|---|
| | | Background level | |
| North Europe | 0.13 | | Reimann et al. (2012) |
| South Europe | 0.22 | | Reimann et al. (2012) |
| USA | 0.36 | | Bradford et al. (1996) |
| China | 0.119 | | Chen et al. (2004) |
| | | Industrial area | |
| France | 44.6 (15.3–8.6) | Area directly adjacent to a Pb and Zn smelter | Tête et al. (2015) |
| Italy | 1.0–4.16 | Home garden on areas near ferroalloy activity | Ferri et al. (2015) |
| Poland | 34.1 (5.24–51.91) | Area directly adjacent to mine and metallurgical plant | Baran et al. (2015) |
| Slovenia | 73.4 (53.6–86.0) | Area surrounding the former lead smelter | Al Sayegh Petkovšek et al. (2014) |
| Spain | 5.0–218 | Near a mineral dressing plant | Álvarez-Ayuso et al. (2013) |
| | | Agricultural area | |
| Croatia | 0.66 | | Romic and Romic (2003) |
| Serbia | 1.4 | | Dragović et al. (2008) |
| Spain | 0.34 (0.10–0.88) | | Micó et al. (2006) |
| USA | 0.27 | | Holmgren et al. (1993) |
| China | 0.27 (0.01–152.95) | | Zhang et al. (2015) |
| | | Urban area | |
| Norway | 0.17 | | Reimann and de Caritat (1998) |
| Spain | 0.32 | | Tume et al. (2011) |
| USA | 0.87 | | McBride et al. (2014) |

The main anthropogenic sources include metal mining and smelting, metallurgic and electronic industries, use of fossil fuels, fertilizing with mineral, and sewage sludge especially including phosphates, as well as municipal and industrial waste. To a certain extent, it also includes warfare and military training. Cadmium penetrates soils in rainwater containing dusts emitted by power plants, local heating plants, road transport, and industrial factories, as well as through direct introduction via fertilizers and pesticides in the past. The physical erosion and chemical decay of mining wastes produced by the exploitation of Zn ores may be a severe threat for ecosystems endangered by the mobilization and emission of toxic elements, including Cd. As noted by Álvarez-Ayuso et al. (2013) in soils of a mining area, the total level of Cd in the soil profile ranged from 5 to 218 mg $kg^{-1}$, while nearby the mining wastes in the superficial layer of the soil, the concentration of cadmium was 0.10–11.0 mg $kg^{-1}$. In most of the studied locations, the amount of Cd was higher than dangerous levels in the biosphere, exceeding 3–8 mg $kg^{-1}$ (Kabata-Pendias and Pendias 1999).

In Europe, the most frequent range of mean Cd levels in superficial soil layers is between 0.1 and 1 mg $kg^{-1}$. Levels below 0.1 mg Cd $kg^{-1}$ had been found in Denmark, Finland, Norway, Sweden, and Portugal, with the highest concentrations found in Switzerland, Greece, Slovenia, and Ireland (Pan et al. 2010). In rural soils it was noted that the background Cd level was <1 mg $kg^{-1}$ and the mean concentration was usually between 0.2 and 0.8 mg $kg^{-1}$ (Adriano 2001). Similar levels of Cd (0.1–0.3 mg $kg^{-1}$) were detected in uncontaminated rural and agricultural areas of NE USA (Holmgren et al. 1993; US EPA 2005). In this matter the situation in some industrialized regions of China seems to be much worse, as the Cd concentration in some areas of agricultural soils exceeded 3 mg $kg^{-1}$, even though the mean concentration of Cd in rural areas of China is the same as in the USA at 0.27 mg $kg^{-1}$ (Zhang et al. 2015).

Phosphate-based fertilizers may contain Cd as a contaminant at levels varying from trace amounts to as much as 300 mg Cd $kg^{-1}$ of dry product and therefore can be considered the main source of cadmium contamination in rural regions (Grant and Sheppard 2008; Six and Smolders 2014). It is estimated that 0.3–10 g Cd $ha^{-1}$ $year^{-1}$ on average is introduced in this manner (Kabata-Pendias and Szteke 2012). An important source of Cd in the soil is sewage sludge used as organic fertilizer. The estimated Cd content in sewage sludges in the UE-27 is 1.8 mg $kg^{-1}$. In the USA, the Cd concentrations ranged from 0.21 to 11.8 mg $kg^{-1}$ (ECB 2005; EPA 2009; Six and Smolders 2014).

Atmospheric transport is probably the most significant vector of Cd distribution in the environment. Once released into the atmosphere, it is diffused, transported by winds, and finally deposited onto the ground and water bodies. This deposition may occur locally (close to the source), regionally, or in distant places from the sources of emission. In some cases, Cd may be transported with air masses for hundreds or thousands of kilometers. The level of atmospheric dispersion and the distance of embedding of this element depend on many factors, including the size of the molecules, height of the emitter (chimney), and meteorological conditions (UNEP 2006). Cadmium present in the atmosphere is adsorbed on the surface of dust particles of different diameter, from <0.1 to 1 μm (Molnár et al. 1995).

The removal of Cd and other dangerous substances on dust particulates happens through wet and dry deposition. Atmospheric deposition of Cd provides increased content in soil in areas located close to the sources of emission, such as power and incineration plants, traffic emitting Cd from combusted fuels, and mechanic attrition of the road surface and building components, but mostly by tire erosion (Fergusson and Kim 1991). For this reason an increased concentration of not only Cd but also other heavy metals is observed next to the roadways. The concentration of Cd in road dust may equal a few to even 11 mg $kg^{-1}$ (Faiz et al. 2009).

The concentration of Cd in the air is strongly diversified. As noted by Reimann and de Caritat (1998), in areas at a distance from the emission sources, the level of Cd ranges from 0.002 to 2.2 ng $m^{-3}$, but in industrialized regions it ranges from 0.5 to even 620 ng $m^{-3}$. Such a great diversity in Cd in the air is mostly the result of the type of activity undertaken in a particular area, the type of installations (including de-dusting modules) and the distance between the point of measure and the source of emission. In urbanized areas, the background levels of Cd ranged from 0.2 to 3 ng $m^{-3}$ (WHO 2007).

In Europe, depending on the region, the concentration of Cd in ambient air is from 0.05 to 0.60 ng $m^{-3}$, reaching the lowest levels in the northern parts (Ilyin et al. 2015). In North America, the concentrations of Cd in ambient air are 1 ng $m^{-3}$ in nonindustrialized areas and 40 ng $m^{-3}$ in urban areas, whereas in Asia the concentration of Cd in ambient air is below the detection range in rural regions up to 12.9 ng $m^{-3}$ in urbanized areas (ATSDR 2012; Shridhar et al. 2010; Islam et al. 2015).

Atmospheric deposition plays an important role in urbanized areas, mostly due to the high concentration of low-height emission sources, car transport and industry. The atmospheric deposition flux of Cd in different regions of the world is strongly diversified, and it is estimated that the dry and wet atmospheric depositions to the soil are <1 and <10 g $ha^{-1}$ $year^{-1}$, respectively (Kabata-Pendias 2001).

Acid rains and soil acidification cause an increase in the geochemical mobility of Cd, which in turn results in an increased concentration of this element in surface and groundwater (Campbell 2006). The range of average concentrations of Cd in the seas and oceans is 0.07–0.11 μg $L^{-1}$ (Kabata-Pendias and Szteke 2012), but some sweetened water bodies with poor connection to oceans, for example, the Baltic Sea, have a concentration reaching 2 μg $L^{-1}$ (Szefer 2002). Much higher concentrations of this element are noted in watercourses, especially in river estuaries (Table 14.3).

The inflow of Cd into rivers includes direct municipal and industrial wastewater discharges, runoff from agricultural areas (anthropogenic sources), and Cd from atmospheric precipitation (natural and anthropogenic sources). Probably about 900 and 3600 tonnes of Cd are deposited into aquatic environments throughout the world through atmospheric deposition of emissions originating from anthropogenic and natural sources. From weathering and erosion, large quantities of Cd (about 15,000 tones) are transported by rivers into the world's oceans (UNEP 2010). Total waterborne Cd input into the Greater North Sea from eight European

**Table 14.3** Cadmium concentrations in river water

| Country | Range (mean) μg $L^{-1}$ | References |
|---|---|---|
| Norway | <0.05–0.09 | Reimann et al. (2009) |
| Belgium | (0.07) | Crévecoeur et al. (2011) |
| France | (0.03) | Idlafkih et al. (1995) |
| Hungary | <0.10–0.60 (0.4) | Bird et al. (2003) |
| Italy | 0.001–0.091 (0.03) | Protano et al. (2014) |
| Latvia | (0.02) | Klavinš et al. (2000) |
| Poland | 0.002–1.090 | Helios-Rybicka et al. (2005) |
| Romania | 0.30–3.00 (1.20) | ICPDR (2002) |
| USA | 1.00–75.0 | Angelo et al. (2007) |

countries, the UK, Sweden, Norway, the Netherlands, Germany, France, Denmark, and Belgium, was evaluated at 23 tonnes $year^{-1}$, with main source of this element from agricultural activity (UNEP 2010).

In water ecosystems, ionic Cd is relatively quickly assimilated by hydrobionts. However the bioavailability may be reduced by the amount of suspended particulate matter (SPM), which is considered as the most important carrier of Cd and other trace metals, and is responsible for the introduction of elements into coastal systems and for the incorporation into sediments. Bottom sediments that are formed as a result of the sedimentation of solid particles become both the reservoir and the source of many pollutants.

## 4 Cadmium Production and Uses

Cadmium is produced as a by-product of processing zinc-bearing ores (80% of production) and refining of lead and copper from sulfide ores (Morrow 2001). This element is collected in the pyrometallurgical recovery and during the recycling of batteries. Recycled Cd is about 18% of the total amount of world Cd supplies (UNEP 2010).

Until the 1980s, the production of Cd in the world was systematically growing, and in 1988 it exceeded 20,000 tonnes $year^{-1}$ and currently is oscillating around this level (Fig. 14.1). Today, China is the world leader in this area and provides about 1/3 of global Cd production. It is estimated that more than 60% of refined Cd production

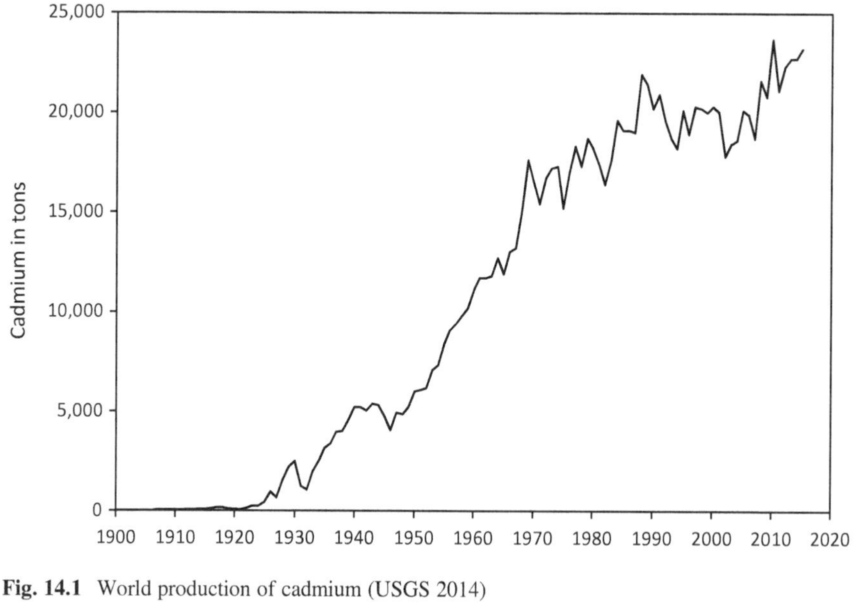


**Fig. 14.1** World production of cadmium (USGS 2014)

is located in Asia and just over a dozen percent in North America (MCS 201CHAPTER 144). Cadmium is used for many purposes, including metal and alloy plating processes due to its anticorrosive properties. Cadmium can also absorb neutrons, so for this reason is used in nuclear reactors (AMAP 2005; UNEP 2006).

Cadmium is used in the production of nickel-cadmium batteries (Ni-Cd), chain reaction controlling rods, and filters to absorb thermal neutrons and pigments (e.g., cadmium yellow) for plastics. The usage of Ni-Cd batteries in consumer electronics has shown a decreasing trend in recent years as a consequence of legal regulations that limit the use of traditional batteries in preference to lithium-ion batteries (Li-Ion) which dominate in laptops, mobile phones, and other small electronic devices due to their better properties. However, the demand for Cd may increase because of a few new possibilities in the use of nickel-cadmium batteries, especially in industry, for storing the energy obtained by some photovoltaic systems (Tolcin 2011).

## *4.1 Cadmium in Nature: Geogenic and Anthropogenic Sources*

Cadmium is emitted into the environment from natural and anthropogenic sources, which mostly have points of distribution. Volcanoes, airborne soil particles, fires, and biogenic release are the main natural sources of Cd emission. The anthropogenic sources include industrial processes, e.g., nonferrous metal production (mining and smelting), stationary fossil fuel combustion, waste incineration, iron and steel production, cement production, and wear of the tread on motor vehicle tires (Fig. 14.2).

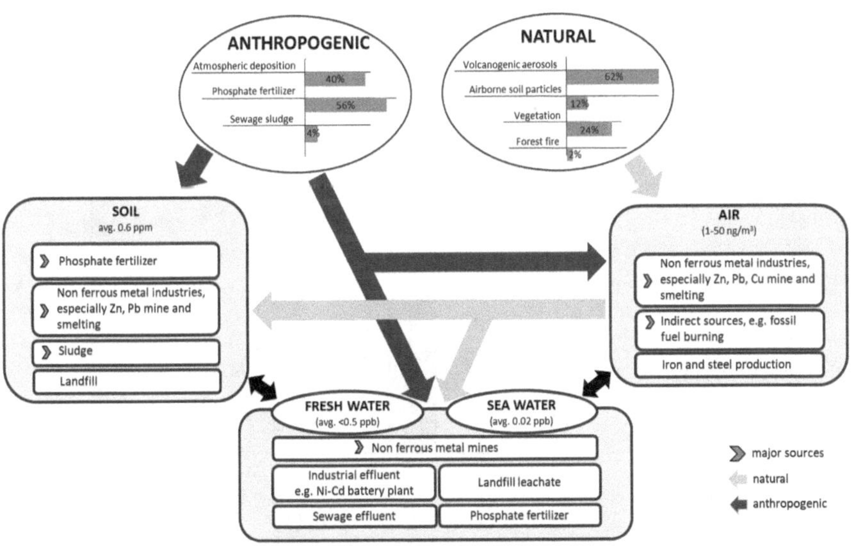


**Fig. 14.2** Natural and anthropogenic sources of cadmium emission

As noted by Henley and Berger (2013), “trace elements in a complex particle—gas mixtures in atmospheric plumes above active volcanoes are enriched by up to several hundred times relative to their parent magmas.” The studies on volcanic materials (lava and gases) performed during the Tolbachik (Kamchatka) eruption in February, April, and May 2013 showed that the concentration of Cd in lava reached about 0.056 ppm, while in emitted gases, it reached 0.60 ppm (Zelenski et al. 2014). Moreover, volcanic gases and dust contributed to significant physicochemical changes in rainwater following eruptions, in the neighborhood of the volcanoes and nearby water reservoirs—they, for example, induced a decrease in the pH and an increase in the level of heavy metals, including Cd (Eriksson et al. 2003).

Forest and peat bog fires are another relevant natural source of Cd and other elements. During such incidents, large amounts of dust transport different organic and inorganic substances, including the toxic metals, into the atmosphere. It was shown that the percentage of metals in the total mass concentration of PM 2.5, emitted during the fire, may range from 1% to 30%, which is subjected mostly to the amount of carbonaceous matter in the dust (See et al. 2007). This fraction of dust coming from the natural fires may contain even about 40 ng Cd $m^{-3}$. It is estimated that the concentration of Cd in such dust may increase by 2–16 times compared to the background (Betha et al. 2013).

It is also estimated that anthropogenic emissions of Cd are a factor of two to three times higher than natural emissions (Pacyna and Pacyna 2001). In 28 EU countries, as reported by the European Environmental Agency (EEA 2015), the 1990–2013 emission of Cd into atmosphere decreased by 75% (from 253 to 63 tonnes $year^{-1}$). Today, China plays the biggest role in global Cd emission from anthropogenic sources. During 1990–2010 the total Cd emission in China increased by about 360% (from 474 tonnes to 2186 tonnes) due to rapid economic growth (Fig. 14.3).

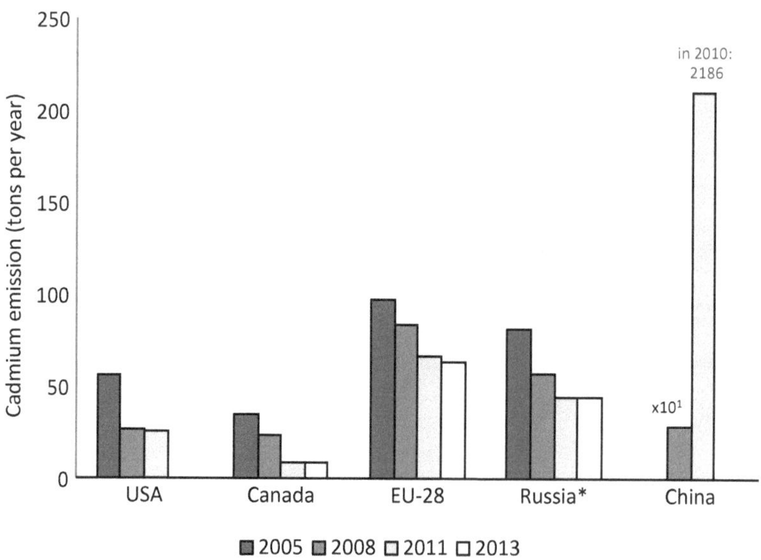


**Fig. 14.3** Total cadmium emission from anthropogenic sources in different countries (asterisk including Asian and European parts of Russia) (Shao et al. 2013; CEIP 2015)

The total atmospheric emission of Cd evaluated for China in 2010 was about 77% and 14% from nonferrous metal smelting and coal combustion, respectively (Shao et al. 2013). Similarly in Canada, the production of nonferrous metals is the main source of atmospheric Cd emission while in the USA fossil fuel combustion (UNEP 2010).

## 5 Biological Status of Cadmium

Bioavailable Cd located in the soil, absorbed by plants and soil invertebrates, continues to appear in the food chain (Fig. 14.4). The invertebrates collect Cd directly from the soil or plants by ingestion, while some are also able to absorb it over the entire surface of the body. Heikens et al. (2001) noted that in most invertebrate taxonomic groups, the concentrations of Cd in the body increased with increasing soil concentrations and could be organized in the order: Lumbricidae > Arachnida > Diplura > Diplopoda > Collembola > Coleoptera. The taxonomic groups for which Cd accumulation is independent of the total soil concentration follow the order: Isopoda > Formicidae > Chilopoda. The authors also observed that internal Cd concentrations were high in Isopoda and low in Coleoptera and Chilopoda. The observed differences in Cd bioaccumulation between the invertebrate species result from differences in morphology, physiology (e.g., regulation capabilities), behavior, habitat, and food preferences. In general, invertebrates that have direct contact with the soil as well as those that consume organic matter rich in Cd accumulate more of this element.

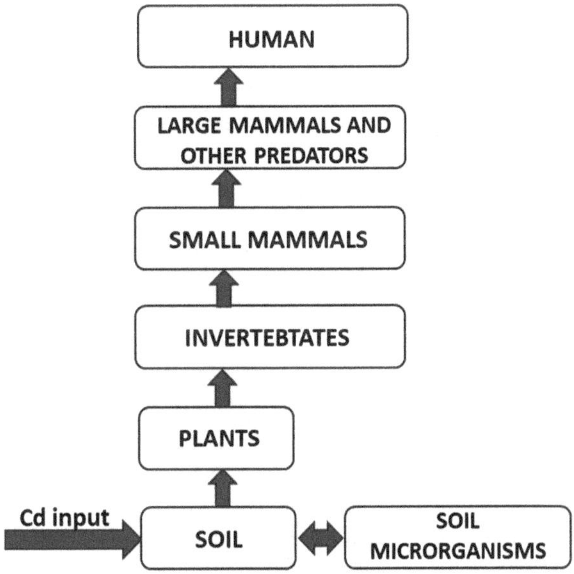


**Fig. 14.4** Translocation of cadmium in the food chain

The amount of Cd accumulated in invertebrates depends also on the effectiveness of detoxification mechanisms. Such detoxification may proceed with the excretion of Cd and/or via storage of Cd in the form of neutral compounds (e.g., metallothionein complexes) and/or by storage in special structures that prevent the transport of Cd to crucial places where it may cause disturbances in biochemical processes (Lindqvist et al. 1995). Earthworms and other invertebrates that are an important link in the food chain accumulate considerable amounts of Cd and mediate its transfer to carnivores, mammals, and birds (Roodbergen et al. 2008; Schipper et al. 2012).

## 5.1 *Toxicity of Cadmium*

Cadmium and its compounds move to the homeothermic organisms of vertebrates via inhalation and ingestion. Via the respiratory system, about 5–50% of inhaled Cd is absorbed, of which 10% (as CdO) accumulates in the lungs, with the remainder entering the bloodstream. Via the digestive system about 1%–10% of Cd is absorbed (ATSDR 2012). Cadmium, an element without any proven biological role, does not have any transporters to facilitate its absorption and distribution in the organism. By using the phenomenon of "ionic mimicry," the absorbed Cd connects with transporters specific for other divalent ions such as $Fe^{2+}$, $Zn^{2+}$, and $Ca^{2+}$. Moreover, Cd can pass through the membranes of erythrocytes with thiol group complexes, including L-cysteine and glutathione, GSH (Bridges and Zalups 2005).

In the distribution of Cd, an important role is played by albumin and metallothionein (MT). Four types of MT are defined, by which MT-1 and MT2 are induced by Cd and other metal ions or stress. In turn, MT-3 and MT-4 are not inducible by Cd (see review by Nordberg 2009). Cadmium complexes with serum proteins are transported to the liver, where the Cd induces the synthesis of new molecules of MT. In hepatocytes, the bigger share of MT binds Cd, and some is released into the bloodstream where it complexes with Cd (CdMT) and in this form moves to the kidneys. CdMT is easily filtered in the renal glomeruli and reabsorbed in the proximal channels. After degradation of CdMT, the $Cd^{2+}$ ions are released and start to act toxically to this organ, leading to disorders (Klassen et al. 1999).

In the case of respiratory exposure, the target organs for Cd are the kidneys and the lungs, while for digestive exposure—the kidneys and the bones (ATSDR 2012). However, it has been shown that the form of this element and its dose also play a role in the transport and deposition of Cd. For instance, Groten et al. (1994) have found that in the oral administration of Cd as $CdCl_2$ and CdMT to rats at a dose of 30 and 90 mg $kg^{-1}$ body weight (mg $kg^{-1}$ bw), higher concentrations of $CdCl_2$ were noted in the liver, while in the kidneys, CdMT accumulated to a higher extent. At smaller doses (0.3 and 3 mg $kg^{-1}$ bw), no such differences were observed.

Cadmium is an element identified with carcinogenic and teratogenic activity. It is found to be a modulator of endocrinal activity, negatively affecting the reproductive process. Also, the cardiotoxicity, hepatotoxicity, and neurotoxicity of Cd have been demonstrated (Limaye and Shaikh 1999; Kim et al. 2013; Xu et al. 2013; Wang et al. 2015). One of the most important aspects of cadmium toxicity is its ability to induce (indirectly) oxidative stress in cells (Stohs et al. 2001; Wang et al. 2015).

In 1993, the International Agency for Research on Cancer (IARC) defined Cd as a carcinogenic substance and qualified it to category I (IARC 1993). The carcinogenic activity of Cd is more related to the oxidative stress this element causes than to its mutagenic activity. Joseph (2009) defined four main mechanisms of Cd activity in the process of carcinogenesis: (1) induction of oxidative stress, (2) aberrant gene expression, (3) blocking DNA repair mechanisms, and (4) blocking apoptosis. The other features of Cd that affect the carcinogenic potential include the ability to cause the aberrant DNA methylation, cell proliferation, and endocrine disruptions. The mechanism of carcinogenic activity of Cd may also concern the disruption of intercellular communication causing changes in protein adhesion, which directly translates into the growth, differentiation, and migration of cells. Between the other tumors caused by the exposure to different Cd compounds, we can include adenocarcinomas of the lung, tumors of the pancreas and the testes, sarcomas, and mammary gland tumors in rats and mice (Garcia-Morales et al. 1994; Waalkes et al. 2000; Goyer et al. 2004; Murphy et al. 2012).

Experimental studies on animals have shown that Cd directly acts on bone cells to decrease bone formation and increase bone resorption. The exposure to even environmental levels of Cd causes a decrease in mineral bone density (demineralization, osteomalation, and osteoporosis) and extends the risk of bone fractures. This results from disturbances in calcium (Ca) metabolism, as well as other bioelements including copper (Cu). It has been demonstrated that oral administration of Cd to laboratory mammals causes a release of Ca from the bones and an increase in Ca in urinary excretion (Wilson and Bhattacharyya 1997). However, in other studies, no changes in the concentrations of circulating 1.25$(OH)_2$ vitamin D, parathormone, nor calcitonin were found, although Cd causes hormonal disturbances (Sacco-Gibson et al. 1992). Bone mechanical strength depends not only on the level of bone mineralization but also on the organic matrix, constructed mostly from collagen. Iguchi and Sano (1982) showed that Cd disturbs the maturation of collagen via inhibiting the Cu-dependent enzyme—lysyl oxidase—that catalyzes the collagen cross-linking.

Cadmium was also defined as an endocrine activity modulator that shows estrogenic and antiandrogenic potential. By disturbing the functions of the hypothalamus-pituitary-gonad axis, Cd disrupts the work of the reproductive system and normal development (Lafuente et al. 2001).

On the basis of former studies, Cd has been defined as a gonadotoxin responsible for decreasing fertility in men and other mammals. Experiments by Haffor and Abou-Tarboush (2004) showed that Cd was a testicular toxicant in mice, causing damage in the endothelium and Sertoli cells. In studies on rats, impairments in spermato- and spermiogenesis were also observed (Hew et al. 1993) as well as a decrease in sperm motility to a total decay in the group with the biggest exposure (0.8 mg $kg^{-1}$ bw) (Xu et al. 2001). In rat females under Cd treatment, a disrupted ovarian histoarchitecture, an extended estrous cycle, and delayed pubertal onset were noted (Samuel et al. 2011).

Studies of the embryotoxicity of Cd in Wistar rats (administration: 20 mg Cd $kg^{-1}$ bw) have demonstrated the occurrence of external malformations and an increase in the number of different anomalies in fetuses with reduced metacarpus ossification, cleft palate, and right or left renal cavitation (Salvatori et al. 2004). Also Aprioku et al. (2014) found many abnormalities in Cd-exposed pregnant female Wistar rats and their offspring. The females had received 4 and 8 mg of Cd $kg^{-1}$ $day^{-1}$ (from conception to gestation), which resulted in a disturbance in the growth and developments of the fetuses, as an effect of improper functioning of the ovaries and placenta. Moreover the authors noted reductions in fetal body weights and limb bone lengths, inhibition of weight gain, and miscarriages. At 4 mg $kg^{-1}$ $day^{-1}$, the number of fetuses in the uterus was significantly lower in comparison to the control group, and at 8 mg $kg^{-1}$ $day^{-1}$, all pregnancies were aborted.

In birds, exposure to Cd causes similar morphological and physiological changes as in mammals. Li et al. (2010), studying testicular toxicity induced by dietary Cd (150 mg $kg^{-1}$ diet/60 days) in cocks, noted a decrease in antioxidative enzyme activity (SOD and GSHPx) in the testicular tissue and a decrease in serum testosterone levels, with a simultaneous increase in the amount of lipid peroxidation and the number of apoptotic cells in the testes. Also, testicular damage was noted, such as edematous testes, severe necrosis, and degeneration of seminiferous tubules, as well as spermatogenesis inhibition. Dietary treatment with Cd resulted in reduced egg production and eggshell thickness in laying hens fed diets supplemented with 3, 12, and 48 mg $kg^{-1}$ of Cd for 12 weeks (Leach et al. 1979).

## 5.2 *Toxicokinetics and the Effects of Cadmium in Wildlife*

Cadmium, in both metallic and saline forms, is poorly absorbed in mammals. It is estimated that via the respiratory, ingestive, and dermal paths, about 25%, 1–10%, and <1% of a dose are absorbed (ATSDR 2012). After absorption, Cd is widely distributed throughout the body and accumulates in many organs, including the liver, kidneys, pancreas, testes, and intestines with the highest concentration found in the liver and kidneys. Although the kidneys are pointed out as the main target for Cd, sometimes the liver shows higher concentrations of Cd. This depends on the form in which the cadmium occurs. Studies by Cherian (1983), aiming to assess the absorption and tissue distribution of Cd in mice (C57BL/6J) repeatedly fed with $^{109}CdCl_2$ or $^{109}$Cd-MT (20 μg Cd once a week for 5 weeks), have shown differences in the distribution of this element in the organs. In the group of mice treated with $^{109}CdCl_2$, the location of Cd between the organs was as follows: liver (73.63%) > kidney (15.62%) > stomach (2.92%) > pancreas (2.52%) > small intestine (2.04%) > cecum (0.41%) > lung (0.41%) > heart (0.39%) > spleen (0.28%) > blood (0.23%) > testes (0.14%) > bone (0.10%). In the group treated with $^{109}$Cd-MT, the ratios were kidney (69.8%) > liver (18.01%) > small intestine (4.24%) > cecum (1.76%) > stomach (1.07%) > pancreas (0.81%) > spleen (0.35%) > lung (0.28%) > heart, blood (0.24%) > testes (0.22%) > bone (0.11%). Moreover these studies revealed a similar body retention of $^{109}CdCl_2$ and $^{109}$Cd-Mt for just the first 3 weeks. In the following weeks, the mice treated with $^{109}CdCl_2$ continued to retain more Cd. In rats it was demonstrated that if fodder contained low levels of Cd (0.11 and 1.1 ppm), its concentration in the liver increased throughout all the experiment (until the eighth month), while in greater concentrations (5–40 ppm) it reached a plateau in the liver in the fourth month. However, in the kidneys, regardless of the dose, the level of Cd constantly increased until the end of the experiment (Hiratsuka et al. 1999).

Studies on birds by García-Fernandez et al. (1995) indicated that bird kidneys are the primary organ for Cd accumulation (61%), followed by the liver (31%), brain (4%), bones (3.5%), and blood (0.5%). Also the amount of absorbed Cd affects the distribution of Cd in the organs of the birds. It was shown that in lower doses, Cd accumulates mostly in the kidneys and, in higher doses, in the liver (Lehman and Klaassen 1986).

In homeothermic vertebrates, exposure to Cd runs in many ways, but the digestive tract has the biggest meaning for environmental exposure to this element. Once absorbed, Cd remains in the organism for a very long time. The biological half-life ($T_{1/2}$) of this metal is very diversified in particular tissues and organs. Based on the Nordberg-Kjellström model, $T_{1/2}$ of Cd was estimated in the human liver and kidneys at 4–19 years and 6–38 years, respectively. For the other tissues, this period has been assessed at 9–47 years (ATSDR 2012). In mouse organs, $T_{1/2}$ is 1.2 years for the liver, 11.8 years for the kidneys, ~1 year for the spleen, and 58 days for the gonads (Matsubara-Khan 1974). Such a long $T_{1/2}$ is caused by the fact that the absorbed Cd is excreted very slowly. It is removed from the organism mainly through the wall of the gastrointestinal tract and to some extent by the hepatobiliary system (Cikrt and Tichý 1974). When Cd enters the organism through ingestion, then the fecal and urinary excretions are approximately equal, and daily excretion is estimated to be 0.007% and 0.009% of body burden, respectively (Kjellström and Nordberg 1985). With respiratory exposure, Cd is excreted mostly with feces and to a lesser extent with urine (Rusch et al. 1986).

In birds, it is considered that they can eliminate Cd and other metals through excretion and depositing Cd in the feathers and eggshell. The most discussed is the role of the eggs in the deposition and elimination of Cd. Eggs, especially their shells, provide an effective method of excretion of Cd. Depending on the species of the bird, the eggshell may contain different amounts of Cd, for example, only 5% of the body burden in roseate terns (*Sterna dougallii*) and 29% for herring gulls (*Larus argentatus*) (Burger 1994). However, other studies suggest that the transfer of Cd to the eggshell is very low and the share in the excretion of Cd from the organism of female birds is low (Leach et al. 1979). Comparative studies by Burger and Gochfeld (1991) on the content of Cd in the feathers and eggs of the common tern (*Sterna hirundo*) have shown that the feathers may contain even more than 12 times more Cd than eggs (0.05 and 0.004 mg $kg^{-1}$, respectively). Due to the relatively low content of Cd in the feathers, the cited authors suppose that during the molting period, the share of feathers in Cd elimination is significant.

Environmental exposure to Cd produces a wide variety of chronic and less frequent acute effects in wildlife species, including mammals and birds, which are often similar to those seen in humans. To estimate the ecological risk for terrestrial wildlife, a toxicity reference value (TRV) is used, which indicates the exposure dose/ concentration above which ecologically significant effects may occur in wildlife species and below which it is expected that such effects will not occur. In the USA, in 2009, the Biological Technical Assistance Group (BTAG) updated the avian cadmium TRV-low (NOAEL, no-observed-adverse-effect level) from 0.08 to 0.07 mg $kg^{-1}$ bw $day^{-1}$, based mainly on data concerning the renal effects of Cd in wood ducks *Aix sponsa* described in the work by Mayack et al. (1981) (DTSC 2009). On the other hand, the lowest-observed-adverse-effect level (LOAEL) was established at 1.0 mg $kg^{-1}$ bw $day^{-1}$ (based on kidney nephrosis in mallards *Anas platyrhynchos*, Cain et al. 1983). The TRV-high as a mid-range adverse effect level was established at 10.43 mg $kg^{-1}$ bw $day^{-1}$ based on reproductive and multiple systemic effects observed in Japanese quail *Coturnix japonica*, described by Richardson and Fox (1974). In reference to mammals, the low and high TRV for Cd were established at 0.06 and 2.64 mg $kg^{-1}$ bw $day^{-1}$) (DTSC 2009). To establish these values, data on the influence of Cd exposure to the reproductive potential of mice were used, as cited in the papers by Webster (1988) and Schroeder and Mitchener (1971).

In ecotoxicological studies on homoeothermic mammals, Cd is primarily measured in the liver and kidneys and less often in muscles (mostly in game animals due to the potential risk of human intoxication). The other tissues are occasionally analyzed in this matter. Scheuhammer (1987) suggested that Cd levels below 3 mg $kg^{-1}$ dw in the liver and 8 mg $kg^{-1}$ dw in the kidneys of adult nonaquatic birds reflect the background levels and above those values indicate an increase in environmental exposure to the element. Adverse effects are expected when Cd concentrations in the liver and kidney of the birds exceed 40 and 100 mg $kg^{-1}$ ww, respectively (Furness 1996). In the case of mammals, acceptable Cd concentrations in the whole kidney were established at $<100$ mg $kg^{-1}$ ww (or $<350$ mg $kg^{-1}$ dw) and in the kidney cortex at 150 mg $kg^{-1}$ ww. Adverse effects on the kidney such as cellular damage are expected at 105 mg Cd $kg^{-1}$ dw and above (Shore and Douben 1994).

## 5.3 *Cadmium Bioaccumulation*

The process of heavy metal accumulation is affected by factors dependent on both the organism (mostly physiological) and the environment. Cadmium is characterized by a high potential of accumulation due to the relatively high mobility and bioavailability for plant and soil invertebrates, mostly earthworms. Al Sayegh Petkovšek et al. (2015) showed that the bio-concentration factor (BCF) of cadmium in soil earthworms ranged from 1.8 to over 29, which indicates that earthworms could be an important source of Cd exposure for animals placed higher in the food chain, such as some mammals (rodents, insectivores, wild boar *Sus scrofa*, badger *Meles meles*) and birds. Despite data indicating an increased concentration of Cd in the tissues of animals from the higher trophic levels (Lazarus et al. 2008; Durkalec et al. 2015), it has not been unambiguously proven if Cd undergoes biomagnification or not (Mann et al. 2011).

### 5.3.1 Cadmium Bioaccumulation in Small- and Medium-Sized Mammals

Small mammals (or micromammals, mainly insectivores like shrews, moles, and rodents), due to the small size of the body and high metabolic rate, are more exposed to the accumulation of environmental pollutants than large mammals. They assimilate Cd mostly through ingestion and to a smaller extent through inhalation or the skin. In the case of fetuses, Cd transfers to their organisms via the placenta and the maternal blood. Similarly, the assimilation of Cd by small mammals depends primarily on the pollution level—mostly of the soil and the type of diet, the way of feeding, trophic level, season, age, and sex. The average concentrations of Cd in this group of mammals, in the case of the liver and kidneys, range from <0.05 to 163.2 and <0.50–99.6 mg kg$^{-1}$ dw (Table 14.4). Analysis of the type of diet and trophic level in relation to the level of Cd in micromammal tissues has shown that omnivorous mammals and carnivores accumulate more Cd in their tissues than herbivores. Hamers et al. (2006) have demonstrated that an estimated daily dietary intake of Cd in the predatory common shrew *Sorex araneus* was about three orders of magnitude higher than in the herbivorous bank vole *Myodes glareolus* (4.6–5.9 vs 0.006–0.007 mg Cd kg$^{-1}$ bw day$^{-1}$, respectively). This reflects in the accumulation of Cd in the kidneys of these animals as the common shrew presented from 5 to even 10 times higher concentrations of Cd than the bank vole. As was presented in the studies by Smith and Rongstad (1982), omnivorous mammal like deer mice *Peromyscus maniculatus* contained more Cd than the herbivorous meadow vole *Microtus pennsylvanicus*. The common shrew, in a diet of which earthworms constituted a significant part, was characterized by a higher concentration of Cd than the bank vole (Veltman et al. 2007a).

Fritsch et al. (2010) in his studies on a multispecific group of small mammals (Rodentia, wood mouse *Apodemus sylvaticus*, bank vole *Myodes glareolus*, European pine vole *Microtus subterraneus*, field vole *Microtus agrestis*; Insectivora, common shrew *Sorex araneus*, pygmy shrew *Sorex minutus*, greater white-toothed shrew *Crocidura russula*) living at diverse distances from a former French smelter have concluded that Cd accumulation in both the liver and the kidneys followed in the animals in the descending order: common and pygmy shrew > wood mouse > voles. It also should be mentioned that in mammals from an area of low metal contamination, the median values of hepatic Cd concentration in insectivorous species ranged from 126 to 163 mg kg$^{-1}$ dw (with a maximum level of 274 mg kg$^{-1}$ dw), and in herbivorous rodents from the same area, medians were between 1.3 and 3.6 mg kg$^{-1}$ dw (with maximum level of 23 mg kg$^{-1}$ dw). The mentioned authors have demonstrated statistically significant differences in the concentration of Cd among rodent species and between rodents and shrews. An analogical regularity in

**Table 14.4** Cadmium concentration (mg $kg^{-1}$ dw) in tissues of small and medium size mammals

| Species | Localization | Concentration | References |
|---|---|---|---|
| *Liver* | | | |
| Yellow-necked mouse *Apodemus flavicollis* | Slovenia | | Al Sayegh Petkovšek et al. (2014) |
| | Control site | 0.17 (0.03–0.36) | |
| | Former lead smelter | 17.9 (0.89–57.1) | |
| | Thermal power plant | 0.37 (0.03–1.80) | |
| | Slovakia | 0.10 (0.03–0.20) | Kramárová et al. (2005) |
| Wood mouse *Apodemus sylvaticus* | Slovakia | 0.23 (0.10–0.30) | Kramárová et al. (2005) |
| | France | | Tête et al. (2015) |
| | Control site | 1.4 (0.3–4.3) | |
| | Former Pb/Zn smelter | 5.6 (1.4–18.6) | |
| | Portugal | | Lourenço et al. (2013) |
| | Control site | 0.30 ± 0.08 (±SD) | |
| | Uranium mining site | 0.44 ± 0.15 (±SD) | |
| Bank vole *Myodes glareolus* | Slovenia | | Al Sayegh Petkovšek et al. (2014) |
| | Lead smelter | 162.7 (12.5–445.5) | |
| | Thermal power plant | 0.56 (0.30–0.86) | |
| | The reference area | 0.63 (0.10–1.82) | |
| | France | | Fritsch et al. (2011) |
| | Smelter-impacted area | 11 (0.13–16) | |
| | Poland | | Włostowski et al. (2009) |
| | Spring | 3.42 | |
| | Autumn | 1.98 | |
| | UK, Wales | | Milton et al. (2003) |
| | Reference area | 0.10 ± 0.03 (±SE) | |
| | Former lead mine | 0.30 ± 0.10 (±SE) | |
| Wild rat *Rattus rattus* | Portugal | | Pereira et al. (2006) |
| | Sulfur mill (SM) | 0.320 ± 0.140 (±SE) | |
| | ~5 km from the SM | 0.381 ± 0.025 (±SE) | |
| Greater white-toothed shrew *Crocidura russula* | France | | Fritsch et al. (2011) |
| | Smelter-impacted area | 72 (3–741) | |

(continued)

Table 14.4 (continued)

| Species | Localization | Concentration | References |
|---|---|---|---|
| Common shrew *Sorex araneus* | France | | Fritsch et al. (2010) |
| | Low contaminated area near former smelter | 163.2 (18.6–267) | |
| Brown hare *Lepus europaeus* | Croatia | | Linšak et al. (2014) |
| | Reference area | 2.52 (1.70–2.98) | |
| | Agricultural area | 3.48 (1.48–8.10) | |
| | Serbia | 0.43 (0.03–1.35) | Petrović et al. (2014) |
| Arctic hare *Lepus arcticus* | Canada | | Pedersen and Lierhagen (2006) |
| | Adult | 4.48 (1.68–10.90) | |
| | Juvenile | 0.196 (0.02–0.44) | |
| | Portugal | 0.23 (0.10–0.73) | Eira et al. (2005) |
| European polecat *Mustela putorius* | Italy | 0.26 | Alleva et al. (2006) |
| American mink *Neovison vison* | USA | 0.33 (0.20–0.71) GM | Mayack (2012) |
| Stone marten *Martes foina* | Italy | 3.17 (<DL–30.06) | Alleva et al. (2006) |
| *Kidney* | | | |
| Yellow-necked mouse *Apodemus flavicollis* | Slovakia | 0.84 (0.28–1.68) | Kramárová et al. (2005) |
| Wood mouse *Apodemus sylvaticus* | Slovakia | 2.08 (0.60–3.88) | Kramárová et al. (2005) |
| | France | | Tête et al. (2015) |
| | Control site | 5.1 (0.3–17.8) | |
| | Former Pb/Zn smelter | 16.5 (4.3–40.1) | |
| | Netherland | | van den Brink et al. (2010) |
| | Floodplain area | 0.18 (0.06–1.81) | |
| | Sandy soil | 9.85 (0.82–86.6) | |
| | Portugal | | Lourenço et al. (2013) |
| | Control site | 0.7 ± 0.27 (±SD) | |
| | Uranium mining site | 1.35 ± 0.49 (±SD) | |

(continued)

**Table 14.4** (continued)

| Species | Localization | Concentration | References |
|---|---|---|---|
| Bank vole *Myodes glareolus* | Poland | | Włostowski et al. (2009) |
| | Spring | 3.36 | |
| | Autumn | 1.88 | |
| | UK, Wales | | Milton et al. (2003) |
| | Reference area | 0.3 ± 0.1 (±SE) | |
| | Former lead mine | 1.9 ± 0.6 (±SE) | |
| | Netherlands | 1.7 | Hamers et al. (2002) |
| Wild rat *Rattus rattus* | Portugal | | Pereira et al. (2006) |
| | Sulfur mill (SM) | 1.28 | |
| | ~5 km from the SM | 0.41 | |
| Common shrew *Sorex araneus* | Netherlands | 7.2 | Hamers et al. (2002) |
| Common shrew *Sorex araneus* | France | | Fritsch et al. (2010) |
| | Low contaminated area near former smelter | 99.6 (4.2–325.7) | |
| Pygmy shrew *Sorex minutus* | France | | Fritsch et al. (2010) |
| | Low contaminated area near former smelter | 42.1 (22.6–626.5) | |
| Brown hare *Lepus europaeus* | Croatia | | Linšak et al. (2014) |
| | Reference area | 12.4 (11.8–20.2) | |
| | Agricultural area | 66.4 (44.8–160.2) | |
| | Slovakia | 5.18 (<DL–15.58) | Kramárová et al. (2005) |
| | Poland | 15.9 (median) | Mysłek and Kalisińska (2006) |
| | Serbia | 4.19 (0.20–17.69) | Petrović et al. (2014) |
| European rabbit *Oryctolagus cuniculus* | Portugal | 3.37 (0.76–12.54) | Eira et al. (2005) |
| Arctic hare *Lepus arcticus* | Canada | | Pedersen and Lierhagen (2006) |
| | Adult | 106.6 (52.2–219.9) | |
| | Juvenile | 1.73 (0.29–4.41) | |
| Pine marten *Martes martes* | Croatia | 4.24 (1.14–19.32) GM | Bilandžić et al. (2012a) |
| Eurasian otters *Lutra lutra* | Korea | 0.339 | Kang et al. (2015) |
| Mink *Neovison vison* | Canada, Yukon | 0.81 | Gamberg et al. (2005) |

(continued)

**Table 14.4** (continued)

| Species | Localization | Concentration | References |
|---|---|---|---|
| *Muscle* | | | |
| Bank vole *Myodes glareolus* | UK, Wales | | Milton et al. (2003) |
| | Reference area | 0.03 ± 0.01 | |
| | Former lead mine | 0.04 ± 0.01 | |
| Wild rat *Rattus rattus* | Spain, Canary Archipelago | 0.0032 | Torres et al. (2011) |
| Brown hare *Lepus europaeus* | Slovakia | 0.16 | Kottferová and Koréneková (2000) |
| | Croatia | | Linšak et al. (2014) |
| | Reference area | 0.13 (0.07–0.19) | |
| | Agricultural area | 0.15 (0.09–1.04) | |
| Arctic hare *Lepus arcticus* | Canada | | Pedersen and Lierhagen (2006) |
| | Adult | 0.082 (0.035–0.201) | |
| | Juvenile | 0.005 (0.002–0.009) | |
| | Canada | 0.20 | Mallory et al. (2004) |
| | Baffin Island | | |
| European rabbit *Oryctolagus cuniculus* | Portugal | 0.16 (0.12–0.16) | Eira et al. (2005) |

Mean and range in parentheses
*GM* geometric mean, *SD* standard deviation, *SE* standard error

the diversification of Cd level in the tissues of micromammals with a different diet (shrews > voles > wood mouse) was noted also by other researchers (Gall et al. 2015). The interspecific differences observed in the concentration of Cd in tissues may also have a physiological ground. Hunter et al. (1987) suggest that the field vole and wood mouse appeared to regulate metal accumulation, while common shrew showed considerable accumulation of Cd. The examples of concentration of Cd in the tissues of small and medium size are shown in Table 14.4.

In the case of insectivorous shrews, it was found that the concentration of Cd in the liver and in the whole organism is closely correlated to the content of Cd in the soil, while in the case of herbivorous voles, no such relation was noted (Veltman et al. 2007b). Van den Brink et al. (2011), in studies concerning not only soil properties on Cd accumulation in small mammals but also species traits and habitat, found that Cd accumulation patterns differed between habitats and were affected by species traits. The authors indicated that local soil properties were most important for predicting Cd accumulation in the specific feeding nonmobile species like the common vole. In the case of the mobile species (wood mice) with a variable diet, the most important factor was diet composition. And for intermediately mobile species (bank vole) with a variable diet, both diet and soil properties were of significance.

Studies on Cd bioaccumulation concerning both sex and age of the micro-Mammalia do not always confirm a connection between the concentration of this metal in tissues and diet and soil properties or even results opposite to each other. For instance, Sánchez-Chardi et al. (2007) concluded that in insectivorous shrews, the content of Cd was rising with age. Similarly, Hunter et al. (1989) noted that between three species of micromammals, wood mouse, field vole, and common shrew, only in the shrew did Cd accumulate significantly positively with age, which corresponds to the results of Hunter et al. (1989). However Fritsch et al. (2010) stated that Cd concentrations in the liver and kidneys increased with age in herbivorous (wood mouse, bank vole, field vole, European pine vole) and insectivorous (common shrew, pygmy shrew, greater white-toothed shrew) micromammals. Blagojević et al. (2012) noted that in the black-striped field mouse *Apodemus agrarius* from Lešnica, Serbia (unpolluted area), the highest accumulations of Cd were characteristic for the young individuals, and not the adults. These authors explain these findings twofold: (1) the higher Cd accumulation rate may be caused by the higher metabolic rate of the juveniles, which translates into a higher uptake of food and accordingly Cd, and (2) in adult individuals a decrease in intestinal absorption of Cd takes place. Due to the fact that the authors have not found any differences between Cd accumulation in young individuals from polluted and not polluted areas, they suggest that in neonates and juveniles, the bioaccumulation of Cd and other heavy metals is more likely related to the high rate of metabolism than the degree of pollution of the environment. In carnivores like the American mink *Neovision vision*, European otter *Lutra lutra*, and polecat *Mustela putorius*, Cd concentrations usually increased with age (Hyvärinen et al. 2003; Grove and Henny 2008; Mayack 2012). Also in herbivorous species of the Leporidae family, an analogical regularity was observed, where in brown hares *Lepus europaeus* aged 3–6 months, the concentration of Cd in the kidneys was 0.39 mg $kg^{-1}$ ww and was about one order of magnitude lower than in individuals aged 24–36 months (2.37 mg $kg^{-1}$ ww). In the case of the liver, the oldest brown hare individuals had about five times higher concentrations of Cd than the youngest (Petrović et al. 2014).

The relations between sex and Cd bioaccumulation in small mammals are rarely analyzed, and studies in this area usually do not confirm any differences between males and females (Hyvärinen et al. 2003; Lemarchand et al. 2010). If they are found, usually females have higher levels of Cd than males (Scheirs et al. 2006). Mayack (2012) explained these with gender differences in relative growth.

Another factor that may affect the accumulation of Cd in animal organisms is the seasonal rhythm of changes and the diversity in diet composition related to it. Hunter et al. (1987) observed differences in the level of Cd in the field vole, wood mouse, and common shrew, which were the effect of seasonal changes in the diet of each. These results correspond to an analysis by Włostowski et al. (2009), who found that the accumulation of Cd in the liver and kidneys was about 70% higher in the bank voles caught in March than in November.

Fernández et al. (2012) in studies on the variability in the transfer and bioaccumulation of heavy metals, including Cd, in trophic compartments of terrestrial ecosystems in Spain, observed a common accumulation pattern: secondary consumers (shrews) > detritivores (slugs) > primary consumers (wood mouse) > productors (oaks), yet without biomagnification of Cd in the woodlands.

Among the semiaquatic carnivore species, some differences in the concentration of Cd were found that were due to different compositions in diet. As noted by Mayack (2012), Cd accumulation in American mink was greater than in the river otter *Lontra canadensis.* The typical terrestrial prey of the mink are small mammals, especially those dependent on detritus-based food chains that may contain increased Cd levels, synonymous with transferring a higher Cd load to the organism of the mink. In contrast to the mink, the river otter feeds mainly on fish. The author suggests that biomagnification of Cd in land trophic chains may affect the higher and more diversified concentrations of Cd in minks than otters.

The accumulation of Cd in particular organs of animals depends on their level of exposure. Hunter et al. (1989) have observed that in the common shrew, with low exposure, accumulation was located in the kidneys and liver, while with higher exposure (at ingestion rates of 25 mg $kg^{-1}$ $day^{-1}$), Cd accumulated mostly in the liver. In studies on Cd in the brown hare, it was demonstrated that the main organs of accumulation in both the contaminated and reference areas were the kidneys. However, the ratio between Cd levels in the kidneys and the concentration in the liver ($Cd_{kidney}$:$Cd_{liver}$) of hares from the polluted area was a few times higher than in individuals from the reference area, at 15.7 and 4, respectively. Moreover, differences in the accumulation of Cd in the muscles and brain between the studied animals were found. In the polluted area, the concentration of Cd in the brain of the hare was higher than in the muscles, while in the reference area, more Cd was found in the muscles than in the brain (Linšak et al. 2014). High values of the ratio $Cd_{kidney}$:$Cd_{liver}$ (from 9 to 30) in hares were also reported by Petrović et al. (2014), wherein they indicated that this ratio was highest in older individuals.

### 5.3.2 Bioaccumulation in Large Mammals

In Europe and North America, among the herbivorous ungulates, most data concerns the concentration of Cd in cervids like roe deer *Capreolus capreolus*, red deer *Cervus elaphus*, moose *Alces alces*, white-tailed deer *Odocoileus virginianus*, and reindeer *Rangifer tarandus* (reindeer in Europe, caribou in North America). Omnivorous and carnivorous animals are less intensively studied in this respect. In the liver and kidneys of large herbivorous mammals (cervids) from different regions of the world, mean concentrations of Cd ranged from 0.19 to 32 and from 3 to 172 mg $kg^{-1}$, respectively, and in Leporidae from 0.2 to 4.5 and from 1.5 to 107 mg $kg^{-1}$ dw (Tables 14.4 and 14.5), respectively.

In omnivores and carnivores, their hepatic and renal concentrations ranged from <DL to 20 and from <0.8 to almost 200 mg $kg^{-1}$ dw, respectively (Tables 14.6 and 14.7). The degree of environmental contamination had a significant effect on Cd bioaccumulation in cervids from polluted areas that lived in close proximity to mines or ironworks—those animals had a few or even dozens of times higher concentrations of Cd in comparison to individuals of the same species from uncontaminated reference areas. For example, studies in Poland have demonstrated that mean

**Table 14.5** Cadmium concentration (mean and range in parentheses; mg $kg^{-1}$ dw) in tissues of large herbivores

| Species | Localization | Concentration | References |
|---|---|---|---|
| *Liver* | | | |
| Red deer *Cervus elaphus* | Croatia | 0.54 | Lazarus et al. (2014) |
| | Poland | 0.42 | Wieczorek-Dąbrowska et al. (2013) |
| | Slovakia | 0.93 (0.25–3.11) | Kramárová et al. (2005) |
| | Spain | | Rodríguez-Estival et al. (2011) |
| | Reference area | 0.197 (0.066–0.616) | |
| | Mines | 0.275 (0.099–1.162) | |
| Reindeer *Rangifer tarandus* | Norway | 2.33 (0.621–7.804) | Hassan et al. (2012) |
| Caribou *Rangifer tarandus* | Canada, Québec | | Robillard et al. (2002) |
| | | 3.57 immature | |
| | | 3.78 adult | |
| Moose *Alces alces* | Canada, Yukon | 17.49 | Gamberg et al. (2005) |
| | Canada, Nova Scotia | 5.8 (<0.05–51.9) | Pollock (2005) |
| | USA, Alaska | 3.64 (0.21–32.13) | Arnold et al. (2006) |
| | Sweden | 5.93 (0.86–21.96) | Frank et al. (2000) |
| Roe deer *Capreolus capreolus* | Croatia | 1.824 | Lazarus et al. (2014) |
| | Poland | | Durkalec et al. (2015) |
| | Polluted area | 22.97 (3.35–73.58) | |
| | Control area | 0.41 (0.05–1.83) | |
| | Czech | 0.79 | Čelechovská et al. (2008) |
| | Slovakia | 0.96 | Kottferová and Koréneková (2000) |
| Fallow deer *Dama dama* | Croatia | 0.45 | Lazarus et al. (2014) |
| | Czech | 0.25 | Čelechovská et al. (2008) |
| White-tailed deer *Odocoileus virginianus* | Canada, Nova Scotia | 1.1 (0.05–28.1) | Pollock (2005) |
| European bison *Bison bonasus* | Poland | 1.61 (1.11–2.07) | Włostowski et al. (2006) |
| | Poland | 2.96 | Kośla et al. (2008) |
| Kidney | | | |
| Red deer *Cervus elaphus* | Croatia | Whole kidney, 9.44 | Lazarus et al. (2008) |
| | | Cortex, 21.57 | |

(continued)

**Table 14.5** (continued)

| Species | Localization | Concentration | References |
|---|---|---|---|
| | Slovakia | 10.40 (1.40–25.06) | Kramárová et al. (2005) |
| Roe deer *Capreolus capreolus* | Poland | | Durkalec et al. (2015) |
| | Polluted area | 172 (0.05–390.98) | |
| | Control area | 6.88 (1.01–39.26) | |
| | Czech | 2.99 | Čelechovská et al. (2008) |
| | Slovakia | 8.09 | Kottferová and Koréneková (2000) |
| Moose *Alces alces* | Canada, Yukon | 122.24 | Gamberg et al. (2005) |
| | Canada, Nova Scotia | 60.4 (14.3–346.1) | Pollock (2005) |
| | Alaska | Cortex: 38.19 | Arnold et al. (2006) |
| | | (0.435–285.80) | |
| | Sweden | 41.76 (3.74–107.88) | Frank et al. (2000) |
| Caribou *Rangifer tarandus* | Canada, Québec | | Robillard et al. (2002) |
| | | 23.19 immature | |
| | | 30.80 adult | |
| | Canada | 28.28 (6.53–191.4) GM | Pollock et al. (2009) |
| Fallow deer *Dama dama* | Czech | 3.01 | Čelechovská et al. (2008) |
| European bison *Bison bonasus* | Poland | Cortex: 12.14 | Włostowski et al. (2006) |
| | | (8.48–15.31) | |
| | Poland | 6.28 | Kośla et al. (2008) |
| Muscle | | | |
| Red deer *Cervus elaphus* | Spain | | Taggart et al. (2011) |
| | Control area | 0.119 (0.007–0.867) | |
| | Mined area | 0.112 (0.007–1.211) | |
| Red deer | Croatia | 0.21 | Lazarus et al. (2014) |
| Reindeer *Rangifer tarandus* | Norway | 0.007 (0.004–0.025) | Hassan et al. (2012) |
| Caribou *Rangifer tarandus* | Canada, Québec | | Robillard et al. (2002) |
| | | 0.046 immature | |
| | | 0.046 adult | |
| Roe deer *Capreolus capreolus* | Croatia | 0.039 | Lazarus et al. (2014) |
| | Poland | | Durkalec et al. (2015) |
| | Polluted area | 0.15 (0.011–0.539) | |
| | Control area | 0.025 (<DL–0.142) | |
| | Czech | 0.028 | Čelechovská et al. (2008) |
| | Slovakia | 0.143 | Kottferová and Koréneková (2000) |

(continued)

**Table 14.5** (continued)

| Species | Localization | Concentration | References |
|---|---|---|---|
| Fallow deer *Dama dama* | Croatia | 0.086 | Lazarus et al. (2014) |
| | Czech | 0.021 | Čelechovská et al. (2008) |
| Moose *Alces alces* | Canada, Yukon | 0.107 | Gamberg et al. (2005) |

*GM* geometric mean, *DL* detection limit

**Table 14.6** Cadmium concentration (mean and range in parentheses; mg kg$^{-1}$ dw) in different tissues of carnivorous mammals

| Species | Localization | Concentration | References |
|---|---|---|---|
| *Liver* | | | |
| Red fox *Vulpes vulpes* | Switzerland | 1.72 | Dip et al. (2001) |
| | Italy | 0.33 (<DL–2.54) | Alleva et al. (2006) |
| | Spain | 0.113 (<DL–1.425) GM | Millán et al. (2008) |
| | Slovakia | 0.69 (0.40–1.25) | Piskorová et al. (2003) |
| Iberian wolf *Canis lupus signatus* | Spain | 0.53 (<DL–1.55) | Hernández-Moreno et al. (2013) |
| Gray wolf *Canis lupus* | Croatia | 0.055 Med | Vihnanek Lazarus et al. (2013) |
| | Russia | <DL | Shore et al. (2001) |
| | Canada | 1.42 | Gamberg and Braune (1999) |
| Iberian lynx *Lynx pardinus* | Spain | 0.09 (0.037–0.254) GM | Millán et al. (2008) |
| Eurasian lynx *Lynx lynx* | Croatia | 1.07 (0.828–1.393) GM | Bilandžić et al. (2012a) |
| *Kidneys* | | | |
| Red fox *Vulpes vulpes* | Switzerland | 5.37 | Dip et al. (2001) |
| | Slovakia | 0.925 (0.74–1.776) | Piskorová et al. (2003) |
| Iberian wolf *Canis lupus signatus* | Spain | 2.69 (0.079–5.14) | Hernández-Moreno et al. (2013) |
| Gray wolf *Canis lupus* | Canada | 5.93 | Gamberg and Braune (1999) |
| | Croatia | 0.925 (0.033–4.477) GM | Bilandžić et al. (2012a) |
| *Muscle* | | | |
| Gray wolf *Canis lupus* | Croatia | 0.037 (0.011–0.525) GM | Bilandžić et al. (2012a) |
| Red fox *Vulpes vulpes* | Spain | 0.007 (<DL–0.047) GM | Millán et al. (2008) |
| | Slovakia | 0.111 (0.074–0.259) | Piskorová et al. (2003) |
| Iberian lynx *Lynx pardinus* | Spain | 0.004 (<DL–0.009) GM | Millán et al. (2008) |
| Eurasian lynx *Lynx lynx* | Croatia | 0.011 (0.007–0.015) GM | Bilandžić et al. (2012a) |

*DL* detection limit, *GM* geometric mean, *Med* median

**Table 14.7** Cadmium concentration (mean and range in parentheses; mg $kg^{-1}$ dw) in tissues of omnivorous mammals

| Species | Localization | Concentration | References |
|---|---|---|---|
| *Liver* | | | |
| Wild boar *Sus scrofa* | Croatia | 1.40 | Lazarus et al. (2014) |
| | | 1.17 | Bilandžić et al. (2012b) |
| | Poland | | Durkalec et al. (2015) |
| | Polluted area | 19.90 (2.74–141.37) | |
| | Control area | 0.67 (0.104–3.37) | |
| | Spain | | Taggart et al. (2011) |
| | Control area | 0.119 (0.007–0.867) | |
| | Mined area | 0.112 (0.007–1.211) | |
| | Italy | 0.303 (0.029–1.360) | Amici et al. (2012) |
| | Slovakia | 1.00 (0.43–3.36) | Piskorová et al. (2003) |
| European badger *Meles meles* | Croatia | 1.92 (1.107–3.713) GM | Bilandžić et al. (2012a) |
| | Italy | 2.39 (<DL–17.32) | Alleva et al. (2006) |
| | Czech | 3.89 | Bukovjan et al. (2014) |
| | Spain | 0.017 GM | Millán et al. (2008) |
| Brown bear *Ursus arctos* | Croatia | 1.239 (1.035–1.499) GM | Bilandžić et al. (2012a) |
| | Croatia | 3.891 | Lazarus et al. (2014) |
| | Slovakia | 2.959 (0.893–4.427) | Čelechovská et al. (2006) |
| *Muscle* | | | |
| Wild boar *Sus scrofa* | Croatia | 0.208 | Lazarus et al. (2014) |
| | Poland | | Durkalec et al. (2015) |
| | Polluted area | 0.16 (0.032–0.560) | |
| | Control area | 0.016 (<0.001–0.088) | |
| | Spain | | Taggart et al. (2011) |
| | Mined area | 0.41 (<DL–1.566) | |
| | Control area | 0.11 (<DL–0.456) | |
| | Italy | 0.316 (0.124–1.524) | Amici et al. (2012) |
| | Slovakia | 0.16 (0.08–0.68) | Piskorová et al. (2003) |
| Eurasian badger *Meles meles* | Spain | 0.003 GM | Millán et al. (2008) |
| European badger *Meles meles* | Croatia | 0.036 (0.024–0.08) GM | Bilandžić et al. (2012a) |
| | Czech | 0.40 | Bukovjan et al. (2014) |
| Brown bear *Ursus arctos* | Croatia | 0.032 | Lazarus et al. (2014) |
| | Croatia | 0.016 (0.012–0.020) GM | Bilandžić et al. (2012a) |
| | Slovakia | 0.068 (0.036–0.0.168) | Čelechovská et al. (2006) |
| *Kidney* | | | |
| Wild boar *Sus scrofa* | Italy | 4.208 (0.064–12.32) | Amici et al. (2012) |
| | Slovakia | 2.24 (0.56–109.28) | Piskorová et al. (2003) |
| Brown bear *Ursus arctos* | Slovakia | 69.44 (33.60–4.588) | Čelechovská et al. (2006) |
| | Croatia | Cortex, 66.0 | Lazarus et al. (2014) |
| European badger *Meles meles* | Croatia | 11.285 (6.475–30.44) GM | Bilandžić et al. (2012a) |
| | Czech | 8.24 | Bukovjan et al. (2014) |

*DL* detection limit, *GM* geometric mean

concentrations of Cd in the kidneys of roe deer from highly industrialized and from uncontaminated areas were 172 and less than 7 mg $kg^{-1}$ dw, respectively. An analogous difference was found in the kidneys of the omnivorous wild boar *Sus scrofa* 198 vs ~8 mg $kg^{-1}$ dw (Durkalec et al. 2015). In large carnivorous mammals like the wolf *Canis lupus*, a high diversity in tissue bioaccumulation of Cd was observed, but still not as high as in ungulates. The highest concentration of this metal was found in the livers of wolves from Yukon, Canada (1.42 mg $kg^{-1}$ dw), and the lowest in the European wolves from Spain and Croatia, respectively, 0.53 and 0.05 mg $kg^{-1}$ dw (Gamberg and Braune 1999; Bilandžić et al. 2012a; Hernández-Moreno et al. 2013). At the same time, no Cd at all was found in the livers of wolves from Northwest Russia. Differences in the presented data regarding the high hepatic Cd concentration in the wolves from Yukon are probably brought about by the abundance of Cd in the geological base. Cadmium penetrates plants, including the *Salix* type, characterized by a high ability to accumulate Cd, and an important ingredient in the diets of caribou and moose, the main prey of wolves in that area (Elkin and Bethke 1995; Gamberg et al. 2005).

As in present studies, the interspecific diversification in Cd content in analogous tissues of animals located in the same area results from their food preferences to a high extent (Alleva et al. 2006; Bilandžić et al. 2012a; Mayack 2012; Durkalec et al. 2015). Among carnivores living near Pesaro (Italy), the highest Cd levels in the liver were found in the omnivorous badger *Meles meles* and stone marten *Martes foina*, at 9 and 6 times more than in the red fox *Vulpes vulpes* with a diet consisting mainly of small mammals (Alleva et al. 2006).

Many researchers have indicated that a higher position in a trophic chain is not necessarily related to the level of Cd in the organism. Metcheva et al. (2003) observed that the contamination observed in the examined mammalian species depended not only on the trophic chain position but also on the food composition and lifestyle. The highest concentrations of Cd were found in the tissues of large and small mammals with a diet rich in earthworms, as these accumulate the highest amounts of Cd and other metals (Scheifler et al. 2006; Roodbergen et al. 2008).

Due to the slow elimination of Cd from the organism and its long half-life, a higher concentration of this element is usually observed in older animals in comparison to young individuals. Dip et al. (2001) demonstrated that in the red fox from urban and suburban areas, the concentration of Cd in the kidneys was significantly higher in older animals than in younger ones (e.g., in adult and young animals from suburban areas, the median concentrations of Cd were 1.82 and 0.73 mg $kg^{-1}$ ww, respectively). An age-dependent accumulation of Cd was also found in the liver of the fox. In turn, in wolves, this relation was observed only in the kidneys (Gamberg and Braune 1999). A positive relation between the concentration of Cd in the kidneys and the age of the animals was also found in wild ruminants (Arnold et al. 2006; Lazarus et al. 2008).

The bioaccumulation of Cd in large- and medium-sized mammals depends on many biological factors, such as the quality and amount of consumed food, size of the body, and age of the organism. Sometimes differences in Cd accumulation between males and females have been observed, which may have resulted from the different metabolic rates of the gender, as well as the hormonal and physiological status of the body (Burger et al. 2007). Sex-related differences could also be associated with differential feeding habits and feeding areas between males and females. However, diversity in Cd accumulation between the sexes is rarely described and is usually, as in small mammals, not confirmed statistically, as presented in the moose, red fox, and Egyptian mongoose *Herpestes ichneumon* by Custer et al. (2004) and Millán et al. (2008), or a higher concentration is found in females, as was shown in the wolf by Gamberg and Braune (1999) and Hernández-Moreno et al. (2013).

In herbivores, an important role in Cd bioaccumulation is played by the season, which is related not only to the availability of particular species of plants in the foraging area but also to changes in Cd assimilation by plants and translocation, which both directly affect the Cd level in particular parts of the plant. As was reported by Brekken and Steinnes (2004), the annual fluctuations of Cd concentration in plants may range from 2% to 82%. In general, the highest Cd concentration in plants is noted in spring and then in autumn and the lowest in summer. The drop in Cd level in plants in summer was explained by the so-called dilution effect (Brekken and Steinnes, 2004).

From an ecotoxicological point of view, as well as human health, the evaluation of Cd bioaccumulation is particularly relevant in the consumable parts of game animals. In the European Union, the maximum acceptable concentrations of this metal in the muscles, liver, and kidneys are 0.05, 0.5, and 1.0 mg $kg^{-1}$ ww, respectively (EC 1881/2006). As noted by Piskorová et al. (2003), wild boar from Slovakia exceeded the norms for Cd content in 13.3% of muscle samples, 20% of liver samples, and 26.6% of kidney samples in this omnivorous mammal. Such over maximal Cd levels were also found in typically herbivorous cervids (roe deer, red deer), European hares, and wild boar in Eastern Slovakia (Kottferová and Koréneková 2000). It is worth indicating that free-living ruminants assimilate more cadmium in their tissues than domesticated ruminants (Kramárová et al. 2005). Włostowski et al. (2006) have also observed that free-ranging European bison accumulate more Cd in the liver and kidneys than domestic animals. For example, mean concentrations of this element in the liver of European bison and cattle were 0.45 and 0.20 mg $kg^{-1}$ ww, respectively.

## 5.4 Bioaccumulation in Birds

According to the trophic position, age, environment, and Cd content in the diet, very diverse concentrations of Cd have been observed in the tissues of wild birds (Table 14.8). From recently published data about hepatic and nephric Cd concentrations in terrestrial birds, values range widely, from <DL to 17.8 and from <DL to >56.5 mg $kg^{-1}$ dw, respectively (Levengood 2003; Alleva et al. 2006; Orłowski et al. 2012; Carneiro et al. 2014; Binkowski and Sawicka-Kapusta 2015). The concentrations of Cd in the livers of birds indicate elevated values to the background level (>3 mg $kg^{-1}$ dw) (Dauwe et al. 2005; Orłowski et al. 2012). They can occur,

**Table 14.8** Concentration of cadmium (mean and range in parentheses; mg kg$^{-1}$ dw) in the tissues of bird species representing different taxonomic and trophic groups

| Species and trophic group | | Localization | Concentration | References |
|---|---|---|---|---|
| *Liver* | | | | |
| House sparrow *Passer domesticus* | G | Italy | 0.27 | Alleva et al. (2006) |
| | | Kosovo | 1.318 (0.075) | Millaku et al. (2015) |
| Great tit *Parus major* | I | Belgium | 2.72 (0.56–11.8) | Dauwe et al. (2005) |
| | | China | 0.68 ± 0.10 | Deng et al. (2007) |
| Red crossbill *Loxia curvirostra* | G | Italy | 0.58 (0.30–0.91) | Alleva et al. (2006) |
| Greenfinch *Carduelis chloris* | G | Italy | 0.55 (0.30–0.91) | Alleva et al. (2006) |
| | | China | 0.56 ± 0.09 | Deng et al. (2007) |
| European blackbird *Turdus merula* | O | Italy | 0.58 (<DL–1.49) | Alleva et al. (2006) |
| Rook *Corvus frugilegus* | O | Poland | 17.2 (16.3–17.8) | Orłowski et al. (2012) |
| Common pheasant *Phasianus colchicus* | O | Italy | 1.03 | Alleva et al. (2006) |
| | | Czech | 1.00 ± 0.024 | Čelechovská et al. (2008) |
| Greater sage-grouse *Centrocercus urophasianus* | H | USA, Wyoming and Montana | 4.75 (0.20–48.5) | Dailey et al. (2008) |
| Eurasian sparrow hawk *Accipiter nisus* | C | Italy | 0.52 (<DL–2.33) | Alleva et al. (2006) |
| Common buzzard *Buteo buteo* | C | Italy | 0.49 (<DL–1.58) | Alleva et al. (2006) |
| Common kestrel *Falco tinnunculus* | C | Italy | 0.09 (<DL–0.33) | Alleva et al. (2006) |
| | | Poland/Czech | 0.3 | Kalisińska et al. (2009) |
| Bald eagle *Haliaeetus leucocephalus* | P | USA, Maine | 0.35 (0.04–1.95) | Mierzykowski and Todd (2012) |
| Brown owl *Strix aluco* | C | Italy | 1.30 (0.18–2.49) | Alleva et al. (2006) |
| | | Spain | 5.52 ± 8.33 | Pérez-López et al. (2008) |
| | | Poland/Czech | 0.04 | Kalisińska et al. (2009) |
| Barn owl *Tyto alba* | C | Italy | 0.18 (<DL–1.67) | Alleva et al. (2006) |
| Little owl *Athene noctua* | I | Italy | 0.66 (<DL–5.36) | Alleva et al. (2006) |
| | | Spain | 0.60 ± 0.57 | Pérez-López et al. (2008) |
| | | Spain | 1.39 ± 0.87 | Pérez-López et al. (2008) |
| | | Italy | 0.64 | Naccari et al. (2009) |
| | | Poland/Czech | 0.9 | Kalisińska et al. (2009) |
| | | Portugal | 0.184 (<DL–1.801) | Carneiro et al. (2014) |

(continued)

**Table 14.8** (continued)

| Species and trophic group | | Localization | Concentration | References |
|---|---|---|---|---|
| Long-eared owl *Asio otus* | C | Italy | 0.82 (<DL–6.48) | Alleva et al. (2006) |
| | | Spain | 1.24 ± 0.75 | Pérez-López et al. (2008) |
| Canada geese *Branta canadensis* | H | USA | 0.60 | Tsipoura et al. (2011) |
| Mallard *Anas platyrhynchos* | H | USA | <2.54–2.94 | Levengood (2003) |
| | | Poland | 0.85 | Kalisińska et al. (2004) |
| | | Japan | 4.82 (0.94–14.45) | Mochizuki et al. (2002) |
| Wood duck *Aix sponsa* | H | USA | <2.54–6.06 | Levengood (2003) |
| Gray heron *Ardea cinerea* | P | Italy | 0.06 (<DL–0.21) | Alleva et al. (2006) |
| | | Korea | 0.24 (0.19–0.29) | Kim and Oh (2015) |
| Black-crowned night herons | P | Korea | 0.22 (0.15–0.30) | Kim and Oh (2015) |
| | | Italy | 0.17 (<DL–0.37) | Alleva et al. (2006) |
| *Kidney* | | | | |
| House sparrow *Passer domesticus* | G | Kosovo | 1.318 ± 0.075 | Millaku et al. (2015) |
| Great tit *Parus major* | I–G | China | 1.32 ± 0.25 | Deng et al. (2007) |
| | | Belgium | 14.1 (4.1–28.3) | Dauwe et al. (2005) |
| Rooks *Corvus frugilegus* | O | Poland | 17.0 (15.5–17.7) | Orłowski et al. (2012) |
| Common pheasant *Phasianus colchicus* | O | Czech | 0.42 | Čelechovská et al. (2008) |
| Brown owl *Strix aluco* | C | Poland/Czech | 14 | Kalisińska et al. (2009) |
| Common buzzard *Buteo buteo* | C | Netherlands | 7.41 ± 0.90 | Jager et al. (1996) |
| | | Spain | 0.27 | García-Fernandez et al. (1995) |
| | | Italy | 2.09 | Naccari et al. (2009) |
| | | Poland/Czech | 1.2 | Kalisińska et al. (2009) |
| | | Portugal | 0.865 (0.033–8.344) | Carneiro et al. (2014) |
| Mallard *Anas platyrhynchos* | | Poland | 2.588 ± 1.819 | Kalisińska et al. (2004) |
| | | Japan | 15.4 (4.68–38.07) | Mochizuki et al. (2002) |
| *Muscle* | | | | |
| Canada geese *Branta canadensis* | H | USA | 0.020 | Tsipoura et al. (2011) |

(continued)

**Table 14.8** (continued)

| Species and trophic group | | Localization | Concentration | References |
|---|---|---|---|---|
| Mallard *Anas platyrhynchos* | H | Poland | 0.27 (0.08–0.0.71) | Szymczyk and Zalewski (2003) |
| Rook *Corvus frugilegus* | O | Poland | 17.2 (15.6–17.8) | Orłowski et al. (2012) |
| | | Italy | 0.09 (0.03–0.18) | Naccari et al. (2009) |
| Common pheasant *Phasianus colchicus* | O | Czech | 0.010 | Čelechovská et al. (2008) |
| *Feathers* | | | | |
| Great tit *Parus major* | I–G | Belgium | 11.6 ± 1.5[a] | Dauwe et al. (2002) |
| | | Belgium | 0.93 ± 0.07 | Dauwe et al. (2002) |
| | | China | 0.11 ± 0.09 | Deng et al. (2007) |
| | | Portugal | 0.10 ± 0.03 | Costa et al. (2013) |
| | | Slovenia | 0.07 ± 0.11 | Al Sayegh Petkovšek et al. (2015) |
| Blue tit *Cyanistes caeruleus* | I–G | Belgium | 8.0 ± 1.4[a] | Dauwe et al. (2002) |
| | | Belgium | 1.4 ± 0.4 | Dauwe et al. (2002) |
| Greenfinch *Carduelis chloris* | H | China | 0.001 | Deng et al. (2007) |
| Canada geese *Branta canadensis* | H | USA | 0.086 ± 0.010 | Tsipoura et al. (2011) |
| Mallard *Anas platyrhynchos* | H | Poland | <DL–0.04 | Binkowski and Sawicka-Kapusta (2015) |
| Common buzzard *Buteo buteo* | C | Italy | 0.11 (0.04–0.18) | Naccari et al. (2009) |

*G* granivores, *I* insectivores, *H* herbivores, *C* carnivores, *P* piscivores, *O* omnivores, *DL* detection limit
[a]Polluted area

for example, in birds from rural areas in which phosphate and organic (manure and slurry) fertilizers containing large amounts of Cd are used (Orłowski et al. 2012).

One of the most important factors regulating the bioaccumulation of Cd in birds is the level of contamination of the bird habitat by this metal. Moreover, other important roles are played by factors affecting the bioavailability and transfer of Cd, such as the characteristics of the soil and landscape composition. Fritsch et al. (2012), in studies on trace metal transfer to European blackbirds *Turdus merula*, observed that Cd in the blood was weak correlated with soil pollution. For this reason the authors think that in this case, ecological processes like prey availability like feeding behavior are major indicators of metal transfer than the availability of

Cd in the soil. Moreover, Fritsch et al. (2012) suggest that "food chain effects" of Cd transfer result more from the preferential use of selected types of habitats by birds than from their general heterogeneity. In studies on the impact of the diversified diet of the small owl *Athene noctua* on the level of Cd accumulation, it was shown that it is strongly and positively correlated with the number of earthworms in the diet. Also it was concluded that the affinity of the common vole to a diet of earthworms assists in lowering the exposure of this bird species to Cd (Schipper et al. 2012). Extremely high concentrations of Cd were noted in the kidneys and bones of a particular type of ptarmigan (white-tailed ptarmigan *Lagopus leucurus*) living in the Colorado Rocky Mountains (USA) (Larison 2002), the diet of which included the willow, assimilating Cd in an order of magnitude higher amounts than other plants.

Interspecies differences in Cd bioaccumulation are explained mostly by the qualitative and quantitative differences in the diets of the birds (Berglund et al. 2011). For this reason exposure to Cd can differ between different bird species in the same habitat. For example, birds whose diet includes mostly arthropods (such as the pied flycatcher *Ficedula hypoleuca*) accumulate higher amounts of Cd in their tissues than birds that feed on seeds, fruits, and other plant elements, like the great tit *Parus major* (Berglund et al. 2011). Alleva et al. (2006) have also made a comparison of Cd concentrations between particular trophic groups, which demonstrated that the highest concentrations of Cd were typical in the liver for omnivorous birds (0.46 mg $kg^{-1}$ ww) and for herbivorous birds (0.01 mg $kg^{-1}$ ww). In general, the order of particular trophic groups according to the hepatic Cd concentration is as follows: omnivorous > frugivorous > insectivorous, granivorous > carnivorous (small mammals) > carnivorous (aquatic invertebrates) > piscivorous > carnivorous (bird-eating) > herbivorous. Moreover, the interspecies differences in Cd accumulation in avian bodies may also be the result of different genetic and biochemical features, including the ability to synthesize specific types and amounts of metallothionein (MT), which take part in Cd detoxification, which is most intensive in the liver and kidneys. In studies in this area, it was shown that at the same level of exposure to Cd in different types of ducks, the highest concentration of Cd is usually found in mallard *Anas platyrhynchos* (Levengood 2003; Szymczyk and Zalewski 2003; Lucia et al. 2010). The results of studies by Lucia et al. (2010) indicate that mallards are characterized by higher basal MT protein levels in the liver and kidneys than the Muscovy duck *Cairina moschata*, which is why a higher accumulation of Cd is found in mallard organs than in the Muscovy duck. Moreover, they observed that in the case of mallards exposed to greater doses of Cd, induction of the MT gene arose after 10 days and at lower doses after 40 days, while in Muscovy duck, induction of the MT gene appeared after 40 days for both levels of exposure to Cd.

Usually, in birds, no differences in tissue concentration of Cd are observed (Kalisińska et al. 2004; Deng et al. 2007; Carneiro et al. 2014; Binkowski and Sawicka-Kapusta 2015). Such differences do however occur between birds from different age categories, while usually higher concentrations of Cd are noted in the tissues of older individuals (García-Fernández et al. 1996; Larison 2002; Kalisińska et al. 2004; Berglund et al. 2011; Carneiro et al. 2014). For example, in the Portuguese common buzzard *Buteo buteo*, Carneiro et al. (2014) showed that young individuals were characterized by twice lower concentrations of Cd in the liver (0.209 vs 0.409 mg $kg^{-1}$ dw) and a three times lower level of Cd in the kidneys (0.698 vs 2.165 mg $kg^{-1}$ dw) than adult individuals. Still, Naccari et al. (2009), in studies on the same species in Sicily, found a higher concentration of Cd in the kidneys of adult animals in comparison to young. Also in the case of the white-tailed ptarmigan *Lagopus leucurus*, the concentration of Cd in the kidneys increases with age at a rate of approximately 0.5 μg Cd $day^{-1}$ of exposure (Larison 2002).

However, there are also some reports indicating that higher concentrations of Cd in most of the internal organs are noted for immature birds, especially chicks (Fritsch et al. 2012). This phenomenon may be related to diversity between the diet of the adult and young birds as well as the quicker rate of metabolic change in the intensively growing chicks, which strongly affects Cd absorption from the food. It was shown that the young individuals of both granivorous and omnivorous bird species feeding by parents with mostly soil invertebrates such as the ground-dwelling Coleoptera and earthworms accumulate considerable amounts of Cd. With such a diet, chicks assimilate large amounts of this element, which accumulate in their organisms (Fritsch et al. 2012; Orłowski et al. 2012).

## 5.5 Bioindicators and Biomarkers in Ecotoxicological Studies

One of the most commonly used methods for monitoring contamination of the environment by heavy metals is evaluation of their bioaccumulation in the tissues of different species of animals used as bioindicators. In biomonitoring and ecotoxicological studies, wild (including micromammals and game animals) and domesticated mammals and birds from the human neighborhood are used.

Small land mammals are found to be very good bioindicators, mostly due to the widespread occurrence, limited home range, generalized feeding habits, short life, high reproductive indexes and metabolic rate, as well as simplicity in obtaining them for research. They are an important part of the ecosystems as they are part of many trophic chains and for this reason play a great role in the circulation and transfer of many elements, including Cd. Ecotoxicological studies are usually performed with wood mice (Sánchez-Chardi et al. 2007; van den Brink et al. 2010; Lourenço et al. 2013; Tête et al. 2015), field mice, yellow-necked mice, voles, moles, and shrews (Damek-Poprawa and Sawicka-Kapusta 2003; Veltman et al. 2007a; Fritsch et al. 2011; Nesterkova et al. 2014) and the slightly bigger wild rat (Pereira et al. 2006). Many researchers consider mice and rats to be particularly useful in bioindication, as they pass many of the important criteria for good indicators: (1) their populations are properly large; (2) they have a wide geographic distribution; (3) they occur in both polluted and nonpolluted areas; (4) they can live in human neighborhoods and consume human food, which make mice and rats good indicators of human exposure; (5) they have a small home range, so they may be treated as site-specific indicators; and finally (6) they are located in a mid-position in many trophic chains (Pereira et al. 2006).

Some researchers in studies on rodents indicate that some species are particularly sensitive and useful in ecotoxicological evaluations. Martiniakova et al. (2012), based on studies of the accumulation of Cd and other elements in the kidney, liver, testis, uterus, and bones of free-living wild rodents from a polluted area in Slovakia, indicate that bank voles are more sensitive heavy metal-loaded bioindicators than yellow-necked mice. In turn Braeuer et al. (2015) observed that between four examined species of small mammals (*Apodemus sylvaticus*, *Microtus arvalis*, *Myodes glareolus*, and *Sorex araneus*), the most sensitive bioindicator was the common shrew.

Another important group in biomonitoring and ecotoxicological studies is the herbivorous Leporidae, wild rabbits and hares, as they are an important edible game species, and moreover they are important prey for several predator species (Lazarus et al. 2014). The concentration of Cd measured in their tissues gives indirect information not only of the degree of pollution in the environment they live in but also about the risks of consuming them. Eira et al. (2005) reported that the mean Cd concentration in wild rabbit kidneys from Portugal was 1.02 mg $kg^{-1}$ ww and that, in 30% of kidney samples, Cd levels exceeded the threshold value (1.0 mg $kg^{-1}$ ww, the maximum value set for the kidneys of cattle and domestic poultry for human consumption in the EU).

In ecotoxicological studies on Cd, carnivores are also considered, mostly from the Canidae, Mustelidae, and Procyonidae families. The diverse food preferences of the particular species allow them to provide information on the level of contamination in the different habitats. Among the mostly used species, the polecat (*Mustela putorius*) and marten (*Martes martes* and *M. foina*) seem to be the best bioindicators of land ecosystem pollution, as their diet is less related to water ecosystems and, as cited by Ryšavá-Nováková and Koubek (2009), is dominated by mammals and birds. Also, the American mink (*Neovison vison*), which is a subject of numerous ecotoxicological and biomonitoring studies, is used as a sensitive bioindicator of pollution of terrestrial ecosystems, even though the proportion of fish in their diet is considerable (Krawczyk et al. 2013). Thereby, the load of Cd in the organism of a mink in part comes from a water habitat. Otters in turn are mostly connected with water ecosystems, and for this reason, their role as a bioindicator of pollution of land ecosystems is low.

Game animals are good bioindicators as they provide information not only about the quality of the ecosystem of which they are part of but also about the potential threat to venison consumers. The most common studies are on wild ruminants, such as roe deer, red deer, fallow deer, and in northern countries the moose and reindeer.

The wild boar also plays a significant role in ecotoxicological studies, whose role as an indicator of environmental pollution is largely in relation to xenobiotics, which are accumulated in the soil. Also animals that live in human neighborhoods (e.g., the fox) are relevant for delivering information about the level of contamination of the human habitat. Dip et al. (2001) indicate that foxes living in urban and suburban areas may be valuable bioindicators of Cd contamination of urbanized areas.

Widely spread species of birds and mammals may also be considered especially valuable biomonitors, allowing comparisons of the level of Cd accumulation within a single continent or between them (Eurasia and North America), as well as species with similar biology, for example, the erne *Haliaeetus albicilla* vs American erne *H. leucocephalus* or reindeer vs caribou.

Usually the concentration of Cd in the organs of mammals follows the descending sequence, kidney > liver > muscle, caused by the fact that the kidneys are the most important organ in the accumulation of this element. According to the fact that the affinity of Cd to the bones is much less than, for example, Pb, cadmium is measured in bone tissue less often. Such studies were undertaken by Łanocha et al. (2013), who observed similar concentrations of Cd in the bones of Canidae (foxes, dogs) and humans, which indicate the usefulness of these animals to evaluate the human threat of exposure to this element. As noted by Herber (2004), only in large exposures to Cd is this element stored in the liver, while in small doses, Cd accumulates mostly in the kidneys.

The concentration of Cd is most frequently evaluated in the kidneys and liver; however the research material is diverse and may differ in particular groups of animals. In murine, the concentration of Cd is measured in the tissues and organs (mostly in the blood, liver, and kidneys and occasionally in the spleen, brain, and muscles) as well as in the whole body. Also the tail of these animals is used (Ferreira et al. 2015). In bigger mammals, studies are performed usually on the liver, kidneys, muscles, as well as fur and bones. Blood may be used as an indicator of current exposure to Cd while hair and bones as indicators of long-term exposure.

Wild birds are often used in biomonitoring and ecotoxicological studies. Usually these studies focus on predatory birds, located at the top of the trophic chains. Pérez-López et al. (2008) indicated that the local- and upper-trophic-level species are believed to be especially vulnerable to metals which explains their important role as environmental contamination indicators. Diurnal species, e.g., the common buzzard (*Buteo buteo*), northern goshawk (*Accipiter gentilis*), common kestrel (*Falco tinnunculus*), Eurasian sparrow hawk (*Accipiter nisus*), as well as nocturnal species, e.g., the long-eared owl (*Asio otus*), tawny owl (*Strix aluco*), and barn owl (*Tyto alba*), are found to be useful species for the biomonitoring of metals and local monitors of contaminant levels (Pérez-López et al. 2008). These birds are usually sedentary, and their wide diet includes small mammals, birds, amphibians, reptiles, and insects.

Many works have concerned smaller birds from other trophic levels and typical for particular ecosystems, as well as those living in human neighborhoods. Although Cd and other metals accumulate better in the organs of longer-lived animals, it seems that relatively short-living passerines, e.g., the house sparrow, great tit, and blue tit, are good biomonitors of Cd levels (Markowski et al. 2014). In studies on the level of

Cd in tissues of the house sparrow (*Passer domesticus*) from locations with different degrees of contamination, Millaku et al. (2015) stated that the concentration of Cd was significantly higher in the tissues of sparrows from polluted areas in comparison to those living in the reference region. This indicates that the house sparrow could be considered as an indicator for environmental contamination. Moreover, passerines are numerous, have a wide range of presence and a small home range, and are relatively highly located in the food chain. Another advantage of these animals is the fact that the biology of this species is well known (which simplifies the interpretation of results) as is their ecotoxicological use (which in turn simplifies comparisons of results from different research teams, regions, and countries). Additionally in the case of the great tit and blue tit, it is easy to monitor their population especially during the breeding season, as they often use nesting boxes made by people. This favors noninvasive sample collections, for example, in the case of studies on the natural mortality of chicks, as well as in observations of possible reproductive or behavioral aberrances in the polluted areas. A particularly important feature of passerines in their use in biomonitoring and ecotoxicological studies is their foraging in small home ranges, making them suitable as biomonitors for point source contamination (Markowski et al. 2014).

The pied flycatcher (*Ficedula hypoleuca*) is commonly used in ecotoxicological studies, and especially the breeding population (Berglund et al. 2011; Eeva and Lehikoinen 2015), even though it is a migratory bird. This species spends winter in Central Africa and visits its breeding areas from the middle of April. Nevertheless it is a species sensitive to environmental pollution (reductions in the hatchability of eggs, increases in nesting mortality) (Berglund and Nyholm 2011), which predisposes it for ecotoxicology. Also, soil invertebrate-feeding birds, like the rook (*Corvus frugilegus*), seem to be good biomonitors, which similar to predators from the top levels of trophic levels are vulnerable to high doses of heavy metals in their diet. Until now this species had been frequently passed by in biomonitoring and ecotoxicological studies. Only a few works about the levels of environmental contamination in the tissues of this bird are available (Orłowski et al. 2012, 2014), with most dated in the 1960s and 1970s.

Costa et al. (2013) suggest the use of chicks in nesting boxes in the evaluation of local environmental contamination. The authors explain that this may simplify material and data collection for a defined area and time period. Additionally, the chicks in nesting boxes are exposed to a lesser extent than the adult individuals to external airborne deposition from industrial sources. This allows assessment of the effects of diet in the concentration of Cd in the tissues of chicks.

In biomonitoring and ecotoxicological studies in birds, Cd is mostly measured in the liver, kidneys, and pectoral muscle, rarely in the blood, and only occasionally in other types of sample. The content of Cd in the liver and kidneys is a reflection of the temporal exposure, while blood levels illustrate the current exposure. The ratio of Cd concentration in the liver to the kidney may be used to evaluate the length and dose of exposure. A factor lower than 1 suggests long exposure to small doses (Scheuhammer 1987). Data about the concentration of Cd in the liver and kidneys may also be used to settle regression lines in the CSRL (Cd standard regression line) model developed by Mochizuki et al. (2008), which allows the definition of whether animals are intoxicated with Cd.

More and more often, researchers decide to use material which may be obtained in a noninvasive way, such as feathers, eggs, or eggshells (Deng et al. 2007; Costa et al. 2013; Markowski et al. 2014); however in some studies, a low usefulness in the assessment of Cd environmental pollution has been demonstrated with those sources (Binkowski and Sawicka-Kapusta 2015). This is caused by the fact that the content of Cd in feathers is not always correlated with the level of this element in the blood and organs nor with the pollution of the environment (e.g., in the soil) (Tsipoura et al. 2011). Due to the fact that during the molting season birds are able to eliminate large amounts of metals, in such studies, feathers formed during or just after the breeding period should be chosen. Additionally Costa et al. (2013) have observed that the feathers of chicks differ in the content of metals from the feathers of fully grown birds, which in part is related to the constant growth of the feathers in chicks. For this reason, they suggest that during interpretation of the results, the stage of feather development should be taken into account.

It seems that the use of eggshells in biomonitoring is controversial, as the location of the habitat (e.g., big and small cities, countryside) did not affect the Cd content in them, as was shown by Orłowski et al. (2014). This observation concerns also whole eggs, in which the concentration of Cd did not always reflects the environmental levels of contamination by Cd and was often below the detection limit (Roodbergen et al. 2008). This indicates that the concentration of Cd, in contrast to the other heavy metals, is probably physiologically regulated in eggs.

Despite the fact that feces are more often used as an object of studies on the evaluation of environmental contamination and Cd exposure (Berglund and Nyholm 2011; Berglund et al. 2011), Binkowski and Sawicka-Kapusta (2015) have stated that feces do not seem to be a useful material to study as they do not present internally accumulated concentrations.

In ecotoxicological studies, *biomarkers* are widely used as biological indicators, allowing for qualitative or quantitative evaluation of interactions between an organism and chemical, physical, and biological factors of the environment. Biomarkers have been divided into three groups—biomarkers of (1) exposure, (2) effect, and (3) susceptibility (Mussali-Galante et al. 2013)—of which the first two play the biggest role in the assessment of the risk of Cd exposure. Among the exposure biomarkers, the most often used are the biomarkers of the internal dose that show the presence of a particular toxic factor in the organism, with the tissues and organs treated as internal dosimeters. To a lesser extent, the biomarkers of an effective biological dose are used. The biomarkers of the effects are DNA single- and double-strand break (earthworms; Fourie et al. 2007), chromosome aberrations (dipterans; Michailova et al. 2000), micronuclei (wood mouse; Sánchez-Chardi et al. 2007), and sister chromatid exchange (Algerian mice; Tapisso et al. 2009). Among the biochemical biomarkers of the effects of Cd exposure, the most often used are metallothionein which increase in concentration at the moment of a direct threat (great tit, Vanparys et al. 2008). There are also some studies that confirm the usefulness of *N*-acetyl-β-D-glucosaminidase activity (NAG) as a biomarker of the effects of exposure (pheasants; Zielińska et al. 2015).

# 6 Conclusions

Cadmium is a toxic element, which naturally occurs in the environment. Despite that the global Cd emission is decreasing in the last few decades, its presence in the environment is still a severe ecological problem. Animals that live in polluted areas may contain extremely high concentrations of Cd in their tissues. Long exposure even to low doses of Cd leads not only to many disorders within the organism itself but may also result in changes in the whole ecosystem.

# References


Adriano DC (2001) Trace elements in terrestrial environments: biogeochemistry, bioavailability and risk of metals. Springer, New York

Al Sayegh Petkovšek S, Kopušar N, Kryštufek B (2014) Small mammals as biomonitors of metal pollution: a case study in Slovenia. Environ Monit Assess 186:4261–4274

Al Sayegh Petkovšek S, Kopušar N, Tome D, Kryštufek B (2015) Risk assessment of metals and PAHs for receptor organisms in differently polluted areas in Slovenia. Sci Total Environ 532: 404–414

Alleva E, Francia N, Pandolfi M, De Marinis AM, Chiarotti F, Santucci D (2006) Organochlorine and heavy-metal contaminants in wild mammals and birds of Urbino-Pesaro province, Italy: an analytic overview for potential bioindicators. Arch Environ Contam Toxicol 51:123–134

Alloway BJ (1995) Soil processes and the behaviour of metals. In: Alloway BJ (ed) Heavy metals in soils. Blackie Academic and Professional, London, pp 11–37

Álvarez-Ayuso E, Otones V, Murciego A, García-Sánchez A, Santa Regina I (2013) Zinc, cadmium and thallium distribution in soils and plants of an area impacted by sphalerite-bearing mine wastes. Geoderma 207–208:25–34

AMAP (2005) Arctic Monitoring and Assessment Programme. AMAP Assessment 2002: heavy metals in the Arctic, Oslo, p 265

Amici A, Danieli P, Russo C, Primi R, Ronchi B (2012) Concentration of some toxic and trace elements in wild boar (Sus scrofa) organs and tissue in different areas of the province of Viterbo, Central Italy. Ital J Anim Sci 11:e65

Angelo RT, Cringan MS, Chamberlain DL, Stahl AJ, Haslouer SG, Goodrich CA (2007) Residual effects of lead and zinc mining on freshwater mussels in the Spring River Basin (Kansas, Missouri, and Oklahoma, USA). Sci Total Environ 384:467–496

Aprioku JS, Ebenezer B, Ijomah MA (2014) Toxicological effects of cadmium during pregnancy in Wistar albino rats. Toxicol Environ Health Sci 6:16–24

Arnold S, Zarnke R, Lynn TV, Chimonas MA, Frank A (2006) Public health evaluation of cadmium concentrations in liver and kidney of moose (*Alces alces*) from four areas of Alaska. Sci Total Environ 357:103–111

ATSDR (2012) Toxicological profile for cadmium. U.S. Department of Health and Human Services, Public Health Service, Agency for Toxic Substances and Disease Registry

Baran A, Czech T, Wieczorek J (2015) Chemical properties and toxicity of soils contaminated by mining activity. Ecotoxicology 23:1234–1244

Berglund Å, Nyholm N (2011) Slow improvements of metal exposure, health- and breeding conditions of pied flycatchers (*Ficedula hypoleuca*) after decreased industrial heavy metal emissions. Sci Total Environ 409:4326–4334

Berglund Å, Koivula M, Eeva T (2011) Species- and age-related variation in metal exposure and accumulation of two passerine bird species. Environ Pollut 159:2368–2374

Betha R, Pradani M, Lestari P, Joshi U, Reid J, Balasubramanian R (2013) Chemical speciation of trace metals emitted from Indonesian peat fires for health risk assessment. Atmos Res 122: 571–578

Bilandžić N, Deždek D, Sedak M, Dokić M, Simić B, Rudan N et al (2012a) Trace elements in tissues of wild carnivores and omnivores in Croatia. Bull Environ Contam Toxicol 88:94–99

Bilandžić N, Sedak M, Dokić M, Šimić B (2012b) Heavy metal concentration in tissues of wild boar of continental Croatia. Int J Environ Prot 2:6–9

Binkowski Ł, Sawicka-Kapusta K (2015) Cadmium concentrations and their implications in Mallard and Coot from fish pond areas. Chemosphere 119:620–625

Bird G, Brewer P, Macklin M, Balteanu D, Driga B, Serban M, Zaharia S (2003) The impact and significance of metal mining activities on the environmental quality of Romanian river systems. In: CCMESI (ed) Proceedings of the First International Conference on Environmental Research and Assessment, Bucharest, Romania. University of Bucharest, pp 316–332

Blagojević J, Jovanović V, Stamenković G, Jojić V, Bugarski-Stanojević V, Adnađević T, Vujošević M (2012) Age differences in bioaccumulation of heavy metals in populations of the black-striped field mouse, *Apodemus agrarius* (Rodentia, Mammalia). Int J Environ Res 6:1045–1052

Bojakowska I (2009) Kadm w surowcach mineralnych polski i jego potencjalna emisja do środowiska. Ochrona Środowiska i Zasobów Naturalnych 40:22–30

Bradford G, Change A, Page A, Bakhtar D, Frampton JA, Wright H (1996) Background concentrations of trace and major elements in California soils. Kearney Foundation of Soil Science, Division of Agriculture and Natural Resources, University of California, Riverside, pp 1–32

Braeuer S, Čadková Z, Száková J, Tlustoš P, Goessler W (2015, 22–26 Febuary) Wild small terrestrial mammals as possible bioindicators for trace element levels in the environment. In: Abstracts of the European Winet Conference on Plasma Spectrochemistry, Münster

Brekken A, Steinnes E (2004) Seasonal concentrations of cadmium and zinc in native pasture plants: consequences for grazing animals. Sci Total Environ 326:181–195

Bridges C, Zalups R (2005) Molecular and ionic mimicry and the transport of toxic metals. Toxicol Appl Pharmacol 204:274–308

Bukovjan K, Toman A, Kutlvašr K, Marada P, Kodet R, Sláma P, Křikava L (2014) Contents of chemical elements in tissues of European badger (*Meles meles*) affected by ovarian tumour: a case report. Acta Vet Brno 83:139–143

Burger J (1994) Heavy metals in avian eggshells: another excretion method. J Toxicol Environ Health 41(2):207–220

Burger J, Gochfeld M (1991) Cadmium and lead in common terns (Aves: *Sterna hirundo*): Relationship between levels in parents and eggs. Environ Monit Assess 16:253–258

Burger J, Fossi C, McClellan-Green P, Orlando EF (2007) Methodologies, bioindicators, and biomarkers for assessing gender-related differences in wildlife exposed to environmental chemicals. Environ Res 104:135–152

Butterman WC, Plachy J (2002) Mineral Commodity Profiles: Cadmium. USGS Open-File Report 2002-238. pubs.usgs.gov/of/2002/of02-238/of02-238.pdf

Cain BW, Sileo L, Franson L, Moore J (1983) Effects of dietary cadmium on mallard ducklings. Environ Res 32:286–297

Campbell PGC (2006) Cadmium – a priority pollutant. Environ Chem 3:387–388

Carneiro M, Colaço B, Brandão R, Ferreira C, Santos N, Soeiro V et al (2014) Biomonitoring of heavy metals (Cd, Hg, and Pb) and metalloid (As) with the Portuguese common buzzard (Buteo buteo). Environ Monit Assess 186:7011–7021

CEIP (2015) Centre on Emission Inventories and Projections. Trend tables. http://www.ceip.at/ms/ceip_home1/ceip_home/status_reporting/2015_submissions/

Čelechovská O, Literák I, Ondruš S, Pospíšil Z (2006) Heavy metals in brown bears from the Central European Carpathians. Acta Vet Brno 75:501–506

Čelechovská O, Malota L, Zima S (2008) Entry of heavy metals into food chains: a 20-year comparison study in Northern Moravia (Czech Republic). Acta Vet Brno 77:645–652

Chen TB, Zheng YM, Chen H, Zheng GD (2004) Background concentrations of soil heavy metals in Beijing. Huan Jing Ke Xue 25:117–122

Cherian MG (1983) Absorption and tissue distribution of cadmium in mice after chronic feeding with cadmium chloride and cadmium-metallothionein. Bull Environ Contam Toxicol 30:33–36

Cikrt M, Tichý M (1974) Excretion of cadmium through bile and intestinal wall in rats. Br J Ind Med 31:134–139

Costa RA, Eeva T, Eira C, Vaqueiro J, Vingada JV (2013) Assessing heavy metal pollution using Great Tits (*Parus major*): feathers and excrements from nestlings and adults. Environ Monit Assess 185:5339–5344

Crévecoeur S, Debacker V, Joaquim-Justo C, Gobert S, Scippo M-L, Dejonghe W et al (2011) Groundwater quality assessment of one former industrial site in Belgium using a TRIAD-like approach. Environ Poll 159:2461–2466

Custer T, Cox E, Gray B (2004) Trace elements in moose (*Alces alces*) found dead in Northwestern Minnesota, USA. Sci Total Environ 330:81–87

Dailey RN, Raisbeck MF, Siemion RS, Cornish TE (2008) Liver metal concentrations in Greater - Sage-grouse (*Centrocercus urophasianus*). J Wildl Dis 44:494–498

Damek-Poprawa M, Sawicka-Kapusta K (2003) Damage to the liver, kidney, and testis with reference to burden of heavy metals in yellow-necked mice from areas around steelworks and zinc smelters in Poland. Toxicology:126:1–126:12610

Dauwe T, Bervoets L, Janssens E, Pinxten R, Blust R, Eens M (2002) Great and blue tit feathers as biomonitors for heavy metal pollution. Ecol Indic 1:227–234

Dauwe T, Janssens E, Bervoets L, Blust R, Eens M (2005) Heavy-metal concentrations in female laying great tits (*Parus major*) and their clutches. Arch Environ Contam Toxicol 49:249–256

De Vos W, Tarvainen T, Salminen R, Reeder S, De Vivo B, Demetriades A et al (2006) Geochemical atlas of Europe. Part 2, Geological Survey of Finland, Espoo

Deng H, Zhang Z, Chang C, Wang Y (2007) Trace metal concentration in Great Tit (Parus major) and Greenfinch (*Carduelis sinica*) at the Western Mountains of Beijing, China. Environ Pollut 148:620–626

Dip R, Stieger C, Deplazes P, Hegglin D, Müller U, Dafflon O et al (2001) Comparison of heavy metal concentrations in tissues of red foxes from adjacent urban, suburban, and rural areas. Arch Environ Contam Toxicol 40:551–556

Dragović S, Mihailović N, Gajić B (2008) Heavy metals in soils: distribution, relationship with soil characteristics and radionuclides and multivariate assessment of contamination sources. Chemosphere 72:491–495

DTSC (2009, February 24) HERD EcoNOTE 6: Cadmium – Currently Recommended U.S. Environmental Protection Agency Region 9 Biological Technical Assistance Group (BTAG) Mammalian and Avian Toxicity Reference Values (TRVs)

Durkalec M, Szkoda J, Kołacz R, Opalinski S, Nawrocka A, Zmudzki J (2015) Bioaccumulation of lead, cadmium and mercury in roe deer and wild boars from areas with different levels of toxic metal pollution. Int J Environ Res 9:205–212

EC 1881/2006 Commission Regulation (CE) No 1881/2006 of 19 December 2006. Setting maximum levels for certain contaminants in foodstuffs. http://eurlex.europa.eu/LexUriServ/LexUriServ.do?uri=OJ:L:2006:364:0005:0024:EN:PDF

ECB (2005) Risk assessment: cadmium metal/cadmium oxide. Final, but not adopted version of December 2005. European Chemicals Bureau, Ispra, Italy

EEA (2015) European Union emission inventory report 1990–2013 under the UNECE convention on Long-range Transboundary Air Pollution (LRTAP). Copenhagen, Denmark

Eeva T, Lehikoinen E (2015) Long-term recovery of clutch size and egg shell quality of the pied flycatcher (*Ficedula hypoleuca*) in a metal polluted area. Environ Pollut 201:26–33
Eira C, Torres J, Vingadab J, Miquela J (2005) Concentration of some toxic elements in *Oryctolagus cuniculus* and in its intestinal cestode *Mosgovoyia ctenoides*, in Dunas de Mira (Portugal). Sci Total Environ 346:81–86
Elkin BT, Bethke RW (1995) Environmental contaminants in caribou in the northwest territories, Canada. Sci Total Environ 160-161:307–321
EPA (2009) Cadmium Compounds. U.S. Environmental Protection Agency, Washington, DC. http://www.epa.gov/ttn/atw/hlthef/cadmium.html
Eriksson H, Edberg F, Borg H (2003) Effects of forest fire and fire-fighting operations on water chemistry in Tyresta National Park, Stockholm, Sweden. J Phys IV 107(I):427–430. https://doi.org/10.1051/jp4:20030332
Faiz Y, Tufail M, Tayyeb Javeda M, Chaudhrya MM, Siddique N (2009) Road dust pollution of Cd, Cu, Ni, Pb and Zn along Islamabad Expressway, Pakistan. Microchem J 92:186–192
Fergusson JE, Kim ND (1991) Trace elements in street and house dusts: sources and speciation. Sci Total Environ 100:125–150
Fernández AJ, Aboal JR, González XI, Carballeira A (2012) Transfer and bioaccumulation variability of Cd, Co, Cr, Hg, Ni and Pb in TROPHIC compartments of terrestrial ecosystems in Northern Spain. Fresenius Environ Bull 21:3527–3532
Ferreira A, Garcia P, Camarinho R, dos Santos Rodrigues A (2015) Volcanogenic pollution and testicular damage in wild mice. Chemosphere 132:135–141
Ferri R, Hashim D, Smith DR, Guazzetti S, Donna F, Ferretti E et al (2015) Metal contamination of home garden soils and cultivated vegetables in the province of Brescia, Italy: implications for human exposure. Sci Total Environ 518–519C:507–517
Finkelman RB (1993) Trace and minor elements in coal. In: Engel MH, Macko SA (eds) Organic geochemistry. Plenum, New York, pp 593–607
Fourie F, Reinecke SA, Reinecke AJ (2007) The determination of earthworm species sensitivity differences to cadmium genotoxicity using the comet assay. Ecotoxicol Environ Saf 67:361–368
Frank A, Danielsson UR, Jones B (2000) The 'mysterious' disease in Swedish moose. Concentrations of trace elements in liver and kidneys and clinical chemistry. Comparison with experimental molybdenosis and copper deficiency in the goat. Sci Total Environ 249:107–122
Fritsch C, Cosson RP, Cœurdassier M, Raoul F, Giraudoux P, Crini N et al (2010) Responses of wild small mammals to a pollution gradient: host factors influence metal and metallothionein levels. Environ Pollut 158:827–840
Fritsch C, Cœurdassier M, Giraudoux P, Raoul F, Douay F, Rieffel D et al (2011) Spatially explicit analysis of metal transfer to biota: influence of soil contamination and landscape. PLoS One 6: e20682
Fritsch C, Coeurdassier M, Faivre B, Baurand PE, Giraudoux P, van den Brink NW, Scheifler R (2012) Influence of landscape composition and diversity on contaminant flux in terrestrial food webs: a case study of trace metal transfer to European blackbirds *Turdus merula*. Sci Total Environ 432:275–287
Furness RW (1996) Cadmium in birds. In: Beyer WN, Heinz GH, Redmon-Norwood AW (eds) Environmental contaminants in wildlife: interpreting tissue concentrations. CRC Press, Boca Raton, FL, pp 389–404
Gall JE, Boyd RS, Rajakarunaet N (2015) Transfer of heavy metals through terrestrial food webs: a review. Environ Monit Assess 187:201
Gamberg M, Braune BM (1999) Contaminant residue levels in arctic wolves (*Canis lupus*) from the Yukon Territory, Canada. Sci Total Environ 243–244:329–338
Gamberg M, Boila G, Stern G, Roach P (2005) Cadmium, mercury and selenium concentrations in mink (Mustela vison) from Yukon, Canada. Sci Total Environ 351–352:523–529
García-Fernandez AJ, Sanchez-Garcia JA, Jimenez-Montalban P, Luna A (1995) Lead and cadmium in wild birds in southeastern Spain. Environ Toxicol Chem 14:2049–2058

García-Fernández AJ, Sanchez-Garcia JA, Gomez-Zapata M, Luna A (1996) Distribution of cadmium in blood and tissues of wild birds. Arch Environ Contam Toxicol 30:252–258

Garcia-Morales P, Saceda M, Kenney N, Kim N, Salomon DS, Gottardis MM et al (1994) Effect of cadmium on estrogen receptor levels and estrogen-induced responses in human breast cancer cells. J Biol Chem 269:16896–16901

Goyer RA, Liu J, Waalkes MP (2004) Cadmium and cancer of prostate and testis. Biometals 17: 555–558

Grant CA, Sheppard SC (2008) Fertilizer impacts on cadmium availability in agricultural soils and crops. Hum Ecol Risk Assess 14:210–228

Groten JP, Koeman JH, van Nesselrooij JH, Luten JB, Fentener van Vlissingen JM, Stenhuis WS, van Bladeren PJ (1994) Comparison of renal toxicity after long-term oral administration of cadmium chloride and cadmium-metallothionein in rats. Fundam Appl Toxicol 23:544–552

Grove RA, Henny CJ (2008) Environmental contaminants in male river otters from Oregon and Washington, USA, 1994–1999. Environ Monit Assess 145:49–73

Haffor AS, Abou-Tarboush FM (2004) Testicular cellular toxicity of cadmium: transmission electron microscopy examination. J Environ Biol 3:251–258

Hamers T, Smit LAM, Bosveld ATC, van den Berg JH, Koeman JH, van Schooten FJ, Murk AJ (2002) Lack of a distinct gradient in biomarker responses in small mammals collected at different distances from a highway. Arch Environ Contam Toxicol 43:345–355

Hamers T, van den Berg JHJ, van Gestel CAM et al (2006) Risk assessment of metals and organic pollutants for herbivorous and carnivorous small mammal food chains in a polluted floodplain (Biesbosch, The Netherlands). Environ Pollut 144:581–595

Hassan AA, Rylander C, Brustad M, Sandanger TM (2012) Level of selected toxic elements in meat, liver, tallow and bone marrow of young semi-dometicated reindeer (*Rangifer tarandus tarandus*) from Northern Norway. Int J Circumpolar Health 71:1–7

Heikens A, Peijnenburgb W, Hendriks A (2001) Bioaccumulation of heavy metals in terrestrial invertebrates. Environ Pollut 113:385–393

Helios-Rybicka E, Adamiec E, Aleksander-Kwaterczak U (2005) Distribution of trace metals in the Odra River system: water–suspended matter–sediments. Limnol Ecol Manag Inl Wat (3): 185–198

Henley R, Berger B (2013) Nature's refineries—metals and metalloids in arc volcanoes. Earth-Sci Rev 125:146–170

Herber R (2004) Cadmium. In: Merian E, Anke M, Ihnat M, Stoeppler M (eds) Elements and their compounds in the environment. Wiley, Weinheim, pp 689–708

Hernández-Moreno D, de la Casa Resino I, Fidalgo L, Llaneza L, Soler Rodríguez F, Pérez-López M, López-Beceiro A (2013) Noninvasive heavy metal pollution assessment by means of Iberian wolf (*Canis lupus signatus*) hair from Galicia (NW Spain): a comparison with invasive samples. Environ Monit Assess 185:10421–10430

Hew KW, Ericson WA, Welsh MJ (1993) A single low cadmium dose causes failure of spermiation in the rat. Toxicol Appl Pharmacol 121:15–21

Hiratsuka H, Satoh S, Satoh M, Katsuki Y, Suzuki I, Nakagawa J et al (1999) Tissue distribution of cadmium in rats given minimum amounts of cadmium-polluted rice or cadmium chloride for 8 months. Toxicol Appl Pharmacol 160:183–191

Holmgren GGS, Meyer MW, Chaney RL, Daniels RB (1993) Cadmium, lead, zinc, copper and nickel in agricultural soils of the United States of America. J Environ Qual 22:335–348

Hunter BA, Johnson MS, Thompson DJ (1987) Ecotoxicology of copper and cadmium in a contaminated grassland ecosystem. III. Small mammals. J Appl Ecol 24:601–614

Hunter BA, Johnson MS, Thompson DJ (1989) Ecotoxicology of copper and cadmium in a contaminated grassland ecosystem. IV. Tissue distribution and age accumulation in small mammals. J Appl Ecol 26(1):89–99

Hyvärinen H, Tyni P, Nieminen P (2003) Effects of Moult, age, and sex on the accumulation of heavy metals in the Otter (*Lutra lutra*) in Finland. Bull Environ Contam Toxicol 70:278–284

IARC (1993) Cadmium and certain cadmium compounds. In: IARC monographs on the evaluation of the carcinogenic risk of chemicals to humans. Beryllium, cadmium, mercury and exposures in the glass manufacturing industry. IARC monographs, vol 58. World Health Organization, International Agency for Research on Cancer, Lyon, France, pp 119–236

ICPDR (2002) Joint Danube survey, final report (p 261). International Commission for the Protection of the Danube River, Vienna

Idlafkih Z, Cossa D, Meybeck M (1995) Comportements des contaminants en trace dissous et particulaires (As, Cd, Cu, Hg, Pb, Zn) dans la Seine. Hydroecol Appl 7:127–150

Iguchi H, Sano S (1982) Effect of cadmium on the bone collagen metabolism of rat. Toxicol Appl Pharmacol 62:126–136

Ilyin I, Rozovskaya O, Travnikov O, M. Varygina (2015) Heavy metals: analysis of long-term trends, country-specific research and progress in mercury regional and global modeling. EMEP Status Report 2/2015. http://www.msceast.org/index.php/reports

Islam F, Majumder SS, Al Mamun A, Khan B, Rahman MA, Salam A (2015) Trace metals concentrations at the atmosphere particulate matters in the Southeast Asian Mega City (Dhaka, Bangladesh). Open J Air Pollut 4:86–98

Jager LP, Rijnierse VJ, Esselink H, Baars AJ (1996) Biomonitoring with the buzzard *Buteo buteo* in the Netherlands: heavy metals and sources of variation. J Ornithol 137:295–318

Joseph P (2009) Mechanisms of cadmium carcinogenesis. Toxicol Appl Pharmacol 238:272–279

Kabata-Pendias H (2001) Trace elements in soils and plants. CRC Press, Boca Raton, FL

Kabata-Pendias A, Pendias H (1999) Biogeochemia pierwiastków śladowych, PWN Warszawa

Kabata-Pendias A, Szteke B (2012) Pierwiastki śladowe w geo- i biosferze. JUNG-BIB, Puławy, pp 91–100

Kalisińska E, Salicki W, Mysłek P, Kavestka K, Jackowski A (2004) Using the Mallard to biomonitor heavy metal contamination of wetlands in north-western Poland. Sci Total Environ 320:145–161

Kalisińska E, Budis H, Wilk A, Łanocha N, Jackowski A (2009) Lead and cadmium in kidney and liver of nocturnal and diurnal raptors. Ochr Śr Zasobów Nat 41:102–110

Kang S, Kang JH, Kim S, Lee SH, Lee S, HJ Y et al (2015) Trace element analysis of three tissues from Eurasian otters (*Lutra lutra*) in South Korea. Ecotoxicology 24:1064–1072

Kim J, Oh JM (2015) Comparison of trace element concentrations in grey heron and black-crowned night heron chicks. Environ Monit Assess 187:4124

Kim S, Cheon HS, Kim SY, Juhnn Y, Kim YY (2013) Cadmium induces neuronal cell death through reactive oxygen species activated by GADD153. BMC Cell Biol 14:4

Kjellström T, Nordberg GF (1985) Kinetic model of cadmium metabolism. In: Friberg L, Elinder CG, Kjellström T et al (eds) Cadmium and health: a toxicological and epidemiological appraisal, Exposure, dose and metabolism, vol I. CRC Press, Boca Raton, FL, pp 179–197

Klassen CD, Liu J, Choudhuri S (1999) Metallothionein: an intracelluar protein to protect against cadmium toxicity. Ann Rev Pharmacol Toxicol 39:267–294

Klavinš M, Briede A, Rodinov V, Kokorite I, Parele E, Kļaviņa I (2000) Heavy metals in rivers of Latvia. Sci Total Environ 262:175–183

Kortenski J, Sotirov A (2002) Occurrence and distribution of environmentally hazardous elements in the Katrishte lignite bed, Strouma-Mesta Province, Bulgaria. Environ Geosci 9:191–199

Kośla T, Skibniewska EM, Skibniewski M (2008) Evaluation of cadmium content in the kidneys and liver of European bisons from the Białowieża forest. Med Weter 64:1129–1131

Kottferová J, Koréneková B (2000) Game as an indicator of environmental pollution by cadmium and lead. J Trace Microprobe Tech 18:571–575

Kramárová M, Massányi P, Slamecka J, Tataruch F, Jancová A, Gasparik J et al (2005) Distribution of cadmium and lead in liver and kidney of some wild animals in Slovakia. J Environ Sci Health Part A 40:593–600

Krawczyk AJ, Bogdziewicz M, Czyż M (2013) Diet of the American mink Neovison vison in an agricultural landscape in western Poland. Folia Zool 62:303–309

Lafuente A, Márquez N, Pérez-Lorenzo M, Pazo D, Esquifino AI (2001) Cadmium effects on hypothalamic-pituitary-testicular axis in male rats. Exp Biol Med (Maywood) 226:605–611

Łanocha N, Kalisińska E, Kosik-Bogacka D, Budis H, Sokolowski S, Bohatyrewicz A (2013) Comparison of metal concentrations in bones of long-living mammals. Biol Trace Elem Res 152:195–203

Larison JR (2002) Effects of cadmium on white-tailed ptarmigan in Colorado. http://fs.ogm.utah.gov/pub/MINES/AMR_Related/NAAMLP/BioVeg/Larison.pdf

Lazarus M, Orct T, Blanuša M, Vickovic I, Sostarić B (2008) Toxic and essential metal concentrations in four tissues of red deer (*Cervus elaphus*) from Baranja, Croatia. Food Add Contam 25:270–283

Lazarus M, Crnić AP, Bilandžić N, Kusak J, Reljić S (2014) Cadmium, lead, and mercury exposure assessment among Croatian consumers of free-living game. Arh Hig Rada Toksikol 65:281–292

Leach RM Jr, Wang KW, Baker DE (1979) Cadmium and the food chain: the effect of dietary cadmium on tissue composition in chicks and laying hens. J Nutr 109:437–443

Lehman LD, Klaassen CD (1986) Dosage-dependent disposition of cadmium administered orally to rats. Toxicol Appl Pharmacol 84:159–167

Lemarchand C, Rosoux R, Berny P (2010) Organochlorine pesticides, PCBs, heavy metals and anticoagulant rodenticides in tissues of Eurasian otters (*Lutra lutra*) from upper Loire River catchment (France). Chemosphere 80:1120–1124

Levengood JM (2003) Cadmium and lead in tissues of Mallards (*Anas platyrhynchos*) and Wood Ducks (*Aix sponsa*) using the Illinois River (USA). Environ Pollut 122:177–181

Li JL, Gao R, Li S, Wang JT, Wang JT, Tang ZX, Xu SW (2010) Testicular toxicity induced by dietary cadmium in cocks and ameliorative effect by selenium. Biometals 23:695–705

Limaye DA, Shaikh ZA (1999) Cytotoxicity of cadmium and characteristics of its transport in cardiomyocytes. Toxicol Appl Pharmacol 154:59–66

Lindqvist L, Block M, Tjälve H (1995) Distribution and excretion of Cd, Hg, methyl-Hg and Zn in the predatory beetle *Pterostichus niger* (Coleoptera: Carabidae). Environ Toxicol Chem 14: 1195–1201

Linšak DT, Linšak Z, Spirić Z, Srebočan E, Glad M, Cenov A (2014) Influence of cadmium on metallothionein expression and products of lipid peroxidation in the organs of hares (*Lepus europaeus* Pallas). J Appl Toxicol 34:289–295

Lourenço J, Pereira R, Gonçalves F, Mendo S (2013) Metal bioaccumulation, genotoxicity and gene expression in the European wood mouse (*Apodemus sylvaticus*) inhabiting an abandoned uranium mining area. Sci Total Environ 443:673–680

Lucia M, Andre JM, Gonzalez P, Baudrimont M, Bernadet MD, Gontier K (2010) Effects of dietary cadmium contamination on bird *Anas platyrhynchos*—comparison with species *Cairina moschata*. Ecotoxicol Environ Saf 73:2010–2016

Mallory ML, Wayland M, Braune BM, Drouillard KG (2004) Trace elements in marine birds, arctic hare and ringed seals breeding near Qikiqtarjuaq, Nunavut, Canada. Mar Pollut Bull 49:119–141

Mann RM, Vijver MG, Peijnenburg W (2011) Metals and metalloids in terrestrial systems: bioaccumulation, biomagnification and subsequent adverse effects. In: Sánchez-Bayo F, van den Brink P, Mann R (eds) Ecological impacts of toxic chemicals, Bentham Science Publishers Ltd, Beijing, China, pp 43-62

Markowski M, Bańbura M, Kaliński A, Markowski J, Skwarska J, Wawrzyniak J et al (2014) Spatial and temporal variation of lead, cadmium, and zinc in feathers of great tit and blue tit nestlings in central Poland. Arch Environ Contam Toxicol 67:507–518

Martiniakova M, Omelka R, Jancova A, Formicki G, Stawarz R, Bauerova M (2012) Accumulation of risk elements in kidney, liver, testis, uterus and bone of free-living wild rodents from a polluted area in Slovak. J Environ Sci Health Part A 47:1202–1206

Matsubara-Khan J (1974) Compartmental analysis for the evaluation of biological half-lives of cadmium and mercury in mouse organs. Environ Res 7:54–67

Mayack LA (2012) Hepatic mercury, cadmium and lead in mink and otter from New York State: monitoring environmental contamination. Environ Monit Assess 184:2497–2516

Mayack LA, Bush P, Fetcher O, Page RK, Fendley TT (1981) Tissues residues of dietary cadmium in wood ducks. Arch Environ Contam Toxicol 10:637–645

McBride MB, Shayler HA, Spliethoff HM, Mitchell RG, Marquez-Bravo LG, Ferenz GS et al (2014) Concentrations of lead, cadmium and barium in urban garden-grown vegetables: the impact of soil variables. Environ Pollut 194:254–261

Metcheva R, Teodorova S, Topashka-Ancheva M (2003) A comparative analysis of the heavy metal loading of small mammals in different regions of Bulgaria I: monitoring points and bioaccumulation features. Ecotoxicol Environ Saf 54:176–187

Michailova P, Petrova N, Bovero S, Caviccioli O, Ramella L, Sella G (2000) Effect of environmental pollution on the chromosomal variability of *Chironomus riparius* Meigen (Diptera, Chironomidae) larvae from two Piedmont stations. Genetica 108:171–180

Micó C, Recatalá L, Peris M, Sánchez J (2006) Assessing heavy metal sources in agricultural soils of an European Mediterranean area by multivariate analysis. Chemosphere 65:863–872

Mierzykowski SE, Todd CS (2012) Environmental contaminants in a crossed bill bald eagle recovered in Maine. USFWS. Special Project Report, FY12-MEFO-3-EC. Maine Field Office. Orono, ME, 28 pp

Millaku L, Imeri R, Trebicka A (2015) Bioaccumulation of heavy metals in tissues of house sparrow (*Passer domesticus*). Res J Environ Toxicol (2):107–112

Millán J, Mateo R, Taggart MA, López-Bao JV, Viota M, Monsalve L et al (2008) Levels of heavy metals and metalloids in critically endangered Iberian lynx and other wild carnivores from Southern Spain. Sci Total Environ 399:193–201

Milton A, Cooke JA, Johnson MS (2003) Accumulation of lead, zinc, and cadmium in a wild population of *Clethrionomys glareolus* from an abandoned lead mine. Arch Environ Contam Toxicol 44:405–411

Mochizuki M, Hondo R, Kumon K, Sasaki R, Matsuba H, Ueda F (2002) Cadmium contamination in wild birds as an indicator of environmental pollution. Environ Monit Assess 73:229–235

Mochizuki M, Mori M, Hondo R, Ueda F (2008) A new index for evaluation of cadmium pollution in birds ans mammals. Environ Monit Assess 137:35–49

Molnár A, Mészáros E, Polyák K, Borbély-Kiss I, Koltay E, Szabó G, Horváth Z (1995) Atmospheric budget of different elements in aerosol particles over Hungary. Atmos Environ 29: 1821–1828

Morrow H (2001) Cadmium and cadmium alloys. In: Kirk-Othmer encyclopedia of chemical technology. John Wiley & Sons, Inc, pp 471-507

Murphy RD, Smith AL, Todd SL, Elpers JT, Radde BN, Klinge CM et al (2012) The endocrine disruptors cadmium chloride and sodium arsenate induce human lung adenocarcinoma cell proliferation by activating the estrogen receptor-mediated signaling pathway. FASEB J 26(suppl 1):765.3

Mussali-Galante P, Tovar-Sánchez E, Valverde M, Rojas Del Castillo E (2013) Biomarkers of exposure for assessing environmental metal pollution: from molecules to ecosystems. Rev Int Contam Ambient 29:117–140

Mysłek P, Kalisińska E (2006) Contents of selected heavy metals in the liver, kidneys and abdominal muscle of the brown hare (*Lepus europaeus* Pallas) in Central Pomerania, Poland. Pol J Vet Sci 9:31–41

Naccari C, Cristani M, Cimino F, Arcoraci T, Trombetta D (2009) Common buzzards (*Buteo buteo*) bioindicator of heavy metals pollution in Sicily (Italy). Environ Int 35:594–598

Nesterkova DV, Vorobeichik EL, Reznichenko IS (2014) The effect of heavy metals on the soil-earthworm-European mole food chain under the conditions of environmental pollution caused by the emissions of a copper smelting plant. Contemp Probl Ecol 7:587–596

Nordberg FG (2009) Historical perspectives on cadmium toxicology. Toxicol Appl Pharmacol 238:192–200

Orłowski G, Kamiński P, Kasprzykowski Z, Zawada Z, Koim-Puchowska B, Szady-Grad M, Klawe JJ (2012) Essential and nonessential elements in nestling rooks *Corvus frugilegus* from eastern Poland with a special emphasis on their high cadmium contamination. Arch Environ Contam Toxicol 63:601–611

Orłowski G, Kasprzykowski Z, Dobicki W, Pokorny P, Wuczyński A, Polechoński R, Mazgajski TD (2014) Residues of chromium, nickel, cadmium and lead in rook *Corvus frugilegus* eggshells from urban and rural areas of Poland. Sci Total Environ 490:1057–1064

Pacyna JM, Pacyna EG (2001) An assessment of global and regional emissions of trace metals to the atmosphere from anthropogenic sources worldwide. Environ Rev 9:269–298

Pan J, Plant J, Voulvoulis N, Oates CJ, Ihlenfeld C (2010) Cadmium levels in Europe implications for human health. Environ Geochem Health 32:1–12

Pedersen S, Lierhagen S (2006) Heavy metal accumulation in arctic hares (*Lepus arcticus*) in Nunavut, Canada. Sci Total Environ 368:951–955

Pereira R, Pereira ML, Ribeiro R, Gonçalves F (2006) Tissue and hair residues and histopathology in wild rats (*Rattus rattus* L.) and Algerian mice (*Mus spretus* Lataste) from an abandoned mine area (Southeast Portugal). Environ Pollut 139:561–575

Pérez-López M, Hermoso De Mendoza M, Soler Rodríguez F (2008) Heavy metal (Cd, Pb, Zn) and metalloid (As) content in raptor species from Galicia (NW Spain). Ecotoxicol Environ Saf 70: 154–162

Peterson JM (2001) Cadmium. http://hpschapters.org/northcarolina/NSDS/cadmium.pdf. Accessed 22 Aug 2015

Petrović Z, Teodorović V, Djurić S, Milićević D, Vranić D, Lukić M (2014) Cadmium and mercury accumulation in European hare (*Lepus europaeus*): age-dependent relationships in renal and hepatic tissue. Environ Sci Pollut Res 21:14058–14068

Piskorová L, Vasilková Z, Krupicer I (2003) Heavy metal residues in tissues of wild boar (*Sus scrofa*) and red fox (*Vulpes vulpes*) in the Central Zemplin region of the Slovak Republic. Czech J Anim Sci 48:134–138

Pollock B (2005) Trace elements status of white-tailed deer (*Odocoileus virginianus*) and moose (*Alces alces*) in Nova Scotia. Wildlife Damage Management, Internet Center for Canadian Cooperative Wildlife Health Centre: Newsletters & Publications, University of Nebraska–Lincoln

Pollock B, Penashue B, Scott MV, Vanleeuwen J, Daoust PY, Burgess NM, Tasker AR (2009) Liver parasites and body condition in relation to environmental contaminants in caribou (*Rangifer tarandus*) from Labrador, Canada. Arctic 62:1–12

Protano CL, Zinna S, Giampaoli V, Romano Spica V, Chiavarini S, Vitali M (2014) Heavy metal pollution and potential ecological risks in rivers: a case study from southern Italy. Bull Environ Contam Toxicol 92:75–80

Querol X, Cabrera LI, Pickel W, López-Solera A, Hagemannc HW, Fernández-Turiela JL (1996) Geological controls on the coal quality of the Mequinenza subbituminous coal deposit, northeast Spain. Int J Coal Geol 29:67–91

Reimann C, de Caritat P (1998) Chemical elements in the environment. Factsheets for the Geochemist and Environmental Scientist. Springer, Berlin, Heidelberg, 398 pp

Reimann C, Finne TE, Nordgulen R, Sæthera OM, Arnoldussen A, Banks D (2009) The influence of geology and land-use on inorganic stream water quality in the Oslo region, Norway. Appl Geochem 24:1862–1874

Reimann C, de Caritat P, GEMAS Project Team, NGSA Project Team (2012) New soil composition data for Europe and Australia: demonstrating comparability, identifying continental-scale processes and learning lessons for global geochemical mapping. Sci Total Environ 416:239–252

Richardson ME, Fox MR (1974) Dietary cadmium and enteropathy in the Japanese quail: histochemical and ultrastructural studies. Lab Invest 31:722–731

Robillard S, Beauchamp G, Paillard G, Bélanger D (2002) Levels of cadmium, lead, mercury and $^{137}$caesium in caribou (*Rangifer tarandus*) tissues from Northern Québec. Arctic 55:1–9

Rodríguez-Estival J, Taggart MA, Mateo R (2011) Alterations in vitamin A and E levels in liver and testis of wild ungulates from a lead mining area. Arch Environ Contam Toxicol 60:361–371

Romic M, Romic D (2003) Heavy metals distribution in agricultural topsoils in urban area. Environ Geol 43:795–805

Roodbergen M, Klok C, van der Hout A (2008) Transfer of heavy metals in the food chain earthworm Black-tailed godwit (*Limosa limosa*): comparison of a polluted and a reference site in The Netherlands. Sci Total Environ 406:407–412

Rusch GM, O'Grodnick JS, Rinehart WE (1986) Acute inhalation study in rat of comparative uptake, distribution and excretion of different cadmium containing materials. Am Ind Hyg Assoc J 47:754–763

Ryšavá-Nováková M, Koubek P (2009) Feeding habits of two sympatric mustelid species, European polecat *Mustela putorius* and stone marten *Martes foina*, in the Czech Republic. Folia Zool 58:66–75

Sacco-Gibson N, Chaudhry S, Brock A, Sickles AB, Patel B, Hegstad R et al (1992) Cadmium effects on bone metabolism: accelerated resorption in ovariectomized, aged beagles. Toxicol Appl Pharmacol 113:274–283

Salvatori F, Talassi CB, Salzgeber SA, Spinosa HS, Bernardi MM (2004) Embryotoxic and long-term effects of cadmium exposure during embryogenesis in rats. Neurotoxicol Teratol 26: 673–680

Samuel JB, Stanley JA, Princess RA, Shanthi P, Sebastian MS (2011) Gestational cadmium exposure-induced ovotoxicity delays puberty through oxidative stress and impaired steroid hormone levels. J Med Toxicol 7:195–204

Sánchez Chardi A, Peñarroja-Matutano C, Oliveira Ribeiro CA, Nadal J (2007) Bioaccumulation of metals and effects of a landfill in small mammals. Part II. The wood mouse, *Apodemus sylvaticus*. Chemosphere 70:101–109

Scheifler R, De Vaufleury A, Coeurdassier M, Crini N, Badot PM (2006) Transfer of Cd, Cu, Ni, Pb, and Zn in a soil-plant-invertebrate food chain: a microcosm study. Environ Toxicol Chem 25:815–822

Scheirs J, De Coen A, Covaci A, Beernaert J, Kayawe VM, Caturla M et al (2006) Genotoxicity in wood mice (*Apodemus sylvaticus*) along a pollution gradient: exposure-, age-, and gender-related effects. Environ Toxicol Chem 25:2154–2162

Scheuhammer AM (1987) The chronic toxicity of aluminum, cadmium, mercury, and lead in birds: a review. Environ Pollut 46:263–295

Schipper AM, Wijnhoven S, Baveco H, van den Brink NW (2012) Contaminant exposure in relation to spatio-temporal variation in diet composition: a case study of the little owl (*Athene noctua*). Environ Pollut 163:109–116

Schroeder HA, Mitchener M (1971) Toxic effects of trace elements on the reproduction of mice and rats. Arch Environ Health 23:102–106

See SW, Balasubramanian R, Rianawati E, Karthikeyan S, Streets DG (2007) Characterization and source apportionment of particulate matter $\leq$2.5 μm in Sumatra, Indonesia during a recent peat fire episode. Environ Sci Technol 41:3488–3494

Shao X, Cheng H, Li Q, Lin C (2013) Anthropogenic atmospheric emissions of cadmium in China. Atmos Environ 79:155–160

Shore RF, Douben PET (1994) The ecological significance of cadmium intake and residues in terrestrial small mammals. Ecotoxicol Environ Saf 29:101–112

Shore RF, Casulli A, Bologov V, Wienburg CL, Afsar A, Toyne P, Dell'Omo G (2001) Organochlorine pesticide, polychlorinated biphenyl and heavy metal concentrations in wolves (*Canis lupus* L. 1758) from north-west Russia. Sci Total Environ 280:45–54

Shridhar V, Khillare PS, Agarwal T, Ray S (2010) Metallic species in ambient particulate matter at rural and urban location of Delhi. J Hazard Mater 175:600–607

Six L, Smolders E (2014) Future trends in soil cadmium concentration under current cadmium fluxes to European agricultural soils. Sci Total Environ 485-486:319–328

Smith GJ, Rongstad OJ (1982) Small mammal heavy metal concentrations from mined and controlled sites. Environ Pollut 28:121–134

Stepanova VA, Pokrovsky OS, Viers J, Mironycheva-Tokareva NP, Kosykh NP, Vishnyakova EK (2015) Elemental composition of peat profiles in western Siberia: effect of the micro-landscape, latitude position and permafrost coverage. Appl Geochem 53:53–70

Stohs SJ, Bagchi D, Hassoun E, Bagchi M (2001) Oxidative mechanisms in the toxicity of chromium and cadmium ions. J Environ Pathol Toxicol Oncol 20:77–88

Stricker GD, Flores RM, Trippi MH, Ellis MS, Olson CM, Sullivan JE et al (2007) Coal quality and major, minor, and trace elements in the Powder River, Green River, and Williston basins, Wyoming and North Dakota: U.S. Geological Survey Open-File Report 2007-1116, 31 pp

Szefer P (2002) Metals, metalloids and radionuclides in the Baltic Sea Ecosystem. Trace metals in the environment. Elsevier, Amsterdam, Germany

Szymczyk K, Zalewski K (2003) Copper, zinc, lead and cadmium content in liver and muscles of mallards (*Anas platyrhychnos*) and other hunting fowl species in Warmia and Mazury in 1999-2000. Pol J Environ Stud 12:381–386

Taggart M, Reglero M, Camarero P, Mateo R (2011) Should legislation regarding maximum Pb and C levels in human food also cover large game meat? Environ Int 37:11–25

Tapisso J, Marques C, Mathias M Ramalhinho L (2009) Induction of micronuclei and sister chromatid exchange in bone-marrow cells and abnormalities in sperm of Algerian mice (*Mus spretus*) exposed to cadmium, lead and zinc. Mutat Res 678:59–64

Tête N, Afonso E, Bouguerra G, Scheifler R (2015) Blood parameters as biomarkers of cadmium and lead exposure and effects in wild wood mice (*Apodemus sylvaticus*) living along a pollution gradient. Chemosphere 138:940–946

Tolcin A (2011) Cadmium. In: 2011 Mineral Yearbook

Torres J, Eira C, Miquel J, Foronda P, Feliu C (2011) Cadmium and lead concentrations in *Moniliformis moniliformis* (Acanthocephala) and *Rodentolepis microstoma* (Cestoda), and in their definitive hosts, *Rattus rattus* and *Mus domesticus* in El Hierro (Canary Archipelago, Spain). Acta Parasitol 56:320–324

Tsipoura N, Burger J, Newhouse M, Jeitner C, Gochfeld M, Mizrahi D (2011) Lead, mercury, cadmium, chromium, and arsenic levels in eggs, feathers, and tissues of Canada geese of the New Jersey Meadowlands. Environ Res 111:775–778

Tume P, Bech J, Reverter F, Bech J, Longan L, Tume L, Sepúlveda B (2011) Concentration and distribution of twelve metals in Central Catalonia surface soils. J Geochem Exp 109:92–103

UNEP (2006) http://www.chem.unep.ch/Pb_and_Cd/WG/WG-meeting-documents.htm

UNEP (2010) Final review of scientific information on cadmium. http://www.unep.org/chemicalsandwaste/Portals/9/Lead_Cadmium/docs/Interim_reviews/UNEP_GC26_INF_11_Add_2_Final_UNEP_Cadmium_review_and_apppendix_Dec_2010.pdf

US EPA (2005) Ecological soil screening levels for cadmium. interim final. OSWER Directive 92857e65. US EPA office of Solid Waste and Emergency Response, Washington, DC, 20460

USGS (2014) Cadmium. In: Kelly TD, Matos GR (eds) Historical statistics for mineral and material commodities in the United States (2014 version): U.S. Geological Survey Data Series 140, http://minerals.usgs.gov/minerals/pubs/historical-statistics/

Van den Brink N, Lammertsma D, Dimmers W, Boerwinkel MC, van der Hout A (2010) Effects of soil properties on food web accumulation of heavy metals to the wood mouse (*Apodemus sylvaticus*). Environ Pollut 158:245–251

Van den Brink N, Lammertsma D, Dimmers W, Boerwinkel MC (2011) Cadmium accumulation in small mammals: species traits, soil properties, and spatial habitat use. Environ Sci Technol 45: 7497–7502

Vanparys C, Dauwe T, Campenhout K, Bervoets L, De Coen W, Blust R, Eens M (2008) Metallothioneins and aminolevulinic acid dehydratase (ALAd) as biomarkers of metal pollution in great tits (*Parus major*) along a pollution gradient. Sci Total Environ 401:184–193

Veltman K, Huijbregts MAJ, Hamers T, Wijnhoven S, Hendriks AJ (2007a) Cadmium accumulation in herbivorous and carnivorous small mammals: meta-analysis of field data and validation of the bioaccumulation model optimal modeling for ecotoxicological applications. Environ Toxicol Chem 26:1488–1496

Veltman K, Huijbregts MAJ, Hendriks AJ (2007b) Cadmium bioaccumulation factors for terrestrial species: application of the mechanistic bioaccumulation model OMEGA to explain field data. Sci Total Environ 406:413–418
Vihnanek Lazarus M, Sekovanić A, Kljaković-Gašpić Z, Orct T, Jurasović J, Kusak J et al (2013) Cadmium and lead in grey wolf liver samples: optimisation of a microwave-assisted digestion method. Arh Hig Rada Toksikol 64:395–403
Waalkes MP, Rehm S, Cherian MG (2000) Repeated cadmium exposures enhance the malignant progression of ensuing tumors in rats. Toxicol Sci 54:110–120
Wang J, Hao M, Liua C, Liu R (2015) Cadmium induced apoptosis in mouse primary hepatocytes: the role of oxidative stress-mediated ERK pathway activation and the involvement of histone H3 phosphorylation. RSC Adv 5:31798–31806
Webster WS (1988) Chronic cadmium exposure during pregnancy in the mouse: influence of exposure levels on the fetal and maternal uptake. J Toxicol Environ Health 74:183–192
WHO (2007) Health risks of heavy metals from long-range transboundary air pollution. Joint WHO/Convention Task Force on the Health Aspects of Air Pollution. www.euro.who.int/_data/assets/pdf.../E91044.pdf. Accessed 18 Dec 2007
Wieczorek-Dąbrowska M, Tomza-Marciniak A, Pilarczyk B, Balicka-Ramisz A (2013) Roe and red deer as bioindicators of heavy metals contamination in north-western Poland. Chem Ecol 29: 100–110
Wilson AK, Bhattacharyya MH (1997) Effects of cadmium on bone: an in vivo model for the early response. Toxicol Appl Pharmacol 145:68–74
Włostowski T, Bonda E, Krasowska A (2006) Free-ranging European bisons accumulate more cadmium in the liver and kidneys than domestic cattle in north-eastern Poland. Sci Total Environ 364:295–300
Włostowski T, Krasowska A, Salińska A, Włostowska M (2009) Seasonal changes of body iron status determine cadmium accumulation in the wild bank voles. Biol Trace Elem Res 131: 291–297
Xu LC, Wang SY, Yang XF, Wang XR (2001) Effects of cadmium on rat sperm motility evaluated with computer assisted sperm analysis. Biomed Environ Sci 4:312–317
Xu S, Pi H, Chen Y, Zhang N, Guo P, Lu Y et al (2013) Cadmium induced Drp1-dependent mitochondrial fragmentation by disturbing calcium homeostasis in its hepatotoxicity. Cell Death Dis 4:e540
Zhang X, Chen D, Zhong T, Zhang X, Cheng M, Li X (2015) Assessment of cadmium (cd) concentration in arable soil in China. Environ Sci Pollut Res 22:4932–4941
Zelenski M, Malik N, Taran Y (2014) Emissions of trace elements during the 2012–2013 effusive eruption of Tolbachik volcano, Kamchatka: enrichment factors, partition coefficients and aerosol contribution. J Volc Geoth Res 285:136–149
Zielińska S, Pieniążek M, Dżugan M (2015) Changes in activity of acid hydrolases in tissues of wild pheasants induced by heavy metals. Ecol Chem Eng 22:93–101

# Chapter 15
# Fluorine, F

**Izabela Gutowska, Monika Rać, and Dariusz Chlubek**

**Abstract** We summarize literature data on the concentrations of fluorine (F) in various tissues of terrestrial mammals and birds. The results of various studies indicate the usefulness of the determination of fluoride in hard tissues of wildlife for monitoring the environmental F contamination, as well as its role in the assessment of the effectiveness of measures taken to reduce F emissions, e.g., from industrial sources.

## 1 Introduction

Atmospheric air is contaminated with fluorine compounds not only as the result of various human economic activities but also by natural factors such as aeolian dust, volcanic emissions, evaporation of chemical components from sea and ocean surfaces, cosmic dust, and natural disasters (Francisca and Carro Perez 2009; Kabata-Pendias 2011). Changes in the chemical composition of the environment disrupt the homeostasis of ecosystems, an essential factor for their natural existence (Kabata-Pendias 2011). The accumulation of fluoride in soils is particularly dangerous, as the soil is the first barrier regulating its transfer into the food chain, and any changes in the natural fluoride concentration in the soil, such as an increase in levels or changes in quantitative proportions, cause adverse ecological and nutritional effects (Flueck and Smith-Flueck 2013; Kabata-Pendias 2011).

I. Gutowska (✉)
Department of Biochemistry and Human Nutrition, Pomeranian Medical University in Szczecin, Szczecin, Poland
e-mail: gutowska@pum.edu.pl

M. Rać · D. Chlubek
Department of Biochemistry and Medical Chemistry, Pomeranian Medical University in Szczecin, Szczecin, Poland



E. Kalisińska (ed.), *Mammals and Birds as Bioindicators of Trace Element Contaminations in Terrestrial Environments*,
https://doi.org/10.1007/978-3-030-00121-6_15

## 2 General Properties

At room temperature, fluorine is a pale yellow-green gas (Kabata-Pendias 2011) with a penetrating odor, strongly affecting the mucous membrane. As one of the most electronegative and reactive of all elements (Kolditz 1994), at room or elevated temperatures, fluorine binds directly to all elements apart from oxygen and nitrogen and therefore readily reacts with the most organic compounds. Fluorine ions have a high ability to form complexes with heavy metal ions in aqueous solutions (Kabata-Pendias 2011). Fluorine also creates compounds with nonmetallic elements, comprising covalent bonds (Kabata-Pendias 2011). Thanks to its high electronegativity, fluorine oxidizes many elements to the highest degrees, and the small size of its atom leads to high coordination numbers (Kolditz 1994).

Assessments of exposure to fluorine and its compounds can be made by determinations in different parts of the environment. The maximum allowable fluoride level in the air at workplaces is 0.05 mg $m^{-3}$, with an allowable momentary level of 0.4 mg $m^{-3}$. Acceptable concentrations of fluorine in air emissions are 0.03 mg $m^{-3}$ (for 30 min), 0.01 mg $m^{-3}$ (average daily concentration), and 0.0016 mg $m^{-3}$ (average annual concentration) (Węglarz and Michalski 1998).

## 3 Fluorine Minerals, Production, and Uses

Fluorine constitutes about 0.06–0.09% of elements in the earth's crust, i.e., 625 mg $kg^{-1}$ on average (Kabata-Pendias 2011). Due to the fact that fluorine is one of the most reactive elements, in the environment, it creates common and poorly soluble mineral complexes (WHO, 2006), which include fluorspar, rock phosphate, cryolite, apatite, mica, hornblende, and others (Murray 1986). Igneous and sedimentary rocks are the richest in the poorly soluble calcium fluoride (fluorite, $CaF_2$). The highest F levels are found in alkaline to acidic igneous rocks, while the lowest levels are found in ultra-alkaline rocks (Kabata-Pendias 2011). The average F level in rocks ranges from 0.1 to 1.0 g $kg^{-1}$ (Indulski 1989). Fluorine exhibits lipophilic and oxyphilic tendencies (Kabata-Pendias 2011).

Fluorspar is the most common commercial source of F. Global mining production of fluorspar, apart from the United States (data not available), was estimated in 2008 at about 5840 kt and rose in 2014 to about 6850 kt (US Geological Survey 2015).

Other minerals used in the industry include phosphate rock applied in the manufacture of phosphate fertilizers and cryolite used in the production of aluminum (Murray 1986) and as a pesticide. The by-products of the conversion of phosphate rock into phosphate fertilizers are fluorosilicates, which after purification are a source of fluoride added to drinking water in various countries in the past or currently to assist in preventing dental caries (Reeves 1986).

# 4 Fluorine in Nature: Geogenic and Anthropogenic Sources

## *4.1 Water*

Fluoride is naturally present in various water bodies, ranging from 1 mg $L^{-1}$ in the sea to approx. 0.5 mg $L^{-1}$ (or less) in rivers and lakes. The concentration in groundwater depends mainly on the structure of the mineral rock and the degree of solubility of fluorine compounds (Hem 1989). In addition, calcium and sodium ions can reduce or increase the content of fluoride in water: the more calcium, the lower the concentration of fluoride; the reverse is true for sodium (Edmunds and Smedley 1996).

High concentrations of fluoride in groundwater resulting from the construction of mineral substrates have been noted in areas of India, Pakistan, West and South Africa, Sri Lanka, Thailand, in all provinces and autonomous regions of China, and in South America (WHO 2006). Central European countries that have elevated levels of fluoride in groundwater include Ukraine, Moldova, Hungary, and Slovakia (Fordyce et al. 2007).

Particularly high concentrations of fluorine compounds in water can be observed in areas of high volcanic activity (Allibone et al. 2012; Edmunds and Smedley 1996). Therefore, thermal waters—especially those with a high pH—are rich in fluorine compounds (Edmunds and Smedley 1996). Many lakes in Sudan, Ethiopia, Uganda, Kenya, and Tanzania have a very high content of fluorine ranging from 690 mg $L^{-1}$ in Tanzania to 2800 mg $L^{-1}$ in Kenya. Also groundwaters in those areas are characterized by high levels of fluoride ranging from 30 to 50 mg $L^{-1}$. Samples of groundwater from bores drilled to source drinking water exceeded the safety threshold of 1–1.5 mg $L^{-1}$ (WHO 2006), which correlated with a high rate of fluorosis in those areas (Manji and Kapila 1986). Many scholars have also indicated a relationship between volcanic activity and the increased incidence of fluorosis among people and animals, even at large distances from the sources of emissions (Allibone et al. 2012; Cronin and Sharp 2002; Flueck and Smith-Flueck 2013; Francisca and Carro Perez 2009).

The countries of South America, especially Chile, Peru, Argentina, and Paraguay, are particularly susceptible to high concentrations of fluoride in the groundwater because of the geographical location of those countries, bordering the Ring of Fire, a seismically active region. Research conducted by Francisca and Carro Perez (2009) showed elevated concentrations of fluoride in groundwater samples in Argentina (0–13.5 mg F $L^{-1}$) and directly linked to recent volcanic eruptions. After the eruption of the Puyehue-Cordon Caulle in 2011 (PCCVE, June 2011), Argentina and Chile saw an increase in the incidence of fluorosis in humans, confirming the impact of the eruption on the content of fluoride in the environment (Flueck and Smith-Flueck 2013). Similar observations were made by Oruc (2008) after analyzing data from areas located near Tendurek Volcano in Eastern Turkey.

## 4.2 Air

Many researchers believe that only the areas around industrial plants are environmentally contaminated, endangering only local fauna and flora. However, research in areas located far from the emitters has shown that these are also ecologically threatened. The spread of pollutants greatly depends on the directions of winds. The distribution of winds, climatic conditions, and the level of emissions are factors determining the extent and spread of industrial emissions (Machoy et al. 2001, 2002).

Fluorine does not occur in the atmosphere in a free state. Fluoride is emitted into the air in the form of gas and solids. The greatest spread of fluoride in the air can be observed close to active volcanoes (Edmunds and Smedley 1996). Another natural source of atmospheric fluoride in the soil is in seawater spray carried by the wind. In urban areas, most of the fluoride found in the air is the result of human activity. The aluminum industry is responsible for the emission of 10% of the total amount of atmospheric fluoride. Ironworks, superphosphate factories, brickyards, glassworks, coal-fired power plants, and petroleum refineries also pollute the atmosphere with fluoride (Machoy et al. 2002).

# 5 Biological Status of Fluorine

Fluoride may be present in inorganic and organic compounds. Individual forms are soluble to varying degrees and then absorbed by plants, animals, and humans, exhibiting diverse levels of toxicity (Kabata-Pendias 2011). This is significant in the case of fluoride—an element that acts slowly, not giving symptoms for many years. In particular, this applies to the skeletal system because the physiological role of fluorine is mainly related to the processes of bone mineralization (Machoy-Mokrzyńska 2000).

The affinity of fluorine to metals, in particular divalent ones (Ca, Mg, Mn, Fe, Cu, Zn), promotes the formation of different compounds, which change the physiological effects of these cations. Precipitated magnesium and calcium fluorides are deposited in different, mainly hard, tissues (Gutowska et al. 2005). Fluorine, when ingested in large quantities, can be harmful to human health and animals, particularly taking into account the very narrow margin of safety between a tolerable and a toxic dose (Giżewska and Machoy 1988; Machoy-Mokrzyńska 2000).

Acute fluoride poisoning affects virtually all organs in the human or animal body and can be fatal due to blockages in cell metabolism. Fluoride inhibits enzymatic processes and may even halt them. In particular, this applies to metalloenzymes. Also significant is the disruption of calcium-dependent essential functions of the body. Other metal ions may also be bound by fluoride, which then causes blockages of various biochemical reactions (Barbier et al. 2010). The toxicity of fluoride may be enhanced by the reaction of the environment. The acidic pH of the stomach, the oral cavity, and the final sections of the urinary tract favors the formation of highly toxic hydrogen fluoride (Giżewska and Machoy 1988; Jędrzejuk and Milewicz 1996).

The main route of gaseous and particulate fluoride absorption is the respiratory system (Barbier et al. 2010). In the gastrointestinal tract, fluorine is easily absorbed by simple diffusion (Giżewska and Machoy 1988). It permeates rapidly through cell membranes and is fed to all tissues (Barbier et al. 2010). In direct contact, fluoride can also be absorbed through the skin (Giżewska and Machoy 1988) and penetrates skin appendages—for example, both the organic and inorganic parts of nails. In addition to transepidermal absorption, fluorine exhibits a potential for transfollicular absorption via hair follicles and sebaceous glands. However, this route is less effective due to the small area of absorption (Giżewska and Machoy 1988).

Absorbed fluoride spreads throughout the human or animal body (Barbier et al. 2010; Giżewska and Machoy 1988). It can be found in all tissues, with the greatest accumulation in bones and teeth (Giżewska and Machoy 1988; Kabata-Pendias 2011). Fluoride ions are rapidly taken up by the bone, which results in the replacement of hydroxyl ions included in hydroxyapatite and creation of fluorhydroxyapatite or fluorapatite (Machoy 1990; Gutowska et al. 2005).

Fluoride retained in skeletal bone may be gradually released to the blood and urine. Blood acts as a means of fluoride transport in the body. Approximately 75% of fluoride in the blood is located in the plasma; the remainder is inside the red blood cells or on their surface. Unlike in bones, soft tissue concentrations do not increase with age or the duration of exposure (Indulski 1989).

The major route of excretion of fluoride is the urinary tract, although small amounts may also be excreted in sweat, feces, and saliva. Elimination of fluoride takes place in two phases: a rapid phase taking place over a matter of weeks and probably related to the ion exchange in the hydration shell, followed by a second slower phase with an average half-life lasting about 8 years, resulting from the resorption by osteoclast cells. Fluoride is excreted faster from trabecular bone than from compact bone. Approximately half of the absorbed fluoride is excreted in the urine via glomerular filtration. Fluoride may undergo resorption in the renal tubules, depending on the pH of the urine. In urine, fluoride is mainly present as $F^-$ and in small amounts as HF (Indulski 1989).

## 6 Toxicity of Fluorine in Homeothermic Animals

### *6.1 Effect on Bones and Teeth*

In the case of long-term absorption of fluoride into the body, a negative two-way action can be observed. Fluoride damages bone matrix, affecting biosynthesis of collagen by cells producing intercellular substance, via alteration of their activity, damage, or destruction (Yan et al. 2015). Secondly, it can form sparingly soluble calcium and magnesium fluorides, which lead to hypermineralization (Yan et al.

2015). The excessive accumulation of fluoride in bones can interrupt the formation of the crystal nucleation process. Crystallization occurs continuously toward the c-axis at the peripheral area, while the central area remains amorphous, resulting in crystal defects (Kakei et al. 2007). In addition, in trabecular bone, excessive accumulation of fluoride may increase bone volume and trabecular thickness without a concomitant increase in trabecular connectivity, which may reduce bone quality despite the increase in bone mass (Aaron et al. 1991). In addition, severe fluorosis leads to chronic osteomyelitis (Schultz et al. 1998).

Teeth with fluorosis are characterized by varying degrees of discoloration of the enamel, hypoplasia, and posteruptive surface defects. Subsurface layers of enamel exhibit hypomineralization, accentuated striae of Retzius, and wide hypomineralized incremental ranges of abnormal enamel structures, which may result from the negative effect of fluoride on the secretory activity of ameloblasts and enamel maturation (Kierdorf et al. 1996c, 2000b; Richter et al. 2010). Chronic exposure to fluoride also causes defects in the mineralization of the dentine, manifested by hypomineralization or interglobular dentin. However, according to one of the hypotheses, in certain stages of life, this structure includes mechanisms that protect it from exposure to high doses of fluorine. Those mechanisms include the placental barrier in the fetal period and the blood-milk barrier in the breast-feeding period; the uptake of most serum fluoride by the developing skeletal system occurs in this period of rapid growth of young individuals (Richter et al. 2010).

Severe fluorosis causes pathologically increased wear and fracture of teeth and periodontal bone lesions caused by inflammation of the periosteum. The result of the strong abrasion of the occlusal surface of teeth is opening of the pulp chamber, formation of periapical ulcers, and—eventually—tooth loss (Schultz et al. 1998).

## 6.2 *Neurotoxicity of Fluorine Compounds*

Fluoride has a negative effect on many metabolic processes in the human body. It is also a potent neurotoxin, resulting in the degeneration of structures such as the hippocampus, cerebral cortex, and cerebellum (Shivarajashankara et al. 2002). Exposure to fluoride inhibits receptors in the brain and reduces the production of neurotransmitters (Yu et al. 2008). Moreover, chronic intake of fluoride during the development of the brain may lead to reduced intelligence and disorders of the processes associated with memory and learning (Xiang et al. 2003), which may be related to changes in the structure and function of synapses (Celio and Blümcke 1994; Ziemiańska et al. 2012). At high concentrations, fluorine is an inhibitor of the metalloproteinases MMP2 and MMP9 (Kato et al. 2014), enzymes responsible for proteolysis in the perineuronal network that perform a protective function against inflammatory lesions and degeneration of nerve cells and participate in the stabilization of synapses and matured neurons (Celio and Blümcke 1994).

Research conducted by Basha et al. (2011) in rats dosed with 100 and 200 mg $L^{-1}$ NaF in drinking water showed histopathological changes in discrete brain regions and decreased learning and memory abilities in multigenerational rats. The presence of eosinophilic Purkinje cells, degenerating neurons, decreased granular cells, and vacuolations was noted in discrete brain regions of the fluoride-treated group. In the T-maze experiments, rats treated with fluoride showed worse results in comparison to the control group (poor acquisition and retention and higher latency), and the observed changes deepened in the subsequent generations of the rats (Basha et al. 2011), which may indicate a negative cumulative impact of fluorine on the developing brain. The observed changes may be caused by a growing oxidative stress induced by both the increase in the concentration of free oxygen radicals (Eisenbrandt and Nitschke 1989) and inhibited action of antioxidant enzymes by fluorine (Wang et al. 2004). At the same time, abnormal production of thyroid hormones induced by fluorine compounds also contributed to the antioxidant imbalance (Basha et al. 2011).

Piscivorous birds are particularly vulnerable to fluoride compounds. Their brain fluoride concentrations are significantly higher than in mammals, which results from the cumulative properties of fluorine originating from their prey (Kalisińska et al. 2014).

## *6.3 Effects of Fluorine on Glucose Metabolism*

There exist only a few studies on the effects of fluoride on the muscle metabolism of glycogen, a very important source of glucose and energy for muscles (McGown and Suttie 1977; Dost et al. 1977; Shashi et al. 1988). Waldbott et al. (1978), after examining about 500 people with chronic fluorosis, compiled a list of clinical features, which included chronic fatigue that is not relieved by rest, weakness, and involuntary muscle contractions. Their observations were later confirmed by Susheela (2001).

The regulation of blood glucose levels is dependent on the concentration of insulin in the blood. Diurnal variation in the concentration of this hormone depends on melatonin secretion by the pineal gland (Peschke et al. 2013). Studies on birds (Kalisińska et al. 2014) show that this gland accumulates large amounts of fluorine, compared to the brain and bone, which significantly interferes with the metabolism of glucose in the pineal gland, leading to the formation of paracrystalline structures containing large amounts of glycogen. Deficiency of this hormone can have significant consequences on the metabolism of birds and seasonal migrations (Kalisińska et al. 2014).

## *6.4 The Role of Fluorine in the Development of Inflammation*

The role of fluorine in the development of the inflammatory process is poorly defined (Schamschula and Barmes 1981). It is known that prolonged exposure to fluoride stimulates oxidative stress, accelerating pathological changes within cells (Das et al. 2006; Guan et al. 2000). NaF increases the formation of ROS in macrophages (Goldman et al. 1995) and activates a signaling cascade in response to stress, which involves the activation of MAP kinase in vascular endothelial cells. In addition, fluoride lowers the activity of antioxidant enzymes (glutathione peroxidase, superoxide dismutase, glutathione reductase, and catalase) in macrophages, which can indirectly contribute to an increase in ROS levels observed in these cells. Higher production of ROS increases lipid peroxidation, which in turn leads to apoptosis (Gutowska et al. 2010).

NaF in in vitro cultures reduces the intracellular concentration of ATP and the loss of mitochondrial transmembrane potential (Gutowska et al. 2010), suggesting that the multidirectional effect of NaF on cells can be caused by limiting the availability of ATP in these cells, a substance necessary for protein phosphorylation rather than by G protein activation (Goldman et al. 1995).

Prolonged exposure to fluoride leads to changes in the efficiency and catalytic activity of enzymes involved in the metabolism of eicosanoids, inflammatory processes, vascular remodeling, coronary heart disease, myocardial infarction, etc. (Goldman et al. 1995, 1997; Gutowska et al. 2012; Wessel et al. 1989). Eicosanoid synthesis in macrophages is controlled by the availability of free arachidonic acid. Fluoride increases arachidonic acid release in the cell membranes of macrophages, which increases the activity of cytosolic phospholipase A2 ($cPLA_2$) and the synthesis of prostaglandins in macrophages (Gutowska et al. 2011, 2012; Schulze-Specking et al. 1991; Wessel et al. 1989). It is believed that by stimulating G protein (guanine nucleotide binding), fluoride increases the amount of thromboxane $B_2$ ($TXB_2$), 6-ketoprostaglandin F1 alpha, and prostaglandin F2 alpha (PGF2α) in the serum of people exposed to NaF (Dodam and Olson 1995), which has been confirmed by in vitro studies (Coffee et al. 1992).

By changing the intracellular calcium concentration (Murao et al. 2000; Xu et al. 2007), fluoride can cause translocation of protein kinase C (PKC) from the cytosol to the membrane, where the increase in PKC activity has a positive effect on the release of arachidonic acid from the membrane and its availability for the synthesis of prostaglandins (Garcia et al. 1992). In addition, the increase in intracellular calcium concentration and an increase in PKC activity are factors stimulating $PLA_2$ and PLC (Garcia et al. 1992). cAMP plays an important role in regulating many processes in macrophages, i.e., phagocytosis or migration. High intracellular concentrations of cAMP are negatively correlated with the activity of these processes, and incubation of macrophages with sodium fluoride ions increases the levels of intracellular cAMP (Houdijk et al. 1991).

Chronic fluorosis also leads to a change in the composition and structure of cell membranes (Wang et al. 2000). These changes include changing the content of neutral lipids and phospholipids (Guan et al. 2000). In studies on macrophages, an increase in diacylglycerol synthesis was observed following exposure to fluoride (Dieter and Fitzke 1993; Wessel et al. 1989), which may have taken place via the hydrolysis of other phospholipids (Dieter and Fitzke 1993). Literature data suggest that activation of $PLA_2$ and increased eicosanoid synthesis are caused by fluoride-induced synthesis of diacylglycerols (Wessel et al. 1989).

# 7 Fluoride in Mammals

Ecotoxicological research designed to indirectly evaluate environmental pollution by various substances, for example, by fluorine compounds, is focused on the determination of concentrations in the bodies of wild animals. Primarily, concentrations are determined in the organs responsible for detoxification in mammals and birds, namely, the liver and kidneys. However, some elements, including fluorine, accumulate in increasing quantities in hard tissues. Therefore, determinations for long-term pollution and exposure of living organisms to fluoride are much more frequently based on levels in the hard tissues building bones, teeth, and antlers (Bezerra de Menezes et al. 2003).

## *7.1 Farm Animals*

The first alarming symptoms of the negative impact of pollution with fluoride on living organisms were observed in farm animals, such as cattle (Burns, 1969; Dale and Crampton 1955; Filippovskiĭ 1969; Green 1946; Gründer 1972; Murray 1967; Obel and Erne 1971; Schmid 1956; Udall and Keller 1952) and sheep (Burns 1969). Research conducted over many years and in many countries has shown a link between emissions from certain industries and fluorosis in cattle (Gründer 1972; Krook and Maylin 1979; Choubisa 2015). It was noted that cattle grazing on pastures near aluminum smelters or heat and power plants ate grasses growing in the contaminated area and thus accumulated fluoride in their bodies, negatively affecting their health and thus farm productivity and profits (Bunce 1985; Krook and Maylin 1979).

However, not only atmospheric emissions may increase fluoride levels in the body. Long-term (30 years) fertilization of soil with phosphorus fertilizers contaminated with fluorine compounds may result in increased consumption of this element by animals (Grace et al. 2008). Areas that naturally contain high concentrations of fluoride in the soil and waters (India, China) also demonstrate the occurrence of fluorosis and numerous disturbances in the normal functioning of the body and the retention of a number of elements in farm animals in numerous studies (Wang et al. 1995; Choubisa et al. 2012; Narwaria and Saksena 2012; Choubisa 2014; Khandare et al. 2015; Choubisa and Choubisa 2016).

## 7.2 Ungulates

The most common materials in research on environmental pollution with fluorine compounds are the antlers and mandibles of deer, due to the cumulative capacity of F in hard tissues of the body (Shupe et al. 1984; Jelenko and Pokorny 2010). However, some studies have also used hair and hooves (Zakrzewska et al. 2004). Antlers have a well-known annual growth cycle and therefore enable the determination of fluoride pollution in the previous year. The rapid growth of antlers results in the accumulation of particularly large amount of fluoride during antlerogenesis and allows using antlers as a model to study the impact of large amounts of fluoride on bone formation. Furthermore, analysis of antler samples and pedicle bone permits a noninvasive monitoring of environmental pollution with fluoride (Kierdorf et al. 1997, 2000a, b). The mandible accumulates fluoride during the entire life of the organism and thus reflects the cumulative effect of fluoride on the body (Kierdorf et al. 1989; Gutowska et al. 2004). Fluoride is captured faster during the growth of the bone than later in life, when accumulation occurs during normal bone remodeling (Kierdorf et al. 1995). Both tissues are easily accessible. Mandibles are often systematically collected by hunting associations in order to control the number and health of wild game, while antlers (as hunting trophies) can come from private collections (Jelenko and Pokorny 2010).

Antlers and bones of deer are excellent materials from which to study large-scale environmental contamination with fluorides in areas inhabited by these animals, due to the large numbers of animals and the high degree of adaptation to the conditions in a given area (Shupe et al. 1984; Vikøren and Stuve 1996b; Machoy et al. 1991; Kierdorf et al. 2000a, b, 2012; Piotrowska et al. 2006). In addition, antlers are highly useful in the analysis of the direction and extent of contamination with fluoride from emission sources such as aluminum smelters (Kierdorf and Kierdorf 2002; Kierdorf et al. 2012) and iron and steel smelters (Kierdorf and Kierdorf 2003).

Jelenko and Pokorny (2010), after examining 141 antlers and 220 mandibles of roe deer (*Capreolus capreolus*) derived from animals hunted between 1960 and 2007 (antlers) and 1997 and 2009 (mandibles) in the area of the largest Slovene thermal power plant of Šoštanj (STPP), showed a high correlation between annual emissions from the plant and the average annual contents of fluoride in the antlers and mandibles. In addition, by comparing the results obtained for samples from different years, they noted a significant reduction in the emission of pollutants containing fluorine compounds from the plant after 1995 and 2000, which was connected with the introduction of equipment purifying exhaust gases. Kierdorf and Kierdorf (2000), after examining roe deer (*Capreolus capreolus*) antlers between 1932 and 1998 also showed seasonal changes in the emissions of pollutants containing fluoride in the eastern suburbs of Cologne (an area of 800 $km^2$, Germany). The fluoride content in the antlers dropped significantly in the 1980s and 1990s, which indicated a reduction in fluoride emissions (Kierdorf and Kierdorf 2000). Also, research conducted by the same researchers (Kierdorf and Kierdorf 2001) in the period of 1951–1999 in the industrialized Ruhr area (Western Germany) showed a correlation between the concentration of fluoride in the environment and its content in the tested antlers.

The results showed a gradual reduction in the amount of fluoride in the atmosphere in the study area after 1980, which probably resulted from the use of effective measures to control emissions in Germany and neighboring countries (Kierdorf and Kierdorf 2001).

Research conducted by Newman and Yu (1976) on the black-tailed deer (*Odocoileus hemionus columbianus*) from an industrial area northwest of Washington showed dental abnormalities and abnormal patterns of tooth wear associated with the occurrence of fluorosis in these animals. The levels of fluoride in the bones were 10–30 times higher than the concentration of this element in the bones of animals from reference areas (Newman and Yu 1976). Shupe et al. (1984), after examining the bones and teeth of deer (*Odocoileus hemionus columbianus*), moose (*Alces alces*), and bison (*Bison bison*), from Utah, Idaho, Montana, and Wyoming, observed changes in the bones and teeth of those animals, correlating with the amount of fluoride pollution in their habitats (industry and water with high fluoride levels).

Zakrzewska et al. (2005) conducted a study on F in the bones of red deer (*Cervus elaphus*) from an area of northwestern Poland in the 1990s, a period covering the opening of the Police chemical plant near the city of Szczecin (Poland) (Zakrzewska et al. 2005). Bone samples collected from animals in the area more exposed to F compounds (areas of Szczecin and Police) contained an average of 50% more F compared to samples from areas distant from the source of emission. Similar conclusions were reached by other authors who examined the mandibles of deer coming from areas exposed to the emissions of fluorine compounds and in the areas without any major industrial plants (Poland) (Machoy et al. 1995; Gutowska et al. 2004). Dąbkowska et al. (1995a, b) examined animal bones collected between 1982 and 1990 and showed that the reduction in production volume and modernization of the chemical plant in Police resulted in a reduction of emissions into the environment, which in turn resulted in lower fluoride content in the bones of animals in 1990 compared to samples from 1982 (Dąbkowska et al. 1995a, b).

Teeth are also recommended as an excellent material to study the level of intake of fluoride and a very good indicator of chronic exposure to this element (Kierdorf et al. 1993, 1996b, 1999; Vikøren and Stuve 1996a, b; Richter et al. 2011). In addition, damage to the teeth significantly correlates with the degree of damage to forests in which the tested animals dwelled (Zemek et al. 2006). The teeth with fluorosis were characterized by an opaque enamel, and in cases of severe fluorosis, the enamel became damaged or its surface reduced, and occlusal surfaces were subject to greater wear. This led to hypermineralization and ameloblast dysfunction and throughout the enamel striae of Retzius became strengthened (Schultz et al. 1998; Appleton et al. 2000; Kierdorf et al. 1993, 1996b, 2000a, b). Studies on the teeth of wild boar (*Sus scrofa*) originating from areas of the Czech Republic (Bohemia) and Germany (Saxony) showed a significantly higher content of fluoride in comparison with samples taken from animals originating from areas of western Germany (Kierdorf et al. 2000a, b). Also, teeth of the red and roe deer (*Capreolus*

*capreolus*) proved to be an excellent material for the analysis of environmental exposure to fluoride. Extending the research to teeth with dentine enabled a more precise determination of specific periods of exposure to fluoride during the life of the individual (Kierdorf et al. 1999; Richter et al. 2010). Schultz et al. (1998), after examining 545 red deer mandibles, showed the occurrence of severe fluorosis in more than 11% of the animals, accompanied by pathologically increased wear and fractures of the teeth, periodontal bone lesions caused by inflammation of the periosteum, and chronic osteomyelitis in the mandible. The strong occlusal abrasions resulted in the opening of the pulp chamber, forming periapical ulcers and eventually tooth loss, which according to the authors is an important factor reducing the lifespan of the animals (Schultz et al. 1998).

In addition to human activities, volcanic eruptions are a significant factor introducing fluorine compounds to the environment. This source of fluorine may affect organisms over much larger areas and at larger distances. Flueck and Smith-Flueck (2013) decided to examine the impact of these emissions using animals as bioindicators. The study involved red deer from areas about 100 km from the Puyehue-Cordon Caulle volcano a year after an eruption. The researchers recorded a level of fluoride in bones exceeding 5000 mg $kg^{-1}$ and the occurrence of severe dental fluorosis, manifested in enamel hypoplasia, breakages, pitting, mottling, and extremely rapid ablation of entire crowns down to the underlying pulp cavities. Tooth loss reduced the physical condition of animals, and although a preliminary analysis of the water and volcanic ash showed no danger for living organisms, a study on ruminants as indicators of the contaminated ecosystem clearly demonstrated a 38-fold increase in the level of fluoride in bones during the first 15 months of exposure after the volcano eruption (Flueck and Smith-Flueck 2013) (Table 1).

## 7.3 Canids

Biomonitoring of environmental risks associated with fluoride pollution is usually based on samples of bones, teeth, and antlers from ungulates. However, medium-sized omnivorous mammals seem to be more suitable for this type of research, because their type of diet and longevity make them more similar to humans (Kay et al. 1975). Such animals include the red fox (*Vulpes vulpes*) and raccoon dog (*Nyctereutes procyonoides*). The red fox fulfills the conditions established for good bioindicators (Ellenberg 1991), which include large geographical coverage, but not a too large range, stable local population, fixed position in the food chain, and the ease of specimen collection via hunting. For these reasons, this species is often used in studies of various types of environmental pollution, including fluorine compounds (Palczewska-Komsa et al. 2014). Although both red fox and raccoon dog have many features that make them potentially ideal bioindicators (Apostoli 1992), there is little data on the concentration of F in the hard tissues of red foxes (Kalisińska and Palczewska 2007; Kalisińska and Palczewska-Komsa 2011; Palczewska-Komsa et al. 2014) and raccoon dogs (Palczewska-Komsa et al. 2014).

**Table 1** Concentrations of fluoride in hard tissues of ungulates

| Species | Tissue | Year | Fluoride (mg kg$^{-1}$ ) | Place of collection | Reference |
|---|---|---|---|---|---|
| Roe deer *Capreolus capreolus* | Mandible $n = 112$ | 1986 | 43–901 | 5 localities in England and Scotland | Walton and Ackroyd (1988) |
| | Antler $n = 10$ | | 1220–2010 | | |
| | Mandible $n = 39$ | 1985–1993 | 208–1026 (dw) | Harz mountains (Germany) | Kierdorf et al. (1995) |
| Moose *Alces alces* | Mandible $n = 1104$ | 1990–1993 | >8000 | The vicinity of seven Norwegian aluminum smelters (Norway) | Vikøren and Stuve (1996b) |
| Roe deer *Capreolus capreolus* | Mandible $n = 147$ | 1990–1993 | >8000 | Vicinity of seven Norwegian aluminum smelters (Norway) | Vikøren and Stuve (1996b) |
| Red deer *Cervus elaphus* | Mandible $n = 24$ (examined group) | 1985–1993 | (dw) 948–4680 | N-Bohemian brown coal belt (the vicinity of the two towns: Karlovy Vary and Chomutov—examined group; Harz mountains, State of Lower Saxony—control group) | Kierdorf et al. (1996a) |
| | $n = 39$ (control) | | 208–1026 | | |
| Red deer (*Cervus elaphus*) | Mandible $n = 27$, Karlovy Vary $n = 18$, Nejdek $n = 15$, forest district Eibenstock $n = 39$, control | 1986–1993 1982–1990 1988, 1989 1985–1993 | 2754 ± 1088 1244 ± 523 883 ± 444 540 ± 227 | The North-Bohemian brown coal belt (the vicinity of the two towns (Karlovy Vary and Chomutov) and Nejdek and forest district Eibenstock—examined groups; Harz mountains, State of Lower Saxony—control group) | Kierdorf et al. (1996b) |
| Red deer (*Cervus elaphus*) | Antler and pedicle $n = 18$ | | Antler (ba): Bohemia 845 ± 257 Control 206 ± 124 Pedicle (ba): Bohemia 1448 ± 461 Control 322 ± 157 | North Bohemia (Czech Republic) and two uncontaminated areas in West Germany | Kierdorf et al. (1997) |
| Red deer *Cervus elaphus* | Mandible $n = 61$ | 1985–1993 | >4000 (dw) | Ore mountains and their southern foreland, Czech-German border region | Schultz et al. (1998) |

(continued)

**Table 1** (continued)

| Species | Tissue | Year | Fluoride (mg kg$^{-1}$ ) | Place of collection | Reference |
|---|---|---|---|---|---|
| Roe deer *Capreolus capreolus* | Mandible Ruhr area $n = 76$ | 1955–1998 | (dw) 150 (2-year-old specimen taken in 1997) 5724 (10-year-old specimen taken in 1957) | The federal state of North Rhine-Westphalia, Germany | Kierdorf and Kierdorf (2000b) |
| | Cologne area $n = 81$ Age range 1–11 years | 1983–1998 | | | |
| Roe deer (*Capreolus capreolus*) | Antlers | 1932–1998 | 158–3713 dw | The eastern suburbs of Cologne, Germany | Kierdorf and Kierdorf (2000) |
| Wild boars *Sus scrofa* | Mandible and teeth $n = 47$ | 1995–1997 | Mandible (dw): Bohemia 754.3 ± 149.6 Saxony 490.8 ± 135.1 Control 304.7 ± 91.0 Teeth (dw): Bohemia 382.1 ± 165.2 Saxony 125.0 ± 38.3 Control 33.6 ± 26.7 | Fluoride-polluted areas in Bohemia (Czech Republic) and Saxony, Germany | Kierdorf et al. (2000a) |
| Roe deer *Capreolus capreolus* | Antlers $n = 167$ | 1951–1999 | 110–8178 (ba) | Industrialized Ruhr area, W Germany | Kierdorf and Kierdorf (2001) |
| Roe deer (*Capreolus capreolus*) | Antlers $n = 188$ | 1990–1999 | 113–11, 995 (ba) | 14 areas of North Rhine-Westphalia, Germany | Kierdorf and Kierdorf (2002) |
| Roe deer *Capreolus capreolus* | Antlers $n = 116$ | 1948–2000 | 118–5428 (ba) | Industrialized area of Siegen, W Germany | Kierdorf and Kierdorf (2003) |
| Red deer *Cervus elaphus* Roe deer *Capreolus capreolus* | Mandible $n = 51$ $n = 175$ | Hunting season 1998/1999 | 55–273 171–430 | NW Poland | Gutowska et al. (2004) |
| Roe deer *Capreolus capreolus* | Mandible $n = 7$ $n = 7$ | Early 1990s | (dw) 1374–3790 1719–3411 | Fluoride-polluted area along Czech-German border | Richter et al. (2010) |

(continued)

**Table 1** (continued)

| Species | Tissue | Year | Fluoride (mg kg$^{-1}$ ) | Place of collection | Reference |
|---|---|---|---|---|---|
| Red deer *Cervus elaphus* | | | | | |
| Roe deer *Capreolus capreolus* | Antler $n = 141$ | 1960–2007 | 110–1210 (yearlings) 130–2340 (young adults) 250–2590 (older adults) | Area of Slovene thermal power plant of Šoštanj (Slovenia) | Jelenko and Pokorny (2010) |
| | Mandible $n = 220$ | 1997–2009 | 30.0–227 (fawns) 33.8–383 (yearlings) 61.5–1020 (adults) | | |
| Roe deer *Capreolus capreolus* | Mandible $n = 157$ $n = 117$ | 1996–1997, 2009 | Median (dw) 3147 350 | Five counties in the northwestern part of the Czech Republic | Kierdorf et al. (2012) |
| Red deer *Cervus elaphus* | $n = 127$ $n = 72$ | 1996–1997, 2009 | 1263 288 | | |
| Red deer *Cervus elaphus* | Bone $n = 26$ | 2012 | >5175 (dw) | 100 km from the volcano: the Puyehue-Cordon Caulle (Chile) | Flueck and Smith-Flueck (2013) |

*dw* dry weight, *ba* bone ash

One such study is the work of Kay et al. (1975), in which the analysis of the bones of coyotes (*Canis latrans*) from uncontaminated areas of North America showed an almost two times lower content of fluoride in samples when compared to those from femoral wild canids (32 red foxes and 18 raccoon dogs) from contaminated sites in northwestern Poland (Palczewska-Komsa et al. 2014). These studies confirm the results obtained by other authors pointing to increased accumulation of fluoride in the hard tissues of red foxes (Walton 1984) living in areas surrounding industrial plants emitting fluorine compounds into the atmosphere (Table 2).

## 7.4 *Small Wild Mammals*

A large number of biomonitoring studies on small mammals have demonstrated the importance of eating habits in selecting species for this type of research (Talmage and Walton 1991). Environmental studies on exposure to fluorine compounds have

**Table 2** Concentrations of fluoride in hard tissues of selected canid species

| Species | Tissue | Year | Fluoride (mg kg$^{-1}$) | Place of collection | Reference |
|---|---|---|---|---|---|
| Coyote *Canis latrans* | Bone $n = 2$ | 1975 | 321 (dry weight) | Montana (USA) | Kay et al. (1975) |
| Red fox *Vulpes vulpes* | Mandible $n = 230$ | – | 283 (Aberdeen in Scotland) 1650 (aluminum plant, Anglesey) (dry weight) | Several areas within the United Kingdom (areas not contaminated and areas near aluminum plant, Anglesey) | Walton (1984) |
| Red fox (*Vulpes vulpes*) | – | Hunting seasons 2004/2005 2005/2006 | 514 (group I) 389 (group II) (dry weight) | West (group I) and north (group II) of Szczecin, Western Pomerania (Poland) | Kalisińska and Palczewska (2007) |
| Red fox (*Vulpes vulpes*) | Teeth (first molars of the permanent teeth) $n = 35$ | The hunting seasons 2004/2005 2005/2006 | 297 (6–12 months, $n = 11$), 385 (12–20 months, $n = 10$), 654 (>20 months, $n = 14$) (bone ash) | West Pomeranian and Pomeranian Voivodeships (Poland) | Kalisińska and Palczewska-Komsa (2011) |
| Red fox *Vulpes vulpes* | Femur bone $n = 32$ | The hunting seasons 2008/2009, 2011/2012, 2012/2013 | 175.9–3668.1 (dry weight) | Area of the north-western Poland | Palczewska-Komsa et al. (2014) |
| Raccoon dog *Nyctereutes procyonoides* | $n = 18$ | 2009/2010, 2011/2012, 2012/2013 | 83.7–1190.3 (dry weight) | Area of Warta Mouth National Park | |

Concentrations in mg kg$^{-1}$ bone ash or dry weight, maximum value or range reported in respective references

been carried out on small wild mammals such as wood mice (*Apodemus sylvaticus*), shrew (*Sorex araneus*), and field vole (*Microtus agrestis*) (Shore 1995).

Boulton et al. (1997) conducted a study on the impact of fluoride on the appearance of the teeth of the short-tailed field vole from fields affected by different levels of industrial fluoride pollution. On the incisors of the voles consuming relatively low doses of fluoride, the changes were largely limited to the distortion of enamel pigmentation, as in the case of molars. In areas with a high degree of fluoride pollution, changes were observed in the composition and strength of both the enamel and dentin; incisors showed hypoplasia and enamel pigmentation, while the occlusal surfaces of the molars exhibited severe erosion and exposed dentin (Boulton et al. 1997). Also in this case, the amount and rate of accumulation of fluorine in hard tissue were dependent on the age of the tested animals. After birth, the young did not show any change in the construction of the incisors, but after weaning, teeth showed significant morphological changes and serious damage. The rate of accumulation of fluoride in bones showed age-related changes rapidly in young individuals, whereas in adults, it reached a relative equilibrium, due to the slowdown in the animal's growth (Boulton et al. 1994).

## 8 Fluoride in Wild and Domestic Birds

Due to increasing anthropogenic pollution, it is very important to be able to identify areas contaminated with fluoride through biomonitoring based on bird bones. This type of research reflects the natural geochemical background and the effects of human activity. However, as most birds are migratory animals, living in different environments and with different feeding behaviors, it is difficult to use these species as indicators. The concentrations of fluoride in the bones of wild birds are characterized by a very large interindividual and interspecies variability.

About 99% of fluorine present in the bird's body is stored in the bones. The amount of fluorine deposited in bones is age and diet dependent. In the skeleton of domestic ducks (*Anas platyrhynchos f. dom.*), 28.7% of fluorine is deposited in the skull and 69.8% in other bones. Birds seem to be able to tolerate fairly high doses of fluoride, which has been proven experimentally in American kestrel's (*Falco sparverius*) fed doses of 1120 and 2240 mg $kg^{-1}$ $F^-$. Those doses did not interfere with the development of internal organs, including the brain, although decreased bone resistance to fractures (Bird et al. 1992). For this reason, the birds could be suitable for the biomonitoring of fluoride contamination.

Fluoride absorption and incorporation into the bone are strongest during the rapid growth phase of the birds. The normal concentration of fluoride in the bones of chickens (*Gallus gallus f. dom.*) and other birds can range from 500 to 1000 mg $F^-$ $kg^{-1}$ dry weight (Xie and Sun 2003; Committee on toxicology 1993). Higher concentrations induce bone fluorosis. However, studies on the bones of the black heron (*Egretta ardesiaca*) living in the vicinity of a phosphate processing complex in the United States showed a level of 1700 mg $F^-$ $kg^{-1}$ dw (Henny and Burke 1990). Gulls (Larinae) are vulnerable to exposure from high doses of fluoride when they reside in the vicinity of aluminum smelters. A study on migratory species of gulls has shown that measurements of F concentration in eggshells are useful for monitoring the local exposure of wild birds to fluoride. In contrast, the volume of eggs, shell thickness, and percentage of fertilized eggs are not good indicators of the exposure of seagulls to fluorine compounds (Vikøren and Stuve 1996a, b). Interestingly, a study on the tissues of the Adélie penguin (*Pygoscelis adeliae*) living on an island in Ardley Maxwell Bay showed an extremely high content of $F^-$ in the range from 832 to 7187 mg $kg^{-1}$, while radiographs provided no evidence of skeletal fluorosis. Very high concentrations of fluoride in the bones of penguins can be explained by the diet of these birds, with an 80% share of krill in which the average fluoride concentration in soft tissue is more than 1200 mg $kg^{-1}$, while in the shell, it is even higher than 3800 mg $kg^{-1}$. It is likely that penguins and other seabirds that live in the waters of Antarctica are extremely resistant to high concentrations of fluoride in their tissues, as part of the adaptation of these animals to an environment with a high concentration of fluoride (Xie and Sun 2003). Moreover, in the white-faced heron (*Egretta novaehollandiae*) and in different species of cormorants (*Phalacrocoracidae*) living in industrially uncontaminated regions of New Zealand, the average fluoride concentration is up to 35 mg $kg^{-1}$ of ash (670–1580 mg $kg^{-1}$ dw) (Robertson and Lock 1994).

Studies show that treating chickens with sodium fluoride containing water for several weeks at doses ranging from a few to several mmol $L^{-1}$ increases their rate of bone maturation and secondary mineralization. Such changes can have a significant impact on bone density and fragility (Lundy et al. 1992). A postmortem study showed that the supplementation of drinking water with fluoride at a concentration of 6–20 mg $L^{-1}$ for 17 weeks of growth, and then for 57 further weeks, resulted in a linear increase in weight gain and a fluoride content in the femur of laying hens. In that study, the addition of fluoride to drinking water had a significant effect on the production of eggs, but not the strength of the eggshell. Histopathological examination showed no changes in the liver or kidneys. Data from that study indicated that laying hens tolerated a fluoride intake of 4.453 mg day $^{1}$ for up to 74 weeks (Coetzee et al. 1997). Other authors have demonstrated that a long-term intake of fluorine at up to 300 mg $kg^{-1}$ body weight had no negative effects on the morphology nor mechanical properties of bones (Suttie et al. 1984) and that fluorine had a positive effect on bone strength at lower doses, with an adverse influence only at higher doses (Turner et al. 1992).

There is an interesting study showing that in chickens fed F for 50 days at a concentration of 110 mg $kg^{-1}$ body weight, there were significant and strongly age-related differences between the content of fluoride in the bone tissues. The compact bone fluorine level was 0.192 mg $g^{-1}$ and increased with age up to 0.336 mg $g^{-1}$. Cancellous bone fluoride levels rose with age, from 0.174 to 0.224 mg $g^{-1}$. Bone marrow fluorine remained between 0.009 and 0.012 mg $g^{-1}$, with the lowest concentrations in the cartilage; at 0.005 mg $g^{-1}$, this value was decreasing during the study. The authors noted that the concentration of fluoride in the cartilage negatively correlated with the level in the cancellous bone and marrow (Dołegowska 2002; Dołegowska et al. 2003). In summary, the increase in the content of fluorine in both cortical and cancellous bone correlated with intensive processes of bone formation and remodeling in young chickens. The slow decline of the fluorine content in the articular cartilage is probably due to an age-related decrease in intensity and specificity of calcification processes within the cartilage.

The main problem in cage poultry breeding is the development of dyschondroplasia, which greatly reduces the economic results of farms. Tibial dyschondroplasia changes the cartilage at the base of the tibia of fast-growing broilers. Soft bones can thicken at the base, deform, and break up. Inflammation of the tibia can lead to problems with mobility, dehydration, and death. Prevention of the disease consists in slowing down the growth of chickens at the beginning of breeding.

Treatment with sodium fluoride increases the strength of the humeral bones and in tibial fractures, which is a consequence of the increase in the marrow ash fluoride levels. At the same time, this supplementation did not result in a decreased quality and efficiency of egg laying (Merkley 1981). Even small concentrations of F (e.g., 0.46 mg $kg^{-1}$ dw) are likely to increase the hardness of the bone, and in addition the chicks grow better if a feed containing sodium fluoride is also added with the appropriate proportions of calcium phosphate (Shim et al. 2011). On the other hand, the toxic effects of fluoride in the diet lead to the inhibition of growth of chickens due to a decreased appetite and reduced feed intake, which has a potential impact on the growth and health of the animals (Choubisa 2010). It has been found that doses of 1000 and 1300 mg $kg^{-1}$ dw fluoride decreased feed intake, weight gain, and efficiency of egg production. Long-term administration of high doses of sodium fluoride, however, does not lead to a permanent loss of value of production, because the changes related to the lack of appetite subside within 6 months of cessation of exposure. An increase in fluoride in the diet is accompanied by a reduction in the volume of eggs with a tendency for a better quality eggshell (resistance to deformation and fracture) and also a drop in the retention of phosphorus, magnesium, and calcium levels, resulting in less calcium available for bone formation. There is also a slowdown in yolk synthesis (Guenter and Hahn 1986).

Experimental studies show that differences in the content of fluoride in the bones of poultry are gender related due to physiological factors associated with the production of eggs. Bone fluoride levels are higher in hens than cocks, due to an increased metabolism of minerals in the bones of females during egg production. Calcium is removed from bones to create eggshells. The increased absorption of calcium is also accompanied by increased fluoride absorption. Probably, fluoride moves with calcium to the bone, but when the calcium is removed from bones to form the eggshell, fluoride remains, and the concentration proportionally increases with each cycle of egg production.

A 2-year experiment proved that the concentration of fluoride in the bones of chickens is significantly higher after puberty in relation to young chickens (Michel et al., 1984). Higher concentrations of fluoride were recorded in the bones of females of other birds, e.g., owls, in the breeding season with a diet containing 200 mg $F^{-}$ kg $^{-1}$ dw (Pattee et al. 1988), and gulls during the breeding season living near an aluminum smelter (Vikøren and Stuve 1996a, b). Such dependences were not observed in free-living urban pigeons (Salicki and Kalisinska 2006).

There are significant differences in bone fluoride levels in birds depending on the type of the bone. It was shown that in spinal and cranial bones, fluoride concentrations are higher than in the long bones and higher in the compact bone than in cancellous bone (Xie and Sun 2003). One should not forget that the amount of fluoride accumulation in the bones of birds is also a function of age (Henny and Burke 1990).

Most fluoride accumulated in the body of the birds is found in the bones. However, fluorides are also present in the blood and soft tissues and then excreted in the feces and to a lesser extent in the mucus of salt glands located above the eyes of certain birds, especially in seabirds (Culik 1987).

In total, there are two ways to eliminate fluoride from circulation: excretion by the kidneys and deposition in the bones and eggshells. Interestingly, in chickens, even at a concentration of $F^-$ in the diet reaching 1300 mg $kg^{-1}$, the amount of fluoride in the egg white did not exceed 1 mg $kg^{-1}$.

The addition of aluminum salts in the diet causes a reduction in fluoride levels in the liver and kidney, but not in the pectoral and tibial muscle in birds (Hahn and Guenter 1986). The results of studies on impaired reproductive efficiency of owls (*Otus asio*) in areas contaminated with fluoride showed that despite having no significant differences in the hematocrit value, hemoglobin, plasma calcium levels, nor alkaline phosphatase, between control and treatment groups, the addition of 200 mg $kg^{-1}$ of sodium fluoride to the diet of the birds produced a significant reduction in the volume and weight of the eggs. One-day-old owls weighed 10% less, and their tibiotarsus was significantly shorter in comparison to the control group. In addition, in the group of owls fed with sodium fluoride at 40 mg $kg^{-1}$, there were significantly higher serum phosphate levels compared to the control group (Hoffman et al. 1985).

In another study on 7-day-old chicks of American kestrels (*Falco sparverius*) fed daily with fluoride supplementation at concentration of 0 (control), 1120, and 2240 mg $kg^{-1}$, there were no significant differences between the three groups in the length of the duodenum, the lengths of the jejunum and the ileum, and the weight of the adrenal glands, brain, stomach, spleen, heart, kidney, liver, pancreas, and pectoral muscle. Treatment with NaF resulted in a significant reduction in the resistance of bones to fractures (Bird et al. 1992). On the other hand, other authors report pathological effects of fluoride intake on the renal function (Bai et al. 2010) and thymus (Chen et al. 2010) in chickens.

In China, chicken farms suffered considerable economic losses due to fluorosis associated with the use of a feed additive containing about 300 mg $kg^{-1}$ fluorine and in some cases up to 2000 mg $kg^{-1}$ (Liu et al. 2003). This has resulted in numerous reports by local researchers to help in understanding the cellular mechanism of fluoride toxicity. Some authors indicated a decrease in the percentage of T lymphocytes caused by oxidative damage and apoptosis resulting from the toxic effects of fluoride on the lymphoid tissue in the avian caecum (Chen et al. 2009; Liu et al. 2012). Others showed that feeding broilers with feed containing NaF at 800 and 1200 mg $kg^{-1}$ induce apoptosis of lymphocytes via DNA damage mechanisms and also decrease the synthesis of Bcl-2, an integral membrane protein on the outer membrane of mitochondria. An overexpression protects cells from apoptosis in response to different stimuli, since the role of Bcl-2 is to block the release of cytochrome c from mitochondria (Liu et al. 2013).

The results of yet another work proved that high doses of fluoride caused severe oxidative stress and damage to the spleen cells of broiler chickens on a diet containing 800 and 1200 mg $F^-$ $kg^{-1}$ for 42 days. Those chickens experienced an increase in plasma concentrations of fluoride and damage to the mitochondria, a decrease in superoxide dismutase (SOD) and glutathione peroxidase (GSH-Px), and an increased level of malondialdehyde as a product of lipid peroxidation in the spleen (Chen et al. 2011). In contrast to those results, other experiments showed that a 3-week diet containing fluoride at 300 mg $kg^{-1}$ had no effect on the activity of L-gulonolactone oxidase, the resources of ascorbic acid in the selected tissues of chickens (plasma, kidney, liver, muscle, and adrenal gland), and the resistance to stressors (Maurice et al. 2002).

Birds have an incomplete blood-brain barrier, which increases the likelihood of penetration of various substances into the brain (Kuenzel et al. 1997). It appears that an excess of fluoride is deposited in soft tissues, especially in the brain of birds; it constitutes a permanent biologically inactive deposit that the body is not able to remove.

There are no known adaptive mechanisms to prevent or reduce the toxicity of fluoride in birds. One can only presume the important role of calcium ions, which may bind F and form insoluble salts, such as $CaF_2$ (Monsour and Kruger 1985). This was confirmed by studies that found a surprisingly high concentration of F in the brains of common merganser (*Mergus merganser*) in northwestern Poland. The highest mean concentration of fluoride in this species was observed in bones (430 mg $kg^{-1}$ dw) and the pineal gland (780 mg $kg^{-1}$ dw) and the lowest in the brain (170 mg $kg^{-1}$ dw), in both adult and young mergansers living in areas located between a fertilizer production plant and a power plant (Kalisińska et al. 2014).

Pathological changes caused by the toxic effects of fluoride on the brains of migratory birds may have serious consequences for their populations. Experiments have shown that fluoride has neurotoxic properties and adversely affects the functioning of the brain. Even at low doses, fluoride contributes to the induction of apoptosis of neurons and formation of oxidative stress, increasing the amount of free radicals and lipid peroxidation in the brain (Choi et al. 2012). On the other hand, in laying hens, phenylmethylsulfonyl fluoride (PMSF)—an inhibitor of serine proteases—reduces the risk of organophosphorus neuropathy which leads to severe paralysis of the peripheral nerves induced by toxic effects of tricresyl phosphate (TOCP), a pesticide that may be present in feed (Song et al. 2009; Mangas et al. 2014). In addition, using a feed with a combination of fluoride at a concentration of 10 mg $kg^{-1}$ accompanied by vitamin K and calcium also helped to achieve a 20% reduction in the loss of bone structures, thus preventing osteoporosis in laying hens (Fleming et al. 2003).

## 9 Summary

Fluoride is ubiquitous in the environment as a particulate in the air and in ionic forms in surface and groundwaters, soil, and sediments. Its natural circulation in nature is accompanied by processes related to the economic activities of man (Shaw 2012). In light of the data we have presented in this chapter, it seems necessary to monitor the presence of fluoride in the environment, as its uncontrolled intake may pose a serious risk for human and animal health (Machoy et al. 2002). The results of various studies indicate the usefulness of the determination of F in hard tissues of wildlife for monitoring the environmental F contamination, as well as its role in the assessment of the effectiveness of measures taken to reduce F emissions, e.g., from industrial sources (Kierdorf and Kierdorf 2000b).

## References


Aaron JE, de Vernejoul MC, Kanis JA (1991) The effect of sodium fluoride on trabecular architecture. Bone 12:307–310

Allibone R, Cronin SJ, Charley DT, Neall VE, Stewart RB, Oppenheimer C (2012) Dental fluorosis linked to degassing of Ambrym volcano, Vanuatu: a novel exposure pathway. Environ Geochem Health 34:155–170

Apostoli P (1992) Criteria for the definition of reference values for toxic metals. Sci Total Environ 120:23–37

Appleton J, Chesters J, Kierdorf U, Kierdorf H (2000) Changes in the structure of dentine from cheek teeth of deer chronically exposed to high levels of environmental fluoride. Cells Tissues Organs 167:266–272

Bai CM, Chen T, Cui Y, Gong T, Peng X, Cui HM (2010) Effect of high fluoride on the cell cycle and apoptosis of renal cells in chickens. Biol Trace Elem Res 138:173–180

Barbier O, Arreola-Mendoza L, Del Razo LM (2010) Molecular mechanisms of fluoride toxicity. Chem Biol Interact 188:319–333

Basha PM, Rai P, Begum S (2011) Fluoride toxicity and status of serum thyroid hormones, brain histopathology, and learning memory in rats: a multigenerational assessment. Biol Trace Elem Res 144:1083–1094

Bezerra de Menezes LM, Volpato MC, Rosalen PL, Cury JA (2003) Bone as a biomarker of acute fluoride toxicity. Forensic Sci Int 137:209–214

Bird DM, Carriere D, Lacombe D (1992) The effect of dietary sodium fluoride on internal organs, breast muscle, and bones in captive American kestrels (*Falco sparverius*). Arch Environ Contam Toxicol 22:242–246

Boulton IC, Cooke JA, Johnson MS (1994) Age-accumulation of fluoride in an experimental population of short-tailed field voles (*Microtus agrestis* L.). Sci Total Environ 154:29–37

Boulton IC, Cooke JA, Johnson MS (1997) Fluoride-induced lesions in the teeth of the short-tailed field vole (*Microtus agrestis*): a description of the dental pathology. J Morphol 232:155–167

Bunce HW (1985) Fluoride in air, grass, and cattle. J Dairy Sci 68:1706–1711

Burns KN (1969) Dental fluorosis and some other dental disorders in cattle and sheep. Proc R Soc Med 62:1297–1300

Celio MR, Blümcke I (1994) Perineuronal nets—a specialized form of extracellular matrix in the adult nervous system. Brain Res Brain Res Rev 19:128–145

Chen T, Cui Y, Bai CM, Gong T, Peng X, Cui H (2009) Decreased percentages of the peripheral blood T-cell subsets and the serum IL-2 contents in chickens fed on diets excess in fluorine. Biol Trace Elem Res 132:122–128

Chen T, Cui HM, Cui Y, Bai C, Gong T, Peng X (2010) Cell-cycle blockage associated with increased apoptotic cells in the thymus of chickens fed on diets high in fluorine. Hum Exp Toxicol 30:685–692

Chen T, Cui H, Cui Y, Bai C, Gong T (2011) Decreased antioxidase activities and oxidative stress in the spleen of chickens fed on high-fluorine diets. Hum Exp Toxicol 30:1282–1286

Choi AL, Sun G, Zhang Y, Grandjean P (2012) Developmental fluoride neurotoxicity: a systematic review and meta-analysis. Environ Health Perspect 120:1362–1368

Choubisa SL (2010) Osteo-dental fluorosis in domestic horses and donkeys in Rajasthan, India. Fluoride 43:5–12

Choubisa SL (2014) Bovine calves as ideal bio-indicators for fluoridated drinking water and endemic osteodental fluorosis. Environ Monit Assess 186(7):4493–4498. https://doi.org/10.1007/s10661-014-3713-x

Choubisa SL (2015) Industrial fluorosis in domestic goats (*Capra hircus*), Rajasthan, India. Fluoride 48:105–112

Choubisa SL, Choubisa D (2016) Status of industrial fluoride pollution and its diverse adverse health effects in man and domestic animals in India. Environ Sci Pollut Res Int 23(8):7244–7254. https://doi.org/10.1007/s11356-016-6319-8

Choubisa SL, Modasiya V, Bahura CK, Sheikhc Z (2012) Toxicity of fluoride in cattle of the Indian Thar desert, Rajasthan, India. Fluoride 45:371–376

Committee on toxicology (1993) National Research. http://www.fluoridation.com/skeletal.htm

Coetzee CB, Casey NH, Meyer JA (1997) Fluoride tolerance of laying hens. Br Poult Sci 38:597–602

Coffee KA, Halushka PV, Ashton SH, Tempel GE, Wise WC, Cook JA (1992) Endotoxin tolerance is associated with altered GTP-binding protein function. J Appl Physiol 73:1008–1013

Cronin SJ, Sharp DS (2002) Environmental impacts on health from continuous volcanic activity at Yasur (Tanna) and Ambrym, Vanuatu. Int J Environ Health Res 12:109–123

Culik B (1987) Fluoride turnover in Adelie Penguins (*Pygoscelis adeliae*) and other bird species. Polar Biol 7:179–187

Dąbkowska E, Machoy-Mokrzyńska A, Straszko J, Machoy Z, Samujło D (1995a) Temporal changes in the fluoride levels of jaws of European deer in industrial regions of Western Pomerania, Poland. Environ Geochem Health 17:155–158

Dąbkowska E, Machaliński B, Chlubek D, Noceń I, Machoy Z, Ogoński T et al (1995b) Zawartość głównych składników mineralnych zębów łosia europejskiego. Czas Stomatol 9:576–581 [in Polish]

Dale DG, Crampton EW (1955) Observations on chronic fluorosis in dairy cattle. Can J Comp Med Vet Sci 19:6–16

Das SS, Maiti R, Ghosh D (2006) Fluoride-induced immunotoxicity in adult male albino rat: a correlative approach to oxidative stress. J Immunotoxicol 3:49–55

Dieter P, Fitzke E (1993) Formation of diacylglycerol, inositol phosphates, arachidonic acid and its metabolites in macrophages. Eur J Biochem 218(2):753–758

Dodam JR, Olson NC (1995) Effect of fluoride on cardiopulmonary function and release of eicosanoids in pigs. J Appl Physiol 78:569–577

Dołegowska B (2002) Evaluation of fatty acid levels and selected bioelements in femoral bones of chicks depending on age. Ann Acad Med Stetin 48:45–58

Dołegowska B, Machoy Z, Chlubek D (2003) Changes in the content of zinc and fluoride during growth of the femur in chicken. Biol Trace Elem Res 91:67–76

Dost FN, Knaus RM, Johnson DE, Wang CH (1977) Fluoride impairment of glucose utilization: nature of effect in rats during and after continuous NaF infusion. Toxicol Appl Pharmacol 4:451–458

Edmunds WM, Smedley PL (1996) Groundwater geochemistry and health: an overview. In: Appleton JD, Fuge R, GJH MC (eds) Environmental geochemistry and health, vol 113. Geological Society Special Publication, London, pp 91–105

Eisenbrandt DL, Nitschke KD (1989) Inhalation toxicity of sulfuryl fluoride in rats and rabbits. Fundam Appl Toxicol 12:540–557

Ellenberg H (1991) Bioindicator and biological monitoring. In: Biological monitoring. Signal from the environment. Gate/GTZ, Braunschweig, pp 13–74

Filippovskiĭ TP (1969) Enzootic diseases of the teeth of cattle. Veterinariia 46:68–69

Fleming RH, McCormack HA, McTeir L, Whitehead CC (2003) Effects of dietary particulate limestone, vitamin K3 and fluoride and photostimulation on skeletal morphology and osteoporosis in laying hens. Br Poult Sci 44:683–689

Flueck WT, Smith-Flueck JA (2013) Severe dental fluorosis in juvenile deer linked to a recent volcanic eruption in Patagonia. J Wildl Dis 49:355–366

Fordyce FM, Vrana K, Zhovinsky E, Povoroznuk V, Toth G, Hope BC et al (2007) health risk assessment for fluoride in Central Europe. Environ Geochem Health 29:83–102

Francisca FM, Carro Perez ME (2009) Assessment of natural arsenic in groundwater in Cordoba Province, Argentina. Environ Geochem Health 31:673–682

Garcia JG, Stasek J, Natarajan V, Patterson CE, Dominguez J (1992) Role of protein kinase C in regulation of prostaglandin synthesis in human endothelium. Am J Respir Cell Mol Biol 6:315–325

Giżewska M, Machoy Z (1988) Czynniki wpływające na wchłanianie i wydalanie fluoru u ludzi i zwierząt. Czas Stomatol 41:603–609 [in Polish]

Goldman R, Granot Y, Zor U (1995) A pleiotropic effect of fluoride on signal transduction in macrophages: is it mediated by GPT-binding proteins? J Basic Clin Physiol Pharmacol 6:79–94

Goldman R, Ferber E, Zor U (1997) Involvement of reactive oxygen species in phospholipase A2 activation: inhibition of protein tyrosine phosphatases and activation of protein kinases. Adv Exp Med Biol 400A:25–30

Grace ND, Loganathan P, Hedley MJ (2008) The effect of age on the fluoride concentration in the metacarpus of grazing sheep in New Zealand. N Z Vet J 56:115–119

Green HH (1946) An outbreak of industrial fluorosis in cattle. Proc R Soc Med 39:795–796

Gründer HD (1972) Effects of fluorine emission on cattle. Comprehensive results of studies conducted over several years in the neighborhood of a hydrofluoric acid and aluminum factory. Zentralbl Veterinarmed A 19:265–309

Guan ZZ, Xiao KQ, Zeng XY, Long YG, Cheng YH, Jiang SF et al (2000) Changed cellular membrane lipid composition and lipid peroxidation of kidney in rats with chronic fluorosis. Arch Toxicol 74:602–608

Guenter W, Hahn PH (1986) Fluorine toxicity and laying hen performance. Poult Sci 65:769–778

Gutowska I, Machoy Z, Machaliński B, Chlubek D (2004) Living conditions of deer in the provinces of Western Pomerania and Lubuskie as revealed by mandibular content of fluoride, calcium, and magnesium. 1. Inter-relations between fluoride, calcium, and magnesium content in mandible. Ann Acad Med Stetin Suppl 50:42–46

Gutowska I, Machoy Z, Machaliński B (2005) The role of bivalent metals in hydroxyapatite structures as revealed by molecular modeling with the HyperChem software. J Biomed Mater Res 75A(4):788–793

Gutowska I, Baranowska-Bosiacka I, Baśkiewicz M, Millo B, Siennicka A, Marchlewicz M et al (2010) Fluoride as a pro-inflammatory factor and inhibitor of ATP bioavailability in differentiated human THP1 monocytic cells. Toxicol Lett 96:74–79

Gutowska I, Baranowska-Bosiacka I, Siennicka A, Baskiewicz M, Chlubek D, Machaliński B et al (2011) Fluoride and generation of pro-inflammatory factors in human macrophages. Fluoride 44:125–134

Gutowska I, Baranowska-Bosiacka I, Siennicka A, Telesiński A, Stańczyk-Dunaj M, Wesołowska T et al (2012) Activation of phospholipase A(2) by low levels of fluoride in THP1 macrophages via altered Ca(2+) and cAMP concentration. Prostaglandins Leukot Essent Fatty Acids 86:99–105

Hahn PH, Guenter W (1986) Effect of dietary fluoride and aluminum on laying hen performance and fluoride concentration in blood, soft tissue, bone, and egg. Poult Sci 65:1343–1349

Hem JD (1989) Study and interpretation of the chemical characteristics of natural water. Water Supply Paper 2254, 3rd edn. US Geological Survey, Washington DC, 263 p

Henny CJ, Burke PM (1990) Fluoride accumulation and bone strength in wild black-crowned night-herons. Arch Environ Contam Toxicol 19:132–137

Houdijk AP, Van Leeuwen PA, Adolfs MJ, Bonta IL (1991) GTP-related difference in cyclic AMP production between resident and inflammatory human peritoneal macrophages. Int J Tissue React 13:279–285

Hoffman DJ, Pattee OH, Wiemeyer SN (1985) Effects of fluoride on screech owl reproduction: teratological evaluation, growth, and blood chemistry in hatchlings. Toxicol Lett 26(1):19–24

Indulski J (1989) Fluorine and fluoride. PZWL, Warsaw [in Polish]

Jędrzejuk D, Milewicz A (1996) Fluorine toxicology. Bromat Chem Toksykol 3:205–211 [in Polish]

Jelenko I, Pokorny B (2010) Historical biomonitoring of fluoride pollution by determining fluoride contents in roe deer (*Capreolus capreolus* L.) antlers and mandibles in the vicinity of the largest Slovene thermal power plant. Sci Total Environ 409:430–438

Kabata-Pendias A (2011) Trace elements in soil and plants. Taylor & Francis, New York, pp 385–391

Kakei M, Sakae T, Yoshikawa M, Tamura N (2007) Effect of fluoride ions on apatite crystal formation in rat hard tissues. Fluoride 40:198–204

Kalisińska E, Palczewska M (2007) Fluoride in the teeth of the red fox *Vulpes vulpes* from western Pomerania. Ochr Srod Zas Nat 31:428–433

Kalisińska E, Palczewska-Komsa M (2011) Teeth of the red fox *Vulpes vulpes* (L., 1758) as a bioindicator in studies on fluoride pollution. Acta Theriol 56:343–351

Kalisińska E, Bosiacka-Baranowska I, Lanocha N, Kosik-Bogacka D, Krolaczyk K, Wilk A, Kavetska K, Budis H, Gutowska I, Chlubek D (2014) Fluoride concentrations in the pineal gland, brain and bone of goosander (*Mergus merganser*) and its prey in Odra River estuary in Poland. Environ Geochem Health 36:1063–1077

Kato MT, Bolanho A, Zarella BL, Salo T, Tjäderhane L, Buzalaf MA (2014) Sodium fluoride inhibits MMP-2 and MMP-9. J Dent Res 93:74–77

Kay CE, Tourangeau PC, Gordon CC (1975) Fluoride levels in indigenous animals and plants collected from uncontaminated ecosystems. Fluoride 8:125–133

Khandare AL, Validandi V, Rao S, Nagalla B (2015) Effects of strontium and fluoride ions on bone mechanical and biochemical indices in guinea pigs (*Cavia porcellus*). Fluoride 48:149–159

Kierdorf U, Kierdorf H, Erdelen M, Machoy Z (1995) Mandibular bone fluoride accumulation in wild red deer (Cervus elaphus L.) of known age. Comp Biochem Physiol A Physiol 110(4):299–302

Kierdorf H, Kierdorf U (2000a) Roe deer antlers as monitoring units for assessing temporal changes in environmental pollution by fluoride and lead in a German forest area over a 67-year period. Arch Environ Contam Toxicol 39:1–6

Kierdorf U, Kierdorf H (2000b) Temporal and geographical variation in skeletal fluoride content of roe deer (*Capreolus capreolus*) from industrialized areas in Germany. Comp Biochem Physiol C Toxicol Pharmacol 126:61–68

Kierdorf U, Kierdorf H (2001) Fluoride concentrations in antler bone of roe deer (*Capreolus capreolus*) indicate decreasing fluoride pollution in an industrialized area of western Germany. Environ Toxicol Chem 20:1507–1510

Kierdorf U, Kierdorf H (2002) Assessing regional variation of environmental fluoride concentrations in western Germany by analysis of antler fluoride content in roe deer (Capreolus capreolus). Arch Environ Contam Toxicol 42:99–104

Kierdorf U, Kierdorf H (2003) Temporal variation of fluoride concentration in antlers of roe deer (*Capreolus capreolus*) living in an area exposed to emissions from iron and steel industry, 1948-2000. Chemosphere 52:1677–1681

Kierdorf U, Kierdorf H, Erdelen M, Korsch JP (1989) Mandibular fluoride concentration and its relation to age in roe deer (*Capreolus capreolus* L.). Comp Biochem Physiol A Comp Physiol 94:783–785

Kierdorf U, Kierdorf H, Fejerskov O (1993) Fluoride-induced developmental changes in enamel and dentine of European roe deer (*Capreolus capreolus* L.) as a result of environmental pollution. Arch Oral Biol 38:1071–1081

Kierdorf U, Kierdorf H, Erdelen M, Machoy Z (1996a) Mandibular bone fluoride accumulation in wild red deer (*Cervus elaphus* L.) of known age. Comp Biochem Physiol A Physiol 110:299–302

Kierdorf H, Kierdorf U, Sedlacek F, Erdelen M (1996b) Mandibular bone fluoride levels and occurrence of fluoride induced dental lesions in populations of wild red deer (*Cervus elaphus*) from Central Europe. Environ Pollut 93:75–81

Kierdorf U, Kierdorf H, Sedlacek F, Fejerskov O (1996c) Structural changes in fluorosed dental enamel of red deer (*Cervus elaphus* L.) from a region with severe environmental pollution by fluorides. J Anat 188:183–195

Kierdorf U, Richards A, Sedlacek F, Kierdorf H (1997) Fluoride content and mineralization of red deer (*Cervus elaphus*) antlers and pedicles from fluoride polluted and uncontaminated regions. Arch Environ Contam Toxicol 32:222–227

Kierdorf H, Kierdorf U, Sedlacek F (1999) Monitoring regional fluoride pollution in the Saxonian Ore mountains (Germany) using the biomarker dental fluorosis in roe deer (*Capreolus capreolus* L.). Sci Total Environ 232:159–168

Kierdorf H, Kierdorf U, Richards A, Sedlacek F (2000a) Disturbed enamel formation in wild boars (*Sus scrofa* L.) from fluoride polluted areas in Central Europe. Anat Rec 259:12–24

Kierdorf U, Kierdorf H, Boyde A (2000b) Structure and mineralisation density of antler and pedicle bone in red deer (*Cervus elaphus* L.) exposed to different levels of environmental fluoride: a quantitative backscattered electron imaging study. J Anat 196:71–83

Kierdorf U, Bahelková P, Sedláček F, Kierdorf H (2012) Pronounced reduction of fluoride exposure in free-ranging deer in North Bohemia (Czech Republic) as indicated by the biomarkers skeletal fluoride content and dental fluorosis. Sci Total Environ 414:686–695

Kolditz L (1994) Inorganic chemistry. Part I and II. PWN, Warszawa

Krook L, Maylin GA (1979) Industrial fluoride pollution. Chronic fluoride poisoning in Cornwall Island cattle. Cornell Vet 69(Suppl 8):1–70

Kuenzel WJ, McCune SK, Talbot RT, Sharp PJ, Hill JH (1997) Sites of gene expression for vasoactive intestinal polypeptide throughout the brain of the chick (*Gallus domesticus*). J Comp Neurol 381:101–118

Liu G, Chai C, Cui L (2003) Fluoride causing abnormally elevated serum nitric oxide levels in chicks. Environ Toxicol Pharmacol 13:199–204

Liu J, Cui HM, Peng X, Fang J, Zuo Z, Wang H et al (2012) Decreased percentages of T cell subsets and IL-2 contents in the cecal tonsil of broilers fed diets high in fluorine. Fluoride 45:53–57

Liu J, Cui H, Peng X, Fang J, Zuo Z, Wang H et al (2013) Dietary high fluorine induces apoptosis and alters Bcl-2, Bax, and caspase-3 protein expression in the cecal tonsil lymphocytes of broilers. Biol Trace Elem Res 152:25–30

Lundy MW, Russell JE, Avery J, Wergedal JE, Baylink DJ (1992) Effect of sodium fluoride on bone density in chickens. Calcif Tissue Int 50:420–426

Machoy Z (1990) Different mechanisms of fluoride binding by bones and teeth. Czas Stomatol 43:689–692 [in Polish]

Machoy Z, Dabkowska E, Nowicka W (1991) Increased fluoride content in mandibular bones of deer living in industrialised regions of Poland. Environ Geochem Health 13:161–163

Machoy Z, Dabkowska E, Samujło D, Ogoński T, Raczyński J, Gebczynska Z (1995) Relationship between fluoride content in bones and the age in European elk (*Alces alces* L.). Comp Biochem Physiol C Pharmacol Toxicol Endocrinol 111:117–120

Machoy Z, Straszko J, Dziedziejko V, Gutowska I, Zakrzewska H (2001) Statistical neutral network analysis of fluoride bioaccumulation in deer bones from Polish industrial areas. Fluoride 34:197–198

Machoy Z, Straszko J, Dziedziejko V, Gutowska I (2002) Estimation of influence of industrial emissions on fluoride accumulation in deers jaws by neural network. Environ Sci 30:289–300

Machoy-Mokrzyńska A (2000) The importance of inhalation toxicity of fluorine compounds in Poland. Bromat Chem Toksykol 333:133–136

Mangas I, Vilanova E, Estévez J (2014) Kinetic interactions of a neuropathy potentiator (phenylmethylsulfonyl fluoride) with the neuropathy target esterase and other membrane bound esterases. Arch Toxicol 88:355–366

Manji F, Kapila S (1986) Fluorides and fluorosis in Kenya. Part 1. The occurrence of fluorides. Odontostomatol Trop 9:15–20

Maurice DV, Lightsey SF, Abudabos A, Toler JE (2002) Factors affecting ascorbic acid biosynthesis in chickens: III. Effect of dietary fluoride on L-gulonolactone oxidase activity and tissue ascorbic acid (AsA) concentration. J Anim Physiol Anim Nutr (Berl) 86:383–388
McGown EL, Suttie JW (1977) Mechanism of fluoride-induced hyperglycemia in the rat. Toxicol App Pharmacol 40:83–90
McRae ME, US Geological Survey (2015) Mineral commodity summaries, pp 56–57
Merkley JW (1981) The effect of sodium fluoride on egg production, egg quality, and bone strength of caged layers. Poult Sci 60:771–776
Michel JN, Suttie JW, Sunde ML (1984) Fluorine deposition in bone as related to physiological state. Poult Sci 63:1407–1411
Monsour PA, Kruger BJ (1985) Effect of fluoride on soft tissues in vertebrates: a review. Fluoride 8:53–61
Murao H, Sakagami N, Iguchi T, Murakami T, Suketa Y (2000) Sodium fluoride increases intracellular calcium in rat renal epithelial cell line NRK-52E. Biol Pharm Bull 23:581–584
Murray M (1967) Fluorosis in a herd of cattle in Kenya. Bull Epizoot Dis Afr 15:259–262
Murray JJ (ed) (1986) Appropriate use of fluorides for human health. WHO, Geneva
Narwaria YS, Saksena DN (2012) Incidence of dental fluorosis in domestic animals of Shivpuri, Madhya Pradesh, India. J Environ Res Dev 7:426–430
Newman JP, Yu MH (1976) Fluorosis in black-tailed deer. J Wildl Dis 12:39–41
Obel AL, Erne K (1971) Bovine fluorosis in Sweden. Acta Vet Scand 12:164–184
Oruc N (2008) Occurrence and problems of high fluoride waters in Turkey: an overview. Environ Geochem Health 30:315–323
Palczewska-Komsa M, Kalisińska E, Kosik-Bogacka DI, Lanocha N, Budis H, Baranowska-Bosiacka I et al (2014) Fluoride in the bones of foxes (*Vulpes vulpes* Linnaeus, 1758) and raccoon dogs (*Nyctereutes procyonoides* Gray, 1834) from North-Western Poland. Biol Trace Elem Res 160:24–31
Pattee OH, Wiemeyer SN, Swineford DM (1988) Effects of dietary fluoride on reproduction in Eastern Screech-owls. Arch Environ Contam Toxicol 17:213–218
Peschke E, Bahr I, Mühlbauer E (2013) Melatonin and pancreatic islets: interrelationships between melatonin, insulin and glucagon. Int J Mol Sci 14:6981–7015
Piotrowska S, Machoy Z, Chlubek D (2006) Calcium, magnesium, fluoride, and phosphorus content in antlers and skull bones of roe deer (*Capreolus capreolus*) in relation to age. Ann Acad Med Stetin 52(Suppl 1):83–87
Reeves TG (1986) Water fluoridation. A manual for engineers and technicians. US Department of Health and Human Services, Centres for Disease Control and Prevention, Atlanta, GA, 138 pp
Richter H, Kierdorf U, Richards A, Kierdorf H (2010) Dentin abnormalities in cheek teeth of wild red deer and roe deer from a fluoride-polluted area in Central Europe. Ann Anat 20(192):86–95
Richter H, Kierdorf U, Richards A, Melcher F, Kierdorf H (2011) Fluoride concentration in dentine as a biomarker of fluoride intake in European roe deer (*Capreolus capreolus*)—an electron-microprobe study. Arch Oral Biol 56:785–792
Robertson CJR, Lock JW (1994) Fluoride in New Zealand birds: a review. Sci Res Ser 71:1–28
Salicki W, Kalisinska E (2006) Fluorine and calcium concentrations in bones of the wood pigeon from the environs of Szczecin: a 2002–2004 study. Ann Acad Med Stetin 52:89–95
Schamschula RG, Barmes DE (1981) Fluoride and health: dental caries, osteoporosis, and cardiovascular disease. Annu Rev Nutr 1:427–435
Schmid G (1956) Fluorosis in cattle. Bull Schweiz Akad Med Wiss 12:397–418
Schultz M, Kierdorf U, Sedlacek F, Kierdorf H (1998) Pathological bone changes in the mandibles of wild red deer (*Cervus elaphus* L) exposed to high environmental levels of fluoride. J Anat 193:431–442
Schulze-Specking A, Duyster J, Gebicke-Haerter PJ, Wurster S, Dieter P (1991) Effect of fluoride, pertussis and cholera toxin on the release of arachidonic acid and the formation of prostaglandin E2, D2, superoxide and inositol phosphates in rat liver macrophages. Cell Signal 3:599–606

Shashi JP, Singh, Thapar SP (1988) Changes in glycogen content in some tissues during fluorosis an experimental study on rabbits. Fluoride 21:82–86

Shaw D (2012) Weeping and wailing and gnashing of teeth: the legal fiction of water fluoridation. Med Law Int 12:11–27

Shim MY, Parr C, Pesti GM (2011) The effects of dietary fluoride on growth and bone mineralization in broiler chicks. Poult Sci 90:1967–1974

Shivarajashankara YM, Shivashankara AR, Bhat PG, Rao SM, Rao SH (2002) Histological changes in the brain of Young fluoride-intoxicated rats. Fluoride 35:12–21

Shore RF (1995) Predicting cadmium, lead and fluoride levels in small mammals from soil residues and by species-species extrapolation. Environ Pollut 88:333–340

Shupe JL, Olson AE, Peterson HB, Low JB (1984) Fluoride toxicosis in wild ungulates. J Am Vet Med Assoc 185(11):1295–1300

Song F, Yan Y, Zhao X, Dou D, Zhang C, Xie K (2009) Phenylmethylsulfonyl fluoride protects against the degradation of neurofilaments in tri-ortho-cresyl phosphate (TOCP) induced delayed neuropathy. Toxicology 262:258–264

Susheela AK (2001) A treatise on fluorosis. Fluorosis Research and Rural Development Foundation, Delhi, India, pp 53–60, 78–79

Suttie JW, Kolstad DL, Sunde ML (1984) Fluoride tolerance of the young chick and turkey poult. Poult Sci 63:738–743

Talmage SS, Walton BT (1991) Small mammals as monitors of environmental contaminants. Rev Environ Contam Toxicol 119:47–145

Turner CH, Akhter MP, Heaney RP (1992) The effects of fluoridated water on bone strength. J Orthop Res 10:581–587

Udall DH, Keller KP (1952) A report on fluorosis in cattle in the Columbia River Valley. Cornell Vet 42:159–184

Vikøren T, Stuve G (1996a) Fluoride exposure and selected characteristics of eggs and bones of the herring gull (*Larus argentatus*) and the common gull (*Larus canus*). J Wildl Dis 32:190–198

Vikøren T, Stuve G (1996b) Fluoride exposure in cervids inhabiting areas adjacent to aluminum smelters in Norway. II. Fluorosis. J Wildl Dis 32:181–189

Waldbott GL, Burgstahler AW, McKinney HL (1978) Fluoridation: the great dilemma. Coronado, Lawrence, KS

Walton KC (1984) Fluoride in fox bone near an aluminium reduction plant in Anglesey, Wales and elsewhere in the United Kingdom. Environ Pollut B 7:273–280

Walton KC, Ackroyd S (1988) Fluoride in mandibles and antlers of roe and red deer from different areas of England and Scotland. Environ Pollut 54:17–27

Wang JD, Hong JP, Li JX (1995) Studies on alleviation of industrial fluorosis in Baotou goats. Fluoride 28:131–134

Wang YN, Xiao KQ, Liu JL, Dallner G, Guan ZZ (2000) Effect of long term fluoride exposure on lipid composition in rat liver. Toxicology 146:161–169

Wang J, Ge Y, Ning H, Wang S (2004) Effects of high fluoride and low iodine on oxidative stress and antioxidant defense of the brain in offspring rats. Fluoride 37:264–270

Węglarz A, Michalski R (1998) Methods to determine fluorine and its compounds. Air protection and waste problems 1

Wessel K, Resch K, Kaever V (1989) Aluminum fluoride enhances phospholipase A2 activity and eicosanoid synthesis in macrophages. Eicosanoids 2:223–227

WHO (2006) Fluoride in drinking-water. Fawell J, Bailey K, Chilton J, Dahi E, Fewtrell L, Magara Y. IWA, London

Xiang Q, Liang Y, Wang C, Chen B, Chen X, Zhou M (2003) Effect of fluoride in drinking water on children's intelligence. Fluoride 36:84–94

Xie Z, Sun L (2003) Fluoride content in bones of Adelie penguins and environmental media in Antarctica. Environ Geochem Health 25:483–490

Xu H, Zhou YL, Zhang JM, Liu H, Jing L, Li GS (2007) Effects of fluoride on the intracellular free Ca2+ and Ca2+-ATPase of kidney. Biol Trace Elem Res 116:279–288

Yan X, Hao X, Nie Q, Feng C, Wang H, Sun Z et al (2015) Effects of fluoride on the ultrastructure and expression of Type I collagen in rat hard tissue. Chemosphere 128:36–41

Yu Y, Yang W, Dong Z, Wan C, Zhang J, Liu J et al (2008) Neurotransmitter and receptor changes in the brains of fetuses from areas of endemic fluorosis. Fluoride 41:134–138

Zakrzewska H, Brzezińska M, Orowicz W, Samujło D, Wójcik A, Kolanus A (2004) Content of fluoride in hair and hoofs of wild boars and deer from Western Pomerania as a bioindicator of environmental pollution. Ann Acad Med Stetin 50:100–103

Zakrzewska H, Machoy-Mokrzyńska A, Materny M, Gutowska I, Machoy Z (2005) Estimation of fluoride distribution in the mandible and teeth of the red deer (*Cervus elaphus* L.) from industrially polluted areas in Poland. Arch Oral Biol 50:309–316

Zemek F, Herman M, Kierdorf H, Kierdorf U, Sedlácek F (2006) Spatial distribution of dental fluorosis in roe deer (*Capreolus capreolus*) from North Bohemia (Czech Republic) and its relationships with environmental factors. Sci Total Environ 370:491–505

Ziemiańska K, Konopka A, Wilczyński G (2012) The role of extracellular proteolysis of synaptic plasticity in the CNS. Postepy Hig Med Dosw 66:959–975

# Chapter 16
# Lead, Pb

**Irena Baranowska-Bosiacka, Jan Korbecki, and Mariola Marchlewicz**

**Abstract** We present literature data on the concentrations of lead (Pb) in various organs of terrestrial mammals and birds. The presented data have been summarized in terms of animal species, organs, and the approximate year of sampling and area. Studies on Pb in the muscles of various mammalian species indicate that Pb levels in the environment are not as high as 25 years ago, and in most of the non-contaminated areas they do not exceed 1 mg $kg^{-1}$ dw in the muscle of herbivorous or predatory mammals. Bioindication research helps to determine not only the history and the present state of lead pollution but also allows an understanding of the phenomenon of the circulation of lead in nature, as well as in prediction of the risk of its presence in the trophic chain. Long-term bioindication research conducted on the bones and soft tissues of mammals and birds, as well as the eggs and feathers, indicates the usefulness of this type of material for evaluating the state of the environment.

## 1 Introduction

Lead (Lat. plumbum, Pb) is believed to be the first metal smelted by humans (Lessler 1988). From the very beginnings of lead smelting, anthropogenic lead dust has spread in the atmosphere by air currents, reaching areas as distant as Greenland and leaving clear traces in the environment. The isotopic composition of lead particles preserved under the layers of ice in Greenland confirms their anthropogenic nature and indicates their place of origin, thanks to the unique combinations of lead isotopes in each lead ore (Gale and Stos-Gale 1981). Only the first of the isotopes, $^{204}$

I. Baranowska-Bosiacka (✉) · J. Korbecki
Department of Biochemistry and Medical Chemistry, Pomeranian Medical University in Szczecin, Szczecin, Poland
e-mail: irena.bosiacka@pum.edu.pl

M. Marchlewicz
Department of Aesthetic Dermatology, Pomeranian Medical University in Szczecin, Szczecin, Poland



E. Kalisińska (ed.), *Mammals and Birds as Bioindicators of Trace Element Contaminations in Terrestrial Environments*,
https://doi.org/10.1007/978-3-030-00121-6_16

Pb, is always present in the same amount, while the levels of the remaining three isotopes vary: $^{206}Pb$ (1.35–1.5%), $^{207}Pb$ (23.5–27%), and $^{208}Pb$ (20.5–23%) (ATSDR 2013).

After the introduction of silver coin minting in the Mediterranean region (approximately 680 BC), relying on the use of litharge (lead oxide) in silver smelting (cupellation), the levels of anthropogenic lead began to clearly rise in the successive ice layers of Greenland (Nriagu 1983; Rosman et al. 1997). Lead also became commonly used in various areas of daily life. For centuries, lead acetate served as a sweetener, wine preservative, an inhibitor of cider fermentation, an anti-inflammatory agent, and an abortifacient. Other lead compounds, such as galena (lead sulfide), cerussite (lead carbonate), and litharge, were used as drug ingredients. Metallic lead, thanks to its plasticity and softness, was used for making pipes, roofs (lead sheets are still being used for this purpose), pots, coins, bullets, weights, and many other common objects.

The toxic effects of lead have been known for centuries, with the first description of lead poisoning found in "Alexipharmaca," a poem by Nikander from the second century BC (Waldron 1973). Despite the accumulated knowledge about its harmful effects, lead was responsible for the longest-lasting environmental poisoning epidemic in human history. In the twentieth century, the discovery of anti-knock properties of tetraethyl lead and the use of leaded petrol-fueled engines resulted in a rapid increase in the concentration of lead in the atmosphere. Although leaded petrol has been removed from use in many countries (e.g., in 1976–1986 in the United States, in 2005 in the European Union (EU)), lead compounds are still being used in aviation fuels. In addition, although the sale of lead paints was banned in the United States in the late 1970s, and in the EU in 1992, they are still used in the restoration and preservation of historic buildings (including the interiors) and art (Directive EP 1998; Statutory Instruments 2005, 2009; European Commission 2008).

The current presence and toxicity of lead makes it the second most dangerous environmental poison according to ATSDR Priority Substance List (2013). No lowest safe concentration exists for lead, which contributes to 0.6% of the global burden of disease (WHO 2009). Although the carcinogenic action of lead compounds has not yet been fully proven (Mccabe et al. 2001), it has been shown that lead can contribute to the induction of neoplastic processes by the inhibition of DNA repair enzymes (Rajaraman et al. 2006). The International Agency for Research on Cancer (IARC) has classified inorganic lead compounds to Group 2A: Probably carcinogenic to humans and lead to Group 2B: Possibly carcinogenic to humans (IARC 2016).

## 2 General Properties of Lead

Lead is a carbon group element with atomic number 82, group IVa, atomic weight 207.19, density 11.340 g $cm^{-3}$ at 20°C, and melting point 327.46°C. This element can occur in oxidation states II and IV. It is a soft metal (hardness 1.5 Mohs), with a bluish-white color and easily oxidizes in the air, which results in its surface being coated with hydroxide and carbonate that prevent further oxidation (US DHHS 1999).

## 3 Lead in Nature

Environmental lead can be divided into primary and secondary deposits. The former became the component of rocks at the time of their formation, while the secondary lead (or radiogenic) derived from the radioactive decay of uranium and thorium (Kabata-Pendias and Pendias 1999). Lead occurs naturally in igneous rocks and clay (at 10–40 mg $kg^{-1}$ (or ppm, part per million)) and at much lower levels in alkaline igneous rocks and carbonate sediments (0.1–10 ppm) (Kabata-Pendias and Pendias 1999). Feldspar and mica are richest in this element, due to lead's ability to replace potassium in those formations. The most important minerals of lead include galena (PbS), anglesite ($PbSO_4$), cerussite ($PbCO_3$), pyromorphite ($Pb_5(PO_4)_3Cl$), and mimetite ($Pb_5(AsO_4)_3Cl$). Lead can be found in the deposits of many metals, including zinc, silver, and gold. Most commonly, it is derived from lead sulfide.

Lead has potent chalcophilic properties. The average lead concentration in coal is 25 ppm, with the maximum content in ash following combustion exceeds 2000 ppm. Ash from oil may contain 500 ppm of lead on average (Kabata-Pendias and Pendias 1999). Lead exhibits low migration into soils; hence its distribution there reflects its content in rocks and anthropogenic influences in surface layers (Adriano 2001; Hettiarachchi and Pierzynski 2004). The common occurrence of zinc and lead ores, as well as their sulfides and galena, have resulted in the formation of joint deposits via post-magmatic processes. Metasomatic deposits found in limestone and dolomite are the most important category of zinc and lead deposits. These include mesothermal deposits whose galena is often argentiferous and may also contain gold. Many of these fields are substantial and contain significant resources (from 10 to 15 million tons), including those found in Leadville, Colorado (USA), Broken Hill in Northern Rhodesia (Africa), and Nerchinsky District in Zabaykalsky Krai (Russia). Telethermal deposits are characterized by ore mineralization occurring usually in certain layers that are themselves layered. The largest deposits of this kind in the world can be found in the mainly zinc-containing "Joplin District" in the USA (parts of Missouri, Oklahoma, and Kansas) and the mainly lead-containing deposits in southeastern Missouri. The largest European deposits are found in Poland, in the Silesian-Cracow area (US Geological Survey 2015).

In 2014, the global mine production of lead was expected to be about 5.50 million tons, with production increases in Australia, China, and the United States. The International Lead and Zinc Study Group (ILZSG) forecast global refined lead production to increase slightly from that in 2013, to 11.3 million tons, primarily driven by increases in Australia, Belgium, China, India, Italy, and South Korea. ILZSG projected global lead consumption to increase slightly in 2014 from that in 2013, to 11.3 million tons, partially owing to an increase in China, and that global refined lead production would exceed consumption by 38,000 tons (U.S. Geological Survey 2015).

The lead-acid battery industry accounted for about 90% of the reported US lead consumption during 2014. Lead-acid batteries were primarily used as starting-lighting-ignition (SLI) batteries for automobiles and trucks and as industrial-type batteries for standby power for computer and telecommunications networks and for motive power. During the first 9 months of 2014, 93.5 million lead-acid automotive batteries were shipped by North American producers (U.S. Geological Survey 2015).

Atmospheric lead is a result of industrial and traffic-related emissions of dust, including significant quantities from coal combustion (WHO 1995; Marcus and McBratney 2011). According to WHO (2000), the average concentration of lead in the air in and around crop fields reaches 0.1–0.3 μg $m^{-3}$ and about 0.5 μg $m^{-3}$ in the vicinity of villages and towns. Near large European cities, it ranges from 0.5 to 3.0 μg $m^{-3}$. The most lead-polluted air is found around mines and metal smelters, where it can reach a dozen or so μg $m^{-3}$ (WHO 2000). Although the use of tetraethyl lead—the most prevalent compound of this metal in the atmosphere—had been eliminated from fuels with the introduction of the US ban on the sale of leaded petrol in 1986, the concentration of this element in the atmosphere continues to grow, particularly in countries with developed metallurgical and mining industries, as well as from the related recycling of various products, e.g., batteries (Statutory Instruments 2005, 2009). In Europe, the ban on the sale of lead-containing fuel has been in force since 2005, based on the Directive of the European Parliament relating to the quality of petrol and diesel fuels (Directive EP 1998). An additional source of lead in the atmosphere is secondary dust coming from asphalted streets, parking lots, and sports fields with surfaces made of polyvinyl chloride.

Lead is also released from natural sources as a result of natural mobilization during volcanic activity and the weathering of rocks. The major natural sources of emissions into the air are volcanoes, airborne soil particles, sea spray, biogenic material, and forest fires. A recent study estimates emissions from natural sources between 0.22 and 4.9 million tons per year (UNEP 2010). The total emissions and distribution by sources vary considerably among countries. Overall emissions of lead decreased by about 95% over the 21-year period from 1982 to 2002, falling from about 54,500 tons per year in 1982 to about 1550 tons in 2002. The significant reduction in lead emissions was mainly due to restrictions and bans on the use of leaded petrol for vehicles and also implementation of improved air pollution controls (UNEP 2010).

Along with other air pollutants, lead falls on the surface layer of the ground and from there is absorbed into the deeper layers together with rainwater. It accumulates in soils, and due to the lack of biodegradation, this is where it reaches the highest environmental concentrations. Lead-contaminated soil becomes an indirect source of lead for humans, via plants growing near roads with heavy traffic or in the vicinity of current or former mines and smelters (Marcus and McBratney 2011). The bioavailability of lead in the soil depends on its pH; acidic soil increases the bioavailability of lead and its penetration into plants, while neutral soil pH reduces the absorption of heavy metals by flora. In addition, the bioavailability of lead is raised by high zinc concentrations, triggered by low soil pH (Alvarenga et al. 2014; Wuana and Okieimen 2011).

The major source of direct lead release to soils is the use of ammunition. In 2003 the total global consumption of lead for ammunition was about 120,000 tons. Ammunition is partly used for hunting and partly in shooting ranges, where the lead is either accumulated at the range or collected for recycling. Moreover, lost lead shots may poison waterfowl and other birds ingesting the shots. Terrestrial environments in various countries are also exposed to lead present in paints, balancing weights for vehicles, sheathing of cables left in the ground, and lead batteries. Large amounts of lead are directed to landfills and waste dumps with discarded products and residues from mining and base metal production (UNEP 2010).

Lead may be present on the surfaces of leaves and other plant parts due to deposition from the atmosphere. Plants can also take up lead from contaminated soils and retain most of it in the roots or transport it to other parts. The bioavailability of lead in soils is limited due to the strong adsorption of lead by soil organic matter; however, this availability may increase with a decrease in pH and organic matter. It also depends on soil moisture and type of soil amendments. For example, the addition of calcium and phosphorus to the soil reduces the bioavailability of lead, while the addition of $K_2EDTA$ may increase lead uptake. Finally, some plant species have a high sensitivity to lead, while others exhibit a high tolerance. The potential use of plants to remediate contaminated soil has recently received a great deal of interest, since heavy metal contamination of soil is a serious problem in industrial and postindustrial areas (Henry 2000).

Lichens and funghi are excellent bioacumulators of trace elements, since the concentrations found in their thalli can be directly correlated with those in the environment. Some of them are able to accumulate heavy metals at very high concentrations without showing visible pathological symptoms (Baranowska-Bosiacka et al. 2001).

Plants can accumulate lead in those parts considered edible for herbivores and humans, which is a serious problem in many parts of the world. Dietary exposure to lead for adult consumers in 19 European countries ranges from 0.36 to 1.24 μg $kg^{-1}$ body weight (bw) per day (lower bound for a country with lowest average exposure—upper bound for a country with highest average exposure) and from 0.73 to 2.43 μg $kg^{-1}$ bw per day for high consumers, respectively. Overall, cereals and vegetables (potatoes) are the most significant sources of lead exposure in the general European population (EFSA 2010).

Water, both on the surface and in the ground, is a source of lead mainly as a result of human industrial activities. Surface waters may accumulate lead from precipitation in air-polluted areas or when it is washed out with rain from the soil, but the most lead-contaminated waters are found near industrial plants releasing polluted wastewater or in areas of crops fertilized with nitrogen agents contaminated with lead (WHO 1995). Drinking water also may contain trace amounts of lead due to outdated water supply systems, where pipes of internal water supply systems, welds, valves, and house connections may contain lead (WHO 1995). In many European countries, lead systems are used in many buildings, especially in the older districts of large cities (Hayes and Skubala 2009). In recent years, numerous studies have shown that exposure to even relatively low concentrations of lead in the water, within the limits of 25–50 μg $L^{-1}$, previously regarded as harmless to human health, do increase the concentration of lead in the blood. As a result, the experts of the World Health Organization have concluded that concentrations of lead in water intended for drinking should not exceed 10 μg $L^{-1}$ (WHO 2011).

## 4 Lead Accumulation and Toxicity in Humans

Lead enters the human body primarily through the gastrointestinal tract and respiratory system and is then transported to all tissues via the blood. For this reason, lead poisoning and determination of exposure in general and occupationally exposed populations are measured by the whole blood lead levels (Pb-B) (WHO 1995). However, this good indicator of current exposure to lead does not directly reflect levels in the brain (Lidsky and Schneider 2003; White et al. 2007; Baranowska-Bosiacka et al. 2011). Lead easily penetrates the blood-brain barrier, reaching much higher concentrations in the cerebrospinal fluid than in the whole blood. This is due to difference in half-life period between the places of deposition; in the blood it is about 35 days, while in the brain it is about 2 years, and in bones from 1 to 30 years (Conti et al. 2012a). In addition to the skeletal system, background levels of lead accumulate in the liver (1 mg $kg^{-1}$), kidney (0.8 mg $kg^{-1}$ in the cortex and 0.5 mg $kg^{-1}$ in the core), and then in the ovaries and pancreas (0.4 mg $kg^{-1}$), spleen (0.3 mg $kg^{-1}$), prostate and adrenal glands (0.2 mg $kg^{-1}$), brain and adipose tissues (0.1 mg $kg^{-1}$), testes (0.08 mg $kg^{-1}$), heart (0.07 mg $kg^{-1}$) and skeletal muscle (0.05 mg $kg^{-1}$) (Giel-Pietraszuk et al. 2012).

Inorganic lead is not transformed in the body, while tetraalkyl compounds are dealkylated in the liver. Two thirds of inorganic lead is excreted with urine and in 1/3 via bile into the intestine and further excreted in the feces. Small amounts, physiologically irrelevant for the body, are excreted in sweat, milk, and saliva and accumulate in the hair and nails. Di- and mono-alkyl derivatives, resulting from the dealkylation of tetraalkyl lead compounds, are excreted primarily in the urine. Inhaled lead is either exhaled or moved with discharge into the gastrointestinal tract, where the particles not absorbed in the intestines are excreted in the feces (Baranowska-Bosiacka and Chlubek 2006).

Currently, researchers focus particularly on the neurotoxic actions of lead (Rao Barkur and Bairy 2015; Sanders et al. 2015). These studies tend to propose an ever lower "safe threshold concentration" of this element. Although in 1991 the American Center for Disease Control (CDC, United States Department of Health and Human Services) lowered the safe threshold concentration of Pb-B to 10 μg $dL^{-1}$ (WHO 1995), followed by 5 μg $dL^{-1}$ for children and pregnant women (CDC 2012), a lot of data indicates that even lower concentrations pose a threat to health (CDC 2004, 2005, 2007). Some researchers argue that there is no such thing as "the safe level of lead." This means that at each concentration of lead in the body, we may expect irregularities in the biochemical processes in many organs, particularly in the nervous tissue (Koller et al. 2004). In addition, this metal has been found to accumulate in some tissues/organs from which it may be released under certain conditions (pregnancy, osteoporosis, hormonal disorders). This may lead to increased blood lead levels and consequently higher brain levels. Even the bone pool of lead, previously considered a permanent place of deposition, can be mobilized according to a recent study (Conti et al. 2012b; CDC 2004).

For lead risk assessment and follow-up of time trends, there is a need for adequate information on exposure. The neurotoxic impact of very low levels of prenatal lead exposure (below 5 μg $dL^{-1}$) has been observed in infants and very young children, which suggests a revision of established health guidelines for prenatal exposure to lead (CDC 2012).

Acute lead poisonings, such as those resulting in encephalopathy (70–100 μg $dL^{-1}$), are rare nowadays. However, as recently as 2009–2010, 400 children died from lead poisoning in two villages in the province of Zamfara in northwestern Nigeria in Africa. Thousands of local children and adults are still seriously ill due to lead contamination from illegal gold mining, undertaken with the simplest primitive methods, with no safety precautions against the relatively high proportion of lead in gold ores (Moszynski 2010; Dooyema et al. 2012; Plumlee et al. 2013).

The toxicity of lead affects many organs, the most sensitive being the brain, blood, and kidneys (Baranowska-Bosiacka and Chlubek 2006). Its hematoxicity is associated with the inhibition of heme synthesis, leading to anemia; here, lead inhibits the activity of δ-aminolevulinic acid dehydratase (ALAD) and ferrochelatase, which catalyzes the insertion of ferrous iron into protoporphyrin IX, yielding heme. A reduction in their activity leads to the accumulation of δ-aminolevulinic acid (ALA) and erythrocyte protoporphyrin in the blood and an increase in the level of porphyrins in the urine and feces (Giel-Pietraszuk et al. 2012). The blood lead level, which triggers ALAD inhibition and an increase in the concentration of free erythrocyte protoporphyrin, is about 20–30 μg $dL^{-1}$. The first symptoms of anemia occur at concentrations higher than 50 μg $dL^{-1}$. Nephropathy, with a typical atherosclerosis, glomerular atrophy, and interstitial fibrosis, develops at a concentration greater than 60 μg $dL^{-1}$ (Giel-Pietraszuk et al. 2012).

## 5 Lead Accumulation and Toxicity in Birds and Mammals

Many factors, such as age, sex, physiological status, diet, dose, and exposure time, affect the absorption of lead in birds. Lead levels in tissues causing physiological dysfunction, the clinical symptoms of poisoning, and death vary between species. A comprehensive study on interpreting tissue concentration of environmental contaminants (Nelson Beyer and Meador 2011) indicates the background concentration of lead in birds to be <20 μg $dL^{-1}$ in blood, <2 mg $kg^{-1}$ wet weight (ww) in the liver and kidneys, and <10 mg $kg^{-1}$ dry weight (dw) of the bone in birds.

For birds in general, liver lead concentrations within the clinical poisoning range (≥6 mg $kg^{-1}$ ww) suggest lead poisoning. Bone lead concentrations of ≥20 mg $kg^{-1}$ dw are considered to suggest excessive exposure. Because of the rapid uptake and slow release of lead from the bone, bone concentration can be used to determine the geographical patterns of poisoning in populations. The suggested threshold of increasing severity of effects for Anseriformes are 20–50 μg $dL^{-1}$ in blood and two < 6 mg $kg^{-1}$ ww in the liver and kidney (subclinical poisoning); 50–100 μg $dL^{-1}$ in blood, 6–10 mg $kg^{-1}$ ww in the liver, and 6–15 mg $kg^{-1}$ ww in the kidneys (clinical poisoning); and >100 μg $dL^{-1}$ in blood, 10 mg $kg^{-1}$ ww in the liver, and >5 mg $kg^{-1}$ ww in the kidneys (severe clinical poisoning). Suggested thresholds for pigeons and doves (ordo Columbiformes) are 20 < 200 μg $dL^{-1}$ in blood, two < 6 mg $kg^{-1}$ ww in the liver, and two < 15 mg $kg^{-1}$ ww in the kidneys (subclinical poisoning); 200–300 μg $dL^{-1}$ in blood, 6–15 mg $kg^{-1}$ ww in the liver, and 15–30 mg $kg^{-1}$ ww in the kidneys (clinical poisoning); and >300 μg $dL^{-1}$ in blood, >15 mg $kg^{-1}$ ww in the liver, and >30 mg $kg^{-1}$ ww in the kidneys (severe clinical poisoning). Lead concentrations in mammalian blood and soft tissue are useful as a biomarker of acute lead poisoning, whereas lead in the bone is more relevant for evaluating health effects over a long period. Tables 1 and 2 present

**Table 1** Toxic and nontoxic levels of lead in the blood and urine of various animals according to the Normal Range Values for WVDL Toxicology (μg $dL^{-1}$) (WVDL 2015)

| Animal group | Normal | High | Toxic |
|---|---|---|---|
| *Blood* | | | |
| Avian | 0.02–0.2 | >0.6 | 1–30 |
| Canine | 0.01–0.1 | 0.3–0.8 | 0.6–7.4 |
| Caprine | <0.2 | 0.5–0.9 | >0.9 |
| Equine | 0.04–0.25 | 0.3–0.6 | 0.33–2.5 |
| Feline | <0.1 | 0.7 | >1 |
| Lapine | 0.02–0.3 | 0.3–1.4 | |
| Ovine | 0.02–0.25 | 0.7–0.9 | 1-5 |
| Porcine | 0.04–0.1 | 0.3–3.0 | >1.2 |
| *Urine* | | | |
| Bovine | <0.05 | | 0.2–1.0 |
| Canine | <0.05 | | >0.075 |
| Feline | 0.05–0.2 | | 0.5–1.8 |
| Bovine | <0.05 | | 0.2–1.0 |

**Table 2** Toxic and nontoxic lead levels (mg kg$^{-1}$ ww) in the soft tissues of various animals, according to the Normal Range Values for WVDL Toxicology (WVDL 2015)

| Animal group | Normal | High | Toxic |
|---|---|---|---|
| *Brain* | | | |
| Bovine | 0.25–0.50 | 1.5–4.5 | 3.5–7.0 |
| *Kidney* | | | |
| Avian | 0.4–4.0 | 4–48 | 32–6400 |
| Bovine | 0.08–8.0 | 12–80 | 20–2800 |
| Camelid | <8 | | |
| Canine | 0.4–10.0 | 20–40 | 40–200 |
| Caprine | 2.0–4.0 | 15–20 | >40 |
| Cervid | <4 | | |
| Equine | 0.12–5.2 | 12–20 | 20–800 |
| Feline | <4.0 | | |
| Lapine | 0.4–4.0 | | >40 |
| Ovine | 0.4–3.2 | 20–400 | 20–800 |
| Porcine | <2.8 | 20–100 | |
| Ursine | 0.8–8.0 | | |
| *Liver* | | | |
| Avian | 0.3–1.6 | 3–33 | 20–466 |
| Bovine | 0.3–3.3 | 7–33 | 16–1000 |
| Camelid | <7 | | |
| Canine | 0.3–11.7 | 12–17 | 167–667 |
| Caprine | 1.7–3.3 | 10–17 | >33 |
| Cervid | <3 | | |
| Equine | 0.27–4.7 | 10–17 | 13–1667 |
| Feline | <3 | | 33–243 |
| Lapine | 0.7–2.0 | 10–19.0 | >33 |
| Mammals | <3 | 7–17 | >17 |
| Ovine | 0.10–2.7 | 17–83 | 33–333 |
| Porcine | <2.3 | 16–83 | |
| Ursine | 0.20–6.7 | | |

nontoxic and toxic Pb levels in the blood, urine, and tissues of various animals (WVDL 2015).

There is no evidence of a difference in susceptibility between mammals and humans, suggesting a predictiveness of their lead dose-effect relations. Blood lead levels >5 μg dL$^{-1}$ are associated with neurobehavioral deficits and neurotoxicic effects. Levels >20 μg Pb dL$^{-1}$ are associated with adverse reproductive effects, and levels >40 μg dL$^{-1}$ result in nephrotoxic and hematological changes. Blood lead levels >80 μg dL$^{-1}$ cause death (Nelson Beyer and Meador 2011).

## 6 Bioaccumulation of Lead in Wildlife

Lead found in wild animals can come from both natural and anthropogenic sources. Anthropogenic pollution contributes to an increase in the concentration of Pb in the air, water, and food chain (Fig. 1). Ingestion and inhalation are the most significant routes of exposure to Pb in terrestrial animals. Currently, there exist two major anthropogenic sources of this toxic metal—ammunition and non-ammunition-based lead. Ecotoxicological studies on Pb are usually performed on bones (from recently caught or dead animals or those stored in museums) and in the liver and kidneys (as organs responsible for the detoxification of the body). Many studies also analyze Pb levels in the meat of hunted animals, due to its possible effects on human health. In contrast, much less frequently studied are animal nervous system tissues, including the brain, even though the central nervous system is the target destination of Pb toxicity in warm-blooded vertebrates. This clearly smaller group of works (compared to studies on the bone, liver, and kidney) is partly due to the difficulty and time required to perform trepanation (Scheuhammer and Norris 1996; Hunt et al. 2009a; Lazarus et al. 2014; Legagneux et al. 2014).

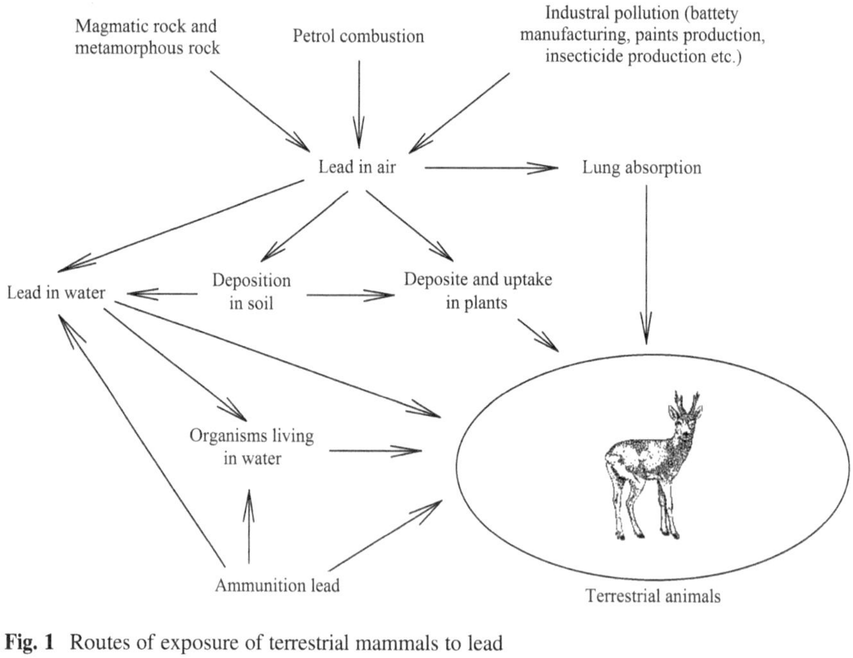


**Fig. 1** Routes of exposure of terrestrial mammals to lead

## 6.1 *Lead in Mammalian Tissues*

Animals intake lead primarily orally, with approximately 40% of lead being absorbed from the gastrointestinal tract into the bloodstream (Smith et al. 2008). Studies in rats, orally given lead acetate (PbAc), at 5 mg $kg^{-1}$ of body weight per day for 6 weeks, show that Pb accumulated in all tissues, although most abundantly in the long bones (Senapati et al. 2001). After 6 weeks of exposure, Pb increased five times from 15.52 to 80.2 mg $kg^{-1}$ dw. However, in the soft tissues, such as the liver and kidneys, the increase in Pb was much lower: three times in the liver (from 3.05 to 9.80 mg $kg^{-1}$ dw), more in the kidneys (from 4.97 to 19.1 mg $kg^{-1}$ dw), and considerably less in the whole blood (0.31–0.96 mg $kg^{-1}$ dw). In the brain there was a near twofold increase in Pb (from 2.89 to 5.09 mg $kg^{-1}$ dw) (Senapati et al. 2001), although Pb levels in different brain structures varied depending on the brain structure and exposure period (Klein and Koch 1981).

### 6.1.1 Industrial Activities of Man and Lead in Mammalian Tissues

Pb pollution is a major problem in industrialized countries, with the highest levels observed near mines, metal smelters, industrial areas, and large cities. Lead levels in bones and soft tissues of small mammals are a good indicator of Pb contamination in these areas. A UK study conducted in the 1990s at a site close to a lead mine showed that the concentration of this element in the femur of the bank vole (*Clethrionomys glareolus*) was more than 60 times higher than in the bones of the same species from not-polluted areas, 203.0 ± 13.0 vs. 3.2 ± 0.8 mg $kg^{-1}$ dw (Milton et al. 2003). In specimens living in the vicinity of the mine, muscle Pb was 36 times higher than in those trapped in uncontaminated areas (3.6 ± 0.6 vs. 0.10 ± 0.02 mg $kg^{-1}$ dw), 13 times higher in the liver, and 50 times higher in the kidneys.

Research on bone Pb in roe deer found near defunct lead mines in other European regions conducted in 2004–2006 (Sierra Madrona Mountains and the valley of Alcudia in Southern Spain) also showed a significantly higher content of Pb (averaging 3.53 ± 0.92 mg $kg^{-1}$ dw) than in roe deer living in non-polluted areas of southern Spain and forest areas in Europe in the 1970s (0.93 ± 0.10 mg $kg^{-1}$ dw on average) (Fig. 2) (Kierdorf and Kierdorf 2000; Reglero et al. 2008).

Comparative test results on Pb in the tissues of the European hare (*Lepus europaeus*) living in Finland also provided evidence of the relationship between human industrial activity and increased levels of Pb in the soft tissues of the animals (Venäläinen et al. 1996). In the early 1980s, liver Pb concentrations in the European hare living in the industrial areas of southern Finland (3.4 ± 2.9 mg $kg^{-1}$ dw) were significantly higher than in non-polluted areas of Finland where they did not exceed 1.60 mg $kg^{-1}$ dw (1.57 ± 0.47 mg $kg^{-1}$ dw) (Fig. 3).

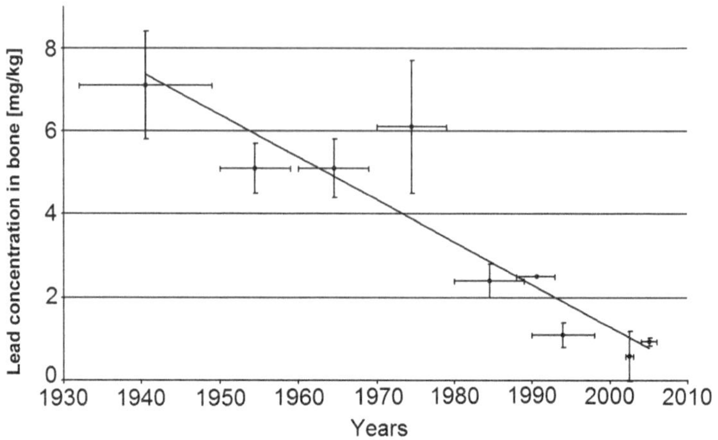


**Fig. 2** Lead levels in cervids 1930–2005

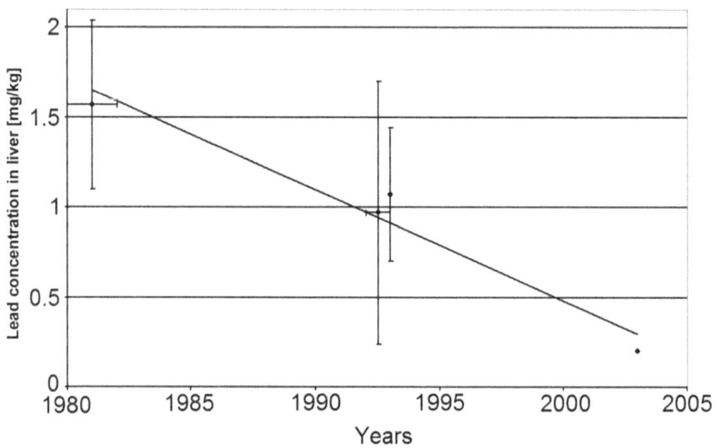


**Fig. 3** Lead levels in the liver of rabbits over the years 1980–2003

### 6.1.2 Combustion of Leaded Gasoline and Lead in Mammalian Tissues

Long-term use of gasoline with tetraethyl lead has strongly influenced the levels of Pb near busy roads, which is reflected in levels in the tissues of small mammals living in the vicinity. An example of this research is the bank vole, carried out in the UK in 1972. In the kidneys of the animals living in the nearby roadsides, Pb levels were several times higher (an average of 13.0 mg $kg^{-1}$ dw) than in specimens captured 0.5 mile from the road (an average of 5 mg $kg^{-1}$ dw) (Williamson and Evans 1972). Also, studies conducted in the United States on the tissues of the

Norway rat (*Rattus norvegicus*) caught near roads in the city have shown several times higher liver Pb concentrations than in specimens caught in suburban areas away from busy roads ($3.34 \pm 0.45$ vs. $0.44 \pm 0.09$ mg $kg^{-1}$ dw) (Mouw et al. 1975). Another study conducted in Texas in the late 1970s showed that in all the tissues of urban Norway rats, Pb concentrations were significantly elevated compared to individuals caught in non-contaminated suburban areas. In the bones of urban rats from Houston (USA), Pb concentrations were dozens of times higher than in individuals living in the outer suburbs, 146.0 vs. 8.8 mg $kg^{-1}$ dw (Way and Schroder 1982), and their liver and kidney Pb levels were about three times higher compared to rats from the outer suburbs. Research conducted in the early 1990s in Europe (Vienna, Austria) also showed elevated concentrations of Pb in the femur of the European mole (*Talpa europaea*) living near busy roads, averaging $29.8 \pm 11.6$ mg $kg^{-1}$ dw (Komarnicki 2000).

### 6.1.3 Ammunition and Fishing Sinkers as Sources of Lead Poisoning in Wildlife

At first, studies of exposure of wildlife to lead were mainly ascribed to industrial activities, such as metallurgy, the use of lead paint and gasoline with tetraethyl lead (Blus et al. 1999; Hernberg 2000; Finkelstein et al. 2003). More recent papers also mention contamination caused by ammunition and fishing sinkers because of their widespread recreational and subsistence use in wildlife habitats (Haig et al. 2014). Currently, larger mammals and birds are hunted to control the number of wild animals and as a hobby. In the northern hemisphere, in particular, this refers to deer, lagomorphs, wild boar, foxes, and birds such as pheasants, partridges, and ducks. Lead ammunition (bullets and pellets) often remains in the environment and pollute woodlands and meadows. This poses a direct threat to birds, which find and consume the lead pellets (Pain 1990). It is estimated that this causes Pb poisoning in approximately 4% of all large birds (Pain et al. 2015). Birds are also an important link in the food chain, which results in elevated levels of Pb in the bodies of carnivores (Pain et al. 2010; Rogers et al. 2012). Lead can also enter the body directly, if the hunted animal is not killed during a hunt and the ammunition remains in the body for a long time. Lead poisoning related to lead pellets was observed in 24% of bald eagles (*Haliaeetus leucocephalus*) in the state of Wyoming, USA (Bedrosian et al. 2012). In addition, in the state of Iowa (USA), more than 50% of bald eagles had high Pb levels in the body due to the presence of lead pellets in their bodies (Neumann 2009). Lead pellets and bullets remain in the flesh of many hunted animals (Hunt et al. 2009b; Pain et al. 2010). Consumption of that meat and offal by people increases the risk of increased blood Pb levels. According to the Regulation of the European Commission in 2006, the permissible Pb concentration in consumed meat must not exceed 0.10 and 0.50 mg $kg^{-1}$ ww in offal (Commission Regulation EC 2006).

Fishing results in the pollution of waterway beds due to lost lead sinkers. Small lead weights can be eaten by fish and shellfish feeding on the bottom, and thus Pb from this source becomes included in the trophic chain. Such weights may also be swallowed by waterfowl, including dabbling and diving ducks, swans, and other birds feeding at the bottom (Haig et al. 2014). As shown by a recent study, among the toxicologically significant sources, Pb-based paints, mining, metallurgy, leaded gasoline, lead pellet, and fishing sinker are primary exposure pathways for birds in terrestrial and aquatic systems (Beyer et al. 2013; Haig et al. 2014). For this reason, regulators worldwide have introduced proposals to ban lead ammunition and hardware in hunting and fishing (Haig et al. 2014).

### 6.1.4 Lead in the Muscles of Mammals

In the next section, we present literature data on the concentrations of lead in various organs of terrestrial mammals, including muscle (Table 3). To facilitate comparison, the presented data have been summarized in terms of animal species, organs, and the approximate year of sampling. Primarily, we cite literature data on the concentrations of lead in the muscles of mammals, an important component of the diet for carnivorous animals and humans. Therefore, this is an important indicator of exposure to lead for animals at the apex of the trophic chain, indicating the accumulation of lead in the entire food chain.

Studies on Pb in the muscles of various mammalian species indicate that Pb levels in the environment are not as high as 25 years ago, and in most of the non-contaminated areas they do not exceed 1 mg $kg^{-1}$ dw in the muscle of herbivorous or predatory mammals. Pb in the muscle of moose (*Alces alces*) inhabiting the forests of Canada, conducted in 2000, averaged 0.1 $\pm$ 0.3 mg $kg^{-1}$ dw (Gamberg et al. 2005). Muscle Pb in this species caught in the forests of Finland was very low and amounted to 0.066 mg $kg^{-1}$ dw (Venäläinen et al. 2005). In reindeer (*Rangifer tarandus*) caught in Norway in 2005, muscle Pb was 0.026 mg $kg^{-1}$ dw (Hassan et al. 2012). In red deer (*Cervus elaphus*) from the forests of Slovakia and Poland, investigated in the late 1990s and 2000s, muscle Pb was 0.30 $\pm$ 0.07 and 0.6 $\pm$ 1.1 mg $kg^{-1}$ dw, respectively (Kottferová and Koréneková 1998; Jarzyńska and Falandysz 2011). In the Arctic hare (*Lepus arcticus*) living in non-polluted forest areas of Canada, it was 0.011 mg $kg^{-1}$ dw (Pedersen and Lierhagen 2006). Muscle Pb in European hares (*Lepus europaeus*) harvested in the early 1980s in Finland averaged 0.43 mg $kg^{-1}$ dw, while in a study conducted in the early 1990s it had dropped to 0.166 mg $kg^{-1}$ dw, indicating a significant improvement in the environmental Pb contamination (Venäläinen et al. 1996). However, muscle Pb in European hares inhabiting non-polluted forests in Slovakia examined in the early 1990s were significantly higher with an average 1.2 $\pm$ 1.6 mg $kg^{-1}$ dw (Kottferová and Koréneková 1998).

Studies on carnivorous mammals show significantly more diverse concentrations of Pb, depending on the location and species. In a Croatian population of gray wolf (*Canis lupus*) hunted in 2009–2010, muscle Pb concentrations averaged

**Table 3** Arithmetic mean lead levels (mg Pb $kg^{-1}$ dw) in the muscles of various mammalian species

| Trophic group and species | Mean level | SD | Years | Country | References |
|---|---|---|---|---|---|
| Herbivores<br>Micromammals | | | | | |
| Bank vole *Clethrionomys glareolus* | 0.100 | 0.020 | 1995 | Britain | Milton et al. (2003) |
| Wood mouse *Apodemus sylvaticus* | 0.051 | 0.043 | 2001 | Spain | Torres et al. (2004) |
| Medium and large size | | | | | |
| European hare *Lepus europaeus* | 0.167 | | 1992–1993 | Finland | Venäläinen et al. (1996) |
| European hare *Lepus europaeus* | 1.12 | 1.20 | 1998–2001 | Poland | Mysłek and Kalisińska (2006) |
| European hare *Lepus europaeus* | 1.170 | 1.600 | 1993 | Slovakia | Kottferová and Koréneková (1998) |
| Arctic hare *Lepus arcticus* | 0.011 | | 2003 | Canada | Pedersen and Lierhagen (2006) |
| Red deer *Cervus elaphus* | 0.030 | 0.070 | 1993 | Slovakia | Kottferová and Koréneková, (1998) |
| Moose *Alces alces* | 0.100 | 0.300 | 1994–2001 | Canada | Gamberg et al. (2005) |
| Reindeer *Rangifer tarandus* | 0.033 | 0.013 | 1990–1991 | Finland | Rintala et al. (1995) |
| Omnivores | | | | | |
| Wild boar *Sus scrofa* | 0.410 | 0.090 | 2005–2006 | Italy | Danieli et al. (2012) |
| Insectivore | | | | | |
| Mouse-eared bat *Myotis myotis* | 0.670 | 0.030 | 2007 | Czech Rep. | Pikula et al. (2010) |
| Carnivores | | | | | |
| Brown bear *Ursus arctos* | 0.010 | | 2009–2010 | Croatia | Bilandžić et al. (2012) |
| Eurasian badger *Meles meles* | 0.260 | | 2009–2010 | Croatia | Bilandžić et al. (2012) |
| Eurasian lynx *Lynx lynx* | 0.017 | | 2009–2010 | Croatia | Bilandžić et al. (2012) |
| Gray wolf *Canis lupus* | 0.070 | | 2009–2010 | Croatia | Bilandžić et al. (2012) |
| Red fox *Vulpes vulpes* | 0.133 | 0.030 | 2010 | Italy | Naccari et al. (2013) |
| Pine marten *Martes martes* | 0.013 | | 2009–2010 | Croatia | Bilandžić et al. (2012) |
| Stone marten *Martes foina* | 0.100 | 0.260 | 2008–2009 | Croatia | Bilandžić et al. (2010) |

*SD* standard deviation

0.133 ± 0.830 mg kg$^{-1}$ dw (Bilandžić et al. 2010, 2012), like in foxes (*Vulpes vulpes*) from Italian forests examined at the same time, 0.133 ± 0.03 mg kg$^{-1}$ dw (Naccari et al. 2013).

In the muscles of large omnivorous mammals, e.g., brown bears (*Ursus arctos*), hunted in the forests of Croatia in 2010, Pb concentrations were very low at 0.01 mg kg$^{-1}$ dw (Bilandžić et al. 2012). In wild boar coming from forested areas in Italy at the same time, muscle Pb was 0.414 ± 0.093 mg kg$^{-1}$ dw (Danieli et al. 2012). In the muscles of smaller predators, such as the stone marten (*Martes foina*) caught in forested areas in Croatia in 2008–2009, Pb level was 0.10 ± 0.26 mg kg$^{-1}$ dw (Bilandžić et al. 2010). Although in a study conducted in Croatia, muscle Pb in Eurasian lynx (*Lynx lynx*) was much lower, at 0.0166 mg kg$^{-1}$ dw (Bilandžić et al. 2012), similar to pine marten (*Martes martes*) 0.013 mg kg$^{-1}$ dw (Bilandžić et al. 2012), Eurasian badgers (*Meles meles*) from Croatia studied in 2009–2010 had muscle Pb of 0.25 mg kg$^{-1}$ dw (Bilandžić et al. 2012).

In a recent study in the Czech Republic, muscle Pb in very small mammals, such as mouse-eared bat (*Myotis myotis*) and pipistrelle bats (*Pipistrellus pipistrellus*), were 0.67 ± 0.03 and 0.67 ± 0.07 mg kg$^{-1}$ dw, respectively (Pikula et al. 2010). In a research conducted in 2001, the content of Pb in the muscles of wood mice (*Apodemus sylvaticus*) living in non-polluted areas of Spain was much lower (0.051 ± 0.043 mg kg$^{-1}$ dw, Torres et al. 2004) than in mentioned insectivorous bats from polluted Czech Republic.

### 6.1.5 Lead in Mammalian Bones

Animals may periodically experience lead poisoning associated with the ingestion of lead shot or fishing sinkers or after being shot by hunters. Such periodic fluctuations in the concentration of lead in the blood are reflected in the concentration of lead in soft tissues. In contrast, an animal's liftetime exposure to lead can be shown by bone lead levels. Hence, data presented later in this section can be used for a relatively thorough analysis of environmental pollution by lead in the investigated area.

Over the last 10 years, research on the bones of herbivorous mammals, such as the red deer (*Cervus elaphus*) living in the forests of Spain and Croatia, showed Pb concentrations at 0.93 ± 0.10 and 0.58 ± 0.60 mg kg$^{-1}$ dw, respectively (Reglero et al. 2008; Lazarus et al. 2005). Deers belong to the most significant group of game animals and therefore are a good material for bioindicative research. Long-term studies show that Pb concentrations in the tissues of these animals continue to decline. This is primarily related to the prohibition on leaded gasoline and paints containing lead and also to a growing use of non-lead bullets and pellets. A long-term study on the bones of deer living in the forests bordering the eastern suburbs of Cologne (Germany) showed that in roe deer (*Capreolus capreolus*) hunted between 1932 and 1949 (including World War II), bone Pb (average of 7.1 ± 1.3 mg kg$^{-1}$ dw) was high compared to animals living in the same area but examined in the following decades. From 1950 to 1969 and 1970 to 1979, the average bone Pb levels were 6.1 ± 1.6 and 5.1 ± 0.6 mg kg$^{-1}$ dw, respectively (Kierdorf and Kierdorf 2000).

After the ban on the use of Pb in gasoline and paints in many European countries, by as early as the 1980s–1990s, bone Pb had fallen steadily in the deer. In the period 1980–1989 and in the early 1990s, bone Pb in cervids in Germany and the Netherlands had dropped by half compared to the 1970s, 2.4 ± 0.4 and 2.5 mg $kg^{-1}$ dw, respectively (Kuiters 1996; Kierdorf and Kierdorf 2000). At present, the concentration of Pb in the bones of roe deer inhabiting Germany and the Netherlands has fallen below 1.0 mg $kg^{-1}$ dw.

In carnivorous mammals Pb content in the bones is varied. In foxes (*Vulpes vulpes*) living in non-polluted regions of Poland in the late 2000s, it was 0.98 ± 1.15 mg $kg^{-1}$ dw (Lanocha et al. 2012). In the 1990s, bone Pb in the gray wolf (*Canis lupus*) from the forests of North America (Canada) was 0.40 ± 0.86 mg $kg^{-1}$ dw (Gamberg and Braune 1999). In the Dutch omnivorous wild boars (*Sus scrofa*) examined in the late 1980s and 1990s, bone Pb averaged 4.8 mg $kg^{-1}$ dw (Kuiters 1996), and this value is much higher than in mentioned carnivorous and herbivorous species.

### 6.1.6 Lead in the Mammalian Brain

Little is known about the subclinical signs, biochemical and behavioral changes and corresponding concentrations of brain Pb in animals (ATSDR 2007). In control mammalian groups used in laboratory experiments and small mammals from reference area in field studies, brain Pb levels generally ranged from <0.10 to 0.50 mg $kg^{-1}$ dw (Stowe et al. 1973; Mierau and Favara 1975; Yoo et al. 2002; Mari et al. 2014; Kalisinska et al. 2016). A concentration of <0.50 mg Pb $kg^{-1}$ dw in the brain can therefore be considered as reflecting the background level.

## *6.2 Lead in Tissues in Birds*

Birds are sensitive to Pb exposure, leading to apparent sublethal or lethal toxic responses, with exposure documented in 20 species (reviewed by Tranel and Kimmel 2009). Lead shot pellets or bullet fragments are commonly ingested by many avian species and may cause lethal poisoning and sublethal effects, with population level implication for conservation of some species (Bellinger et al. 2013).

After permeating the bird's body, Pb is transported by the blood to all tissues. A high concentration of this metal in the blood and soft tissues indicates a recent exposure to Pb prior to sampling. In contrast, increased bone Pb concentrations indicate chronic exposure or short-term exposure to high concentrations of lead. Importantly, bone Pb may be released to the blood due to bone remodeling. Nevertheless, this process takes a long time and bone Pb is generally assumed to reflect the average lifelong exposure. Levels greater than 5 mg $kg^{-1}$ in the bone are thought to indicate increased environmental pollution with lead (Orłowski et al. 2012). In Table 4 are gathered data on lead in urban pigeons.

**Table 4** Arithmetic mean lead levels (mg Pb $kg^{-1}$ dw) in the bones of pigeons living in urban areas

| Species | Mean level | SD | Years | Country | Notice | References |
|---|---|---|---|---|---|---|
| Pigeon | 42 | 21 | 1970s | Japan | Tokyo, central city | Ohi et al. (1974) |
| Pigeon | 2.67 | 4.00 | 1970s | Japan | Tokyo, suburban city | Ohi et al. (1974) |
| Feral pigeon *Columba livia f. domestica* | 40.9 | 25.6 | 1983–1985 | Slovakia | Bratislava | Janiga and Zemberyová (1998) |
| Feral pigeon | 29.9 | 10.4 | 1983–1985 | Slovakia | Bratislava | Janiga and Zemberyová (1998) |
| Feral pigeon | 39.3 | 28.1 | 2000–2001 | Korea | Seoul, urban area | Nam and Lee (2006a) |
| Feral pigeon | 14.0 | 6.3 | 2000–2001 | Korea | Ansan, industrial area | Nam and Lee (2006a) |
| Feral pigeon | 31.7 | 18.3 | 2000–2001 | Korea | Busan, industrial area | Nam and Lee (2006a) |
| Feral pigeon | 2.40 | 1.15 | 2000–2001 | Korea | Duckjuk Island, rural | Nam and Lee (2006a) |

*SD* standard deviation

In contrast, feathers are low in Pb compared to the soft tissues and bones (Ek et al. 2004). This is due to the fact that Pb is transported from the blood to the feathers only during the short period of their formation. As a result, Pb does not accumulate in the feathers. Eggs also contain little Pb (Ek et al. 2004). Similar to feathers, this is related to the fact that Pb penetrates in the short period of egg development in the breeding season. Bioindication research sometimes uses bird droppings, which have been shown to have 3× higher Pb levels than soft tissues (Ek et al. 2004).

### 6.2.1 The Impact of Environmental Pollution on Pb Concentration in the Avian Eggs

Eggshells are good bioindicative material due to the relative ease of sampling and the noninvasive character, as well as their ability to accumulate Pb and relative durability. This makes it possible to perform comparative studies of Pb pollution in the area over many years. Bird eggs are also an important object of environmental studies due to the strictly defined time and place of nesting. Pb from the blood of a female bird reaches the eggs during their development in the period preceding the laying of eggs in a nest. Therefore, the amount of Pb in eggs correlates with the level of Pb in the blood of birds during the breeding season. This, in turn, correlates with the content of Pb in the diet of the birds near the nesting site (Ek et al. 2004).

Due to their position in the food chain, birds of prey play an important part in research on the bioindication of environmental pollution with heavy metals. An example of this is the american kestrel (*Falco sparverius*), whose eggshells collected in the area of New York (USA) in the early 1970s showed average Pb levels of 0.908 ± 0.073 mg $kg^{-1}$ dw (Lincer and McDuffie 1974). In the eggshells of double-crested cormorant (*Phalacrocorax auritus*) in Minnesota (USA), Pb concentrations in the eggs were 0.128 ± 0.390 mg $kg^{-1}$ dw (Burger and Gochfeld 1996). However, in the eggs of osprey (*Pandion haliaetus*) collected in New Jersey, USA, concentrations of Pb in 1989 averaged 0.03 mg $kg^{-1}$ dw (Clark et al. 2001) and had increased in studies conducted 10 years later, to 0.87 mg $kg^{-1}$ dw (Clark et al. 2001).

Pb pollution in cities may also be indicated by the levels determined in the eggshells of birds that live permanently in cities, e.g., doves. In Seoul and Ansan (Korea), eggshells of doves (*Columba livia*) showed very high Pb levels. In Seoul, eggshell Pb in the early twenty-first century averaged 4.8 ± 0.6 mg $kg^{-1}$ dw (Nam and Lee 2006b), while in Ansan it was 3.3 ± 0.3 mg $kg^{-1}$ dw. In a study conducted at the same time in Europe, in a highly industrialized area in Antwerp (Belgium), the eggshells and egg contents of the great tit (*Parus major*) also showed high concentrations of Pb, 1.505 ± 0.597 and 2.18 ± 0.72 mg $kg^{-1}$ dw, respectively (Dauwe et al. 2005).

### 6.2.2 Lead in Bird Feathers

Feathers are an important bioindicative material because of the possibility of intravital sampling, which is important in the study of birds threatened with extinction. Due to the high content of keratin, rich in sulfur amino acids, feathers easily bind Pb, which has a high affinity for –SH groups in the protein. Although air pollution does not directly affect Pb levels in feathers, lead dust can cover the outer surface of feathers. Pb is incorporated in the feathers via the blood during their formation (Lodenius and Solonen 2013). Therefore, feathers are a valuable indicator of Pb contamination in the breeding and wintering areas, and this biological material is often used in ecotoxicology (Table 5). It is understood that a Pb concentration greater than 4 mg $kg^{-1}$ dw indicates sublethal poisoning, which negatively affects the reproductive functions of birds (Norouzi et al. 2012).

Due to the accumulation of Pb in the food chain, research on birds of prey gives valuable information about lead contamination over large areas. Studies conducted in the late 1990s in common buzzards (*Buteo buteo*) from non-contaminated areas of Italy showed feather Pb to average 1.48 mg $kg^{-1}$ dw (Battaglia et al. 2005). In southeastern Spain, common buzzard feathers from environmentally clean areas had lower Pb levels, averaging 1.01 ± 0.20 mg $kg^{-1}$ dw (Martínez-López et al. 2004). Feather Pb in the booted eagle (*Hieraaetus pennatus*), also from southeastern Spain, and the Spanish imperial eagle (*Aquila adalberti*) from the south-central part of the country caught in recent years (Martínez-López et al. 2004; Rodriguez-Ramos Fernandez et al. 2011) were 0.72 ± 0.31 and 0.043 ± 0.078 mg $kg^{-1}$ dw. In northern goshawk (*Accipiter gentilis*), caught in the same area of southeastern Spain, feather

**Table 5** Arithmetic mean lead levels (mg Pb $kg^{-1}$ dw) in the feathers of diurnal and nocturnal avian predators

| Species | Sex | Mean level | SD | Years | Country | References |
|---|---|---|---|---|---|---|
| Common buzzard *Buteo buteo* | | 1.48 | | 1998–1999 | Italy | Battaglia et al. (2005) |
| Common buzzard | | 1.01 | 0.20 | 1999–2000 | Spain | Martínez-López et al. (2004) |
| Common buzzard | M | 0.95 | 1.00 | 1997–2007 | Spain | Castro et al. (2011) |
| | F | 1.75 | 1.19 | 1997–2008 | Spain | Castro et al. (2011) |
| Booted eagle *Hieraaetus pennatus* | | 0.72 | 0.31 | 1999–2000 | Spain | Martínez-López et al. (2004) |
| Spanish imperial eagle *Aquila adalberti* | | 0.043 | 0.078 | 1997–2008 | Spain | Rodriguez-Ramos Fernandez et al. (2011) |
| Golden eagle *Aquila chrysaetos* | | 0.68 | 0.09 | 2008–2010 | USA, Montana | Harmata and Restani (2013) |
| Northern goshawk *Accipiter gentilis* | | 0.98 | 0.29 | 1999–2000 | Spain | Martínez-López et al. (2004) |
| Northern goshawk | M | 1.55 | 0.77 | 1997–2005 | Spain | Castro et al. (2011) |
| | F | 2.24 | 1.86 | 1997–2006 | Spain | |
| Osprey *Pandion haliaetus* | | 0.80 | 2.20 | 2000–2001 | USA, Florida | Lounsbury-Billie et al. (2008) |
| Tawny owl *Strix aluco* | | 0.38<br>0.024 | | 1986<br>2005 | Norway | Bustnes et al. (2013) |
| Little owl *Athene noctua* | | 2 | | 1998–1999 | Italy | Battaglia et al. (2005) |
| Brown owl *Strix aluco* | M<br>F | 1.84<br>3.47 | 8.38<br>14.45 | 1997–2009<br>1997–2010 | Spain | Castro et al. (2011) |

*SD* standard deviation, *F* female, *M* male

Pb was 0.98 ± 0.29 mg $kg^{-1}$ dw. In North America (USA, Montana), feather Pb in golden eagles (*Aquila chrysaetos*) from ecologically clean areas was 0.68 ± 0.09 mg $kg^{-1}$ dw (Harmata and Restani 2013). Feather Pb in osprey from the cleaner areas of Florida Bay (USA) did not differ from the content in feathers of other birds of prey, at 0.80 ± 2.20 mg $kg^{-1}$ dw (Lounsbury-Billie et al. 2008).

Feather Pb levels in tawny owls (*Strix aluco*) have shown that Pb contamination in Norway has significantly reduced over the last 25 years. In 1986, feather Pb in this species was 0.36 mg $kg^{-1}$ dw, while in 2005 it was only 0.024 mg $kg^{-1}$ dw (Bustnes et al. 2013). In contrast, recent studies on feather Pb levels in the little owl (*Athene noctua*) living in the woods in northern Italy point to a persistent contamination of the environment with lead, showing levels as high as 2 mg $kg^{-1}$ dw (Battaglia et al. 2005).

In lead biomonitoring of urban ecosystems, feathers of some species are used. Due to the constant presence of pigeons in large cities, these birds are an important indicator of urban pollution by heavy metals. Research on pigeon feathers shows very high pollution in urban areas and indicates a high tolerance of this species to Pb poisoning. In studies conducted in recent years in Greater Paris (France), Pb concentrations in the feathers of feral pigeons were as high as 13.82 mg $kg^{-1}$ dw (Frantz et al. 2012). In Iraq, in the Hormod Protected Area, feather Pb in the rock dove was 7.7 mg $kg^{-1}$ dw (Norouzi et al. 2012). Similarly, high concentrations of Pb were found in the feathers of feral pigeons living in the cities of Seoul and Ansan, Korea, 4.99 $\pm$ 1.21 and 9.01 $\pm$ 3.00 mg $kg^{-1}$ dw (Nam et al. 2004). In comparison, in the late 1990s the concentration of Pb in the feathers of the mourning dove (*Zenaida macroura*) from the nuclear reservation Savannah River Site in Par Pond, South Carolina, USA, were much lower, at 0.466 $\pm$ 0.132 mg $kg^{-1}$ dw (Burger et al. 1997).

### 6.2.3 Lead in the Liver of Birds

The liver is involved in many physiological functions, including blood plasma protein synthesis or metabolism of glucose. Therefore, studies on the concentrations of harmful substances in this organ allow to assess the health status of birds in areas contaminated with lead. It is understood that liver Pb $<$ 2 mg $kg^{-1}$ dw is the concentration resulting from the natural occurrence of Pb in nature (Kim and Oh 2013), not affecting its normal functions. The range of 6.0–30.0 mg $kg^{-1}$ dw is considered toxic, disturbing normal function of the body and indicating lead poisoning. The concentration of Pb in the liver of birds above 30.0 mg $kg^{-1}$ dw reflects acute poisoning which leads to death.

The following literature data show the concentration of lead in the liver of birds of prey, situated top of the food chain and therefore at a higher risk of Pb accumulation in the liver than other birds. Because of the rapid elimination of lead from soft tissues, the collected data show only the general condition of the biota in a specific area within a specified time. Results vary significantly due to the randomness of serious lead poisoning associated with swallowing lead pellets or ingestion of a shot animal.

According to a recent study, in the white-tailed eagle (*Haliaeetus albicilla*) from Korea, liver Pb was 4 mg $kg^{-1}$ dw (Nam and Lee 2011). The moderate pollution with Pb in the south Baltic coast in Poland is indicated by liver Pb levels in the white-tailed eagles averaging 1.10 $\pm$ 0.60 mg $kg^{-1}$ dw (Falandysz et al. 2001).

Studies on another group of predators, owls (*Strigiformes*), living in industrial areas in Korea, confirmed the contamination of those areas with Pb. Liver Pb levels of the Eurasian eagle-owl (*Bubo bubo*) averaged 3.6 mg kg$^{-1}$ dw (Nam and Lee 2011). In Europe, increased levels of pollution with Pb were recorded in northern Italy, and accordingly, liver Pb in little owls from Bologna and Parma Pb were 0.667 ± 0.037 and 0.773 ± 1.267 mg kg$^{-1}$ dw (Zaccaroni et al. 2003). In Galicia (Spain) Pb concentrations in the liver of the barn owl (*Tyto alba*) and the little owl were 3.12 ± 2.21 and 4.00 ± 0.26 mg kg$^{-1}$ dw, respectively (Pérez-López et al. 2008). In the long-eared owl (*Asio otus*) and tawny owl, respectively, liver Pb levels were 4.09 ± 2.56 and 2.75 ± 2.65 mg kg$^{-1}$ dw (Pérez-López et al. 2008). In Sicily, southern Italy, the livers of common buzzard had very high concentrations of Pb, on average 48.4 ± 12.5 mg kg$^{-1}$ dw (Naccari et al. 2009).

Passerine birds commonly occur on all continents of the northern hemisphere and constitute a valuable reference for lead pollution. Due to the similar location of passerines in the food chain, the literature data presented below allow to compare the state of the environment between the selected areas. The levels of this heavy metal in the liver of these birds in non-contaminated areas are less than 1 mg kg$^{-1}$ dw. An example of this is the recent study of the northern cardinal population (*Cardinalis cardinalis*) from the state of Missouri in the United States, in which the level of Pb in the liver equaled 0.45 ± 0.16 mg kg$^{-1}$ dw (Beyer et al. 2013). In the hooded crow (*Corvus cornix*) from the province of Cuneo in Italy, liver Pb was 0.3 ± 0.9 mg kg$^{-1}$ dw (Giammarino et al. 2014). In the great tit from Belgium, Pb levels in the liver were also low and averaged 0.81 mg kg$^{-1}$ dw (Dauwe et al. 2005).

Liver Pb levels in industrial areas related to the extraction of metals, can significantly exceed 1 mg kg$^{-1}$ dw (Table 6). In the vicinity of Southeast Missouri (USA) Lead Mining District, liver Pb levels in northern cardinal (*Cardinalis cardinalis*) and the American robin (*Turdus migratorius*) were 10.5 ± 2.3 and 33.6 ± 10.0 mg kg$^{-1}$ dw, respectively (Beyer et al. 2013). In China, in the province of Beijing in the vicinity of the Capital Steel Company, Pb level in the Eurasian tree sparrow (*Passer montanus*) was 10.68 ± 5.27 mg kg$^{-1}$ dw in females and 13.26 ± 1.67 mg kg$^{-1}$ dw in males (Pan et al. 2008). Some areas of India are also highly polluted with lead, and in the industrial Nilgiris District, Pb concentrations in the liver of jungle babbler (*Turdoides striatus*) and common myna (*Acridotheres tristis*) were as high as 14.77 ± 4.93 and 11.233 ± 0.867 mg kg$^{-1}$ dw, respectively (Jayakumar and Muralidharan 2011). Some areas of Europe, which are not associated with the ore mining, can also be heavily contaminated with lead. As demonstrated in a study on rural areas in northeastern Poland, in rook chicks (*Corvus frugilegus*) Pb levels in the liver were as high as 5.0 ± 0.8 mg kg$^{-1}$ dw (Orłowski et al. 2012).

**Table 6** Arithmetic mean lead levels (mg Pb $kg^{-1}$ dw) in passerine livers

| Species | Mean level | SD | Years | Place | Notice | References |
|---|---|---|---|---|---|---|
| American robin *Turdus migratorius* | 3.41 | 1.10 | 2009–2010 | USA, Missouri | | Beyer et al. (2013) |
| American robin *Turdus migratorius* | 33.6 | 10.0 | 2009–2010 | USA, Missouri | Mining area | Beyer et al. (2013) |
| Common myna *Acridotheres tristis* | 1.23 | 0.87 | 1998–1999 | India | | Jayakumar and Muralidharan (2011) |
| Eurasian tree sparrow *Passer montanus* | 7.25 | 8.90 | 2002 | China, Beijing | Adult | Pan et al. (2008) |
| Great tit *Parus major* | 0.81 | | 2000 | Belgium | Female | Dauwe et al. (2005) |
| Great tit *Parus major* | 0.64 | 0.15 | 2004 | China, Beijing | | Deng et al. (2007) |
| Greenfinch *Chloris chloris* | 0.45 | 0.06 | 2004 | China, Beijing | | Deng et al. (2007) |
| Hooded crow *Corvus cornix* | 0.30 | 0.90 | 2005–2006 | Italy | | Giammarino et al. (2014) |
| Jungle babbler *Turdoides striatus* | 14.77 | 4.93 | 1998–1999 | India | | Jayakumar and Muralidharan (2011) |
| Northern cardinal *Cardinalis cardinalis* | 0.45 | 0.16 | 2009–2010 | USA, Missouri | | Beyer et al. (2013) |
| Northern cardinal *Cardinalis cardinalis* | 10.50 | 2.33 | 2009–2010 | USA, Missouri | Mining area | Beyer et al. (2013) |
| Northern cardinal *Cardinalis cardinalis* | 8.40 | | 1999 | USA, Georgia | Firearms training facility | Lewis et al. (2001) |
| Rook *Corvus frugilegus* | 5.0 | 0.8 | 2005 | Poland | | Orłowski et al. (2012) |

*SD* standard deviation, *F* female

# 7 Conclusions

Over the last several decades, there has been increased awareness of the dangers to living organisms and ecosystems from lead pollution. Gradual understanding of the mechanisms of lead toxicity has helped to reduce its use. These actions have been taken largely in economically developed countries, yet the global anthropogenic emissions of lead into the atmosphere still remains at a high level. Bioindication research helps to determine not only the history and the present state of lead pollution but also allows an understanding of the phenomenon of the circulation of lead in nature, as well as in prediction of the risk of its presence in the trophic chain.

Long-term bioindication research conducted on the bones and soft tissues of mammals and birds, as well as the eggs and feathers, indicates the usefulness of this type of material for evaluating the state of the environment.

# References

Adriano DC (2001) Trace elements in terrestrial environments: biogeochemistry, bioavailability and risks of metals, 2nd edn. Springer, New York

Alvarenga P, Simões I, Palma P, Amaral O, Matos JX (2014) Field study on the accumulation of trace elements by vegetables produced in the vicinity of abandoned pyrite mines. Sci Total Environ 470–471:1233–1242

ATSDR (2007) Toxicological profile for lead. Agency for Toxic Substances and Disease Registry, Department of Health and Human Services, Atlanta, GA

ATSDR (2013) Substance priority list. Agency for Toxic Substances and Disease Registry. Available: http://www.atsdr.cdc.gov/SPL/index.html

Baranowska-Bosiacka I, Chlubek D (2006) Biochemical mechanisms of neurotoxic lead activity. Posteps Biochem 52:320–329

Baranowska-Bosiacka I, Pieńkowski P, Bosiacka B (2001) Content and localisation of heavy metals in thalli of hemerophilous lichens. Pol J Environ Stud 10:213–216

Baranowska-Bosiacka I, Gutowska I, Marchetti C, Rutkowska M, Marchlewicz M, Kolasa A et al (2011) Altered energy status of primary cerebellar granule neuronal cultures from rats exposed to lead in the pre- and neonatal period. Toxicology 280:24–32

Battaglia A, Ghidini S, Campanini G, Spaggiari R (2005) Heavy metal contamination in little owl (*Athene noctua*) and common buzzard (*Buteo buteo*) from northern Italy. Ecotoxicol Environ Saf 60:61–66

Bedrosian B, Craighead D, Crandall R (2012) Lead exposure in bald eagles from big game hunting, the continental implications and successful mitigation efforts. PLoS One 7:e51978

Bellinger DCJ, Burger TJ, Cade DA, Cory-Slechta M, Finkelstein H, Hu M et al (2013) Health risks from lead-based ammunition in the environment. Environ Health Perspect 121:A178–A179

Beyer WN, Franson JC, French JB, May T, Rattner BA, Shearn-Bochsler VI et al (2013) Toxic exposure of songbirds to lead in the Southeast Missouri Lead Mining District. Arch Environ Contam Toxicol 65:598–610

Bilandžić N, Dežđek D, Sedak M, Dokić M, Solomun B, Varenina I et al (2010) Concentrations of trace elements in tissues of red fox (*Vulpes vulpes*) and stone marten (*Martes foina*) from suburban and rural areas in Croatia. Bull Environ Contam Toxicol 85:486–491

Bilandžić N, Dežđek D, Sedak M, Dokić M, Simić B, Rudan N et al (2012) Trace elements in tissues of wild carnivores and omnivores in Croatia. Bull Environ Contam Toxicol 88:94–99

Blus LJ, Henny CJ, Hoffman DJ, Sileo L, Audet DJ (1999) Persistence of high lead concentrations and associated effects in tundra swans captured near a mining and smelting complex in northern Idaho. Ecotoxicology 8:125–132

Burger J, Gochfeld M (1996) Heavy metal and selenium levels in birds at Agassiz National Wildlife Refuge, Minnesota: food chain differences. Environ Monit Assess 43:267–282

Burger J, Kennamer RA, Brisbin IL Jr, Gochfeld M (1997) Metal levels in mourning doves from South Carolina: potential hazards to doves and hunters. Environ Res 75:173–186

Bustnes JO, Bårdsen BJ, Bangjord G, Lierhagen S, Yoccoz NG (2013) Temporal trends (1986-2005) of essential and non-essential elements in a terrestrial raptor in northern Europe. Sci Total Environ 458-460:101–106

Castro I, Aboal JR, Fernández JA, Carballeira A (2011) Use of raptors for biomonitoring of heavy metals: gender, age and tissue selection. Bull Environ Contam Toxicol 86:347–351

CDC (2004) United States Department of Health and Human Services, Atlanta. Available: http://www.cdc.gov/nceh/lead/ACCLPP/meetingMinutes/lessThan10MtgMAR04.pdf

CDC (2005) National Center for Environmental Health. Preventing lead poisoning in young children, Atlanta

CDC (2007) Interpreting and managing blood lead levels <10 μg/dL in children and reducing childhood exposures to lead: recommendations of CDC's Advisory Committee on Childhood Lead Poisoning Prevention. Centers for Disease Control and Prevention. MMWR Recomm Rep 56:1–16 Available: http://www.cdc.gov/mmwr/preview/mmwrhtml/rr5608a1.htm

CDC (2012) Report of the Advisory Committee on Childhood Lead Poisoning Prevention of the Centers for Disease Control and Prevention. Low level lead exposure harms children: a renewed call for primary prevention. Available: http://www.cdc.gov/nceh/lead/acclpp/final_document_030712.pdf

Clark KE, Stansley W, Niles LJ (2001) Changes in contaminant levels in New Jersey osprey eggs and prey, 1989 to 1998. Arch Environ Contam Toxicol 40:277–284

Commission Regulation EC (2006) Commission Regulation (EC) No 1881/2006 of 19 December 2006 setting maximum levels for certain contaminants in foodstuffs. Off J EU L, 364/5

Conti MI, Bozzini C, Facorro GB, Lee CM, Mandalunis PM, Piehl LL et al (2012a) Lead bone toxicity in growing rats exposed to chronic intermittent hypoxia. Bull Environ Contam Toxicol 89:693–698

Conti MI, Terrizzi AR, Lee CM, Mandalunis PM, Bozzini C, Piñeiro AE et al (2012b) Effects of lead exposure on growth and bone biology in growing rats exposed to simulated high altitude. Bull Environ Contam Toxicol 88:1033–1037

Danieli PP, Serrani F, Primi R, Ponzetta MP, Ronchi B, Amici A (2012) Cadmium, lead, and chromium in large game: a local-scale exposure assessment for hunters consuming meat and liver of wild boar. Arch Environ Contam Toxicol 63:612–627

Dauwe T, Janssens E, Bervoets L, Blust R, Eens M (2005) Heavy-metal concentrations in female laying great tits (*Parus major*) and their clutches. Arch Environ Contam Toxicol 49:249–256

Deng H, Zhang Z, Chang C, Wang Y (2007) Trace metal concentration in great tit (*Parus major*) and greenfinch (*Carduelis sinica*) at the Western Mountains of Beijing, China. Environ Pollut 148:620–626

Directive EP (1998) Directive 98/70/EC of the European Parliament and of the Council of 13 October 1998 relating to the quality of petrol and diesel fuels and amending Council Directive 93/12/EEC. Off J EC L 350/58

Dooyema CA, Neri A, Lo YC, Durant J, Dargan PI, Swarthout T et al (2012) Outbreak of fatal childhood lead poisoning related to artisanal gold mining in northwestern Nigeria. Environ. Health Perspect 120:601–607

EFSA (2010) Scientific opinion on lead in food. European Food Safety Authority, EFSA. EFSA J 8 (4):1570 http://www.efsa.europa.eu/sites/default/files/scientific_output/files/main_documents/1570.pdf

Ek KH, Morrison GM, Lindberg P, Rauch S (2004) Comparative tissue distribution of metals in birds in Sweden using ICP-MS and laser ablation ICP-MS. Arch Environ Contam Toxicol 47:259–269

European Commission (2008) Opinion of the TC NES on the Industry Voluntary Risk Assessments (Human Health) on Lead and Lead compounds. European Commission, Directorate-General JRC, Institute for Health and Consumer Protection Toxicology and Chemical Substances (& ECB), pp 7. http://echa.europa.eu/doc/trd_substances/VRAR/Lead/tcnes_opinion/tcnes_opinion_env.pdf

Falandysz J, Ichihashi H, Szymczyk K, Yamasaki S, Mizera T (2001) Metallic elements and metal poisoning among white-tailed sea eagles from the Baltic south coast. Mar Pollut Bull 42:1190–1193

Finkelstein ME, Gwiazda RH, Smith DR (2003) Lead poisoning of seabirds: environmental risks from leaded paint at a decommissioned military base. Environ Sci Technol 37(15):3256–3260

Frantz A, Pottier MA, Karimi B, Corbel H, Aubry E, Haussy C et al (2012) Contrasting levels of heavy metals in the feathers of urban pigeons from close habitats suggest limited movements at a restricted scale. Environ Pollut 168:23–28

Gale NH, Stos-Gale ZA (1981) Cycladic lead and silver metallurgy. ABSA 76:169–224

Gamberg M, Braune BM (1999) Contaminant residue levels in arctic wolves (*Canis lupus*) from the Yukon Territory, Canada. Sci Total Environ 243–244:329–338

Gamberg M, Palmer M, Roach P (2005) Temporal and geographic trends in trace element concentrations in moose from Yukon, Canada. Sci Total Environ 351–352:5300–5308

Giammarino M, Quatto P, Squadrone S, Abete MC (2014) The hooded crow (*Corvus cornix*) as an environmental bioindicator species of heavy metal contamination. Bull Environ Contam Toxicol 93:410–416

Giel-Pietraszuk M, Hybza K, Chełchowska M, Barciszewski J (2012) Mechanizmy toksycznośc ołowiu. Post Biol Kom 39:217–248

Haig SM, D'Elia J, Eagles-Smith C, Fair JM, Gervais J, Herring G, Rivers JW, Schulz JH (2014) The persistent problem of lead poisoning in birds from ammunition and fishing tackle. Condor 116(3):408–428

Harmata AR, Restani M (2013) Lead, mercury, selenium, and other trace elements in tissues of golden eagles from southwestern Montana, USA. J Wildl Dis 49:114–124

Hassan AA, Rylander C, Brustad M, Sandanger TM (2012) Level of selected toxic elements in meat, liver, tallow and bone marrow of young semi-domesticated reindeer (*Rangifer tarandus tarandus* L.) from Northern Norway. Int J Circumpolar Health 71:1–7

Hayes CR, Skubala ND (2009) Is there still a problem with lead in drinking water in the European Union? J Water Health 7:569–580

Henry JR (2000) National Network of Environmental Management Studies (NNEMS). Fellow May–August 2000. An overview of the phytoremediation of lead and mercury. Prepared For U.S. Environmental Protection Agency Office of Solid Waste and Emergency Response Technology Innovation Office Washington, DC. http://clu-in.org

Hernberg S (2000) Lead poisoning in a historical perspective. Am J Ind Med 38(3):244–254

Hettiarachchi GM, Pierzynski GM (2004) Soil lead bioavailability and in situ remediation of lead-contaminated soils: a review. Environ Prog Sustain Energy 23:78–93

Hunt WG, Parish CN, Orr K, Aguilar RF (2009a) Lead poisoning and the reintroduction of the California condor in northern Arizona. J Avian Med Surg 23:145–150

Hunt WG, Watson RT, Oaks JL, Parish CN, Burnham KK, Tucker RL et al (2009b) Lead bullet fragments in venison from rifle-killed deer: potential for human dietary exposure. PLoS One 4: e5330

IARC (2016) International Agency for Research on Cancer, IARC, Definition and list of compounds. Agents classified by the IARC Monographs, vols 1–112. http://monographs.iarc.fr/ENG/Classification/latest_classif.php

Janiga M, Zemberyová M (1998) Lead concentration in the bones of the feral pigeons (*Columba livia*): sources of variation relating to body condition and death. Arch Environ Contam Toxicol 35:70–74

Jarzyńska G, Falandysz J (2011) Selenium and 17 other largely essential and toxic metals in muscle and organ meats of ted deer (*Cervus elaphus*)—consequences to human health. Environ Int 37:882–888

Jayakumar R, Muralidharan S (2011) Metal contamination in select species of birds in Nilgiris district, Tamil Nadu, India. Bull Environ Contam Toxicol 87:166–170

Kabata-Pendias A, Pendias H (1999) Biogeochemistry of trace elements. Warszawa PWN (in Polish)

Kalisinska E, Lanocha-Arendarczyk N, Kosik-Bogacka D, Budis H, Podlasinska J, Popiolek M, Pirog A, Jedrzejewska E, Freedman JH (2016) Brains of native and alien mesocarnivores in biomonitoring of toxic metals in Europe. PLoS One 11(8):e0159935

Kierdorf H, Kierdorf U (2000) Roe deer antlers as monitoring units for assessing temporal changes in environmental pollution by fluoride and lead in a German forest area over a 67-year period. Arch Environ Contam Toxicol 39:1–6

Kim J, Oh JM (2013) Assessment of trace metals in four bird species from Korea. Environ Monit Assess 185:6847–6854

Klein AW, Koch TR (1981) Lead accumulations in brain, blood, and liver after low dosing of neonatal rats. Arch Toxicol 47:257–262

Koller K, Brown T, Spurgeon A, Levy L (2004) Recent developments in low-level lead exposure and intellectual impairment in children. Environ Health Perspect 112:987–994

Komarnicki GJ (2000) Tissue, sex and age specific accumulation of heavy metals (Zn, Cu, Pb, Cd) by populations of the mole (*Talpa europaea* L.) in a central urban area. Chemosphere 41:1593–1602

Kottferová J, Koréneková B (1998) Distribution of Cd and Pb in the tissues and organs of free-living animals in the territory of Slovakia. Bull Environ Contam Toxicol 60:171–176

Kuiters AT (1996) Accumulation of cadmium and lead in red deer and wild boar at the Veluwe, The Netherlands. Vet Q 18(Suppl 3):S134–S135

Lanocha N, Kalisinska E, Kosik-Bogacka DI, Budis H, Noga-Deren K (2012) Trace metals and micronutrients in bone tissues of the red fox *Vulpes vulpes* (L., 1758). Acta Theriol 57:233–244

Lazarus M, Vicković I, Sostarić B, Blanusai M (2005) Heavy metal levels in tissues of red deer (*Cervus elaphus*) from Eastern Croatia. Arh Hig Rada Toksikol 56:233–240

Lazarus M, Prevendar Crnić A, Bilandžić N, Kusak J, Reljić S (2014) Cadmium, lead, and mercury exposure assessment among Croatian consumers of free-living game. Arh Hig Rada Toksikol 65:281–292

Legagneux P, Suffice P, Messier JS, Lelievre F, Tremblay JA, Maisonneuve C et al (2014) High risk of lead contamination for scavengers in an area with high moose hunting success. PLoS One 9:e111546

Lessler MA (1988) Lead and lead poisoning from antiquity to modern times. Ohio J Sci 88:78–84

Lewis LA, Poppenga RJ, Davidson WR, Fischer JR, Morgan KA (2001) Lead toxicosis and trace element levels in wild birds and mammals at a firearms training facility. Arch Environ Contam Toxicol 41:208–214

Lidsky TI, Schneider JS (2003) Lead neurotoxicity in children: basic mechanisms and clinical correlates. Brain 126:5–19

Lincer JL, McDuffie B (1974) Heavy metal residues in the eggs of wild American kestrels (*Falco sparverius* linn). Bull Environ Contam Toxicol 12:227–232

Lodenius M, Solonen T (2013) The use of feathers of birds of prey as indicators of metal pollution. Ecotoxicology 22:1319–1334

Lounsbury-Billie MJ, Rand GM, Cai Y, Bass OL Jr (2008) Metal concentrations in osprey (*Pandion haliaetus*) populations in the Florida Bay estuary. Ecotoxicology 17:616–622

Marcus J, McBratney A (2011) A review of the contamination of soil with lead II. Spatial distribution and risk assessment of soil lead. Environ Int 27:399–411

Mari M, Nadal M, Schuhmacher M, Barbería E, García F, Domingo JL (2014) Human exposure to metals: levels in autopsy tissues of individuals living near a hazardous waste incinerator. Biol Trace Elem Res 159:15–21

Martínez-López E, Martínez JE, María-Mojica P, Peñalver J, Pulido M, Calvo JF et al (2004) Lead in feathers and delta-aminolevulinic acid dehydratase activity in three raptor species from an unpolluted Mediterranean forest (Southeastern Spain). Arch Environ Contam Toxicol 47:270–275

Mccabe MJ Jr, Singh KP, Reiners JJ Jr (2001) Low level lead exposure in vitro stimulates the proliferation and expansion of alloantigen-reactive CD4 (high) T cells. Toxicol Appl Pharmacol 177:219–231

Mierau GW, Favara BE (1975) Lead poisoning in roadside populations of deer mice. Environ Pollut 8:55–64

Milton A, Cooke JA, Johnson MS (2003) Accumulation of lead, zinc, and cadmium in a wild population of Clethrionomys glareolus from an abandoned lead mine. Arch Environ Contam Toxicol 44:405–411

Moszynski P (2010) Lead poisoning in Nigeria causes "unprecedented" emergency. BMJ 341: c4031

Mouw D, Kalitis K, Anver M, Schwartz J, Constan A, Hartung R et al (1975) Lead. Possible toxicity in urban vs rural rats. Arch Environ Health 30:276–280

Mysłek P, Kalisińska E (2006) Contents of selected heavy metals in the liver, kidneys and abdominal muscle of the brown hare (Lepus europaeus Pallas, 1778) in Central Pomerania, Poland. Pol J Vet Sci 9:31–41

Naccari C, Cristani M, Cimino F, Arcoraci T, Trombetta D (2009) Common buzzards (*Buteo buteo*) bio-indicators of heavy metals pollution in Sicily (Italy). Environ Int 35:594–598

Naccari C, Giangrosso G, Macaluso A, Billone E, Cicero A, D'Ascenzi C et al (2013) Red foxes (*Vulpes vulpes*) bioindicator of lead and copper pollution in Sicily (Italy). Ecotoxicol Environ Saf 90:41–45

Nam DH, Lee DP (2006a) Monitoring for Pb and Cd pollution using feral pigeons in rural, urban, and industrial environments of Korea. Sci Total Environ 357:288–295

Nam DH, Lee DP (2006b) Reproductive effects of heavy metal accumulation on breeding feral pigeons (*Columba livia*). Sci Total Environ 366:682–687

Nam DH, Lee DP (2011) Mortality factors and lead contamination of wild birds from Korea. Environ Monit Assess 178:161–169

Nam DH, Lee DP, Koo TH (2004) Monitoring for lead pollution using feathers of feral pigeons (*Columba livia*) from Korea. Environ Monit Assess 95:13–22

Nelson Beyer W, Meador JP (eds) (2011) Environmental contaminants in biota. Interpreting tissue concentrations, 2nd edn. CRC, Boca Raton, FL

Neumann K (2009) Bald eagle lead poisoning in winter. In: Watson RT, Fuller M, Pokras M, Hunt WG (eds) Ingestion of lead from spent ammunition: implications for wildlife and humans. The Peregrine Fund, Boise, ID, pp 210–218

Norouzi M, Mansouri B, Hamidian AH, Ebrahimi T, Kardoni F (2012) Comparison of the metal concentrations in the feathers of three bird species from southern Iran. Bull Environ Contam Toxicol 89:1082–1086

Nriagu JO (1983) Occupational exposure to lead in ancient times. Sci Total Environ 31:105–116

Ohi G, Seki H, Akiyama K, Yagyu H (1974) The pigeon, a sensor of lead pollution. Bull Environ Contam Toxicol 12:92–98

Orłowski G, Kamiński P, Kasprzykowski Z, Zawada Z, Koim-Puchowska B, Szady-Grad M et al (2012) Essential and nonessential elements in nestling rooks *Corvus frugilegus* from eastern Poland with a special emphasis on their high cadmium contamination. Arch Environ Contam Toxicol 63:601–611

Pain DJ (1990) Lead shot ingestion by waterbirds in the Camargue, France: an investigation of levels and interspecific differences. Environ Pollut 66:273–285

Pain DJ, Cromie RL, Newth J, Brown MJ, Crutcher E, Hardman P et al (2010) Potential hazard to human health from exposure to fragments of lead bullets and shot in the tissues of game animals. PLoS One 5:e10315

Pain DJ, Cromie R, Green RE (2015) Poisoning of birds and other wildlife from ammunition-derived lead in the UK. In: Delahay RJ Spray CJ (eds) Proceedings of the Oxford lead symposium. Lead ammunition: understanding and minimising the risks to human and environmental health. Edward Grey Institute, The University of Oxford, UK, pp 58–84

Pan C, Zheng G, Zhang Y (2008) Concentrations of metals in liver, muscle and feathers of tree sparrow: age, inter-clutch variability, gender, and species differences. Bull Environ Contam Toxicol 81:558–560

Pedersen S, Lierhagen S (2006) Heavy metal accumulation in arctic hares (*Lepus arcticus*) in Nunavut, Canada. Sci Total Environ 368:951–955

Pérez-López M, Hermoso de Mendoza M, López Beceiro A, Soler Rodríguez F (2008) Heavy metal (Cd, Pb, Zn) and metalloid (As) content in raptor species from Galicia (NW Spain). Ecotoxicol Environ Saf 70:154–162

Pikula J, Zukal J, Adam V, Bandouchova H, Beklova M, Hajkova P et al (2010) Heavy metals and metallothionein in vespertilionid bats foraging over aquatic habitats in the Czech Republic. Environ Toxicol Chem 29:501–506

Plumlee GS, Durant JT, Morman SA, Neri A, Wolf RE, Dooyema CA (2013) Linking geological and health sciences to assess childhood lead poisoning from artisanal gold mining in Nigeria. Environ Health Perspect 121:744–750

Rajaraman P, Stewart PA, Samet JM, Schwartz BS, Linet MS, Zahm SH et al (2006) Lead, genetic susceptibility, and risk of adult brain tumors. Cancer Epidemiol Biomarkers Prev 15:2514–2520

Rao Barkur R, Bairy LK (2015) Evaluation of passive avoidance learning and spatial memory in rats exposed to low levels of lead during specific periods of early brain development. Int J Occup Med Environ Health 28:533–544

Reglero MM, Monsalve-González L, Taggart MA, Mateo R (2008) Transfer of metals to plants and red deer in an old lead mining area in Spain. Sci Total Environ 406:287–397

Rintala R, Venäläinen E-R, Hirvi T (1995) Heavy metals in muscle, liver, and kidney from finnish reindeer in 1990-91 and 1991-92. Bull Environ Contam Toxicol 54(1):158–165

Rodriguez-Ramos Fernandez J, Höfle U, Mateo R, Nicolas de Francisco O, Abbott R, Acevedo P et al (2011) Assessment of lead exposure in Spanish imperial eagle (*Aquila adalberti*) from spent ammunition in central Spain. Ecotoxicology 20:670–681

Rogers TA, Bedrosian B, Graham J, Foresman KR (2012) Lead exposure in large carnivores in the greater Yellowstone ecosystem. J Wildl Manag 76:575–582

Rosman KJR, Chisholm W, Hong S, Candelone JP, Boutron CF (1997) Lead from Carthaginian and Roman Spanish mines isotopically identified in Greenland ice dated from 600 B.C. to 300 A.D. Environ Sci Technol 31:3413–3416

Sanders AP, Claus Henn B, Wright RO (2015) Perinatal and childhood exposure to cadmium, manganese, and metal mixtures and effects on cognition and behavior: a review of recent literature. Curr Environ Health Rep 2:284–294

Scheuhammer AM, Norris SL (1996) The ecotoxicology of lead shot and lead fishing weights. Ecotoxicology 5:279–295

Senapati SK, Dey S, Dwivedi SK, Swarup D (2001) Effect of garlic (*Allium sativum* L.) extract on tissue lead level in rats. J Ethnopharmacol 76:229–232

Smith DM Jr, Mielke HW, Heneghan JB (2008) Subchronic lead feeding study in male rats. Arch Environ Contam Toxicol 55:518–528

Statutory Instruments (2005) The restriction of the use of certain hazardous substances in electrical and electronic equipment regulations. Environmental Protection, EP, No. 2748. http://www.legislation.gov.uk/uksi/2005/2748/pdfs/uksi_20052748_en.pdf

Statutory Instruments (2009) The restriction of the use of certain hazardous substances in electrical and electronic equipment (amendment) regulations. Environmental Protection, EP, No. 581. http://www.legislation.gov.uk/uksi/2009/581/pdfs/uksi_20090581_en.pdf

Stowe HD, Goyer RA, Krigman MR (1973) Experimental oral lead toxicity in young dogs. Arch Pathol 95:106–116
Torres J, de Lapuente J, Eira C, Nadal J (2004) Cadmium and lead concentrations in *Gallegoides arfaai* (Cestoda: Anoplocephalidae) and *Apodemus sylvaticus* (Rodentia: Muridae) from Spain. Parasitol Res 94:468–470
Tranel MA, Kimmel RO (2009) Impacts of lead ammunition on wildlife, the environment, and human health—a literature review and implications for Minnesota. In: Watson RT, Fuller M, Pokras M, Hunt WG (eds) Ingestion of lead from spent ammunition: implications for wildlife and humans. The Peregrine Fund, Boise, ID, pp 318–337
UNEP (2010) Final review of scientific information on lead. United Nations Environment Programme Chemicals Branch, DTIE. http://www.unep.org/hazardoussubstances/Portals/9/Lead_Cadmium/docs/Interim_reviews/UNEP_GC26_INF_11_Add_1_Final_UNEP_Lead_review_and_apppendix_Dec_2010.pdf
US DHHS (1999) Toxicological profile for lead, United States Department of Health and Human Services, Atlanta, GA, USA
U.S. Geological Survey (2015) Mineral commodity summaries 2015. U.S. Geological Survey, p 196. https://doi.org/10.3133/70140094
Venäläinen ER, Niemi A, Hirvi T (1996) Heavy metals in tissues of hares in Finland, 1980-82 and 1992-93. Bull Environ Contam Toxicol 56:251–258
Venäläinen ER, Anttila M, Peltonen K (2005) Heavy metals in tissue samples of Finnish moose, Alces alces. Bull Environ Contam Toxicol 74:526–536
Waldron HA (1973) Lead poisoning in the ancient world. Med Hist 17:391–399
Way CA, Schroder GD (1982) Accumulation of lead and cadmium in wild populations of the commensal rat, Rattus norvegicus. Arch Environ Contam Toxicol 11:407–417
White LD, Cory-Slechta DA, Gilbert ME, Tiffany-Castiglioni E, Zawia NH, Virgolini M et al (2007) New and evolving concepts in the neurotoxicology of lead. Toxicol Appl Pharmacol 225:1–27
WHO (1995) Environmental Health Criteria 165, Inorganic lead, Geneva
WHO (2000) Regional publications, European series. Inorganic pollutants. Lead 91:1–17
WHO (2009) Global health risks: mortality and burden of disease attributable to selected major risks, Geneva. http://www.who.int/healthinfo/global_burden_disease/GlobalHealthRisks_report_full.pdf
WHO (2011) Guidelines for drinking water quality, 4th edn. Geneva, pp 383–385
Williamson P, Evans PR (1972) Lead: levels in roadside invertebrates and small mammals. Bull Environ Contam Toxicol 8:280–288
Wuana RA, Okieimen FE (2011) Heavy metals in contaminated soils: a review of sources, chemistry, risks and best available strategies for remediation. ISRN Ecol 2011:Article ID 402647
WVDL (2015) Normal range values for WVDL toxicology. www.wvdl.wisc.edu/wp-content/uploads/2013/06/WVDL.Info_.Toxicology_Normal_Ranges.pdf. Accessed 28 Apr 2015
Yoo YC, Lee SK, Yang JY, In SW, Kim KW, Chung KH et al (2002) Organ distribution of heavy metals in autopsy material from normal Korean. J Health Sci 48:186–194
Zaccaroni A, Amorena M, Naso B, Castellani G, Lucisano A, Stracciari GL (2003) Cadmium, chromium and lead contamination of Athene noctua, the little owl, of Bologna and Parma, Italy. Chemosphere 52:1251–1258

# Chapter 17
# Mercury, Hg

**Elżbieta Kalisińska, Natalia Łanocha-Arendarczyk, and Danuta I. Kosik-Bogacka**

**Abstract** In nature, mercury (Hg) occurs in the elemental form ($Hg^0$), as well as in inorganic (InHg) and organic (OrgHg) compounds. It is the only heavy metal that is liquid at room temperature and easily turns into a gas. Mercury vapours can be transported with air masses for hundreds and thousands of kilometres and—after falling down—contribute to the pollution of land and waters. In aquatic environments biogeochemical processes promote the natural microbial conversion of InHg to methylmercury (MeHg), the most bioavailable form of Hg.

Human activities have increased atmospheric Hg concentrations 3–5 times over the past 150 years, mainly as a result of the combustion of fossil fuels. It is believed that all forms of Hg are toxic to endothermic animals and humans, but MeHg is particularly dangerous because of its neurotoxic and teratogenic effects as well as negative influence on reproduction. Moreover, in nature MeHg is biomagnified, and its concentration reaches the highest levels in top predators, especially piscivorous species. For several decades, there have also been reports documenting the local occurrence of dangerously high concentrations of Hg in organisms living in terrestrial ecosystems (including spiders, insects and songbirds feeding on them) in areas, which had been subject to anthropogenic Hg pollution many decades ago. Studies on inland aquatic and terrestrial ecosystems have indicated the long-term persistence of Hg introduced into the environment and the complexity of its transformations and circulation in nature. A better understanding of these processes requires further research, including the issue of bioaccumulation and biomagnification of MeHg in various ecosystems.

E. Kalisińska (✉) · N. Łanocha-Arendarczyk · D. I. Kosik-Bogacka
Department of Biology and Medical Parasitology, Pomeranian Medical University in Szczecin, Szczecin, Poland
e-mail: ekalist@pum.edu.pl



E. Kalisińska (ed.), *Mammals and Birds as Bioindicators of Trace Element Contaminations in Terrestrial Environments*,
https://doi.org/10.1007/978-3-030-00121-6_17

# 1 Introduction

Mercury causes many environmental and health problems. Together with lead and cadmium, it belongs to the group of particularly toxic metals, which do not have any physiological functions in warm-blooded vertebrates (including humans), and therefore even small amounts of absorbed mercury result in the disruption of biochemical processes in the body. Its elevated concentrations in birds and mammals lead to the development of many diseases (mainly in the nervous and excretory systems) and death (Clarkson and Magos 2006).

# 2 General Properties

Mercury (Hg from "hydrargyrum", i.e. "liquid silver" from Greek "hydr-" for water and "argyros" for silver) is a heavy metal with a density of 13.55 g $cm^{-3}$. It is the only metal which occurs in a liquid form at room temperature; its freezing point is –38.83°C and boiling point is 356.73°C. It has good electrical conductivity and high volatility, reaching a vapour pressure of $1.22 \times 10^{-3}$ mm at 20°C ($2.8 \times 10^{-3}$ mm at 30°C). Its solubility in water is $6 \times 10^{-6}$ $g^{-1}$ 100 ml (25°C). In the atomic table of elements, mercury is located in the group IIB, with atomic number 80 and atomic mass 200.59. There are 33 known isotopes of Hg, of which 7 are stable. The general pool of Hg is dominated by three isotopes: $^{199}Hg$, $^{200}Hg$ and $^{202}Hg$. at 16.9%, 23.1% and 29.7%, respectively (Blum 2011). In the environment, mercury exists in the elemental form ($Hg^0$) and in compounds with I (mercurous or $Hg^+$) and II oxidation states (mercuric or $Hg^{2+}$). Elemental mercury is an extremely good "solvent" for gold, silver and many other metals (except iron) via the formation of amalgams (alloys). It forms both inorganic and organic compounds, with the latter known as organomercurials. Common mercury salts contain halides (fluorine, chlorine, bromine and iodine) and sulphur (HgS). Organic compounds occur as $R_2Hg$ and RHgX, where R represents a simple alkyl group such as methyl ($CH_3^-$) and X represents atoms or groups such as chlorine, bromine, iodine, cyanide and hydroxyl. Two of the organic compounds are monomethyl mercury $CH_3HgX$ (methylmercury, MeHg) and dimethyl mercury $(CH_3)_2Hg$, the most important chemical forms of Hg with respect to environmental impact assessments (National Research Council 2000; Scoullos et al. 2001).

# 3 Mercury in Nature

Mercury is a natural component of the Earth's crust, occurring in soil, water and air where it penetrates into living organisms. In the environment it is found in an elemental form or in inorganic and organic compounds with varying degrees of toxicity to plants and animals, including vertebrates.

## 3.1 Mercury in the Abiotic Environment

It is estimated that Hg constitutes only $0.083 \times 10^{-4}\%$ of the Earth's crust and is in the 63rd position in terms of percentage share in the lithosphere. Mercury is present in the Earth's upper crust at a mean concentration of ~0.05 mg $kg^{-1}$ (ppm). Its abundance in igneous rocks is lower than in sedimentary rocks (0.004–0.08 and 0.01–0.40 ppm, respectively) and is mainly concentrated in argillaceous sediments. As a chalcophile element, this metal exhibits high affinity for sulphur and low to oxygen. Mercury occurs mainly in minerals containing sulphides and sulpho-salts and accompanies the ores of many metals (including copper, silver, zinc and lead). Generally, Hg is considered to be a rare element and extensively dispersed in the lithosphere (Yaroshevsky 2006; Kabata-Pendias and Mukherjee 2007; Kabata-Pendias 2011).

About 90 Hg minerals have been described, including cinnabar (HgS) and calomel ($Hg_2Cl_2$). Various ores generally contain from 0.1 to 2.5% Hg and occasionally >7% Hg. In some parts of the world there are geological anomalies with very high accumulations of minerals rich in Hg. Geologists have described more than 2200 sites where ores not only contain significant amounts of mercury but where also the soil, deposits of coal and oil and inland waters are characterized by elevated Hg content. Most of these sites are located within three transcontinental belts, usually with significant volcanic activity. The first belt (Mediterraneo-Himalayan) runs from the Iberian Peninsula in Europe to the Himalayas in Asia, the second covers the area lying along the west coast of the Pacific, and the third runs through the western areas of the Americas, together with the Pacific Ocean adjoining them; therefore, the Pacific is surrounded by the zone naturally high in mercury (Rytuba 2003; AMAP/UNEP 2013). Ores containing cinnabar, the most widespread of natural mercury-containing minerals, are present in approximately 60 countries. Five of the richest deposits of Hg include three European sites (Almaden in Spain, Monte Amiata in Central Italy, Idrija in Slovenia) and one located in North America (including New Almaden and New Idria in California, USA) and in South America (Huancavelica in Peru). These deposits were exploited for hundreds of years but eventually were closed in the period 1982–2002 (Ferrara et al. 1999; Gnamuš and Horvat 1999; US GS 2016a). In addition to those already mentioned, areas particularly rich in mercury can be found in China and Kyrgyzstan (Scoullos et al. 2001; Hylander and Meili 2003; Gómez et al. 2007).

The concentration of Hg in environmental samples is generally low outside of these geological anomalies and areas anthropogenically contaminated by this element. Hg levels in the air in Greenland range between 0.01 and 0.06 ng $m^{-3}$ and in snow and rainwater do not normally exceed 0.2 μg $L^{-1}$. In inland surface waters, the concentration of Hg ranges from 0.2 to 1.0 μg $L^{-1}$, and it is typically lower in rivers than in lakes (Adriano 2001). Globally, the average concentration of mercury in soils assumes is about 0.16 mg $kg^{-1}$ dry weight, (range 0.06–0.20 mg $kg^{-1}$ dw), but in European agricultural soils, it is markedly less and does not exceed 0.04 mg $kg^{-1}$ dw (Adriano 2001; De Vos et al. 2006). Much higher values are listed in soils of volcanic origin, where the concentration of Hg can exceed 7 mg $kg^{-1}$ (Kabata-Pendias 2011). From the environmental and economic points of view, the mercury content in mined and processed raw materials is most essential. These are mainly ores of mercury and other metals, which are accompanied by mercury, rocks used in the cement industry and fossil fuels (Table 17.1). To obtain mercury on an industrial scale, ore with an average content of 0.6–3.2% is exploited, while there are also deposits in Almaden (Spain) which comprise 8% Hg or 80,000 ppm. In addition, some local rocks there contain small drops of native mercury (Kim et al. 2004; Gómez et al. 2007). Most mercury mines in the world have been closed, with those remaining open are located mainly in Asia.

## *3.2 Mercury Production and Uses*

Due to its unique properties, mercury and its compounds have been used in a variety of applications since ancient times. Over the centuries, cinnabar (vermilion) with a characteristic vivid red colour was widely used as a pigment in art, wall decorations, cosmetics and some medicines in Rome, mediaeval Europe, Egypt, India and China. Even in the twenty-first century, it is used in some ritualistic and spiritual practices. Up to now mercury was extracted in poor countries by heating cinnabar in a current of air and condensing the vapour. By 500 BC, mercury was used to make amalgams with other metals. This property of mercury to form alloys is still widely used, particularly in obtaining precious metals and the preparation of dental amalgams ("silver fillings"). Such cheap and permanent fillings have been used in dentistry since the nineteenth century. Since the last century, mercury has been used on a large scale in the chemical and electrochemical industries for electrical and electronic applications (among others in switches, batteries, fluorescent lamps and energy-saving light bulbs). It is also found in some control devices (thermometers, barometers and manometers) and some pesticides, although developed countries significantly reduced the use of mercury in various products and processes due to its high toxicity and environmental hazard (Caley 1928; Parsons and Percival 2005; Masur 2011; Teaf and Garber 2012).

The world's richest source of cinnabar and quicksilver in Almaden (Spain) was operated for over 2000 years, with about 7 million tons of Hg extracted (Tejero et al. 2015). For comparison, from 1500 to 2000, the entire world production of Hg was less than 1 million tons, of which Almaden accounted for ~33% (Gómez et al. 2007; Hylander and Meili 2003; Tejero et al. 2015). As late as in 1971–1980, world production of Hg was very large, with an estimated production of 81,925 tons Hg, of which the former Soviet Union (including Ukraine, Russia, Kyrgyzstan and Tajikistan) accounted for 26%, Spain (Almaden) for 18% and the United States (California and Nevada) for 10.2% (Hylander and Meili 2003). Since then, the global excavation of Hg has dropped more than five times, and in the decade from 2001 to 2010, it amounted to a total of 16,310 tons (US GS 2001–2011). Table 17.2

**Table 17.1** Mercury concentrations in raw materials for industry (dw, dry weight)

| Material | Mean value | Range | Source |
|---|---|---|---|
| Rocks | | | |
| Raw materials for metallurgy, mg $kg^{-1}$ | | | |
| Mercury ores | | 188–22310 | Kim et al. (2004) |
| Zinc and lead ores | | 27–1198 | Rytuba (2003) |
| Raw materials for cement industry, μg $kg^{-1}$ dw | | | |
| Limestone | 17 | <1–391 | Hills and Stevenson (2006) |
| Sand | 29 | <1–556 | Hills and Stevenson (2006) |
| Clay | 52 | 2–270 | Hills and Stevenson (2006) |
| Shale | 57 | 2–436 | Hills and Stevenson (2006) |
| Fossil fuels | | | |
| Peat, μg $kg^{-1}$ dw | | | |
| World | | 60–300 | Adriano (2001) |
| Brown coal, μg $kg^{-1}$ dw | | | |
| North America | 130 | 7–1200 | Toole-O'Neil et al. (1999) |
| Europe | | 20–1500 | Glodek and Pacyna (2009) |
| Coal, μg $kg^{-1}$ dw | | | |
| World | 217 | 10–1780 | Pye et al. (2006) |
| United States | 170 | | Mukherjee et al. (2008) |
| Canada | 60–140 | | Pye et al. (2006) |
| Europe | | 10–1500 | Glodek and Pacyna (2009) |
| Ukraine | 390 | | Pye et al. (2006) |
| China | 30–340 | 20–1590 | Pye et al. (2006) |
| India | | 110–800 | Mukherjee et al. (2008) |
| Crude oil, μg $kg^{-1}$ dw | | | |
| World | 7.3 | 0.5–600 | Wilhelm et al. (2007) |
| North America | 5.3 | | Wilhelm et al. (2007) |
| Asia | 220 | | Wilhelm et al. (2007) |
| Thailand | 593 | | Wilhelm et al. (2007) |
| Near East | 0.8 | | Wilhelm et al. (2007) |
| Europe | 8.7 | | Wilhelm et al. (2007) |
| Russia | | 3.6–19.5 | Lang et al. (2012) |
| Natural gas, μg $Nm^{-3}$ | | | |
| World | | 1–1000 | Visvanathan (2003) |
| North America | | 1–20 | Eckersley (2010) |
| Europe | | 1–50 | Eckersley (2010) |
| East Europe | | 1–2000 | Lang et al. (2012) |
| Asia SE | | 10–2000 | Eckersley (2010) |
| Near East | | 1–10 | Eckersley (2010) |
| Biomass, μg $kg^{-1}$ dw | | | |
| | | 30–80 | Pye et al. (2006) |

**Table 17.2** Three countries with the highest mercury production in selected years (metric tons)

| Year | Three countries with the highest mercury production | Source |
|---|---|---|
| 1980 | USSR (2139), Spain (1697), Yugoslavia (1058) | US Bureau of Mines (1981) |
| 1985 | USSR (2240), Spain (1554), Algeria (793) | US Bureau of Mines (1986) |
| 1990 | USSR (2100), China (800), Mexico (735) | US Bureau of Mines (1991) |
| 1995 | Spain (1700), China (500), Algeria (450) | US GS (1996) |
| 2000 | Spain (600), Kyrgyzstan (600), Algeria (200) | US GS (2001) |
| 2005 | China (500), Kyrgyzstan (300), Spain (150) | US GS (2006) |
| 2010 | China (1400), Kyrgyzstan (250), Chile (150, Hg as a by-product) | US GS (2011) |
| 2015 | China (1600), Mexico—exports (500), Kyrgyzstan (70), Russia (50) | US GS (2016b) |

shows the three countries with the highest production of Hg in the period 1980–2015.

By the end of 1970, mining mercury in European mines in Almaden, Monte Amiata, Idrija and North America, and Hg use in various sectors of the economy in those parts of the world, was significantly higher compared to developing countries in Europe and Asia. Over time, it changed significantly, and since 2005 China has been the world leader in extraction (Table 17.2). In 2015, 1600 tons Hg was excavated in China, accounting for nearly 70% of global production (US GS 2016b). It is estimated that 80% of the world's mercury reserves have already been processed through human products (Meinert et al. 2016).

Before 1980 metallic mercury had been used in significant quantities, mainly for the extraction of gold and silver (for centuries), in dental amalgam fillings, as a catalyst in the chlor-alkali industry (where liquid Hg is the cathode, and this is one of three chlorine production technologies) and production of vinyl chloride monomer (VCM) used to synthesize polyvinyl chloride, PVC, to produce tubes, bottles, window frames and many articles. Moreover, metallic Hg is used in measuring devices, in electrical and electronic switches as well as in fluorescent lamps. Inorganic mercury compounds were used, among others, in Hg-oxide batteries, as pigments and dyes and as antiseptics in pharmaceuticals, while organic compounds of Hg (including alkyl forms) were used mainly as effective biocides in the paper industry and were added as an antifouling agent to paints and as fungicide to protect seeds and plants from fungal diseases (Hylander and Meili 2003). Due to the strong toxicity of mercury, already well documented in medical and ecotoxicological studies from 1950 to 1980, and focusing on the protection of health and care for the quality of the environment, regulations limiting economic exploitation, mining and trade of mercury have been gradually introduced in the European Union (EU) and North America. In those parts of the world, mercury mines had been shut before 2002. The consequence of the aforementioned actions was a drastic reduction in demand for mercury and a drop in its prices (Hylander and Meili 2003; Parsons and Percival 2005; Mohapatra and Mitchell 2009; UNEP 2013).

World mercury mining in 1980 was still relatively high at 6811 tons, but in 2005 it fell to 1520 tons. At that time, production and consumption of mercury shifted significantly from Europe and North America to Asia (US GS 1981, 2006). In 2005, including in Asia, Europe and North America, various sectors of the economy consumed 3188 tons Hg, of which Asia accounted for almost 67%, Europe (EU25 + CIS and other European countries) for 22.5% and North America 10.8% (AMAP/UNEP 2008). In Asia, most mercury is used in VCM and battery production (750 and 280 tons, respectively), in EU25 in mercury-cell chlor-alkali production (175 tons) and dental amalgam production (95 tons) and in North America in mercury-cell chlor-alkali production (60 tons) and production of measuring and control devices (48 tons). Several years later (in 2011), the global demand for mercury had dropped to 1930 tons, and the dominant recipient of this metal was chemical manufacturing (including 15% of the chlor-alkali industry and 21% of vinyl chloride monomer production) and artisanal and small-scale gold mining ASGM (24%) and batteries (13%), and further positions were dental amalgams 8%, measuring and control devices 7%, electrical and electronic devices 7% and fluorescent lighting 4% (UNEP 2013). According to a report by the United Nations Environmental Programme (UNEP) Global Mercury Partnership and its mercury-cell chlor-alkali production partnership area, this industry saw a very noticeable reduction in global demand for mercury. Between the base year 2005 and 2015, the consumption of mercury in the chlor-alkali industry fell by 50%, from 500 to 250 tons, resulting from the reduction in the number of plants that uses mercury in the production of chlorine and alkalis, through their closure or a shift into mercury-free technology (UNEP 2016), especially in this regard in the EU, where the use of mercury in chlor-alkali industry will have ceased in 2017 (Eurochlor 2016).

Although between 1980 and 2007 the global demand for mercury fell dramatically, and its production decreased almost six times (from 6811 to 1170 tons according to US GS in 1981, 2008), in recent years this downward trend has unfortunately changed, caused by the global economic crisis in 2008. For comparison, in 2008 and 2015, the global production of mercury was, respectively, 1320 and 2340 tons, significantly higher than in 2007 (US GS 2010, 2016b). The current increase in demand for mercury is significantly associated with an increased demand for gold, as its acquisition by the inexpensive method of amalgamation requires Hg. This method is mainly used in ASGM in developing countries (UNEP 2013).

### 3.2.1 Emission Sources of Mercury

Hg is released from natural (geogenic) and anthropogenic sources, including intentional (Hg acquisition from its ores, meeting the needs of certain sectors of the economy) and unintentional, that accompany various production and energy processes. Geogenic sources of mercury in nature include volcanic eruptions, weathering of rocks, natural forest fires and steppes and evaporation of the seas and oceans. Partially, these also include areas around active and abandoned Hg mines (with the deposited waste), often with significant levels of that element.

Terrestrial sources and the oceans are credited with 48 and 52% of total annual emissions of mercury into the air. Researchers that from 80 to 600 tons of Hg reach from the land to the air, with the geogenic emissions mainly caused by mass burning (13%) and metal release from the desert, metalliferous and non-vegetated zones (10%), as well as some biomes such as tundra, grassland, savannah, prairie and chaparral (9%) (Pirrone et al. 2010; AMAP/UNEP 2013). In 2010 oceanic sources accounted for up to 2900 tons of Hg released into the global atmosphere, including the contribution from re-emission processes, which are emissions of previously deposited Hg originating from anthropogenic and natural sources, and primary emissions from natural reservoirs (AMAP/UNEP 2013).

Over the past few decades, the major sources of anthropogenic mercury unintentionally released into the air are the combustion of fossil fuels, mining and the processing of non-ferrous ores, cement production, natural gas cleaning, recycling and government stockpiles and incineration of sludge from biological treatment (Mohapatra and Mitchell 2009). Fossil fuels and various industrial raw materials usually contain small quantities of Hg (Table 17.1), but given the huge amounts used by man, their contribution to environmental pollution with Hg is a key position in its biogeochemical cycle. However, in 2010 it was recognized that global anthropogenic emissions of mercury to the air are mainly based on artisanal and small-scale gold mining (ASGM), before the process of burning coal for the needs of electro-energy (AMAP/UNEP 2013). It is estimated that in 2010, Hg from anthropogenic sources amounted to about 2000 tons, and another 1000 tons was released into waters, wherein the emission of water is much less recognized and evaluated in comparison to the atmospheric release. It is believed that chlor-alkali plants, paper pulp factories and mine wastes have been the major industrial sources that discharge mercury waste into water bodies (Mohapatra and Mitchell 2009; AMAP/UNEP 2013; UNEP 2013). In 2010, global atmospheric mercury emissions totalled 8900 tons, of which the current emissions from natural and anthropogenic sources account for 80–600 tons and about 2000 tons. The remaining amount of Hg (60%) in the annual amount came from re-emission, with the terrestrial and oceanic volumes estimated to be 1700–2800 and 2000–2950 tons, respectively (AMAP/UNEP 2013).

For about 200 years, we have seen a significant increase in the quantity of mercury circulating in nature. This is indicated by comparative studies of lake bottom sediments, peat deposits and core glaciers (Schuster et al. 2002; Allan et al. 2013). It is estimated that, compared to pre-industrial times, the concentration of Hg in the atmosphere and in the geochemical background has increased at least three times and probably 5–10 times in relation to the natural level (Mason et al. 2012; Horowitz et al. 2014). On a global scale, in the period 1850–2010, unintentional anthropogenic sources (from "by-product" sectors including fossil fuel combustion) issued to the atmosphere 215,000 tons of mercury. During that time, a further 540,000 tons of mercury was introduced into the environment from intentional commercial Hg uses and nonatmospheric releases from chlor-alkali plants and mining processes. From this very large pool, 20% reached the air, 30% waters, 30% soils and 20% landfill wastes. Some of this mercury remains in landfills or is associated with bottom sediments, but a significant quantity (310,000 tons) actively participates in the geochemical cycle (Horowitz et al. 2014).

Emissions of mercury into the environment have clearly differed between the Northern and Southern Hemispheres, where human economic activity releases 70% and 30% Hg, respectively (Pacyna et al. 2006; Selin et al. 2008; Pirrone et al. 2010; AMAP/UNEP 2013). This disparity in the emissions of Hg between the two hemispheres has historical, economic and demographic reasons.

Mercury released from natural and anthropogenic sources circulates in nature for a long time and is transmitted over long distances by strong atmospheric and ocean currents. Probably, it will take about a thousand years before mercury is released from stable formations in the lithosphere and circulating in the air-water-soil system, settles on the ocean floor and is permanently bound by mineral deposits in the rock formation processes (Mason et al. 2012; Horowitz et al. 2014).

Between 1980 and 2007, the mining of mercury decreased almost six times, which was driven by the results of numerous studies and regulations for the protection of health and the environment. Scientific studies provide ample evidence of the strong toxicity of Hg (especially MeHg) on humans and other warm-blooded vertebrates and document a dramatic increase in the amount of anthropogenic environment (Hylander and Meili 2003; Clarkson and Magos 2006; Horowitz et al. 2014). Out of many disasters caused by environmental Hg poisoning, the best known are the tragic events from the Japanese Minamata Bay from the 1950s, with the mass Hg poisoning of residents, their cats and wild birds, via the fish and seafood consumed. The primary source of mercury was wastewater from chemical plants discharging into the bay. The increasing awareness of risks arising from the increase in the amount of anthropogenic Hg in the environment has led to the introduction of regulations aimed at limiting the extraction, use and trade of Hg and consequently a reduction in the release of mercury into the air, water and soil from anthropogenic sources. Such pro-health and pro-environmental legislative action were taken earliest in the well-developed countries of the EU, North America and Japan, but globally more important will be the implementation of the provisions of the Minamata Convention, adopted on 10 October 2013 at a diplomatic conference held in Kumamoto, Japan. The convention entered into force on 16 August 2017 (www.mercuryconvention.org).

## 3.3 *Biological Status of Mercury*

According to current knowledge, mercury does not have any physiological function in eukaryotic and in most prokaryotic organisms. Its accumulation results in various life-threatening disorders and can lead to fatal poisoning (Clarkson 1992; Barkay and Wagner-Döbler 2005; Scheuhammer et al. 2015). Recently, Gregoire and Poulain (2016) showed a peculiar exception among prokaryotes: photosynthetic microorganisms from the group of purple non-sulphur bacteria (representing genera

*Rhodobacter* and *Rhodopseudomonas*) are able to use Hg as an electron acceptor during photosynthesis.

Mercury was identified thousands of years ago and is one of the oldest toxicants known. The three forms of Hg, i.e. elemental, inorganic and organic mercury (especially $CH_3Hg$-R; methyl-Hg or MeHg), have different toxicological properties. Mercury can occur in compounds either in +1 or +2 oxidation state, i.e. in mercurous (I) and mercuric(II) compounds, respectively. In nature, inorganic divalent Hg (II) compounds predominate, with relatively few monovalent Hg(I) compounds. Monovalent Hg compounds are less toxic than Hg(II) compounds as they are less soluble in water (WHO 2003; Park and Zheng 2012).

The biogeochemical cycle of Hg and toxicity involve bacteria that produce MeHg. In the environment some anaerobic sulphate- and iron-reducing bacteria can methylate oxidized mercury ($Hg^{2+}$) and to a smaller degree $Hg^0$, thus generating MeHg (Hu et al. 2013; Li and Cai 2013). Biologically mediated production of MeHg predominantly occurs under anaerobic conditions in sediments of inland waters, nearshore and oceanic sea floors, as well as in peatlands, wetland soils and some rice paddy fields, for example, in China (Zhang et al. 2010; Gu et al. 2011; Windham-Myers et al. 2014; Zhao et al. 2016). MeHg is also present in most if not all aquatic organisms. Methylation of InHg to MeHg and demethylation of MeHg are the two most important processes in the cycling of MeHg, determining the levels of MeHg in aquatic and terrestrial ecosystems. Aerobic bacteria have evolved an efficient strategy of eliminating mercuric ($Hg^{2+}$) and organic mercury compounds (including MeHg) from the environment through the reduction of $Hg^{2+}$ to $Hg^0$ (Li and Cai 2013).

Methylation and biomagnification of Hg have been well researched in aquatic ecosystems due to the consumption of Hg-contaminated fish, crayfish and molluscs, which may lead to poisoning of humans and other warm-blooded vertebrates. By contrast, studies on Hg and especially MeHg in terrestrial ecosystems are few (Clarkson 1992; Larosa and Allen-Gil 1995; Wolfe et al. 1998; Jackson et al. 2011; Douglas et al. 2012; Kalisinska et al. 2012a; Rieder et al. 2013; Scheuhammer et al. 2015). Since MeHg in aquatic ecosystems is subject to biomagnification, Hg reaches its highest levels in predatory fish, piscivorous birds and marine and semiaquatic mammals. Mercury concentrations in those biotas can be many millions of times greater than in the waters which serve as their aquatic habitat or food source (Lavoie et al. 2013; Finley et al. 2016). The greatest increase in MeHg concentration occurs in the trophic step between water and algae. It is estimated that the biomagnification factor (BMF) between water and seston often ranges from $\sim 10^5$ to $\sim 10^6$ with the BMF of MeHg concentrations between successive trophic levels above algae generally less than $10^1$ (Wolfe et al. 2007). In terrestrial ecosystems, biomagnification of MeHg also occurs, yet this phenomenon has been much less researched (Rimmer et al. 2010; Rieder et al. 2013; Osborn et al. 2011; Jackson et al. 2015; Abeysinghe et al. 2017).

## 3.4 Mercury Toxicity

In the 1950s, dramatic events took place in the Japanese Bay of Minamata with many lethal mercury poisonings in humans, cats and wild birds. Over 3000 brain-damaged victims were diagnosed with "Minamata disease", and veterinary medicine introduced the term "dancing cats" to describe the neurological symptoms observed in cats. Both "Minamata disease" and "dancing cats" were the result of Hg poisoning accompanied by other contaminants spilled into the gulf from a nearby chemical factory. In the gulf's sediments, bacteria transformed inorganic mercury into MeHg, whose levels progressively increased in organisms from successive trophic levels. Large amounts of MeHg in fish, crustaceans and mussels were consumed by humans and animals inhabiting those areas, resulting in diseases and fatal poisonings (D'Itri 1991; Aronson 2005; Hachiya 2006; Ekino et al. 2007; Grandjean et al. 2010). Also in the 1950s, MeHg toxicity in the developing brain was first recognized in cases of congenital Minamata disease among newborns and children. At the same time, it was noted that the mothers had no symptoms of Hg toxicity or were minimal (Clarkson and Magos 2006; Ekino et al. 2007).

A few later studies from the 1960s to 1970s were conducted by Swedish naturalists on birds and rodents feeding on grains and on predators feeding on these granivores. They showed that Hg poisoning can also occur in terrestrial environments, not just aquatic environments. Inorganic and organic Hg compounds (including MeHg) were then common components of pesticides (fungicides) serving as seed dressing. Large quantities of Hg from the fungicides were detected in granivores and even larger levels in predatory birds and mammals preying on the passerines and rodents (Borg et al. 1969; Johnles and Westermark 1969). From 1960 to 1990, Hg-containing fungicides had been banned in Northern Hemisphere countries with highly developed agriculture (UNEP 2002). After all those years, it is very difficult to determine how much of the Hg pesticides has been introduced into the environment since the usage (launched in the first quarter of the twentieth century) lasted dozens of years. In the United States, Sweden and Japan, it is estimated that 800, 600 and 1600 tons of Hg fungicides were sprayed each year in rural areas of those countries (with Japan being more than 20 times smaller in area than in the United States) (Smart 1968; Kiesling and Lloyd 1971). Currently, agricultural soils are also being contaminated with anthropogenic Hg due to fertilization with sewage sludge, but this process is much less intense. It is estimated that in the EU, the Hg concentrations in sewage sludge recycled to agriculture vary among its member states from 0.2 to 4.6 mg $kg^{-1}$ dw (Milieu Ltd. WRc and RPA 2010). In the 2000s the amount of mercury introduced into agricultural soils in the 27 EU countries probably exceeded 4 tons per year (AMAP/UNEP 2013). Total Hg from atmospheric deposition (derived from natural and anthropogenic sources) of agricultural origin and released from soil rocks contributes to pollution of the terrestrial environment. Mercury is washed away from these areas and is transported to various waters bodies where it is methylated and (partly as $Hg^0$) is released into the atmosphere and transported over considerable distances. In addition, soils in river valleys are exposed to various forms of Hg during periodic inundations. However, in aquatic environments, as compared to land, Hg is to a much greater degree integrated into food chains, and aquatic food can be a significant threat to the health of humans and wildlife. Generally free-living terrestrial animals are chronically exposed to low concentrations of Hg contained in food, water and ambient air. Mercury toxicity has been studied at the levels of molecules, cells, tissues, organisms, species and ecosystems (Borg et al. 1969; Wren 1984; Scheuhammer et al. 1998a, b; Aschner 2000; Schurz et al. 2000; Silva-Pereira et al. 2005; Wolfe et al. 2007).

The toxicity of mercury has been attributed to its high affinity to protein-containing sulfhydryl (thiol) groups (–SH). These groups are especially abundant in proteins containing cysteine and methionine, which are sulphur amino acids. Proteins rich in cysteine include glutathione peroxidase (GSH-Px), metallothioneins (MTs) and keratins. GSH-Px belongs to the family of very important antioxidant enzymes, which also contain selenium (Se) (Clarkson and Magos 2006). MTs and keratin structures (including hair and feathers) may contain up to 30% and 26% of cysteine, respectively (Clarkson and Magos 2006; Agarwal and Behari 2007; Greenwold and Sawyer 2013). The MTs are low-molecular-weight proteins and are present in various cells (especially in the liver and kidneys) and serum of vertebrates, but they were also discovered in invertebrates. MTs have a few main hypothesized functions: homeostasis of essential metals such as zinc (Zn) and copper (Cu), detoxification of non-essential Hg and cadmium (Cd), protection against oxidative damage and free radical scavenging (Isani and Carpenè 2014).

All mercury species are accumulated by eukaryotic organisms. Vertebrates can uptake toxic mercury from the environment through the lungs, gills, skin and from the digestive tract. In wildlife the alimentary tract plays the most important route. From avian and mammalian gastrointestinal tracts, MeHg is most effectively absorbed at a rate over 90%. InHg is absorbed from the diet, at most at a rate of a few to a dozen percent, and $Hg^0$ at <0.01% (Serafin 1984; Clarkson and Magos 2006; Park and Zheng 2012; Ye et al. 2016). Inhaled $Hg^0$ vapour in the lungs of mammals is absorbed at up to 85%, as demonstrated by experimental research on mammals and epidemiological studies of humans occupationally exposed to mercury vapour (Pendergrass et al. 1997; Falnoga et al. 2000; Bose-O'Reilly et al. 2010; Bernhoft 2012).

Mercury toxicity studies have taken into account many factors, including the physico-chemical properties of this element. Mercury is classified as a chalcophile element (alongside Se, Cd and Pb), with a typically higher affinity to sulphur (S) and a lower affinity to oxygen (O) than iron (Fe). In living organisms, Hg is highly competitive in relation to essential metals, mainly Zn and Cu, which are displaced from the S binding sites in cysteine to be replaced by $Hg^{+2}$ and/or $MeHg^+$. Sulphur amino acids (cysteine, Cys, and methionine, Met) are constituents of enzyme, transport and structural proteins, which after binding to Hg change their properties and structure (Grosicki and Kowalski 2002; Fraga 2005; García-Barrera et al. 2012; Dobrakowski et al. 2013). In the case of Cys, over the course of evolution, S has been replaced by Se to form the 21st amino acid, selenocysteine (SeCys). It is a natural component of selenoproteins in all animal kingdoms including vertebrates

(Lu and Holmgren 2009). From this group of proteins, the most important are enzymes such as GSH-Px, thioredoxin reductase and iodothyronine deiodinase. These proteins participate in the antioxidant protection of cells and the metabolism of thyroid hormones and of immunological processes. Selenoproteins may contain from 1 to 15 SeCys per protein subunit (Ralston et al. 2008; Mehdi et al. 2013). $MeHg^+$ ions possess electrophilic properties, and they interact with and oxidize nucleophilic groups of various biomolecules, especially those containing sulfhydryl groups. Besides proteins (i.e. antioxidant enzymes, neurotransmitter receptors, transporters), sulphydryl groups contain nonprotein thiols such as cysteine and glutathione, GSH (Farina et al. 2013). GSH is an important antioxidant in animals, preventing damage to cellular components caused by reactive oxygen species and other factors including $Hg^{+2}$ and $MeHg^+$ (Schurz et al. 2000; Pompella et al. 2003; Clarkson and Magos 2006; Wolfe et al. 2007).

As the binding affinity of Hg for Se is up to a million times higher than for S, Hg (especially $Hg^{2+}$ and $MeHg^+$) inexorably sequesters Se, directly impairing selenoenzyme activity and synthesis. At the same time, Se compounds are able to decrease the toxicity of Hg, which has been established in all investigated species of mammals, birds and fish (Dietz et al. 2000; Belzile et al. 2009; Ralston and Raymond 2010).

### 3.4.1 Mercury Cytotoxicity, Genotoxicity, Cancerogenicity and Teratogenicity

The cytotoxicity and genotoxicity of the various forms of Hg are evaluated mainly in vitro assays on human and non-human cell lines (De Flora et al. 1994; Silva-Pereira et al. 2005; Robinson et al. 2010; Polunas et al. 2011; Fernandes Azevedo et al. 2012; Roy et al. 2013; Wang et al. 2013, 2016). The results of in vivo Hg genotoxicity tests (based mostly on leucocytes) that assessed the damage of nuclear genetic material (comet assay, micronucleus test, chromosome aberration tests) do not always confirm differences between the material obtained from warm-blooded vertebrates exposed to Hg and from control/comparison groups (Hansteen et al. 1993; Rozgaj et al. 2005; Kenow et al. 2008; Crespo-López et al. 2009). Various ions of Hg exhibit a high ability to bind –SH groups of protein and nonprotein compounds, and on this ground a number of hypotheses have been formulated about molecular mechanisms of Hg genotoxicity. In this respect, the most commonly mentioned are four mechanisms: oxidative stress connected with generation of free radicals, effect on microtubules, influence on DNA repair mechanisms and direct interaction with DNA molecules (De Flora et al. 1994; Myers and Davidson 1998; Burbacher et al. 1990; Crespo-López et al. 2009).

In vitro cytotoxicity studies have shown that in various human and animal cell lines, both InHg and MeHg induce numerous adverse changes. These changes mainly rely on altering mitochondrial function and raising oxidative stress by generating free radicals or by interacting with sulfhydryl groups (Polunas et al. 2011; Farina et al. 2013; Agrawal et al. 2015; Wang et al. 2013, 2016). Additionally, in human embryonic neural progenitor cells, MeHg induces oxidative damage to mitochondrial DNA (Wang et al. 2016).

The carcinogenetic potential of InHg is still being debated. In the 1990s the International Agency for Research on Cancer (IARC 1993) classified MeHg compounds as possibly carcinogenic to humans (Group 2B), but metallic mercury and InHg compounds were not classifiable as carcinogenic to humans (Group 3). The classifications of MeHg, $Hg^0$ and InHg are still in use by the IARC (2017). Drasch et al. (2004) reviewed papers concerning the influence of Hg on laboratory rats and mice and revealed that male rats receiving extremely high oral doses of $HgCl_2$ or MeHgCl had an increased number of renal tubule tumours. These compounds caused severe nephropathy in the rodents. It is likely that dietary MeHgCl may act in mice as a tumour promoter. However, the connection between Hg exposure and carcinogenesis remains controversial (Drasch et al. 2004; Crespo-López et al. 2009).

Methylmercury is known as an embryotoxic and teratogenic agent. The teratogenicity of MeHg is well documented in fish, birds, mammals and humans. This compound especially affects normal development of the central nervous system (Domingo 1994; Samson and Shenker 2000; Schurz et al. 2000; Heinz et al. 2011; Gandhi et al. 2013). In addition to the already mentioned adverse effects of Hg on wildlife, its effects on endocrine and immune systems are also important (Zhu et al. 2000; Kenow et al. 2007; Tan et al. 2009). Tan et al. (2009) listed five main endocrine-related mechanisms of Hg across these systems: (a) accumulation in the endocrine system, (b) specific cytotoxicity in endocrine tissues, (c) changes in hormone concentrations, (d) interactions with sex hormones and (e) upregulation or downregulation of enzymes within the steroidogenesis pathway. However, disorder and impairment of endocrine and immune systems by Hg and the net effects on the demography of wild animals are poorly understood (Kenow et al. 2007; Frederick and Jayasena 2011).

### 3.4.2 Mercury Neurotoxicity and Lethal Levels of Total Mercury in Soft Tissues

All three Hg species may occur in the brain, including elemental Hg. A certain part of inhaled $Hg^0$ is deposited in the brain as demonstrated in humans and laboratory animals (Warfvinge et al. 1992; Tjälve and Henriksson 1999; Bose-O'Reilly et al. 2010; Park and Zheng 2012). Also $Hg^0$ can be transported through the olfactory pathway to the olfactory bulbs and later into other brain areas (Galić et al. 1999; Tjälve and Henriksson 1999; Park and Zheng 2012). As $Hg^0$ is lipid soluble and highly diffusible, it can cross the blood-brain barrier and other cellular and intracellular membranes (Park and Zheng 2012). In humans inhalation of $Hg^0$ vapour can cause acute and chronic intoxication. Typical symptoms connecting with this include airway symptoms and many neurological problems (tremor, ataxia, coordination disturbances, abnormal reflexes, polyneuropathy with sensation difficulties, loss of memory, neurocognitive disorders) as well as kidney problems such as proteinuria (Bose-O'Reilly et al. 2010). In laboratory animals, the neurological symptoms following exposure to $Hg^0$ are poorly understood, but in rats a significant increase in Hg concentrations in different parts of the brain (primarily in the neocortex, in the basal nuclei and in the cerebellar granule cells) and in the kidneys were shown in comparison to the control group (Warfvinge et al. 1992; Galić et al. 1999). Unlike elemental and organic mercury species, the oxidized Hg forms ($Hg^{2+}$) are not able to effectively cross the blood-brain barrier, but such process could not be excluded (Park and Zheng 2012). Organic Hg compounds, especially MeHg, can easy cross the blood-brain barrier (however, less efficiently as $Hg^0$) and are accumulated in vertebrate brains. The probable active transport of MeHg (via neutral amino acid transporters) into the brain is preceded by the formation of MeHg-cysteine complexes (ADSTR 1999; Clarkson and Magos 2006). MeHg does not uniformly affect the nervous system, and Hg concentration in the brain varies between the compartments (Eto et al. 1999, 2010; Farina et al. 2013).

Speciation analyses of brain Hg in vertebrates show that a much larger proportion of THg is present in the form of MeHg (typically >80%) and a small fraction as InHg. Depending on the degree and duration of exposure to MeHg, the percentage of brain THg may change over time and varies greatly between individuals of the same species and between various species. In extreme cases in some people exposed to MeHg in childhood and for more than 20 years, up to 80% of brain THg may be InHg (Farina et al. 2013). Most wildlife is exposed to long-term exposure to small amounts of MeHg contained in the diet, with the exception of long-living piscivorous species. MeHg, which has penetrated into the brain, is gradually demethylated and transformed into InHg. The demethylation of MeHg has been found in the brain of humans and several wild species of birds and mammals from inland environments (Eto et al. 1999; Gnamuš and Horvat 1999; Scheuhammer et al. 2008; Strom 2008; Eagles-Smith et al. 2009; Haines et al. 2010; Basu 2012; Kalisinska et al. 2014a; Jo et al. 2015). Presumably, the remaining part of brain InHg can occur in insoluble and biologically inert compounds with selenium such as tiemannite (HgSe) (Whanger 2001; Nakazawa et al. 2011). In long-lived animals and humans, the half-life for MeHg in the brain is determined in days or months, but for InHg it is many years (Vahter et al. 1994; ADSTR 1999; Rice et al. 2014). Until recently it had been assumed that MeHg that gets into the brain did not leave, similar to InHg produced by MeHg demethylation or oxidation of $Hg^0$. However, works by Lohren et al. (2015, 2016), investigating MeHg and InHg transfer across the blood-brain barrier in a primary in vitro porcine model, may lead to the revision of this view. In the latter paper, Lohren et al. (2016), the researchers documented the transport of MeHg via the barrier in both directions, with diffusion as the transfer mechanism. Additionally for $HgCl_2$, their data delivered evidence that the blood-brain barrier transfers InHg out of the brain.

Lethal brain levels of Hg have not yet been established for most mammals and birds. In literature, in the brains of piscivorous mammals experimentally intoxicated with MeHg, river otter *Lontra canadensis* and American mink *Neovison vison* (previously *Mustela vison*), Hg levels were 23.7 and 19.9 mg $kg^{-1}$ ww and 11.9 mg $kg^{-1}$ ww, respectively (Aulerich et al. 1974; Wobeser et al. 1976; O'Connor and Nielsen 1981). In field studies in North America, brain Hg in single dead or dying river otter and mink were ~30 and 13.4 mg $kg^{-1}$ ww, respectively (Wren 1985; Sleeman et al. 2010; Wobeser and Swift 1976). A lower range was shown by THg concentrations (8.1–18.6 mg $kg^{-1}$ ww) in experimentally and non-intentionally intoxicated domestic cats from Japan and Northwestern Ontario Reserve (Canada), which revealed neurological symptoms typical for Minamata disease (Takeuchi et al. 1977). Shore et al. (2011) defined >10 mg THg $kg^{-1}$ ww as a lower indicative value in mammalian brains, which may be associated with adverse effects on survival and resulting in death. Krey et al. (2015) analysed a large number of reports on mammalian brain Hg concentrations and proposed a THg threshold concentrations for toxic endpoints: clinical symptoms >6.75 mg $kg^{-1}$ ww (29 mg $kg^{-1}$ dw), neuropathological signs >4 mg $kg^{-1}$ ww (17.2 mg $kg^{-1}$ dw), neurochemical changes >0.4 mg $kg^{-1}$ ww (1.72 mg $kg^{-1}$ dw) and neurobehavioral changes >0.1 mg $kg^{-1}$ ww (0.43 mg $kg^{-1}$ dw).

In adult passerines (starling *Sturnus vulgaris*, grackle *Quiscalus quiscula*, red-winged blackbird *Agelaius phoeniceus*, brown-headed cowbird *Molothrus ater*, zebra finch *Poephila guttata* and piscivorous great egret *Ardea albus*), which were experimentally intoxicated with MeHg, the concentration of brain THg was in the range of 20–45 mg $kg^{-1}$ ww (Finley et al. 1979; Scheuhammer 1988; Spalding et al. 2000). The highest THg residues in brains among wild birds found dead in fields were within the range of 13–14 mg $kg^{-1}$ ww: in tawny owl *Strix aluco* from Norway feeding on small rodents, piscivorous white-tailed eagle *Haliaeetus albicilla* from Sweden and common loon *Gavia immer* from Canada (Jensen et al. 1972; Holt et al. 1979; Scheuhammer et al. 2008). The values found in wild birds were clearly smaller than in experimental studies, but free-living animals are more exposed to various adverse environmental factors, including food shortages, than captive ones (Van der Molen et al. 1982; Wiener et al. 2003). A combination of the environmental factors can cause premature death before brain THg levels in birds reach ≥20 mg $kg^{-1}$ ww, established as lethal in laboratory conditions. In addition, dead animals are quickly eaten by scavengers, which is why they are rarely obtained for analysis. It has been experimentally shown that chicks are more sensitive to the toxic effects of Hg than adult birds. Data presented by Heinz and Locke (1976) indicate that lethal brain THg levels can be as low as 3–7 ppm in mallard ducklings. Shore et al. (2011) suggested indicative values of THg concentrations for avian brains of non-marine species, which may be associated with bird deaths at >15 mg $kg^{-1}$ ww and >3 mg $kg^{-1}$ ww in adults and developing youngs, respectively, and correspond well to data from Jensen et al. (1972), Holt et al. (1979), Scheuhammer et al. (2008) and Heinz and Locke (1976). Neurological symptoms (e.g. trembling) have been observed in 1 hatch-year osprey with 1.2 mg $kg^{-1}$ ww THg in the brain (or 6.2 mg $kg^{-1}$ dw) (Hopkins et al. 2007). However, THg threshold concentrations for toxic endpoints analogous to those proposed for mammals have not been developed, i.e. ones that would include subclinical, neuropathological, neurochemical and neurobehavioral changes, although some attempts have been made in this regard (Scheuhammer et al. 2008; Rutkiewicz et al. 2011; Rutkiewicz 2012).

Mercury is not only neurotoxic but also nephrotoxic for elemental and inorganic mercury species. The kidney is a major repository of InHg in warm-blooded vertebrates. Within the kidney divalent Hg accumulates primarily in the cortex and outer stripe of the outer medulla (Aschner and Aschner 1990; Wolfe et al. 1998; Bridges and Zalups 2010). It should be underlined that birds differ from mammals in having a renal portal system. In birds the venous blood from the terminal portion of the digestive tract flows to the kidney rather than the liver, as in mammals. This may make the avian kidney more vulnerable than the mammalian (Wolfe et al. 1998). Indicative values of THg concentrations in mammalian kidney associated with death, as proposed by Shore et al. (2011), are lower than in avian species, >25–30 mg $kg^{-1}$ ww compared to >40 mg $kg^{-1}$ ww. Also THg indicative value estimated for the adult mammalian brain are lower than in the analogous avian organ. However, in the case of the liver, the indicative value is higher in mammals than birds: >25–30 THg $kg^{-1}$ versus >20 mg THg $kg^{-1}$ (Shore et al. 2011).

Lethal concentrations of THg in the soft tissues of mammals and birds are most commonly determined in the liver and kidney, followed by the brain. Muscles are rarely taken into consideration, although they constitute a large part of the body weight of the animals, and the collection of muscle samples is easy when compared to the brain (Finley et al. 1979; O'Connor and Nielsen 1981; Wren et al. 1987; Farrar et al. 1994; Thompson 1996; Shore et al. 2011; WVDL 2015). In addition, the efficient functioning and coordination of skeletal muscles play a key role, especially in predators, because they co-determine the effectiveness of hunting. Based on relatively scarce data concerning THg concentrations in tissue pairs: liver (L)–muscle (M) and muscle–brain (B) in adult individuals of wild animals and birds, and those experimentally intoxicated with organomercury, correlation coefficients (*r*) and the values of two indices $M_{THg}/L_{THg}$ and $B_{THg}/M_{THg}$ were calculated, and potentially lethal muscle THg concentrations were estimated. In both animal groups, an increasing hepatic THg concentration was initially accompanied by a marked increase in muscle levels (Fig. 17.1, panels a and b). After exceeding ~10 mg $kg^{-1}$ ww in the muscle, the increase in THg slowed down and stabilized at 25–35 and 25–40 mg $kg^{-1}$ ww in mammals and birds, respectively, while the hepatic THg significantly exceeded 100 mg $kg^{-1}$ ww over time in some cases. Among inland mammalian and avian species, the highest hepatic THg levels were detected in river otter (96 mg $kg^{-1}$ ww) and common loon (200 mg $kg^{-1}$ and 370 mg $kg^{-1}$ ww) (Wren 1985; Stone and Okoniewski 2001; Scheuhammer et al. 2008). In the livers of marine mammals and birds, levels exceeding 1000 and 200 mg THg $kg^{-1}$ ww, respectively, were found in some cases (Kim et al. 1996; Storelli et al. 1999; Pompe-Gotal et al. 2009).

In the multispecies groups of mammals and birds, the correlation coefficient between the concentration of THg in the liver and muscle exceeded 0.95, and the values of *r* were, respectively, 0.928 and 0.964 (Fig. 17.1, panels a and b). Using the appropriate equations from panels A and B, we calculated THg concentrations for avian and mammalian muscle when the concentration of hepatic THg reached the lower limit values of the estimated lethal range (25 and 20 mg $kg^{-1}$ ww, respectively) (Shore et al. 2011). At these hepatic THg concentrations in mammalian and avian muscle, potentially lethal values were 9.8 and 7.3 mg $kg^{-1}$ ww. Other researchers had also found a significant correlation (*r* ranging from 0.60 to 0.98) between muscle

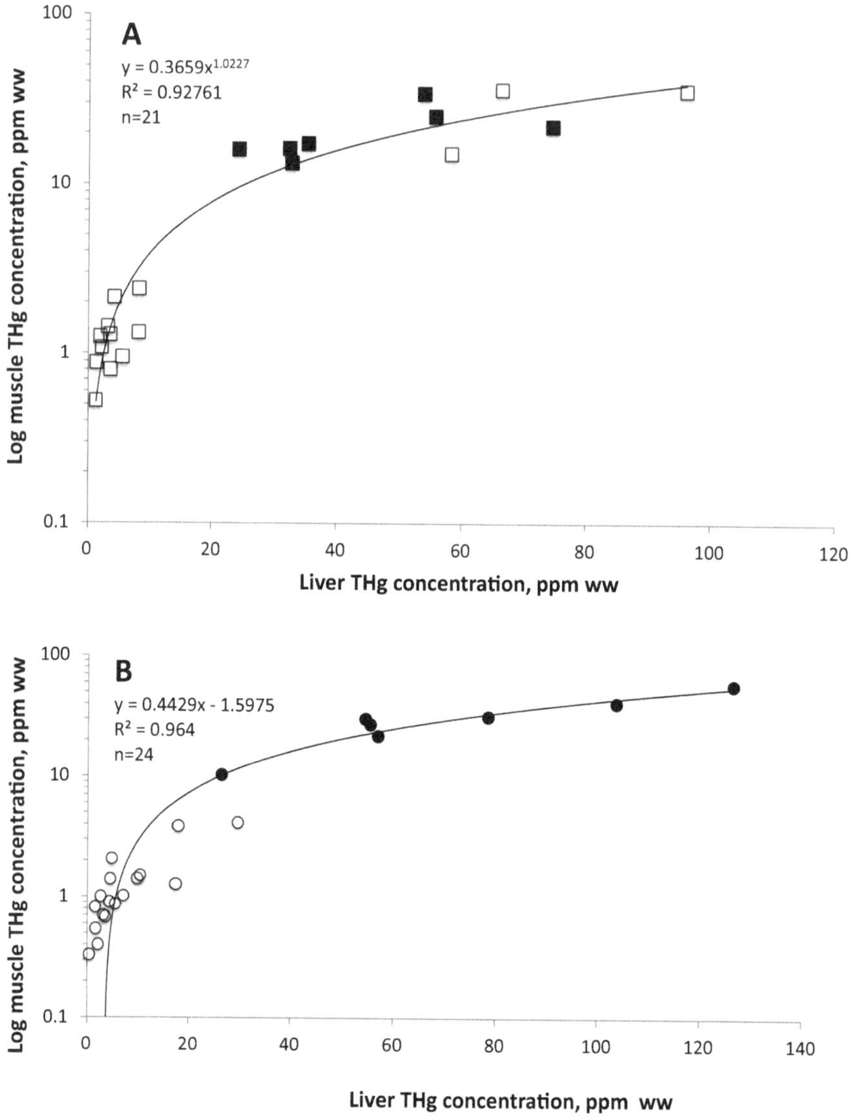


**Fig. 17.1** Relationship between total mercury (THg) concentrations (mg $kg^{-1}$ = ppm ww, wet weight) in liver and muscle, muscle and brain in terrestrial mammals and birds. Panel (**A**) Used data of intoxicated mammals such as river otter *Lontra canadensis* (O'Connor and Nielsen 1981), American mink *Neovison vison* (Aulerich et al. 1974; Wobeser et al. 1976), cat *Felis catus* (Charbonneau et al. 1974), ferret *Mustela putorius furo* (Hanko et al. 1970) and wild animals such as river otter (Sheffy and St Amant 1982; Wren 1985; Langlois and Langis 1995; Fortin et al. 2001; Strom 2008; Sellers 2010; Sleeman et al. 2010; Dornbos et al. 2013), Eurasian otter *Lutra lutra* (Hernández et al. 1985; Hyvärinen et al. 2003; Lodenius et al. 2014), American mink *Neovison vison* (Sheffy and St Amant 1982; Langlois and Langis 1995; Fortin et al. 2001; Wobeser and Swift 1976). Panel (**B**) Used data of intoxicated birds such as cowbird *Molothrus ater* (Finley et al. 1979), redwing *Agelaius phoeniceus* (Finley et al. 1979), starling *Sturnus vulgaris*

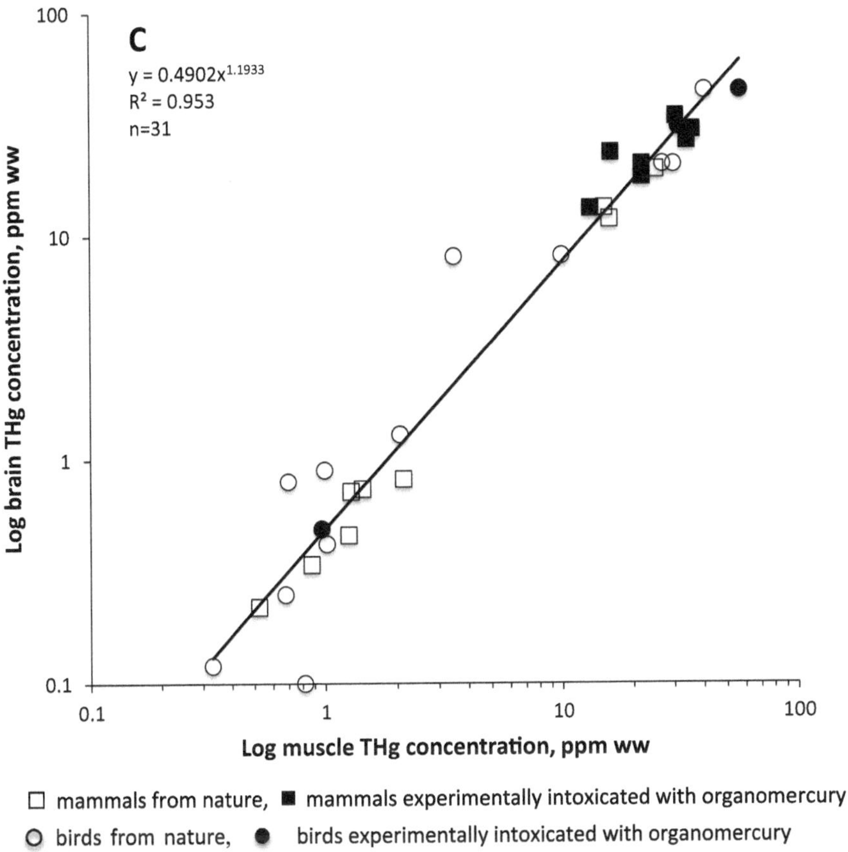


**Fig. 17.1** (continued) (Finley et al. 1979), grackle *Quiscalus quiscula* (Finley et al. 1979), American kestrel *Falco sparverius* (Bennett et al. 2009), mallard *Anas platyrhynchos* (Hough and Zabik 1972) and wild birds from nature such as osprey *Pandion haliaetus* (Holt et al. 1979; Norheim and Frøslie 1978; Evers et al. 2005; Hopkins et al. 2007; Kalisinska et al. 2014a), bald eagle *Haliaeetus leucocephalus* (Evers et al. 2005; Evans 1993), white-tailed eagle *H. albicilla* (Norheim and Frøslie 1978; Kalisinska et al. 2014a; Henriksson et al. 1966; Falandysz et al. 2000), common loon *Gavia immer* (Evers et al. 2005; Scheuhammer et al. 1998b), common merganser *Mergus merganser* (Langlois and Langis 1995; Scheuhammer et al. 1998b; Kalisinska et al. 2010). Panel (**C**) Used data of intoxicated mammals: river otter (O'Connor and Nielsen 1981), American mink (Aulerich et al. 1974; Wobeser et al. 1976), cat (Charbonneau et al. 1974), ferret (Hanko et al. 1970) and wild mammals from nature such as river otter (Sheffy and St Amant 1982; Wren 1985; Fortin et al. 2001; Strom 2008; Sleeman et al. 2010; Dornbos et al. 2013), American mink (Sheffy and St Amant 1982; Fortin et al. 2001; Wobeser and Swift 1976), Eurasian otter (Kalisinska et al. 2016, 2017); intoxicated birds, cowbird (Finley et al. 1979), redwing (Finley et al. 1979), starling (Finley et al. 1979), grackle (Finley et al. 1979), mallard (Hough and Zabik 1972); and wild birds from nature, osprey (Holt et al. 1979; Hopkins et al. 2007; Kalisinska et al. 2014a), bald eagle (Evans 1993), white-tailed eagle (Holt et al. 1979; Kalisinska et al. 2014a; Henriksson et al. 1966; Jensen et al. 1972), common merganser (Kalisinska et al. 2010)

and hepatic THg concentrations in inland mammals (Lord et al. 2002; Millan et al. 2008; Strom 2008; Kalisinska et al. 2009; Lodenius et al. 2014) and birds (Hopkins et al. 2007; Eagles-Smith et al. 2008), although not always (Halbrook et al. 1994; Kalisinska et al. 2010; Lanocha et al. 2014). These ambiguous results may be related to the large variations of hepatic THg concentration and $M_{THg}/L_{THg}$ index in endothermic animals exposed to Hg. The mean value of the index is statistically higher in mammals than birds (0.42 versus 0.31, $t = 2.34$; $p < 0.03$). Wolfe et al. (2007) emphasized a poor correlation between liver THg concentration and its effects. Unlike the liver, the muscle THg concentration is more representative of brain THg concentration and correlates better with its effect. Moreover, MeHg is a dominant species of Hg in the brain and muscle tissues. These suggestions are supported by our analysis of data on THg concentration in the muscle and brain of mammals and birds combined into one group (Fig. 17.1, panel c). The correlation coefficient for this relationship exceeded 0.97, and values of the index $B_{THg}/M_{THg}$ for mammals and birds were close, at 0.73 and 0.82, respectively. In another study, Shore et al. (2011) suggested that the lethal concentrations of THg in the brain of mammals and birds are >10 and >15 mg $kg^{-1}$ ww, respectively. Taking into consideration the equation from panel c, it may be assumed that the lethal THg level in muscle is about 13 mg $kg^{-1}$ ww for mammals and 18 mg $kg^{-1}$ ww for birds. On the basis of equations from Fig. 17.1, it can be assumed that the lethal concentration of THg in the muscles of mammals and birds is in the range 10–13 mg $kg^{-1}$ ww and 7–18 mg $kg^{-1}$ ww, respectively. Heinz (1996), based on literature data, estimated that muscle Hg concentrations associated with harmful Hg exposure in adult birds ranged from 15 to 30 mg $kg^{-1}$ ww. In the context of our analysis of avian muscle, it seems that the lower value suggested by Heinz (1996) is more likely.

### 3.4.3 Inorganic and Organic Mercury Distribution in Bodies of Mammals and Birds

The three forms of Hg (elemental, inorganic and organic) that penetrate the organisms of vertebrates differ with respect to their toxicokinetics regarding absorption, distribution and accumulation. In laboratory studies, the influence of MeHg (in MeHgCl form) and mercury compounds of Hg(II) (especially $HgCl_2$) are most frequently investigated. Mercurous mercury Hg(I), for example, in the form of mercurous chloride ($Hg_2Cl_2$), is little absorbed in the body. This compound readily dissociates in body fluids where, from $Hg_2Cl_2$, double atom cations of $Hg_2{}^{2+}$ are realized and from this is formed one atom of divalent $Hg^{2+}$ and another of elemental mercury ($Hg^0$). Elemental mercury from this unimportant source and the vapour of this metal from inhaled air are oxidized into the mercuric form ($Hg^{2+}$) in erythrocytes and tissues. Both inorganic and organic Hg species are excreted primarily in faeces. Absorption of MeHg from the digestive tract in warm-blooded vertebrates is very high (about 90%), with a great amount of it excreted in faeces (about 85–90%) and 5% with urine. Scientists have estimated that only up to 15% of absorbed MeHg is incorporated in various tissues and organs. Fur or hair in mammals as well as feather in birds are also an important route of excretion, especially MeHg (Farris et al. 1993; Clarkson and Magos 2006; Wolfe et al. 2007).

Mammalian pelt and avian plumage sometimes incorporates even >80% of THg in the body. MeHg is permanently built into hair and feathers during their growth. It is a dominant species of Hg in these tissues and becomes biologically inactive there, as confirmed in studies on experimental animals and wildlife from inland ecosystems. After long exposure to MeHg in laboratory experiments and chronic exposure of wildlife, MeHg and/or THg concentrations in these keratin skin structures usually reach the highest values in comparison to liver, kidney, brain and muscle THg (Thomas et al. 1988; Farris et al. 1993; Wood et al. 1996; DesGranges et al. 1998; Mierle et al. 2000; Hyvärinen et al. 2003; Bennett et al. 2009; Lieske et al. 2011; Nam et al. 2012; Wang et al. 2014). However, THg and/or MeHg are rarely assayed in all of the mentioned tissues in the same individuals. Eventually, MeHg is removed from mammalian and avian bodies during moulting, and therefore hair and feathers are also an important additional route of Hg excretion (Honda et al. 1986; Farris et al. 1993; Clarkson and Magos 2006; Wolfe et al. 2007; Wang et al. 2014; Evans et al. 2016). After Hg in fur and feathers, the second largest Hg pool can be found in skeletal muscles, with up to 50% of the remaining MeHg in the body (Farris et al. 1993; DesGranges et al. 1998; Saeki et al. 2000; Nam et al. 2005) from the large proportion of skeletal muscles in the body mass of vertebrates and their vasculature. For example, in the body of predatory mammals and birds, these muscles represent on average 50–55% and 30–40% of body mass, respectively (Honda et al. 1986; Biewener 2011; Muchlinski et al. 2012), and in the case of fish, it is up to 70% of their body weight (Kisia 1996). In the muscles of warm-blooded vertebrates from inland ecosystems, Hg occurs mainly in the form of MeHg (70–95% of THg), and the concentration is usually low ($<0.50$ mg $kg^{-1}$ ww), with the exception of the muscles of fish species near the top of a food web and piscivorous wildlife (Wren et al. 1980; Mason et al. 1986; Rothschild and Duffy 2005; Kinghorn et al. 2007; Strom 2008; Ruelas-Inzunza et al. 2009; Chumchal et al. 2011; Burger et al. 2013; Hall et al. 2014; Kalisinska et al. 2014a, b, 2017; Wentz et al. 2014). Observed transient storage of large amounts of MeHg in the muscle may protect other tissues against MeHg toxicity.

Because of the large proportion of muscles in body weight and easily digestible MeHg contained in them, they play an important role in the transfer of this substance from freshwater invertebrates and fish to semiaquatic piscivores and benthophages and from carrion of these animals to terrestrial scavengers (Sheffy and St Amant 1982; Halbrook et al. 1994; Langlois and Langis 1995; Fortin et al. 2001; Evers et al. 2005; Chumchal et al. 2011; Kalisinska et al. 2009, 2016). However, mercury, especially MeHg, is rarely assayed in the muscles of warm-blooded vertebrates. Among the tissues of terrestrial vertebrates, Hg achieves the highest concentration in the liver and kidneys, so THg is most frequently analysed in these organs, although in total they account for no more than 4–6% of the body weight of mammals and birds (Fischer and Bartlett 1957; Holliday et al. 1967; Hughes 1970; Kruska and Schreiber 1999; Lanszki et al. 2008; Balk et al. 2009; Kalisinska et al. 2010). In the kidney and livers of many fish-eating mammalian and avian species, the percentage of MeHg decreases as THg concentration increases in the organs (Norheim and Frøslie 1978; Wiener et al. 2003; Gamberg et al. 2005a). The liver and kidney have been suggested as one of the major sites of MeHg demethylation in mammals and birds. Above the threshold value of 10 mg THg $kg^{-1}$ dw (~3 mg THg $kg^{-1}$ ww), hepatic %MeHg declines rapidly from a high value (~90%) (Eagles-Smith et al. 2009). However, interspecies differences are observed in this respect, and hepatic intensification of MeHg demethylation in birds can occur already in the range of 5–7 mg THg $kg^{-1}$ dw, because then %MeHg in THg falls below 70% (Scheuhammer et al. 1998b; Dietz et al. 2013; Kalisinska et al. 2014c). Some researchers (Gamberg et al. 2005a; Martin et al. 2011) suggest that in piscivorous mammals (such as mink), the demethylation process of hepatic MeHg is activated well below the 30 mg THg $kg^{-1}$ dw threshold (10 mg $kg^{-1}$ ww) suggested by Wiener et al. (2003). Energy costs of MeHg demethylation in avian and mammalian livers are probably high but to date have not been estimated (Eagles-Smith et al. 2009; Dietz et al. 2013; Kalisinska et al. 2014c). Methylmercury demethylation is observed in tissues other than the liver and kidney but at a lower intensity and efficiency. This process is well documented in the brain of a number of mammals and birds, including terrestrial species. However, species of endothermic animals differ in the proportion of brain MeHg to THg (Vahter et al. 1994; Farina et al. 2003; Scheuhammer et al. 2008, 2015). It is generally assumed that demethylation of MeHg in fish and other vertebrate muscles does not occur or is negligible, with the percentage of MeHg in THg usually exceeding 80–90% (O'Connor and Nielsen 1981; Houserova et al. 2006; Strom 2008; George et al. 2011; Kalisinska et al. 2014b; Harley et al. 2015; Scheuhammer et al. 2015). However, in a few papers concerning the muscle of fish, birds and mammals, we can find data indicating that %MeHg can be $\leq$70%, especially in cases where THg $< 0.5$ mg $kg^{-1}$ ww. Pal et al. (2012) and Park et al. (2010) found in 8 out of 13 (8/13) and 5/13 investigated Asian freshwater fish species (generally with muscle THg 0.05–0.45 mg $kg^{-1}$ ww) mean values of % MeHg were in the range 50–69%. Sometimes in predatory freshwater fish (such as *Elops machnata* and *Pelates quadrilineatus* from Taiwan), whose muscles contained >1 mg THg $kg^{-1}$ ww, MeHg did not exceed 70% of THg (Huang et al. 2008). In three aquatic birds species from Mexico THg muscle levels varied from 0.32 to 0.45 mg $kg^{-1}$ ww, and the %MeHg was in the range of 26–61% (Ruelas-Inzunza et al. 2009). In two populations of white-tailed eagle from Europe, the share of MeHg reached 45 and 58% when the mean THg in the eagle muscle was just 4.8 and 0.46 mg $kg^{-1}$ ww, respectively (Norheim and Frøslie 1978; Kalisinska et al. 2014a). In the muscle of the piscivorous river otter mean, 72% MeHg of THg was sporadically revealed (THg $= 0.89$ mg $kg^{-1}$ ww, Wren et al. 1980), but in marine cetaceans a value <70% was very often noticed. In 11 out of 16 studied species, the means were in the range 36–67%, and THg concentrations varied from 1.0 to 39.5 mg $kg^{-1}$ ww (Endo et al. 2005). The data quoted above may indicate that MeHg demethylation in vertebrate muscles does occur, although this process requires further investigation and collection of more data. On the basis of comparative studies of two populations of blue shark (*Prionace glauca*) from the Azores and the Canary Islands, Branco et al. (2004) speculated that the diet of migrating animals may differ significantly in MeHg content due to differences in exposure to Hg at different locations. Periodic stays in areas where prey contains less MeHg promote gradual demethylation and elimination of MeHg already accumulated in the muscles of sharks, and at the same time the supply of new portions of MeHg with food to their organisms is then lower. Branco et al. (2004) found in the shark's muscle from the Canaries %MeHg much lower than in sharks from the Azores 55–70% and >80%, respectively, although muscle THg concentrations were similar.

## *3.5 Mercury in Elements of Inland Food Chains*

Food is the main source of Hg for humans and wildlife, but its absorption from digestive tract is strictly dependent on the chemical form and amount of Hg in various diets. Mercury concentrations increased from autotrophic organisms to herbivores < detritivores < omnivores < carnivores (Rimmer et al. 2010). For terrestrial herbivorous and omnivorous animals, plant, fungi and invertebrates are the most important components of their food. Soil invertebrates, insects, spiders and other arthropods or small- and medium-sized birds and mammals are eaten by different predators depending on their body size and food preferences. Some carnivore mammals and aquatic birds of inland habitats are highly specialized in catching fish. Below are presented some aspects of Hg transfer between different environmental components, including soil and plants, plants and warm-blooded herbivores as well as preys and predators.

### 3.5.1 Mercury in Plants and Mushrooms

The amount of MeHg in soils is low relative to THg, and the dominant form in soils is InHg (Burton et al. 2006). Bioavailability of soil InHg for plants is very low. A significant part of the InHg taken from the soil is retained in the roots, which are a barrier to mercury uptake. There is a positive correlation between the concentration of InHg in the soil and roots, but it does not occur between soil Hg and its content in shoots and leaves, which are about ten times lower than in the roots, and probably the transport of Hg from the roots to the stems either does not occur or is a very slow process (Wang and Greger 2004; Tomiyasu et al. 2005). The main soil factors affecting the collection of this toxic metal by plants include organic matter, oxygen and carbon, redox potential, Hg species and their concentrations and the presence of other metals in the soil solution (Tomiyasu et al. 2005; Patra and Sharma 2000; Azevedo and Rodriguez 2012). In plants, the dominant form is InHg, which is >97% THg (Mailman and Bodaly 2005; Dombaiová 2005). In unpolluted areas, THg concentration in leaves is negligible and is characterized by considerable variability, ranging from several to several dozen micrograms per kilogram of dry matter ($\mu g\ kg^{-1}$ dw). In addition to the species diversity of plants, it is related to seasonal variation. In young leaves, compared to older ones, at the end of the growing season, the concentration of mercury is an order of magnitude smaller. The mercury detected in the leaves basically comes from the surrounding air, most likely $Hg^0$, entering through the stomata, and probably leaf uptake of Hg is irreversible (Bushey et al. 2008; Laacouri et al. 2013). In areas where Hg was mined (e.g. Almaden zone in Spain), Hg concentration in soils is many thousand mg $kg^{-1}$ dw, and in some herbal plant species, it reaches 7–23 mg THg $kg^{-1}$ dw (or 7000–23,000 μg $kg^{-1}$ dw), thousands of times greater than in plants in uncontaminated areas (Moreno-Jimenez et al. 2006; Laacouri et al. 2013). In contrast to InHg, which is absorbed by the root system and kept there, in wetlands MeHg enters more efficiently both to water plants and through the roots to the aerial parts of plants (Patra and Sharma 2000; Windham-Myers et al. 2014). This organic species of mercury in plants from paddy fields may reach levels up to 63 μg $kg^{-1}$ dw in rice grain and pose a significant health risk to people, as has been demonstrated for rice grown on soils with a high concentration of Hg in Asia (Qiu et al. 2012; Rothenberg et al. 2014). Probably, due to the consumption of rice grain from such areas, not only humans but also grain-feeding animals (especially granivorous birds) are at risk of MeHg intoxication.

Of the nonanimal inland organisms, fungi are considered the greatest accumulator of Hg from the soil (Falandysz and Borovička 2013). Usually, higher Hg concentrations are detected in these than in their substrates, and fungi accumulate especially high levels in the areas of geochemical anomalies such as the mercuriferous Eurasian belt (including Almaden in Spain, Monte Amiata in Italy and Chinese Yunnan Province). In the mushrooms found there, the average concentration of THg varies from 1 to 100 mg $kg^{-1}$ dw (Bargali and Baldi 1984; Esbri et al. 2011; Falandysz et al. 2015). The concentration of MeHg in mushrooms is generally low and ranges between 0.01 and 3.7 mg $kg^{-1}$ dw, with the proportion of MeHg in THg not exceeding 5% (Bargali and Baldi 1984; Rieder et al. 2011).

### 3.5.2 Mercury in Earthworms

For some terrestrial invertebrates and vertebrates, the source of mercury may be soil contaminated with Hg. It is the essential food of earthworms or is an admixture for the intaken plant and animal foods of soil invertebrates, birds and mammals (Hargreaves et al. 2011; Rieder et al. 2013). In soils, over 90% of the invertebrate biomass may consists of earthworms. That is why, they are a significant object in ecotoxicological studies on Hg (Zhang et al. 2009; Teršič and Gosar 2012; Rieder et al. 2011; Abeysinghe et al. 2017). Concentrations of Hg in earthworm bodies depend on animal species and various soil conditions such as Hg forms and their amount, content of organic matter, pH and oxygen availability (Zhang et al. 2009; Rieder et al. 2013; Abeysinghe et al. 2017). Additionally, Rieder et al. (2011) demonstrated that earthworms inhabiting topsoils (endogenic) contained the highest concentrations THg and MeHg, followed by deep-burrowing earthworms (anecic) and litter-inhabiting organisms (epigeic). Methylated organic Hg species bioaccumulate more readily, and much higher bioconcentration factors (BCFs) from soil to earthworms are reported for MeHg than for THg (BCFs are calculated as THg or MeHg concentrations in the organisms divided by the corresponding concentrations in the soils). For example, in earthworms from Swiss forest, soils non-contaminated with Hg (mean THg level at 0.18 mg $kg^{-1}$ dw) mean THg and MeHg in all investigated earthworm groups were 1.04 and 0.09 mg $kg^{-1}$ dw, respectively. The share of MeHg in THg did not exceed 9%. BCF for THg and MeHg differed significantly: 7.2 vs. 83.1 (Rieder et al. 2011).

Analogical data has also been presented for earthworms living in rice paddy soils (Abeysinghe et al. 2017) sampled at various distances from abandoned mercury mines in Guizhou (China) and at control sites without a history of Hg mining. The highest mean THg concentrations were detected in the soil near the mines (80–125 mg $kg^{-1}$ dw). However, even at sites distant from a mine (7–8 km) and in control samples THg levels were quite high (~20 and ~0.6 mg $kg^{-1}$ dw, respectively). On the other hand, in those samples the concentration of MeHg was negligible and did not exceed 0.001 mg $kg^{-1}$ dw, with the proportion of MeHg in THg estimated at $\leq$0.01%. In earthworm bodies, mean concentrations of THg and MeHg decreased with the increasing distance from the mine. In the animal samples at sites distant 7–8 km from the mine and at control areas, the average THg concentrations were approximately 10 and 0.60 mg $kg^{-1}$ dw and for MeHg 0.10 and 0.05 mg $kg^{-1}$ dw, respectively. Share of MeHg in THg in the two group of earthworms was ~8 and ~2%, similar to levels reported by other researchers (Zhang et al. 2009; Rieder et al. 2011). In the study, BCFs for THg and especially for MeHg increased with distance from the mine. In earthworms sampled at 7–8 km from mines and at reference sites mean values of BCFs for THg were in the range 0.5–1.0 and almost three orders of magnitude lower than BCFs calculated for MeHg. Mean values of BCF for MeHg at control sites and 7–8 km from mines were about 900 and 300, respectively. Abeysinghe et al. (2017) suggested that specific soil conditions in rice paddies may make the earthworms important biomagnifiers of MeHg. Such large differences observed between the BCF for THg and BCF for MeHg in the case of earthworms and soils (even with negligible Hg contamination) are influenced by very high absorption of lipophilic MeHg from their intestine compared to InHg. This is probably due to the methylation of InHg occurring in their digestive tract due to the activity of the microbiota. At least two arguments for this are given by Rieder et al. (2013) on the basis of their experimental studies. Firstly, earthworms contained about six times higher concentrations of MeHg if they grew in soils treated with InHg than in soils without Hg. Secondly, the concentrations of MeHg in earthworm casts and in the soils were similar and did not change over time.

### 3.5.3 Mercury in Spiders and Insects

Studies of MeHg contamination of food webs have historically focused on aquatic organisms including those inhabiting inland reservoirs. However, recent reports have shown that terrestrial organisms such as songbirds, bats and reptiles can exhibit elevated Hg burden by feeding on MeHg-contaminated spiders and insects (Cristol et al. 2008; Jackson et al. 2011; Drewett et al. 2013; Yates et al. 2014; Gann et al.

2015). Studies in this field are mainly conducted in floodplains, riparian and wetland ecosystems of North America, which have documented historical influence of Hg pollutants. It has been shown that in such areas (especially not too distant from Hg point sources) MeHg in terrestrial predatory spiders from the Lycosidae family reach high concentrations, in the range 0.60–1.29 mg $kg^{-1}$ dw, which may be comparable or greater than in fish from neighbouring waters (Cristol et al. 2008; Speir et al. 2014; Gann et al. 2015; Standish 2016). In areas with negligible contamination or unpolluted with Hg, average concentration of MeHg in Lycosidae varies from 0.06 to 0.15 mg $kg^{-1}$ dw (Bartrons et al. 2015; Gann et al. 2015; Tavshunsky et al. 2017). Depending on trophic position (which can be derived from $\delta^{15}N$), other arthropods in the areas with the historically proven exposure to Hg may exhibit MeHg concentrations from 0.02 mg $kg^{-1}$ dw in herbivorous leafhoppers to 1.18 mg $kg^{-1}$ dw in detritofagous isopods (Cristol et al. 2008; Standish 2016).

Long-lived cicadas are another example of increased concentrations of MeHg in arthropods. The larvae of these insects live in the ground (2–17 years) and feed on root juice. In the Hg-contaminated soils, the effective absorption of MeHg occurs through the roots from where it can be taken up by cicadas. In Huludao City (NE China), with a chlor-alkali plant and two zinc smelters (industrial sources of Hg), its soils contained on average 4.08 mg THg $kg^{-1}$ dw and 0.009 mg MeHg $kg^{-1}$ dw. Cicadas *Cryptotympana atrata* from such soils accumulated in their bodies on average 0.124 mg MeHg $kg^{-1}$ dw, in a range from 0.021 to 0.319 mg MeHg $kg^{-1}$ dw (Zheng et al. 2010). Thus, these insects, although not associated with aquatic food chains, may constitute an important local source of MeHg intoxication for predatory arthropods, insectivorous birds, bats and other animals. The number of studies on bioaccumulation and biomagnification of MeHg in terrestrial food webs is gradually increasing, which should result in a better understanding and explanation of these processes. Importantly, this requires close cooperation between specialists in various fields, including zoology, ecology, toxicology of animals, plants and soils.

### 3.5.4 Transfer of Mercury from Inland Aquatic Ecosystems to Terrestrial Vertebrates

Compared to the Hg transfer between the links of the aforementioned food chains, much more data has been gathered on predatory warm-blooded vertebrates (including semiaquatic mammals and aquatic birds) that inhabit inland areas and feed on aquatic food, especially fish. Studies on the relationships between these organisms, taking into account Hg forms and their concentration levels, have been conducted at least since the mid-twentieth century. Their initiation was closely related to the dramatic events in the Gulf of Minamata and documented the neurotoxic and disruptive effects of Hg on reproductive processes in humans and other homeothermic animals. Fish (and in less degree shellfish) are considered most significant source of MeHg exposure for humans and wildlife. Therefore, many countries have set standards to protect humans from Hg in food. For example, in the EU the limit for Hg in freshwater fish for humans is 0.5 mg $kg^{-1}$ ww or 500 μg $kg^{-1}$ ww

(1000 μg $kg^{-1}$ ww for pike *Esox lucius* and eel *Anguilla anguilla*) (Commission Regulation, EC 2006), and in the United States 300 μg MeHg $kg^{-1}$ ww is recommended (US EPA 2001, 2010). According to the EU Water Framework Directive, Environmental Quality Standards (EQS) for some chemicals in biota have been set, with mercury being defined as a priority hazardous substance (Directive 2008/105/EC). EQS are intended to protect top predators against secondary poisoning and refer to THg; for freshwater fish, the EQS for Hg (EQS/Hg) is at 20 μg $kg^{-1}$ ww. Apart from the EU, only Canada has a standard designed of Hg (MeHg) to protect fish-eating animals at 33 μg $kg^{-1}$ ww (Canadian Environmental Quality Guidelines 2000). The Canadian standard concerning Hg in freshwater fish is 65% higher than the European EQS/Hg. In North America the value of 100 μg $kg^{-1}$ ww in fish is of concern for the protection of piscivorous mammals, including mink and otters (Scudder et al. 2009). However, robust data on the dietary Hg exposure thresholds that result in deleterious effects, including disturbances in reproduction, exist only for very few bird species. Typical range of Hg effect thresholds are approximately from 200 to over 1400 μg $kg^{-1}$ ww in natural and/or experimental diets (Fuchsman et al. 2017). In North America, the piscivorous common loon has been intensively studied in field and laboratory settings (Evers et al. 2003; Kenow et al. 2008; Scheuhammer et al. 2008; Kenow et al. 2011). The dietary screening benchmark of 180 μg Hg $kg^{-1}$ ww in whole body prey fish was established for this species, characterized as moderately sensitive to Hg intoxication (Heinz et al. 2009; Depew et al. 2012).

The concentration of Hg in fish depends on the degree of environmental pollution with this metal, the intensity of Hg methylation, the size of fish (closely correlated with their age) and their trophic level (Depew et al. 2013a; Eagles-Smith et al. 2014). Because of the higher cost of MeHg analysis (2–3 times greater than that for Hg analysis), THg in various animal tissues is assayed in the most investigations, including monitoring studies. It is generally accepted that the Hg in fish muscle occurs in the form of MeHg, which accounts for ~90% of THg (US EPA 2010). Concentration of Hg in freshwater fish in various parts of the world varies considerably. The United States and Canada have very large databases on Hg concentration in many species of fish. These data (after appropriate selection, standardization and statistical processing) allow an estimate of Hg concentrations in prey ($Hg_{PREY}$) of piscivorous fish and wildlife and evaluate their potential. In North America ecological monitoring of Hg depends crucially on top piscivorous fish such as walleye *Sander vitreus* and northern pike *Esox lucius* and among fish from lower trophic levels—yellow perch *Perca flavescens* and largemouth bass *Micropterus salmoides*. Among piscivorous wildlife, Hg monitoring uses common loon, bald eagle (to a smaller extent), mink and river otter (Evers and Clair 2005; Evers et al. 2007; Depew et al. 2013a). The United States Geological Survey (USGS) developed the National Descriptive Model for Mercury in Fish (NDMMF, http://emmma.usgs.gov; Wente 2004), which was later adopted in Canadian reports (Depew et al. 2013b). For example, in standardized fish (collected in 1998–2005) coming from streams across the United States, fish Hg concentrations at 27% of sampled sites exceeded the US EPA human health criterion (300 μg $kg^{-1}$ ww). However, THg concentrations in fish from >66% of the sites exceeded the value of 100 μg $kg^{-1}$ ww that is of concern for the protection of piscivorous mammals. The highest mean Hg concentrations (between 1800 and 1950 μg $kg^{-1}$ ww) were noticed in fish from blackwater coastal-plain streams draining forests or wetlands in eastern and south-eastern part of the United States as well as from streams draining gold- or Hg-mined basins in the Western United States (Scudder et al. 2009). Clearly lower concentrations of Hg were found in fish living in 21 national parks in the Western United States, with average value at ~78 μg $kg^{-1}$ ww (Eagles-Smith et al. 2014). According to Depew et al. (2013b) Hg concentration in Canadian fish (gathered in years 1967–2010) averaged 370 μg $kg^{-1}$ ww (from below detection to 10,430 μg $kg^{-1}$ ww). In fish from years 1990–2010 estimated $Hg_{PREY}$ ranged from 10 to 960 μg $kg^{-1}$ ww with a mean of 90 μg $kg^{-1}$ ww, decreasing westwards. This is consistent with spatio-temporal tendency in the United States of a decrease in $Hg_{PREY}$ from east to west (Evers et al. 2007). This situation is closely related to the strong industrialization of the south-eastern regions of Canada and the Northeastern United States, where winds carry air masses anthropogenically contaminated with Hg. Mercury contamination is gradually deposited westwards, but the influence of mercury from bedrocks cannot be ruled out either (Page and Murphy 2005; Evers et al. 2007; Wentz et al. 2014).

The European Union as a whole lacks a common database on Hg concentration in fish that could be comparable to the North American one. In Scandinavian countries Hg concentrations in freshwater fish have been reported regularly since the late 1960s and early 1970s. The most data was collected for top predator northern pike followed by Eurasian perch *Perca fluviatilis*, which represents a lower trophic level (Munthe et al. 2007; Danielsson et al. 2011; Akerblom et al. 2014). Munthe et al. (2007) took into account all lacustrine data for Sweden, Norway and Finland from 1965 to 2004. In a standardized size of pike and perch, they found mean Hg concentrations of 730 and 400 μg $kg^{-1}$ ww, respectively, in the three Scandinavian countries. Importantly, mean value of Hg in "standard fish" (1 kg pike or 0.3 kg perch or 3.2 kg brown trout *Salmo trutta* or 1.4 kg Arctic char *Salvelinus alpinus*) was estimated to be as high as 690 μg $kg^{-1}$ ww (Munthe et al. 2007). The authors of that report stated that the data from Scandinavia show some similarity with data from a large survey in NE North America, when considering the mean values for various fish species. In the recent past in both regions, similar levels of atmospheric Hg pollution were noticed, and the geographic characteristics of bedrock and soils exhibit many analogies. For Eurasian perch and North American yellow perch, the mean concentrations were comparable: 400 versus 440 μg $kg^{-1}$ ww. However, an important difference was observed between these two regions, with Hg concentration in the pike from Scandinavia higher than in NE North America: 730 vs. 640 μg $kg^{-1}$ ww (Munthe et al. 2007). Miller et al. (2013) analysed data concerning Eurasian perch from Sweden and Finland covering the period 1974–2005. Swedish data from a later period (post-1996) show that in the fish from 22 and 72% lakes Hg concentrations were as high as >500 μg $kg^{-1}$ ww and between 200 and 500 μg $kg^{-1}$ ww, respectively. By contrast, after 1996 more lakes in Finland showed Hg concentrations in fish greater than 500 μg $kg^{-1}$ ww (31%), while fewer lakes had fish Hg concentrations below 500 μg $kg^{-1}$ ww (68%). Despite considerable reductions in Hg use and production as well as lower Hg atmospheric deposition in these countries, Miller et al. (2013) indicated that Hg concentrations in the fish exceeded the EQS/Hg (and EQS/Hg for the Nordic region was 200–250 μg $kg^{-1}$ ww). Moreover, in both Finland and Sweden, the perch from over 90% lakes exhibited Hg concentration exceeding 100 μg $kg^{-1}$ ww, which in North America is a level of concern for the protection of piscivorous mammals. One of the probable reasons for the persistence of elevated Hg concentrations in fish may be significantly lower selenium concentration in the Scandinavian environment (similar to Poland and eastern Germany). The deficiency of this element in the diet of vertebrates is accompanied by an increased accumulation of Hg, and in the case of fish from Scandinavia and Poland, this was indicated by Julshamn et al. (1986), Lindqvist et al. (1991), Hultberg (2002) and Kalisinska et al. (2017). In addition to the aforementioned species, bream *Abramis brama* is used to assess the quality of the aquatic environments in Europe, a common benthofagous species sampled in the German Environmental Specimen Bank (Wellmitz 2010). German biomonitoring research from the years 1994–2009 showed that on average Hg concentrations changed from ~100 to 350 μg $kg^{-1}$ ww and exceeded the EQS/Hg in all analysed years and all 17 sites from which breams came from: rivers Rhine, Danube, Saar, Elbe and its tributaries Saale and Mulde (Wellmitz 2010). Between 2007 and 2013, Hg levels were analysed in breams from five riverine places in France, Netherlands, Sweden and United Kingdom and one German lake as reference site (Nguetseng et al. 2015). Means of Hg concentration ranged from 18 to 246 μg $kg^{-1}$ ww. However, the EQS/Hg was exceeded in all years and at all riverine sites including the reference site except for the year 2012. The available data show that in Europe, the areas with not exceeded EQS/Hg in various fish species (even in non-piscivorous breams) are not very often reported; exceptions include some freshwater aquifers in Poland and Croatia (Zrncic et al. 2013; Szkoda et al. 2014).

In Asia, several year-long and systematic biomonitoring of Hg in freshwater fish has only been conducted in South Korea. In other countries occasional research has usually concerned individual species and reservoirs (Jin et al. 2006; Kim et al. 2012; Pal et al. 2012; Zhu et al. 2012). In 2006–2008 in Korea, analysis covered 55 species of wild freshwater fish, among which seven species predominated. The most numerous of them were two piscivores (largemouth bass *Micropterus salmoides* and Far Eastern catfish *Silurus asotus*) and five omnivores (steed barbell *Hemibarbus labeo*, Korean bullhead *Pseudobagrus fulvidraco*, pale chub *Zacco platypus*, crucian carp *Carassius auratus*, carp *Cyprinus carpio*). Each freshwater fish species was assigned to an appropriate trophic level (piscivore, carnivore, omnivore, planktivore). The piscivores had the highest median Hg concentration (148 μg $kg^{-1}$ ww) than carnivores and omnivores (83 μg $kg^{-1}$ ww and 68 μg $kg^{-1}$ ww, respectively). The median in planktivores was the lowest, at 30 μg $kg^{-1}$ ww. In most piscivorous species (including largemouth bass) from 12 sites Hg level exceeded 500 μg $kg^{-1}$ ww, which is recommended by the Korea Food and Drug Administration and the World Health Organization to protect human health (Kim et al. 2012).

The fish bioaccumulation factor (BAF), which expresses the ratio of THg (or MeHg) concentration in fish to the concentration in ambient water, depends on many factors including trophic position and fish size (US EPA 2000; Yu et al. 2011). BAF is mainly presented in a logarithmic form (log10), and in freshwater prey fish and larger predatory fish, it is usually in the range from 5.9 to 6.6 (Yu et al. 2011; Scudder Eikenberry et al. 2015; Wu 2017). In ecotoxicological studies, analysis of MeHg biomagnification is very important, including indicators of changes in concentration between different trophic levels (TMF, trophic magnification factor). Extensive analysis of Lavoie et al. (2013) shows that in freshwater food webs MeHg levels increase by a factor of 8.1 per trophic level. In addition, they stated that TMF is higher in lentic than lotic waters (7.6 vs. 9.8), and values of this factor increase from tropical via temperate to polar climatic zones (TMF, 3.9, 7.5 and 12, respectively). Finally, biomagnification factors (BMFs) are also estimated within food web, and the factor concerning MeHg (or THg) is expressed as the ratio of concentration in animal bodies to the concentration in their food (in ppm or ppb) (Rolfhus et al. 2011). BAF and BMF are very seldom presented in piscivorous inland birds and mammals, which participate in transport of Hg from aquatic to terrestrial environments. BAF and BMF calculated for piscivorous birds take into account Hg concentrations in water, fish and avian blood, feathers or eggs but rarely in soft tissues including muscle (Henny et al. 2009; Yu et al. 2011; Falkowska et al. 2013). In the transmission of Hg (especially MeHg) from freshwater fish to piscivorous wildlife of inland ecosystems muscle tissue plays important role for at least two reasons. Firstly, among soft digestible tissues, skeletal muscles represent the largest percentage of body weight, and secondly Hg present in them is almost all in MeHg form, which is easily absorbed. Therefore, it seems reasonable to analyse BMF using Hg concentrations in fish and wildlife muscle (not fish muscle and indigestible fur or feathers). For example (based on muscle tissue), in two pairs American mink—fish and Eurasian otter—fish from Western Poland BMFs were 27.3 and 10.9, respectively (Kalisinska et al. 2017). In addition, they found that these mammals quite often die on the roads and are later eaten by scavengers, thus contributing to the further transmission of Hg in the local terrestrial food web.

## 3.6 *Mercury Concentrations in Soft Tissues in Various Groups of Inland Wildlife*

The literature available in English includes many publications on the concentration of THg and much fewer investigate MeHg in soft and hard tissues of wild animals, especially in North America and Europe. In spite of this, there are basically no studies which estimate the average concentrations of THg representing the main ecotrophic groups. In order to characterize and compare the concentration of THg in wild terrestrial mammals and birds of the Northern Hemisphere, 140 studies from the years 1973–2017 were selected, including data on at least one of four soft tissues: liver, kidney, muscle and brain. The average concentrations of THg (mainly arithmetic means) from these reports concerned a minimum of three specimens of an individual species. If several groups of animals of the same species were included in the study (due to sex, age, temporal or territorial division), mean THg concentrations selected for the analysis referred to the largest number of individuals, preferring adults. Since the concentrations in soft tissues were given in mg kg$^{-1}$ in dry or wet weight, we made appropriate calculations, and the final results are presented in mg kg$^{-1}$ ww. Mammalian livers, kidneys, muscles and brains contain on average 70%, 75%, 75% and 80% of water, respectively, as calculated on the basis of several works (Weiner 1973, Blus and Henny 1990, Reinoso et al. 1997; Gamberg et al. 2005a, b; Sleeman et al. 2010; Kalisinska et al. 2012a, b). In the case of birds, it was assumed that their liver, kidney, muscle and brain contain 70%, 75%, 70% and 80% water, respectively (mean values were calculated based on the work of Cosson et al. 1988; Cosson 1989; Binkowski et al. 2013; Kalisinska et al. 2010, 2014a).

Data on mammals and birds were grouped according to their ecotrophic category. Among mammals, three groups were identified: Herb-M (predominantly herbivorous), Carn-M (terrestrial carnivores) and SemCarn-M (semiaquatic carnivores). Four groups were distinguished among the birds: TerrOmn-B (terrestrial omnivores and herbivores), TerrPred-B (diurnal and nocturnal predators), W-B (non-piscivorous waterfowl) and Pisc-B (piscivores). Groups, names of species and data sources are given in Table 17.3.

The analysis excluded cases indicating very high THg concentrations in the liver and kidneys, which were recorded in warm-blooded vertebrates living in areas heavily contaminated with mercury. For carnivorous mammals and those who prefer a different diet, excessive concentrations of mercury in liver and/or kidney were assumed to be above 16.5 and 12.5 mg kg$^{-1}$ ww, respectively, i.e. two-thirds and one-half of the value associated with mortality of mammals (lower value of the range: <25–30 mg kg$^{-1}$ ww), which was reported by Shore et al. (2011). The data of piscivorous birds and other ecotrophic groups that indicated very high Hg exposure were not included in this analysis. The threshold levels in the livers and kidneys in Pisc-B were over 2/3 of the levels, shown by Shore et al. (2011) to be associated with the death of non-marine birds (20 mg kg$^{-1}$ ww and >40 mg kg$^{-1}$ ww, respectively). Therefore, cases were excluded from statistical calculations when hepatic and nephric THg concentrations were higher than 13.2 and 26.4 mg kg$^{-1}$ ww. In relation to other ecotrophic bird groups, the exclusion limit was 1/2 of the levels indicated for avian liver and kidney by Shore et al. (2011), i.e. 10 and 20 mg kg$^{-1}$ ww.

Figure 17.2 shows the mean concentrations of THg in soft tissues of the various ecotrophic groups of birds and mammals inhabiting the inland areas in the Northern Hemisphere. Many species included in Table 17.3 occur in both Eurasia and North America. Some of them are native species on both continents (such as common loon, mallard, osprey, Eurasian elk/moose, reindeer/caribou), but some of them have been introduced, for example, fallow deer from Europe to North America and American mink and raccoon from North America to Europe (Genovesi et al. 2012; Bradley et al. 2014). Belonging to the same species and/or genus, occurrence on both continents and the large biological similarity (e.g. bald eagle and white-tailed eagle) are a justification for using their THg data in joint analyses.

**Table 17.3** Analysed ecotrophic groups of mammals and birds, species names and source of data

| Group | Species | Source of data |
|---|---|---|
| *Mammals* | | |
| Herb-M, predominantly herbivore | Ungulates<br>Red deer *Cervus elaphus*, roe deer *Capreolus capreolus*, fallow deer *Dama dama*, Japanese serow *Capricornis crispus*, mule deer *Odocoileus hemionus*, white-tailed deer *O. virginianus*, Eurasian elk/moose *Alces alces*, reindeer/caribou *Rangifer tarandus*, wild boar *Sus scrofa*<br>Others<br>European hare *Lepus europaeus*, snowshoe hare *L. americanus*, common rabbit *Oryctolagus cuniculus*, Eurasian beaver *Castor fiber*, American beaver *C. canadensis*, muskrat *Ondatra zibethicus* | Europe<br>Aastrup et al. (2000); Albinska et al. (2011); Berzas Nevado et al. (2012); Bilandžić et al. (2010a); Celechovska et al. (2008); Dobrowolska and Melosik (2002); Eira et al. (2005); Falandysz (1994); Gasparik et al. (2012); Giżejewska et al. (2014); Gnamuš and Horvat (1999); Kalas et al. (1995); Krynski et al. (1982); Lazarus et al. (2008); Piskorova et al. (2003); Rudy (2010); Sobanska (2005); Spiric et al. (2012); Srebocan et al. (2011); Suran et al. (2013); Szkoda et al. (2012); Zarski et al. (1995)<br>North America<br>Gamberg et al. (2005b, c); Keeyask Hyd Ltd (2012); Khan and Forester (1995); Langlois and Langis (1995); Robillard et al. (2002); Smith and Armstrong (1975); Wren (1986); Wren et al. (1980)<br>Asia<br>Honda et al. (1987) |
| Carn-M, terrestrial carnivore | Canids<br>red fox *Vulpes vulpes*, Arctic fox *Alopex lagopus*, raccoon dog *Nyctereutes procyonoides*, wolf *Canis lupus*<br>Others<br>Pine marten *Martes martes*, beech marten *M. foina*, American marten *M. americana*, European polecat *Mustela putorius*, Egyptian mongoose *Herpestes ichneumon*, common ganet *Ganetta ganetta*, Iberian lynx *Lynx pardinus*, brown bear *Ursus arctos* | Europe<br>Alleva et al. (2006); Bilandžić et al. (2010b); Corsolini et al. (1999); Kalisinska et al. (2009, 2012a); Komov et al. (2016); Lodenius et al. (2014); Millan et al. (2008); Piskorova et al. (2003); Prestrud et al. (1994); Zilincar et al. (1992)<br>North America<br>Dehn et al. (2006); Gamberg and Braune (1999); Hoekstra et al. (2003); Langlois and Langis (1995); Sheffy and St Amant (1982) |
| SemCarn-M, semiaquatic carnivore | Eurasian otter *Lutra lutra*, river otter *Lontra canadensis*, American mink *Neovison vison* (previously *Mustela* vison), raccoon *Procyon lotor* | Europe<br>Brzezinski et al. (2014); Gutleb et al. (1998); Hyvärinen et al. (2003); Kalisinska et al. (2012b, 2016); Kalisinska unpbl. data; Kruuk et al. (1997); Lanocha et al. (2014); Lanszki et al. (2009); Lemarchand et al. (2010); Ljungvall et al. (2017); Lodenius et al. (2014); Mason and Madsen (1992); Mason et al. (1986); |

(continued)

**Table 17.3** (continued)

| Group | Species | Source of data |
|---|---|---|
| | | Norheim et al. (1984)<br>North America<br>Bank et al. (2007); Bowman et al. (2012); Carmichael and Baker (1989); Dornbos et al. (2013); Evans et al. (2000); Fortin et al. (2001); Gamberg et al. (2005a); Halbrook et al. (1994); Harding et al. (1998); Hernandez et al. (2017); Keeyask Hyd Ltd (2012); Khan et al. (1995); Klenavic et al. (2008); Kucera (1983); Langlois and Langis (1995); Lord et al. (2002); Martin et al. (2011); Mayack (2012); Mierle et al. (2000); Poole and Elkin (1992); Poole et al. (1995); Ropek and Neely (1993); Sellers (2010); Sheffy and St Amant (1982); Souza et al. (2013); Stansley et al. (2010); Strom (2008); Wren (1986); Wren et al. (1980); Yates et al. (2005) |
| *Birds* | | |
| TerrOmn–B, terrestrial herbivore and omnivore | Galliformes<br>willow ptarmigan *Lagopus lagopus*, rock ptarmigan *L. muta*, ring-necked pheasant *Phasianus colchicus*<br>Passeriformes<br>Black-capped chickade *Poecile atricapillus*, House Wren *Troglodytes aedon*, great tit *Parus major*, song thrush *Turdus philomelos*, hooded raven *Corvus cornix*<br>Others<br>Song thrush *Gallinago gallinago* | Europe<br>Celechovska et al. (2008); Dauwe et al. (2005); Holt et al. (1979); Kalas et al. (1995); Zarski et al. (2015)<br>North America<br>Ackerman et al. (2016); Braune and Malone (2006a) |
| TerrPred–B, diurnal and nocturnal predators | Falconiformes<br>Eurasian sparrowhawk *Accipiter nisus*, Eurasian sparrowhawk *A. gentilis*, red kite *Milvus migrans*, golden eagle *Aquila chrysaetos*<br>Strigiformes<br>Eurasian eagle-owl *Bubo bubo*, tawny owl *Strix aluco* | Europe<br>Holt et al. (1979); Houserova et al. (2005); Kenntner et al. (2003); Kenntner et al. (2007); Kitowski et al. (2015); Norheim and Frøslie (1978); Walker et al. (2016)<br>Asia<br>Honda et al. (1986) |
| W–B, non-piscivore waterfowl | Anseriformes<br>Mallard *Anas platyrhynchos*, gadwall *A. strepera*, green-winged teal *A. crecca*, American black duck *A. rubripes*, northern pintail *A. acuta*, northern shoveler *A. clypeata*, wood duck *Aix sponsa*, canvasback *Aythya* | Europe<br>Falandysz et al. (1988); Florijančić et al. (2016); Kalisinska et al. (2013); Kitowski et al. (2015); Parslow et al. (1982)<br>North America<br>Ackerman et al. (2016); Braune and |

(continued)

**Table 17.3** (continued)

| Group | Species | Source of data |
|---|---|---|
| | *valisineria*, lesser scaup *A. affinis*, Canada goose *Branta canadensis*, brant *B. bernicla*, white-fronted goose *Anser albifrons*, mute swan *Cygnus olor* | Malone 2006b; Burger and Gochfeld (1985); Champoux et al. (1999); Cristol et al. (2012); Gerstenberger (2004); Hughes et al. (2014); Lindsay and Dimmick (1983); Petrie et al. (2007); Pollock and Machin (2008); Rothschild and Duffy (2005); Stickel et al. (1977); Tsipoura et al. (2011)<br>Asia<br>Saeki et al. (2000); Zamani-Ahmadmahmoodi et al. (2010) |
| Pisc–B, piscivore | Falconiformes<br>Osprey *Pandion haliaetus*, white-tailed eagle *Haliaeetus albicilla*, bald eagle *H. leucocephalus*,<br>Anseriformes<br>common merganser *Mergus merganser*<br>Suliformes<br>Great cormorant *Phalacrocorax carbo*, double-crested cormorant *P. auritus*<br>Podicipediformes<br>Great crested grebe *Podiceps cristatus*<br>Gaviiformes<br>common loon *Gavia immer*<br>Pelecaniformes<br>Grey heron *Ardea cinerea*, great blue heron *A. herodias*, American white pelican *Pelecanus erythrorhynchos* | Europe<br>Falandysz et al. (2000); Holt et al. (1979); Houserova et al. (2007); Kalisinska et al. (2010, 2014a); Kenntner et al. (2001); Kitowski et al. (2015); Krone et al. (2004, 2006); Lemarchand et al. (2012); Norheim and Frøslie (1978)<br>North America<br>Evers et al. (2005); Greichus et al. (1973); Hopkins et al. (2007); Langlois and Langis (1995); Mierzykowski et al. (2011, 2013); Rutkiewicz et al. (2011); Scheuhammer et al. (1998b); Sepúlveda et al. (1998); Stone and Okoniewski (2001); Stout and Trust (2002); Weech et al. (2003); Wolfe and Norman (1998); Wood et al. (1996)<br>Asia<br>Mazloomi et al. (2008); Nam et al. (2005) |

### 3.6.1 Mercury Concentrations in Mammalian Soft Tissues

The low concentration of THg in the aboveground parts of plants (with the predominant share of InHg poorly absorbed in the gastrointestinal tracts of mammals and birds) results in a negligible exposure of most herbivorous animals to this toxic metal. In the three groups of mammals we distinguished above, THg concentrations were the smallest in Herb-M, but they can be arranged in the following ascending order: muscle < brain < liver < kidney (0.015, 0.026, 0.056 and 0.173 mg $kg^{-1}$ ww). According to Wisconsin Veterinary Diagnostic Laboratory (WVDL 2015), normal THg concentration in cervid kidney and liver does not exceed 0.1 mg $kg^{-1}$ ww, which only in the case of liver is consistent with the level established for the

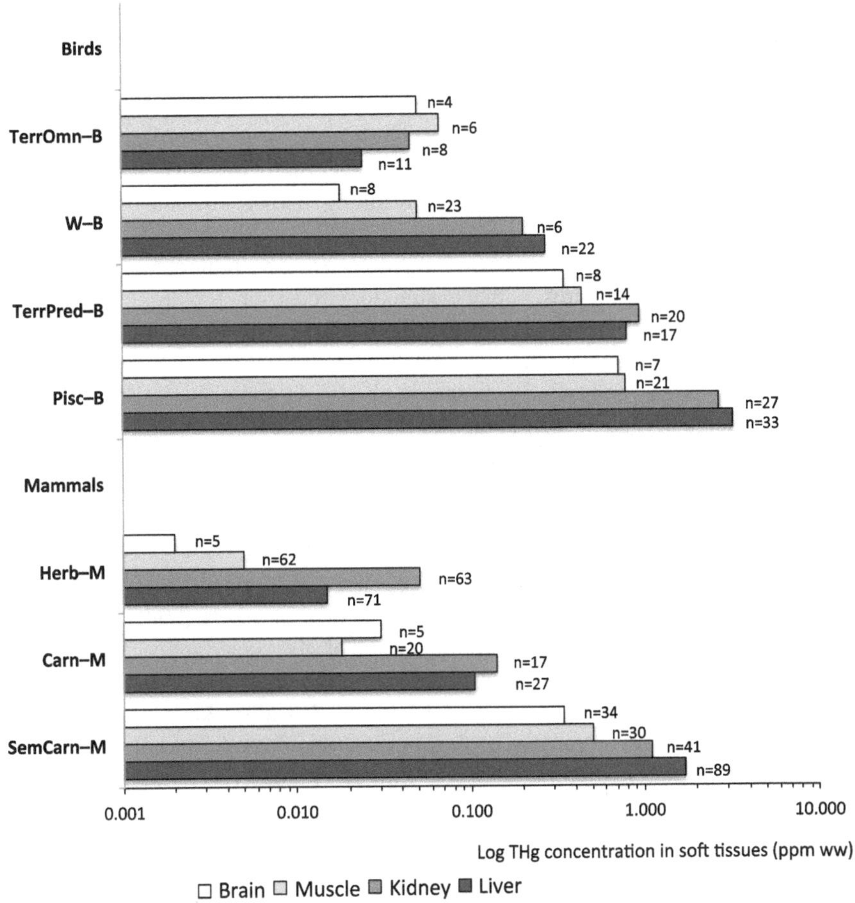


**Fig. 17.2** Medians of total mercury concentrations (ppm = mg $kg^{-1}$; ww, wet weight) in soft tissues of inland mammals (Herb–M, predominantly herbivore; Carn–M, terrestrial carnivore; SemCarn–M, semiaquatic carnivore) and birds (TerrOmn–B, terrestrial omnivore and herbivore; TerrPred–B, diurnal and nocturnal predators; W–B, non-piscivore waterfowl; Pisc–B, piscivore birds) from the Northern Hemisphere (for more details see Table 17.3)

multispecies group of Herb-M. Two reports on special cases were excluded from this group, which indicated that increased concentrations of THg may be found even among herbivores. In the early 1990s among herbivorous ungulate mammals, there were exceptions such as roe deer from zones of a mercury mine in Idrija (Slovenia), which was active in the 1990s, and caribou from Canadian Arctic, Northwest Territories and Nunavut (Gnamuš and Horvat 1999; Gamberg et al. 2005b). In the roe deer, liver and kidney Hg levels were 0.64 and 15.56 mg $kg^{-1}$ ww. Hepatic and nephric tissues of the caribou contained 2.04 and 12.80 mg Hg $kg^{-1}$ ww, respectively. In both cases, the main reason for such a high concentration of Hg was the specific diet of these animals, containing large amounts of Hg. In the aboveground parts of plants from a smelter area of Idrija, average Hg was ~50 mg $kg^{-1}$ dw. Leaves of plants in those areas intensively absorbed $Hg^0$ released during the roasting of ores containing this metal (Gnamuš and Horvat 1999). Caribou in the far north, on the other hand, mainly feeds on mosses and lichens, perennial plants which lack root systems and absorb contaminants (including Hg), along with their nutrients, from atmospheric deposition. In addition, high Hg levels in detoxification organs were related to the caribou weight loss in spring, resulting in lower absolute organ weights (Gamberg et al. 2005b).

The diet of carnivores (Carn-M) is very diverse. Their prey consists mostly of rodents, lagomorphs and birds, with the admixture of carrion, reptiles, frogs, insects, fruits and other parts of plants. In these predators, the average THg concentration in the liver and kidneys was similar and amounted to 0.105 mg $kg^{-1}$ ww and 0.140 mg $kg^{-1}$ ww, respectively. Farrar et al. (1994) argue that in the liver and kidneys of the dog, the concentration of THg usually does not exceed 0.1 mg $kg^{-1}$ ww, and in the WVDL list for canids from 2015 (including domestic dog), the normal Hg concentration in tissues is <0.1 mg $kg^{-1}$ ww and <0.200 mg $kg^{-1}$ ww, respectively. In both cases, these values do not differ from those calculated by us for the multispecies Carn-M group. In their muscles and brain, the THg concentration was an order of magnitude lower than in the liver and kidneys, and they did not exceed 0.018 mg $kg^{-1}$ ww and 0.030 mg $kg^{-1}$ ww, respectively (Fig. 17.2).

SemCarn-M group represents four species of the superfamily Musteloidea, including piscivorous mustelids. Among them, the diet of otters is 90% fish, American mink 60%, and raccoon 30% (Table 17.3; Kalisinska et al. 2017). This mammalian group is characterized by the largest body of data (especially with regard to the liver and kidney). Median THg concentration in the liver, kidney, muscle and brain of SemCarn-M were, respectively, 1.70, 1.09, 0.51 and 0.34 mg $kg^{-1}$ ww. Comparisons of median hepatic and nephric THg concentrations between Eurasian otter (liver $n = 12$, 2.57 mg $kg^{-1}$ ww; kidney $n = 7$, 1.30 mg $kg^{-1}$ ww) and river otter from North America (liver $n = 25$, 1.78 mg $kg^{-1}$ ww; kidney $n = 11$, 1.42 mg $kg^{-1}$ ww) showed no significant differences.

According to WVDL (2015), normal THg concentrations in the liver and kidneys of mustelids are <0.20–0.70 mg $kg^{-1}$ ww and <1.0 mg $kg^{-1}$ ww, respectively, much lower than our results. In North American studies from the 1980s, when Hg intoxication of otters and minks was much more frequent, background hepatic THg in those piscivorous species was indicated as <4–5 mg $kg^{-1}$ ww and ~2 mg $kg^{-1}$ ww, respectively (Wren 1986; O'Connor and Nielsen 1981; Carmichael and Baker 1989). In the light of the quoted papers from 1980s and our statistical analysis (taking into account European and North American reports from 1981 to 2017), it can be assumed that currently the values of hepatic background level for otters and American mink are <3.0 mg $kg^{-1}$ ww and <1.5 mg $kg^{-1}$ ww, respectively. According to our analysis, THg levels in the kidney, muscle and brain of the piscivorous mammals are <1.5, 1.0–1.3 and 0.3–0.6 mg $kg^{-1}$ ww, respectively.

Comparisons of hepatic THg concentration showed statistically confirmed differences between all three ecotrophic groups, and their values can be arranged in a decreasing series of SemCarn-M > Carn-M > Herb-M (1.700 > 0.105 >

0.015 mg kg$^{-1}$ ww). In relation to SemCarn-M, the concentrations of THg in the kidneys, muscles and brain of Carn-M were about an order of magnitude lower, and in the Herb-M groups, it was two orders of magnitude lower. No significant differences were found in kidney and brain THg between Carn-M and Herb-M (Fig. 17.2). In muscle, the concentrations of THg in Carn-M and Herb-M were more than 28 and 100 times lower than in SemCarn-M. In comparison to Carn-M, Herb-M had a 3.6 times lower level of muscle THg. In the muscles, similar to the liver, the concentrations could be arranged in a descending order (0.508 > 0.018 > 0.005 mg kg$^{-1}$ ww), and all intergroup differences were statistically significant.

### 3.6.2 Mercury Concentrations in Avian Soft Tissues

In birds, the lowest levels of THg in the liver, kidneys, muscles and brain occurred in the TerrOmn-B group, and their medians ranged from 0.024 to 0.067 mg kg$^{-1}$ ww. In some reports, especially in the case of muscles and the brain, the concentrations were very low, below the limit of detection, but increased levels (≥0.10 mg kg$^{-1}$ ww) were found in tissues of granivorous birds from Scandinavia in the 1970s, when large amounts of organic Hg fungicides were used in agriculture in that part of Europe (Holt et al. 1979).

Although in the liver and kidneys of waterfowl (W-B group) we found an order of magnitude higher THg concentration than in the TerrOmn-B group, the differences between these groups were not statistically significant. The largest number of differences were recorded between piscivorous birds (Pisc-B) and other analysed groups of birds. Pisc-B had the highest concentration of THg in the liver, kidneys, muscles and brain (3.21, 2.69, 0.78, 0.72 mg kg$^{-1}$ ww, respectively) and significantly differed in this regard from TerrOmn-B and W-B. Pisc-B and TerrPred-B showed no statistically confirmed difference in muscle THg (0.78 vs. 0.44 mg kg$^{-1}$ ww) and brain THg (0.72 vs. 0.35 mg kg$^{-1}$ ww).

In contrast to mammals, the WVDL list (2015) does not include the normal THg level for or different systematic groups of birds. Normal concentration is proposed of avian liver in the range of 0.01–0.10 mg kg$^{-1}$ ww and for the kidney at <0.02–0.30 mg kg$^{-1}$ ww. Puls (1988) suggested that normal concentrations of THg in the liver, kidney, muscle and brain of poultry were 0.01–0.10, 0.05–0.30, 0.008–0.100 and 0.10 mg kg$^{-1}$ ww, respectively. These levels coincide with those we calculated for TerrOmn–B, i.e. typical terrestrial birds, including galliformes. Other researchers, investigating various wild water birds, argue that hepatic and renal THg residues represent background concentrations when they are 0.3–3.0 mg kg$^{-1}$ ww (Ohlendorf 1993; Badzinski et al. 2009). This range includes median hepatic THg concentrations calculated by us for two groups of bird: W-B and TerrPred-B (Fig. 17.2). In the group of piscivorous birds (Pisc–B), hepatic THg exceeds 3.0 mg kg$^{-1}$ ww (3.21), but for the kidneys it is lower (2.69 mg kg$^{-1}$ ww).

Mammals and birds are characterized by different sensitivity to Hg, and depending on the type of intaken food, they accumulate different amounts of this toxic element. Significantly, the lowest adverse effect level (LOAEL) has not been established for most wild endothermic animals. In the common loon, in the case of the liver, kidney, breast muscle and brain, LOAEL values do not exceed 4.0, 2.3, 1.2 and 0.80 mg $kg^{-1}$ ww, respectively (Zhang et al. 2013). The quoted values coincide with the THg levels proposed by us for piscivorous birds, with the exception of muscle THg, which we estimated to be ~0.80 mg $kg^{-1}$ ww.

## 3.7 Mercury Concentrations in Hair and Feathers of Inland Wildlife

Hair (fur) and feathers are often used in ecotoxicological studies because they can be taken from living individuals. Here, the dominant form of Hg is MeHg, which reaches hair/fur/feathers in the period of their growth and reflects only that period. As in the case of soft tissues, the concentration of Hg in fur/feathers is closely related to the ecotrophic association of wildlife and Hg contamination of habitats. Sheffy and St Amant (1982) based on various furbearers from Wisconsin (USA) considered that Hg 1–5 mg $kg^{-1}$ ww (ppm dw) in hair to be normal background levels. In herbivorous mammals (such as American beaver, muskrat, white-tailed deer and lagomorphs), the average concentration of hair Hg does not exceed 0.3 ppm dw, and in many individuals it is below the limit of detection (Cumbie and Jenkins 1975; Sheffy and St Amant 1982; Stevens et al. 1997; Lourenco et al. 2011). In omnivorous mammals, such as common opossum *Didelphis marsupialis*, average hair Hg, depending on the environmental Hg pollution, ranged from 1.3 to 44 ppm dw (Cumbie and Jenkins 1975). Until recently, piscivorous mammals were thought to have the highest hair Hg concentrations among terrestrial mammals. The average hair Hg concentrations in these mammals from North America in the twentieth and twenty-first centuries usually exceeded 5, and sometimes 15 ppm dw (Sheffy and St Amant 1982; Stevens et al. 1997; Wolfe & Norman 1998; Mierle et al. 2000; Yates et al. 2005; Strom 2008). The maximum values in river otter and mink from the United States (Maine) reached 234 and 68.5 ppm dw, respectively (Yates et al. 2005). However, several years ago even greater concentrations were detected in the hair of insectivorous bats from Virginia (the South River, USA): little brown bat *Myotis lucifugus* and big brown bat *Eptesicus fuscus*, at 274 and 65.4 mg $kg^{-1}$ dw, respectively (Wada et al. 2010; Nam et al. 2012). Probably, the concentrations greater than 30 ppm dw in fur are associated with the clinical neurological effects, or they may be lethal (Wobeser and Swift 1976; Evers 2005; Basu et al. 2007), but there is little data on wild mammals in this respect.

In monitoring programs, feathers have low priority status for several reasons. Feather Hg concentration is characterized by high variability even in the same individual (depending, among others, on the type and location of feathers). Moreover, it relatively weakly correlates with the Hg concentration in soft tissues. Usually, the times of moulting and replacement of certain types of feathers are not known for most species, and it is even more complicated for migratory birds. In addition, the period of feather growth is accompanied by the redistribution of Hg in internal organs and its increased transport to feathers, both in chickens and older individuals (Honda et al. 1986; Eagles-Smith et al. 2008; Ackerman et al. 2011, 2016; Odsjo et al. 2012). In general, bird feathers have average Hg concentrations in the range of 0.1–5 ppm dw (Lodenius and Solonen 2013), but in some European herbivores, such as wood pigeon *Columba palumbus* and red-legged partridge *Alectoris rufa*, it may be <0.1 ppm (Hahn et al. 1993; Lourenco et al. 2011). Natural background levels of Hg in feathers of non-piscivorous raptorial birds are in the range 1–5 ppm dw (Scheuhammer 1991). Among adult piscivorous birds, it is estimated that this level for bald eagle in North America is much higher and in some regions ~20 ppm dw (DeSorbo et al. 2008). Among piscivorous birds the maximum concentration of Hg in feathers sometimes exceeds 190 ppm, for example, in osprey from Canada (DesGranges et al. 1998) and white-tailed eagle from Germany (Niecke et al. 1998). Mercury levels in feathers that are associated with adverse effects in birds are 5 ppm fresh weight or 7.5 mg $kg^{-1}$ dw. Concentrations of 15 ppm are required for adverse effects of mercury in some predatory birds (Burger and Gochfeld 2009). In raptorial birds concentrations >20 ppm may be connected with toxic effects, but in bald eagle it is probably >60 ppm (Scheuhammer 1991; DeSorbo et al. 2008).

Despite the large number of works with Hg concentration in mammalian fur and bird feathers, huge species and ecological diversity of wildlife make interpretation of results difficult, especially since the correlation between Hg concentration in these tissues and concentration in soft tissues in general are usually very weak or non-existent. Therefore, information on Hg obtained from fur and feather samples is not sufficient to clearly assess Hg intoxication of wildlife and their habitats.

## 4 Conclusions

Long-term studies of the abiotic environment, human toxicology and the ecotoxicology of Hg hold major gaps in knowledge on the behaviour of Hg in nature (including MeHg biomagnification) and subsequent long-term ignoring of the evidence of the negative effects of this metal on humans and other vertebrates. Maintaining the functioning of the various economic sectors based on Hg and coal-based energy has led to a dramatic increase in the environmental problems associated with Hg. Currently, the most important way to reduce anthropogenic Hg emissions and to reduce the health risks to humans and ecosystems globally is to act in international agreements. The first formal and very important preventive action was the signing of the Minamata Convention on Mercury in October 2013 (Kessler 2013; Larson 2014). However, its ratification, implementation and raising of awareness of entire societies and individuals will determine not only the health condition of this and future generations and the different environments and also the survival of many sensitive species, especially those directly or indirectly dependent on aquatic food chains. This requires, among other things, control of the presence of Hg in abiotic and biotic environments, including biomonitoring.

## References

Aastrup P, Riget F, Dietz R, Asmund D (2000) Lead, zinc, cadmium, mercury, selenium and copper in Greenland caribou and reindeer (*Rangifer tarandus*). Sci Total Environ 245:149–159

Abeysinghe KS, Yang XD, Goodale E, Anderson CWN, Bishop K, Cao A et al (2017) Total mercury and methylmercury concentrations over a gradient of contamination in earthworms living in rice paddy soil. Environ Toxicol Chem 36:1202–1210

Ackerman JT, Eagles-Smith CA, Herzog MP (2011) Bird mercury concentrations change rapidly as chicks age: toxicological risk is highest at hatching and fledging. Environ Sci Technol 45:5418–5425

Ackerman JT, Eagles-Smith CA, Herzog MO, Hartman CA, Peterson SH, Evers DC et al (2016) Avian mercury exposure and toxicological risk across western North America: a synthesis. Sci Total Environ 568:749–769

Adriano DC (2001) Trace elements in terrestrial environments. Biogeochemistry, BIOAVAILABILITY, AND RISK OF METALS. Springer, New York, pp 411–458

ADSTR (1999) Toxicological profile for mercury. Agency for Toxic Substances and Disease Registry, US Department of Health and Human Services, Public Health Service, Atlanta, pp 600

Agarwal R, Behari JR (2007) Role of selenium in mercury intoxication in mice. Ind Health 45:388–395

Agrawal H, Bhatnagar P, Flora SJS (2015) Changes in tissue oxidative stress, brain biogenic amines and acetylcholinesterase following co-exposure to lead, arsenic and mercury in rats. Food Chem Toxicol 86:208–216

Akerblom S, Bignert A, Meili M, Sonesten L, Sundbom M (2014) Half a century of changing mercury levels in Swedish freshwater fish. Ambio 43(Suppl 1):91–103

Albinska J, Góralski J, Szynkowska MI, Leśniewska E, Paryjczak T (2011) Mercury in carcasses of wild animals hunted in the province of Lodz [in Polish]. Rocz Ochr Środ 13:525–540

Allan M, Le Roux G, Sonke JE, Piotrowska N, Streel M, Fagel N (2013) Reconstructing historical atmospheric mercury deposition in Western Europe using: Misten peat bog cores, Belgium. Sci Total Environ 442:209–301

Alleva E, Francia N, Pandolfi M, De Marinis AM, Chiarotti F, Santucci D (2006) Organochlorine and heavy-metal contaminants in wild mammals and birds of Urbino-Pesaro province, Italy: an analytic overview for potential bioindicators. Arch Environ Conatm Toxicol 51:123–134

AMAP/UNEP (2008) Technical background report to the global atmospheric mercury assessment. Arctic Monitoring and Assessment Programme/UNEP Chemicals Branch, pp 159

AMAP/UNEP (2013) Technical background report for the global mercury assessment 2013. Arctic Monitoring and Assessment Programme, Oslo, Norway/UNEP Chemicals Branch, Geneva, Switzerland. vi + 263 p

Aronson SM (2005) The dancing cats of Minamata Bay. Med Health R I 88:209

Aschner JL (2000) Possible mechanisms of methylmercury cytotoxicity. Mol Biol Today 1:43–48

Aschner M, Aschner JL (1990) Mercury neurotoxicity: mechanisms of blood–brain barrier transport. Neurosci Biobehav Rev 14:169–176

Aulerich RJ, Ringer RK, Iwamoto S (1974) Effects of dietary mercury on mink. Arch Environ Contam Toxicol 2:43–50

Azevedo R, Rodriguez E (2012) Phytotoxicity of mercury in plants: a review. J Bot 2012, Article ID 848614, pp 6

Badzinski SS, Gorman KB, Petrie SA (2009) Relationships between hepatic trace element concentrations, reproductive status, and body condition of female greater scaup. Environ Pollut 157:1886–1893

Balk L, Hägerroth PA, Akerman G, Hanson M, Tjärnlund U, Hansson T et al (2009) Wild birds of declining European species are dying from a thiamine deficiency syndrome. Proc Natl Acad Sci USA 106:12001–12006

Bank MS, Burgess JR, Evers DC, Loftin CS (2007) Mercury contamination of biota from Acadia National Park, Maine: a review. Environ Monit Assess 126:105–115

Bargali R, Baldi F (1984) Mercury and methyl mercury in higher fungi and their relation with the substrata in a cinnabar mining area. Chemosphere 13:1059–1071

Barkay T, Wagner-Döbler I (2005) Microbial transformations of mercury: potentials, challenges, and achievements in controlling mercury toxicity in the environment. Adv Appl Microbiol 57:1–52

Bartrons M, Gratton C, Spiesman BJ, Vander Zanden MJ (2015) Taking the trophic bypass: aquatic-terrestrial linkage reduces methylmercury in a terrestrial food web. Ecol Appl 25:151–159

Basu N (2012) Piscivorous mammalian wildlife as sentinels of methylmercury exposure and neurotoxicity in humans. In: Ceccatelli S, Aschner M (eds) Methylmercury and neurotoxicity, Current topics in neurotoxicity, vol 2. Springer, Boston, MA, pp 357–370

Basu N, Scheuhammer AM, Bursian SJ, Elliott J, Rouvinen-Watt K, Chan HM (2007) Mink as a sentinel species in environmental health. Environ Res 103:130–144

Belzile N, Chen YW, Yang DY, Truong YTH, Zhao QX (2009) Selenium bioaccumulation in freshwater organisms and antagonistic effect against mercury assimilation. Environ Bioindic 4:203–221

Bennett RS, French JB, Rossmann R, Haebler R (2009) Dietary toxicity and tissue accumulation of methylmercury in American kestrels. Arch Environ Contam Toxicol 56:149–156

Bernhoft RA (2012) Mercury toxicity and treatment: a review of the literature. J Environ Publ Health 2012:460508

Berzas Nevado JJ, Rodríguez Martín-Doimeadios RC, Mateo R, Rodríguez Fariñas N, Rodríguez-Estival J, Patiño Ropero MJ (2012) Mercury exposure and mechanism of response in large game using the Almaden mercury mining area (Spain) as a case study. Environ Res 112:58–66

Biewener AA (2011) Muscle function in avian flight: achieving power and control. Philos Trans R Soc B 366:1496–1506

Bilandžić N, Sedak M, Dokić M, Simic B (2010a) Wild boar tissue levels of cadmium, lead and mercury in seven regions of continental Croatia. Bull Environ Contam Toxicol 84:738–743

Bilandžić N, Deždek D, Sedak M, Dokić M, Solomun B, Verenina I et al (2010b) Concentrations of trace elements in tissues of red fox (*Vulpes vulpes*) and stone marten (*Martes foina*) from suburban and rural areas in Croatia. Bull Environ Contam Toxicol 85:486–491

Binkowski ŁJ, Sawicka-Kapusta K, Szarek J, Strzyżewska E, Felsmann M (2013) Histopathology of liver and kidneys of wild living mallards *Anas platyrhynchos* and coots *Fulica atra* with considerable concentrations of lead and cadmium. Sci Total Environ 450–451:326–333

Blum JD (2011) Applications of stable mercury isotopes to biogeochemistry. In: Baskaran M (ed) Handbook of environmental isotope geochemistry, Advances in isotope geochemistry. Springer, Berlin, pp 229–245

Blus LJ, Henny CJ (1990) Lead and cadmium concentrations in mink from northern Idaho. Northwest Sci 64:219–223

Borg K, Wanntrop H, Erne K, Hanko E (1969) Alkyl mercury poisoning in terrestrial Swedish wildlife. Viltrevy 6:301–379

Bose-O'Reilly S, McCarty KM, Steckling N, Lettmeier B (2010) Mercury exposure and children's health. Curr Probl Pediatr Adolesc Health Care 40:186–215

Bowman J, Kidd AG, Martin PA, McDaniel TV, Nituch LA, Schulte-Hostedde AI (2012) Testing for bias in a sentinel species: contaminants in free-ranging domestic, wild, and hybrid mink. Environ Res 112:77–82

Bradley RD, Ammerman LK, Baker RJ, Bradley LC, Cook JA, Dowler RC et al (2014) Revised checklist of North American mammals north of Mexico. Occas Pap Mus Tex Tech Univ 327:1–27

Branco V, Canario J, Vale C, Raimundo J, Reis C (2004) Total and organic mercury concentrations in muscle tissue of the blue shark (*Prionace glauca* L.1758) from the Northeast Atlantic. Mar Pollut Bull 49:854–874

Braune BM, Malone BJ (2006a) Organochlorines and trace elements in upland birds harvested in Canada. Sci Total Environ 363:60–69

Braune BM, Malone BJ (2006b) Mercury and selenium in livers of waterfowl harvested in northern Canada. Arch Environ Contam Toxicol 50:284–289
Bridges CC, Zalups RK (2010) Transport of inorganic mercury and methylmercury in target tissues and organs. J Toxicol Environ Health B 13:385–410
Brzezinski M, Zalewski A, Niemczynowicz A, Jarzyna I, Suska-Maławska M (2014) The use of chemical markers for the identification of farm escapees in feral mink populations. Ecotoxicology 23:767–778
Burbacher TM, Rodier PM, Weiss B (1990) Methylmercury developmental neurotoxicity: a comparison of effects in human and animals. Neurotoxicol Teratol 12:191–202
Burger J, Gochfeld M (1985) Comparisons of nine heavy metals in salt gland and liver of greater scaup (Aythya marila), black duck (Anas rubripes) and mallard (A. platyrhynchos). Comp Biochem Physiol Part C Comp Pharmacol 81(2):287–292
Burger J, Gochfeld M (2009) Comparison of arsenic, cadmium, chromium, lead, manganese, mercury and selenium in feathers in bald eagle (*Haliaeetus leucocephalus*), and comparison with common eider (*Somateria mollissima*), glaucous-winged gull (*Larus glaucescens*), pigeon guillemot (*Cepphus columba*), and tufted puffin (*Fratercula cirrhata*) from the Aleutian Chain of Alaska. Environ Monit Assess 152:357–367
Burger J, Jehl JR, Gochfeld M (2013) Selenium:mercury molar ratio in eared grebes (*Podiceps nigricollis*) as a possible biomarker of exposure. Ecol Indic 34:60–68
Burton DT, Turley SD, Fisher DJ, Green DJ, Shedd TR (2006) Bioaccumulation of total mercury and monomethylmercury in the earthworm *Eisenia fetida*. Water Air Soil Pollut 170:37–54
Bushey JT, Nallana AG, Montesdeoca MR, Driscoll CT (2008) Mercury dynamics of a northern hardwood canopy. Atmos Environ 42:6905–6914
Caley ER (1928) Mercury and its compounds in ancient times. J Chem Educ 5:419–424
Canadian Environmental Quality Guidelines (2000) Canadian tissue residue guidelines for the protection of wildlife consumers of aquatic biota: methylmercury. Canadian Council of Ministers of the Environment Winnipeg, pp 7. http://ceqg-rcqe.ccme.ca/download/en/294, accessed 20.04.2014
Carmichael DB, Baker OE (1989) Pesticide, PCB and heavy metal residues in South Carolina mink. Proc Annu Conf Southeast Assoc Fish Wildl Agen 43:444–451
Celechovska O, Malota L, Zima S (2008) Entry of heavy metals into food chains: a 20-year comparison study in Northern Moravia (Czech Republic). Acta Vet Brno 77:645–652
Champoux L, Rodrigue J, Braune B, Leclair D (1999) Contaminants in Northern Québec wildlife. In: Jensen J (ed) Synopsis of research conducted under the 1997-1998 Northern Contaminants Program. Department of Indian Affairs and Northern Development, Ottawa, Canada, pp 109–116
Charbonneau SM, Munro IC, Nera EA, Willes RF, Kuiper-Goodman T, Iverson F et al (1974) Subacute toxicity of methylmercury in the adult cat. Toxicol Appl Pharmacol 27:569–581
Chumchal MM, Rainwater TR, Osborn SC, Roberts AP, Abel MT, Cobb GP et al (2011) Mercury speciation and biomagnification in the food web of Caddo Lake, Texas and Louisiana, USA, a subtropical freshwater ecosystem, USA, a subtropical freshwater ecosystem. Environ Chem 30:1153–1162
Clarkson TW (1992) Mercury: major issues in environmental health. Environ Health Perspect 100:31–38
Clarkson TW, Magos L (2006) The toxicology of mercury and its chemical compounds. Crit Rev Toxicol 36:609–662
Commission Regulation, EC (2006) Commission Regulation No 1881/2006 setting maximum levels for certain contaminants in foodstuffs. OJ EU L364/5
Corsolini S, Focardi S, Leonzio C, Lovari S, Monaci F, Romeo G (1999) Heavy metals and chlorinated hydrocarbon concentrations in the red fox in relation to some biological parameters. Environ Monit Assess 54:87–100

Cosson RP (1989) Relationships between heavy metal and metallothionein-like protein levels in the liver and kidney of two birds: the greater flamingo and the little egret. Comp Biochem Physiol 94C:243–248

Cosson RP, Amiard JC, Amiard-Triquet C (1988) Trace elements in little egrets and flamingos of Camargue, France. Ecotoxicol Environ Saf 15:107–116 (Hg, Cd, Pb, Se)

Crespo-López ME, Macêdo GL, Pereira SI, Arrifano GP, Picanço-Diniz DL, Nascimento JL et al (2009) Mercury and human genotoxicity: critical considerations and possible molecular mechanisms. Pharmacol Res 60:212–220

Cristol DA, Brasso RL, Condon AM, Fovargue RE, Friedman SL, Hallinger KK et al (2008) The movement of aquatic mercury through terrestrial food webs. Science 320:335

Cristol DA, Savoy L, Evers DC, Perkins C, Taylor R, Varian-Ramos CW (2012) Mercury in waterfowl from a contaminated river in Virginia. J Wildl Manag 76:1617–1624

Cumbie PM, Jenkins JH (1975) Mercury accumulation in native mammals of the southeast. Proc Annu Conf Southeast Assoc Game Fish Comm 28:639–648

Danielsson S, Hedman J, Miller A, Bignert A (2011) Mercury in perch from Norway, Sweden and Finland—Geographical patterns and temporal trends. Swedish Museum of Natural History, Stockholm, Report nr 8:2011, pp 22

Dauwe T, Janssens E, Bervoets L, Blust R, Eens M (2005) Heavy-metal concentrations in female laying great tits (*Parus major*) and their clutches. Arch Environ Contam Toxicol 49:249–256

De Flora S, Bennicelli C, Bagnasco M (1994) Genotoxicity of mercury compounds. A review. Mutat Res 317:57–79

Dehn LA, Follmann EH, Thomas DL, Sheffield GG, Rosa C, Duffy LK et al (2006) Trophic relationships in an Arctic food web and implications for trace metal transfer. Sci Total Environ 362:103–123

Depew DC, Basu N, Burgess NM, Campbell LM, Evers DC, Grasman KA et al (2012) Derivation of screening benchmarks for dietary methylmercury exposure for the common loon (*Gavia immer*): rationale for use in ecological risk assessment. Toxicol Chem 31:2399–2407

Depew DC, Burgess NM, Anderson MR, Baker R, Bhavsar SP, Bodaly RA et al (2013a) An overview of mercury concentrations in freshwater fish species: a national fish mercury dataset for Canada. Can J Fish Aqua Sci 70:436–451

Depew DC, Burgess NM, Campbell LM (2013b) Modelling mercury concentrations in prey fish: derivation of a national-scale common indicator of dietary mercury exposure for piscivorous fish and wildlife. Environ Pollut 176:234–243

DesGranges JL, Rodrigue J, Tardif B, Laperle M (1998) Mercury accumulation and biomagnification in ospreys (*Pandion haliaetus*) in the James Bay and Hudson Bay regions of Québec. Arch Environ Contam Toxicol 35:330–341

DeSorbo CR, Nye PE, Loukmas JJ, Evers DC (2008) Assessing mercury exposure and spatial patterns in adult and nestling bald eagles in New York State, with an emphasis on the Catskill Region. Report BRI 2008-06 Submitted to The Nature Conservancy, Albany, New York. BioDiversity Research Institute, Gorham, Maine, pp 1–34

De Vos W, Tarvainen T, Salminen R, Reeder S, De Vivo B, Demetriades A et al (2006) Geochemical Atlas of Europe. Part 2. Interpretation of geochemical maps, additional tables, figures, maps and related publications. Geological Survey, Finland

Dietz R, Riget F, Born EW (2000) An assessment of selenium to mercury in Greenland marine animals. Sci Total Environ 245:15–24

Dietz R, Sonne C, Basu N, Braune B, O'Hara T, Letcher RJ et al (2013) What are the toxicological effects of mercury in Arctic biota? Sci Total Environ 443:775–790

Directive 2008/105/EC (2008) Directive of the European Parliament and of the Council on environmental quality standards in the field of water policy, amending and subsequently repealing Council Directives 82/176/EEC, 83/513/EEC, 84/156/EEC, 84/491/EEC, 86/280/EEC and amending Directive 2000/60/EC of the European Parliament and of the Council. OJ EU L348/84

D'Itri FM (1991) Mercury contamination—what we have learned since Minamata. Environ Monit Assess 19:165–182

Dobrakowski M, Kiełtucki J, Wyparło-Wszelaki M, Kasperczyk S (2013) Effects of a chronic lead intoxication on the pathophysiological changes in the digestive system and interactions of lead with trace elements. Med Środ 16:42–46 [in Polish]

Dobrowolska A, Melosik M (2002) Mercury contents in liver and kidneys wild boar (*Sus scrofa*) and red deer (*Cervus elaphus*). Z Jagdwiss 48:156–160

Dombaiová R (2005) Mercury and methylmercury in plants from differently contaminated sites in Slovakia. Plant Soil Environ 51:456–463

Domingo JL (1994) Metal-induced developmental toxicity in mammals: a review. J Toxicol Environ Health 42:123–141

Dornbos P, Strom S, Basu N (2013) Mercury exposure and neurochemical biomarkers in multiple brain regions of Wisconsin river otters (*Lontra canadensis*). Ecotoxicology 22:469–475

Douglas TA, Loseto LL, Macdonald RW, Outridge P, Dommergue A, Poulain A et al (2012) The fate of mercury in Arctic terrestrial and aquatic ecosystems, a review. Environ Chem 9:321–355

Drasch G, Horvat M, Stoeppler M (2004) Mercury. In: Merian E, Anke M, Ihnat M, Stoepper M (eds) Elements and their compounds in the environment. WILEY-VCH, Weinheim, pp 931–1005

Drewett DVV, Willson JD, Cristol DA, Chin SY, Hopkins WA (2013) Inter- and intraspecific variation in mercury bioaccumulation by snakes inhabiting a contaminated river floodplain. Environ Toxicol Chem 32:1178–1186

Eagles-Smith CA, Ackerman JT, Yee J, Adelsbach TL, Takekawa JY, Miles AK et al (2008) Mercury correlations among six tissues for four waterbird species breeding in San Francisco Bay, California, USA. Environ Toxicol Chem 27:2136–2153

Eagles-Smith CA, Ackerman JT, Yee J, Adelsbach TL (2009) Mercury demethylation in livers of four waterbird species: evidence for dose-response thresholds with liver total mercury. Environ Toxicol Chem 28:568–577

Eagles-Smith CA, Willacker JJ, Flanagan Pritz CM (2014) Mercury in fishes from 21 national parks in the Western United States—inter- and intra-park variation in concentrations and ecological risk. U.S. Geological Survey Open-File Report 2014-1051, pp 54

Eckersley N (2010) Advanced mercury removal technologies. Hydrocarbon Proc 89:29–35

Eira C, Torres J, Vingada J, Miquel J (2005) Concentration of some toxic elements in *Oryctolagus cuniculus* and in its intestinal cestode *Mosgovoyia ctenoides*, in Dunas de Mira (Portugal). Sci Total Environ 346:81–86

Ekino S, Susa M, Ninomiya T, Imamura K, Kitamura T (2007) Minamata disease revisited: an update on the acute and chronic manifestations of methyl mercury poisoning. J Neurol Sci 262:131–144

Endo T, Haraguchi K, Hotta Y, Hisamichi Y, Lavery S, Dalebout ML et al (2005) Total mercury, methyl mercury, and selenium levels in the red meat of small cetaceans sold for human consumption in Japan. Environ Sci Technol 39:5703–5708

Esbri JM, Lopez-Berdonzes MA, Higueras P, Gonzalez-Pavon A (2011) Mercury bioaccumulation in wild fungi from Almaden mining district (Spain). Geophys Res Abstr 13:EGU2011-12550-1

Eto K, Takizawa Y, Akagi H, Haraguchi K, Asano S, Takahata N et al (1999) Differential diagnosis between organic and inorganic mercury poisoning in human cases—the pathologic point of view. Toxicol Pathol 27:664–671

Eto K, Marumoto M, Takeya M (2010) The pathology of methylmercury poisoning (Minamata disease). Neuropathology 30:471–479

Eurochlor (2016) Chlor-alkali industry needs permanent disposal solutions and welcomes the proposed EU Mercury Regulation (2016/0023). Eurochlor 17, 7 Mar 2016

Evans ED (1993) Mercury and other metals in bald eagle feathers and other tissues from Michigan, nearby areas of Minnesota, Wisconsin, Ohio, Ontario and Alaska 1985-1989. Wildlife Division Report No. 3200, Michigan Department of Natural Resources, Lansing, pp 57

Evans RD, Addison EM, Villeneuve JY, MacDonald KS, Joachim DG (2000) Distribution of inorganic and methylmercury among tissues in mink (*Mustela vison*) and otter (*Lutra canadensis*). Environ Res 84:133–139

Evans RD, Hickie B, Rouvinen-Watt K, Wang W (2016) Partitioning and kinetics of methylmercury among organs in captive mink (*Neovison vison*): a stable isotope tracer study. Environ Toxicol Pharmacol 42:163–169
Evers DC (2005) Mercury connections: the extent and effects of mercury pollution in northeastern North America. BioDiversity Research Institute, Gorham, ME pp 28
Evers DC, Clair T (2005) Mercury in northeastern North America: a synthesis of existing databases. Ecotoxicology 14:7–14
Evers DC, Taylor KM, Major A, Taylor RJ, Poppenga R, Scheuhammer AM (2003) Common loon eggs as indicators of methylmercury availability in North America. Ecotoxicology 12:69–81
Evers DC, Burgess NM, Champoux L, Hoskins B, Major A, Goodale WM et al (2005) Patterns and interpretation of mercury exposure in freshwater avian communities in northeastern North America. Ecotoxicology 14:193–221
Evers DC, Han YJ, Driscoll CT, Kamman NC, Goodale MW, Lambert KF et al (2007) Biological mercury hotspots in the northeastern United States and southeastern Canada. BioScience 57:29–43
Falandysz J (1994) Some toxic and trace metals in big game hunted in the northern part of Poland in 1987–1991. Sci Total Environ 141:59–73
Falandysz J, Borovička J (2013) Macro and trace mineral constituents and radionuclides in mushrooms—health benefits and risks. Appl Microbiol Biotechnol 97:477–501
Falandysz J, Jakuczun B, Mizera T (1988) Metals and organochlorines in four female white-tailed eagles. Marine Pollut Bull 19:521–526
Falandysz J, Ichihashi H, Mizera T, Yamasaki S (2000) Mineral composition of selected tissues and organs of white-tailed eagle. Rocz PZH 51:1–5 (in Polish)
Falandysz J, Zhang J, Wang Y-Z, Saba M, Krasinska G, Wiejak A et al (2015) Evaluation of mercury contamination in fungi *Boletus* species from latosols, lateritic red earths, and red and yellow earths in the circum-Pacific mercuriferous belt of southwestern China. PLoS One 10(11): e0143608
Falkowska L, Reindl AR, Szumiło E, Kwaśniak J, Staniszewska M, Bełdowska M et al (2013) Mercury and chlorinated pesticides on the highest level of the food web as exemplified by herring from the Southern Baltic and African penguins from the Zoo. Water Air Soil Pollut 224:1549
Falnoga I, Tusek-Znidaric M, Horvat M, Stegnar P (2000) Mercury, selenium, and cadmium in human autopsy samples from Idrija residents and mercury mine workers. Environ Res 84:211–218
Farina M, Dahm KC, Schwalm FD, Brusque AM, Frizzo ME, Zeni G et al (2003) Methylmercury increases glutamate release from brain synaptosomes and glutamate uptake by cortical slices from suckling rat pups: modulatory effect of ebselen. Toxicol Sci 73:135–140
Farina M, Avila DS, da Rocha JBT, Aschner M (2013) Metals, oxidative stress and neurodegeneration: a focus on iron, manganese and mercury. Neurochem Int 62:575–594
Farrar WP, Edwards JF, Willard MD (1994) Pathology in a dog associated with elevated tissue mercury concentrations. J Vet Diagn Invest 6:511–514
Farris FF, Dedrick RL, Allen PV, Smith JC (1993) Physiological model for the pharmacokinetics of methyl mercury in the growing rat. Appl Pharm 119:74–90
Fernandes Azevedo B, Barros Furieri L, Peçanha FM, Wiggers GA, Vassalio PF, Simones MR et al (2012) Toxic effects of mercury on the cardiovascular and central nervous systems. J Biomed Biotechnol 2012:article ID: 949048 pp 11
Ferrara R, Maserti BE, Mazzolai B, Di Francesco F, Eijner H, Svanberg S et al (1999) Atmospheric mercury in abandoned mine structures and restored mine buildings at Mt. Amiata, Italy. In: Ebinghaus R, Turner RR, de Lacerda LDD, Vasiliev O, Salomons W (eds) Mercury contaminated sites: characterization, risk assessment, and remediation. Springer, Berlin, pp 249–257
Finley MT, Stickel WH, Christensen RE (1979) Mercury residues in tissues of dead and surviving birds fed methylmercury. Bull Environ Contam Toxicol 21:105–110

Finley MLD, Kidd KA, Curry RA, Lescord GL, Clayden MG, O'Driscoll NJ (2016) A comparison of mercury biomagnification through lacustrine food webs supporting brook trout (*Salvelinus fontinalis*) and other salmonid fishes. Front Environ Sci 4:23
Fischer HI, Bartlett LM (1957) Diurnal cycles in liver weights in birds. Condor 59:364–372
Florijančić T, Opačak A, BoŠković I, Jelkić D, Ozimec SŠ, Bogdanović T, ListeŠ I, Škrivanko M, PuŠkadija Z (2016) Heavy metal concentrations in the liver of two wild duck species: influence of species and gender. Ital J Anim Sci 8(sup3):222–224
Fortin C, Beauchamp G, Dansereau M, Larivière N, Bélanger D (2001) Spatial variation in mercury concentrations in wild mink and river otter carcasses from the James Bay territory, Quebec, Canada. Arch Environ Contam Toxicol 40:121–127
Fraga CG (2005) Relevance, essentiality and toxicity of trace elements in human health. Mol Aspects Med 26:235–244
Frederick P, Jayasena N (2011) Altered pairing behaviour and reproductive success in white ibises exposed to environmentally relevant concentrations of methylmercury. Proc Biol Sci 278:1851–1857
Fuchsman PC, Brown LE, Henning MH, Bock MJ, Magar VS (2017) Toxicity reference values for methylmercury effects on avian reproduction: critical review and analysis. Environ Toxicol Chem 36:294–319
Galić N, Prpic-Mehicic G, Prester L, Blanusa M, Krnic Z, Ferencic Z (1999) Dental amalgam mercury exposure in rats. Biometals 12:227–231
Gamberg M, Braune BM (1999) Contaminant residue levels in arctic wolves (*Canis lupus*) from the Yukon Territory, Canada. Sci Total Environ 243–244:329–338
Gamberg M, Boila G, Stern G, Roach P (2005a) Cadmium, mercury and selenium concentrations in mink (*Mustela vison*) from Yukon, Canada. Sci Total Environ 351–352:523–529
Gamberg M, Braune BM, Davey E, Elkin B, Hoekstra PF, Kennedy D et al (2005b) Spatial and temporal trends of contaminants in terrestrial biota from the Canadian Arctic. Sci Total Environ 351–352:148–164
Gamberg M, Palmer M, Roach P (2005c) Temporal and geographic trends in trace element concentrations in moose from Yukon, Canada. Sci Total Environ 351–352:530–538
Gandhi DN, Panchal GM, Dhull DK (2013) Influence of gestational exposure on the effects of prenatal exposure to methyl mercury on postnatal development in rats. Cent Eur J Public Health 21:30–35
Gann GL, Powell CH, Chumchal MM, Drenner RW (2015) Hg-contaminated terrestrial spiders pose a potential risk to songbirds at Caddo Lake (Texas/Louisiana, USA). Environ Toxicol Chem 34:303–306
García-Barrera T, Gómez-Ariza JL, González-Fernández M, Moreno F, García-Sevillano MA, Gómez-Jacinto V (2012) Biological responses related to agonistic, antagonistic and synergistic interactions of chemical species. Anal Bioanal Chem 403:2237–2225
Gasparik J, Dobias M, Capcarova M, Smehyl P, Slamecka J, Bujko J et al (2012) Concentration of cadmium, mercury, zinc, copper and cobalt in the tissues of wild boar (*Sus scrofa*) hunted in the western Slovakia. J Environ Sci Health A Tox Hazard Subst Environ Eng 47:1212–1216
Genovesi P, Carnevali L, Alonzi A, Scalera R (2012) Alien mammals in Europe: updated numbers and trends, and assessment of the effects on biodiversity. Integr Zool 7:247–253
George GN, MacDonald TC, Korbas M, Singh SP, Myers GJ, Watson GE et al (2011) The chemical forms of mercury and selenium in whale skeletal muscle. Metallomics 3:1232–1237
Gerstenberger SL (2004) Mercury concentrations in migratory waterfowl harvested from Southern Nevada Wildlife Management areas, USA. Environ Toxicol 19:35–44
Giżejewska A, Spodniewska A, Barski D (2014) Concentration of lead, cadmium, and mercury in tissues of European beaver (*Castor fiber*) from the north-eastern Poland. Bull Vet Inst Pulawy 58:77–80
Glodek A, Pacyna JM (2009) Mercury emission from coal-fired power plants in Poland. Atmos Environ 43:5668–5673

Gnamuš A, Horvat M (1999) Mercury in the terrestrial food web of the Idrija mining area. In: Ebinghaus R, Turner RR, de Lacerda LDD, Vasiliev O, Salomons W (eds) Mercury contaminated sites: characterization, risk assessment, and remediation. Springer, Berlin, pp 281-317
Gómez MG, Klink JDC, Boffetta P, Español S, Sällsten G, Quintana JG (2007) Exposure to mercury in the mine of Almadén. Occup Environ Med 64:389–395
Grandjean P, Satoh H, Murata K, Eto K (2010) Adverse effects of methylmercury: environmental health research implications. Environ Health Perspect 118:1137–1145
Greenwold MJ, Sawyer RH (2013) Molecular evolution and expression of archosaurian β-keratins: diversification and expansion of archosaurian β-keratins and the origin of feather β-keratins. J Exp Zool (Mol Dev Evol) 9999:1–13
Gregoire DS, Poulain AJ (2016) A physiological role for Hg during phototrophic growth. Nat Geosci 9:121–125
Greichus YA, Greichus A, Emerick RJ (1973) Insecticides, polychlorinated biphenyls and mercury in wild cormorants, pelicans, their eggs, food and environment. Bull Environ Contam Toxicol 9:321–328
Grosicki A, Kowalski B (2002) Lead, cadmium and mercury influence on selenium fate in rats. Bull Vet Inst Pulawy 46:337–343
Gu B, Bian Y, Miller CL, Dong W, Jiang X, Liang L (2011) Mercury reduction and complexation by natural organic matter in anoxic environments. Proc Natl Acad Sci USA 108:1479–1483
Gutleb AC, Kranz A, Nechay G, Toman A (1998) Heavy metal concentrations in livers and kidneys of the otter (*Lutra lutra*) from central Europe. Bull Environ Contam Toxicol 60:273–279
Hachiya N (2006) The history and the present of Mina mata disease. JMAJ 49:112–118
Hahn E, Hahn K, Stoeppler M (1993) Bird feathers as bioindicators in areas of the German environmental specimen bank—bioaccumulation of mercury in food-chains and exogenous deposition of atmospheric pollution with lead and cadmium. Sci Total Environ 140:259–270
Haines KJR, Evans RD, O'Brien M, Evans HE (2010) Accumulation of mercury and selenium in the brain of river otters (*Lontra canadensis*) and wild mink (*Mustela vison*) from Nova Scotia, Canada. Sci Total Environ 408:537–542
Halbrook RS, Jenkins JH, Bush PB, Seabolt ND (1994) Sublethal concentrations of mercury in river otters: monitoring environmental contamination. Arch Environ Contam Toxicol 27:306–310
Hall BD, Doucette JL, Bates LM, Bugajski A, Niyogi S, Somers CM (2014) Differential trends in mercury concentrations in double-crested cormorant populations of the Canadian Prairies. Ecotoxicology 23:419–428
Hanko E, Erne K, Wanntorp H, Borg K (1970) Poisoning in ferrets by tissues of alkyl mercury-fed chickens. Acta Vet Scand 11:268–282
Hansteen H, Ellingsen DG, Clausen KO, Kjuus H (1993) Chromo some aberrations in chloralkali workers previously exposed to mercury vapour. Scand J Work Environ Health 19:375–381
Harding L, Harris M, Elliott J (1998) Heavy and trace metals in wild mink (*Mustela vison*) and river otter (*Lontra canadensis*) captured on rivers receiving metals discharges. Bull Environ Contam Toxicol 61:600–607
Hargreaves AL, Whiteside DP, Gilchrist G (2011) Concentrations of 17 elements, including mercury, in the tissues, food and abiotic environment of Arctic shorebirds. Sci Total Environ 409:3757–3770
Harley J, Lieske C, Bhojwani S, Castellini JM, López JA, O'Hara TM (2015) Mercury and methylmercury distribution in tissues of sculpins from the Bering Sea. Polar Biol 38:1535–1543
Heinz GH (1996) Mercury poisoning in wildlife. In: Faibrother A, Locke LN, Hoff GL (eds) Non-infectious diseases of wildlife. The Iowa State University Press, Ames, IA, pp 118–127
Heinz GH, Locke LN (1976) Brain lesions in mallard ducklings from parents fed methylmercury. Avian Dis 20:9–17
Heinz GH, Hoffman DJ, Klimstra JD, Stebbnis KR, Kondrad SL, Erwin CA (2009) Species differences in the sensitivity of avian embryos to methylmercury. Arch Environ Contam Toxicol 56:129–138

Heinz GH, Hoffman DJ, Klimstra JD, Stebbins KR, Kondrad SL, Erwin CA (2011) Teratogenic effects of injected methylmercury on avian embryos. Environ Toxicol Chem 30:1593–1598

Henny CJ, Kaiser JL, Grove RA (2009) PCDDs, PCDFs, PCBs, OC pesticides and mercury in fish and osprey eggs from Willamette River, Oregon (1993, 2001 and 2006) with calculated biomagnification factors. Ecotoxicology 18:151–173

Henriksson K, Karppanen E, Helminen M (1966) High residue of mercury in Finnish white-tailed eagles. Ornis Fennica 43:38–45

Hernández LM, González MJ, Rico MC, Fernández MA, Baluja G (1985) Presence and biomagnification of organochlorine pollutants and heavy metals in mammals of Doñana National Park (Spain), 1982-1983. J Environ Sci Health B 20:633–650

Hernandez F, Oldenkamp RE, Webster S, Beasley JC, Farina LL, Wisely SM (2017) Raccoons (*Procyon lotor*) as sentinels of trace element contamination and physiological effects of exposure to coal fly ash. Arch Environ Contam Toxicol 72:235–246

Hills LM, Stevenson RW (2006) Mercury and lead content in raw materials. PCA R&D Serial No. 2888 (www.cement.org, 12.03.2012)

Hoekstra PF, Braune BM, Elkin B, Armstrong FAJ, Muir DCG (2003) Concentrations of selected essential and non-essential elements in arctic fox (*Alopex lagopus*) and wolverines (*Gulo gulo*) from the Canadian Arctic. Sci Total Environ 309:81–92

Holliday MA, Potter D, Jarrah A, Bearg S (1967) The relation of metabolic rate to body weight and organ size. Pediatr Res 1:185–195

Holt G, Frøslie A, Norheim G (1979) Mercury, DDE, and PCB in the avian fauna in Norway 1965-1976. Acta Vet Scand (Suppl) 70:1–28

Honda K, Nasu T, Tatsukawa R (1986) Seasonal changes in mercury accumulation in the black-eared kite, *Milvus migrans lineatus*. Environ Pollut 42A:325–334

Honda K, Ichihashi H, Tatsukawa R (1987) Tissue distribution of heavy metals and their variations with age, sex, and habitat in Japanese serows (*Capricornis crispus*). Arch Environ Conatm Toxicol 16:551–561

Hopkins WA, Hopkins LB, Unrine JM, Snodgrass J, Elliot JD (2007) Mercury concentrations in tissues of osprey from the Carolinas, USA. J Wildl Manag 71:1819–1829

Horowitz HM, Jacob DJ, Amos HM, Streets DG, Sunderland EM (2014) Historical mercury releases from commercial products: global environmental implications. Environ Sci Technol 48:10242–10250

Hough EJ, Zabik ME (1972) Distribution of mercury in organs of McGraw-mallard ducks given methyl mercury chloride. Poult Sci 51:2101–2103

Houserova P, Hedbavny J, Matejicek D, Kràčmar S, Sitko J, Kubàň V (2005) Determination of total mercury in muscle, intestines, liver and kidney tissues of cormorant (*Phalacrocorax carbo*), great crested grebe (*Podiceps cristatus*) and Eurasian buzzard (*Buteo buteo*). Vet Med Czech 50:61–68

Houserova P, Kubàň V, Spurny P, Habarata P (2006) Determination of total mercury and mercury species in fish and aquatic ecosystems of Moravian rivers. Vet Med 51:101–110

Houserova P, Kubàň V, Kràčmar S, Sitko J (2007) Total mercury and mercury species in birds and fish in an aquatic ecosystem in the Czech Republic. Environ Pollut 145:185–194

Hu H, Lin H, Zheng W, Tomanicek SJ, Johs A, Feng X et al (2013) Oxidation and methylation of dissolved elemental mercury by anaerobic bacteria. Nat Geosci 6:751–754

Huang SW, Chen CY, Chen MH (2008) Total and organic hg in fish from the reservoir of a chlor-alkali plant in Tainan, Taiwan. J Food Drug Anal 16:75–80

Hughes MR (1970) Relative kidney size in nonpasserine birds with functional salt glands. Condor 72:164–168

Hughes KD, Martin PA, de Solla SR (2014) Contaminants in overwintering canvasbacks (*Aythya valisineria*) and resident mallards (*Anas platyrhynchos*) in the Lake St. Clair/St. Clair River Area. Environment Canada, Ecotoxicology and Wildlife Health Division, pp 21

Hultberg H (2002) Treatment of lakes and storage reservoirs with very low dosages of selenium to reduce methyl mercury in fish. Report, IVL Swedish Environmental Research Institute Ltd, pp 38
Hylander LD, Meili M (2003) 500 years of mercury production: global annual inventory by region until 2000 and associated emissions. Sci Total Environ 304:13–27
Hyvärinen H, Tyni P, Nieminen P (2003) Effects of moult, age, and sex on the accumulation of heavy metals in the otter (*Lutra lutra*) in Finland. Bull Environ Conatm Toxicol 70:278–284
IARC (1993) International Agency for Research on Cancer Monographs on the Evaluation of Carcinogenic Risks to Humans, vol 58, WHO
IARC (2017) International Agency for Research on Cancer Monographs on the Evaluation of Carcinogenic Risks to Humans, vol 58, WHO. http://monographs.iarc.fr/ENG/Classification/latest_classif.php
Isani G, Carpenè E (2014) Metallothioneins, unconventional proteins from unconventional animals: a long journey from nematodes to mammals. Biomolecules 4:435–457
Jackson AK, Evers DC, Folsom SB, Condon AM, Diener J, Goodrick LF et al (2011) Mercury exposure in terrestrial birds far downstream of an historical point source. Environ Pollut 159:3302–3308
Jackson AK, Evers DC, Adams EM, Cristol DA, Eagles-Smith C, Edmonds ST et al (2015) Songbirds as sentinels of mercury in terrestrial habitats of eastern North America. Ecotoxicology 24:453–467
Jensen S, Johnels AG, Olsson M, Westermark T (1972) The avifauna of Sweden as indicators of environmental contamination with mercury and chlorinated hydrocarbons. In: Brill EJ (ed) Proceedings of the 15th international ornithological congress, Hague, The Netherlands, pp 455–465
Jin L, Liang L, Jiang G, Ying Xu Y (2006) Methylmercury, total mercury and total selenium in four common freshwater fish species from Ya-Er Lake, China. Environ Geochem Health 28:401–407
Jo S, Woo HD, Kwon HJ, Oh SY, Park JD, Hong YS et al (2015) Estimation of the biological half-life of methylmercury using a population toxicokinetic model. Int J Environ Res Public Health 12:9054–9067
Johnles A, Westermark T (1969) Mercury contamination of the environment in Sweden. In: Miller MW, Berg GG (eds) Chemical fallout. Charles C. Thomas, Springfield, IL, pp 221–239
Julshamn K, Ringdal O, Haugsnes J (1986) Minerals and trace elements in fillets of nine freshwater fishes from Norway. Fisk Dir Skr Ser Ernering 2:185–191
Kabata-Pendias A (2011) Trace elements in soils and plants, 4th edn. CRC, Boca Raton, FL
Kabata-Pendias A, Mukherjee AB (2007) Trace elements from soil to human. Springer, Berlin
Kalas JA, Ringsby TH, Lierhagen S (1995) Metals and selenium in wild animals from Norwegian areas close to Russian nickel smelters. Environ Monit Assess 36:251–270
Kalisinska E, Lisowski P, Salicki W, Kucharska T, Kavetska K (2009) Mercury in wild terrestrial carnivorous mammals from north-western Poland and unusual fish diet of red fox. Acta Theriol 54:345–356
Kalisinska E, Budis H, Podlasińska J, Łanocha N, Kavetska KM (2010) Body condition and mercury concentration in apparently healthy goosander (*Mergus merganser*) wintering in the Odra estuary, Poland. Ecotoxicology 19:1382–1399
Kalisinska E, Lisowski P, Kosik-Bogacka DI (2012a) Red fox *Vulpes vulpes* (L., 1758) as a bioindicator of mercury contamination in terrestrial ecosystems of north-western Poland. Biol Trace Elem Res 145:172–180
Kalisinska E, Budis H, Łanocha N, Podlasińska J, Baraniewicz E (2012b) Comparison of hepatic and nephric concentrations of mercury between feral and ranch American mink (*Neovison vison*) from NW Poland. Bull Environ Contam Toxicol 88:802–806
Kalisinska E, Kosik-Bogacka DI, Lisowski P, Lanocha N, Jackowski A (2013) Mercury in the body of the most commonly occurring European game duck, the mallard (Anas platyrhynchos L. 1758), from northwestern poland. Arch Environ Contam Toxicol 64(4):583–593
Kalisinska E, Gorecki J, Lanocha N, Okonska A, Melgarejo JB, Budis H et al (2014a) Total and methyl mercury in soft tissues of white-tailed eagle (*Haliaeetus albicilla*) and osprey (*Pandion haliaetus*) collected in Poland. AMBIO 43:858–870

Kalisinska E, Gorecki J, Okonska A, Pilarczyk B, Tomza-Marciniak A, Budis H et al (2014b) Mercury and selenium in the muscle of piscivorous common mergansers (*Mergus merganser*) from a selenium-deficient European country. Ecotoxicol Environ Saf 101:107–115
Kalisinska E, Gorecki J, Okonska A, Pilarczyk B, Tomza-Marciniak A, Budis H et al (2014c) Hepatic and nephric mercury and selenium concentration in common merganser *Mergus merganser* from Baltic Region, Europe. Environ Toxicol Chem 33:421–340
Kalisinska E, Kosik-Bogacka DI, Lanocha-Arendarczyk N, Budis H, Podlasinska J, Popiolek M et al (2016) Brains of native and alien mesocarnivores in biomonitoring of toxic metals in Europe. PLoS One 11(8):e0159935
Kalisinska E, Lanocha-Arendarczyk N, Kosik-Bogacka DI, Budis H, Pilarczyk B, Tomza-Marciniak A et al (2017) Muscle mercury and selenium in fishes and semiaquatic mammals from a selenium-deficient area. Ecotoxicol Environ Saf 136:24–30
Keeyask Hyd Ltd (2012) Keeyask Generation Project environmental impact stetement. Supporting volume terrestrial environment, pp 75. http://keeyask.com/wp/wp-content/uploads/2012/07/Section-8-Wildlife-and-Mercury.pdf
Kenntner N, Tataruch F, Krone O (2001) Heavy metals in soft tissue of white-tailed eagles found dead or moribund in Germany and Austria from 1993 to 2000. Environ Toxicol Chem 20:1831–1837
Kenntner N, Krone O, Altenkamp R, Tataruch F (2003) Environmental contaminants in liver and kidney of free-ranging northern goshawks (*Accipiter gentilis*) from three regions of Germany. Arch Environ Contam Toxicol 45:128–135
Kenntner N, Crettenand Y, Funfstuck HJ, Janovsky M, Tataruch F (2007) Lead poisoning and heavy metal exposure of golden eagles (*Aquila chrysaetos*) from the European Alps. J Ornithol 148:173–177
Kenow KP, Grasman KA, Hines RK, Meyer MW, Gendron-Fitzpatrick A, Spalding MG et al (2007) Effects of methylmercury exposure on the immune function of juvenile common loons (*Gavia immer*). Environ Toxicol Chem 26:1460–1469
Kenow KP, Hoffman DJ, Hines RK, Meyer MW, Bickham JW, Matson CW et al (2008) Effects of methylmercury exposure on glutathione metabolism, oxidative stress, and chromosomal damage in captive-reared common loon (*Gavia immer*) chicks. Environ Pollut 156:732–738
Kenow KP, Meyer MW, Rossmann R, Gendron-Fitzpatrick A, Gray BR (2011) Effects of injected methylmercury on the hatching of common loon (*Gavia immer*) eggs. Ecotoxicology 20:1684–1693
Kessler M (2013) Minamata Convention on Mercury. A first step towards protecting future generations. Environ Health Perspect 121:A304–A309
Khan AT, Forester DM (1995) Mercury in white-tailed deer forage in Russell Plantation, Macon County, Alabama. Vet Hum Toxicol 37:45–46
Khan AT, Thompson SJ, Mielke HW (1995) Lead and mercury levels in raccoons from Macon County, Alabama. Bull Environ Contam Toxicol 54:812–816
Kiesling RL, Lloyd EH (1971) Chemicals: fungicide uses and problems in North Dakota. Farm Res 28:29–31
Kim EY, Saeki K, Tanabe S, Tanaka H, Tatsukawa R (1996) Specific accumulation of mercury and selenium in seabirds. Environ Pollut 94:261–265
Kim CS, Rytuba JJ, Brown GE (2004) Geological and anthropogenic factors influencing mercury speciation in mine wastes: an EXAFS spectroscopy study. Appl Geochem 19:379–393
Kim CK, Lee TW, Lee KT, Lee JH, Lee CB (2012) Nationwide monitoring of mercury in wild and farmed fish from fresh and coastal waters of Korea. Chemosphere 89:1360–1368
Kinghorn A, Solomon P, Chan HM (2007) Temporal and spatial trends of mercury in fish collected in the English-Wabigoon river system in Ontario, Canada. Sci Total Environ 372:615–623
Kisia SM (1996) Structure of fish locomotory muscle. In: Datta-Munshi JS, Gutta HM (eds) Fish morphology—horizon of new research. Science, pp 169–178
Kitowski I, Kowalski R, Komosa A, Sujak A (2015) Total mercury concentration in the kidneys of birds from Poland. Turk J Zool 39:1–9

Klenavic K, Champoux L, O'Brien M, Daoust PY, Evans RD, Evans HE (2008) Mercury concentration in wild mink (*Mustela vison*) and river otters (*Lontra canadensis*) collected from eastern and Atlantic Canada: relationship to age and parasitism. Environ Pollut 156:359–366

Komov VT, Ivanova ES, Gremyachikh VA, Poddubnaya NY (2016) Mercury content in organs and tissues of indigenous (*Vulpes vulpes* L.) and invasive (*Nyctereutes procyonoides* Gray) species of canids from areas near Cherepovets (North-Western Industrial Region, Russia). Bull Environ Contam Toxicol 97:480–485

Krey A, Kwan M, Chan HM (2015) Mercury speciation in brain tissue of polar bears (*Ursus maritimus*) from the Canadian Arctic. Environ Res 114:24–30

Krone O, Willie F, Kenntner N, Boertmann D, Tataruch F (2004) Mortality factors, environmental contaminants, and parasites of white-tailed sea eagles from Greenland. Avian Dis 48:417–424

Krone O, Stjernberg T, Kenntner N, Tataruch F, Koivusaari J, Nuuja I (2006) Mortality factors, helminth burden, and contaminant residues in white-tailed sea eagles (Haliaeetus albicilla) from Finland. AMBIO J Hum Environ 35(3):98–104

Kruska D, Schreiber A (1999) Comparative morphometrical and biochemical–genetic investigations in wild and ranch mink (*Mustela vison*: Carnivora: Mammalia). Acta Theriol 44:377–382

Kruuk H, Conroy JWH, Webb A (1997) Concentration of mercury in otters (*Lutra lutra*) in Scotland in relation to rainfall. Environ Pollut 96:13–18

Krynski A, Kałużynski J, Wlazełko M, Adamowski A (1982) Contamination of roe deer by mercury compounds. Acta Theriol 27:499–507

Kucera E (1983) Mink and otter as indicators of mercury in Manitoba waters. Can J Zool 61:2250–2256

Laacouri A, Nater EA, Kolka RK (2013) Distribution and uptake dynamics of mercury in leaves of common deciduous tree species in Minnesota, U.S.A. Environ Sci Technol 47:10462–10470

Lang D, Holmes J, Gardner M (2012) Mercury arising from oil and gas production in the United Kingdom and UK continental shelf. IKIMP, Mercury Knowledge Exchange, University of Oxford, Oxford, pp 42

Langlois C, Langis R (1995) Presence of airborne contaminants in the wildlife of northern Québec. Sci Total Environ 160(161):391–402

Lanocha N, Kalisinska E, Kosik-Bogacka DI, Budis H, Podlasinska J, Jedrzejewska E (2014) Mercury levels in raccoons (*Procyon lotor*) from the Warta Mouth National Park, north-western Poland. Biol Trace Elem Res 159:152–160

Lanszki J, Sugár L, Orosz E, Nagy D (2008) Biological data from post mortem analysis of otters in Hungary. Acta Zool Acad Sci Hung 54:201–212

Lanszki J, Orosz E, Sugar L (2009) Metal levels in tissues of Eurasian otters (*Lutra lutra*) from Hungary: variation with sex, age, condition and location. Chemosphere 74:741–743

Larosa B, Allen-Gil S (1995) The methylmercury to total mercury ratio in selected marine, freshwater, and terrestrial organism. Water Air Soil Pollut 80:905–913

Larson H (2014) The Minamata Convention on Mercury: risk in perspective. Lancet 383:198–199

Lavoie RA, Jardine TD, Chumchal MM, Kidd KA, Campbell LM (2013) Biomagnification of mercury in aquatic food webs: a worldwide meta-analysis. Environ Sci Technol 47:13385–13394

Lazarus M, Orct T, Blanusa M, Vickovic I, Sostarić B (2008) Toxic and essential metal concentrations in four tissues of red deer (*Cervus elaphus*) from Baranja, Croatia. Food Addit Contam A Chem Anal Control Expo Risk Assess 25:270–283

Lemarchand C, Rosoux R, Berny P (2010) Organochlorine pesticides, PCBs, heavy metals and anticoagulant rodenticides in tissues of Eurasian otters (*Lutra lutra*) from upper Loire River catchment (France). Chemosphere 80:1120–1124

Lemarchand C, Rosoux R, Penide ME, Berny P (2012) Tissue concentrations of pesticides, PCBs and metals among ospreys, *Pandion haliaetus*, collected in France. Bull Environ Contam Toxicol 88:89–93

Li YB, Cai Y (2013) Progress in the study of mercury methylation and demethylation in aquatic environments. Chin Sci Bull 58:177–185
Lieske CL, Moses SK, Castellini JM, Klejka J, Hueffer K, O'Hara TM (2011) Toxicokinetics of mercury in blood compartments and hair of fish-fed sled dogs. Acta Vet Scand 53:66
Lindqvist O, Johansson K, Bringmark L, Timm B, Aastrup M, Andersson A et al (1991) Mercury in the Swedish environment—recent research on causes, consequences and corrective methods. Water Air Soil Pollut 55:1–261
Lindsay RC, Dimmick RW (1983) Mercury residues in wood ducks and wood duck foods in eastern Tennessee. J Wildl Dis 19:114–117
Ljungvall K, Magnusson U, Korvela M, Norrby M, Bergquist J, Persson S (2017) Heavy metal concentrations in female wild mink (*Neovison vison*) in Sweden: sources of variation and associations with internal organ weights. Environ Toxicol Chem 36:2030–2035
Lodenius M, Solonen T (2013) The use of feathers of birds of prey as indicators of metal pollution. Ecotoxicology 22:1319–1334
Lodenius M, Skaren U, Hellstedt P, Tulisalo E (2014) Mercury in various tissues of three mustelid and other trace metals in liver o European otter from eastern Finland. Environ Monit Assess 186:325–333
Lohren H, Bornhorst J, Galla H-J, Schwerdtle T (2015) The blood–cerebrospinal fluid barrier—first evidence for an active transport of organic mercury compounds out of the brain. Metallomics 7:1420
Lohren H, Bornhorst J, Fitkau R, Pohl G, Galla H-J, Schwerdtle T (2016) Effects on and transfer across the blood-brain barrier in vitro—Comparison of organic and inorganic mercury species. BMC Pharmacol Toxicol 17:63
Lord CG, Gaines KF, Boring CS, Brisbin IL, Gochfeld M Jr, Burger J (2002) Raccoon (*Procyon lotor*) as a bioindicator of mercury contamination at the U.S. Department of Energy's Savannah River Site. Arch Environ Contam Toxicol 43:356–363
Lourenco R, Tavares PC, Degaldo MM, Rabaca JE, Penteriani V (2011) Superpredation increases mercury levels in a generalist top predator, the eagle owl. Ecotoxicology 20:635–642
Lu J, Holmgren A (2009) Selenoproteins. J Biol Chem 284:723–727
Mailman M, Bodaly RA (2005) Total mercury, methyl mercury, and carbon in fresh and burned plants and soil in Northwestern Ontario. Environ Pollut 138:161–166
Martin PA, McDaniel TV, Hughes KD, Hunter B (2011) Mercury and other heavy metals in free-ranging mink of the lower Great Lakes basin, Canada, 1998–2006. Ecotoxicology 20:1701–1712
Mason CF, Madsen AB (1992) Mercury in Danish otters (*Lutra lutra*). Chemosphere 25:865–867
Mason CF, Last NI, Macdonald SM (1986) Mercury, cadmium, and lead in British otters. Bull Environ Contam Toxicol 37:844–849
Mason RP, Choi AL, Fitzgerald WF, Hammerschmidt CR, Lamborg CH, Soerensen AL et al (2012) Mercury biogeochemical cycling in the ocean and policy implications. Environ Res 119:101–117
Masur LC (2011) A review of the use of mercury in historic and current ritualistic and spiritual practices. Altern Med Rev 16:314–320
Mayack DT (2012) Hepatic mercury, cadmium, and lead in mink and otter from New York State: monitoring environmental contamination. Environ Monit Assess 184:2497–2516
Mazloomi SA, Esmaeili SM, Ghasempoori SM, Omidi A (2008) Mercury distribution in liver, kidney, and feathers of Caspian Sea common cormorant (*Phalacrocorax carbo*). Res J Environ Sci 2:433–437
Mehdi Y, Hornick JL, Istasse L, Dufranse I (2013) Selenium in the environment, metabolism and involvement in body functions. Molecules 18:3292–3311
Meinert LD, Robinson GR, Nassar NT (2016) Mineral resources: reserves, peak production and the future. Resources 5:14
Mierle G, Addison EM, MacDonald KS, Joachim DG (2000) Mercury levels in tissues of otters from Ontario, Canada: variation with age, sex, and location. Environ Toxicol Chem 19:3044–3051

Mierzykowski SE, Smith JEM, Todd CS, Kusnierz D, DeSorbo CR (2011) Liver contaminants in bald eagle carcasses from Maine. USFWS Spec Proj Rep FY09-MEFO-6-EC, Maine Field Office, Orono, ME, pp 53

Mierzykowski SE, Todd CS, Pokras MA, Oliveira RD (2013) Lead and mercury levels in livers of bald eagles recovered in New England. USFWS. Spec Proj Rep FY13-MEFO-2-EC, Maine Field Office, Orono, ME, pp 26

Milieu Ltd (2010) Environmental, economic and social impacts of the use of sewage sludge on land. Part II: Report on Options and Impacts. Report prepared for the European Commission under Study Contract DG ENV.G.4/ETU/2008/0076r

Millan J, Mateo R, Taggart MA, López-Bao JV, Viota M, Monsalve L et al (2008) Levels of heavy metals and metalloids in critically endangered Iberian lynx and other wild carnivores from southern Spain. Sci Total Environ 399:193–201

Miller A, Bignert A, Porvari P, Danielsson S, Verta M (2013) Mercury in perch (*Perca fluviatilis*) from Sweden and Finland. Water Air Soil Pollut 224:1472

Mohapatra SP, Mitchell A (2009) Mercury trade in globalizing world. In: Watanabe Y, Yamashita H (eds) Trade policy in globalizing world. Nova, New York, pp 141–150

Moreno-Jimenez E, Gamarra R, Carpena-Ruiz RO, Millan R, Penalosa JM, Esteban E (2006) Mercury bioaccumulation and phytotoxicity in two wild plant species of Almaden area. Chemosphere 63:1969–1973

Muchlinski MN, Snodgrass JJ, Terranova CJ (2012) Muscle mass scaling in primates: an energetic and ecological perspective. Am J Primatol 74:395–407

Mukherjee AB, Zevenhoven R, Bhattacharya P, Sajwan KS, Kikuchi R (2008) Mercury flow via coal and coal utilization by-products: a global perspective. Resour Conser Recycl 52:571–591

Munthe J, Wängberg I, Rognerud S, Fjeld E, Verta M, Porvari P et al (2007) Mercury in Nordic ecosystems. IVL Report B1761

Myers GJ, Davidson PW (1998) Prenatal methylmercury exposure and children: neurologic, developmental, and behavioral research. Environ Health Perspect 106(Suppl 3):841–847

Nakazawa E, Ikemoto T, Hokura A, Terada Y, Kunito T, Tanabe S et al (2011) The presence of mercury selenide in various tissues of the striped dolphin: evidence from μ-XRF-XRD and XAFS analyses. Metallomics 3:719–725

Nam DH, Anan Y, Ikemoto T, Okabe Y, Kim EY, Subramanian A et al (2005) Specific accumulation of 20 trace elements in great cormorants (*Phalacrocorax carbo*) from Japan. Environ Pollut 134:503–514

Nam DH, Yates D, Ardapple P, Evers DC, Schmerfeld J, Basu N (2012) Elevated mercury exposure and neurochemical alterations in little brown bats (*Myotis lucifugus*) from a site with historical mercury contamination. Ecotoxicology 21:1094–1101

National Research Council (2000) *Toxicological effects of methylmercury*. The National Academies Press, Washington, DC. https://doi.org/10.17226/9899

Nguetseng R, Fliedner A, Knopf B, Lebreton B, Quack M, Rüdel H (2015) Retrospective monitoring of mercury in fish from selected European freshwater and estuary sites. Chemosphere 134:427–434

Niecke M, Kruger A, Hauff P, Ellenberg H, Labes R, Niecke S (1998) Quecksilber in Seeadlerfedern aus Mecklenburg-Vorpommern mit Hilfe der Hamburger Protonenmikrosonde. Z Umweltchem Okotox 10:3–14 (in German)

Norheim G, Frøslie A (1978) The degree of methylation and organ distribution of mercury in some birds of prey in Norway. Acta Pharmacol Toxicol 43:196–204

Norheim G, Sivertsen T, Brevik EM, Frøslie A (1984) Mercury and selenium in wild mink (*Mustela vision*) from Norway. Nord Vet Med 36:43–48 (in Norwegian)

O'Connor DJ, Nielsen SW (1981) Environmental survey of methylmercury levels in wild mink (*Mustela vison*) and otter (*Lutra canadensis*) from the northeastern United States and experimental pathology of methylmercurialism in the otter. In: Chapman JA, Pursley D (eds) Worldwide furbearer conference proceedings, 3–11 Aug 1980, Frostburg, MD, pp 1728–1745

Odsjo T, Raikkonen J, Bignert A (2012) Time trends of metals in liver and muscle of reindeer (*Rangifer tarandus*) from northern and central Lapland, Sweden, 1983-2005. Swedish monitoring programme in terrestrial biota. Swedish Museum of Natural History, Stockholm, p 33

Ohlendorf HM (1993) Marine birds and trace elements in the temperate North Pacific. In: Vermeer K, Briggs KT, Morgan KH, Siegel-Causey D (eds) The status, ecology, and conservation of marine birds of the North Pacific. Canadian Wildlife Service Special Publication, Ottawa, pp 232–240

Osborn CE, Evers DC, Duron M, Schoch N, Yates D, Buck D et al (2011) Mercury contamination within terrestrial ecosystems in New England and Mid-Atlantic states: profiles of soil, invertebrates, songbirds, and bats. Report BRI 2011-09. Submitted to the Nature Conservancy—Eastern New York Chapter. Biodiversity Research Institute, Gorham, ME, pp 100

Pacyna EG, Pacyna JM, Steenhuisen F, Wilson S (2006) Global anthropogenic mercury emission inventory for 2000. Atmos Environ 40:4048–4063

Page KD, Murphy JB (2005) Mercury concentrations in the bedrock of southwestern Nova Scotia: a reconnaissance study. Atl Geol 40:31–40

Pal M, Ghosh S, Mukhopadhyay M, Ghosh M (2012) Methyl mercury in fish—a case study on various samples collected from Ganges River at West Bengal. Environ Monit Assess 184:3407–3414

Park JD, Zheng W (2012) Human exposure and health effects of inorganic and elemental mercury. J Prev Med Public Health 45:344–352

Park JS, Lee JS, Kim GB, Cha JS, Shin SK, Kang HG et al (2010) Mercury and methylmercury in freshwater fish and sediments in South Korea using newly adopted purge and trap GC-MS detection method. Water Air Soil Pollut 207:391–401

Parslow JLF, Thomas GJ, Williams TD (1982) Heavy metals in the livers of waterfowl from the ouse washes, England. Environ Pollut Ser A, Ecol Biol 29(4):317–327

Parsons MB, Percival JB (2005) A brief history of mercury and its environmental impact. In: Parsons MB, Percival JB (eds) Mercury: sources, measurements, cycles and effects. Mineralogical Association of Canada, Halifax, Nova Scotia pp 20

Patra M, Sharma A (2000) Mercury toxicity in plants. Bot Rev 66:379–422

Pendergrass JC, Haley BE, Vimy MJ, Winfield SA, Lorscheider FL (1997) Mercury vapor inhalation inhibits binding of GTP to tubulin in rat brain: similarity to a molecular lesion in Alzheimer diseased brain. Neurotoxicology 18:315–324

Petrie SA, Badzinski SS, Drouillard KG (2007) Contaminants in lesser and greater scaup staging on the lower Great Lakes. Arch Environ Contam Toxicol 52:580–589

Pirrone N, Cinnirella S, Feng X, Finkelman RB, Friedli HR, Leaner J et al (2010) Global mercury emissions to the atmosphere from anthropogenic and natural sources. Atmos Chem Phys 10:5951–5964

Piskorova L, Vasilkova Z, Krupicer I (2003) Heavy metals residues in tissues of wild boar (*Sus scrofa*) and red fox (*Vulpes vulpes*) in the Central Zemplin region of the Slovak Republik. Czech J Anim Sci 48:134–138

Pollock B, Machin KL (2008) Effects of cadmium, mercury, and selenium on reproductive indices in male lesser scaup (Aythya affinis) in the western Boreal forest. Arch Environ Contam Toxicol 54:730–739

Polunas M, Halladay A, Tjalkens RB, Philbert MA, Lowndes H, Reuhl K (2011) Role of oxidative stress and the mitochondrial permeability transition in methylmercury cytotoxicity. Neurotoxicology 32:526–534

Pompe-Gotal J, Srebocan E, Gomercic H, Prevendar Crinic A (2009) Mercury concentrations in the tissues of bottlenose dolphins (Tursiops truncatus) and striped dolphins (*Stenella coeruloalba*) stranded on the Croatian Adriatic coast. Vet Med 54:598–606

Pompella A, Visvikis A, Paolicchi A, De Tata V, Casini AF (2003) The changing faces of glutathione, a cellular protagonist. Biochem Pharmacol 66:1499–1503

Poole KG, Elkin B (1992) Environmental contaminants, population structure, and biological condition of harvested mink in the Western Northwest Territories, 1991–92. Department of

Renewable Resources Government of the Northwest Territories Yellowknife, NWT., Report No 66

Poole KG, Elkin BT, Bethke RW (1995) Environmental contaminants in wild mink in the Northwest Territories, Canada. Sci Total Environ 160(161):473–786

Prestrud P, Norheim G, Sivertsen T, Daae HL (1994) Levels of toxic and essential elements in arctic fox in Svalbard. Polar Biol 14:155–159

Puls R (1988) Mineral levels in animal health. Sherpa, Clearbrook, BC

Pye S, Jones G, Stewart R, Woodfield M, Kubica K, Kubica R, et al (2006) Costs and environmental effectiveness of options for reducing mercury emissions to air from small-scale combustion installations. AEAT/ED48706/Final Report, AEA Technology Environment, Harwell, Oxon, UK, pp 122

Qiu G, Feng X, Meng B, Wang X (2012) Methylmercury in rice (*Oryza sativa* L.) grown from the Xunyang Hg mining area, Shaanxi province, northwestern China. Pure Appl Chem 84:281–289

Ralston NV, Raymond LJ (2010) Dietary selenium's protective effects against methylmercury toxicity. Toxicology 278:112–123

Ralston NVC, Ralston CR, Blackwell JL, Raymond LJ (2008) Dietary and tissue selenium in relation to methylmercury toxicity. Neurotoxicology 29:802–811

Reinoso RF, Telfer BA, Rowland M (1997) Tissue water content in rats measured by desiccation. J Pharmacol Toxicol Methods 38:87–92

Rice KM, Walker EM, Wu M, Gillette C, Blough ER (2014) Environmental mercury and its toxic effects. J Prev Med Public Health 47:74–83

Rieder SR, Brunner I, Horvat M, Jacobs A, Frey B (2011) Accumulation of mercury and methylmercury by mushrooms and earthworms from forest soils. Environ Pollut 159:2861–2869

Rieder SR, Brunner I, Daniel O, Liu B, Frey B (2013) Methylation of mercury in earthworms and the effect of mercury on the associated bacterial communities. PLoS One 8:e61215

Rimmer CC, Miller EK, McFarland KP, Taylor RJ, Faccio SD (2010) Mercury bioaccumulation and trophic transfer in the terrestrial food web of a montane forest. Ecotoxicology 19:697–709

Robillard S, Beauchamp G, Paillard G, Bélanger D (2002) Levels of cadmium, lead, mercury and $^{137}$caesium in caribou (*Rangifer tarandus*) tissues from northern Québec. Arctic 55:1–9

Robinson JF, Guerrette Z, Yu X, Hong S, Faustman EM (2010) A systems-based approach to investigate dose- and time-dependent methylmercury-induced gene expression response in C57BL/6 mouse embryos undergoing neurulation. Birth Defects Res B Dev Reprod 89:188–200

Rolfhus KR, Hall BD, Monson BA, Paterson MJ, Jeremiason JD (2011) Assessment of mercury bioaccumulation within the pelagic food web of lakes in the western Great Lakes region. Ecotoxicology 20:1520–1529

Ropek RM, Neely RK (1993) Mercury levels in Michigan river otters, *Lutra canadensis*. J Freshwat Ecol 8:141–147

Rothenberg SE, Windham-Myers L, Creswell JE (2014) Rice methylmercury exposure and mitigation: a comprehensive review. Environ Res 133:407–423

Rothschild RFN, Duffy LK (2005) Mercury concentrations in muscle, brain and bone of Western Alaskan waterfowl. Sci Total Environ 349:277–283

Roy A, Dey SK, Saha C (2013) Modification of cyto- and genotoxicity of mercury and lead by antioxidant on human lymphocytes *in vitro*. Curr Sci 104:224–228

Rozgaj R, Kasuba V, Blanusa M (2005) Mercury chloride genotoxicity in rats following oral exposure, evaluated by comet assay and micronucleus test. Arh Hig Rada Toksikol 56:9–15

Rudy M (2010) Chemical composition of wild boar meat and relationship between age and bioaccumulation of heavy metals in muscle and liver tissue. Food Addit Contam A Chem Anal Control Expos Risk Assess 27:464–472

Ruelas-Inzunza J, Hernández-Osuna J, Páez-Osuna F (2009) Organic and total mercury in muscle tissue of five aquatic birds with different feeding habits from the SE Gulf of California, Mexico. Chemosphere 76:415–418

Rutkiewicz JM (2012) Neurochemical biomarkers to assess mercury's health impacts in birds. PhD thesis, University of Michigan, Ann Arbor, MI, pp 200

Rutkiewicz J, Nam DH, Cooley T, Neumann K, Padilla IB, Route W et al (2011) Mercury exposure and neurochemical impacts in bald eagles across several Great Lakes states. Ecotoxicology 20:1669–1676

Rytuba JJ (2003) Mercury from mineral deposits and potential environmental impact. Environ Geol 43:326–338

Saeki K, Okabe Y, Kim E, Tanabe S, Fukuda M, Tatsukawa R (2000) Mercury and cadmium in common cormorants (*Phalacrocorax carbo*). Environ Pollut 108:249–255

Samson JC, Shenker J (2000) The teratogenic effects of methylmercury on early development of the zebrafish, *Danio rerio*. Aqua Toxicol 48:343–354

Scheuhammer AM (1988) Chronic dietary toxicity of methylmercury in the zebra Finch, *Poephila guttata*. Bull Environ Contam Toxicol 40:123–130

Scheuhammer AM (1991) Effects of acidification on the availability of toxic metals and calcium to wild birds and mammals. Environ Pollut 71:329–375

Scheuhammer AM, Atchison CM, Wong AHK, Evers DC (1998a) Mercury exposure in breeding common loons (*Gavia immer*) in central Ontario, Canada. Environ Toxicol Chem 17:191–196

Scheuhammer AM, Wong AH, Bond D (1998b) Mercury and selenium accumulation in common loons (*Gavia immer*) and common mergansers (*Mergus merganser*) from eastern Canada. Environ Toxicol Chem 17:197–201

Scheuhammer AM, Basu N, Burgess NM, Elliott JE, Campbell GD, Wayland M et al (2008) Relationships among mercury, selenium, and neurochemical parameters in common loons (*Gavia immer*) and bald eagles (*Haliaeetus leucocephalus*). Ecotoxicology 17:93–101

Scheuhammer AM, Braune B, Chan HM, Frouin H, Krey A, Letcher R et al (2015) Recent progress on our understanding of the biological effects of mercury in fish and wildlife. Sci Total Environ 509-510:91–103

Schurz F, Sabater-Vilar M, Fink-Gremmels J (2000) Mutagenicity of mercury chloride and mechanisms of cellular defence: the role of metal-binding proteins. Mutagenesis 15:525–530

Schuster PF, Krabbenhoft DP, Naftz DL, Cecil LD, Olson ML, Dewild JF et al (2002) Atmospheric mercury deposition during the last 270 years: a glacial ice core record of natural and anthropogenic sources. Environ Sci Technol 36:2303–2310

Scoullos M, Vonkeman GH, Thorton I, Makuch Z (2001) Mercury. In: Scoullos M, Vonkeman GH, Thorton I, Makuch Z (eds) Mercury—cadmium—lead handbook for sustainable heavy metals policy and regulation. Kluwer Academic, Dordrecht, pp 11–68

Scudder Eikenberry BC, Riva-Murray K, Knightes CD, Journey CA, Chasar LC, Brigham ME et al (2015) Optimizing fish sampling for fish-mercury bioaccumulation factors. Chemosphere 135:467–473

Scudder BC, Chasar LC, Wentz DA, Bauch NJ, Brigham ME, Moran PW et al (2009) Mercury in fish, bed sediment, and water from streams across the United States, 1998–2005. U.S. Geological Survey Scientific Investigations Report 2009–5109, pp74

Selin NE, Jackob DJ, Yantosca RM, Strode S, Jaegle L, Sunderland EM (2008) Global 3-D land-ocean-atmosphere model for mercury: present-day versus preindustrial cycles and anthropogenic enrichment factors for deposition. Glob Biogeochem Cycle 22:GB2011

Sellers P (2010) A *survey* of *chemical contaminants* in *wild meat harvested* from the *traditional territories* of Wabauskang *First Nation* (Wabauskang), Asubpeeschoseewagong Netum Anishinabek (Grassy Narrows), and Wabaseemoong Independent Nation (Whitedog). First Nations Environmental *Contaminants* Program (National) as Partial fulfillment of Project No. HQ0900055, pp 65

Sepúlveda MS, Poppenga RH, Arregis JJ, Quinn LB (1998) Concentrations of mercury and selenium in tissues of double-crested cormorants (*Phalacrocorax auritus*) from southern Florida. Colon Waterbirds 21:35–42

Serafin JA (1984) Avian species differences in the intestinal absorption of xenobiotics (PCB, dieldrin, $Hg^{2+}$). Comp Biochem Physiol C 78:4910–4496

Sheffy TB, St Amant JR (1982) Mercury burdens in furbearers in Wisconsin. J Wildl Manage 46:1117–1120
Shore RF, Pereira MG, Walker LA, Thompson DR (2011) Mercury in nonmarine birds and mammals. In: Beyer WN, Meador JP (eds) Environmental contaminants in biota. CRC, Boca Raton, FL, pp 609–642
Silva-Pereira LC, Cardoso PCS, Leite DS, Bahia MO, Bastos WR, Smith MAC et al (2005) Cytotoxicity and genotoxicity of low doses of mercury chloride and methylmercury chloride on human lymphocytes *in vitro*. Braz J Med Biol Res 38:901–907
Sleeman JM, Cristol DA, White AE, Evers DC, Gerhold RW, Keel MK (2010) Mercury poisoning in free-living northern river otter (*Lontra canadensis*). J Wildl Dis 46:1035–1039
Smart NA (1968) Use and residues of mercury compounds in agriculture. In: Gunther FA (ed) Residue review. Springer, New York, p 36
Smith TG, Armstrong FAJ (1975) Mercury in seals, terrestrial carnivores, and principal food items of the Inuit from Holman, N.W.T. J Fish Res Board Can 32:795–801
Sobanska MA (2005) Wild boar hair *Sus scrofa* as a non-invasive indicator of mercury pollution. Sci Total Environ 339:81–88
Souza MJ, Donnell R, Ramsay E (2013) Metal accumulation and health effects in raccoons (*Procyon lotor*) associated with coal fly ash exposure. Arch Environ Contam Toxicol 64:529–536
Spalding MG, Frederick PC, McGill HC, Bouton SN, McDowell LR (2000) Methylmercury accumulation in tissues and its effects on growth and appetite in captive great egrets. J Wildl Dis 36:411–422
Speir SL, Chumchal MM, Drenner RW, Cocke WG, Lewis ME, Whitt HJ (2014) Methyl mercury and stable isotopes of nitrogen reveal that a terrestrial spider has a diet of emergent aquatic insects. Environ Toxicol Chem 33:2506–2509
Spiric Z, Srebocan E, Crnic AP (2012) Mercury in hares organs (*Lepus europaeus* Pallas) in the vicinity of the mercury-contaminated natural gas treatment plant in Croatia. J Environ Sci Health A Tox Hazard Subst Environ Eng 47:77–83
Srebocan E, Prevendar Crnić A, Ekert-Kabalin AM, Lazarus M, Jurasović J, Tomljanović K et al (2011) Cadmium, lead, and mercury concentrations in tissues of roe deer (*Capreolus capreolus* L.) and wild boar (*Sus scrofa* L.) from lowland Croatia. Czech J Food 29:624–633
Standish CL (2016) Evaluation of total mercury and methylmercury concentrations of terrestrial invertebrates along Lower East Fork Poplar Creek in Oak Ridge, Tennessee. Master's thesis, University of Tennessee, pp 117. http://trace.tennessee.edu/utk_gradthes/4078
Stansley W, Velinsky D, Thomas R (2010) Mercury and halogenated organic contaminants in river otters (*Lontra canadensis*) in New Jersey, USA. Environ Toxicol Chem 29:2235–2242
Stevens RT, Ashwood TL, Sleeman JM (1997) Mercury in hair of muskrats (*Ondatra zibethicus*) and mink (*Mustela vison*) from the U. S. Department of Energy Oak Ridge Reservation. Bull Environ Contam Toxicol 58:720–725
Stickel LF, Stickel WH, McLanc MAR, Bruns M (1977) Prolonged retention of methyl mercury by mallard drakes. Bull Environ Contam Toxicol 18:393–400
Stone WB, Okoniewski JC (2001) Necropsy findings and environmental contaminants in common loons from New York. J Wildl Dis 37:178–184
Storelli MM, Zizzo N, Marcotrigiano GO (1999) Heavy metals and methylmercury in tissues of Risso's dolphin (*Grampus griseus*) and Cuvier's beaked whale (*Ziphius cavirostris*) stranded in Italy (South Adriatic Sea). Bull Environ Contam Toxicol 63:703–710
Stout JH, Trust KA (2002) Elemental and organochlorine residues in bald eagles from Adak Island, Alaska. J Wildl Dis 38:511–517
Strom SM (2008) Total mercury and methylmercury residues in river otters (*Lutra canadensis*) from Wisconsin. Arch Environ Contam Toxicol 54:546–554
Suran J, Prisc M, Rasic R, Srebocan E, Crnic AP (2013) Malondialdehyde and heavy metal concentrations in tissues of wild boar (*Sus scrofa* L.) from central Croatia. J Environ Sci Health B 48:147–152

Szkoda J, Durkalec M, Kołacz R, Opaliński S, Żmudzki J (2012) Content of cadmium, lead and mercury in the tissues of game animals. Med Weter 68:689–692 (in Polish)

Szkoda J, Zmudzki J, Nawrocka A, Kmieciak M (2014) Toxic elements in free-living freshwater fish, water and sediments in Poland. Bull Vet Inst Pulawy 58:589–595

Takeuchi T, D'Itri FM, Fischer PV, Annett CS, Okabe M (1977) The outbreak of Minamata disease (methyl mercury poisoning) in cats on Northwestern Ontario Reserves. Environ Res 13:215–228

Tan SW, Meiller JC, Mahaffey KR (2009) The endocrine effects of mercury in humans and wildlife. Crit Rev Toxicol 39:228–269

Tavshunsky I, Eggert SL, Mitchell CPJ (2017) Accumulation of methylmercury in invertebrates and masked shrews (*Sorex cinereus*) at an Upland Forest-Peatland Interface in Northern Minnesota, USA. Bull Environ Contam Toxicol 99:673–678

Teaf CM, Garber M (2012) Mercury exposure considerations: evaluating the chemical form and activities of the individual. In: Proceedings of the annual international conference on soils, sediments, water and energy, vol 17, pp 25–42

Tejero J, Higueras PL, Garrido I, Esbrí JM, Oyarzun R, Español S (2015) An estimation of mercury concentrations in the local atmosphere of Almadén (Ciudad Real Province, South Central Spain) during the twentieth century. Environ Sci Pollut Res 22:4833–4841

Teršič T, Gosar M (2012) Comparison of elemental contents in earthworm cast and soil from a mercury-contaminated site (Idrija area, Slovenia). Sci Total Environ 430:28–33

Thomas DJ, Fisher HL, Sumler MR, Hall LL, Mushak P (1988) Distribution and retention of organic and inorganic mercury in methyl mercury-treated neonatal rats. Environ Res 47:59–71

Thompson DR (1996) Mercury in birds and terrestrial mammals. In: Beyer WN, Heinz GH, Redmon-Norwood AW (eds) Environmental contaminants in wildlife: interpreting tissue concentrations. Lewis, Boca Raton, FL, pp 341–356

Tjälve H, Henriksson J (1999) Uptake of metals in the brain via olfactory pathways. Neurotoxicology 20:181–195

Tomiyasu T, Matsuo T, Miyamoto J, Imura R, Anazawa K, Sakamoto H (2005) Low level mercury uptake by plants from natural environments—mercury distribution in *Solidago altissima* L. Environ Sci 12:231–238

Toole-O'Neil B, Tewalt SJ, Finkelmanb RB, Akers DJ (1999) Mercury concentration in coal—unraveling the puzzle. Fuel 78:47–54

Tsipoura N, Burger J, Newhouse M, Mizrahi D (2011) Lead, mercury, cadmium, chromium, and arsenic levels in eggs, feathers and tissues of Canada geese of the New Jersey Meadowlands. Environ Res 111:775–784

UNEP (2002) Chemicals. Global mercury assessment. Report no. 54790-01. Geneva, Switzerland, pp 258. http://www.chem.unep.ch

UNEP (2013) Mercury: time to act. Technical report. Chemicals Branch, Division of Technology, Industry and Economics, United Nations Environment Programme, UNEP, Geneva, pp 1–44. http://www.unep.org/PDF/PressReleases/Mercury_TimeToAct.pdf/

UNEP (2016) Business plan of the mercury cell chlor-alkali production partnership area. http://www.unep.org/chemicalsandwaste/Portals/9/Mercury/Chloralkali/Chlor-alkali%20business%20plan%2002_2016.pdf

US Bureau of Mines (1981) Mercury. In: Bureau of Mines Minerals Yearbook. US Bureau of Mines, Washington, DC, pp 585–591

US Bureau of Mines (1986) Mercury. In: Bureau of Mines Minerals Yearbook. US Bureau of Mines, Washington, DC, pp 659–665

US Bureau of Mines (1991) Mercury. In: Bureau of Mines Minerals Yearbook. US Bureau of Mines, Washington, DC, pp 989–995

US EPA (2000) Bioaccumulation testing and interpretation for the purpose of sediment quality assessment. Status and needs. United States Environmental Protection Agency, Bioaccumulation Analysis Workgroup, Washington, DC, EPA-823-R-00-001, pp 136

US EPA (2001) Water quality criterion for the protection of human health: methylmercury. US Environmental Protection Agency EPA-823-R-01-001. Office of Water, Washington, DC. http://water.epa.gov/scitech/swguidance/standards/criteria/aqlife/methylmercury/upload/2009_01_15_criteria_methylmercury_mercury-criterion.pdf
US EPA (2010) Guidance for Implementing the January 2001 Methylmercury Water Quality Criterion. EPA 823-R-10-001. U.S. Environmental Protection Agency, Office of Water, Washington, DC
US GS (1981) Minerals yearbook. Mercury. US Department of the U.S. Geological Survey, pp 585–591
US GS (1996) Mercury. In: Mineral commodity summaries. US Geological Survey, Washington, DC, pp 106–107
US GS (2001) Mercury. In: Mineral Commodity Summaries. US Geological Survey, Washington, DC, pp 104–105
US GS (2006) Mercury. In: Mineral commodity summaries. US Geological Survey, Washington, DC, pp 108–109
US GS (2010) Mercury. In: Mineral commodity summaries. US Geological Survey, Washington, DC, p 101
US GS (2011) Mercury. In: Mineral commodity summaries. US Geological Survey, Washington, DC, pp 102–103
US GS (2016a) 2014 Minerals yearbook. Mercury. US Department of the US Geological Survey, pp 48.1–48.5. https://minerals.usgs.gov/minerals/pubs/commodity/mercury/myb1-2014-mercu.pdf
US GS (2016b) Mercury. In: Mineral commodity summaries. US Geological Survey, Washington, DC, pp 108–109
Vahter M, Mottet NK, Friberg L, Lind B, Shen D, Burbacher T (1994) Speciation of mercury in the primate blood and brain following long-term exposure to methylmercury. Toxicol Appl Pharmacol 124:221–229
Van der Molen EJ, Blok AA, de Graaf GJ (1982) Winter starvation and mercury intoxication in grey herons (*Ardea cinerea*) in the Netherlands. Ardea 70:173–184
Visvanathan C (2003) Treatment and disposal of mercury contaminated waste from oil and gas exploration facilities. In: International environmental disaster and emergency response conference, 13–14 Nov 2003, Yunlin, Taiwan, pp 11
Wada H, Yates DE, Evers DC, Taylor RJ, Hopkins WA (2010) Tissue mercury concentrations and adrenocortical responses of female big brown bats (*Eptesicus fuscus*) near a contaminated river. Ecotoxicology 19:1277–1284
Walker LA, Chaplow JS, Grant HK, Lawlor AJ, Pereira MG, Potter ED et al (2016) Mercury (Hg) concentrations in predatory bird livers and eggs as an indicator of changing environmental concentrations: a Predatory Bird Monitoring Scheme (PBMS) report. Centre for Ecology & Hydrology, Lancaster, UK, pp 23
Wang Y, Greger M (2004) Clonal differences in mercury tolerance, accumulation, and distribution in willow. J Environ Qual 33:1779–1785
Wang H, Tong J, Bi Y, Wang C, Guo L, Lu Y (2013) Evaluation of mercury mediated *in vitro* cytotoxicity among cell lines established from green sea turtles. Toxicol In Vitro 27:1025–1030
Wang W, Evans D, Hickie BE, Rouvinen-Watt K, Evans HE (2014) Methylmercury accumulation and elimination in mink (*Neovison vison*) hair and blood: results of a controlled feeding experiment using stable isotope tracers. Environ Toxicol Chem 33:2873–2880
Wang X, Yan M, Zhao L, Wu Q, Wu C, Chang X et al (2016) Low-dose methylmercury-induced apoptosis and mitochondrial DNA mutation in human embryonic neural progenitor cells. Oxid Med Cell Longev 2016:article ID 5137042
Warfvinge K, Hua J, Berlin M (1992) Mercury distribution in the rat brain after mercury vapor exposure. Toxicol Appl Pharmacol 117:46–52

Weech SA, Wilson LK, Langelier KM, Elliott JE (2003) Mercury residues in livers of bald eagles (*Haliaeetus leucocephalus*) found dead or dying in British Columbia, Canada (1987–1994). Arch Environ Contam Toxicol 45:562–569
Weiner J (1973) Dressing percentage, gross body composition and caloric value of the roe-deer. Acta Theriol 18:209–222
Wellmitz J (2010) Mercury levels and trends in fish and mussels from German surface waters—comparison with the EQS as specified in Directive 2008/105/EC. German Federal Environment Agency, Sec II 2.5, pp 26. www.umweltprobenbank.de
Wente SP (2004) A statistical model and national data set for partitioning fish-tissue mercury concentration variation between spatiotemporal and sample characteristic effects. US Geological Survey Scientific Investigation Report 2004-5199, pp 15
Wentz DA, Brigham ME, Chasar LC, Lutz MA, Krabbenhoft DP (2014) Mercury in the Nation's streams— Levels, trends, and implications: U.S. Geological Survey Circular 1395, pp 90. https://doi.org/10.3133/cir1395
Whanger PD (2001) Selenium and the brain: a review. Nutr Neurosci 4:81–97
WHO (2003) Elemental mercury and inorganic mercury compounds: human health aspects. http://www.who.int/ipcs/publications/cicad/en/cicad50.pdf
Wiener JG, Krabbenhoft DP, Heinz GH, Scheuhammer AM (2003) Ecotoxicology of mercury. In: Hoffman DJ, Rattner BA, Burton GA, Cairns J (eds) Handbook of ecotoxicology, 2nd edn. CRC, Boca Raton, FL, pp 409–463
Wilhelm SM, Liang L, Cussen D, Kirchgessener DA (2007) Mercury in crude oil processed in the United States (2004). Environ Sci Technol 41:4509–5414
Windham-Myers L, Marvin-DiPasquale M, Kakouros E, Agee JL, Kieu le H, Stricker CA et al (2014) Mercury cycling in agricultural and managed wetlands of California, USA: seasonal influences of vegetation on mercury methylation, storage, and transport. Sci Total Environ 484:308–318
Wobeser G, Swift M (1976) Mercury poisoning in a wild mink. J Wildl Dis 12:335–340
Wobeser G, Nielsen NO, Schiefer B (1976) Mercury and mink. II. Experimental methyl mercury intoxication. Can J Comp 40:34–45
Wolfe M, Norman D (1998) Effects of waterborne mercury on terrestrial wildlife at Clear Lake: evaluation and testing of a predictive model. Environ Toxicol Chem 17:214–227
Wolfe MF, Schwarzbach S, Sulaiman RA (1998) Effects of mercury on wildlife: a comprehensive review. Environ Toxicol Chem 17:146–160
Wolfe MF, Atkeson T, Bowerman W, Burger J, Evers DC, Murray MW et al (2007) Wildlife Indicators. In: Harris R, Krabbenhoft DP, Mason R, Murray MW, Reash RJ, Saltman T (eds) Ecosystem responses to mercury contamination: indicators of change. SETAC books. CRC, Boca Raton, FL, pp 123–189
Wood PB, White JH, Steffer A, Wood JM, Facemire CF, Percival HF (1996) Mercury concentrations in tissues of Florida bald eagle. J Wildl Manage 60:178–185
Wren CD (1984) Distribution of metals in tissues of beaver, raccoon and otter from Ontario, Canada. Sci Total Environ 34:177–184
Wren CD (1985) A probable case of mercury poisoning in a wild otter (*Lutra canadensis*) from north-western Ontario. Can Field Nat 99:112–114
Wren CD (1986) A review of metal accumulation and toxicity in wild mammals. I. Mercury. Environ Res 40:210–244
Wren CD, MacCrimmon H, Frank R, Suda P (1980) Total methylmercury levels in wild mammals from the Precambrian shield area of south central Ontario, Canada. Bull Environ Contam Toxicol 25:100–105
Wren CD, Hunter DB, Leatherland JE, Stokes PM (1987) The effects of polychlorinated biphenyls and methylmercury, singly and in combination, on mink. I. Uptake and toxic responses. Arch Environ Contam Toxicol 16:441–447
Wu P (2017) Methylmercury in boreal freshwater food webs. PhD thesis, Swedish University of Agricultural Sciences University, Uppsala, pp 67
WVDL (2015) Normal range values for WVDL toxicology. accessed 28 Apr 2015

Yaroshevsky AA (2006) Abundances of chemical elements in the Earth's crust. Geochem Int 44:48–55

Yates DE, Mayack DT, Munney K, Evers DC, Major A, Kaur T, Taylor RJ (2005) Mercury levels in mink (*Mustela vison*) and river otter (*Lontra canadensis*) from northeastern North America. Ecotoxicology 14:263–274

Yates DE, Adams EM, Angelo SE, Evers DC, Schmerfeld J, Moore MS et al (2014) Mercury in bats from the northeastern United States. Ecotoxicology 23:45–55

Ye B-J, Kim B-G, Jeon MJ, Kim S-Y, Kim HC, Jang T-W et al (2016) Evaluation of mercury exposure level, clinical diagnosis and treatment for mercury intoxication. Ann Occup Environ Med 28:5

Yu X, Driscoll CT, Montesdeoca M, Evers D, Duron M, Williams K et al (2011) Spatial patterns of mercury in biota of Adirondack, New York lakes. Ecotoxicology 20:1543–1554

Zamani-Ahmadmahmoodi R, Esmaili-Sari A, Savabieasfahani M, Ghasempouri SM, Bahramifar N (2010) Mercury pollution in three species of waders from Shadegan Wetlands at the head of the Persian Gulf. Bull Environ Contam Toxicol 84(3):326–330

Zarski TP, Debski B, Samek M (1995) Relation between selenium and mercury concentrations in tissues of hares (*Lepus europaeus* Pall.) from regions with various environmental contaminations. Ekologia (Bratislava) 14:93–97

Zarski TP, Rejt L, Zarska H, Jarmul J (2015) Investigation on the distribution of mercury in tissues and organs of wild birds obtained from the area covered by Greater Warsaw. J Elem 20:247–254

Zhang ZS, Zheng DM, Wang QC, Lv XV (2009) Bioaccumulation of total and methyl mercury in three earthworm species (*Drawida* sp., *Allolobophora* sp., and *Limnodrilus* sp.). Bull Environ Contam Toxicol 83:937–942

Zhang H, Feng X, Larssen T, Shang L, Li P (2010) Bioaccumulation of methylmercury versus inorganic mercury in rice (*Oryza sativa* L.) grain. Environ Sci Technol 44:4499–4504

Zhang R, Wu F, Li H, Guo G, Feng C, Giesy JP, Chang H (2013) Toxicity reference values and tissue residue criteria for protecting avian wildlife exposed to methylmercury in China. Rev Environ Contam Toxicol 223:53–80

Zhao L, Anderson WNC, Qiu G, Meng B, Wang D, Feng X (2016) Mercury methylation in paddy soil—source and distribution of mercury species at a Hg mining area, Guizhou Province, China. Biogeosciences 13:2429–2440

Zheng D, Zhang Z, Wang Q (2010) Total and methyl mercury contents and distribution characteristics in cicada, *Cryptotympana atrata* (Fabricius). Bull Environ Contam Toxicol 84:749–753

Zhu X, Kusaka Y, Sato K, Zhang Q (2000) The endocrine disruptive effects of mercury. Environ Health Prev Med 4:174–183

Zhu H, Yan B, Cao H, Wang L (2012) Risk assessment for methylmercury in fish from the Songhua River, China: 30 years after mercury-containing wastewater outfalls were eliminated. Environ Monit Assess 184:77–88

Zilincar VJ, Bystrica B, Zvada P, Kubin D, Hell P (1992) Die Schwermeallbelastung bei den Braunbaren in den Westkarpten. Z Jagdwiss 38:235–243 (in German)

Zrncic S, Oraic D, Caleta M, Mihaljevic Z, Zanella D, Bilandzic N (2013) Biomonitoring of heavy metals in fish from the Danube River. Environ Monit Assess 185:1189–1119

# Chapter 18
# Silver, Ag

**Lidia Strużyńska**

**Abstract** The profile of silver (Ag) emission to the environment has changed significantly in recent decades. Although the photography industry has lost its importance, global Ag production continues, together with rising demand for nanosilver (nano-Ag), which is widely used in many products. Hence, increasing volumes of Ag waste are expected to be released into waters and on land. The results of experimental studies has demonstrated the hazardous impact of Ag/nano-Ag on mammalian organisms. However, the belief exists that Ag is extremely toxic for aquatic species only, so studies conducted on terrestrial wildlife are scarce. The data presented here indicate that Ag accumulates in high amounts in the soft tissues and bones of the investigated mammals and birds. It also appears to be present in higher concentrations in liver and brain, which could, hence, be pointed to as target organs. In addition, birds inhabiting aquatic areas and those connected with aquatic food chains seem to be suitable bioindicators of water-body contamination, whereas small birds, such as the great tit, may be useful indicators of urban pollution. Consideration should also be given to the value of current observations concerning the close relation between the presence of Ag in organisms and their environments. It should also be highlighted that detailed studies on avian and mammalian terrestrial wildlife merits high priority in order to evaluate the possible hazardous impact of increased environmental emission of Ag and nano-Ag.

## 1 Introduction

Metals have become increasingly important as a group of environmental pollutants and potentially dangerous toxins. Major advances have been made in recent decades in our understanding of the biological effects of silver (chemical symbol: Ag, from

L. Strużyńska (✉)
Laboratory of Pathoneurochemistry, Department of Neurochemistry, Mossakowski Medical Research Centre, Polish Academy of Sciences, Warsaw, Poland
e-mail: lidkas@imdik.pan.pl



E. Kalisińska (ed.), *Mammals and Birds as Bioindicators of Trace Element Contaminations in Terrestrial Environments*,
https://doi.org/10.1007/978-3-030-00121-6_18

the Latin *argentum*) in experimental animals, but we still do not know enough about its environmental levels, bioavailability, bioaccumulation, and ecotoxicology. Silver is a nonessential element. It is known to bioaccumulate in and to be highly toxic to aquatic organisms (Luoma 2008). Since in terrestrial mammals this metal is not extensively metabolized and its absorption is low, the belief exists about its relatively low toxicity to this group of organisms. However, it should be remembered that metals, because of their persistent nature in the environment, may generally be more dangerous when accumulated over a long period of time. The industrial pattern of Ag use has changed significantly over the years. Parallel to the reduced importance of analog photography, the manufacture of Ag-containing electrical goods has increased (Eckelman and Graedel 2007). Moreover, the widespread use of Ag in its particulate form has been observed recently. The expanding development of nanotechnologies has led to increased production of engineered silver nanoparticles (AgNPs) utilized in many medical and consumer products (Rejeski 2011). Nano-Ag-based textiles, plastics, and cosmetics with antimicrobial effects are enjoying increasing popularity in Asia, North America, and Europe. Their growing production and use represent a significant source of Ag emissions and enhance the likelihood of environmental exposure to AgNPs (Eckelman and Graedel 2007). Thus, environmental concerns about Ag have currently switched from its ecotoxicological to its nanoecotoxicological impacts (Kahru and Dubourguier 2010). The increasing economic importance of nanomaterials has not been accompanied by appropriate safety regulations (Bondarenko et al. 2013). This creates an increasing need to assess the potential adverse health effects of AgNPs in animals and humans. Although recently progress has been made in research on experimental nano-Ag toxicity, there are still few data concerning its ecotoxicological effects. Much remains to learn about nano-Ag behavior in environmental media, its bioavailability in wildlife, and its interactions at the molecular level. Current data concerning nanotoxicity, including AgNPs toxicity, are incomplete and inconsistent owing to differences in experimental approaches and should be extended to those test species and biological endpoints that are usually used in environmental hazard assessment (Bondarenko et al. 2013). All these issues have become a challenge for nanoecotoxicological research, which is expected to contribute to new regulations being issued by European Commission and US authorities (Kahru and Dubourguier 2010).

## 2 General Properties of Silver

Silver is a chemical element with atomic number 47 that belongs to the group of transition metals. In the periodic table, Ag is situated in period 5, group 11, and d-bloc. It has an atomic mass of 107.868, a density of 10.49 g $cm^{-3}$, and melting and boiling points of 961.78°C and 2162°C, respectively (Wieser et al. 2013; http://periodictable.com/Properties/A/CrustAbundance.html). It possesses the lowest contact resistance and the highest electrical and thermal conductivity of metals. It is water-insoluble but dissolves in nitric acid. Elemental silver Ag(0) and monovalent silver Ag(I) are the most common oxidation states (ATSDR 1990). Other oxidation states, Ag(II) and Ag(III), although rare, are also possible. Silver ions are dissociated from particulate Ag and from different salts. There are 2 natural stable isotopes of Ag, $^{107}Ag$ and $^{109}Ag$, and 20 artificial radioisotopes, of which the radioisotope $^{110}Ag$ has the longest physical half-life (253 days). None of the radioisotopes occurs naturally (Eisler 1996; WHO 2002). Several compounds of Ag possess explosive properties: silver oxalate ($Ag_2C_2O_4$) explodes when heated; silver acetylide ($Ag_2C_2$) detonates on contact; and silver azide ($AgN_3$) detonates spontaneously under certain circumstances (WHO 2002).

Silver occurs relatively rarely in the Earth's crust, appearing in the 65th position in the natural elements (Yaroshevsky 2006). The mean concentration of Ag in the upper continental crust is estimated to be between 0.053 and 0.070 mg $kg^{-1}$ (Rudnick and Gao 2003; Yaroshevsky 2006).

## 3 Silver in Nature: Geogenic Sources of Metal

Crustal Ag occurs naturally as native Ag, alone or with other metals, frequently with gold, although Ag deposits tend to be of a rather polymetallic nature. It is predominantly concentrated in basalt (0.10 mg $kg^{-1}$) and igneous rocks (0.07 mg $kg^{-1}$) in the form of minerals: acanthite ($Ag_2S$) and chlorargyrite (AgCl). Chlorargyrite, Ag chloride, is also known as horn Ag when weathered by air. Acanthite, the sole form of Ag sulfide stable in normal air temperature, transforms to argentite above 173°C. It is the main ore from which Ag is extracted. Important sources of the metal are also pyrargyrite ($Ag_3SbS_3$), known as dark red Ag, and the corresponding proustite ($Ag_3AsS_3$), or light red Ag (US EPA 2006). From these natural sources Ag is released into the environment through the weathering of rocks and erosion of soils. Some areas in the world are rich in Ag, including the Cordilleras region of the Americas, Abitibi and Cobalt areas on the Canadian Shield, certain mountain ranges of Europe (Cornwall, Central Massive, Harz, Internal Carpathians), the Caucasus, Altai Mountains, Chukotka, Japan, and Eastern Australia. Fine large and pure crystals are found in Saxony and Harz Mountains (Germany), in Joachimsthal (Czech Republic), in the Calstock, Cambourne, and St. Just districts of England, and in Cornwall. In the USA, Ag ores occur in large masses in the Lake Valley district, in the states of New Mexico, Arizona, and Nevada (Palache et al. 1951; Klein and Hurlbut 1985).

The average content of Ag in rocks has been determined to be 0.05 mg $kg^{-1}$ in acidic and ultraalkaline rocks, and about 0.1 mg $kg^{-1}$ in alkaline rocks. In abiotic elements of the environment, its concentrations are also naturally elevated in crude oil (as high as 100 mg $kg^{-1}$) (Eisler 1996; WHO 2002). Among other energy resources, Ag content of coal is about 0.011 mg $kg^{-1}$ (Schweinfurt 2009).

The background Ag in soils is of lithogenic origin and range in concentration from 0.4 to 0.8 mg $kg^{-1}$ dry weight (dw), as reported for both eastern and western regions of the USA (US EPA 2003). According to Kabata-Pendias (2011) average background Ag content in continental soil is 0.13 mg $kg^{-1}$. Understanding background concentrations of metals in soil is important for interpreting the toxicity-derived so-called ecological soil screening levels (Eco-SSLs).

High Ag content is exhibited by some geothermal water reservoirs, for example, Salton Lake (California, USA) and Taupo Lake (New Zealand): 0.8–2.0 and 2–7 mg $L^{-1}$, respectively. Benthic salt muds of the Red Sea exhibit 18 mg Ag $L^{-1}$, whereas uncontaminated sea and ocean waters have scant levels of Ag and contain on average 0.15–0.29 ng $L^{-1}$ or even less, 0.03–0.10 ng $L^{-1}$ (WHO 2002; Luoma 2008). Average Ag concentrations in natural inland waters like rivers and lakes are in the range of 0.2–0.3 μg $L^{-1}$ (US EPA 1980). The measurements of Ag in freshwater reservoirs show levels even below 0.01 μg $L^{-1}$ in unpolluted areas and much lower concentration ranges are observed for Ag content in air (ng $m^{-3}$) (US EPA 2003).

Another form of the metal is nanoparticulate Ag (nano-Ag), which is generally a product of anthropogenic activity. However, it was reported that AgNPs may be formed via the reduction of $Ag^+$ in the presence of humic acid, which is the element of water sediments and soils. This suggests that nano-Ag might be not only of anthropogenic origin but could be naturally formed under the proper environmental conditions as well (Akaighe et al. 2011).

## 4 Silver Production and Uses. Anthropogenic Sources of Silver

Silver was discovered around 4000 BC and has been used for millennia for currency and jewelry. Owing to the strong antibacterial properties of its ionized form, it was also used in food service and water disinfection in ancient times (Barillo and Marx 2014). Its medicinal application in wound care has been known for 200 years. Silver compounds have also been used in the treatment of mental illness, epilepsy, and infections like syphilis and gonorrhea (Marshall and Shneider 1977). In crafts and industry, it was used for many centuries in mirror production and photographic materials. Other traditional industrial applications of metallic Ag included coin and metal fabrication, electrical and electronic components, soldering, and plating. As an antibacterial agent it has been used in medical instruments and materials like central venous and bladder catheters, endotracheal tubes lowering the risk of ventilator-associated pneumonia, wound dressing, and gels.

The rediscovery of the biocidal properties of Ag for medicinal and many other uses occurred owing to the development of nanotechnologies (dos Santos et al. 2014). The twenty-first century has been a time of growing importance of nanoparticles (NPs) in general and AgNPs in particular. Engineered NPs are defined as materials having less than 100 nm in size at least in one dimension. Thanks to their large surface area per unit mass, NPs display enhanced properties compared with microsized materials, including a high reactivity both in chemical and biological systems (Nel et al. 2006). Furthermore, the unique physicochemical properties of NPs may be modified by coating their surface with different substances and chemical groups (Mody et al. 2010). Interesting properties of the nanometer-sized form of Ag that make it different from the bulk form have allowed for its extensive use in medicine and medical devices, as well as in optics, sensing, and painting, and its further extension to many other consumer products. The market for commercially available nano-Ag-containing products continues to expand. Currently, about 30% of all NP-based products contain AgNPs (WWC 2013; Rejeski 2011). Because of their potent antimicrobial effect, exceeding that observed for ionic Ag, AgNPs have found application in various home appliances like refrigerators, dishwashers, washing machines, and water filters. The antibacterial properties of nano-Ag-coated textiles are also popular in the fields of sports and medicine. They are also utilized wherever hygienic conditions and sterile procedures are particularly important, for example, in hospitals, kitchens and food preparation, sanitary facilities, air conditioning and ventilation systems, and material packaging (Bondarenko et al. 2013; Fewtrell 2014; Gaillet and Rouanet 2015). Important fields of nano-Ag applications are sensors (25%), antimicrobial agents (19%), catalysis (13%), optics (8%), and sensing devices (7%) (Bondarenko et al. 2013).

In conventional medicine, the coating of various medical devices, such as catheters, stents, implants, or prostheses with nano-Ag, significantly minimizes the risk of infection during surgical procedures (Ge et al. 2014). Apart from that, AgNPs are also very important tools in next-generation medicine as vehicles for gene therapy and targeted drug delivery systems, as well as biomarkers and elements of imaging systems (Leite et al. 2015).

Because of the global development of nanoindustry, nanoproducts have found application in almost all areas of life. As a result, the growing importance of nano-Ag relative to its metallic or ionic forms is currently observed.

There is a long list of countries around the world where Ag is mined. In the twentieth century most of the world's Ag in the Northern Hemisphere (75%) was mined and produced in the Mexico, USA, Canada, Japan, and Russia; among these countries the USA produced about 50% of the world's Ag (Eisler 1996). Up to 1970 the USA produced less than 15% of the world's Ag and consumed more than 60%, exceeding extractions from ores (Eisler 1996). World production of Ag increased over the years from 7700 tons in 1964 to 12,700 tons in 1984. At the end of the last century, the estimated world mine production of Ag had increased to 14,200 tons (Table 18.1). The major producers of the metal were Mexico (17% of the total), the USA (14%), Peru (12%), the former Soviet Union (10%), and Canada (9%) (Eisler 1996; http://minerals.usgs.gov/minerals/pubs/commodity/myb/). A sharp increase in Ag mining has occurred in the current century. Global mine production of Ag during the period 2004–2014 increased from about 19,800 to 27,200 tons. In 2010, 13,800 tons of Ag were used in industrial applications, including 4734 and 2860 tons used for Ag jewelry and coins/medals, respectively (https://www.silverinstitute.org). In 2013 global Ag use increased by 13% compared with 2012, with a concomitant decreased use for electric, electronic, and photographic applications (http://minerals.usgs.gov/minerals/pubs/commodity/myb/).

**Table 18.1** Global mine production of silver over last 50 years (1964–2014)

| Year | Tons of silver |
|---|---|
| 1964 | 7700 |
| 1974 | 9060 |
| 1984 | 12,700 |
| 1994 | 14,200 |
| 2004 | 19,800 |
| 2014 | 27,200 |

According to U.S. Geological Survey, http://minerals.usgs.gov/minerals/pubs/commodity/myb/

**Table 18.2** Silver production in top ten countries, 2000–2013

| Country | Silver production (tons) in selected years | | |
|---|---|---|---|
| | 2000[a] | 2010[a] | 2013[b] |
| Mexico | 2620 | 4411 | 4892 |
| Peru | 2145 | 3640 | 3342 |
| China | 1600 | 3500 | 3339 |
| Australia | 2060 | 1864 | 1675 |
| Russia | 370 | 1545 | 1273 |
| Boliwia | 434 | 1259 | 1165 |
| Chile | 1242 | 1287 | 1109 |
| Poland | 1100 | 1181 | 1064 |
| USA | 1980 | 1280 | 990 |
| Argentina | 78 | 723 | 699 |

According to data supplied by [a]U.S. Geological Survey (http://minerals.usgs.gov/minerals/pubs/commodity/myb/) and [b]Silver Institute (https://www.silverinstitute.org)

In recent years Mexico has been the dominant player in Ag production, at nearly 4900 tons; Peru, China, Australia, and Russia are the world's other leading producers (Table 18.2).

China's Ag production, including mined, byproduct output, and recycled material, grew by almost 15% every year from 1990 to 2009. By 2010, China was producing over 3000 tons of Ag per year, making it the world's third largest producer of mined Ag and transforming it from a net exporter to a net importer of the metal (https://www.silverinstitute.org/site/supply-demand/silver-production/). Along with China and Mexico, Russia, with 1273 tons of Ag, was responsible for much of the global increase in the metal's supply in 2013. Poland is at the forefront of Ag-producing countries, posssessing the second-largest Ag reserves in the world, estimated at 85,000 tons.

Apart from being exploited from its natural deposits, Ag is frequently obtained as a byproduct in the process of copper, gold, nickel, and lead-zinc ore refining (Eisler 1996). Other important anthropogenic sources of Ag in the biosphere, except mines, are smelters, manufacture of photographic and electrical supplies, coal combustion, and cloud seeding with Ag iodide (Eisler 1996). Silver is emitted into the atmosphere mainly in the forms of Ag sulfide, Ag sulfate, Ag carbonate, Ag halides, and metallic

Ag (Smith and Carson 1977), of which about 50% is transported more than 100 km and is eventually deposited in precipitation (US PHS 1990). It may be found in wastewater from mines and various industries but is also emitted in fallout from coal-fired power plants, from where it can pass into surface waters or accumulate in the soil (Nordberg and Gerhardsson 1988). It has been reported that liquid effluents from the nuclear industry often contain significant quantities of radionuclide $^{110}$Ag (Eisler 1996). In the late 1970s it was estimated that Ag released into the environment from industrial sources was approx. 2500 tons per year just in the USA. Nowadays, the amount of Ag waste entering terrestrial or aquatic ecosystems each year worldwide is 300 tons (Wijnhoven et al. 2009). Almost all (about 95%) of the Ag emitted from anthropogenic sources remains in the soil and wastewater compartments (Shafer et al. 1998). As much as 80,000 tons of Ag per year is used on agricultural lands and ends up in industrial sewage sludge; 150 tons of Ag enters aquatic environments every year from the photography industry, mine tailings, and electroplating. Measurements of Ag in rivers and lakes show levels of about 0.1 μg $L^{-1}$ in urban and industrialized areas (WHO 2002). The atmosphere receives 300 tons of Ag each year from a variety of sources. Maximum concentrations of total Ag recorded during the 1970s and 1980s in selected areas were as follows: 36.5 ng $m^{-3}$ in air near smelters; 2.0 μg $m^{-3}$ in atmospheric dust; 6.0 μg $L^{-1}$ in groundwater near hazardous waste sites; 260 μg $L^{-1}$ near photographic manufacturing waste discharge; and as much as 150 mg $kg^{-1}$ in river sediments (WHO 2002).

Regionally, Asia emits high amounts of Ag from landfills directly to the land. Asian countries, such as China, India, and Indonesia, emit significant amounts of the metal into water (Eckelman and Graedel 2007). A survey conducted in North Pacific waters in 2002 found significant (1.2 ng $L^{-1}$) Ag contamination, the most likely source of which were emissions from coal-burning and coastal waters, especially in Asia. Levels of silver were about 50 times over baseline levels found in uncontaminated waters of the Atlantic Ocean. Increased concentrations of the metal in the open ocean suggest Asia may be a so-called hotspot where Ag contamination is high (Stephens 2005; http://currents.ucsc.edu/04-05/03-14/silver.asp).

Engineered nano-Ag may be considered a significant anthropogenic source of Ag potentially affecting the environment. The issue of AgNPs released into the environment and a potential risk of contamination of natural water systems and aquatic organisms is currently under debate (Blaser et al. 2008; Fabrega et al. 2011). It was estimated that currently about 320 tons per year of nano-Ag are produced and used worldwide (Gottschalk et al. 2010). In Europe , up to 2010, approximately 110–230 tons of Ag was used in the form of nano-Ag-containing biocidal products, where Germany alone used about 8 tons, mainly for water purification purposes (Blaser et al. 2008). In China, the market demand for AgNPs used as antibacterial agents was estimated to have increased to 366 tons by 2014 compared to 45 tons in 2008 (Gao et al. 2013). Worldwide annual production of Ag-containing nanomaterials is reported to be in the range of 5.5–550 tons per year (Piccinno et al. 2012). According to Massarsky et al. (2014), who reviewed data on worldwide Ag production, the amount of nano-Ag increased from 0.4–46 tons per year in 2006 to 7.2–716 tons per year in 2014. Assuming that increase, environmental concentrations of nano-Ag predicted for 2016 in surface water may even range between 163 and 1306 ng $L^{-1}$, reaching a maximum value of 57.1 mg $kg^{-1}$ in sediments.

The first relevant processes leading to the mass flow of nano-Ag into terrestrial systems are the disposal of sewage sludge and solid waste management (EEA 2001). Hence, the concentration of nano-Ag in soil will likely continue to rise. The future emission scenario envisions increasing Ag use for the production of different categories of biocidal products and, therefore, increasing flow of Ag into the environment. Silver sulfide concentrations expected in the wastewaters of the European Union in 2015 were estimated to be at 15.7 μg $L^{-1}$ (Blaser et al. 2008).

Silver bioavailability depends on water characteristics such as hardness, natural organic matter (NOM), the presence of chloride, sulfides, and sulfates, all of which lower Ag toxicity. Part of wastewater Ag enters natural freshwater reservoirs where it settles in sediment or is kept in solution in colloidal or complexed form. When associated with colloidal sediment fractions, Ag may naturally occur in the form of NP clusters (Luther and Rickard 2005). To form complexes, it tends to react with available inorganic anionic ligands, mostly with chloride ($Cl^-$), sulfide ($S^{2-}$), thiosulfate ($S_2O_2{}^{-3}$), or organic thiolate (e.g., glutathionate, cysteinate) (Hiriart-Baer et al. 2006). The formation of complexes increases together with enhanced salinity, so that in marine environments ionic Ag binding in complexes with chloride predominates (Ratte 1999). In freshwater and soils, the primary Ag compounds under oxidizing conditions are bromides, chlorides, and iodides; under reducing conditions the free metal and Ag sulfide predominate (US PHS 1990). The opinion also exists that under environmental conditions, neither AgCl (seawater) nor Ag-NOM (freshwater) dominates in Ag speciation, but Ag sulfides are expected to be the predominant forms of the metal (Kramer et al. 2002). Hence, in the current risk assessment, Ag sulfides are considered to be the environmentally relevant Ag compounds.

## 5 Biological Status of Silver

The global biogeochemical cycle of Ag in nature occurs by its release into the atmosphere, water, and land from natural and, mainly, anthropogenic sources, its long-range transport in the atmosphere, wet and dry deposition, and absorption in soils and sediments (US PHS 1990). Although anthropogenic emission of Ag is still increasing, the mechanisms of its environmental behavior are poorly understood, because no urgent environmental problems have been caused by Ag, in contrast to other metals like mercury, cadmium, or lead.

In soil, Ag is generally highly immobile, relative to other metals, due to precipitation into insoluble salts. Manganese and iron oxides have a strong affinity for binding to it. Silver present in ionic form or in the form of organic complexes in soil solutions is considered the most bioavailable fraction. The amount of soluble Ag (I) depends on dissolved organic carbon (DOC) concentrations, which strongly binds to Ag in soil water extracts. It is also related to the pH, redox conditions, and strength of binding and complexation with dissolved organic matter (US EPA 2006; Settimio et al. 2015). Microorganisms are very important biocomponents of soils because they are involved in the cycling of chemical elements that are responsible for their mobilization and accumulation in soils (Kabata-Pendias 2011). Since Ag exhibits very high toxicity toward microorganisms, soils contaminated with Ag may be deficient in microbiota, which is of importance in the case of cultivated soils.

Eco-SSLs are concentrations of contaminants in soil that are protective of biota living in or on soil. These values are applied at the screening stage of an ecological risk assessment, which are assumed to provide adequate protection of terrestrial ecosystems. The silver Eco-SSL value calculated for plants is 560 mg $kg^{-1}$ dw in soil, while values for avian and mammalian wildlife range from 4.2 to 14 mg $kg^{-1}$ dw, respectively (US EPA 2006). These concentrations significantly exceed background concentrations in soil in the USA (US EPA 2003).

As a nonessential metal, Ag fulfills no biochemical or physiological roles in tissues.

Bioaccumulation of this metal may occur via the body surface and food in aquatic organisms (bioconcentration) or mainly via the alimentary tract in terrestrial invertebrates and vertebrates. In microorganisms, the bioconcentration of Ag takes place by adsorption on the surface. In terrestrial plants bioaccumulation through the leaves and roots predominates (Ratte 1999).

Microorganisms take up Ag by the adsorption process, that is, by binding on the surface by covalent, electrostatic bonds. In higher organisms uptake of Ag requires energy-dependent active transport with a macromolecular carrier (Ratte 1999). In marine invertebrates, Ag accumulates more intensively than other metals like cadmium, indicating a specific salinity-dependent process (Luoma et al. 1995).

Since Ag is highly toxic to microbes, mechanisms of bioconcentration can be investigated in species with high tolerance. Several bacteria from the genus *Pseudomonas* were found to exhibit extremely high tolerance followed by an accumulation rate over 300 g Ag $kg^{-1}$ dw (Charley and Bull 1979). The strong bactericidal properties of Ag influence bacteria, which fix nitrogen and break down organic matter in soil ecosystems. It was demonstrated that Ag inhibits the growth of heterotrophic and chemolithotrophic bacteria in soil even in concentrations well below those characteristic of heavy metals (Throbäck et al. 2007). Soil organic matter binds Ag strongly, limiting its absorption.

Regarding plants, the absorption of Ag from soils by terrestrial plants is generally low, even if the soil is amended with Ag-containing sewage sludge (WHO 2002). It accumulates mainly in the root systems of plants occurring in the vicinity of mines or in areas contaminated with metal-containing wastewaters. This concerns both trees and other plants, including agricultural crops, grasses, or fungi. Suitable for the detection of soil contamination by Ag is the earthworm *Lumbricus terrestris*. Exposed to soil with increasing concentrations of $Ag_2S$, it exhibited low bioaccumulation of Ag but reduced growth due to the contact toxicity (for review see: Ratte 1999).

While the biological status of ionic Ag is well established, what happens with AgNPs in both environmental and biological media remains under investigation. It is important to understand the mechanisms of their biotransformation to assess their potential environmental impact. It has been found that NPs lose surface-coating agents such as citric acid, sodium dodecyl sulfate, and amino acids, for example, which are added to prevent the agglomeration of NPs and thereby protect their dispersed state. The result of this process is the unstable state of nano-Ag, leading to aggregation and agglomeration (Li et al. 2013; McShan et al. 2014). Another mechanism involved in biotransformation is the oxidation of NPs' surface to Ag oxide, resulting in the release of ions after interaction with redox-active compounds (Liu et al. 2010). This mechanism of AgNP transformation occurs in environmental and biological media, as it does inside cells. It has also been pointed out that the concentration of sulfur ions, dissolved oxygen, chlorine ions, and biological macromolecules (mainly proteins) is of importance for interactions with AgNPs (McShan et al. 2014). In the environment, AgNPs should be released and concentrated in water systems. In natural water systems, as well as in plants treated with Ag-containing wastewaters, sulfidation of AgNPs has been observed (Choi et al. 2009; Kim et al. 2010a). The lower solubility of nano-Ag sulfide results in a significantly reduced toxicity (Levard et al. 2012). NOM formed from degraded plants and animals is ubiquitous in natural water systems. It was reported that large polymeric NOM molecules, such as humic acid and fulvic acid, might adsorb NPs, increasing their stability (Sharma et al. 2014). The stable state facilitates dispersion, allowing NPs to remain longer in the water system (Tripathy 2008; Liu et al. 2010).

Nano-Ag affects denitrifying bacteria, disrupting denitrification processes at very low concentrations (0.14 mg $L^{-1}$). This is important because it can result in eutrophication of rivers, lakes, and marine ecosystems. Nitrifying species, which are used for wastewater treatment, are especially susceptible to nano-Ag inhibitory action (Punita 2012; Choi et al. 2009).

Information on Ag toxicity to organisms living in soil and sediments is limited. However, like ionic and metallic forms, AgNPs could have a toxic effect on bacteria that are essential for soil formation. It seems that Ag can disrupt soil microbial species, inhibiting the growth of bacteria well below concentrations of other toxic metals (Murata et al. 2007). As in waters, toxicity depends on physicochemical soil and sediments properties.

Heterotrophic ammonifying bacteria and chemolithotrophic bacteria that consume inorganic material and are essential in soil formation are especially vulnerable to nano-Ag. Nematodes, which are widely found in soils and play an important role in the production, decomposition, and cycling of nutrients, and are food components of many higher animals, serve in natural ecosystems as useful indicators of the presence of soil pollutants. It was found that both AgNPs and bulk Ag are toxic to nematodes (including *Caenorhabditis elegans*), impairing their growth and reproduction (Wang et al. 2009).

The environmental toxicity risks of nano-Ag are poorly understood. No studies have examined the mechanisms of biotransformation and transmission of Ag to plants and further into the food chain. Mammals overexposed to Ag exhibit a characteristic gray or blue-gray and irreversible skin discoloration known as argyria. Silver accumulation in the skin leads to argyria when the body burden is $>1$ g, which is an effect of the precipitation of insoluble Ag salts such as Ag chloride and Ag phosphate (Stokinger 1981; ATSDR 1990). Silver granules are mostly deposited in both pigmented and unpigmented skin together with sulfur and selenium in lysosomal fractions (Lansdown 2007). Skin discoloration may be caused by a photoreduction of Ag chloride to metallic Ag, which is further oxidized in the tissue to black Ag sulfide (Danscher 1981). Argyria was suggested to be a mechanism of Ag detoxification in tissues (Venugopal and Luckey 1978).

With respect to nano-Ag, recent studies have revealed that besides absorption from nanotechnology products, it might be formed in organisms following exposure to an ionic form of Ag (Liu et al. 2012). The chemical mechanisms of nano-Ag biotransformation are virtually unknown. Many of them have not yet been explored, like interactions with selenium and photoreaction of nano-Ag-biocomplexes in the skin (Cheng et al. 2011). Others, like mechanisms of dissolution in the gastrointestinal tract, are still unclear. Studies by Liu et al. (2012) on nano-Ag reactions in biological media mimicking the composition of fluids in different compartments of the human (mammalian) body suggest the existence of an argyrialike pathway of transformation. In addition, it seems that Ag particles deposited in the skin or other tissues have a similar composition and contain sulfur and selenium regardless of whether ionic or nanoparticulate Ag was administered (Hadrup and Lam 2014).

Investigations of the possible environmental fate of nano-Ag under experimental conditions mimicking estuarine mesocosms revealed a high rate of adsorption and bioaccumulation in different biota, such as biomagnification via trophic transfer from sand to clams (Cleveland et al. 2012). According to Dehn et al. (2006) trophic transfer of Ag may occur also in the benthic food chain. Since Ag has a high affinity to sulfur ligands in water and sediments, it may accumulate in the benthic food web (cephalopods, bivalves, crustaceans). In turn, the results of a study by Yoo-iam et al. (2014) revealed no evidence of biomagnification of either $Ag^+$ or AgNPs in the food chain, although aquatic organisms in lower trophic groups such as phytoplankton and zooplankton accumulated higher concentrations of $Ag^+$ and nano-Ag than animals in a higher trophic group (fish). Generally, a few existing studies on the bioaccumulation and biomagnification of Ag have been carried out mainly using aquatic ecosystems. The potential of Ag bioconcentration and biomagnification in terrestrial ecosystems remains unclear. Based on research conducted by Dauwe et al. (2004), one may suspect food chain exposure in certain bird species. Silver concentrations in food (vegetative plants and caterpillars) and excreta of great tit (*Parus major*) from polluted areas were positively correlated. On the other hand, magnification was insignificant for higher organisms analyzed as components of the Arctic food web (Dehn et al. 2006).

## *5.1 Toxicity of Various Forms of Silver in Homeothermic Animals*

Silver has no known physiological function. It has been considered for a long time to be nontoxic to humans and terrestrial animals, although it is known to be, after mercury, the most toxic metal for aquatic organisms and is classified as a priority pollutant of surface waters (US EPA 1980; Luoma 2008). However, regarding the fact that in all major Ag-producing countries emissions in tailings and landfills have the greatest environmental impact, the question regarding Ag toxicity has arisen in connection with the contamination of terrestrial ecosystems (Eckelman and Graedel 2007).

Years of research on Ag toxicity have shown that it depends on the specific Ag compound rather than on total Ag concentration. The metallic form of Ag is less toxic than the ionic form, of which the most toxic is Ag nitrate solution. This compound has been reported to be highly irritating to the skin, mucous membranes, and eyes, causing ocular damage. Applied to the gingival mucosa may result in necrotizing, ulcerative gingivitis (Stokinger 1981).

Silver toxicity to wildlife is of importance mainly in aquatic environments, where the probability of ion formation is high. In surface waters it is classified as a pollutant that is extremely harmful to freshwater fish and invertebrates, with lethal concentrations for some sensitive species as low as 1–5 $\mu g\ L^{-1}$ (US EPA 1980; CEC 1996; Wijnhoven et al. 2009). According to Eisler (1996) and the World Health Organization's (WHO) "Assessment of Environmental Aspects of Silver and Silver Compounds," no data are available on the effects of Ag on wild birds or mammals (WHO 2002). In general, there are very few reports concerning Ag concentration in wildlife-derived samples. In terrestrial animals, the toxicity of Ag ions and AgNPs has been investigated predominantly in laboratory conditions using rodents, most frequently rats and mice (Matuk et al. 1981; Rungby and Danscher 1984; Skalska et al. 2015; El Mahdy et al. 2015).

### 5.1.1 Acute and Chronic Toxicity

Acute toxicity through inhalation exposure to Ag(0) or Ag compounds causes both upper and lower respiratory tract irritation, where the action of Ag nitrate is the most intense likely due to the corrosive effect of the nitrate itself (Rosenman et al. 1979). The inhalation of an aerosol containing colloidal Ag was shown to lead to ultrastructural damage to the epithelial cells of the trachea in rabbits (ATSDR 1990). In humans, accidental ingestion of large doses of Ag nitrate has led to corrosive damage to the gastrointestinal tract, abdominal pain, diarrhea, vomiting, shock, convulsions, and death (US EPA 1985).

Chronic toxicity is connected to exposure to metallic Ag and Ag compounds in small doses over periods of months or years, commonly resulting in argyria, regardless of the route of exposure (Gulbranson et al. 2000). Argyria is recognized as the most common indicator of long-term exposure to Ag or Ag compounds in humans. However, rats experimentally exposed to 222 mg Ag $kg^{-1}$ b.w. $day^{-1}$ for 37 weeks also exhibited granular Ag deposits in the eyes, apart from growth depression and shortened lifespan (Matuk et al. 1981). Long-term oral exposure of experimental animals resulted in a thickening of the basement membranes of the renal glomeruli and granular Ag-containing deposits in skin, eyes, and internal organs, as well as as cardiac enlargement and anemia (Matuk et al. 1981; Drake and Hazelwood 2005). Mild allergic responses have been reported in connection with prolonged dermal exposure to Ag (ATSDR 1990). Oral administration of repeated doses of nano-Ag to mice induced an inflammatory response and significant increase in liver enzymes, indicating hepatotoxicity. Additionally, histopathological changes in kidney have been reported (Park et al. 2010).

### 5.1.2 Genotoxicity

The genotoxicity of metallic/ionic Ag was not confirmed in experimental studies using in vivo models of exposure. However, the combined effects of Ag and other environmental factor, such as ultraviolet (UVB) radiation, were investigated in vitro using human keratinocyte cell line HaCaT. It was found that Ag(I) interacts with cellular DNA and spurs the formation of pyrimidine dimers in both cellular and isolated DNA in previously radiated cells (Zhao et al. 2014). Similarly, the genotoxic effect of nano-Ag was not confirmed in exposed animals (Kim et al. 2011), although in vitro studies in mammalian cells demonstrated DNA damage. The induction of DNA strand break formation, the inhibition of enzymes required for DNA repair, and the upregulation of DNA damage repair proteins have been reported (Ahamed et al. 2008; AshaRani et al. 2009), just as chromosomal aberrations and sister-chromatid exchanges were noted in Chinese hamster fibroblast cell lines (Ahlberg et al. 2014).

### 5.1.3 Carcinogenicity and Teratogenicity

There are no data indicating the carcinogenicity of Ag to humans or animals irrespective of route of exposure (Furst and Schlauder 1978). Silver is placed in the group of substances not classifiable as being carcinogenic to humans (US EPA 1997). There are also no studies indicating that Ag, in either ionic or nano form, can act as a teratogen (ATSDR 1990).

### 5.1.4 Endocrine System Toxicity

The issue of Ag-induced endocrine disruption has received limited attention. Available data mostly originate from in vitro studies. Scant in vivo results were obtained using aquatic species, wherein increasing levels of the stress hormone cortisol and estrogenic effects were shown, signaling the endocrine-disrupting potential of nano-Ag (for a review see Iavicoli et al. 2013).

### 5.1.5 Reproductive Toxicity

Exposure of male Wistar rats to nano-Ag damaged DNA in germ cells and decreased epididymal sperm count (Gromadzka-Ostrowska et al. 2012). In addition, disturbances in the proliferation signaling cascade of spermatogonial stem cells were observed, indicating an inhibitory effect on reproductive processes (Braydich-Stolle et al. 2010).

Additionally, relevant rodent studies on nano-Ag indicate its harmful effect on fetal development. Developing embryos of mice treated with nano-Ag had a high concentration of the metal in the liver, spleen, and visceral yolk sac (Austin et al. 2012). Rat fetuses obtained from mothers exposed during pregnancy to different doses of nano-Ag (0.4–0.8 mg $kg^{-1}$ b.w.) exhibited significantly reduced weights and lengths. Placental weight and volume were also lowered relative to control animals (Mahabady 2012).

### 5.1.6 Neurotoxicity

In long-term studies in orally exposed rodents, hypoactivity and granular Ag-containing deposits in the central nervous system (CNS) were observed, mainly in areas responsible for motor control (Rungby and Danscher 1984). The cellular distribution pattern indicated large motor neurons and protoplasmic astrocytes as a place of preference (Rungby 1986). However, it is unclear whether Ag possesses neurotoxic properties. Although some experimental data demonstrate that Ag ions penetrate the blood–brain barrier (BBB) and accumulate in glial cells and neurons in hippocampus and pons (Rungby and Danscher 1983), others are contradictory (Scott and Norman 1980). It is claimed that Ag does not meet the criteria for neurotoxin, and pathological or behavioral changes induced by this metal are somewhat rarely observed (Lansdown 2007). However, recent data obtained by the author's team indicate that prolonged exposure of rats to low doses (0.2 mg $kg^{-1}$ b.w. $day^{-1}$) of ionic Ag leads to ultrastructural and biochemical changes in brain synapses (Skalska et al. 2015).

Things are different in the case of Ag nanoformulations. Although few investigations have been conducted into Ag's neurotoxic effects, interest in such studies is on the rise. The ability of nano-Ag to freely cross cell membranes is of great concern in neuroscience due to the development of new therapeutic nanotechnologies, wherein AgNPs are applied as drug carriers or elements of drug-carrier systems (Yang et al. 2010; Leite et al. 2015). Unlike the ionic form, nanoformulation of Ag not only penetrates the brain but has been shown to cause destruction of the BBB, astrocyte swelling, and neuronal degeneration (Tang et al. 2009; Yang et al. 2010). Orally administered, it affects neurotransmitters (5-HT and dopamine) concentration in rat brain (Hadrup et al. 2012). Moreover, while accumulating in brain parenchyma, it aggravated existing brain pathologies. Administration of nano-Ag to rats prior to their being subjected to heat stress led to greater BBB disruption, edema formation, impairment of cognitive and motor functions, and brain damage (Sharma and Sharma 2007).

Experimental studies of prolonged oral exposure of rats to small (10 nm) commercial-grade AgNPs revealed their efficient absorption into the blood and induction of changes in nerve endings, even at very low concentrations (Skalska et al. 2015). This experimental design reflects a possible profile of toxic events in animals exposed to low doses of nano-Ag that is relevant to events occurring environmentally.

Exposure to even low concentrations of toxic substances may have a detrimental effect on fetal development. Hence, there is also concern about the toxic effect of nano-Ag on the developing CNS. Processes ongoing in the developing CNS during embryogenesis, such as cell division, differentiation, and migration, are extremely responsive to even weak stimuli derived from the intra- or extracellular environment. Since nano-Ag was reported to interfere in vitro with the growth of human embryonic neural precursor cells, further research is needed to assess its adverse effects on the CNS during embryonic development (Soderstjerna et al. 2013).

### 5.1.7 Cellular Mechanisms of Ionic Silver/Nanosilver Toxicity at Molecular and Biochemical Levels

Ionic Ag exhibits a high affinity for the thiol groups (–SH) of cellular compounds, binding mainly to the reduced glutathione (Baldi et al. 1988) and thus depleting its pool for other biochemical pathways. A strong binding to –SH groups in the collagen of connective tissues and basement membranes has also been observed (Lansdown 2007). It forms complexes with proteins and nucleic acids as well by binding to sulfhydryl, amino, carboxyl, and phosphate groups.

Moreover, exposure to Ag induces the expression of metallothioneins (MTs), which serve a protective function in cells (Bremner and Beattie 1990). These cystein-rich proteins bind Ag absorbed by tissue. In parallel, zinc ions may be liberated from zinc-MT complexes, leading to an excessive amount of this metal first in intracellular and then in extracellular compartments (Lansdown 2007). This mechanism may be of importance in the case of the neuronal form of MT, MT-III, due to the neurotoxic effects of excessive zinc (Koh et al. 1996).

Molecular mechanisms of nano-Ag toxicity are currently under intensive investigation. Unique physical properties of nano forms include a higher reactivity, which is inversely proportional to the particle size and directly proportional to the toxicity (Christian et al. 2008). Substantial evidence exists for and against the different toxic mechanisms of ionic and nano-Ag (for a review see Hadrup and Lam 2014). Nevertheless, the toxicity of AgNPs seems to be mediated by two combined mechanisms: Ag ions released from their surface and unique features of nano formulations (Pratsinis et al. 2013; Hadrup and Lam 2014; Ziemińska et al. 2014;

Skalska et al. 2015). Studies using a variety of experimental animal models have shown that they perturb the function of mitochondria, increasing free radical production and subsequently leading to oxidative stress (Rahman et al. 2009; Strużyński et al. 2014; Wu and Zhou 2013). The generation of excessive reactive oxygen species (ROSs) results in cell death through apoptosis (Piao et al. 2011; Yin et al. 2013). Recently, excitotoxicity via the activation of glutaminergic receptors, followed by calcium imbalance, destabilization of mitochondrial function, and ROS production, was suggested as a mechanism involved specifically in nano-Ag neurotoxicity toward cultured cerebellar granule cells (CGCs) (Ziemińska et al. 2014). Moreover, under nano-Ag exposure an imbalance between extracellular and intracellular zinc levels was also noticed. Zinc supplementation positively influenced nano-Ag-induced changes in CGCs, which was presumed to be due to an inhibitory effect on NMDA-sensitive calcium channels (Ziemińska and Strużyńska 2016).

## 5.2 *Toxicokinetics of Silver and Its Effects in Homeothermic Animals*

Due to overexposure, Ag may accumulate mainly in the skin, corneas, gingivae, mucose membranes of the body, liver, kidneys, and spleen. Since it is thought not to evoke serious toxic risks to terrestrial (but not aquatic) organisms, safe reference values are difficult to assess. Studies on the toxic effects of Ag in wild animals are very limited. However, it is evident that the concentration of Ag measured in tissues of exposed organisms is not strictly related to its toxicity. Vertebrates may be exposed to Ag via different routes: orally, by inhalation, or by dermal contact. The results of studies in humans and experimental animals indicate that Ag compounds are mainly absorbed orally, by inhalation, and, to a lesser extent, through the skin (ATSDR 1990).

Absorption, mainly investigated in experimental rodents, was found to be relatively low, much like in domestic birds fed Ag-containing food (for a review see Ratte 1999). It was estimated that about 10% of the initial dose of Ag is absorbed by an animal's body, and of this 4% is retained in tissues. The biological half-life of Ag in human liver ranges from several to 50 days (Nordberg and Gerhardsson 1988). The liver is also one of the main organs absorbing Ag and involved in its excretion with bile in the feces (US EPA 1980). Fecal excretion values reported by Furchner et al. (1968) were 99.6% for mice, 98% for rats, 90% for dogs, and 94% for monkeys; limited amounts were excreted in urine, regardless of the animal species examined. Following oral administration, Ag is absorbed in the digestive tract into portal venous circulation, passes through the liver, and is partially excreted in bile. The remaining portion is distributed to tissues by systemic circulation (ATSDR 1990). In addition to the high concentrations of Ag in rat liver, high concentrations were observed in spleen, bone marrow, lymph nodes, kidney, bladder, and all parts of the gastrointestinal tract (Olcott 1948; Rungby and Danscher 1983; Loeschner et al.

2011; Hadrup and Lam 2014). Some authors also mention muscles and brain as locations of Ag deposition (Rungby and Danscher 1983; Fung and Bowen 1996). After inhalation of metallic Ag by dogs, the largest part of the initial dose was found in the lungs (96.9%); the rest was deposited in the liver, gallbladder, intestines, and stomach (Phalen and Morrow 1973). Intravenous injection of radio-labeled Ag leads to its absorption mainly by the liver and of lesser amounts in testes, where it was found in all cell types of spermatogenesis and in the Sartoli cells (Ernst et al. 1991). In animals exposed experimentally via inhalation, 90–99% of absorbed Ag was excreted in the feces (Phalen and Morrow 1973; US EPA 1985).

The toxicokinetics of nano-Ag is generally similar to that observed in its ionic form. Following oral gavage, it was found at the highest concentrations in the liver, kidney, and small intestine, with the lowest observed in the lungs and brain of rats. Excretion in the feces was estimated to be 63% $\pm$ 23% of a daily dose (Loeschner et al. 2011). Rats exposed chronically (90 days) by inhalation showed enhanced Ag concentration in blood, liver, olfactory bulb, brain, and kidney (Sung et al. 2009), whereas in mice 4% of the total Ag dose retention was observed in the lungs (Stebounova et al. 2011). According to the most recent studies, comparison of the bioavailability of different forms of Ag after oral administration revealed a lower level for nanoparticulate than ionic Ag, based on higher fecal excretion (68% vs. 49%) and lower absolute levels in organs (Loeschner et al. 2011; van der Zande et al. 2012; Hadrup and Lam 2014). These data are inconsistent with earlier results indicating high fecal excretion of ionic Ag. However, measurement of Ag concentration in the blood of rats following 2-week oral exposure to ionic and nano-Ag (0.2 mg $kg^{-1}$ b.w. $day^{-1}$) revealed equal absorption into the serum compartment regardless of the form of Ag (Skalska et al. 2015).

Studies on oral exposure using Ag nitrate, Ag oxide, or Ag chloride in various animal species have reported lethal doses in ranges that are considered indicative of slight to moderate toxicity. Lethal doses for Ag ($LD_{50}$) were estimated to be 280 mg $kg^{-1}$ b.w. for rats and 800 mg $kg^{-1}$ b.w. for rabbits (Tamimi et al. 1998). $LD_{50}$ values for reported for mice oral colloidal Ag or Ag nitrate are 100 and 129 mg $kg^{-1}$ b.w., respectively. The experimentally assessed no-observed-adverse-effect level (NOAEL) for $AgNO_3$ in test conditions of lethal effect was 181.2 mg $kg^{-1}$ b.w. for orally exposed rats and 137.13 mg $kg^{-1}$ b.w. in guinea pigs following dermal exposure. NOAEL assessed for neuronal effects in mice exposed chronically to $AgNO_3$ or AgCl was 18.1 mg $kg^{-1}$ b.w. (Ratte 1999). In mice NOAEL for nano-Ag administered orally was 0.50 mg $kg^{-1}$ b.w. $day^{-1}$ based on the hepatotoxicity and histopathological changes visible in kidneys at the highest dose examined, or 30 mg $kg^{-1}$ b.w. $day^{-1}$ based on the increased Ag concentration in kidney with no adverse effect observed (Kim et al. 2009; Park et al. 2010). A similar NOAEL value (30 mg $kg^{-1}$ b.w. $day^{-1}$) was established for rats in a 90-day repeated-dose oral study based on decreased kidney weight (Kim et al. 2009). The lowest-observed-adverse-effect level (LOAEL), the lowest concentration of nano-Ag that causes observed adverse changes in morphology, function, growth, or development of a target organism, was 125 mg $kg^{-1}$ b.w. $day^{-1}$ based on morphological changes in the liver, bile duct, and intestine (Kim et al. 2010b).

The lungs and liver were determined to be the main target organs of rats chronically exposed to nano-Ag (20 nm) by inhalation (Sung et al. 2009). The authors determined a NOAEC (no-observed-adverse-effect concentration) of 0.1 mg $m^{-3}$ based on histopathological changes in the bile duct and features of inflammation in the lungs.

The highest available dose in the literature of nano-Ag tested was 5000 mg $kg^{-1}$ b.w. $day^{-1}$ and was not lethal to guinea pigs (Maneewattanapinyo et al. 2011).

## *5.3 Bioaccumulation of Silver in Mammalian and Avian Species*

The uptake of this trace element in environmental conditions mainly depends on dietary levels, digestion, and absorption from the gastrointestinal tract. It might be suspected that it occurs to a marginal extent via damaged skin. Silver absorbed into the blood is initially deposited in soft tissues, where the levels mirror the recent exposure.

Of all the tissues, bones reflect long-term exposure to metals. From the research of Hamilton et al. (1972) it is apparent that in human Ag achieves the highest concentrations in bones, i.e., 1.1 mg $kg^{-1}$ in ashed bones, which is 0.67 mg $kg^{-1}$ dw assuming a ratio of ash to organic matter in human ribs of 61–29% (Table 18.3) (Call et al. 1965). In different regions of the world and different decades of the 20th and twenty-first centuries, the concentration of Ag in the bones of modern humans has changed within a considerable range (from <0.01 to >2.0 mg $kg^{-1}$ dw), mainly depending on environmental conditions. In the period 1970–2000 it generally did not exceed 0.5 mg $kg^{-1}$ dw of Ag in Europeans, whereas it was ten times higher in old inhabitants of industrialized Taiwan in Asia (Table 18.3). Samples collected recently from residents of the city of Obinsk (central Russia) showed that the concentration of Ag in intact bones is very small and does not exceed 0.003 mg $kg^{-1}$ dw, which can be considered the value reflecting the geochemical background (Zaichick and Zaichick 2015).

Information concerning Ag values in human samples was included in the current meta-analysis since data derived from wild mammals are scant. Unpublished data of Kalisinska and coworkers concerning Ag content in mammal bones of predatory semiaquatic racoon *Procyon lotor* from Poland, which is an alien species in the European fauna originating from North America, revealed accumulation of the metal one to two orders of magnitude higher than the typical value (0.01 mg $kg^{-1}$ dw) in mammalian bones reported by Gough et al. (1979). The fact that samples were collected from the "Warta-Mouth" National Park in Poland, located in the vicinity of the copper mining region of Lower Silesia, may explain the enhanced Ag levels.

Research conducted in the 1970s on birds, especially those connected to aquatic food chains and inhabiting water reservoirs contaminated with metals, including Ag,

**Table 18.3** Silver concentration in mammalian and avian bones

| Species | Location and time period | Concentration mg $kg^{-1}$ dw | Remarks | References |
|---|---|---|---|---|
| *Mammals* | | | | |
| Humans<br>*Homo sapiens* | Taiwan, <2000 | 2.8 ± 3.5<br>2.32 ± 1.93<br>3.06 ± 3.61<br>2.23 ± 1.19<br>5.07 ± 8.54 | *n* = 70, all groups<br><40 years<br>41–60 years<br>61–80 years<br>>80 years | Kuo et al. (2000) Age differences among bone Ag levels NS |
| *Homo sapiens* | Verona, Italy, 14th century | 0.01 | *n* = 1; ad | Apostoli et al. (2009) |
| *Homo sapiens*<br>North American Pecos Indian | North America, ~14th century | *n* = 14, ad + im<br>0.15 ± 0.15<br>0.039–0.560<br>ad: 0.066–0.560<br>im: 0.039–0.13 | Inner part of bone, ad<br>*n* = 10<br>im *n* = 4 | Ericson et al. (1991) |
| *Homo sapiens* | Europe | 0.01–0.44 | Modern people | Ericson et al. (1991) |
| *Homo sapiens* | Russia, <2015 | 0.0027 ± 0.0015<br>0.00026–0.0047<br>0.0074 ± 0.0188<br>0.00064–0.0967 | intact bone, *n* = 27<br>osteogenic bone, *n* = 27 | Zaichick and Zaichick (2015) |
| *Homo sapiens* | United Kingdom, <1970 | 1.1<br>0.67 | *n* = 22, rib, ash[a]<br>*n* = 22, rib, dw | Hamilton et al. (1972) |
| *Phocoenoides dalli*<br>Dall's porpoise | Japan, 2000 | 0.001 | *n* = 1, ad male | Yang et al. (2006) |
| Raccoon<br>*Procyon lotor* | Poland, polluted area, 2009–2012 | 0.325 ± 0.047<br>0.244–0.421 | *n* = 14, ad | Kalisinska unpubl. Data |
| *Birds* | | | | |
| *Larus crassirostris*<br>Black-tailed gull | Japan, Rishiri Island, 1999–2001 | 0.006 ± 0.004 | *n* = 4 | Agusa et al. (2005) |
| Great cormorant<br>*Phalacrocorax carbo* | Japan, 2003 | <DL | *n* = 4, ad | Nam et al. (2005a) |
| Great tit<br>*Parus major* | Belgium, 2000 | 0.018<br><DL - 0.154 | *n* = 10, ad<br><DL one tit | Dauwe et al. (2005) |
| Brown pelican<br>*Pelecanus occidentalis* | USA, Florida, 1969 | 2.51<br>2.03–3.12 | *n* = 5 | Connors et al. (1972) |
| | USA, California, 1969–1971 | 2.32<br>1.95–2.75 | *n* = 5 | |
| White pelican<br>*Pelecanus erythrorhynchos* | USA, California, 1969–1971 | 1.92 | *n* = 1 | Connors et al. (1972) |

*n* number, *ad* adult, *im* immature, *DL* detection limit
[a]61% ash in rib, see Call et al. (1965)

identified >2 mg kg$^{-1}$ dw of Ag in bones (Connors et al. 1972; Lande 1977; Ohlendorf et al. 1986). The values correspond to those noted in the inhabitants of industrialized Taiwan (Table 18.3).

The observed decrease of Ag concentrations in the bones of homeothermic vertebrates (including humans) within a few years could be associated, at least in part, with the development of digital photography and the subsequent reduction in metal emissions into the environment. However, it should be kept in mind that at the same time, the mining, production, and use of Ag in other industrial sectors is continuously increasing. Nevertheless, a growing awareness of the necessity to protect the environment reflects the activities undertaken for this purpose.

Experimental studies on Ag toxicokinetics indicate that Ag absorbed into the blood of orally exposed animals is initially deposited in soft tissues such as liver, kidneys, intestines, muscles, and brain (Sect. 5.2). Preferentially it accumulates in the liver, in nuclear, lysosomal, and mitochondrial fractions. Among soft tissues in human tissues, the highest Ag concentrations were reported in liver and brain, 0.020 and 0.018 mg kg$^{-1}$ dw, respectively (Hamilton et al. 1972). The content of Ag in other human tissues such as kidney, lung, muscles, and gonads ranged between 0.010 and 0.007 mg kg$^{-1}$ dw.

Fragmentary studies on wildlife species also demonstrated bioconcentration of Ag within liver tissue (Table 18.4). The concentration was usually in a range of 0.013–4.3 mg kg$^{-1}$ dw, but it varied by location of sample origin. The values were higher in the liver of Javan mongoose *Herpestes javanicus* from Japan than in the liver of Arctic fox *Alopex lagopus* from Alaska (0.043–4.300 vs. < detection limit – 0.100 mg kg$^{-1}$ dw). In turn, samples derived from the same species were similar regardless of place of collection. In Canadian polar bear *Ursus maritimus* and in polar bear from Alaska, Ag concentrations reached on average 0.65 and 0.53 mg kg$^{-1}$ dw, respectively (Dehn et al. 2006; Rush et al. 2008).

For different avian species concentrations of metal found in liver usually ranged between 0.010 and 1.000 mg kg$^{-1}$ dw, with extreme values below (< detection limit) and above (44.000 mg kg$^{-1}$ dw) the range. As was reported, the major part of Ag in the body of adult birds accumulates in the liver (~60%) (Nam et al. 2005a). Previously, apart from the liver, from which a sample was usually collected, kidneys, muscles, and brain were also frequently utilized for analysis. In the most recent studies, eggs and feathers are also used, owing to the fact that birds are frequently protected species. In the liver and kidney of the white-tailed eagle, *Haliaeetus albicilla*, from breeding sites on the Polish coast of the Baltic Sea, Ag was found to be present in rather low concentrations of 0.056 and 0.037 mg kg$^{-1}$ dw, respectively, with maximum values of 0.23 and 0.13 mg kg$^{-1}$ dw (Falandysz et al. 2000, 2001). Apparently resulting from background exposure, these concentrations are rather low and do not induce any acute toxic effects. Much higher concentrations of metal were reported for the bald eagle, *Haliaeetus leucocephalus*, from the US state of Maine (Mierzykowski et al. 2011), which is is heavy metal-burdened area (http://www.nrcm.org/projects-hot-issues/healthy-waters/open-pit-metal-mining-in-maine/). In the liver of this species, 0.67 mg kg$^{-1}$ dw (range 0.21–3.40 mg kg$^{-1}$ dw) of Ag was found, and this value was an order of magnitude greater than that presented by

**Table 18.4** Silver concentration (mg kg$^{-1}$ dw) in soft tissues of wild terrestrial and semiaquatic mammals and birds. Conversion of wet weight (ww) on a dry weight basis are given in parentheses (calculations assume that the liver, kidney, and brain of mammals and birds contain an average of 70%, 75%, 80% water, respectively)

| Species | Place and years | Liver | Kidney | Muscle | Brain | Remarks | References |
|---|---|---|---|---|---|---|---|
| *Mammals* | | | | | | | |
| Hedgehog, T *Erinaceus europaeus* | Belgium, 2002–2003 | 0.12 ± 0.02<br><DL–0.48 | 0.04 ± 0.02<br><DL–0.62 | <LD ± 0.002<br><DL–0.057 | | Liver, *n* = 43<br>Kidney, *n* = 44<br>Muscle, *n* = 44 | D'Havé et al. (2006) |
| Javan mongoose, T *Herpestes javanicus* | Japan, 2004–2005 | 0.143 ± 0.194<br>0.013–1.290<br>(0.477 ± 0.647)<br>(0.043–4.300) | 0.002 ± 0.002<br><DL–0.008<br>(0.008 ± 0.008)<br>(<DL–0.032) | | 0.011 ± 0.003<br><DL–0.014<br>(0.055 ± 0.015)<br>(<DL–0.070) | ww<br>Liver, *n* = 53<br>Kidney, *n* = 30<br>Brain, *n* = 4 | Horai et al. (2006) |
| Arctic fox, T *Alopex lagopus* | USA, Alaska 1999–2000 | 0.01 ± 0.01<br><DL–0.03<br>(0.033 ± 0.033)<br>(<DL–0.100) | | | | ww, *n* = 27 | Dehn et al. (2006) |
| Polar bear, S *Ursus maritimus* | USA, Alaska 1999–2000 | 0.16 ± 0.08<br>0.05–0.35<br>(0.53 ± 0.26)<br>(0.17–1.17) | | | | ww, *n* = 23 | Dehn et al. (2006) |
| Polar bear, S *Ursus maritimus* | Canada, 2002 | | | | | | Rush et al. (2008) |
| | East Baffin Island | 0.13 ± 0.11<br>(0.43 ± 0.34) | | | | ww, *n* = 13 | |
| | Northern Baffin Island | 0.21 ± 0.11<br>(0.70 ± 0.34) | | | | ww, *n* = 13 | |
| | Lancaster Sound | 0.28 ± 0.11<br>(0.93 ± 0.34) | | | | ww, *n* = 13 | |
| | Alaska, 1994–1999 | 0.16 ± 0.10<br>(0.53 ± 0.33) | | | | ww, *n* = 6 | |

(continued)

**Table 18.4** (continued)

| Species | Place and years | Liver | Kidney | Muscle | Brain | Remarks | References |
|---|---|---|---|---|---|---|---|
| *Birds* | | | | | | | |
| Great tit, T *Parus major* | Belgium, 2000 | 0.023<br>0.004–0.150 | 0.008–0.008<br><DL ($n = 3$) | 0.003–0.012<br><DL ($n = 2$) | 0.013<br>0.007–0.026 | Median; ad, F, $n = 10$ | Dauwe et al. (2005) |
| White-winged dove, T *Zenaida asiatica* | USA, Texas, 2003 | <DL<br><DL–22.2 | | | | <DL 69 out of 70 | Fredricks et al. (2009) |
| Black-tailed gull, W *Larus crassirostris* | Japan, Rishiri Island, 1999–2001 | 0.019 ± 0.008 | 0.028 ± 0.027 | 0.006 ± 0.005<br><DL $n = 1$ | 0.030 ± 0.010 | ad, $n = 5$ | Agusa et al. (2005) |
| Great cormorant, W *Phalacrocorax carbo* | Japan, Lakes Biwa and Mie, 1993 and 2003 | 0.084 ± 0.021 | 0.028 ± 0.019 | 0.001 ± 0.001 | 0.036 ± 0.002 | Liver 1993<br>0.081 ± 0.043<br>0.021–0.160<br>Liver 2003<br>0.047 ± 0.024<br>0.022–0.092 | Nam et al. (2005a) |
| Great cormorant, W *Phalacrocorax carbo* | Japan, Lakes Biwa and Mie, 2003 | 0.117 ± 0.179 | 0.011 ± 0.020 | <0.001 | | $n = 17$; ad+juv | Nam et al. (2005b) |
| Mallard, W *Anas platyrhynchos* | Japan, Izumi, 2003 | 0.190 ± 0.241 | 0.014 ± 0.009 | <0.001 | | $n = 13$ | Nam et al. (2005b) |

| | | | | | | | |
|---|---|---|---|---|---|---|---|
| Cammon teal, W *Anas crecca* | Japan, Izumi, 2003 | 0.083 | 0.003 | <0.001 | | $n = 2$, ad | Nam et al. (2005b) |
| Pintail, W *Anas acuta* | Japan, Izumi, 2003 | 0.010 | 0.010 | <0.001 | | $n = 2$, ad | Nam et al. (2005b) |
| Bald eagle, W/T *Haliaeetus leucocephalus* | USA, Maine, 2001–2007 | 0.96<br>0.21–3.40 | | | | $n = 47$ | Mierzykowski et al. (2011) |
| White-tailed eagle, W/T *Haliaeetus albicilla* | Poland, 1991–1995 | 0.056<br>0.002–0.230 | 0.037<br>0.003–0.130 | 0.010 | | $n = 10–12$ | Falandysz et al. (2000) |
| Greater scaup, W *Aythya marila* | USA, San Francisco Bay, 1982 | 1.050 ± 0.146<br>0.390–3.100 | | | | $n = 18$, ad (16 M+2F); 94% with residue; contaminated area | Ohlendorf et al. (1986) |
| Surf scoter, W *Melanitta perspicillata* | USA, San Francisco Bay, 1982 | 0.900 ± 0.179<br>0.330–3.700 | | | | $n = 22$, M (21 ad+1 juv); 91% with residue; contaminated area | Ohlendorf et al. (1986) |
| Greater scaup, W *Aythya marila* | Canada, British Columbia, 1976 | | | | | | Vermeer and Peakall (1979) |
| | Iona Island | 0.32 ± 0.08<br>(1.04 ± 0.27) | | | | ww, $n = 10$ | |
| | Roberts Bank | 0.04 ± 0.004<br>(0.13 ± 0.013) | | | | ww, $n = 10$ | |

(continued)

**Table 18.4** (continued)

| Species | Place and years | Liver | Kidney | Muscle | Brain | Remarks | References |
|---|---|---|---|---|---|---|---|
| Surf scoter, W *Melanitta perspicillata* | Canada, British Columbia, 1976 | | | | | | Vermeer and Peakall (1979) |
| | Iona Island | 0.14 ± 0.03 (0.47 ± 0.10) | | | | ww, $n = 10$ | |
| | Roberts Bank | 0.03 ± 0.004 (0.10 ± 0.013) | | | | ww, $n = 10$ | |
| Common eider, W *Somateria mollissima* | Norway, 1972–1973 | 44.00 | 7.00 | 2.00 | | ad, F, $n = 6$ polluted area | Lande (1977) |
| Lesser black-backed gull, W *Larus fuscus* | Norway, 1972–1973 | 2.00 | 1.00 | 3.00 | | ad, $n = 6$ polluted area | Lande (1977) |

*T* terrestrial, *S* semiaquatic, *W* waterbird, *DL* detection limit, *F* female, *M* male, *ad* adult, *juv* juvenile

Falandysz et al. (2001) for white-tailed eagle, whose biology is very similar to that of the North American bald eagle.

In avian species inhabiting inland water bodies or coastal areas, Ag concentration in liver and other soft tissues varies to a large extent, depending on the degree of pollution and the metal content in the diet. It seems that exposure to elevated concentrations of Ag contained in seafood is a source of high body burden of metal in birds breeding in coastal locations. In the great cormorant, *Phalacrocorax carbo*, feeding in the large freshwater reservoir contaminated with heavy metals in Japan, high concentrations of Ag were noted in liver and brain (0.084 and 0.036 mg $kg^{-1}$ dw, respectively), just like in kidney (0.028 mg $kg^{-1}$ dw), whereas the concentration of Ag in feathers of this bird did not exceed 0.01 mg $kg^{-1}$ dw (Nam et al. 2005a, b). However, the greatest Ag abundance was detected in certain duck species (common eider, *Somateria mollissma*) inhabiting Norwegian fjords highly polluted with heavy metals (Lande 1977). In the liver, kidney, and muscle, average concentrations of Ag was 44.0, 7.0, and 1.0 mg $kg^{-1}$ dw, respectively. This seems to be the largest average value of the metal estimated in the avian liver. Simultaneously, it should be highlighted that environmental Ag contamination, mainly of anthropogenic origin, results in an increased accumulation of the metal in the bodies of homeothermic vertebrates, especially those associated with aqueous food chains, including aquatic birds. This is confirmed by both the earlier research conducted during the period of the environment burdening with uncontrolled discharge of pollutants and works carried out in the twenty-first century, while in Europe, the USA, and Canada strict regulations on environmental protection were introduced (Lande 1977; Ohlendorf et al. 1986; Mierzykowski et al. 2011).

Silver is incorporated to a lesser extent into avian eggs compared with internal organs. The levels of this element seem to be low in eggshell and egg content: <0.001–0.012 and 0.004–0.013 mg $kg^{-1}$ dw, respectively (Agusa et al. 2005; Ikemoto et al. 2005). However, higher Ag levels were observed in eggs taken from birds originating from heavy-metal-polluted areas in Norway (common eider, *Somateria mollissima*, and lesser black-backed gull, *Larus fuscus*) (Lande 1977) and Belgium (great tit) (Dauwe et al. 1999, 2005).

It is also noteworthy that the distribution of Ag is very large in feathers. Studies by Nam et al. (2005a) showed that adult birds accumulated about 30% of absorbed Ag in their feathers. Among birds from various trophic and taxonomic groups, Ag content in feathers generally was in a narrow range of 0.010–0.094 mg $kg^{-1}$ dw (Table 18.5) or even undetectable (Scanlon et al. 1980), although feathers sampled from great tit living near smelters exhibited much higher Ag concentrations (over 3.5 mg $kg^{-1}$ dw in adult specimens) relative to that from unpolluted areas (0.13 mg $kg^{-1}$ dw). In addition, tit nestlings from polluted areas had higher Ag concentrations in feathers compared with nestlings from referenced area: 0.020 vs. 0.001 mg $kg^{-1}$ dw (Janssens et al. 2001).

**Table 18.5** Silver concentration (mg kg$^{-1}$ dw) in feathers of various bird species

| Species | Location and time period | Concentration | Reference |
|---|---|---|---|
| *Unpolluted area* | | | |
| Ruffed grouse *Bonassa umbellus* | USA, Virginia, 1977–1979 | < 0.010 | Scanlon et al. (1980) |
| Black-tailed gull *Larus crassirostris* | Japan, Rishiri Island, 1999–2001 | 0.019 ± 0.003 | Agusa et al. (2005) |
| Great cormorant *Phalacrocorax carbo* | Japan, Lake Biwa and Mie, 2003 | 0.010 ± 0.004 | Nam et al. (2005a) |
| Great tit *Parus major* | Belgium, Brasschaat, 2000 | 0.130 ± 0.070 | Janssens et al. (2001) |
| Sparrowhawk *Accipiter nisus* | Belgium, Flanders, 2001 | 0.023 ± 0.012 (0.013–0.040) | Dauwe et al. (2003) |
| Little owl *Athene noctua* | Belgium, Flanders, 2001 | 0.018 ± 0.009 (0.013–0.025) | Dauwe et al. (2003) |
| Barn owl *Tyto alba* | Belgium, Flanders, 2001 | 0.021 ± 0.005 (0.018–0.027) | Dauwe et al. (2003) |
| Tawny owl *Strix aluco* | Belgium, Flanders, 2001 | 0.061 ± 0.009 (0.029–0.094) | Dauwe et al. (2003) |
| *Polluted area* | | | |
| Great tit *Parus major* | Belgium, Antwerp, 1997–1998 | 3.590 ± 0.600 | Janssens et al. (2001) |

## *5.4 Ecological Effects of Silver*

As discussed earlier, free Ag ions are extremely toxic to aquatic organisms and even lethal to certain sensitive species of invertebrates and fish at concentrations of 1.2–4.9 μg L$^{-1}$. Available data on Ag toxicity concern mainly its effects on small laboratory mammals and poultry. Toxic effects on avian or mammalian wildlife have not been extensively studied since it was believed to be an element of low toxicity. Many of the review papers on the presence and toxicity of Ag in wild species highlight the fact that there are no data concerning the ecological effects of this metal in terrestrial avian and especially mammalian species (Eisler 1996; WHO 2002).

For mammals, no data are available to predict the ecotoxicological effects of Ag. Some of the species studied include polar bear (semiaquatic species from an ecological point of view), Arctic fox, and Javan mongoose (Table 18.4). Reported Ag concentrations in liver fell within a relatively high range (from below the detection limit to 4.30 mg kg$^{-1}$ dw). Despite this, liver may be considered a location of the highest Ag accumulation. The limited types of samples prevent identifying another target organ. However, samples derived from Javan mongoose (carnivore) indicate that Ag concentration in soft tissues reaches its highest levels in liver and then in the brain and kidney (Horai et al. 2006).

Birds, while affected environmentally, usually exhibit the highest bioconcentrations of metal in liver and brain, the lowest in muscles, and intermediate levels in the kidneys. Concentrations in avian tissues were frequently found to be elevated mainly in the vicinity of metal-contaminated areas in Europe and the USA and were higher in liver compared to other examined tissues like kidney or muscles (Lande 1977; Ohlendorf et al. 1986). Its presence was especially observed in birds in connection with aquatic systems. Maximum concentrations of total Ag, recorded in field collections of living organisms, were found in the liver of four species: lesser black-backed gull, surf scoter (*Melanitta perspicillata*), greater scaup (*Aythya marila*), and common eider. Whole tissue concentrations ranged in these species between 0.9 and 44.0 mg $kg^{-1}$ dw (Lande 1977), whereas in conspecifics from areas remote from anthropogenic contamination, Ag concentrations were usually much lower, within a range of 0.010–0.190 mg $kg^{-1}$ dw (Agusa et al. 2005; Nam et al. 2005a, b).

Aquatic and other birds may be exposed to Ag mainly via their diet or accidentally swallowed small Ag-containing things like gastroliths, although contamination by respiration cannot be excluded in industrial areas (Agusa et al. 2005; Dauwe et al. 2005; Fredricks et al. 2009). The latter publication reports extremely high concentrations of Ag found in the liver of one white-winged dove, *Zenaida asiatica*, from Texas (Fredricks et al. 2009).

Data collected in Table 18.4 indicate that Ag may be preferentially enriched in some avian tissues, mainly in liver and brain. The same may be concluded based on the results obtained from human and mammalian specimens. Thus, these organs should be considered targets for Ag. As has been suggested, lying eggs may constitute the specific mechanism for Ag excretion by female birds. However, a decrease in the Ag body burden by its transfer to eggs is limited (Dauwe et al. 2005). Although avian eggs have been widely used as indicators of heavy metal exposure, they seem not to be suitable in the case of exposure to Ag. However, the negative influence of accumulated metal on embryonic development cannot be excluded and may be relevant in the context of ecotoxicological effects (Dauwe et al. 2005).

Recent studies emphasize the need to develop safe methods of biomonitoring that will make it possible to assess the bioconcentration of various elements in living organisms. Bird feathers are good for this purpose because they are easy to collect noninvasively and repeatedly without affecting investigated individuals. Silver shows high affinity for the sulfhydryl groups (−SH) of keratin. Thus, it accumulates in feathers, which can be considered an important excretory pathway for this metal (Agusa et al. 2005). Moreover, birds eliminate heavy metals from tissues by sequestering them into plumage during the molting period. Since birds are in many cases protected, noninvasive techniques of sample collection are desirable. Feathers provide an alternative to internal organs and have proven to be suitable biomonitors for Ag pollution (Dauwe et al. 2003). Feathers of species that inhabit uncontaminated or relatively lightly contaminated environments and are nonmigratory, are presumably the best source of background concentrations because such concentrations are observable in the case of ruffed grouse, *Bonassa umbellus*, from forested areas in the vicinity of the US state of Virginia (Scanlon et al. 1980). Moreover, external contamination may have an important impact on the level of Ag detected in feathers, resulting in higher concentrations of metal in feathers most exposed to external atmospheric conditions or preening (Dauwe et al. 2003).

The present collection of data provides some information on the abundance of Ag in various avian and mammalian species. First of all, ranges for Ag concentrations, regardless of the analyzed tissue, can be related to their local environment and to the pollution status of ecosystems. However, it is essential to realize that this collection may be of limited value. The dearth of research on Ag toxicity in the wild, together with the recognition that individuals may differ in their responses to Ag, may cause difficulties in predicting the exact ecotoxicological effects of Ag. However, these data in conjunction with knowledge of local pollution levels show that Ag present in the environment is capable of entering both avian and mammalian species (including humans). Additonal issues emerge from the data collected in Table 18.4. First, the research does not cover the last 10 years and, second, it comes mainly from the USA and Europe. There are almost no data relating to individuals inhabiting large territories of Asia (except Japan). This is a significant consideration in light of the fact that Asia emits substantial amounts of Ag to the land and into water bodies (Eckelman and Graedel 2007) and is regarded as a hotspot of Ag/nano-Ag pollution. We do not know exactly yet how nanotechnological advances may change human and animal environments. Drastically increasing both the production and abundance of nano-Ag in the environment may spur increases in toxicity for wildlife than has been estimated so far.

Based on current information on the fate of nano-Ag in the environment, it may have ecotoxicological effects, particularly after being discharged into water. Flowing into water bodies, nanometer-sized Ag has a variety of physiological effects on living organisms, including fish and invertebrates (Fabrega et al. 2011; Schirmer et al. 2013). Therefore, the ecotoxicological effects of nano-Ag (similarly to ionic Ag) could be expected to be closely linked with the aquatic environment and affect species inhabiting water ecosystems.

## 5.5 *Bioindicators and Biomarkers for Silver in Ecotoxicological Studies*

Since the problem of Ag pollution of the environment is predicted to increase, it is important to look for organisms that might be useful as bioindicators of Ag contamination. Species to be used as biological indicators of Ag contamination should reflect the level of environmental pollution. Birds are often used as bioindicators since they are particularly well-known organisms with well-established behavior and biology and have relatively long life spans, which makes it possible to assess the long-term effects of exposure (Furness 1993). The food items and metal concentrations in the diet of birds may vary considerably in different areas. Analysis of the data collected in Table 18.4 indicates that birds feeding in differently contaminated zones reveal an interdependence between metal concentrations in tissues and the higher level expected in their habitat. Bioaccumulation of Ag in sea ducks—surf scoter and greater scaup—feeding exclusively in the marine and estuarine habitat of Iona Island polluted by wastewater from Vancouver (Canada) was dependent on the level of contamination of their feeding areas as well as prey (Vermeer and Peakall 1979). Moreover, samples of aquatic birds from extensively polluted areas such as San Fransico Bay in California (Ohlendorf et al. 1986) or Trondheimsfjorden, Norway (Lande 1977), reflected well the high concentrations of Ag, regardless of the investigated species of surf scoter/greater scoup and common eider/lesser black-backed gull, respectively. Aquatic, especially coastal/marine, ecosystems are of concern in the case of Ag contamination since Ag is known to bioaccumulate and to be highly toxic to aquatic organisms. Current meta-analysis performed on the basis of limited studies suggested that aquatic birds may be treated as potential bioindicators of Ag pollution in costal/estuarine areas since high concentrations of metal have been encountered in tissues of many species. In analyzing other avian species, strong evidence can be found confirming relationships between Ag concentrations in tissues versus those in more typical habitats. Birds such as the great tit, which inhabit urban areas, exhibited higher Ag levels in tissues (Dauwe et al. 2005) relative to migratory species like the white-winged dove (Fredricks et al. 2009). This implies that food preferences may significantly contribute to the bioaccumulation of Ag in birds. The rate at which the diet passes through the gastrointestinal tract and dietary composition may affect metal absorption and increase its levels in avian tissues. Primary determinants of food retention are food characteristics and the digestive anatomy of the avian species (Nam et al. 2005a, b). Because birds may ingest toxicants through food or water, soil contamination may represent a significant hazard for birds. Airborne deposition of Ag on feathers may also be a significant source of exposure during bird self-grooming. In turn, female birds may eliminate pollutants by depositing them in eggs. During the breeding period, the great tit collects food in its territory, so the metal content in the eggs is derived to a large extent from local sources, making the eggs suitable bioindicators of local pollution (Dauwe et al. 2005). However, the egg load factor of Ag may have limited value. According to Dauwe et al. (1999), metal concentrations in egg contents and eggshells, although high, were poorly correlated with metal concentrations in internal tissues and feathers. Collected data indicate that Ag concentrations in the egg content of black-tailed gull and great tit are 0.008 mg $kg^{-1}$ dw (Agusa et al. 2005) and 0.012 mg $kg^{-1}$ dw (Dauwe et al. 2005), which is comparable to concentrations in the muscles of those species (Table 18.4).

Generally metals, including Ag, are incorporated into the keratin structure of a feather during the growth period, when the feathers receive a supply of blood. However, Ag may also be deposited onto the surface of feathers by airborne dust. Thus, feathers may reflect both endogenous contaminations originating from bird diet and exogenous adsorption. It is stressed that analyzing the level of different metals present in feathers is a method of monitoring the ecological consequences of environmental metal pollution. Consequently, feathers can be used as bioindicators (Markowski et al. 2013), including of Ag contamination. In feathers of ruffed grouse from forested areas of the US state of Virginia, Ag was not detected at sensitivity levels (Scanlon et al. 1980). In contrast, in feathers of the great tit from polluted areas of Antwerp, Belgium, Ag levels were very high (Janssens et al. 2001).

Silver concentrations in feathers of birds of prey are significantly lower than in the great tit (Table 18.5). One might think, then, that birds of prey would be useful for assessing environmental Ag contamination because of their high position at the top of the food chain. However, to date, there is no convincing evidence of Ag biomagnification in trophic chains, especially in the higher trophic groups. Currently available data do not confirm an increased concentration of Ag in comparisons of levels in tissues of predators and their prey. Moreover, the extended home range and feeding area of birds of prey make it difficult to determine the exact location of their exposure. For these reasons, resident passerine birds like the great tit seem to be better suited for the biomonitoring of Ag ecotoxicological effects within limited regions such as urban areas. As a resident species in a relatively small home range, the great tit feeds within a limited area, and the content of Ag present in organisms may reflect local contamination. Feathers of the great tit, both adult and nestlings, were shown previously to reflect well the profile of Ag pollution in urban areas (Janssens et al. 2001; Dauwe et al. 2004).

However, it should be kept in mind that the aforementioned conclusions are limited by insufficient studies focusing on terrestrial organisms, especially semi-aquatic mammals from inland ecosystems.

## 6 Conclusions

To understand the bioavailability and uptake of Ag within natural ecosystems, efforts have focused on wildlife. Concentrations of many pollutants may increase with trophic level. Thus, studies should begin with the transfer of Ag contaminating terrestrial environments in species within a food chain to assess the risks posed to upper-trophic-level organisms, especially mammals and birds. This is very important given the increasing usage of nano-Ag-containing products. Particular attention should be directed toward understanding natural nano-Ag cycling and predicting the risk of environmental hazards.

## References

Agusa T, Matsumoto T, Ikemoto T, Anan Y, Kubota R, Yasunaga G et al (2005) Body distribution of trace elements in black-tailed gulls from Rishiri Island, Japan: age-dependent accumulation and transfer to feathers and eggs. Environ Toxicol Chem 24:2107–2120

Ahamed M, Karns M, Goodson M, Rowe J, Hussain SM, Schlager JJ et al (2008) DNA damage response to different surface chemistry of Ag nanoparticles in mammalian cells. Toxicol Appl Pharmacol 233:404–410

Ahlberg S, Antonopulos A, Diendorf J, Dringen R, Epple M, Flöck R et al (2014) PVP-coated, negatively charged silver nanoparticles: a multi-center study of their physicochemical characteristics, cell culture and in vivo experiments. Beilstein J Nanotechnol 5:1944–1965

Akaighe N, MacCuspie RI, Navarro DA, Aga DS, Banerjee S, Sohn M et al (2011) Humic acid-induced silver nanoparticle formation under environmentally relevant conditions. Environ Sci Technol 45:3895–3901

Apostoli P, De Palma G, Catalani S, Bortolotti F, Tagliaro F (2009) Multielemental analysis of tissues from Cangrande della Scala, Prince of Verona, in the 14th Century. J Anal Toxicol 33:322–327

AshaRani PV, Low Kah Mun G, Hande MP, Valiyaveettil S (2009) Cytotoxicity and genotoxicity of silver nanoparticles in human cells. ACS Nano 3:279–290

ATSDR (1990) Toxicological profile for silver. US Public Health Service, Agency for Toxic Substances and Disease Registry, Atlanta, GA, (TP-90-24) pp 145

Austin CA, Umbreit TH, Brown KM, Barber DS, Dair BJ, Francke-Carroll S et al (2012) Distribution of silver nanoparticles in pregnant mice and developing embryos. Nanotoxicology 6:912–922

Barillo DJ, Marx DE (2014) Silver in medicine: a brief history BC 335 to present. Burns 40(suppl): S3–S8

Baldi C, Minoia C, Di Nucci A, Capodaglio E, Manzo L (1988) Effects of silver in isolated rat hepatocytes. Toxicol Lett 41:261–268

Blaser SA, Sheringer M, Macleod M, Hungerbuhler K (2008) Estimation of cumulative aquatic exposure and risk due to silver: contribution of nanofunctionalized plastics and textiles. Sci Total Environ 390:396–409

Bondarenko O, Juganson K, Ivask A, Kasemets K, Mortimer M, Kahru A (2013) Toxicity of Ag, CuO and ZnO nanoparticles to selected environmentally relevant test organisms and mammalian cells in vitro: a critical review. Arch Toxicol 87:1181–1200

Braydich-Stolle LK, Lucas B, Schrand A, Murdock RC, Lee T, Schlager J et al (2010) Silver nanoparticles disrupt GDNF/Fyn kinase signalling in spermatogonial stem cells. Toxicol Sci 116:577–589

Bremner I, Beattie JH (1990) Metallothionein and the trace minerals. Annu Rev Nutr 10:63–83

Call RA, Greenwood DA, Lecheminant WH, Shupe JL, Nielsen HM, Olson LE et al (1965) Histological and chemical studies in man on effects of fluoride. Public Health Rep 80:529–538

CEC (1996) Commission of the European Communities. Technical guidance document in support of commission directive 93/67/EEC on risk assessment for new notified substances. Part II, Environmental Risk Assessment, Luxembourg

Charley RC, Bull AT (1979) Bioaccumulation of silver by multispecies community of bacteria. Arch Microbiol 123:239–244

Cheng Y, Yin L, Lin S, Wiesner M, Bernhardt E, Liu J (2011) Toxicity reduction of polymer-stabilized silver nanoparticles by sunlight. J Phys Chem C 115:4425–4432

Choi O, Cleuenger TE, Deng BL, Surampalli RY, Ross L, Hu ZQ (2009) Role of sulfide and ligand strength in controlling nanosilver toxicity. Water Res 43:1879–1886

Christian P, Von der Kammer F, Baalousha M, Hofmann TH (2008) Nanoparticles: structure, properties preparation and behaviour in environmental media. Ecotoxicology 17:326–343

Cleveland D, Long SE, Pennington PL, Cooper E, Fulton MH, Scott GI et al (2012) Pilot estuarine mesocosm study on the environmental fate of silver nanomaterials leached from consumer products. Sci Tot Environ 421–422:267–272

Connors PG, Anderlini VC, Risebrough RW, Martin JH, Schroeiber RW, Anderson DW (1972) Heavy metal concentrations in brown pelicans from Florida and California. Cal-Neva Wildlife 56–64

Danscher G (1981) Light and electron microscopic localization of silver in biological tissue. Histochemistry 71:177–186

Dauwe T, Bervoets L, Blust R, Pinxten R, Eens M (1999) Are eggshell and egg contents of great and blue tits suitable indicators of heavy metal pollution? Belg J Zool 129:439–447

Dauwe T, Bervoets L, Pinxten R, Blust R, Eens M (2003) Variation of heavy metals within and among feathers of birds of prey: effects of molt and external contamination. Environ Pollut 124:429–436
Dauwe T, Bervoets L, Pinxten R, Blust R, Eens M (2004) Relationships between metal concentrations in great tit nestlings and their environment and food. Environ Pollut 131:373–380
Dauwe T, Janssens E, Bervoets L, Blust R, Eens M (2005) Heavy-metal concentrations in female laying great tits (*Parus major*) and their clutches. Arch Environ Contam Toxicol 49:249–256
Dehn LA, Follmann EH, Thomas DL, Sheffield GG, Rosa C, Duffy LK, O'Hara TM (2006) Trophic relationships in an Arctic food web and implications for trace metal transfer. Sci Total Environ 362:103–123
D'Havé H, Scheirs J, Mubiana VK, Verhagen R, Blust R, De Coen W (2006) Non-destructive pollution exposure assessment in the European hedgehog (*Erinaceus europaeus*). II. Hair and spines as indicators of endogenous metal and as concentrations. Environ Pollut 142:438–448
Dos Santos C, Seckler MM, Ingle AP, Gupta I, Galdiero S, Galdiero M et al (2014) Silver nanoparticles: therapeutical uses, toxicity, and safety issues. J Pharm Sci 103:1931–1944
Drake PL, Hazelwood KJ (2005) Exposure-related health effects of silver and silver compounds: a review. Ann Occup Hyg 49:575–585
Eckelman MJ, Graedel TE (2007) Silver emissions and their environmental impacts: a multilevel assessment. Environ Sci Technol 41:6283–6289
EEA (2001) Sewage sludge—a future waste problem? European Environmental Agency, Indicator Fact Sheet Signals, EEA, Copenhagen. http://themes.eea.europa.eu/Environmental_issues/waste/indicators/sewage/w5_sludge.pdf
Eisler R (1996) Silver hazards to fish, wildlife, and invertebrates: a synoptic review. US National Biological Service Biological Report 32
El Mahdy MM, Eldin TAS, Aly HS, Mohammed FF, Shaalan MI (2015) Evaluation of hepatotoxic and genotoxic potential of silver nanoparticles in albino rats. Exp Toxicol Pathol 46:21–29
Ericson JE, Smith DR, Flegal AR (1991) Skeletal concentrations of lead, cadmium, zinc, and silver in ancient North American Pecos Indians. Environ Health Perspect 93:217–223
Ernst E, Rungby J, Baatrup E (1991) Ultrastructural localization of silver in rat testis and organs distribution of radioactive silver in the rat. J Appl Toxicol 11:317–321
Fabrega J, Luoma SN, Tyler CR, Galloway TS, Lead JR (2011) Silver nanoparticles: behaviour and effects in the aquatic environment. Environ Int 37:517–531
Falandysz J, Ichihashi H, Mizera T, Yamasaki S (2000) Mineral composition of selected tissues and organs of white-tailed eagle. Rocz PZH 51:1–5 (in Polish)
Falandysz J, Ichihashi H, Szymczyk K, Yamasaki S, Mizera T (2001) Metallic elements and metal poisoning among white-tailed sea eagles from the Baltic South coast. Mar Pollut Bull 42:1190–1193
Fewtrell L (2014) Silver: water disinfection and toxicity. Centre for Research into Environment and Health, 55 pp
Fredricks TB, Fedynich AM, Benn S, Ford L (2009) Environmental contaminants in white-winged doves (*Zenaida asiatica asiatica*) from the Lower Rio Grande Valley of Texas, USA. Arch Environ Contam Toxicol 57:387–396
Fung MC, Bowen DL (1996) Silver products for medical indications: risk-benefit assessment. J Toxicol Clin Toxicol 34:119–126
Furchner JE, Richmond CR, Drake GA (1968) Comparative metabolism of radionuclide in mammals-IV. Retention of silver-110m in the mouse, rat, monkey and dog. Health Phys 15:505–514
Furness RW (1993) Birds as monitors of pollutants. In: Furness RW, Greenwood JJD (eds) Birds as monitors of environmental change. Chapman and Hall, London, pp 86–143
Furst A, Schlauder MC (1978) Inactivity of two noble metals as carcinogens. J Environ Pathol Toxicol 1:51–57
Gaillet S, Rouanet JM (2015) Silver nanoparticles: their potential toxic effects after oral exposure and underlying mechanisms—a review. Food Chem Toxicol 77:58–63

Gao Y, Luo Z, He N, Wang MK (2013) Metallic nanoparticle production and consumption in China between 2000 and 2010 and associative aquatic environmental risk assessment. J Nanopart Res 15:168

Ge L, Li Q, Wang M, Ouyang J, Li X, Xing MM (2014) Nanosilver particles in medical applications: synthesis, performance, and toxicity. Int J Nanomed 9:2399–2407

Gottschalk F, Scholz RW, Nowack B (2010) Probabilistic material flow modeling for assessing the environmental exposure to compounds: methodology and an application to engineered nano-TiO2 particles. Environ Model Softw 25:320–332

Gough LP, Shacklette HT, Case AA (1979) Element concentrations toxic to plants, animals, and man. In: Geological Survey Bulletin 1466, An appraisal of the toxicity hazard to plants, animals, and man from natural and manmade element concentrations of environmental concern. United States Government Printing Office, Washington, DC, pp 44–46

Gromadzka-Ostrowska J, Dziendzikowska K, Lankoff A, Dobrzyńska M, Instanes C, Brunborg G et al (2012) Silver nanoparticles effects on epididymal sperm in rats. Toxicol Lett 214:251–258

Gulbranson SH, Hud JA, Hansen RC (2000) Argyria following the use of dietary supplements containing colloidal silver protein. Cutis 66:373–376

Hadrup N, Lam HR (2014) Oral toxicity of silver ions, silver nanoparticles and colloidal silver—a review. Regul Toxicol Pharmacol 68:1–7

Hadrup N, Loeschner K, Mortensen A, Sharma AK, Qvortrup K, Larsen EH et al (2012) The similar neurotoxic effects of nanoparticulate and ionic silver in vivo and in vitro. Neurotoxicology 33:416–423

Hamilton EI, Minski MJ, Cleary JJ (1972) The concentration and distribution of some stable elements in healthy human tissues from the United Kingdom. Sci Total Environ 1:341–374

Hiriart-Baer VP, Fortin C, Lee DY, Campbell PG (2006) Toxicity of silver to two freshwater algae Chlamydomonas reinhardtii and Pseudokirchneriella subcapitata, grown under continuous culture conditions: influence of thiosulphate. Aquat Toxicol 78:136–148

Horai S, Minagawa M, Ozaki H, Watanabe I, Takeda Y, Yamada K et al (2006) Accumulation of Hg and other heavy metals in the Javan mongoose (*Herpestes javanicus*) captured on Amamioshima Island, Japan. Chemosphere 65:657–665

Iavicoli I, Fontana L, Leso V, Bergamaschi A (2013) The effects of nanomaterials as endocrine disruptors. Int J Mol Sci 14:16732–16801

Ikemoto T, Kunito T, Tanabe S, Tsurumi M, Sato F, Oka N (2005) Non-destructive monitoring of trace element levels in short-tailed albatrosses (Phoebastria albatrus) and black-footed albatrosses (*Phoebastria nigripes*) from Torishima Island, Japan using eggs and blood. Mar Poll Bull 51:889–895

Janssens E, Dauwe T, Bervoets L, Eens M (2001) Heavy metals and selenium in feathers of great tits (*Parus major*) along a pollution gradient. Environ Toxicol Chem 20:2815–2820

Kabata-Pendias A (2011) Trace elements in soils and plants, 4th edn. CRC, Boca Raton, FL

Kahru A, Dubourguier HC (2010) From ecotoxicology to nanoecotoxicology. Toxicology 269:105–119

Kim WY, Kim J, Park JD, Ryu HY, Yu IJ (2009) Histological study of gender differences in accumulation of silver nanoparticles in kidneys of Fischer 344 rats. J Toxicol Environ Health A 72:1279–1284

Kim B, Park C-S, Murayama M, Hochella MF (2010a) Discovery and characterization of silver sulfide nanoparticles in final sewage sludge products. Environ Sci Technol 44:7509–7514

Kim YS, Song MY, Park JD, Song KS, Ryu HR, Chung YH et al (2010b) Subchronic oral toxicity of silver nanoparticles. Part Fibre Toxicol 7:20

Kim JS, Sung JH, Ji JH, Song KS, Lee JH, Kang CS et al (2011) In vivo genotoxicity of silver nanoparticles after 90-day silver nanoparticle inhalation exposure. Saf Health Work 2:34–38

Klein C, Hurlbut CS (1985) Manual of mineralogy, 20th edn. Wiley, New York, pp 271–272

Koh JY, Suh SW, Gwag BJ, He YY, Hsu CY, Choi DW (1996) The role of zinc in selective neuronal death after transient global cerebral ischemia. Science 272:1013–1016

Kramer JR, Benoit G, Bowles KC, DiToro DM, Herrin RT, Luther GW III et al (2002) Environmental chemistry of silver. In: Andren AW, Bober TW (eds) Silver in the environment: transport, fate, and effects. SETAC, Pensacola, FL
Kuo H-W, Kuo S-M, Chou C-H, Lee T-C (2000) Determination of 14 elements in Taiwanese bones. SciTotal Environ 255:45–54
Lande E (1977) Heavy metal pollution in Trondheimsfjorden, Norway, and the recorded effects on the fauna and flora. Environ Pollut 12:187–198
Lansdown AB (2007) Critical observations on the neurotoxicity of silver. Crit Rev Toxicol 37:237–250
Leite PE, Pereira MR, Granjeiro JM (2015) Hazard effects of nanoparticles in central nervous system: searching for biocompatible nanomaterials for drug delivery. Toxicol In Vitro 29:1653–1660
Levard C, Hotze EM, Lowry GV, Brown GE Jr (2012) Environmental transformations of silver nanoparticles: impact on stability and toxicity. Environ Sci Technol 46:6900–6914
Li Y, Zhang W, Niu J, Chen Y (2013) Surface-coating-dependent dissolution, aggregation, and reactive oxygen species (ROS) generation of silver nanoparticles under different irradiation conditions. Environ Sci Technol 47:10293–10301
Liu J, Sonshine DA, Shervani S, Hurt RH (2010) Controlled release of biologically active silver from nanosilver surfaces. ACS Nano 4:6903–6913
Liu J, Wang Z, Liu FD, Kane AB, Hurt RH (2012) Chemical transformations of nanosilver in biological environments. ACS Nano 6:9887–9899
Loeschner K, Hadrup N, Qvortrup K, Larsen A, Gao X, Vogel U (2011) Distribution of silver in rats following 28 days of repeated oral exposure to silver nanoparticles or silver acetate. Part Fibre Toxicol 8:18
Luoma SN (2008) Silver nanotechnologies and the environment: old problems or new challenges? Woodrow Wilson International Center for Scholars, Project on Emerging Nanotechnologies and the PEW Charitable Trusts, Washington, DC
Luoma SN, Ho YB, Bryan GW (1995) Fate, bioavailability and toxicity of silver in estuarine environment. Mar Poll Bull 31:44–54
Luther GW, Rickard DT (2005) Metal sulfide cluster complexes and their biogeochemical importance in the environment. J Nanopart Res 7:389–407
Mahabady MK (2012) The evaluation of teratogenicity of nanosilver on skeletal system and placenta of rat foetuses in prenatal period. Afr J Pharm Pharmacol 6:419–424
Maneewattanapinyo P, Banlunara W, Thammacharoen C, Ekgasit S, Kaewamatawong T (2011) An evaluation of acute toxicity of colloidal silver nanoparticles. J Vet Med Sci 73:1417–1423
Markowski M, Kaliński A, Skwarska J, Wawrzyniak J, Bańbura M, Markowski J et al (2013) Avian Feathers as bioindicators of the exposure to heavy metal contamination of food. Bull Environ Contam Toxicol 91:302–305
Marshall JP, Shneider RP (1977) Systemic argyria secondary to topical silver nitrate. Arch Dermatol 113:1077–1079
Massarsky A, Trudeau VL, Moon TW (2014) Predicting the environmental impact of nanosilver. Environ Toxicol Pharmacol 38:861–873
Matuk Y, Gosh M, McCulloch C (1981) Distribution of silver in the eyes and plasma proteins of the albino rat. Can J Ophthalmol 16:145–150
McShan D, Ray PC, Yu H (2014) Molecular toxicity mechanism of nanosilver. J Food Drug Anal 22:116–127
Mierzykowski SE, Smith JEM, Todd CS, Kusnierz D, DeSorbo CR (2011) Liver contaminants in bald eagle carcasses from Maine. USFWS. Spec. Proj. Rep. FY09-MEFO-6-EC. Maine Field Office, Orono, ME, 53 pp
Mody W, Siwale R, Singh A, Mody HR (2010) Introduction to metallic nanoparticles. J Pharm Bioallied Sci 2:282–289

Murata T, Kanao-Koshikawa M, Takamatsu T (2007) Effects of Pb, Cu, Sb, In and Ag contamination on the proliferation of soil bacterial colonies, soil dehydrogenase activity, and phospholipid fatty acid profiles of soil microbial communities. Water Air Soil Pollut 164:103–118

Nam D-H, Anan Y, Ikemoto T, Okabe Y, Kim E-Y, Subramanian A et al (2005a) Specific accumulation of 20 trace elements in great cormorants (*Phalacrocorax carbo*) from Japan. Environ Pollut 134:503–514

Nam D-H, Anan Y, Ikemoto T, Tanabe S (2005b) Multielemental accumulation and its intracellular distribution in tissues of some aquatic birds. Mar Pollut Bull 50:1347–1362

Nel A, Xia T, Madler L, Li N (2006) Toxic potential of materials at the nanolevel. Science 311:622–662

Nordberg GF, Gerhardsson L (1988) Silver. In: Seiler HG, Sigel H, Sigel A (eds) Handbook on toxicity of inorganic compounds. Marcel Dekker, New York, pp 619–623

Ohlendorf HM, Lowe RW, Kelly PR, Harvey TE (1986) Selenium and heavy metals in San Francisco Bay diving ducks. J Wildl Manage 50:64–71

Olcott CT (1948) Experimental argyrosis. Morphologic changes in the experimental animal. Am J Pathol 24:813–833

Palache C, Berman H, Frondel C (1951) Dana's system of mineralogy, vol II. Wiley, New York, pp 11–15

Park EJ, Bae E, Yi J, Kim Y, Choi K, Lee SH, Yoon J, Lee BC, Park K (2010) Repeated-dose toxicity and inflammatory responses in mice by oral administration of silver nanoparticles. Environ Toxicol Pharmacol 30:162–168

Phalen RF, Morrow PE (1973) Experimental inhalation of metallic silver. Health Phys 24:509–518

Piao MJ, Kang KA, Lee IK, Kim HS, Kim S, Choi JY et al (2011) Silver nanoparticles induce oxidative cell damage in human liver cells through inhibition of reduced glutathione and induction of mitochondria-involved apoptosis. Toxicol Lett 201:92–100

Piccinno F, Gottschalk F, Seeger S, Nowack B (2012) Industrial production quantities and uses of ten engineered nanomaterials in Europe and the world. J Nanopart Res 14:1109

Pratsinis A, Hervella P, Leroux JC, Pratsinis SE, Sotiriou GA (2013) Toxicity of silver nanoparticles in macrophages. Small 9:2576–2584

Rahman MF, Wang J, Patterson TA, Saini UT, Robinson BL, Newport GD et al (2009) Expression of genes related to oxidative stress in the mouse brain after exposure to silver-25 nanoparticles. Toxicol Lett 187:15–21

Ratte HT (1999) Bioaccumulation and toxicity of silver compounds. A review. Environ Toxicol Chem 18:89–108

Rejeski D (2011) Project on Emerging Nanotechnologies. Woodrow Wilson International Center for Scholars, Washington DC

Rosenman KD, Moss A, Kon S (1979) Argyria: clinical implications of exposure to silver nitrate and silver oxide. J Occup Med 21:430–435

Rudnick RL, Gao S (2003) Composition of the continental crust. In: Rudnick RL (ed) The crust, treatise on geochemistry, vol 3. Elsevier, Oxford, pp 1–70

Rungby J (1986) Exogenous silver in dorsal root ganglia, peripheral nerve, enteric ganglia, and adrenal medulla. Acta Neuropathol 69:45–53

Rungby J, Danscher G (1983) Localization of exogenous silver in brain and spinal cord of silver exposed rats. Acta Neuropathol 60:92–98

Rungby J, Danscher G (1984) Hypoactivity in silver exposed mice. Acta Pharmacol Toxicol 55:398–401

Rush SA, Borga K, Dietz R, Born EW, Sonne C, Evans T et al (2008) Geographic distribution of selected elements in the livers of polar bears from Greenland, Canada and the United States. Environ Poll 153:618–626

Scanlon PF, Oderwald RG, Dietrick TJ, Coggin JL (1980) Heavy metal concentrations in feathers of ruffed grouse shot by Virginia hunters. Bull Environ Contam Toxicol 25:947–949

Schirmer K, Behra R, Sigg L, Suter MJ-F (2013) Ecotoxicological aspects of nanomaterials in the aquatic environment. In: Luther W, Zweck A (eds) Safety aspects of engineered nanomaterials. Pan Stanford, Singapore, pp 141–162
Schweinfurt SP (2009) An introduction to coal quality. In: Pierce BS, Dennen KO (eds) The National Coal Resource Assessment Overview. U.S. Geological Survey, Reston, VA, pp 1–16
Scott T, Norman PM (1980) A silver deposition in arteriolar basal laminae in the cerebral cortex of argyric rats. Acta Neuropathol 52:243–246
Settimio L, McLaughlin MJ, Kirby JK, Langdon KA, Janik L, Smith S (2015) Complexation of silver and dissolved organic matter in soil water extracts. Environ Pollut 199:174–184
Shafer MM, Overdier JT, Armstong DE (1998) Removal, partitioning, and fate of silver and other metals in wastewater treatment plants and effluent-receiving streams. Environ Toxicol Chem 17:630–641
Sharma HS, Sharma A (2007) Nanoparticles aggravate heat stress induced cognitive deficits, blood-brain barrier disruption, edema formation and brain pathology. Prog Brain Res 162:245–273
Sharma VK, Siskova KM, Zboril R, Gardea-Torresdey JL (2014) Organic-coated silver nanoparticles in biological and environmental conditions: fate, stability and toxicity. Adv Colloid Interface Sci 204:15–34
Skalska J, Frontczak-Baniewicz M, Strużyńska L (2015) Synaptic degeneration in rat brain after prolonged oral exposure to silver nanoparticles. Neurotoxicology 46:145–154
Smith IC, Carson BL (1977) Trace metals in the environment. V. 2: Silver. Ann Arbor Science, Ann Arbor, MI pp 469
Soderstjerna E, Johansson F, Klefbohm B, Englund Johansson U (2013) Gold- and silver nanoparticles affect the growth characteristics of human embryonic neural precursor cells. PLoS One 8:e58211
Stebounova LV, Adamcakova-Dodd A, Kim JS, Park H, O'Shaughnessy PT, Grassian VH, Thorne PS (2011) Nanosilver induces minimal lung toxicity or inflammation in a subacute murine inhalation model. Part Fibre Toxicol 8:5
Stephens T (2005) Survey finds silver contamination in North Pacific waters, probably from industrial emissions in Asia. UC Santa Cruz Curr 9(28):14–20
Stokinger HE (1981) Silver. In: Clayton E, Clayton P (eds) Industrial hygiene and toxicology, vol 2A. Wiley, New York, pp 1881–1894
Strużyński W, Dąbrowska-Bouta B, Grygorowicz T, Zieminska E, Strużynska L (2014) Markers of oxidative stress in hepatopancreas of crayfish (*Orconectes limosus*, Raf.) experimentally exposed to nanosilver. EnvironToxicol 29:1283–1291
Sung JH, Ji JH, Park JD, Yoon JU, Kim DS, Jeon KS et al (2009) Subchronic inhalation toxicity of silver nanoparticles. Toxicol Sci 108:452–461
Tamimi SO, Zmeili SM, Gharaibeh MN, Shubair MS, Salhab AS (1998) Toxicity of a new antismoking mouthwash 881010 in rats and rabbits. J Toxicol Environ Health A 53:47–60
Tang J, Xiong L, Wang S, Wang J, Liu L, Li J, Yuan F, Xi T (2009) Distribution, translocation and accumulation of silver nanoparticles in rats. J Nanosci Nanotechnol 9:4924–4932
Throbäck IN, Johansson M, Rosenquist M, Pell M, Hansson M, Hallin S (2007) Silver (Ag+) reduces denitrification and induces enrichment of novel nirK genotypes in soil. FEMS Microbiol Lett 270:189–194
Tripathy SK (2008) Nanophotothermolysis of poly-(vinyl) alcohol capped silver particles. Nanoscale Res Lett 3:164–167
US EPA (1980) Ambient water quality criteria for silver. Environmental Protection Agency
US EPA (1985) Drinking water criteria document for silver. Environmental Criteria and Assessment Office, Cincinnati, OH. ECAO-CIN-026, PB86-118288
US EPA (1997) Silver, CASRN 7440-22-4. Integrated Risk Information System, IRI
US EPA (2003) Guidance for developing ecological soil screening levels (Eco-SSLs). Review of background concentrations for metals. Attachment 1–4
US EPA (2006) Environmental Protection Agency Office of Solid Waste and Emergency Response Ecological Soil Screening Levels for Silver. OSWER Directive 9285.7-77. Washington, DC

US PHS (1990) Toxicological profile for silver. Agency for Toxic Substances and Disease Registry Public Health Service, pp 1–157

Van der Zande M, Vandebriel RJ, Van DE, Kramer E, Herrera Rivera Z, Serrano-Rojero CS et al (2012) Distribution, elimination, and toxicity of silver nanoparticles and silver ions in rats after 28-day oral exposure. ACS Nano 6:7427–7442

Venugopal B, Luckey TD (1978) Metal toxicity in mammals. Vol. 2: chemical toxicity of metals and metalloids. Plenum Press, New York

Vermeer K, Peakall DB (1979) Trace metals in seaducks of the Fraser River delta intertidal area, British Columbia. Mar Pollut Bull 10:189–191

Wang H, Wick RL, Xing B (2009) Toxicity of nanoparticulate and bulk ZnO, Al2O3 and TiO2 to the nematode *Caenorhabditis elegans*. Environ Poll 157:1171–1177

WHO (2002) Concise International Chemical Assessment Document 44. Silver and silver compounds: environmental aspects. World Health Organization, Geneva

Wieser ME, Holden N, Coplen TB, Böhlke JK, Berglund M, Brand WA et al (2013) Atomic weights of the elements 2011 (IUPAC Technical Report). Pure Appl Chem 85:1047–1078

Wijnhoven SWP, Peijnenburg WJGM, Herberts CA, Hagens WI, Oomen AG, Heugens EHW et al (2009) Nanosilver—a review of available data and knowledge gaps in human and environmental risk assessment. Nanotoxicology 3:109–138

Wu Y, Zhou Q (2013) Silver nanoparticles cause oxidative damage and histological changes in medaka (*Oryzias latipes*) after 14 days of exposure. Environ Toxicol Chem 32:165–173

WWC (2013) Consumer products inventory: an inventory of nanotechnology-based consumer products introduced on the market. Woodrow Wilson Center: Project on Nanotechnology, Washington, DC. http://www.nanotechproject.org/cpi

Yang J, Miyazaki N, Kunito T, Tanabe S (2006) Trace elements and butyltins in a Dall's porpoise (*Phocoenoides dalli*) from the Sanriku coast of Japan. Chemosphere 63:449–457

Yang Z, Liu ZW, Allaker RP, Reip P, Oxford J, Ahmad Z et al (2010) A review of nanoparticle functionality and toxicity on the central nervous system. J R Soc Interface 7(Suppl 4):S411–S422

Yaroshevsky AA (2006) Abundances of chemical elements in the Earth's crust. Geochem Int 44:48–55

Yin N, Liu Q, Liu J, He B, Cui L, Li Z et al (2013) Silver nanoparticle exposure attenuates the viability of rat cerebellum granule cells through apoptosis coupled to oxidative stress. Small 9:1831–1841

Yoo-iam M, Chaichana R, Satapanajaru T (2014) Toxicity, bioaccumulation and biomagnification of silver nanoparticles in green algae (*Chlorella* sp.), water flea (*Moina macrocopa*), blood worm (*Chironomus* spp.) and silver barb (*Barbonymus gonionotus*). Chem Spec Bioavil 26:257–265

Zaichick S, Zaichick V (2015) The content of silver, cobalt, chromium, iron, mercury, rubidium, antimony, selenium, and zinc in osteogenic sarcoma. J Cancer Ther 6:493–503

Zhao X, Toyooka T, Ibuki Y (2014) Silver ions enhance UVB-induced phosphorylation of histone H2AX. Environ Mol Mutagen 55:556–565

Ziemińska E, Strużyńska L (2016) Zinc modulates nanosilver-induced toxicity in primary neuronal cultures. Neurotox Res 29:325–343

Ziemińska E, Stafiej A, Strużyńska L (2014) The role of the glutamatergic NMDA receptor in nanosilver-evoked neurotoxicity in primary cultures of cerebellar granule cells. Toxicology 315:38–48

# Chapter 19
# Tin, Sn

**Agnieszka Tomza-Marciniak, Bogumiła Pilarczyk, Andrzej Marciniak, Renata Pilarczyk, and Małgorzata Bąkowska**

**Abstract** The harmfulness of tin (Sn) on the environment depends on the chemical form in which it occurs. In general, organic Sn compounds are more toxic than metallic tin and inorganic tin compounds. Some studies suggest that tin is an essential trace element for animals and perhaps for humans, but no consensus exists in this regard. Concentrations of inorganic tin in the air, soil, and water are usually low, apart from those areas with naturally high Sn content and regions surrounding tin processing plants. The toxic activity of Sn, caused by environmental exposure to tin, has not been reported in plants, animals, or humans. From an ecotoxicological point of view, the most important compounds are the organotins, mostly due to their androgenic activity and contribution to the increasing number of imposex individuals between marine vertebrates and invertebrates. Literature data about the bioaccumulation of inorganic tin in land ecosystems is very limited, especially in relation to mammals. Also, most of the data concerning the aptitude of some species of animals and biological parameters to be used as bioindicators and biomarkers of environmental exposure to tin usually relate to marine habitats and organic forms of this element. It seems that the problem of land habitat pollution with tin is not well elucidated.

A. Tomza-Marciniak (✉) · B. Pilarczyk · M. Bąkowska
Department of Animal Reproduction Biotechnology and Environmental Hygiene, West Pomeranian University of Technology in Szczecin, Szczecin, Poland
e-mail: agnieszka.tomza-marciniak@zut.edu.pl

A. Marciniak
Department of Animal Reproduction Biotechnology and Environmental Hygiene, West Pomeranian University of Technology in Szczecin, Szczecin, Poland

R. Pilarczyk
Laboratory of Biostatistics, West Pomeranian University of Technology in Szczecin, Szczecin, Poland



E. Kalisińska (ed.), *Mammals and Birds as Bioindicators of Trace Element Contaminations in Terrestrial Environments*,
https://doi.org/10.1007/978-3-030-00121-6_19

# 1 Introduction

Metallic tin (Sn) is a natural compound of the earth's crust, in concentrations reaching about 2–3 mg $kg^{-1}$ (Budavari 2001). Tin is released into the environment from both natural and anthropogenic sources, but for the organic forms of this element, the anthropogenic sources are dominant. Tin reaches the atmosphere with soil and road dust and agricultural activity. Fires and volcanic emissions are also sources of tin, but their significance is quite low. Industrial emissions of tin are mostly related to smelting and refining, the industrial use of tin, and combustion of waste and fossil fuels (WHO 2005).

Pesticides, landfilling with Sn-containing wastes, and the application of pretreated municipal sludge and urban refuse as soil amendments are sources of tin in soils (ATSDR 2005). The concentration of tin in European soils ranges from <2 to 106 mg $kg^{-1}$, while in unpolluted soils it usually does not exceed 5 mg $kg^{-1}$ (De Vos et al. 2006). In natural water reservoirs (rivers, lakes, estuaries, and oceans), tin is present in trace amounts and usually does not exceed 5 ng $L^{-1}$ (WHO 2004).

Tin is an essential trace element for plants and fungi. Some studies suggest that tin is an essential trace element for animals and perhaps for humans, but there is no consensus on this issue. The toxicity of tin to animals depends on the form in which this element is present. Inorganic tin, due to its low solubility, poor absorption, and small retention in tissues, is relatively harmless (Johnson and Greger 1982). Some organotin compounds show a high biological activity, as used in the production of biocides as pesticides with antifouling agents (WHO 2005).

Literature data on the bioaccumulation of inorganic tin in land ecosystems is very limited, especially in relations to mammals. No reports concerning the transfer of this element along land trophic chains are available. A little information in this regard was provided in the study by Hsu et al. (2006); however, the results were insufficient to completely understand this issue.

Bioindicators and biomarkers are important tools in ecotoxicological studies and in evaluation of the risk of exposure. In highly toxic compounds, scientific literature provides abundant information on the usefulness of different species and biological parameters used in evaluations. However, in the case of metallic tin or its inorganic compounds, almost no data is available. A slightly better situation exists in regard to organic forms of tin, due to their higher toxicity, but most of the reports concern aquatic ecosystems, particularly marine habitats.

In relations to land habitats, some interesting information was provided in two publications by Mizukawa et al. (2009) and Miedico et al. (2016). The first indicates the usefulness of bird feathers as a noninvasive research material to monitor the level of organotins, while the second indicates ruminants (particularly sheep) as good biomonitors of environmental pollution of metallic tin, mostly in the neighborhood of its emitters.

## 2 General Properties

Tin (Sn, atomic weight 118.7) is a metallic element located in the 14th group of the periodic table, showing oxyphilic and chalcophylic properties. It is a silvery white, soft, malleable metal with +2 and +4 degrees of oxidation. Tin is found in the two allotropic forms: β form, white tin (density 7.31 g $cm^{-3}$); and α form, gray tin (density 5.85 g $cm^{-3}$). The melting point of Sn is 231.9 °C, and boiling temperature ranges between 2260 °C and 2270 °C. Cassiterite ($SnO_2$) is the main mineral of tin and the only one with commercial significance. The other minerals are stannin ($Cu_2FeSnS_4$) and teallite ($Cu_2SnS_4$).

The average concentration of tin in the earth's crust is 2.5 mg $kg^{-1}$. The highest concentrations are found in loamy rocks at 6–10 mg $kg^{-1}$, magma rocks at 0.3-- 3.6 mg $kg^{-1}$, and carbonaceous and arenaceous rocks at 0.5 mg $kg^{-1}$ (Kabata-Pendias and Pendias 1999). Among the rock-formed minerals, the highest content of Sn is typically in biotite and muscovite. During the erosion process, Sn is released probably as $Sn^{4+}$, composing hydrolyzates (Migaszewski and Gałuszka 2007). Form $Sn^{4+}$ is absorbed by loamy materials, iron and aluminum oxides, and organic matter, leading to a secondary accumulation of Sn in sediments and coal, in which the concentration may reach even 2.3 mg $kg^{-1}$ (Llorens et al. 2000).

In Europe, soils contain from <2 to 106 mg $kg^{-1}$ of tin (De Vos et al. 2006). The mean content of tin in uncontaminated soils ranges between <0.1 and 4.0 mg $kg^{-1}$, with the lowest amounts found in sandy soils and rendzina and the highest in heavy clay soils (Kabata-Pendias and Szteke 2012) (Table 19.1).

Geochemical background levels for Sn in the soils in Western European and Scandinavian countries are 1–2 mg $kg^{-1}$ (De Vos et al. 2006), in Slovakia 5 mg $kg^{-1}$ (Curlik and Šefeik 1999), and in Lithuania 2.1 mg $kg^{-1}$ (Kadûnas et al. 1999). Acceptable Sn concentration in soils is 50 mg $kg^{-1}$, while in heavily polluted soils, even 800 mg Sn $kg^{-1}$ is found (Kabata-Pendias and Pendias 1999).

**Table 19.1** Content of tin in soil

| Country | Layer of soil (cm) | Mean level (mg $kg^{-1}$) | Range (mg $kg^{-1}$) | References |
|---|---|---|---|---|
| Cyprus | 0–25 | 0.8 | <0.2–96.3 | Cohen et al. (2012) |
| | 50–75 | 0.6 | <0.2–52.8 | |
| Italy | 0–20 | 5.0[a] | 2.1–12.8 | Adamo et al. (2014) |
| | 30–40 | 5.0[a] | 2.1–13.0 | |
| Greece | 0–10 | 5.5[b] | 0.6–156 | Argyraki and Kelepertzis (2014) |
| Norway | 14 | 0.749 | <0.15–10 | Reimann et al. (2015) |
| | 30 (10–80) | 0.443 | <0.1–4.28 | |
| Spain | | 0.23[b] | – | Peña-Fernández et al. (2015) |
| | | 0.21[c] | – | |
| Poland | 0–20 | 0.4 | 0.1–2.6 | Pasieczna (2012) |

[a]Polluted agricultural lands
[b]Urban soil
[c]Industrial soil

**Table 19.2** Production and reserves of Sn in selected countries (in metric tons)

| Country | Refined tin production[a] | Mine production[b] | Reserves[b] |
|---|---|---|---|
| Australia | nd | 5900 | 240.000 |
| Bolivia | 12.106 | 18.000 | 400.000 |
| Brazil | nd | 11.900 | 700.000 |
| China | 114.200 | 100.000 | 1.500.000 |
| Indonesia | 27.431 | 40.000 | 800.000 |
| Malaysia | 30.260 | 3.700 | 250.000 |
| Peru | 20.224 | 26.100 | 91.000 |
| Russia | nd | 300 | 350.000 |
| Thailand | 10.502 | 300 | 170.000 |
| World total | nd | 230.000 | 4.700.000 |

*nd* no data
[a]ITRI 2016 (data for 2015)
[b]MSC 2014 (data for 2013)

## 3 Tin Uses and Production

Tin is used mostly in metallurgy and also in the paint and enamel industry and in the production of plastics and biocides. It is estimated that about 80% of tin production is used in the production of tinplate, bronze, and solder (Adriano 2001). In the USA, the major uses for tin are cans and containers 23%, construction 18%, transportation 17%, electrical 12%, and others 30% (MSC 2014). Butyltin compounds such as monobutyltin (MBT), dibutyltin (DBT), and tributyltin (TBT) are used mostly as biocides (antifouling and anti-mollusk agents, fungicides, insecticides, rodenticides, and acaricides), as well as stabilizers in many branches of industry. Organic compounds of tin are used to stabilize polyvinyl chloride (PVC) to protect it from thermal and photochemical degradation, as well as catalyzers in the synthesis of silicon rubbers and polyurethane foams and as coolers in transformers (Falandysz 2003). Other uses include production of lithium-ion batteries, combustibility reducing agents, glass packages, and TFS conversion (ITRI 2012).

Global reserves of tin are located mostly in western Africa, southeastern Asia, Australia, Bolivia, Brazil, China, Indonesia, and Russia. Mine production and reserves of tin are presented in Table 19.2.

## 4 Tin in Nature: Geogenic and Anthropogenic Sources

Tin is released into the environment from both natural and anthropogenic sources. Metallic tin is a natural compound of the earth's crust, where it is present at concentrations between 2 and 3 mg $kg^{-1}$ (Budavari 2001). Cassiterite ($SnO_2$) is the main ore of Sn; the other significant minerals of Sn are stannite ($Cu_2SnFeS_4$) and montesita ($PbSn_4S_5$) (Alloway 1990).

Inorganic tin may originate from natural and anthropogenic sources; organotin compounds are emitted mostly as the effect of human activity. Relatively low amounts of organic Sn forms arise in chemical and biochemical methylation reactions where inorganic Sn compounds are transformed into a methyltin form.

Tin is a compound present in dusts from soils, roads, agriculture, and industry. Small amounts of Sn are also released into the environment from fires or volcanic eruptions. Anthropogenic Sn sources play a main role in contaminating the environment. Emission takes place during the production, use, storage, and recycling of Sn, as well as during waste combustion, such as for municipal waste. The concentration of Sn in volatile dusts may reach 8.7 mg $kg^{-1}$ (Llorens et al. 2000).

Organic Sn compounds may be released into the air by spraying of fertilizers and antifouling agents; evaporation; incineration of materials treated with organotins or stabilized with organotin compounds; and the processes of glass coating. It has been said that the evaporation of organotin compounds into the air is not a significant source due to their low vapor pressures and rapid photodegradation (Fent 1996).

Natural concentrations of tin in the atmospheric air range between 0.01 and 0.1 ng $m^{-3}$ (Kabata-Pendias and Pendias 1999). Tin and its compounds in environmental conditions are considered nonvolatile and related mainly to atmospheric dust. Atmospheric transport of tin depends therefore mostly on the size and weight of dust particles and the meteorological conditions (Senesi et al. 1999). In the air, tin largely associates with small respirable particles from 1–3 μm diameter (WHO 1980). Huang and Klemm (2004) reported that organotins such as butyltins, methyltins, and octyltins are found in the gas phase (<100 pg $m^{-3}$) with a dominance of tri- and di- substituted organotins and octyltins.

In general, the concentration of tin in the atmospheric air is quite low, apart from the areas in which sources of tin emissions are located. In urbanized areas, the concentration of tin in the air is <6 ng $m^{-3}$ (e.g., Barcelona 2.3, Athens 1.1, Oporto 5.9, Zurich 5 ng $m^{-3}$), while in rural regions it is <1 ng $m^{-3}$ (Minguillóna et al. 2012; Martins et al. 2016).

The soil environment absorbs tin from organotin-containing pesticides, landfilling of Sn-containing wastes, or the application of pretreated municipal sludge and urban refuse as soil amendments (ATSDR 2005). Sewage sediments are particularly rich in tin at 40–700 mg Sn $kg^{-1}$ dw. Much smaller amounts of Sn are found in manure and poultry wastes, 3.7–7.4 and 2.0–4.1 mg Sn $kg^{-1}$ dw, respectively (Senesi et al. 1999; ATSDR 2005).

In natural water reservoirs, tin is present in trace amounts. The concentration of tin in rivers, estuaries, and oceans is generally <5 ng $L^{-1}$ (WHO 2004). The main sources of inorganic forms of tin are ground surface flows from agricultural and industrial areas, whereas for organic forms of tin, mostly triorganotin compounds, their use in antifouling paints is considered to be the main source.

# 5 Biological Status of Tin

## *5.1 Toxicity of Tin*

Although Sn is an essential trace element for plants and fungi, in the case of animal organisms, no unambiguous opinion about its necessity has been agreed. There are however some studies indicating that this element may be an essential ultratrace element for rats and maybe for animals generally. Yokoi et al. (1990) observed a weaker growth of Wistar rats fed with fodder containing 17 ng Sn $g^{-1}$, in comparison to rats fed with fodder supplemented with 1.99 mg $kg^{-1}$, as well as hair loss and lowered response to sound. Additionally, the authors observed significant differences in mineral concentrations in tissues between the two groups. They noted, for example, that the Sn-deficient group was characterized by a higher concentration of Ca and lower Mg in the lungs, a lower level of Cu and Zn in the heart muscle, and a higher content of Fe in the spleen and kidneys, in comparison with the group fed with the abundant Sn dose.

The toxicity of Sn to animals depends on the form in which the element exists. Inorganic Sn, due to its low solubility, poor absorption, and low retention in tissues, is not relatively harmful (Johnson and Greger 1982). On the other hand, organotins show a high biological activity. For plants, the most toxic are the alkyl Sn compounds which inhibit cellular proliferation in plants (Radecki et al. 1989).

The toxicity of inorganic Sn to animals is mostly related to its ability to interfere with the activity of some enzymes and the metabolism of some crucial elements like Zn, Cu, Ca, and Fe. Still, organotin compounds are much more toxic and—as lipophilic substances—mostly attack the central nervous system; myelopathy and spongiform encephalopathy are observed (Nath 2000).

Typical for abiotic habitats, the degradation processes of organotins involving the removal of subsequent alkyl (dealkylation) or aryl (dearylation) groups lead to inorganic Sn according to the following scheme: $R_4Sn \rightarrow R_3Sn^+ \rightarrow R_2Sn^{2+} \rightarrow RSn^{3+} \rightarrow Sn$ (IV) (Maguire et al. 1986). This process results in a decrease in toxicity of tin compound, which, from the ecotoxicological point of view, is very advantageous for the environment. Unfortunately, the process of biomethylation may run in parallel, leading to more toxic organotins (Ostrakhovitch 2013).

The toxicity of the various forms of tin, including the organic compounds in relation to marine vertebrates and invertebrates, has been widely described in literature (see reviews: Okoro 2011; Graceli et al. 2013; Schilithz et al. 2013). Still, no data concerning the harmful effects of Sn compounds in terrestrial wildlife are available. Studies on laboratory animal models have been a good source of knowledge in this area.

### 5.1.1 Inorganic Tin

Exposure to Sn causes hematological changes in animals (ATSDR 2005). Inorganic tin compounds, similarly to Pb, disturb the process of heme synthesis and contribute to hemolytic anemia caused by an improper use of Fe (Chmielnicka et al. 1993; Chmielnicka 2006), replacing Zn in δ-aminolevulinic acid dehydratase (ALADH), thus decreasing the activity of the enzyme. The effect of Sn on heme biosynthesis was shown to depend on the concentration of Zn (Chmielnicka et al. 1992). It has been shown that exposure to Sn results in a disturbance in the metabolism of many elements. Reicks and Rader (1990) observed that the uptake and metabolism of Cu are negatively influenced by dietary Sn, resulting in a Cu deficiency in animals. Other studies have shown that Sn negatively affects the metabolism of Ca and P (and in consequence the bone mineralization process; Yamaguchi et al. 1981) and also inhibits the synthesis of collagen (Yamaguchi et al. 1982). In the bones of the golden retriever and Labrador retriever dogs suffering from osteoarthritis, a significantly higher content of Sn in the bones was found in comparison to the healthy dogs (Nganvongpanit et al. 2016).

Reicks and Rader (1990) demonstrated that exposure to Sn leads to a reduction in liver antioxidant protection, mostly due to a decrease in the activity of numerous antioxidative enzymes like liver glutathione peroxidase and superoxide dismutase.

### 5.1.2 Organic Tin Compounds (OTs)

Organotins are characterized by the presence of a single or multiple numbers of covalent bonds between the atoms of Sn and C and are depicted by the general formula $R_nSnX_{4-n}$ (R, an alkyl or aryl group; X, an anionic species; $n = 1,\ldots,4$). The differing chemical structures of the compounds cause particular organotins to differ in the physicochemical properties translating into biological availability, tissue distribution, and the biological activity and toxicity of the organotins which depend mostly on the number and type of organic group bonds with the central atom of Sn, with activity decreasing accordingly: (tri) $R_3SnX$ > (di) $R_2SnX_2$ > (mono) $RSnX_3$.

The relatively high biological activity of triorganotin most likely results from the ability to bind to some proteins; however the locations in which these binding arise are not well known. It has been observed that trimethyltin compounds express a high toxicity to fungi and insects and triethyltin to mammals. In turn, tripropyl- and tributyltin show a higher toxicity to fungi, mollusks, fish, bacteria, and plants (de Carvalho and Santelli 2010).

Organic Sn compounds inhibit oxidative phosphorylation in mitochondria, leading to discrepancies in mitochondrial oxidation and damage to the mitochondria themselves. It was shown that triethyltin may build complexes with hemoglobin. Some organotins have the ability to interfere with the transport of Ca, Mg, K, and Na ions through cellular membranes, which in turn may result in blocking ATP-ases in the brain (Chmielnicka 2006).

Organic compounds, especially TBT, show an androgenic activity and are a reason for an increased frequency of imposex occurrence among aquatic vertebrates and invertebrates (Shimasaki et al. 2003; Horiguchi et al. 2004). The additional development of masculine sex organs in females results in masculinization of a population. No such data referring to terrestrial organisms has been noted.

## 5.2 *Bioaccumulation*

### 5.2.1 Inorganic Sn

Bioaccumulation happens when the rate of absorption of any substance clearly exceeds the potential to expel it from the organism. In the case of inorganic tin, it was found that the alimentary tract absorbed only a small percentage of ingested Sn (3–5%; Johnson and Greger 1982), with the absorption of divalent tin higher than tetravalent (Hiles 1974). In studies on rats, it was demonstrated that after oral administration of Sn compounds (in fruit juice), almost 99% of the ingested tin was passed from the organism in the feces over the next 24 h (Benoy et al. 1971). Therefore inorganic Sn is characterized by a low potential to accumulate.

It was also shown experimentally that the amount of absorbed tin depends on the size of the dose. Johnson and Greger (1982) stated that at a daily dietary intake of 49.7 mg Sn, retention in the organism was 1.3 ± 1.5 mg $day^{-1}$ (about 3%), while at lower doses like 0.11 mg $day^{-1}$, the retention and absorption of Sn were 0.05 ± 0.03 mg $day^{-1}$, about 50%.

The small amounts of Sn, which remain in an organism, accumulate mostly in the bones, thymus, lungs, and muscles, wherein Sn (IV) shows a higher bone uptake and less soft tissue accumulation than Sn (II). Generally, the descending concentrations of Sn in particular organs are as follows: bones > lymphatic gland > lungs > muscles > liver > kidneys > brain (Chmielnicka 2006). With chronic exposure, bones are the main place of Sn deposition. It is estimated that its biological half-life in bone ranges from 34 to 100 days, depending on the species of animal, while in the liver and kidneys 10–20 days (Hiles 1974; Chmielnicka 2006).

Plants are a very important link in the trophic chains of terrestrial habitats that determine the intake of various elements by animals. Tin is an immobile element in arable soil, especially at neutral pH, which causes a low intake by plants. A decrease in soil pH results in a higher content of soluble forms of Sn and in consequence a higher Sn absorption by plants. Usually the concentration in plants ranges from <0.10 to 3.0 mg $kg^{-1}$. Plants are able to accumulate Sn in the roots, with only a small amount translocated to the foliage. Transfer coefficients for Sn in a soil-plant system are between 0.01 and 0.10 (Kloke et al. 1984). This means that plants may be a source of tin for animals, but only for those that consume whole plants or just the underground parts.

Literature data on the bioaccumulation of inorganic tin in terrestrial ecosystems is very limited, especially in reference to mammals. Only a few studies on the concentrations of Sn in avian and mammalian tissues are available, with those usually have a fragmentary character, originating from the 1980s to 1990s (Tables 19.3 and 19.4).

**Table 19.3** Butyltin residues in avian liver and kidney of piscivores

| Species | Country | Residues (μg $kg^{-1}$ ww) | | References |
|---|---|---|---|---|
| | | Liver | Kidney | |
| Red-throated diver *Gavia stellata* | Poland | 610 | nd | Kannan and Falandysz (1997) |
| Razorbill *Alca torda* | Poland | 330 | nd | Kannan and Falandysz (1997) |
| Great crested grebe *Podiceps cristatus* | Poland | 540 | nd | Kannan and Falandysz (1997) |
| Great cormorant *Phalacrocorax carbo* | Poland | 870 | nd | Kannan and Falandysz (1997) |
| | Japan | 270 | 290 | Guruge et al. (1996) |
| | Japan | 385 | 370 | Mizukawa et al. (2009) |
| Long-tailed duck *Clangula hyemalis* | Poland | 4600, female<br>280, male | nd | Kannan and Falandysz (1997) |
| White-tailed eagles *Haliaeetus albicilla* | Poland | 35, female | nd | Kannan and Falandysz (1997) |
| Guillemot *Uria aalge* | Poland | 500 | nd | Kannan and Falandysz (1997) |
| Surf scoter *Melanitta perspicillata* | Canada | 41 | nd | Elliott et al. (2007) |

*nd* no data

**Table 19.4** Metallic Sn and organotin concentration in mammals

| Species | Country | Residues | References |
|---|---|---|---|
| Japanese macaque *Macaca fuscata* | Japan | Liver, μg $kg^{-1}$ ww<br>MBT: <4.0<br>DBT: <3.0<br>TBT: <2.0–2.7 | Takahashi et al. (1997) |
| Raccoon dog *Nyctereutes procyonoides* | Japan | Liver, μg $kg^{-1}$ ww<br>MBT: 9–120<br>DBT: 18–280<br>TBT: 3–10 | Takahashi et al. (1997) |
| American mink *Neovison vison* | Canada | Liver, mg $kg^{-1}$ dw<br>Sn: 5.4<br>Kidney, mg $kg^{-1}$ dw<br>Sn: 5.9 | Harding et al. (1998) |
| Otter *Lontra canadensis* | Canada | Liver, mg $kg^{-1}$ dw<br>Sn: 3.3 | Harding et al. (1998) |
| Polar bear *Ursus maritimus* | USA, Alaska | Liver, mg $kg^{-1}$ dw<br>Sn: 0.071 | Kannan et al. (2007) |

*MBT* monobutyltin, *DBT* dibutyltin, *TBT* tributyltin

The most complete elaboration is likely the study by Hsu et al. (2006), showing the results of heavy metal bioaccumulation (including tin) in the ecosystem of Kenting National Park (Taiwan). The studies included animals such as insects (six specimens), earthworms, snails, crabs, amphibians, lizards, snakes, fish, and bats (whole organism) as well as soil samples, plants and fungi. The highest mean concentrations of Sn were noted in snails and plants, reaching, respectively, 16.8 ± 21.6 and 11.91 ± 33.6 mg $kg^{-1}$ dw, followed by insects, snakes, earthworms, bats, lizards, and amphibians, in which the mean concentrations of Sn ranged from 10.6 to 6.94 mg $kg^{-1}$ dw. In the soil and in fungi, the Sn content was <1 mg $kg^{-1}$ dw. Bioconcentration factors (BCF) calculated from the collected data were between 0.58 and 19.44 and were arranged in the following descending order: snail > plant > insect > snake > earthworm > bat > lizard > amphibian > crab > fungi > fish. An interesting fact arose where, although in the land organisms quite high values of BCF were found, the soil itself still did not contain an elevated level of Sn and the enrichment factor was low (Sn in soil/Sn in upper continental crust = 0.16).

Data about the content of Sn in the kidneys of American mink (*Neovison vison*) and livers of the mink and river otter (*Lontra canadensis*) from Canada by Harding et al. (1998) are presented (Table 19.4). The examined animals were collected from several regions: otters, from upper Fraser River, lower Fraser River, upper Columbia River, lower Columbia River, and Kootenay River; and minks, from upper and lower Fraser River (Canada). The concentrations of Sn in the livers and kidneys of the mink were comparable at an average 5.2–6.3 mg $kg^{-1}$ dw, while in the livers of the otters, the concentration of Sn was lower, usually <4 mg $kg^{-1}$ dw. The authors did not find any differences in Sn concentrations between the collection areas, sexes, or species.

One of the most recent works is Miedico's et al. (2016) in Italy. While studying the accumulation of Sn in animals (bovine and ovine species) raised in an area surrounding oil wells in Italy, the authors observed that the concentrations of Sn in the organs were arranged as follows: lungs > liver > kidneys (which indicates that the main pathway of exposure was the respiratory tract). The noted concentrations of Sn in these organs (for both species together) were 0.081 ± 0.145, 0.055 ± 0.128, and 0.022 ± 0.045 mg $kg^{-1}$ ww, respectively.

There are many works confirming the ability of organic Sn compounds to bioconcentrate and bioaccumulate in aquatic ecosystems (especially marine) (Harino et al. 2000; Zhang et al. 2003; Strand et al. 2005), yet almost no works studying terrestrial habitats are available. Usually such studies are limited to animals which do not in fact live in an aquatic ecosystem but still are strongly associated with it, for example, with some trophic relationships. An example of such work is Lilley et al. (2013), in which the authors made an evaluation of the accumulation of TBT in Daubenton's bat (*Myotis daubentonii*), considered to be very vulnerable to numerous environmental pollutants due to their long life expectancy and high position in the trophic chains. These animals prefer to feed above the water surface on newly hatched adult chironomids, which are effective vectors for the transport of OTC from aquatic to terrestrial ecosystems (Laws et al. 2016). The authors noted only trace amounts of TBT in bat fur samples. A significant positive correlation between sampling site sediment TBT concentrations and bat fur DBT concentrations was found. The analysis of results with regard to selected biological factors showed a comparable concentration of TBT in male and female bats (8.2 vs. 7.77 mg $kg^{-1}$)

and a higher level of TBT artifacts in adult individuals in comparison to juveniles (8.95 vs. 3.49 mg $kg^{-1}$), although in the last example the observed differences were not confirmed statistically.

In the case of wild migratory birds, the accumulation of butyltins (BTs) such as monobutyltin (MBT), dibutyltin (DBT), and tributyltin (TBT) was reported in soft tissues, muscle, and feathers (Guruge et al. 1996; Senthilkumar et al. 1998). In this study it was shown that of the BTs, MBT was the predominant compound retained in the birds, with concentrations recorded in the tail feathers at 73–360 μg $kg^{-1}$ ww, while DBT and TBT concentrations were 29–56 and 37–67 μg $kg^{-1}$, respectively.

Guruge et al. (1996) while studying the remains of BTs in common cormorants (*Phalacrocorax carbo*) from Lake Biwa (Japan) found that the accumulation of BTs in cormorant bodies followed the orders MBT > DBT > TBT and muscle ≥ feathers > skin > liver > rest of the tissues and organs. The authors suggest that the higher levels of BT residues in the feathers occurred due to excretion of BT (about 25% of body burden) during a complete molting cycle and that it may be a natural mechanism of organism detoxication.

In birds, differently than in mammals, concentrations of BTs in the kidney and fat are usually comparable to those in the liver (Guruge et al. 1996; Mizukawa et al. 2009), where it suggests that this may result from the presence of a specific protein in birds which is able to bind BTs in the kidneys, a weak binding affinity of BTs to proteins in the liver and/or different metabolic and elimination pathways for BTs. In the case of fish and marine mammals, TBT accumulates mostly in fats and muscle and DBT in the liver and kidney (Guruge et al. 1996; Kannan et al. 1996). Unfortunately, no such information is available in reference to land mammals.

Kannan et al. (1998) have stated that habitat and feed preferences affect the levels of BT artifacts in the liver of birds. The aforementioned authors observed that concentrations of BTs in the liver of birds collected from areas surrounding a lake were ≤27 μg $kg^{-1}$ ww, whereas those from marine coastal areas contained amounts more than three times higher, and saltwater mollusk-feeding ducks accumulated greater concentrations of BTs than predatory birds feeding on fish, other birds, or small mammals.

An attempt to evaluate biomagnification in piscivorous birds was undertaken by Guruge et al. (1996). The authors assessed the biomagnification factor for the birds-fish level at 1.1–4.1. Unfortunately, there is no such information available on the potential transfer of inorganic tin or organotin compounds from lower trophic levels to higher levels, especially in terrestrial habitats.

### 5.3 Bioindicators and Biomarkers of Tin in Ecotoxicological Studies

Bioindicators and biomarkers play an important role in monitoring and evaluation of the status of an environment. In relation to the toxic substances of particular interest to ecotoxicologists, many publications provide considerations on the usefulness of various animal species in the monitoring of land habitat quality and the health status of particular ecosystems. In the case of metallic tin or its inorganic compounds, not many such works are available, mostly as a result of its low toxicity. Moreover, as the main target of studies itself, tin is examined rarely, usually in combination with other more harmful elements.

In evaluation of the status of environmental pollution by tin in which atmospheric deposition plays an important role in the neighborhood of tin emitters, herbivorous animals seem to be the best bioindicators as they consume only what is deposited on the surfaces of the plants from the atmosphere and what penetrates an organism via the respiratory tract. Tin has a low absorption rate from the soil by plants, as well as a low translocation from the roots to the foliage, leaving the main vector as direct atmospheric deposition. In these conditions, tin intake occurs more via the respiratory tract than the digestive tract. Although wild animals are usually used in ecotoxicological studies (e.g., cervids), Miedico et al. (2016) point out that farm animals may also turn out to be useful, especially ruminants in extensive farming systems without any additional in fodder. The authors have stated that for exposures to metal (including tin), sheep are more useful as bioindicators/biomonitors than cattle, which are often raised under controlled feeding conditions, which have overly complex interactions between the elements to be representative of a contaminated area.

In more complex studies regarding bioaccumulation and tin transfer via the trophic chain, animals from the top of a selected trophic chain may be useful. Unfortunately in relation to terrestrial ecosystems, no such studies have been performed. Little data about the transfer of pollutants (Sn) from water to land habitats have been presented. Such studies are usually performed using water birds and also bats and mustelids (mink, otter), which are treated as a species equivalent to avian predators (Harding et al. 1998; Lilley et al. 2013).

The most common biomarker of exposure is the presence of a selected xenobiotic, in this case Sn or its metabolite(s) or products of Sn interactions with some target molecules (e.g., adducts) in tissues, body fluids, or secretions/excrements of an exposed organism. Although experimental studies suggest that the bones are a potential location of Sn accumulation, many literature references study Sn levels in just the liver and kidneys. This applies to both the organic and inorganic forms of Sn.

In some avian species, the feathers may be a useful research material to monitor the organic forms of Sn. Mizukawa et al. (2009), in their studies on the distribution of organic Sn compounds (OTs) in the tissues of cormorants, noted a significant correlation between the concentrations of DBT, DBT + TBT, and TPT (triphenyltin) in the ventral feathers and the liver and have stated that there is a possibility of using ventral feather samples as a nondestructive indicator in OT monitoring. The authors have developed equations (presented below) that may be advantageous in evaluating the hepatic levels of OTs in cormorants based on the concentrations of OTs in the ventral feathers:

$$DBT_{LIVER} = 1.0 \times \text{ventral feather concentration} + 110$$

$$DBT_{LIVER} + TBT_{LIVER} = 1.0 \times \text{ventral feather concentration} + 110$$

$$TPT_{LIVER} = 0.80 \times \text{ventral feather concentration} + 50$$

According to literature data, feathers may not be useful in every avian species to evaluate hepatic concentration of OTs. For example, in the study by Elliot et al. (2007) aimed at assessing residual BTs in the tissues of surf scoter (*Melanitta perspicillata*) wintering on the south coast of British Columbia in Canada, no artifacts of BTs were noted in feather samples taken from the breast, despite the fact that these compounds were found in the liver at a mean level of 41 μg Sn $kg^{-1}$.

An attempt to use the fur of bats in biomonitoring studies was undertaken by Lilley et al. (2013). However, the authors noted some limitations in using such material due to the fact that bats molt annually, so the measured concentrations would reflect relatively close exposure to BTs, and moreover, regarding the storing of these compounds in the fur as a detoxication mechanism, the results may also vary between individuals. The authors stated that liver samples would in that case provide more detailed information than fur samples.

There are no specific biomarkers of the effects of inorganic tin compounds. Still, the following biomarkers used in the evaluation of marine organism exposure to TBT are mentioned: metallothionein induction, acetyl cholinesterase inhibition, imposex, lysosomal enlargement, lysosomal membrane destabilization, peroxisome proliferation, lysosomal activity, genetic or molecular biomarkers, TBT-sensitive immunological biomarkers, apoptosis induction, phagocytic index, and amoebocytic index (Okoro 2011). Perhaps some of those would be useful in ecotoxicological studies on terrestrial animals.

## 6 Conclusion

After an analysis of literature data, it can be concluded that the problem of terrestrial habitat pollution with different forms of tin (Sn) is not yet well understood. The data concerning bioaccumulation of inorganic tin is very limited in reference to mammals and also to other groups of animals. Also it lacks data from which it would be possible to indicate the species that would be useful as bioindicators/biomonitors in ecotoxicological studies and in the evaluation of environmental exposure to Sn compounds. Data in this area would certainly be very useful for further studies.

## References

Adamo P, Iavazzo P, Albanese S et al (2014) Bioavailability and soil-to-plant transfer factors as indicators of potentially toxic element contamination in agricultural soils. Sci Total Environ 500–501:11–22

Adriano DC (2001) Trace elements in terrestrial environments: biogeochemistry, bioavailability and risk of metals. Springer, New York

Alloway BJ (1990) Heavy metals in soils. Wiley, New York

Argyraki A, Kelepertzis E (2014) Urban soil geochemistry in Athens, Greece: the importance of local geology in controlling the distribution of potentially harmful trace elements. Sci Total Environ 482–483:366–377

ATSDR (2005) Toxicological profile for tin and tin compounds. U.S. Department of Health and Human Services, Public Health Service, Agency for Toxic Substances and Disease Registry

Benoy CJ, Hooper PA, Schneider R (1971) The toxicity of tin in canned fruit juices and solid foods. Food Cosmet Toxicol 9(5):645–656

Budavari S (2001) Tin. In: Budavari S (ed) The Merck index: an encyclopedia of chemicals, drugs, and biologicals, 13th edn. Wiley, New York, p 1685

Chmielnicka J (2006) Tin. In: Senczuk W (ed) Contemporary toxicology. PZWL, Warsaw, pp 382–386

Chmielnicka J, Zareba G, Grabowska U (1992) Protective effect of zinc on heme biosynthesis disturbances in rabbits after administration per os of tin. Ecotoxicol Environ Saf 24:266–274

Chmielnicka J, Zareba G, Polkowska-Kulesza E, Najder M, Korycka A (1993) Comparison of tin and lead toxic action on erythropoietic system in blood and bone marrow of rabbits. Biol Trace Elem Res 36:73–87

Cohen DR, Rutherford NF, Morisseau E, Zissimos AM (2012) Geochemical patterns in the soils of Cyprus. Sci Total Environ 420:250–262

Curlik J, Šefeik P (1999) Geochemical atlas of the Slovak Republic. Soils. Ministry of the Environment of the Slovak Republic

de Carvalho OR, Santelli RE (2010) Occurrence and chemical speciation analysis of organotin compounds in the environment: a review. Talanta 82:9–24

De Vos W, Tarvainen T, Salminen R et al (2006) Geochemical atlas of Europe. Part 2, Geological Survey of Finland, Espoo

Elliott JE, Harris ML, Wilson LK, Smith BD, Batchelor SP, Maguire J (2007) Butyltins, trace metals and morphological variables in surf scoter (*Melanitta perspicillata*) wintering on the south coast of British Columbia, Canada. Environ Pollut 149:114–124

Falandysz J (2003) Butyltin and its degradation products in the aspects of food toxicology. Rocz Państw Zakl Hig 54:13–23

Fent K (1996) Ecotoxicology of organotin compounds. Crit Rev Toxicol 26:1–117

Graceli JB, Sena GC, Lopes PF, Zamprogno GC, da Costa MB, Godoi AF et al (2013) Organotins: a review of their reproductive toxicity, biochemistry, and environmental fate. Reprod Toxicol 36:40–52

Guruge KS, Tanabe S, Iwata H, Taksukawa R, Yamagishi S (1996) Distribution, biomagnification, and elimination of butyltin compound residues in common cormorants (*Phalacrocorax carbo*) from Lake Biwa, Japan. Arch Environ Contam Toxicol 31:210–217

Harding LE, Harris ML, Elliott JE (1998) Heavy and trace metals in wild mink (*Mustela vison*) and river otter (*Lontra canadensis*) captured on rivers receiving metals discharges. Bull Environ Contam Toxicol 61:600–607

Harino H, Fukushima MH, Kawai S (2000) Accumulation of butyltin and phenyltin compounds in various fish species. Arch Environ Contam Toxicol 39:13–19

Hiles R (1974) Absorption, distribution and excretion of inorganic tin in rats. Toxicol Appl Pharmacol 27:366–379

Horiguchi T, Li Z, Uno S, Shimizu M, Shiraishi H, Morita M et al (2004) Contamination of organotin compounds and imposex in molluscs from Vancouver, Canada. Mar Environ Res 57:75–88

Hsu MJ, Selvaraj K, Agoramoorthy G (2006) Taiwan's industrial heavy metal pollution threatens terrestrial biota. Environ Pollut 143:327–333

Huang J-H, Klemm O (2004) Atmospheric speciation of ionic organotin, organolead and organomercury compounds in NE Bavaria (Germany). Atmos Environ 38:5013–5023

ITRI (2012) Tin for tomorrow. Contributing to global sustainable development. Available http://www.itri.co.uk
ITRI (2016) The top 10 refined tin producers of 2015. Available https://www.itri.co.uk
Johnson MA, Greger JL (1982) Effects of dietary tin on tin and calcium metabolism of adult males. Am J Clin Nutr 35:655–660
Kabata-Pendias A, Pendias H (1999) Biochemistry of trace elements. PWN, Warszawa
Kabata-Pendias A, Szteke B (2012) Trace elements in geo- and biosphere. IUNG-PIB, Puławy, pp 91–100
Kadûnas V, Budavicius R, Gregorauskiene V, Katinas V, Klaugiene E, Radzevicius A et al (1999) Geochemical atlas of Lithuania. Geological Survey of Lithuania. Geology Institute, Vilnius
Kannan K, Falandysz J (1997) Butyltin residues in sediment, fish, fish-eating birds, harbour porpoise and human tissues from the Polish coast of the Baltic Sea. Mar Pollut Bull 34:203–207
Kannan K, Corsolini S, Focardi S, Tanabe S, Tatsukawa R (1996) Accumulation pattern of butyltin compounds in dolphin, tuna and shark collected from the Italian coastal waters. Arch Environ Contam Toxicol 31:19–23
Kannan K, Senthilkumar K, Elliott JE, Feyk LA, Giesy JP (1998) Occurrence of butyltin compounds in tissues of water birds and sea ducks from the United States and Canada. Arch Environ Contam Toxicol 35:64–69
Kannan K, Agusa T, Evans TJ, Tanabe S (2007) Trace element concentrations in livers of polar bears from two populations in northern and western Alaska. Arch Environ Contam Toxicol 53:473–482
Kloke A, Sauerbeck D, Vetter H (1984) The contamination of plants and soil with heavy metals and the transport of metals in terrestrial food chain. In: Nriagu JO (ed) Changing metal cycles and human health. Springer, Berlin, pp 113–141
Laws J, Heppell K, Sheahan D, Liu CF, Grey J (2016) No such thing as a free meal: organotin transfer across the freshwater–terrestrial interface. Freshw Biol 61:2051–2062
Lilley TM, Ruokolainen L, Meierjohann A, Kanerva M, Stauffer J, Laine VN et al (2013) Resistance to oxidative damage but not immunosuppression by organic tin compounds in natural populations of Daubenton's bats (*Myotis daubentonii*). Comp Biochem Physiol Part C: Toxicol Pharmacol 157:298–305
Llorens JF, Fernandez JL, Querol X (2000) The fate of trace elements in a large coal-fired power plant. Environ Geol 40:409–416
Maguire RJ, Tkacz RJ, Chau YK et al (1986) Occurrence of organotin compounds in water and sediment in Canada. Chemosphere 15:253–274
Martins V, Moreno T, Mendes L, Eleftheriadis K, Diapouli E, Alves CA et al (2016) Factors controlling air quality in different European subway systems. Environ Res 146:35–46
Miedico O, Iammarino M, Paglia G, Tarallo M, Mangiacotti M, Chiaravalle AE (2016) Environmental monitoring of the area surrounding oil wells in Val d'Agri (Italy): element accumulation in bovine and ovine organs. Environ Monit Assess 188:338
Migaszewski Z, Gałuszka A (2007) Fundamentals of environmental geochemistry. WNT, Warszawa
Minguillóna MC, Querolb X, Baltenspergera U, Prevot AS (2012) Fine and coarse PM composition and sources in rural and urban sites in Switzerland: local or regional pollution? Sci Total Environ 427–428:191–202
Mizukawa H, Takahashi S, Nakayama K, Sudo A, Tanabe S (2009) Contamination and accumulation feature of organotin compounds in common cormorants (*Phalacrocorax carbo*) from Lake Biwa, Japan. In: Obayashi Y, Isobe T, Subramanian A, Suzuki S, Tanabe S (eds) Interdisciplinary studies on environmental chemistry – environmental research in Asia. Terrapub, Tokyo, pp 153–161
MSC (2014) Mineral commodity summaries 2014. U.S. Geological Survey, p 196. ISBN 978-1-4113-3765-7

Nath R (2000) Tin. In: Nath R (ed) Health and disease role of micronutrients and trace elements: recent advances in the assessment of micronutrients and trace elements deficiency in humans. APH Publishing, New Delhi, pp 385–389

Nganvongpanit K, Buddhachat K, Brown JL (2016) Comparison of bone tissue elements between normal and osteoarthritic pelvic bones in dogs. Biol Trace Elem Res 171:344–353

Okoro HK (2011) Sources, environmental levels and toxicity of organotin in marine environment: a review. Asian J Chem 23(2):473–482

Ostrakhovitch EA (2013) Tin. In: Nordberg GF, Fowler BA, Nordberg M (eds) Handbook on the toxicology of metals, 4th edn, pp 1241–1286

Pasieczna A (2012) Molybdenum and tin in soils of Poland. Biul Państw Inst Geol 450:75–82

Peña-Fernández A, Lobo-Bedmar MC, González-Muñoz MJ (2015) Annual and seasonal variability of metals and metalloids in urban and industrial soils in Alcalá de Henares (Spain). Environ Res 136:40–46

Radecki J, Banaszkiewicz T, Klasa A (1989) The effect of different tin compounds on the mitotic activity of maize root tip cells. Acta Physiol Plant 4:359

Reicks M, Rader JI (1990) Effects of dietary tin and copper on rat hepatocellular antioxidant protection. Proc Soc Exp Biol Med 195:123–128

Reimann C, Fabian K, Schilling J, Roberts D, Englmaier P (2015) A strong enrichment of potentially toxic elements (PTEs) in Nord-Trøndelag (central Norway) forest soil. Sci Total Environ 536:130–141

Schilithz PF, Dorneles PR, Lailson-Brito J (2013) Cetacean exposure to butyltin compounds: a review. Oecol Aust 17:411–423

Senesi GS, Baldassarre G, Senesi N, Radina B (1999) Trace element inputs into soils by anthropogenic activities and implications for human health. Chemosphere 39:343–377

Senthilkumar K, Kannan K, Tanabe S, Prudente M (1998) Butyltin compounds in resident and migrant birds collected from Philippines. Fresenius Environ Bull 7:561–571

Shimasaki Y, Kitano T, Oshima Y, Inoue S, Imada N, Honjo T (2003) Tributyltin causes masculinization in fish. Environ Toxicol Chem 22:141–144

Strand J, Larsen MM, Lockyer C (2005) Accumulation of organotin compounds and mercury in harbor porpoises (*Phocoena phocoena*) from the Danish waters and West Greenland. Sci Total Environ 350:59–71

Takahashi H (1997) Huddling relationships in night sleeping groups among wild Japanese macaques in Kinkazan Island during winter. Primates 38:57–68

WHO (1980) World Health Organization: International Programme on Chemical Safety (WHO/IPCS). Environmental Health Criteria 15

WHO (2004) Inorganic tin in drinking-water. Background document for development of WHO Guidelines for Drinking-water Quality, Geneva, World Health Organization (WHO/SDE/WSH/03.04/115)

WHO (2005) Tin and inorganic tin compounds. World Health Organization, Geneva. Available http://www.who.int/ipcs/publications/cicad/cicad_65_web_version.pdf

Yamaguchi M, Sugii K, Okada S (1981) Action of inorganic tin on bone metabolism in rats – decreases in calcium content and phosphatase activity. J Toxicol Sci 6:238–239

Yamaguchi M, Sugii K, Okada S (1982) Inhibition of collagen synthesis in the femur of rats orally administered stannous chloride. J Pharmacobiodyn 5:388–393

Yokoi K, Kimura M, Itokawa Y (1990) Effect of dietary tin deficiency on growth and mineral status in rats. Biol Trace Elem Res 24:223–231

Zhang G, Yan J, Fu JM, Parker A, Li XD, Wang ZS (2003) Butyltins in sediments and biota from the Pearl River Delta, South China. Chem Spec Bioavailab 14:35–42

FSC
www.fsc.org
MIX
Papier | Fördert gute Waldnutzung
FSC® C083411

Kolibri 360 GmbH
Ferdinand-Jühlke-Straße 7
99095 Erfurt, Deutschland
produktsicherheit@kolibri360.de